英和対訳

化学品の分類および表示に関する世界調和システム（GHS）

改訂10版

Globally Harmonized System of
Classification and Labelling of
Chemicals(GHS)

Tenth revised edition

GHS 関係省庁連絡会議 仮訳

ST/SG/AC.10/30/Rev.10

GLOBALLY HARMONIZED SYSTEM OF CLASSIFICATION AND LABELLING OF CHEMICALS (GHS)

Tenth revised edition

ST/SG/AC.10/30/Rev.10

化学品の分類および表示に関する世界調和システム（GHS）

改訂 10 版

国際連合
ニューヨークおよびジュネーブ、2023

© 2023 United Nations
All rights reserved worldwide

Requests to reproduce excerpts or to photocopy should be addressed to the Copyright Clearance Center at copyright.com

All other queries on rights and licenses, including subsidiary rights, should be addressed to: United Nations Publications, 405 East 42nd Street, S-09FW001, New York, NY 10017, United States of America. Email: permissions@un.org; website: https://shop.un.org

The designations employed and the presentation of the material in this publication do not imply the expression of any opinion whatsoever on the part of the Secretariat of the United Nations concerning the legal status of any country, territory, city or area, or of its authorities, or concerning the delimitation of its frontiers or boundaries.

United Nations publication issued by the United Nations Economic Commission for Europe.

ST/SG/AC.10/30/Rev.10

ISBN: 978-92-1-117304-8
eISBN: 978-92-1-001907-1

ISSN: 2412-155X
eISSN: 2412-1576

Sales number: E.23.II.E.1

© 2023 for the English edition
© 2025 United Nations for the bilingual English-Japanese edition
All rights reserved worldwide
英語版については© 2023
本対訳版については© 2025 国際連合
全世界的に全ての版権を有する

抜粋の複製または複写の要請はcopyright.comの著作権料清算センターに問い合わせること。

補助的な権利を含む権利とライセンスに関する質問は以下に問い合わせること：United Nations Publications, 405 East 42nd Street, S-09FW001, New York, NY 10017, United States of America. Email: permissions@un.org; website: https://shop.un.org

この出版物において使用している呼称および文章の表現は、国家、領土、市、もしくは地域、またはその行政機関の法的な位置づけ、あるいはその国境や領域に関して、国際連合事務局としてのいかなる見解をも意味するものではない。

国際連合欧州経済委員会による国際連合出版物である。

国際連合との契約に基づく声明

The work is published for and on behalf of the United Nations.
The present work is an unofficial translation for which the publisher accepts full responsibility.
本対訳版は，国際連合に代わり発行するものである。
非公式な翻訳であり，出版社の責任において発行するものである。

ご利用に関する注意事項

この邦訳（日本語訳）は，国際連合の許諾を得て，GHS関係省庁連絡会議監修の下，日本規格協会が発行・複製販売するものです。

著作権に触れるような複製又は利用は固く禁止されています。

また邦訳は，技術的内容を考慮して作成しましたが，原本の利用に際しての情報提供を目的としたものであり，原文と同じ効力を認められたものではありません。翻訳文に疑義がある時は原文に準拠してください。原文のみが有効であり，邦訳のみを使用して生じた不都合な事態に関しては，当協会は一切責任を負うものではありません。

　　　　　　　　　　　　　　　一般財団法人　日本規格協会

FOREWORD

1. The Globally Harmonized System of Classification and Labelling of Chemicals (GHS) is the culmination of more than a decade of work. There were many individuals involved, from a multitude of countries, international organizations, and stakeholder organizations. Their work spanned a wide range of expertise, from toxicology to fire protection, and ultimately required extensive goodwill and the willingness to compromise, in order to achieve this system.

2. The work began with the premise that existing systems should be harmonized in order to develop a single, globally harmonized system to address classification of chemicals, labels, and safety data sheets. This was not a totally novel concept since harmonization of classification and labelling was already largely in place for physical hazards and acute toxicity in the transport sector, based on the work of the United Nations Economic and Social Council's Committee of Experts on the Transport of Dangerous Goods. Harmonization had not been achieved in the workplace or consumer sectors, however, and transport requirements in countries were often not harmonized with those of other sectors.

3. The international mandate that provided the impetus for completing this work was adopted at the 1992 United Nations Conference on Environment and Development (UNCED), as reflected in Agenda 21, para.19.27:

"A globally harmonized hazard classification and compatible labelling system, including material safety data sheets and easily understandable symbols, should be available, if feasible, by the year 2000".

4. The work was coordinated and managed under the auspices of the Interorganization Programme for the Sound Management of Chemicals (IOMC) Coordinating Group for the Harmonization of Chemical Classification Systems (CG/HCCS). The technical focal points for completing the work were the International Labour Organization (ILO); the Organisation for Economic Co-operation and Development (OECD); and the United Nations Economic and Social Council's Sub-Committee of Experts on the Transport of Dangerous Goods.

5. Once completed in 2001, the work was transmitted by the IOMC to the new United Nations Economic and Social Council's Sub-Committee of Experts on the Globally Harmonized System of Classification and Labelling of Chemicals (GHS Sub-Committee). The Sub-Committee was established by Council resolution 1999/65 of 26 October 1999 as a subsidiary body of the former Committee of Experts on the Transport of Dangerous Goods, which was reconfigured and renamed on the same occasion "Committee of Experts on the Transport of Dangerous Goods and on the Globally Harmonized System of Classification and Labelling of Chemicals" (hereafter referred to as "the Committee"). The Committee and its sub-committees work on a biennial basis. Secretariat services are provided by the Sustainable Transport Division of the United Nations Economic Commission for Europe (UNECE).

6. The GHS Sub-Committee is responsible for maintaining the GHS, promoting its implementation and providing additional guidance as needs arise, while maintaining stability in the system to encourage its adoption. Under its auspices, the document is regularly revised and updated to reflect national, regional and international experiences in implementing its requirements into national, regional and international laws, as well as the experiences of those doing the classification and labelling.

7. The first task of the GHS Sub-Committee was to make the GHS available for worldwide use and application. The first version of the document, which was intended to serve as the initial basis for the global implementation of the system, was approved by the Committee at its first session (11-13 December 2002) and published in 2003 under the symbol ST/SG/AC.10/30. Since then, the secretariat has been updating and preparing consolidated revised editions of the GHS every two years, following the decisions taken by the Committee.

8. At its eleventh session (9 December 2022), the Committee adopted a set of amendments to the ninth revised edition of the GHS addressing, *inter alia*, the classification procedure for desensitized explosives (chapter 2.17); the use of non-animal testing methods for classification of health hazards (in particular, skin corrosion/irritation (chapter 3.2), serious eye damage/irritation (chapter 3.3) and respiratory or skin sensitization (chapter 3.4)); further rationalization of precautionary statements to improve users' comprehensibility while taking into account usability for labelling practitioners; and the review of annexes 9 and 10 to ensure alignment of the classification strategy, guidance and tools on metals and metal compounds with the provisions for long-term aquatic classification toxicity in chapter 4.1. The tenth revised edition of the GHS takes account of these amendments which were circulated as document ST/SG/AC.10/50/Add.3.

9. At its eleventh session, the Committee also adopted a set of amendments to the *Manual of Tests and Criteria* (ST/SG/AC.10/50/Add.2). The amendments adopted by the Committee will be reflected in the eigth revised edition of the Manual (ST/SG/AC.10/11/Rev.8).

序文

1. 化学品の分類および表示に関する世界調和システム（The Globally Harmonized System of Classification and Labelling of Chemicals）(GHS)は、10 年以上にわたる作業の成果である。多数の国々、国際機関および関係団体から多くの人々が関与してきた。このシステムを完成させるために、作業は毒物学から消防まで広範囲の専門分野にわたり、また最終的には妥協するための広範な親善と意志が必要とされた。

2. この作業は、化学品の分類、表示および安全データシートの統一的な世界調和システムを開発するためには、既存のシステムを調和させるべきであるということから始まった。国際連合経済社会理事会の危険物輸送専門家委員会（UNCETDG）の作業に基づいて輸送部門における物理化学的危険性と急性毒性の分類と表示の調和は既に広く実施されていたので、このシステムは全く新しい概念というわけではなかった。しかし、作業場や消費部門における調和はまだなされておらず、また各国の輸送に係る要求事項も、他の部門における要求事項と調和していないことも多かった。

3. 国際的な取り決めである、1992 年の国際連合環境開発会議（UNCED）において採択されたアジェンダ 21、第 19 章、第 27 項が、この作業を完成させるための推進力となった。

    「*安全データシートおよび容易に理解できるシンボルも含めた、世界的に調和された危険有害性に関する分類および表示システムを、可能であれば西暦 2000 年までに利用できるようにするべきである。*」

4. 作業の調整および管理は、化学品の適正管理のための国際機関間プログラム（IOMC）の化学品分類システムの調和のための調整グループ（CG/HCCS）が行った。作業を完成させるための技術的な活動の中心は、国際労働機関（ILO）、経済協力開発機構（OECD）、国際連合経済社会理事会の危険物輸送に関する専門家小委員会であった。

5. 作業は 2001 年にいったん終了した後、IOMC から国際連合経済社会理事会の新しい委員会である化学品の分類および表示に関する世界調和システムに関する専門家小委員会（GHS 小委員会）に引き継がれた。この小委員会は、1999 年 10 月 26 日の理事会決議 1999/65 に基づき設立されたもので、「危険物輸送ならびに化学品の分類および表示に関する世界調和システムに関する専門家委員会」（以後「委員会」と呼ぶ）と改組・改名された旧危険物輸送に関する専門家委員会の下部組織である。委員会およびこの小委員会は 2 年間単位で作業を行う。事務局作業は国際連合欧州経済委員会（UNECE）の持続可能な輸送部門により提供されている。

6. GHS 小委員会は、GHS の維持、その実施の促進および必要に応じて追加的なガイダンスを提供する一方、その導入を促進するためにシステムの安定を確保する責任がある。この組織の下で、本文書は、国、地域および国際法へ適用する際に得られた、それぞれの経験ならびに分類および表示を行っている者の経験を反映させるために、改訂および更新がなされている。

7. GHS 小委員会が最初に取り組んだ課題は、GHS の世界的な利用と適用を可能にすることであった。GHS の初版は、このシステムの最初の実施に供されることを目的として、委員会の最初の会合（2002 年 12 月 11－13 日）で承認され、ST/SG/AC.10/30 として 2003 年に出版された。これ以降、委員会の決定にしたがって、事務局が 2 年ごとに GHS の改訂版を更新し、作成してきた。

8. 委員会は第 11 回会議（2022 年 12 月 9 日）において、GHS 改訂 9 版に対する以下の修正を採択した。*特に*、鈍性化爆発物の分類手順（第 2.17 章）；健康有害性の分類に対する動物試験によらない方法の利用（特に、皮膚腐食性/刺激性（第 3.2 章）、眼に対する重篤な損傷性/眼刺激性（第 3.3 章）および呼吸器または皮膚感作性（第 3.4 章））；表示実務者の使いやすさを考慮しつつ、利用者の理解しやすさを向上させるための注意書きのさらなる合理化；金属および金属化合物に関する分類戦略、手引きおよびツールと第 4.1 章の長期水生有害性分類に関する規定との整合性を確保するための附属書 9 および 10 の見直し。GHS 改訂 10 版は、ST/SG/AC.10/50/Add.3 文書として出されたこれらの修正を反映させたものである。

9. 第 11 回会議において、委員会はまた*試験方法および判定基準のマニュアル*（ST/SG/AC.10/50/Add.2）に対する修正も採択した。委員会で採択された修正はマニュアルの改訂 8 版に反映されるであろう（ST/SG/AC.10/11/Rev.8）。

10. While Governments, regional institutions and international organizations are the primary audiences for the GHS, it also contains sufficient context and guidance for those in industry who will ultimately be implementing the national requirements which are adopted. Availability of information about chemicals, their hazards, and ways to protect people, will provide the foundation for national programmes for the safe management of chemicals. Widespread management of chemicals in countries around the world will lead to safer conditions for the global population and the environment, while allowing the benefits of chemical use to continue. Harmonization will also have benefits in terms of facilitating international trade, by promoting greater consistency in the national requirements for chemical hazard classification and communication that companies engaged in international trade must meet.

11. In paragraph 23 (c) of its Plan of Implementation adopted in Johannesburg on 4 September 2002, the World Summit on Sustainable Development (WSSD) encouraged countries to implement the GHS as soon as possible with a view to having the system fully operational by 2008. Subsequently, the United Nations Economic and Social Council invited Governments that had not yet done so, to take the necessary steps, through appropriate national procedures and/or legislation, to implement the GHS as recommended in the WSSD Plan of Implementation[1]. It also reiterated its invitation to the regional commissions, United Nations programmes, specialized agencies and other organizations concerned, to promote the implementation of the GHS and, where relevant, to amend their international legal instruments addressing transport safety, workplace safety, consumer protection or the protection of the environment so as to give effect to the GHS through such instruments. Information about the status of implementation may be found on the UNECE Sustainable Transport Division website[2].

12. Additional information on the work of the Committee and its two sub-committees, as well as corrigenda which may be issued after publication of this document, can be found on the UNECE Sustainable Transport Division website[3].

---

[1] *Resolutions 2003/64 of 25 July 2003, 2005/53 of 27 July 2005, 2007/6 of 23 July 2007, 2009/19 of 29 July 2009, 2011/25 of 27 July 2011, 2013/25 of 25 July 2013, 2015/7 of 8 June 2015, 2017/13 of 8 June 2017, 2019/7 of 6 June 2019 and 2021/13 of 8 June 2021.*

[2] *https://unece.org/es/node/9225.*

[3] *https://unece.org/transport/dangerous-goods and https://unece.org/about-ghs.*

10.　各国政府、地域政府および国際機関をGHSの主要な対象者とするが、各国で採用されている国内の要求事項を最終的に実行する産業界の関係者のための十分な内容およびガイダンスも含んでいる。化学品とその危険有害性および人々を保護する方法に関する情報が利用可能になれば、化学品の安全管理に係る国家プログラムの基礎ができるであろう。世界中の国々における化学品管理の拡大は、化学品の利用による便益を得ながら、世界の人々と環境をより安全な状態に導くであろう。化学品の分類および表示に関する世界調和は、貿易を行う企業が守らなければならない化学品の危険有害性に関する分類および情報の伝達に関する各国の要求事項がより一貫性をもつことから、国際貿易の促進にも役に立つであろう。

11.　持続可能な開発に関する世界首脳サミットは2002年9月4日にヨハネスブルグで採択した行動計画23(c)において、2008年までにGHSという新しいシステムを完全に実施することを目指して、各国ができる限り早期にGHSを実施するよう奨励した。その後、国際連合経済社会理事会はまだ実施していない政府に対し、WSSDの実施計画にあるようにGHSを実施するために、行政手続きや法令を整備すること等により、必要な手段を講じるよう促した[1]。国連経済社会理事会はまた、地域共同体、国連計画、特定の官庁やGHSを推進するその他の関係組織に対し、GHSを効果的にするために輸送安全、労働安全、消費者保護や環境保護に関する国際関連法令を修正することを求めた。実施状況に関する情報はUNECE輸送部門のウェブサイト[2]で見ることができる。

12.　委員会および二つの小委員会の作業に関する追加情報は、この文書が発行された後に出されるかもしれない正誤表と同様、UNECE持続可能な輸送部門のウェブサイト[3]で見つけることができる。

---

[1] *Resolutions 2003/64 of 25 July 2003, 2005/53 of 27 July 2005, 2007/6 of 23 July 2007, 2009/19 of 29 July 2009, 2011/25 of 27 July 2011, 2013/25 of 25 July 2013, 2015/7 of 8 June 2015, 2017/13 of 8 June 2017, 2019/7 of 6 June 2019 and 2021/13 of 8 June 2021.*

[2] *https://unece.org/es/node/9225.*

[3] *https://unece.org/trans/dangerous-goods and https//unece.org/about-ghs.*

TABLE OF CONTENTS

		Page
Part 1. INTRODUCTION		1
Chapter 1.1	Purpose, scope and application of the GHS	3
Chapter 1.2	Definitions and abbreviations	11
Chapter 1.3	Classification of hazardous substances and mixtures	17
Chapter 1.4	Hazard communication: Labelling	23
Chapter 1.5	Hazard communication: Safety Data Sheets (SDS)	35
Part 2. PHYSICAL HAZARDS		39
Chapter 2.1	Explosives	41
Chapter 2.2	Flammable gases	51
Chapter 2.3	Aerosols and chemicals under pressure	57
Chapter 2.4	Oxidizing gases	65
Chapter 2.5	Gases under pressure	67
Chapter 2.6	Flammable liquids	71
Chapter 2.7	Flammable solids	75
Chapter 2.8	Self-reactive substances and mixtures	77
Chapter 2.9	Pyrophoric liquids	83
Chapter 2.10	Pyrophoric solids	85
Chapter 2.11	Self-heating substances and mixtures	87
Chapter 2.12	Substances and mixtures which, in contact with water, emit flammable gases	91
Chapter 2.13	Oxidizing liquids	93
Chapter 2.14	Oxidizing solids	95
Chapter 2.15	Organic peroxides	99
Chapter 2.16	Corrosive to metals	105
Chapter 2.17	Desensitized explosives	107
Part 3. HEALTH HAZARDS		113
Chapter 3.1	Acute toxicity	115
Chapter 3.2	Skin corrosion/irritation	127
Chapter 3.3	Serious eye damage/eye irritation	145
Chapter 3.4	Respiratory or skin sensitization	169
Chapter 3.5	Germ cell mutagenicity	191
Chapter 3.6	Carcinogenicity	197
Chapter 3.7	Reproductive toxicity	207
Chapter 3.8	Specific target organ toxicity – Single exposure	217
Chapter 3.9	Specific target organ toxicity – Repeated exposure	227
Chapter 3.10	Aspiration hazard	237

目 次

	頁
第1部　序	1
第1.1章　GHSの目的、範囲および適用	3
第1.2章　定義および略語	11
第1.3章　危険有害性のある物質および混合物の分類	17
第1.4章　危険有害性に関する情報の伝達：表示	23
第1.5章　危険有害性に関する情報の伝達：安全データシート（SDS）	35
第2部　物理化学的危険性	39
第2.1章　爆発物	41
第2.2章　可燃性ガス	51
第2.3章　エアゾールおよび加圧下化学品	57
第2.4章　酸化性ガス	65
第2.5章　高圧ガス	67
第2.6章　引火性液体	71
第2.7章　可燃性固体	75
第2.8章　自己反応性物質および混合物	77
第2.9章　自然発火性液体	83
第2.10章　自然発火性固体	85
第2.11章　自己発熱性物質および混合物	87
第2.12章　水反応可燃性物質および混合物	91
第2.13章　酸化性液体	93
第2.14章　酸化性固体	95
第2.15章　有機過酸化物	99
第2.16章　金属腐食性	105
第2.17章　鈍性化爆発物	107
第3部　健康に対する有害性	113
第3.1章　急性毒性	115
第3.2章　皮膚腐食性/刺激性	127
第3.3章　眼に対する重篤な損傷性/眼刺激性	145
第3.4章　呼吸器感作性または皮膚感作性	169
第3.5章　生殖細胞変異原性	191
第3.6章　発がん性	197
第3.7章　生殖毒性	207
第3.8章　特定標的臓器毒性－単回ばく露	217
第3.9章　特定標的臓器毒性－反復ばく露	227
第3.10章　誤えん有害性	237

TABLE OF CONTENTS *(cont'd)*

			Page
Part 4.	**ENVIRONMENTAL HAZARDS**		243
	Chapter 4.1	Hazardous to the aquatic environment	245
	Chapter 4.2	Hazardous to the ozone layer	267
ANNEXES			269
	Annex 1	Classification and labelling summary tables	271
	Annex 2	*(Reserved)*	289
	Annex 3	Codification of hazard statements, codification and use of precautionary statements, codification of hazard pictograms and examples of precautionary pictograms	291
	Annex 4	Guidance on the preparation of Safety Data Sheets (SDS)	407
	Annex 5	Consumer product labelling based on the likelihood of injury	431
	Annex 6	Comprehensibility testing methodology	437
	Annex 7	Examples of arrangements of the GHS label elements	449
	Annex 8	An example of classification in the Globally Harmonized System	471
	Annex 9	Guidance on hazards to the aquatic environment	479
	Annex 10	Guidance on transformation/dissolution of metals and metal compounds in aqueous media	561
	Annex 11	Guidance on other hazards not resulting in classification	573

目　次（続き）

頁

第4部	環境に対する有害性	243
第4.1章	水生環境有害性	245
第4.2章	オゾン層への有害性	267
附属書		269
附属書1	分類および表示のまとめ	271
附属書2	（保留）	289
附属書3	危険有害性情報のコード、注意書きのコードと使用法 絵表示のコードおよび注意絵表示の例	291
附属書4	安全データシート（SDS）作成指針	407
附属書5	危害の可能性に基づく消費者製品の表示	431
附属書6	理解度に関する試験方法	437
附属書7	GHSラベル要素の配置例	449
附属書8	世界調和システムにおける分類例	471
附属書9	水生環境有害性に関する手引き	479
附属書10	水性媒体中の金属および金属化合物の変化/溶解に関する手引き	561
附属書11	分類に結びつかない他の危険有害性に関する手引き	573

PART 1

INTRODUCTION

第1部

序

第１集

中

CHAPTER 1.1

PURPOSE, SCOPE AND APPLICATION OF THE GLOBALLY HARMONIZED SYSTEM OF CLASSIFICATION AND LABELLING OF CHEMICALS (GHS)

1.1.1 Purpose

1.1.1.1 The use of chemicals to enhance and improve life is a widespread practice worldwide. But alongside the benefits of these products, there is also the potential for adverse effects to people or the environment. As a result, a number of countries or organizations have developed laws or regulations over the years that require information to be prepared and transmitted to those using chemicals, through labels or safety data sheets (SDS). Given the large number of chemicals available, individual regulation of all of them is simply not possible for any entity. Provision of information gives those using chemicals the identities and hazards of these chemicals and allows the appropriate protective measures to be implemented in the local use settings.

1.1.1.2 While these existing laws or regulations are similar in many respects, their differences are significant enough to result in different labels or SDS for the same chemical in different countries. Through variations in definitions of hazards, a chemical may be considered flammable in one country, but not another. Or it may be considered to cause cancer in one country, but not another. Decisions on when or how to communicate hazards on a label or SDS thus vary around the world, and companies wishing to be involved in international trade must have large staffs of experts who can follow the changes in these laws and regulations and prepare different labels and SDS. In addition, given the complexity of developing and maintaining a comprehensive system for classifying and labelling chemicals, many countries have no system at all.

1.1.1.3 Given the reality of the extensive global trade in chemicals, and the need to develop national programs to ensure their safe use, transport, and disposal, it was recognized that an internationally-harmonized approach to classification and labelling would provide the foundation for such programs. Once countries have consistent and appropriate information on the chemicals they import or produce in their own countries, the infrastructure to control chemical exposures and protect people and the environment can be established in a comprehensive manner.

1.1.1.4 Thus the reasons for setting the objective of harmonization were many. It is anticipated that, when implemented, the GHS will:

(a) enhance the protection of human health and the environment by providing an internationally comprehensible system for hazard communication;

(b) provide a recognized framework for those countries without an existing system;

(c) reduce the need for testing and evaluation of chemicals; and

(d) facilitate international trade in chemicals whose hazards have been properly assessed and identified on an international basis.

1.1.1.5 The work began with examination of existing systems, and determination of the scope of the work. While many countries had some requirements, the following systems were deemed to be the "major" existing systems and were used as the primary basis for the elaboration of the GHS:

(a) Requirements of systems in the United States of America for the workplace, consumers and pesticides;

(b) Requirements of Canada for the workplace, consumers and pesticides;

(c) European Union directives for classification and labelling of substances and preparations;

(d) The *UN Model Regulations*.

第 1.1 章

化学品の分類および表示に関する
世界調和システム（GHS）の目的、範囲および適用

1.1.1 目的

1.1.1.1　化学品は、生活を向上させ改善するため、全世界で広く利用されている。しかし、こうした製品はその利点に加え、人や環境に対して悪影響をもたらす可能性がある。その結果、数多くの国々または機関は、近年、ラベルや安全データシート（SDS）を通じて化学品を使用する側に向けた情報の作成と伝達を求める法律や規則を定めるにいたっている。利用可能な化学品の膨大さを考えれば、そのすべてについて個々に規制することはいずれの機関にとっても不可能である。情報提供により、化学品の利用者は個々の化学品を特定してその危険有害性を知り、各地域の状況に応じた適正な防護対策を実施することができる。

1.1.1.2　こうした既存の法律または規則は多くの点で相互に似ているものの、その相異もまた大きいため、結果として同一化学品に対するラベルまたは SDS が国ごとに異なっている。危険有害性の定義が様々なために、ある化学品がある国では可燃性物質とみなされ、他の国ではそうならないことがある。また、ある国では発がん物質とみなされても、他の国ではそうでないかもしれない。ラベルまたは SDS についてどの段階で、どのように情報提供を行うかに関する決定は世界中で異なり、国際貿易を行おうとする企業は、そうした法律および規則に関する相異に対応し、様々なラベルおよび SDS を作成できる大規模な専門家集団を抱えなければならない。さらに、化学品の分類と表示のための包括的なシステムを開発し、維持することは面倒であるために、多くの国々にそのようなシステムはない。

1.1.1.3　化学品の国際貿易が広く行われているという現実、およびその安全な使用、輸送、廃棄を確実に行うための国内計画策定の必要性を考慮すると、国際的に調和された分類および表示方法がそうした計画の基礎となるであろうとの認識がなされた。国内に輸入されたり、または国内で生産される化学品に関して、各国が一貫性のある適切な情報を得られれば、化学品へのばく露を管理し、人々と環境を保護するための基盤を包括的に確立することができる。

1.1.1.4　このように、世界調和を目標に定める理由は数多くある。GHS の実施により以下の点が期待される：

(a) 危険有害性の情報伝達に関する国際的に理解されやすいシステムの導入によって、人の健康と環境の保護が強化される；

(b) 既存のシステムを持たない国々に対し国際的に承認された枠組みが提供される；

(c) 化学品の試験および評価の必要性が減少する；さらに、

(d) 危険有害性が国際的に適正に評価され確認された化学品の国際取引が促進される。

1.1.1.5　作業は、既存システムの検討と、作業の範囲を定めることから始められた。多くの国々が一定の要求事項を設けていたが、中でも以下のシステムが既存の「主要」システムであるとみなされ、GHS 策定の基礎となった：

(a) 米国における作業場、消費者および農薬に関する制度の要件；

(b) カナダにおける作業場、消費者および農薬に関する制度の要件；

(c) 物質および混合物の分類および表示のための EU 指令；

(d) *UN モデル規則*。

1.1.1.6   The requirements of other countries were also examined as the work developed, but the primary task was to find ways to adopt the best aspects of these existing systems and develop a harmonized approach. This work was done based on agreed principles of harmonization that were adopted early in the process:

(a) the level of protection offered to workers, consumers, the general public and the environment should not be reduced as a result of harmonizing the classification and labelling systems;

(b) the hazard classification process refers principally to the hazards arising from the intrinsic properties of substances and mixtures, whether natural or synthetic[1];

(c) harmonization means establishing a common and coherent basis for chemical hazard classification and communication, from which the appropriate elements relevant to means of transport, consumer, worker and environment protection can be selected;

(d) the scope of harmonization includes both hazard classification criteria and hazard communication tools, e.g. labelling and safety data sheets, taking into account especially the four existing systems identified in the ILO report[2];

(e) changes in all these systems will be required to achieve a single globally harmonized system; transitional measures should be included in the process of moving to the new system;

(f) the involvement of concerned international organizations of employers, workers, consumers, and other relevant organizations in the process of harmonization should be ensured;

(g) the comprehension of chemical hazard information, by the target audience, e.g. workers, consumers and the general public should be addressed;

(h) validated data already generated for the classification of chemicals under the existing systems should be accepted when reclassifying these chemicals under the harmonized system;

(i) a new harmonized classification system may require adaptation of existing methods for testing of chemicals;

(j) in relation to chemical hazard communication, the safety and health of workers, consumers and the public in general, as well as the protection of the environment, should be ensured while protecting confidential business information, as prescribed by the competent authorities.

1.1.2   Scope

1.1.2.1   The GHS includes the following elements:

(a) harmonized criteria for classifying substances and mixtures according to their health, environmental and physical hazards; and

(b) harmonized hazard communication elements, including requirements for labelling and safety data sheets.

1.1.2.2   This document describes the classification criteria and the hazard communication elements by type of hazard (e.g. acute toxicity; flammability). In addition, decision logics for each hazard have been developed. Some examples of classification of chemicals in the text, as well as in annex 8, illustrate how to apply the criteria. There is also some discussion about issues that were raised during the development of the system where additional guidance was thought to be necessary to implement the system.

1.1.2.3   The scope of the GHS is based on the mandate from the 1992 United Nations Conference on Environment and Development (UNCED) for development of such a system as stated in paragraphs 26 and 27 of the Agenda 21, Chapter 19, Programme Area B, reproduced below:

---

[1]   *In some cases it is necessary also to take into account hazards arising from other properties, such as the physical state of the substance or mixture (e.g. pressure and temperature) or properties of substances produced by certain chemical reactions (e.g. flammability of gases produced by contact with water).*

[2]   *1992 ILO Report on the size of the task of harmonizing existing systems of classification and labelling for hazardous chemicals.*

1.1.1.6 こうした作業を続ける中、この他の国々における要求事項についても検討が行われたが、第一の課題は、こうした既存システムの最も良い点を取り入れ、調和のとれる手法を見出すことであった。この作業は、その初期に採択し合意した以下の調和原則に基づいて行われた：

(a) 分類および表示システムを調和させることにより、労働者、消費者、一般市民および環境に対する保護レベルを低下させるべきでない；

(b) 危険有害性分類は、原則として、天然、人工の別を問わず、物質および混合物に固有の性質に由来する危険有害性について行う[1]；

(c) 調和とは、化学品の危険有害性の分類および情報の伝達を目的とした共通の一貫した基盤を確立することを意味し、この中から輸送手段、消費者、労働者および環境の保護の点から該当する要素を選択できるようにする；

(d) 調和の対象範囲は、危険有害性の分類の基準と危険有害性に関する情報の伝達手段（表示および安全データシート等）の双方を含んでおり、特に ILO の報告書において認められた 4 つの既存システムを考慮にいれる[2]；

(e) 世界的に調和のとれた単一のシステムを導入するには、すべての既存システムで変更の必要が生じるであろう。したがって、新システムへの移行過程には暫定措置を設けるべきである；

(f) 調和の過程においては、雇用者、労働者および消費者に関係する国際機関、ならびにその他関係機関の参加を確保するべきである；

(g) 化学品の危険有害性に関する情報は、対象となる労働者、消費者および一般市民等に理解されやすいものとなるよう配慮するべきである；

(h) 調和された新たなシステムの下で再分類を行う場合には、既存のシステムの下で化学品の分類のために既に得られた有効なデータを受け入れるべきである；

(i) 調和された新たな分類システムは、化学品の試験のために既存の方法の採用を求めてもよい。

(j) 化学品の危険有害性に関する情報の伝達にあたっては、労働者、消費者および一般市民の健康と安全ならびに環境保護を図ると同様に、所管官庁の定めに従って、企業の営業秘密情報の保護を保証するべきである。

1.1.2 範囲

1.1.2.1 GHS は以下の項目を含む：

(a) 物質および混合物を、健康、環境、および物理化学的危険有害性に応じて分類するために調和された判定基準；および

(b) 表示および安全データシートの要求事項を含む、調和された危険有害性に関する情報の伝達に関する事項。

1.1.2.2 本文書は、危険有害性の種類（例えば急性毒性や引火性）別に分類基準および危険有害性に関する情報の伝達に関する事項を記載している。また、各危険有害性についての判定の手順を策定した。判定基準の適用方法を説明する目的で、化学品の分類例を本文および附属書 8 に示した。さらに、GHS の策定段階で、その実施のために追加の指針が必要と考えられる部分について提起された問題もある。

1.1.2.3 GHS の対象とする範囲は、次に示す 1992 年の国連環境開発会議（UNCED）のアジェンダ 21 第 19 章プログラム分野 B の 26、27 項に記されている、当該システムの開発に向けた指示事項に基づくものである。

---

[1] 物質あるいは混合物の物理的状態（例えば圧力や温度）またはある種の化学反応（例えば、水との接触により可燃性ガスを発生する）により生じる物質の性質に起因する危険性を考慮する必要がある場合もある。

[2] 1992 年の「危険有害化学品の分類および表示の既存システム間における調和作業の規模に関する *ILO 報告書*」

> *"26. Globally harmonized hazard classification and labelling systems are not yet available to promote the safe use of chemicals, inter alia, at the workplace or in the home. Classification of chemicals can be made for different purposes and is a particularly important tool in establishing labelling systems. There is a need to develop harmonized hazard classification and labelling systems, building on ongoing work;*
>
> *27. A globally harmonized hazard classification and compatible labelling system, including material safety data sheets and easily understandable symbols, should be available, if feasible, by the year 2000."*

1.1.2.4 This mandate was later analysed and refined in the harmonization process to identify the parameters of the GHS. As a result, the following clarification was adopted by the Interorganization Programme for the Sound Management of Chemicals (IOMC) Coordinating Group to ensure that participants were aware of the scope of the effort:

> *"The work on harmonization of hazard classification and labelling focuses on a harmonized system for all chemicals, and mixtures of chemicals. The application of the components of the system may vary by type of product or stage of the life cycle. Once a chemical is classified, the likelihood of adverse effects may be considered in deciding what informational or other steps should be taken for a given product or use setting. Pharmaceuticals, food additives, cosmetics, and pesticide residues in food will not be covered by the GHS in terms of labelling at the point of intentional intake. However, these types of chemicals would be covered where workers may be exposed, and, in transport if potential exposure warrants. The Coordinating Group for the Harmonization of Chemical Classification Systems (CG/HCCS) recognizes that further discussion will be required to address specific application issues for some product use categories which may require the use of specialized expertise."*[3]

1.1.2.5 In developing this clarification, the CG/HCCS carefully considered many different issues with regard to the possible application of the GHS. There were concerns raised about whether certain sectors or products should be exempted, for example, or about whether or not the system would be applied at all stages of the life cycle of a chemical. Three parameters were agreed in this discussion and are critical to application of the system in a country or region. These are described below:

(a) Parameter 1:	**The GHS covers all hazardous chemicals. The mode of application of the hazard communication elements of the GHS (e.g. labels, safety data sheets) may vary by product category or stage in the life cycle. Target audiences for the GHS include consumers, workers, transport workers, and emergency responders.**

(i) Existing hazard classification and labelling systems address potential exposures to all potentially hazardous chemicals in all types of use situations, including production, storage, transport, workplace use, consumer use, and presence in the environment. They are intended to protect people, facilities, and the environment. The most widely applied requirements in terms of chemicals covered are generally found in the parts of existing systems that apply to the workplace or transport. It should be noted that the term chemical is used broadly in the UNCED agreements and subsequent documents to include substances, products, mixtures, preparations, or any other terms that may be used in existing systems to denote coverage.

(ii) Since all chemicals in commerce are made in a workplace (including consumer products), handled during shipment and transport by workers, and often used by workers, there are no complete exemptions from the scope of the GHS for any particular type of chemical or product. In some countries, for example, pharmaceuticals are currently covered by workplace and transport requirements in the manufacturing, storage, and transport stages of the life cycle. Workplace requirements may also be applied to employees involved in the administration of some drugs, or clean-up of spills and other types of potential exposures in health care settings. SDS's and training must be available for these employees under some systems. It is anticipated that the GHS would be applied to pharmaceuticals in a similar fashion.

---

[3] *IOMC Description and further clarification of the anticipated application of the Globally Harmonized System (GHS), IFCS/ISG3/98.32B*

> 「26 項　現在のところ、化学品の安全な利用を促すための世界的に調和された危険有害性に関する分類および表示システムは、特に作業場および家庭においては依然として利用できない状況にある。化学品の分類は様々な目的で行われるが、表示システムの確立にあたっては特に重要なものである。したがって、現在構築中の調和された危険有害性に関する分類および表示システムを確立する必要がある；
>
> 27 項　安全データシートおよび容易に理解できるシンボルも含めた、世界的に調和された危険有害性に関する分類および表示システムを、可能であれば西暦 2000 年までに利用できるようにするべきである。」

1.1.2.4　この指示事項は調和作業の過程で検討され、さらに熟考されて、GHS に含めるべき要素が特定された。その結果、関係者がその範囲について確実に認識できるように、次のような説明が化学品の適正管理のための機関間プログラム（IOMC）調整グループ（Coordinating Group）によって採択された：

> 「危険有害性の分類および表示の調和に関する作業は、すべての化学品およびその混合物に対して調和されたシステムという点に主眼を置く。GHS の構成要素の適用は、製品の種類またはライフサイクルの段階によって異なってもよい。一旦ある化学品を分類すれば、起こりうる影響を考慮して特定の製品または利用状況において必要な情報やその他の対策を決定する事が可能になる。医薬品、食品添加物、化粧品、あるいは食物中の残留農薬は、意図的な摂取という理由からラベルの範囲とはしない。しかし、このような種類の化学品に労働者がばく露される可能性のある場所、およびばく露の可能性がある輸送の際には GHS が適用されるであろう。化学品分類システムの調和のための調整グループ（CG/HCCS）は、専門知識を必要とする一部の製品の使用に関する個別の問題については、さらなる議論が必要になることを認めている。」[3]

1.1.2.5　この内容を具体化するにあたり、CG/HCCS は GHS の適用可能性に関係する数多くの様々な問題について慎重に検討を行った。例えば、特定の部門や製品を除外すべきかどうか、あるいは GHS を化学品のライフサイクルの全段階に適用するどうか、などが関心事項となった。検討の中で 3 つの要素について合意されたが、これらの要素は各国または各地域での GHS の適用に際して非常に重要なものである。これらを以下に示す。

> (a)　要素 1：GHS はすべての危険有害な化学品に適用される。GHS の危険有害性に関する情報の伝達要素（例えばラベル、安全データシート）の適用方法は、製品の種類やライフサイクルにおける段階によって異なってもよい。GHS の対象者には、消費者、労働者、輸送担当者、緊急時対応者が含まれる。

    (i)　既存の危険有害性分類および表示システムは、生産、貯蔵、輸送、作業場での利用、消費者の利用、環境中での存在等あらゆる利用状況下において、潜在的に危険有害性を有する化学品すべてに対するばく露の可能性を想定している。これらは、人、施設、環境を保護するためのものである。化学品について最も広く適用されている要求事項は、作業場や輸送段階で適用されている既存のシステムの中に見られる。UNCED 合意およびそれに続く文書においては、化学品という語が、既存システムにおいて物質、製品、混合物、調剤、またはその他の適用範囲を示すあらゆる語を含む形で広く用いられている点に注意するべきである。

    (ii)　取引されるすべての化学品は（消費者製品を含めて）作業場で製造され、労働者の手により出荷、輸送され、また労働者によってよく利用されるため、特定の化学品や製品が GHS の適用範囲から完全に除外されることはありえない。例えばある国では、医薬品は、そのライフサイクルにおける製造、貯蔵、輸送段階で作業場と輸送に関する要件の適用を受けている。作業場における要件を、一部薬品の投与や汚染の浄化など潜在的にばく露の可能性がある医療現場における職員に適用してもよい。そうした職員に対して SDS および訓練を利用できるようにすることを義務付けているシステムもある。GHS も同じように、医薬品に適用されることが期待される。

---

[3]　*IOMC による世界調和システム（GHS）の予想される適用範囲とその明確化 IFCS/ISG3/98.32B*

(iii) At other stages of the life cycle for these same chemicals, the GHS may not be applied at all. For example, at the point of intentional human intake or ingestion, or intentional application to animals, products such as human or veterinary pharmaceuticals are generally not subject to hazard labelling under existing systems. Such requirements would not normally be applied to these products as a result of the GHS. (It should be noted that the risks to subjects associated with the medical use of human or veterinary pharmaceuticals are generally addressed in package inserts and are not part of this harmonization process.) Similarly, products such as foods that may have trace amounts of food additives or pesticides in them are not currently labelled to indicate the presence or hazard of those materials. It is anticipated that application of the GHS would not require them to be labelled as such.

(b) Parameter 2: **The mandate for development of a GHS does not include establishment of uniform test methods or promotion of further testing to address adverse health outcomes.**

(i) Tests that determine hazardous properties, which are conducted according to internationally recognized scientific principles, can be used for purposes of a hazard determination for health and environmental hazards. The GHS criteria for determining health and environmental hazards are test method neutral, allowing different approaches as long as they are scientifically sound and validated according to international procedures and criteria already referred to in existing systems for the hazard class of concern and produce mutually acceptable data. While OECD is the lead organization for development of harmonized health hazard criteria, the GHS is not tied to the OECD Test Guidelines Program. For example, drugs are tested according to agreed criteria developed under the auspices of the World Health Organization (WHO). Data generated in accordance with these tests would be acceptable under the GHS. Criteria for physical hazards under the UNSCETDG are linked to specific test methods for hazard classes such as flammability and explosivity.

(ii) The GHS is based on currently available data. Since the harmonized classification criteria are developed on the basis of existing data, compliance with these criteria will not require retesting of chemicals for which accepted test data already exists.

(c) Parameter 3: **In addition to animal data and valid in vitro testing, human experience, epidemiological data, and clinical testing provide important information that should be considered in application of the GHS.**

(i) Most of the current systems acknowledge and make use of ethically obtained human data or available human experience. Application of the GHS should not prevent the use of such data, and the GHS explicitly acknowledges the existence and use of all appropriate and relevant information concerning hazards or the likelihood of harmful effects (i.e. risk).

1.1.2.6   *Other scope limitations*

1.1.2.6.1   The GHS is not intended to harmonize risk assessment procedures or risk management decisions (such as establishment of a permissible exposure limit for employee exposure), which generally require some risk assessment in addition to hazard classification. However, information on risk management is occasionally provided in the GHS on a case-by-case basis for guidance purposes. Competent authorities are best placed to determine in regulations or standards the appropriate risk assessment procedures and risk management measures. In addition, chemical inventory requirements in various countries are not related to the GHS[3].

1.1.2.6.2   *Hazard versus risk*

1.1.2.6.2.1   Each hazard classification and communication system (workplace, consumer, transport) begins coverage with an assessment of the hazards posed by the chemical involved. The degree of its capacity to harm depends on its intrinsic properties, i.e. its capacity to interfere with normal biological processes, and its capacity to burn, explode, corrode, etc. This is based primarily on a review of the scientific studies available. The concept of risk or the likelihood of harm occurring, and subsequently communication of that information, is introduced when exposure is considered in

---

[3]   *IOMC Description and further clarification of the anticipated application of the Globally Harmonized System (GHS),* IFCS/ISC3/98.32B.

(iii) 同じ化学品のライフサイクルにおいても、段階によっては、GHS がまったく適用されない場合もある。例えば、一般に既存システムでは、ヒトまたは動物用の医薬品のような製品には、ヒトが意図的に摂取する、または動物に対して意図的に投与する時点において、危険有害性に関する表示義務はない。通常これらの製品に GHS のための表示の要件が適用されることはないであろう。(ヒトまたは動物用医薬品を医療において使用する者に対する危険性については、一般に包装内の説明書による対応がなされており、これは調和とは関係ないということに注意するべきである。) 同様に、微量の食品添加物または農薬を含む可能性のある食品等の製品は、現在そうした物質の存在または危険有害性を示す表示がなされていない。これらの製品に GHS の適用による表示を義務付けることにはならないであろう。

(b) 要素2：GHS の指示事項には、健康に対する悪影響に対応するための統一的な試験方法の確立または追加試験を促す項目は含まれていない。

(i) 危険有害性を特定するための、国際的に認められた科学的原則に従って実施される試験は、健康および環境に対する有害性の特定に利用できる。健康および環境に対する有害性を特定するための GHS の判定基準は、中立的な評価方法である。すなわち、既存システムで既に参照されている国際的な手順および判定基準に従って有効性が確認され、相互に受け入れ可能なデータが得られている限り、それらの方法も受け入れる。調和された健康有害性の判定基準に関しては OECD が主導的な組織となっているが、GHS は OECD のテストガイドラインプログラムに連動するものではない。例えば、医薬品は世界保健機関（WHO）の支援により策定され、合意された判定基準に従って試験されている。こうした試験によって作成されたデータは、GHS の下でも受け入れられるものである。UNSCETDG の物理化学的な危険性の判定基準は、引火性や爆発性といった危険性の種類により決められた方法に連動するものである。

(ii) GHS は現時点で利用可能なデータに基づく。調和された分類基準は既存データに基づいて策定されており、既に認められた試験データがある化学品については、この基準を満足させるための再試験は必要ない。

(c) 要素3：GHS の適用にあたっては、動物試験データおよび有効な in vitro 試験に加え、重要な情報を提供するヒトによる経験、疫学データ、臨床試験も考慮するべきである。

(i) 現在のシステムの大半は、倫理的に問題なく得られたヒトのデータまたは利用可能なヒトによる経験を認め、利用している。GHS の適用に際してもこうしたデータの利用を妨げるべきでなく、また GHS は、危険有害性または有害な影響の可能性（すなわちリスク）に関係した、すべての該当する適切な情報の存在とこれの利用を認める。

1.1.2.6 適用範囲に関するその他の制約

1.1.2.6.1 GHS は、一般に危険有害性分類に加えて一定のリスク評価を要するような、リスク評価手続またはリスクマネジメントに係る決定（作業者に対するばく露許容限度の設定等）の調和を図ることを意図するものではない。しかしながら、リスクマネジメントに関する情報が、手引きを目的としてケースバイケースで、GHS において提供されることもある。所管官庁は、規制または基準のなかで適切なリスクアセスメントの手順やリスクマネジメントの対策を決定する最適の立場にある。さらに各国の化学品インベントリーに係る要求事項も GHS に関係するものではない[3]。

1.1.2.6.2 危険有害性とリスク

1.1.2.6.2.1 各危険有害性の分類および情報の伝達システム（作業場、消費者、輸送）では、まず関連する化学品がもたらす危険有害性の評価を行う。危害を与える能力の程度は、固有の性質、すなわち正常な生物学的活動を妨げる能力および燃焼、爆発、腐食などの能力に依存する。これらの能力は、主として利用

---

[3] IOMC の記述および世界調和システム（GHS）の予想される適用範囲とその明確化、IFCS/ISG3/98.32B

conjunction with the data regarding potential hazards. The basic approach to risk assessment is characterized by the simple formula:

$$\text{hazard} \times \text{exposure} = \text{risk}$$

1.1.2.6.2.2   Thus if you can minimize either hazard or exposure, you minimize the risk or likelihood of harm. Successful hazard communication alerts the user to the presence of a hazard and the need to minimize exposures and the resulting risks.

1.1.2.6.2.3   All of the systems for conveying information (workplace, consumer, transport) include both hazard and risk in some form. They vary in where and how they provide the information, and the level of detail they have regarding potential exposures. For example, exposure of the consumer to pharmaceuticals comprises a specific dose that is prescribed by the physician to address a certain condition. The exposure is intentional. Therefore, a determination has been made by a drug regulatory agency that for the consumer, an acceptable level of risk accompanies the specific dosage provided. Information that is provided to the person taking the pharmaceutical conveys the risks assessed by the drug regulatory agency rather than addressing the intrinsic hazards of the pharmaceutical or its components.

1.1.3   Application of the GHS

1.1.3.1   Harmonization of the application of the GHS

1.1.3.1.1   The goal of the GHS is to identify the intrinsic hazards found in substances and mixtures and to convey hazard information about these hazards. The criteria for hazard classification are harmonized. Hazard statements, symbols and signal words have been standardized and harmonized and now form an integrated hazard communication system. The GHS will allow the hazard communication elements of the existing systems to converge. Competent authorities will decide how to apply the various elements of the GHS based on the needs of the competent authority and the target audience. (See also chapter 1.4, *Hazard Communication: Labelling,* (paragraph 1.4.10.5.4.2) and annex 5 *Consumer Product Labelling Based on the Likelihood of Injury)*.

1.1.3.1.2   For transport, it is expected that application of the GHS will be similar to application of current transport requirements. Containers of dangerous goods will be marked with pictograms that address acute toxicity, physical hazards, and environmental hazards. As is true for workers in other sectors, workers in the transport sector will be trained. The elements of the GHS that address such elements as signal words and hazard statements are not expected to be adopted in the transport sector.

1.1.3.1.3   In the workplace, it is expected that all of the GHS elements will be adopted, including labels that have the harmonized core information under the GHS, and safety data sheets. It is also anticipated that this will be supplemented by employee training to help ensure effective communication.

1.1.3.1.4   For the consumer sector, it is expected that labels will be the primary focus of GHS application. These labels will include the core elements of the GHS, subject to some sector-specific considerations in certain systems. (See also chapter 1.4 *Hazard Communication: Labelling* (paragraph 1.4.10.5.4.2) and annex 5 *Consumer Product Labelling Based on the Likelihood of Injury)*.

1.1.3.1.5   *Building block approach*

1.1.3.1.5.1   Consistent with the building block approach, countries are free to determine which of the *building blocks* will be applied in different parts of their systems. However, where a system covers something that is in the GHS, and implements the GHS, that coverage should be consistent. For example, if a system covers the carcinogenicity of a chemical, it should follow the harmonized classification scheme and the harmonized label elements.

1.1.3.1.5.2   In examining the requirements of existing systems, it was noted that coverage of hazards may vary by the perceived needs of the target audience for information. In particular, the transport sector focuses on acute health effects and physical hazards but has not to date covered chronic effects due to the types of exposures expected to be encountered in that setting. But there may be other differences as well, with countries choosing not to cover all of the effects addressed by the GHS in each use setting.

1.1.3.1.5.3   The harmonized elements of the GHS may thus be seen as a collection of building blocks from which to form a regulatory approach. While the full range is available to everyone and should be used if a country or organization chooses to cover a certain effect when it adopts the GHS, the full range does not have to be adopted. While physical hazards are important in the workplace and transport sectors, consumers may not need to know some of the specific physical hazards in the type of use they have for a product. As long as the hazards covered by a sector or system are

可能な科学的研究結果についての文献調査に基づく。ばく露が潜在的危険有害性に関するデータと関連づけられた時、リスクの概念すなわち危害が生じる可能性およびこれらの情報伝達が導入される。リスク評価の基本的アプローチは、以下の公式で定義される。

$$危険有害性 \times ばく露 = リスク$$

1.1.2.6.2.2　したがって、危険有害性またはばく露を最小にすることができれば、リスクすなわち危害の可能性は最小となる。適切な危険有害性に関する情報の伝達により、使用者は危険有害性の存在およびばく露とその結果生じるリスクを最小にする必要性に対して、注意を喚起される。

1.1.2.6.2.3　すべての情報伝達のためのシステム（作業場、消費者、輸送）には、何らかの形式での危険有害性とリスクの双方が含まれる。これらは情報提供を行うべき場所と方法、そしてばく露可能性の程度によって異なる。例えば、医薬品に対する消費者のばく露の程度は、ある状況に対処するために医師が処方する投与量によって決まる。ばく露は意図的である。したがって医薬品管理機関は、消費者にとって受容可能なレベルのリスクで医薬品の投与量を定めている。医薬品の投与を受ける人に提供される情報は、医薬品やその成分に固有の有害性ではなく、そうした医薬品管理機関が評価したリスクを伝える。

1.1.3　GHS の適用

1.1.3.1　*GHS 適用方法の調和*

1.1.3.1.1　GHS の目的は、物質および混合物に固有な危険有害性を特定し、そうした危険有害性に関する情報を伝えることである。危険有害性の分類に関する判定基準が調和され、危険有害性情報、シンボルや注意喚起語が標準化・調和されて、危険有害性に関して統合された情報伝達の仕組みとなった。GHS は既存システムの危険有害性に関する情報の項目をまとめることになるであろう。所管官庁は、各関連所管官庁と対象者のニーズに基づいて GHS の様々な要素を適用する方法を決定するであろう。(1.4 章 *危険有害性に関する情報の伝達：表示*（1.4.10.5.4.2）および附属書 5 *危害の可能性に基づく消費者製品の表示*を参照。)

1.1.3.1.2　輸送については、GHS の適用は現行の輸送に係る要求事項と同様になると予想される。危険物の容器には急性毒性、物理化学的危険性、環境有害性を示した絵表示が記載されるであろう。他の部門の労働者と同様、輸送部門の労働者も訓練が必要であろう。注意喚起語や危険有害性情報などの GHS の要素は、輸送部門には採用されないと予想される。

1.1.3.1.3　作業場においては、GHS で調和された必須な情報についての表示および安全データシートを含むすべての GHS の要素が採用されるものと期待される。また、有効な情報伝達を確実に行うために従業員の訓練を行うことが期待される。

1.1.3.1.4　消費者部門については、表示が GHS の中心となるであろう。これらのラベルでは、部門に特異な点も考慮した上で GHS に必須な要素を含むことになるであろう。(1.4 章 *危険有害性に関する情報の伝達：表示*（1.4.10.5.4.2）および附属書 5 *危害の可能性に基づく消費者製品の表示*を参照。)

1.1.3.1.5　選択可能方式（Building block approach）

1.1.3.1.5.1　選択可能方式によって、各国はそれぞれのシステムにどのような*部分*を当てはめるかを自由に決めることができる。しかし、あるシステムが GHS の一部を含み、かつそのシステムにより GHS を実施する場合には、その適用範囲には一貫性を持たせるべきである。例えば、あるシステムが化学品の発がん性を対象にするならば、調和された分類体系と表示項目に従うべきである。

1.1.3.1.5.2　既存のシステムの要求事項について調査したところ、危険有害性の範囲が、対象者の情報に対するニーズによって異なることが指摘された。特に、輸送部門では急性の健康影響と物理化学的危険性に重点を置いているが、輸送で起こりうるばく露の形態を考慮し、まだ慢性影響については扱っていない。また、GHS が扱う影響のすべてには対応しないという選択を行った国々においては、それぞれの部門でこの他にも相違は存在するであろう。

1.1.3.1.5.3　このように、GHS において調和された要素群は、規制方法を形成する単位の集合体と見なすことができる。誰でも GHS 全体を利用することが可能であるが、GHS を導入する国や組織がある影響のみに対処する目的でこれを利用する場合には、その全体を採り入れる必要はない。物理化学的危険性は作業場や輸送部門において重要であるが、消費者はその製品の使い方によっては物理化学的危険性について

covered consistently with the GHS criteria and requirements, it will be considered appropriate implementation of the GHS. Notwithstanding the fact that an exporter needs to comply with importing countries' requirements for GHS implementation, it is hoped that the application of the GHS worldwide will eventually lead to a fully harmonized situation.

1.1.3.1.5.4 Guidance on the interpretation of the building block approach

(a) Hazard classes are building blocks:

Within their jurisdiction and keeping in mind the goal of full harmonization as well as international conventions, competent authorities may decide which hazard classes they apply;

(b) Within a hazard class, each hazard category can be seen as a building block:

For a given hazard class, competent authorities have the possibility not to apply all categories. Nevertheless, in order to preserve consistency, some restrictions to this principle should be set, as follows:

(i) The classification criteria such as the cut-off values or concentration limits for adopted hazard categories should not be altered. However, adjacent sub-categories (e.g. carcinogenicity Categories 1A and 1B) may be merged into one category. Nevertheless, adjacent hazard categories should not be merged if it results in renumbering the remaining hazard categories. Furthermore, where sub-categories are merged, the names or numbers of the original GHS sub-categories should be retained (e.g. carcinogenicity Category 1 or 1A/B) to facilitate hazard communication;

(ii) Where a competent authority adopts a hazard category, it should also adopt all the categories for higher hazard levels in that class. As a consequence, when a competent authority adopts a hazard class, it will always adopt at least the highest hazard category (Category 1), and, where more than one hazard category is adopted, these hazard categories will form an unbroken sequence.

NOTE 1: *Some hazard classes contain additional categories that can be considered on a stand-alone basis, for example, Category 3 "transient target organ effects" for the hazard class "Specific target organ toxicity" (chapter 3.8), and hazard category "Effects on or via lactation" for the hazard class "reproductive toxicity" (chapter 3.7).*

NOTE 2: *It is noted, however, that the goal of the GHS is to achieve worldwide harmonization (see 1.1.2.3). Therefore, while differences between sectors may persist, the use of an identical set of categories at a worldwide level within each sector should be encouraged.*

1.1.3.2 *Implementation and maintenance of the GHS*

1.1.3.2.1 For the purposes of implementing the GHS, the United Nations Economic and Social Council (ECOSOC) reconfigured the UN Committee of Experts on the Transport of Dangerous Goods by resolution 1999/65 of 26 October 1999. The new Committee of Experts on the Transport of Dangerous Goods and the Globally Harmonized System of Classification and Labelling of Chemicals (UNCETDG/GHS), maintains its Sub-Committee of Experts on the Transport of Dangerous Goods (UNSCETDG) and a new subsidiary body, the Sub-Committee of Experts on the Globally Harmonized System of Classification and Labelling of Chemicals (UNSCEGHS), has been created. The UNSCEGHS has the following functions:

(a) To act as custodian of the GHS, managing and giving direction to the harmonization process;

(b) To keep the GHS system up to date as necessary, considering the need to introduce changes, ensure its continued relevance and practical utility, and determining the need for and timing of the updating of technical criteria, working with existing bodies as appropriate;

(c) To promote understanding and use of the GHS and to encourage feedback;

(d) To make the GHS available for worldwide use and application;

知る必要はないであろう。ある部門またはシステムが対象とする危険有害性について、GHSの判定基準および要求事項と矛盾することがない限り、それはGHSの適切な実施とみなされる。輸出者が輸入国のGHS実施のための要求事項を遵守する必要があるという事実があったとしても、最終的には世界的なGHSの適用により、完全に調和された状況になることが望まれる。

1.1.3.1.5.4　選択可能方式の解釈ガイダンス

(a) 危険有害性クラスは選択可能：

国際的な協約と同様に、完全に調和することを念頭に、所管官庁はそれぞれの法規のなかで、どの危険有害性クラスを適用するかを決めることができる；

(b) ある危険有害性クラスのなかで、それぞれの区分は選択可能としてもよい：

ある危険有害性クラスに対して、所管官庁が必ずしも全ての区分を適用しないこともあろう。しかしながら一貫性を維持するためには以下のようないくつかの制限が必要である：

(i) 適用する危険有害性区分のカットオフ値や濃度限界のような分類基準をかえるべきではない。しかし隣同士の細区分（例、発がん性の区分1Aと1B）は1つの区分にすることも可能であろう。しかしながら、残りの危険有害性区分の番号を変更せざるを得ないような区分の統合はすべきではない。さらに、細区分を統一した場合、危険有害性情報の伝達を容易にするために、もとのGHS細区分の名前や番号は保持するべきである（例、発がん性区分1あるいは1A/B）；

(ii) 所管官庁がある危険有害性の区分を適用する場合、その危険有害性クラスにおける他のすべてのより危険性の高い区分も採用しなければならない。したがって所管官庁がある危険有害性を採用するときは、常に少なくとも最も危険有害性の高い（区分1）区分を採用することになり、さらに1つ以上の危険有害性区分を採用する場合には、これらの区分は分断のない一続きのものとなろう。

注記1：*いくつかの危険有害性クラスは、独立したものと考えてもよい付加的な区分、例えば、「特定標的臓器毒性」（3.8章）における区分3「一過性の標的臓器への影響」、および「生殖毒性」（3.7章）における区分「授乳に対する、または授乳を介した影響」を含んでいる。*

注記2：*GHSの最終目標は世界的な調和を成し遂げるということである（1.1.2.3参照）。したがって分野間での相違が続くとしても、それぞれの分野で世界的に同一の区分の使用が促進されるべきである。*

1.1.3.2　*GHSの実施と維持*

1.1.3.2.1　GHSの実施を目的として、国連経済社会理事会（ECOSOC）は1999年10月26日付の決議1999/65に基づき、危険物輸送に関する専門家委員会を再編した。これにより、「危険物輸送ならびに化学品の分類および表示に関する世界調和システムに関する専門家委員会（UNCETDG/GHS）」が新設され、従来からの「危険物輸送に関する専門家小委員会（UNSCETDG）」と新たに設けられた「化学品の分類および表示に関する世界調和システムに関する専門家小委員会（UNSCEGHS）」は、その下部組織となった。UNSCEGHSの役割は以下のとおりである：

(a) GHSの管理機関として活動し、調和の手続に関する管理を行い、方向性を与える；

(b) 変更を行う必要性を考慮し、GHSの継続性と実践での有用性を確保し、技術基準の更新に対する必要性およびその時期を決定し、担当する機関と協力しながらGHSシステムを最新のものにする；

(c) GHSの理解と利用を促進し、フィードバックを促す；

(d) GHSを世界的に利用、適用できるようにする；

(e) To make guidance available on the application of the GHS, and on the interpretation and use of technical criteria to support consistency of application; and

(f) To prepare work programmes and submit recommendations to the committee.

1.1.3.2.2　　The UNSCEGHS and the UNSCETDG, both operate under the parent committee with responsibility for these two areas. The Committee is responsible for strategic issues rather than technical issues. It is not envisaged that it would review, change or revisit technical recommendations of the sub-committees. Accordingly, its main functions are:

(a) To approve the work programmes for the sub-committees in the light of available resources;

(b) To coordinate strategic and policy directions in areas of shared interests and overlap;

(c) To give formal endorsement to the recommendations of the sub-committees and provide the mechanism for channelling these to ECOSOC; and

(d) To facilitate and coordinate the smooth running of the sub-committees.

1.1.4　　The GHS document

1.1.4.1　　This document describes the GHS. It contains harmonized classification criteria and hazard communication elements. In addition, guidance is included in the document to assist countries and organizations in the development of tools for implementation of the GHS. The GHS is designed to permit self-classification. The provisions for implementation of the GHS allow the uniform development of national policies, while remaining flexible enough to accommodate any special requirements that might have to be met. Furthermore, the GHS is intended to create user-friendly approach, to facilitate the work of enforcement bodies and to reduce the administrative burden.

1.1.4.2　　While this document provides the primary basis for the description of the GHS, it is anticipated that technical assistance tools will be made available as well to assist and promote implementation.

(e) GHS の適用に関する指針および適用における一貫性を確保するための技術基準の解釈と利用に関する指針を策定する；そして

(f) 作業計画を準備し、委員会に勧告書を提出する。

1.1.3.2.2　UNSCEGHS と UNSCETDG の 2 つの小委員会は、ともに親委員会の下で 2 部門について責任をもって活動を行う。親委員会は、技術的な問題よりも戦略的な問題について責任を有する。親委員会は、小委員会の技術面での勧告について検討し、変更または再審査を行うことは目的としていない。したがって、その主たる機能は以下のとおりである：

(a) 利用可能な資源に照らして、小委員会の作業計画を承認する；

(b) 利害が共通する分野および重複する分野において戦略および政策方針を調整する；

(c) 小委員会の勧告に正式な承認を与え、それらを ECOSOC に伝える役割を果たす；そして

(d) 各小委員会の円滑な運営を促進し、調整を行う。

1.1.4　GHS 文書

1.1.4.1　本文書は GHS について解説している。ここには調和のとれた分類基準と危険有害性に関する情報の伝達の要素が含まれる。加えて、指針には、GHS を実施するためのツールを開発する国や機関を支援する文書が含まれている。GHS は、自主的な分類ができるように策定されている。GHS 実施のための規定は、個々の国の国家政策の統一的な発展を可能にする一方で、遵守を求められるいかなる要求事項にも適応できるよう十分な柔軟性も保持している。さらに、GHS は、利用者にとって使いやすいものであると同時に、行政機関の活動を円滑化し、かつ行政上の負担を軽減することを目指している。

1.1.4.2　本文書は GHS についての基本的な事項を規定しているが、技術的な支援ツールとして利用され、実施を支援、促進することも期待されている。

CHAPTER 1.2

DEFINITIONS AND ABBREVIATIONS

This chapter provides definitions and abbreviations of general applicability that are used in the GHS. Additional definitions of the individual hazard classes are presented in the relevant chapters.

For the purposes of the GHS:

ADN means the "European Agreement concerning the International Transport of Dangerous Goods by Inland Waterways", as amended;

ADR means the "Agreement concerning the International Carriage of Dangerous Goods by Road", as amended;

Alloy means a metallic material, homogeneous on a macroscopic scale, consisting of two or more elements so combined that they cannot be readily separated by mechanical means. Alloys are considered to be mixtures for the purpose of classification under the GHS;

Aspiration means the entry of a liquid or solid chemical into the trachea and lower respiratory system directly through the oral or nasal cavity, or indirectly from vomiting;

ASTM means the "American Society of Testing and Materials";

BCF means "bioconcentration factor";

BOD/COD means "biochemical oxygen demand/chemical oxygen demand";

Carcinogen means a substance or a mixture which induce cancer or increase its incidence;

CAS means "Chemical Abstract Service";

Chemical identity means a name that will uniquely identify a chemical. This can be a name that is in accordance with the nomenclature systems of the International Union of Pure and Applied Chemistry (IUPAC) or the Chemical Abstracts Service (CAS), or a technical name;

Competent authority means any national body(ies) or authority(ies) designated or otherwise recognized as such in connection with the Globally Harmonized System of Classification and Labelling of Chemicals (GHS);

Critical temperature means the temperature above which a pure gas cannot be liquefied, regardless of the degree of compression;

Defined approach means an approach to testing and assessment that consists of a fixed data interpretation procedure used to interpret data generated with a defined set of information sources, that can either be used on its own, or together with other information sources within an overall weight of evidence, to satisfy a specific regulatory need;

Dust means solid particles of a substance or mixture suspended in a gas (usually air);

EC$_{50}$ means the effective concentration of substance that causes 50 % of the maximum response;

EC Number or (ECN) is a reference number used by the European Communities to identify dangerous substances, in particular those registered under EINECS;

ECOSOC means the Economic and Social Council of the United Nations;

ECx means the concentration associated with x % response;

EGC Code means the Code for Existing Ships Carrying Liquefied Gases in Bulk;

EINECS means "European Inventory of Existing Commercial Chemical Substances";

第 1.2 章

定義および略語

本章では GHS で使用される定義及び略語の一般的な適用性について記載している。それぞれの危険有害性クラスに関する追加的な定義については該当する章に示されている。

GHS の目的において：

ADN とは、「内陸水路による危険物の国際輸送に関する欧州協定」（European Agreement concerning the International Transport of Dangerous Goods by Inland Waterways）改訂版をいう。

ADR とは、道路での危険物の国際輸送に関する欧州協定（European Agreement concerning the International Carriage of Dangerous Goods by Road）改訂版をいう。

合金（Alloy）とは、機械的手段で容易に分離できないように結合した 2 つ以上の元素から成る巨視的にみて均質な金属体をいう。合金は、GHS による分類では混合物とみなされる。

誤えん（aspiration）とは、液体または固体の化学品が口または鼻腔から直接、または嘔吐によって間接的に、気管および下気道へ侵入することをいう。

ASTM とは、「米国材料試験協会」（American Society of Testing and Material）をいう。

BCF とは、「生物濃縮係数」（bioconcentration factor）をいう。

BOD/COD とは、「生物化学的酸素要求量/化学的酸素要求量」（biochemical oxygen demand/chemical oxygen demand）をいう。

発がん性物質（Carcinogen）とは、がんを誘発し、またはその発生頻度を増大させる物質または混合物をいう。

CAS とは、「ケミカル・アブストラクツ・サービス」（Chemical Abstract Service）をいう。

化学的特定名（Chemical identity）とは、化学品を一義的に識別する名称をいう。これは、国際純正応用化学連合（IUPAC）またはケミカル・アブストラクツ・サービス（CAS)の命名法に従う名称、あるいは専門名を用いることができる。

所管官庁（Competent authority）とは、化学品の分類および表示に関する世界調和システム（GHS)に関連して、所管機関として指定または認定された国家機関、またはその他の機関をいう。

臨界温度（Critical temperature）とは、その温度を超えると圧縮の程度に関係なく、純粋なガスを液化できない温度をいう。

ディファインドアプローチ（Defined Approach）とは、予め定められた情報源を用いてもたらされたデータを解釈するために使われる、予め決められたデータ解釈手順による試験および評価手法をいい、具体的な規制上の必要性を満たすために、単独で、又は他の情報源とともに、包括的な証拠の重み付けの中で使用することができる。

粉塵（Dust）とは、ガス（通常空気）の中に浮遊する物質または混合物の固体の粒子をいう。

EC₅₀ とは、ある反応を最大時の 50%に減少させる物質の濃度をいう。

EC 番号または(ECN) とは、特に、EINECS に登録された危険有害物質を特定するために、欧州委員会により用いられる参照番号をいう。

ECOSOC とは、「国連経済社会理事会」（Economic and Social Council of the United Nations）をいう。

ECx とは、x%の反応を示す濃度をいう。

EGC Code とは、「液化ガスのばら積み輸送のための既存船舶コード」（Code for Existing Ships Carrying Liquefied Gases in Bulk）をいう。

EINECS とは、「欧州既存商業化学物質インベントリー」（European Inventory of Existing Commercial Chemical Substances）をいう。

***ErC*₅₀** means EC$_{50}$ in terms of reduction of growth rate;

EU means the "European Union";

Flash point means the lowest temperature (corrected to a standard pressure of 101.3 kPa) at which the application of an ignition source causes the vapours of a liquid to ignite under specified test conditions;

FAO means the "Food and Agriculture Organization of the United Nations";

Gas means a substance which (i) at 50°C has a vapour pressure greater than 300 kPa (absolute); or (ii) is completely gaseous at 20 °C at a standard pressure of 101.3 kPa;

GC Code means the "Code for the Construction and Equipment of Ships Carrying Liquefied Gases in Bulk" (Gas Carrier Code);

GESAMP means the "Joint Group of Experts on the Scientific Aspects of Marine Environmental Protection of IMO/FAO/UNESCO/WMO/WHO/IAEA/UN/UNEP";

GHS means the "Globally Harmonized System of Classification and Labelling of Chemicals";

Hazard category means the division of criteria within each hazard class, e.g. oral acute toxicity includes five hazard categories and flammable liquids includes four hazard categories. These categories compare hazard severity within a hazard class and should not be taken as a comparison of hazard categories more generally;

Hazard class means the nature of the physical, health or environmental hazard, e.g. flammable solid, carcinogen, oral acute toxicity;

Hazard statement means a statement assigned to a hazard class and category that describes the nature of the hazards of a hazardous product, including, where appropriate, the degree of hazard;

IAEA means the "International Atomic Energy Agency";

IARC means the "International Agency for the Research on Cancer";

IATA means "Integrated Approach on Testing and Assessment";

IBC Code means the "International Code for the Construction and Equipment of Ships carrying Dangerous Chemicals in Bulk" (International Bulk Chemical Code).

IGC Code means the International Code for the Construction and Equipment of Ships Carrying Liquefied Gases in Bulk, including applicable amendments to which the vessel has been certified;

ILO means the "International Labour Organization";

IMDG Code means the "International Maritime Dangerous Goods Code", as amended;

IMO means the "International Maritime Organization";

IMSBC Code means the "International Maritime Solid Bulk Cargoes Code", as amended;

Initial boiling point means the temperature of a liquid at which its vapour pressure is equal to the standard pressure (101.3 kPa), i.e. the first gas bubble appears;

IOMC means the "Inter-organization Programme on the Sound Management of Chemicals";

IPCS means the "International Programme on Chemical Safety";

ISO means the "International Organization for Standardization";

IUPAC means the "International Union of Pure and Applied Chemistry";

ErC₅₀ とは、生長阻害の観点から見た EC₅₀ をいう。

EU とは、「欧州連合」（European Union）をいう。

引火点（Flash point）とは、一定の試験条件の下で任意の液体の蒸気が発火源により発火する最低温度をいう（標準気圧 101.3kPa での温度に換算）。

FAO とは、国連食糧農業機関（Food and Agriculture Organization of the United Nations）をいう。

ガス(Gas)とは、(i) 50℃で 300kPa（絶対圧）を超える蒸気圧を有する物質、または (ii) 101.3kPa の標準気圧、20℃において完全にガス状である物質をいう。

GC Code とは、「液化ガスのばら積み輸送のための船舶の構造及び設備に関するコード」（Code for the Construction and Equipment of Ships Carrying Liquefied Gases in Bulk, Gas Carrier Code）をいう。

GESAMP とは、IMO/FAO/UNESCO/WMO/WHO/IAEA/UN/UNEP の「海洋環境保護の科学的事項に関する専門家合同グループ」（Joint Group of Experts on the Scientific Aspects of Marine Environmental Protection of IMO/FAO/UNESCO/WMO/WHO/IAEA/UN/UNEP）をいう。

GHS とは、「化学品の分類および表示に関する世界調和システム」（Globally Harmonized System of Classification and Labelling of Chemicals）をいう。

危険有害性区分（Hazard category）とは、各危険有害性クラス内の判定基準の区分けをいう。例えば、経口急性毒性には 5 つの有害性区分があり、引火性液体には 4 つの危険性区分がある。これらの区分は危険有害性クラス内で危険有害性の強度により相対的に区分されるもので、より一般的な危険有害性区分の比較とみなすべきでない。

危険有害性クラス（Hazard class）とは、可燃性固体、発がん性物質、経口急性毒性のような、物理化学的危険性、健康または環境有害性の種類をいう。

危険有害性情報（Hazard statement）とは、危険有害性クラスおよび危険有害性区分に割り当てられた文言であって、危険有害な製品の危険有害性の性質を、該当する程度も含めて記述する文言をいう。

IAEA とは、「国際原子力機関」（International Atomic Energy Agency）をいう。

IARC とは、「国際がん研究機関」（International Agency for the Research on Cancer）をいう。

IATA とは、「試験および評価に関する統合的アプローチ」（Integrated Approach on Testing and Assessment）を意味する。

IBC Code とは、「危険化学品のばら積み輸送のための船舶の構造及び設備に関する国際規則」（International Code for the Construction and Equipment of Ships carrying Dangerous Chemicals in Bulk, International Bulk Chemical Code）をいう。

IGC Code とは、証明された船舶に対する適用可能な修正を含む「液化ガスのばら積み輸送のための船舶の構造及び設備に関する国際規則」（International Code of the Construction and Equipment of Ships Carrying Liquefied Gases in Bulk）をいう。

ILO とは、「国際労働機関」（International Labour Organization）をいう。

IMDG Code とは、「国際海上危険物規程」（International Maritime Dangerous Goods Code）改訂版をいう。

IMO とは、「国際海事機関」（International Maritime Organization）をいう。

IMSBC Code とは、「国際海上固体ばら積み貨物コード」（International Maritime Solid Bulk Cargoes Code）改訂版をいう。

初留点（Initial boiling point）とは、ある液体の蒸気圧が標準気圧（101.3kPa）に等しくなる、すなわち最初にガスの泡が発生する時点での液体の温度をいう。

IOMC とは、「化学品の適正な管理に関する国際機関間プログラム」（Inter-organization Programme on the Sound Management of Chemicals）をいう。

IPCS とは、「国際化学品安全性計画」（International Programme on Chemical Safety）をいう。

ISO とは、「国際標準化機構」（International Organization for Standardization）をいう。

IUPAC とは、「国際純正・応用化学連合」（International Union of Pure and Applied Chemistry）をいう。

Label means an appropriate group of written, printed or graphic information elements concerning a hazardous product, selected as relevant to the target sector(s), that is affixed to, printed on, or attached to the immediate container of a hazardous product, or to the outside packaging of a hazardous product;

Label element means one type of information that has been harmonized for use in a label, e.g. pictogram, signal word;

LC$_{50}$ (50 % lethal concentration) means the concentration of a chemical in air or of a chemical in water which causes the death of 50 % (one half) of a group of test animals;

LD$_{50}$ means the amount of a chemical, given all at once, which causes the death of 50 % (one half) of a group of test animals;

L(E)C$_{50}$ means LC$_{50}$ or EC$_{50}$;

Liquid means a substance or mixture which at 50°C has a vapour pressure of not more than 300 kPa (3 bar), which is not completely gaseous at 20 °C and at a standard pressure of 101.3 kPa, and which has a melting point or initial melting point of 20 °C or less at a standard pressure of 101.3 kPa. A viscous substance or mixture for which a specific melting point cannot be determined shall be subjected to the ASTM D 4359-90 test; or to the test for determining fluidity (penetrometer test) prescribed in section 2.3.4 of Annex A of the Agreement concerning the International Carriage of Dangerous Goods by Road (ADR);

Manual of Tests and Criteria means the latest revised edition of the United Nations publication bearing this title, and any published amendment thereto;

MARPOL means the International Convention for the Prevention of Pollution from Ships, 1973, as modified by the Protocol of 1978 relating thereto, as amended;

Mist means liquid droplets of a substance or mixture suspended in a gas (usually air);

Mixture means a mixture or a solution composed of two or more substances in which they do not react;

Montreal Protocol means the Montreal Protocol on Substances that Deplete the Ozone Layer as either adjusted and/or amended by the Parties to the Protocol;

Mutagen means an agent giving rise to an increased occurrence of mutations in populations of cells and /or organisms;

Mutation means a permanent change in the amount or structure of the genetic material in a cell;

NGO means "non-governmental organization";

NOEC (no observed effect concentration) means the test concentration immediately below the lowest tested concentration with statistically significant adverse effect. The NOEC has no statistically significant adverse effect compared to the control;

OECD means the "Organization for Economic Cooperation and Development";

Ozone Depleting Potential (ODP) means an integrative quantity, distinct for each halocarbon source species, that represents the extent of ozone depletion in the stratosphere expected from the halocarbon on a mass-for-mass basis relative to CFC-11. The formal definition of ODP is the ratio of integrated perturbations to total ozone, for a differential mass emission of a particular compound relative to an equal emission of CFC-11;

Pictogram means a graphical composition that may include a symbol plus other graphic elements, such as a border, background pattern or colour that is intended to convey specific information;

Precautionary statement means a phrase (and/or pictogram) that describes recommended measures that should be taken to minimize or prevent adverse effects resulting from exposure to a hazardous product, or improper storage or handling of a hazardous product;

Product identifier means the name or number used for a hazardous product on a label or in the SDS. It provides a unique means by which the product user can identify the substance or mixture within the particular use setting e.g. transport, consumer or workplace;

ラベル（Label）とは、危険有害な製品に関する書面、印刷またはグラフィックによる情報要素のまとまりであって、目的とする部門に対して関連するものが選択されており、危険有害性のある物質の容器に直接、あるいはその外部梱包に貼付、印刷または添付されるものをいう。

ラベル要素（Label element）とは、ラベル中で使用するために国際的に調和されている情報、例えば、絵表示や注意喚起語をいう。

LC_{50}（50%致死濃度）とは、試験動物の50%を死亡させる大気中または水中における試験物質濃度をいう。

LD_{50}とは、一度に投与した場合、試験動物の50%を死亡させる化学品の量をいう。

$L(E)C_{50}$とは、LC_{50}またはEC_{50}をいう。

液体（Liquid）とは、50℃において300kPa（3bar）以下の蒸気圧を有し、20℃、標準気圧101.3kPaでは完全にガス状ではなく、かつ、標準気圧101.3kPaにおいて融点または融解が始まる温度が20℃以下の物質をいう。固有の融点が特定できない粘性の大きい物質または混合物は、ASTMのD4359-90試験を行うか、または危険物の国際道路輸送に関する欧州協定（ADR）の附属文書Aの2.3.4節に定められている流動性特定のための（針入度計）試験を行わなければならない。

試験方法および判定基準のマニュアル（Manual of Tests and Criteria）とは、このタイトルを持つ国際連合の出版物の最新改訂版および公表されたこれへの修正をいう。

MARPOLとは、「1978年の議定書によって修正された1973年の船舶による汚染の防止のための国際条約」（International Convention for the Prevention of Pollution from Ships, 1973, as modified by the Protocol of 1978 relating thereto, as amended.）改訂版をいう。

ミスト（Mist）とは、ガス（通常空気）の中に浮遊する物質または混合物の液滴をいう。

混合物（Mixture）とは、2つ以上の物質で構成される反応を起こさない混合物または溶液をいう。

モントリオール議定書（Montreal Protocol）とは、議定書の締約国によって調整または修正された、オゾン層破壊物質に関するモントリオール議定書をいう。

変異原性物質（Mutagen）とは、細胞の集団または生物体に突然変異を発生する頻度を増大させる物質をいう。

突然変異（Mutation）とは、細胞内の遺伝物質の量または構造における恒久的な変化をいう。

NGOとは、「非政府組織」（non-governmental organization）をいう。

NOEC「無影響濃度」（no observed effect concentration）とは、統計的に有意な悪影響を示す最低の試験濃度直下の試験濃度をいう。NOECではコントロール群と比べて有意な悪影響は見られない。

OECDとは、「経済協力開発機構」（Organization for Economic Cooperation and Development）をいう。

オゾン層破壊係数（ODP）とは、ハロカーボンによって見込まれる成層圏オゾンの破壊の程度を、CFC-11に対して質量ベースで相対的に表した積算量であり、ハロカーボンの種類ごとに異なるものである。ODPの正式な定義は、等量のCFC-11排出量を基準にした、特定の化合物の排出に伴う総オゾンの擾乱量の積算値の比の値である。

絵表示（Pictogram）とは、特定の情報を伝達することを意図したシンボルと境界線、背景のパターンまたは色のような図的要素から構成されるものをいう。

注意書き（Precautionary statement）とは、危険有害性のある製品へのばく露あるいは危険有害性のある製品の不適切な貯蔵または取扱いから生じる有害影響を最小にするため、または予防するために取るべき推奨措置を記述した文言（または絵表示）をいう。

製品特定名（Product identifier）とは、ラベルまたはSDSにおいて危険有害性のある製品に使用される名称または番号をいう。これは、製品使用者が特定の使用状況、例えば輸送、消費者、あるいは作業場の中で物質または混合物を確認することができる一義的な手段となる。

UN Model Regulations means the Model Regulations annexed to the latest revised edition of the Recommendations on the Transport of Dangerous Goods published by the United Nations;

QSAR means "quantitative structure-activity relationship";

Respiratory sensitizer means a substance or mixture that induces hypersensitivity of the airways occurring after inhalation of the substance" or mixture;

RID means the "Regulations concerning the International Carriage of Dangerous Goods by Rail" [Annex 1 to Appendix B (Uniform Rules concerning the Contract for International Carriage of Goods by Rail) (CIM) of COTIF (Convention concerning international carriage by rail)], as amended;

Rotterdam Convention means the "Rotterdam Convention on the Prior Informed Consent Procedure for Certain Hazardous Chemicals and Pesticides in International Trade";

SAR means "Structure Activity Relationship";

SDS means "Safety Data Sheet";

Self-accelerating decomposition temperature (SADT) means the lowest temperature at which self-accelerating decomposition may occur with substance as packaged;

Signal word means a word used to indicate the relative level of severity of hazard and alert the reader to a potential hazard on the label. The GHS uses "Danger" and "Warning" as signal words;

Skin sensitizer means a substance or mixture that will induce an allergic response following skin contact;

SOLAS means the "International Convention for the Safety of Life at Sea", 1974, as amended;

Solid means a substance or mixture which does not meet the definitions of liquid or gas;

Stockholm Convention means the "Stockholm Convention on Persistent Organic Pollutants";

Substance means chemical elements and their compounds in the natural state or obtained by any production process, including any additive necessary to preserve the stability of the product and any impurities deriving from the process used, but excluding any solvent which may be separated without affecting the stability of the substance or changing its composition;

Supplemental label element means any additional non-harmonized type of information supplied on the container of a hazardous product that is not required or specified under the GHS. In some cases. this information may be required by other competent authorities or it may be additional information provided at the discretion of the manufacturer/distributor;

Symbol means a graphical element intended to succinctly convey information;

Technical name means a name that is generally used in commerce, regulations and codes to identify a substance or mixture, other than the IUPAC or CAS name, and that is recognized by the scientific community. Examples of technical names include those used for complex mixtures (e.g. petroleum fractions or natural products), pesticides (e.g. ISO or ANSI systems), dyestuffs (Colour Index system) and minerals;

UNCED means the "United Nations Conference on Environment and Development";

UNCETDG/GHS means the "United Nations Committee of Experts on the Transport of Dangerous Goods and on the Globally Harmonized System of Classification and Labelling of Chemicals";

UN means the "United Nations";

UNEP means the "United Nations Environment Programme";

UNESCO means the "United Nations Educational, Scientific and Cultural Organization";

UNITAR means the "United Nations Institute for Training and Research";

UN モデル規則 (UN Model Regulations) とは、国際連合（United Nations）から出版されている「危険物輸送に関する勧告」（Recommendations on the Transport of Dangerous Goods）の最新改訂版の付属書となっているモデル規則をいう。

QSAR とは、「定量的構造活性相関」（quantitative structure-activity relationship）を意味する。

呼吸器感作性物質（Respiratory sensitizer）とは、物質または混合物の吸入後に起きる気道の過敏反応を誘発する物質または混合物をいう。

RID とは、「鉄道による危険物の国際輸送に関する規則」（The Regulations concerning the International Carriage of Dangerous Goods by Rail）をいう。[COTIF（鉄道による国際輸送に関する条約）の付録 B 附属書 1（鉄道による貨物の国際輸送に関する統一規則）（CIM）]

*ロッテルダム条約*とは、「国際貿易の対象となる特定の有害な化学物質及び駆除剤についての事前のかつ情報に基づく同意の手続に関するロッテルダム条約」（Rotterdam Convention on the Prior Informed Consent Procedure for Certain Hazardous Chemicals and Pesticides in International Trade）をいう。

SAR とは、「構造活性相関」（Structure Activity Relationship）をいう。

SDS とは、「安全データシート」（Safety Data Sheet）をいう。

自己加速分解温度（SADT ; Self-Accelerating Decomposition Temperature）とは、密封状態において物質に自己加速分解が起こる最低温度をいう。

注意喚起語（Signal Word）とは、ラベル上で危険有害性の重大さの相対レベルを示し、利用者に潜在的な危険有害性を警告するために用いられる言葉をいう。GHS では、「危険（Danger）」や「警告（Warning）」を注意喚起語として用いている。

皮膚感作性物質（Skin sensitizer）とは、皮膚への接触によりアレルギー反応を誘発する物質または混合物をいう。

SOLAS とは、「海上における人命の安全のための国際条約 1974」（International Convention for the Safety of Life at Sea, 1974）改訂版をいう。

固体（Solid）とは、液体または気体の定義に当てはまらない物質または混合物をいう。

*ストックホルム条約*とは、「残留性有機汚染物質に関するストックホルム条約」（Stockholm Convention on Persistent Organic Pollutants）をいう。

物質（Substance）とは、自然状態にあるか、または任意の製造過程において得られる化学元素およびその化合物をいう。製品の安定性を保つ上で必要な添加物や用いられる工程に由来する不純物も含むが、当該物質の安定性に影響せず、またその組成を変化させることなく分離することが可能な溶媒は除く。

補助的ラベル要素（Supplemental label element）とは、危険有害性のある製品の容器に付される情報であって、GHS において要求または指定されていない追加情報をいう。こうした情報は、他の所管官庁による要求事項であることもあれば、製造者/流通業者の自由裁量で提供される追加情報のこともある。

シンボル（Symbol）とは、情報を簡潔に伝達するように意図された画像要素をいう。

専門名（Technical name）とは、IUPAC または CAS 名以外の名称であって、物質または混合物を特定するために商業、法規制、規格等で一般に使用され科学者・専門家に認められた名称をいう。専門名の例には、複雑な混合物（例：石油留分や天然産物）、農薬（例：ISO や ANSI システム）、染料（カラーインデックスシステム）、鉱物などに使用されるものがある。

UNCED とは、「国連環境開発会議」（United Nations Conference on Environment and Development）をいう。

UNCETDG/GHS とは、「国連危険物輸送ならびに化学品の分類および表示に関する世界調和システムに関する専門家委員会」（United Nations Committee of Experts on the TDG and on the GHS）をいう。

UN とは、「国際連合」（United Nations）をいう。

UNEP とは、「国連環境計画」（United Nations Environment Programme）をいう。

UNESCO とは、「国連教育科学文化機構」（United Nations Educational, Scientific and Cultural Organization）をいう。

UNITAR とは、「国連訓練調査研究所」（United Nations Institute for Training and Research）をいう。

UNSCEGHS means the "United Nations Sub-Committee of Experts on the Globally Harmonized System of Classification and Labelling of Chemicals";

UNSCETDG means the "United Nations Sub-Committee of Experts on the Transport of Dangerous Goods";

Vapour means the gaseous form of a substance or mixture released from its liquid or solid state;

VDI means the "Association of German Engineers" (*Verein Deutscher Ingenieure*);

WHO means the "World Health Organization";

WMO means the "World Meteorological Organization".

UNSCEGHS とは、「国連化学品の分類および表示に関する世界調和システムに関する専門家小委員会」（United Nations Sub-Committee of Experts on the GHS）をいう。

UNSCETDG とは、「国連危険物輸送に関する専門家小委員会」（United Nations Sub-Committee of Experts on the Transport of Dangerous Goods）をいう。

蒸気（Vapour）とは、液体または固体の状態から放出されたガス状の物質または混合物をいう。

VDI とは、「ドイツ技術者協会」（Verein Deutscher Ingenieure）をいう。

WHO とは、「世界保健機関」（World Health Organization）をいう。

WMO とは、「世界気象機関」（World Meteorological Organization）をいう。

CHAPTER 1.3

CLASSIFICATION OF HAZARDOUS SUBSTANCES AND MIXTURES

1.3.1   Introduction

Development of the GHS began with the work on classification criteria by the OECD Task Force on Harmonization of Classification and Labelling (Task Force on HCL) for health and environmental hazards, and by the UNCETDG/ILO Working Group for Physical Hazards.

1.3.1.1   *Health and environmental hazard classes: OECD Task Force on Harmonization of Classification and Labelling (OECD Task Force on HCL)*

1.3.1.1.1   The work of the OECD Task Force on HCL was generally of three related kinds:

(a) Comparison of the major classification systems, identification of similar or identical elements and, for the elements which were dissimilar, development of a consensus on a compromise;

(b) Examination of the scientific basis for the criteria which define the hazard class of concern (e.g. acute toxicity, carcinogenicity), gaining expert consensus on the test methods, data interpretation and level of concern, and then seeking consensus on the criteria. For some hazard classes, the existing schemes had no criteria and the relevant criteria were developed by the Task Force;

(c) Where there was a decision-tree approach (e.g. irritation) or where there were dependent criteria in the classification scheme (acute aquatic toxicity), development of consensus on the process or the scheme for using the criteria.

1.3.1.1.2   The OECD Task Force on HCL proceeded stepwise in developing its harmonized classification criteria. For each hazard class the following steps were undertaken:

(a) Step 1:   A thorough analysis of existing classification systems, including the scientific basis for the system and its criteria, its rationale and an explanation of how it is used. Step 1 documents were prepared and amended as required after discussion by the OECD Task Force on HCL for the following hazard classes: eye irritation/serious eye damage, skin irritation/corrosion, sensitizing substances, germ cell mutagenicity, reproductive toxicity, specific target organ toxicity, and chemical mixtures;

(b) Step 2:   A proposal for a harmonized classification system and criteria for each hazard class and category was developed. A Step 2 document was prepared and amended as required after discussion by the OECD Task Force on HCL;

(c) Step 3:

  (i) OECD Task Force on HCL reached consensus on the revised Step 2 proposal; or

  (ii) If attempts at consensus building failed, the OECD Task Force on HCL identified specific "non-consensus" items as alternatives in a revised Step 2 proposal for further discussion and resolution.

(d) Step 4:   Final proposals were submitted to the OECD Joint Meeting of the Chemicals Committee and the Working Party on Chemicals, Pesticides and Biotechnology for approval and subsequently to the IOMC CG-HCCS for incorporation into the GHS.

1.3.1.2   *UNCETDG/ILO working group on physical hazards*

The UNCETDG/ILO working group on physical hazards used a similar process to the OECD Task Force on HCL. The work involved a comparison of the major classification systems, identification of similar or identical elements, and for the elements which were dissimilar, development of a consensus on a compromise. For physical hazards, however, the transport definitions, test methods and classification criteria were used as a basis for the work since they were already substantially harmonized. The work proceeded through examination of the scientific basis for the criteria,

第1.3章

危険有害性のある物質および混合物の分類

1.3.1 序文

GHS の策定は、分類および表示の調和に関する OECD タスクフォース（HCL に関するタスクフォース）による健康と環境有害性に対する分類基準および UNCETDG/ILO の作業グループによる物理化学的危険性に関する分類基準の作業から開始された。

1.3.1.1 *健康と環境に対する危険有害性クラス：分類および表示の調和に関する OECD タスクフォース（HCL に関する OECD タスクフォース）*

1.3.1.1.1 OECD の HCL タスクフォースの作業は、相互に関連する以下の3種類であった：

(a) 主要な分類システムの比較検討、類似または同一の要素の特定、ならびに異なる要素に関する妥協案についての合意形成；

(b) 懸念される危険有害性クラス（例えば急性毒性や発がん性）を定義する判定基準についての科学的根拠の調査、試験方法、データの解釈、ならびに有害性の程度に関する専門家の合意、その上での基準に関する合意形成。一部の危険有害性クラスについては、既存の判定基準がなく、同タスクフォースが判定基準を策定した；

(c) 枝分かれ図による手法を用いたもの（例えば刺激性）または分類において依拠する判定基準があったもの（急性水生環境毒性）については、その手順または判定基準の用い方に関する合意の形成。

1.3.1.1.2 HCL に関する OECD タスクフォースは、段階的にその調和分類基準の策定を行った。危険有害性クラスごとに、以下の手順がとられた：

(a) 第1段階：システムとその判定基準の科学的根拠、その理論的解釈および使用方法の説明等、既存の分類システムの徹底的な分析。第1段階の文書は、以下の危険有害性クラスおよび混合物について HCL に関する OECD のタスクフォースの検討を経て作成され、必要に応じて修正された。有害性クラス：眼に対する重篤な損傷性/眼刺激性、皮膚腐食性/刺激性、感作性物質、生殖細胞変異原性、生殖毒性、特定標的臓器毒性；

(b) 第2段階：各危険有害性クラスと区分に対して、調和分類システムおよび判定基準の案が策定された。第2段階の文書は、HCL に関する OECD タスクフォースの検討を経て作成され、必要に応じて修正された；

(c) 第3段階：

　(i) HCL に関する OECD タスクフォースは、修正された第2段階の案について合意した；または

　(ii) 合意に至らなかった場合、HCL に関する OECD タスクフォースが「合意していない」項目を確認し、更なる検討と決定を行うため、第2段階への提案課題とした。

(d) 第4段階：最終提案への承認を求めるため、OECD の化学品委員会と化学品、農薬、およびバイオテクノロジーに関する作業部会による合同会議に同案を提出し、その後、GHS で使用するために IOMC CG - HCCS に提出した。

1.3.1.2 *UNCETDG/ILO の物理化学的危険性に関する作業グループ*

UNCETDG/ILO の物理化学的危険性に関する作業グループは、HCL に関する OECD タスクフォースと同様の手順を用いた。作業は主要分類システムの比較検討、類似または同一要素の特定および異なる要素に関する妥協案をめぐる合意の形成についてなされた。物理化学的危険性に関しては、既に輸送部門に

gaining consensus on the test methods, data interpretation and on the criteria. For most hazard classes, the existing schemes were already in place and being used by the transport sector. On this basis, a portion of the work focused on ensuring that workplace, environment and consumer safety issues were adequately addressed.

1.3.2　General considerations on the GHS

1.3.2.1　*Scope of the system*

1.3.2.1.1　The GHS applies to pure substances and their dilute solutions and to mixtures. "Articles" as defined in the Hazard Communication Standard (29 CFR 1910.1200) of the Occupational Safety and Health Administration of the United States of America, or by similar definition, are outside the scope of the system.

1.3.2.1.2　One objective of the GHS is for it to be simple and transparent with a clear distinction between classes and categories in order to allow for "self-classification" as far as possible. For many hazard classes the criteria are semi-quantitative or qualitative and expert judgement is required to interpret the data for classification purposes. Furthermore, for some hazard classes (e.g. eye irritation, explosives or self-reactive substances) a decision tree approach is provided to enhance ease of use.

1.3.2.2　*Concept of "classification"*

1.3.2.2.1　The GHS uses the term "hazard classification" to indicate that only the intrinsic hazardous properties of substances or mixtures are considered.

1.3.2.2.2　Hazard classification incorporates only three steps, i.e.:

(a) identification of relevant data regarding the hazards of a substance or mixture;

(b) subsequent review of those data to ascertain the hazards associated with the substance or mixture; and

(c) a decision on whether the substance or mixture will be classified as a hazardous substance or mixture and the degree of hazard, where appropriate, by comparison of the data with agreed hazard classification criteria.

1.3.2.2.3　As noted in IOMC Description and further clarification of the anticipated application of the GHS in the *Purpose, scope and application* (chapter 1.1, paragraph 1.1.2.4), it is recognized that once a chemical is classified, the likelihood of adverse effects may be considered in deciding what informational or other steps should be taken for a given product or use setting.

1.3.2.3　*Classification criteria*

1.3.2.3.1　The classification criteria for substances and mixtures are presented in Parts 2, 3 and 4 of this document, each of which is for a specific hazard class or a group of closely related hazard classes. For most hazard classes, the recommended process of classification of mixtures is based on the following sequence:

(a) Where test data are available for the complete mixture, the classification of the mixture will always be based on that data;

(b) Where test data are not available for the mixture itself, then bridging principles included and explained in each specific chapter should be considered to see whether they permit classification of the mixture. Bridging may also be applied when test data conclusively show that no classification is warranted;

In addition, for health and environmental hazards,

(c) If (i) test data are not available for the mixture itself, and (ii) the available information is not sufficient to allow application of the above mentioned bridging principles, then the agreed method(s) described in each chapter for estimating the hazards based on the information known will be applied to classify the mixture.

おいて定義、試験方法、分類基準が実質的に調和されていたので、これを作業の基礎として用いることができた。作業は科学的根拠に関する調査を通じ、試験方法、データの解釈、判定基準に関する合意について進められた。大半の危険有害性クラスに関しては、輸送部門において既に体系が整えられ、用いられていた。これを基礎として、作業場、環境、消費者の安全に関する問題について適正に対処することに重点を置いた。

1.3.2　GHSに関する一般事項

1.3.2.1　システムの範囲

1.3.2.1.1　GHSは、純粋な物質とその希釈溶液および混合物に適用する。米国労働安全衛生局（Occupational Safety and Health Administration）の危険有害性周知基準（29CFR1910.1200）および同様の定義項目に定められている「物品（Article）」は、本システムの範囲から除外される。

1.3.2.1.2　GHSの1つの目標は、可能な限り「自主的な分類」ができるよう、本システムを簡潔にし、かつ透明性を持たせ、危険有害性クラスや区分間に明確な区別を設けるようにすることである。多くの危険有害性クラスについて、判定基準は半定量的または定性的であり、分類目的でデータの解釈を行うためには専門家の判断が必要である。さらに、一部の危険有害性クラス（例えば眼刺激性、爆発物、自己反応性物質）については、枝分かれ図による手法を取り入れ、簡単に使えるようにした。

1.3.2.2　「分類」の概念

1.3.2.2.1　GHSでは、物質または混合物の固有な危険有害性のみに着目していることを示すために「危険有害性の分類」という語を用いている。

1.3.2.2.2　危険有害性の分類は3つの手順から成る：

(a)　物質または混合物についての関連するデータの特定；

(b)　物質または混合物のもつ危険有害性を確認する目的での上記データの検討；

(c)　合意された危険有害性の分類基準とデータとの比較検討に基づく、物質または混合物の該当する危険有害性クラスおよび区分についての決定。

1.3.2.2.3　効果、範囲および適用（第1.1章1.1.2.4）にある、GHSに関する指示事項のIOMCによる説明文書で確認されているように、いったんある化学品を分類すれば、起こりうる影響を考慮して特定の製品または利用状況において必要な情報やその他の対策を決定することが可能になる。

1.3.2.3　分類基準

1.3.2.3.1　物質および混合物の分類基準は本文書の第2、第3および第4部に示すが、そこでは特定の危険有害性クラスまたは密接に関連しあった危険有害性クラスについて記載してある。ほとんどの危険有害性クラスに関して、混合物の分類について推奨する手順は次のとおりである：

(a) 混合物そのものの試験データが利用できる場合、混合物の分類は常にそのデータに基づいて行う；

(b) 混合物そのものの試験データが利用できない場合には、混合物の分類が可能かどうかについて、それぞれの章で説明されているつなぎの原則（bridging principle）を考慮するべきである。つなぎの原則は、区分に該当しないことが保証されることを試験データが決定的に示している場合に適用されてもよい。

さらに、健康および環境に対する危険有害性クラスに関しては、

(c) もし(ⅰ)混合物そのものの試験データが利用できず、(ⅱ)利用可能な情報が不十分でつなぎの原則が適用できなければ、既知の情報に基づいて危険有害性を推定するためにそれぞれの章に記述されている承認された方法を適用して、混合物を分類する。

1.3.2.3.2   In most cases, it is not anticipated that reliable data for complete mixtures will be available for germ cell mutagenicity, carcinogenicity, and reproductive toxicity hazard classes. Therefore, for these hazard classes, mixtures will generally be classified based on the available information for the individual ingredients of the mixtures, using the cut-off values/concentration limit methods in each chapter. The classification may be modified on a case-by-case basis based on available test data for the complete mixture, if such data are conclusive as described in each chapter.

1.3.2.4   *Available data, test methods and test data quality*

1.3.2.4.1   The GHS itself does not include requirements for testing substances or mixtures. Therefore, there is no requirement under the GHS to generate test data for any hazard class. It is recognized that some parts of regulatory systems do require data to be generated (e.g. pesticides), but these requirements are not related specifically to the GHS. The criteria established for classifying a mixture will allow the use of available data for the mixture itself and/or similar mixtures and/or data for ingredients of the mixture.

1.3.2.4.2   The classification of a substance or mixture depends both on the criteria and on the reliability of the test methods underpinning the criteria. In some cases, the classification is determined by a pass or fail of a specific test, (e.g. the ready biodegradation test for substances or ingredients of mixtures), while in other cases, interpretations are made from dose/response curves and observations during testing. In all cases, the test conditions need to be standardized so that the results are reproducible with a given substance and the standardized test yields "valid" data for defining the hazard class of concern. In this context, validation is the process by which the reliability and the relevance of a procedure are established for a particular purpose.

1.3.2.4.3   Tests that determine hazardous properties, which are conducted according to internationally recognized scientific principles, can be used for purposes of a hazard determination for health and environmental hazards. The GHS criteria for determining health and environmental hazards are test method neutral, allowing different approaches as long as they are scientifically sound and validated according to international procedures and criteria already referred to in existing systems for the hazard of concern and produce mutually acceptable data. Test methods for determining physical hazards are generally more clear-cut and are specified in the GHS.

1.3.2.4.4   *Previously classified chemicals*

One of the general principles established by the IOMC-CG-HCCS states that test data already generated for the classification of chemicals under the existing systems should be accepted when classifying these chemicals under the harmonized system thereby avoiding duplicative testing and the unnecessary use of test animals. This policy has important implications in those cases where the criteria in the GHS are different from those in an existing system. In some cases, it may be difficult to determine the quality of existing data from older studies. In such cases, expert judgement will be needed.

1.3.2.4.5   *Substances/mixtures posing special problems*

1.3.2.4.5.1   The effect of a substance or mixture on biological and environmental systems is influenced, among other factors, by the physico-chemical properties of the substance or mixture and/or ingredients of the mixture and the way in which ingredient substances are biologically available. Some groups of substances may present special problems in this respect, for example, some polymers and metals. A substance or mixture need not be classified when it can be shown by conclusive experimental data from internationally acceptable test methods that the substance or mixture is not biologically available. Similarly, bioavailability data on ingredients of a mixture should be used where appropriate in conjunction with the harmonized classification criteria when classifying mixtures.

1.3.2.4.5.2   Certain physical hazards (e.g. due to explosive or oxidizing properties) may be altered by dilution, as is the case for desensitized explosives, by inclusion in a mixture or article, packaging or other factors. Classification procedures for specific sectors (e.g. storage) should take experience and expertise into account.

1.3.2.4.6   *Animal welfare*

The welfare of experimental animals is a concern. This ethical concern includes not only the alleviation of stress and suffering but also, in some countries, the use and consumption of test animals. Where possible and appropriate, tests and experiments that do not require the use of live animals are preferred to those using sentient live experimental animals. To that end, for certain hazards non-animal observations/measurements are included as part of the classification system. Additionally, alternative animal tests, using fewer animals or causing less suffering are internationally accepted and should be preferred.

1.3.2.3.2　多くの場合、生殖細胞変異原性、発がん性そして生殖毒性の有害性クラスに関して混合物全体としての信頼すべきデータは期待できない。そこで混合物は、これらの有害性クラスに関してそれぞれの章にあるカットオフ値/濃度限界を用いて、個々の成分に関して入手できる情報に基づいて分類される。混合物全体としてのデータが各章で記述されているように決定的である場合には、混合物の分類はそのデータに基づいてケースバイケースで修正されてもよい。

1.3.2.4　利用可能なデータ、試験方法および試験データの質

1.3.2.4.1　GHS自体では、物質や混合物の試験は要求されていない。つまりどの危険有害性クラスについてもGHSのために試験データを取る必要はない。既存の規制システムの中にもデータの取得を要求するものがある（例えば農薬）ことはよく知られているが、この要求はGHSとは直接関係ない。混合物の分類のための判定基準では、混合物そのもの/または類似の混合物/または混合物の成分のデータを利用することが可能である。

1.3.2.4.2　物質や混合物の分類は、判定基準および判定基準の基礎となる試験の信頼性の両方に依存している。分類が特定の試験の合否によって決定される例（例えば、易生分解性試験）もあり、また、量-反応曲線および試験中の所見から解釈を行う例もある。いずれの場合も、試験条件を標準化して、所定の物質について再現性のある結果が得られ、標準化された試験から、懸念される危険有害性クラスを決定するための「有効」なデータが得られるようにする必要がある。この意味では、有効性の検証は、特定の目的を達成するための信頼性および妥当性を確立する過程である。

1.3.2.4.3　危険有害性を特定するための、国際的に認められた科学的原則に従って実施される試験は、健康および環境に対する有害性の特定に利用できる。健康および環境に対する有害性を特定するためのGHS判定基準は、中立的な評価方法であり、既存システムで既に参照されている国際的手順および判定基準に従って有効性が確認され、相互に受け入れ可能なデータが得られている限り、異なった方法も受け入れる。物理化学的危険性を決定する試験方法は、一般的により明確であり、GHSにおいても具体的に記述されている。

1.3.2.4.4　既に分類されている化学品

　IOMC-CG-HCCSにより策定された一般原則の1つによれば、化学品を調和されたシステムに従って分類する際には、試験の重複および試験動物の不必要な使用を避けるために、化学品分類のための既存システムにより得られている試験データを受け入れるべきであるとしている。この原則には、GHSにおける判定基準が既存システムの判定基準と異なっているような状況では重要な意味がある。ずっと以前の試験で得た既存データの質を決定することが困難な状況もある。そのような場合には専門家の判断が必要となる。

1.3.2.4.5　特殊な問題のある物質/混合物

1.3.2.4.5.1　生物系および環境系への物質または混合物の影響は、とりわけ物質または混合物および/または混合物中の成分の物理化学的性質と、成分が生物学的にどのように利用されるかに左右される。一部の物質、例えばある種のポリマーや金属では、この点に関して特殊な問題が生じる。国際的に認められている試験方法による決定的な実験データによって、物質または混合物が生物学的に利用されないことが示されるならば、それらを分類する必要はない。同様に、混合物の成分に関するこのような生物学的利用性についてのデータは、これらの混合物を分類するときに、該当する調和された分類基準と共に使用するべきである。

1.3.2.4.5.2　ある種の物理的危険性（例えば爆発性や酸化性）は、鈍性化爆発物の例に見られるように、混合物や物品に含まれたり包装されたりあるいは他の要因によって、希釈され変化するであろう。特定の分野（例えば貯蔵）に対する分類手順では経験や専門性を考慮しなければならない。

1.3.2.4.6　動物愛護

　実験動物の愛護は懸案事項である。この倫理的問題には、ストレスや痛みの緩和だけでなく、国によっては試験動物の使用および消費も含まれる。可能で適切であるならば、生きた動物を必要としない試験および実験が、生きて感覚を持つ実験動物を用いる試験よりも望ましい。そのために、ある有害性については、動物を用いない観察/測定が分類システムの中に含まれている。さらに、動物数を少なくした、または痛みを軽減させた動物試験代替法が国際的に受け入れられており、それらが優先されるべきである。

1.3.2.4.7   *Evidence from humans*

For classification purposes, reliable epidemiological data and experience on the effects of chemicals on humans (e.g. occupational data, data from accident databases) should be taken into account in the evaluation of human health hazards of a chemical. Testing on humans solely for hazard identification purposes is generally not acceptable.

1.3.2.4.8   *Expert judgement*

The approach to classifying mixtures includes the application of expert judgement in a number of areas in order to ensure existing information can be used for as many mixtures as possible to provide protection for human health and the environment. Expert judgement may also be required in interpreting data for hazard classification of substances, especially where weight of evidence assessments are needed.

1.3.2.4.9   *Weight of evidence*

1.3.2.4.9.1   For some hazard classes, classification results directly when the data satisfy the criteria. For others, classification of a substance or a mixture is made on the basis of the total weight of evidence. This means that all available information bearing on the determination of toxicity is considered together, including the results of valid in vitro tests, relevant animal data, and human experience such as epidemiological and clinical studies and well-documented case reports and observations.

1.3.2.4.9.2   The quality and consistency of the data are important. Evaluation of substances or mixtures related to the material being classified should be included, as should site of action and mechanism or mode of action study results. Both positive and negative results are assembled together in a single weight of evidence assessment.

1.3.2.4.9.3   Positive effects which are consistent with the criteria for classification in each chapter, whether seen in humans or animals, will normally justify classification. Where evidence is available from both sources and there is a conflict between the findings, the quality and reliability of the evidence from both sources must be assessed in order to resolve the question of classification. Generally, data of good quality and reliability in humans will have precedence over other data. However, even well-designed and conducted epidemiological studies may lack sufficient numbers of subjects to detect relatively rare but still significant effects, or to assess potentially confounding factors. Positive results from well-conducted animal studies are not necessarily negated by the lack of positive human experience but require an assessment of the robustness and quality of both the human and animal data relative to the expected frequency of occurrence of effects and the impact of potentially confounding factors.

1.3.2.4.9.4   Route of exposure, mechanistic information and metabolism studies are pertinent to determining the relevance of an effect in humans. When such information raises doubt about relevance in humans, a lower classification may be warranted. When it is clear that the mechanism or mode of action is not relevant to humans, the substance or mixture should not be classified.

1.3.2.4.9.5   Both positive and negative results are assembled together in the weight of evidence assessment. However, a single positive study performed according to good scientific principles and with statistically and biologically significant positive results may justify classification.

1.3.3   Specific considerations for the classification of mixtures

1.3.3.1   *Definitions*

1.3.3.1.1   In order to ensure a full understanding of the provisions for classifying mixtures, definitions of certain terms are required. These definitions are for the purpose of evaluating or determining the hazards of a product for classification and labelling and are not intended to be applied in other situations such as inventory reporting. The intent of the definitions as drawn is to ensure that:

(a) all products within the scope of the Globally Harmonized System are evaluated to determine their hazards, and are subsequently classified according to the GHS criteria as appropriate; and

(b) the evaluation is based on the actual product involved, i.e. on a stable product. If a reaction occurs during manufacture and a new product results, a new hazard evaluation and classification must take place to apply the GHS to the new product.

1.3.2.4.7 *ヒトより得られた証拠*

　分類を目的として化学品のヒトの健康に対する有害性評価を行う際は、ヒトに対する化学品の作用に関する信頼できる疫学的データおよび経験（例：職業に関するデータ、事故のデータベースからのデータ）を考慮するべきである。有害性の特定のためだけにヒトで試験することは、一般に認められない。

1.3.2.4.8 *専門家の判断*

　混合物の分類にあたっては、ヒトの健康と環境を保護するためにできるだけ多くの混合物について既存の情報を確実に使用できるように、多くの領域で専門家の判断の活用も必要であろう。また、特に証拠の重み付け評価が必要な場合には、物質の有害性分類でのデータの解釈に専門家の判断を要するであろう。

1.3.2.4.9 *証拠の重み付け*

1.3.2.4.9.1　危険有害性クラスによっては、データが判定基準を満たした場合に直ちに分類されるものもある。また、総合的な証拠の重み付けにより物質または混合物が分類される場合もある。これは、有効なin vitro 試験の結果や、関連する動物データ、疫学的調査や臨床研究、記録の確かな症例報告および所見等のヒトでの経験など、毒性の決定に関するあらゆる利用可能な情報をすべて考慮するということである。

1.3.2.4.9.2　データの質および一貫性は重要である。作用部位および作用機序や作用様式についての研究結果と同様に、調査物質に関連した物質または混合物の評価も加えるべきである。陽性結果と陰性結果の両方を組み合わせて証拠の重み付け評価を行う。

1.3.2.4.9.3　ヒトのデータでも、動物のデータでも、各章に示されている判定基準と一致する陽性の作用は、分類を裏付けるものであろう。2 つの情報源から証拠が得られ、その知見が矛盾している場合には、分類の問題を解決するために、それらの情報源から得られる証拠の質および信頼性を評価しなければならない。一般的に、質および信頼性に優れたヒトのデータは、他のデータより優先される。ただし、適切に計画され実施された疫学的調査であっても、対象数が少ないために、比較的まれなしかし重要な影響を検出できないとか、あるいは潜在的交絡要因を推定できないということもありうる。適切に実施された動物試験から陽性の結果が得られたならば、ヒトで陽性の経験が得られていなくとも、その結果を否定しなくともよいが、むしろ予測される影響の発生率および潜在的交絡要因の影響に関する、ヒトおよび動物における両方のデータの頑健性および質についての評価が求められる。

1.3.2.4.9.4　ばく露経路、機序に関する情報および代謝に関する研究は、ある影響がヒトに現れるかどうかを決定する際に有用である。そのような情報からヒトへの適用について疑問が生じたときは、低い方の分類が適当な場合もある。作用機序または作用様式がヒトに該当しないことが明らかであるならば、その物質または混合物はその影響について有害であると分類されるべきでない。

1.3.2.4.9.5　陽性結果と陰性結果の両方を組み合わせて証拠の重み付け評価を行う。しかし、優れた科学的原則に従って行われており、統計学的および生物学的に有意な結果が得られているならば、1 つの陽性結果を示す研究からでも危険有害性の分類は可能であろう。

1.3.3　混合物の分類のための特別に留意すべき事項

1.3.3.1 *定義*

1.3.3.1.1　混合物を分類する規定の理解を確実にするためには、用語の定義が必要である。これらの定義は、分類と表示に向けて製品の危険有害性を評価または決定する目的のためのものであり、インベントリー報告などの他の状況で適用するためのものではない。定義の意図は、次のことを確実にすることである：

(a) GHS の対象範囲内のすべての製品がそれらの危険有害性を決定するために評価され、そして該当する GHS 判定基準に従って分類されること；および

(b) 評価は、実際の製品、すなわち安定した製品に基づくこと。もし製造中に反応が起こり、新しい生成物が生ずる場合には、GHS を適用するため、その生成物に対して新たに危険有害性についての評価および分類を行わなければならない。

1.3.3.1.2 Working definitions have been accepted for the following terms: substance, mixture, alloy (see chapter 1.2 for other definitions and abbreviations used in the GHS).

> Substance: Chemical elements and their compounds in the natural state or obtained by any production process, including any additive necessary to preserve the stability of the product and any impurities deriving from the process used, but excluding any solvent which may be separated without affecting the stability of the substance or changing its composition.
>
> Mixture: Mixtures or solutions composed of two or more substances in which they do not react.
>
> Alloy: An alloy is a metallic material, homogeneous on a macroscopic scale, consisting of two or more elements so combined that they cannot be readily separated by mechanical means. Alloys are considered to be mixtures for the purpose of classification under the GHS.

1.3.3.1.3 These definitions should be used to maintain consistency when classifying substances and mixtures in the GHS. Note also that where impurities, additives or individual constituents of a substance or mixture have been identified and are themselves classified, they should be taken into account during classification if they exceed the cut-off value/concentration limit for a given hazard class.

1.3.3.1.4 It is recognized, as a practical matter, that some substances may react slowly with atmospheric gases, e.g. oxygen, carbon dioxide, water vapour, to form different substances; or they may react very slowly with other ingredients of a mixture to form different substances; or they may self-polymerise to form oligomers or polymers. However, the concentrations of different substances produced by such reactions are typically considered to be sufficiently low that they do not affect the hazard classification of the mixture.

1.3.3.2 *Use of cut-off values/concentration limits*

1.3.3.2.1 When classifying an untested mixture based on the hazards of its ingredients, generic cut-off values or concentration limits for the classified ingredients of the mixture are used for several hazard classes in the GHS[1]. While the adopted cut-off values/concentration limits adequately identify the hazard for most mixtures, there may be some that contain hazardous ingredients at lower concentrations than the harmonized cut-off value/concentration limit that still pose an identifiable hazard. There may also be cases where the harmonized cut-off value/concentration limit is considerably lower than could be expected on the basis of an established non-hazardous level for an ingredient.

1.3.3.2.2 Normally, the generic cut-off values/concentration limits adopted in the GHS should be applied uniformly in all jurisdictions and for all sectors. However, if the classifier has information that the hazard of an ingredient will be evident below the generic cut-off values/concentration limits, the mixture containing that ingredient should be classified accordingly.

1.3.3.2.3 On occasion, conclusive data may show that the hazard of an ingredient will not be evident when present at a level above the generic GHS cut-off value(s)/concentration limit(s). In these cases the mixture could be classified according to those data. The data should exclude the possibility that the ingredient would behave in the mixture in a manner that would increase the hazard over that of the pure substance. Furthermore, the mixture should not contain ingredients that would affect that determination.

1.3.3.2.4 Adequate documentation supporting the use of any values other than the generic cut-off values/concentration limits should be retained and made available for review on request.

1.3.3.3 *Synergistic or antagonistic effects*

When performing an assessment in accordance with the GHS requirements, the evaluator must take into account all available information about the potential occurrence of synergistic effects among the ingredients of the mixture. Lowering classification of a mixture to a less hazardous category on the basis of antagonistic effects may be done only if the determination is supported by sufficient data.

---

[1] *For the purposes of the GHS, the terms "cut-off value" and "concentration limit" are equivalent and are meant to be used interchangeably. Competent authorities may choose whether to use either term to define thresholds that trigger classification.*

1.3.3.1.2　物質、混合物、合金について、次の定義(working definitions)が採用された（GHSで用いられる他の定義および略語については第1.2章参照）。

　　物質：自然状態にあるか、または任意の製造過程において得られる化学元素およびその化合物をいう。製品の安定性を保つ上で必要な添加物や用いられる工程に由来する不純物を含むが、当該物質の安定性に影響せず、またその組成を変化させることなく分離することが可能な溶媒は除く。

　　混合物：2つ以上の物質で構成される反応を起こさない混合物または溶液をいう。

　　合金：機械的手段で容易に分離できないように結合した2つ以上の元素から成る巨視的にみて均質な金属体をいう。合金は、GHSによる分類では混合物とみなされる。

1.3.3.1.3　GHSで物質および混合物の分類を一貫して行うためは、これらの定義を用いるべきである。また、不純物、添加物、または物質もしくは混合物の成分が特定されてその各々が分類され、ある危険有害性クラスについてカットオフ値/濃度限界を超える場合は、これらも分類の際に考慮に入れるべきである。

1.3.3.1.4　実際には、物質によっては、大気中の気体、例えば、酸素、二酸化炭素、水蒸気などとゆっくり反応して、異なる物質を形成するものがあるかもしれず、また、混合物の他の成分と極めてゆっくり反応して、異なる物質を形成するものがあるかもしれないし、あるいは自己重合して、オリゴマーやポリマーを形成するものがあるかもしれない。しかし、このような反応によって生成する物質の濃度は、一般的に十分低いと考えられるので、混合物の危険有害性分類に影響しない。

1.3.3.2　*カットオフ値/濃度限界の使用*

1.3.3.2.1　未試験の混合物を成分の危険有害性に基づいて分類する場合、GHS[1] では、ある危険有害性クラスについて、混合物の分類された成分に対して統一的なカットオフ値または濃度限界が使用される。採用されたカットオフ値/濃度限界で、ほとんどの混合物について危険有害性が適切に特定されるが、カットオフ値/濃度限界以下の濃度でもその成分が特定可能な危険有害性を呈する場合がある。また、カットオフ値/濃度限界が、その成分が危険有害性を示さないと予想される濃度よりも、かなり低い場合もある。

1.3.3.2.2　通常、GHSで採用されたカットオフ値/濃度限界は、全ての管轄分野および全ての部門で一様に適用するべきである。しかし、分類する者が、ある成分が統一的なカットオフ値/濃度限界以下でも危険有害性を有することが明白であるという情報を持つ場合には、その成分を含む混合物はその情報に従って分類するべきである。

1.3.3.2.3　ある成分が統一的なGHSのカットオフ値/濃度限界以上の濃度で存在していても、危険有害性が顕在化しないという明確なデータが示される場合がある。この場合、混合物は、そのデータに従って分類できる。データにより、ある成分が単独で存在する場合よりも、混合物中でより危険有害性が増すという可能性が除外されるべきである。さらに、混合物には、その決定に影響を与える他の成分を含むべきではない。

1.3.3.2.4　統一的なGHSのカットオフ値/濃度限界以外の値を利用する理由を示した十分な書類は保管し、後で要求があった場合に審理に利用できるようにするべきである。

1.3.3.3　*相乗または拮抗作用*

　GHSの要求事項に従って評価を行う場合、評価者は、混合物成分間の潜在的相乗作用についてのあらゆる情報を考慮に入れなければならない。拮抗作用に基づいて混合物の分類をより低位の区分に下げることは、その決定が十分なデータによって裏付けされる場合に限る。

---

[1]　*GHSでは、「カットオフ値」および「濃度限界」は同意義であり、どちらを使用してもよい。所管官庁は分類を行う境界を定義するために、どちらかの用語を使用するかどうか選択してもよい。*

CHAPTER 1.4

HAZARD COMMUNICATION: LABELLING

1.4.1 Objectives, scope and application

1.4.1.1　　One of the objectives of the work on the Globally Harmonized System (GHS) has been the development of a harmonized hazard communication system, including labelling, safety data sheets and easily understandable symbols, based on the classification criteria developed for the GHS. This work was carried out under the auspices of the ILO, by the ILO working group on hazard communication using the same 3-step procedure outlined for the harmonization of classification in *Classification of hazardous substances and mixtures* (chapter 1.3, paragraph 1.3.1.1.2).

1.4.1.2　　The harmonized system for hazard communication includes the appropriate labelling tools to convey information about each of the hazard classes and categories in the GHS. The use of symbols, signal words or hazard statements other than those which have been assigned to each of the GHS hazard classes and categories, would be contrary to harmonization.

1.4.1.3　　The ILO working group considered the application of the general principles described in the IOMC CG/HCCS terms of reference[1] as they apply to hazard communication and recognized that there will be circumstances where the demands and rationale of systems may warrant some flexibility in whether to incorporate certain hazard classes and categories for certain target audiences.

1.4.1.4　　For example, the scope of the *UN Model Regulations*, encompasses only the most severe hazard categories of the acute toxicity hazard class. This system would not label substances or mixtures falling within the scope of the less severe hazard categories (e.g. those falling within the oral range > 300 mg/kg). However, should the scope of that system be amended to incorporate substances and mixtures falling in these less severe hazard categories, they should be labelled with the appropriate GHS labelling tools. The use of different cut-off values to determine which products are labelled in a hazard category would be contrary to harmonization.

1.4.1.5　　It is recognized that the *UN Model Regulations* provide label information primarily in a graphic form because of the needs of the target audiences. Therefore, the UN Sub-Committee of Experts on the Transport of Dangerous Goods may choose not to include signal words and hazard statements as part of the information provided on the label under the Model Regulations.

1.4.2 Terminology

1.4.2.1　　A description of common terms and definitions related to hazard communication is included in chapter 1.2 *Definitions and abbreviations*.

1.4.3 Target audiences

1.4.3.1　　The needs of the target audiences that will be the primary end-users of the harmonized hazard communication scheme have been identified. Particular attention was given to a discussion of the manner in which these target audiences will receive and use the information conveyed about hazardous chemicals. Factors discussed include the potential use of products, availability of information other than the label and the availability of training.

1.4.3.2　　It was recognized that it is difficult to completely separate the needs of different target audiences. For example, both workers and emergency responders use labels in storage facilities, and products such as paints and solvents are used both by consumers and in workplaces. In addition, pesticides can be used in consumer settings (e.g. lawn and garden products) and workplaces (e.g. pesticides used to treat seed in seed treatment plants). That said, there are certain characteristics which are particular to the different target audiences. The following paragraphs in this section consider the target audiences and the type of information they need.

1.4.3.3　　*Workplace*: Employers and workers need to know the hazards specific to the chemicals used and or handled in the workplace, as well as information about the specific protective measures required to avoid the adverse effects that might be caused by those hazards. In the case of storage of chemicals, potential hazards are minimized by the containment (packaging) of the chemical, but in the case of an accident, workers and emergency responders need to know

---

[1] *IOMC, Coordinating group for the harmonization of chemical classification systems, revised terms of reference and work programme (IOMC/HCS/95 – 14 January 1996).*

第 1.4 章

危険有害性に関する情報の伝達：表示

1.4.1 目的、範囲および適用

1.4.1.1 世界調和システム（GHS）の作業における目的のひとつは、GHS のために策定された分類の判定基準に基づいた表示、安全データシート、容易に理解できるシンボルを含む、調和された危険有害性に関する情報の伝達のシステムを確立することにあった。この作業は、ILO の支援の下、危険有害性に関する情報の伝達に関する ILO 作業グループによって、*危険有害性物質および混合物の分類*（第 1.3 章 1.3.1.1.2）における分類の調和で示したものと同じ 3 段階の手続で行われた。

1.4.1.2 危険有害性に関する情報の伝達に関する調和システムは、GHS での各危険有害性クラスおよび区分に関する情報を伝達するためにそれぞれに該当する表示要素を含む。GHS の各危険有害性クラスおよび区分に割り当てられたシンボル、注意喚起語、危険有害性情報以外のものを使用することは、調和の取り組みに反するものである。

1.4.1.3 ILO 作業グループは、危険有害性に関する情報の伝達についても IOMC CG/HCCS の委任事項[1]に記載されている一般原則の適用を考慮し、また、特定の危険有害性クラスおよび区分を特定の対象者に当てはめるか否かに関して、システムの要求事項および原則にある程度柔軟性が必要となる状況があることを認めた。

1.4.1.4 例えば、*UN モデル規則*は、急性毒性でも最も厳しい有害性区分のみを対象としている。このシステムでは、より緩い有害性区分の範囲内（例えば、経口摂取量が 300mg/kg より多い範囲内）にある物質または混合物については表示を行わない。しかし、同システムの適用範囲が変更され、こうした比較的低い危険有害性区分に収まる物質および混合物も組み入れることになれば、これらは該当する GHS の表示要素により表示を行うべきである。製品の危険有害性に関する表示を決定するために、異なるカットオフ値を用いることは調和に反する。

1.4.1.5 *UN モデル規則*では、その対象者のニーズから、主として図形で表示情報を提示することが認められている。したがって国連の危険物輸送に関する専門家小委員会は、モデル規則の下で、表示に注意喚起語と危険有害性情報を含めないという選択が可能である。

1.4.2 専門用語

1.4.2.1 危険有害性に関する情報の伝達に関する共通の用語および定義は、第 1.2 章 *定義および略語* に含まれる。

1.4.3 対象者

1.4.3.1 調和された危険有害性に関する情報の伝達システムの主な末端利用者となる対象者のニーズが確認された。特に、これらの対象者が危険有害性のある化学品についての情報を受け取り、利用する方法について集中的に議論が行われた。製品の予想される用途、ラベル以外の情報の利用可能性および訓練の利用可能性等について議論された。

1.4.3.2 異なる対象者のニーズを完全に分離することは困難であることがわかった。例えば、作業者と緊急時対応者の両方が貯蔵施設でラベルを利用するし、塗料や溶剤などの製品は、消費者と作業場の両方で使用される。さらに、農薬は、消費者部門でも（例えば芝や園芸品など）作業場でも（例えば種子の処理施設において使用される農薬）使用される。これは、対象者によってはそれぞれの特徴があるということである。この節の以下の段落では、対象者と彼らが必要とする情報の種類を検討する。

1.4.3.3 *作業場*：事業主と作業者は、作業場で使用または取り扱われる化学品に特有の危険有害性とそれによる悪影響を避けるために必要な防護対策に関する情報を知っている必要がある。化学品の貯蔵においては、潜在的な危険有害性は化学品の容器（包装）により最小限に抑えられているが、事故が起きた場合

---

[1] IOMC、化学品の分類システムの調和のための調整グループ、委任事項および作業プログラム改訂版（*IOMC/HCS/95-1996 年 1 月 14 日*）

what mitigation measures are appropriate. Here they may require information which can be read at a distance. The label, however, is not the sole source of this information, which is also available through the SDS and workplace risk management system. The latter should also provide for training in hazard identification and prevention. The nature of training provided and the accuracy, comprehensibility and completeness of the information provided on the SDS may vary. However, compared to consumers for example, workers can develop a more in depth understanding of symbols and other types of information.

1.4.3.4   *Consumers:* The label in most cases is likely to be the sole source of information readily available to the consumer. The label, therefore, will need to be sufficiently detailed and relevant to the use of the product. There are considerable philosophical differences on the approach to providing information to consumers. Labelling based on the likelihood of injury (i.e. risk communication) is considered to be an effective approach in this respect by some consumer labelling systems, whilst others take account of the "right to know" principle in providing information to consumers which is solely based on the product's hazards. Consumer education is more difficult and less efficient than education for other audiences. Providing sufficient information to consumers in the simplest and most easily understandable terms presents a considerable challenge. The issue of comprehensibility is of particular importance for this target audience, since consumers may rely solely on label information.

1.4.3.5   *Emergency responders:* Emergency responders need information on a range of levels. To facilitate immediate responses, they need accurate, detailed and sufficiently clear information. This applies in the event of an accident during transportation, in storage facilities or at workplaces. Fire fighters and those first at the scene of an accident for example, need information that can be distinguished and interpreted at a distance. Such personnel are highly trained in the use of graphical and coded information. However, emergency responders also require more detailed information about hazards and response techniques, which they obtain from a range of sources. The information needs of medical personnel responsible for treating the victims of an accident or emergency may differ from those of fire fighters.

1.4.3.6   *Transport:* The *UN Model Regulations*, cater for a wide range of target audiences although transport workers and emergency responders are the principal ones. Others include employers, those who offer or accept dangerous goods for transport or load or unload packages of dangerous goods into or from transport vehicles, or freight containers. All need information concerning general safe practices that are appropriate for all transport situations. For example, a driver will have to know what has to be done in case of an accident irrespective of the substance transported: (e.g. report the accident to authorities, keep the shipping documents in a given place, etc.). Drivers may only require limited information concerning specific hazards, unless they also load and unload packages or fill tanks, etc. Workers who might come into direct contact with dangerous goods, for example on board ships, require more detailed information.

1.4.4   Comprehensibility

1.4.4.1   Comprehensibility of the information provided has been one of the most important issues addressed in the development of the hazard communication system (see annex 6, *Comprehensibility testing methodology*). The aim of the harmonized system is to present the information in a manner that the intended audience can easily understand. The GHS identifies some guiding principles to assist this process:

(a) Information should be conveyed in more than one way;

(b) The comprehensibility of the components of the system should take account of existing studies and literature as well as any evidence gained from testing;

(c) The phrases used to indicate degree (severity) of hazard should be consistent across different hazard types.

1.4.4.2   The latter point was subject to some debate concerning the comparison of severity between long-term effects such as carcinogenicity and physical hazards such as flammability. Whilst it might not be possible to directly compare physical hazards to health hazards, it may be possible to provide target audiences with a means of putting the degree of hazard into context and therefore convey the same degree of concern about the hazard.

1.4.4.3   *Comprehensibility testing methodology*

A preliminary review of the literature undertaken by the University of Maryland indicated that common principles related to comprehensibility could be applied to the development of the harmonized hazard communication scheme. The University of Cape Town developed these into a comprehensive testing methodology to assess the comprehensibility of the hazard communication system (see annex 6). In addition to testing individual label elements, this methodology considers the comprehensibility of label elements in combination. This was considered particularly important to assess the comprehensibility of warning messages for consumers where there is less reliance on training to

には、作業者と緊急時対応者は災害を小さくする適切な方法を知る必要がある。事故の場合、ある程度離れていても読むことができる情報が必要であろう。しかし、ラベルは唯一の情報源ではなく、SDSや作業場のリスク管理システムを通しても情報は入手できる。リスク管理システムは危険有害性の特定および防止に関する訓練についても規定するべきである。行われる訓練の内容およびSDSで提供される情報の正確さ、分かりやすさ、完成度は様々であろう。とはいっても、例えば消費者と比較して、作業者はシンボルや他の種類の情報をより深く理解することができる。

1.4.3.4　*消費者*：大抵の場合、ラベルは消費者にとって容易に入手できる唯一の情報源である。そのため、ラベルはその製品の使用について、十分詳細かつ適切であることが必要となる。消費者への情報提供に関して、大きな基本的考え方の相異があった。障害の可能性に基づいた表示（すなわちリスクコミュニケーション）は、ある消費者表示システムにおいては有効な手法と考えられるが、一方で、「知る権利」の原則を考慮し製品の危険有害性だけに基づいた消費者への情報提供を行うシステムもある。消費者教育は他の対象者教育より困難で効率が悪い。消費者に最も簡単で最も容易に理解できる用語で十分な情報を提供するのは、かなりの難題である。消費者はラベル情報だけに頼るであろうから、分かりやすさの問題は特に重要である。

1.4.3.5　*緊急時対応者*：緊急時対応者は、広範囲なレベルについて情報を必要とする。また、緊急対応を容易にするために、正確かつ詳細で十分に明確な情報を必要とする。これは、輸送中、貯蔵施設、または作業場の事故の場合に当てはまる。例えば、消防士や最初に事故現場にいる者は、ある程度離れていてもはっきりしていて意味のわかる情報を必要とする。このような作業者は、図および記号化された情報の使用について高度に訓練されている。さらに、緊急時対応者は危険有害性と対応策についてより詳細な情報を必要とし、彼らはこれを広範囲な情報源から入手している。事故または緊急時の被害者を治療する医療従事者が必要とする情報は、消防士のものとは異なるであろう。

1.4.3.6　*輸送*：UNモデル規則は、輸送従事者と緊急時対応者が主対象であるが、より広範囲の対象者に使用されている。事業主、輸送委託者もしくは受託者、または車両もしくは貨物コンテナでの輸送物の荷役従事者なども関係する。これらの全員が、あらゆる輸送状況に対応した一般安全慣行に関する情報を必要とする。例えば、運転者は輸送する物質にかかわらず、事故の場合に何をすべきかを知らなければならない（例えば事故を所管官庁に報告する、船積み書類を所定場所に保管するなど）。運転者が包装品の積み卸しやタンクへの充填などを行わない場合は、彼らは特定の危険有害性に関する限られた情報だけを必要とするであろう。乗船する作業者等、危険物に直接接触する可能性がある作業者は、より詳細な情報を必要とする。

1.4.4　理解度

1.4.4.1　提供される情報の分かりやすさは、危険有害性に関する情報の伝達システムを策定する際の最も重要な課題のひとつであった（附属書6 *理解度に関する試験方法*を参照）。調和されたシステムの目的は、対象者が容易に理解できるように情報を提示することである。GHSでは、この理解の促進のため、以下の原則を確認した：

(a)　情報は複数の方法で伝達するべきである；

(b)　システムの構成要素の分かりやすさは、試験から得られた証拠だけでなく、既存の研究と文献を考慮するべきである；

(c)　危険有害性の程度（重大さ）を示すために用いられる用語は、異なる危険有害性の種類にわたって一貫しているべきである。

1.4.4.2　最後の点に関しては、発がん性などの長期的影響と可燃性などの物理化学的危険性との間の重大さの比較に関して議論がなされた。物理化学的危険性をヒトの健康に対する有害性と直接比較することは可能ではないかも知れないが、危険有害性の程度を対象者に示すことで、危険有害性について同程度の懸念を伝達することは可能であろう。

1.4.4.3　*理解度に関する試験方法*

　メリーランド大学が行った予備的な文献調査により、理解度に関係した一般原則は、調和された危険有害性に関する情報の伝達システムの策定に適用できることが示された。ケープタウン大学はこれを発展させ、危険有害性に関する情報の伝達システムの理解度を評価する試験方法にした（附属書6参照）。個々のラベル要素の試験に加え、この方法では、ラベル要素を組み合わせた時の理解度も考慮している。これは、理解力を高める訓練にそれほど頼れない消費者に対する警告メッセージの理解度を評価する際に特に重要

aid understandability. The testing methodology also includes a means of assessing SDS comprehensibility. A summary description of this methodology is provided in annex 6.

1.4.5 Translation

Options for the use of textual information present an additional challenge for comprehensibility. Clearly words and phrases need to retain their comprehensibility when translated, whilst conveying the same meaning. The IPCS chemical safety card programme has gained experience of this in translating standard phrases in a wide variety of languages. The EU also has experience of translating terms to ensure the same message is conveyed in multiple languages e.g. hazard, risk etc. Similar experience has been gained in North America where the North American Emergency Response Guidebook, which uses key phrases, is available in a number of languages.

1.4.6 Standardization

1.4.6.1 To fulfil the goal of having as many countries as possible adopt the system, much of the GHS is based on standardized approaches to make it easier for companies to comply with and for countries to implement the system. Standardisation can be applied to certain label elements (symbols, signal words, statements of hazard, precautionary statements) and to label format and colour and to SDS format.

1.4.6.2 *Application of standardization in the harmonized system*

For labels, the hazard symbols, signal words and hazard statements have all been standardized and assigned to each of the hazard categories. These standardized elements should not be subject to variation and should appear on the GHS label as indicated in the chapters for each hazard class in this document. For safety data sheets, chapter 1.5 *Hazard communication: Safety Data Sheets* provides a standardized format for the presentation of information. Although precautionary statements have not been fully harmonized in the current GHS, annex 3 provides guidance to aid in the selection of appropriate statements. Additional work to achieve greater standardization in this area may be undertaken in the future, once countries have gained experience with the system.

1.4.6.3 *Use of non-standardized or supplemental information*

1.4.6.3.1 There are many other label elements which may appear on a label which have not been standardized in the harmonized system. Some of these clearly need to be included on the label, for example precautionary statements. Competent authorities may require additional information, or suppliers may choose to add supplementary information on their own initiative. In order to ensure that the use of non-standardized information does not lead to unnecessarily wide variation in information or undermine GHS information, the use of supplementary information should be limited to the following circumstances:

(a) the supplementary information provides further detail and does not contradict or cast doubt on the validity of the standardized hazard information; or

(b) the supplementary information provides information about hazards not yet incorporated into the GHS.

In either instance, the supplementary information should not lower standards of protection.

1.4.6.3.2 The labeller should have the option of providing supplementary information related to the hazard, such as physical state or route of exposure, with the hazard statement rather than in the supplementary information section on the label, see also 1.4.10.5.4.1.

1.4.7 Updating information

1.4.7.1 All systems should specify a means of responding in an appropriate and timely manner to new information and updating labels and SDS information accordingly. The following are examples of how this could be achieved.

1.4.7.2 *General guidance on updating of information*

1.4.7.2.1 Suppliers should respond to "new and significant" information they receive about a chemical hazard by updating the label and safety data sheet for that chemical. New and significant information is any information that changes the GHS classification of the substance or mixture and leads to a resulting change in the information provided on the label or any information concerning the chemical and appropriate control measures that may affect the SDS. This could include,

と考えられた。この試験方法は、SDS の理解度を評価する手段も含んでいる。この方法の概説は、附属書 6 に示した。

1.4.5 翻訳

　文言の使い方で理解度が異なる。翻訳する際に分かりやすさを保ちつつ、同じ意味を伝達しなければならない。例えば、IPCS 化学品安全カードプログラム（Chemical Card Programme）は、標準的な文言の多種多様な言語への翻訳でこの種の経験を積んでいる。欧州連合も例えば、危険有害性やリスクなど、同じメッセージを多数の言語で伝達するという翻訳経験を持っている。キーフレーズを用いている北米の緊急時対応ガイドブック（North American Emergency Response Guidebook）でも同様の試みがなされており、多くの言語に翻訳したものを利用することができる。

1.4.6 標準化

1.4.6.1.　できるだけ多くの国にシステムを導入させるために、GHS は、企業がシステムを遵守しやすく、また国がシステムを実行しやすいように、システムの大部分を標準化した手順に基づいたものにした。標準化は、特定のラベル要素（シンボル、注意喚起語、危険有害性情報、注意書き）およびラベルの書式と色、そして SDS の書式に適応される。

1.4.6.2 *調和システムにおける標準化の適用*

　ラベルでは、危険有害性シンボル、注意喚起語および危険有害性情報はすべて標準化され、各危険有害性区分に割り当てられている。これらの標準化された要素は変更されるべきでなく、本文書の危険有害性クラスに関する各章に示されたとおり、GHS ラベル上に記載されるべきである。安全データシートについては、*危険有害性に関する情報の伝達：安全データシート*（第 1.5 章）に、情報提示の方法について標準化した様式を示した。注意書きは、現行の GHS では完全に調和されていないが、本文書の附属書 3 は、適切な文言を選択する際の助けとなるよう手引きを示している。国々が、このシステムに経験を積めば、この分野において、さらに標準化を達成するための追加作業が将来着手されるかも知れない。

1.4.6.3 *標準化されていない情報または補足情報の使用*

1.4.6.3.1　調和されたシステムで標準化されていないラベルに記載される他の多くの要素がある。これらの一部は明らかに、注意書き等としてラベルに含める必要がある。追加情報は所管官庁が要求する場合もあるであろうし、また供給者が自主的に補足情報を加えることもできる。標準化されていない情報を使用することにより、不必要な情報が増加したり、GHS 情報が軽視されることにつながらないようにするために、補足情報の使用は次のような場合に限定するべきである：

　（a）　補足情報はより詳細な情報を提供するものであり、標準化された危険有害性に関する情報の妥当性に矛盾したり、疑いを生じさせたりしないこと；または

　（b）　補足情報により、GHS にまだ取り入れられていない危険有害性に関する情報が提供されること。

　いずれの場合でも、補足情報により保護されるレベルを低下させるべきではない。

1.4.6.3.2　表示を行う者は、物理的状態やばく露経路など、危険有害性に関する補足情報については、ラベル上の補足情報の部分に示すのではなく、危険有害性情報と共に示すべきである。1.4.10.5.4.1 も参照のこと。

1.4.7 情報の更新

1.4.7.1　すべてのシステムは、新しい情報に適切かつ適時に対応し、それに応じたラベルと SDS 情報を更新する手段を定めるべきである。例を以下に示す。

1.4.7.2 *情報更新の全般的指針*

1.4.7.2.1　供給者は、化学品の危険有害性について入手した「新しくかつ重要な」情報に対応し、その物質に関するラベルおよび安全データシートを更新するべきである。新しくかつ重要な情報とは、物質また

for example, new information on the potential adverse chronic health effects of exposure as a result of recently published documentation or test results, even if a change in classification may not yet be triggered.

1.4.7.2.2   Updating should be carried out promptly on receipt of the information that necessitates the revision. The competent authority may choose to specify a time limit within which the information should be revised. This applies only to labels and SDS for products that are not subject to an approval mechanism such as pesticides. In pesticide labelling systems, where the label is part of the product approval mechanism, suppliers cannot update the supply label on their own initiative. However, when the products are subject to the transport of dangerous goods requirements, the label used should be updated on receipt of the new information, as above.

1.4.7.2.3   Suppliers should also periodically review the information on which the label and safety data sheet for a substance or mixture is based, even if no new and significant information has been provided to them in respect of that substance or mixture. This will require e.g. a search of chemical hazard databases for new information. The competent authority may choose to specify a time (typically 3 – 5 years) from the date of original preparation, within which suppliers should review the labels and SDS information.

1.4.8   Confidential business information

1.4.8.1   Systems adopting the GHS should consider what provisions may be appropriate for the protection of confidential business information. Such provisions should not compromise the health and safety of workers or consumers, or the protection of the environment. As with other parts of the GHS, the rules of the importing country should apply with respect to confidential business information claims for imported substances and mixtures.

1.4.8.2   Where a system chooses to provide for protection of confidential business information, competent authorities should establish appropriate mechanisms, in accordance with national law and practice, and consider:

（a） whether the inclusion of certain chemicals or classes of chemicals in the arrangements is appropriate to the needs of the system;

（b） what definition of "confidential business information" should apply, taking account of factors such as the accessibility of the information by competitors, intellectual property rights and the potential harm disclosure would cause to the employer or supplier's business; and

（c） appropriate procedures for the disclosure of confidential business information, where necessary to protect the health and safety of workers or consumers, or to protect the environment, and measures to prevent further disclosure.

1.4.8.3   Specific provisions for the protection of confidential business information may differ among systems in accordance with national law and practice. However, they should be consistent with the following general principles:

（a） For information otherwise required on labels or safety data sheets, confidential business information claims should be limited to the names of substances, and their concentrations in mixtures. All other information should be disclosed on the label and/or safety data sheet, as required;

（b） Where confidential business information has been withheld, the label or safety data sheet should so indicate;

（c） Confidential business information should be disclosed to the competent authority upon request. The competent authority should protect the confidentiality of the information in accordance with applicable law and practice;

（d） Where a medical professional determines that a medical emergency exists due to exposure to a hazardous substance or mixture, mechanisms should be in place to ensure timely disclosure by the supplier or employer or competent authority of any specific confidential information necessary for treatment. The medical professional should maintain the confidentiality of the information;

（e） For non-emergency situations, the supplier or employer should ensure disclosure of confidential information to a safety or health professional providing medical or other safety and health services to exposed workers or consumers, and to workers or workers' representatives. Persons requesting the information should provide specific reasons for the disclosure, and should agree

は混合物に関する GHS の分類の変更と、ラベルに記載すべき情報またはその化学品に関するあらゆる情報および SDS に影響する適切な予防対策の変更につながるものをさす。例えば、分類の変更にはすぐに至らないが、最近公表された文書または試験の結果から、ばく露による潜在的な慢性的健康影響に関する新たな情報が明らかになったような場合がこれにあたる。

1.4.7.2.2 情報の更新は、変更を必要とする情報を入手し次第、迅速に行うべきである。所管官庁は情報を改訂するまでの時間的期限を定めてもよい。これは、農薬で行われるような認可手続を伴わない製品の表示や SDS にのみ適用される。表示が製品認可手続の一部であるような農薬の表示システムでは、供給者が供給品の表示を自発的に更新することはできない。しかし、製品が危険物の輸送に関する要求事項の適用を受ける場合は、輸送に用いられる表示については、上記のとおり新情報の入手時に更新するべきである。

1.4.7.2.3 また供給者は、たとえ新しく重要な情報がなかったとしても、物質または混合物の表示および安全データシートの基礎となる情報について定期的に見直しを行うべきである。これには例えば、化学品の危険有害性のデータベースにおける新情報の検索が必要となろう。所管官庁は、当初の作成期日から起算した期限（通常 3～5 年）を定め、その期間内に供給者が関連の表示および SDS 情報の見直しを行うようにしてもよい。

1.4.8　営業秘密情報

1.4.8.1　GHS を採用しているシステムでは、どのような規定が営業秘密情報の保護に適切かを考慮するべきである。このような規定によって、作業者や消費者の健康と安全、または環境保護を危うくするべきではない。GHS の他の部分と同様、輸入される物質または混合物の営業秘密情報の申請については、輸入国の規則を適用するべきである。

1.4.8.2　システムで営業秘密情報の保護を規定することに決めた場合、所管官庁は国の法律と慣行に従い、適切なメカニズムを確立し、以下を考慮するべきである：

(a) ある特定の化学品または化学品の危険有害性クラスを含めることが、システムの要求事項に合っているかどうか；

(b) 競合相手が情報を入手してしまう可能性や、知的所有権などの要因、潜在的危険有害性の開示が事業主または供給者の事業に与える要因を考慮して、どのような「営業秘密情報」の定義を適用するべきか；および

(c) 作業者や消費者の健康と安全を保護するあるいは環境を保護する必要がある場合、営業秘密情報の開示の適切な手順、および追加の開示を防止する措置。

1.4.8.3　営業秘密情報の保護に関する規定は、国の法律と慣行により、システム間で異なる場合がある。しかし、これらは次の一般原則と一致させるべきである：

(a) ラベルまたは安全データシートで要求される情報については、営業秘密情報の申請は物質の名前と混合物中の濃度に制限するべきである。他のすべての情報は、要求どおり、ラベルまたは安全データシートで開示するべきである；

(b) 営業秘密情報がある場合は、ラベルまたは安全データシートでその事実を示すべきである；

(c) 営業秘密情報は要請に応じて、所管官庁に開示するべきである。所管官庁は適用される法律と慣行に従い、情報の機密性を保護するべきである；

(d) 危険有害性のある物質または混合物へのばく露による緊急事態であると医療関係者が決定した場合、供給者または事業主あるいは所管官庁が治療に必要な特定の秘密情報を適時に開示する手段を確保するべきである。医療関係者は情報の機密性を保持するべきである；

(e) 緊急事態でない場合には、供給者または事業主は、ばく露した作業者または消費者に医療や他

to use the information only for the purpose of consumer or worker protection, and to otherwise maintain its confidentiality;

(f) Where non-disclosure of confidential business information is challenged, the competent authority should address such challenges or provide for an alternative process for challenges. The supplier or employer should be responsible for supporting the assertion that the withheld information qualifies for confidential business information protection.

1.4.9 Training

Training users of hazard information is an integral part of hazard communication. Systems should identify the appropriate education and training for GHS target audiences who are required to interpret label and/or SDS information and to take appropriate action in response to chemical hazards. Training requirements should be appropriate for and commensurate with the nature of the work or exposure. Key target audiences for training include workers, emergency responders, and those involved in the preparation of labels, SDS and hazard communication strategies as part of risk management systems. Others involved in the transport and supply of hazardous chemicals also require training to varying degrees. In addition, systems should consider strategies required for educating consumers in interpreting label information on products that they use.

1.4.10 Labelling procedures

1.4.10.1 *Scope*

The following sections describe the procedures for preparing labels in the GHS, comprising the following:

(a) Allocation of label elements;

(b) Reproduction of the symbol;

(c) Reproduction of the hazard pictogram;

(d) Signal words;

(e) Hazard statements;

(f) Precautionary statements and pictograms;

(g) Product and supplier identification;

(h) Multiple hazards and precedence of information;

(i) Arrangements for presenting the GHS label elements;

(j) Special labelling arrangements.

1.4.10.2 *Label elements*

The tables in the individual chapters for each hazard class detail the label elements (symbol, signal word, hazard statement) that have been assigned to each of the hazard categories of the GHS. Hazard categories reflect the harmonized classification criteria. A summary of the allocation of label elements is provided in annex 1. Special arrangements to take into account the information needs of different target audiences are further described in 1.4.10.5.4.

1.4.10.3 *Reproduction of the symbol*

The following hazard symbols are the standard symbols which should be used in the GHS. With the exception of the new symbol which will be used for certain health hazards and the exclamation mark, they are part of the standard symbol set used in the *UN Model Regulations*.

の安全衛生サービスを提供する安全衛生の専門家、および作業者または作業者の代表者への秘密情報の開示を保証すべきである。情報を要求する者は、開示の理由を示し、消費者または作業者保護の目的でのみ情報を使用し、他の目的に使用しないことに同意するべきである；

(f) 営業秘密情報の非開示が要求された場合、所管官庁はこのような要求に対応するか、あるいは要求に対する代替の方法を規定するべきである。供給者または事業主は、保留された情報が営業秘密情報保護の対象になるという主張に対して責任を持つべきである。

1.4.9 訓練

危険有害性に関する情報の使用者に対する訓練は、情報伝達の重要な部分である。システムでは、GHS対象者はラベルまたは SDS 情報を解釈し、化学品の危険有害性に対応して適切な措置をとることが要求されるので、GHS の対象者に対する適切な教育と訓練の内容が明らかにされるべきである。訓練規定は、作業またはばく露の内容に見合った適切なものとすべきである。訓練の主な対象者は、作業者、緊急時対応者、ならびにリスクマネージメントシステムの一環としてラベル、SDS および危険有害性に関する情報の伝達方策の立案に関係する者を含む。危険有害性のある化学品の輸送と供給に関係する他の者も、様々なレベルで訓練を必要とする。加えて、システムでは、使用する製品のラベル情報の解釈に関する消費者の教育に必要な方策も考慮するべきである。

1.4.10 表示手順

1.4.10.1 *範囲*

以降の節では、GHS における表示の準備のための手順を説明する。その手順は以下の項目からなる：

(a) ラベル要素の割り当て；

(b) シンボルの記載；

(c) 危険有害性の絵表示の記載；

(d) 注意喚起語；

(e) 危険有害性情報；

(f) 注意書きおよび絵表示；

(g) 製品および供給者の特定；

(h) 複数の危険有害性および危険有害性に関する情報の優先順位；

(i) GHS ラベル要素の配置方法；

(j) ラベルに関する特別な取決め。

1.4.10.2 *ラベル要素*

各章の表には、GHS のそれぞれの危険有害性クラスに割り当てられたラベル要素（シンボル、注意喚起語、危険有害性情報）が列挙されており、これらは、GHS の危険有害性判定基準を反映している。ラベル要素の割り当てに関しては、附属書1にまとめられている。対象者別に必要な情報について考慮した特別の取決めについては、1.4.10.5.4 で詳述する。

1.4.10.3 *危険有害性シンボルの記載*

次の危険有害性シンボルは、GHS で使用すべき標準シンボルである。健康有害性に使用される新しいシンボル、感嘆符を除き、*UN モデル規則*で使用される標準シンボルが用いられている。

Flame	Flame over circle	Exploding bomb
🔥	🔥⭕	💥
Corrosion	Gas cylinder	Skull and crossbones
🧪	⬛	☠️
Exclamation mark	Environment	Health hazard
❗	🌳🐟	👤✴️

1.4.10.4 *Pictograms and reproduction of the hazard pictograms*

1.4.10.4.1 A pictogram means a graphical composition that may include a symbol plus other graphic elements, such as a border, background pattern or colour that is intended to convey specific information.

1.4.10.4.2 *Shape and colour*

1.4.10.4.2.1 All hazard pictograms used in the GHS should be in the shape of a square set at a point.

1.4.10.4.2.2 For transport, the pictograms (commonly referred to as labels in transport regulations) prescribed by the *UN Model Regulations* should be used. The *UN Model Regulations* prescribe transport pictogram specifications including colour, symbols, size, background contrast, additional safety information (e.g. hazard class) and general format. Transport pictograms are required to have minimum dimensions of 100 mm by 100 mm, with some exceptions for allowing smaller pictograms for very small packagings and for gas cylinders. Transport pictograms include the symbol in the upper half of the label. The *UN Model Regulations* require that transport pictograms be printed or affixed to a packaging on a background of contrasting colour. An example showing a typical label for a flammable liquid hazard according to the *UN Model Regulations* is provided below:

Pictogram for flammable liquid in the *UN Model Regulations* (Symbol: Flame: black or white;
Background: red; Figure 3 in bottom corner; minimum dimensions 100 mm × 100 mm)

1.4.10.4.2.3 Pictograms prescribed by the GHS but not the *UN Model Regulations*, should have a black symbol on a white background with a red frame sufficiently wide to be clearly visible. However, when such a pictogram appears on a label for a package which will not be exported, the competent authority may choose to give suppliers and employers discretion to use a black border. In addition, competent authorities may allow the use of *UN Model Regulations* pictograms in other use settings where the package is not covered by the Model Regulations. An example of a GHS pictogram used for a skin irritant is provided below.

炎	円上の炎	爆弾の爆発
🔥	🔥	💥
腐食性	ガスボンベ	どくろ
🧪	🛢	☠
感嘆符	環境	健康有害性
❗	🌳🐟	⭐

1.4.10.4　危険有害性を表す絵表示の記載

1.4.10.4.1　絵表示とは、ある情報を伝達することを意図した、シンボルと境界線、背景のパターンまたは色などの図的要素から構成されるものをいう。

1.4.10.4.2　*形と色*

1.4.10.4.2.1　GHS で使用されるすべての危険有害性を示す絵表示は、1 つの頂点で正立させた正方形の中に書かれるべきである。

1.4.10.4.2.2　輸送に対しては、*UN モデル規則*で指定された絵表示（一般に、輸送の規則における標札と呼ばれる）を用いるべきである。*UN モデル規則*は、色、シンボル、サイズ、背景の濃淡、および追加的な安全情報（例：危険有害性クラス）および様式を含む輸送の絵表示を規定している。輸送の絵表示は、最小でも 100mm 角の大きさが要求されているが、非常に小さい包装の場合、またはガスシリンダーに対しては、より小さな絵表示を例外として認めている。また、輸送の絵表示では標札の上半分にシンボルを置く。*UN モデル規則*では、輸送の絵表示は、コントラストのある色を背景として、包装の上に、印刷するか、または貼付する。引火性液体について *UN モデル規則*で使用する標札の例を下に示す：

*UN モデル規則*の　引火性液体の絵表示（シンボル：炎：黒または白、
背景：赤、下部の隅に数字の 3、最小寸法　100mm×100mm）

1.4.10.4.2.3　GHS で規定されているが、*UN モデル規則*では規定されていない絵表示は、白い背景の上に黒いシンボルを置き、はっきり見えるように十分に幅広い赤い枠で囲むべきである。しかし、輸出されない包装品のラベルにこのような絵表示を用いるときは、所管官庁は、供給者および事業主に黒い境界線を使用する許可を与えることができる。さらに、所管官庁は、包装品が *UN モデル規則*の対象とならない他の部門でも、*UN モデル規則*の絵表示の使用を許可することができる。皮膚刺激性物質に使用されるGHS 絵表示の例を下に示す。

Pictogram for skin irritant

1.4.10.4.3 *Codification*

Pictograms prescribed by the GHS for sectors other than transport, and a code uniquely identifying each one, are listed in section 4 of annex 3. The pictogram code is intended to be used for reference purposes only. It is not part of the pictogram and should not appear on labels or in section 2 of the safety data sheet.

1.4.10.4.4 *Use of GHS pictograms in transport*

In transport, a GHS pictogram not required by the *UN Model Regulations* should only appear as part of a complete GHS label (see 1.4.10.5.4.1) and not independently.

1.4.10.5 *Allocation of label elements*

1.4.10.5.1 *Information required for packages covered by the UN Model Regulations*

Where a *UN Model Regulations* pictogram appears on a label, a GHS pictogram for the same hazard should not appear. The GHS pictograms not required for the transport of dangerous goods should not be displayed on freight containers, road vehicles or railway wagons/tanks.

1.4.10.5.2 *Information required on a GHS label*

(a) Signal words

A signal word means a word used to indicate the relative level of severity of hazard and alert the reader to a potential hazard on the label. The signal words used in the GHS are "Danger" and "Warning". "Danger" is mostly used for the more severe hazard categories (i.e. in the main for hazard categories 1 and 2), while "Warning" is mostly used for the less severe. The tables in the individual chapters for each hazard class detail the signal words that have been assigned to each of the hazard categories of the GHS.

(b) Hazard statements

(i) A hazard statement means a phrase assigned to a hazard class and category that describes the nature of the hazards of a hazardous product, including, where appropriate, the degree of hazard. The tables of label elements in the individual chapters for each hazard class detail the hazard statements that have been assigned to each of the hazard categories of the GHS;

(ii) Hazard statements and a code uniquely identifying each one are listed in section 1 of annex 3. The hazard statement code is intended to be used for reference purposes. It is not part of the hazard statement text and should not be used to replace it.

(c) Precautionary statements and pictograms

(i) A precautionary statement means a phrase (and/or pictogram) that describes recommended measures that should be taken to minimise or prevent adverse effects resulting from exposure to a hazardous product, or improper storage or handling of a hazardous product. The GHS label should include appropriate precautionary information, the choice of which is with the labeller or the competent authority. Annex 3 contains examples of precautionary statements, which can be used, and also examples of precautionary pictograms, which can be used where allowed by the competent authority;

皮膚刺激性の絵表示

1.4.10.4.3　コード化

　輸送以外の分野に対する GHS で規定されている絵表示およびそれぞれを一意的に認識できるコードは附属書 3 の第 4 節に記載した。絵表示コードは参照のためだけに使用されるものである。これは絵表示の一部ではなく、ラベルや安全データシート第 2 節に記載するべきではない。

1.4.10.4.4　輸送における GHS 絵表示の使用

　輸送においては、UN モデル規則で要求されていない GHS の絵表示は完全な GHS ラベルの一部としてのみ表示しなければならない（1.4.10.5.4.1 を参照）し、また単独で用いるべきではない。

1.4.10.5　ラベル要素の配置

1.4.10.5.1　UN モデル規則による包装に必要な情報

　UN モデル規則の絵表示をラベルに使用する場合には、同じ危険有害性に関する GHS の絵表示を使用すべきでない。また、危険物輸送に要求されない GHS 絵表示は、貨物輸送用コンテナ、道路車両または鉄道貨車/タンクに付けるべきでない。

1.4.10.5.2　GHS ラベルに必要な情報

(a) 注意喚起語

　注意喚起語とは、危険有害性の重大性の相対的レベルを示し、利用者に対して潜在的な危険有害性について警告するための語句を意味する。GHS で用いられる注意喚起語は、「危険 (Danger)」と「警告 (Warning)」である。「危険」は多くの場合、より厳しい危険有害性区分に用いられ（主として危険有害性の区分 1 と 2）、「警告」は多くの場合より緩い区分に用いられる。GHS の各危険有害性の区分に割り当てられた注意喚起語は、各章のそれぞれの危険有害性クラスに関する表に示されている。

(b) 危険有害性情報

(i) 危険有害性情報とは、各危険有害性クラスおよび区分に割り当てられた文言で、該当製品の危険有害性の性質と該当する場合はその程度を示すものである。GHS の各危険有害性区分に割り当てられた危険有害性情報は、各章のそれぞれの危険有害性クラスに関する表に示されている；

(ii) 危険有害性情報およびそれらを特定するコードは附属書 3 の第 1 節に記載されている。危険有害性情報のコードは参照するためのものである。コードは危険有害性情報の文言の一部ではないので、文言の代わりに用いることはできない。

(c) 注意書きおよび絵表示

(i) 注意書きは、危険有害性をもつ製品へのばく露、または、その不適切な貯蔵や取扱いから生じる被害を防止し、または最小にするために取るべき推奨措置について記述した文言（または絵表示）を意味する。GHS ラベルは適切な注意書きを含むべきであるが、その選択は表示者または所管官庁が行う。附属書 3 では使用できる注意書きの例、および所管官庁が許可した場合に使用できる予防策を表す絵表示の例を示す；

(ii) Precautionary statements and a code uniquely identifying each one are listed in section 2 of annex 3. The precautionary statement code is intended to be used for reference purposes. It is not part of the precautionary statement text and should not be used to replace it.

(d) Product identifier

(i) A product identifier should be used on a GHS label and it should match the product identifier used on the SDS. Where a substance or mixture is covered by the *UN Model Regulations*, the UN proper shipping name should also be used on the package;

(ii) The label for a substance should include the chemical identity of the substance. For mixtures or alloys, the label should include the chemical identities of all ingredients or alloying elements that contribute to acute toxicity, skin corrosion or serious eye damage, germ cell mutagenicity, carcinogenicity, reproductive toxicity, skin or respiratory sensitization, or specific target organ toxicity (STOT), when these hazards appear on the label. Alternatively, the competent authority may require the inclusion of all ingredients or alloying elements that contribute to the hazard of the mixture or alloy;

(iii) Where a substance or mixture is supplied exclusively for workplace use, the competent authority may choose to give suppliers discretion to include chemical identities on the SDS, in lieu of including them on labels;

(iv) The competent authority rules for confidential business information take priority over the rules for product identification. This means that where an ingredient would normally be included on the label, if it meets the competent authority criteria for confidential business information, its identity does not have to be included on the label.

(e) Supplier identification

The name, address and telephone number of the manufacturer or supplier of the substance or mixture should be provided on the label.

1.4.10.5.3 *Multiple hazards and precedence of hazard information*

The following arrangements apply where a substance or mixture presents more than one GHS hazard. It is without prejudice to the building block principle described in the *Purpose, scope and application* (chapter 1.1). Therefore, where a system does not provide information on the label for a particular hazard, the application of the arrangements should be modified accordingly.

1.4.10.5.3.1 Precedence for the allocation of symbols

For substances and mixtures covered by the *UN Model Regulations*, the precedence of symbols for physical hazards should follow the rules of the *UN Model Regulations*. In workplace situations, the competent authority may require all symbols for physical hazards to be used. For health hazards the following principles of precedence apply:

(a) if the skull and crossbones applies, the exclamation mark should not appear;

(b) if the corrosive symbol applies, the exclamation mark should not appear where it is used for skin or eye irritation;

(c) if the health hazard symbol appears for respiratory sensitization, the exclamation mark should not appear where it is used for skin sensitization or for skin or eye irritation.

1.4.10.5.3.2 Precedence for allocation of signal words

If the signal word "Danger" applies, the signal word "Warning" should not appear.

1.4.10.5.3.3 Precedence for allocation of hazard statements

All assigned hazard statements should appear on the label, except where otherwise provided in this subsection. The competent authority may specify the order in which they appear.

(ii) 注意書きおよびそれらを特定するコードは附属書 3 の第 2 節に記載されている。注意書きのコードは参照するためのものである。コードは注意書きの文言の一部ではないので、文言の代わりに用いることはできない。

(d) 製品特定名

(i) 製品特定名は、GHS ラベルに使用されるべきであるが、これは SDS で使用した製品特定名と一致させるべきである。当該物質または混合物に *UN モデル規則* が適応される場合は、包装品に国連品名も記載するべきである；

(ii) 物質用のラベルは、物質の化学的特定名を含むべきである。混合物または合金であって、急性毒性、皮膚腐食性または眼に対する重篤な損傷性、生殖細胞変異原性、発がん性、生殖毒性、皮膚感作性または呼吸器感作性、あるいは特定標的臓器毒性（STOT）の有害性がラベルに示される場合、これらに関与するすべての成分または合金元素の物質の化学的特定名をラベルに示すべきである。また、所管官庁は、混合物または合金の上記以外の健康有害性に関与するすべての成分または合金元素についてもラベルに記すよう要求することができる；

(iii) 物質または混合物が作業場での使用のためだけに供給される場合には、所管官庁は、物質の化学的特定名をラベルではなく SDS に記載する裁量を供給者に与えることができる；

(iv) 営業秘密情報に関する所管官庁の規則は製品の特定名の規則よりも優先される。つまり、通常であれば成分がラベルに記載される場合でも、その成分が営業秘密情報に関する所管官庁の判断基準を満たす場合は、その特定名をラベルに記載しなくてもよい。

(e) 供給者の特定

物質または混合物の製造者、または供給者の名前、住所および電話番号をラベルに記載すべきである。

1.4.10.5.3 *複数の危険有害性および危険有害性に関する情報の優先順位*

物質または混合物が複数の GHS 危険有害性を示す場合には以下のように取り扱う。これは、*目的、範囲、適用*（第 1.1 章）に記述されている選択可能方式の原則を侵すものではない。したがって、このシステムで、ある危険有害性に関する情報をラベルに記載しない場合には、以下の取決めはそれに応じて変更するべきである。

1.4.10.5.3.1 シンボルの割当てに関する優先順位

UN モデル規則 が適用される物質および混合物については、物理化学的危険性のシンボルの優先順位は *UN モデル規則* に従うべきである。作業場については、所管官庁は物理化学的危険性のすべてのシンボルの使用を要求してもよい。健康に対する有害性については、次の優先順位の原則が適用される：

(a) どくろを適用する場合、感嘆符を使用するべきではない；

(b) 腐食性シンボルを適用する場合、皮膚または眼刺激性を表す感嘆符を使用するべきではない；

(c) 呼吸器感作性に関する健康有害性シンボルを使用する場合、皮膚感作性または皮膚/眼刺激性を表す感嘆符を使用するべきではない。

1.4.10.5.3.2 注意喚起語の割り当てに関する優先順位

注意喚起語「危険」を適用する場合、注意喚起語「警告」を使用するべきではない。

1.4.10.5.3.3 危険有害性情報の割当てに関する優先順位

ラベルには、本節で定められた他の方法を除いて、割り当てられたすべての危険有害性情報を記載するべきである。所管官庁は、それらを示す順序を指定してもよい。

However, to avoid evident duplication or redundancy in the information conveyed by hazard statements, the following precedence rules may be applied:

(a) If the statement H410 "Very toxic to aquatic life with long lasting effects" is assigned, the statement H400 "Very toxic to aquatic life" may be omitted;

(b) If the statement H411 "Toxic to aquatic life with long lasting effects" is assigned, the statement H401 "Toxic to aquatic life" may be omitted;

(c) If the statement H412 "Harmful to aquatic life with long lasting effects" is assigned, the statement H402 "Harmful to aquatic life" may be omitted;

(d) If the statement H314 "Causes severe skin burns and eye damage" is assigned, the statement H318 "Causes serious eye damage" may be omitted.

Competent authorities may decide whether to require use of the above precedence rules, or to leave the choice to the manufacturer/supplier.

Table A3.1.2 in annex 3 includes specified combinations of hazard statements. Where a combined hazard statement is indicated, the competent authority may specify whether the combined hazard statement or the corresponding individual statements should appear on the label or may leave the choice to the manufacturer/supplier.

1.4.10.5.4 *Arrangements for presenting the GHS label elements*

1.4.10.5.4.1 Location of GHS information on the label

The GHS hazard pictograms, signal word and hazard statements should be located together on the label. The competent authority may choose to provide a specified layout for the presentation of these and for the presentation of precautionary information or allow supplier discretion. Specific guidance and examples are provided in the chapters on individual hazard classes.

There have been some concerns about how the label elements should appear on different packagings. Specific examples are provided in annex 7.

1.4.10.5.4.2 Supplemental information

The competent authority has the discretion to allow the use of supplemental information subject to the parameters outlined in 1.4.6.3. The competent authority may choose to specify where this information should appear on the label or allow supplier discretion. In either approach, the placement of supplemental information should not impede identification of GHS information.

1.4.10.5.4.3 Use of colour outside pictograms

In addition to its use in pictograms, colour can be used on other areas of the label to implement special labelling requirements such as the use of the pesticide bands in the FAO Labelling Guide, for signal words and hazard statements or as background to them, or as otherwise provided for by the competent authority.

1.4.10.5.4.4 Labelling of small packagings

The general principles that should underpin labelling of small packagings are:

(a) All the applicable GHS label elements should appear on the immediate container of a hazardous substance or mixture where possible;

(b) Where it is impossible to put all the applicable label elements on the immediate container itself, other methods of providing the full hazard information should be used in accordance with the definition of "Label" in the GHS. Factors influencing this include inter alia:

(i) the shape, form or size of the immediate container;

(ii) the number of label elements to be included, particularly where the substance or mixture meets the classification criteria for multiple hazard classes;

しかし、危険有害性情報における明らかな重複や冗長を避けるために、次のような優先に関する決まりを適用してもよい：

(a) H410「長期継続的影響により水生生物に非常に強い毒性」が割り当てられた場合、H400「水生生物に非常に強い毒性」は省略することができる；

(b) H411「長期継続的影響により水生生物に毒性」が割り当てられた場合、H401「水生生物に毒性」は省略することができる；

(c) H412「長期継続的影響により水生生物に有害」が割り当てられた場合、H402「水生生物に有害」は省略することができる；

(d) H314「重篤な皮膚の薬傷・眼の損傷」が割り当てられた場合、H318「重篤な眼の損傷」は省略することができる。

所管官庁は上記の優先に関する決まりを要求するか、あるいはその選択を製造者/供給者に委ねるか決めることができる。

附属書3の表A3.1.2には危険有害性情報の特別な組み合わせが示してある。組み合わせられた危険有害性情報が示されたところに関しては、所管官庁は組み合わせられた危険有害性情報かまたはそれぞれ個々の危険有害性情報のどちらをラベルに記載するか、あるいはその選択を製造者/供給者に委ねるか決めることができる

1.4.10.5.4 *GHS ラベル要素を提示する際の取決め*

1.4.10.5.4.1 ラベル上の GHS 情報の配置

GHS の危険有害性を表す絵表示、注意喚起語および危険有害性情報はラベル上に一緒に配置するべきである。所管官庁は、これらの記載および注意書きの記載について配置を指定するか、または供給者の自由裁量に任せることができる。各章の危険有害性クラスのところにガイダンスと例が示されている。

ラベル要素を種々の包装にどのように表示すべきかについての関心が示されてきた。特定の例が附属書7に示されている。

1.4.10.5.4.2 補足情報

所管官庁は、1.4.6.3 で概説された事項に従った補足情報の使用を許可する裁量を有する。所管官庁は、この情報のラベルの記載すべき場所を指定しても、または選択に任せてもよい。いずれの場合においても、補足情報の配置が GHS で定められている情報を妨げるべきでない。

1.4.10.5.4.3 絵表示外での色の使用

色は、絵表示で使用するほか、特別なラベルの要件を満たすためにラベルの他の領域で使用することができる。例えば、FAO 表示ガイドにおける農薬標識への使用、注意喚起語や危険有害性情報、またはそれらの背景、あるいは所管官庁による他の規定での使用などがある。

1.4.10.5.4.4 小さな包装のラベル

小さな包装のラベルについて勘案されなければならない原則は以下のとおりである：

(a) 可能であれば、すべての適用される GHS ラベル要素は危険有害な物質あるいは混合物が直接入っている容器に記載されていなければならない；

(b) すべての適用されるラベル要素が直接の容器に記載できない場合には、GHS の「ラベル」の定義にしたがって、すべての危険有害性情報を示す他の方法が用いられなければならない。これに影響する要素には特に次のようなものがある：

　　(i) 直接容器の形やサイズ；

　　(ii) 含まれるべきラベル要素の数、特に物質や混合物が多くの危険有害性クラスに対して判定基準が当てはまる場合；

(iii) the need for label elements to appear in more than one official language.

(c) Where the volume of a hazardous substance or mixture is so low and the supplier has data demonstrating, and the competent authority has determined, that there is no likelihood of harm to human health and/or the environment, then the label elements may be omitted from the immediate container;

(d) Competent authorities may allow certain label elements to be omitted from the immediate container for certain hazard classes/categories where the volume of the substance or mixture is below a certain amount;

(e) Some labelling elements on the immediate container may need to be accessible throughout the life of the product, e.g. for continuous use by workers or consumers.

1.4.10.5.5   *Special labelling arrangements*

The competent authority may choose to allow communication of certain hazard information for carcinogens, reproductive toxicity and specific target organ toxicity through repeated exposure on the label and on the SDS, or through the SDS alone (see specific chapters for details of relevant cut-offs for these classes).

Similarly, for metals and alloys, the competent authority may choose to allow communication of the hazard information through the SDS alone when they are supplied in the massive, non-dispersible, form.

Where a substance or mixture is classified as corrosive to metals but not corrosive to skin and/or eyes, the competent authority may choose to allow the hazard pictogram linked to "corrosive to metals" to be omitted from the label of such substances or mixtures which are in the finished state as packaged for consumer use.

1.4.10.5.5.1   Workplace labelling

Products falling within the scope of the GHS will carry the GHS label at the point where they are supplied to the workplace, and that label should be maintained on the supplied container in the workplace. The GHS label or label elements should also be used for workplace containers. However, the competent authority can allow employers to use alternative means of giving workers the same information in a different written or displayed format when such a format is more appropriate to the workplace and communicates the information as effectively as the GHS label. For example, label information could be displayed in the work area, rather than on the individual containers.

Alternative means of providing workers with the information contained in GHS labels are needed usually where hazardous chemicals are transferred from an original supplier container into a workplace container or system, or where chemicals are produced in a workplace but are not packaged in containers intended for sale or supply. Chemicals that are produced in a workplace may be contained or stored in many different ways such as: small samples collected for testing or analysis, piping systems including valves, process or reaction vessels, ore cars, conveyer systems or free-standing bulk storage of solids. In batch manufacturing processes, one mixing vessel may be used to contain a number of different mixtures.

In many situations, it is impractical to produce a complete GHS label and attach it to the container, due, for example, to container size limitations or lack of access to a process container. Some examples of workplace situations where chemicals may be transferred from supplier containers include: containers for laboratory testing or analysis, storage vessels, piping or process reaction systems or temporary containers where the chemical will be used by one worker within a short timeframe. Decanted chemicals intended for immediate use could be labelled with the product identifier and directly refer the user to the supplier label information and SDS.

All such systems should ensure that there is clear hazard communication. Workers should be trained to understand the specific communication methods used in a workplace. Examples of alternative methods include: use of product identifiers together with GHS symbols and other pictograms to describe precautionary measures; use of process flow charts for complex systems to identify chemicals contained in pipes and vessels with links to the appropriate SDS; use of displays with GHS symbols, colour and signal words in piping systems and processing equipment; use of permanent placarding for fixed piping; use of batch tickets or recipes for labelling batch mixing vessels and use of piping bands with hazard symbols and product identifiers.

(iii) 1つの公用語以上でラベルに記載する必要がある場合。

(c) 物質あるいは混合物の容量が非常に少なくて、供給者がヒトの健康や環境への害がなさそうであることを示し、所管官庁が決定した場合には、ラベル要素は直接容器から省略することができる：

(d) 所管官庁は、物質や混合物の容量がある量よりも少ない場合には決められた危険有害性のクラスや区分を直接容器から省略することを認めてもよい：

(e) 直接容器上のラベル表示要素のいくつかは、製品のライフサイクルを通じて利用可能とする必要があろう、例えば労働者や消費者によって続けて使用されるものなど。

1.4.10.5.5　ラベルに関する特別な取決め

所管官庁は、発がん性物質、生殖毒性および特定標的臓器毒性反復ばく露に関する特定の危険有害性に関する情報については、ラベルおよび SDS、または SDS のみにより、情報伝達を行う場合がある（これらの危険有害性クラスに関連したカットオフの詳細については各章を参照すること）。

同様に、金属と合金が大量かつ散逸しない状態で供給されるときには、所管官庁は SDS だけで危険有害性に関する情報の伝達を行うことを許可することもある。

所管官庁は、物質または混合物が金属に対して腐食性であるが皮膚および/または眼に対しては腐食性でない場合には、消費者製品として包装され完成しているそのような物質または混合物のラベルから「金属腐食性」に関連した絵表示の削除を許可することを選択してもよい。

1.4.10.5.5.1　作業場用の表示

GHS の対象となる製品には、作業場に供給される時点で GHS のラベルが付けられるが、そのラベルは、作業場においてもその供給された容器にずっと付けておくべきである。また、GHS のラベルあるいはラベル要素は作業場の容器にも使用されるべきである。所管官庁は同じ情報を作業者に伝える代替手段として、事業主が、異なる記述あるいは表示様式を用いることを許可することができる。ただし、このような様式は作業場において、より適切で、必要な情報が GHS ラベルと同様に有効に伝達される場合に限る。例えば、ラベル情報を個々の容器上に付すのではなく、作業区域内に表示することもできる。

労働者に対して GHS ラベルに含まれる情報を示すための代替手段は、通常、危険有害性を有する化学品が供給者の容器から作業場の容器もしくはシステムに移し替えられる場合や、化学品が作業場で製造され、販売もしくは供給用の容器に収納されない場合に必要となる。作業場で製造される化学品は、様々な方法で容器に投入あるいは貯蔵される。例えば試験もしくは分析用に集められた少量の試料や、弁、処理工程もしくは反応容器を含む配管、鉱石運搬車、コンベアシステム、ばら積などがあげられる。バッチ式製造工程においては、様々な混合物を入れるのに1つの混合容器が用いられる場合もある。

多くの状況において、完全な GHS のラベルを作成し、それを容器に貼付することは、容器のサイズによる制約や工程用の容器に近づけないなどの理由から現実的ではない。化学品が供給用容器から移し替えられるような作業場としては、例えば、研究所での試験または分析用容器、貯蔵容器、パイプまたは反応システム、1人の作業者が化学品を短時間だけ利用するための一時的な容器などがある。すぐ利用するために分取した化学品の製品特定名についてラベルで示し、使用者に供給者のラベル情報と SDS を直接参照させることが必要となろう。

このすべてのシステムにおいて、危険有害性に関する明確な情報の伝達が保証されるべきである。労働者には作業場で用いられる情報伝達の方法について理解できるような訓練をするべきである。代替手段の例としては、GHS シンボルおよびその他の予防対策を表した絵表示とともに製品の特定名を用いる、パイプや容器に含まれる化学品の識別を行うために SDS とともに複雑なシステムの工程にはフローチャートを用いる、配管および工程の設備に GHS のシンボル、色、注意喚起語を使った表示を行う、固定配管には恒久的な掲示を行う、バッチ式混合容器の表示にバッチ表示や配合表を用いる、危険有害性シンボルおよび製品の特定名を示す配管標識を用いる、などがある。

1.4.10.5.5.2   Consumer product labelling based on the likelihood of injury

All systems should use the GHS classification criteria based on hazard, however competent authorities may authorize consumer labelling systems providing information based on the likelihood of harm (risk-based labelling). In the latter case the competent authority would establish procedures for determining the potential exposure and risk for the use of the product. Labels based on this approach provide targeted information on identified risks but may not include certain information on chronic health effects (e.g. specific target organ toxicity (STOT)) following repeated exposure, reproductive toxicity and carcinogenicity), that would appear on a label based on hazard alone. A general explanation of the broad principles of risk-based labelling is contained in annex 5.

1.4.10.5.5.3   Tactile warnings

If tactile warnings are used, the technical specifications should conform with ISO 11683:1997 "Tactile warnings of danger: Requirements".

1.4.10.5.5.2　危害の可能性に基づく消費者製品の表示

　すべてのシステムは、GHS 分類基準を使用するべきである。しかし、所管官庁は、障害の可能性に基づいて情報を提供する消費者表示システムを認可することができる（リスクに基づくラベル）。その場合、所管官庁は製品使用に対する潜在的ばく露およびリスクを決定する手順を確立することとなる。この方法に基づくラベルでは、特定されたリスクに関して目的とされる情報を提供するが、有害性だけに基づくラベルで示される慢性健康影響（例えば、反復ばく露による特定標的臓器毒性（STOT）、生殖毒性、発がん性）に関する情報を含まない場合がある。リスクに基づくラベル表示に関する大まかな原則の説明を、附属書 5 に示す。

1.4.10.5.5.3　触覚による警告

　触覚による警告が使用される場合、技術仕様は「触覚による危険の警告：要求事項」に関する ISO 規格 11683（1997 年版）に従うべきである。

CHAPTER 1.5

HAZARD COMMUNICATION: SAFETY DATA SHEETS (SDS)

1.5.1 **The role of the safety data sheet (SDS) in the harmonized system**

1.5.1.1　　The SDS should provide comprehensive information about a substance or mixture for use in workplace chemical control regulatory frameworks. Both employers and workers use it as a source of information about hazards, including environmental hazards, and to obtain advice on safety precautions. The information acts as a reference source for the management of hazardous chemicals in the workplace. The SDS is product related and, usually, is not able to provide specific information that is relevant for any given workplace where the product may finally be used, although where products have specialized end uses the SDS information may be more workplace-specific. The information therefore enables the employer (a) to develop an active programme of worker protection measures, including training, which is specific to the individual workplace; and (b) to consider any measures which may be necessary to protect the environment.

1.5.1.2　　In addition, the SDS provides an important source of information for other target audiences in the GHS. So certain elements of information may be used by those involved with the transport of dangerous goods, emergency responders (including poison centers), those involved in the professional use of pesticides and consumers. However, these audiences receive additional information from a variety of other sources such as the *UN Model Regulations* document and package inserts for consumers and will continue to do so. The introduction of a harmonized labelling system therefore, is not intended to affect the primary use of the SDS which is for workplace users.

1.5.2 **Criteria for determining whether an SDS should be produced**

An SDS should be produced for all substances and mixtures which meet the harmonized criteria for physical, health or environmental hazards under the GHS and for all mixtures which contain ingredients that meet the criteria for carcinogenic, toxic to reproduction or specific target organ toxicity in concentrations exceeding the cut-off limits for SDS specified by the criteria for mixtures (see 1.5.3.1). The competent authority may also require SDS's for mixtures not meeting the criteria for classification as hazardous, but which contain hazardous ingredients in certain concentrations (see 1.5.3.1).

1.5.3 **General guidance for compiling a safety data sheet**

1.5.3.1 *Cut-off values/concentration limits*

1.5.3.1.1　　An SDS should be provided based on the generic cut-off values/concentration limits indicated in table 1.5.1:

Table 1.5.1: Cut-off values/concentration limits for each health and environmental hazard class

Hazard class	Cut-off value/concentration limit
Acute toxicity	$\geq 1.0\%$
Skin corrosion/Irritation	$\geq 1.0\%$
Serious eye damage/eye irritation	$\geq 1.0\%$
Respiratory/Skin sensitization	$\geq 0.1\%$
Germ cell mutagenicity (Category 1)	$\geq 0.1\%$
Germ cell mutagenicity (Category 2)	$\geq 1.0\%$
Carcinogenicity	$\geq 0.1\%$
Reproductive toxicity	$\geq 0.1\%$
Specific target organ toxicity (single exposure)	$\geq 1.0\%$
Specific target organ toxicity (repeated exposure)	$\geq 1.0\%$
Aspiration hazard (Category 1)	$\geq 1.0\%$
Aspiration hazard (Category 2)	$\geq 1.0\%$
Hazardous to the aquatic environment	$\geq 1.0\%$

第 1.5 章

危険有害性に関する情報の伝達：安全データシート（SDS）

1.5.1 調和システムにおける 安全データシート（SDS）の役割

1.5.1.1 SDS は、作業場の化学品管理規制の枠組みの中で使用するために、物質または混合物に関する包括的な情報を提供するべきである。事業主と作業者の両者は、環境に対する危険有害性も含めた危険有害性に関する情報源として、また、安全対策に関する助言を得るために、これを使用する。この情報は、作業場で使用する危険有害性のある化学品を管理するための情報源としての役割を果たす。製品に特殊な最終用途がある場合には、SDS 情報はより作業場に特化したものとなることがあるが、通常は、SDS は製品に関連したものであり、製品が最終的に使用される特定の作業場に関連した特殊な情報を提供することはできない。したがって、その情報によって、事業主は、(a) 個々の作業場に特化した訓練などの、作業者保護対策の活動プログラムを開発し、(b) 環境の保護に必要な対策を考慮することができる。

1.5.1.2 また、SDS は GHS の他の対象者にとって重要な情報源となる。したがって、情報の一部分が、危険物輸送従事者、緊急時対応者（毒物管理センターを含む）、農薬の専門的使用者、および消費者によって使用されることもある。これらの対象者は、一方で *UN モデル規則* や消費者向けの包装内の説明書き等様々な他の情報源から追加情報を受けており、また引き続きこれらの情報を受けることになろう。調和した表示システムの導入が、作業場の使用者に向けた SDS の基本的な使用に影響を与えることはない。

1.5.2 SDS を作成するべきかどうかの判断基準

SDS は、GHS に基づく物理化学的な危険性や、ヒトの健康または環境に対する有害性に関する調和された判定基準を満たすすべての物質および混合物について作成されるべきである。また、混合物に対する判定基準で指定されたカットオフ限界（1.5.3.1 参照）を超える濃度の発がん性、生殖毒性、特定標的臓器毒性のある成分を含むすべての混合物についても作成されるべきである。所管官庁は、危険有害性として分類される判定基準に合致しなくても、危険有害な成分を一定濃度以上含む混合物に対して SDS を要求することができる（1.5.3.1 参照）。

1.5.3 安全データシート作成のための全般的指針

1.5.3.1 *カットオフ値/濃度限界*

1.5.3.1.1 SDS は、表 1.5.1 に示した統一的なカットオフ値/濃度限界に基づいて作成されるべきである。

表 1.5.1：健康および環境の各危険有害性クラスに対するカットオフ値/濃度限界

危険有害性クラス	カットオフ値/濃度限界
急性毒性	1.0%以上
皮膚腐食性/刺激性	1.0%以上
眼に対する重篤な損傷性/眼刺激性	1.0%以上
呼吸器感作性または皮膚感作性	0.1%以上
生殖細胞変異原性（区分1）	0.1%以上
生殖細胞変異原性（区分2）	1.0%以上
発がん性	0.1%以上
生殖毒性	0.1%以上
特定標的臓器毒性（単回ばく露）	1.0%以上
特定標的臓器毒性（反復ばく露）	1.0%以上
誤えん有害性（区分1）	1.0%以上
誤えん有害性（区分2）	1.0%以上
水生環境有害性	1.0%以上

1.5.3.1.2　　As noted in the *Classification of hazardous substances and mixtures* (see chapter 1.3), there may be some cases when the available hazard data may justify classification on the basis of other cut-off values/concentration limits than the generic ones specified in the health and environment hazard class chapters (chapters 3.2 to 3.10 and 4.1). When such specific cut-off values are used for classification, they should also apply to the obligation to compile an SDS.

1.5.3.1.3　　Some competent authorities may require SDS's to be compiled for mixtures which are not classified for acute toxicity or aquatic toxicity as a result of application of the additivity formula, but which contain acutely toxic or toxic to the aquatic environment ingredients in concentrations equal to or greater than 1 %[1].

1.5.3.1.4　　In accordance with the building block approach, some competent authorities may choose not to regulate certain categories within a hazard class. In such situations, there would be no obligation to compile an SDS.

1.5.3.1.5　　Once it is clear that an SDS is required for a substance or a mixture then the information required to be included in the SDS should in all cases be provided in accordance with GHS requirements.

1.5.3.2　　*SDS format*

1.5.3.2.1　　The information in the SDS should be presented using the following 16 headings in the order given below:

1. Identification
2. Hazard(s) identification
3. Composition/information on ingredients
4. First-aid measures
5. Fire-fighting measures
6. Accidental release measures
7. Handling and storage
8. Exposure controls/personal protection
9. Physical and chemical properties
10. Stability and reactivity
11. Toxicological information
12. Ecological information
13. Disposal considerations
14. Transport information
15. Regulatory information
16. Other information.

1.5.3.3　　*SDS content*

1.5.3.3.1　　The SDS should provide a clear description of the data used to identify the hazards. The minimum information in table 1.5.2 should be included, where applicable and available, on the SDS under the relevant headings[2]. If specific information is not applicable or not available under a particular subheading, the SDS should clearly state this. Additional information may be required by competent authorities. Guidance on the preparation of SDS's under the requirements of the GHS can be found in annex 4.

1.5.3.3.2　　Some subheadings relate to information that is national or regional in nature, for example "EC number" and "occupational exposure limits". Suppliers or employers should include information under such SDS subheadings that is appropriate and relevant to the countries or regions for which the SDS is intended and into which the product is being supplied.

---

[1] *The cut-off values for classification of mixtures are normally specified by concentrations expressed as % of the ingredients. In some cases, for example acute toxicity (human health), the cut-off values are expressed as acute toxicity values (ATE). The classification of a mixture is determined by additivity calculation based on acute toxicity values (see chapter 3.1) and concentrations of ingredients. Similarly, acute aquatic toxicity classification may be calculated on the basis of acute aquatic toxicity values (see chapter 4.1) and where appropriate, corrosion/irritation by adding up concentrations of ingredients (see chapters 3.2 and 3.3). Ingredients are taken into consideration for application of the formula when the concentration is equal to or greater than 1 %. Some competent authorities may use this cut-off as a basis of obligation to compile an SDS.*

[2] *Where "applicable" means where the information is applicable to the specific product covered by the SDS. Where "available" means where the information is available to the supplier or other entity that is preparing the SDS.*

1.5.3.1.2　危険有害性物質および混合物の分類（第 1.3 章参照）で述べたように、利用可能な有害性データがある場合には、ヒトの健康および環境に対する危険有害性クラスについての章（第 3.2 章～第 3.10 章および第 4.1 章）で指定されている統一的なカットオフ値/濃度限界以外の値に基づく分類が妥当なこともある。このような特別のカットオフ値を分類に用いる場合、それらは SDS を作成する場合にも適用するべきである。

1.5.3.1.3　所管官庁は、加算式を適用した結果として急性毒性または水生環境有害性とは分類されないが、急性毒性物質または水生生物への有害性を有する物質を 1%以上の濃度で含む混合物について、SDS を作成するよう求めてもよい[1]。

1.5.3.1.4　所管官庁は、選択可能方式の原則に従い、ある危険有害性クラスにおける区分に関して規制をしなくてもよい。この場合、SDS にこの区分について記載する義務はないであろう。

1.5.3.1.5　ある物質または混合物に関して SDS が必要となることが明らかになった場合、SDS に含めるべき情報は、GHS の要求事項に従って提供するべきである。

1.5.3.2　*SDS の様式*

1.5.3.2.1　SDS の情報は、次の 16 項目を使用し、下に示す順序で記載するべきである：
1. 特定情報
2. 危険有害性の要約
3. 組成および成分情報
4. 応急措置
5. 火災時の措置
6. 漏出時の措置
7. 取扱いおよび保管上の注意
8. ばく露防止および保護措置
9. 物理的および化学的性質
10. 安定性および反応性
11. 有害性情報
12. 環境影響情報
13. 廃棄上の注意
14. 輸送上の注意
15. 適用法令
16. その他の情報。

1.5.3.3　*SDS の内容*

1.5.3.3.1　SDS は、関係する危険有害性を特定するのに用いられたデータを明確に記載するべきである。表 1.5.2 に示した最低限の情報は、該当する場合であってかつ入手可能な場合において、SDS の関連する項目に含めるべきである[2]。小項目に該当する特定の情報がない、または入手不能である場合は、SDS にその事実を明示するべきである。所管官庁は追加情報を要求してもよい。GHS で要求される SDS の作成ガイダンスは附属書 4 にある。

1.5.3.3.2　一部の小項目は、例えば「EC 番号」や「職業ばく露限界」などの国内または地域的な情報に関係するものである。供給者または事業者は、これらの SDS 小項目に、その SDS が用いられ、その製品が供給される国または地域に該当し関連する情報を盛り込むべきである。

---

[1]　混合物の分類のためのカットオフ値は、通常、成分物質の%濃度で定められる。急性毒性（人の健康）等一部の事例では、上限値が急性毒性推定値（ATE）として表される。混合物の分類は、急性毒性値と成分物質の濃度に基づく加算的な計算によって決定される（第 3.1 章参照）。同様に、急性水生環境有害性の分類も急性水生毒性値（第 4.1 章参照）に基づいて、また、腐食性/刺激性も該当する場合は個々の物質の濃度を加算して算定することができる（第 3.2 章、第 3.3 章を参照）。成分物質の濃度が 1%以上になった場合に算定式の適用が考慮される。所管官庁は、このカットオフ値に基づき SDS への記載を求めてもよい

[2]　「該当する」場合とは、関係の情報が SDS の対象とする個々の製品に適用される場合をいう。「利用可能」な場合とは、情報が供給者またはその他 SDS の作成を行う組織にとって入手可能なものである場合をいう。

1.5.3.3.3   Additional safety and environmental information is required to address the needs of seafarers and other transport workers in the bulk transport of dangerous goods in sea-going or inland navigation bulk carriers or tank-vessels subject to IMO or national regulations. Paragraph A4.3.14.7 of annex 4 recommends the inclusion of basic classification information when such cargoes are transported in bulk according to IMO instruments. In addition, ships carrying oil or oil fuel, as defined in Annex I of MARPOL, in bulk or bunkering oil fuel are required before loading to be provided with a "material safety data sheet" in accordance with the IMO's Maritime Safety Committee (MSC) resolution "Recommendations for Material Safety Data Sheets (MSDS) for MARPOL Annex I Oil Cargo and Oil Fuel" (MSC.286(86)). Therefore, in order to have one harmonized SDS for maritime and non-maritime use, the additional provisions of Resolution MSC.286(86) may be included in the GHS SDS, where appropriate, for marine transport of MARPOL Annex I cargoes and marine fuel oils.

Table 1.5.2: Minimum information for an SDS

1.	**Identification of the substance or mixture and of the supplier**	(a) GHS product identifier; (b) Other means of identification; (c) Recommended use of the chemical and restrictions on use; (d) Supplier's details (including name, address, phone number etc.); (e) Emergency phone number.
2.	**Hazards identification**	(a) GHS classification of the substance/mixture and any national or regional information; (b) GHS label elements, including precautionary statements. (Hazard symbols may be provided as a graphical reproduction of the symbols in black and white or the name of the symbol e.g. "flame", "skull and crossbones"); (c) Other hazards which do not result in classification (e.g. "dust explosion hazard") or are not covered by the GHS.
3.	**Composition/ information on ingredients**	**Substance** (a) Chemical identity; (b) Common name, synonyms, etc.; (c) CAS number and other unique identifiers; (d) Impurities and stabilizing additives which are themselves classified and which contribute to the classification of the substance. **Mixture** The chemical identity and concentration or concentration ranges of all ingredients which are hazardous within the meaning of the GHS and are present above their cut-off levels. *NOTE: For information on ingredients, the competent authority rules for confidential business information take priority over the rules for product identification.*
4.	**First-aid measures**	(a) Description of necessary measures, subdivided according to the different routes of exposure, i.e. inhalation, skin and eye contact and ingestion; (b) Most important symptoms/effects, acute and delayed. (c) Indication of immediate medical attention and special treatment needed, if necessary.
5.	**Fire-fighting measures**	(a) Suitable (and unsuitable) extinguishing media. (b) Specific hazards arising from the chemical (e.g. nature of any hazardous combustion products). (c) Special protective equipment and precautions for fire-fighters.
6.	**Accidental release measures**	(a) Personal precautions, protective equipment and emergency procedures. (b) Environmental precautions. (c) Methods and materials for containment and cleaning up.
7.	**Handling and storage**	(a) Precautions for safe handling. (b) Conditions for safe storage, including any incompatibilities.
8.	**Exposure controls/personal protection**	(a) Control parameters e.g. occupational exposure limit values or biological limit values. (b) Appropriate engineering controls. (c) Individual protection measures, such as personal protective equipment.

(Cont'd on next page)

1.5.3.3.3　IMO あるいは国の規制を受ける航海や内陸航行のばら積み船あるいはタンク船での危険物のばら積み輸送における船員や他の輸送労働者の要求に応えるために、追加的な安全および環境情報が必要とされている。該当する荷が IMO 文書にしたがって液体ばら積みで輸送されるときは、基本的な分類情報を含めるように附属書 4 のパラグラフ A4.3.14.7 で推奨している。さらに油や燃料油を、MARPOL 附属書 I で定義されている、ばら積みあるいはバンカー燃料油で運んでいる船舶は、IMO 海上安全委員会（MSC）決議「MARPOL 附属書 I 油及び燃料油に対する物質安全データシート（MSDS）についての勧告」(MSC.286(86)) にしたがって、積込みの前に「安全データシート」を備えることが要求されている。それゆえ、海上および非海上の使用で 1 つの調和された SDS にするために、可能であれば、MARPOL 附属書 I 油及び燃料油の海上輸送に関して MSC.286(86) 決議の追加規定を GHS SDS に含めてもよい。

表 1.5.2：SDS の必要最少情報

1.	物質または混合物および会社情報	(a) GHS の製品特定手段 (b) 他の特定手段 (c) 化学品の推奨用途と使用上の制限 (d) 供給者の詳細（社名、住所、電話番号など） (e) 緊急時の電話番号
2.	危険有害性の要約	(a) 物質/混合物の GHS 分類と国/地域情報 (b) 注意書きも含む GHS ラベル要素。（危険有害性シンボルは、黒と白を用いたシンボルの図による記載またはシンボルの名前、例えば、「炎」、「どくろ」などとして示される場合がある） (c) 分類に関係しない（例「粉じん爆発危険性」）または GHS で扱われない他の危険有害性
3.	組成および成分情報	**物質** (a) 化学的特定名 (b) 慣用名、別名など (c) CAS 番号およびその他の特定名 (d) それ自体が分類され、物質の分類に寄与する不純物および安定化添加物 **混合物** GHS 対象の危険有害性があり、カットオフ値以上で存在するすべての成分の化学名と濃度または濃度範囲 *注記*：成分に関する情報については、製品の特定規則より営業秘密情報に関する所管官庁の規則が優先される。
4.	応急措置	(a) 異なるばく露経路、すなわち吸入、皮膚や眼との接触、および経口摂取に従って細分された必要な措置の記述 (b) 急性および遅延性の最も重要な症状/影響 (c) 必要な場合、応急処置および必要とされる特別な処置の指示
5.	火災時の措置	(a) 適切な（および不適切な）消火剤 (b) 化学品から生じる特定の危険有害性（例えば、「有害燃焼生成物の性質」） (c) 消火作業者用の特別な保護具と予防措置
6.	漏出時の措置	(a) 人体に対する予防措置、保護具および緊急時措置 (b) 環境に対する予防措置 (c) 封じ込めおよび浄化方法と機材
7.	取扱いおよび保管上の注意	(a) 安全な取扱いのための予防措置 (b) 混触危険性等、安全な保管条件
8.	ばく露防止および保護措置	(a) 職業ばく露限界値、生物学的限界値等の管理指標 (b) 適切な工学的管理 (c) 個人用保護具などの個人保護措置

(次ページに続く)

Table 1.5.2: Minimum information for an SDS *(cont'd)*

9.	Physical and chemical properties	Physical state; Colour; Odour; Melting point/freezing point; Boiling point or initial boiling point and boiling range; Flammability; Lower and upper explosion limit/flammability limit; Flash point; Auto-ignition temperature; Decomposition temperature; pH; Kinematic viscosity; Solubility; Partition coefficient: n-octanol/water (log value); Vapour pressure; Density and/or relative density; Relative vapour density; Particle characteristics.
10.	Stability and reactivity	(a) Reactivity (b) Chemical stability; (c) Possibility of hazardous reactions; (d) Conditions to avoid (e.g. static discharge, shock or vibration); (e) Incompatible materials; (f) Hazardous decomposition products.
11.	Toxicological information	Concise but complete and comprehensible description of the various toxicological (health) effects and the available data used to identify those effects, including: (a) information on the likely routes of exposure (inhalation, ingestion, skin and eye contact); (b) Symptoms related to the physical, chemical and toxicological characteristics; (c) Delayed and immediate effects and also chronic effects from short- and long-term exposure; (d) Numerical measures of toxicity (such as acute toxicity estimates).
12.	Ecological information	(a) Ecotoxicity (aquatic and terrestrial, where available); (b) Persistence and degradability; (c) Bioaccumulative potential; (d) Mobility in soil; (e) Other adverse effects.
13.	Disposal considerations	Description of waste residues and information on their safe handling and methods of disposal, including the disposal of any contaminated packaging.
14.	Transport information	(a) UN number; (b) UN proper shipping name; (c) Transport hazard class(es); (d) Packing group, if applicable; (e) Environmental hazards (e.g. Marine pollutant (Yes/No)); (f) Transport in bulk according to IMO instruments; (g) Special precautions which a user needs to be aware of, or needs to comply with, in connection with transport or conveyance either within or outside their premises.
15.	Regulatory information	Safety, health and environmental regulations specific for the product in question.
16.	Other information including information on preparation and revision of the SDS	

NOTE: *The order of the physical and chemical properties presented in section 9 may be followed on the SDS as shown in this table but is not mandatory. The competent authority may decide to prescribe an order for section 9 of the SDS, or they may leave it to the preparer of the SDS to re-order the properties, if deemed appropriate.*

表 1.5.2：SDS の必要最少情報 *(続き)*

9.	物理的および化学的性質	物理状態； 色； 臭い； 融点/凝固点； 沸点または初留点および沸点範囲； 燃焼性； 爆発下限および上限/引火限界； 引火点； 自然発火温度； 分解温度； pH； 動粘性率； 溶解度； 分配係数：n-オクタノール/水（log 値）； 蒸気圧； 密度および/または比重； 蒸気比重； 粒子特性。
10.	安定性および反応性	(a) 反応性 (b) 化学的安定性 (c) 危険有害反応性の可能性 (d) 避けるべき条件（静電放電、衝撃、振動等） (e) 混触危険物質 (f) 危険有害性のある分解生成物
11.	有害性情報	種々の毒性学的（健康）影響の簡潔だが完全かつ包括的な記述および次のような影響の特定に使用される利用可能なデータ： (a) 可能性の高いばく露経路（吸入、経口摂取、皮膚および眼接触）に関する情報 (b) 物理的、化学的および毒性学的特性に関係した症状 (c) 短期および長期ばく露による遅発および即時影響、ならびに慢性影響 (d) 毒性の数値的尺度（急性毒性推定値など）
12.	環境影響情報	(a) 生態毒性（利用可能な場合、水生および陸生） (b) 残留性と分解性 (c) 生物蓄積性 (d) 土壌中の移動度 (e) 他の有害影響
13.	廃棄上の注意	廃棄残留物の記述とその安全な取扱いに関する情報、汚染容器包装の廃棄方法を含む
14.	輸送上の注意	(a) 国連番号 (b) 国連品名 (c) 輸送における危険有害性クラス (d) 容器等級（該当する場合） (e) 環境有害性（例：海洋汚染物質（該当/非該当）） (f) IMO 文書に基づいたばら積み輸送 (g) 使用者が構内もしくは構外の輸送または輸送手段に関連して知る必要がある、または従う必要がある特別の安全対策
15.	適用法令	当該製品に特有の安全、健康および環境に関する規則
16.	SDS の作成と改訂に関する情報を含むその他の情報	

注記：*9 節に示されている物理的および化学的性質の SDS における順番は、本表にしたがって良いが、強制ではない。所管官庁は SDS の 9 節における順番について規定してもよいし、適当であれば、並べ替えを SDS の作成者に委ねてもよい。*

PART 2

PHYSICAL HAZARDS

第2部

物理化学的危険性

CHAPTER 2.1

EXPLOSIVES

2.1.1 Definitions and general considerations

2.1.1.1 *Definitions*

An *explosive substance or mixture* is a solid or liquid substance or mixture which is in itself capable by chemical reaction of producing gas at such a temperature and pressure and at such a speed as to cause damage to the surroundings. Pyrotechnic substances and mixtures are included even when they do not evolve gases.

A *pyrotechnic substance or mixture* is an explosive substance or mixture that is designed to produce an effect by heat, light, sound, gas or smoke or a combination of these as the result of non-detonative self-sustaining exothermic chemical reactions.

An *explosive article* is an article containing one or more explosive substances or mixtures.

Explosive or pyrotechnic effect in the context of 2.1.1.2.1 (c) means an effect produced by self-sustaining exothermic chemical reactions including shock, blast, fragmentation, projection, heat, light, sound, gas and smoke.

Division means the classification of an explosive substance, mixture or article according to Part I of the Manual of Tests and Criteria and relates to it being in a certain configuration.

Primary packaging means the minimum level of packaging of a configuration assigned a division, in which the explosive substance, mixture or article is intended to be retained until use.

NOTE: *Divisions are generally assigned for the purpose of transport and may be subject to further packaging specifications according to the UN Model Regulations to be valid.*

2.1.1.2 *Scope*

2.1.1.2.1 Except as provided in 2.1.1.2.2, the class of explosives comprises:

(a) Explosive substances and mixtures;

(b) Explosive articles, except devices containing explosive substances or mixtures in such quantity or of such a character that their inadvertent or accidental ignition or initiation shall not cause any effect external to the device either by projection, fire, smoke, heat or loud noise; and

(c) Substances, mixtures and articles not mentioned under (a) and (b) above which are manufactured with the view to producing a practical explosive or pyrotechnic effect.

2.1.1.2.2 The following substances and mixtures are excluded from the class of explosives:

(a) Ammonium nitrate-based emulsions, suspensions or gels which meet the criteria of test series 8 of the *Manual of Tests and Criteria* for classification as ANEs of Category 2 oxidizing liquids (chapter 2.13) or Category 2 oxidizing solids (chapter 2.14);

(b) Substances and mixtures which meet the criteria for classification as desensitized explosives according to the criteria of chapter 2.17;

(c) Substances and mixtures which have not been manufactured with the view to producing, in themselves, an explosive or pyrotechnic effect and which;

(i) are self-reactive substances and mixtures according to the criteria of chapter 2.8; or

(ii) are organic peroxides according to the criteria of chapter 2.15; or

第 2.1 章

爆発物

2.1.1 定義および一般事項

2.1.1.1 *定義*

*爆発性物質*または*混合物*は、それ自体の化学反応により、周囲環境に損害を及ぼすような温度および圧力ならびに速度でガスを発生する能力のある固体または液体の物質または混合物をいう。火工物質および混合物は、たとえガスを発生しない場合でも含まれる。

*火工物質*または*混合物*は、非爆轟性で自己持続的発熱化学反応により、熱、光、音、ガスまたは煙若しくはこれらの組み合わせの効果を生じるよう作られた爆発性物質または混合物である。

*爆発性物品*は、爆発性物質または混合物を一種類以上含む物品である。

爆発または火工効果（2.1.1.2.1(c) の文脈における）は、衝撃、爆風、破砕、飛散、熱、光、音、ガス、および煙を含む自己持続的発熱化学反応によって生じる効果をいう。

等級（*Division*）とは、試験方法及び評価基準のマニュアル第Ｉ部にしたがった、爆発性物質、混合物または物品の分類をいい、ある特定の構成にあることに関係する。（訳者注：UN モデル規則では Division を「区分」と訳しているが、GHS では爆発物の Division 1.1～1.6 についてのみ「等級」を用いた。）

*一次包装*とは、爆発性物質、混合物または物品を使用する前まで保持することを目的とした、区分に割り当てられた構成の最低レベルの包装をいう。

注記：等級は一般に輸送の目的で割り当てられ、有効にするために UN モデル規則に基づくさらなる包装要件にしたがってもよい。

2.1.1.2 *適用範囲*

2.1.1.2.1　2.1.1.2.2 に挙げられているものを除き、次のものが爆発物に分類される：

(a) 爆発性物質および混合物；

(b) 爆発性物品、ただし不注意または偶発的な点火または起爆によって、飛散物、火炎、発煙、発熱または大音響のいずれかによって装置の外側に対し何ら影響を及ぼさない程度の量またはそのような特性の爆発性物質または混合物を含む装置を除く；および

(c) 上記(a)および(b)以外の物質、混合物および物品であって、実質的な爆発または火工効果を目的として製造されたもの。

2.1.1.2.2　以下の物質および混合物は爆発物からは除外される：

(a) 硝酸アンモニウムエマルション、サスペンションまたはゲル（ANE）であって、*試験方法及び評価基準のマニュアル*の試験シリーズ 8 により、酸化性液体区分 2（第 2.13 章）または酸化性固体区分 2（第 2.14 章）に適合するもの；

(b) 第 2.17 章の判定基準にしたがって鈍性化爆発物として分類されるための判定基準を満たす物質および混合物；

(c) それら自体が、爆発または火工効果を目的として製造されていない物質および混合物で以下のもの；

  (i) 第 2.8 章の判定基準にしたがった自己反応性物質および混合物；または

  (ii) 第 2.15 章の判定基準にしたがった有機過酸化物；または

(iii) are deemed not to have explosive properties on basis of the screening procedures in appendix 6 of the *Manual of Tests and Criteria*; or

(iv) are too insensitive for inclusion in the hazard class according to test series 2 of the *Manual of Tests and Criteria*; or

(v) are excluded from assignment within Class 1 of the *UN Model Regulations* based on results in test series 6 of the *Manual of Tests and Criteria*.

NOTE: *Performing test series 2 requires a substantial amount of material, which may not be available in the initial stages of research and development. Substances and mixtures in the research and development phase for which not enough material exists to perform test series 2 of the Manual of Tests and Criteria may, for the purpose of further scientific characterisation, be regarded as self-reactive substances and mixtures Type C (see chapter 2.8), provided that:*

(a) *The substance or mixture is not manufactured with the view to producing an explosive or pyrotechnic effect; and*

(b) *The decomposition energy of the substance or mixture is less than 2000 J/g; and*

(c) *The result in test 3(a) and test 3(b) of the Manual of Tests and Criteria is negative; and*

(d) *The result in test 2(b) of the Manual of Tests and Criteria is "no explosion" at an orifice diameter of 6 mm; and*

(e) *The expansion of the lead block in Test F.3 of the Manual of Tests and Criteria is less than 100 ml per 10 g substance or mixture.*

2.1.1.2.3　　For explosive articles that are assigned a specific UN number in a class other than Class 1 according to the Dangerous Goods List of the *UN Model Regulations*, the following applies.

2.1.1.2.3.1　　Explosive articles that are assigned a specific UN number in Class 2, 3, 4 or 5 are classified in the GHS hazard class and, where available, category corresponding to the transport classification, and excluded from the hazard class explosives, provided that:

(a) they are in the transport configuration; or

(b) the transport classification does not depend on a particular configuration; or

(c) they are in use, see 2.1.1.3.4.

2.1.1.2.3.2　　Explosive articles that are assigned a specific UN number in Class 9 are classified as explosives in Sub-category 2C, provided that:

(a) they are in the transport configuration; or

(b) the transport classification does not depend on a particular configuration; or

(c) they are in use, see 2.1.1.3.4.

NOTE 1: *Subject to approval from the competent authority, explosive articles that are assigned a specific UN number in division 6.1 within Class 6 or in Class 8 may be classified in the GHS hazard class and, where available, category corresponding to the transport classification, and excluded from the hazard class explosives, provided that conditions (a) to (c) of 2.1.1.2.3.1 are met.*

NOTE 2: *According to the UN Model Regulations, articles are normally not assigned packing groups and hence a category within the corresponding GHS hazard class cannot always be assigned on this basis. Expert judgement should be used to assign an appropriate category in these cases, taking into account the GHS classification of the substances or mixtures contained.*

(iii) 試験方法及び評価基準のマニュアルの付録 6 のスクリーニング手順に基づき爆発性はないとされたもの；または

(iv) 試験方法及び評価基準のマニュアルの試験シリーズ 2 にしたがうと危険有害性クラスに含めるにはあまりに鈍感であるもの；または

(v) 試験方法及び評価基準のマニュアルの試験シリーズ 6 の結果に基づき UN モデル規則のクラス 1 の中には割り当てられないとされたもの。

注記：試験シリーズ 2 を実行するにはかなりの量の試料が必要であるが、研究および開発の初期段階では利用できない場合もある。試験方法及び評価基準のマニュアルの試験シリーズ 2 を実行するための十分な量の試料が存在しない研究や開発段階物質や混合物は、さらなる科学的特性評価の目的で、以下の条件で、自己反応性物質および混合物タイプ C（第 2.8 章参照）とみなしてもよい：

(a) 物質または混合物は爆発または火工効果を目的として製造されていない；および

(b) 物質および混合物の分解エネルギーは 2000J/g 未満である；および

(c) 試験方法及び評価基準のマニュアルの試験 3(a)および 3(b)の結果が「－」である；および

(d) 試験方法及び評価基準のマニュアルの試験 2(b)の結果が、6mm のオリフィス径において「爆発なし」である；および

(e) 試験方法及び評価基準のマニュアルの試験 F.3 における鉛ブロックの膨張が、物質または混合物 10 g あたり 100ml 未満である。

2.1.1.2.3　UN モデル規則の危険物リストにおいてクラス 1 以外のクラスで特定の国連番号が割り当てられている爆発性物品については、以下が適用される：

2.1.1.2.3.1　クラス 2、3、4 または 5 で特定の国連番号が割り当てられている爆発性物品は、輸送分類に対応する GHS の危険有害性クラスおよび、可能な場合には危険有害性区分に分類され、以下の条件で危険有害性クラス爆発物からは除外される：

(a) それらは輸送の構成になっている；または

(b) 輸送の分類は特定の構成に依存していない；または

(c) それらは使用されている、2.1.1.3.4 参照。

2.1.1.2.3.2　クラス 9 で特定の国連番号が割り当てられている爆発性物品は以下の条件で危険有害性細区分 2C に分類される：

(a) それらは輸送の構成になっている；または

(b) 輸送の分類は特定の構成に依存していない；または

(c) それらは使用中である、2.1.1.3.4 参照。

注記1：　所管官庁の承認を受ける場合、クラス 6 の区分 6.1 またはクラス 8 で特定の国連番号が割り当てられた爆発性物品は、輸送分類に対応する GHS の危険有害性クラスおよび、可能な場合には、危険有害性区分に分類され、危険有害性クラス爆発物からは除外されてもよい、ただし 2.1.1.2.3.1 の(a)から(c)の条件を満たさなければならない。

注記2：　UN モデル規則によると、通常、物品には容器等級が割り当てられないため、対応する GHS 危険有害性クラス内の危険有害性区分が常に割り当てられるとは限らない。これらの場合、含まれる物質または混合物の GHS 分類を考慮して、専門家の判断を利用して適切な危険有害性区分を割り当てるべきである。

2.1.1.3 *Other considerations*

2.1.1.3.1 *The relation to the classification according to the UN Model Regulations*

The GHS classification of substances, mixtures and articles as explosives builds largely on the classification used for transport according to the *UN Model Regulations*. Information on their transport division and, when available, some of the underlying test results according to Part I of the *Manual of Tests and Criteria*, is therefore relevant for the GHS classification. Test data is not required when classification using expert judgement is possible based on available information from previous testing and characterization. Where appropriate, analogy to tested explosives may be used, taking into consideration whether changes to the configuration may affect the hazard posed compared to the tested configuration. While the transport divisions are designed for the purpose of safe transport of explosives, the GHS classification draws from this classification to ensure appropriate hazard communication in other sectors, in particular supply and use. In doing this, any mitigating effects of the transport configuration on the explosive behaviour, such as a particular packaging, are evaluated as they may not be present in sectors outside of transport.

2.1.1.3.2 *The configuration dependence of the division*

Entry into the hazard class of explosives is based on the intrinsic explosive properties of substances and mixtures. The assignment to a division, however, is also dependent on the configuration using packaging, and the incorporation into articles of such substances and mixtures. The division is the relevant level of classification when the explosive is in the configuration to which the division was assigned, e.g. when transported or stored, and may form the basis for explosives licencing and safety measures such as distance requirements. The hazard categories, on the other hand, are the relevant level of classification for the safe handling.

2.1.1.3.3 *The hierarchy of the categories*

Category 2 only contains explosives which have been assigned to a division and corresponds to Class 1 of the *UN Model Regulations*. The sub-categories within Category 2 classify explosives on basis of the hazardous behaviour of the explosive in its primary packaging or, where applicable, of the explosive article alone. An explosive that has not been assigned a division is classified in Category 1 of the hazard class of explosives. This may be because it is considered too dangerous to be assigned a division, or because it is not (yet) in a suitable configuration to assign it to a division. Explosives in Category 1 are therefore not necessarily more hazardous than explosives in Category 2.

2.1.1.3.4 *Change of classification over the life cycle*

As the assignment to a division depends on the configuration, the classification of an explosive may change over its life cycle as a result of reconfiguration. An explosive that was assigned a division in a certain configuration, and hence classified in a sub-category within Category 2, may no longer retain that division when out of that configuration. If assigned to another division in the new configuration, it may need to be classified in another sub-category within Category 2, and if not assigned a division it should be classified in Category 1. However, the use of an explosive, meaning the preparation and intentional functioning, including removal from the primary packaging for functioning or installation or deployment in readiness for functioning, is not intended to require such re-classification.

2.1.1.3.5 *Exclusions from the hazard class*

Some substances, mixtures and articles that have explosive properties are excluded from the hazard class of explosives because they are not considered sensitive enough or because they do not present a significant explosion hazard in a particular configuration. The safety data sheet is an appropriate means to convey information on explosive properties for such substances and mixtures, and the explosion hazards of such articles (see chapter 1.4).

2.1.1.3 その他の考慮事項

2.1.1.3.1 UN モデル規則にしたがった分類との関係

爆発物としての物質、混合物および物品の GHS 分類は、主に UN モデル規則にしたがって輸送に使用される分類に基づいている。したがって、輸送区分に関する情報および、利用可能な場合は、試験方法及び評価基準のマニュアルの第 I 部にしたがった基礎となる試験結果の一部は GHS 分類に関連している。以前の試験および特性評価から利用可能な情報に基づいて専門家の判断による分類が可能な場合、試験データは必要ない。必要に応じて、構成の変更が試験された構成と比較してもたらされる危険に影響を与える可能性があるかどうかを考慮して、試験された爆発物との類似性を使用してもよい。輸送区分は爆発物の安全な輸送を目的として設計されている一方、GHS 分類はこの輸送分類に基づき、他の分野、特に供給と使用における適切な危険有害性の情報伝達を確実にする。この際に、特定の包装など輸送構成による爆発挙動の軽減効果は、輸送以外の分野には存在しない可能性があるので評価される。

2.1.1.3.2 輸送区分の構成依存性

爆発物の危険有害性クラスへの分類は、物質および混合物固有の爆発特性に基づいている。しかし、等級への割り当ては包装を使用した構成およびそのような物質や混合物の物品への組み込みにも依存する。等級は、爆発物が等級に割り当てられた構成の中にある場合、例えば、輸送または保管されている場合に適切な分類レベルであり、爆発物の認可および保安距離設定などの安全対策の基礎をなすであろう。一方、危険有害性区分は安全な取扱いに関する分類レベルである。

2.1.1.3.3 危険有害性区分の階層

区分 2 はある等級が割り当てられた爆発物のみが含まれ、UN モデル規則のクラス 1 に相当する。区分 2 内の細区分は、一次包装内の爆発物または、該当する場合は、爆発性物品のみの危険挙動に基づいて爆発物を分類する。等級が割り当てられていない爆発物は、危険有害性区分 1 の爆発物に分類される。これは等級を割り当てるには危険すぎるとみなされているか、等級を割り当てるのに（まだ）適切な構成になっていないことが原因である可能性がある。したがって区分 1 の爆発物は必ずしも区分 2 の爆発物よりもより危険であるとは限らない。

2.1.1.3.4 ライフサイクルによる分類の変化

等級への割り当ては構成に依存するため、爆発物の分類は、構成変更の結果としてそのライフサイクルにわたって変化する可能性がある。特定の構成で区分が割り当てられ、区分 2 内の細区分に分類された爆発物は、その構成から外れると、その区分のままではなくなる可能性がある。新しい構成で別の等級に割り当てられた場合は、区分 2 内の別の細区分に分類する必要があるかもしれないし、等級が割り当てられない場合は、区分 1 に分類する必要がある。ただし、爆発物の使用、つまり作動するための一次包装からの取り出しや設置、または作動準備のための配置を含む、調製および意図した作動では、そのような再分類を要求してはいない。

2.1.1.3.5 危険有害性クラスからの除外

爆発性を有する一部の物質、混合物および物品は、十分な感度がないとみなされるため、または特定の構成で重大な爆発の危険性を示さないために、爆発物の危険有害性クラスからは除外される。安全データシートは、そのような物質および混合物の爆発特性、およびそのような物品の爆発の危険性に関する情報を伝達するための適切な手段である（第 1.4 章参照）。

2.1.2   Classification criteria

2.1.2.1   Substances, mixtures and articles of this class are classified into one of two categories, and for Category 2 into one of three sub-categories according to the following table:

Table 2.1.1: Criteria for explosives

Category	Sub-category	Criteria

1		Explosive substances, mixtures and articles which
		(a) have not been assigned a division and which
		    (i) are manufactured with the view of producing an explosive or pyrotechnic effect; or
		    (ii) are substances or mixtures which show positive results when tested in Test series 2 of the *Manual of Tests and Criteria*
		or
		(b) are out of the primary packaging of the configuration to which a division was assigned[a], unless they are explosive articles assigned a division:
		    (i) without a primary packaging; or
		    (ii) in a primary packaging that does not attenuate the explosive effect, taking into account also intervening packaging material, spacing or critical orientation.

2	2A	Explosive substances, mixtures and articles which have been assigned (a) Division 1.1, 1.2, 1.3, 1.5 or 1.6; or (b) Division 1.4 and are not meeting the criteria for sub-category 2B or 2C.[b]
2	2B	Explosive substances, mixtures and articles which have been assigned Division 1.4 and a compatibility group other than S, and which: (a) do not detonate and disintegrate when functioned as intended; and (b) exhibit no high hazard event[c] in test 6 (a) or 6 (b) of the *Manual of Tests and Criteria*; and (c) do not require attenuating features, other than that which may be provided by a primary packaging, to mitigate a high hazard event[c].
2	2C	Explosive substances, mixtures and articles which have been assigned Division 1.4 compatibility group S, and which: (a) do not detonate and disintegrate when functioned as intended; and (b) exhibit no high hazard event[c] in test 6 (a) or 6 (b), or in the absence of these test results, similar results in test 6 (d) of the *Manual of Tests and Criteria*; and (c) do not require attenuating features, other than that which may be provided by a primary packaging, to mitigate a high hazard event[c].

[a]   *Explosives in Category 2 that are removed from their primary packaging for use remain classified in Category 2, see 2.1.1.3.4.*

[b]   *The manufacturer, supplier or competent authority may classify an explosive of Division 1.4 as sub-category 2A on basis of data or other considerations even if it meets the technical criteria for sub-category 2B or 2C.*

[c]   *A high hazard event is exhibited when performing test 6(a) or 6(b), according to the Manual of Tests and Criteria, by:*

   (i) *a significant change in the witness plate shape, such as perforation, gouge, substantial dent or bowing; or*

   (ii) *instantaneous scattering of most of the confining material.*

2.1.2 分類基準

2.1.2.1 下表にしたがい、このクラスに分類される物質、混合物および物品は、2つの区分のうちの1つに、さらに区分2では3つの細区分のうちの1つに分類される：

表 2.1.1：爆発物の判定基準

区分	細区分	判定基準
1		爆発性物質、混合物および物品で： (a) 等級が割り当てられておらず、 　(i) 爆発または火工効果を目的として製造されたもの；または 　(ii) 試験方法及び評価基準のマニュアルの試験シリーズ2で試験されたとき「＋」の結果を示す物質または混合物 または (b) 等級が割り当てられた構成の一次包装の外にあるもの[a]、ただし等級が割り当てられた爆発性物品ではないこと： 　(i) 一次包装がないもの；または 　(ii) 爆発効果を減衰させない一次包装の中にあるもの、包装内の介在する梱包材料、物品の間隔または物品の危険な方向性も考慮する。
2	2A	爆発性物質、混合物および物品で以下が割り当てられているもの (a) 等級 1.1、1.2、1.3、1.5 または 1.6；または (b) 等級 1.4 が割り当てられており、細区分 2B または 2C の判定基準に合致しないもの[b]。
2	2B	爆発性物質、混合物および物品で、等級 1.4 および S 以外の隔離区分が割り当てられている、以下のもの： (a) 意図したとおりに作動したときに爆轟および崩壊しないもの；および (b) *試験方法及び評価基準のマニュアル*の試験 6(a)または 6(b)で危険性の高い事象[c]を示さないもの；および (c) 危険性の高い事象[c]を軽減させるために、一次包装により提供されるであろう減衰機能以外に減衰機能を必要としないもの[c]。
2	2C	爆発性物質、混合物および物品で等級 1.4 隔離区分 S が割り当てられている、以下のもの： (a) 意図したとおりに作動したときに爆轟および崩壊しないもの；および (b) *試験方法及び評価基準のマニュアル*の試験 6(a)または 6(b)で危険性の高い事象[c]を示さないもの、またはこれらの試験結果がなく、試験 6(d)で同様の結果；および (c) 危険性の高い事象を軽減させるために、一次包装により提供されるであろう減衰機能以外に減衰機能を必要としないもの[c]。

[a] 区分 2 の爆発物は、使用するために一次包装から取り出されても区分 2 に分類されたままである、2.1.1.3.4 参照。

[b] 製造者、供給者または所管官庁は、たとえ細区分 2B または 2C の技術的な基準に合ったとしても、データまたは他の検討に基づいて、等級 1.4 の爆発物を細区分 2A に分類してもよい。

[c] *試験方法及び評価基準のマニュアル*にしたがった試験 6(a)または 6(b)を実行して、以下の危険性の高い事象が示されるとは：

(i) 貫通孔、溝、相当なへこみまたは曲がりなど証拠板における相当な変化；または

(ii) 閉じ込め材料の瞬間的な飛散。

2.1.2.2 The divisions are as follows:

(a) Division 1.1: Substances, mixtures and articles which have a mass explosion hazard (a mass explosion is one which affects almost the entire quantity present virtually instantaneously);

(b) Division 1.2: Substances, mixtures and articles which have a projection hazard but not a mass explosion hazard;

(c) Division 1.3: Substances, mixtures and articles which have a fire hazard and either a minor blast hazard or a minor projection hazard or both, but not a mass explosion hazard:

  (i) combustion of which gives rise to considerable radiant heat; or

  (ii) which burn one after another, producing minor blast or projection effects or both;

(d) Division 1.4: Substances, mixtures and articles which present no significant hazard: substances, mixtures and articles which present only a small hazard in the event of ignition or initiation. The effects are largely confined to the package and no projection of fragments of appreciable size or range is to be expected. An external fire shall not cause virtually instantaneous explosion of almost the entire contents of the package;

(e) Division 1.4 compatibility group S: Substances, mixtures and articles so packed or designed that any hazardous effects arising from accidental functioning are confined within the package unless the package has been degraded by fire, in which case all blast or projection effects are limited to the extent that they do not significantly hinder fire-fighting or other emergency response efforts in the immediate vicinity of the package;

(f) Division 1.5: Very insensitive substances or mixtures which have a mass explosion hazard: substances and mixtures which have a mass explosion hazard but are so insensitive that there is very little probability of initiation or of transition from burning to detonation under normal conditions. The probability of transition from burning to detonation is greater when large quantities are present;

(g) Division 1.6: Extremely insensitive articles which do not have a mass explosion hazard: articles which predominantly contain extremely insensitive substances or mixtures and which demonstrate a negligible probability of accidental initiation or propagation. The hazard from articles of Division 1.6 is limited to the explosion of a single article.

NOTE 1: *For some regulatory purposes, the divisions are further subdivided into compatibility groups which identify the kinds of explosives that are deemed to be compatible (see chapter 2.1, section 2.1.2 of the UN Model Regulations).*

NOTE 2: *While Division 1.4 compatibility group S is not a division of its own, this classification corresponds to a separate division based on additional criteria.*

NOTE 3: *For classification tests on explosive substances or mixtures, the tests should be performed on the substance or mixture as presented. If for example, for the purposes of supply or transport, the same substance or mixture is to be presented in a physical form different from that which was tested and which is considered likely to materially alter its performance in a classification test, the substance or mixture must also be tested in the new form.*

2.1.3 Hazard communication

General and specific considerations concerning labelling requirements are provided in *Hazard communication: Labelling* (chapter 1.4). Annex 1 contains summary tables about classification and labelling. Annex 3 contains examples of precautionary statements and pictograms which can be used where allowed by the competent authority. Table 2.1.2 presents specific label elements for substances and mixtures classified into this hazard class based on the criteria in this chapter.

2.1.2.2 等級（輸送区分）は以下の通りである：

(a) 等級 1.1：大量爆発の危険性を持つ物質、混合物および物品（大量爆発とは、ほぼ瞬時にほとんど全量に影響が及ぶような爆発をいう）；

(b) 等級 1.2：大量爆発の危険性はないが、飛散物の危険性を有する物質、混合物および物品；

(c) 等級 1.3：大量爆発の危険性はないが、火災の危険性を有し、かつ弱い爆風の危険性または僅かな飛散物の危険性のいずれか、若しくは両方を持っている物質、混合物および物品：

 (i) その燃焼により大量の放射熱を放出するもの；または

 (ii) 弱い爆風または飛散物のいずれか若しくは両方の効果を発生しながら次々に燃焼するもの；

(d) 等級 1.4：重大な危険性の認められない物質、混合物および物品：点火または起爆した場合にも僅かな危険性しか示さない物質、混合物および物品。その影響はほとんどが包装内に限られ、ある程度以上の大きさと飛散距離を持つ破片の飛散は想定されない。外部火災が原因となり包装物のほとんどすべての内容物がほぼ瞬時に爆発を起こしてはならない；

(e) 等級 1.4 隔離区分 S：包装が火災で劣化した場合を除き、偶発的な作用から生じる危険な影響が包装内に限定されるように梱包または設計された物質、混合物および物品。火災の場合、全ての爆風または飛散物効果は、包装のすぐ近くでの消火活動またはその他の緊急対応の取組を著しく妨げない範囲に限られる；

(f) 等級 1.5：大量爆発の危険性を有するが、非常に鈍感な物質：大量爆発の危険性を持っているが、非常に鈍感で、通常の条件では、起爆の確率あるいは燃焼から爆轟に転移する確率が極めて小さい物質および混合物。燃焼から爆轟に転移する確率は、大量に存在する場合には大きくなる；

(g) 等級 1.6：大量爆発の危険性を有しない極めて鈍感な物品：主としてきわめて鈍感な物質または混合物を含む物品で、偶発的な起爆または伝播の確率をほとんど無視できるようなものである。等級 1.6 の物品による危険性は単一の物品の爆発に限られる。

注記 1：一部の規制目的では、さらに適合性があるとみなすことができる爆発物の種類を識別する隔離区分に細区分される（UN モデル規則第 2.1 章 2.1.2 参照）。

注記 2：等級 1.4 隔離区分 S はそれ自体区分ではないが、これは追加の判定基準に基づく別の等級に相当する。

注記 3：爆発性物質または混合物の分類試験では、物質または混合物のあり姿で試験されるべきである。例えば、供給または輸送の目的で、同じ物質または混合物が、試験されたものとは異なる物理的形態で提示され、かつ分類試験でその結果を大幅に変える可能性が高いと考えられる場合には、その物質または混合物は新たな形態でも試験されなければならない。

2.1.3　危険有害性情報の伝達

表示要件に関する通則および細則は、危険有害性に関する情報の伝達：表示（第 1.4 章）に定める。附属書 1 に分類と表示に関する概要表を示す。附属書 3 には、注意書きおよび所管官庁が許可した場合に使用可能な注意絵表示の例を示す。表 2.1.2 は本章の判定基準に基づいて、この危険有害性クラスに分類される物質および混合物のラベル要素を示す。

Table 2.1.2: Label elements for explosives

Category	1	2		
Sub-category	Not applicable	2A	2B	2C
Symbol[a]	Exploding bomb	Exploding bomb	Exploding bomb	Exclamation mark
Signal word	Danger	Danger	Warning	Warning
Hazard statement	Explosive	Explosive	Fire or projection hazard	Fire or projection hazard
Additional hazard statement	Very sensitive[b] __or__ May be sensitive[c]	Not applicable	Not applicable	Not applicable

[a] For Divisions 1.4, 1.5 and 1.6 no symbol appears on the label for transport, according to the UN Model Regulations.

[b] To be assigned additionally to explosives that are sensitive to initiation as determined by test series 3 or 4 of the Manual of Tests and Criteria. May also be applied to explosives sensitive to other stimuli, e.g. electrostatic discharge.

[c] To be assigned additionally to explosives for which sufficient information on their sensitivity to initiation is not available.

NOTE: *Substances and mixtures excluded by 2.1.1.2.2 (c) (v) still have explosive properties. The user should be informed of these intrinsic explosive properties because they have to be considered for handling – especially if the substance or mixture is removed from its packaging or is repackaged – and for storage. For this reason, the explosive properties of the substance or mixture should be communicated in subsection 2.3 (other hazards that do not result in classification) and section 9 (Physical and chemical properties) or 10 (Stability and reactivity) of the Safety Data Sheet in accordance with table 1.5.2, and other sections of the Safety Data Sheet, as appropriate.*

2.1.4 Decision logic and guidance

The decision logic and guidance, which follow, are not part of the harmonized classification system, but have been provided here as additional guidance. It is strongly recommended that the person responsible for classification studies the criteria before and during use of the decision logic.

表 2.1.2：爆発物のラベル要素

区分	1	2		
細区分	適用なし	2A	2B	2C
シンボル a	爆弾の爆発	爆弾の爆発	爆弾の爆発	感嘆符
注意喚起語	危険	危険	警告	警告
危険有害性情報	爆発物	爆発物	火災または飛散危険性	火災または飛散危険性
追加的危険有害性情報	非常に敏感 b **または** 敏感である可能性 c	適用なし	適用なし	適用なし

a UN モデル規則にしたがった輸送用のラベルでは、等級 1.4、1.5 および 1.6 に対してシンボルはない。
b 試験方法及び評価基準のマニュアルの試験シリーズ 3 または 4 によって決定される起爆に敏感な爆発物に追加で割り当てられる。他の刺激、例えば静電放電、に敏感な爆発物にも適用される。
c 起爆に対する敏感さの十分な情報が利用できない爆発物に追加で割り当てられる。

注記： 2.1.1.2.2.(c)(v)により除外されている物質や混合物は依然として爆発特性を持つ。これらの固有の爆発特性は、取扱い―特に物質または混合物が包装から取り出されまたは再包装された場合―および保管のために考慮されなければならないので、使用者に伝えられるべきである。このため物質または混合物の爆発特性は、表 1.5.2 にしたがった安全データシートの 2.3 小節（分類に関係しない他の危険有害性）および 9 節（物理的および科学的性質）または 10 節（安定性および反応性）、および他の安全データシートの節にしたがって適切に伝えられるべきである。

2.1.4 判定論理および手引き

　以下の判定論理および手引きは、調和分類システムには含まれないが、追加的な手引きとしてここに示す。分類の責任者に対し、この判定論理を使用する前および使用する際に判定基準についてよく調べ理解することを強く勧める。

2.1.4.1 *Decision logics*

Decision logic 2.1 (a) for categories of explosives

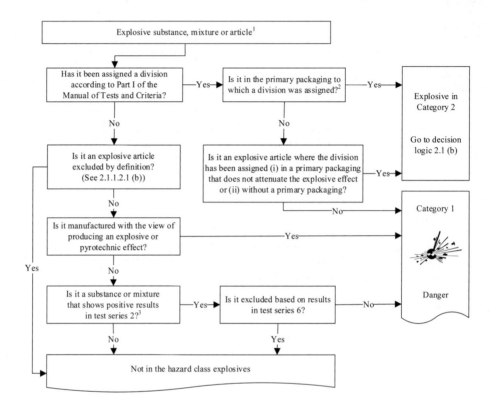

---

[1] *ANEs, desensitized explosives, organic peroxides and self-reactive substances and mixtures are classified in other hazard classes, see 2.1.1.2.2.*

[2] *Unless it is for use, see 2.1.1.3.4.*

[3] *Screening procedures may be used to avoid testing, see 2.1.1.2.2.*

2.1.4.1 判定論理

爆発物の分類に関する判定論理 2.1(a)

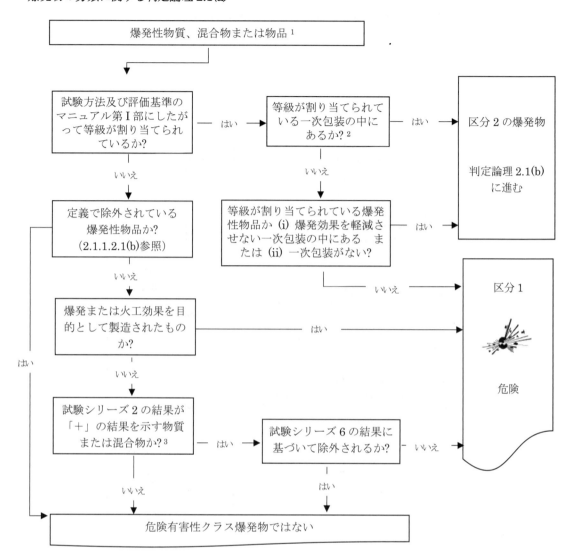

---
[1] ANSs、鈍性化爆発物、有機過酸化物及び自己反応性物質および混合物は他の危険有害性クラスに分類される、2.1.1.2.2 参照。
[2] 使用しない限り、2.1.1.3.4 参照。
[3] 試験を避けるためにスクリーニング手順を使用してもよい、2.1.1.2.2 参照。

Decision logic 2.1 (b) for sub-categories of explosives

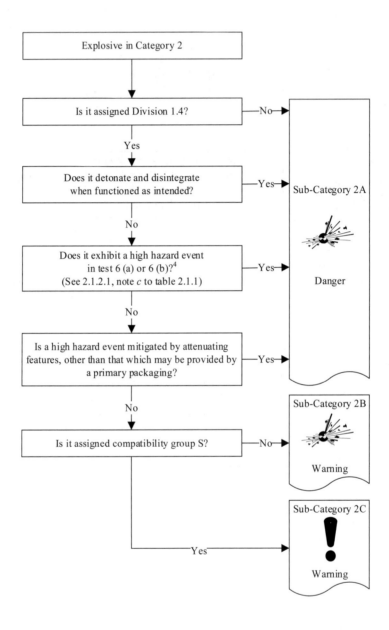

---

[4] *In the absence of results from test 6 (a) or 6 (b), results from test 6 (d) may be used to assess whether there was a high hazard event, see 2.1.2.1. If the configuration includes attenuating features that are likely to mitigate a high hazard event, such as spacing or a specific orientation of explosive articles, sub-category 2A may be assigned without the need to assess test data.*

判定論理 2.1(b) 爆発物の細区分

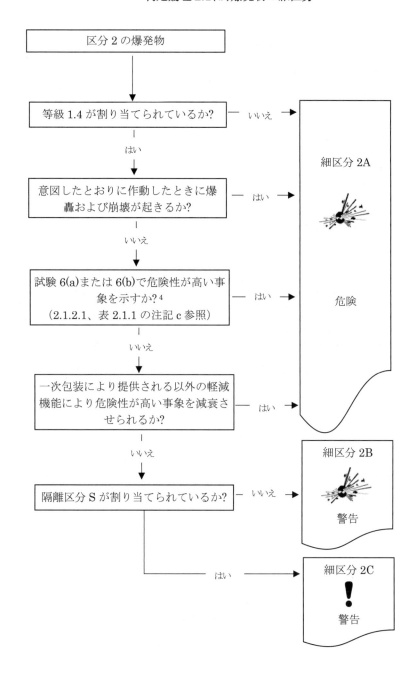

---

[4] 試験 6(a)または 6(b)の結果がない場合、試験 6(d)の結果を使用して、危険性の高い事象が発生したかどうかを評価してもよい、2.1.2.1 参照。構成に爆発性物品の間隔や特定の方向性など、危険性が高い事象を減衰する可能性のある軽減特性が含まれている場合、試験データを評価する必要なしに細区分 2A を割り当ててもよい。

2.1.4.2 Description of explosion hazard levels

Sub-category	Explosion hazard level
2A	Sub-category 2A represents a high explosion hazard. An explosive in this sub-category has the potential to cause complete destruction of objects and lethal or very severe injuries to persons.
2B	Sub-category 2B represents a medium explosion hazard. An explosive in this sub-category has the potential to cause serious damage to objects and serious injuries to persons. Injuries may result in permanent impairment.
2C	Sub-category 2C represents a low explosion hazard. An explosive in this sub-category can cause minor damage to objects and moderate injuries to persons. Injuries would not normally result in permanent impairment.

2.1.4.3 Principles of explosives classification

2.1.4.3.1 *Assigning explosives to divisions by testing*

2.1.4.3.1.1   Explosives are assigned divisions based on testing of specific configurations, which quantifies levels of blast, projections and fire. Formation of a configuration provides a level of protection from outside stimuli and fixes the sensitivity and hazard magnitude, which enables the assignment to a division. The divisions therefore describe the explosive behaviour in the particular configuration. Such descriptions reflect attenuating properties of the packaging and article, which may include spacing, or specific orientations of explosive articles to mitigate an explosive effect. The configuration is further controlled by design and packaging requirements specified in the *UN Model Regulations*.

2.1.4.3.1.2   Although divisions are not valid outside of the configurations to which they were assigned, they may still be used as a basis for regulatory measures in storage and handling when these configurations are modified. This normally presumes that additional safety measures are taken to account for the modified configurations, e.g. aggregate quantity limits and protective building designs.

2.1.4.3.2 *Assigning explosives to divisions based on analogy*

While classification in a division or a sub-category is based on testing in accordance with Part I of the *Manual of Tests and Criteria*, similar explosives configurations may be classified without testing, where appropriate, based on analogy to tested explosives. The use of analogy should take into consideration whether changes to the configuration may affect the hazard posed compared to the tested configuration, and is narrowly limited according to the quantity, packaging and design of the explosive.

2.1.4.3.3 *Assigning explosives to sub-categories*

2.1.4.3.3.1   Assignment to sub-categories within Category 2 builds on the information provided by the division to better reflect the hazard of the explosive in its primary packaging, which is intended to be retained until use. The primary packaging is all or part of the original tested configuration. It is normally the immediate container or the innermost packaging layer and may include attenuating properties which mitigate hazardous effects. However, only flexible inner packaging such as a thin-wall plastic bag or other unsubstantial material which provides negligible attenuation of explosive effects should not be considered the primary packaging. As explosives are unpackaged from their primary packaging they may present greater sensitivity or blast, projection or fire hazards. Retaining the primary packaging until use and limiting the amount of unpacked explosives are therefore generally important safety measures when handling explosives. When an explosive is installed or deployed and is later removed from use without initiation, it should be replaced in its primary packaging or an identical primary packaging.

2.1.4.3.3.2   Multiple explosive articles may sometimes be supplied where they are in direct contact without any intervening packaging material or spacing, or critical orientation. Provided all applicable classification evaluation occurred in this configuration, their primary packaging can be discarded without affecting the classification.

2.1.4.3.3.3   Occasionally, larger explosive articles are supplied without any packaging, e.g. in a handling device such as a cradle. In these cases, there may be no primary packaging, i.e. the classification is of the article as such. Handling devices that do not affect the classification can be discarded.

2.1.4.2 *爆発危険性のレベル*

細区分	爆発危険性レベル
2A	細区分 2A は高い爆発危険性を示す。この細区分の爆発物は物体を完全に破壊し、人に致命的または非常に重篤な傷害を負わせる可能性がある。
2B	細区分 2B は中程度の爆発危険性を示す。この細区分の爆発物は物体に重大な損傷を与え、人に重大な傷害を引き起こす可能性がある。傷害は永久的な障害になるかもしれない。
2C	細区分 2C は低い爆発危険性を示す。この細区分の爆発物は物体に軽微な損傷を与え、人に中程度の傷害を引き起こす。傷害は通常、永久的な障害にはならないであろう。

2.1.4.3 *爆発物分類の基本*

2.1.4.3.1 *試験による爆発物の等級の割り当て*

2.1.4.3.1.1 爆発物は、爆風、飛散物および火災のレベルを定量化する特定の構成に対する試験に基づいて等級が割り当てられる。構成のつくりは外部刺激からの保護レベルを提供し、感度と危険性の大きさを確定し、これにより等級への割り当てが可能になる。したがって等級は特定の構成での爆発挙動を説明している。そのような説明は、爆発の影響を軽減させるための爆発性物品の間隔または特定の方向性を含む、包装および物品の減衰特性を反映している。さらに構成は *UN モデル規則* で指定されている設計および包装要件によって管理されている。

2.1.4.3.1.2 等級は割り当てられた構成以外では有効ではないが、これらの構成が変更された場合でも、保管および取扱いにおける規制措置の基礎として使用してもよい。これは通常、変更された構成を考慮して追加の安全対策、例えば総量制限および保護的な建物の設計、が講じられていることを前提としている。

2.1.4.3.2 *類推に基づく爆発物の等級への割り当て*

等級または細区分の分類は*試験方法及び評価基準のマニュアル第 I 部*にしたがった試験に基づいているが、同様の爆発物構成は、必要に応じて試験された爆発物との類似性に基づいて、試験なしで分類してもよい。類推の使用は、構成の変更が試験された構成と比較してもたらされる危険性に影響を与える可能性があるかどうかを考慮すべきであり、爆発物の量、包装および設計によって狭く制限されている。

2.1.4.3.3 *爆発物の細区分への割り当て*

2.1.4.3.3.1 区分 2 内の細区分への割り当ては、使用するまで保持することを目的とした一次包装の爆発物の危険性をより適切に反映するために、等級から提供される情報に基づいている。一次包装は元の試験済み構成の全部または一部である。これは通常、直接の容器または最も内側の包装層であり、危険な影響を軽減する減衰特性が含まれる場合がある。ただし爆発の影響を無視できる程度しか減衰しない薄いプラスチック袋または他の支持力のない材料などの柔軟な内部包装のみを一次包装とみなすべきではない。爆発物が一次包装から開梱された際、感度が高くなるまたは爆風、飛散物あるいは火災の危険性が生じる可能性がある。したがって使用するまで一次包装を保持し、開梱された爆発物の量を制限することは、爆発物の取り扱う際の一般的に重要な安全対策である。爆発物が設置または配備され、後で起爆せずに使用を中止する場合には、元の一次包装にまたは同等の一次包装に戻すべきである。

2.1.4.3.3.2 複数の爆発性物品が、包装内の介在する梱包材料、物品の間隔または危険な方向性を考慮せずに直接接触して供給されることがある。該当する全ての分類評価がこの構成についてなされた場合、それらの一次包装は分類に影響を与えることなく無視できる。

2.1.4.3.3.3 時折、より大きな爆発性物品は包装なしで供給される、例えばクレイドルなどの取扱い装置。これらの場合、一次包装がない可能性がある、つまり分類は物品に対するようなものとなる。分類に影響を与えない取扱い装置は無視してもよい。

2.1.4.3.4　　　　*Classification of explosives in situations where they cannot be assigned a division*

2.1.4.3.4.1　　　Explosives in manufacturing processing and otherwise unfinished stages cannot be assigned a division until configured for transport, and hence are assigned to Category 1. Similarly, explosives assigned to Category 2 when taken out of their primary packaging for purposes other than use, are re-assigned to Category 1 (unless their primary packaging can be discarded, see 2.1.4.3.3).

2.1.4.3.4.2　　　The sensitivity and hazard severity of unpackaged explosives is dependent on non-intrinsic parameters related to the methods used, including quantity, depth, confinement, initiation stimulus, composition, physical state such as particle size, etc. The hazards posed by explosives in Category 1 thus vary extensively and may also vary dynamically as they flow through a process. For these reasons, the hazard communication for Category 1 cannot provide any details regarding the explosive behaviour. Process hazards analysis and risk management principles should be applied in these cases to identify and manage the risk of processes in accordance with best practices and applicable regulations.

2.1.4.3.5　　　　*Safety related to explosives failing test series 3 or 4*

Category 1 also includes explosives that fail test series 3 or test series 4 as configured, having an unacceptable level of sensitivity to stimuli encountered during transport. The thresholds of these tests may not be representative of the energy levels encountered during explosives processing and manufacturing. Furthermore, these tests do not include all types of stimuli that may be encountered, such as electrostatic discharge. Additional investigations of the properties of the explosive at hand may thus be needed for safe processing and handling.

2.1.4.3.4　等級を割り当てることができない状況での爆発物の分類

2.1.4.3.4.1　製造工程およびその他の未完成段階の爆発物は、輸送用に構成されるまで等級を割り当てることができないため区分1が割り当てられる。同様に、区分2に割り当てられた爆発物が使用以外の目的で一次包装から取り出されたときは、新たに区分1に割り当てられる（一次包装が廃棄されない場合は、2.1.4.3.3参照）。

2.1.4.3.4.2　包装されていない爆発物の感度と危険度は、量、深さ、閉じ込め、起爆刺激、組成、粒子サイズのような物理的状態など、使用される方法に関連する非固有パラメータに依存する。したがって区分1の爆発物によってもたらされる危険性は大きく異なり、プロセスを流れる時に動的に変化することもある。これらの理由により、区分1に関する危険性情報伝達においては爆発の挙動に関する詳細は提供できない。これらの場合、ベストプラクティスおよび適用規制にしたがって、プロセスリスクの同定と管理に、プロセスにおける危険性分析およびリスク管理の原則を適用するべきである。

2.1.4.3.5　試験シリーズ3または4に失敗した爆発物に関係する安全性

区分1には構成されたものとしての試験シリーズ3または4に失敗し、輸送中に遭遇する刺激に対して許容できないレベルの感度を持つ爆発物も含まれる。これらの試験のしきい値は、爆発物の処理および製造中に遭遇するエネルギーレベルを表していない可能性がある。さらに、これらの試験には、静電放電など、遭遇する可能性のあるすべてのタイプの刺激が含まれているわけではない。したがって安全な処理と取扱いのために、手元にある爆発物の特性の追加調査が必要になる場合がある。

CHAPTER 2.2

FLAMMABLE GASES

2.2.1 Definitions

2.2.1.1 A *flammable gas* is a gas having a flammable range with air at 20 °C and a standard pressure of 101.3 kPa.

2.2.1.2 A *pyrophoric gas* is a flammable gas that is liable to ignite spontaneously in air at a temperature of 54 °C or below.

2.2.1.3 A *chemically unstable gas* is a flammable gas that is able to react explosively even in the absence of air or oxygen.

2.2.2 Classification criteria

2.2.2.1 A flammable gas is classified in Category 1A, 1B or 2 according to the following table. Flammable gases that are pyrophoric and/or chemically unstable are always classified in Category 1A.

Table 2.2.1: Criteria for categorization of flammable gases

Category			Criteria
1A	Flammable gas		Gases, which at 20 °C and a standard pressure of 101.3 kPa: (a) are ignitable when in a mixture of 13 % or less by volume in air; or (b) have a flammable range with air of at least 12 percentage points regardless of the lower flammability limit unless data show they meet the criteria for Category 1B
	Pyrophoric gas		Flammable gases that ignite spontaneously in air at a temperature of 54°C or below
	Chemically unstable gas	A	Flammable gases which are chemically unstable at 20 °C and a standard pressure of 101.3 kPa
		B	Flammable gases which are chemically unstable at a temperature greater than 20 °C and/or a pressure greater than 101.3 kPa
1B	Flammable gas		Gases which meet the flammability criteria for Category 1A, but which are not pyrophoric, nor chemically unstable, and which have at least either: (a) a lower flammability limit of more than 6 % by volume in air; or (b) a fundamental burning velocity of less than 10 cm/s;
2	Flammable gas		Gases, other than those of Category 1A or 1B, which, at 20 °C and a standard pressure of 101.3 kPa, have a flammable range while mixed in air

NOTE 1: Ammonia and methyl bromide may be regarded as special cases for some regulatory purposes.

NOTE 2: Aerosols should not be classified as flammable gases. See chapter 2.3.

NOTE 3: In the absence of data allowing classification into Category 1B, a flammable gas that meets the criteria for Category 1A is classified per default in Category 1A.

NOTE 4: Spontaneous ignition for pyrophoric gases is not always immediate, and there may be a delay.

NOTE 5: In the absence of data on its pyrophoricity, a flammable gas mixture should be classified as a pyrophoric gas if it contains more than 1 % (by volume) of pyrophoric component(s).

第 2.2 章

可燃性ガス

2.2.1 定義

2.2.1.1　可燃性ガスとは、標準気圧 101.3kPa で 20℃において、空気との混合気が燃焼範囲を有するガスをいう。

2.2.1.2　自然発火性ガスとは、54 ℃ 以下の空気中で自然発火しやすいような可燃性ガスをいう。

2.2.1.3　化学的に不安定なガスとは、空気や酸素が無い状態でも爆発的に反応しうる可燃性ガスをいう。

2.2.2 分類基準

2.2.2.1　可燃性ガスは、次表にしたがって区分 1A、1B または 2 のいずれかに分類される。自然発火性および/または化学的に不安定な可燃性ガスは、つねに区分 1A に分類される。

表 2.2.1：可燃性ガスの判定基準

区分			判定基準
1A	可燃性ガス		標準気圧 101.3kPa で 20℃において以下の性状を有するガス： (a) 空気中の容積で 13%以下の混合気が可燃性であるもの；または (b) 燃焼（爆発）下限界に関係なく空気との混合気の燃焼範囲（爆発範囲）が 12%以上のもの 区分 1B の判定基準に合致した場合を除く
	自然発火性ガス		54℃以下の空気中で自然発火する可燃性ガス
	化学的に不安定なガス	A	標準気圧 101.3kPa で 20℃において化学的に不安定である可燃性ガス
		B	気圧 101.3kPa 超および/または 20℃超において化学的に不安定である可燃性ガス
1B	可燃性ガス		区分 1A の可燃性ガスの判定基準を満たし、自然発火性ガスでも化学的に不安定なガスでもなく、少なくとも以下のどちらかの条件を満たすもの： (a) 燃焼下限が空気中の容積で 6%を超える；または (b) 基本的な燃焼速度が 10 cm/s 未満；
2	可燃性ガス		区分 1A または 1B 以外のガスで、標準気圧 101.3kPa、20℃においてガスであり、空気との混合気が燃焼範囲を有するもの

注記 1：アンモニアおよび臭化メチルは、規制目的によっては特殊例と見なされる。

注記 2：エアゾールは可燃性ガスと分類するべきではない、第 2.3 章参照。

注記 3：区分 1B に分類するための十分なデータがない場合には、区分 1A の判定基準を満たす可燃性ガスは自動的に区分 1A とする。

注記 4：自然発火性ガスの自然発火は常に直ちに起こるとは限らず、遅れることもある。

注記 5：可燃性ガスの混合物で、自然発火性に関するデータがなく、1%を超える（容量）自然発火性成分を含む場合には自然発火性ガスに分類するべきである。

2.2.3 **Hazard communication**

2.2.3.1 General and specific considerations concerning labelling requirements are provided in Hazard communication: Labelling (chapter 1.4). Annex 1 contains summary tables about classification and labelling. Annex 3 contains examples of precautionary statements and pictograms which can be used where allowed by the competent authority. Table 2.2.2 presents specific label elements for substances and mixtures classified into this hazard class based on the criteria in this chapter.

Table 2.2.2: Label elements for flammable gases

	Category 1A	Gases categorized as 1A by meeting pyrophoric or unstable gas A/B criteria			Category 1B	Category 2
		Pyrophoric gas	Chemically unstable gas			
			Category A	Category B		
Symbol	Flame	Flame	Flame	Flame	Flame	*No symbol*
Signal word	Danger	Danger	Danger	Danger	Danger	Warning
Hazard statement	Extremely flammable gas	Extremely flammable gas. May ignite spontaneously if exposed to air	Extremely flammable gas. May react explosively even in the absence of air	Extremely flammable gas. May react explosively even in the absence of air at elevated pressure and/or temperature	Flammable gas	Flammable gas

2.2.3.2 If a flammable gas or gas mixture is classified as pyrophoric and/or chemically unstable, then all relevant classification(s) should be communicated on the safety data sheet as specified in annex 4, and the relevant hazard communication elements included on the label.

2.2.4 **Decision logic and guidance**

The decision logic and guidance, which follow, are not part of the harmonized classification system, but have been provided here as additional guidance. It is strongly recommended that the person responsible for classification studies the criteria before and during use of the decision logic.

2.2.4.1 *Decision logic for flammable gases*

To classify a flammable gas, data on its flammability, on its ability to ignite in air and on its chemical instability are required. In case of categorization in Category 1B, data on its lower flammability limit or its fundamental burning velocity are required. The classification is according to decision logic 2.2.

2.2.3 危険有害性情報の伝達

2.2.3.1 表示要件に関する通則および細則は、*危険有害性に関する情報の伝達：表示（第1.4章）*に定める。附属書1に分類と表示に関する概要表を示す。附属書3には、注意書きおよび所管官庁が許可した場合に使用可能な注意絵表示の例を示す。表2.2.2は本章の判定基準に基づいて、この危険有害性クラスに分類される物質および混合物のラベル要素を示す。

表2.2.2：可燃性ガスのラベル要素

| | 区分1A | 自然発火性ガスまたは化学的に不安定なガスA/Bの判定基準を満たす、1Aガスの区分 ||| 区分1B | 区分2 |
| | | 自然発火性ガス | 化学的に不安定なガス || | |
			区分A	区分B		
シンボル	炎	炎	炎	炎	炎	なし
注意喚起語	危険	危険	危険	危険	危険	警告
危険有害性情報	極めて可燃性の高いガス	極めて可燃性の高いガス 空気に触れると自然発火するおそれ	極めて可燃性の高いガス 空気が無くても爆発的に反応するおそれ	極めて可燃性の高いガス 圧力および/または温度が上昇した場合、空気が無くても爆発的に反応するおそれ	可燃性ガス	可燃性ガス

2.2.3.2 可燃性ガスやガスの混合物が自然発火性および/または化学的に不安定に分類された場合、すべての関連する分類は附属書4で定められているように安全データシートにおいて伝達されるべきで、関連する危険有害性情報の要素はラベルに含まれるべきである。

2.2.4 判定論理および手引き

以下の判定論理および手引きは、調和分類システムには含まれないが、追加的な手引きとしてここに示す。分類の責任者に対し、この判定論理を使用する前および使用する際に判定基準についてよく調べ理解することを強く勧める。

2.2.4.1 *可燃性ガスの判定論理*

可燃性ガスの分類には、その可燃性、空気中での可燃能力および化学的不安定性に関するデータが求められる。区分1Bにおける分類では、燃焼下限や基本的な燃焼速度に関するデータが求められる。分類は、判定論理2.2に従う。

Decision logic 2.2

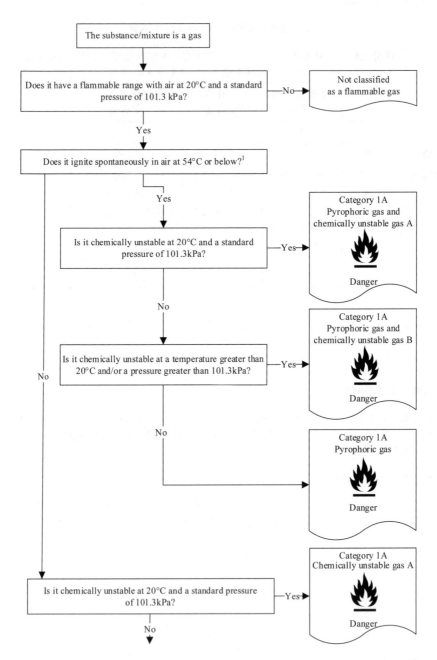

(cont'd on next page)

---

[1] *In the absence of data on its pyrophoricity, a flammable gas mixture should be classified as a pyrophoric gas if it contains more than 1% (by volume) of pyrophoric component(s).*

判定論理 2.2

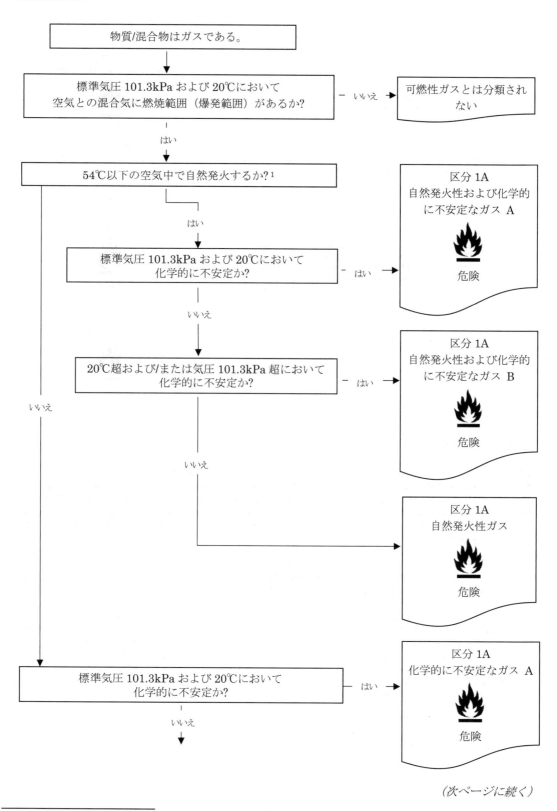

1 可燃性ガスの混合物で、自然発火性に関するデータがなく、1%を超える（容量）自然発火性成分を含む場合には自然発火性ガスに分類するべきである。

- 53 -

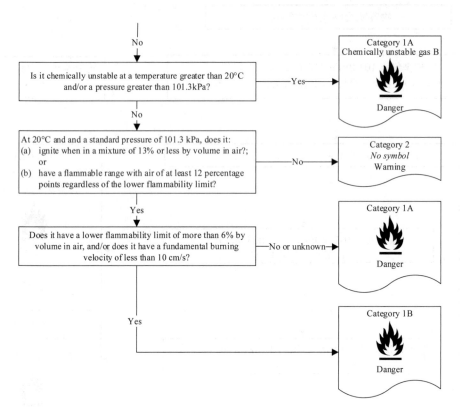

2.2.4.2 *Guidance*

2.2.4.2.1 Flammability should be determined by tests or by calculation in accordance with methods adopted by ISO (see ISO 10156:2017 "Gases and gas mixtures – Determination of fire potential and oxidizing ability for the selection of cylinder valve outlets" and, if using fundamental burning velocity for Category 1B, see ISO 817:2014 "Refrigerants-Designation and safety classification, Annex C: Method of test for burning velocity measurement of flammable gases"). Where insufficient data are available to use these methods, tests by a comparable method recognized by the competent authority may be used.

2.2.4.2.2 Pyrophoricity should be determined at 54°C in accordance with either IEC 60079-20-1 ed1.0 (2010-01) "Explosive atmospheres – Part 20-1: Material characteristics for gas and vapour classification – Test methods and data" or DIN 51794 "Determining the ignition temperature of petroleum products".

2.2.4.2.3 The classification procedure for pyrophoric gases need not be applied when experience in production or handling shows that the substance does not ignite spontaneously on coming into contact with air at a temperature of 54 °C or below. Flammable gas mixtures, which have not been tested for pyrophoricity and contain more than one percent pyrophoric components, should be classified as a pyrophoric gas. Expert judgement on the properties and physical hazards of pyrophoric gases and their mixtures should be used in assessing the need for classification of flammable gas mixtures containing one percent or less pyrophoric components. In this case, testing need only be considered if expert judgement indicates a need for additional data to support the classification process.

2.2.4.2.4 Chemical instability should be determined in accordance with the method described in Part III of the *Manual of Tests and Criteria*. If the calculations in accordance with ISO 10156:2017 show that a gas mixture is not flammable it is not necessary to carry out the tests for determining chemical instability for classification purposes.

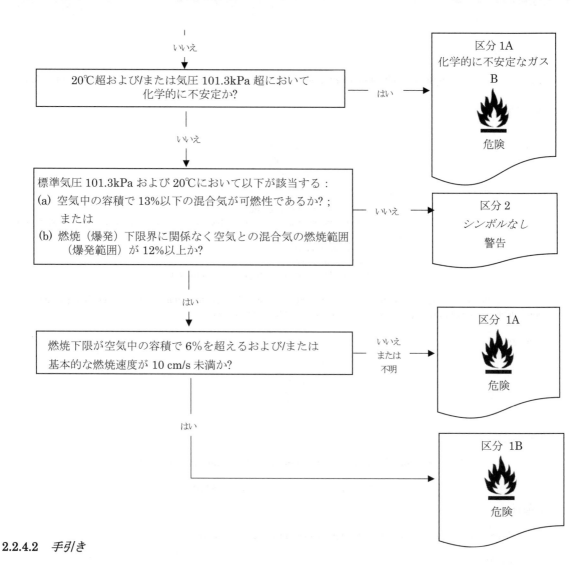

2.2.4.2　手引き

2.2.4.2.1　可燃性は ISO の採択する方法にしたがって、試験または計算により決定すべきである（ISO 10156:2017「ガスおよびガス混合物-シリンダー放出弁の選択のための着火および酸化能力の決定」参照、もし区分 1B に関して基本的な燃焼速度を用いるのであれば ISO 817:2014「冷媒－呼称と安全区分」附属書 C: 可燃性ガスの燃焼速度に関する試験方法）を参照）。これらの方法を利用するための十分なデータがない場合には、所管官庁が認める類似の方法による試験を用いることができる。

2.2.4.2.2　自然発火性は、54℃ において IEC 60079-20-1 ed1.0（2010-01）「爆発雰囲気 – Part 20-1：ガスおよび蒸気の分類に関する材料の特性– 試験方法及びデータ」または DIN 51794「石油製品の発火温度の測定」のいずれかの方法により測定する。

2.2.4.2.3　生産または取扱いにおける経験で物質が 54 ℃ 以下で空気に接触しても自然に発火しないことがわかっている場合には、自然発火性ガスの分類手順を適用する必要はない。自然発火性が試験されておらずしかも 1%を超える自然発火性成分を含む可燃性ガスの混合物は自然発火性ガスと分類されるべきである。自然発火性ガスおよびその混合物の性質や物理的危険性に関する専門家の判断は、1%以下の自然発火性成分を含む可燃性ガスの混合物の分類が必要な場合に行われるべきである。この場合、専門家の判断が分類をおこなうために追加的なデータを必要としているとした場合のみ、試験が検討される必要がある。

2.2.4.2.4　化学的不安定性は*試験方法及び判定基準のマニュアルの第 III 部に記載されている方法*にしたがって決定される。ISO 10156:2017 にしたがった計算でガス混合物が可燃性とならなかった場合には、分類のための化学的不安定性を見る試験を行う必要はない。

2.2.5 Example: Classification of a flammable gas mixture by calculation according to ISO 10156:2017

Formula

$$\sum_{i}^{n} \frac{V_i\%}{T_{ci}}$$

where:

$V_i\%$	=	the equivalent flammable gas content
T_{ci}	=	the maximum concentration of a flammable gas in nitrogen at which the mixture is still not flammable in air
i	=	the first gas in the mixture
n	=	the nth gas in the mixture
K_i	=	the equivalency factor for an inert gas versus nitrogen

Where a gas mixture contains an inert diluent other than nitrogen, the volume of this diluent is adjusted to the equivalent volume of nitrogen using the equivalency factor for the inert gas (K_i).

Criterion:

$$\sum_{i}^{n} \frac{V_i\%}{T_{ci}} > 1$$

Gas mixture

For the purpose of this example the following is the gas mixture to be used

$$2\,\%\,(H_2) + 6\,\%(CH_4) + 27\,\%(Ar) + 65\,\%(He)$$

Calculation

1. Ascertain the equivalency factors (K_i) for the inert gases versus nitrogen:

   K_i (Ar) = 0.55
   K_i (He) = 0.9

2. Calculate the equivalent mixture with nitrogen as balance gas using the K_i figures for the inert gases:

   $2\,\%(H_2) + 6\,\%(CH_4) + [27\,\% \times 0.55 + 65\,\% \times 0.9](N_2) = 2\,\%(H_2) + 6\,\%(CH_4) + 73.35\,\%(N_2) = 81.35\,\%$

3. Adjust the sum of the contents to 100 %:

   $\dfrac{100}{81.35} \times [2\,\%(H_2) + 6\,\%(CH_4) + 73.35\,\%(N_2)] = 2.46\,\%(H_2) + 7.37\,\%(CH_4) + 90.17\,\%(N_2)$

4. Ascertain the T_{ci} coefficients for the flammable gases:

   $T_{ci}\ H_2$ = 5.5 %
   $T_{ci}\ CH_4$ = 8.7 %

5. Calculate the flammability of the equivalent mixture using the formula:

$$\sum_{i}^{n} \frac{V_i\%}{T_{ci}} = \frac{2.46}{5.5} + \frac{7.37}{8.7} = 1.29 \qquad\qquad 1.29 > 1$$

Therefore the mixture is <u>flammable</u> in air.

2.2.5 例：ISO 10156: 2017 に従った計算による可燃性ガス混合物の分類

公式

$$\sum_i^n \frac{V_i\%}{T_{ci}}$$

ここで：
- $V_i\%$ ＝ 可燃性ガスの含量
- T_{ci} ＝ 混合物が空気中ではまだ可燃性とならない窒素中の可燃性ガス最大濃度
- i ＝ 混合物のi番目のガス
- n ＝ 混合物中のn番目のガス
- K_i ＝ 不活性ガス対窒素に関する等価係数

ガス混合物に窒素以外の不活性希釈ガスが含まれる場合、この希釈ガスの体積はその不活性ガスの等価係数（K_i）を用いて補正し窒素の等価体積とする。

判定基準

$$\sum_i^n \frac{V_i\%}{T_{ci}} \geq 1$$

ガス混合物

この例においては、次式のガス混合物を用いる。

$$2\% (H_2) + 6\% (CH_4) + 27\% (Ar) + 65\% (He)$$

計算

1. 窒素に対するこれら不活性ガスの各等価係数（K_i）を確認する：

    $K_i (Ar) = 0.55$
    $K_i (He) = 0.9$

2. 不活性ガスの K_i 値を用いて窒素をバランスガスとして等価の混合物を計算する：
    $2\% (H_2) + 6\% (CH_4) + [27\% \times 0.55 + 65\% \times 0.9] (N_2) = 2\% (H_2) + 6\% (CH_4) + 73.35\% (N_2) = 81.35\%$

3. 含量合計を補正して 100%とする：

    $(100/81.35) \times [2\% (H_2) + 6\% (CH_4) + 73.35\% (N_2)] = 2.46\% (H_2) + 7.37\% (CH_4) + 90.17\% (N_2)$

4. これらの可燃性ガスの T_{ci} 係数を確かめる：

    $T_{ci} H_2 = 5.5\%$
    $T_{ci} CH_4 = 8.7\%$

5. 次式を用いて等価の混合物の可燃性を計算する：

    $$\sum_i^n \frac{V_i\%}{T_{ci}} = 2.46/5.5 + 7.37/8.7 = 1.29 \qquad \mathbf{1.29 > 1}$$

したがってこの混合物は空気中で<u>可燃性</u>である。

CHAPTER 2.3

AEROSOLS AND CHEMICALS UNDER PRESSURE

2.3.0 **Introduction**

This chapter contains the definitions, classification criteria, hazard communication elements, decision logics and guidance for aerosols and chemicals under pressure. Although they present similar hazards, aerosols and chemicals under pressure are separate hazard classes and are covered in separate sections. While the hazards are similar and the classification is based on flammable properties and heat of combustion, they are presented in two different sections due to allowable pressure, capacity and construction of the two kinds of receptacles. A substance or mixture is classified as either an aerosol in accordance with 2.3.1 or a chemical under pressure in accordance with 2.3.2.

2.3.1 **Aerosols**

2.3.1.1 *Definition*

Aerosols, this means aerosol dispensers, are any non-refillable receptacles made of metal, glass or plastics and containing a gas compressed, liquefied or dissolved under pressure, with or without a liquid, paste or powder, and fitted with a release device allowing the contents to be ejected as solid or liquid particles in suspension in a gas, as a foam, paste or powder or in a liquid state or in a gaseous state.

2.3.1.2 *Classification criteria*

2.3.1.2.1 Aerosols are classified in one of the three categories of this hazard class, in accordance with table 2.3.1, depending on:

- their flammable properties;

- their heat of combustion; and

- if applicable, test results from the ignition distance test, the enclosed space ignition test and the aerosol foam flammability test, performed in accordance with subsections 31.4, 31.5 and 31.6 of the *Manual of Tests and Criteria*.

They should be considered for classification in Category 1 or 2 if they contain more than 1 % components (by mass) which are classified as flammable according to the GHS criteria, i.e.:

- Flammable gases (see chapter 2.2);

- Flammable liquids (see chapter 2.6);

- Flammable solids (see chapter 2.7);

or if their heat of combustion is at least 20 kJ/g.

第 2.3 章

エアゾールおよび加圧下化学品

2.3.0　序

この章には、エアゾールおよび加圧下化学品に関する定義、分類判定基準、危険性情報要素、判定論理および手引きを含む。これらは類似の危険性を示すが、エアゾールおよび加圧下化学品は別の危険有害性クラスであり、異なる節で扱われる。危険性は似ておりその分類は可燃特性および燃焼熱に基づいているが、2 種類の容器の許容圧力、容量および構造にしたがって異なる節で取り扱われる。物質または混合物は、2.3.1 にしたがってエアゾールとして、あるいは 2.3.2 にしたがって加圧下化学品として、それぞれ分類される。

2.3.1　エアゾール

2.3.1.1　定義

エアゾール、すなわちエアゾール噴霧器とは、圧縮ガス、液化ガスまたは溶解ガス（液状、ペースト状または粉末を含む場合もある）を内蔵する金属製、ガラス製またはプラスチック製の再充填不能な容器に、内容物をガス中に浮遊する固体もしくは液体の粒子として、または液体中またはガス中に泡状、ペースト状もしくは粉状として噴霧する噴射装置を取り付けたものをいう。

2.3.1.2　分類基準

2.3.1.2.1　エアゾールは、表 2.3.1 にしたがって以下の条件により、この危険有害性クラスの 3 つの区分のうちの 1 つに分類される：

- 燃焼特性；
- 燃焼熱；および
- 可能であれば、試験方法及び判定基準のマニュアル 31.4、31.5 および 31.6 にしたがって実施する着火距離試験、密閉空間発火試験および泡状エアゾールの可燃性試験の結果。

次の GHS 判定基準にしたがった可燃性/引火性に分類される成分（質量）を 1%超含むエアゾールの分類は、区分 1 あるいは 2 を検討すべきである：

GHS 判定基準：

- 可燃性ガス（第 2.2 章参照）；
- 引火性液体（第 2.6 章参照）；
- 可燃性固体（第 2.7 章参照）；

または燃焼熱量が少なくとも 20kJ/g であるエアゾール。

Table 2.3.1: Criteria for aerosols

Category	Criteria

1	(a) Any aerosol that contains ≥ 85 % flammable components (by mass) and has a heat of combustion of ≥ 30 kJ/g;
	 (b) Any aerosol that dispenses a spray that, in the ignition distance test, has an ignition distance of ≥ 75 cm; or
	 (c) Any aerosol that dispenses a foam that, in the foam flammability test, has:
	   (i) a flame height of ≥ 20 cm and a flame duration of ≥ 2 s; or
	   (ii) a flame height of ≥ 4 cm and a flame duration of ≥ 7 s.

2	(a) Any aerosol that dispenses a spray that, based on the results of the ignition distance test, does not meet the criteria for Category 1, and which has:
	   (i) a heat of combustion of ≥ 20 kJ/g;
	   (ii) a heat of combustion of < 20 kJ/g along with an ignition distance of ≥ 15 cm; or
	   (iii) a heat of combustion of < 20 kJ/g and an ignition distance of < 15 cm along with either, in the enclosed space ignition test:
	     - a time equivalent of ≤ 300 s/m^3; or
	     - a deflagration density of ≤ 300 g/m^3; or
	 (b) Any aerosol that dispenses a foam that, based on the results of the aerosol foam flammability test, does not meet the criteria for Category 1, and which has a flame height of ≥ 4 cm and a flame duration of ≥ 2 s.

3	(a) Any aerosol that contains ≤ 1 % flammable components (by mass) and that has a heat of combustion < 20 kJ/g; or (b) Any aerosol that contains > 1 % (by mass) flammable components or which has a heat of combustion of ≥ 20 kJ/g but which, based on the results of the ignition distance test, the enclosed space ignition test or the aerosol foam flammability test, does not meet the criteria for Category 1 or Category 2.

NOTE 1: *Flammable components do not cover pyrophoric, self-heating or water-reactive substances and mixtures because such components are never used as aerosol contents.*

NOTE 2: *Aerosols containing more than 1 % flammable components or with a heat of combustion of at least 20 kJ/g, which are not submitted to the flammability classification procedures in this chapter should be classified as aerosols, Category 1.*

NOTE 3: *Aerosols do not fall additionally within the scope of chapter 2.2 (flammable gases), section 2.3.2 (chemicals under pressure), chapters 2.5 (gases under pressure), 2.6 (flammable liquids) and 2.7 (flammable solids). Depending on their contents, aerosols may however fall within the scope of other hazard classes, including their labelling elements.*

2.3.1.3 *Hazard communication*

General and specific considerations concerning labelling requirements are provided in *Hazard communication: Labelling* (chapter 1.4). Annex 1 contains summary tables about classification and labelling. Annex 3 contains examples of precautionary statements and pictograms which can be used where allowed by the competent authority. Table 2.3.2 presents specific label elements for substances and mixtures classified into this hazard class based on the criteria in this chapter.

表 2.3.1：エアゾールの判定基準

区分	判定基準

1	(a) 85%以上（質量）の可燃性/引火性成分を含有し、かつ 30kJ/g 以上の燃焼熱を有するすべてのエアゾール；
	
	 (b) 着火距離試験で着火距離が 75cm 以上のスプレーを出すすべてのエアゾール；または
	
	 (c) 泡の可燃性試験の結果が以下のような、泡を出すすべてのエアゾール：
	    (i) 炎の高さが 20cm 以上かつ炎持続時間が 2 秒以上；または
	    (ii) 炎の高さが 4cm 以上かつ炎持続時間が 7 秒以上。

2	(a) 着火距離試験の結果が区分 1 の判定基準には該当せず、以下の条件を満たすスプレーを出すすべてのエアゾール：
	    (i) 燃焼熱が 20kJ/g 以上；
	    (ii) 燃焼熱が 20kJ/g 未満で着火距離が 15cm 以上；または
	    (iii) 燃焼熱が 20kJ/g 未満で着火距離が 15cm 未満かつ以下の密閉空間発火試験結果：
	      − 時間等量が 300s/m^3 以下；または
	      − 爆燃密度が 300g/m^3 以下；または
	
	 (b) 泡状エアゾール可燃性試験の結果が区分 1 の判定基準には該当せず、炎の高さが 4cm 以上かつ炎持続時間が 2 秒以上のすべてのエアゾール。

3	(a) 1%以下（質量）の可燃性成分を有し、かつ燃焼熱が 20kJ/g 未満のすべてのエアゾール；または (b) 1%超（質量）の可燃性成分を有するかまたは燃焼熱が 20kJ/g 以上であるが、着火距離試験、密閉空間発火試験または泡状エアゾール可燃性試験の結果が区分 1 または区分 2 の判定基準に該当しないすべてのエアゾール。

注記 1： *可燃性/引火性成分には自然発火性物質、自己発熱性物質または水反応性物質は含まない。なぜならば、これらの物質はエアゾール内容物として用いられることはないためである。*

注記 2： *本章で可燃性/引火性の分類の手順を踏まない、1%超の可燃性/引火性成分を含むまたは燃焼熱が少なくとも 20kJ/g のエアゾールは、区分 1 に分類するべきである。*

注記 3： *エアゾールが追加的に第2.2章（可燃性ガス）、2.3.2（加圧下化学品）、第2.5章（高圧ガス）、第2.6章（引火性液体）および第2.7章（可燃性固体）の範疇で分類されることはない。しかし成分により、ラベル要素も含め、エアゾールが他の危険有害性クラスの範疇に分類されることはありうる。*

2.3.1.3 *危険有害性情報の伝達*

表示要件に関する通則および細則は、*危険有害性に関する情報の伝達：表示*（第 1.4 章）に定める。附属書1に分類と表示に関する概要表を示す。附属書3には、注意書きおよび所管官庁が許可した場合に使用可能な注意絵表示の例を示す。表 2.3.2 は本章の判定基準に基づいて、この危険有害性クラスに分類される物質および混合物のラベル要素を示す。

Table 2.3.2: Label elements for aerosols

	Category 1	**Category 2**	**Category 3**
Symbol	Flame	Flame	*No symbol*
Signal word	Danger	Warning	Warning
Hazard statement	Extremely flammable aerosol Pressurized container: May burst if heated	Flammable aerosol Pressurized container: May burst if heated	Pressurized container: May burst if heated

2.3.1.4   *Decision logic*

2.3.1.4.1   The decision logic, which follows, has been provided here as additional guidance. It is strongly recommended that the person responsible for classification studies the criteria before and during use of the decision logic.

2.3.1.4.2   To classify an aerosol data on its flammable components, on its chemical heat of combustion and, if applicable, the results of the ignition distance test and enclosed space test (for spray aerosols) and of the foam test (for foam aerosols) are required. Classification should be made according to decision logics 2.3.1 (a) to 2.3.1 (c).

Decision logic 2.3.1 (a) for aerosols

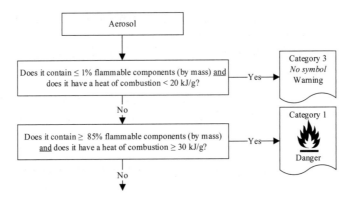

For spray aerosols, go to decision logic 2.3.1 (b);
For foam aerosols, go to decision logic 2.3.1 (c);

表 2.3.2：エアゾールのラベル要素

	区分1	区分2	区分3
シンボル	炎	炎	シンボルなし
注意喚起語	危険	警告	警告
危険有害性情報	極めて可燃性の高いエアゾール高圧容器：熱すると破裂のおそれ	可燃性エアゾール高圧容器：熱すると破裂のおそれ	高圧容器：熱すると破裂のおそれ

2.3.1.4 判定論理

2.3.1.4.1 以下の判定論理は、追加的な手引きとしてここに示す。分類の責任者に対し、この判定論理を使用する前および使用する際に判定基準についてよく調べ理解することを強く勧める。

2.3.1.4.2 エアゾールを分類するには、その可燃性/引火性成分、その化学燃焼熱、および該当する場合には火炎長（着火距離）試験および密閉空間試験（噴射式エアゾールの場合）ならびに泡試験（泡エアゾールの場合）に関するデータが求められる。分類は 2.3.1(a)から 2.3.1(c)の判定論理に従うべきである。

判定論理 2.3.1(a)　エアゾール

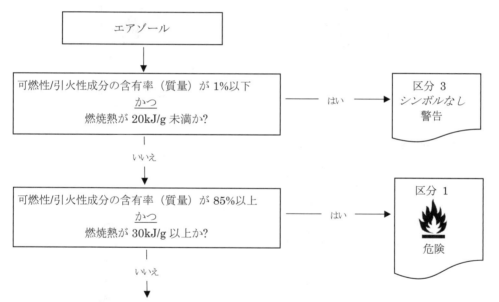

Decision logic 2.3.1 (b) for spray aerosols

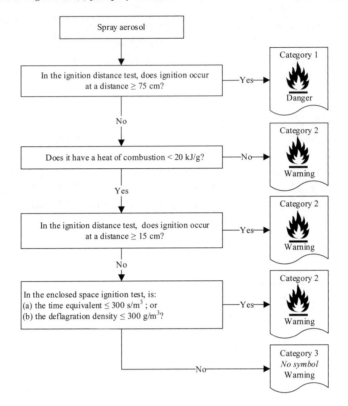

Decision logic 2.3.1 (c) for foam aerosols

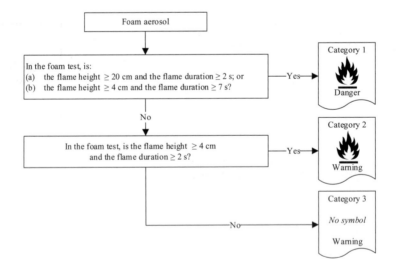

判定論理 2.3.1(b)　噴射式エアゾール

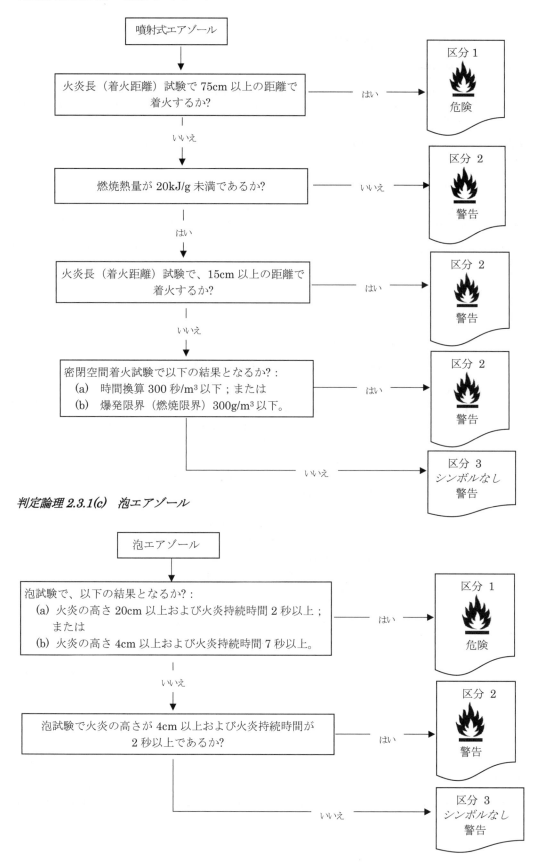

判定論理 2.3.1(c)　泡エアゾール

- 60 -

2.3.2 Chemicals under pressure

2.3.2.1 *Definition*

Chemicals under pressure are liquids or solids (e.g. pastes or powders), pressurized with a gas at a pressure of 200 kPa (gauge) or more at 20 °C in pressure receptacles other than aerosol dispensers and which are not classified as gases under pressure.

NOTE: Chemicals under pressure typically contain 50 % or more by mass of liquids or solids whereas mixtures containing more than 50 % gases are typically considered as gases under pressure.

2.3.2.2 *Classification criteria*

2.3.2.2.1 Chemicals under pressure are classified in one of three categories of this hazard class, in accordance with table 2.3.3, depending on their content of flammable components and their heat of combustion (see 2.3.2.4.1).

2.3.2.2.2 Flammable components are components which are classified as flammable in accordance with the GHS criteria, i.e.:

- Flammable gases (see chapter 2.2);
- Flammable liquids (see chapter 2.6);
- Flammable solids (see chapter 2.7).

Table 2.3.3: Criteria for chemicals under pressure

Category	Criteria
1	Any chemical under pressure that: (a) contains ≥ 85 % flammable components (by mass); and (b) has a heat of combustion of ≥ 20 kJ/g.
2	Any chemical under pressure that: (a) contains > 1 % flammable components (by mass); and (b) has a heat of combustion < 20 kJ/g; or that: (a) contains < 85 % flammable components (by mass); and (b) has a heat of combustion ≥ 20 kJ/g.
3	Any chemical under pressure that: (a) contains ≤ 1% flammable components (by mass); and (b) has a heat of combustion of < 20 kJ/g.

NOTE 1: The flammable components in a chemical under pressure do not include pyrophoric, self-heating or water-reactive, substances and mixtures because such components are not allowed in chemicals under pressure in accordance with the UN Model Regulations.

NOTE 2: Chemicals under pressure do not fall additionally within the scope of section 2.3.1 (aerosols), chapters 2.2 (flammable gases), 2.5 (gases under pressure), 2.6 (flammable liquids) and 2.7 (flammable solids). Depending on their contents, chemicals under pressure may however fall within the scope of other hazard classes, including their labelling elements.

2.3.2.3 *Hazard communication*

General and specific considerations concerning labelling requirements are provided in *Hazard communication: Labelling* (chapter 1.4). Annex 1 contains summary tables about classification and labelling. Annex 3 contains examples of precautionary statements and pictograms which can be used where allowed by the competent authority. Table 2.3.4 presents specific label elements for substances and mixtures classified into this hazard class based on the criteria in this chapter.

2.3.2 加圧下化学品

2.3.2.1 定義

加圧下化学品とは、エアゾール噴霧器ではなく、かつ高圧ガスとは分類されない、圧力容器中で20℃において200kPa以上（ゲージ圧）の圧力でガスにより加圧された液体または固体（例えばペーストまたは粉体）をいう。

注記：加圧下化学品は一般に質量で50％以上の液体または固体を含むが、50％以上のガスを含む混合物は一般に高圧ガスと考えられる。

2.3.2.2 分類基準

2.3.2.2.1 加圧下化学品は、可燃性/引火性成分の量およびそれらの燃焼熱によって、表2.3.2にしたがって、この危険有害性クラスの3つの区分のうちの1つに分類される（2.3.2.4.1参照）。

2.3.2.2.2 可燃性/引火性成分とは以下のGHSの判定基準にしたがって可燃性/引火性と分類された成分のことである、すなわち：

— 可燃性ガス（第2.2章参照）；

— 引火性液体（第2.6章参照）；

— 可燃性固体（第2.7章参照）。

表2.3.3：加圧下化学品の判定基準

区分	判定基準
1	以下のようなすべての加圧下化学品： (a) 85％以上（質量）の可燃性/引火性成分を含み；かつ (b) 燃焼熱が20kJ/g以上。
2	以下のようなすべての加圧下化学品： (a) 1％超（質量）の可燃性/引火性成分を含み；かつ (b) 燃焼熱が20kJ/g未満； または： (a) 85％未満（質量）の可燃性/引火性成分を含み；かつ (b) 燃焼熱が20kJ/g以上。
3	以下のようなすべての加圧下化学品： (a) 1％以下（質量）の可燃性/引火性成分を含み；かつ (b) 燃焼熱が20kJ/g未満。

注記1：加圧下化学品の可燃性/引火性成分には、自然発火性、自己発熱性または水反応性の物質や混合物は含まれない。それらの成分はUNモデル規則により加圧下化学品として認められていないからである。

注記2：加圧下化学品が追加的に2.3.1（エアゾール）、第2.2章（可燃性ガス）、第2.5章（高圧ガス）、第2.6章（引火性液体）および第2.7章（可燃性固体）の範疇で分類されることはない。しかし成分により、ラベル要素も含め、加圧下化学品が他の危険有害性クラスの範疇に分類されることはありうる。

2.3.2.3 危険有害性情報の伝達

表示要件に関する通則および細則は、*危険有害性に関する情報の伝達：表示*（第1.4章）に定める。附属書1に分類と表示に関する概要表を示す。附属書3には、注意書きおよび所管官庁が許可した場合に使用可能な注意絵表示の例を示す。表2.3.4は本章の判定基準に基づいて、この危険有害性クラスに分類される物質および混合物のラベル要素を示す。

Table 2.3.4: Label elements for chemicals under pressure

	Category 1	Category 2	Category 3
Symbol	Flame Gas cylinder	Flame Gas cylinder	Gas cylinder
Signal word	Danger	Warning	Warning
Hazard statement	Extremely flammable chemical under pressure: May explode if heated	Flammable chemical under pressure: May explode if heated	Chemical under pressure: May explode if heated

2.3.2.4   *Decision logic*

The decision logic 2.3.2 has been provided as additional guidance. It is strongly recommended that the person responsible for classification studies the criteria before and during use of the decision logic.

2.3.2.4.1   *Decision logic*

To classify a mixture as chemicals under pressure, data on its pressure, its flammable components, and on its specific heat of combustion are required. Classification should be made in accordance with decision logic 2.3.2.

Decision logic 2.3.2 for chemicals under pressure

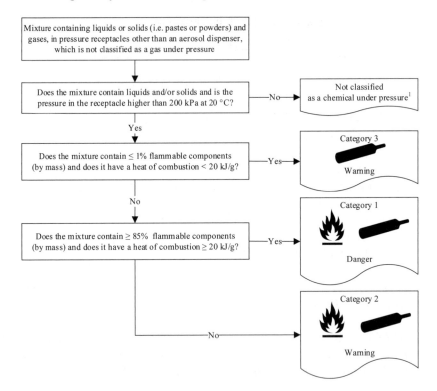

---

[1]   *Should be considered for classification in other physical hazard classes as appropriate.*

表 2.3.4：加圧下化学品のラベル要素

	区分 1	区分 2	区分 3
シンボル	炎 ガスボンベ	炎 ガスボンベ	ガスボンベ
注意喚起語	危険	警告	警告
危険有害性情報	極めて可燃性の高い加圧下化学品：熱すると爆発のおそれ	可燃性の加圧下化学品：熱すると爆発のおそれ	加圧下化学品：熱すると爆発のおそれ

2.3.2.4 判定論理

判定論理 2.3.2 を追加的な手引きとして示す。分類責任者に対し、この判定論理を使用する前および使用する際に判定基準についてよく調べ理解することを強く勧める。

2.3.2.4.1 判定論理

加圧下化学品として混合物を分類するためには、その圧力、可燃性/引火性成分およびその燃焼熱に関するデータが必要である。分類は判定論理 2.3.2 にしたがって行われるべきである。

判定論理 2.3.2　加圧下化学品

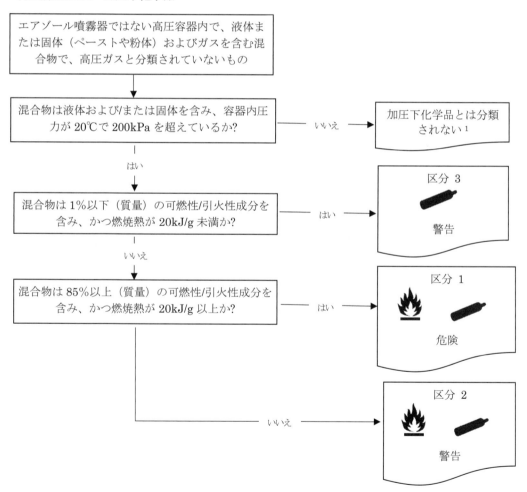

---

[1] *必要に応じて他の物理化学的危険性に関する分類を考慮する。*

2.3.3 **Guidance on specific heat of combustion**

2.3.3.1 For a composite formulation, the specific heat of combustion of the product is the summation of the weighted specific heats of combustion for the individual components, as follows:

$$\Delta H_c(\text{product}) = \sum_{i}^{n} [\, w(i) \times \Delta H_c(i)\,]$$

where:

$\Delta H_c(\text{product})$ = specific heat of combustion (kJ/g) of the product;
$\Delta H_c(i)$ = specific heat of combustion (kJ/g) of component i in the product;
$w(i)$ = mass fraction of component i in the product;
n = total number of components in the product.

The specific heats of combustion, which are given in kilojoules per gram (kJ/g), can be found in the scientific literature, calculated or determined by tests (see ASTM D 240 and NFPA 30B). Note that experimentally measured heats of combustion usually differ from the corresponding theoretical heats of combustion, since the combustion efficiency normally is less than 100 % (a typical combustion efficiency is 95 %).

2.3.3 燃焼熱に関する手引き

2.3.3.1 混合物においては、製品の燃焼熱は以下のようにそれぞれの成分に対して重み付けした燃焼熱の合計である：

$$\Delta H_c (製品) = \sum_{i}^{n} [\, w(i) \times \Delta H_c(i) \,]$$

ここで：
 ΔH_c（製品） ＝ 製品の燃焼熱（kJ/g）；
 $\Delta H_c(i)$ ＝ 製品を構成する成分 i の燃焼熱（kJ/g）；
 $w(i)$ ＝ 製品を構成する成分 i の質量百分率；
 n ＝ 製品の成分数。

燃焼熱は、グラム当たりのキロジュール（kJ/g）で与えられ、文献報告値、計算値または試験（ASTM D 240 および NFPA 30B）による測定値でもよい。試験的に測定された燃焼熱は、燃焼効率が普通100％未満（典型的な燃焼効率は95％である）なので、対応する理論的な燃焼熱とは通常異なることに注意が必要である。

CHAPTER 2.4

OXIDIZING GASES

2.4.1 **Definition**

An *oxidizing gas* is any gas which may, generally by providing oxygen, cause or contribute to the combustion of other material more than air does.

NOTE: *"Gases which cause or contribute to the combustion of other material more than air does" means pure gases or gas mixtures with an oxidizing power greater than 23.5 % as determined by a method specified in ISO 10156:2017.*

2.4.2 **Classification criteria**

An oxidizing gas is classified in a single category for this class according to the following table:

Table 2.4.1: Criteria for oxidizing gases

Category	Criteria
1	Any gas which may, generally by providing oxygen, cause or contribute to the combustion of other material more than air does.

2.4.3 **Hazard communication**

General and specific considerations concerning labelling requirements are provided in *Hazard communication: Labelling* (chapter 1.4). Annex 1 contains summary tables about classification and labelling. Annex 3 contains examples of precautionary statements and pictograms which can be used where allowed by the competent authority. Table 2.4.2 presents specific label elements for substances and mixtures classified into this hazard class based on the criteria in this chapter.

Table 2.4.2: Label elements for oxidizing gases

	Category 1
Symbol	Flame over circle
Signal word	Danger
Hazard statement	May cause or intensify fire; oxidizer

2.4.4 **Decision logic and guidance**

The decision logic and guidance, which follow, are not part of the harmonized classification system, but have been provided here as additional guidance. It is strongly recommended that the person responsible for classification studies the criteria before and during use of the decision logic.

2.4.4.1 *Decision logic*

To classify an oxidizing gas, tests or calculation methods as described in ISO 10156:2017 "Gases and gas mixtures – Determination of fire potential and oxidizing ability for the selection of cylinder valve outlets" should be performed.

第2.4章

酸化性ガス

2.4.1 定義

酸化性ガスとは、一般的には酸素を供給することにより、空気以上に他の物質の燃焼を引き起こす、または燃焼を助けるガスをいう。

注記：「*空気以上に他の物質の燃焼を引き起こすガス*」*とは、ISO 10156:2017 により定められる方法によって測定された 23.5%以上の酸化能力を持つ純粋ガスあるいは混合ガスをいう。*

2.4.2 分類基準

酸化性ガスは、次表にしたがってこのクラスにおける単一の区分に分類される。

表 2.4.1：酸化性ガスの判定基準

区分	判定基準
1	一般的には酸素を供給することにより、空気以上に他の物質の燃焼を引き起こす、または燃焼を助けるガス

2.4.3 危険有害性情報の伝達

表示要件に関する通則および細則は、*危険有害性に関する情報の伝達：表示*（第 1.4 章）に定める。附属書 1 に分類と表示に関する概要表を示す。附属書 3 には、注意書きおよび所管官庁が許可した場合に使用可能な注意絵表示の例を示す。表 2.4.2 は本章の判定基準に基づいて、この危険有害性クラスに分類される物質および混合物のラベル要素を示す。

表 2.4.2：酸化性ガスのラベル要素

	区分1
シンボル	円上の炎
注意喚起語	危険
危険有害性情報	発火または火災助長のおそれ；酸化性物質

2.4.4 判定論理および手引き

以下の判定論理および手引きは、調和分類システムには含まれないが、追加的な手引きとしてここに示す。分類の責任者に対し、この判定論理を使用する前および使用する際に判定基準についてよく調べ理解することを強く勧める。

2.4.4.1 *判定論理*

酸化性ガスの分類には、ISO 10156:2017「ガスおよびガス混合物-シリンダー放出弁の選択のための着火および酸化能力の決定」に記載された試験または計算方法を実施するべきである。

Decision logic 2.4 for oxidizing gases

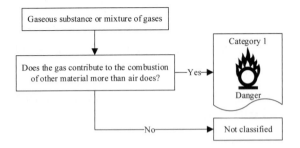

2.4.4.2   *Guidance*

Example of the classification of an oxidizing gas mixture by calculation according to ISO 10156:2017.

The classification method described in ISO 10156 uses the criterion that a gas mixture should be considered as more oxidizing than air if the oxidizing power of the gas mixture is higher than 0.235 (23.5 %).

The oxidizing power (OP) is calculated as follows:

$$OP = \frac{\sum_{i=1}^{n} x_i C_i}{\sum_{i=1}^{n} x_i + \sum_{k=1}^{p} K_k B_k}$$

Where:

- x_i = molar fraction of the i:th oxidizing gas in the mixture;
- C_i = coefficient of oxygen equivalency of the i:th oxidizing gas in the mixture;
- K_k = coefficient of equivalency of the inert gas k compared to nitrogen;
- B_k = molar fraction of the k:th inert gas in the mixture;
- n = total number of oxidizing gases in the mixture;
- p = total number of inert gases in the mixture;

Example mixture: 9 % (O_2) + 16 % (N_2O) + 75 % (He)

Calculation steps

Step 1:

Ascertain the coefficient of oxygen equivalency (Ci) for the oxidizing gases in the mixture and the nitrogen equivalency factors (Kk) for the non-flammable, non-oxidizing gases.

C_i (N_2O)   =   0.6 (nitrous oxide)

C_i (O_2) =   1 (oxygen)

K_k (He)=   0.9 (helium)

Step 2:

Calculate the oxidizing power of the gas mixture

$$OP = \frac{\sum_{i=1}^{n} x_i C_i}{\sum_{i=1}^{n} x_i + \sum_{k=1}^{p} K_k B_k} = \frac{0.09 \times 1 + 0.16 \times 0.6}{0.09 + 0.16 + 0.75 \times 0.9} = 0.201 \qquad 20.1 < 23.5$$

Therefore the mixture is <u>not</u> considered as <u>an oxidizing gas.</u>

判定論理 2.4　酸化性ガス

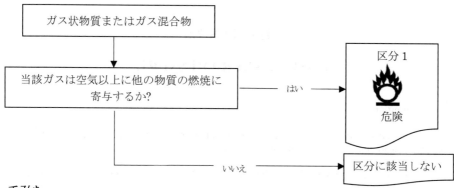

2.4.4.2　手引き

ISO-10156:2017 に従った計算による酸化性ガス混合物分類の例

ISO-10156 に記載されている分類方法では、ガス混合物の酸化力が 0.235（23.5%）を超える場合にガス混合物は空気よりもより酸化力が高いとみなされるべきである、という判定基準を採用している。

酸化力（oxidizing power: OP）は以下のように計算される：

$$OP = \frac{\sum_{i=1}^{n} x_i C_i}{\sum_{i=1}^{n} x_i + \sum_{k=1}^{p} K_k B_k}$$

ここで、
- X_i = 混合物中 i 番目の酸化性ガスのモル分率
- C_i = 混合物中 i 番目の酸化性ガス酸素当量係数
- K_k = 窒素と比較した非活性ガス k の当量係数
- B_k = 混合物中 k 番目の非活性ガスのモル分率
- n = 混合物中の酸化性ガスの総数
- p = 混合物中の非活性ガスの総数

混合物例：　9% (O_2) + 16% (N_2O) + 75% (He)

計算手順

ステップ1：

当該混合物中の酸化性ガスの酸素当量（C_i）係数および非可燃性、非酸化性ガスの窒素当量係数（K_k）を確認する。

C_i (N_2O) = 0.6（亜酸化窒素）
C_i (O_2) = 1（酸素）
K_k(He) = 0.9（ヘリウム）

ステップ2：

ガス混合物の酸化力を計算する

$$OP = \frac{\sum_{i=1}^{n} x_i C_i}{\sum_{i=1}^{n} x_i + \sum_{k=1}^{p} K_k B_k} = \frac{0.09 \times 1 + 0.16 \times 0.6}{0.09 + 0.16 + 0.75 \times 0.9} = 0.201 \qquad 20.1 < 23.5$$

したがって混合物は<u>酸化性ガス</u>とはみなされ<u>ない</u>。

CHAPTER 2.5

GASES UNDER PRESSURE

2.5.1 Definition

Gases under pressure are gases which are contained in a receptacle at a pressure of 200 kPa (gauge) or more at 20 °C, or which are liquefied or liquefied and refrigerated.

They comprise compressed gases, liquefied gases, dissolved gases and refrigerated liquefied gases.

2.5.2 Classification criteria

2.5.2.1 Gases under pressure are classified, according to their physical state when packaged, in one of four groups in the following table:

Table 2.5.1: Criteria for gases under pressure

Group	Criteria
Compressed gas	A gas which when packaged under pressure is entirely gaseous at -50°C; including all gases with a critical temperature ≤ -50°C.
Liquefied gas	A gas which when packaged under pressure, is partially liquid at temperatures above -50°C. A distinction is made between: (a) High pressure liquefied gas: a gas with a critical temperature between -50°C and +65°C; and (b) Low pressure liquefied gas: a gas with a critical temperature above +65°C.
Refrigerated liquefied gas	A gas which when packaged is made partially liquid because of its low temperature.
Dissolved gas	A gas which when packaged under pressure is dissolved in a liquid phase solvent.

The critical temperature is the temperature above which a pure gas cannot be liquefied, regardless of the degree of compression.

NOTE: *Aerosols and chemicals under pressure should not be classified as gases under pressure. See chapter 2.3.*

2.5.3 Hazard communication

General and specific considerations concerning labelling requirements are provided in *Hazard communication: Labelling* (chapter 1.4). Annex 1 contains summary tables about classification and labelling. Annex 3 contains examples of precautionary statements and pictograms which can be used where allowed by the competent authority. Table 2.5.2 presents specific label elements for substances and mixtures classified into this hazard class based on the criteria in this chapter.

Table 2.5.2: Label elements for gases under pressure

	Compressed gas	Liquefied gas	Refrigerated liquefied gas	Dissolved gas
Symbol	Gas cylinder	Gas cylinder	Gas cylinder	Gas cylinder
Signal word	Warning	Warning	Warning	Warning
Hazard statement	Contains gas under pressure; may explode if heated	Contains gas under pressure; may explode if heated	Contains refrigerated gas; may cause cryogenic burns or injury	Contains gas under pressure; may explode if heated

第 2.5 章

高圧ガス

2.5.1 定義

高圧ガスとは、20℃、200kPa（ゲージ圧）以上の圧力の下で容器に充填されているガスまたは液化または深冷液化されているガスをいう。

高圧ガスには、圧縮ガス、液化ガス、溶解ガスおよび深冷液化ガスが含まれる。

2.5.2 分類基準

2.5.2.1 高圧ガスは、充填された時の物理的状態によって、次表の4つのグループのいずれかに分類される。

表 2.5.1：高圧ガスの判定基準

グループ	判定基準
圧縮ガス	加圧して容器に充填した時に、−50℃で完全にガス状であるガス；臨界温度−50℃以下のすべてのガスを含む。
液化ガス	加圧して容器に充填した時に−50℃を超える温度において部分的に液体であるガス。次の2つに分けられる： （a）高圧液化ガス：臨界温度が−50℃と+65℃の間にあるガス；および （b）低圧液化ガス：臨界温度が+65℃を超えるガス。
深冷液化ガス	容器に充填したガスが低温のために部分的に液体であるガス。
溶解ガス	加圧して容器に充填したガスが液相溶媒に溶解しているガス。

臨界温度とは、その温度を超えると圧縮の程度に関係なく純粋ガスが液化されない温度をいう。

注記：エアゾールおよび加圧下化学品は高圧ガスとして分類するべきではない。第 2.3 章参照。

2.5.3 危険有害性情報の伝達

表示要件に関する通則および細則は、*危険有害性に関する情報の伝達：表示（第 1.4 章）*に定める。附属書1に分類と表示に関する概要表を示す。附属書3には、注意書きおよび所管官庁が許可した場合に使用可能な注意絵表示の例を示す。表 2.5.2 は本章の判定基準に基づいて、この危険有害性クラスに分類される物質および混合物のラベル要素を示す。

表 2.5.2：高圧ガスのラベル要素

	圧縮ガス	液化ガス	深冷液化ガス	溶解ガス
シンボル	ガスボンベ	ガスボンベ	ガスボンベ	ガスボンベ
注意喚起語	警告	警告	警告	警告
危険有害性情報	高圧ガス；熱すると爆発するおそれ	高圧ガス；熱すると爆発するおそれ	深冷液化ガス；凍傷または傷害のおそれ	高圧ガス；熱すると爆発するおそれ

2.5.4 **Decision logic and guidance**

The decision logic and guidance, which follow, are not part of the harmonized classification system, but have been provided here as additional guidance. It is strongly recommended that the person responsible for classification studies the criteria before and during use of the decision logic.

2.5.4.1 *Decision logic*

Classification can be made according to decision logic 2.5.

Decision logic 2.5 for gases under pressure

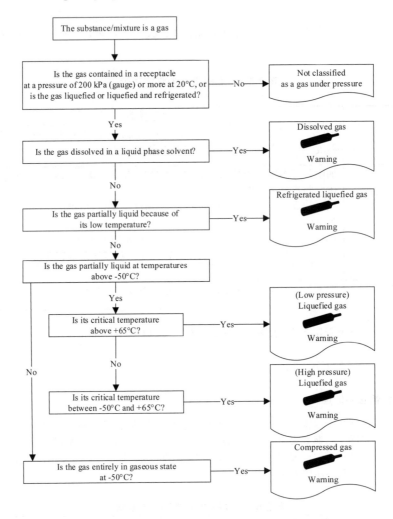

2.5.4 判定論理および手引き

以下の判定論理および手引きは、調和分類システムには含まれないが、追加的な手引きとしてここに示す。分類の責任者に対し、この判定論理を使用する前および使用する際に判定基準についてよく調べ理解することを強く勧める。

2.5.4.1 判定論理

分類は判定論理 2.5 にしたがって行う事ができる。

判定論理 2.5　高圧ガス

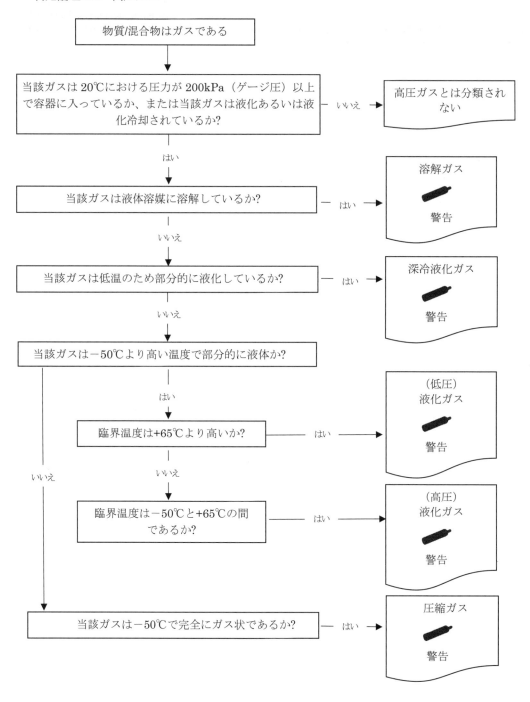

2.5.4.2 *Guidance*

For this group of gases, the following information is required to be known:

(a) The vapour pressure at 50°C;

(b) The physical state at 20 °C at standard ambient pressure;

(c) The critical temperature.

In order to classify a gas, the above data are needed. Data can be found in literature, calculated or determined by testing. Most pure gases are already classified in the *UN Model Regulations*. Most-off mixtures require additional calculations that can be very complex.

2.5.4.2 *手引き*

このガスのグループには次の既知情報が必要である：

(a) 50℃における蒸気圧；

(b) 20℃および標準気圧における物理的性状；

(c) 臨界温度。

ガスの分類には、上記のデータが必要である。データは文献、計算または試験測定で得られる。ほとんどの純粋ガスは *UN モデル規則* ですでに分類されている。ほとんどの 1 回限りの混合物は非常に複雑な追加計算が必要となる。

CHAPTER 2.6

FLAMMABLE LIQUIDS

2.6.1 Definition

A *flammable liquid* means a liquid having a flash point of not more than 93 °C.

2.6.2 Classification criteria

A flammable liquid is classified in one of the four categories for this class according to the following table:

Table 2.6.1: Criteria for flammable liquids

Category	Criteria
1	Flash point < 23 °C and initial boiling point ≤ 35 °C
2	Flash point < 23 °C and initial boiling point > 35 °C
3	Flash point ≥ 23 °C and ≤ 60 °C
4	Flash point > 60 °C and ≤ 93 °C

NOTE 1: *Gas oils, diesel and light heating oils in the flash point range of 55°C to 75°C may be regarded as a special group for some regulatory purposes.*

NOTE 2: *Liquids with a flash point of more than 35 °C and not more than 60 °C may be regarded as non-flammable liquids for some regulatory purposes (e.g. transport) if negative results have been obtained in the sustained combustibility test L.2 of Part III, section 32 of the Manual of Tests and Criteria.*

NOTE 3: *Viscous flammable liquids such as paints, enamels, lacquers, varnishes, adhesives and polishes may be regarded as a special group for some regulatory purposes (e.g. transport). The classification or the decision to consider these liquids as non-flammable may be determined by the pertinent regulation or competent authority.*

NOTE 4: *Aerosols should not be classified as flammable liquids. See chapter 2.3.*

2.6.3 Hazard communication

General and specific considerations concerning labelling requirements are provided in *Hazard communication: Labelling* (chapter 1.4). Annex 1 contains summary tables about classification and labelling. Annex 3 contains examples of precautionary statements and pictograms which can be used where allowed by the competent authority. Table 2.6.2 presents specific label elements for substances and mixtures classified into this hazard class based on the criteria in this chapter.

Table 2.6.2: Label elements for flammable liquids

	Category 1	Category 2	Category 3	Category 4
Symbol	Flame	Flame	Flame	*No symbol*
Signal word	Danger	Danger	Warning	Warning
Hazard statement	Extremely flammable liquid and vapour	Highly flammable liquid and vapour	Flammable liquid and vapour	Combustible liquid

第 2.6 章

引火性液体

2.6.1 定義

*引火性液体*とは、引火点が 93℃以下の液体をいう。

2.6.2 分類基準

引火性液体は、次表にしたがってこのクラスにおける 4 つの区分のいずれかに分類される：

表 2.6.1：引火性液体の判定基準

区分	判定基準
1	引火点＜ 23℃および初留点≦35℃
2	引火点＜ 23℃および初留点＞35℃
3	引火点≧23℃および≦60℃
4	引火点＞ 60℃および≦93℃

注記 1：引火点が 55℃から 75℃の範囲内にある軽油類、ディーゼル油および軽加熱油は、規制目的によっては 1 つの特殊グループとされることがある。

注記 2：引火点が 35℃を超え 60℃を超えない液体は、試験方法及び判定基準のマニュアル第 III 部、32 節の燃焼持続試験 L.2 において否の結果が得られている場合は、規制目的（輸送など）によっては引火性液体とされないことがある。

注記 3：ペイント、エナメル、ラッカー、ワニス、接着剤、つや出し剤等の粘性の引火性液体は、規制目的（輸送など）によっては 1 つの特殊グループとされることがある。この分類またはこれらの液体を非引火性とすることは、関連法規または所管官庁により決定することができる。

注記 4：エアゾールは引火性液体と分類すべきではない、第 2.3 章参照。

2.6.3 危険有害性情報の伝達

表示要件に関する通則および細則は、*危険有害性に関する情報の伝達：表示*（第 1.4 章）に定める。附属書 1 に分類と表示に関する概要表を示す。附属書 3 には、注意書きおよび所管官庁が許可した場合に使用可能な注意絵表示の例を示す。表 2.6.2 は本章の判定基準に基づいて、この危険有害性クラスに分類される物質および混合物のラベル要素を示す。

表 2.6.2：引火性液体のラベル要素

	区分 1	区分 2	区分 3	区分 4
シンボル	炎	炎	炎	シンボルなし
注意喚起語	危険	危険	警告	警告
危険有害性情報	極めて引火性の高い液体および蒸気	引火性の高い液体および蒸気	引火性液体および蒸気	可燃性液体

2.6.4   **Decision logic and guidance**

The decision logic and guidance, which follow, are not part of the harmonized classification system, but have been provided here as additional guidance. It is strongly recommended that the person responsible for classification studies the criteria before and during use of the decision logic.

2.6.4.1   *Decision logic*

Once the flash point and the initial boiling point are known, the classification of the substance or mixture and the relevant harmonized label information can be obtained according to decision logic 2.6.

Decision logic 2.6 for flammable liquids

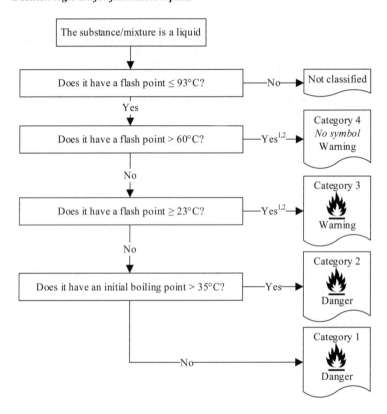

---

[1] *Gas oils, diesel and light heating oils in the flash point range of 55 °C to 75 °C may be regarded as a special group for some regulatory purposes as these hydrocarbons mixtures have varying flash point in that range. Thus, classification of these products in Category 3 or 4 may be determined by the pertinent regulation or competent authority.*

[2] *Liquids with a flash point of more than 35 °C and not more than 60 °C may be regarded as non-flammable liquids for some regulatory purposes (e.g. transport) if negative results have been obtained in the sustained combustibility test L.2 of Part III, section 32 of the Manual of Tests and Criteria.*

2.6.4　判定論理および手引き

　以下の判定論理および手引きは、調和分類システムには含まれないが、追加的な手引きとしてここに示す。分類の責任者に対し、この判定論理を使用する前および使用する際に判定基準についてよく調べ理解することを強く勧める。

2.6.4.1　判定論理

　引火点および初留点が既知の場合は、その物質または混合物の分類および調和された関連表示情報は次の枝分かれ図から得られる。

判定論理 2.6　引火性液体

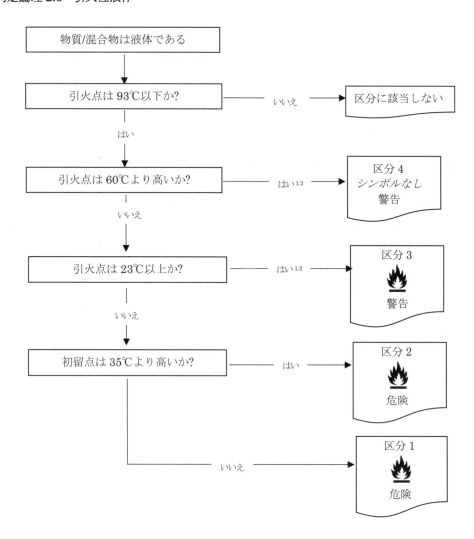

---

[1]　引火点が55℃から75℃の範囲内にある軽油類、ディーゼル油および軽加熱油は、規制目的によっては1つの特殊グループと見なされる。なぜならば、これらの炭化水素類の混合物はこの範囲で引火点が変わるためである。したがって、これらの製品の区分3または区分4への分類は、関連法規または所管官庁が判断することができる。

[2]　引火点が35℃より高く60℃を超えない液体は、試験方法及び判定基準のマニュアル第III部、32節の燃焼持続性試験 L.2 において否の結果が得られている場合には、規制目的（輸送など）によっては引火性液体とされないことがある。

2.6.4.2 *Guidance*

2.6.4.2.1 In order to classify a flammable liquid, data on its flash point and initial boiling point are needed. Data can be determined by testing, found in literature or calculated.

2.6.4.2.2 In the case of mixtures[3] containing known flammable liquids in defined concentrations, although they may contain non-volatile ingredients e.g. polymers, additives, the flash point need not be determined experimentally if the calculated flash point of the mixture, using the method given in 2.6.4.2.3 below, is at least 5°C[4] greater than the relevant classification criterion and provided that:

    (a)   The composition of the mixture is accurately known (if the material has a specified range of composition, the composition with the lowest calculated flash point should be selected for assessment);

    (b)   The lower explosion limit of each ingredient is known (an appropriate correlation has to be applied when these data are extrapolated to other temperatures than test conditions) as well as a method for calculating the lower explosion limit of the mixture;

    (c)   The temperature dependence of the saturated vapour pressure and of the activity coefficient is known for each ingredient as present in the mixture;

    (d)   The liquid phase is homogeneous.

2.6.4.2.3 A suitable method is described in Gmehling and Rasmussen (Ind. Eng. Chem. Fundament, 21, 186, (1982)). For a mixture containing non-volatile ingredients, e.g. polymers or additives, the flash point is calculated from the volatile ingredients. It is considered that a non-volatile ingredient only slightly decreases the partial pressure of the solvents and the calculated flash point is only slightly below the measured value.

2.6.4.2.4 If data are not available, the flash point and the initial boiling point should be determined through testing. The flash point should be determined by closed-cup test method. Open-cup tests are acceptable for liquids which cannot be tested in closed-cup test methods (e.g. due to their viscosity) or when open-cup test data is already available. In these cases, 5.6 °C should be subtracted from the measured value because open-cup test methods generally result in higher values than closed-cup methods.

2.6.4.2.5 The following methods for determining the flash point of flammable liquids should be used:

International standards:

    ISO 1516
    ISO 1523
    ISO 2719
    ISO 13736
    ISO 3679
    ISO 3680

National standards:

American Society for Testing Materials International, 100Barr Harbor Drive, PO Box C 700, West Conshohocken, Pennsylvania, USA 19428-2959:

    ASTM D3828-07a, "Standard Test Methods for Flash Point by Small Scale Closed Cup Tester"
    ASTM D56-05, "Standard Test Method for Flash Point by Tag Closed Cup Tester"
    ASTM D3278-96(2004)e1, "Standard Test Methods for Flash Point of Liquids by Small Scale Closed Cup Apparatus"

---

[3] *Up to now, the calculation method is validated for mixtures containing up to six volatile components. These components may be flammable liquids like hydrocarbons, ethers, alcohols, esters (except acrylates), and water. It is however not yet validated for mixtures containing halogenated, sulphurous, and/or phosphoric compounds as well as reactive acrylates.*

[4] *If the calculated flash point is less than 5°C greater than the relevant classification criterion, the calculation method may not be used and the flash point should be determined experimentally.*

2.6.4.2　手引き

2.6.4.2.1　引火性液体を分類するには、その引火点および初留点に関するデータが必要である。データは試験結果、文献報告値または計算により決定できる。

2.6.4.2.2　混合物[3]を構成している既知の引火性液体の濃度がわかっている場合、その混合物が例えば高分子や添加剤などの非揮発性成分を含んでいたとしても、もし下記2.6.4.2.3に示す方法で当該混合物の引火点計算値が、関連する分類基準より5℃以上[4]高い場合には、次の各項を満たすことを条件にその引火点を実験で測定する必要はない：

(a) 混合物を構成する成分が正確にわかっている（その材料の組成範囲が特定されているならば、引火点計算値が最も低くなる組成を選択して評価すべきである）；

(b) 混合物の爆発下限界の計算方法と各成分の爆発下限界がわかっている（こうしたデータを試験条件以外の別の温度に換算する場合には、該当する補正を行わなければならない）；

(c) 混合物中に存在する状態での各成分の飽和蒸気圧および活量係数の温度依存性がわかっている；

(d) 液相が均一である。

2.6.4.2.3　これに適する方法は Gmehling and Rasmussen (Ind. Eng. Chem. Fundament, 21, 186, (1982)) に報告されている。例えば高分子または添加剤等の非揮発性成分を含む混合物では、引火点は揮発性成分から算出する。非揮発性成分は、その溶媒の分圧を僅か低下させるだけであり、引火点計算値は測定値より僅かに低いだけであると考えられている。

2.6.4.2.4　データが利用できない場合には、引火点および初留点は試験をして決定すべきである。引火点は密閉式試験法で測定すべきである。開放式試験法は特殊な場合に限って適用される。開放式試験法は、密閉式試験法で試験できない液体（例えば、その粘度を理由に）または開放式試験データが既に利用可能である場合に許容される。これらの場合、開放式試験法は一般的に密閉式試験法よりも高い値をもたらすため、測定値から5.6℃を差し引く必要がある。

2.6.4.2.5　以下の引火性液体の引火点測定方法を使用すべきである。

国際規格：

ISO 1516
ISO 1523
ISO 2719
ISO 13736
ISO 3679
ISO 3680

各国標準：

米国材料試験協会、100 Barr Harbor Drive, P.O.Box C700, West Conshohocken, Pennsylvania, USA 19428-2959:

ASTM D 3828-07a、「小規模密閉式試験器による引火点標準試験法」
ASTM D 56-05、「タグ密閉式試験器による引火点標準試験法」
ASTM D 3278-96(2004)e1、「小規模密閉式試験装置による液体の引火点標準試験法」

---

[3] これまでのところ計算方法は6つの揮発性成分を含む混合物まで有効であると確認されている。これらの成分としては炭化水素、エーテル、アルコール、エステル（アクリレートを除く）のような引火性液体および水である。しかし反応性に富むアクリレートと同様にハロゲン、硫黄、リン等の化合物を含む混合物に対しては有効性が確認されていない。

[4] 計算した引火点が相当する判定基準よりは大きいもののその差が5℃未満である場合には、計算結果は使用せず、引火点は実験的に求めるべきであろう。

ASTM D93-08, "Standard Test Methods for Flash Point by Pensky-Martens Closed Cup Tester"

Association française de normalisation, AFNOR, 11, rue de Pressensé. 93571 La Plaine Saint-Denis Cedex":

    French Standard NF M 07 - 019
    French Standards NF M 07 - 011 / NF T 30 - 050 / NF T 66 - 009
    French Standard NF M 07 - 036

Deutsches Institut für Normung, Burggrafenstr. 6, D-10787 Berlin:

    Standard DIN 51755 (flash points below 65°C)

State Committee of the Council of Ministers for Standardization, 113813, GSP, Moscow, M-49 Leninsky Prospect, 9:

    GOST 12.1.044-84

2.6.4.2.6   The following methods for determining the initial boiling point of flammable liquids should be used:

International standards:

    ISO 3924
    ISO 4626
    ISO 3405

National standards:

    American Society for Testing Materials International, 100 Barr Harbor Drive, PO Box C700, West Conshohocken, Pennsylvania, USA 19428-2959:

        ASTM D86-07a, "Standard Test Method for Distillation of Petroleum Products at Atmospheric Pressure"
        ASTM D1078-05, "Standard Test Method for Distillation Range of Volatile Organic Liquids"

Further acceptable methods:

    Method A.2 as described in Part A of the Annex to Commission Regulation (EC) No.440/2008[5].

---

[5] *Commission Regulation (EC) No 440/2008 of 30 May 2008 laying down test methods pursuant to Regulation (EC) No 1907/2006 of the European Parliament and of the Council on the Registration, Evaluation, Authorisation and Restriction of Chemicals (REACH) (Official Journal of the European Union, No. L142 of 31.05.2008, p1-739 and No. L143 of 03.06.2008, p.55)*.

ASTM D 93-08、「Pensky-Martens 密閉式試験器による引火点標準試験法」

フランス規格協会、*AFNOR,* 11, rue de Pressense. 93571 La Plaine Saint-Denis Cedex：

 フランス規格 NF M 07-019
 フランス規格 NF M 07-011/NF T 30-050/NF T 66-009
 フランス規格 NF M 07-036

ドイツ規格協会、Burggrafenstr. 6, D-10787 Berlin:

 DIN 51755（引火点 65℃以下）

ロシア連邦閣僚会議国家標準委員会、113813, GSP, Moscow, M-49 Leninsky Prospect, 9

 GOST 12.1.044-84

2.6.4.2.6　以下の引火性液体の初留点測定方法を使用すべきである：

国際規格：

 ISO 3924
 ISO 4626
 ISO 3405

各国標準：

 米国材料試験協会、100 Barr Harbor Drive, P.O.Box C700, West Conshohocken, Pennsylvania, USA 19428-2959
  ASTM D86-07a　大気圧下での石油製品蒸留標準試験法
  ASTM D1078-05　揮発性有機液体の蒸留範囲に関する標準試験法

他の好ましい方法：

 委員会規則（EC）No440/2008[5]の付属書 A に記載されている方法 A.2

---

[5] 欧州議会および理事会規則*(EC)No1907/2006* にしたがって試験方法を定めた、登録、評価、認可および制限（*REACH*）に関する *2008 年 5 月 30 日の委員会規則(EC)No440/2008*　（欧州連合広報 *No.L142. 31.05.2008, p1-739* および *L 143、03.06.2008, p.55*）

CHAPTER 2.7

FLAMMABLE SOLIDS

2.7.1 Definitions

A *flammable solid* is a solid which is readily combustible or may cause or contribute to fire through friction.

Readily combustible solids are powdered, granular, or pasty substances which are dangerous if they can be easily ignited by brief contact with an ignition source, such as a burning match, and if the flame spreads rapidly.

Metal powders are powders of metals or metal alloys.

2.7.2 Classification criteria

2.7.2.1 Powdered, granular or pasty substances or mixtures shall be classified as readily combustible solids when the time of burning of one or more of the test runs, performed in accordance with the test method described in the *Manual of Tests and Criteria*, Part III, subsection 33.2, is less than 45 s or the rate of burning is more than 2.2 mm/s.

2.7.2.2 Metal powders shall be classified as flammable solids when they can be ignited and the reaction spreads over the whole length of the sample (100 mm) in 10 min or less.

2.7.2.3 Solids which may cause fire through friction shall be classified in this class by analogy with existing entries (e.g. matches) until definitive criteria are established.

2.7.2.4 A flammable solid is classified in one of the two categories for this class using method N.1 as described in Part III, subsection 33.2 of the *Manual of Tests and Criteria,* according to the following table:

Table 2.7.1: Criteria for flammable solids

Category	Criteria
1	Burning rate test: Substances or mixtures other than metal powders: (a) wetted zone does not stop fire; and (b) burning time < 45 s or burning rate > 2.2 mm/s Metal powders: burning time ≤ 5 min
2	Burning rate test: Substances or mixtures other than metal powders: (a) wetted zone stops the fire for at least 4 min; and (b) burning time < 45 s or burning rate > 2.2 mm/s Metal powders: burning time > 5 min and ≤ 10 min

NOTE 1: *For classification tests on solid substances or mixtures, the tests should be performed on the substance or mixture as presented. If for example, for the purposes of supply or transport, the same chemical is to be presented in a physical form different from that which was tested and which is considered likely to materially alter its performance in a classification test, the substance must also be tested in the new form.*

NOTE 2: *Aerosols should not be classified as flammable solids. See chapter 2.3.*

2.7.3 Hazard communication

General and specific considerations concerning labelling requirements are provided in *Hazard communication: Labelling* (chapter 1.4). Annex 1 contains summary tables about classification and labelling. Annex 3 contains examples of precautionary statements and pictograms which can be used where allowed by the competent authority. Table 2.7.2 presents specific label elements for substances and mixtures classified into this hazard class based on the criteria in this chapter.

第 2.7 章

可燃性固体

2.7.1 定義

*可燃性固体*とは、易燃性を有する、または摩擦により発火あるいは発火を助長するおそれのある固体をいう。

*易燃性固体*とは、粉末状、顆粒状、またはペースト状の物質で、燃えているマッチ等の発火源と短時間の接触で容易に発火しうる、また、炎が急速に拡散する危険なものをいう。

*金属粉末*とは、金属または金属合金の粉末をいう。

2.7.2 分類基準

2.7.2.1 粉末状、顆粒状またはペースト状の物質あるいは混合物は、*試験方法及び判定基準のマニュアル*の第 III 部、33.2.1 にしたがって 1 種以上の試験を実施し、その燃焼時間が 45 秒未満か、または燃焼速度が 2.2mm/秒より速い場合には、易燃性固体として分類される。

2.7.2.2 金属粉末は、発火し、その反応がサンプルの全長 (100mm) にわたって 10 分間以内に拡散する場合、可燃性固体として分類される。

2.7.2.3 摩擦によって火が出る固体は、確定的な判定基準が確立されるまでは、既存のもの(マッチなど)との類推によって、このクラスに分類される。

2.7.2.4 可燃性固体は、*試験方法及び判定基準のマニュアル*の第 III 部、33.2.1 に示すように、試験方法 N.1 を用いて、下記の表にしたがってこのクラスにおける 2 つの区分のいずれかに分類される。

表 2.7.1：可燃性固体の判定基準

区分	判定基準
1	燃焼速度試験： 　金属粉末以外の物質または混合物： 　　(a) 火が湿潤部分を越える；および 　　(b) 燃焼時間＜45 秒、または燃焼速度＞2.2mm/秒 　金属粉末：燃焼時間≦5 分
2	燃焼速度試験： 　金属粉末以外の物質または混合物： 　　(a) 火が湿潤部分で少なくとも 4 分間以上止まる；および 　　(b) 燃焼時間＜45 秒、または燃焼速度＞2.2mm/秒 　金属粉末：燃焼時間＞5 分 および 燃焼時間≦10 分

注記 1：固体物質または混合物の分類試験では、当該物質または混合物は提供された形態で試験を実施すること。例えば、供給または輸送が目的で、同じ物質が、試験したときとは異なった物理的形態で、しかも評価試験を著しく変える可能性が高いと考えられる形態で提供されるとすると、そうした物質もまたその新たな形態で試験されなければならない。

注記 2：エアゾールは可燃性固体と分類すべきではない。*第 2.3 章参照。*

2.7.3 危険有害性情報の伝達

表示要件に関する通則および細則は、*危険有害性に関する情報の伝達：表示*（第 1.4 章）に定める。附属書 1 に分類とラベル表示に関する概要表を示す。附属書 3 には、注意書きおよび所管官庁が許可した場合に使用可能な注意絵表示の例を示す。表 2.7.2 は本章の判定基準に基づいて、この危険有害性クラスに分類される物質および混合物のラベル要素を示す。

Table 2.7.2: Label elements for flammable solids

	Category 1	Category 2
Symbol	Flame	Flame
Signal word	Danger	Warning
Hazard statement	Flammable solid	Flammable solid

2.7.4 Decision logic

The decision logic which follows, is not part of the harmonized classification system, but has been provided here as additional guidance. It is strongly recommended that the person responsible for classification studies the criteria before and during use of the decision logic.

To classify a flammable solid, the test method N.1 as described in Part III, subsection 33.2 of the *Manual of Tests and Criteria* should be performed. The procedure consists of two tests: a preliminary screening test and a burning rate test. Classification is according to decision logic 2.7.

Decision logic 2.7 for flammable solids

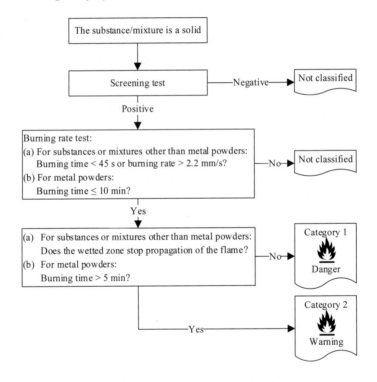

- 76 -

表 2.7.2：可燃性固体のラベル要素

	区分 1	区分 2
シンボル	炎	炎
注意喚起語	危険	警告
危険有害性情報	可燃性固体	可燃性固体

2.7.4 判定論理

以下の判定論理は、調和分類システムには含まれないが、追加的な手引きとしてここに示す。分類の責任者に対し、この判定論理を使用する前および使用する際に判定基準についてよく調べ理解することを強く勧める。

可燃性固体の分類には、*試験方法及び判定基準のマニュアル*の第 III 部、33.2.1 にしたがって試験方法 N.1 を実施すること。この手順は、予備スクリーニング試験および燃焼速度試験の 2 つの試験から構成されている。分類は、判定論理 2.7 に従う。

判定論理 2.7　可燃性固体

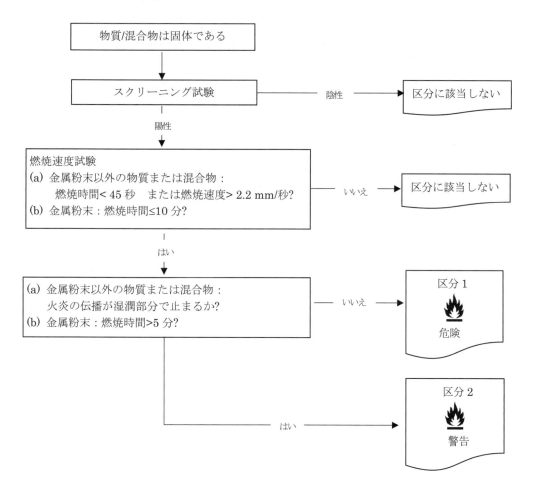

- 76 -

CHAPTER 2.8

SELF-REACTIVE SUBSTANCES AND MIXTURES

2.8.1 Definitions

2.8.1.1 *Self-reactive substances or mixtures* are thermally unstable liquid or solid substances or mixtures liable to undergo a strongly exothermic decomposition even without participation of oxygen (air). This definition excludes substances and mixtures classified under the GHS as explosives, organic peroxides or as oxidizing.

2.8.1.2 A self-reactive substance or mixture is regarded as possessing explosive properties when in laboratory testing the formulation is liable to detonate, to deflagrate rapidly or to show a violent effect when heated under confinement.

2.8.2 Classification criteria

2.8.2.1 Any self-reactive substance or mixture should be considered for classification in this class unless:

(a) They are explosives, according to the GHS criteria of chapter 2.1;

(b) They are oxidizing liquids or solids, according to the criteria of chapters 2.13 or 2.14, except that mixtures of oxidizing substances which contain 5 % or more of combustible organic substances shall be classified as self-reactive substances according to the procedure defined in the note below;

(c) They are organic peroxides, according to the GHS criteria of chapter 2.15;

(d) Their heat of decomposition is less than 300 J/g; or

(e) Their self-accelerating decomposition temperature (SADT) is greater than 75 °C for a 50 kg package.

NOTE: *Mixtures of oxidizing substances, meeting the criteria for classification as oxidizing substances, which contain 5.0 % or more of combustible organic substances and which do not meet the criteria mentioned in (a), (c), (d) or (e) above, shall be subjected to the self-reactive substances classification procedure;*

Such a mixture showing the properties of a self-reactive substance type B to F (see 2.8.2.2) shall be classified as a self-reactive substance.

2.8.2.2 Self-reactive substances and mixtures are classified in one of the seven categories of "types A to G" for this class, according to the following principles:

(a) Any self-reactive substance or mixture which can detonate or deflagrate rapidly, as packaged, will be defined as **self-reactive substance TYPE A**;

(b) Any self-reactive substance or mixture possessing explosive properties and which, as packaged, neither detonates nor deflagrates rapidly, but is liable to undergo a thermal explosion in that package will be defined as **self-reactive substance TYPE B**;

(c) Any self-reactive substance or mixture possessing explosive properties when the substance or mixture as packaged cannot detonate or deflagrate rapidly or undergo a thermal explosion will be defined as **self-reactive substance TYPE C**;

(d) Any self-reactive substance or mixture which in laboratory testing:

(i) detonates partially, does not deflagrate rapidly and shows no violent effect when heated under confinement; or

(ii) does not detonate at all, deflagrates slowly and shows no violent effect when heated under confinement; or

第2.8章

自己反応性物質および混合物

2.8.1 定義

2.8.1.1 *自己反応性物質または混合物*は、熱的に不安定で、酸素（空気）がなくとも強い発熱分解を起し易い液体または固体の物質あるいは混合物である。GHSのもとで、爆発物、有機過酸化物または酸化性物質として分類されている物質および混合物は、この定義から除外される。

2.8.1.2 自己反応性物質または混合物は、実験室の試験において組成物が密封下の加熱で爆轟、急速な爆燃または激しい反応を起こす場合には、爆発性の性状を有すると見なされる。

2.8.2 分類基準

2.8.2.1 自己反応性物質または混合物は、このクラスでの分類を検討すること。ただし下記の場合を除く：

(a) 第2.1章のGHS判定基準に従い、爆発物である；

(b) 第2.13章または第2.14章の判定基準に基づく酸化性液体または酸化性固体、ただし、5%以上有機可燃性物質を含有する酸化性物質の混合物は注記に規定する手順により自己反応性物質に分類しなければならない；

(c) 第2.15章のGHS判定基準に従い、有機過酸化物である；

(d) 分解熱が300J/gより低い；または

(e) 50kgの包装物の自己加速分解温度（SADT）が75℃を超えるもの。

注記：*酸化性物質の分類の判定基準に適合し、かつ5%以上有機可燃性物質を含有する酸化性物質の混合物であって、上記(a)、(c)、(d)または(e)の基準に適合しないものは自己反応性物質の分類手順に拠らなければならない；*

自己反応性物質タイプBからFの性状（2.8.2.2参照）を有する混合物は、自己反応性物質に分類しなければならない。

2.8.2.2 自己反応性物質および混合物は、下記の原則にしたがって、このクラスにおける「タイプAからG」の7種類の区分のいずれかに分類される：

(a) 包装された状態で爆轟しまたは急速に爆燃し得る自己反応性物質または混合物は**自己反応性物質タイプA**と定義される；

(b) 爆発性を有するが、包装された状態で、爆轟も急速な爆燃もしないが、その包装物内で熱爆発を起こす傾向を有する自己反応性物質または混合物は**自己反応性物質タイプB**として定義される；

(c) 爆発性を有するが、包装された状態で、爆轟も急速な爆燃も熱爆発も起こすことのない自己反応性物質または混合物は**自己反応性物質タイプC**として定義される；

(d) 実験室の試験で以下のような性状の自己反応性物質または混合物は**自己反応性物質タイプD**として定義される：

    (i) 爆轟は部分的であり、急速に爆燃することなく、密封下の加熱で激しい反応を起こさない；または

    (ii) 全く爆轟せず、緩やかに爆燃し、密封下の加熱で激しい反応を起こさない；または

(iii) does not detonate or deflagrate at all and shows a medium effect when heated under confinement;

will be defined as self-reactive substance TYPE D;

(e) Any self-reactive substance or mixture which, in laboratory testing, neither detonates nor deflagrates at all and shows low or no effect when heated under confinement will be defined as **self-reactive substance TYPE E**;

(f) Any self-reactive substance or mixture which, in laboratory testing, neither detonates in the cavitated state nor deflagrates at all and shows only a low or no effect when heated under confinement as well as low or no explosive power will be defined as **self-reactive substance TYPE F**;

(g) Any self-reactive substance or mixture which, in laboratory testing, neither detonates in the cavitated state nor deflagrates at all and shows no effect when heated under confinement nor any explosive power, provided that it is thermally stable (self-accelerating decomposition temperature is 60 °C to 75°C for a 50 kg package), and, for liquid mixtures, a diluent having a boiling point greater than or equal to 150°C is used for desensitization will be defined as **self-reactive substance TYPE G**. If the mixture is not thermally stable or a diluent having a boiling point less than 150°C is used for desensitization, the mixture shall be defined as self-reactive substance TYPE F.

NOTE 1: *Type G has no hazard communication elements assigned but should be considered for properties belonging to other hazard classes.*

NOTE 2: *Types A to G may not be necessary for all systems.*

2.8.2.3 *Criteria for temperature control*

Self-reactive substances need to be subjected to temperature control if their self-accelerating decomposition temperature (SADT) is less than or equal to 55°C. Test methods for determining the SADT as well as the derivation of control and emergency temperatures are given in the *Manual of Tests and Criteria*, Part II, section 28. The test selected shall be conducted in a manner which is representative, both in size and material, of the package.

2.8.3 Hazard communication

General and specific considerations concerning labelling requirements are provided in *Hazard communication: Labelling* (chapter 1.4). Annex 1 contains summary tables about classification and labelling. Annex 3 contains examples of precautionary statements and pictograms which can be used where allowed by the competent authority. Table 2.8.1 presents specific label elements for substances and mixtures classified into this hazard class based on the criteria in this chapter.

Table 2.8.1: Label elements for self-reactive substances and mixtures

	Type A	Type B	Type C and D	Type E and F	Type G[a]
Symbol	Exploding bomb	Exploding bomb and flame	Flame	Flame	*There are no label elements allocated to this hazard category*
Signal word	Danger	Danger	Danger	Warning	
Hazard statement	Heating may cause an explosion	Heating may cause a fire or explosion	Heating may cause a fire	Heating may cause a fire	

[a] *Type G has no hazard communication elements assigned but should be considered for properties belonging to other hazard classes.*

(iii) 全く爆轟も爆燃もせず、密封下の加熱では中程度の反応を起こす；

(e) 実験室の試験で、全く爆轟も爆燃もせず、かつ密封下の加熱で反応が弱いかまたは無いと判断される自己反応性物質または混合物は、**自己反応性物質タイプ E** として定義される；

(f) 実験室の試験で、空気泡の存在下で全く爆轟せず、また全く爆燃もすることなくかつ、密封下の加熱でも爆発力の試験でも、反応が弱いかまたは無いと判断される自己反応性物質または混合物は、**自己反応性物質タイプ F** として定義される；

(g) 実験室の試験で、空気泡の存在下で全く爆轟せず、また全く爆燃もすることなく、かつ、密封下の加熱でも爆発力の試験でも反応を起こさない自己反応性物質または混合物は、**自己反応性物質タイプ G** として定義される。ただし、熱的に安定である（SADT が 50kg の包装物では 60℃ から 75℃）、および液体混合物の場合には沸点が 150℃以上の希釈剤で鈍性化されていることを前提とする。混合物が熱的に安定でない、または沸点が 150℃未満の希釈剤で鈍性化されている場合、その混合物は自己反応性物質タイプ F として定義すること。

注記 1：タイプ G には危険有害性情報の伝達要素の指定はないが、別の危険有害性クラスに該当する特性があるかどうか考慮する必要がある。

注記 2：タイプ A からタイプ G はすべてのシステムに必要というわけではない。

2.8.2.3 *温度管理基準*

自己加速分解温度（SADT）が 55℃以下の自己反応性物質は、温度管理が必要である。SADT 決定のための試験法並びに管理温度および緊急対応温度の判定は*試験方法及び判定基準のマニュアル*の第 II 部、28 節に規定されている。選択された試験は、包装物の寸法および材質のそれぞれに対する方法ついて実施しなければならない。

2.8.3 危険有害性情報の伝達

表示要件に関する通則および細則は、*危険有害性に関する情報の伝達：表示*（第 1.4 章）に定める。附属書 1 に分類と表示に関する概要表を示す。附属書 3 には、注意書きおよび所管官庁が許可した場合に使用可能な注意絵表示の例を示す。表 2.8.1 は本章の判定基準に基づいて、この危険有害性クラスに分類される物質および混合物のラベル要素を示す。

表 2.8.1：自己反応性物質および混合物のラベル要素

	タイプ A	タイプ B	タイプ C&D	タイプ E&F	タイプ G[a]
シンボル	爆弾の爆発	爆弾の爆発と炎	炎	炎	この危険性区分にはラベル表示要素の指定はない
注意喚起語	危険	危険	危険	警告	
危険有害性情報	熱すると爆発のおそれ	熱すると火災または爆発のおそれ	熱すると火災のおそれ	熱すると火災のおそれ	

[a] タイプ G には危険有害性情報の伝達要素は指定されてはいないが、別の危険有害性クラスに該当する特性があるかどうか考慮する必要がある。

2.8.4 **Decision logic and guidance**

The decision logic and guidance which follow, are not part of the harmonized classification system, but have been provided here as additional guidance. It is strongly recommended that the person responsible for classification studies the criteria before and during use of the decision logic.

2.8.4.1 *Decision logic*

To classify a self-reactive substance or mixture test series A to H as described in Part II of the *Manual of Tests and Criteria* should be performed. Classification is according to decision logic 2.8.

The properties of self-reactive substances or mixtures which are decisive for their classification should be determined experimentally. Test methods with pertinent evaluation criteria are given in the *Manual of Tests and Criteria*, Part II (test series A to H).

2.8.4 判定論理および手引き

以下の判定論理および手引きは、調和分類システムには含まれないが、追加的な手引きとしてここに示す。分類の責任者に対し、この判定論理を使用する前および使用する際に判定基準についてよく調べ理解することを強く勧める。

2.8.4.1 *判定論理*

自己反応性物質または混合物を分類するには、*試験方法及び判定基準のマニュアル*の第 II 部に記載された試験シリーズ A から H を実施すること。分類は、判定論理 2.8 に従う。

自己反応性物質または混合物の分類に決定的な特性は、実験によって判定すること。試験法および関連する評価判定基準は、*試験方法及び判定基準のマニュアル*の第 II 部(試験シリーズ A～H)に記載されている。

Decision logic 2.8 for self-reactive substances and mixtures

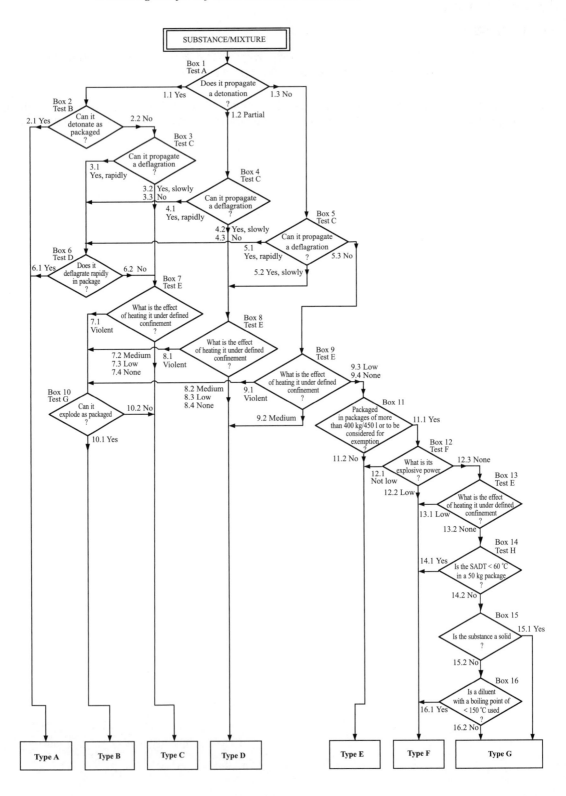

判定論理 2.8　自己反応性物質および混合物

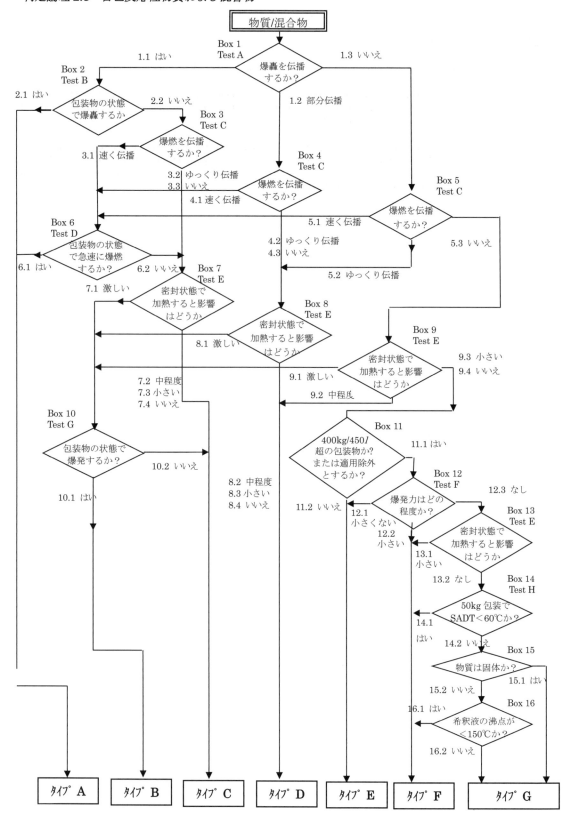

2.8.4.2   *Guidance*

The classification procedures for self-reactive substances and mixtures need not be applied if:

(a)   There are no chemical groups present in the molecule associated with explosive or self-reactive properties; examples of such groups are given in tables A6.1 and A6.2 in the appendix 6 of the *Manual of Tests and Criteria*; or

(b)   For a single organic substance or a homogeneous mixture of organic substances, the estimated SADT is greater than 75°C or the exothermic decomposition energy is less than 300 J/g. The onset temperature and decomposition energy may be estimated using a suitable calorimetric technique (see 20.3.3.3 in Part II of the *Manual of Tests and Criteria*).

2.8.4.2 *手引き*

以下の場合、自己反応性物質および混合物の分類手順を適用する必要はない。

(a) その分子内に爆発性または自己反応性に関連する官能基が存在しない。そのような官能基の例は*試験方法及び判定基準のマニュアル*の付録 6、表 A6.1 および表 A6.2 に示されている。または

(b) 単一有機物質または有機物質の均一な混合物では、SADT 推定値が 75℃より高いか、または発熱分解エネルギーが 300J/g 未満である。分解開始温度および分解エネルギーは、適切な熱量測定法により推定してもよい（*試験方法及び判定基準のマニュアル*の第Ⅱ部、20.3.3.3 参照）。

CHAPTER 2.9

PYROPHORIC LIQUIDS

2.9.1 Definition

A *pyrophoric liquid* is a liquid which, even in small quantities, is liable to ignite within five minutes after coming into contact with air.

2.9.2 Classification criteria

A pyrophoric liquid is classified in a single category for this class using test N.3 in Part III, subsection 33.4.5 of the *Manual of Tests and Criteria*, according to the following table:

Table 2.9.1: Criteria for pyrophoric liquids

Category	Criteria
1	The liquid ignites within 5 min when added to an inert carrier and exposed to air, or it ignites or chars a filter paper on contact with air within 5 min.

2.9.3 Hazard communication

General and specific considerations concerning labelling requirements are provided in *Hazard communication: Labelling* (chapter 1.4). Annex 1 contains summary tables about classification and labelling. Annex 3 contains examples of precautionary statements and pictograms which can be used where allowed by the competent authority. Table 2.9.2 presents specific label elements for substances and mixtures classified into this hazard class based on the criteria in this chapter.

Table 2.9.2: Label elements for pyrophoric liquids

	Category 1
Symbol	Flame
Signal word	Danger
Hazard statement	Catches fire spontaneously if exposed to air

2.9.4 Decision logic and guidance

The decision logic and guidance which follow, are not part of the harmonized classification system, but have been provided here as additional guidance. It is strongly recommended that the person responsible for classification studies the criteria before and during use of the decision logic.

2.9.4.1 *Decision logic*

To classify a pyrophoric liquid, the test method N.3 as described in Part III, subsection 33.4.5 of the *Manual of Tests and Criteria* should be performed. The procedure consists of two steps. Classification is according to decision logic 2.9.

第 2.9 章

自然発火性液体

2.9.1 定義

*自然発火性液体*とは、たとえ少量であっても、空気と接触すると 5 分以内に発火しやすい液体をいう。

2.9.2 分類基準

自然発火性液体は、*試験方法及び判定基準のマニュアル*の第Ⅲ部、33.4.5 の試験 N.3 により、下記の表にしたがってこのクラスの単一の区分に分類される：

表 2.9.1：自然発火性液体の判定基準

区分	判定基準
1	液体を不活性担体に浸けて空気に接触させると 5 分以内に発火する、または液体を空気に接触させると 5 分以内にろ紙を発火させるか、ろ紙を焦がす。

2.9.3 危険有害性情報の伝達

表示要件に関する通則および細則は、*危険有害性に関する情報の伝達：表示*（第 1.4 章）に定める。附属書 1 に分類と表示に関する概要表を示す。附属書 3 には、注意書きおよび所管官庁が許可した場合に使用可能な注意絵表示の例を示す。表 2.9.2 は本章の判定基準に基づいて、この危険有害性クラスに分類される物質および混合物のラベル要素を示す。

表 2.9.2：自然発火性液体のラベル要素

	区分 1
シンボル	炎
注意喚起語	危険
危険有害性情報	空気に触れると自然発火

2.9.4 判定論理および手引き

以下の判定論理および手引きは、調和分類システムには含まれないが、追加的な手引きとしてここに示す。分類の責任者に対し、この判定論理を使用する前および使用する際に判定基準についてよく調べ理解することを強く勧める。

2.9.4.1 *判定論理*

自然発火性液体を分類するには、*試験方法及び判定基準のマニュアル*の第Ⅲ部、33.4.5 の試験 N.3 を実施すること。分類手順は二段階となっている。分類は、判定論理 2.9 に従う。

Decision logic 2.9 for pyrophoric liquids

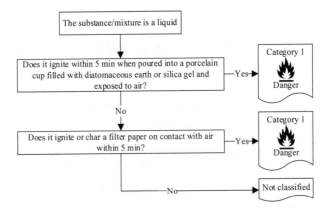

2.9.4.2   *Guidance*

The classification procedure for pyrophoric liquids need not be applied when experience in production or handling shows that the substance or mixture does not ignite spontaneously on coming into contact with air at normal temperatures (i.e. the substance is known to be stable at room temperature for prolonged periods of time (days)).

判定論理 2.9　自然発火性液体

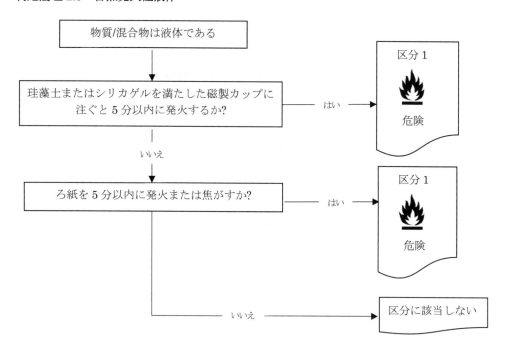

2.9.4.2　手引き

　製造または取扱時の経験から、当該物質または混合物が、常温で空気と接触しても自然発火しないことが認められている（すなわち、当該物質が室温で長期間（日単位）にわたり安定であることが既知である）ならば、自然発火性液体の分類手順を適用する必要はない。

CHAPTER 2.10

PYROPHORIC SOLIDS

2.10.1 Definition

A *pyrophoric solid* is a solid which, even in small quantities, is liable to ignite within five minutes after coming into contact with air.

2.10.2 Classification criteria

A pyrophoric solid is classified in a single category for this class using test N.2 in Part III, subsection 33.4.4 of the *Manual of Tests and Criteria* according to the following table:

Table 2.10.1: Criteria for pyrophoric solids

Category	Criteria
1	The solid ignites within 5 min of coming into contact with air.

NOTE: *For classification tests on solid substances or mixtures, the tests should be performed on the substance or mixture as presented. If for example, for the purposes of supply or transport, the same chemical is to be presented in a physical form different from that which was tested and which is considered likely to materially alter its performance in a classification test, the substance or mixture must also be tested in the new form.*

2.10.3 Hazard communication

General and specific considerations concerning labelling requirements are provided in *Hazard communication: Labelling* (chapter 1.4). Annex 1 contains summary tables about classification and labelling. Annex 3 contains examples of precautionary statements and pictograms which can be used where allowed by the competent authority. Table 2.10.2 presents specific label elements for substances and mixtures classified into this hazard class based on the criteria in this chapter.

Table 2.10.2: Label elements for pyrophoric solids

	Category 1
Symbol	Flame
Signal word	Danger
Hazard statement	Catches fire spontaneously if exposed to air

2.10.4 Decision logic and guidance

The decision logic and guidance which follow, are not part of the harmonized classification system, but have been provided here as additional guidance. It is strongly recommended that the person responsible for classification studies the criteria before and during use of the decision logic.

2.10.4.1 *Decision logic*

To classify a pyrophoric solid, the test method N.2 as described in Part III, subsection 33.4.4 of the *Manual of Tests and Criteria* should be performed. Classification is according to decision logic 2.10.

第 2.10 章

自然発火性固体

2.10.1 定義

*自然発火性固体*とは、たとえ少量であっても、空気と接触すると 5 分以内に発火しやすい固体をいう。

2.10.2 分類基準

自然発火性固体は、*試験方法及び判定基準のマニュアル*の第Ⅲ部、33.4.4 の試験 N.2 により、以下の表にしたがって、このクラスの単一の区分に分類される。

表 2.10.1：自然発火性固体の判定基準

区分	判定基準
1	固体が空気と接触すると5分以内に発火する。

注記：固体物質または混合物の分類試験では、当該物質または混合物は実際に提供される形態で試験を実施すること。例えば、供給または輸送が目的で、同じ物質が、試験したときとは異なった物理的形態で、しかも評価試験結果を著しく変える可能性が高いと考えられる形態で提供されるとすると、そうした物質もまたその新たな形態で試験されなければならない。

2.10.3 危険有害性情報の伝達

表示要件に関する通則および細則は、*危険有害性に関する情報の伝達：表示*（第 1.4 章）に定める。附属書 1 に分類と表示に関する概要表を示す。附属書 3 には、注意書きおよび所管官庁が許可した場合に使用可能な注意絵表示の例を示す。表 2.10.2 は本章の判定基準に基づいて、この危険有害性クラスに分類される物質および混合物のラベル要素を示す。

表 2.10.2：自然発火性固体のラベル要素

	区分1
シンボル	炎
注意喚起語	危険
危険有害性情報	空気に触れると自然発火

2.10.4 判定論理および手引き

以下の判定論理および手引きは、調和分類システムには含まれないが、追加的な手引きとしてここに示す。分類の責任者に対し、この判定論理を使用する前および使用する際に判定基準についてよく調べ理解することを強く勧める。

2.10.4.1 *判定論理*

自然発火性固体を分類するには、*試験方法及び判定基準のマニュアル*の第Ⅲ部、33.4.4 の試験 N.2 を実施すること。分類は、判定論理 2.10 に従う。

Decision logic 2.10 for pyrophoric solids

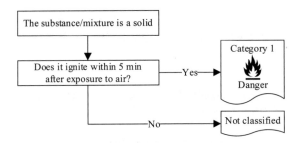

2.10.4.2 *Guidance*

The classification procedure for pyrophoric solids need not be applied when experience in production or handling shows that the substance or mixture does not ignite spontaneously on coming into contact with air at normal temperatures (i.e. the substance or mixture is known to be stable at room temperature for prolonged periods of time (days)).

判定論理 2.10　自然発火性固体

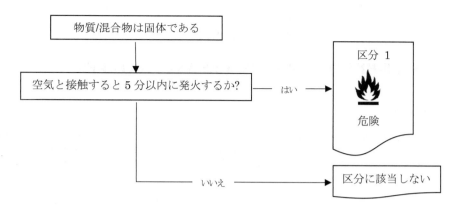

2.10.4.2　手引き

製造または取扱時の経験から、当該物質または混合物が、常温で空気と接触しても自然発火しないことが認められている（すなわち、当該物質または混合物は室温で長期間（日単位）にわたり安定であることが既知である）ならば、自然発火性固体の分類手順を適用する必要はない。

CHAPTER 2.11

SELF-HEATING SUBSTANCES AND MIXTURES

2.11.1 Definition

A *self-heating substance or mixture* is a solid or liquid substance or mixture, other than a pyrophoric liquid or solid, which, by reaction with air and without energy supply, is liable to self-heat; this substance or mixture differs from a pyrophoric liquid or solid in that it will ignite only when in large amounts (kilograms) and after long periods of time (hours or days).

NOTE: *Self-heating of a substance or mixtures is a process where the gradual reaction of that substance or mixture with oxygen (in air) generates heat. If the rate of heat production exceeds the rate of heat loss, then the temperature of the substance or mixture will rise which, after an induction time, may lead to self-ignition and combustion.*

2.11.2 Classification criteria

2.11.2.1 A substance or mixture shall be classified as a self-heating substance of this class, if in tests performed in accordance with the test method given in the *Manual of Tests and Criteria*, Part III, subsection 33.4.6:

(a) A positive result is obtained using a 25 mm cube sample at 140 °C;

(b) A positive result is obtained in a test using a 100 mm sample cube at 140 °C and a negative result is obtained in a test using a 100 mm cube sample at 120 °C <u>and</u> the substance or mixture is to be packed in packages with a volume of more than 3 m^3;

(c) A positive result is obtained in a test using a 100 mm sample cube at 140 °C and a negative result is obtained in a test using a 100 mm cube sample at 100°C <u>and</u> the substance or mixture is to be packed in packages with a volume of more than 450 litres;

(d) A positive result is obtained in a test using a 100 mm sample cube at 140 °C <u>and</u> a positive result is obtained using a 100 mm cube sample at 100°C.

2.11.2.2 A self-heating substance or mixture is classified in one of the two categories for this class if, in test performed in accordance with test method N.4 in Part III, subsection 33.4.6 of the *Manual of Tests and Criteria*, the result meets the criteria shown in table 2.11.1.

Table 2.11.1: Criteria for self-heating substances and mixtures

Category	Criteria
1	A positive result is obtained in a test using a 25 mm sample cube at 140 °C
2	(a) A positive result is obtained in a test using a 100 mm sample cube at 140 °C and a negative result is obtained in a test using a 25 mm cube sample at 140 °C <u>and</u> the substance or mixture is to be packed in packages with a volume of more than 3 m^3; or
	(b) A positive result is obtained in a test using a 100 mm sample cube at 140 °C and a negative result is obtained in a test using a 25 mm cube sample at 140 °C, a positive result is obtained in a test using a 100 mm cube sample at 120 °C <u>and</u> the substance or mixture is to be packed in packages with a volume of more than 450 litres; or
	(c) A positive result is obtained in a test using a 100 mm sample cube at 140 °C and a negative result is obtained in a test using a 25 mm cube sample at 140 °C and a positive result is obtained in a test using a 100 mm cube sample at 100°C.

NOTE 1: *For classification tests on solid substances or mixtures, the tests should be performed on the substance or mixture as presented. If for example, for the purposes of supply or transport, the same chemical is to be presented in a physical form different from that which was tested and which is considered likely to materially alter its performance in a classification test, the substance or mixture must also be tested in the new form.*

NOTE 2: *The criteria are based on the self-ignition temperature of charcoal, which is 50°C for a sample cube of 27 m^3. Substances and mixtures with a temperature of spontaneous combustion higher than 50°C for a volume of 27 m^3*

第 2.11 章

自己発熱性物質および混合物

2.11.1 定義

　*自己発熱性物質または混合物*とは、自然発火性液体または自然発火性固体以外の固体物質または混合物で、空気との接触によりエネルギー供給がなくとも、自己発熱しやすいものをいう。この物質または混合物が自然発火性液体または自然発火性固体と異なるのは、それが大量（キログラム単位）にあり、かつ長期間（数時間または数日間）経過後に限って発火する点にある。

注記：物質あるいは混合物の自己発熱は、それらが酸素（空気中）と徐々に反応し発熱する過程である。発熱の速度が熱損失の速度を超えると物質あるいは混合物の温度は上昇し、ある誘導時間を経て、自己発火や燃焼となる。

2.11.2 分類基準

2.11.2.1　試験方法及び判定基準のマニュアルの第 III 部の 33.4.6 に示される試験法にしたがって試験し、以下の結果となった場合、物質または混合物はこのクラスの自己発熱性物質に分類される：

(a) 25mm 立方体のサンプルを用いて 140℃で肯定的結果が得られる；

(b) 100mm 立方体のサンプルを用いて 140℃で肯定的結果が得られ、かつ 100mm 立方体サンプルを用いて 120℃で否定的結果が得られ、かつ、当該物質または混合物が $3m^3$ より大きい容積のパッケージとして包装される；

(c) 100mm 立方体のサンプルを用いて 140℃で肯定的結果が得られ、かつ 100mm 立方体サンプルを用いて 100℃で否定的結果が得られ、かつ、当該物質または混合物が 450 リットルより大きい容積のパッケージとして包装される；

(d) 100mm 立方体のサンプルを用いて 140℃で肯定的結果が得られ、かつ、100mm 立方体サンプルを用いて 100℃で肯定的結果が得られる。

2.11.2.2　自己発熱性物質または混合物は、*試験方法及び判定基準のマニュアルの第 III 部の 33.4.6* に示される試験 N.4 にしたがって実施された試験で得られた結果が表 2.11.1 の判定基準に適合するならば、このクラスにおける 2 つの区分のいずれかに分類される。

表 2.11.1：自己発熱性物質および混合物の判定基準

区分	判定基準
1	25mm 立方体サンプルを用いて 140℃における試験で肯定的結果が得られる
2	(a) 100mm 立方体のサンプルを用いて 140℃で肯定的結果が得られ、かつ 25mm 立方体サンプルを用いて 140℃で否定的結果が得られ、かつ、当該物質または混合物が $3m^3$ より大きい容積パッケージとして包装される；または (b) 100mm 立方体のサンプルを用いて 140℃で肯定的結果が得られ、かつ 25mm 立方体サンプルを用いて 140℃で否定的結果が得られ、100mm 立方体のサンプルを用いて 120℃で肯定的結果が得られ、かつ、当該物質または混合物が 450 リットルより大きい容積のパッケージとして包装される；または (c) 100mm 立方体のサンプルを用いて 140℃で肯定的結果が得られ、かつ 25mm 立方体サンプルを用いて 140℃で否定的結果が得られ、かつ 100mm 立方体のサンプルを用いて 100℃で肯定的結果が得られる。

注記 1：固体物質または混合物の分類試験では、当該物質または混合物は提供された形態で試験を実施すること。例えば、供給または輸送が目的で、同じ物質が、試験したときとは異なった物理的形態で、しかも評価試験結果を著しく変える可能性が高いと考えられる形態で提供されるとすると、そうした物質もまたその新たな形態で試験されなければならない。

注記 2：この判断基準は、$27m^3$ の立方体サンプルの自己発火温度が 50℃である木炭の例をもとにしている。$27m^3$ の容積の自然燃焼温度が 50℃より高い物質および混合物はこの危険有害性クラスに指定される

should not be assigned to this hazard class. Substances and mixtures with a self-ignition temperature higher than 50°C for a volume of 450 litres should not be assigned to hazard Category 1 of this hazard class.

2.11.3   Hazard communication

General and specific considerations concerning labelling requirements are provided in *Hazard communication: Labelling* (chapter 1.4). Annex 1 contains summary tables about classification and labelling. Annex 3 contains examples of precautionary statements and pictograms which can be used where allowed by the competent authority. Table 2.11.2 presents specific label elements for substances and mixtures classified into this hazard class based on the criteria in this chapter.

Table 2.11.2: Label elements for self-heating substances and mixtures

	Category 1	Category 2
Symbol	Flame	Flame
Signal word	Danger	Warning
Hazard statement	Self-heating; may catch fire	Self-heating in large quantities; may catch fire

2.11.4   Decision logic and guidance

The decision logic and guidance which follow, are not part of the harmonized classification system, but have been provided here as additional guidance. It is strongly recommended that the person responsible for classification studies the criteria before and during use of the decision logic.

2.11.4.1   *Decision logic*

To classify a self-heating substance or mixture, test method N.4, as described in Part III, subsection 33.4.6 of the *Manual of Tests and Criteria*, should be performed. Classification is according to decision logic 2.11.

べきでない。容積450リットルの自己発火温度が50℃より高い物質および混合物は、この危険有害性クラスの区分1に指定すべきでない。

2.11.3 危険有害性情報の伝達

表示要件に関する通則および細則は、*危険有害性に関する情報の伝達：表示*（第1.4章）に定める。附属書1に分類と表示に関する概要表を示す。附属書3には、注意書きおよび所管官庁が許可した場合に使用可能な注意絵表示の例を記載する。表2.11.2は本章の判定基準に基づいて、この危険有害性クラスに分類される物質および混合物のラベル要素を示す。

表2.11.2：自己発熱性物質および混合物のラベル要素

	区分1	区分2
シンボル	炎	炎
注意喚起語	危険	警告
危険有害性情報	自己発熱；火災のおそれ	大量の場合自己発熱；火災のおそれ

2.11.4 判定論理および手引き

以下の判定論理および手引きは、調和分類システムには含まれないが、追加的な手引きとしてここに示す。分類の責任者に対し、この判定論理を使用する前および使用する際に判定基準についてよく調べ理解することを強く勧める。

2.11.4.1 *判定論理*

自己発熱性物質を分類するには、*試験方法及び判定基準のマニュアル*の第Ⅲ部、33.4.6の試験N.4を実施すること。分類は、判定論理2.11に従う。

Decision logic 2.11 for self-heating substances and mixtures

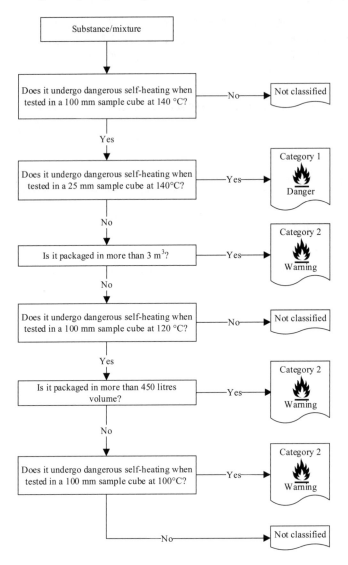

2.11.4.2   Guidance

The classification procedure for self-heating substances or mixtures need not be applied if the results of a screening test can be adequately correlated with the classification test and an appropriate safety margin is applied. Examples of screening tests are:

(a)   The Grewer Oven test (VDI guideline 2263, part 1, 1990, Test methods for the Determination of the Safety Characteristics of Dusts) with an onset temperature 80 K above the reference temperature for a volume of 1 l;

(b)   The Bulk Powder Screening Test (Gibson, N. Harper, D. J. Rogers, R. Evaluation of the fire and explosion risks in drying powders, Plant Operations Progress, 4 (3), 181-189, 1985) with an onset temperature 60 K above the reference temperature for a volume of 1 l.

判定論理 2.11　自己発熱性物質および混合物

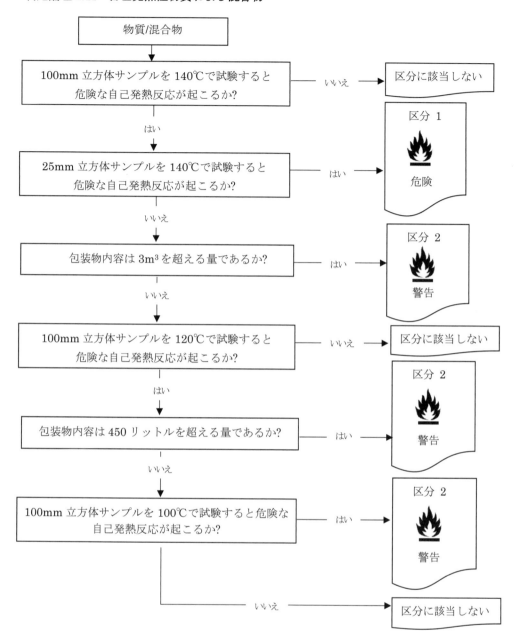

2.11.4.2　手引き

スクリーニング試験の結果と分類試験の結果にある程度の相関が認められ、かつ適切な安全範囲が適用されるならば、自己発熱性物質の分類手順を適用する必要はない。スクリーニング試験には以下のような例がある：

(a)　Grewer Oven 試験（VDI ガイドライン 2263, Part 1, 1990, 粉塵の安全特性判定試験法）で、容積 1 リットルにつき開始温度が標準温度より 80K 高い；

(b)　Bulk Powder Screening 試験（Gibson, N. Harper, D.J. Rogers, Evaluation of fire and explosion risks in drying powders, Plant Operation Progress, 4(3), 181-189, 1985）で、容積 1 リットルにつき開始温度が標準温度より 60K 高い。

CHAPTER 2.12

SUBSTANCES AND MIXTURES WHICH, IN CONTACT WITH WATER, EMIT FLAMMABLE GASES

2.12.1 Definition

Substances or mixtures which, in contact with water, emit flammable gases are solid or liquid substances or mixtures which, by interaction with water, are liable to become spontaneously flammable or to give off flammable gases in dangerous quantities.

2.12.2 Classification criteria

A substance or mixture which, in contact with water, emit flammable gases is classified in one of the three categories for this class, using test N.5 in Part III, subsection 33.5.4 of the *Manual of Tests and Criteria*, according to the following table:

Table 2.12.1: Criteria for substances and mixtures which, in contact with water, emit flammable gases

Category	Criteria
1	Any substance or mixture which reacts vigorously with water at ambient temperatures and demonstrates generally a tendency for the gas produced to ignite spontaneously, or which reacts readily with water at ambient temperatures such that the rate of evolution of flammable gas is equal to or greater than 10 litres per kilogram of substance over any one minute.
2	Any substance or mixture which reacts readily with water at ambient temperatures such that the maximum rate of evolution of flammable gas is equal to or greater than 20 litres per kilogram of substance per hour, and which does not meet the criteria for Category 1.
3	Any substance or mixture which reacts slowly with water at ambient temperatures such that the maximum rate of evolution of flammable gas is greater than 1 litre per kilogram of substance per hour, and which does not meet the criteria for categories 1 and 2.

NOTE 1*:   A substance or mixture is classified as a substance which, in contact with water, emits flammable gases if spontaneous ignition takes place in any step of the test procedure.*

NOTE 2*:   For classification tests on solid substances or mixtures, the tests should be performed on the substance or mixture as presented. If for example, for the purposes of supply or transport, the same chemical is to be presented in a physical form different from that which was tested and which is considered likely to materially alter its performance in a classification test, the substance or mixture must also be tested in the new form.*

2.12.3 Hazard communication

General and specific considerations concerning labelling requirements are provided in *Hazard communication: Labelling* (chapter 1.4). Annex 1 contains summary tables about classification and labelling. Annex 3 contains examples of precautionary statements and pictograms which can be used where allowed by the competent authority. Table 2.12.2 presents specific label elements for substances and mixtures classified into this hazard class based on the criteria in this chapter.

Table 2.12.2: Label elements for substances and mixtures, which in contact with water, emit flammable gases

	Category 1	Category 2	Category 3
Symbol	Flame	Flame	Flame
Signal word	Danger	Danger	Warning
Hazard statement	In contact with water releases flammable gases which may ignite spontaneously	In contact with water releases flammable gases	In contact with water releases flammable gases

第 2.12 章

水反応可燃性物質および混合物

2.12.1 定義

水と接触して可燃性ガスを発生する物質または混合物とは、水との相互作用により、自然発火性となるか、または可燃性ガスを危険となる量発生する固体または液体の物質あるいは混合物をいう。

2.12.2 分類基準

水と接触して可燃性ガスを発生する物質または混合物は、試験方法及び判定基準のマニュアルの第III部、33.5.4 の試験 N.5 により、下記の表にしたがって、このクラスにおける 3 つの区分のいずれかに分類される:

表 2.12.1：水と接触して可燃性ガスを発生する物質または混合物の判定基準

区分	判定基準
1	室温で水と激しく反応し、自然発火性のガスを生じる傾向が全般的に認められる物質または混合物、または室温で水と激しく反応し、その際の可燃性ガスの発生速度は、どの 1 分間をとっても物質 1kg につき 10 リットル以上であるような物質または混合物。
2	室温で水と急速に反応し、可燃性ガスの最大発生速度が 1 時間あたり物質 1kg につき 20 リットル以上であり、かつ区分 1 に適合しない物質または混合物。
3	室温では水と穏やかに反応し、可燃性ガスの最大発生速度が 1 時間あたり物質 1kg につき 1 リットルを超えて、かつ区分 1 や区分 2 に適合しない物質または混合物。

注記 1：試験手順のどの段階であっても自然発火する物質または混合物は、水と接触して可燃性ガスを発生する物質として分類される。

注記 2：固体物質または固体混合物を分類する試験では、その物質または混合物が提示されている形態で試験を実施する必要がある。例えば同一化学品でも、供給または輸送のために、試験が実施された形態とは異なる、および分類試験におけるその試験結果を著しく変更する可能性が高いと思われる物理的形態として提示されるような場合、その物質または混合物はその新たな形態でも試験されなければならない。

2.12.3 危険有害性情報の伝達

表示要件に関する通則および細則は、*危険有害性に関する情報の伝達：表示（第 1.4 章）*に定める。附属書 1 に分類と表示に関する概要表を示す。附属書 3 には、注意書きおよび所管官庁が許可した場合に使用可能な注意絵表示の例を示す。表 2.12.2 は本章の判定基準に基づいて、この危険有害性クラスに分類される物質および混合物のラベル要素を示す。

表 2.12.2：水反応可燃性物質および混合物のラベル要素

	区分 1	区分 2	区分 3
シンボル	炎	炎	炎
注意喚起語	危険	危険	警告
危険有害性情報	水に触れると自然発火するおそれのある可燃性ガスを発生	水に触れると可燃性ガスを発生	水に触れると可燃性ガスを発生

2.12.4 **Decision logic and guidance**

The decision logic and guidance which follow, are not part of the harmonized classification system, but have been provided here as additional guidance. It is strongly recommended that the person responsible for classification studies the criteria before and during use of the decision logic.

2.12.4.1 *Decision logic*

To classify a substance or mixture which, in contact with water emits flammable gases, test N.5 as described in Part III, subsection 33.5.4 of the *Manual of Tests and Criteria,* should be performed. Classification is according to decision logic 2.12.

Decision logic 2.12 for substances and mixtures which, in contact with water, emit flammable gases

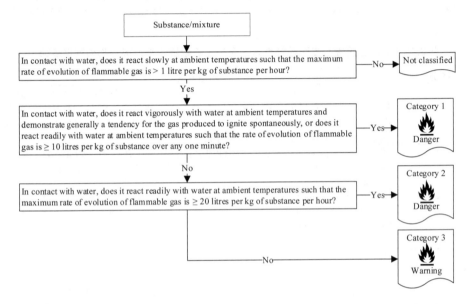

2.12.4.2 *Guidance*

The classification procedure for this class need not be applied if:

(a) The chemical structure of the substance or mixture does not contain metals or metalloids;

(b) Experience in production or handling shows that the substance or mixture does not react with water, e.g. the substance is manufactured with water or washed with water; or

(c) The substance or mixture is known to be soluble in water to form a stable mixture.

2.12.4 判定論理および手引き

以下の判定論理および手引きは、調和分類システムには含まれないが、追加的な手引きとしてここに示す。分類の責任者に対し、この判定論理を使用する前および使用する際に判定基準についてよく調べ理解することを強く勧める。

2.12.4.1 *判定論理*

水と接触して可燃性ガスを発生する物質および混合物を分類するには、*試験方法及び判定基準のマニュアル*の第III部、33.5.4の試験N.5を実施すること。分類は、判定論理2.12に従う。

判定論理2.12 水反応可燃性物質および混合物

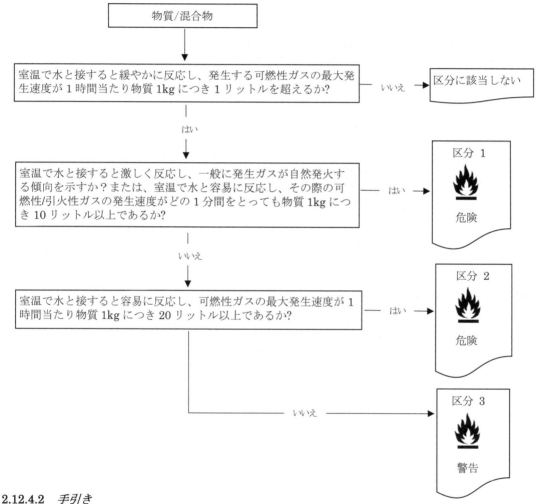

2.12.4.2 *手引き*

以下の場合、このクラスへの分類手順を適用する必要はない：

(a) 当該物質または混合物の化学構造に金属または半金属（metalloids）が含まれていない；

(b) 製造または取扱の経験上、当該物質または混合物は水と反応しないことが認められている、例えば当該物質は水を用いて製造されたか、または水で洗浄しているなど；

または

(c) 当該物質または混合物は水に溶解して安定な混合物となることがわかっている。

CHAPTER 2.13

OXIDIZING LIQUIDS

2.13.1 Definition

An *oxidizing liquid* is a liquid which, while in itself not necessarily combustible, may, generally by yielding oxygen, cause, or contribute to, the combustion of other material.

2.13.2 Classification criteria

An oxidizing liquid is classified in one of the three categories for this class using test O.2 in Part III, subsection 34.4.2 of the *Manual of Tests and Criteria,* according to the following table:

Table 2.13.1: Criteria for oxidizing liquids

Category	Criteria
1	Any substance or mixture which, in the 1:1 mixture, by mass, of substance (or mixture) and cellulose tested, spontaneously ignites; or the mean pressure rise time of a 1:1 mixture, by mass, of substance and cellulose is less than that of a 1:1 mixture, by mass, of 50 % perchloric acid and cellulose;
2	Any substance or mixture which, in the 1:1 mixture, by mass, of substance (or mixture) and cellulose tested, exhibits a mean pressure rise time less than or equal to the mean pressure rise time of a 1:1 mixture, by mass, of 40 % aqueous sodium chlorate solution and cellulose; and the criteria for Category 1 are not met;
3	Any substance or mixture which, in the 1:1 mixture, by mass, of substance (or mixture) and cellulose tested, exhibits a mean pressure rise time less than or equal to the mean pressure rise time of a 1:1 mixture, by mass, of 65 % aqueous nitric acid and cellulose; and the criteria for categories 1 and 2 are not met.

2.13.3 Hazard communication

General and specific considerations concerning labelling requirements are provided in *Hazard communication: Labelling* (chapter 1.4). Annex 1 contains summary tables about classification and labelling. Annex 3 contains examples of precautionary statements and pictograms which can be used where allowed by the competent authority. Table 2.13.2 presents specific label elements for substances and mixtures classified into this hazard class based on the criteria in this chapter.

Table 2.13.2: Label elements for oxidizing liquids

	Category 1	Category 2	Category 3
Symbol	Flame over circle	Flame over circle	Flame over circle
Signal word	Danger	Danger	Warning
Hazard statement	May cause fire or explosion; strong oxidizer	May intensify fire; oxidizer	May intensify fire; oxidizer

2.13.4 Decision logic and guidance

The decision logic and guidance which follow, are not part of the harmonized classification system, but have been provided here as additional guidance. It is strongly recommended that the person responsible for classification studies the criteria before and during use of the decision logic.

第 2.13 章

酸化性液体

2.13.1 定義

*酸化性液体*とは、それ自体は必ずしも可燃性を有しないが、一般的には酸素の発生により、他の物質を燃焼させまたは助長するおそれのある液体をいう。

2.13.2 分類基準

酸化性液体は、*試験方法及び判定基準のマニュアル*の第 III 部、34.4.2 の試験 O.2 により、下記の表にしたがって、このクラスにおける 3 つの区分のいずれかに分類される:

表 2.13.1：酸化性液体の判定基準

区分	判定基準
1	物質（または混合物）をセルロースとの重量比 1:1 の混合物として試験した場合に自然発火する、または物質とセルロースの重量比 1:1 の混合物の平均昇圧時間が、50%過塩素酸とセルロースの重量比 1:1 の混合物より短い物質または混合物；
2	物質（または混合物）をセルロースとの重量比 1:1 の混合物として試験した場合の平均昇圧時間が、塩素酸ナトリウム 40%水溶液とセルロースの重量比 1:1 の混合物の平均昇圧時間以下である、および区分 1 の判定基準が適合しない物質または混合物；
3	物質（または混合物）をセルロースとの重量比 1:1 の混合物として試験した場合の平均昇圧時間が、硝酸 65%水溶液とセルロースの重量比 1:1 の混合物の平均昇圧時間以下である、および区分 1 および 2 の判定基準が適合しない物質または混合物。

2.13.3 危険有害性情報の伝達

表示要件に関する通則および細則は、*危険有害性に関する情報の伝達：表示*（第 1.4 章）に定める。附属書 1 に分類と表示に関する概要表を示す。附属書 3 には、注意書きおよび所管官庁が許可した場合に使用可能な注意絵表示の例を記載する。表 2.13.2 は本章の判定基準に基づいて、この危険有害性クラスに分類される物質および混合物のラベル要素を示す。

表 2.13.2：酸化性液体のラベル要素

	区分 1	区分 2	区分 3
シンボル	円上の炎	円上の炎	円上の炎
注意喚起語	危険	危険	警告
危険有害性情報	火災または爆発のおそれ；強酸化性物質	火災助長のおそれ；酸化性物質	火災助長のおそれ；酸化性物質

2.13.4 判定論理および手引き

以下の判定論理および手引きは、調和分類システムには含まれないが、追加的な手引きとしてここに示す。分類の責任者に対し、この判定論理を使用する前および使用する際に判定基準についてよく調べ理解することを強く勧める。

2.13.4.1   *Decision logic*

To classify an oxidizing liquid test method O.2 as described in Part III, subsection 34.4.2 of the *Manual of Tests and Criteria* should be performed. Classification is according to decision logic 2.13.

Decision logic 2.13 for oxidizing liquids

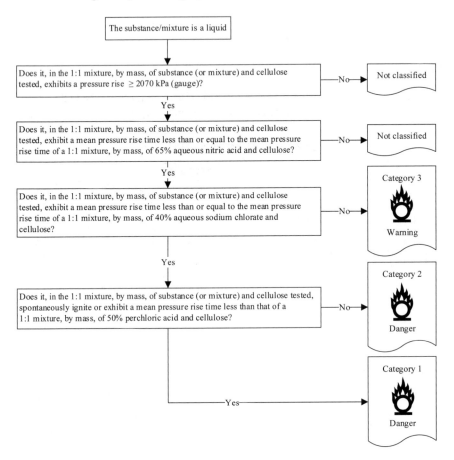

2.13.4.2   *Guidance*

2.13.4.2.1   Experience in the handling and use of substances or mixtures which shows them to be oxidizing is an important additional factor in considering classification in this class. In the event of divergence between tests results and known experience, judgement based on known experience should take precedence over test results.

2.13.4.2.2   In some cases, substances or mixtures may generate a pressure rise (too high or too low), caused by chemical reactions not characterising the oxidizing properties of the substance or mixture. In these cases, it may be necessary to repeat the test described in Part III, subsection 34.4.2 of the *Manual of Tests and Criteria* with an inert substance, e.g. diatomite (kieselguhr), in place of the cellulose in order to clarify the nature of the reaction.

2.13.4.2.3   For organic substances or mixtures the classification procedure for this class need not be applied if:

(a)   The substance or mixture does not contain oxygen, fluorine or chlorine; or

(b)   The substance or mixture contains oxygen, fluorine or chlorine and these elements are chemically bonded only to carbon or hydrogen.

2.13.4.2.4   For inorganic substances or mixtures, the classification procedure for this class need not be applied if they do not contain oxygen or halogen atoms.

2.13.4.1　判定論理

　酸化性液体を分類するには、試験方法及び判定基準のマニュアルの第Ⅲ部、34.4.2 の試験 O.2 を実施すること。分類は、判定論理 2.13 に従う。

判定論理 2.13　酸化性液体

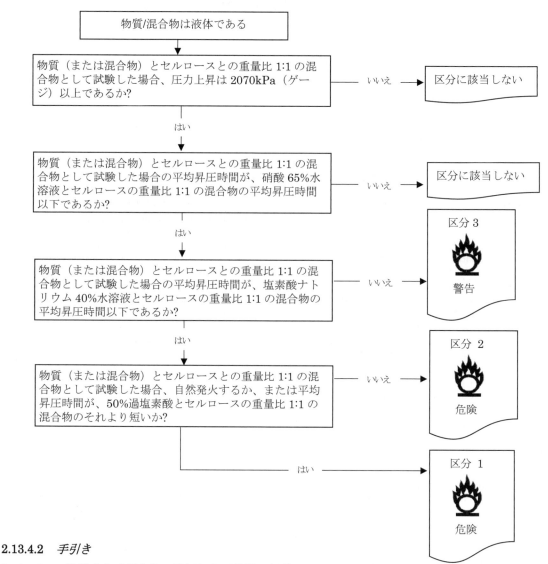

2.13.4.2　手引き

2.13.4.2.1　物質または混合物の取扱および使用の経験からこれらが酸化性であることが認められるような場合、このことはこのクラスへの分類を検討する上で重要な追加要因となる。試験結果と既知の経験に相違が見られるようであったならば、既知の経験を試験結果より優先させること。

2.13.4.2.2　物質または混合物が、その物質または混合物の酸化性を特徴づけていない化学反応によって圧力上昇（高すぎる、または低すぎる）を生じることもある。そのような場合には、その反応の性質を明らかにするために、セルロースの代わりに不活性物質、例えば珪藻土などを用いて試験方法及び判定基準のマニュアルの第Ⅲ部、34.4.2 の試験を繰返して実施する必要があることもある。

2.13.4.2.3　有機物質または混合物は、以下の場合にはこのクラスへの分類手順を適用する必要はない：

　　（a）　物質または混合物は、酸素、フッ素または塩素を含まない；または

　　（b）　物質または混合物は、酸素、フッ素または塩素を含み、これらの元素が炭素または水素にだけ化学結合している。

2.13.4.2.4　無機物質または混合物は、酸素原子またはハロゲン原子を含まないならば、このクラスへの分類手順を適用する必要はない。

CHAPTER 2.14

OXIDIZING SOLIDS

2.14.1 Definition

An *oxidizing solid* is a solid which, while in itself is not necessarily combustible, may, generally by yielding oxygen, cause, or contribute to, the combustion of other material.

2.14.2 Classification criteria

An oxidizing solid is classified in one of the three categories for this class using test O.1 in Part III, subsection 34.4.1 or test O.3 in Part III, subsection 34.4.3, of the *Manual of Tests and Criteria,* according to the following table:

Table 2.14.1: Criteria for oxidizing solids

Category	Criteria using test O.1	Criteria using test O.3
1	Any substance or mixture which, in the 4:1 or 1:1 sample-to-cellulose ratio (by mass) tested, exhibits a mean burning time less than the mean burning time of a 3:2 mixture, (by mass), of potassium bromate and cellulose.	Any substance or mixture which, in the 4:1 or 1:1 sample-to-cellulose ratio (by mass) tested, exhibits a mean burning rate greater than the mean burning rate of a 3:1 mixture (by mass) of calcium peroxide and cellulose.
2	Any substance or mixture which, in the 4:1 or 1:1 sample-to-cellulose ratio (by mass) tested, exhibits a mean burning time equal to or less than the mean burning time of a 2:3 mixture (by mass) of potassium bromate and cellulose and the criteria for Category 1 are not met.	Any substance or mixture which, in the 4:1 or 1:1 sample-to-cellulose ratio (by mass) tested, exhibits a mean burning rate equal to or greater than the mean burning rate of a 1:1 mixture (by mass) of calcium peroxide and cellulose and the criteria for Category 1 are not met.
3	Any substance or mixture which, in the 4:1 or 1:1 sample-to-cellulose ratio (by mass) tested, exhibits a mean burning time equal to or less than the mean burning time of a 3:7 mixture (by mass) of potassium bromate and cellulose and the criteria for categories 1 and 2 are not met.	Any substance or mixture which, in the 4:1 or 1:1 sample-to-cellulose ratio (by mass) tested, exhibits a mean burning rate equal to or greater than the mean burning rate of a 1:2 mixture (by mass) of calcium peroxide and cellulose and the criteria for categories 1 and 2 are not met.

NOTE 1: *Some oxidizing solids may also present explosion hazards under certain conditions (e.g. when stored in large quantities). For example, some types of ammonium nitrate may give rise to an explosion hazard under extreme conditions and the "Resistance to detonation test" (IMSBC Code, Appendix 2, Section 5) may be used to assess this hazard. Appropriate comments should be made in the Safety Data Sheet.*

NOTE 2: *For classification tests on solid substances or mixtures, the tests should be performed on the substance or mixture as presented. If for example, for the purposes of supply or transport, the same chemical is to be presented in a physical form different from that which was tested and which is considered likely to materially alter its performance in a classification test, the substance or mixture must also be tested in the new form.*

2.14.3 Hazard communication

General and specific considerations concerning labelling requirements are provided in *Hazard communication: Labelling* (chapter 1.4). Annex 1 contains summary tables about classification and labelling. Annex 3 contains examples of precautionary statements and pictograms which can be used where allowed by the competent authority. Table 2.14.2 presents specific label elements for substances and mixtures classified into this hazard class based on the criteria in this chapter.

第2.14章

酸化性固体

2.14.1 定義

*酸化性固体*とは、それ自体は必ずしも可燃性を有しないが、一般的には酸素の発生により、他の物質を燃焼させまたは助長するおそれのある固体をいう。

2.14.2 分類基準

酸化性固体は、*試験方法及び判定基準のマニュアル*の第III部、34.4.1の試験O.1または第III部、34.4.3の試験O.3を用いて、下記の表にしたがってこのクラスにおける3つの区分のいずれかに分類される:

表2.14.1:酸化性固体の判定基準

区分	O.1による判定基準	O.3による判定基準
1	サンプルとセルロースの重量比 4:1 または 1:1 の混合物として試験した場合、その平均燃焼時間が臭素酸カリウムとセルロースの重量比 3:2 の混合物の平均燃焼時間より短い物質または混合物。	サンプルとセルロースの重量比 4:1 または 1:1 の混合物として試験した場合、その平均燃焼速度が過酸化カルシウムとセルロースの重量比 3:1 の混合物の平均燃焼速度より大きい物質または混合物。
2	サンプルとセルロースの重量比 4:1 または 1:1 の混合物として試験した場合、その平均燃焼時間が臭素酸カリウムとセルロースの重量比 2:3 の混合物の平均燃焼時間以下であり、かつ区分1の判断基準に適合しない物質または混合物。	サンプルとセルロースの重量比 4:1 または 1:1 の混合物として試験した場合、その平均燃焼速度が過酸化カルシウムとセルロースの重量比 1:1 の混合物の平均燃焼速度以上であり、かつ区分1の判定基準に適合しない物質または混合物。
3	サンプルとセルロースの重量比 4:1 または 1:1 の混合物として試験した場合、その平均燃焼時間が臭素酸カリウムとセルロースの重量比 3:7 の混合物の平均燃焼時間以下であり、かつ区分1および2の判断基準に適合しない物質または混合物。	サンプルとセルロースの重量比 4:1 または 1:1 の混合物として試験した場合、その平均燃焼速度が過酸化カルシウムとセルロースの重量比 1:2 の混合物の平均燃焼速度以上であり、かつ区分1および2の判断基準に適合しない物質または混合物。

注記1:一部の酸化性固体はある条件下で爆発危険性を持つことがある(大量に貯蔵しているような場合)。例えば、一部の硝酸アンモニウムは厳しい条件下で爆発する可能性があり、この危険性の評価には「爆発抵抗試験」(IMSBCコード、付録2、第5節)が使用できるであろう。適切なコメントを安全データシートに記載すべきである。

注記2:固体物質または混合物の分類試験では、当該物質または混合物は提供された形態で試験を実施すること。例えば、供給または輸送が目的で、同じ物質が、試験したときとは異なった物理的形態で、しかも評価試験を著しく変える可能性が高いと考えられる形態で提供されるとすると、そうした物質もまたその新たな形態で試験されなければならない。

2.14.3 危険有害性情報の伝達

表示要件に関する通則および細則は、*危険有害性に関する情報の伝達:表示*(第1.4章)に定める。附属書1に分類と表示に関する概要表を示す。附属書3には、注意書きおよび所管官庁が許可した場合に使用可能な注意絵表示の例を示す。表2.14.2は本章の判定基準に基づいて、この危険有害性クラスに分類される物質および混合物のラベル要素を示す。

Table 2.14.2: Label elements for oxidizing solids

	Category 1	Category 2	Category 3
Symbol	Flame over circle	Flame over circle	Flame over circle
Signal word	Danger	Danger	Warning
Hazard statement	May cause fire or explosion; strong oxidizer	May intensify fire; oxidizer	May intensify fire; oxidizer

2.14.4 Decision logic and guidance

The decision logic and guidance which follow, are not part of the harmonized classification system, but have been provided here as additional guidance. It is strongly recommended that the person responsible for classification studies the criteria before and during use of the decision logic.

2.14.4.1 *Decision logic*

To classify an oxidizing solid test method O.1 as described in Part III, subsection 34.4.1 or test method O.3 as described in Part III, subsection 34.4.3 of the *Manual of Tests and Criteria,* should be performed. Classification is according to decision logic 2.14.

Decision logic 2.14 for oxidizing solids

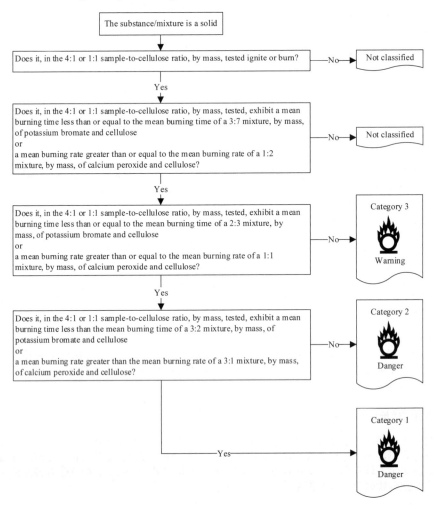

- 96 -

表 2.14.2：酸化性固体のラベル要素

	区分 1	区分 2	区分 3
シンボル	円上の炎	円上の炎	円上の炎
注意喚起語	危険	危険	警告
危険有害性情報	火災または爆発のおそれ；強酸化性物質	火災助長のおそれ；酸化性物質	火災助長のおそれ；酸化性物質

2.14.4　判定論理および手引き

以下の判定論理および手引きは、調和分類システムには含まれないが、追加的な手引きとしてここに示す。分類の責任者に対し、この判定論理を使用する前および使用する際に判定基準についてよく調べ理解することを強く勧める。

2.14.4.1　判定論理

酸化性固体を分類するには、*試験方法及び判定基準のマニュアル*の第 III 部、34.4.1 の試験 O.1 または第 III 部、34.4.3 の試験 O.3 を実施すること。分類は、判定論理 2.14 に従う。

判定論理 2.14　酸化性固体

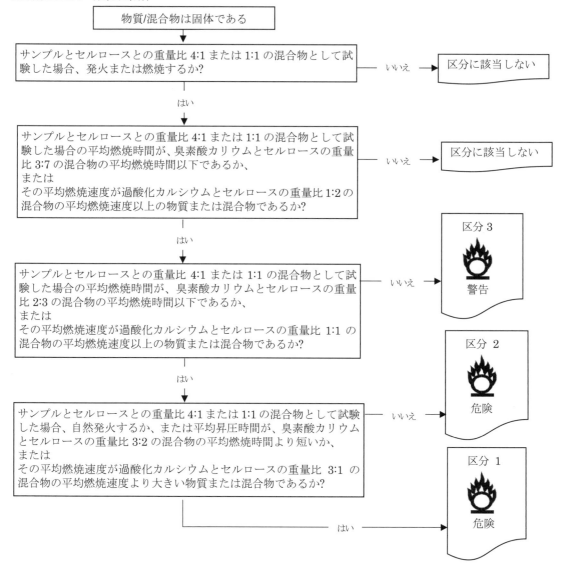

2.14.4.2 *Guidance*

2.14.4.2.1　　Experience in the handling and use of substances or mixtures which shows them to be oxidizing is an important additional factor in considering classification in this class. In the event of divergence between tests results and known experience, judgement based on known experience should take precedence over test results.

2.14.4.2.2　　The classification procedure for this class need not be applied to organic substances or mixtures if:

    (a)　　The substance or mixture does not contain oxygen, fluorine or chlorine; or

    (b)　　The substance or mixture contains oxygen, fluorine or chlorine and these elements are chemically bonded only to carbon or hydrogen.

2.14.4.2.3　　The classification procedure for this class need not be applied to inorganic substances or mixtures if they do not contain oxygen or halogen atoms.

2.14.4.2 *手引き*

2.14.4.2.1　物質または混合物の取扱いおよび使用の経験から、これら物質が酸化性があることが認められるような場合、このことはこのクラスへの分類を検討する上で重要な追加要因となる。試験結果と既知の経験に相違が見られるようであったならば、既知の経験を試験結果より優先させること。

2.14.4.2.2　有機物質または混合物は、以下の場合にはこのクラスへの分類手順を適用する必要はない：

(a)　物質または混合物は、酸素、フッ素または塩素を含まない；または

(b)　物質または混合物は、酸素、フッ素または塩素を含み、これらの元素が炭素または水素にだけ化学結合している。

2.14.4.2.3　無機物質または混合物は、酸素原子またはハロゲン原子を含まないならば、このクラスへの分類手順を適用する必要はない。

CHAPTER 2.15

ORGANIC PEROXIDES

2.15.1 Definition

2.15.1.1 *Organic peroxides* are liquid or solid organic substances which contain the bivalent -O-O- structure and may be considered derivatives of hydrogen peroxide, where one or both of the hydrogen atoms have been replaced by organic radicals. The term also includes organic peroxide formulations (mixtures). Organic peroxides are thermally unstable substances or mixtures, which may undergo exothermic self-accelerating decomposition. In addition, they may have one or more of the following properties:

    (a) be liable to explosive decomposition;

    (b) burn rapidly;

    (c) be sensitive to impact or friction;

    (d) react dangerously with other substances.

2.15.1.2 An organic peroxide is regarded as possessing explosive properties when in laboratory testing the formulation is liable to detonate, to deflagrate rapidly or to show a violent effect when heated under confinement.

2.15.2 Classification criteria

2.15.2.1 Any organic peroxide shall be considered for classification in this class, unless it contains:

    (a) not more than 1.0 % available oxygen from the organic peroxides when containing not more than 1.0 % hydrogen peroxide; or

    (b) not more than 0.5 % available oxygen from the organic peroxides when containing more than 1.0 % but not more than 7.0 % hydrogen peroxide.

NOTE: The available oxygen content (%) of an organic peroxide mixture is given by the formula:

$$16 \times \sum_{i}^{n} \left(\frac{n_i \times c_i}{m_i} \right)$$

where:

n_i = *number of peroxygen groups per molecule of organic peroxide i;*

c_i = *concentration (mass %) of organic peroxide i;*

m_i = *molecular mass of organic peroxide i.*

2.15.2.2 Organic peroxides are classified in one of the seven categories of "Types A to G" for this class, according to the following principles:

    (a) Any organic peroxide which, as packaged, can detonate or deflagrate rapidly will be defined as **organic peroxide TYPE A**;

    (b) Any organic peroxide possessing explosive properties and which, as packaged, neither detonates nor deflagrates rapidly, but is liable to undergo a thermal explosion in that package will be defined as **organic peroxide TYPE B**;

    (c) Any organic peroxide possessing explosive properties when the substance or mixture as packaged cannot detonate or deflagrate rapidly or undergo a thermal explosion will be defined as **organic peroxide TYPE C**;

第 2.15 章

有機過酸化物

2.15.1 定義

2.15.1.1 *有機過酸化物*とは、2 価の-O-O-構造を有し、1 あるいは 2 個の水素原子が有機ラジカルによって置換されている過酸化水素の誘導体と考えられる、液体または固体有機物質をいう。この用語はまた、有機過酸化物組成物（混合物）も含む。有機過酸化物は熱的に不安定な物質または混合物であり、自己発熱分解を起こすおそれがある。さらに、以下のような特性を 1 つ以上有する：

  (a) 爆発的な分解をしやすい；
  (b) 急速に燃焼する；
  (c) 衝撃または摩擦に敏感である；
  (d) 他の物質と危険な反応をする。

2.15.1.2 有機過酸化物は、実験室の試験でその組成物が爆轟したり、急速に爆燃したり、または密封下の加熱で激しい反応を起こす傾向があるときは、爆発性を有するものと見なされる。

2.15.2 分類基準

2.15.2.1 いかなる有機過酸化物でも、以下を除いて、このクラスへの分類を検討すること：

  (a) 過酸化水素の含有量が 1.0%以下の場合において、有機過酸化物に基づく活性酸素量が 1.0%以下のもの；または
  (b) 過酸化水素の含有量が 1.0%を超え 7.0%以下である場合において、有機過酸化物に基づく活性酸素量が 0.5%以下のもの。

  注記：有機過酸化物混合物の活性酸素量(%)は以下の式で求められる：

$$16 \times \sum_{i}^{n} \left(\frac{n_i \times c_i}{m_i} \right)$$

ここで：

  n_i = *有機過酸化物 i の一分子あたりの過酸基（ペルオキソ基）の数；*
  c_i = *有機過酸化物 i の濃度（重量%）；*
  m_i = *有機過酸化物 i の分子量。*

2.15.2.2 有機過酸化物は、下記の原則にしたがってこのクラスにおける 7 つの区分「タイプ A～タイプ G」のいずれかに分類される：

  (a) 包装された状態で、爆轟しまたは急速に爆燃し得る有機化酸化物は、**有機過酸化物タイプ A**として定義される；
  (b) 爆発性を有するが、包装された状態で爆轟も急速な爆燃もしないが、その包装物内で熱爆発を起こす傾向を有する有機過酸化物は、**有機過酸化物タイプ B**として定義される；
  (c) 爆発性を有するが、包装された状態で爆轟も急速な爆燃も熱爆発も起こすことのない有機過酸化物は、**有機過酸化物タイプ C**として定義される；

(d) Any organic peroxide which in laboratory testing:

(i) detonates partially, does not deflagrate rapidly and shows no violent effect when heated under confinement; or

(ii) does not detonate at all, deflagrates slowly and shows no violent effect when heated under confinement; or

(iii) does not detonate or deflagrate at all and shows a medium effect when heated under confinement;

will be defined as **organic peroxide TYPE D**;

(e) Any organic peroxide which, in laboratory testing, neither detonates nor deflagrates at all and shows low or no effect when heated under confinement will be defined as **organic peroxide TYPE E**;

(f) Any organic peroxide which, in laboratory testing, neither detonates in the cavitated state nor deflagrates at all and shows only a low or no effect when heated under confinement as well as low or no explosive power will be defined as **organic peroxide TYPE F**;

(g) Any organic peroxide which, in laboratory testing, neither detonates in the cavitated state nor deflagrates at all and shows no effect when heated under confinement nor any explosive power, provided that it is thermally stable (self-accelerating decomposition temperature is 60 °C or higher for a 50 kg package), and, for liquid mixtures, a diluent having a boiling point of not less than 150°C is used for desensitization, will be defined as **organic peroxide TYPE G**. If the organic peroxide is not thermally stable or a diluent having a boiling point less than 150°C is used for desensitization, it shall be defined as organic peroxide TYPE F.

NOTE 1: *Type G has no hazard communication elements assigned but should be considered for properties belonging to other hazard classes.*

NOTE 2: *Types A to G may not be necessary for all systems.*

2.15.2.3 *Criteria for temperature control*

The following organic peroxides need to be subjected to temperature control:

(a) Organic peroxide types B and C with an SADT ≤ 50°C;

(b) Organic peroxide type D showing a medium effect when heated under confinement[1] with an SADT ≤ 50°C or showing a low or no effect when heated under confinement with an SADT ≤ 45°C; and

(c) Organic peroxide types E and F with an SADT ≤ 45°C.

Test methods for determining the SADT as well as the derivation of control and emergency temperatures are given in the *Manual of Tests and Criteria*, Part II, section 28. The test selected shall be conducted in a manner which is representative, both in size and material, of the package.

---

[1] *As determined by test series E as prescribed in the Manual of Tests and Criteria, Part II.*

(d) 実験室の試験で以下のような性状の有機過酸化物は**有機過酸化物タイプ D** として定義される:

 (i) 爆轟は部分的であり、急速に爆燃することなく、密閉下の加熱で激しい反応を起こさない；または

 (ii) 全く爆轟せず、緩やかに爆燃し、密閉下の加熱で激しい反応を起こさない；または

 (iii) 全く爆轟も爆燃もせず、密閉下の加熱で中程度の反応を起こす；

(e) 実験室の試験で、全く爆轟も爆燃もせず、かつ密閉下の加熱で反応が弱いか、または無いと判断される有機過酸化物は、**有機過酸化物タイプ E** として定義される；

(f) 実験室の試験で、空気泡の存在下で全く爆轟せず、また全く爆燃もすることなく、また、密閉下の加熱でも、爆発力の試験でも、反応が弱いかまたは無いと判断される有機過酸化物は、**有機過酸化物タイプ F** として定義される；

(g) 実験室の試験で、空気泡の存在下で全く爆轟せず、また全く爆燃することなく、密閉下の加熱でも、爆発力の試験でも、反応を起こさない有機過酸化物は、**有機過酸化物タイプ G** として定義される。ただし熱的に安定である（自己促進分解温度（SADT）が 50kg のパッケージでは 60℃以上）、また液体混合物の場合には沸点が 150℃以上の希釈剤で鈍性化されていることを前提とする。有機過酸化物が熱的に安定でない、または沸点が 150℃未満の希釈剤で鈍性化されている場合、その有機過酸化物は有機過酸化物タイプ F として定義される。

注記1：タイプ G には危険有害性情報の伝達要素は指定されていないが、他の危険有害性クラスに該当する特性があるかどうか検討する必要がある。

注記2：タイプ A から G はすべてのシステムに必要というわけではない。

2.15.2.3 *温度管理基準*

次に掲げる有機過酸化物は、温度管理が必要である：

 (a) SADT が 50℃以下のタイプ B および C の有機過酸化物；

 (b) SADT が 50℃以下であり密閉加熱における試験結果[1]が中程度または SADT が 45℃以下であり密閉加熱における試験結果が低いか若しくは反応なしのタイプ D の有機過酸化物；および

 (c) SADT が 45℃以下のタイプ E および F の有機過酸化物。

SADT 決定のための試験法並びに管理温度および緊急対応温度の判定は、試験方法及び判定基準のマニュアルの第Ⅱ部、28 節に規定されている。選択された試験は、包装物の寸法および材質のそれぞれに対する方法について実施しなければならない。

---

[1] 試験方法及び判定基準のマニュアルの第Ⅱ部に規定する試験シリーズ E により決定される。

2.15.3 Hazard communication

General and specific considerations concerning labelling requirements are provided in *Hazard communication: Labelling* (chapter 1.4). Annex 1 contains summary tables about classification and labelling. Annex 3 contains examples of precautionary statements and pictograms which can be used where allowed by the competent authority. Table 2.15.1 presents specific label elements for substances and mixtures classified into this hazard class based on the criteria in this chapter.

Table 2.15.1: Label elements for organic peroxides

	Type A	Type B	Type C and D	Type E and F	Type G [a]
Symbol	Exploding bomb	Exploding bomb and flame	Flame	Flame	*There are no label elements allocated to this hazard category.*
Signal word	Danger	Danger	Danger	Warning	
Hazard statement	Heating may cause an explosion	Heating may cause a fire or explosion	Heating may cause a fire	Heating may cause a fire	

[a] *Type G has no hazard communication elements assigned but should be considered for properties belonging to other hazard classes.*

2.15.4 Decision logic and guidance

The decision logic and guidance which follow, are not part of the harmonized classification system, but have been provided here as additional guidance. It is strongly recommended that the person responsible for classification studies the criteria before and during use of the decision logic.

2.15.4.1 *Decision logic*

To classify an organic peroxide test series A to H as described in Part II of the *Manual of Tests and Criteria,* should be performed. Classification is according to decision logic 2.15.

2.15.3　危険有害性情報の伝達

表示要件に関する通則および細則は、*危険有害性に関する情報の伝達：表示*（第 1.4 章）に定める。附属書 1 に分類と表示に関する概要表を示す。附属書 3 には、注意書きおよび所管官庁が許可した場合に使用可能な注意絵表示の例を示す。表 2.15.1 は本章の判定基準に基づいて、この危険有害性クラスに分類される物質および混合物のラベル要素を示す。

表 2.15.1：有機過酸化物のラベル要素

	タイプ A	タイプ B	タイプ C&D	タイプ E&F	タイプ G [a]
シンボル	爆弾の爆発	爆弾の爆発と炎	炎	炎	この危険性区分にはラベル表示要素の指定はない
注意喚起語	危険	危険	危険	警告	
危険有害性情報	熱すると爆発のおそれ	熱すると火災または爆発のおそれ	熱すると火災のおそれ	熱すると火災のおそれ	

[a] *TYPE G には危険有害性情報の伝達要素は指定されていないが、他の危険有害性クラスに該当する特性があるかどうか考慮する必要がある。*

2.15.4　判定論理および手引き

以下の判定論理および手引きは、調和分類システムには含まれないが、追加的な手引きとしてここに示す。分類の責任者に対し、この判定論理を使用する前および使用する際に判定基準についてよく調べ理解することを強く勧める。

2.15.4.1　*判定論理*

有機過酸化物を分類するには、*試験方法及び判定基準のマニュアル*の第 II 部に規定されている試験シリーズ A～H を実施すること。分類は、判定論理 2.15 に従う。

Decision logic 2.15 for organic peroxides

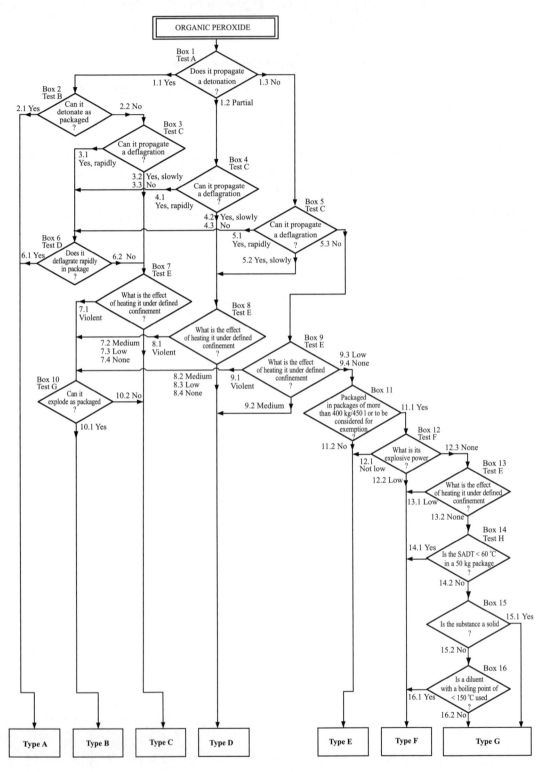

判定論理 2.15　有機過酸化物

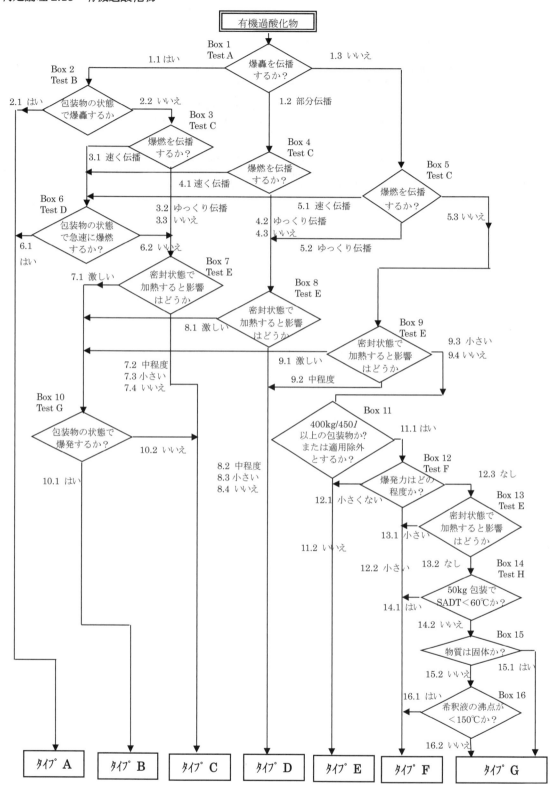

2.15.4.2 *Guidance*

2.15.4.2.1 Organic peroxides are classified by definition based on their chemical structure and on the available oxygen and hydrogen peroxide contents of the mixture (see 2.15.2.1).

2.15.4.2.2 The properties of organic peroxides which are decisive for their classification should be determined experimentally. Test methods with pertinent evaluation criteria are given in the *Manual of Tests and Criteria*, Part II (Test Series A to H).

2.15.4.2.3 Mixtures of organic peroxides may be classified as the same type of organic peroxide as that of the most dangerous ingredient. However, as two stable ingredients can form a thermally less stable mixture, the self-accelerating decomposition temperature (SADT) of the mixture shall be determined.

2.15.4.2　*手引き*

2.15.4.2.1　有機過酸化物は、その化学構造にしたがって、および当該混合物の活性酸素および過酸化水素の含量にしたがって分類される（第 2.15.2.1 参照）。

2.15.4.2.2　有機過酸化物はその分類に決定的な特性については実験的に判定すること。試験方法はこれに関連する評価判断基準と共に*試験方法及び判定基準のマニュアル*の第 II 部（試験シリーズ A~H）に定められている。

2.15.4.2.3　有機過酸化物の混合物は、これを構成する最も危険な成分の有機過酸化物と同じタイプとして分類されることもある。ただし 2 種類の安定な成分でも混合物が熱的に安定でなくなる可能性もあるため、当該混合物の自己加速分解温度（SADT）を測定しておくこと。

CHAPTER 2.16

CORROSIVE TO METALS

2.16.1 Definition

A *substance or a mixture which is corrosive to metals* is a substance or a mixture which by chemical action will materially damage, or even destroy, metals.

2.16.2 Classification criteria

A substance or a mixture which is corrosive to metals is classified in a single category for this class, using the test in Part III, subsection 37.4 of the *Manual of Tests and Criteria*, according to the following table:

Table 2.16.1: Criteria for substances and mixtures corrosive to metal

Category	Criteria
1	Corrosion rate on either steel or aluminium surfaces exceeding 6.25 mm per year at a test temperature of 55°C when tested on both materials.

NOTE: Where an initial test on either steel or aluminium indicates the substance or mixture being tested is corrosive the follow-up test on the other metal is not required.

2.16.3 Hazard communication

General and specific considerations concerning labelling requirements are provided in *Hazard communication: Labelling* (chapter 1.4). Annex 1 contains summary tables about classification and labelling. Annex 3 contains examples of precautionary statements and pictograms which can be used where allowed by the competent authority. Table 2.16.2 presents specific label elements for substances and mixtures classified into this hazard class based on the criteria in this chapter.

Table 2.16.2: Label elements for substances and mixtures corrosive to metals

	Category 1
Symbol	Corrosion
Signal word	Warning
Hazard statement	May be corrosive to metals

NOTE: Where a substance or mixture is classified as corrosive to metals but not corrosive to skin and/or eyes, some competent authorities may allow the labelling provisions described in 1.4.10.5.5.

2.16.4 Decision logic and guidance

The decision logic and guidance which follow, are not part of the harmonized classification system but have been provided here as additional guidance. It is strongly recommended that the person responsible for classification studies the criteria before and during use of the decision logic.

第 2.16 章

金属腐食性

2.16.1 定義

金属に対して腐食性である物質または混合物とは、化学反応によって金属を著しく損傷し、または破壊する物質または混合物をいう。

2.16.2 分類基準

金属に対して腐食性である物質または混合物は、*試験方法及び判定基準のマニュアル*の第 III 部、37.4 を用いて、下記の表にしたがってこのクラスにおける単一の区分に分類される：

表 2.16.1：金属に対して腐食性である物質または混合物の判定基準

区分	判定基準
1	55℃の試験温度で、鋼片およびアルミニウム片の両方で試験されたとき、浸食度がいずれかの金属において年間 6.25mm を超える。

注記：*鋼片またはアルミニウムにおける最初の試験で物質あるいは混合物が腐食性を示したならば、他方の金属による追試をする必要はない。*

2.16.3 危険有害性情報の伝達

表示要件に関する通則および細則は、*危険有害性に関する情報の伝達：表示（第 1.4 章）*に定める。附属書 1 に分類と表示に関する概要表を示す。附属書 3 には、注意書きおよび所管官庁が許可した場合に使用可能な注意絵表示の例を示す。表 2.16.2 は本章の判定基準に基づいて、この危険有害性クラスに分類される物質および混合物のラベル要素を示す。

表 2.16.2：金属に対して腐食性である物質または混合物のラベル要素

	区分 1
シンボル	腐食性
注意喚起語	警告
危険有害性情報	金属腐食のおそれ

注記：*物質または混合物が、金属腐食性があり、皮膚および/または眼には腐食性がないと分類される場合には、所管官庁は 1.4.10.5.5 に記載されているラベルに関する規定を許可してもよい。*

2.16.4 判定論理および手引き

以下の判定論理および手引きは、調和分類システムには含まれないが、追加的な手引きとしてここに示す。分類の責任者に対し、この判定論理を使用する前および使用する際に判定基準についてよく調べ理解することを強く勧める。

2.16.4.1   *Decision logic*

Decision logic 2.16 for substances and mixtures corrosive to metals

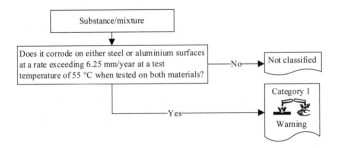

2.16.4.2   *Guidance*

The corrosion rate can be measured according to the test method of Part III, subsection 37.4 of the *Manual of Tests and Criteria*. The specimen to be used for the test should be made of the following materials:

(a)   For the purposes of testing steel, steel types S235JR+CR (1.0037 resp.St 37-2), S275J2G3+CR (1.0144 resp.St 44-3), ISO 3574, Unified Numbering System (UNS) G 10200, or SAE 1020;

(b)   For the purposes of testing aluminium: non-clad types 7075-T6 or AZ5GU-T6.

2.16.4.1　判定論理

判定論理 2.16　金属に対して腐食性である物質または混合物

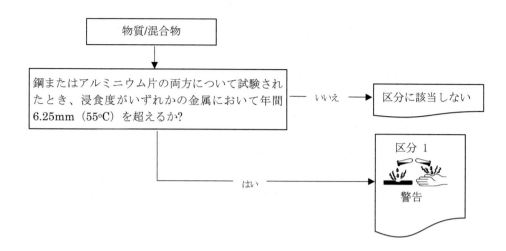

2.16.4.2　手引き

浸食度は、*試験方法及び判定基準のマニュアル*の第III部、37.4 節の試験法で測定可能である。試験で用いられる物質は、下記のものでなされなければならない：

(a)　鋼を用いる試験に対する鋼のタイプ：
S235JR+CR（1.0037 resp.St37-2）
S275J2G3+CR（1.0144 resp.St 44-3）、ISO 3574,米国ナンバリングシステム
（UNS）G10200 または、SAE 1020

(b)　アルミニウム試験：クラッド加工していない 7075-T6 または AZ5GU-T6 のようなタイプ

CHAPTER 2.17

DESENSITIZED EXPLOSIVES

2.17.1 Definitions and general considerations

2.17.1.1　Desensitized explosives are substances and mixtures in the scope of chapter 2.1 which are phlegmatized to suppress their explosive properties in such a manner that they meet the criteria as specified in 2.17.2 and thus may be exempted from the hazard class "Explosives" (see chapter 2.1, paragraph 2.1.1.2.2).

2.17.1.2　The class of desensitized explosives comprises:

    (a)　Solid desensitized explosives: explosive substances or mixtures which are wetted with water or alcohols or are diluted with other substances, to form a homogeneous solid mixture to suppress their explosive properties.

        NOTE: This includes desensitization achieved by formation of hydrates of the substances.

    (b)　Liquid desensitized explosives: explosive substances or mixtures which are dissolved or suspended in water or other liquid substances, to form a homogeneous liquid mixture to suppress their explosive properties.

2.17.2 Classification criteria

2.17.2.1　A phlegmatized explosive should be considered for inclusion in this class if, in that state, the exothermic decomposition energy is ≥ 300 J/g.

NOTE 1:　*The exothermic decomposition energy may be estimated using a suitable calorimetric technique (see section 20, subsection 20.3.3.3 in Part II of the Manual of Tests and Criteria).*

NOTE 2:　*Substances and mixtures with an exothermic decomposition energy < 300 J/g should be considered for inclusion in other physical hazard classes (e.g. as flammable liquids or flammable solids).*

2.17.2.2　A phlegmatized explosive should be considered for inclusion in this class if, in that state, it meets the following criteria:

    (a)　It is not intended to produce a practical explosive or pyrotechnic effect; and

    (b)　it is phlegmatized to an extent that:

        (i)　it has no mass explosion hazard in accordance with test 6 (a) or 6 (b) of the *Manual of Tests and Criteria*; and

        (ii)　it is not too sensitive or thermally unstable in accordance with test series 3 of the *Manual of Tests and Criteria*;

    or that

        (iii)　it is too insensitive for inclusion into in the class of explosives in accordance with test series 2 of the *Manual of Tests and Criteria*; and

    (c)　it presents no mass explosion hazard and has a corrected burning rate ≤ 1200 kg/min in accordance with the burning rate test of subsection 51.4 of the *Manual of Tests and Criteria*.

NOTE:　*Phlegmatized explosives which do not meet the criteria of 2.17.2.2 should be classified as explosives (see chapter 2.1).*

第2.17章

鈍性化爆発物

2.17.1 定義および一般事項

2.17.1.1 鈍性化爆発物とは、2.17.2 に規定された基準を満たすように、爆発性を抑制するために鈍性化され、したがって危険有害性クラス「爆発物」から除外されている、第2.1章の範囲にある物質または混合物をいう（第2.1章；パラグラフ 2.1.2.2 の注記も参照）。

2.17.1.2 鈍性化爆発物のクラスには以下のものを含む：

(a) 固体鈍性化爆発物：水もしくはアルコールで湿性とされるかあるいはその他の物質で希釈されて、均一な固体混合物となり爆発性を抑制されている爆発性物質または混合物。

 注記：これには物質を水和物とすることによる鈍性化も含まれる。

(b) 液体鈍性化爆発物：水もしくは他の液体に溶解または懸濁されて、均一な液体混合物となり爆発性を抑制されている爆発性物質または混合物。

2.17.2 分類基準

2.17.2.1 鈍性化された爆発物は、その状態で発熱分解エネルギーが 300J/g 以上であれば、本分類に含めることを考慮する。

注記1：発熱分解エネルギーは、適当な熱量測定法を用いて推定してもよい（試験方法及び判定基準のマニュアルの第II部20節20.3.3.3参照）。

注記2：発熱分解エネルギーが < 300J/g の物質および混合物は、他の物理化学的危険性クラス（例えば、引火性液体または可燃性固体として）に含めることを考慮すべきである。

2.17.2.2 鈍性化された爆発物は、その状態において、以下の基準を満たす場合、本クラスに含めることを検討すべきである：

(a) 実用的な爆発性または火工効果をもたらすことを意図していない；かつ

(b) それは以下の程度まで鈍性化する：

 (i) 試験方法および判定基準マニュアルの試験6（a）または6（b）にしたがって、大量爆発の危険性がないこと；かつ

 (ii) 試験方法および判定基準マニュアルの試験シリーズ 3 にしたがって、それほど敏感ではなく、熱的に不安定ではないこと；

 または

 (iii) 試験方法および判定基準マニュアルの試験シリーズ 2 にしたがった爆発物のクラスに含めるにはあまりにも鈍感であること；かつ

(c) 大量爆発の危険性がなく、試験方法および判定基準マニュアルの 51.4 項の燃焼速度試験にしたがった補正燃焼速度≦1200 kg/分である。

 注記：2.17.2.2 の基準を満たさない鈍性化された爆発物は、爆発物として分類されるべきである（2.1章参照）。

2.17.2.3   In addition to the criteria in 2.17.2.1 and 2.17.2.2, nitrocellulose should be stable in accordance with appendix 10 of the *Manual of Tests and Criteria* in order to be used in nitrocellulose mixtures considered for this class.

NOTE: *Nitrocellulose mixtures containing no explosives other than nitrocellulose, do not need to meet the criterion of 2.17.2.2 (b) (ii).*

2.17.2.4   Desensitized explosives shall be classified as packaged for supply and use in one of the four categories of this class depending on the corrected burning rate (Ac) determined using the burning rate (external fire) test described in Part V, subsection 51.4 of the *Manual of Tests and Criteria*, according to table 2.17.1:

Table 2.17.1: Criteria for desensitized explosives

Category	Criteria
1	Desensitized explosives with a corrected burning rate (A_C) equal to or greater than 300 kg/min but not more than 1200 kg/min
2	Desensitized explosives with a corrected burning rate (A_C) equal to or greater than 140 kg/min but less than 300 kg/min
3	Desensitized explosives with a corrected burning rate (A_C) equal to or greater than 60 kg/min but less than 140 kg/min
4	Desensitized explosives with a corrected burning rate (A_C) less than 60 kg/min

NOTE 1:   *Desensitized explosives should be prepared so that they remain homogeneous and do not separate during normal storage and handling, particularly if desensitized by wetting. The manufacturer/supplier should give information in the safety data sheet about the shelf-life and instructions on verifying desensitization. Under certain conditions the content of desensitizing agent (e.g. phlegmatizer, wetting agent or treatment) may decrease during supply and use, and thus, the hazard potential of desensitized explosive may increase. In addition, the safety data sheet should include advice on avoiding increased fire, blast or protection hazards when the substance or mixture is not sufficiently desensitized.*

NOTE 2:   *Desensitized explosives may be treated differently for some regulatory purposes (e.g. transport). Classification of solid desensitized explosives for transport purposes is addressed in chapter 2.4, section 2.4.2.4 of the UN Model Regulations. Classification of liquid desensitized explosives is addressed in chapter 2.3, section 2.3.1.4 of the Model Regulations.*

NOTE 3:   *Explosive properties of desensitized explosives should be determined by test series 2 of the Manual of Tests and Criteria and should be communicated in the safety data sheet. For testing of liquid desensitized explosives for transport purposes, refer to section 32, subsection 32.3.2 of the Manual of Tests and Criteria. Testing of solid desensitized explosives for transport purposes is addressed in section 33, subsection 33.2.3 of the Manual of Tests and Criteria.*

NOTE 4:   *For the purposes of storage, supply and use, desensitized explosives do not fall additionally within the scope of chapters 2.1 (explosives), 2.6 (flammable liquids) and 2.7 (flammable solids).*

2.17.3   Hazard communication

General and specific considerations concerning labelling requirements are provided in *Hazard communication: Labelling* (chapter 1.4). Annex 1 contains summary tables about classification and labelling. Annex 3 contains examples of precautionary statements and pictograms which can be used where allowed by the competent authority. Table 2.17.2 presents specific label elements for substances and mixtures classified into this hazard class based on the criteria in this chapter.

2.17.2.3　2.17.2.1 および 2.17.2.2 の基準に加え、このクラスで考慮されるニトロセルロース混合物に使用するために、ニトロセルロースは、試験方法および判定基準マニュアルの付録 10 にしたがって安定でなければならない。

注記：ニトロセルロース以外の爆発物を含まないニトロセルロース混合物は、2.17.2.2 (b) (ii) の基準を満たす必要はない。

2.17.2.4　鈍性化爆発物は、供給と使用のため包装状態で、このクラスの 4 つの区分に分類されなければならない。分類は試験方法及び判定基準のマニュアルの第 V 部 51.4 小節に記載されている「燃焼速度試験（外炎）」を用いて決定された補正燃焼速度（A_c）に基づいて、表 2.17.1 にしたがって行う：

表 2.17.1：鈍性化爆発物の判定基準

区分	判定基準
1	補正燃焼速度（A_c）が 300 kg/min 以上、1200 kg/min を超えない鈍性化爆発物
2	補正燃焼速度（A_c）が 140 kg/min 以上、300 kg/min 未満の鈍性化爆発物
3	補正燃焼速度（A_c）が 60 kg/min 以上、140 kg/min 未満の鈍性化爆発物
4	補正燃焼速度（A_c）が 60 kg/min 未満の鈍性化爆発物

注記 1：鈍性化爆発物は、特に湿性で鈍性化されている場合には、均一性を保ち通常の貯蔵や取扱いで分離しないようにつくられているべきである。製造者・供給者は、鈍性化を確認するための貯蔵期間や手順について安全データシートに情報を提供すべきである。ある状況下では、供給や使用の間に鈍性化剤（例えば、鈍感化剤、湿性剤または処理）が減少し、したがって鈍性化爆発物の危険性が増加する可能性がある。さらに、安全データシートには、物質または混合物が十分に鈍性化されていない時に増大する火災、爆風または飛散危険性を避けるための情報を含めるべきである。

注記 2：鈍性化爆発物は規制の目的（例えば輸送）によって異なる扱いになるであろう。輸送目的の固体の鈍性化爆発物の分類は UN モデル規則の第 2.4 章 2.4.2.4 節で扱われている。液体の鈍性化爆発物の分類はモデル規則第 2.3 章 2.3.1.4 節で扱われている。

注記 3：鈍性化爆発物の爆発性は、試験方法及び判定基準のマニュアルのテストシリーズ 2 によって決定されるべきであり、安全データシートに記載されるべきである。輸送目的での液体鈍性化爆発物の試験は試験方法及び判定基準のマニュアル 32 節、32.3.2 を参照する。輸送目的での固体鈍性化爆発物の試験は、試験方法及び判定基準のマニュアル 33 節 33.2.3 で扱われている。

注記 4：貯蔵、供給および使用の目的では、鈍性化爆発物が追加的に第 2.1 章（爆発物）、第 2.6 章（引火性液体）および第 2.7 章（可燃性固体）になることはない。

2.17.3　危険有害性情報の伝達

表示要件に関する通則および細則は、*危険有害性に関する情報の伝達：表示（第 1.4 章）*に定める。附属書 1 に分類と表示に関する概要表を示す。附属書 3 には、注意書きおよび所管官庁が許可した場合に使用可能な注意絵表示の例を示す。表 2.17.2 は本章の判定基準に基づいて、この危険有害性クラスに分類される物質および混合物のラベル要素を示す。

Table 2.17.2: Label elements for desensitized explosives

	Category 1	Category 2	Category 3	Category 4
Symbol	Flame	Flame	Flame	Flame
Signal word	Danger	Danger	Warning	Warning
Hazard statement	Fire, blast or projection hazard; increased risk of explosion if desensitizing agent is reduced	Fire or projection hazard; increased risk of explosion if desensitizing agent is reduced	Fire or projection hazard; increased risk of explosion if desensitizing agent is reduced	Fire hazard; increased risk of explosion if desensitizing agent is reduced

2.17.4 **Decision logic and guidance**

The decision logic and guidance which follow are not part of the harmonized classification system, but have been provided here as additional guidance. It is strongly recommended that the person responsible for classification studies the criteria before and during use of the decision logic.

2.17.4.1 *Decision logic*

To classify desensitized explosives, data for the sensitivity, thermal stability, explosive potential and the corrected burning rate should be determined as described in Part I and Part V of the *Manual of Tests and Criteria*. Where a mixture contains nitrocellulose, additional data for the stability of the nitrocellulose as described in appendix 10 of the *Manual of Tests and Criteria* are needed in order to be used in nitrocellulose mixtures considered for this class. Classification is according to decision logic 2.17.1.

表 2.17.2 : 鈍性化爆発物のラベル要素

	区分 1	区分 2	区分 3	区分 4
シンボル	炎	炎	炎	炎
注意喚起語	危険	危険	警告	警告
危険有害性情報	火災、爆風または飛散危険性；鈍性化剤が減少した場合には爆発の危険性の増加	火災または飛散危険性；鈍性化剤が減少した場合には爆発の危険性の増加	火災または飛散危険性；鈍性化剤が減少した場合には爆発の危険性の増加	火災危険性；鈍性化剤が減少した場合には爆発の危険性の増加

2.17.4 判定論理および手引き

以下の判定論理および手引きは、調和分類システムには含まれないが、追加的な手引きとしてここに示す。分類の責任者に対し、この判定論理を使用する前および使用する際に判定基準についてよく調べ理解することを強く勧める。

2.17.4.1 *判定論理*

鈍性化爆発物を分類するためには、*試験方法及び判定基準のマニュアル*の第Ⅰ部と第Ⅴ部に記載されているように、感度、熱安定性、爆発可能性および補正燃焼速度のデータを測定するべきである。混合物がニトロセルロースを含む場合、このクラスで考慮されるニトロセルロース混合物に使用するためには、*試験方法および判定基準マニュアル*の付録 10 に記載されているニトロセルロースの安定性に関する追加データが必要である。分類は、判定論理 2.17.1 に従う。

Decision logic 2.17.1 for desensitized explosives

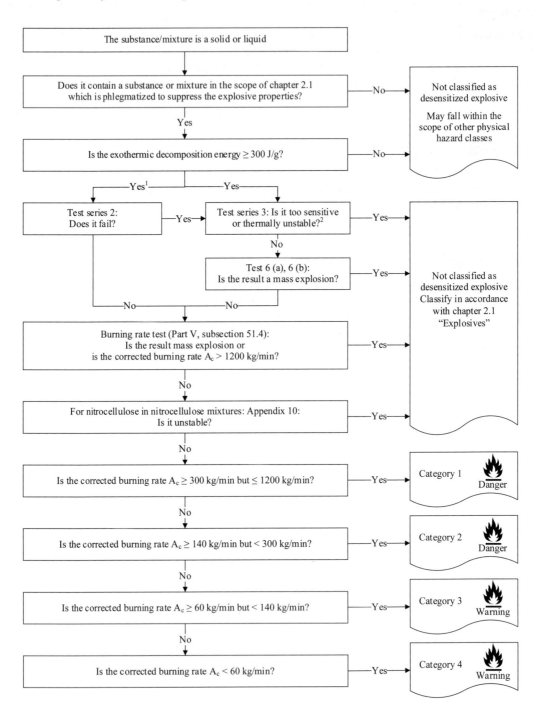

---

[1] Test series 2 is optional. The alternative route (via test 6 (a) and (b) and test series 3) may be taken directly without performing test series 2.

[2] Test series 3 is not applicable to nitrocellulose mixtures containing no explosives other than nitrocellulose.

判定論理 2.17.1　鈍性化爆発物

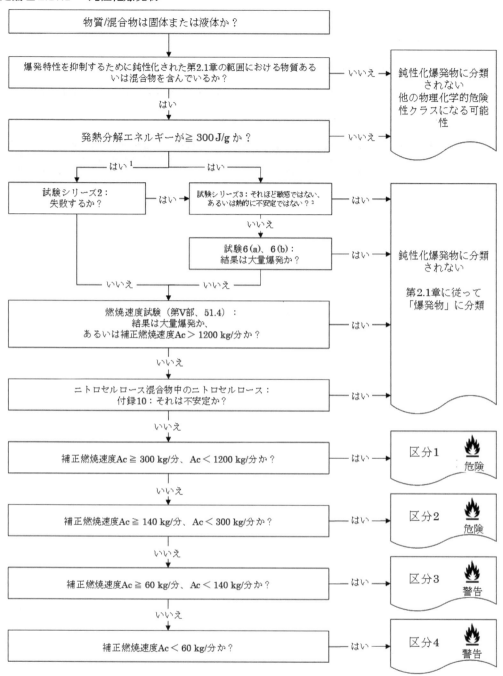

1　試験シリーズ2は任意である。代替ルート（試験6(a)および6(b)および試験シリーズ3）は試験シリーズ2を実施することなしに直接採用されるかもしれない。
2　試験シリーズ3はニトロセルロース以外の爆発物を含まないニトロセルロース混合物には適用されない。

2.17.4.2 *Guidance*

2.17.4.2.1 The classification procedure for desensitized explosives does not apply if:

(a) The substances or mixtures contain no explosives according to the criteria in chapter 2.1; or

(b) The exothermic decomposition energy is less than 300 J/g.

2.17.4.2.2 The exothermic decomposition energy should be determined using the explosive already desensitized (i.e.: the homogenous solid or liquids mixture formed by the explosive and the substance(s) used to suppress its explosive properties). The exothermic decomposition energy may be estimated using a suitable calorimetric technique (see section 20, subsection 20.3.3.3 in Part II of the *Manual of Tests and Criteria*).

2.17.4.2　*手引き*

2.17.4.2.1　以下の場合には鈍性化爆発物の分類手順を適用しない：

(a) 物質または混合物が、第 2.1 章の判定基準に従った爆発物を含まない；または

(b) 発熱分解エネルギーが 300 J/g.未満である。

2.17.4.2.2　発熱分解エネルギーはすでに鈍性化された爆発物を用いて測定されるべきである（すなわち：爆発物および爆発性を抑制するために用いられる物質により構成される均一な固体または液体混合物）。発熱分解エネルギーは、適当な熱量測定法をもちいて推定してもよい（*試験方法及び判定基準のマニュアルの第 II 部 20 節 20.3.3.3 参照*）。

PARTIE 2

ROMANAN HERBETS

PART 3

HEALTH HAZARDS

第3部

健康に対する有害性

CHAPTER 3.1

ACUTE TOXICITY

3.1.1 Definition

Acute toxicity refers to serious adverse health effects (i.e. lethality) occurring after a single or short-term oral, dermal or inhalation exposure to a substance or mixture.

3.1.2 Classification criteria for substances

3.1.2.1 Substances can be allocated to one of five hazard categories based on acute toxicity by the oral, dermal or inhalation route according to the numeric cut-off criteria as shown in the table below. Acute toxicity values are expressed as (approximate) LD_{50} (oral, dermal) or LC_{50} (inhalation) values or as acute toxicity estimates (ATE). While some in vivo methods determine LD_{50}/LC_{50} values directly, other newer in vivo methods (e.g. using fewer animals) consider other indicators of acute toxicity, such as significant clinical signs of toxicity, which are used by reference to assign the hazard category. Explanatory notes are shown following table 3.1.1.

Table 3.1.1: Acute toxicity estimate (ATE) values and criteria for acute toxicity hazard categories

Exposure route	Category 1	Category 2	Category 3	Category 4	Category 5
Oral (mg/kg bodyweight) *See notes (a) and (b)*	ATE ≤ 5	5 < ATE ≤ 50	50 < ATE ≤ 300	300 < ATE ≤ 2000	2000< ATE≤ 5000 *See detailed criteria in note (g)*
Dermal (mg/kg bodyweight) *See notes (a) and (b)*	ATE ≤ 50	50 < ATE ≤ 200	200 < ATE ≤ 1000	1000 < ATE ≤ 2000	
Gases (ppmV) *See notes (a), (b) and (c)*	ATE ≤ 100	100 < ATE ≤ 500	500 < ATE ≤ 2500	2500 < ATE ≤ 20000	*See detailed criteria in note (g)*
Vapours (mg/l) *See notes (a), (b), (c), (d) and (e)*	ATE ≤ 0.5	0.5 < ATE ≤ 2.0	2.0 < ATE ≤ 10.0	10.0 < ATE ≤ 20.0	
Dusts and Mists (mg/l) *See notes (a), (b), (c) and (f)*	ATE ≤ 0.05	0.05 < ATE ≤ 0.5	0.5 < ATE ≤ 1.0	1.0 < ATE ≤ 5.0	

Note: Gas concentrations are expressed in parts per million per volume (ppmV).

Notes to table 3.1.1:

(a) The acute toxicity estimate (ATE) for the classification of a substance is derived using the LD_{50}/LC_{50} where available;

(b) The acute toxicity estimate (ATE) for a substance in a mixture is derived using:

  (i) the LD_{50}/LC_{50} where available; otherwise,

  (ii) the appropriate conversion value from table 3.1.2 that relates to the results of a range test; or

  (iii) the appropriate conversion value from table 3.1.2 that relates to a classification category;

(c) Inhalation cut-off values in the table are based on 4 hour testing exposures. Conversion of existing inhalation toxicity data which has been generated according to 1 hour exposures should be by dividing by a factor of 2 for gases and vapours and 4 for dusts and mists;

(d) It is recognized that saturated vapour concentration may be used as an additional element by some regulatory systems to provide for specific health and safety protection (e.g. UN Model Regulations);

(e) For some substances the test atmosphere will not just be a vapour but will consist of a mixture of liquid and vapour phases. For other substances the test atmosphere may consist of a vapour which is near the

第 3.1 章

急性毒性

3.1.1　定義

　急性毒性とは、物質または混合物への単回または短時間の経口、経皮または吸入ばく露後に生じる健康への重大な有害影響（すなわち致死作用）をさす。

3.1.2　物質の分類基準

3.1.2.1　物質は、経口、経皮および吸入経路による急性毒性に基づいて表に示されるようなカットオフ値の判定基準によって 5 つの有害性区分の 1 つに割当てることができる。急性毒性の値は LD_{50}（経口、経皮）または LC_{50}（吸入）値または、急性毒性推定値（ATE）で表わされる。in vivo 試験により直接的に LD_{50}/LC_{50} が求められる一方、他の新しい in vivo 試験（例、より少ない動物を使用した）では、毒性の明確な臨床徴候など有害性区分の割り当てに参照されるような、急性毒性の他の指標も考慮される。説明のための注記は表 3.1.1 に続いて示されている。

表 3.1.1：急性毒性区分に関する急性毒性推定値（ATE）および判定基準

ばく露経路	区分 1	区分 2	区分 3	区分 4	区分 5
経口 (mg/kg 体重) 注記(a),(b)参照	ATE ≦ 5	5 ＜ ATE ≦ 50	50 ＜ ATE ≦ 300	300 ＜ATE≦ 2000	2000＜ATE≦5000 注記(g)詳細な判定基準参照
経皮 (mg/kg 体重) 注記(a),(b) 参照	ATE ≦ 50	50 ＜ ATE ≦ 200	200＜ATE≦1000	1000 ＜ATE≦2000	
気体(ppmV) 注記(a), (b),(c)参照	ATE ≦100	100＜ATE≦500	500＜ATE≦2500	2500 ＜ATE≦20000	注記(g)詳細な判定基準参照
蒸気（mg/l) 注記(a), (b),(c),(d),(e)参照	ATE ≦ 0.5	0.5 ＜ ATE ≦ 2.0	2.0 ＜ATE≦10.0	10.0 ＜ATE≦20.0	
粉塵およびミスト(mg/l) 注記(a), (b), (c), (f)参照	ATE≦0.05	0.05 ＜ATE≦ 0.5	0.5 ＜ ATE ≦1.0	1.0 ＜ ATE ≦ 5.0	

注記：気体濃度は容積での百万分の一 *(ppmV)* を単位として表されている。

表 3.1.1 への注記：

(a)　物質の分類のための急性毒性推定値*(ATE)*は、利用可能な LD_{50}/LC_{50} から得られる；

(b)　混合物成分の分類のための急性毒性推定値*(ATE)*は、次を用いて得られる：

　(i)　利用可能な LD_{50} / LC_{50}；そのほか

　(ii)　範囲試験の結果に関連した表 3.1.2 からの適切な変換値；または

　(iii)　分類区分に関連した表 3.1.2 からの適切な変換値；

(c)　表中の吸入試験のカットオフ値は 4 時間試験ばく露に基づく。1 時間ばく露で求めた、既存の吸入毒性データを換算するには、気体および蒸気の場合は 2、粉塵およびミストの場合 4 で割る；

(d)　ある規制システムでは、飽和蒸気濃度を追加要素として使用し、特別な健康および安全保護規定を設けている（例：UN モデル規則）；

(e)　物質によっては、試験対象となる物質の状態が蒸気だけでなく、液体相と蒸気相で混成される。また他の化学品では、試験雰囲気が、ほぼ気体相に近い蒸気であることもある。この後者の例で

gaseous phase. In these latter cases, classification should be based on ppmV as follows: Category 1 (100 ppmV), Category 2 (500 ppmV), Category 3 (2500 ppmV), Category 4 (20000 ppmV).

The terms "dust", "mist" and "vapour" are defined as follows:

(i) *Dust*: solid particles of a substance or mixture suspended in a gas (usually air);

(ii) *Mist*: liquid droplets of a substance or mixture suspended in a gas (usually air);

(iii) *Vapour*: the gaseous form of a substance or mixture released from its liquid or solid state.

Dust is generally formed by mechanical processes. Mist is generally formed by condensation of supersaturated vapours or by physical shearing of liquids. Dusts and mists generally have sizes ranging from less than 1 to about 100 μm;

(f) The values for dusts and mists should be reviewed to adapt to any future changes to OECD test guidelines with respect to technical limitation in generating, maintaining and measuring dust and mist concentrations in respirable form;

(g) Criteria for Category 5 are intended to enable the identification of substances which are of relatively low acute toxicity hazard but which under certain circumstances may present a danger to vulnerable populations. These substances are anticipated to have an oral or dermal LD_{50} in the range of 2000-5000 mg/kg bodyweight and equivalent doses for inhalation. The specific criteria for Category 5 are:

(i) The substance is classified in this category if reliable evidence is already available that indicates the LD_{50} (or LC_{50}) to be in the range of Category 5 values or other animal studies or toxic effects in humans indicate a concern for human health of an acute nature.

(ii) The substance is classified in this category, through extrapolation, estimation or measurement of data, if assignment to a more hazardous category is not warranted, and:

- reliable information is available indicating significant toxic effects in humans; or

- any mortality is observed when tested up to Category 4 values by the oral, inhalation, or dermal routes; or

- where expert judgement confirms significant clinical signs of toxicity, when tested up to Category 4 values, except for diarrhoea, piloerection or an ungroomed appearance; or

- where expert judgement confirms reliable information indicating the potential for significant acute effects from other animal studies.

Recognizing the need to protect animal welfare, testing in animals in Category 5 ranges is discouraged and should only be considered when there is a strong likelihood that results of such a test would have a direct relevance for protecting human health.

3.1.2.2 The harmonized classification system for acute toxicity has been developed in such a way as to accommodate the needs of existing systems. A basic principle set by the IOMC Coordinating Group/Harmonization of Chemical Classification Systems (CG/HCCS) is that "harmonization means establishing a common and coherent basis for chemical hazard classification and communication from which the appropriate elements relevant to means of transport, consumer, worker and environment protection can be selected". To that end, five categories have been included in the acute toxicity scheme.

3.1.2.3 The preferred test species for evaluation of acute toxicity by the oral and inhalation routes is the rat, while the rat or rabbit are preferred for evaluation of acute dermal toxicity. Test data already generated for the classification of chemicals under existing systems should be accepted when reclassifying these chemicals under the harmonized system. When experimental data for acute toxicity are available in several animal species, scientific judgement should be used in selecting the most appropriate LD_{50} value from among valid, well-performed tests. In cases where data from human experience (i.e. occupational data, data from accident databases, epidemiology studies, clinical reports) are also available, they should be considered in a weight of evidence assessment consistent with the principles described in 1.3.2.4.9.

3.1.2.4 Category 1, the highest hazard category, has cut-off values (see table 3.1.1) currently used primarily by the transport sector for classification for packing groups.

は、区分 1 (100ppmV)、区分 2 (500ppmV)、区分 3 (2500ppmV)、区分 4 (20000ppmV) のように、ppmV 濃度により分類すべきである；

「粉塵」、「ミスト」および「蒸気」という用語は以下のとおり定義される：

(i) <u>粉塵</u>: 気体（通常空気）の中に浮遊する物質または混合物の固体の粒子；

(ii) <u>ミスト</u>: 気体（通常空気）の中に浮遊する物質または混合物の液滴；

(iii) <u>蒸気</u>: 液体または固体の状態から放出された気体状の物質または混合物。

一般に粉塵は、機械的な工程で形成される。一般にミストは、過飽和蒸気の凝縮または液体の物理的な剪断で形成される。粉塵およびミストの大きさは、一般に 1μm 未満からおよそ 100μm までである；

(f) 粉塵およびミストの数値については、今後 OECD テストガイドラインが、吸入可能な形態での粉塵およびミストの発生、維持および濃度測定の技術的限界のために変更された場合、これらに適合できるよう見直すべきである；

(g) 区分 5 の判定基準は、急性毒性の有害性は比較的弱いが、ある状況下では高感受性集団に対して危険を及ぼすような物質を識別できるようにすることを目的としている。こうした物質は、経口または経皮 LD_{50} 値が 2000-5000mg／kg、また吸入で同程度の投与量であると推定されている。区分 5 に対する特定の判定基準は：

(i) LD_{50}（または LC_{50}）が区分 5 の範囲内にあることを示す信頼できる証拠がすでに得られている場合、またはその他の動物試験あるいはヒトにおける毒性作用から、ヒトの健康に対する急性的な懸念が示唆される場合、その物質は区分 5 に分類される。

(ii) より危険性の高い区分へ分類されないことが確かな場合、データの外挿、推定または測定により、および下記の場合に、その物質は区分 5 に分類される：

- ヒトにおける明確な毒性影響を示唆する信頼できる情報が得られている；または

- 経口、吸入または経皮により区分 4 の数値に至るまで試験した場合に 1 匹でも死亡が認められた場合；または

- 区分 4 の数値に至るまで試験した場合に、専門家の判断により毒性の明確な臨床徴候（下痢、立毛、不十分な毛繕いは除く）が確認された場合；または

- 専門家の判断により、その他の動物試験から明確な急性影響の可能性を示す信頼できる情報があると確認された場合。

動物愛護の必要性を認識した上で、区分 5 の範囲での動物の試験は必要ないと考えられ、動物試験結果からヒトの健康保護に関する直接的関連性が得られる可能性が高い場合にのみ検討されるべきである。

3.1.2.2 急性毒性に関する調和分類システムは、既存システムの要求と合致するように策定されている。IOMC CG/HCCS（Coordinating Group/Harmonization of Chemical Classification Systems）の定めた基本原則では「調和とは、化学品の有害性の分類および情報伝達のための共通かつ首尾一貫した基盤を確立することを意味する。これより輸送手段、消費者、労働者および環境保護に関連する適切な条項の選択が可能である」としている。このために、急性毒性の体系には 5 つの分類区分が含まれている。

3.1.2.3 経口および吸入経路による急性毒性評価のために望ましい試験動物種はラットであり、急性経皮毒性評価にはラットおよびウサギが望ましい。既存システムのもとで化学品の分類のためにすでに得られた試験データは、これらの化学品を調和システムにしたがって再分類する際に受け入れられるべきである。複数種の動物での急性毒性実験データが利用可能である場合には、有効であり、適切に実施された試験の中から、最もふさわしい LD_{50} 値を選択する際に科学的判断を行うべきである。またヒトの経験に基づいたデータ（すなわち職業データ、事故情報データベース、疫学研究、臨床報告）が利用可能な時には、これらは 1.3.2.4.9 に記載されている原則に従った証拠の重み付け評価により検討されるべきである。

3.1.2.4 区分 1 は、最も有害性が強い区分であり、そのカットオフ値（表 3.1.1 参照）は、主として輸送分野で容器等級の分類に採用されている。

3.1.2.5   Category 5 is for substances which are of relatively low acute toxicity but which, under certain circumstances, may pose a hazard to vulnerable populations. Criteria for identifying substances in Category 5 are provided in addition to the table. These substances are anticipated to have an oral or dermal LD_{50} value in the range 2000 - 5000 mg/kg bodyweight and equivalent doses for inhalation exposure[1]. In light of animal welfare considerations, testing in animals in Category 5 ranges is discouraged and should only be considered when there is a strong likelihood that results of such testing would have a direct relevance to the protection of human health.

3.1.2.6   *Specific considerations for inhalation toxicity*

3.1.2.6.1   Values for inhalation toxicity are based on 4 hours tests in laboratory animals. When experimental values are taken from tests using a 1 hour exposure, they can be converted to a 4 hour equivalent by dividing the 1 hour value by a factor of 2 for gases and vapours and 4 for dusts and mists. Guidance on the conversion of experimental values for times other than a 1-hour exposure is provided in 3.1.5.3.

3.1.2.6.2   Units for inhalation toxicity are a function of the form of the inhaled material. Values for dusts and mists are expressed in mg/l. Values for gases are expressed in ppmV. Acknowledging the difficulties in testing vapours, some of which consist of mixtures of liquid and vapour phases, the table provides values in units of mg/l. However, for those vapours which are near the gaseous phase, classification should be based on ppmV. As inhalation test methods are updated, the OECD and other test guideline programs will need to define vapours in relation to mists for greater clarity.

3.1.2.6.3   Vapour inhalation values are intended for use in classification of acute toxicity for all sectors. It is also recognized that the saturated vapour concentration of a chemical is used by the transport sector as an additional element in classifying chemicals for packing groups.

3.1.2.6.4   Of particular importance is the use of well articulated values in the highest hazard categories for dusts and mists. Inhaled particles between 1 and 4 microns mean mass aerodynamic diameter (MMAD) will deposit in all regions of the rat respiratory tract. This particle size range corresponds to a maximum dose of about 2 mg/l. In order to achieve applicability of animal experiments to human exposure, dusts and mists would ideally be tested in this range in rats. The cut-off values in the table for dusts and mists allow clear distinctions to be made for materials with a wide range of toxicities measured under varying test conditions. The values for dusts and mists should be reviewed in the future to adapt to any future changes in OECD or other test guidelines with respect to technical limitations in generating, maintaining, and measuring dust and mist concentrations in respirable form.

3.1.2.6.5   In addition to classification for inhalation toxicity, if data are available that indicates that the mechanism of toxicity was corrosivity of the substance or mixture, certain authorities may also choose to label it as *corrosive to the respiratory tract*. Corrosion of the respiratory tract is defined by destruction of the respiratory tract tissue after a single, limited period of exposure analogous to skin corrosion; this includes destruction of the mucosa. The corrosivity evaluation could be based on expert judgment using such evidence as: human and animal experience, existing (in vitro) data, pH values, information from similar substances or any other pertinent data.

3.1.3   **Classification criteria for mixtures**

3.1.3.1   The criteria for substances classify acute toxicity by use of lethal dose data (tested or derived). For mixtures, it is necessary to obtain or derive information that allows the criteria to be applied to the mixture for the purpose of classification. The approach to classification for acute toxicity is tiered and is dependent upon the amount of information available for the mixture itself and for its ingredients. The flow chart of figure 3.1.1 below outlines the process to be followed:

---

[1]   *Guidance on Category 5 inhalation values: The OECD Task Force on Harmonization of Classification and Labelling (HCL) did not include numerical values in table 3.1.1 above for acute inhalation toxicity Category 5 but instead specified doses "equivalent" to the range of 2000-5000 mg/kg bodyweight by the oral or dermal route (see note (g) to table 3.1.1). In some systems, the competent authority may prescribe values.*

3.1.2.5　区分5は、急性毒性は比較的弱いが、特定条件下で特に高感受性の集団に有害性の可能性がある物質である。区分5に分類される物質を特定するための判定基準を表の追加部分に示す。これらの物質の経口または経皮 LD$_{50}$ 値は 2000～5000mg/kg の範囲内、また吸入経路でもこれに相当する数値であると想定される[1]。動物愛護の観点から、区分5の範囲での動物の試験は必要ないと考えられ、動物試験結果からヒトの健康保護に関する直接的関連性が得られる可能性が高い場合にのみ検討されるべきである。

3.1.2.6　*吸入毒性に関して特別に留意すべき事項*

3.1.2.6.1　*吸入毒性* に関する数値は、4時間の動物試験に基づいている。1時間のばく露試験からの実験値を採用する場合には、1時間での数値を、気体および蒸気の場合は 2 で、粉塵およびミストの場合は 4 で割ることで、4時間に相当する数値に換算できる。1時間ばく露以外の時間の実験値の換算に関する手引きは、3.1.5.3 に示す。

3.1.2.6.2　吸入毒性の単位は吸入された物質の形態によって決定される。粉塵およびミストの場合の数値は mg/l として表示される。気体の場合の数値は ppmV として表示される。液体相および蒸気相で混成されるような蒸気を試験する困難さを認め、表中では単位を mg/l として数値の表示をしている。ただし、気相に近いような蒸気の場合には、分類は ppmV 濃度に基づくべきである。吸入試験方法を更新する場合には、OECD およびその他のテストガイドライン（試験指針）プログラムは、蒸気について、ミストとの関係をより明確にして定義することが必要となろう。

3.1.2.6.3　蒸気吸入の数値は、あらゆる分野での急性毒性分類に採用されることを目的としている。また、化学品の飽和蒸気濃度は輸送分野で、化学品を容器等級で分類する際に追加要素として採用されている。

3.1.2.6.4　特に重要なのは、粉塵およびミストの最も有害性が強い区分において明確な数値を用いることである。空気力学的質量中央径（MMAD）が 1～4 ミクロンの吸入された粒子は、ラットの呼吸器のすべての部分に沈着する。この粒子サイズ範囲で約 2mg/l の最大用量に対応する。動物実験の結果をヒトのばく露に外挿することができるためには、粉塵およびミストはラットにおいてこのサイズで試験することが理想的である。粉塵およびミストの表におけるカットオフ値は、様々な試験条件下で測定された広範囲の毒性をもつ物質に対して明確な区別ができるようになっている。粉塵およびミストに関する値については、将来的に見直しを行い、吸入可能な形態での粉塵とミストの生成、維持、測定の技術的制約に関する OECD や他のテストガイドラインの将来的な変更に対応していくべきである。

3.1.2.6.5　吸入毒性の分類に加えて、物質または混合物の毒性のメカニズムが腐食性であることを示すデータがあれば、所管官庁は気道に対する腐食性を表示する選択をしてもよい。*気道の腐食* は、皮膚の腐食に類似した、一回の限られた時間でのばく露後の気道組織の破壊（粘膜の破壊を含む）として定義される。ヒトおよび動物での経験、既存の（in vitro）データ、pH の値、類似の物質からの情報、他の適切なデータなどの証拠を使用し、専門家の判断に基づいて、腐食性の評価をすることができる。

3.1.3　混合物の分類基準

3.1.3.1　物質に対する判定基準では、致死量データ（試験または予測による）を使用して急性毒性を分類する。混合物については、分類の目的で判定基準を適用するための情報を入手または予測する必要がある。急性毒性の分類方法は、段階的で、混合物そのものとその成分について利用可能な情報の量に依存する。図 3.1.1 のフローチャートに、従うべき手順の概要を示す：

---

[1]　*区分5の吸入値についての指針：分類と表示の調和に関する OECD タスクフォース（HCL）は区分5の急性吸入毒性について上記の 3.1.1 に数値を示さず、かわりに経口あるいは経皮での 2000～5000mg/kg 体重に「相当」する投与量を指定した（表 3.1.1 の(g)参照）。システムによっては、所管官庁が値を規定してもよい。*

Figure 3.1.1: Tiered approach to classification of mixtures for acute toxicity

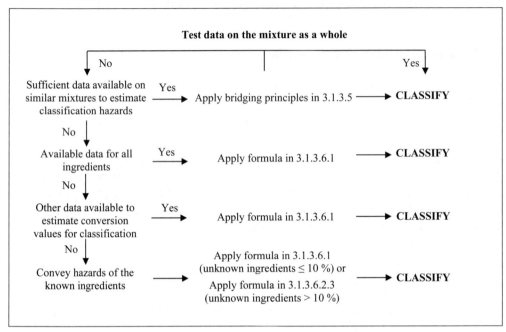

3.1.3.2   Classification of mixtures for acute toxicity can be carried out for each route of exposure, but is only needed for one route of exposure as long as this route is followed (estimated or tested) for all ingredients and there is no relevant evidence to suggest acute toxicity by multiple routes. When there is relevant evidence of toxicity by multiple routes of exposure, classification is to be conducted for all appropriate routes of exposure. All available information should be considered. The pictogram and signal word used should reflect the most severe hazard category and all relevant hazard statements should be used.

3.1.3.3   In order to make use of all available data for purposes of classifying the hazards of mixtures, certain assumptions have been made and are applied where appropriate in the tiered approach:

(a) The "relevant ingredients" of a mixture are those which are present in concentrations $\geq 1\ \%$ (w/w for solids, liquids, dusts, mists and vapours and v/v for gases), unless there is a reason to suspect that an ingredient present at a concentration $< 1\ \%$ is still relevant for classifying the mixture for acute toxicity. This point is particularly relevant when classifying untested mixtures which contain ingredients that are classified in Category 1 and Category 2;

(b) Where a classified mixture is used as an ingredient of another mixture, the actual or derived acute toxicity estimate (ATE) for that mixture may be used when calculating the classification of the new mixture using the formulas in 3.1.3.6.1 and 3.1.3.6.2.3;

(c) If the converted acute toxicity point estimates for all ingredients of a mixture are within the same category, then the mixture should be classified in that category;

(d) When only range data (or acute toxicity hazard category information) are available for ingredients in a mixture, they may be converted to point estimates in accordance with table 3.1.2 when calculating the classification of the new mixture using the formulas in 3.1.3.6.1 and 3.1.3.6.2.3.

図 3.1.1：混合物の急性毒性に関する分類のための段階的なアプローチ

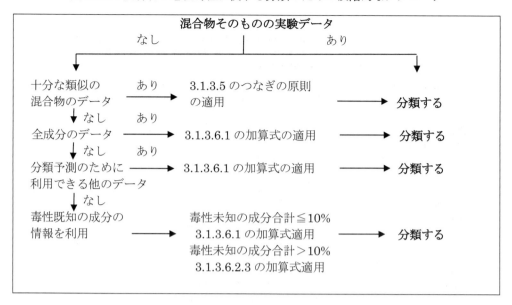

3.1.3.2 急性毒性に関する混合物の分類は、各ばく露経路について行うことができるが、1つのばく露経路だけが全成分について検討（推定または試験）され、複数の経路による急性毒性を示唆する適当な証拠はないとされる場合には、その経路だけが分類される。複数のばく露経路による毒性に関して適当な証拠がある場合には、全経路からのばく露に対しての区分を決める。利用できるすべての情報を考慮すべきである。用いる絵表示や注意喚起語はもっとも厳しい有害性区分を反映させるべきであり、すべての危険有害性情報を記載すべきである。

3.1.3.3 混合物の有害性を分類する目的で利用できるあらゆるデータを使用するために、ある条件が与えられており、該当する段階的アプローチが適用される：

(a) 混合物の「考慮すべき成分」とは、≧1%の濃度（固体、液体、粉塵、ミストおよび蒸気については重量/重量、気体については体積/体積）で存在するものである。ただし＜1%の成分でも、その混合物の急性毒性を分類することに関連すると予想される場合は、この限りではない。これは特に、区分1や区分2に分類される成分を含む未試験の混合物を分類する場合に関係する；

(b) 分類された混合物が別の混合物の成分として使用される場合は、3.1.3.6.1 および 3.1.3.6.2.3 の式を用いて新しい混合物の分類を計算する際に、分類された混合物の実際のあるいは予測される急性毒性推定値(ATE)を使用してもよい；

(c) 混合物のすべての成分に対する変換した急性毒性点推定値が同じ区分にあれば、混合物は同じ区分とするべきである；

(d) 3.1.3.6.1 および 3.1.3.6.2.3 における式を利用して新しい混合物の区分を計算する際に、混合物の成分に関して範囲を示すデータ（または急性毒性の区分に関する情報）のみが利用できるときは、それらを表 3.1.2 にしたがって点推定値に変換する。

Table 3.1.2: Conversion from experimentally obtained acute toxicity range values (or acute toxicity hazard categories) to acute toxicity point estimates for use in the formulas for the classification of mixtures

Exposure routes	Classification category or experimentally obtained acute toxicity range estimate (see note 1)	Converted acute toxicity point estimate (see note 2)
Oral (mg/kg bodyweight)	0 < Category 1 ≤ 5	0.5
	5 < Category 2 ≤ 50	5
	50 < Category 3 ≤ 300	100
	300 < Category 4 ≤ 2000	500
	2000 < Category 5 ≤ 5000	2500
Dermal (mg/kg bodyweight)	0 < Category 1 ≤ 50	5
	50 < Category 2 ≤ 200	50
	200 < Category 3 ≤ 1000	300
	1000 < Category 4 ≤ 2000	1100
	2000 < Category 5 ≤ 5000	2500
Gases (ppmV)	0 < Category 1 ≤ 100	10
	100 < Category 2 ≤ 500	100
	500 < Category 3 ≤ 2500	700
	2500 < Category 4 ≤ 20000	4500
	Category 5 - See footnote to 3.1.2.5.	
Vapours (mg/l)	0 < Category 1 ≤ 0.5	0.05
	0.5 < Category 2 ≤ 2.0	0.5
	2.0 < Category 3 ≤ 10.0	3
	10.0 < Category 4 ≤ 20.0	11
	Category 5 - See footnote to 3.1.2.5.	
Dust/mist (mg/l)	0 < Category 1 ≤ 0.05	0.005
	0.05 < Category 2 ≤ 0.5	0.05
	0.5 < Category 3 ≤ 1.0	0.5
	1.0 < Category 4 ≤ 5.0	1.5
	Category 5 - See footnote to 3.1.2.5.	

Note: Gases concentration are expressed in parts per million per volume (ppmV).

NOTE 1: Category 5 is for mixtures which are of relatively low acute toxicity but which under certain circumstances may pose a hazard to vulnerable populations. These mixtures are anticipated to have an oral or dermal LD_{50} value in the range of 2000-5000 mg/kg bodyweight or equivalent dose for other routes of exposure. In light of animal welfare considerations, testing in animals in Category 5 ranges is discouraged and should only be considered when there is a strong likelihood that results of such testing would have a direct relevance for protecting human health.

NOTE 2: These values are designed to be used in the calculation of the ATE for classification of a mixture based on its ingredients and do not represent test results. The values are conservatively set at the lower end of the range of Categories 1 and 2, and at a point approximately $1/10^{th}$ from the lower end of the range for Categories 3 – 5.

3.1.3.4 *Classification of mixtures where acute toxicity test data are available for the complete mixture*

Where the mixture itself has been tested to determine its acute toxicity, it will be classified according to the same criteria as those used for substances presented in table 3.1.1. If test data for the mixture are not available, the procedures presented below should be followed.

3.1.3.5 *Classification of mixtures where acute toxicity test data are not available for the complete mixture: bridging principles*

3.1.3.5.1 Where the mixture itself has not been tested to determine its acute toxicity, but there are sufficient data on both the individual ingredients and similar tested mixtures to adequately characterize the hazards of the mixture, these data will be used in accordance with the following agreed bridging principles. This ensures that the classification process uses the available data to the greatest extent possible in characterizing the hazards of the mixture without the necessity for additional testing in animals.

表 3.1.2：実験的に得られた急性毒性範囲推定値（または急性毒性区分）から式を利用して
混合物を分類するための急性毒性点推定値への変換

ばく露経路	分類または実験で得られた 急性毒性範囲推定値 （注記1参照）	変換値 (Conversion Value) （注記2参照）
経口 (mg/kg 体重)	0＜ 区分1 ≦5 5＜ 区分2 ≦50 50＜ 区分3 ≦300 300＜ 区分4 ≦2000 *2000＜ 区分5 ≦5000*	0.5 5 100 500 *2500*
経皮 (mg/kg 体重)	0＜ 区分1 ≦50 50＜ 区分2 ≦200 200＜ 区分3 ≦1000 1000＜ 区分4 ≦2000 *2000＜ 区分5 ≦5000*	5 50 300 1100 *2500*
気体 (ppmV)	0＜ 区分1 ≦100 100＜ 区分2 ≦500 500＜ 区分3 ≦2500 2500＜ 区分4 ≦20000 *区分5　3.1.2.5 脚注参照*	10 100 700 4500
蒸気 (mg/l)	0＜ 区分1 ≦0.5 0.5＜ 区分2 ≦2.0 2.0＜ 区分3 ≦10.0 10.0＜ 区分4 ≦20.0 *区分5　3.1.2.5 脚注参照*	0.05 0.5 3 11
粉塵/ミスト (mg/l)	0＜ 区分1 ≦0.05 0.05＜ 区分2 ≦0.5 0.5＜ 区分3 ≦1.0 1.0＜ 区分4 ≦5.0 *区分5　3.1.2.5 脚注参照*	0.005 0.05 0.5 1.5

注記：気体濃度は容積当りの ppm (ppmV) で表される。

注記1：区分5は、急性毒性は比較的弱いが、ある特定の状況で影響を受けやすい集団に有害性を示す可能性がある混合物に対するものである。これらの混合物は、2000～5000mg/kg の範囲の経口または経皮 LD_{50} 値か、または他のばく露経路で同等の急性毒性値をもつものと予想される。動物愛護の観点から、区分5の範囲での動物の試験は必要ないと考えられ、動物試験結果からヒトの健康保護に関する直接的関連性が得られる可能性が高い場合にのみ検討されるべきである。

注記2：変換値は、混合物の各成分の情報に基づき混合物の分類のための ATE 値を計算する目的のためのものであり、試験結果を示すものではない。変換値は、区分1と2では範囲の下限を、区分3から5では、範囲の幅の 1/10 程度下限から上にずらした値で設定されている。

3.1.3.4　混合物そのものの急性毒性試験データが利用できる場合の混合物の分類

　混合物は、その急性毒性を決定するためにそのものが試験されている場合、表 3.1.1 に示した物質についての判定基準にしたがって分類される。混合物に関するこのような試験データが利用できない状況にある場合には、以下に示した手順に従うべきである。

3.1.3.5　*混合物そのものの急性毒性試験データが利用できない場合の混合物の分類：つなぎの原則 (Bridging principles)*

3.1.3.5.1　混合物そのものは急性毒性を決定する試験がなされていないが、当該混合物の有害性を適切に特定するための、個々の成分および類似の試験された混合物の両方に関して十分なデータがある場合、これらのデータは以下の承認されたつなぎの原則にしたがって使用される。これによって、分類手順において動物試験を追加する必要もなく、混合物の有害性の判定に利用可能なデータを可能な限り最大限に用いることができる。

3.1.3.5.2   *Dilution*

If a tested mixture is diluted with a diluent that has an equivalent or lower toxicity classification than the least toxic original ingredient, and which is not expected to affect the toxicity of other ingredients, then the new diluted mixture may be classified as equivalent to the original tested mixture. Alternatively, the formula explained in 3.1.3.6.1 could be applied.

3.1.3.5.3   *Batching*

The toxicity of a tested production batch of a mixture can be assumed to be substantially equivalent to that of another untested production batch of the same commercial product, when produced by or under the control of the same manufacturer, unless there is reason to believe there is significant variation such that the toxicity of the untested batch has changed. If the latter occurs, a new classification is necessary.

3.1.3.5.4   *Concentration of highly toxic mixtures*

If a tested mixture is classified in Category 1, and the concentration of the ingredients of the tested mixture that are in Category 1 is increased, the resulting untested mixture should be classified in Category 1 without additional testing.

3.1.3.5.5   *Interpolation within one hazard category*

For three mixtures (A, B and C) with identical ingredients, where mixtures A and B have been tested and are in the same hazard category, and where untested mixture C has the same toxicologically active ingredients as mixtures A and B but has concentrations of toxicologically active ingredients intermediate to the concentrations in mixtures A and B, then mixture C is assumed to be in the same hazard category as A and B.

3.1.3.5.6   *Substantially similar mixtures*

Given the following:

(a) Two mixtures: (i)   A + B;

     (ii)   C + B;

(b) The concentration of ingredient B is essentially the same in both mixtures;

(c) The concentration of ingredient A in mixture (i) equals that of ingredient C in mixture (ii);

(d) Data on toxicity for A and C are available and substantially equivalent, i.e. they are in the same hazard category and are not expected to affect the toxicity of B;

If mixture (i) or (ii) is already classified based on test data, then the other mixture can be classified in the same hazard category.

3.1.3.5.7   *Aerosols*

An aerosolized form of a mixture may be classified in the same hazard category as the tested, non-aerosolized form of the mixture for oral and dermal toxicity provided the added propellant does not affect the toxicity of the mixture on spraying. Classification of aerosolized mixtures for inhalation toxicity should be considered separately.

3.1.3.6   **Classification of mixtures based on ingredients of the mixture (additivity formula)**

3.1.3.6.1   *Data available for all ingredients*

In order to ensure that classification of the mixture is accurate, and that the calculation need only be performed once for all systems, sectors, and categories, the acute toxicity estimate (ATE) of ingredients should be considered as follows:

(a) Include ingredients with a known acute toxicity, which fall into any of the GHS acute toxicity hazard categories;

3.1.3.5.2　希釈

試験された混合物が毒性の最も弱い成分に比べて同等以下の毒性分類に属する物質で希釈され、その物質が他の成分の毒性に影響を与えないことが予想されれば、新しい希釈された混合物は、試験された元の混合物と同等として分類してもよい。あるいは 3.1.3.6.1 で説明した式も適用できる。

3.1.3.5.3　製造バッチ

混合物の試験されていない製造バッチに毒性があるかどうかは、同じ製造者によって、またはその管理下で生産された同じ商品の試験された別のバッチの毒性と本質的に同等とみなすことができる。ただし、試験されていないバッチの毒性が変化するような有意な変動があると考えられる理由がある場合はこの限りではない。このような場合には、新しい分類が必要である。

3.1.3.5.4　毒性の強い混合物の濃縮

試験された混合物が区分1に分類され、区分1にある試験された混合物の成分の濃度が増加する場合、試験されていない新しい混合物は、追加試験なしで区分1に分類すべきである。

3.1.3.5.5　1つの有害性区分内での内挿

3つの混合物（A、BおよびC）は同じ成分を持ち、AとBは試験され同じ有害性区分にある。試験されていない混合物Cは混合物AおよびBと同じ毒性学的に活性な成分を持ち、毒性学的に活性な成分の濃度が混合物AとBの中間である場合、混合物CはAおよびBと同じ有害性区分にあるとする。

3.1.3.5.6　本質的に類似した混合物

次を仮定する：

(a)　2つの混合物：　　(i)　　A+B；

　　　　　　　　　　　(ii)　　C+B；

(b)　成分Bの濃度は、両方の混合物で本質的に同じである；

(c)　混合物(i)の成分Aの濃度は、混合物(ii)の成分Cの濃度に等しい；

(d)　AとCの毒性に関するデータは利用でき、実質的に同等であり、すなわちAとCは同じ有害性区分に属し、かつ、Bの毒性には影響を与えることはないと判断される；

混合物(i)または(ii)が既に試験データによって分類されている場合には、他方の混合物は同じ有害性区分に分類することができる。

3.1.3.5.7　エアゾール

エアゾール形態の混合物は、添加された噴霧剤が噴霧時に混合物の毒性に影響しないという条件下では、経口および経皮毒性について試験された非エアゾール形態の混合物と同じ有害性区分に分類してよい。エアゾール化された混合物の吸入毒性に関する分類は、個別に考慮するべきである。

3.1.3.6　混合物の成分に基づく混合物の分類（加算式）

3.1.3.6.1　全成分についてデータが利用できる場合

混合物の分類を正確にし、すべてのシステム、部門および区分について計算を一度だけで済むようにするために、成分の急性毒性推定値(ATE)は次のように考えるべきである：

(a)　急性毒性が知られており、GHS急性毒性有害性区分のいずれかに分類される成分を含める；

(b) Ignore ingredients that are presumed not acutely toxic (e.g. water, sugar);

(c) Ignore ingredients if the data available are from a limit dose test (at the upper threshold for Category 4 for the appropriate route of exposure as provided in table 3.1.1) and do not show acute toxicity.

Ingredients that fall within the scope of this paragraph are considered to be ingredients with a known acute toxicity estimate (ATE). See note (b) to table 3.1.1 and paragraph 3.1.3.3 for appropriate application of available data to the equation below, and paragraph 3.1.3.6.2.3.

The ATE of the mixture is determined by calculation from the ATE values for all relevant ingredients according to the following formula below for oral, dermal or inhalation toxicity:

$$\frac{100}{ATEmix} = \sum_n \frac{C_i}{ATE_i}$$

where:

C_i = concentration of ingredient i;

n ingredients and i is running from 1 to n;

ATE_i = Acute toxicity estimate of ingredient i;

3.1.3.6.2 *Data are not available for one or more ingredients of the mixture*

3.1.3.6.2.1 Where an ATE is not available for an individual ingredient of the mixture, but available information such as listed below can provide a derived conversion value, the formula in 3.1.3.6.1 may be applied.

This may include evaluation of:

(a) Extrapolation between oral, dermal and inhalation acute toxicity estimates[2]. Such an evaluation could require appropriate pharmacodynamic and pharmacokinetic data;

(b) Evidence from human exposure that indicates toxic effects but does not provide lethal dose data;

(c) Evidence from any other toxicity tests/assays available on the substance that indicates toxic acute effects but does not necessarily provide lethal dose data; or

(d) Data from closely analogous substances using structure-activity relationships.

This approach generally requires substantial supplemental technical information, and a highly trained and experienced expert, to reliably estimate acute toxicity. If such information is not available, proceed to the provisions of 3.1.3.6.2.3.

3.1.3.6.2.2 In the event that an ingredient without any useable information for classification is used in a mixture at a concentration ≥ 1 %, it is concluded that the mixture cannot be attributed a definitive acute toxicity estimate. In this situation the mixture should be classified based on the known ingredients only, with the additional statement that × percent of the mixture consists of ingredient(s) of unknown acute (oral/dermal/inhalation) toxicity. The competent authority can decide to specify that the additional statement(s) be communicated on the label or on the SDS or both, or to leave the choice of where to place the statement to the manufacturer/supplier.

---

[2] When mixtures contain ingredients that do not have acute toxicity data for each route of exposure, acute toxicity estimates may be extrapolated from the available data and applied to the appropriate routes (see 3.1.3.2). However, competent authorities may require testing for a specific route. In those cases, classification should be performed for that route based upon the competent authority's requirement.

(b) 急性毒性ではないと考えられる成分を無視する（例えば、水、砂糖）；

(c) 限界用量試験（表 3.1.1 における適当なばく露経路に対して区分 4 に相当する上限値）のデータが利用でき、急性毒性を示していない成分を無視する。

これらの範囲内に入る成分を急性毒性推定値（ATE）が既知の成分であると考える。利用できるデータを下記および 3.1.3.6.2.3 の式に適当に当てはめるためには表 3.1.1 注記(b)および 3.1.3.3 を参照。

混合物の ATE 値は、経口、経皮、吸入毒性について、以下の式に従い、すべての関連成分の ATE 値から計算によって決定される：

$$\frac{100}{ATEmix} = \sum_n \frac{Ci}{ATEi}$$

ここで：

Ci = 成分 i の濃度；

成分数 n のとき、i は 1 から n；

ATEi = 成分 i の急性毒性推定値；

3.1.3.6.2　*混合物の 1 つまたは複数の成分についてデータが利用できない場合*

3.1.3.6.2.1　混合物の個々の成分については ATE 値が利用できないが、以下に挙げたような利用できる情報から、予測された変換値が提供される場合には、3.1.3.6.1 の加算式が適用される。

これには次の評価を用いてもよい：

(a) 経口、経皮、および吸入急性毒性推定値間の外挿[2]。このような評価には、適切なファーマコダイナミクスおよびファーマコキネティクスのデータが必要となることがある；

(b) 毒性影響はあるが致死量データのない、ヒトへのばく露からの証拠；

(c) 急性毒性影響はあるが、必ずしも致死量データはない物質に関して利用できる他の毒性試験/分析からの証拠；または

(d) 構造活性相関を用いた極めて類似した物質からのデータ。

この方法は一般に、急性毒性を信頼できる程度に推定するために、多くの補足技術情報と高度に訓練され経験豊かな専門家の能力を必要とする。このような情報が利用できない場合には、3.1.3.6.2.3 の規定に進むこと。

3.1.3.6.2.2　分類のための利用できる情報の全くない成分が混合物中に ≧1%の濃度で使用されている場合には、混合物は明確な急性毒性推定値を割当てることはできないと結論される。この場合には、混合物の x パーセントは急性（経口/経皮/吸入）毒性が未知の成分から成るという追加の記述と共に混合物は既知の成分だけに基づいて分類するべきである。所管官庁はその追加的な記述をラベルまたは SDS あるいはその両方で伝達することを明記するかどうか、またその記述をどこにするかの選択を製造者/供給者に委ねるかどうかを決めることができる。

---

[2]　*混合物が、それぞれのばく露経路について急性毒性のデータがない成分を含む場合には、急性毒性推定値は利用できるデータから外挿して適当な経路に適用する（3.1.3.2 参照）。所管官庁は特定の経路に対して試験を要求してもよい。この場合、分類は所管官庁の要求に基づいた経路に対して行うべきである。*

3.1.3.6.2.3   If the total concentration of the relevant ingredient(s) with unknown acute toxicity is ≤ 10 % then the formula presented in 3.1.3.6.1 should be used. If the total concentration of the relevant ingredient(s) with unknown toxicity is > 10 %, the formula presented in 3.1.3.6.1 should be corrected to adjust for the percentage of the unknown ingredient(s) as follows:

$$\frac{100 - \left(\sum C_{unknown} \text{ if } > 10\%\right)}{ATE_{mix}} = \sum_n \frac{C_i}{ATE_i}$$

3.1.4 Hazard communication

3.1.4.1   General and specific considerations concerning labelling requirements are provided in *Hazard communication: Labelling* (chapter 1.4). Annex 1 contains summary tables about classification and labelling. Annex 3 contains examples of precautionary statements and pictograms which can be used where allowed by the competent authority. Table 3.1.3 presents specific label elements for substances and mixtures classified into this hazard class based on the criteria in this chapter.

Table 3.1.3: Label elements for acute toxicity

	Category 1	Category 2	Category 3	Category 4	Category 5
Symbol	Skull and crossbones	Skull and crossbones	Skull and crossbones	Exclamation mark	No symbol
Signal word	Danger	Danger	Danger	Warning	Warning
Hazard statement:					
--Oral	Fatal if swallowed	Fatal if swallowed	Toxic if swallowed	Harmful if swallowed	May be harmful if swallowed
--Dermal	Fatal in contact with skin	Fatal in contact with skin	Toxic in contact with skin	Harmful in contact with skin	May be harmful in contact with skin
--Inhalation see note	Fatal if inhaled	Fatal if inhaled	Toxic if inhaled	Harmful if inhaled	May be harmful if inhaled

NOTE: *If a substance/mixture is also determined to be corrosive (based on data such as skin or eye data), corrosivity hazard may also be communicated by some authorities as symbol and/or hazard statement. That is, in addition to an appropriate acute toxicity symbol, a corrosivity symbol (used for skin and eye corrosivity) may be added along with a corrosivity hazard statement such as "corrosive" or "corrosive to the respiratory tract".*

3.1.4.2   The acute toxicity hazard statements differentiate the hazard based on the route of exposure. Communication of acute toxicity classification should also reflect this differentiation. For example, acute oral toxicity Category 1, acute dermal toxicity Category 1 and acute inhalation toxicity Category 1. If a substance or mixture is classified for more than one route of exposure then all relevant classifications should be communicated on the safety data sheet as specified in annex 4 and the relevant hazard communication elements included on the label as prescribed in 3.1.3.2. If the statement "x % of the mixture consists of ingredient(s) of unknown acute (oral/dermal/inhalation) toxicity" is communicated, as prescribed in 3.1.3.6.2.2, then it can also be differentiated based on the route of exposure. For example, "x % of the mixture consists of ingredient(s) of unknown acute oral toxicity" and "x % of the mixture consists of ingredient(s) of unknown acute dermal toxicity".

3.1.3.6.2.3　急性毒性が未知の当該成分の全濃度が≦10%の場合には、3.1.3.6.1 に示した式を用いるべきである。毒性が未知の当該成分の全濃度が＞10%の場合には、3.1.3.6.1に示した加算式は、次のように式（未知成分補正）により未知の成分の%について調整するように補正するべきである：

$$\frac{100 - (\sum C_{unknown\ if\ >10\%})}{ATEmix} = \sum_n \frac{C_i}{ATE_i}$$

3.1.4　危険有害性情報の伝達

3.1.4.1　表示要件に関する通則および細則は、*危険有害性に関する情報の伝達：表示*（第 1.4 章）に定める。附属書1に、分類と表示に関する概要表を示す。附属書3には、注意書きおよび所管官庁が許可した場合に使用可能な注意絵表示の例を示す。下の表には、本章で示された判定基準に基づき急性毒性有害性区分1から5に分類された物質および混合物について、そのラベル要素を示す。表 3.1.3 は本章の判定基準に基づいて、この危険有害性クラスに分類される物質および混合物の具体的なラベル要素を示す。

表 3.1.3：急性毒性のラベル要素

	区分1	区分2	区分3	区分4	区分5
シンボル	どくろ	どくろ	どくろ	感嘆符	シンボルなし
注意喚起語	危険	危険	危険	警告	警告
危険有害性情報					
-- 経口	飲み込むと生命に危険	飲み込むと生命に危険	飲み込むと有毒	飲み込むと有害	飲み込むと有害のおそれ
-- 経皮	皮膚に接触すると生命に危険	皮膚に接触すると生命に危険	皮膚に接触すると有毒	皮膚に接触すると有害	皮膚に接触すると有害のおそれ
-- 吸入 注記参照	吸入すると生命に危険	吸入すると生命に危険	吸入すると有毒	吸入すると有害	吸入すると有害のおそれ

注記：*物質/混合物が（皮膚または眼に関するデータに基づき）腐食物であると決定される場合、所管官庁は、腐食性をシンボルまたは危険有害性情報として伝達してもよい。すなわち、適切な急性毒性のシンボルに加えて、「腐食性」あるいは「気道に腐食性」などの腐食性の危険有害性情報とともに腐食性のシンボル（皮膚と眼の腐食性のために用いられる）を追加してもよい。*

3.1.4.2　急性毒性の危険有害性情報はばく露経路による危険有害性を区別している。急性毒性分類の伝達もまたこの区別を反映させるべきである。例えば、急性経口毒性区分1、急性経皮毒性区分1そして急性吸入毒性区分1である。物質あるいは混合物が1つ以上のばく露経路に対して分類される場合は、全ての関連した分類は附属書4に明記されているように SDS で伝達されなければならないし、関連した危険有害性に関する要素は 3.1.3.2 で規定されているようにラベルに含まれなければならない。3.1.3.6.2.2 で規定されているように、「混合物の x %は急性（経口/経皮/吸入）毒性が未知の成分からなる」という記述が伝達される場合、ばく露の経路による区別もまた可能であろう。例えば、「混合物の x %は急性経口毒性が未知の成分からなる」また「混合物の x %は急性経皮毒性が未知の成分からなる」など。

- 122 -

3.1.5 **Decision logic**

The decision logic which follows, is not part of the harmonized classification system but is provided here as additional guidance. It is strongly recommended that the person responsible for classification study the criteria before and during use of the decision logic.

3.1.5.1 *Decision logic 3.1.1 for acute toxicity*

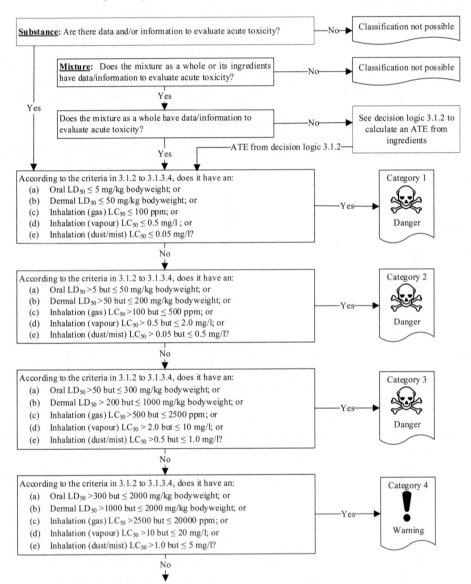

(Cont'd on next page)

3.1.5 判定論理

以下の判定論理は、調和分類システムには含まれないが、追加的な手引きとしてここに示す。分類の責任者に対し、この判定論理を使用する前および使用する際に判定基準についてよく調べ理解することを強く勧める。

3.1.5.1 *判定論理 3.1.1 急性毒性*

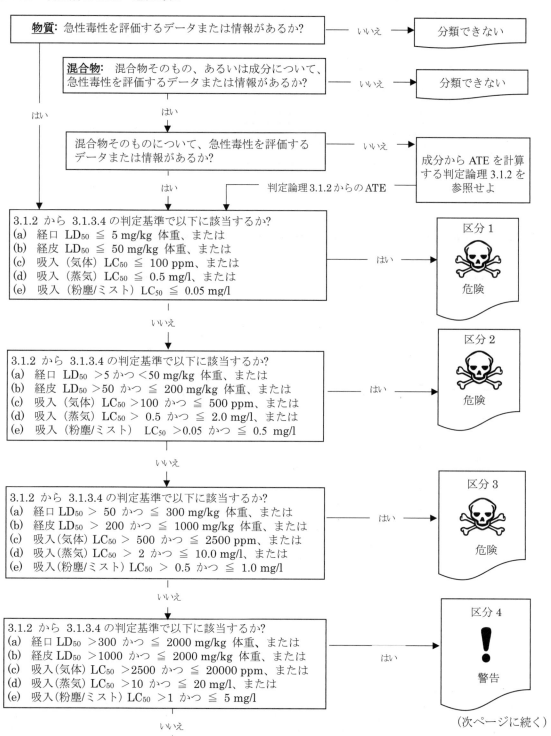

(次ページに続く)

- 123 -

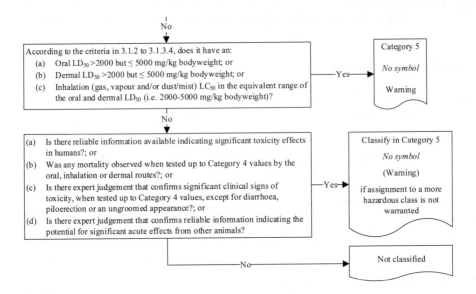

3.1.5.2 *Decision logic 3.1.2 for acute toxicity (see criteria in 3.1.3.5 and 3.1.3.6)*

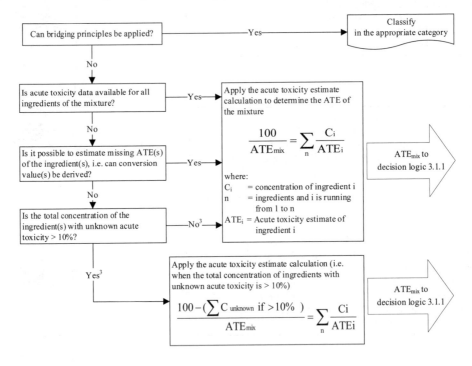

---

[3] *In the event that an ingredient without any useable information is used in a mixture at a concentration ≥ 1%, the classification should be based on the ingredients with the known acute toxicity only, and additional statement(s) should identify the fact that x % of the mixture consists of ingredient(s) of unknown acute (oral/dermal/inhalation) toxicity. The competent authority can decide to specify that the additional statement(s) be communicated on the label or on the SDS or both, or to leave the choice of where to place the statement to the manufacturer/supplier.*

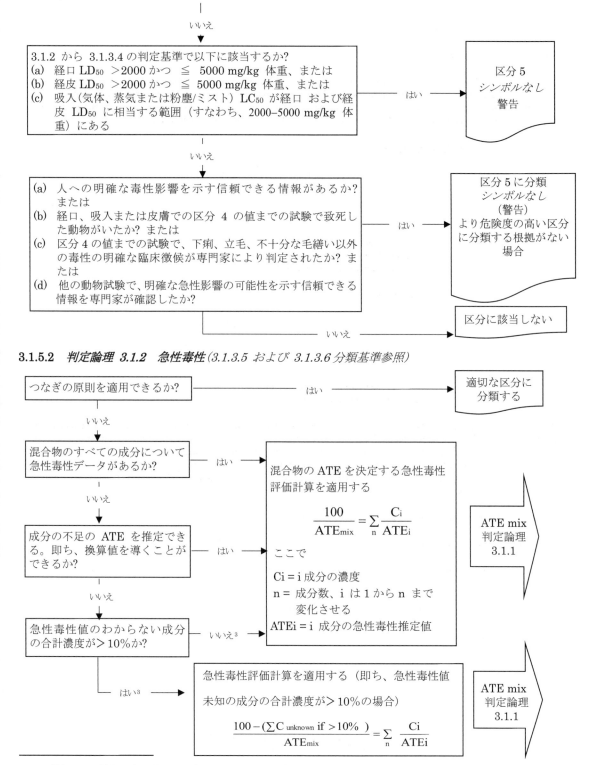

3.1.5.2 判定論理 3.1.2 急性毒性 (3.1.3.5 および 3.1.3.6 分類基準参照)

[3] 利用できる情報がない成分が混合物中に≧1%の濃度で使用されている場合、分類は急性毒性が既知の成分のみに基づいて行われるべきであり、追加の記述で混合物中のx%は急性(経口/経皮/吸入)毒性が未知の成分からなるという事実を明記するべきである。所管官庁はその追加的な記述をラベルまたはSDSあるいはその両方で伝達することを明記するかどうか、またその記述をどこにするかの選択を製造者/供給者に委ねるかどうかを決めることができる。

3.1.5.3 *Guidance*

3.1.5.3.1   The ATE values used for inhalation toxicity classification in table 3.1.1 are based on a 4-hour experimental exposure in laboratory animals (3.1.2.6.1). Existing inhalation LC$_{50}$ values obtained in studies using exposure times other than 1 hour (3.1.2.6.1) can be adjusted to a 4-hour exposure using the ten Berge equation ($C^n \times t = k$) for gases and vapours and Haber's rule ($C \times t = k$) for dusts and mists, as follows:

Formula for gases and vapours

$$LC_{50}(4 \text{ hours}) = \left(\frac{C^n \times t}{4}\right)^{1/n}$$

where:

- C = LC$_{50}$ concentration for exposure duration t
- n = chemical-specific exponent
- t = exposure duration, in hours, for C

Formula for dusts and mists

$$LC_{50}(4 \text{ hours}) = \frac{C \times t}{4}$$

where:

- C = LC$_{50}$ concentration for exposure duration t
- t = exposure duration, in hours, for C

3.1.5.3.2   A default value of 2 is used for n unless additional conclusive information is available to indicate that a different value is more appropriate. The accepted exposure times for conversion are from 30 minutes to 8-hour exposures. A competent authority may decide whether other exposure times are acceptable for conversion. Data from a long-term exposure should not be converted because this hazard class addresses acute toxicity. Guidance on the duration of short-term (i.e. acute) inhalation toxicity exposures can be found in OECD Guidance Document 39 (section 4.1: Outline of the exposure methodology).

Examples: classification using calculated 4-hour LC$_{50}$ values

Example 1: Substance (liquid)

1.   For the purpose of this example the substance has an experimental 6-hour vapour LC$_{50}$ = 13.6 mg/l

2.   No additional information on n is available so the default value (n = 2) will be used.

Criterion:

$$LC_{50}(4 \text{ hours}) = \left(\frac{C^n \times t}{4}\right)^{1/n}$$

Calculation

$$LC_{50}(4 \text{ hours}) = \left(\frac{C^n \times t}{4}\right)^{\frac{1}{n}} = \left(\frac{13.6^2 \times 6}{4}\right)^{\frac{1}{2}} = 16.7$$

3.   Therefore, the substance is classified into Category 4 based on the vapours Category 4 criteria (10.0 < ATE ≤ 20.0) from table 3.1.1.

3.1.5.3 手引き

3.1.5.3.1 表 3.1.1 の吸入毒性の分類に用いられる ATE 値は、実験動物における 4 時間の実験的なばく露に基づいている (3.1.2.6.1)。1 時間以外のばく露時間を用いた試験 (3.1.2.6.1) で得られた既存の吸入の LC_{50} 値は、気体および蒸気については ten Berge の式 ($C^n × t = k$)、粉塵およびミストについては Haber の法則 ($C × t = k$) を用いて、以下のように 4 時間ばく露に調整することができる:

気体と蒸気に対する計算式

$$LC_{50}(4\text{ 時間}) = \left(\frac{C^n × t}{4}\right)^{1/n}$$

ここで:

C = ばく露時間 t に対する LC_{50} 濃度
n = 化学物質固有の指数
t = C に対するばく露時間 (時間単位)

粉塵とミストに対する計算式

$$LC_{50}(4\text{ 時間}) = \frac{C × t}{4}$$

ここで:

C = ばく露時間 t に対する LC_{50} 濃度
t = C に対するばく露時間 (時間単位)

3.1.5.3.2 別の値がより適切であることを示す決定的な追加情報がない限り、n にはデフォルト値 2 が使用される。換算のために許容されるばく露時間は、30 分から 8 時間のばく露である。所管官庁は、他のばく露時間を換算に用いることができるかどうかを決定することができる。この有害性クラスは急性毒性を扱うため、長期ばく露のデータは換算すべきではない。短期 (すなわち急性) 吸入毒性ばく露時間に関するガイダンスは、OECD ガイダンス文書 39 (第 4.1 節:ばく露方法の概要) に記載されている。

例: 計算された 4 時間 LC_{50} 値を用いた分類

例 1: 物質 (液体)

1. この例において、物質の実験値として 6 時間 LC_{50} (蒸気) = 13.6 mg/l である。
2. n に関する追加情報がないため、デフォルト値 (n = 2) が使用される。

   基準:

   $$LC_{50}(4\text{ 時間}) = \left(\frac{C^n × t}{4}\right)^{1/n}$$

   計算

   $$LC_{50}(4\text{ 時間}) = \left(\frac{C^n × t}{4}\right)^{\frac{1}{n}} = \left(\frac{13.6^2 × 6}{4}\right)^{\frac{1}{2}} = \mathbf{16.7}$$

3. したがって、この物質は、表 3.1.1 の蒸気の区分 4 の基準 (10.0 < ATE ≦ 20.0) に基づいて区分 4 に分類される。

Example 2: Substance (solid)

4.   For this example, the substance has an experimental 2-hour dust $LC_{50} = 0.26$ mg/l

Criterion:

$$LC_{50}(4 \text{ hours}) = \frac{C \times t}{4}$$

Calculation

$$LC_{50}(4 \text{ hours}) = \frac{C \times t}{4} = \frac{0.26 \times 2}{4} = 0.13$$

5.   Therefore, the substance is classified into Category 2 based on the dusts and mists Category 2 criteria $(0.05 < ATE \leq 0.5)$ from table 3.1.1.

例 2: 物質 (固体)

4. この例において、物質の実験値として 2 時間 LC_{50} （粉塵）= 0.26 mg/l である。

　　　基準:

$$LC_{50}(4\text{時間}) = \frac{C \times t}{4}$$

　　　計算

$$LC_{50}(4\text{時間}) = \frac{C \times t}{4} = \frac{0.26 \times 2}{4} = 0.13$$

5. したがって、この物質は、表 3.1.1 の粉塵およびミストの区分 2 の基準 (0.05 ＜ ATE ≦ 0.5) に基づいて区分 2 に分類される。

CHAPTER 3.2

SKIN CORROSION/IRRITATION

3.2.1 Definitions and general considerations

3.2.1.1 *Skin corrosion* refers to the production of irreversible damage to the skin; namely, visible necrosis through the epidermis and into the dermis occurring after exposure to a substance or mixture.

Skin irritation refers to the production of reversible damage to the skin occurring after exposure to a substance or mixture.

3.2.1.2 To classify, all available and relevant information on skin corrosion/irritation is collected and its quality in terms of adequacy and reliability is assessed. Classification should be based on mutually acceptable data generated using methods that are validated according to international procedures. These include both OECD guidelines and equivalent methods (see 1.3.2.4.3). Sections 3.2.2.1 to 3.2.2.7 provide classification criteria for the different types of information that may be available.

3.2.1.3 A *tiered approach* (see 3.2.2.8) organizes the available information into levels/tiers and provides for decision-making in a structured and sequential manner. Classification results directly when the information consistently satisfies the criteria. However, where the available information gives inconsistent and/or conflicting results within a tier, classification of a substance or a mixture is made on the basis of the weight of evidence within that tier. In some cases when information from different tiers gives inconsistent and/or conflicting results (see 3.2.2.8.3) or where data individually are insufficient to conclude on the classification, an overall weight of evidence assessment is used (see 1.3.2.4.9, 3.2.2.7 and 3.2.5.3.1).

3.2.1.4 Guidance on the interpretation of criteria and references to relevant guidance documents are provided in 3.2.5.3.

3.2.2 Classification criteria for substances

Substances can be allocated to one of the following three categories within this hazard class:

(a) Category 1 (skin corrosion)

This category may be further divided into up to three sub-categories (1A, 1B and 1C) which can be used by those authorities requiring more than one designation for corrosivity.

Corrosive substances should be classified in Category 1 where subcategorization is not required by a competent authority or where data are not sufficient for subcategorization.

When data are sufficient, and where required by a competent authority, substances may be classified in one of the three sub-categories 1A, 1B or 1C.

(b) Category 2 (skin irritation)

(c) Category 3 (mild skin irritation)

This category is available for those authorities that want to have more than one skin irritation category (e.g. for classifying pesticides).

3.2.2.1 *Classification based on human data (tier 1 in figure 3.2.1)*

Existing reliable and good quality human data on skin corrosion/irritation should be given high weight where relevant for classification (see 3.2.5.3.2) and should be the first line of evaluation, as this gives information directly relevant to effects on the skin. Existing human data could be derived from single or repeated exposure(s), for example in occupational, consumer, transport or emergency response scenarios and epidemiological and clinical studies in well-documented case reports and observations (see 1.1.2.5 (c), 1.3.2.4.7 and 1.3.2.4.9). Although human data from accident or poison centre databases can provide evidence for classification, absence of incidents is not itself evidence for no classification, as exposures are generally unknown or uncertain.

第 3.2 章

皮膚腐食性/刺激性

3.2.1 定義および一般事項

3.2.1.1 *皮膚腐食性*とは皮膚に対する不可逆的な損傷を生じさせることをさす；すなわち物質または混合物へのばく露後に起こる、表皮を貫通して真皮に至る明らかに認められる壊死。

*皮膚刺激性*とは、物質または混合物へのばく露後に起こる、皮膚に対する可逆的な損傷を生じさせることをさす。

3.2.1.2 分類のため、皮膚腐食性/刺激性に関するすべての利用可能な関連する情報は収集され、妥当性と信頼性の観点からその質について評価される。分類は、国際的な手順にしたがって検証された方法により取られた相互に受け入れ可能なデータに基づくべきである。これらには、OECD ガイドラインと同等の方法の両方が含まれる（1.3.2.4.3 参照）。3.2.2.1 から 3.2.2.7 に利用可能な異なるタイプの情報に対する判定基準を示した。

3.2.1.3 *段階的アプローチ*（3.2.2.8 参照）とは、利用可能な情報をレベル/段階に整理して、構造的かつ連続的な方法で意思決定を行うものである。情報が一貫して判定基準を満たしていれば、分類はすぐに終わる。しかし一つの段階において利用可能な情報に一貫性がないおよび/または矛盾する結果が見られる場合には、物質や混合物の分類はその段階のなかで証拠の重み付けに基づいて行われる。異なる段階からの情報に一貫性がないおよび/または矛盾が見られる場合（3.2.2.8.3 参照）または個々のデータが分類を結論付けるには不十分であるような場合には、包括的な証拠の重み付け評価が使用される（1.3.2.4.9、3.2.2.7 および 3.2.5.3.1 参照）。

3.2.1.4 判定基準の解釈に関する手引きおよび関連の手引きへの参照は 3.2.5.3 に記載されている。

3.2.2 物質の分類基準

この有害性クラスでは、物質は以下の 3 つの区分の内の 1 つに割り当てることができる：

(a) 区分 1 （皮膚腐食性）

本区分はさらに、腐食性を複数に分けることを要求する所管官庁が使用することができるように、3 つの細区分（1A、1B および 1C）に分けてもよい。

所管官庁により細区分が要求されていない、またはデータが細区分のためには十分でない場合には、腐食性物質は区分 1 に分類するべきである。

データが十分であり、かつ所管官庁により要求されている場合には、物質は 3 つの細区分 1A、1B または 1C のうちの一つに分類してもよい。

(b) 区分 2 （皮膚刺激性）

(c) 区分 3 （軽度の皮膚刺激性）

本区分は 2 つ以上の皮膚刺激性区分を望む所管官庁に適用される（例えば農薬の分類）。

3.2.2.1 *ヒトのデータに基づく分類（図 3.2.1 の段階 1）*

皮膚腐食性/刺激性に関する既存の信頼できる質の高いヒトのデータは、皮膚への影響に直接的に関係する情報であるので、分類に関連して重視されるべきであり（3.2.5.3.2 参照）、評価の第一番目にあげられるべきである。既存のヒトのデータは、単回または反復ばく露から得ることができる。例えば、職業、消費者、輸送または緊急時対応のシナリオおよびよくまとめられた症例報告や観察による疫学的および臨床的研究などである（1.1.2.5 (c)、1.3.2.4.7 および 1.3.2.4.9 参照）。事故または中毒センターのデータベースから得られるヒトのデータは分類の根拠となり得るが、事例がないこと自体は、一般的にばく露は不明または不確実であるため、区分に該当しない根拠とはならない。

3.2.2.2   *Classification based on standard animal data (tier 1 in figure 3.2.1)*

OECD Test Guideline 404 is the currently available and internationally accepted animal test method for classification as skin corrosive or irritant (see tables 3.2.1 and 3.2.2, respectively) and is the standard animal test. The current version of OECD Test Guideline 404 uses a maximum of 3 animals. Results from animal studies conducted under previous versions of OECD Test Guideline 404 that used more than 3 animals are also considered standard animal tests when interpreted in accordance with 3.2.5.3.3.

3.2.2.2.1   *Skin corrosion*

3.2.2.2.1.1   A substance is corrosive to skin when it produces destruction of skin tissue, namely, visible necrosis through the epidermis and into the dermis, in at least one tested animal after exposure for up to 4 hours.

3.2.2.2.1.2   For those authorities wanting more than one designation for skin corrosion, up to three sub-categories are provided within the corrosion category (Category 1, see table 3.2.1): sub-category 1A, where corrosive responses are noted following up to 3 minutes exposure and up to 1 hour observation; sub-category 1B, where corrosive responses are described following exposure greater than 3 minutes and up to 1 hour and observations up to 14 days; and sub-category 1C, where corrosive responses occur after exposures greater than 1 hour and up to 4 hours and observations up to 14 days.

Table 3.2.1: Skin corrosion category and sub-categories

	Criteria
Category 1	Destruction of skin tissue, namely, visible necrosis through the epidermis and into the dermis, in at least one tested animal after exposure ≤ 4 h
Sub-category 1A	Corrosive responses in at least one animal following exposure ≤ 3 min during an observation period ≤ 1 h
Sub-category 1B	Corrosive responses in at least one animal following exposure > 3 min and ≤ 1 h and observations ≤ 14 days
Sub-category 1C	Corrosive responses in at least one animal after exposures > 1 h and ≤ 4 h and observations ≤ 14 days

3.2.2.2.2   *Skin irritation*

3.2.2.2.2.1   A substance is irritant to skin when it produces reversible damage to the skin following its application for up to 4 hours.

3.2.2.2.2.2   An irritation category (Category 2) is provided that:

(a)   recognizes that some test materials may lead to effects which persist throughout the length of the test; and

(b)   acknowledges that animal responses in a test may be variable.

An additional mild irritation category (Category 3) is available for those authorities that want to have more than one skin irritation category.

3.2.2.2.2.3   Reversibility of skin lesions is another consideration in evaluating irritant responses. When inflammation persists to the end of the observation period in 2 or more test animals, taking into consideration alopecia (limited area), hyperkeratosis, hyperplasia and scaling, then a material should be considered to be an irritant.

3.2.2.2.2.4   Animal irritant responses within a test can be variable, as they are with corrosion. A separate irritant criterion accommodates cases when there is a significant irritant response but less than the mean score criterion for a positive test. For example, a test material might be designated as an irritant if at least 1 of 3 tested animals shows a very elevated mean score throughout the study, including lesions persisting at the end of an observation period of normally 14 days. Other responses could also fulfil this criterion. However, it should be ascertained that the responses are the result of chemical exposure. Addition of this criterion increases the sensitivity of the classification system.

3.2.2.2.2.5   An irritation category (Category 2) is presented in table 3.2.2 using the results of animal testing. Authorities (e.g. for classifying pesticides) also have available a less severe mild irritation category (Category 3). Several criteria distinguish the two categories (table 3.2.2). They mainly differ in the severity of skin reactions. The major criterion

3.2.2.2 標準的動物データに基づく分類 (図3.2.1の段階1)

OECDテストガイドライン404は、現在利用可能で国際的に認められている皮膚腐食性または刺激性の分類に関する動物試験法であり（それぞれ表3.2.1および3.2.2参照）、標準的な動物試験である。現行のOECDテストガイドライン404は最大3匹の動物を使用する。OECDテストガイドライン404の旧版に基づいて実施され、3匹を超える動物を使用した動物試験の結果も、3.2.5.3.3にしたがって解釈する場合には、標準的な動物試験とみなされる。

3.2.2.2.1 皮膚腐食性

3.2.2.2.1.1 物質が皮膚の組織を破壊、すなわち表皮を通して真皮に達する目に見える壊死が、4時間までのばく露後に少なくとも1匹の試験動物で見られた場合に、皮膚腐食性とする。

3.2.2.2.1.2 皮膚腐食性について1つ以上の区分を望む所管官庁のために、腐食性区分(区分1、表3.2.1参照)の中に3つの細区分を与えた。細区分1Aは3分間以内のばく露後、1時間以内の観察期間で腐食性反応が認められる場合、細区分1Bは3分間を超え1時間までのばく露期間後、14日以内の観察期間に腐食性反応が認められる場合、細区分1Cは1時間を超え4時間までのばく露後、14日以内の観察期間に腐食性反応が認められる場合である。

表3.2.1：皮膚腐食性の区分および細区分

	判定基準
区分1	≦4時間のばく露後、少なくとも1匹の試験動物で、皮膚の組織を破壊、すなわち表皮を通して真皮に達する目に見える壊死
細区分 1A	≦3分間のばく露で、少なくとも1匹の動物で、≦1時間の観察により腐食反応
細区分 1B	＞3分間かつ≦1時間のばく露で、少なくとも1匹の動物で、≦14日間の観察により腐食反応
細区分 1C	＞1時間かつ≦4時間のばく露で、少なくとも1匹の動物で、≦14日間の観察により腐食反応

3.2.2.2.2 皮膚刺激性

3.2.2.2.2.1 物質が4時間までのばく露後に皮膚に可逆的な損傷を与えた場合に皮膚刺激性とする。

3.2.2.2.2.2 刺激性区分（区分2）は以下のようになる：

(a) 試験期間全体にわたって継続する作用を生じうる被験物質がある、および

(b) 試験における動物の反応は多様でありうる。

皮膚刺激性物質の区分を1つ以上設けることを望む所管官庁は、さらにもう1つの軽度刺激性物質の区分（区分3）を利用できる。

3.2.2.2.2.3 皮膚病変の可逆性は、刺激性反応評価において考慮すべきもう1つの事項である。試験動物2匹以上で炎症が試験期間終了時まで継続する場合には、脱毛（限定領域）、過角化症、過形成および落屑を考慮に入れて、試料を刺激性物質であると考えるべきである。

3.2.2.2.2.4 試験中の動物の刺激性反応は、腐食性の場合と同様に多様である。有意な刺激性反応はあるが、陽性試験の平均スコア基準値よりも低いような例も加えられるようにするために、別の刺激性の判定基準も加えるべきである。例えば、試験動物3匹中1匹で、通常14日間の観察期間終了時においてもまだ病変が認められるなど、試験期間中を通じて平均スコアがきわめて上昇しているのが認められたならば、被験試料は刺激性物質としてよいかもしれない。他の反応でもこの判定基準が充足されることがある。ただし、その反応は化学品へのばく露によるものであることを確認すべきである。この判定基準を加えれば、本分類システムの精度は高くなる。

3.2.2.2.2.5 動物試験結果から刺激性区分(区分2)が表3.2.2に示されている。所管官庁（例：農薬の分類）によっては、軽度の刺激性区分(区分3)も利用できる。数種類の判定基準によって、この2種類の区分が区別されている（表3.2.2）。これらの区分は主として皮膚反応の重篤度に違いがある。刺激性区分の

for the irritation category is that at least 2 of 3 tested animals have a mean score of ≥ 2.3 and ≤ 4.0. For the mild irritation category, the mean score cut-off values are ≥ 1.5 and < 2.3 for at least 2 of 3 tested animals. Test materials in the irritation category are excluded from the mild irritation category.

Table 3.2.2: Skin irritation categories [a,b]

Categories	Criteria
Irritation (Category 2) (applies to all authorities)	(a) Mean score of ≥ 2.3 and ≤ 4.0 for erythema/eschar or for oedema in at least 2 of 3 tested animals from gradings at 24, 48 and 72 hours after patch removal or, if reactions are delayed, from grades on 3 consecutive days after the onset of skin reactions; or (b) Inflammation that persists to the end of the observation period normally 14 days in at least 2 animals, particularly taking into account alopecia (limited area), hyperkeratosis, hyperplasia, and scaling; or (c) In some cases where there is pronounced variability of response among animals, with very definite positive effects related to chemical exposure in a single animal but less than the criteria above.
Mild irritation (Category 3) (applies to only some authorities)	Mean score of ≥ 1.5 and < 2.3 for erythema/eschar or for oedema in at least 2 of 3 tested animals from gradings at 24, 48 and 72 hours or, if reactions are delayed, from grades on 3 consecutive days after the onset of skin reactions (when not included in the irritant category above).

[a] *Grading criteria are understood as described in OECD Test Guideline 404.*
[b] *Evaluation of a 4, 5 or 6-animal study should follow the criteria given in 3.2.5.3.3.*

3.2.2.3   *Classification based on in vitro/ex vivo data (tier 2 in figure 3.2.1)*

3.2.2.3.1   The currently available individual in vitro/*ex vivo* test methods address either skin irritation or skin corrosion, but do not address both endpoints in one single test. Therefore, classification based solely on in vitro/*ex vivo* test results may require data from more than one method. For authorities implementing Category 3 it is important to note that the currently available internationally validated and accepted in vitro/*ex vivo* test methods do not allow identification of substances classified as Category 3.

3.2.2.3.2   The classification criteria for the currently available in vitro/*ex vivo* test methods adopted by OECD in test guidelines 430, 431, 435 and 439 are described in tables 3.2.6 and 3.2.7 (see 3.2.5.3.4). Other validated in vitro/*ex vivo* test methods accepted by some competent authorities may also be considered. A competent authority may decide which classification criteria, if any, should be applied for other test methods to conclude on classification, including that a substance is not classified for effects on the skin.

3.2.2.3.3   *In vitro/ex vivo* data can only be used for classification when the tested substance is within the applicability domain of the test method(s) used. Additional limitations described in the published literature should also be taken into consideration.

3.2.2.3.4   *Skin corrosion*

3.2.2.3.4.1   Where tests have been undertaken in accordance with OECD Test Guideline 430, 431, or 435, a substance is classified for skin corrosion in category 1 (and, where possible and required into sub-categories 1A, 1B or 1C) based on the criteria in table 3.2.6 (see 3.2.5.3.4).

3.2.2.3.4.2   Some in vitro/*ex vivo* methods do not allow differentiation between sub-categories 1B and 1C (see table 3.2.6). Where sub-categories are required by competent authorities and existing in vitro/*ex vivo* data cannot distinguish between the sub-categories, additional information has to be taken into account to differentiate between these two sub-categories. Where no or insufficient additional information is available, Category 1 is applied.

3.2.2.3.4.3   A substance identified as not corrosive should be considered for classification as skin irritant.

3.2.2.3.5   *Skin irritation*

3.2.2.3.5.1   Where classification for corrosivity can be excluded and where tests have been undertaken in accordance with OECD Test Guideline 439, a substance should be considered for classification as skin irritant in Category 2 based on the criteria in table 3.2.7 (see 3.2.5.3.4).

主な分類基準は、試験動物のうち少なくとも3匹のうち2匹で平均スコアが≧2.3 かつ≦4.0 となることである。軽度刺激性の区分では、少なくとも動物3匹のうち2匹で平均スコア・カットオフ値が≧1.5 かつ<2.3 となることである。刺激性区分に分類されている試験試料は軽度刺激性区分への分類からは除外される。

表 3.2.2：皮膚刺激性の区分 [a, b]

区分	判定基準
刺激性 （区分2） （すべての所管官庁に適用）	(a) 試験動物3匹のうち少なくとも2匹で、パッチ除去後 24、48 および 72 時間における評価または反応が遅発性の場合には皮膚反応発生後3日間連続しての評価で、紅斑/痂皮または浮腫の平均スコアが≧2.3 かつ≦4.0 である、または (b) 少なくとも2匹の動物で、通常 14 日間の観察期間終了時まで炎症が残る、特に脱毛（限定領域内）、過角化症、過形成および落屑を考慮する、または (c) 動物間にかなりの反応の差があり、動物1匹で化学品ばく露に関してきわめて決定的な陽性作用が見られるが、上述の判定基準ほどではないような例もある。
軽度刺激性 （区分3） （限られた所管官庁のみに適用）	試験動物3匹のうち少なくとも2匹で、パッチ除去後 24、48 および 72 時間における評価または反応が遅発性の場合には皮膚反応発生後3日間連続しての評価で、紅斑/痂皮または浮腫の平均スコアが≧1.5 かつ<2.3 である（上述の刺激性区分には分類されない場合）。

[a] *評価基準はOECDテストガイドライン404に記載されている。*
[b] *4, 5 または6匹の動物実験の評価は 3.2.5.3.3 にある判定基準にしたがうべきである。*

3.2.2.3 *in vitro（試験管内）/ex vivo（生体外）のデータに基づく分類（図 3.2.1 の段階2）*

3.2.2.3.1 現在利用可能なそれぞれの in vitro/ex vivo 試験方法は皮膚刺激性または皮膚腐食性のどちらかを評価するが、一つの試験で両方の影響は評価しない。したがって in vitro/ex vivo 試験結果のみに基づいた分類は二つ以上の方法で得られたデータを必要とするであろう。区分3を導入する所管官庁は、現在利用可能な国際的に検証され受け入れられている in vitro/ex vivo 試験方法では、区分3と分類される物質を識別することはできないことを認識することが重要である。

3.2.2.3.2 テストガイドライン 430、431、435、および 439 として OECD によって採用されている現在利用可能な in vitro/ex vivo 試験法に対する分類基準は、表 3.2.6 および 3.2.7 に記載されている（3.2.5.3.4 参照）。一部の所管官庁によって承認された、他の有効な in vitro/ex vivo 試験方法も考慮される場合がある。所管官庁は、物質が皮膚への影響について分類されないことを含め、分類を結論付けるために、他の試験方法について分類基準がある場合、それを適用すべきかを決定することができる。

3.2.2.3.3 試験物質が用いられた試験方法の適用範囲内にある場合にのみ、*in vitro/ex vivo* のデータは分類に使用することができる。公表された文献に記載されている追加的な制限も考慮されるべきである。

3.2.2.3.4 *皮膚腐食性*

3.2.2.3.4.1 OECD テストガイドライン 430、431 または 435 にしたがって試験が実施された場合、表 3.2.6 の判定基準（3.2.5.3.4 参照）に基づいて、物質は皮膚腐食性区分1（また可能であり要求されていれば細区分 1A、1B または 1C）に分類される。

3.2.2.3.4.2 いくつかの in vitro/ex vivo 試験方法では細区分 1B および 1C（表 3.2.6 参照）を区別することはできない。細区分が所管官庁によって要求されており、既存の in vitro/ex vivo データが細区分を区別することができない場合には、これら二つの細区分を区別するための追加的な情報を考慮しなければならない。追加的な情報がないまたは不十分である場合には、区分1が適用される。

3.2.2.3.4.3 腐食性とは同定されない物質は皮膚刺激性としての分類が検討されるべきである。

3.2.2.3.5 *皮膚刺激性*

3.2.2.3.5.1 腐食性に関する分類が除外され、しかも OECD テストガイドライン 439 にしたがった試験が実施された場合には、物質は表 3.2.7 の判定基準（3.2.5.3.4 参照）に基づいて皮膚刺激性区分2としての分類が検討されるべきである。

3.2.2.3.5.2 Where competent authorities adopt category 3, it is important to note that currently available in vitro/*ex vivo* test methods for skin irritation (e.g. OECD Test Guideline 439) do not allow for classification of substances in Category 3.

3.2.2.3.6 *No classification for effect on the skin*

Where competent authorities do not adopt Category 3, a negative result in an in vitro/*ex vivo* test method for skin irritation that is validated according to international procedures, e.g. OECD Test Guideline 439, can be used to conclude as not classified for skin irritation. Where competent authorities adopt Category 3, additional information is required to differentiate between Category 3 and no classification.

3.2.2.4 *Classification based on other existing animal skin data (tier 3 in figure 3.2.1)*

Other existing skin data in animals may be used for classification, but there may be limitations regarding the conclusions that can be drawn (see 3.2.5.3.5). If a substance is highly toxic via the dermal route, an in vivo skin corrosion/irritation study may not have been conducted since the amount of test substance to be applied would considerably exceed the toxic dose and, consequently, would result in the death of the animals. When observations of skin corrosion/irritation in acute toxicity studies are made, these data may be used for classification, provided that the dilutions used and species tested are relevant. Solid substances (powders) may become corrosive or irritant when moistened or in contact with moist skin or mucous membranes. This is generally indicated in the standardised test methods. Guidance on the use of other existing skin data in animals including acute and repeated dose toxicity tests as well as other tests is provided in 3.2.5.3.5.

3.2.2.5 *Classification based on extreme pH (pH ≤ 2 or ≥ 11.5) and acid/alkaline reserve (tier 4 in figure 3.2.1)*

In general, substances with an extreme pH (pH ≤ 2 or ≥ 11.5) are expected to cause significant skin effects, especially when associated with significant acid/alkaline reserve. A substance with pH ≤ 2 or ≥ 11.5 is therefore considered to cause skin corrosion (Category 1) in this tier if it has a significant acid/alkaline reserve or if no data for acid/alkaline reserve are available. However, if consideration of acid/alkaline reserve suggests the substance may not be corrosive despite the extreme pH value, the result is considered inconclusive within this tier (see figure 3.2.1). A pH > 2 and < 11.5 is considered inconclusive and cannot be used for classification purposes. Acid/alkaline reserve and pH can be determined by different methods including those described in OECD Test Guideline 122 and Young et al. (1988), acknowledging that there are some differences between these methods (see 3.2.5.3.6). A competent authority may decide which criteria for significant acid/alkaline reserve can be applied.

3.2.2.6 *Classification based on non-test methods (tier 5 in figure 3.2.1)*

3.2.2.6.1 Classification, including the conclusion not classified, can be based on non-test methods, with due consideration of reliability and applicability, on a case-by-case basis. Non-test methods include computer models predicting qualitative structure-activity relationships (structural alerts, SAR) or quantitative structure-activity relationships (QSARs), computer expert systems, and read-across using analogue and category approaches.

3.2.2.6.2 Read-across using analogue or category approaches requires sufficiently reliable test data on similar substance(s) and justification of the similarity of the tested substance(s) with the substance(s) to be classified. Where adequate justification of the read-across approach is provided, it has in general higher weight than (Q)SARs.

3.2.2.6.3 Classification based on (Q)SARs requires sufficient data and validation of the model. The validity of the computer models and the prediction should be assessed using internationally recognized principles for the validation of (Q)SARs. With respect to reliability, lack of alerts in a SAR or expert system is not sufficient evidence for no classification.

3.2.2.6.4 For conclusions on no classification from read-across and (Q)SARs the adequacy and robustness of the scientific reasoning and of the supporting evidence should be well substantiated and normally requires multiple negative substances with good structural and physical (related to toxicokinetics) similarity to the substance being classified, as well as a clear absence of positive substances with good structural and physical similarity to the substance being classified.

3.2.2.7 *Classification based on an overall weight of evidence assessment (tier 6 in figure 3.2.1)*

3.2.2.7.1 An overall weight of evidence assessment using expert judgement is indicated where none of the previous tiers resulted in a definitive conclusion on classification. In some cases, where the classification decision was postponed until the overall weight of evidence, but no further data are available, a classification may still be possible.

3.2.2.3.5.2　所管官庁が区分3を採用している場合、現在利用可能な皮膚刺激性に関する in vitro/ex vivo 試験方法（例えば OECD テストガイドライン 439）では、物質を区分3に分類することはできないことを認識することが重要である。

3.2.2.3.6　皮膚への影響について分類されない場合

　所管官庁が区分3を採用しない場合、皮膚刺激性に関して国際的な手順にしたがって有効とされた in vitro/ex vivo 試験方法（例えば OECD テストガイドライン 439）における陰性結果を用いて、皮膚刺激性に関して区分に該当しない（分類されない）と結論付けることができる。所管官庁が区分3を採用する場合、区分3と区分に該当しない（分類されない）を区別するための追加情報が必要である。

3.2.2.4　動物におけるその他の既存の皮膚データに基づく分類（図 3.2.1 の段階 3）

　動物における他の既存の皮膚データは分類に使用することもできるが、導き出される結論に関しては制限がある場合がある（3.2.5.3.5参照）。物質が経皮的に強い毒性を持つ場合、適用される試験物質の量がかなりの毒性量を超え、動物の死に結び付くこともあるので、in vivo 皮膚腐食性/刺激性は実施されていないこともありうる。急性毒性試験において皮膚腐食性/刺激性の観察が行われていれば、使用された希釈液や試験された種が適当であるという条件で、これらのデータは分類に使用できるであろう。固体物質（粉体）は、湿った状態あるいは湿った皮膚または粘膜に接触した場合、腐食性または刺激性となることがある。これは一般に標準化された試験方法に示されている。急性および反復ばく露毒性試験を含む、動物における他の既存の皮膚データの使用に関する手引きは 3.2.5.3.5 に示されている。

3.2.2.5　極端なpH (pH≦2 または≧11.5) および酸/アルカリ予備に基づく分類（図 3.2.1 の段階 4）

　一般に、極端な pH (pH≦2 または≧11.5) の物質は、特に相当量の酸/アルカリ予備がある場合に、皮膚に明確な影響を引き起こすことが予想される。したがって、pH≦2 または≧11.5 の物質は、相当量の酸/アルカリ予備がある場合、または酸/アルカリ予備のデータが利用できない場合、この段階で皮膚腐食性（区分1）を示すと見なされる。ただし、酸/アルカリ予備を考慮して、極端な pH 値にもかかわらず物質が腐食性ではない可能性があることを示唆している場合、その結果はこの段階では決定的ではないと見なされる（図 3.2.1 参照）。2 < pH < 11.5 のときは決定的でないと見なされ、分類目的には使用できない。酸/アルカリ予備および pH は、OECD テストガイドライン 122 および Young ら (1988) を含むいくつかの異なる方法で決定され、これらの方法にはいくつかの違いがあることが認められている（3.2.5.3.6参照）。所管官庁は、相当量の酸/アルカリ予備がある場合に対してどの基準を適用できるかを決定することができる。

3.2.2.6　試験によらない方法に基づく分類（図 3.2.1 の段階 5）

3.2.2.6.1　ケースバイケースで信頼性および適用性を十分に考慮した上で、区分に該当しないという結論も含めて、試験によらない方法に基づいて分類を行うことができる。試験によらない方法には、定性的構造活性相関（構造アラート、SAR）や定量的構造活性相関（QSARs）のコンピューターモデル、コンピューターエキスパートシステム、および、類似物質およびカテゴリーアプローチを用いたリード・アクロスがある。

3.2.2.6.2　類似物質やカテゴリーアプローチを用いたリード・アクロスでは、類似物質に関する十分に信頼できる試験データとそして、試験物質と分類対象物質の類似性の正当化が必要である。リード・アクロス手法に関して十分な正当化がなされている場合には、一般には(Q)SARs よりも高い重み付けがなされる。

3.2.2.6.3　(Q)SARs に基づく分類では十分なデータとモデルの検証が必要である。コンピューターモデルとその予測の検証は、(Q)SARs の検証に関する国際的に認知された原則を用いて評価されるべきである。信頼性に関して、SAR における構造アラートまたはエキスパートシステムがないことは、区分に該当しない十分な証拠とはならない。

3.2.2.6.4　リード・アクロスおよび (Q)SAR から得られた区分に該当しない（分類されない）という結論については、科学的推論およびそれを裏付ける証拠の妥当性と頑健性が十分に実証されている必要があり、分類対象物質との構造的および物理的類似性が高い陽性物質が明らかに存在しないことを示すのと同様に、通常は分類対象物質との構造的および物理的（関連するトキシコキネティクスに）類似性が高い複数の陰性物質が必要である。

3.2.2.7　包括的な証拠の重み付け評価に基づく分類（図 3.2.1 の段階 6）

3.2.2.7.1　専門家の判断を使用した包括的な証拠の重み付け評価は、これより前の段階のいずれにおいても分類に関する最終的な結論が得られなかった場合に示される。段階5までのプロセスで分類の決定が判断できず、それ以上のデータが利用できないケースでもまだ分類が可能な場合がある。

3.2.2.7.2　　　　A substance with an extreme pH (pH ≤ 2 or ≥ 11.5) and non-significant acid/alkaline reserve (result considered inconclusive in tier 4; see 3.2.2.5) and for which no other information is available, should be classified as skin corrosion Category 1 in this tier. If inconclusive information is also available from other tiers but the overall weight of evidence assessment remains inconclusive, the extreme pH (pH ≤ 2 or ≥ 11.5) result should take precedence and the substance should be classified as skin corrosion Category 1 in this tier independently of its acid/alkaline reserve. For mixtures, the approach is different and is detailed in 3.2.3.1.3.

3.2.2.8　　　*Classification in a tiered approach (figure 3.2.1)*

3.2.2.8.1　　　　A tiered approach to the evaluation of information should be considered, where applicable (figure 3.2.1), recognizing that not all tiers as well as information within a tier may be relevant. However, all available and relevant information of sufficient quality needs to be examined for consistency with respect to the resulting classification.

3.2.2.8.2　　　　In the tiered approach (figure 3.2.1), existing human and animal data form the highest tier, followed by in vitro/*ex vivo* data, other existing animal skin data, extreme pH and acid/alkaline reserve, and finally non-test methods. Where information from data within the same tier is inconsistent and/or conflicting, the conclusion from that tier is determined by a weight of evidence assessment.

3.2.2.8.3　　　　Where information from several tiers is inconsistent and/or conflicting with respect to the resulting classification, information of sufficient quality from a higher tier is generally given a higher weight than information from a lower tier. However, when information from a lower tier would result in a stricter classification than information from a higher tier and there is concern for misclassification, then classification is determined by an overall weight of evidence assessment. For example, having consulted the guidance in 3.2.5.3 as appropriate, classifiers concerned with a negative result for skin corrosion in an in vitro/*ex vivo* study when there is a positive result for skin corrosion in other existing skin data in animals would utilise an overall weight of evidence assessment. The same would apply in the case where there is human data indicating skin irritation but positive results from an in vitro/*ex vivo* test for corrosion are also available.

3.2.2.7.2　極端な pH (pH≦2 または≧11.5) でかつ酸/アルカリ予備はそれほどあるわけではなく (段階 4 で「inconclusive (決定的でない)」と見なされた結果; 3.2.2.5 参照)、さらに他の情報が利用できない物質は、この段階で皮膚腐食性区分 1 に分類されるべきである。 他の段階でも決定的でない情報が利用可能であるが、包括的な証拠の重み付け評価で結論ができない場合、極端な pH (pH≦2 または≧11.5) の結果が優先され、その物質は酸/アルカリ予備とは無関係にこの段階で皮膚腐食性区分 1 に分類されるべきである。混合物の場合、アプローチは異なり、3.2.3.1.3 に詳述されている。

3.2.2.8　*段階的アプローチによる分類（図 3.2.1）*

3.2.2.8.1　すべての段階および段階内の情報が関連するとは限らないことを認識した上で、該当する場合、情報を評価するための段階的アプローチを考慮するものとする (図 3.2.1)。ただし、十分な質の利用可能な関連情報はすべて、分類結果に関して一貫性があるかどうかを精査される必要がある。

3.2.2.8.2　段階的アプローチ (図 3.2.1) においては、既存のヒトおよび標準的な動物データが最上位にあり、次いで in vitro/*ex vivo* データ、その他の既存の動物皮膚データ、極端な pH および酸/アルカリ予備、そして最後に試験によらない方法となる。同じ段階のデータからの情報に一貫性がないおよび/または矛盾している場合には、この段階での結論は証拠の重み付け評価によって決定される。

3.2.2.8.3　複数の段階からの情報が、分類結果に対して一貫性がないおよび/または矛盾している場合には、一般により上位の段階での十分な質の情報は、下位の段階での情報よりも、高い重みづけがなされる。しかし、下位の段階からの情報がより上位の段階からの情報よりも厳しい分類結果となり、誤分類の懸念がある場合は、分類は包括的な証拠の重み付け評価によって決定される。例えば、3.2.5.3 における手引きを適宜参照し、in vitro/*ex vivo* 試験での皮膚腐食性が陰性であるにもかかわらず、他の既存の動物における皮膚のデータで皮膚腐食性の結果が陽性であることに懸念を持つ分類者は、包括的な証拠の重み付け評価を活用するであろう。また、皮膚刺激性を示すヒトのデータがあるが、in vitro/*ex vivo* 試験では腐食性の陽性結果が得られている場合にも同じことが当てはまる。

Figure 3.2.1: Application of the tiered approach for skin corrosion and irritation[a]

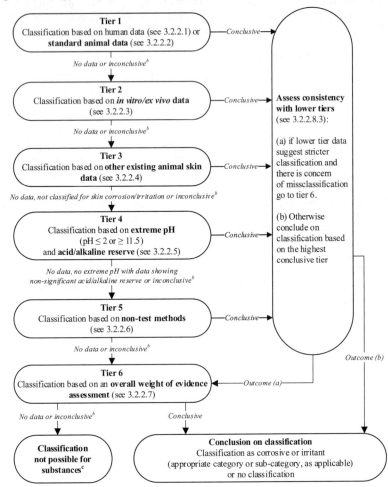

[a] Before applying the approach, the explanatory text in 3.2.2.8 as well as the guidance in 3.2.5.3 should be consulted. Only adequate and reliable data of sufficient quality should be included in applying the tiered approach.

[b] Information may be inconclusive for various reasons, e.g.:
- The available data may be of insufficient quality, or otherwise insufficient/inadequate for the purpose of classification, e.g. due to quality issues related to experimental design and/or reporting;
- The available data may be insufficient to conclude on the classification, e.g. they might be adequate to demonstrate irritancy, but inadequate to demonstrate absence of corrosivity;
- Where competent authorities make use of the mild skin irritation Category 3, the available data may not be capable of distinguishing between Category 3 and Category 2, or between Category 3 and no classification;
- The method used to generate the available data may not be suitable for concluding on no classification (see 3.2.2. and 3.2.5.3 for details). Specifically, in vitro/ex vivo and non-test methods need to be validated explicitly for this purpose.

[c] For mixtures, the flow chart in figure 3.2.2 should be followed.

図 3.2.1：皮膚腐食性および刺激性に関する段階的アプローチの適用 [a]

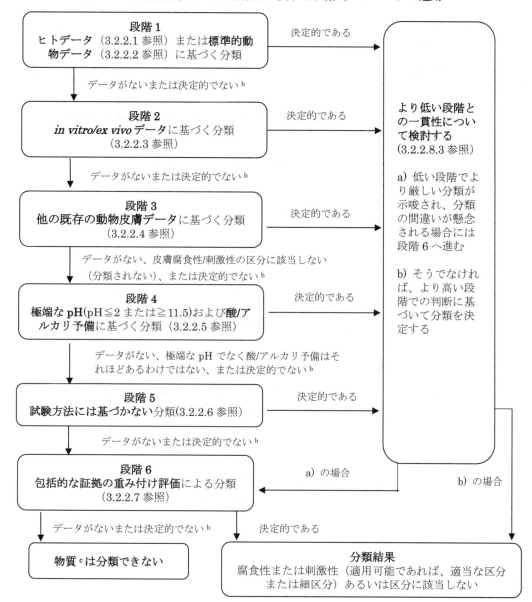

[a] アプローチを適用する前に、手引き 3.2.5.3 と同様に 3.2.2.8 にある説明文章を参照するべきである。満足できる質の適切かつ信頼できるデータのみが段階的アプローチに使用されるべきである。

[b] 情報が決定的でない理由はさまざまあろう、例えば；
- 利用可能なデータの質が十分ではない、すなわち分類の目的には不十分/不適切かもしれない、例えば実験デザインおよび/または報告に関連した質の問題；
- 利用可能なデータが分類を決定するのに不十分なこともあろう、例えば刺激性は十分に示しているが、腐食性がないことを示すには不十分である；
- 所管官庁が皮膚軽度刺激性区分 3 を利用する場合、利用可能なデータは区分 3 および区分 2 を、あるいは区分 3 および区分に該当しないを区別できないかもしれない。
- 利用可能なデータを採取した方法が、区分に該当しないという決定には適当ではないかもしれない（詳細は 3.2.2 および 3.2.5.3 参照）。特に in vitro/ex vivo および試験方法によらない場合には、この目的に関して明確に検証される必要がある。

[c] 混合物の場合、図 3.2.2 のフローチャートに従うこと。

3.2.3 Classification criteria for mixtures

The approach to classification for skin corrosion/irritation is tiered and is dependent upon the amount of information available for the mixture itself and for its ingredients. The flow chart of figure 3.2.2 below outlines the process to be followed.

Figure 3.2.2: Tiered approach to classification of mixtures for skin corrosion/irritation

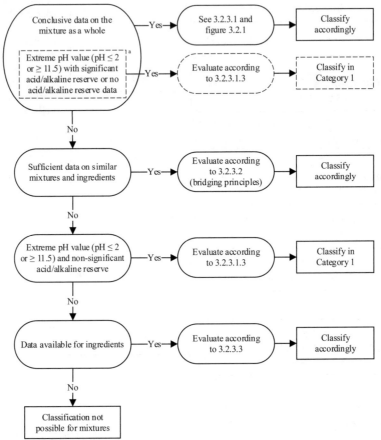

[a] The dashed boxes represent an individual tier within conclusive data on the mixture as a whole. However, in contrast to substances, mixtures having an extreme pH value (pH ≤ 2 or ≥ 11.5) and non-significant acid/alkaline reserve but no other conclusive data on the mixture as a whole, or no conclusive weight of evidence assessment from all available data on the mixture as a whole, are not conclusive within the tiers for conclusive data on the mixture as a whole. Such mixtures should be first evaluated according to the bridging principles before the extreme pH value is considered as conclusive for classification.

3.2.3 混合物の分類基準

皮膚腐食性/刺激性の分類方法は段階的であり、混合物自体とその成分について利用可能な情報の量に依存する。以下の図 3.2.2 のフローチャートは、従うべきプロセスの概要を示す。

図 3.2.2: 皮膚腐食性/刺激性に対する混合物の分類の段階的アプローチ

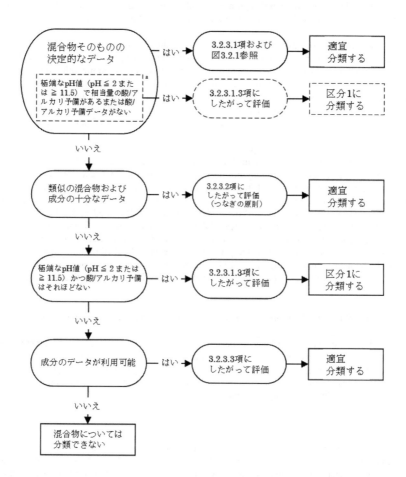

[a] 破線のボックスは、混合物そのものの決定的なデータのうちの個々の段階を表す。ただし、物質とは対照的に、極端なpH値（pH ≦ 2 または ≧ 11.5）かつ酸/アルカリ予備はそれほどないが、混合物そのものに関する他の決定的なデータがない、または混合物そのものに関するすべての利用可能なデータから決定的な証拠の重み付け評価がない混合物は、混合物そのものに関する決定的なデータの段階においては決定的であるとは言えない。そのような混合物は、極端なpH値が分類の決定的なものと見なされる前に、まずつなぎの原則に従って評価されなければならない。

- 133 -

3.2.3.1 *Classification of mixtures when data are available for the complete mixture*

3.2.3.1.1 In general, the mixture should be classified using the criteria for substances, taking into account the tiered approach to evaluate data for this hazard class (as illustrated in figure 3.2.1) and 3.2.3.1.2 and 3.2.3.1.3 below. If classification is not possible using the tiered approach, then the approach described in 3.2.3.2 (bridging principles), or, if that is not applicable 3.2.3.3 (classification based on ingredients) should be followed.

3.2.3.1.2 In vitro/*ex vivo* test methods validated according to international procedures may not have been validated using mixtures; although these methods are considered broadly applicable to mixtures, they can only be used for classification of mixtures when all ingredients of the mixture fall within the applicability domain of the test method(s) used. Specific limitations regarding applicability domains are described in the respective test methods and should be taken into consideration as well as any further information on such limitations from the published literature. Where there is reason to assume or evidence indicating that the applicability domain of a particular test method is limited, data interpretation should be exercised with caution, or the results should be considered not applicable.

3.2.3.1.3 A mixture with an extreme pH (pH ≤ 2 or ≥ 11.5) is considered corrosive (Category 1) in tier 4 if it has a significant acid/alkaline reserve or if no data for acid/alkaline reserve are available. However, if consideration of acid/alkaline reserve suggests the mixture may not be corrosive despite the extreme pH value, the result is considered inconclusive within tier 4 (see figure 3.2.1). If the overall weight of evidence assessment remains inconclusive or no data other than pH and acid/alkaline reserve are available, mixtures with an extreme pH (pH ≤ 2 or ≥ 11.5) and non-significant acid/alkaline reserve should be assessed using the bridging principles described in 3.2.3.2. If the bridging principles cannot be applied, mixtures with an extreme pH (pH ≤ 2 or ≥ 11.5) and non-significant acid/alkaline reserve should be classified as skin Category 1 (see figure 3.2.2). A pH > 2 and < 11.5 is considered inconclusive and cannot be used for classification purposes. Acid/alkaline reserve and pH can be determined by different methods including those described in OECD Test Guideline 122 and Young et al. (1988), acknowledging that there are some differences between these methods (see 3.2.5.3.6). A competent authority may decide which criteria for significant acid/alkaline reserve can be applied.

3.2.3.2 *Classification of mixtures when data are not available for the complete mixture: bridging principles*

3.2.3.2.1 Where the mixture itself has not been tested to determine its skin corrosion/ irritation potential, but there are sufficient data on both the individual ingredients and similar tested mixtures to adequately characterize the hazards of the mixture, these data will be used in accordance with the following agreed bridging principles. This ensures that the classification process uses the available data to the greatest extent possible in characterizing the hazards of the mixture without the necessity for additional testing in animals.

3.2.3.2.2 *Dilution*

If a tested mixture is diluted with a diluent which has an equivalent or lower skin corrosivity/irritancy classification than the least skin corrosive/irritant original ingredient and which is not expected to affect the skin corrosivity/irritancy of other ingredients, then the new diluted mixture may be classified as equivalent to the original tested mixture. Alternatively, the method explained in 3.2.3.3 could be applied.

3.2.3.2.3 *Batching*

The skin corrosion/irritation potential of a tested production batch of a mixture can be assumed to be substantially equivalent to that of another untested production batch of the same commercial product when produced by or under the control of the same manufacturer, unless there is reason to believe there is significant variation such that the skin corrosion/irritation potential of the untested batch has changed. If the latter occurs, a new *classification is necessary.*

3.2.3.2.4 *Concentration* of mixtures of the highest corrosion/irritation category

If a tested mixture classified in the highest sub-category for skin corrosion is concentrated, the more concentrated untested mixture should be classified in the highest corrosion sub-category without additional testing. If a tested mixture classified for skin irritation (Category 2) is concentrated and does not contain skin corrosive ingredients, the more concentrated untested mixture should be classified for skin irritation (Category 2) without additional testing.

3.2.3.2.5 *Interpolation within one hazard category*

For three mixtures (A, B and C) with identical ingredients, where mixtures A and B have been tested and are in the same skin corrosion/irritation hazard category, and where untested mixture C has the same toxicologically active ingredients as mixtures A and B but has concentrations of toxicologically active ingredients intermediate to the

3.2.3.1 混合物そのもののデータが利用できる場合の混合物の分類

3.2.3.1.1　一般に混合物は、物質に関する判定基準を用いて、本有害性クラスに関するデータを評価するための段階的アプローチ（図 3.2.1 に示されている）および下記の 3.2.3.1.2 および 3.2.3.1.3 を考慮に入れて分類されるべきである。段階的アプローチを使用して分類ができない場合には、3.2.3.2（つなぎの原則）に記載されている方法、またはこれが適用できない場合には、3.2.3.3.（成分に基づく分類）にしたがうべきである。

3.2.3.1.2　国際的な手順にしたがって検証された in vitro/ex vivo 試験法は混合物の使用では検証されていないかもしれない；これらの方法は混合物に広く適用可能であると考えられているが、混合物のすべての成分が使用された試験方法の適用範囲に該当する場合にのみ、これらは混合物の分類に使用することができる。それぞれの試験方法において適用範囲に関して明確な制限が記載されており、公表されている文献でのそのような制限に関するさらなる情報と同様に、これらは考慮されるべきである。特定の試験方法の適用範囲を制限していることを示す前提または証拠を示す理由がある場合には、データの解釈には注意を払わなければならず、あるいは結果は適用できないと考えるべきである。

3.2.3.1.3　極端な pH（pH≦2 または≧11.5）の混合物は、相当量の酸/アルカリ予備がある場合、または酸/アルカリ予備のデータが利用できない場合、段階 4 で腐食性（区分 1）と見なされる。ただし、酸/アルカリ予備を考慮して、極端な pH 値にもかかわらず混合物が腐食性ではない可能性があることを示唆している場合、結果は段階 4 において決定的でないと見なされる（図 3.2.1 参照）。包括的な証拠の重み付け評価が決定的でない場合、または pH および酸/アルカリ予備以外のデータが利用できない場合、極端な pH（pH≦2 または≧11.5）かつ酸/アルカリ予備はそれほどない混合物は、3.2.3.2 で説明されているつなぎの原則を使用して評価する必要がある。つなぎの原則を適用できない場合、極端な pH（pH≦2 または≧11.5）かつ酸/アルカリ予備はそれほどない混合物は、皮膚区分 1 に分類する必要がある（図 3.2.2 参照）。2 < pH < 11.5 の場合は決定的でないと見なされ、分類目的には使用できない。酸/アルカリ予備および pH は、OECD テストガイドライン 122 および Young ら（1988）に記載されている方法を含む異なる方法で測定することができるが、これらの方法の間にいくつかの相違があることを認めている（3.2.5.3.6 参照）。所管官庁は、相当量の酸/アルカリ予備がある場合にどの基準を適用できるかを決定することができる。

3.2.3.2 混合物そのものについてデータが利用できない場合の混合物の分類：つなぎの原則（bridging principles）

3.2.3.2.1　混合物そのものは皮膚の腐食性/刺激性があるかどうかを決定する試験がなされていないが、当該混合物の有害性を適切に特定するための、個々の成分および類似の試験された混合物の両方に関して十分なデータがある場合、これらのデータは以下の合意されたつなぎの原則にしたがって利用される。これによって分類手順において、動物試験を追加する必要もなく、混合物の有害性判定に利用可能なデータを可能な限り最大限に用いられるようになる。

3.2.3.2.2　希釈

試験された混合物が皮膚腐食性/刺激性の最も弱い元の成分に比べて同等以下の皮膚腐食性/刺激性分類に属する物質で希釈され、その物質が他の成分の皮膚腐食性/刺激性に影響を与えないことが予想されれば、新しい希釈された混合物は試験された元の混合物と同等として分類してもよい。あるいは、3.2.3.3 節で説明する方法も適用できる。

3.2.3.2.3　製造バッチ

混合物の試験されていない製造バッチに皮膚の腐食性/刺激性があるかどうかは、同じ製造者によって、またはその管理下で生産された同じ商品の試験された別のバッチの毒性と本質的に同等とみなすことができる。ただし、試験されていないバッチの皮膚腐食性/刺激性が変化するような有意な変動があると考えられる理由がある場合はこの限りではない。このような場合には、新しい分類が必要である。

3.2.3.2.4　最も強い腐食性/刺激性区分の混合物の濃縮

皮膚腐食性について最も強い細区分に分類された試験混合物が濃縮された場合には、より濃度が高い試験されていない混合物は追加試験なしで最も強い腐食性の細区分に分類するべきである。皮膚刺激性に分類された試験混合物が濃縮され、皮膚腐食性成分を含まなければ、より濃度が高い試験されていない混合物は追加試験なしで刺激性区分（区分 2）に分類するべきである。

3.2.3.2.5　1 つの有害性区分の中での内挿

3 つの混合物（A、B および C）は同じ成分を持ち、A と B は試験され同じ皮膚腐食性/刺激性の区分にある。試験されていない混合物 C は混合物 A および B と同じ毒性学的に活性な成分を持ち、毒性学的に

concentrations in mixtures A and B, then mixture C is assumed to be in the same skin corrosion/irritation category as A and B.

3.2.3.2.6   *Substantially similar mixtures*

Given the following:

(a)　　Two mixtures:　　(i)　　A + B;
　　　　　　　　　　　　(ii)　　C + B;

(b)   The concentration of ingredient B is essentially the same in both mixtures;

(c)   The concentration of ingredient A in mixture (i) equals that of ingredient C in mixture (ii);

(d)   Data on skin corrosion/irritation for A and C are available and substantially equivalent, i.e. they are in the same hazard category and are not expected to affect the skin corrosion/irritation potential of B.

If mixture (i) or (ii) is already classified based on test data, then the other mixture can be classified in the same hazard category.

3.2.3.2.7   *Aerosols*

An aerosolized form of a mixture may be classified in the same hazard category as the tested non-aerosolized form of the mixture provided that the added propellant does not affect the skin corrosion/irritation properties of the mixture upon spraying.

3.2.3.3   *Classification of mixtures when data are available for all ingredients or only for some ingredients of the mixture*

3.2.3.3.1   In order to make use of all available data for purposes of classifying the skin corrosion/irritation hazards of mixtures, the following assumption has been made and is applied where appropriate in the tiered approach for mixtures (see 1.3.2.3):

The "relevant ingredients" of a mixture are those which are present in concentrations ≥ 1 % (w/w for solids, liquids, dusts, mists and vapours and v/v for gases), unless there is a presumption (e.g. in the case of corrosive ingredients) that an ingredient present at a concentration < 1 % can still be relevant for classifying the mixture for skin corrosion/irritation.

3.2.3.3.2   In general, the approach to classification of mixtures as corrosive or irritant to skin when data are available on the ingredients, but not on the mixture as a whole, is based on the theory of additivity, such that each skin corrosive or irritant ingredient contributes to the overall corrosive or irritant properties of the mixture in proportion to its potency and concentration. A weighting factor of 10 is used for corrosive ingredients when they are present at a concentration below the concentration limit for classification with Category 1 but are at a concentration that will contribute to the classification of the mixture as an irritant. The mixture is classified as corrosive or irritant to skin when the sum of the concentrations of such ingredients exceeds a cut-off value/concentration limit.

3.2.3.3.3   Table 3.2.3 below provides the cut-off value/concentration limits to be used to determine if the mixture is considered to be corrosive or irritant to the skin.

3.2.3.3.4   Particular care must be taken when classifying mixtures containing certain types of substances such as acids and bases, inorganic salts, aldehydes, phenols, and surfactants. The approach explained in 3.2.3.3.1 and 3.2.3.3.2 might not work given that many such substances are corrosive or irritant at concentrations < 1 %. For mixtures containing strong acids or bases the pH should be used as classification criterion (see 3.2.3.1.3) since extreme pH will be a better indicator of corrosion than the concentration limits in table 3.2.3. A mixture containing corrosive or irritant ingredients that cannot be classified based on the additivity approach shown in table 3.2.3, due to chemical characteristics that make this approach unworkable, should be classified as skin corrosion Category 1 if it contains ≥ 1 % of a corrosive ingredient and as skin irritation Category 2 or Category 3 when it contains ≥ 3 % of an irritant ingredient. Classification of mixtures with ingredients for which the approach in table 3.2.3 does not apply is summarized in table 3.2.4 below.

3.2.3.3.5   On occasion, reliable data may show that the skin corrosion/irritation of an ingredient will not be evident when present at a level above the generic cut-off values/concentration limits mentioned in tables 3.2.3 and 3.2.4. In these

活性な成分の濃度が混合物 A と B の中間である場合、混合物 C は、A および B と同じ皮膚腐食性/刺激性の区分であると推定される。

3.2.3.2.6 *本質的に類似した混合物*

次を仮定する：

(a) 2 つの混合物： (i)   A+B
                    (ii)   C+B
(b) 成分 B の濃度は、両方の混合物で本質的に同じである；
(c) 混合物(i)の成分 A の濃度は、混合物(ii)の成分 C の濃度に等しい；
(d) A と C の皮膚腐食性/刺激性に関するデータは利用でき、実質的に同等であり、すなわち A と C は同じ有害性区分に属し、かつ、B の皮膚腐食性/刺激性かどうかには影響を与えることはないと判断される。

混合物(i)または(ii)が既に試験によって分類されている場合には、他方の混合物は同じ有害性区分に分類することができる。

3.2.3.2.7 *エアゾール*

エアゾール形態の混合物は、添加された噴射剤が噴霧時に混合物の皮膚腐食性/刺激性の性質に影響しないという条件下では、試験された非エアゾール形態の混合物と同じ有害性区分に分類してよい。

3.2.3.3 *混合物の全成分についてまたは一部の成分だけについてデータが利用できる場合の混合物の分類*

3.2.3.3.1 混合物の皮膚の皮膚腐食性/刺激性を分類する目的のため利用可能なすべてのデータを使用するために、下記の仮定が設定され、混合物に対する段階的なアプローチ（1.3.2.3 参照）で適宜その仮定が適用される。

混合物の「考慮すべき成分」とは、≧1%の濃度（固体、液体、粉塵、ミストおよび蒸気については重量/重量、気体については体積/体積）で存在するものである。ただし、（例えば腐食性の成分の場合に）＜1%の濃度で存在する成分が、なお皮膚腐食性/刺激性についての分類に関係する可能性があるという前提がある場合はこの限りではない。

3.2.3.3.2 一般的に、各成分のデータは利用可能であるが、混合物そのもののデータがない場合、皮膚への腐食性あるいは刺激性として混合物を分類する方法は加成性の理論に基づいている。すなわち、皮膚への腐食性あるいは刺激性の各成分は、その程度および濃度に応じて、混合物そのものの皮膚への腐食性あるいは刺激性に寄与していると考える。腐食性成分が区分 1 と分類できる濃度以下で、しかし混合物を刺激性に分類するのに寄与する濃度で含まれる場合には、加重係数として 10 を用いる。各成分の濃度の合計が分類基準となるカットオフ値/濃度限界を超えた場合、その混合物は腐食性ないし刺激性として分類される。

3.2.3.3.3 表 3.2.3 に混合物が皮膚に対する腐食性あるいは刺激性に分類されると考えるべきかどうかを決定するために使用されるカットオフ値/濃度限界を示した。

3.2.3.3.4 酸、塩基、無機塩、アルデヒド類、フェノール類および界面活性剤のような特定の種類の物質を含む混合物を分類する場合には特別の注意を払わなければならない。これらの化合物の多くは ＜ 1%の濃度であっても腐食性ないし刺激性を示す場合があるので、3.2.3.3.1 および 3.2.3.3.2 に記述した方法は機能しないであろう。強酸または強アルカリを含む混合物に関して、極端な pH は表 3.2.3 における濃度限界よりも、腐食性のより適した指標であるから、分類基準として pH を使用すべきである（3.2.3.1.3 参照）。また、刺激性あるいは腐食性成分を含む混合物は、化学品の特性により、表 3.2.3 に示された加成方式で分類できない場合で≧1%の腐食性成分を含む場合には、皮膚腐食性区分 1 に、また≧3%の刺激性成分を含む場合は皮膚刺激性区分 2 または区分 3 に分類する。表 3.2.3 における方法が適用できない混合物の分類は表 3.2.4 にまとめられている。

3.2.3.3.5 時には、表 3.2.3 から 3.2.4 に示されている一般的なカットオフ値/濃度限界レベル以上の濃度であっても、成分の皮膚腐食性/刺激性の影響を否定する信頼できるデータがある場合がある。この場合に

cases, the mixture could be classified according to those data (see also 1.3.3.2). On occasion, when it is expected that the skin corrosion/irritation of an ingredient will not be evident when present at a level above the cut-off values/concentration limits mentioned in tables 3.2.3 and 3.2.4, testing of the mixture may be considered.

3.2.3.3.6 If there are data showing that (an) ingredient(s) may be corrosive or irritant to skin at a concentration of < 1 % (corrosive) or < 3 % (irritant), the mixture should be classified accordingly (see also *Classification of hazardous substances and mixtures – Use of cut-off values/Concentration limits* (1.3.3.2)).

Table 3.2.3: Concentration of ingredients of a mixture classified as skin Category 1, 2 or 3 that would trigger classification of the mixture as hazardous to skin (Category 1, 2 or 3)

Sum of ingredients classified as:	Concentration triggering classification of a mixture as:		
	Skin corrosive	Skin irritant	
	Category 1 (see note below)	Category 2	Category 3
Skin Category 1	≥ 5 %	≥ 1 % but < 5 %	
Skin Category 2		≥ 10 %	≥ 1 % but < 10 %
Skin Category 3			≥ 10 %
(10 × skin Category 1) + skin Category 2		≥ 10 %	≥ 1 % but < 10 %
(10 × skin Category 1) + skin Category 2 + skin Category 3			≥ 10 %

NOTE: *Where the sub-categories of skin Category 1 (corrosive) are used, the sum of all ingredients of a mixture classified as sub-category 1A, 1B or 1C respectively, should each be ≥ 5 % in order to classify the mixture as either skin sub-category 1A, 1B or 1C. Where the sum of 1A ingredients is < 5 % but the sum of 1A + 1B ingredients is ≥ 5 %, the mixture should be classified as sub-category 1B. Similarly, where the sum of 1A + 1B ingredients is < 5 % but the sum of 1A + 1B + 1C ingredients is ≥ 5 % the mixture should be classified as sub-category 1C. Where at least one relevant ingredient in a mixture is classified as Category 1 without subcategorization, the mixture should be classified as Category 1 without subcategorization if the sum of all ingredients corrosive to skin is ≥ 5 %.*

Table 3.2.4: Concentration of ingredients of a mixture when the additivity approach does not apply, that would trigger classification of the mixture as hazardous to skin

Ingredient:	Concentration	Mixture classified as: Skin
Acid with pH ≤ 2	≥ 1 %	Category 1
Base with pH ≥ 11.5	≥ 1 %	Category 1
Other corrosive (Category 1) ingredient	≥ 1 %	Category 1
Other irritant (Category 2/3) ingredient, including acids and bases	≥ 3 %	Category 2/3

3.2.4 Hazard communication

General and specific considerations concerning labelling requirements are provided in *Hazard communication: Labelling* (chapter 1.4). Annex 1 contains summary tables about classification and labelling. Annex 3 contains examples of precautionary statements and pictograms which can be used where allowed by the competent authority. Table 3.2.5 presents specific label elements for substances and mixtures classified into this hazard class based on the criteria in this chapter.

は、混合物はそのデータに基づき分類を行う（1.3.3.2 参照）。また表 3.2.3 から 3.2.4 に示されているカットオフ濃度レベル以上の濃度であっても、成分の皮膚腐食性/刺激性がないと予想される場合は、混合物そのものでの試験実施を検討してもよい。

3.2.3.3.6　ある成分に関して腐食性の場合 ＜ 1％、刺激性の場合 ＜ 3％の濃度で皮膚に対して腐食性/刺激性であることを示すデータがある場合には、その混合物はしかるべく分類されるべきである（*危険有害な物質および混合物の分類－カットオフ値/濃度限界の使用（1.3.3.2）参照*）。

表 3.2.3：皮膚区分 1、2 または 3 として分類される混合物成分の濃度、混合物を皮膚有害性と分類する際の基準（区分 1、2 または 3）

各成分の合計による分類	混合物を分類するための成分濃度		
	皮膚腐食性	皮膚刺激性	
	区分1 （下注記参照）	区分2	区分3
皮膚区分1	≧5％	＜5％、≧1％	
皮膚区分2		≧10％	＜10％、≧1％
皮膚区分3			≧10％
(10×皮膚区分1)＋ 皮膚区分2		≧10％	＜10％、≧1％
(10×皮膚区分1)＋ 皮膚区分2＋ 皮膚区分3			≧10％

注記：皮膚区分1（腐食性）の細区分が用いられる場合、混合物を1A、1B、1Cに分類するためには、皮膚区分1A、1B、1Cと分類されている混合物の成分の合計が、各々≧5％であるべきである。1Aの対象成分となる濃度が ＜ 5％の場合で1A＋1Bの濃度が≧5％の場合には、1Bと分類すべきである。同様に1A＋1Bの対象成分となる濃度が ＜ 5％の場合でも 1A＋1B＋1C の合計が≧5％であれば1Cに分類する。混合物の少なくとも1つの成分が細区分なしに区分1に分類されている場合には、皮膚に対して腐食性である成分の合計が≧5％である場合、混合物は細区分なしに区分1と分類されるべきである。

表 3.2.4：加成方式を適用しない混合物成分の濃度、混合物を皮膚有害性と分類する際の基準

成分	濃度	混合物の分類：皮膚
酸　　pH≦2	≧1％	区分1
塩基　pH≧11.5	≧1％	区分1
その他の腐食性（区分1）成分	≧1％	区分1
その他の刺激性（区分2/3）成分 酸、塩基を含む	≧3％	区分2/3

3.2.4　危険有害性情報の伝達

表示要件に関する通則および細則は、*危険有害性に関する情報の伝達：表示*（第1.4章）に定める。附属書1に分類と表示に関する概要表を示す。附属書3には、注意書きおよび所管官庁が許可した場合に使用可能な注意絵表示の例を示す。表 3.2.5 は、本章の判定基準に基づいて、この有害性クラスに分類される物質および混合物の具体的なラベル要素を示す。

Table 3.2.5: Label elements for skin corrosion/irritation

	Category 1			Category 2	Category 3
	1 A	**1 B**	**1 C**		
Symbol	Corrosion	Corrosion	Corrosion	Exclamation mark	*No symbol*
Signal word	Danger	Danger	Danger	Warning	Warning
Hazard statement	Causes severe skin burns and eye damage	Causes severe skin burns and eye damage	Causes severe skin burns and eye damage	Causes skin irritation	Causes mild skin irritation

3.2.5   Decision logics and guidance

The decision logics which follow are not part of the harmonized classification system but are provided here as additional guidance. It is strongly recommended that the person responsible for classification study the criteria before and during use of the decision logics.

3.2.5.1   *Decision logic 3.2.1 for skin corrosion/irritation*

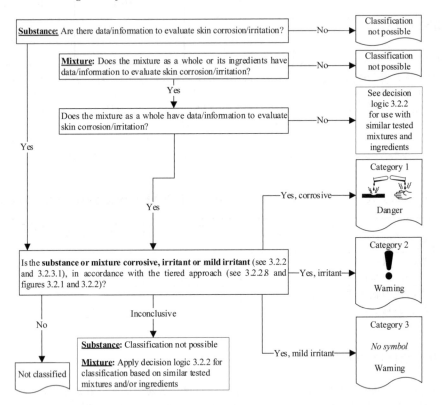

表 3.2.5：皮膚腐食性/刺激性のラベル要素

	区分 1			区分 2	区分 3
	1A	1B	1C		
シンボル	腐食性	腐食性	腐食性	感嘆符	シンボルなし
注意喚起語	危険	危険	危険	警告	警告
危険有害性情報	重篤な皮膚の薬傷および眼の損傷	重篤な皮膚の薬傷および眼の損傷	重篤な皮膚の薬傷および眼の損傷	皮膚刺激	軽度の皮膚刺激

3.2.5 判定論理および手引き

以下の判定論理は、調和分類システムには含まれないが、追加的な手引きとしてここに示す。分類の責任者に対し、この判定論理を使用する前および使用する際に判定基準についてよく調べ理解することを強く勧める。

3.2.5.1　*判定論理 3.2.1　皮膚腐食性/刺激性*

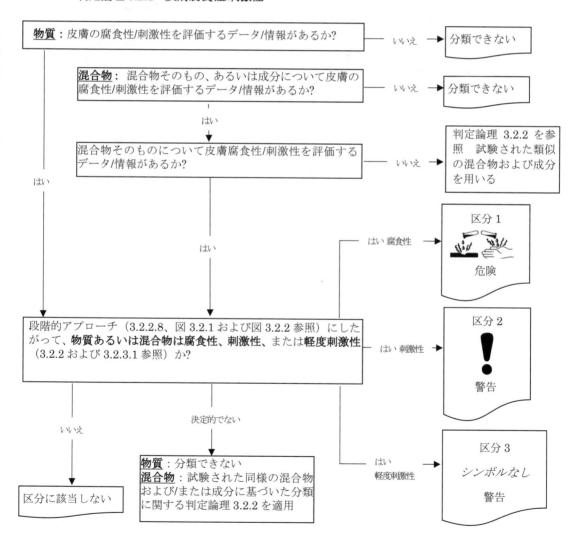

- 137 -

3.2.5.2 *Decision logic 3.2.2 for skin corrosion/irritation*

Classification of mixtures on the basis of information/data on similar tested mixtures and/or ingredients

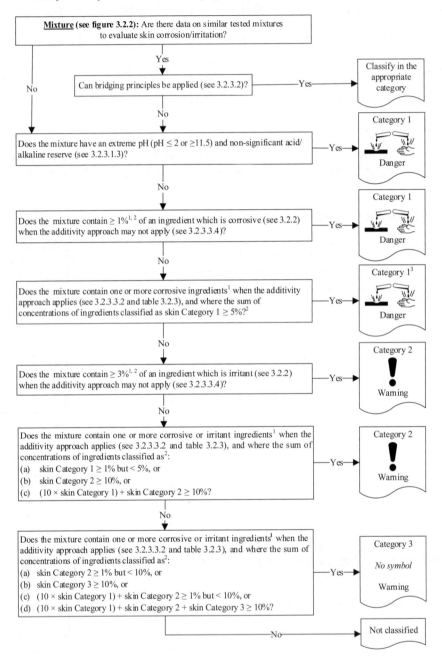

---

[1] *Where relevant < 1%, see 3.2.3.3.1.*

[2] *For specific concentration limits, see 3.2.3.3.5 and 3.2.3.3.6. See also chapter 1.3, paragraph 1.3.3.2 for "Use of cut-off values/concentration limits".*

[3] *See note to table 3.2.3 for details on use of Category 1 sub-categories.*

3.2.5.2　判定論理 3.2.2　皮膚腐食性/刺激性

試験された同様な混合物および/または成分に基づいた情報/データによる混合物の分類

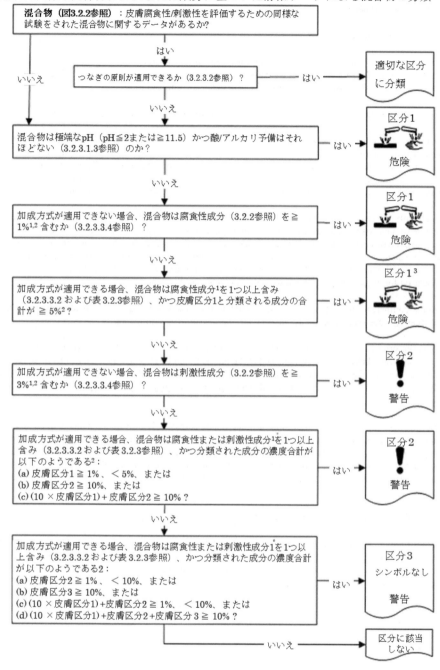

---

[1]　考慮すべき成分 ＜1% の場合、3.2.3.3.1 参照。

[2]　特定の濃度限界については 3.2.3.3.5 および 3.2.3.3.6 参照、「カットオフ値/濃度限界の使用」については第 1.3 章 1.3.3.2 も参照。

[3]　区分 1 の細区分の使用の詳細については表 3.2.3 の注記を参照。

3.2.5.3 *Background guidance*

3.2.5.3.1 *Relevant guidance documents*

Helpful information on the strengths and weaknesses of the different test and non-test methods, as well as useful guidance on how to apply a weight of evidence assessment, is provided in OECD Guidance Document 203 on an integrated approach on testing and assessment (IATA) for skin corrosion and irritation.

3.2.5.3.2 *Guidance on the use of human data for classification as skin corrosion or skin irritation*

3.2.5.3.2.1 Human data generally refers to two types of data: prior human experience (e.g. published case studies from occupational, consumer, transport, emergency response scenarios, epidemiological studies) or from human tests (e.g. clinical trials, dermal patch test). Relevant, reliable and good quality human data is generally given high weight for classification. However, human data may have limitations. Further details on the strengths and limitations of human data for skin irritation/corrosion can be found in OECD Guidance Document 203 (section III. A, Part 1, Module 1).

3.2.5.3.2.2 Generally, Human Patch Tests (HPT) are performed to discriminate between irritant and non-irritant substances. Application of a corrosive substance to human skin is generally avoided. Therefore, another test is normally performed in advance to exclude corrosivity. The HPT alone does not normally discriminate between irritant and corrosive substances. In rare circumstances, there may be HPT data that can be used for classification as corrosive (e.g. application of an HPT after a false negative in vitro test). However, the combination of an HPT and sufficient other information on skin corrosion can be used for classification within a weight of evidence assessment.

3.2.5.3.2.3 Some competent authorities do not allow HPT testing solely for hazard identification (see 1.3.2.4.7) while some competent authorities recognize the use of HPT for classification as skin irritant.

3.2.5.3.2.4 Specific criteria for HPT results leading to classification as Category 2 (skin irritation), Category 3 (mild irritation) or not classified, have not been established at the international level. Therefore, the results of an HPT are generally used within a weight of evidence assessment. However, some competent authorities may provide specific guidance. A clearly negative result in an HPT with a sufficient number of volunteers after exposure to the undiluted substance for 4 hours can justify no classification.

3.2.5.3.2.5 Human case reports may be used for classification as corrosive if irreversible damage to the skin was observed. There are no internationally accepted classification criteria for irritation. Therefore, where competent authorities have not provided specific guidance on this matter, expert judgement may be required to evaluate whether the exposure duration and any available long-term follow-up information are sufficient to allow for a conclusion on classification. Cases resulting in irritation or no effects may not be conclusive on their own but can be used in a weight of evidence assessment.

3.2.5.3.3 *Classification based on standard animal tests with more than 3 animals*

3.2.5.3.3.1 Classification criteria for the skin and eye hazard classes are detailed in the GHS in terms of a 3-animal test. It has been identified that some older test methods may have used up to 6 animals. However, the GHS criteria do not specify how to classify based on existing data from tests with more than 3 animals. Guidance on how to classify based on existing data from studies with 4 or more animals is given in the following paragraphs.

3.2.5.3.3.2 Classification criteria based on a 3-animal test are detailed in 3.2.2.2. Evaluation of a 4, 5 or 6-animal study should follow the criteria in the following paragraphs, depending on the number of animals tested. Scoring for erythema/eschar and oedema should be performed at 24, 48 and 72 hours after exposure or, if reactions are delayed, from grades on 3 consecutive days after the onset of skin reactions.

3.2.5.3.3.3 In the case of a study with 6 animals the following principles apply:

(a) The substance or mixture is classified as skin corrosion Category 1 if destruction of skin tissue (that is, visible necrosis through the epidermis and into the dermis) occurs in at least one animal after exposure up to 4 hours in duration;

(b) The substance or mixture is classified as skin irritation Category 2 if at least 4 out of 6 animals show a mean score per animal of ≥ 2.3 and ≤ 4.0 for erythema/eschar or for oedema;

(c) The substance or mixture is classified as skin irritation Category 3 if at least 4 out of 6 animals show a mean score per animal of ≥ 1.5 and < 2.3 for erythema/eschar or for oedema.

3.2.5.3 基本的手引き

3.2.5.3.1 関連の手引き

証拠の重み付け評価の適用方法に関する有用な手引きと同様、種々の試験の長所および短所さらに試験によらない方法についての有用な情報は、皮膚腐食性および刺激性に関する試験および評価に関する総合的アプローチ（IATA）である OECD ガイダンス文書 203 で提供されている。

3.2.5.3.2 皮膚腐食性または刺激性として分類するためのヒトのデータの使用に関する手引き

3.2.5.3.2.1　一般にヒトのデータには二つのタイプがある：過去のヒトの経験（例えば、職業、消費者、輸送、緊急時対応、疫学的研究による公表されている症例研究）またはヒトでの試験（例えば、臨床試験、皮膚パッチテスト）。関連性があり、信頼できる、質の良いヒトのデータは分類に関して一般に高く評価される。しかしながら、ヒトのデータには限界がある。皮膚刺激性/腐食性に関するヒトのデータの長所および限界についての詳細は OECD ガイダンス文書 203（section III. A, Part 1, Module 1）で見ることができる。

3.2.5.3.2.2　一般にヒトパッチテスト（HPT）は、刺激性および非刺激性物質を識別するために行われる。ヒトの皮膚への腐食性物質の適用は通常避ける。それゆえ腐食性を除外するために、一般的に前もって別の試験が行なわれる。通常は HPT 単独では刺激性および腐食性の識別は行わない。まれには腐食性の分類に使用できる HPT データがあるかもしれない（例えば、偽陰性 *in vitro* 試験後の HPT の適用）。ただし HPT および十分な他の皮膚腐食性に関する情報の組み合わせであれば、証拠の重み付け評価を用いて、分類に使用できる。

3.2.5.3.2.3　HPT 試験単独では有害性の特定を許可しない所管官庁があり（1.3.2.4.7 参照）、一方で皮膚刺激性の分類に HPT の使用を認めている所管官庁がある。

3.2.5.3.2.4　区分 2（刺激性）、区分 3（軽度刺激性）または区分に該当しないという分類に結び付く HPT 結果に関する明確な判定基準は国際的レベルでは確立されていない。したがって、HPT の結果は一般に証拠の重み付け評価のなかで利用される。しかし特定の手引きを提供する所管官庁があるかもしれない。十分な数のボランティアによる非希釈物質への 4 時間のばく露後に、HPT での明らかな陰性結果があれば、区分に該当しないということができる。

3.2.5.3.2.5　皮膚への不可逆的な損傷が観察された場合、ヒトの症例報告は腐食性としての分類に使用されるであろう。刺激性に関して国際的に受け入れられている分類判定基準はない。したがって、所管官庁がこの問題に関する特定の手引きを提供していない場合には、ばく露期間および利用可能な長期間におよぶ観察情報が分類の決定を是認するのに十分かどうか評価するために、専門家判断が必要となるであろう。刺激性または影響なしとなった症例は、それらだけでは決定的とは言えず、証拠の重み付け評価のなかで使用することができる。

3.2.5.3.3 3 匹超の動物による標準的な動物試験に基づいた分類

3.2.5.3.3.1　皮膚および眼の有害性クラスに対する分類判定基準は 3 匹の動物試験として GHS に詳述されている。動物を 6 匹まで使用したとみられる古い試験方法も知られている。しかし、GHS の判定基準では 3 匹を超える動物の試験による既存のデータに基づいた分類方法について明記していない。4 匹以上の動物を用いた試験による既存データに基づいた分類方法に関する手引きを以下のパラグラフに示す。

3.2.5.3.3.2　3 匹の動物試験に基づいた分類判定基準は 3.2.2.2 に詳述されている。4、5 または 6 匹の動物試験の評価は、試験動物の数により、以下のパラグラフによる判定基準にしたがうべきである。紅斑/痂皮および浮腫のスコアは、ばく露後 24、48 および 72 時間後、または反応が遅延している場合、皮膚反応の開始後 3 連続日の進行度から、とるべきである。

3.2.5.3.3.3　6 匹の動物試験の場合、次の原則を適用する：

(a) 皮膚の組織の破壊（すなわち表皮を通して真皮に達する目に見える壊死）が、4 時間以内のばく露後に少なくとも 1 匹の動物で起きた場合に、物質または混合物は皮膚腐食性区分 1 と分類する；

(b) 6 匹中少なくとも 4 匹で、動物ごとの紅斑/痂皮または浮腫に関する平均スコアが ≥ 2.3 および ≤ 4.0 の場合、物質または混合物は皮膚刺激性区分 2 と分類される；

(c) 6 匹中少なくとも 4 匹で、動物ごとの紅斑/痂皮または浮腫に関する平均スコアが ≥ 1.5 および < 2.3 の場合、物質または混合物は皮膚刺激性区分 3 と分類される。

3.2.5.3.3.4 In the case of a study with 5 animals the following principles apply:

    (a) The substance or mixture is classified as skin corrosion Category 1 if destruction of skin tissue (that is, visible necrosis through the epidermis and into the dermis) occurs in at least one animal after exposure up to 4 hours in duration;

    (b) The substance or mixture is classified as skin irritation Category 2 if at least 3 out of 5 animals show a mean score per animal of ≥ 2.3 and ≤ 4.0 for erythema/eschar or for oedema;

    (c) The substance or mixture is classified as skin irritation Category 3 if at least 3 out of 5 animals show a mean score per animal of ≥ 1.5 and < 2.3 for erythema/eschar or for oedema.

3.2.5.3.3.5 In the case of a study with 4 animals the following principles apply:

    (a) The substance or mixture is classified as skin corrosion Category 1 if destruction of skin tissue (that is, visible necrosis through the epidermis and into the dermis) occurs in at least one animal after exposure up to 4 hours in duration;

    (b) The substance or mixture is classified as skin irritation Category 2 if at least 3 out of 4 animals show a mean score per animal of ≥ 2.3 and ≤ 4.0 for erythema/eschar or for oedema;

    (c) The substance or mixture is classified as skin irritation Category 3 if at least 3 out of 4 animals show a mean score per animal of ≥ 1.5 and < 2.3 for erythema/eschar or for oedema.

3.2.5.3.4 *Classification criteria based on in vitro/ex vivo data*

Where in vitro/*ex vivo* tests have been undertaken in accordance with OECD Test Guideline 430, 431, 435 and/or 439, the criteria for classification in Category 1 (and, where possible and required into sub-categories 1A, 1B or 1C) for skin corrosion and in Category 2 for skin irritation are set out in tables 3.2.6 and 3.2.7.

3.2.5.3.3.4 5匹の動物試験の場合、次の原則を適用する：

(a) 皮膚の組織の破壊（すなわち表皮を通して真皮に達する目に見える壊死）が、4時間以内のばく露後に少なくとも1匹の動物で起きた場合に、物質または混合物は皮膚腐食性区分1と分類する；

(b) 5匹中少なくとも3匹で、動物ごとの紅斑/痂皮または浮腫に関する平均スコアが≥2.3 および ≤4.0の場合、物質または混合物は皮膚刺激性区分2と分類される；

(c) 5匹中少なくとも3匹で、動物ごとの紅斑/痂皮または浮腫に関する平均スコアが≥ 1.5 および < 2.3の場合、物質または混合物は皮膚刺激性区分3と分類される。

3.2.5.3.3.5 4匹の動物試験の場合、次の原則を適用する：

(a) 皮膚の組織の破壊（すなわち表皮を通して真皮に達する目に見える壊死）が、4時間以内のばく露後に少なくとも1匹の動物で起きた場合に、物質または混合物は皮膚腐食性区分1と分類する；

(b) 4匹中少なくとも3匹で、動物ごとの紅斑/痂皮または浮腫に関する平均スコアが≥2.3 および ≤4.0の場合、物質または混合物は皮膚刺激性区分2と分類される；

(c) 4匹中少なくとも3匹で、動物ごとの紅斑/痂皮または浮腫に関する平均スコアが≥ 1.5 および < 2.3の場合、物質または混合物は皮膚刺激性区分3と分類される。

3.2.5.3.4 *in vitro/ex vivo データに基づいた分類判定基準*

in vitro/*ex vivo* 試験がOECDテストガイドライン430、431、435および/または439にしたがって行われた場合、皮膚腐食性に対しては区分1（可能でかつ要求されている場合には細区分1A、1Bまたは1C）および皮膚刺激性に対しては区分2に関する判定基準が、表3.2.6および表3.2.7のように設定されている。

Table 3.2.6: Skin corrosion criteria for in vitro/ex vivo methods

Category	OECD Test Guideline 430 (Transcutaneous Electrical Resistance test method)	OECD Test Guideline 431 Reconstructed human Epidermis test methods: Methods 1, 2, 3, 4 and 5 as numbered in Annex 2 of OECD Test Guideline 431			OECD Test Guideline 435 Membrane barrier test method	
	Using rat skin discs corrosive chemicals are identified by their ability to produce a loss of normal *stratum corneum* integrity. Barrier function of the skin is assessed by recording the passage of ions through the skin. The electrical impedance of the skin is measured using transcutaneous electrical resistance (TER). A confirmatory test of positive results using a dye-binding step that assesses if an increase in ionic permeability is performed in case of a reduced TER (less than or around 5 kΩ) due to the physical destruction of the *stratum corneum* is in the absence of obvious damage. The criteria are based on the mean TER value in kΩ and sometimes on dye content.	Four similar methods where the test chemical is applied topically to a three-dimensional reconstructed human epidermis (RhE) which closely mimics the properties of the upper parts of human skin. The test method is based on the premise that corrosive chemicals are able to penetrate the *stratum corneum* by diffusion or erosion and are cytotoxic to the cells in the underlying layers. Tissue viability is assessed by enzymatic conversion of the dye MTT into a blue formazan salt that is quantitatively measured after extraction from the tissues. Corrosive chemicals are identified by their ability to decrease tissue viability below defined threshold values. The criteria are based on the percent tissue viability following a defined exposure period.			An in vitro membrane barrier test method comprising a synthetic macromolecular bio-barrier and a chemical detection system (CDS). Barrier damage is measured after the application of the test chemical to the surface of the synthetic membrane barrier. The criteria are based on the mean penetration/breakthrough time of the chemical through the membrane barrier.	
					Type 1 chemicals (high acid/alkaline reserve)	Type 2 chemicals (low acid/alkaline reserve)
1	(a) mean TER value ≤ 5 kΩ and the skin discs are obviously damaged (e.g. perforated), or (b) mean TER value ≤ 5 kΩ, and (i) the skin discs show no obvious damage (e.g. perforation), but (ii) the subsequent confirmatory testing of positive results using a dye binding step is positive.	Method 1 < 35 % after 3, 60 or 240 min exposure	Methods 2, 3, 4, 5 < 50 % after 3 min exposure; or ≥ 50 % after 3 min exposure and < 15 % after 60 min exposure		≤ 240 min	≤ 60 min
1A	Not applicable	Method 1 < 35 % after 3 min exposure	Method 2 < 25 % after 3 min exposure	Method 3 < 18 % after 3 min exposure	0-3 min.	0-3 min
				Methods 4, 5 < 15 % after 3 min exposure		
1B		≥ 35 % after 3 min exposure and < 35 % after 60 min exposure or ≥ 35 % after 60 min exposure and < 35 % after 240 min exposure	≥ 25 % after 3 min exposure and fulfilling criteria for Category 1	≥ 18 % after 3 min exposure and fulfilling criteria for Category 1	> 3 to 60 min.	> 3 to 30 min
1C				≥ 15 % after 3 min exposure and fulfilling criteria for Category 1	> 60 to 240 min.	> 30 to 60 min
Not classified as skin corrosive	(a) the mean TER value > 5 kΩ, or (b) the mean TER value ≤ 5 kΩ, and (i) the skin discs show no obvious damage (e.g. perforation), and (ii) the subsequent confirmatory testing of positive results using a dye binding step is negative	≥ 35 % after 240 min exposure	≥ 50 % after 3 min exposure and ≥ 15 % after 60 min exposure		> 240 min.	> 60 min

表 3.2.6：in vitro/ex vivo 方法に関する皮膚腐食性判定基準

区分	OECD テストガイドライン 430 (経皮電気抵抗試験)	OECD テストガイドライン 431 再構築ヒト表皮モデル試験：OECD テストガイドライン 431、補遺 2、方法 1、2、3、4 および 5				OECD テストガイドライン 435 膜バリア試験	
	腐食性化学品は、ラット皮膚ディスクを用いて、正常な角質層を破壊する能力で評価される。皮膚のバリア機能は皮膚を透過するイオンの通過を記録することで評価される。皮膚の電気的インピーダンスは経皮電気抵抗(TER)により測定される。明らかな損傷が見られなくてTER が減少(5kΩ付近か未満)した場合には、イオン透過性が増加している恐れがあるため、染料結合試験を用いて、陽性結果の確認試験を実施する。 判定基準は kΩ で示された TER の平均および染料含有量による。	試験化学品を、ヒトの皮膚の上層部に性質が近い 3 次元再構築ヒト表皮 (RhE) に局所的に適用する。腐食性方法は、腐食性化学品は拡散または浸食により角質層を貫通し、かつ下層の細胞に対して毒性があるという前提に基づく。組織の生存率は、MTT 染料の酵素反応による青色ホルマザン塩への変換を、組織から抽出して定量することで評価する。腐食性化学品は、定められた閾値以下の組織の生存率を低下させる能力に基づく。 判定基準は、定義されたばく露期間による組織の生存率に基づく。				in vitro 膜バリア試験方法は合成巨大分子バリアおよび化学品検出システム (CDS) からなる。合成膜バリアは試験化学品の表面に試験化学品を適用後に測定される。 判定基準は化学品が膜バリアを通して貫通/突破する平均時間に基づく。	
		方法 1	方法 2、3、4、5	方法 3	方法 4、5	タイプ 1 化学品 (高酸アルカリ予備)	タイプ 2 化学品 (低酸アルカリ予備)
1	(a) 平均 TER が ≦5kΩ、かつ皮膚ディスクが明らかに損傷 (例えば穿孔)、または (b) 平均 TER が ≦5kΩ、かつ (i) 皮膚ディスクには明らかな損傷 (例えば穿孔) なし、しかし (ii) 陽性結果確認のための染料結合試験が陽性。	3、60 または 240 分ばく露 < 35%	3 分ばく露 < 50%；または 3 分ばく露 ≧50% かつ 60 分ばく露 < 15%	3 分ばく露 < 18% 3 分ばく露 ≧18% かつ 60 分ばく露かつ分 1 の判定基準を満足	3 分ばく露 < 15% 3 分ばく露 ≧15% かつ 60 分ばく露かつ分 1 の判定基準を満足	≦240 分	≦60 分
1A	適用なし	3 分ばく露 < 35%	3 分ばく露 < 25%			0–3 分	0–3 分
1B		3 分ばく露 ≧35% かつ 60 分ばく露 < 35% または 60 分ばく露 ≧35% かつ 240 分ばく露 < 35%	3 分ばく露 ≧25% かつ 60 分ばく露 < 25% かつ分 1 の判定基準を満足			3 分から 60 分	3 分から 30 分
1C		240 分ばく露 ≧35%	3 分ばく露 ≧50%、かつ 60 分ばく露 ≧15%			60 分から 240 分	30 分から 60 分
皮膚腐食性とは分類されない	(a) 平均 TER が > 5kΩ、または (b) 平均 TER が ≦5kΩ、かつ (i) 皮膚ディスクには明らかな損傷 (例えば穿孔) 無し、かつ (ii) 陽性結果確認のための染料結合試験が陰性。		3 分ばく露 ≧50%、かつ 60 分ばく露 ≧15%			> 240 分	> 60 分

- 141 -

Table 3.2.7: Skin irritation criteria for in vitro methods

Category	OECD Test Guideline 439 Reconstructed Human Epidermis test methods
	Four similar methods (1-4) where the test chemical is applied topically to a three-dimensional reconstructed human epidermis (RhE) which closely mimics the properties of the upper parts of human skin. Tissue viability is assessed by enzymatic conversion of the dye MTT into a blue formazan salt that is quantitatively measured after extraction from the tissues. Positive chemicals are identified by their ability to decrease tissue viability below defined threshold levels. The criteria are based on mean percent tissue viability after exposure and post-treatment incubation.
1 or 2	Mean percent tissue viability (≤) 50 %. *Note: The RhE test methods covered by this test guideline cannot resolve between GHS categories 1 and 2. Further information on skin corrosion will be required to decide on its final classification (see also OECD Guidance Document 203).*
2	Mean percent tissue viability ≤ 50 % and the test chemical is found to be noncorrosive (e.g. based on Test Guideline 430, 431 or 435)
Not classified as skin irritant or Category 3	Mean percent-tissue viability > 50 % *Note: The RhE test methods covered by this test guideline cannot resolve between GHS optional Category 3 and not classified as skin irritant. Further information on skin irritation is required for those authorities that want to have more than one skin irritation category.*

表 3.2.7 : in vitro 方法に関する皮膚刺激性判定基準

区分	OECD テストガイドライン 439 再構築ヒト表皮試験方法
	試験化学品を、ヒトの皮膚の上層部に性質がよく似た 3 次元再構築ヒト表皮 (RhE) に局所的に適用する場合の 4 つの方法 (1-4)。組織の生存率は、MTT 染料の酵素反応による青色ホルマザン塩への変換を、青色ホルマザン塩を組織から抽出して定量することで評価する。陽性の化学品は、組織の生存率を定めらえた閾値以下に減少させる能力で決定される。 判定基準は、ばく露処理してインキュベーションした後の組織の平均生存率に基づく。
1 または 2	組織の平均生存率 (≦) 50% 注記：OECD テストガイドラインによる RhE 試験方法では、GHS 区分 1 および 2 を決定することはできない。最終的な分類の決定には皮膚腐食性に関するさらなる情報が必要であろう (OECD ガイダンス文書 203 も参照)。
2	組織の平均生存率≦50%、かつ試験化学品は非腐食性であることが示されている (例えば、TG430, 431 または 435 に基づく)。
皮膚刺激性とは分類されないまたは区分 3	組織の平均生存率 > 50% 注記：OECD テストガイドラインによる RhE 試験方法では、GHS の任意区分 3 および皮膚刺激性区分に該当しない違いを解決することはできない。二つ以上の皮膚刺激性区分を望む所管官庁では、皮膚刺激性に関するさらなる情報が必要である。

3.2.5.3.5 *Guidance on the use of other existing skin data in animals for classification as skin corrosion or skin irritation*

3.2.5.3.5.1 General approach

All existing other animal data should be carefully reviewed and only used if they are conclusive for classification. In evaluating other existing skin data in animals, however, it should be recognized that the reporting of dermal lesions may be incomplete, testing and observations may be made in a species other than the rabbit, and species may differ in sensitivity in their responses. In general skin thickness decreases with body weight. However, other factors also affect species variability. In addition, for most of these tests, irritating and corrosive effects need to be avoided. Therefore, these effects may only be observed in range finding studies using a small number of animals with limited observations and reporting.

3.2.5.3.5.2 Other data limitations and consequences for classification

3.2.5.3.5.2.1 Acute dermal toxicity tests, repeated dose animal studies, skin sensitization studies and skin absorption studies may all differ from the standard in vivo acute dermal irritation/corrosion test (e.g. OECD Test Guideline 404) with regard to exposure duration, area dose, the use of dissolved substances, level of occlusion, patch type, scoring and follow-up of the skin lesions and the test species.

3.2.5.3.5.2.2 Destruction of the skin in any acute dermal toxicity test (e.g. OECD Test Guideline 402) should be considered for classification as corrosive (Category 1 or sub-category 1A, 1B or 1C where possible and required). Skin irritation in an acute dermal study in rabbits fulfilling the criteria in table 3.2.2, should be considered for classification as irritant if the exposure conditions are such that corrosive effects can be excluded. Skin irritation in an acute dermal study in other species should be considered as not conclusive, as these species may be less or more sensitive than rabbits. Such data should be taken into account in a weight-of-evidence assessment. The absence of skin irritation should also be considered as not conclusive and taken into account in a weight-of-evidence assessment.

3.2.5.3.5.2.3 Repeated dose dermal studies (e.g. OECD test guidelines 410 and 411) can be used to classify as corrosive when destruction of the skin is observed after the initial exposures. However, normally such exposures are avoided and corrosive effects may only be observed in the range-finding studies. Moreover, subcategorization for corrosion will rarely be possible due to a longer time period between start of exposure and first observation. The observation of skin irritation or the absence of skin irritating effects should be considered as not conclusive. Skin effects only observed after multiple exposures may indicate skin sensitization rather than skin irritation.

3.2.5.3.5.2.4 In skin sensitization studies in guinea pigs (e.g. OECD Test Guideline 406), severely irritating and corrosive exposure must be avoided. Therefore, such effects are normally only observed in range-finding studies. The range-finding results, with the exception of intradermal exposure in the maximisation test, can be used to classify as corrosive when destruction of the skin is observed. The presence or absence of skin irritation in a skin sensitization study should be considered as not conclusive by itself as the species tested may be more or less sensitive than rabbits, but signs of irritation should be taken into account in a weight of evidence assessment.

3.2.5.3.5.2.5 Irritation data from the Local Lymph Node Assay (e.g. OECD test guidelines 429, 442A and 442B) should normally not be used for classification as the test substance is applied to the dorsum of the ear by open topical application, and in some cases specific vehicles for enhancement of skin penetration are used. Further, due to the proportional increase of skin thickness associated with increased body weight, the skin thickness of mice deviates significantly from that of rabbits and humans.

3.2.5.3.5.2.6 In skin absorption studies (e.g. OECD Test Guideline 427), corrosive exposure conditions are generally avoided as this affects the absorption. Therefore, information on skin effects from these studies does not allow classification directly but may be considered within a weight of evidence assessment. However, information on the dermal absorption may be taken into account in a weight-of-evidence assessment as a high dermal absorption in combination with additional evidence for high cytotoxicity may indicate irritation or corrosivity.

3.2.5.3.6 *Guidance on the use of pH and acid/alkaline reserve for classification as skin corrosion/irritation*

3.2.5.3.6.1 Methods to determine the pH value such as OECD Test Guideline 122 and the method described by Young et al. (1988) differ in the concentration of the substance or mixture for which the pH is determined and include values of 1%, 10% and 100%. These methods also differ in the way the acid/alkaline reserve is determined, namely up to a pH of 7 for both acids and bases (OECD Test Guideline 122) or up to a pH of 4 for acids and a pH of 10 for bases (Young et al., 1988). Furthermore, there are differences between OECD Test Guideline 122 and Young et al. (1988) in the units used to express the acid/alkaline reserve.

3.2.5.3.5 *皮膚腐食性または刺激性とする分類に関する、動物における他の既存皮膚データの使用についての手引き*

3.2.5.3.5.1 一般的アプローチ

　全ての既存の他の動物データは注意深く検討され、それらが分類に関して決定的である場合にのみ使用されるべきである。しかしながら動物における他の既存皮膚データの評価においては、皮膚の損傷に関する報告は不完全で、試験や観察はウサギ以外の種で行われていて、さらに反応における感受性は種により異なっているかもしれないことを認識するべきである。一般的に皮膚の厚さは体重と共に減少する。しかしながら他の要因もまた種の多様性に影響する。加えてこれらの試験の大部分では、刺激および腐食の影響は避ける必要がある。それゆえこれらの影響は、少ない数の動物を使用した濃度範囲を見つけるための試験の限られた観察や報告でしか観察されないであろう。

3.2.5.3.5.2 分類に関する他のデータの制限および結果

3.2.5.3.5.2.1　急性経皮毒性試験、反復ばく露動物試験、皮膚感作性試験および皮膚吸収試験は全て、標準的in vivo急性皮膚刺激性/腐食性試験（例えば、OECDテストガイドライン404）とは、ばく露期間、投与部分、溶解物質の使用、閉塞の程度、パッチタイプ、スコアリングおよび皮膚損傷の管理、さらに試験の種に関して異なる。

3.2.5.3.5.2.2　どのような急性経皮毒性試験（例えば、OECDテストガイドライン402）における皮膚の破壊も分類としては腐食性（区分1または可能であり要求されていれば細区分1A、1Bまたは1C）と考えるべきである。表3.2.2の判定基準を満たすウサギでの急性経皮試験における皮膚刺激性は、ばく露の条件が腐食影響を除外できるようであれば、分類としては刺激性と考えるべきである。他の種による急性経皮試験における皮膚刺激性は、これらの種はウサギよりも感受性が弱いかもしれないしまたは強いかもしれないので、決定的ではないと考えるべきである。そのようなデータは証拠の重み付け評価において考慮されるべきである。皮膚刺激性がないことは決定的ではないと考えるべきであり、証拠の重み付け評価において考慮されるべきである。

3.2.5.3.5.2.3　反復投与経皮毒性試験（例えば、OECDテストガイドライン410および411）は、最初のばく露で皮膚の破壊が観察された場合には、腐食性と分類するために使用することができる。しかしながら通常そのようなばく露は避けられ、腐食の影響は濃度範囲を見つける試験でのみ観察されるであろう。さらにばく露の開始から最初の観察まで長期間かかるので、腐食性の細区分はめったにできないであろう。皮膚刺激性の観察または皮膚刺激性の欠如は決定的と考えるべきではない。複数回のばく露後にのみ観察される皮膚影響は、皮膚刺激性というよりは皮膚感作性を示しているかもしれない。

3.2.5.3.5.2.4　モルモットによる皮膚感作性試験（例えば、OECDテストガイドライン406）においては、重篤な刺激性および腐食性のばく露は避けられなければならない。それゆえ、そのような影響は通常濃度範囲を見つける試験でのみ観察される。Maximisation試験における皮内ばく露を除いて、濃度範囲を見つける試験の結果は、皮膚の破壊が観察された場合、腐食性と分類するために使用することができる。皮膚感作性試験における皮膚刺激性の有無は、試験された種はウサギよりも感受性が弱いかもしれないしまたは強いかもしれないので、それ自身決定的ではないと考えるべきであるが、刺激性の症候は証拠の重み付け評価において考慮されるべきである。

3.2.5.3.5.2.5　局所リンパ節試験（例えば、OECDテストガイドライン429、442Aおよび442B）での刺激性データは、試験物質は耳の背部に局所的に適用されまた皮膚への浸透を強化するための特別な担体が使用されることもあるので、通常は分類に使用すべきではない。さらに、増加する体重と共に皮膚の厚さも増加するために、マウスの皮膚の厚さはウサギやヒトの皮膚の厚さから大きく離れる。

3.2.5.3.5.2.6　皮膚吸収試験（例えばOECDテストガイドライン427）においては、腐食性のばく露条件は、吸収に影響するので、通常回避される。したがってこれらの試験による皮膚影響に関する情報は直接的な分類を許さず、証拠の重み付け評価のなかで検討されるであろう。しかし、強い細胞毒性に関する追加的な証拠を伴った強い皮膚吸収は刺激性または腐食性を示しているかもしれないので、皮膚吸収に関する情報は証拠の重み付け評価のなかで考慮されるであろう。

3.2.5.3.6 *皮膚腐食性/刺激性として分類するためのpHおよび酸/アルカリ予備の使用に関する手引き*

3.2.5.3.6.1　OECDテストガイドライン122やYoungら（1988）によって記述された方法などのpH値を決定する方法は、pHを決定される物質または混合物の濃度が異なり、1%、10%、および100%の値が含まれる。また、これらの方法は、酸/アルカリ予備の決定方法も異なり、つまり酸・塩基ともにpH7まで（OECDテストガイドライン122）、または酸はpH4まで、塩基はpH10まで（Young et al., 1988）である。さらに、OECDテストガイドライン122とYoungら（1988）では、酸/アルカリ予備の表現に使用する単位に違いがある。

3.2.5.3.6.2   Criteria to identify substances and mixtures requiring classification in Category 1 based on pH and acid/alkaline reserve have been developed for effects on the skin (Young et al., 1988). These criteria were developed using a combination of pH and acid/alkaline reserve values that were determined in a specific way (Young et al., 1988). Therefore, these criteria may not be directly applicable when other test concentrations or methods are used to measure pH and acid/alkaline reserve. Furthermore, the calibration and validation of these criteria was based on a limited dataset for effects on the skin. Thus, the predictive value of the combination of pH and acid/alkaline reserve for classification in Category 1 for effects on the skin is limited, especially for substances and mixtures with an extreme pH but a non-significant acid/alkaline reserve. The criteria developed by Young et al. (1988) for classification in Category 1 may be used as a starting point for determining whether a substance or a mixture has a significant acid/alkaline reserve or a non-significant acid/alkaline reserve. A competent authority may decide which criteria for significant acid/alkaline reserve can be applied.

---

*   *References:*

   Young, J.R., M.J. How, A.P. Walker, and W.M. Worth. 1988. Classification as corrosive or irritant to skin of preparations containing acidic or alkaline substances, without testing on animals. Toxicol. In Vitro, 2(1): 19-26. Doi: 10.1016/0887-2333(88)90032-x.

3.2.5.3.6.2　皮膚への影響について、pH および酸/アルカリ予備に基づいて区分 1 に分類する必要がある物質および混合物を特定する基準が策定された（Young et al., 1988）。これらの基準は、具体的な方法で決定された pH と酸/アルカリ予備の値の組み合わせを使用して開発された（Young et al., 1988）。したがって、これらの基準は、他の試験濃度または方法を使用して pH および酸/アルカリ予備を測定した場合、直接適用できない可能性がある。さらに、これらの基準の校正（キャリブレーション）と検証は、皮膚への影響に関する限られたデータセットに基づくものであった。したがって、特に pH が極端だが酸/アルカリ予備がそれほどない物質や混合物については、皮膚への影響に関する区分 1 への分類のための pH と酸/アルカリ予備の組み合わせの予測値は限定的である。物質または混合物が相当量の酸/アルカリ予備があるか、または酸/アルカリ予備はそれほどないかを決定するための出発点として、Young ら（1988）によって区分 1 に分類するために開発された基準を使用することができる。所管官庁は、相当量の酸/アルカリ予備がある場合にどの基準を適用できるかを決定することができる。

参考文献:

Young, J.R., M.J. How, A.P. Walker, and W.M. Worth. 1988. Classification as corrosive or irritant to skin of preparations containing acidic or alkaline substances, without testing on animals. Toxicol. In vitro, 2(1): 19-26. Doi: 10.1016/0887-2333(88)90032-x.".

CHAPTER 3.3

SERIOUS EYE DAMAGE/EYE IRRITATION

3.3.1 Definitions and general considerations

3.3.1.1 *Serious eye damage* refers to the production of tissue damage in the eye, or serious physical decay of vision, which is not fully reversible, occurring after exposure of the eye to a substance or mixture.

Eye irritation refers to the production of changes in the eye, which are fully reversible, occurring after the exposure of the eye to a substance or mixture.

3.3.1.2 To classify, all available and relevant information on serious eye damage/eye irritation is collected and its quality in terms of adequacy and reliability is assessed. Classification should be based on mutually acceptable data/results generated using methods and/or defined approaches[1] that are validated according to international procedures. These include both OECD guidelines and equivalent methods/defined approaches (see 1.3.2.4.3). Sections 3.3.2.1 to 3.3.2.8 provide classification criteria for the different types of information that may be available.

3.3.1.3 A tiered approach (see 3.3.2.10) organizes the available information into levels/tiers and provides for decision-making in a structured and sequential manner. Classification results directly when the information consistently satisfies the criteria. However, where the available information gives inconsistent and/or conflicting results within a tier, classification of a substance or a mixture is made on the basis of the weight of evidence within that tier. In some cases when information from different tiers gives inconsistent and/or conflicting results (see 3.3.2.10.3) or where data individually are insufficient to conclude on the classification, an overall weight of evidence assessment is used (see 1.3.2.4.9, 3.3.2.9 and 3.3.5.3.1).

3.3.1.4 Guidance on the interpretation of criteria and references to relevant guidance documents are provided in 3.3.5.3.

3.3.2 Classification criteria for substances

Substances are allocated to one of the categories within this hazard class, Category 1 (serious eye damage) or Category 2 (eye irritation), as follows:

(a) Category 1 (serious eye damage/irreversible effects on the eye):

substances that have the potential to seriously damage the eyes.

(b) Category 2 (eye irritation/reversible effects on the eye):

substances that have the potential to induce reversible eye irritation.

Those authorities desiring one category for classification of "eye irritation" may use the overall Category 2; others may want to distinguish between Category 2A and Category 2B.

3.3.2.1 *Classification based on human data (tier 1 in figure 3.3.1)*

Existing reliable and good quality human data on serious eye damage/eye irritation should be given high weight where relevant for classification (see 3.3.5.3.2) and should be the first line of evaluation, as this gives information directly relevant to effects on the eye. Existing human data could be derived from single or repeated exposure(s), for example in occupational, consumer, transport or emergency response scenarios and epidemiological and clinical studies in well-documented case reports and observations (see 1.1.2.5 (c), 1.3.2.4.7 and 1.3.2.4.9). Although human data from

---

[1] According to OECD Guidance Document 255 on the reporting of defined approaches to be used within integrated approaches to testing and assessment, a defined approach to testing and assessment consists of a fixed data interpretation procedure (DIP) applied to data generated with a defined set of information sources to derive a result that can either be used on its own, or together with other information sources within an overall weight of evidence assessment, to satisfy a specific regulatory need.

第 3.3 章

眼に対する重篤な損傷性/眼刺激性

3.3.1 定義および一般事項

3.3.1.1 *眼に対する重篤な損傷性*とは、物質または混合物へのばく露後に起こる、眼の組織損傷を生じさせること、すなわち視力の重大な機能低下で、完全には治癒しないものをさす。

*眼刺激性*とは、物質または混合物へのばく露後に起こる、眼に変化を生じさせることで、完全に治癒するものをさす。

3.3.1.2 分類のため、眼に対する重篤な損傷性/眼刺激性に関するすべての利用可能な関連する情報は収集され、妥当性および信頼性の観点からその質について評価される。分類は、国際的な手順にしたがって検証された方法および/またはディファインドアプローチ [1] により取られた相互に受け入れ可能なデータ/結果に基づく必要がある。これらには、OECD ガイドラインと同等の方法/ディファインドアプローチの両方が含まれる（1.3.2.4.3 参照）。3.3.2.1 から 3.3.2.8 に利用可能な異なるタイプの情報に対する分類基準を示した。

3.3.1.3 段階的アプローチ（3.3.2.10 参照）は利用可能な情報をレベル/段階に整理して、構造的かつ連続的な方法で意思決定を行うものである。情報が一貫して判定基準を満たしていれば、分類はすぐに終わる。しかし一つの段階において利用可能な情報に一貫性がないおよび/または矛盾する結果が見られる場合には、物質や混合物の分類はその段階のなかで証拠の重み付けに基づいて行われる。異なる段階からの情報に一貫性がないおよび/または矛盾する結果が見られる場合（3.3.2.10.3 参照）または個々のデータが分類を結論付けるには不十分であるような場合には、包括的な証拠の重み付け評価が使用される（1.3.2.4.9、3.3.2.9 および 3.3.5.3.1 参照）。

3.3.1.4 判定基準の解釈に関する手引きおよび関連の手引きへの参照は 3.3.5.3.に記載されている。

3.3.2 物質の分類基準

物質はこの有害性クラスでは、以下のように区分 1（眼に対する重篤な損傷性）または区分 2（眼刺激）のうちの 1 つに割り当てられる：

(a) 区分 1（眼に対する重篤な損傷性/眼に対する不可逆的作用）：

眼に対して重篤な損傷を与える可能性のある物質。

(b) 区分 2（眼刺激性/眼に対する可逆的影響）：

可逆的な眼刺激作用を起こす可能性のある物質。

「眼刺激」の分類に関して 1 つの区分を望む所管官庁は総合的な区分 2 を使用すればよい；区分 2A および区分 2B を区別したいところもあろう。

3.3.2.1 *ヒトのデータに基づく分類（図 3.3.1 の段階 1）*

眼に対する重篤な損傷性/眼刺激性に関する既存の信頼できる質の高いヒトのデータは、眼に対する影響に直接的に関係する情報であるので、分類に関連して重視されるべきであり(3.3.5.3.2 参照)、評価の第一番目にあげられるべきである。既存のヒトのデータは、単回または反復ばく露から得ることができる。例えば、職業、消費者、輸送または緊急時対応のシナリオおよびよくまとめられた症例報告や観察などによる疫学的および臨床的研究などである。（1.1.2.5 (c)、1.3.2.4.7 および 1.3.2.4.9 参照）。事故または中毒

---

[1] 試験および評価に関する統合的アプローチで使用されるディファインドアプローチの報告に関する OECD ガイダンス文書 255 によれば、ディファインドアプローチは、予め定められた情報源を用いてもたらされたデータを適用された予め決められたデータ解釈手順（DIP）による試験および評価手法をいい、具体的な規制上の必要性を満たすために、それ単独で、あるいは包括的な証拠の重み付け評価の中で他の情報源と組み合わせて結果を導き出すことができる。

accident or poison centre databases can provide evidence for classification, absence of incidents is not itself evidence for no classification, as exposures are generally unknown or uncertain.

3.3.2.2 *Classification based on standard animal data (tier 1 in figure 3.3.1)*

OECD Test Guideline 405 is the currently available and internationally accepted animal test method for classification as serious eye damage or eye irritant (see tables 3.3.1 and 3.3.2, respectively) and is the standard animal test. The current version of OECD Test Guideline 405 uses a maximum of 3 animals. Results from animal studies conducted under previous versions of OECD Test Guideline 405 that used more than 3 animals are also considered standard animal tests when interpreted in accordance with 3.3.5.3.3.

3.3.2.2.1 *Serious eye damage (Category 1)/irreversible effects on the eye*

A single hazard category (Category 1) is adopted for substances that have the potential to seriously damage the eyes. This hazard category includes as criteria the observations listed in table 3.3.1. These observations include animals with grade 4 cornea lesions and other severe reactions (e.g. destruction of cornea) observed at any time during the test, as well as persistent corneal opacity, discoloration of the cornea by a dye substance, adhesion, pannus, and interference with the function of the iris or other effects that impair sight. In this context, persistent lesions are considered those which are not fully reversible within an observation period of normally 21 days. Hazard classification as Category 1 also contains substances fulfilling the criteria of corneal opacity ≥ 3 or iritis > 1.5 observed in at least 2 of 3 tested animals, because severe lesions like these usually do not reverse within a 21 days observation period.

Table 3.3.1: Serious eye damage/Irreversible effects on the eye category [a, b]

	Criteria

Category 1: **Serious eye damage/Irreversible effects on the eye**	A substance that produces:
	 (a) in at least one animal effects on the cornea, iris or conjunctiva that are not expected to reverse or have not fully reversed within an observation period of normally 21 days; and/or
	 (b) in at least 2 of 3 tested animals, a positive response of:
	   (i) corneal opacity ≥ 3; and/or
	   (ii) iritis > 1.5;
	 calculated as the mean scores following grading at 24, 48 and 72 hours after instillation of the test material.

[a] *Grading criteria are understood as described in OECD Test Guideline 405.*

[b] *Evaluation of a 4, 5 or 6-animal study should follow the criteria given in 3.3.5.3.3.*

3.3.2.2.2 *Eye irritation (Category 2)/Reversible effects on the eye*

3.3.2.2.2.1 Substances that have the potential to induce reversible eye irritation should be classified in Category 2 where further categorization into Category 2A and Category 2B is not required by a competent authority or where data are not sufficient for further categorization. When a substance is classified as Category 2, without further categorization, the classification criteria are the same as those for Category 2A.

3.3.2.2.2.2 For those authorities wanting more than one designation for reversible eye irritation, Category 2A and Category 2B are provided:

(a) When data are sufficient and where required by a competent authority, substances may be classified in Category 2A or 2B in accordance with the criteria in table 3.3.2;

(b) For substances inducing eye irritant effects reversing within an observation time of normally 21 days, Category 2A applies. For substances inducing eye irritant effects reversing within an observation time of 7 days, Category 2B applies.

3.3.2.2.2.3 For those substances where there is pronounced variability among animal responses, this information may be taken into account in determining the classification.

センターのデータベースから得られるヒトのデータは分類の根拠となり得るが、一般にばく露は不明または不確実であるため、事例がないこと自体が区分に該当しない根拠とはならない。

3.3.2.2　標準的動物試験データに基づく分類（図 3.3.1 の段階 1）

OECD テストガイドライン 405 は、現在利用可能で国際的に認められている眼に対する重篤な損傷性または眼刺激性の分類に関する動物試験法であり（それぞれ表 3.3.1 および 3.3.2 参照）、標準的な動物試験である。現行の OECD テストガイドライン 405 では、最大 3 匹の動物を使用する。OECD テストガイドライン 405 の旧版に基づいて実施され、3 匹を超える動物を使用した動物試験の結果も、3.3.5.3.3 にしたがって解釈する場合には、標準的な動物試験とみなされる。

3.3.2.2.1　眼に対する重篤な損傷性（区分 1）/眼への不可逆的作用

眼を重篤に損傷する可能性を有する物質には、単一の区分 1 の有害性区分が適用される。この有害性区分には、表 3.3.1 にある判定基準としての観察が含まれている。これらの所見には、試験中のどこかの時点で観察された第 4 段階の角膜病変およびその他の重篤な反応（例：角膜破壊）、持続性の角膜混濁、色素物質による角膜の着色、癒着、角膜の血管増殖、および虹彩機能の障害、または視力を傷害するその他の作用を伴った動物が含まれる。ここで持続性の病変とは、通常 21 日間の観察期間内で完全に可逆的ではない病変をいう。有害性分類区分 1 にはまた、3 匹の試験動物の内少なくとも 2 匹で、角膜混濁≧3、または虹彩炎＞1.5 が観察されるとする判定基準を満たす物質も含まれる。なぜなら、これらのような重篤な病変は、21 日間の観察期間内には通常回復しないからである。

表 3.3.1：眼に対する重篤な損傷性/眼への不可逆的作用区分 [a,b]

	判定基準
区分 1: 眼に対する重篤な損傷性/眼に対する不可逆的作用	以下の作用を示す物質： (a) 少なくとも 1 匹の動物で、角膜、虹彩または結膜に対する、可逆的であると予測されない作用が認められる、または通常 21 日間の観察期間中に完全には回復しない作用が認められる、および/または (b) 試験動物 3 匹中少なくとも 2 匹で、試験物質滴下後 24、48 および 72 時間における評価の平均スコア計算値が 　(i)　角膜混濁　≧3；および/または 　(ii)　虹彩　＞1.5； で陽性反応が得られる。

[a]　評価基準は OECD テストガイドライン 405 に記載されている。
[b]　4, 5 または 6 匹の動物実験の評価は 3.3.5.3.3 にある判定基準にしたがうべきである。

3.3.2.2.2　眼刺激性（区分 2）/眼に関する可逆的作用

3.3.2.2.2.1　所管官庁によりさらに区分 2A および 2B に分類する必要がない、あるいはさらに分類するためのデータが十分でない場合には、可逆的な眼刺激を起こす可能性のある物質は区分 2 に分類するべきである。物質が区分 2 と分類され、さらなる分類がない場合、分類の判定基準は区分 2A とおなじである。

3.3.2.2.2.2　可逆的な眼刺激に対して 2 つ以上の割り当てを望む所管官庁には、区分 2A および区分 2B がある。

(a) データが十分で、所管官庁により要求されている場合には、物質を表 3.3.2 の判定基準にしたがって区分 2A または区分 2B と分類してもよい；

(b) 通常 21 日間の観察期間内に回復する眼刺激作用を起こす物質は区分 2A とする。7 日間の観察期間内に回復する眼刺激作用を起こす物質は区分 2B とする。

3.3.2.2.2.3　動物間で反応にきわめて多様性が認められる化学品に対しては、分類の決定において、その情報を考慮してもよい。

Table 3.3.2: **Reversible effects on the eye categories** [a, b]

	Criteria
	Substances that have the potential to induce reversible eye irritation
Category 2/2A	Substances that produce in at least 2 of 3 tested animals a positive response of: (a) corneal opacity ≥ 1; and/or (b) iritis ≥ 1; and/or (c) conjunctival redness ≥ 2; and/or (d) conjunctival oedema (chemosis) ≥ 2 calculated as the mean scores following grading at 24, 48 and 72 hours after instillation of the test material, and which fully reverses within an observation period of normally 21 days.
Category 2B	Within Category 2A an eye irritant is considered mildly irritating to eyes (Category 2B) when the effects listed above are fully reversible within 7 days of observation.

[a]   Grading criteria are understood as described in OECD Test Guideline 405.
[b]   Evaluation of a 4, 5 or 6-animal study should follow the criteria given in 3.3.5.3.3.

3.3.2.3   *Classification based on defined approaches (tier 2 in figure 3.3.1)*

3.3.2.3.1   Defined approaches consist of a rule-based combination of data obtained from a predefined set of different information sources (e.g. in vitro methods, *ex vivo* methods, physico-chemical properties, non-test methods). It is recognized that most single in vitro/*ex vivo* methods are not able to replace in vivo methods fully for most regulatory endpoints. Thus, defined approaches can be useful strategies of combining data for classifying substances and mixtures. Results obtained with a defined approach validated according to international procedures, such as an OECD defined approach guideline or an equivalent approach, is conclusive for classification for serious eye damage/eye irritation if the criteria of the defined approach are fulfilled (see 3.3.5.3.4)[2]. Data from a defined approach can only be used for classification when the tested substance is within the applicability domain of the defined approach used. Additional limitations described in the published literature should also be taken into consideration.

3.3.2.3.2   Where the results from defined approaches are assigned a level of confidence, a low confidence outcome of a defined approach cannot be used on its own to classify but may be considered in combination with other data.

3.3.2.3.3   Individual evidence used within a defined approach should not also be used outside of that defined approach.

3.3.2.4   *Classification based on in vitro/ex vivo data (tier 2 in figure 3.3.1)*

3.3.2.4.1   The classification criteria for the currently available in vitro/*ex vivo* test methods adopted by OECD in test guidelines 437, 438, 460, 491, 492, 494 and 496 are described in table 3.3.6 (see 3.3.5.3.5.1). When considered individually, these in vitro/*ex vivo* OECD test guidelines address serious eye damage and/or no classification for eye hazard, but do not address eye irritation. Therefore, data from a single in vitro/*ex vivo* OECD test guideline can only be used to conclude on either classification in Category 1 or no classification and cannot be used to conclude on classification in Category 2. When the result of a single in vitro/*ex vivo* method is "no stand-alone prediction can be made" (e.g. see table 3.3.6), a conclusion cannot be drawn on the basis of that single result and further data are necessary for classification (see 3.3.5.3.4.3 and 3.3.5.3.4.4).

3.3.2.4.2   In vitro/*ex vivo* methods in 3.3.2.4.1 with the result "no stand-alone prediction can be made" should within tier 2 only be used in combination with other types of data in defined approaches.

3.3.2.4.3   Other validated in vitro/*ex vivo* test methods accepted by some competent authorities are described in 3.3.5.3.5.2. Some of these in vitro/*ex vivo* test methods may be useful to classify in Category 2. A competent authority may decide which classification criteria, if any, should be applied for these test methods to conclude on classification, including that a substance is not classified for effects on the eye.

---

[2]   *Some defined approaches have been proposed for serious eye damage/eye irritation (Alépée et al., 2019a, b) but no classification criteria have yet been agreed internationally.*

表 3.3.2：可逆的な眼への作用に関する区分 [a,b]

	判定基準
	可逆的な眼刺激作用の可能性を持つ物質
区分 2/2A	試験動物 3 匹中少なくとも 2 匹で以下の陽性反応がえられる。 試験物質滴下後 24、48 および 72 時間における評価の平均スコア計算値が： (a) 角膜混濁 ≥1；および/または (b) 虹彩 ≥1；および/または (c) 結膜発赤 ≥2；および/または (d) 結膜浮腫 ≥2 かつ通常 21 日間の観察期間内で完全に回復する。
区分 2B	区分 2A において、上述の作用が 7 日間の観察期間内に完全に可逆的である場合には、眼刺激性は軽度の眼刺激性（区分 2B）であるとみなされる。

[a] 評価基準は OECD テストガイドライン 405 に記載されている。
[b] 4、5 または 6 匹の動物実験の評価は 3.3.5.3.3 にある判定基準にしたがうべきである。

3.3.2.3 ディファインドアプローチに基づく分類（図 3.3.1 の段階 2）

3.3.2.3.1 ディファインドアプローチは、事前に定められた一連の異なる情報源（in vitro 法、ex vivo 法、物理化学的特性、試験によらない方法等）から得られたデータのルールに基づく組み合わせで構成される。多くの単一での in vitro/ex vivo 法は、ほとんどの規制エンドポイントにおいて in vivo 法に完全に置き換えることができないと認識されている。したがって、ディファインドアプローチは、物質や混合物を分類するためにデータを組み合わせることで有用な戦略となり得る。OECD のディファインドアプローチガイドラインまたは同等のアプローチのような国際的な手順にしたがって検証されたディファインドアプローチで得られた結果は、ディファインドアプローチの基準が満たされている場合、眼に対する重篤な損傷性/眼刺激性の分類において決定的となる（3.3.5.3.4 参照）[2]。ディファインドアプローチのデータは、試験物質が使用されたディファインドアプローチの適用範囲内にある場合にのみ、分類に用いることができる。公表文献に記載されている追加の制限も考慮すべきである。

3.3.2.3.2 ディファインドアプローチの結果に信頼性のレベルが付与されている場合、ディファインドアプローチにおける信頼性が低い結果を単独で分類に用いることはできないが、他のデータと組み合わせて考慮することはできる。

3.3.2.3.3 ディファインドアプローチにおいて使用される個々の証拠は、そのディファインドアプローチ以外では使用されるべきではない。

3.3.2.4 in vitro（試験管内）/ex vivo（生体外）データに基づく分類（図 3.3.1 の段階 2）

3.3.2.4.1 OECD がテストガイドライン 437、438、460、491、492、494 および 496 で採用している、現在利用可能な in vitro/ex vivo 試験法の分類基準を表 3.3.6 に記載する（3.3.5.3.5.1 参照）。個別に検討すると、これらの in vitro/ex vivo 系の OECD テストガイドラインは、眼に対する有害性について、眼に対する重篤な損傷性および/または区分に該当しないの識別は扱うが、眼刺激性は扱っていない。したがって、単一の in vitro/ex vivo 系 OECD テストガイドラインのデータは、区分 1 への分類または分類されないのいずれかを結論づけるためにのみ使用でき、区分 2 への分類を結論づけるために使用することができない。単一の in vitro/ex vivo 法の結果が「単独での予測はできない」（例：表 3.3.6 参照）場合、その単一の結果に基づいて結論を出すことはできず、分類にはさらなるデータが必要となる（3.3.5.3.4.3 および 3.3.5.3.4.4 参照）。

3.3.2.4.2 3.3.2.4.1 で「単独での予測はできない」という結果が得られた in vitro/ex vivo 法は、段階 2 において、ディファインドアプローチの他のタイプのデータとの組み合わせでのみ使用する必要がある。

3.3.2.4.3 一部の所管官庁が認めるその他の有効な in vitro/ex vivo 試験法は 3.3.5.3.5.2 に記載されている。これらの in vitro/ex vivo 試験法の中には、区分 2 に分類するのに有用なものがある。所管官庁は、これらの試験法について、眼に対する影響について物質が分類されないことを含め、分類の結論を出すために適用すべき分類基準がある場合、それを決定することができる。

---

[2] 眼に対する重篤な損傷性/眼刺激性に対しては、いくつかのディファインドアプローチが提案されているが（Alépée et al., 2019a、b）、国際的にはまだ合意された分類基準はない。

3.3.2.4.4   In vitro/*ex vivo* data can only be used for classification when the tested substance is within the applicability domain of the test method(s) used. Additional limitations described in the published literature should also be taken into consideration.

3.3.2.4.5   *Serious eye damage (Category 1)/Irreversible effects on the eye*

3.3.2.4.5.1   Where tests have been undertaken in accordance with OECD test guidelines 437, 438, 460, 491 and/or 496, a substance is classified for serious eye damage in Category 1 based on the criteria in table 3.3.6 (see 3.3.5.3.5.1).

3.3.2.4.5.2   Although the currently available OECD in vitro/*ex vivo* test guidelines and equivalent methods have not been developed to identify substances inducing discolouration of the eye, some comparable effects may be observed in these tests. Therefore, where, after washing, discolouration of the cornea or of the tested cells compared to the control is observed in OECD Test Guideline 437, 438, 492 or 494, or in other equivalent methods, suggesting a permanent effect, a competent authority may require classification of a substance for serious eye damage in Category 1.

3.3.2.4.6   *Eye irritation (Category 2)/Reversible effects on the eye*

3.3.2.4.6.1   A positive result in an in vitro/*ex vivo* test method that is validated according to international procedures for identification of substances inducing eye irritation can be used to classify for eye irritation in Category 2/2A[3].

3.3.2.4.6.2   Where competent authorities adopt Category 2A and Category 2B, it is important to note that the currently validated in vitro/*ex vivo* test methods for effects on the eye do not allow discrimination between these two categories. In this situation, if the criteria for classification in Category 2 have been considered fulfilled, and no other relevant information is available, classification in Category 2/2A should be applied.

3.3.2.4.7   *No classification for effects on the eye*

OECD test guidelines 437, 438, 491, 492, 494 and 496 (see table 3.3.6 in 3.3.5.3.5.1) can be used to conclude that a substance is not classified for effects on the eye.

3.3.2.5   ***Classification based on conclusive human data, standard animal data or in vitro/ex vivo data for skin corrosion (tier 3 in figure 3.3.1)***

Substances classified as corrosive to skin (skin Category 1) based on conclusive human data, standard animal data or in vitro/*ex vivo* data for skin corrosion according to the criteria in chapter 3.2 are also deemed as inducing serious eye damage (eye Category 1). Skin irritation (skin Category 2), mild skin irritation (skin Category 3) and no classification for skin irritation, as well as human patch data (as described in chapter 3.2), cannot be used alone to conclude on eye irritation or no classification for effects on the eye, but may be considered in an overall weight of evidence assessment.

3.3.2.6   ***Classification based on other existing animal skin or eye data (tier 4 in figure 3.3.1)***

Other existing skin or eye data in animals may be used for classification, but there may be limitations regarding the conclusions that can be drawn (see 3.3.5.3.6). Substances classified as corrosive to skin (skin Category 1) based on other existing skin data according to the criteria in chapter 3.2 are also deemed as inducing serious eye damage (eye Category 1). Other existing skin data leading to classification in skin Category 2, 3 or no classification, cannot be used alone to conclude on eye irritation or no classification for effects on the eye, but may be considered in an overall weight of evidence assessment.

3.3.2.7   ***Classification based on extreme pH (pH ≤ 2 or ≥ 11.5) and acid/alkaline reserve (tier 5 in figure 3.3.1)***

In general, substances with an extreme pH (pH ≤ 2 or ≥ 11.5) are expected to cause significant eye effects, especially when associated with significant acid/alkaline reserve. A substance with pH ≤ 2 or ≥ 11.5 is therefore considered to cause serious eye damage (Category 1) in this tier if it has a significant acid/alkaline reserve or if no data for acid/alkaline reserve are available. However, if consideration of acid/alkaline reserve suggests the substance may not cause serious eye damage despite the extreme pH value, the result is considered inconclusive within this tier (see

---

[3]   *Although no classification criteria have yet been agreed internationally for some validated and/or accepted in vitro/ex vivo test methods proposed for identifying substances inducing eye irritation, these test methods may still be accepted by some competent authorities (see 3.3.2.4.2). If a defined approach (see 3.3.2.3) is not available or is not adequate for classification, data from these methods may be considered in a weight of evidence assessment within this tier.*

3.3.2.4.4 In vitro/ex vivo データは、被験物質が使用された試験法の適用範囲内にある場合にのみ分類に使用できる。公表文献に記載されている追加の制限も考慮すべきである。

3.3.2.4.5 *眼に対する重篤な損傷性（区分1）/眼に対する不可逆的な影響*

3.3.2.4.5.1 OECD テストガイドライン 437、438、460、491 および/または 496 にしたがって試験が実施された場合、物質は表 3.3.6 の基準に基づいて眼に対する重篤な損傷性の区分1（かどうか）に分類される（3.3.5.3.5.1 参照）。

3.3.2.4.5.2 現在利用可能な OECD の in vitro/ex vivo 系のテストガイドラインおよび同等の方法は、眼の変色を引き起こす物質を特定する目的で開発されてはいないが、これらの試験でいくつかの類似の影響が観察されることがある。したがって、OECD テストガイドライン 437、438、492 または 494、あるいは他の同等の方法において、洗浄後に対照群と比較して角膜または試験された細胞の変色が観察され、永続的な影響を示唆する場合、所管官庁は、眼に対する重篤な損傷性物質として区分1に分類することを要求できる。

3.3.2.4.6 *眼刺激性 (区分2)/眼に対する可逆的な影響*

3.3.2.4.6.1 眼刺激性物質の同定に関する国際的な手順にしたがって検証された in vitro/ex vivo 試験法での陽性結果は、眼刺激性区分 2/2A[3]に分類するために使用できる。

3.3.2.4.6.2 所管官庁が区分 2A および区分 2B を採用する場合、眼に対する影響に関する現在有効な in vitro/ex vivo 試験法では、これら2つの区分の区別がつかないことに注意することが重要である。この状況において、区分2の分類基準が満たされていると考えられ、他に関連する情報が利用できない場合、分類は区分 2/2A を適用する必要がある。

3.3.2.4.7 *眼に対する影響に関する区分に該当しない*

OECD テストガイドライン 437、438、491、492、494 および 496（3.3.5.3.5.1 の表 3.3.6 参照）は、物質が眼に対する影響に関する区分に該当しないと結論づけるために使用することができる。

3.3.2.5 *皮膚腐食性に関する決定的なヒトのデータ、標準的な動物データ、または in vitro/ex vivo 系データに基づく分類 (図 3.3.1 の段階 3)*

第 3.2 章の基準にしたがって、皮膚腐食性に関する決定的なヒトのデータ、標準的な動物データ、または in vitro/ex vivo 系データに基づいて皮膚腐食性（皮膚区分1）として分類された物質は、眼に対する重篤な損傷性（眼区分1）を引き起こすものともみなされる。皮膚刺激性（皮膚区分2）、軽度の皮膚刺激性（皮膚区分3）、および皮膚刺激性の区分に該当しないという情報、およびヒトパッチテストデータ（第 3.2 章で説明）を単独で使用して、眼刺激性があるまたは眼に対する影響の区分に該当しないと結論付けることはできないが、包括的な証拠の重み付け評価で考慮することはできる。

3.3.2.6 *他の既存の動物の皮膚または眼のデータに基づく分類 (図 3.3.1 の段階 4)*

動物の他の既存の皮膚または眼のデータを分類に使用することもできるが、導き出される結論に関しては制限がある場合がある（3.3.5.3.6 参照）。第 3.2 章の基準にしたがって、他の既存の皮膚データに基づいて皮膚腐食性（皮膚区分1）として分類された物質は、眼に対する重篤な損傷性（眼区分1）も引き起こすものとみなされる。皮膚区分 2、3 または区分に該当しないという他の既存の皮膚データは、単独で使用して眼刺激があるまたは眼に対する影響の区分に該当しないと結論付けることはできないが、包括的な証拠の重み付け評価で考慮することはできる。

3.3.2.7 *極端な pH（pH≦2 または≧11.5）および酸/アルカリ予備に基づく分類 (図 3.3.1 の段階 5)*

一般に、極端な pH（pH≦2 または≧11.5）の物質は、特に相当量の酸/アルカリ予備がある場合、眼に対する明確な影響を引き起こすことが予想される。したがって、pH≦2 または≧11.5 の物質は、相当量の酸/アルカリ予備がある場合、または酸/アルカリ予備のデータが利用できない場合、この段階で眼に対する重篤な損傷性を引き起こす（区分1）と見なされる。ただし、酸/アルカリ予備を考慮して、極端な pH 値にもかかわらず、物質が眼に対する重篤な損傷性を引き起こさない可能性があることを示唆している場合、その結果はこの段階では決定的ではないと見なされる（図 3.3.1 参照）。2 < pH < 11.5 の場合は、

---

[3] 眼刺激性を引き起こす物質を特定するために提案された一部の検証済みおよび/または受け入れられた *in vitro/ex vivo* 試験方法については、まだ分類基準が国際的に合意されていないが、これらの試験方法は一部の所管官庁によって依然として受け入れられている可能性がある（3.3.2.4.2 参照）。ディファインドアプローチ（3.3.2.3 参照）が利用できない場合、または分類に適切でない場合、これらの手法のデータは、この段階内の証拠の重み付け評価で考慮される場合がある。

figure 3.3.1). A pH > 2 and < 11.5 is considered inconclusive and cannot be used for classification purposes. Acid/alkaline reserve and pH can be determined by different methods including those described in OECD Test Guideline 122 and Young et al. (1988), acknowledging that there are some differences between these methods (see 3.3.5.3.7). A competent authority may decide which criteria for significant acid/alkaline reserve can be applied.

3.3.2.8 *Classification based on non-test methods for serious eye damage/eye irritation or for skin corrosion (tier 6 in figure 3.3.1)*

3.3.2.8.1 Classification, including the conclusion not classified, can be based on non-test methods, with due consideration of reliability and applicability, on a case-by-case basis. Non-test methods include computer models predicting qualitative structure-activity relationships (structural alerts, SAR) or quantitative structure-activity relationships (QSARs), computer expert systems, and read-across using analogue and category approaches.

3.3.2.8.2 Read-across using analogue or category approaches requires sufficiently reliable test data on similar substance(s) and justification of the similarity of the tested substance(s) with the substance(s) to be classified. Where adequate justification of the read-across approach is provided, it has in general higher weight than (Q)SARs.

3.3.2.8.3 Classification based on (Q)SARs requires sufficient data and validation of the model. The validity of the computer models and the prediction should be assessed using internationally recognized principles for the validation of (Q)SARs. With respect to reliability, lack of alerts in a SAR or expert system is not sufficient evidence for no classification.

3.3.2.8.4 Conclusive non-test data for skin corrosion may be used for classification for effects on the eye. Thus, substances classified as corrosive to skin (skin Category 1) according to the criteria in chapter 3.2 are also deemed as inducing serious eye damage (eye Category 1). Skin irritation (skin Category 2), mild skin irritation (skin Category 3) and no classification for skin irritation according to chapter 3.2 cannot be used alone to conclude eye irritation or no classification for effects on the eye, but may be considered in an overall weight of evidence assessment.

3.3.2.8.5 For conclusions on no classification from read-across and (Q)SARs the adequacy and robustness of the scientific reasoning and of the supporting evidence should be well substantiated and normally requires multiple negative substances with good structural and physical (related to toxicokinetics) similarity to the substance being classified, as well as a clear absence of positive substances with good structural and physical similarity to the substance being classified.

3.3.2.9 *Classification based on an overall weight of evidence assessment (tier 7 in figure 3.3.1)*

3.3.2.9.1 An overall weight of evidence assessment using expert judgement is indicated where none of the previous tiers resulted in a definitive conclusion on classification. In some cases, where the classification decision was postponed until the overall weight of evidence, but no further data are available, a classification may still be possible.

3.3.2.9.2 A substance with an extreme pH (pH ≤ 2 or ≥ 11.5) and non-significant acid/alkaline reserve (result considered inconclusive in tier 5; see 3.3.2.7) and for which no other information is available, should be classified as serious eye damage Category 1 in this tier. If inconclusive information is also available from other tiers but the overall weight of evidence assessment remains inconclusive, the extreme pH (pH ≤ 2 or ≥ 11.5) result should take precedence and the substance should be classified as serious eye damage Category 1 in this tier independently of its acid/alkaline reserve. For mixtures, the approach is different and is detailed in 3.3.3.1.3.

3.3.2.10 *Classification in a tiered approach (figure 3.3.1)*

3.3.2.10.1 A tiered approach to the evaluation of information should be considered where applicable (figure 3.3.1), recognizing that not all tiers as well as information within a tier may be relevant. However, all available and relevant information of sufficient quality needs to be examined for consistency with respect to the resulting classification.

3.3.2.10.2 In the tiered approach (figure 3.3.1), existing human and standard animal data for eye effects form the highest tier, followed by defined approaches and in vitro/*ex vivo* data for eye effects, existing human/standard animal/in vitro/*ex vivo* data for skin corrosion, other existing animal skin or eye data, extreme pH and acid/alkaline reserve, and finally non-test methods. Where information from data within the same tier is inconsistent and/or conflicting, the conclusion from that tier is determined by a weight of evidence assessment.

3.3.2.10.3 Where information from several tiers is inconsistent and/or conflicting with respect to the resulting classification, information of sufficient quality from a higher tier is generally given a higher weight than information from a lower tier. However, when information from a lower tier would result in a stricter classification than information from a higher tier and there is concern for misclassification, then classification is determined by an overall weight of evidence

決定的ではないと見なされ、分類目的には使用できない。酸／アルカリ予備およびpHは、OECDテストガイドライン 122 および Young ら（1988）を含むいくつかの異なる方法で決定され、これらの方法にはいくつかの違いがあることが認められている（3.3.5.3.7参照）。所管官庁は、相当量の酸/アルカリ予備がある場合にはどの基準を適用できるかを決定することができる。

3.3.2.8 眼に対する重篤な損傷性/眼刺激性または皮膚腐食性の試験によらない方法に基づく分類（図 3.3.1 の段階6）

3.3.2.8.1 ケースバイケースで、信頼性および適用性を十分に考慮した上で、区分に該当しないという結論も含めて、試験によらない方法に基づいて分類を行うことができる。試験によらない方法には、定性的構造活性相関（構造的アラート、SAR）や定量的構造活性相関（QSARs）のコンピューターモデル、コンピューターエキスパートシステム、および、類似物質およびカテゴリーアプローチを用いたリード・アクロスなどがある。

3.3.2.8.2 類似物質やカテゴリーアプローチを用いたリード・アクロスでは、類似物質に関する十分に信頼できる試験データとそして、試験物質と分類対象物質の類似性の正当化が必要である。リード・アクロス手法に関して十分な正当化がなされている場合、一般には(Q)SARs よりも高い重み付けがなされる。

3.3.2.8.3 (Q)SARs に基づく分類では十分なデータとモデルの妥当性検証が必要である。コンピュータモデルとその予測の検証は、(Q)SAR の検証に関する国際的に認知された原則を用いて評価されるべきである。信頼性に関して、SAR における構造アラートまたはエキスパートシステムがないことは、区分に該当しない十分な証拠とはならない。

3.3.2.8.4 皮膚腐食性に関する決定的な試験によらないデータは、眼に対する影響の分類に使用できる。したがって、第 3.2 章の基準にしたがって皮膚腐食性（皮膚区分 1）に分類された物質は、眼に対する重篤な損傷性（眼区分 1）を引き起こすとみなされる。皮膚刺激性（皮膚区分 2）、軽度の皮膚刺激性（皮膚区分 3）および第 3.2 章による皮膚刺激性の区分に該当しないは、単独の使用で眼刺激性または眼に対する影響の区分に該当しないと結論づけることはできないが、包括的な証拠の重み付け評価で考慮することができる。

3.3.2.8.5 リード・アクロスおよび (Q)SAR から得られた区分に該当しない（分類されない）という結論については、科学的推論およびそれを裏付ける証拠の妥当性と頑健性が十分に実証されている必要があり、分類対象物質との構造的および物理的類似性が高い陽性物質が明らかに存在しないことを示すのと同様に、通常は分類対象物質との構造的および物理的(関連するトキシコキネティクスに) 類似性が高い複数の陰性物質が必要である。

3.3.2.9 包括的な証拠の重み付け評価に基づく分類（図 3.3.1 の段階7）

3.3.2.9.1 専門家の判断による包括的な証拠の重み付け評価は、これより前の段階のいずれにおいても分類に関する最終的な結論が得られなかった場合に示される。包括的な証拠の重み付けまで分類の決定が判断できず、それ以上のデータが利用できないケースでもまだ分類が可能な場合がある。

3.3.2.9.2 極端な pH（pH≦2 または≧11.5）かつ酸/アルカリ予備はそれほどなく（段階5で「inconclusive（決定的ではない）」と見なされた結果； 3.3.2.7 参照）、さらに他の情報が利用できない物質は、この段階で眼に対する重篤な損傷性区分 1 に分類されるべきである。他の段階でも結論付けられない情報が利用可能であるが、包括的な証拠の重み付け評価で結論ができない場合、極端な pH（pH≦2 または≧11.5）の結果が優先され、その物質は酸/アルカリ予備とは無関係にこの段階で眼に対する重篤な損傷性区分 1 に分類されるべきである。混合物の場合、アプローチは異なり、3.3.3.1.3.に詳述されている。

3.3.2.10 段階的アプローチによる分類（図 3.3.1）

3.3.2.10.1 すべての段階および段階内の情報が関連するとは限らないことを認識した上で、該当する場合、情報を評価するための段階的アプローチを考慮するものとする（図 3.3.1）。ただし、十分な質の利用可能な関連情報はすべて、結果として得られる分類に関して一貫性があるかどうかを精査される必要がある。

3.3.2.10.2 段階的アプローチ（図 3.3.1）においては、眼に対する影響に関する既存のヒトおよび標準動物のデータが最上位であり、次いで眼への影響に関するディファインドアプローチおよび in vitro/ex vivo データ、皮膚腐食性に関する既存のヒト/標準的な動物/in vitro/ex vivo データ、その他の既存の動物の皮膚または眼のデータ、極端な pH および酸/アルカリ予備、そして最後に試験によらない方法となる。同じ段階のデータからの情報に一貫性がないおよび/または矛盾している場合、この段階での結論は証拠の重み付け評価によって決定される。

3.3.2.10.3 複数の段階からの情報が、分類結果に対して一貫性がないおよび／または矛盾している場合、一般に、より上位の段階での十分な質の情報は、より下位の段階での情報よりも高い重み付けがなされる。しかし、下位の段階の情報が上位の段階の情報よりも厳しい分類結果となり、誤分類の懸念がある場合は、分類は包括的な証拠の重み付け評価によって決定される。例えば、3.3.5.3 の手引きを適宜参照し、in vitro/ex

assessment. For example, having consulted the guidance in 3.3.5.3 as appropriate, classifiers concerned with a negative result for serious eye damage in an in vitro/*ex vivo* study when there is a positive result for serious eye damage in other existing eye data in animals would utilise an overall weight of evidence assessment. The same would apply in the case where there is human data indicating eye irritation but positive results from an in vitro/*ex vivo* test for serious eye damage are also available.

Figure 3.3.1: Application of the tiered approach for serious eye damage/eye irritation[a]

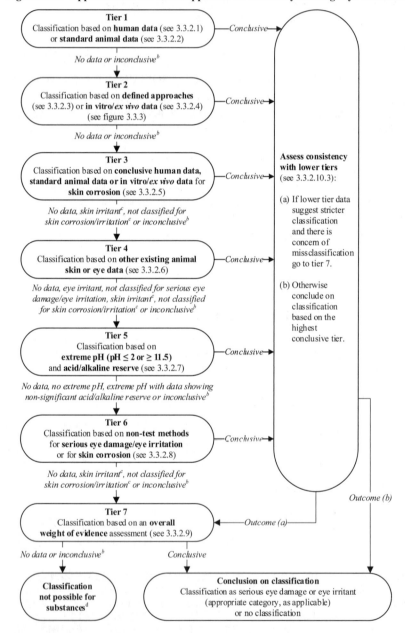

[a] Before applying the approach, the explanatory text in 3.3.2.10 as well as the guidance in 3.3.5.3 should be consulted. Only adequate and reliable data of sufficient quality should be included in applying the tiered approach.

[b] Information may be inconclusive for various reasons, e.g.:
- The available data may be of insufficient quality, or otherwise insufficient/inadequate for the purpose of classification, e.g. due to quality issues related to experimental design and/or reporting;

vivo 試験で眼に対する重篤な損傷性の結果が陰性であるにもかかわらず、他の既存の動物における眼のデータで眼に対する重篤な損傷性の結果が陽性であることを懸念を持つ分類者は、包括的な証拠の重み付け評価を活用するであろう。また、眼刺激性を示すヒトのデータがあるが、in vitro/ex vivo 試験で眼に対する重篤な損傷性の陽性結果が得られている場合にも同じことが当てはまる。

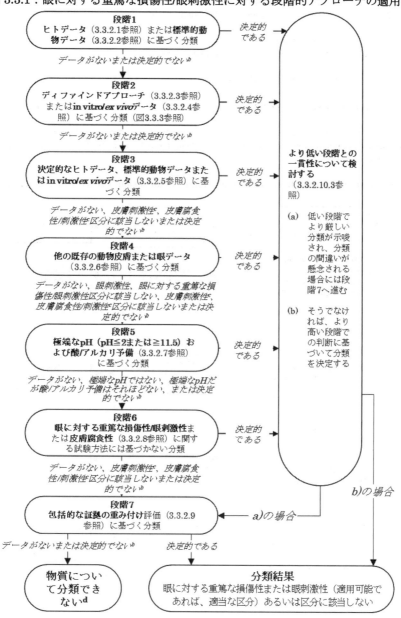

図 3.3.1：眼に対する重篤な損傷性/眼刺激性に対する段階的アプローチの適用 [a]

[a] このアプローチを適用する前に、3.3.2.10 の説明文及び手引き 3.3.5.3 を参照すること。段階的アプローチの適用には、十分な質の信頼できるデータのみが含まれるべきである。

[b] 情報が決定的でなく結論が出ない理由は以下のように様々である：
- 実験計画および／または報告に関連する品質上の問題により、利用可能なデータは質が十分ではない、または分類の目的には不十分／不適切である可能性がある；

- *The available data may be insufficient to conclude on the classification, e.g. they might be indicative for absence of serious eye damage, but inadequate to demonstrate eye irritation;*
- *Where competent authorities make use of the eye irritation categories 2A and 2B, the available data may not be capable of distinguishing between Category 2A and Category 2B.*

c *It is recognized that not all skin irritants are eye irritants and that not all substances that are non-irritant to skin are non-irritant to the eye (see 3.3.2.5, 3.3.2.6, 3.3.2.8.4 and 3.3.2.9.1).*

d *For mixtures, the flow chart in figure 3.3.2 should be followed.*

3.3.3 Classification criteria for mixtures

The approach to classification for serious eye damage/eye irritation is tiered and is dependent upon the amount of information available for the mixture itself and for its ingredients. The flow chart of figure 3.3.2 below outlines the process to be followed.

Figure 3.3.2: Tiered approach to classification of mixtures for serious eye damage/eye irritation

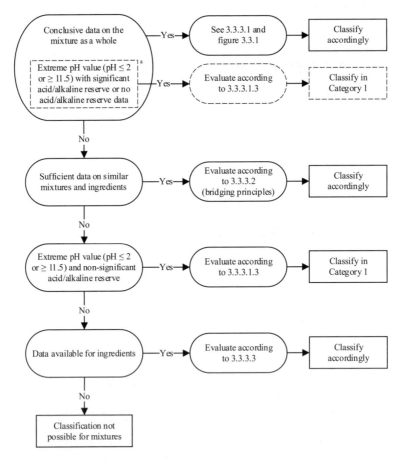

a *The dashed boxes represent an individual tier within conclusive data on the mixture as a whole. However, in contrast to substances, mixtures having an extreme pH value (pH ≤ 2 or ≥ 11.5) and non-significant acid/alkaline reserve but no other conclusive data on the mixture as a whole, or no conclusive weight of evidence assessment from all available data on the mixture as a whole, are not conclusive within the tiers for conclusive data on the mixture as a whole. Such mixtures should be first evaluated according to the bridging principles before the extreme pH value is considered as conclusive for classification.*

- 利用可能なデータは、分類を結論付けるには不十分な可能性がある。例えば、眼に重篤な損傷性がないことを示すかもしれないが、眼刺激性を示すには不十分な可能性がある；
- 所管官庁が眼刺激性区分2Aおよび2Bを使用する場合、利用可能なデータによっては区分2Aと区分2Bを区別することができない可能性がある。

c すべての皮膚刺激性物質が眼刺激性であるとは限らず、また、皮膚に対する刺激性がない物質がすべて眼に対する刺激性もないとは限らないことが認識されている（3.3.2.5、3.3.2.6、3.3.2.8.4 および 3.3.2.9.1 参照）。

d 混合物の場合は、図 3.3.2 のフローチャートに従うこと。

3.3.3 混合物の分類基準

眼に対する重篤な損傷性／眼刺激性の分類方法は段階的であり、混合物自体およびその成分について利用可能な情報量に依存する。以下の図 3.3.2 のフローチャートは、従うべきプロセスの概要を示している。

図 3.3.2：眼に対する重篤な損傷性／眼刺激性の混合物の分類のための段階的アプローチ

```
混合物そのものの              →はい→  3.3.3.1項および   →  適宜
決定的なデータ                        図3.3.1参照          分類する

極端なpH値（pH≦2また          →はい→  3.3.3.1.3項に    →  区分1に
は≧11.5）で相当量の酸/ア              したがって評価      分類する
ルカリ予備があるまたは酸/
アルカリ予備データがない a

    ↓いいえ

類似の混合物および            →はい→  3.3.3.2項に      →  適宜
成分の十分なデータ                    したがって評価      分類する
                                     （つなぎの原則）

    ↓いいえ

極端なpH値（pH≦2または        →はい→  3.3.3.1.3項に    →  区分1に
≧11.5）かつ酸/アルカリ予              したがって評価      分類する
備はそれほどない

    ↓いいえ

成分のデータが利用可能        →はい→  3.3.3.3項に      →  適宜
                                     したがって評価      分類する

    ↓いいえ

混合物については
分類できない
```

a 破線のボックスは、混合物そのものの決定的なデータのうちの個々の段階を表す。ただし、物質とは対照的に、極端なpH値（pH≦2または≧11.5）かつ酸/アルカリ予備はそれほどないが、混合物そのものに関する他の決定的なデータがない、または混合物そのものに関するすべての利用可能なデータから決定的な証拠の重み付け評価がない混合物は、混合物そのものに関する決定的なデータの段階においては決定的であるとは言えない。そのような混合物は、極端なpH値が分類の決定的なものと見なされる前に、まずつなぎの原則に従って評価されなければならない。

3.3.3.1 *Classification of mixtures when data are available for the complete mixture*

3.3.3.1.1 In general, the mixture should be classified using the criteria for substances, taking into account the tiered approach to evaluate data for this hazard class (as illustrated in figure 3.3.1) and 3.3.3.1.2 and 3.3.3.1.3 below. If classification is not possible using the tiered approach, then the approach described in 3.3.3.2 (bridging principles), or, if that is not applicable, 3.3.3.3 (classification based on ingredients) should be followed.

3.3.3.1.2 Defined approaches and/or in vitro/*ex vivo* test methods validated according to international procedures may not have been validated using mixtures; although these approaches/methods are considered broadly applicable to mixtures, they can only be used for classification of mixtures when all ingredients of the mixture fall within the applicability domain of the defined approach or test method(s) used. Specific limitations regarding applicability domains are described in the respective defined approaches and test methods and should be taken into consideration as well as any further information on such limitations from the published literature. Where there is reason to assume or evidence indicating that the applicability domain of a particular defined approach or test method is limited, data interpretation should be exercised with caution, or the results should be considered not applicable.

3.3.3.1.3 A mixture with an extreme pH (pH ≤ 2 or ≥ 11.5) is considered to cause serious eye damage (Category 1) in tier 5 if it has a significant acid/alkaline reserve or if no data for acid/alkaline reserve are available. However, if consideration of acid/alkaline reserve suggests the mixture may not cause serious eye damage despite the extreme pH value, the result is considered inconclusive within tier 5 (see figure 3.3.1). If the overall weight of evidence assessment remains inconclusive or no data other than pH and acid/alkaline reserve are available, mixtures with an extreme pH (pH ≤ 2 or ≥ 11.5) and non-significant acid/alkaline reserve should be assessed using the bridging principles described in 3.3.3.2. If the bridging principles cannot be applied, mixtures with an extreme pH (pH ≤ 2 or ≥ 11.5) and non-significant acid/alkaline reserve should be classified as eye Category 1 (see figure 3.3.2). A pH > 2 and < 11.5 is considered inconclusive and cannot be used for classification purposes. Acid/alkaline reserve and pH can be determined by different methods including those described in OECD Test Guideline 122 and Young et al. (1988), acknowledging that there are some differences between these methods (see 3.3.5.3.7). A competent authority may decide which criteria for significant acid/alkaline reserve can be applied.

3.3.3.2 *Classification of mixtures when data are not available for the complete mixture: bridging principles*

3.3.3.2.1 Where the mixture itself has not been tested to determine its skin corrosivity or potential to cause serious eye damage or eye irritation, but there are sufficient data on both the individual ingredients and similar tested mixtures to adequately characterize the hazards of the mixture, these data will be used in accordance with the following agreed bridging principles. This ensures that the classification process uses the available data to the greatest extent possible in characterizing the hazards of the mixture without the necessity for additional testing in animals.

3.3.3.2.2 *Dilution*

If a tested mixture is diluted with a diluent which has an equivalent or lower classification for serious eye damage/eye irritation than the least seriously eye damaging/eye irritant original ingredient and which is not expected to affect the serious eye damage /eye irritancy of other ingredients, then the new diluted mixture may be classified as equivalent to the original tested mixture. Alternatively, the method explained in 3.3.3.3 could be applied.

3.3.3.2.3 *Batching*

The serious eye damage/eye irritation potential of a tested production batch of a mixture can be assumed to be substantially equivalent to that of another untested production batch of the same commercial product when produced by or under the control of the same manufacturer, unless there is reason to believe there is significant variation such that the serious eye damage/eye irritation potential of the untested batch has changed. If the latter occurs, a new classification is necessary.

3.3.3.2.4 *Concentration of mixtures of the highest serious eye damage/eye irritation category*

If a tested mixture classified for serious eye damage (Category 1) is concentrated, the more concentrated untested mixture should be classified for serious eye damage (Category 1) without additional testing. If a tested mixture classified for eye irritation (Category 2 or 2A) is concentrated and does not contain serious eye damage ingredients, the more concentrated untested mixture should be classified in the same category (Category 2 or 2A) without additional testing.

3.3.3.1　混合物そのもののデータが利用できる場合の混合物の分類

3.3.3.1.1　一般に、混合物は、この危険有害性クラスのデータを評価するための段階的アプローチ（図3.3.1 に示される）および以下の 3.3.3.1.2 および 3.3.3.1.3 を考慮して、物質の基準を用いて分類されるものとする。段階的アプローチで分類できない場合は、3.3.3.2（つなぎの原則）に記載されているアプローチに従うか、それが適用できない場合は 3.3.3.3（成分に基づく分類）に従うべきである。

3.3.3.1.2　ディファインドアプローチおよび／または国際的な手順にしたがって検証された in vitro/ex vivo 試験法は、混合物の使用では検証されていないかもしれない。これらの方式／試験法は、混合物に広く適用可能であると考えられているが、混合物のすべての成分が、使用されたディファインドアプローチまたは試験法の適用範囲に該当する場合にのみ、これらは混合物の分類に使用することができる。それぞれのディファインドアプローチおよび試験法において適用範囲に関して明確な制限が記載されており、公表されている文献でのそのような制限に関するさらなる情報と同様に、これらは考慮されるべきである。具体的なディファインドアプローチまたは試験法の適用範囲を制限していることを示す前提または証拠を示す理由がある場合には、データの解釈に注意を払わなければならず、あるいは結果は適用できないと考えるべきである。

3.3.3.1.3　極端な pH（pH≦2 または≧11.5）の混合物は、相当量の酸/アルカリ予備がある場合、または酸/アルカリ予備のデータが利用できない場合、段階 5 で眼に対する重篤な損傷性（区分 1）があると見なされる。ただし、酸/アルカリ予備を考慮して、極端な pH 値にもかかわらず、混合物が眼に対する重篤な損傷性を引き起こさない可能性があることを示唆している場合、結果は段階 5 において決定的ではないと見なされる（図 3.3.1 参照）。包括的な証拠の重み付け評価で決定的な結論が出ない場合、または pH および酸/アルカリ予備以外のデータが利用できない場合、極端な pH（pH≦2 または≧11.5）かつ酸/アルカリ予備はそれほどない混合物は、3.3.3.2 で説明されているつなぎの原則を使用して評価する必要がある。つなぎの原則を適用できない場合、極端な pH（pH≦2 または≧11.5）かつ酸/アルカリ予備はそれほどない混合物は、眼区分 1 に分類する必要がある（図 3.3.2 参照）。2 < pH < 11.5 の場合は、決定的ではないと見なされ、分類目的には使用できない。酸／アルカリ予備および pH は、OECD テストガイドライン 122 および Young ら（1988）に記載されている方法を含む異なる方法で測定することができるが、これらの方法の間にいくつかの相違があることを認めている（3.3.5.3.7 参照）。所管官庁は、相当量の酸/アルカリ予備がある場合にどの基準を適用できるかを決定することができる。

3.3.3.2　*混合物そのものについてデータが利用できない場合の混合物の分類：つなぎの原則 (bridging principles)*

3.3.3.2.1　混合物そのものは皮膚腐食性、眼に対する重篤な損傷性または眼刺激性を決定する試験がなされていないが、当該混合物の有害性を適切に特定するための、個々の成分および試験された類似の混合物の両方に関して十分なデータがある場合、これらのデータは以下の合意されたつなぎの原則にしたがって利用される。これによって分類手順において、動物試験を追加する必要もなく、混合物の有害性判定に利用可能なデータを可能な限り最大限に用いることができるようになる。

3.3.3.2.2　*希釈*

試験された混合物が眼に対する重篤な損傷性/眼刺激性の最も弱い元の成分に比べて同等以下の眼に対する重篤な損傷性/眼刺激性分類に属する物質で希釈され、その物質が他の成分の眼に対する重篤な損傷性/眼刺激性に影響を与えないことが予想されれば、新しい希釈された混合物は、試験された元の混合物と同等として分類してもよい。あるいは、3.3.3.3 節で説明する方法も適用できる。

3.3.3.2.3　*製造バッチ*

混合物の試験されていない製造バッチに眼に対する重篤な損傷性/眼刺激性があるかどうかは、同じ製造者によって、またはその管理下で生産された同じ商品の試験された別のバッチの毒性と本質的に同等とみなすことができる。ただし、試験されていないバッチの眼に対する重篤な損傷性/眼刺激性の可能性が変化するような有意な変動があると考えられる理由がある場合はこの限りではない。このような場合には、新しい分類が必要である。

3.3.3.2.4　*最も強い眼に対する重篤な損傷性/眼刺激性区分の混合物の濃縮*

眼に対する重篤な損傷性（区分 1）に分類された試験混合物が濃縮された場合には、より濃度が高い試験されていない混合物は追加試験なしで眼に対する重篤な損傷性（区分 1）に分類すべきである。皮膚/眼刺激性について眼刺激性（区分 2 または 2A）に分類された試験混合物が濃縮され、眼に対して重篤な損傷性を引き起こす成分を含まなければ、より濃度が高い試験されていない混合物は追加試験なしで眼刺激性（区分 2 または 2A）に分類すべきである。

3.3.3.2.5   *Interpolation within one hazard category*

For three mixtures (A, B and C) with identical ingredients, where mixtures A and B have been tested and are in the same serious eye damage/eye irritation hazard category, and where untested mixture C has the same toxicologically active ingredients as mixtures A and B but has concentrations of toxicologically active ingredients intermediate to the concentrations in mixtures A and B, then mixture C is assumed to be in the same serious eye damage/eye irritation category as A and B.

3.3.3.2.6   *Substantially similar mixtures*

Given the following:

(a)   Two mixtures: (i)   A +B
                      (ii) C + B;

(b)   The concentration of ingredient B is essentially the same in both mixtures;

(c)   The concentration of ingredient A in mixture (i) equals that of ingredient C in mixture (ii);

(d)   Data on serious eye damage/eye irritation for A and C are available and substantially equivalent, i.e. they are in the same hazard category and are not expected to affect the serious eye damage/eye irritation potential of B.

If mixture (i) or (ii) is already classified based on test data, the other mixture can be classified in the same hazard category.

3.3.3.2.7   *Aerosols*

An aerosolized form of a mixture may be classified in the same hazard category as the tested non-aerosolized form of the mixture provided that the added propellant does not affect the serious eye damage/eye irritation properties of the mixture upon spraying[4].

3.3.3.3   *Classification of mixtures when data are available for all ingredients or only for some ingredients of the mixture*

3.3.3.3.1   In order to make use of all available data for purposes of classifying the serious eye damage/eye irritation hazards of the mixtures, the following assumption has been made and is applied where appropriate in the tiered approach for mixtures (see 1.3.2.3):

The "relevant ingredients" of a mixture are those which are present in concentrations ≥ 1 % (w/w for solids, liquids, dusts, mists and vapours and v/v for gases), unless there is a presumption (e.g. in the case of corrosive ingredients) that an ingredient present at a concentration < 1 % can still be relevant for classifying the mixture for serious eye damage/eye irritation.

3.3.3.3.2   In general, the approach to classification of mixtures as seriously damaging to the eye or eye irritant when data are available on the ingredients, but not on the mixture as a whole, is based on the theory of additivity, such that each corrosive or serious eye damaging/eye irritant ingredient contributes to the overall serious eye damage/eye irritation properties of the mixture in proportion to its potency and concentration. A weighting factor of 10 is used for corrosive and serious eye damaging ingredients when they are present at a concentration below the concentration limit for classification with Category 1, but are at a concentration that will contribute to the classification of the mixture as serious eye damaging/eye irritant. The mixture is classified as seriously damaging to the eye or eye irritant when the sum of the concentrations of such ingredients exceeds a threshold cut-off value/concentration limit.

3.3.3.3.3   Table 3.3.3 provides the cut-off value/concentration limits to be used to determine if the mixture should be classified as seriously damaging to the eye or an eye irritant.

3.3.3.3.4   Particular care must be taken when classifying mixtures containing certain types of substances such as acids and bases, inorganic salts, aldehydes, phenols, and surfactants. The approach explained in 3.3.3.3.1 and 3.3.3.3.2 might not work given that many such substances are seriously damaging to the eye/eye irritating at concentrations < 1 %.

---

[4]   *Bridging principles apply for the intrinsic hazard classification of aerosols, however, the need to evaluate the potential for "mechanical" eye damage from the physical force of the spray is recognized.*

3.3.3.2.5　1つの有害性区分の中での内挿

3つの混合物（A、BおよびC）は同じ成分を持ち、AとBは試験され同じ眼に対する重篤な損傷性/眼刺激性の有害性区分にある。混合物 C は混合物 A および B と同じ毒性学的に活性な成分を持ち、毒性学的に活性な成分の濃度が混合物 A と B の中間である場合、混合物 C は、A および B と同じ眼に対する重篤な損傷性/眼刺激性の区分であると推定される。

3.3.3.2.6　本質的に類似した混合物

次を仮定する：

(a)　2つの混合物：　(i)　　A+B
　　　　　　　　　　(ii)　　C+B

(b)　成分 B の濃度は、両方の混合物で本質的に同じである。

(c)　混合物(i)の成分 A の濃度は、混合物(ii)の成分 C の濃度に等しい。

(d)　A と C の眼に対する重篤な損傷性/眼刺激性に関するデータが利用でき、実質的に同等であり、すなわち A と C は同じ有害性区分に属し、かつ B の眼に対する重篤な損傷性/眼刺激性の可能性には影響を与えることはないと判断される。

混合物(i)または(ii)が既に試験データに基づき分類されている場合には、他方の混合物は同じ有害性区分に分類することができる。

3.3.3.2.7　エアゾール

エアゾール形態の混合物は、添加された噴射剤が噴霧時に混合物の眼に対する重篤な損傷性/眼刺激性に影響しないという条件下では、試験された非エアゾール形態の混合物と同じ有害性区分に分類してよい[4]。

3.3.3.3　混合物の全成分についてまたは一部の成分だけについてデータが利用可能な場合の混合物の分類

3.3.3.3.1　混合物の眼に対する重篤な損傷性/眼刺激性を分類する目的のため利用可能なすべてのデータを使用するために、下記が仮定されており、混合物の段階的なアプローチで適宜その仮定が適用される。

混合物の「考慮すべき成分」とは、≧1%の濃度（固体、液体、粉塵、ミストおよび蒸気については重量/重量、気体については体積/体積）で存在するものである。ただし、（例えば腐食性の成分の場合に）1%より低い濃度で存在する成分が、なお眼に対する重篤な損傷性/眼刺激性についての分類に関係する可能性があるという前提がある場合はこの限りではない。

3.3.3.3.2　一般的に、各成分のデータは利用可能だが、混合物そのもののデータがない場合、眼の重篤な損傷性または眼刺激性として混合物を分類する方法は加成法の理論に基づく。すなわち、腐食性または重篤な損傷性/刺激性の各成分がその程度および濃度に応じて、混合物そのものの眼に対する重篤な損傷性/眼刺激性に寄与しているという理論である。腐食性および眼に対する重篤な損傷性の成分が区分 1 と分類できる濃度以下であるが、混合物を眼に対する重篤な損傷性/刺激性に分類するのに寄与する濃度で含まれる場合には、加重係数として 10 を用いる。各成分の濃度の合計がカットオフ値/濃度限界を超えた場合、その混合物は眼に対する重篤な損傷性または眼刺激性として分類される。

3.3.3.3.3　表 3.3.3 に混合物を眼に対する重篤な損傷性あるいは眼刺激性に分類すべきかを決定するためのカットオフ値/濃度限界を示した。

3.3.3.3.4　酸、塩基、無機塩、アルデヒド、フェノールおよび界面活性剤のようなある具体的な種類の物質を含む混合物を分類する場合には特別の注意を払わなければならない。これらの化合物の多くは＜ 1%の濃度であっても眼に対する重篤な損傷性/眼刺激性を示す場合があるので、3.3.3.3.1 および 3.3.3.3.2 に記述し

---

[4] つなぎの原則はエアゾールの本質的な有害性分類に適用されるが、スプレーの物理的な力による「機械的な」眼損傷の可能性も評価する必要があることが認められている。

For mixtures containing strong acids or bases the pH should be used as the classification criterion (see 3.3.3.1.3) since extreme pH will be a better indicator of serious eye damage than the concentration limits in table 3.3.3. A mixture containing corrosive or serious eye damaging/eye irritating ingredients that cannot be classified based on the additivity approach applied in table 3.3.3 due to chemical characteristics that make this approach unworkable, should be classified as eye Category 1 if it contains ≥ 1 % of a corrosive or serious eye damaging ingredient and as eye Category 2 when it contains ≥ 3 % of an eye irritant ingredient. Classification of mixtures with ingredients for which the approach in table 3.3.3 does not apply is summarized in table 3.3.4.

3.3.3.3.5　　　On occasion, reliable data may show that the serious eye damage/eye irritation of an ingredient will not be evident when present at a level above the generic cut-off values/concentration limits mentioned in tables 3.3.3 and 3.3.4. In these cases, the mixture could be classified according to those data (see also 1.3.3.2). On occasion, when it is expected that the skin corrosion/irritation or the serious eye damage/eye irritation of an ingredient will not be evident when present at a level above the generic cut-off levels/concentration limits mentioned in tables 3.3.3 and 3.3.4, testing of the mixture may be considered.

3.3.3.3.6　　　If there are data showing that (an) ingredient(s) may be corrosive to the skin or seriously damaging to the eye/eye irritating at a concentration of < 1 % (corrosive to the skin or seriously damaging to the eye) or < 3 % (eye irritant), the mixture should be classified accordingly (see also 1.3.3.2 *"Use of cut-off values/concentration limits"*).

Table 3.3.3: Concentration of ingredients of a mixture classified as skin Category 1 and/or eye Category 1 or 2 that would trigger classification of the mixture as hazardous to the eye (Category 1 or 2)

Sum of ingredients classified as	Concentration triggering classification of a mixture as:	
	serious eye damage	eye irritation
	Category 1	Category 2/2A
Skin Category 1 + eye Category 1[a]	≥ 3 %	≥ 1 % but < 3 %
Eye Category 2		≥ 10 %[b]
10 × (skin Category 1 + eye Category 1)[a] + eye Category 2		≥ 10 %

[a]　*If an ingredient is classified as both skin Category 1 and eye Category 1 its concentration is considered only once in the calculation;*

[b]　*A mixture may be classified as eye Category 2B when all relevant ingredients are classified as eye Category 2B.*

Table 3.3.4: Concentration of ingredients of a mixture when the additivity approach does not apply, that would trigger classification of the mixture as hazardous to the eye

Ingredient	Concentration	Mixture classified as: eye
Acid with pH ≤ 2	≥ 1 %	Category 1
Base with pH ≥ 11.5	≥ 1 %	Category 1
Other corrosive (eye Category 1) ingredient	≥ 1 %	Category 1
Other eye irritant (eye Category 2) ingredient	≥ 3 %	Category 2

た方法は機能しないであろう。強酸または強塩基を含む混合物に関して、極端なpHは表 3.3.3 にある濃度限界よりも眼に対する重篤な損傷性のより適切な指標であるから、pHを分類基準として使用すべきである（3.3.3.1.3 参照）。眼に対する重篤な損傷性/眼刺激性の成分を含む混合物で、化学品の特性により、表 3.3.3 に示された加成方式に基づいて分類できない場合、≧1%の腐食性または眼に対する重篤な損傷性の成分を含む場合には、眼区分 1 に分類する。また、≧3%の眼刺激性成分を含む場合は眼区分 2 に分類する。表 3.3.3 の方法が適用できない混合物の分類は表 3.3.4 にまとめられている。

3.3.3.3.5 時には、表 3.3.3 および 3.3.4 に示されている一般的なカットオフ値/濃度限界を超えるレベルで存在するのに、眼に対する重篤な損傷性/眼刺激性を否定する信頼できるデータがある場合がある。この場合には、混合物はそのデータに基づき分類できる（1.3.3.2 参照）。また、ある成分が表 3.3.3 および 3.3.4 に述べる一般的なカットオフ値/濃度限界以上であっても、皮膚の腐食性/刺激性、あるいは眼に対する重篤な損傷性/眼刺激性がないと予想される場合は、混合物そのものでの試験実施を検討してもよい。

3.3.3.3.6 ある成分について、皮膚腐食性ないし眼に対する重篤な損傷性の場合 ＜ 1%、眼刺激性の場合 ＜ 3%の濃度でも、皮膚腐食性ないし眼に対する重篤な損傷性/眼刺激性であることを示すデータがある場合は、混合物はそれにしたがって分類されるべきである（1.3.3.2「カットオフ値/濃度限界の使用」参照）。

表 3.3.3：皮膚区分 1 または眼区分 1、2 として分類される混合物成分の濃度、混合物を眼有害性と分類する際の基準（区分 1 または 2）

各成分の合計による分類	混合物を分類するための成分濃度	
	眼に対する重篤な損傷性	眼刺激性
	区分 1	区分 2/2A
皮膚区分 1 ＋ 眼区分 1 [a]	≧3%	≧1%、＜3%
眼区分 2		≧10% [b]
10×（皮膚区分 1 ＋ 眼区分 1）[a] ＋ 眼区分 2		≧10%

[a] 1つの成分が皮膚区分 1 および眼区分 1 の両方に分類されていた場合、その濃度は計算に一度だけ入れる；
[b] すべての考慮すべき成分が眼区分 2B と分類されている場合、混合物は眼区分 2B と分類してもよい。

表 3.3.4：加成方式を適用しない混合物成分の濃度、混合物を眼有害性と分類する際の基準

成分	濃度	混合物の分類 眼
酸　　pH≦2	≧1%	区分 1
塩基　pH≧11.5	≧1%	区分 1
その他の腐食性（眼区分 1）成分	≧1%	区分 1
その他の眼刺激性（眼区分 2）成分（酸、塩基を含む）	≧3%	区分 2

3.3.4 Hazard communication

General and specific considerations concerning labelling requirements are provided in *Hazard communication: Labelling* (chapter 1.4). Annex 1 contains summary tables about classification and labelling. Annex 3 contains examples of precautionary statements and pictograms which can be used where allowed by the competent authority. Table 3.3.5 presents specific label elements for substances and mixtures classified into this hazard class based on the criteria in this chapter.

Table 3.3.5: Label elements for serious eye damage/eye irritation [a]

	Category 1	**Category 2/2A**	**Category 2B**
Symbol	Corrosion	Exclamation mark	*No symbol*
Signal word	Danger	Warning	Warning
Hazard statement	Causes serious eye damage	Causes serious eye irritation	Causes eye irritation

[a] *Where a chemical is classified as skin Category 1, labelling for serious eye damage/eye irritation may be omitted as this information is already included in the hazard statement for skin Category 1 (Causes severe skin burns and eye damage) (see chapter 1.4, paragraph 1.4.10.5.3.3).*

3.3.5 Decision logics and guidance

The decision logics which follow are not part of the harmonized classification system but are provided here as additional guidance. It is strongly recommended that the person responsible for classification study the criteria before and during use of the decision logics.

3.3.5.1 *Decision logic 3.3.1 for serious eye damage/eye irritation*

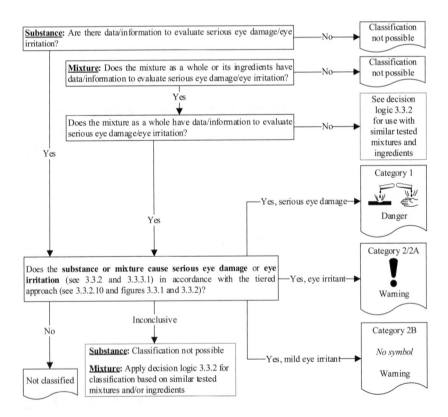

- 155 -

3.3.4 危険有害性情報の伝達

表示要件に関する通則および細則は、*危険有害性に関する情報の伝達：表示*（第1.4章）に定める。附属書1に分類と表示に関する概要表を示す。附属書3には、注意書きおよび所管官庁が許可した場合に使用可能な注意絵表示の例を示す。表3.3.5は、本章の判定基準に基づいて、この有害性クラスに分類される物質および混合物の具体的なラベル要素を示す。

表 3.3.5：眼に対する重篤な損傷性/眼刺激性のラベル要素 [a]

	区分 1	区分 2/2A	区分 2B
シンボル	腐食性	感嘆符	シンボルなし
注意喚起語	危険	警告	警告
危険有害性情報	重篤な眼の損傷	強い眼刺激	眼刺激

[a] 化学品が皮膚区分1と分類されている場合、眼に対する重篤な損傷性/眼刺激の表示は省略しても良い。この情報はすでに皮膚区分1に対する有害性情報の中に含まれている（重篤な皮膚の薬傷・眼の損傷）（第1.4章1.4.10.5.3.3参照）。

3.3.5 判定論理

以下の判定論理は、調和分類システムには含まれないが、追加的な手引きとしてここに示す。分類の責任者に対し、この判定論理を使用する前および使用する際に判定基準についてよく調べ理解することを強く勧める。

3.3.5.1 *判定論理 3.3.1 眼に対する重篤な損傷性/眼刺激性*

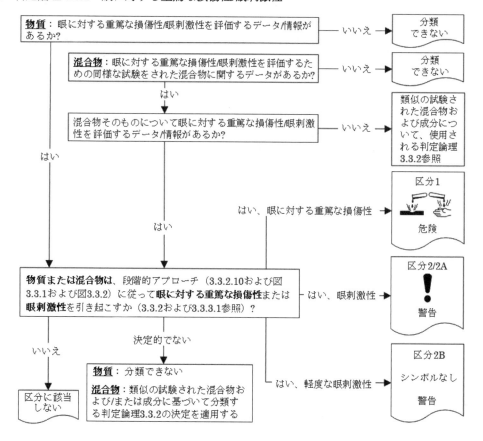

3.3.5.2 *Decision logic 3.3.2 for serious eye damage/eye irritation*

Classification of mixtures on the basis of information/data on similar tested mixtures and ingredients

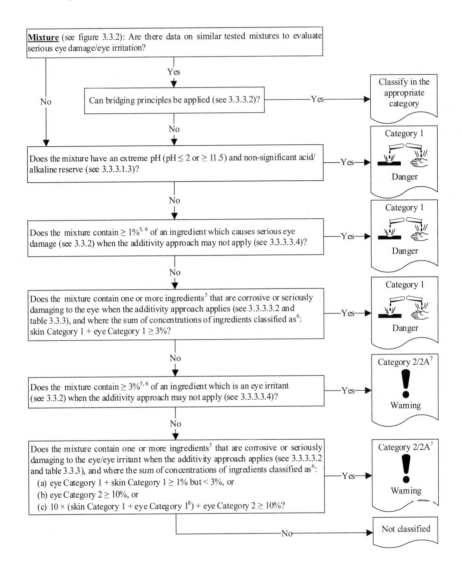

---

[5] Where relevant < 1%, see 3.3.3.3.1.

[6] For specific concentration limits, see 3.3.3.3.5 and 3.3.3.3.6. See also chapter 1.3, paragraph 1.3.3.2 "Use of cut-off values/concentration limits".

[7] A mixture may be classified as eye Category 2B in case all relevant ingredients are classified as eye Category 2B.

[8] If an ingredient is classified as both skin Category 1 and eye Category 1 its concentration is considered only once in the calculation.

- 156 -

3.3.5.2 判定論理 3.3.2 眼に対する重篤な損傷性/眼刺激性

試験された同様な混合物および成分の情報/データに基づいた混合物の分類

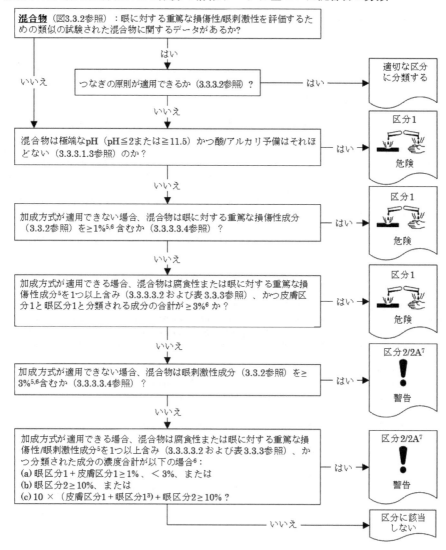

---

5 考慮すべき成分 < 1% の場合、3.3.3.3.1 参照

6 具体的な濃度限界については 3.3.3.3.5 および 3.3.3.3.6 参照、「カットオフ値/濃度限界の使用」については第 1.3 章 1.3.3.2 も参照。

7 すべての考慮すべき成分が眼区分 2B と分類されている場合、混合物は眼区分 2B と分類しても良い。

8 1 つの成分が皮膚区分 1 および眼区分 1 の両方に分類されていた場合、その濃度は計算に一度だけ入れる。

3.3.5.3 *Background guidance*

3.3.5.3.1 *Relevant guidance documents*

Helpful information on the strengths and weaknesses of the different test and non-test methods, as well as useful guidance on how to apply a weight of evidence assessment, is provided in OECD Guidance Document 263 on an integrated approach on testing and assessment (IATA) for serious eye damage and eye irritation.

3.3.5.3.2 *Guidance on the use of human data for classification as serious eye damage/eye irritation*

The availability of human data for serious eye damage/eye irritation is limited and the data available may contain some uncertainty. However, where such data exist, they should be considered based on their quality. Human data may be obtained from epidemiological studies, human experience (e.g. consumer experience), poison control centres, national and international home accident surveillance programs, case studies, or worker experience and accidents. Human case studies may have limited predictive value as often the presence of a substance or mixture in the eye will result in pain and quick washing of the eyes. Therefore, the effects observed may underestimate the intrinsic property of the substance or the mixture to affect the eye without washing. Further details on the strengths and limitations of human data for serious eye damage/eye irritation can be found in OECD Guidance Document 263 (section 4.1. Module 1: Existing human data on serious eye damage and eye irritation).

3.3.5.3.3 *Classification based on standard animal tests with more than 3 animals*

3.3.5.3.3.1 Classification criteria for the skin and eye hazard classes are detailed in the GHS in terms of a 3-animal test. It has been identified that some older test methods may have used up to 6 animals. However, the GHS criteria do not specify how to classify based on existing data from tests with more than 3 animals. Guidance on how to classify based on existing data from studies with 4 or more animals is given in the following paragraphs.

3.3.5.3.3.2 Classification criteria based on a 3-animal test are detailed in 3.3.2.2. Evaluation of a 4, 5 or 6 animal study should follow the criteria in the following paragraphs, depending on the number of animals tested. Scoring should be performed at 24, 48 and 72 hours after instillation of the test material.

3.3.5.3.3.3 In the case of a study with 6 animals the following principles apply:

(a) The substance or mixture is classified as serious eye damage Category 1 if:

(i) at least in one animal effects on the cornea, iris or conjunctiva are not expected to reverse or have not fully reversed within an observation period of normally 21 days; and/or

(ii) at least 4 out of 6 animals show a mean score per animal of ≥ 3 for corneal opacity and/or > 1.5 for iritis.

(b) The substance or mixture is classified as eye irritation Category 2/2A if at least 4 out of 6 animals show a mean score per animal of:

(i) ≥ 1 for corneal opacity; and/or

(ii) ≥ 1 for iritis; and/or

(iii) ≥ 2 for conjunctival redness; and/or

(iv) ≥ 2 for conjunctival oedema (chemosis),

and which fully reverses within an observation period of normally 21 days.

(c) The substance or mixture is classified as irritating to eyes (Category 2B) if the effects listed in sub-paragraph (b) above are fully reversible within 7 days of observation.

3.3.5.3　基本的手引き

3.3.5.3.1　関連の手引き

異なる試験法および試験によらない方法の強みと弱み、ならびに証拠の重み付け評価の適用方法に関する有用な情報は、OECD ガイダンス文書 263 に眼に対する重篤な損傷性および眼刺激性の試験および評価に関する統合的アプローチ（IATA）として記載されている。

3.3.5.3.2　眼に対する重篤な損傷性/眼刺激性の分類のためのヒトのデータの使用に関する手引き

眼に対する重篤な損傷性/眼刺激性に関するヒトのデータの利用可能性は限られており、利用可能なデータには不確実性が含まれる場合がある。しかし、そのようなデータが存在する場合、その質に基づいて検討する必要がある。ヒトのデータは、疫学的研究、ヒトの経験（消費者の経験など）、中毒センター、国内外の家庭事故調査プログラム、ケーススタディ、または労働者の経験や事故から得ることができる。物質や混合物が眼に入ると、痛みを感じたり、すぐに眼を洗ったりすることが多いため、ヒトのケーススタディでは予測値に限界がある場合がある。したがって、観察された影響は、洗浄なしで眼に影響を及ぼす物質または混合物の本質的な特性を過小評価する可能性がある。眼に対する重篤な損傷性/眼刺激性に関するヒトのデータの強みと限界に関するさらなる詳細は、OECD ガイダンス文書 263（第 4.1 項 モジュール 1: 眼に対する重篤な損傷性および眼刺激性に関する既存のヒトのデータ）に記載されている。

3.3.5.3.3　3 匹を超える動物を用いた標準的な動物実験に基づく分類

3.3.5.3.3.1　皮膚および眼の有害性クラスに対する判定基準は 3 匹の動物試験として GHS に詳述されている。動物を 6 匹まで使用してもよいとする古い試験方法も知られている。しかし、GHS の判定基準では 3 匹を超える動物の試験による既存のデータに基づいた分類について明記していない。4 匹以上の動物を用いた試験による既存データに基づいてどのように分類を行うかに関する手引きを以下のパラグラフに示す。

3.3.5.3.3.2　3 匹の動物試験に基づいた分類判定基準は 3.3.2.2 に詳述されている。4、5 または 6 匹の動物試験の評価は、試験動物の数により、以下のパラグラフによる判定基準にしたがうべきである。スコア付けは試験物質の滴下後、24、48 および 72 時間に実施するべきである。

3.3.5.3.3.3　6 匹の動物試験の場合、次の原則を適用する：

(a) 以下のような場合、物質または混合物を眼に対する重篤な損傷性区分 1 と分類する：

　(i) 少なくとも 1 匹の動物で、角膜、虹彩または結膜に対する作用の回復が期待できない、または通常 21 日間の観察期間中に完全には回復しない；および/または

　(ii) 動物 6 匹中少なくとも 4 匹で、動物ごとの平均スコアが角膜混濁 ≥ 3 および/または虹彩炎 > 1.5。

(b) 6 匹の中少なくとも 4 匹で、動物ごとの平均スコアが以下のような場合、物質または混合物を眼刺激区分 2/2A と分類する：

　(i) 角膜混濁 ≥ 1；および/または

　(ii) 虹彩炎 ≥ 1；および/または

　(iii) 結膜発赤 ≥ 2；および/または

　(iv) 結膜浮腫 ≥ 2、

　かつ、通常 21 日間の観察期間に完全に回復する。

(c) 上記(b)に記載された作用が 7 日間の観察で完全に回復した場合には、物質または混合物は眼刺激（区分 2B）と分類する。

3.3.5.3.3.4   In the case of a study with 5 animals the following principles apply:

   (a)   The substance or mixture is classified as serious eye damage Category 1 if:

      (i)   at least in one animal effects on the cornea, iris or conjunctiva are not expected to reverse or have not fully reversed within an observation period of normally 21 days; and/or

      (ii)   at least 3 out of 5 animals show a mean score per animal of ≥ 3 for corneal opacity and/or > 1.5 for iritis.

   (b)   The substance or mixture is classified as eye irritation Category 2/2A if at least 3 out of 5 animals show a mean score per animal of:

      (i) ≥ 1 for corneal opacity; and/or

      (ii) ≥ 1 for iritis; and/or

      (iii) ≥ 2 for conjunctival redness; and/or

      (iv) ≥ 2 for conjunctival oedema (chemosis),

      and which fully reverses within an observation period of normally 21 days.

   (c)   The substance or mixture is classified as irritating to eyes (Category 2B) if the effects listed in sub-paragraph (b) above are fully reversible within 7 days of observation.

3.3.5.3.3.5   In the case of a study with 4 animals the following principles apply:

   (a)   The substance or mixture is classified as serious eye damage Category 1 if:

      (i)   at least in one animal effects on the cornea, iris or conjunctiva are not expected to reverse or have not fully reversed within an observation period of normally 21 days; and/or

      (ii)   at least 3 out of 4 animals show a mean score per animal of ≥ 3 for corneal opacity and/or > 1.5 for iritis.

   (b)   Classification as eye irritation Category 2/2A if at least 3 out of 4 animals show a mean score per animal of:

      (i) ≥ 1 for corneal opacity; and/or

      (ii) ≥ 1 for iritis; and/or

      (iii) ≥ 2 for conjunctival redness; and/or

      (iv) ≥ 2 for conjunctival oedema (chemosis),

      and which fully reverses within an observation period of normally 21 days.

   (c)   The substance or mixture is classified as irritating to eyes (Category 2B) if the effects listed in sub-paragraph (b) above are fully reversible within 7 days of observation.

3.3.5.3.4   *Guidance on the use of defined approaches and/or in vitro/ex vivo data for classification within tier 2 of figure 3.3.1*

3.3.5.3.4.1   Defined approaches consist of a predefined set of different information sources (e.g. in vitro methods, *ex vivo* methods, physico-chemical properties, non-test methods) which, combined together through a fixed Data Interpretation Procedure (DIP) to convert input data into a prediction (or result), can provide a conclusion on the classification of a substance or mixture. A fixed DIP is defined as any fixed algorithm for interpreting data from one or typically several information sources and is rule-based in the sense that it is based, for example on a formula or an algorithm (e.g. decision criteria, rule or set of rules) that do not involve expert judgment. The output of a DIP generally is a prediction of a biological effect of interest or regulatory endpoint. Since in a defined approach the information sources

3.3.5.3.3.4　5匹の動物試験の場合、次の原則を適用する：

(a) 以下のような場合、物質または混合物を眼に対する重篤な損傷性区分1と分類する：

　(i) 少なくとも1匹の動物で、角膜、虹彩または結膜に対する作用の回復が期待できない、または通常21日間の観察期間中に完全には回復しない；および/または

　(ii) 動物5匹中少なくとも3匹で、動物ごとの平均スコアが角膜混濁 ≥3 、および/または虹彩炎 > 1.5。

(b) 5匹の中少なくとも3匹で、動物ごとの平均スコアが以下のような場合、物質または混合物を眼刺激区分2/2Aと分類する：

　(i) 角膜混濁 ≥1；および/または

　(ii) 虹彩炎 ≥1；および/または

　(iii) 結膜発赤 ≥2；および/または

　(iv) 結膜浮腫 ≥2 、

　かつ、通常21日間の観察期間に完全に回復する。

(c) 上記(b)に記載された作用が7日間の観察で完全に回復した場合には、物質または混合物は眼刺激（区分2B）と分類する。

3.3.5.3.3.5　4匹の動物試験の場合、次の原則を適用する：

(a) 以下のような場合、物質または混合物を眼に対する重篤な損傷性区分1と分類する：

　(i) 少なくとも1匹の動物で、角膜、虹彩または結膜に対する作用の回復が期待できない、または通常21日間の観察期間中に完全には回復しない；および/または

　(ii) 動物4匹中少なくとも3匹で、動物ごとの平均スコアが角膜混濁 ≥3 、および/または虹彩炎 > 1.5。

(b) 4匹の中少なくとも3匹で、動物ごとの平均スコアが以下のような場合、物質または混合物を眼刺激区分2/2Aと分類する：

　(i) 角膜混濁 ≥1；および/または

　(ii) 虹彩炎 ≥1；および/または

　(iii) 結膜発赤 ≥2；および/または

　(iv) 結膜浮腫 ≥2 、

　かつ、通常21日間の観察期間に完全に回復する。

(c) 上記(b)に記載された作用が7日間の観察で完全に回復した場合には、物質または混合物は眼刺激性（区分2B）と分類する。

3.3.5.3.4　*図3.3.1の段階2に分類するためのディファインドアプローチおよび／または in vitro/ex vivo データの使用に関する手引き*

3.3.5.3.4.1　ディファインドアプローチは、事前に定められた一連の異なる情報源（in vitro/ex vivo 法、物理化学的特性、試験によらない方法等）に基づき、これらは入力データを予測（または結果）に変換するための予め決められたデータ解釈手順（DIP）により組み合わされ、物質または混合物の分類に関する結論を提供することができる。予め決められたDIPは、1つまたは複数の情報源からのデータを解釈するための予め決められたアルゴリズムとして定められ、専門家の判断を必要としない公式またはアルゴリズム（例えば、決定基準、法則または法則のセット）に基づくという意味でルールベースである。DIPの出力は一般に、興味深い生物学的影響または規制当局のエンドポイントの予測である。ディファインドアプローチでは、情報源が規定され、それらを統合し解釈する方法に関する一連のルールが事前に決められて

are prescribed and the set of rules on how to integrate and interpret them is predetermined, the same conclusion will always be reached by different assessors on the same set of data as there is no room for subjective interpretation. In contrast, in a weight of evidence assessment, expert judgment is applied on an ad hoc basis to the available information, which may lead to different conclusions because there are no fixed rules for interpreting the data.

3.3.5.3.4.2 A stepwise approach to the evaluation of information derived from tier 2 of figure 3.3.1, i.e. defined approaches and/or in vitro/*ex vivo* test methods, should be considered where applicable (figure 3.3.3), recognizing that not all tiers as well as information within a tier may be relevant. However, all available and relevant information of sufficient quality needs to be examined for consistency with respect to the resulting classification. The outcome of a defined approach containing conclusive animal and/or human data may also eventually be considered during the overall weight of evidence in tier 7 (see figure 3.3.1). Where information from several steps is inconsistent and/or conflicting with respect to the resulting classification, information of sufficient quality from a higher step is generally given a higher weight than information from a lower step. However, when information from a lower step would result in a stricter classification than information from a higher step and there is concern for misclassification, then classification is determined by a within-tier weight of evidence assessment. For example, classifiers concerned with a negative result for serious eye damage in a defined approach when there is a positive result for serious eye damage in an in vitro/*ex vivo* method would utilise a within-tier weight of evidence assessment.

3.3.5.3.4.3 Current in vitro/*ex vivo* test methods are not able to distinguish between certain in vivo effects, such as corneal opacity, iritis, conjunctiva redness or conjunctiva chemosis, but they have shown to correctly predict substances inducing serious eye damage/eye irritation independently of the types of ocular effects observed in vivo. Many of the current in vitro/*ex vivo* test methods can thus identify substances or mixtures not requiring classification with high sensitivity but with limited specificity when used to distinguish not classified from classified substances or mixtures. This means that it is reasonably certain that a substance or mixture identified as not requiring classification by OECD Test Guideline 437, 438, 491, 492, 494 or 496 (see table 3.3.6) is indeed not inducing eye effects warranting classification, whereas some substances or mixtures not requiring classification will be over-predicted by these in vitro/*ex vivo* test methods when used in isolation. Furthermore, it should be considered that substances inducing serious eye damage are identified by many of these test methods with a high specificity but a limited sensitivity when used to distinguish Category 1 from Category 2 and not classified. This means that it is reasonably certain that a substance or mixture identified as Category 1 by OECD Test Guideline 437, 438, 460, 491 or 496 (see table 3.3.6) is indeed inducing irreversible eye effects, whereas some substances or mixtures inducing serious eye damage will be under-predicted by these in vitro/*ex vivo* test methods when used in isolation. As a consequence, a single in vitro/*ex vivo* OECD test guideline method is currently sufficient to conclude on either Category 1 or no classification according to the criteria defined in table 3.3.6, but not to conclude Category 2. When the result of an in vitro/*ex vivo* method is "no stand-alone prediction can be made" (e.g. see table 3.3.6), a conclusion cannot be drawn on the basis of that single result and further data are necessary for classification. Some in vitro/*ex vivo* test methods validated according to international procedures but not adopted as OECD test guidelines may be accepted by some competent authorities to classify in Category 2 (see 3.3.5.3.5.2). Moreover, combinations of in vitro/*ex vivo* methods in tiered approaches or their integration in defined approaches (see 3.3.2.3) may reduce the number of false predictions and show adequate performance for classification purposes.

3.3.5.3.4.4 In the absence of an adequate defined approach (see 3.3.2.3) or of conclusive in vitro/*ex vivo* data (see 3.3.2.4.1 and 3.3.2.4.2), a stand-alone prediction is not possible. In such cases, a within-tier weight of evidence assessment of data from more than one method would be needed to classify within tier 2. If a within-tier weight of evidence assessment is still not conclusive, then data from lower tiers may be required to reach a conclusion (see figure 3.3.1).

いるため、主観的な解釈の余地がなく、同じデータセットについて異なる評価者であっても常に同じ結論に達することになる。これに対し、証拠の重み付け評価では、利用可能な情報に対して専門家の判断がその場限りで適用されるが、データを解釈するための予め決められたルールが存在しないため、異なる結論に至る可能性がある。

3.3.5.3.4.2　図 3.3.1 の段階 2 から得られた情報、すなわちディファインドアプローチおよび／または in vitro/*ex vivo* 試験法の評価へのステップワイズアプローチは、すべての段階および段階内の情報が関連するとは限らないことを認識しつつ、該当する場合には考慮すべきである（図 3.3.3）。しかし、十分な質の利用可能な関連情報はすべて、結果として得られる分類との整合性を検討する必要がある。動物および／またはヒトの決定的なデータを含むディファインドアプローチの結果は、最終的に、段階 7 の包括的な証拠の重み付けにおいて考慮されることもある（図 3.3.1 参照）。複数のステップから得られた情報が、結果として得られる分類に関して矛盾および／または相反する場合、一般に、より上位のステップで得られた十分な質の情報が、より下位のステップからの情報よりも高い重みを与えられる。しかし、下位のステップの情報が上位のステップの情報よりも厳しい分類をもたらし、誤分類の懸念がある場合は、証拠の重み付け評価の段階で分類が決定される。例えば、in vitro/*ex vivo* 法で眼に対する重篤な損傷性の陽性結果が出た場合、ディファインドアプローチで眼に対する重篤な損傷性の陰性結果を懸念する分類者は、証拠の重み付け評価の段階を利用することになる。

3.3.5.3.4.3　現在の in vitro/*ex vivo* 試験法は、角膜混濁、虹彩炎、結膜発赤または結膜浮腫などの特定の in vivo 影響を区別することはできないが、in vivo で観察される眼への影響の種類とは別に、眼に対する重篤な損傷性/眼刺激性を引き起こす物質を正しく予測できることが示されている。このように、現在の in vitro/*ex vivo* 試験法の多くは、分類を必要としない物質または混合物を高感度で識別することができるが、分類されていない物質または混合物を分類された物質または混合物と区別するために使用すると、特異性は限られる。これは、OECD テストガイドライン 437、438、491、492、494 または 496（表 3.3.6 参照）による分類を必要としない物質または混合物が、実際に分類を必要とする眼への影響を誘発しないことは合理的に確実であるが、分類を必要としない物質または混合物の中には、単独で使用される場合、これらの in vitro/*ex vivo* 試験法によって過剰予測されるものがあるということである。さらに、眼に対する重篤な損傷性を誘発する物質は、これらの試験法の多くによって高い特異性で同定されるが、区分 1 と区分 2 および区分に該当しないを区別するために使用すると感度が限定されることを考慮すべきである。これは、OECD テストガイドライン 437、438、460、491 または 496（表 3.3.6 参照）で区分 1 とされた物質または混合物が、実際に不可逆的な眼の影響を引き起こすことは合理的に確実であるが、眼に対する重篤な損傷性を引き起こす一部の物質または混合物は、これらの in vitro/*ex vivo* 試験法を単独で使用すると予測値が低くなることを示している。結果として、表 3.3.6 に定義された基準にしたがって区分 1 または区分に該当しないのいずれかに結論づけるには、単一の in vitro/*ex vivo* OECD テストガイドラインの試験法で十分であるが、区分 2 と結論づけることはできない。in vitro/*ex vivo* 法の結果が「単独での予測はできない」（例：表 3.3.6 参照）場合、その単一の結果に基づいて結論を出すことはできず、分類にはさらなるデータが必要である。国際的な手続きにしたがって検証されているが、OECD のテストガイドラインとして採用されていない in vitro/*ex vivo* 試験法の中には、所管官庁によっては区分 2 に分類することが認められるものもある（3.3.5.3.5.2 参照）。さらに、段階的アプローチにおける in vitro/*ex vivo* 法の組み合わせや、ディファインドアプローチにおけるそれらの統合（3.3.2.3 参照）は、誤った予測の数を減らし、分類目的に対して適切な性能を示すことができる。

3.3.5.3.4.4　適切なディファインドアプローチ（3.3.2.3 参照）または決定的な in vitro/*ex vivo* データ（3.3.2.4.1 および 3.3.2.4.2 参照）がない場合、単独での予測は不可能である。このような場合、段階 2 で分類するためには、複数の方法によるデータの段階内での証拠の重み付け評価が必要となる。段階内の重み付け評価でも結論が出ない場合は、結論を出すために下位の段階のデータが必要となる場合がある（図 3.3.1 参照）。

Figure 3.3.3: Classification based on defined approaches and/or in vitro/*ex vivo* data within tier 2 of figure 3.3.1

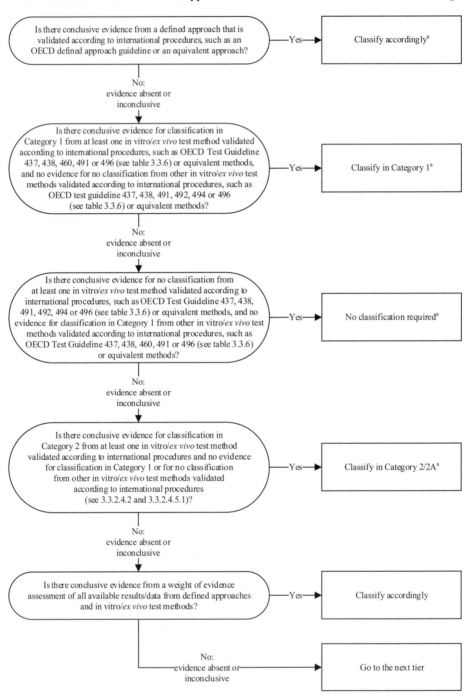

[a] Evidence is considered conclusive if the data fulfil the criteria of the defined approach or of the method and there is no contradicting in vitro/ex vivo information. When information from a lower step would result in a stricter classification than information from a higher step and there is concern for misclassification, then classification is determined by a within-tier weight of evidence assessment.

図 3.3.3　ディファインドアプローチおよび/または図 3.3.1 の段階 2 内の *in vitro/ex vivo* データに基づく分類

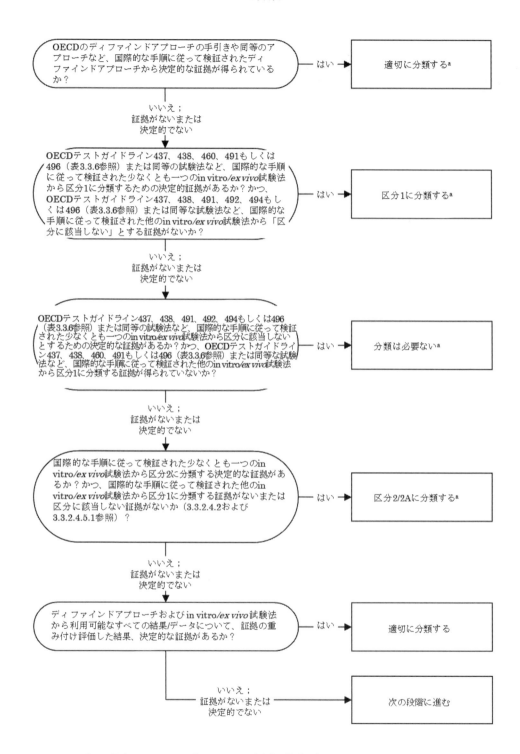

a　データがディファインドアプローチまたは試験法の基準を満たし、in vitro/ex vivo の情報において矛盾がない場合、証拠は決定的であるとみなされる。下位のステップからの情報が上位のステップからの情報よりも厳しい分類となり、誤分類が懸念される場合、分類は証拠の重み付け評価の段階で決定される。

3.3.5.3.5 *Classification criteria based on in vitro/ex vivo data*

3.3.5.3.5.1   Where in vitro/*ex vivo* tests have been undertaken in accordance with OECD test guidelines 437, 438, 460, 491, 492, 494 and/or 496, the criteria for classification in Category 1 for serious eye damage/irreversible effects on the eye and for no classification are set out in table 3.3.6.

3.3.5.3.5　*in vitro/ex vivo* データに基づく分類基準

3.3.5.3.5.1　OECD テストガイドライン 437、438、460、491、492、494 および／または 496 にしたがって in vitro/*ex vivo* 試験が実施された場合、眼に対する重篤な損傷性/不可逆的影響に関する区分 1 への分類および区分に該当しないの基準は、表 3.3.6 に定める

Table 3.3.6: Criteria for serious eye damage/irreversible effects on the eye and for no classification[a] for in vitro/ex vivo methods

Category	OECD Test Guideline 437 Bovine Corneal Opacity and Permeability test method	OECD Test Guideline 438 Isolated Chicken Eye test method	OECD Test Guideline 460 Fluorescein Leakage test method	OECD Test Guideline 491 Short Time Exposure test method	OECD Test Guideline 492 Reconstructed human Cornea-like Epithelium (RhCE)-based test method: Methods 1, 2, 3 and 4 as numbered in Annex II of OECD Test Guideline 492	OECD Test Guideline 494 Vitrigel-Eye Irritancy Test Method	OECD Test Guideline 496 In vitro Macromolecular Test Method (test method 1)
	Organotypic ex vivo assay using isolated corneas from the eyes of freshly slaughtered cattle. Test chemicals are applied to the epithelial surface of the cornea. Damage by the test chemical is assessed by quantitative measurements of: - Corneal opacity changes measured using a light transmission opacitometer (opacitometer 1) or a laserlight-based opacitometer (LLBO, opacitometer 2) - Permeability (sodium fluorescein dye). Both measurements are used to calculate an In Vitro Irritancy Score (IVIS) when using opocitometer 1 or a LLBO Irritancy Score (LIS) when using opacitometer 2. **Criteria based on IVIS or LIS.**	Organotypic ex vivo assay based on the short-term maintenance of chicken eyes in vitro. Test chemicals are applied to the epithelial surface of the cornea. Damage by the test chemical is assessed by (i) a quantitative measurement of increased corneal thickness (swelling), (ii) a qualitative assessment of corneal opacity, (iii) a qualitative assessment of damage to epithelium based on application of fluorescein to the eye, and (iv) a qualitative evaluation of macroscopic morphological damage to the surface. Histopathology can be used to increase the sensitivity of the method for identifying Category 1 non-extreme pH (2 < pH < 11.5) detergents and surfactants.[b] **Criteria based on the scores of corneal swelling, opacity and fluorescein retention, which are used to assign ICE classes (I, II, III or IV) to each endpoint, and on macroscopic and histopathology assessment**[b]	Cytotoxicity and cell-function based in vitro assay that is performed on a confluent monolayer of Madin-Darby Canine Kidney (MDCK) CB997 tubular epithelial cells cultured on permeable inserts. The toxic effects of a test chem cal are measured after a short exposure time (1 minute) by an increase in permeability of sodium fluorescein through the epithelial monolayer of MDCK cells. The amount of fluorescein leakage that occurs is proportional to the chemical-induced damage to the tight junctions, desmosomal junctions and cell membranes, and is used to estimate the ocular toxicity potential of a test chemical. **Criteria based on mean percent fluorescein leakage following a defined exposure period**	Cytotoxicity-based in vitro assay that is performed on a confluent monolayer of Statens Seruminstitut Rabbit Cornea (SIRC) cells. Each test chemical is tested at both 5 % and 0.05 % concentrations. Following five-minute exposure, cell viability is assessed by the enzymatic conversion in viable cells of the vital dye MTT into a blue formazan salt that is quantitatively measured after extraction from cells. **Criteria based on mean percent cell viability following a defined exposure period**	Three-dimensional RhCE tissues are reconstructed from either primary human cells or human immortalised corneal epithelial cells, which have been cultured for several days to form a stratified, highly differentiated squamous epithelium, consisting of at least 3 viable layers of cells and a non-keratinised surface, showing a cornea-like structure morphologically similar to that found in the human cornea. Following exposure and post-treatment incubation (where applicable), tissue viability is assessed by the enzymatic conversion in viable cells of the vital dye MTT into a blue formazan salt that is quantitatively measured after extraction from the tissues. **Criteria based on mean percent tissue viability following defined exposure and post-exposure (where applicable) periods**	In vitro assay using human corneal epithelium models fabricated in a collagen vitrigel membrane (CVM) chamber. The eye irritation potential of the test chemical is predicted by analysing time-dependent changes in transepithelial electrical resistance values using the value Resistance values are measured at intervals of 10 seconds for a period of three minutes after exposure to the test chemical preparation. **Criteria based on the 3 measured indexes: time lag, intensity and plateau level of electrical resistance.**	In vitro assay consisting of a macromolecular plant-based matrix obtained from jack bean *Canavalis ensiformis*. This matrix serves as the target for the test chemical and is composed of a mixture of proteins, glycoproteins, carbohydrates, lipids and low molecular weight components, which form a highly ordered and transparent gel structure upon rehydration. Test chemicals would lead to the disruption and disaggregation of the highly organised macromolecular reagent matrix, and produce turbidity of the macromolecular reagent. Such phenomena is quantified, by measuring changes in light scattering. **Criteria based on a Maximum Qualified Score (MQS) derived from the Optical Density readings at different concentrations, calculated via a software.**

- 162 -

表 3.3.6: *in vitro/ex vivo* 試験法に関する眼に対する重篤な損傷性/不可逆な影響および区分に該当しないの基準

区分	OECD テストガイドライン 437 ウシ角膜を用いる混濁度および透過性試験法	OECD テストガイドライン 438 ニワトリ摘出眼球を用いる試験法	OECD テストガイドライン 460 フルオレセイン漏出試験法	OECD テストガイドライン 491 短時間ばく露試験法	OECD テストガイドライン 492 OECD テストガイドライン 492 の補助 II 試験法 2、3 および 4 として番号を付された再構築ヒト角膜上皮モデル RhCE 法	OECD テストガイドライン 494 Vitrigel-Eye Irritancy Test 法	OECD テストガイドライン 496 In vitro 高分子凝集法（試験法 1）
	屠殺直後の畜牛の眼球から摘出した角膜を用いた器官型 *ex vivo*（生体外）基づく。角膜の上皮表面に試験化学品を塗布する。試験化学品は角膜化学品による損傷は、以下の定量的な測定法によって評価される： ‑ 光透過性オパシトメーター（オパシトメーター1）またはレーザー光シトメーター（LLBO、オパシトメーター2）を用いて測定した角膜の不透明度変化、 ‑ 透過性（フルオレセインナトリウム色素）。 いずれの測定値も、オパシトメーター1を使用する場合は *in vitro* 刺激性スコア（IVIS）、オパシトメーター2を使用する場合はLLBO刺激性スコア（LIS）を計算するために使用される。 IVIS または LIS に基づく基準	ニワトリの眼球を *in vitro*（試験管内）で短期間維持器官（生体）基づく。角膜の上皮表面に試験化学品による損傷を評価する。試験化学品は角膜表面に塗布する。試験化学品による損傷は、以下の定量的な測定法によって評価される：(i)角膜の厚さの増加による損傷、(ii)角膜の定性的評価、(iii)眼内混濁の定性的評価、(iv)眼内のフルオレセインの適用に基づく、(v)表面のフルオレセイン損傷の定性的評価、クロム形態的損傷の定性的評価によって評価される。病理組織検査は、区分1の極端でない pH (2<pH<11.5) の化学剤と界面活性剤を認識する方法の感度を高めるために使用することができる。 各エンドポイントに ICE クラス (I、II、III または IV) を割り当てるための基準 [b]	細胞毒性および細胞機能に基づく *in vitro* アッセイで、透過性シート上で培養した Madin-Darby Canine Kidney (MDCK) CB997 tubular epithelial cells のコンフルエント単層で実施される試験法。試験化学品は、試験化学品による短時間ばく露後、MDCK 細胞の上皮細胞単層を通り抜けるフルオレセインナトリウムの透過増加によって測定される。フルオレセイン漏出量は、密着結合、デスモソーム結合および細胞膜の化学物質の誘発性損傷に比例し、試験物質の眼に対する刺激性を推定するために使用できる。 定義されたばく露期間後のフルオレセイン漏出量の平均パーセントに基づく基準	Statens Seruminstitut Rabbit Cornea (SIRC) 細胞のコンフルエントな単層で実施する細胞毒性に基づく *in vitro* アッセイ。各試験化学品は、5% と 10% の濃度の両方で試験される。5 分間のばく露、細胞生存率は、生細胞内での生体色素である MTT のブルーホルマザン塩への酵素変換によって評価され、定量的に測定される。 定義されたばく露期間後の平均細胞生存率パーセントに基づく基準	3 次元 RhCE 組織は、ヒト初代細胞または死化したヒト角膜上皮細胞のいずれかから再構築され、数日間培養され、少なくとも 3 層の生細胞層および非角化細胞らなる高分化で重層扁平上皮を形成し、ヒト角膜に見られるのと形態的に類似した角膜組織構造を示している。ばく露およびの処理後後に形態的インキュベーション（該当する場合）後、組織の生存率は、組織から抽出したホルマザン塩への生体色素 MTT の生細胞による酵素変換による測定する。 定義されたばく露期間およびばく露後の（該当する場合）の平均組織生存率パーセントに基づく基準	Collagen vitrigel membrane (CVM) チャンバーで作製したヒト角膜上皮モデルを用いた *in vitro* アッセイ。試験物質の眼刺激性は、3 種類の相誘を用いて電気抵抗値の分析時間依存性変化を分析することにより予測される。抵抗値は、試験物質溶液にばく露 3 分間にわたって、露後 10 秒間隔で測定される。 測定された 3 つの指標に基づく基準：電気抵抗のタイムラグ、強度、プラトーレベル	タチナタマメ (*Canavalis ensiformis*) から得られた高分子植物レクチンとのマトリックスからなる *in vitro* アッセイ。このマトリックスは、試験物質のターゲットとなり、タンパク質、糖タンパク質、炭水化物、脂質、低分子量成分の混合物からなり、再水和により高度に秩序立った透明なゲル構造を形成する。眼障害を引き起こす試験化学品は、高度に組織化された高分子量マトリックスの破壊と分解を引き起こし、このような薬に濁りを生じさせる。現象は、光散乱の変化を測定することにより、定量化される。 異なる濃度における光学濃度の測定から、ソフトウェアにより算出される MQS (Maximum Qualified Score) に基づく基準

- 162 -

Table 3.3.6: Criteria for serious eye damage/irreversible effects on the eye and for no classification[a] for in vitro/ex vivo methods (cont'd)

Category	OECD Test Guideline 437 Bovine Corneal Opacity and Permeability test method		OECD Test Guideline 438 Isolated Chicken Eye test method	OECD Test Guideline 460 Fluorescein Leakage test method	OECD Test Guideline 491 Short Time Exposure test method	OECD Test Guideline 492 Reconstructed human Cornea-like Epithelium (RhCE)-based test methods: Methods 1, 2, 3 and 4 as numbered in Annex II of OECD Test Guideline 492				OECD Test Guideline 494 Vitrigel-Eye Irritancy Test Method	OECD Test Guideline 496 In vitro Macromolecular Test Method (test method 1)
1	Opacitometer 1 IVIS > 55	Opacitometer 2 LIS > 30 and lux/7 ≤ 145 and OD490 > 2.5, OR LIS > 30 and lux/7 > 145	At least 2 ICE class IV, OR Corneal opacity = 3 at 30 min (in at least 2 eyes), OR Corneal opacity = 4 at any time point (in at least 2 eyes), OR Severe loosening of the epithelium (in at least 1 eye), OR Certain histopathological effects[b]	Chemical concentration causing 20 % of Fluorescein Leakage (FL$_{20}$) ≤ 100 mg/mL	Viability ≤ 70 % at 5 % and 0.05 %	No stand-alone prediction can be made				No stand-alone prediction can be made	MQS > 30.0
2/2A/2B	No stand-alone prediction can be made.	No stand-alone prediction can be made	No stand-alone prediction can be made	No stand-alone prediction can be made	No stand-alone prediction can be made	No stand-alone prediction can be made				No stand-alone prediction can be made	No stand-alone prediction can be made
Not classified	Opacitometer 1 IVIS ≤ 3	Opacitometer 2 LIS ≤ 30	ICE class I for all 3 endpoints, OR ICE class I for 2 endpoints and ICE class II for the other endpoint, OR ICE class II for 2 endpoints and ICE class I for the other endpoint	No stand-alone prediction can be made	Viability > 70 % at 5 % and 0.05 %	Test method 1 Liquids and Solids: Viability > 60 %	Test method 2 Liquids: Viability > 60 %; Solids: Viability > 50 %	Test method 3 Liquids and Solids: Viability > 40 %	Test method 4 Liquids: Viability > 35 %; Solids: Viability > 60 %	Time lag > 180 seconds and Intensity < 0.05 %/seconds and Plateau level ≤ 5.0 %	MQS ≤ 12.5

[a] Grading criteria are understood as described in OECD test guidelines 437, 438, 460, 491, 492, 494 and 496.
[b] For criteria, please consult OECD Test Guideline 438

表 3.3.6: in vitro/ex vivo 試験法に関する眼に対する重篤な損傷性/不可逆な影響および区分に該当しないの基準 (続き)

区分	OECD テストガイドライン 437 ウシ角膜を用いる混濁度および透過性試験法		OECD テストガイドライン 438 ニワトリ摘出眼球を用いる試験法	OECD テストガイドライン 460 フルオレセイン漏出試験法	OECD テストガイドライン 491 短時間ばく露試験法	OECD テストガイドライン 492 構築ヒト角膜様上皮モデル(RhCE)法 OECD テストガイドライン 492 の補遺 II に試験法 1、2、3 および 4 として番号を付された再				OECD テストガイドライン 494 Vitrigel-Eye Irritancy Test 法	OECD テストガイドライン 496 In vitro 高分子試験法 (試験法 1)
1	オペレーター 1 IVIS > 55	オペレーター 2 IVIS > 30 かつ lux/7 ≤ 145 かつ OD490 > 2.5、または IVIS > 30 かつ lux/7 > 145	ICE クラス IV が 2 つ以上ある、または 30 分後の角膜混濁=3 (少なくとも 2 眼で)、またはいずれかの時点で角膜混濁 = 4 (少なくとも 2 眼で)、または上皮の重度の緩み (少なくとも 1 眼で)、または特定の病理組織検査で見られた影響 b	20%のフルオレセイン漏出を引き起こす化学物質濃度 (FL20) ≤ 100 mg/mL	5%群および 0.05%群における生存率 ≤ 70 %	単独での予測はできない				単独での予測はできない	MQS > 30.0
2/2A/2B	単独での予測はできない		単独での予測はできない	単独での予測はできない	単独での予測はできない	単独での予測はできない				単独での予測はできない	単独での予測はできない
区分に該当しない	オペレーター 1 IVIS ≤ 3	オペレーター 2 LIS ≤ 30	3 つのエンドポイントすべてで ICE クラス I、または 2 つのエンドポイントが ICE クラス I で、もう 1 つのエンドポイントが ICE クラス II、または 2 つのエンドポイントで ICE クラス II で、もう 1 つのエンドポイントで ICE クラス I		5%群および 0.05%群における生存率 > 70 %	試験法 1 液体および固体:生存率 > 60 %	試験法 2 液体:生存率 > 60 %; 固体:生存率 > 50 %	試験法 3 液体および固体:生存率 > 40 %	試験法 4 液体:生存率 > 35 %; 固体:生存率 > 60 %	時間差 > 180 秒 および低下速度 < 0.05 %/秒 および終点での低下率 ≤ 5.0 %	MQS ≤ 12.5

a 判定基準は、OECD テストガイドライン 437、438、460、491、492、494、および 496 に記載されているように理解される。
b 基準については、OECD テストガイドライン 438 を参照のこと。

3.3.5.3.5.2　　A non-exhaustive list of other validated in vitro/*ex vivo* test methods accepted by some competent authorities but not adopted as OECD test guidelines are listed below. A competent authority may decide which classification criteria, if any, should be applied for these test methods:

(a) Time to Toxicity (ET_{50}) tests using the Reconstructed human Cornea-like Epithelia (RhCE) described in OECD Test Guideline 492 (Kandarova et al., 2018; Alépée et al., 2020);

(b) Ex Vivo Eye Irritation Test (EVEIT): an *ex vivo* assay that uses excised rabbit corneal tissues kept in culture for several days and monitors tissue recovery to model both reversible and non-reversible eye effects. Full-thickness tissue recovery is monitored non-invasively using optical coherence tomography (OCT) (Frentz et al., 2008; Spöler et al., 2007; Spöler et al., 2015);

(c) Porcine Ocular Cornea Opacity/Reversibility Assay (PorCORA): an *ex vivo* assay that uses excised porcine corneal tissues kept in culture for up to 21 days and monitors tissue recovery to model both reversible and non-reversible eye effects. The tissues are stained with fluorescent dye and effects on the corneal epithelia are visualised by the retention of fluorescent dye (Piehl et al., 2010; Piehl et al., 2011);

(d) EyeIRR-IS assay: a genomic approach applied to a RhCE model (Cottrez et al., 2021);

(e) In vitro Macromolecular Test Method (test method 2), similar to test method 1 described in OECD Test Guideline 496 (Choksi et al., 2020);

(f) Metabolic activity assay: In vitro assay consisting of measuring changes to metabolic rate in test-material treated L929 cell monolayer (Harbell et al., 1999; EURL ECVAM, 2004a; Hartung et al., 2010; Nash et al., 2014);

(g) Hen's Egg Test on the Chorio-Allantoic Membrane (HET-CAM): an organotypic assay that uses the vascularised membrane of fertile chicken eggs to assess a test material's potential to cause vascular changes (Spielmann et al., 1993; Balls et al., 1995; Spielmann et al., 1996; Brantom et al., 1997; ICCVAM, 2007; ICCVAM, 2010);

(h) Chorio-Allantoic Membrane Vascular Assay (CAMVA): an organotypic assay that uses the vascularised membrane of fertile chicken eggs to assess a test material's potential to cause vascular changes (Bagley et al., 1994; Brantom et al., 1997; Bagley et al., 1999; Donahue et al., 2011);

(i) Neutral Red Release (NRR) assay: In vitro assay that quantitatively measures a substance's ability to induce damage to cell membranes in a monolayer of normal human epidermal keratinocytes (NHEK) (Reader et al. 1989; Reader et al., 1990; Zuang, 2001; EURL ECVAM, 2004b; Settivari et al., 2016); and

(j) Isolated Rabbit Eye (IRE) test, similar to OECD Test Guideline 438 but using isolated rabbit eyes instead of isolated chicken eyes (Burton et al., 1981; Whittle et al. 1992; Balls et al., 1995; Brantom et al., 1997; ICCVAM, 2007; ICCVAM, 2010).

3.3.5.3.6　　*Guidance on the use of other existing skin or eye data in animals for classification as serious eye damage or eye irritation*

3.3.5.3.6.1　　The availability of other animal data for serious eye damage/eye irritation may be limited as tests with the eye as the route of exposure are not normally performed. An exception could be historical data from the Low Volume Eye Test (LVET) that might be used in a weight of evidence assessment. The LVET is a modification of the standard OECD Test Guideline 405 test method.

3.3.5.3.6.2　　Existing data from the LVET test could be considered for the purpose of classification and labelling but must be carefully evaluated. The differences between the LVET and OECD Test Guideline 405 may result in a classification in a lower category (or no classification) based on LVET data, than if the classification was based on data derived from the standard in vivo test (OECD Test Guideline 405). Thus, positive data from the LVET test could be a trigger for considering classification in Category 1 on its own, but data from this test are not conclusive for a Category 2 classification or no classification (ECHA, 2017). Such data may, however, be used in an overall weight of evidence assessment. It is noted that the applicability domain of the LVET is limited to household detergent and cleaning products and their main ingredients (surfactants) (ESAC, 2009).

3.3.5.3.6.3　　Effects on the eyes may be observed in acute or repeated dose inhalation studies with full body exposure. However, normally no scoring according to the Draize criteria is performed and the follow-up period may be shorter than 21 days. Also, the effects on the eyes will likely depend upon the concentration of the substance/mixture and

3.3.5.3.5.2　一部の所管官庁で認められているが、OECDテストガイドラインとして採用されていない、その他の有効なin vitro/ex vivo試験法の非網羅的リストを以下に挙げる。所管官庁は、これらの試験法についてどの分類基準（もしあれば）を適用すべきかを決定することができる。

   (a) OECDテストガイドライン492に記載されている再構築ヒト角膜様上皮モデル（Reconstructed human Cornea-like Epithelia; RhCE）を用いた毒性時間（ET_{50}）試験（Kandarova et al., 2018; Alépée et al., 2020）；

   (b) Ex Vivo Eye Irritation Test（EVEIT）：数日間培養した摘出ウサギ角膜組織を用い、組織の回復をモニターして可逆的および非可逆的な眼の影響をモデル化するex vivoアッセイ。全厚の組織回復を光干渉断層計（OCT）を用いて非侵襲的にモニターする（Frentz et al., 2008; Spöler et al., 2007; Spöler et al., 2015）；

   (c) Porcine Ocular Cornea Opacity/Reversibility Assay (PorCORA)：最大21日間培養した摘出ブタ角膜組織を用い、組織の回復をモニターして可逆的・非可逆的な眼の影響をモデル化するex vivoアッセイ。組織は蛍光色素で染色され、角膜上皮への影響は蛍光色素の保持によって可視化される（Piehl et al., 2010; Piehl et al., 2011）；

   (d) EyeIRR-ISアッセイ：RhCEモデルに適用した遺伝学的アプローチ（Cottrez et al., 2021）；

   (e) In vitro高分子試験法（試験法2）、OECDテストガイドライン496に記載されている試験法1と同様（Choksi et la., 2020）；

   (f) 代謝活性アッセイ：試験物質を処理したL929細胞単層における代謝速度の変化の測定からなるin vitroアッセイ（Harbell et al., 1999; EURL ECVAM, 2004a; Hartung et al., 2010; Nash et al., 2014）；

   (g) 受精鶏卵漿尿膜（Hen's Egg Test on the Chorio-Allantoic Membrane; HET-CAM）：試験物質が血管の変化を引き起こす可能性を評価するために受精卵の血管膜を使用する器官型アッセイ（Spielmann et al., 1993; Balls et al., 1995; Spielmann et al., 1996; Brantom et al., 1997; ICCVAM, 2007; ICCVAM, 2010）；

   (h) 漿尿膜血管アッセイ（Chorio-Allantoic Membrane Vascular Assay; CAMVA）：被験物質が血管の変化を引き起こす可能性を評価する受精卵の血管膜を用いた器官型アッセイ（Bagley et al., 1994; Brantom et al., 1997; Bagley et al., 1999; Donahue et al., 2011）；

   (i) ニュートラルレッド遊離（Neutral Red Release; NRR）アッセイ：正常ヒト表皮ケラチノサイト（NHEK）の単層における細胞膜への損傷を誘発する物質の能力を定量的に測定するin vitroアッセイ（Reader et al. 1989; Reader et al., 1990; Zuang, 2001; EURL ECVAM, 2004b; Settivari et al., 2016）；および

   (j) 単離ウサギ眼（IRE）試験、OECDテストガイドライン438に似ているが、単離鶏眼の代わりに単離ウサギ眼を使用（Burton et al., 1981; Whittle et al., 1992; Balls et al., 1995; Brantom et al., 1997; ICCVAM, 2007; ICCVAM, 2010）

3.3.5.3.6　*眼に対する重篤な損傷性または眼刺激性の分類のための動物における他の既存の皮膚または眼のデータの使用に関する手引き*

3.3.5.3.6.1　眼をばく露経路とする試験は通常行われないため、眼に対する重篤な損傷性／眼刺激性に関する他の動物データの利用可能性は限られるかもしれない。例外として、LVET（Low Volume Eye Test）の背景データを証拠の重み付け評価に使用することができるかもしれない。LVETは、標準的なOECDテストガイドライン405の試験法を修正したものである。

3.3.5.3.6.2　LVET試験の既存データは、分類と表示の目的で考慮することができるが、慎重に評価する必要がある。LVETとOECDテストガイドライン405の違いは、分類が標準的なin vivo試験（OECDテストガイドライン405）から得られたデータに基づく場合よりも、LVETデータに基づく方が低くなる（緩くなる）（または区分に該当しない）可能性がある。したがって、LVET試験の陽性データは、それだけで区分1への分類を検討するきっかけとなり得るが、この試験のデータは、区分2への分類または区分に該当しないことを決定するものではない（ECHA, 2017）。ただし、このようなデータは、包括的な証拠の重み付け評価で使用することができる。LVETの適用領域は、家庭用洗剤および洗浄剤とその主成分（界面活性剤）に限定されていることに留意されたい（ESAC, 2009）。

3.3.5.3.6.3　眼に対する影響は、全身へのばく露を伴う急性または反復投与吸入試験で観察されることがある。しかし、通常、Draize基準によるスコアリングは行われず、追跡調査期間も21日間より短くなる可能性がある。また、眼への影響は、物質/混合物の濃度およびばく露期間に依存すると考えられる。最

the exposure duration. As there are no criteria for minimal concentration and duration, the absence of effects on the eyes or eye irritation may not be conclusive for the absence of serious eye damage. The presence of irreversible effects on the eye should be considered within a weight of evidence assessment.

3.3.5.3.7 *Guidance on the use of pH and acid/alkaline reserve for classification as serious eye damage*

3.3.5.3.7.1 Methods to determine the pH value such as OECD Test Guideline 122 and the method described by Young et al. (1988) differ in the concentration of the substance or mixture for which the pH is determined and include values of 1%, 10% and 100%. These methods also differ in the way the acid/alkaline reserve is determined, namely up to a pH of 7 for both acids and bases (OECD Test Guideline 122) or up to a pH of 4 for acids and a pH of 10 for bases (Young et al., 1988). Furthermore, there are differences between OECD Test Guideline 122 and Young et al. (1988) in the units used to express the acid/alkaline reserve.

3.3.5.3.7.2 Criteria to identify substances and mixtures requiring classification in Category 1 based on pH and acid/alkaline reserve have been developed for effects on the skin (Young et al., 1988) and the same criteria are applied for effects on the eye. These criteria were developed using a combination of pH and acid/alkaline reserve values that were determined in a specific way (Young et al., 1988). Therefore, these criteria may not be directly applicable when other test concentrations or methods are used to measure pH and acid/alkaline reserve. Furthermore, the calibration and validation of these criteria was based on a limited dataset for effects on the skin. Thus, the predictive value of the combination of pH and acid/alkaline reserve for classification in Category 1 for effects on the eye is limited, especially for substances and mixtures with an extreme pH but a non-significant acid/alkaline reserve. The criteria developed by Young et al. (1988) for classification in Category 1 may be used as a starting point for determining whether a substance or a mixture has a significant acid/alkaline reserve or a non-significant acid/alkaline reserve. A competent authority may decide which criteria for significant acid/alkaline reserve can be applied.

---

* *References:*

Alépée, N., E. Adriaens, T. Abo, D. Bagley, B. Desprez, J. Hibatallah, K. Mewes, U. Pfannenbecker, À. Sala, A.R. Van Rompay, S. Verstraelen, and P. McNamee. 2019a. Development of a defined approach for eye irritation or serious eye damage for liquids, neat and in dilution, based on Cosmetics Europe analysis of in vitro STE and BCOP test methods. Toxicol. In Vitro, 57: 154-163. Doi: 10.1016/j.tiv.2019.02.019.

Alépée, N., E. Adriaens, T. Abo, D. Bagley, B. Desprez, J. Hibatallah, K. Mewes, U. Pfannenbecker, À. Sala, A.R. Van Rompay, S. Verstraelen, and P. McNamee. 2019b. Development of a defined approach for eye irritation or serious eye damage for neat liquids based on Cosmetics Europe analysis of in vitro RhCE and BCOP test methods. Toxicol. In Vitro, 59: 100-114. Doi: 10.1016/j.tiv.2019.04.011.

Alépée, N., V. Leblanc, M.H. Grandidier, S. Teluob, V. Tagliati, E. Adriaens, and V. Michaut. 2020. Development of the SkinEthic HCE Time-to-Toxicity test method for identifying liquid chemicals not requiring classification and labelling and liquids inducing serious eye damage and eye irritation. Toxicol. In Vitro, 69: 104960. Doi: 10.1016/j.tiv.2020.104960.

Bagley, D.M., D. Waters, and B.M. Kong. 1994. Development of a 10-day chorioallantoic membrane vascular assay as an alternative to the Draize rabbit eye irritation test. Food Chem. Toxicol., 32(12): 1155-1160. Doi: 10.1016/0278-6915(94)90131-7.

Bagley, D.M., D. Cerven, and J. Harbell. 1999. Assessment of the chorioallantoic membrane vascular assay (CAMVA) in the COLIPA in vitro eye irritation validation study. Toxicol. In Vitro, 13(2): 285-293. Doi: 10.1016/s0887-2333(98)00089-7.

Balls, M., P.A. Botham, L.H. Bruner, and H. Spielmann. 1995. The EC/HO international validation study on alternatives to the draize eye irritation test. Toxicol. In Vitro, 9(6): 871-929. Doi: 10.1016/0887-2333(95)00092-5.

Brantom, P.G., L.H. Bruner, M. Chamberlain, O. De Silva, J. Dupuis, L.K. Earl, D.P. Lovell, W.J. Pape, M. Uttley, D.M. Bagley, F.W. Baker, M. Bracher, P. Courtellemont, L. Declercq, S. Freeman, W. Steiling, A.P. Walker, G.J. Carr, N. Dami, G. Thomas, J. Harbell, P.A. Jones, U. Pfannenbecker, J.A. Southee, M. Tcheng, H. Argembeaux, D. Castelli, R. Clothier, D.J. Esdaile, H. Itigaki, K. Jung, Y. Kasai, H. Kojima, U. Kristen, M. Larnicol, R.W. Lewis, K. Marenus, O. Moreno, A. Peterson, E.S. Rasmussen, C. Robles, and M. Stern. 1997. A summary report of the COLIPA international validation study on alternatives to the draize rabbit eye irritation test. Toxicol. In Vitro, 11: 141-179. Doi:10.1016/S0887-2333(96)00069-0.

小限の濃度や期間についての基準がないため、眼への影響や眼刺激性がないことが、眼に対する重篤な損傷性がないことの決め手とならない場合がある。眼に対する不可逆的な影響の有無は、証拠の重み付け評価の中で考慮されるべきである。

3.3.5.3.7 眼に対する重篤な損傷性として分類するための pH および酸/アルカリ予備の使用に関する手引き

3.3.5.3.7.1　OECD テストガイドライン 122 や Young ら（1988）が記載する方法のような pH 値を決定する方法は、pH を決定する物質または混合物の濃度が異なり、1%、10%、100%の値を含む。また、これらの方法は、酸/アルカリ予備の決定方法にも違いがあり、酸および塩基ともに pH7 まで（OECD テストガイドライン 122）、酸は pH4 まで、塩基は pH10 まで（Young et al., 1988）である。さらに、OECD テストガイドライン 122 と Young ら（1988）では、酸/アルカリ予備を表すために使用する単位に違いがある。

3.3.5.3.7.2　pH と酸/アルカリ予備に基づいて区分 1 の分類が必要な物質と混合物を特定する基準は、皮膚への影響について開発されており（Young et al., 1988）、眼への影響についても同じ基準が適用される。これらの基準は、具体的な方法で決定された pH と酸/アルカリ予備の値の組み合わせを使用して開発された（Young et al., 1988）。したがって、他の試験濃度や方法を用いて pH や酸/アルカリ予備を測定した場合、これらの基準をそのまま適用できない可能性がある。さらに、これらの基準の校正と検証は、皮膚への影響に関する限られたデータセットに基づくものであった。したがって、特に pH が極端だが酸/アルカリ予備がそれほどない物質や混合物については、眼への影響に関する区分 1 の分類のための pH と酸/アルカリ予備の組み合わせの予測値は限定的である。物質または混合物が相当量の酸/アルカリ予備があるか、または酸/アルカリ予備はそれほどないかを決定するための出発点として、Young ら（1988）によって区分 1 に分類するために開発された基準を使用することができる。所管官庁は、相当量の酸/アルカリ予備がある場合にどの基準を適用できるかを決定することができる。

---

*　　　*References:*

Alépée, N., E. Adriaens, T. Abo, D. Bagley, B. Desprez, J. Hibatallah, K. Mewes, U. Pfannenbecker, À. Sala, A.R. Van Rompay, S. Verstraelen, and P. McNamee. 2019a. Development of a defined approach for eye irritation or serious eye damage for liquids, neat and in dilution, based on Cosmetics Europe analysis of in vitro STE and BCOP test methods. Toxicol. In vitro, 57: 154-163. Doi: 10.1016/j.tiv.2019.02.019.

Alépée, N., E. Adriaens, T. Abo, D. Bagley, B. Desprez, J. Hibatallah, K. Mewes, U. Pfannenbecker, À. Sala, A.R. Van Rompay, S. Verstraelen, and P. McNamee. 2019b. Development of a defined approach for eye irritation or serious eye damage for neat liquids based on Cosmetics Europe analysis of in vitro RhCE and BCOP test methods. Toxicol. In vitro, 59: 100-114. Doi: 10.1016/j.tiv.2019.04.011.

Alépée, N., V. Leblanc, M.H. Grandidier, S. Teluob, V. Tagliati, E. Adriaens, and V. Michaut. 2020. Development of the SkinEthic HCE Time-to-Toxicity test method for identifying liquid chemicals not requiring classification and labelling and liquids inducing serious eye damage and eye irritation. Toxicol. In vitro, 69: 104960. Doi: 10.1016/j.tiv.2020.104960.

Bagley, D.M., D. Waters, and B.M. Kong. 1994. Development of a 10-day chorioallantoic membrane vascular assay as an alternative to the Draize rabbit eye irritation test. Food Chem. Toxicol., 32(12): 1155-1160. Doi: 10.1016/0278-6915(94)90131-7.

Bagley, D.M., D. Cerven, and J. Harbell. 1999. Assessment of the chorioallantoic membrane vascular assay (CAMVA) in the COLIPA in vitro eye irritation validation study. Toxicol. In vitro, 13(2): 285-293. Doi: 10.1016/s0887-2333(98)00089-7.

Balls, M., P.A. Botham, L.H. Bruner, and H. Spielmann. 1995. The EC/HO international validation study on alternatives to the draize eye irritation test. Toxicol. In vitro, 9(6): 871-929. Doi: 10.1016/0887-2333(95)00092-5.

Brantom, P.G., L.H. Bruner, M. Chamberlain, O. De Silva, J. Dupuis, L.K. Earl, D.P. Lovell, W.J. Pape, M. Uttley, D.M. Bagley, F.W. Baker, M. Bracher, P. Courtellemont, L. Declercq, S. Freeman, W. Steiling, A.P. Walker, G.J. Carr, N. Dami, G. Thomas, J. Harbell, P.A. Jones, U. Pfannenbecker, J.A. Southee, M. Tcheng, H. Argembeaux, D. Castelli, R. Clothier, D.J. Esdaile, H. Itigaki, K. Jung, Y. Kasai, H. Kojima, U. Kristen, M. Larnicol, R.W. Lewis, K. Marenus, O. Moreno, A. Peterson, E.S. Rasmussen, C. Robles, and M. Stern. 1997. A summary report of the COLIPA international validation study on alternatives to the draize rabbit eye irritation test. Toxicol. In vitro, 11: 141-179. Doi:10.1016/S0887-2333(96)00069-0.

Burton, A.B., M. York, and R.S. Lawrence. 1981. The in vitro assessment of severe eye irritants. Food Cosmet. Toxicol., 19(4): 471-480. Doi: 10.1016/0015-6264(81)90452-1.

Choksi, N., S. Lebrun, M. Nguyen, A. Daniel, G. DeGeorge, J. Willoughby, A. Layton, D. Lowther, J. Merrill, J. Matheson, J. Barroso, K. Yozzo, W. Casey, and D. Allen. 2020. Validation of the OptiSafe™ eye irritation test. Cutan. Ocul. Toxicol., 39(3): 180-192. Doi: 10.1080/15569527.2020.1787431.

Cottrez, F., V. Leblanc, E. Boitel, H. Groux, and N. Alépée. 2021. The EyeIRR-IS assay: Development and evaluation of an in vitro assay to measure the eye irritation subcategorization of liquid chemicals. Toxicol. In Vitro, 71: 105072. Doi: 10.1016/j.tiv.2020.105072.

Donahue, D.A., L.E. Kaufman, J. Avalos, F.A. Simion, and D.R Cerven. 2011. Survey of ocular irritation predictive capacity using Chorioallantoic Membrane Vascular Assay (CAMVA) and Bovine Corneal Opacity and Permeability (BCOP) test historical data for 319 personal care products over fourteen years. Toxicol. In Vitro, 25(2): 563-572. Doi: 10.1016/j.tiv.2010.12.003.

ECHA. 2017. Guidance on the Application of the CLP Criteria. Version 5.0. Reference ECHA-17-G-21-EN. doi: 10.2823/124801. Available at: https://echa.europa.eu/guidance-documents/guidance-on-clp.

ESAC. 2019. Statement on the use of existing low volume eye test (LVET) data for weight of evidence decisions on classification and labelling of cleaning products and their main ingredients. Statement of the ECVAM Scientific Advisory Committee (ESAC) of 9th July 2009. Available at: https://ec.europa.eu/jrc/sites/jrcsh/files/esac31_lvet_20090922.pdf.

EURL ECAM. 2004a. Tracking System for Alternative Methods Towards Regulatory Acceptance (TSAR). Method TM2004-01. The cytosensor microphysiometer toxicity test. Available at: https://tsar.jrc.ec.europa.eu/test-method/tm2004-01.

EURL ECAM. 2004b. Tracking System for Alternative Methods Towards Regulatory Acceptance (TSAR). Method TM2004-03. Neutral Red Release Assay. Available at: https://tsar.jrc.ec.europa.eu/test-method/tm2004-03.

Frentz, M., M. Goss, M. Reim, and N.F. Schrage. 2008. Repeated exposure to benzalkonium chloride in the Ex Vivo Eye Irritation Test (EVEIT): observation of isolated corneal damage and healing. Altern. Lab. Anim., 36(1): 25-32. Doi: 10.1177/026119290803600105.

Harbell, J.W., R. Osborne, G.J. Carr, and A. Peterson. 1999. Assessment of the Cytosensor Microphysiometer Assay in the COLIPA In Vitro Eye Irritation Validation Study. Toxicol. In Vitro, 13(2): 313-323. Doi: 10.1016/s0887-2333(98)00090-3.

Hartung, T., L. Bruner, R. Curren, C. Eskes, A. Goldberg, P. McNamee, L. Scott, and V. Zuang. 2010. First alternative method validated by a retrospective weight of evidence approach to replace the Draize eye test for the identification of non-irritant substances for a defined applicability domain. ALTEX, 27(1): 43-51. Doi: 10.14573/altex.2010.1.43.

ICCVAM. 2007. ICCVAM test method evaluation report: in vitro ocular toxicity test methods for identifying ocular severe irritants and corrosives. NIH Publication No. 07–4517. National institute of environmental health sciences, research Triangle Park, North Carolina, USA.

ICCVAM. 2010. ICCVAM test method evaluation report: current validation status of in vitro test methods proposed for identifying eye injury hazard potential of chemicals and products. NIH Publication No. 10-7553. National Institute of Environmental Health Sciences, Research Triangle Park, North Carolina, USA.

Kandarova, H., S. Letasiova, E. Adriaens, R. Guest, J.A. Willoughby Sr., A. Drzewiecka, K. Gruszka, N. Alépée, S. Verstraelen, and A.R. Van Rompay. 2018. CON4EI: CONsortium for in vitro Eye Irritation testing strategy - EpiOcular™ time-to-toxicity (EpiOcular ET-50) protocols for hazard identification and labelling of eye irritating chemicals. Toxicol. In Vitro, 49: 34-52. Doi: 10.1016/j.tiv.2017.08.019.

Nash, J.R., G. Mun, H.A. Raabe, and R. Curren. 2014. Using the cytosensor microphysiometer to assess ocular toxicity. Curr. Protoc. Toxicol. 61: 1.13.1-11. Doi: 10.1002/0471140856.tx0113s61.

Piehl, M., A. Gilotti, A. Donovan, G. DeGeorge, and D. Cerven. 2010. Novel cultured porcine corneal irritancy assay with reversibility endpoint. Toxicol. In Vitro 24: 231-239. Doi:10.1016/j.tiv.2009.08.033.

Burton, A.B., M. York, and R.S. Lawrence. 1981. The in vitro assessment of severe eye irritants. Food Cosmet. Toxicol., 19(4): 471-480. Doi: 10.1016/0015-6264(81)90452-1.

Choksi, N., S. Lebrun, M. Nguyen, A. Daniel, G. DeGeorge, J. Willoughby, A. Layton, D. Lowther, J. Merrill, J. Matheson, J. Barroso, K. Yozzo, W. Casey, and D. Allen. 2020. Validation of the OptiSafe™ eye irritation test. Cutan. Ocul. Toxicol., 39(3): 180-192. Doi: 10.1080/15569527.2020.1787431.

Cottrez, F., V. Leblanc, E. Boitel, H. Groux, and N. Alépée. 2021. The EyeIRR-IS assay: Development and evaluation of an in vitro assay to measure the eye irritation subcategorization of liquid chemicals. Toxicol. In vitro, 71: 105072. Doi: 10.1016/j.tiv.2020.105072.

Donahue, D.A., L.E. Kaufman, J. Avalos, F.A. Simion, and D.R Cerven. 2011. Survey of ocular irritation predictive capacity using Chorioallantoic Membrane Vascular Assay (CAMVA) and Bovine Corneal Opacity and Permeability (BCOP) test historical data for 319 personal care products over fourteen years. Toxicol. In vitro, 25(2): 563-572. Doi: 10.1016/j.tiv.2010.12.003.

ECHA. 2017. Guidance on the Application of the CLP Criteria. Version 5.0. Reference ECHA-17-G-21-EN. doi: 10.2823/124801. Available at: https://echa.europa.eu/guidance-documents/guidance-on-clp.

ESAC. 2019. Statement on the use of existing low volume eye test (LVET) data for weight of evidence decisions on classification and labelling of cleaning products and their main ingredients. Statement of the ECVAM Scientific Advisory Committee (ESAC) of 9th July 2009. Available at: https://ec.europa.eu/jrc/sites/jrcsh/files/esac31_lvet_20090922.pdf.

EURL ECAM. 2004a. Tracking System for Alternative Methods Towards Regulatory Acceptance (TSAR). Method TM2004-01. The cytosensor microphysiometer toxicity test. Available at: https://tsar.jrc.ec.europa.eu/test-method/tm2004-01.

EURL ECAM. 2004b. Tracking System for Alternative Methods Towards Regulatory Acceptance (TSAR). Method TM2004-03. Neutral Red Release Assay. Available at: https://tsar.jrc.ec.europa.eu/test-method/tm2004-03.

Frentz, M., M. Goss, M. Reim, and N.F. Schrage. 2008. Repeated exposure to benzalkonium chloride in the Ex Vivo Eye Irritation Test (EVEIT): observation of isolated corneal damage and healing. Altern. Lab. Anim., 36(1): 25-32. Doi: 10.1177/026119290803600105.

Harbell, J.W., R. Osborne, G.J. Carr, and A. Peterson. 1999. Assessment of the Cytosensor Microphysiometer Assay in the COLIPA In vitro Eye Irritation Validation Study. Toxicol. In vitro, 13(2): 313-323. Doi: 10.1016/s0887-2333(98)00090-3.

Hartung, T., L. Bruner, R. Curren, C. Eskes, A. Goldberg, P. McNamee, L. Scott, and V. Zuang. 2010. First alternative method validated by a retrospective weight of evidence approach to replace the Draize eye test for the identification of non-irritant substances for a defined applicability domain. ALTEX, 27(1): 43-51. Doi: 10.14573/altex.2010.1.43.

ICCVAM. 2007. ICCVAM test method evaluation report: in vitro ocular toxicity test methods for identifying ocular severe irritants and corrosives. NIH Publication No. 07–4517. National institute of environmental health sciences, research Triangle Park, North Carolina, USA.

ICCVAM. 2010. ICCVAM test method evaluation report: current validation status of in vitro test methods proposed for identifying eye injury hazard potential of chemicals and products. NIH Publication No. 10-7553. National Institute of Environmental Health Sciences, Research Triangle Park, North Carolina, USA.

Kandarova, H., S. Letasiova, E. Adriaens, R. Guest, J.A. Willoughby Sr., A. Drzewiecka, K. Gruszka, N. Alépée, S. Verstraelen, and A.R. Van Rompay. 2018. CON4EI: CONsortium for in vitro Eye Irritation testing strategy - EpiOcular™ time-to-toxicity (EpiOcular ET-50) protocols for hazard identification and labelling of eye irritating chemicals. Toxicol. In vitro, 49: 34-52. Doi: 10.1016/j.tiv.2017.08.019.

Nash, J.R., G. Mun, H.A. Raabe, and R. Curren. 2014. Using the cytosensor microphysiometer to assess ocular toxicity. Curr. Protoc. Toxicol. 61: 1.13.1-11. Doi: 10.1002/0471140856.tx0113s61.

Piehl, M., A. Gilotti, A. Donovan, G. DeGeorge, and D. Cerven. 2010. Novel cultured porcine corneal irritancy assay with reversibility endpoint. Toxicol. In vitro 24: 231-239. Doi:10.1016/j.tiv.2009.08.033.

Piehl, M., M. Carathers, R. Soda, D. Cerven, and G. DeGeorge. 2011. Porcine corneal ocular reversibility assay (PorCORA) predicts ocular damage and recovery for global regulatory agency hazard categories. Toxicol. In Vitro, 25: 1912-1918. Doi:10.1016/j.tiv.2011.06.008.

Reader, S.J., V. Blackwell, R. O'Hara, R.H. Clothier, G. Griffin, and M. Balls. 1989. A vital dye release method for assessing the short-term cytotoxic effects of chemicals and formulations. Altern. Lab. Anim., 17: 28-33. Doi: 10.1177/026119298901700106.

Reader, S.J., V. Blackwell, R. O'Hara, R.H. Clothier, G. Griffin, and M. Balls. 1990. Neutral red release from pre-loaded cells as an in vitro approach to testing for eye irritancy potential. Toxicol. In Vitro, 4(4-5): 264-266. Doi: 10.1016/0887-2333(90)90060-7.

Settivari, R.S., R.A. Amado, M. Corvaro, N.R. Visconti, L. Kan, E.W. Carney, D.R. Boverhof, and S.C. Gehen. 2016. Tiered application of the neutral red release and EpiOcular™ assays for evaluating the eye irritation potential of agrochemical formulations. Regul. Toxicol. Pharmacol., 81: 407-420. Doi: 10.1016/j.yrtph.2016.09.028.

Spielmann, H., S. Kalweit, M. Liebsch, T. Wirnsberger, I. Gerner, E. Bertram-Neis, K. Krauser, R. Kreiling, H.G. Miltenburger, W. Pape, and W. Steiling. 1993. Validation study of alternatives to the Draize eye irritation test in Germany: Cytotoxicity testing and HET-CAM test with 136 industrial chemicals. Toxicol. In Vitro, 7(4): 505-510. Doi: 10.1016/0887-2333(93)90055-a.

Spielmann, H., M. Liebsch, S. Kalweit, F. Moldenhauer, T. Wirnsberger, H.-G. Holzhütter, B. Schneider, S. Glaser, I. Gerner, W.J.W. Pape, R. Kreiling, K. Krauser, H.G. Miltenburger, W. Steiling, N.P. Luepke, N. Müller, H. Kreuzer, P. Mürmann, J. Spengler, E. Bertram-Neis, B. Siegemund, and F.J. Wiebel. 1996. Results of a validation study in Germany on two in vitro alternatives to the Draize eye irritation test, HET-CAM test and the 3T3 NRU cytotoxicity test. Altern. Lab. Anim., 24: 741-858.

Spöler, F., M. Först, H. Kurz, M. Frentz, and N.F. Schrage. 2007. Dynamic analysis of chemical eye burns using high-resolution optical coherence tomography. J. Biomed. Opt., 12: 041203. doi:10.1117/1.2768018.

Spöler, F., O. Kray, S. Kray, C. Panfil, and N.F. Schrage. 2015. The Ex Vivo Eye Irritation Test as an alternative test method for serious eye damage/eye irritation. Altern. Lab. Anim., 43(3): 163-179. Doi: 10.1177/026119291504300306.

Whittle, E., D. Basketter, M. York, L. Kelly, T. Hall, J. McCall, P. Botham, D. Esdaile, and J. Gardner. 1992. Findings of an interlaboratory trial of the enucleated eye method as an alternative eye irritation test. Toxicol. Mech. Methods., 2: 30-41.

Young, J.R., M.J. How, A.P. Walker, and W.M. Worth. 1988. Classification as corrosive or irritant to skin of preparations containing acidic or alkaline substances, without testing on animals. Toxicol. In Vitro, 2(1): 19-26. Doi: 10.1016/0887-2333(88)90032-x.

Zuang, V. 2001. The neutral red release assay: a review. Altern. Lab. Anim., 29(5): 575-599. Doi: 10.1177/026119290102900513.

Piehl, M., M. Carathers, R. Soda, D. Cerven, and G. DeGeorge. 2011. Porcine corneal ocular reversibility assay (PorCORA) predicts ocular damage and recovery for global regulatory agency hazard categories. Toxicol. In vitro, 25: 1912-1918. Doi:10.1016/j.tiv.2011.06.008.

Reader, S.J., V. Blackwell, R. O'Hara, R.H. Clothier, G. Griffin, and M. Balls. 1989. A vital dye release method for assessing the short-term cytotoxic effects of chemicals and formulations. Altern. Lab. Anim., 17: 28-33. Doi: 10.1177/026119298901700106.

Reader, S.J., V. Blackwell, R. O'Hara, R.H. Clothier, G. Griffin, and M. Balls. 1990. Neutral red release from pre-loaded cells as an in vitro approach to testing for eye irritancy potential. Toxicol. In vitro, 4(4-5): 264-266. Doi: 10.1016/0887-2333(90)90060-7.

Settivari, R.S., R.A. Amado, M. Corvaro, N.R. Visconti, L. Kan, E.W. Carney, D.R. Boverhof, and S.C. Gehen. 2016. Tiered application of the neutral red release and EpiOcular™ assays for evaluating the eye irritation potential of agrochemical formulations. Regul. Toxicol. Pharmacol., 81: 407-420. Doi: 10.1016/j.yrtph.2016.09.028.

Spielmann, H., S. Kalweit, M. Liebsch, T. Wirnsberger, I. Gerner, E. Bertram-Neis, K. Krauser, R. Kreiling, H.G. Miltenburger, W. Pape, and W. Steiling. 1993. Validation study of alternatives to the Draize eye irritation test in Germany: Cytotoxicity testing and HET-CAM test with 136 industrial chemicals. Toxicol. In vitro, 7(4): 505-510. Doi: 10.1016/0887-2333(93)90055-a.

Spielmann, H., M. Liebsch, S. Kalweit, F. Moldenhauer, T. Wirnsberger, H.-G. Holzhütter, B. Schneider, S. Glaser, I. Gerner, W.J.W. Pape, R. Kreiling, K. Krauser, H.G. Miltenburger, W. Steiling, N.P. Luepke, N. Müller, H. Kreuzer, P. Mürmann, J. Spengler, E. Bertram-Neis, B. Siegemund, and F.J. Wiebel. 1996. Results of a validation study in Germany on two in vitro alternatives to the Draize eye irritation test, HET-CAM test and the 3T3 NRU cytotoxicity test. Altern. Lab. Anim., 24: 741-858.

Spöler, F., M. Först, H. Kurz, M. Frentz, and N.F. Schrage. 2007. Dynamic analysis of chemical eye burns using high-resolution optical coherence tomography. J. Biomed. Opt., 12: 041203. doi:10.1117/1.2768018.

Spöler, F., O. Kray, S. Kray, C. Panfil, and N.F. Schrage. 2015. The Ex Vivo Eye Irritation Test as an alternative test method for serious eye damage/eye irritation. Altern. Lab. Anim., 43(3): 163-179. Doi: 10.1177/026119291504300306.

Whittle, E., D. Basketter, M. York, L. Kelly, T. Hall, J. McCall, P. Botham, D. Esdaile, and J. Gardner. 1992. Findings of an interlaboratory trial of the enucleated eye method as an alternative eye irritation test. Toxicol. Mech. Methods., 2: 30-41.

Young, J.R., M.J. How, A.P. Walker, and W.M. Worth. 1988. Classification as corrosive or irritant to skin of preparations containing acidic or alkaline substances, without testing on animals. Toxicol. In vitro, 2(1): 19-26. Doi: 10.1016/0887-2333(88)90032-x.

Zuang, V. 2001. The neutral red release assay: a review. Altern. Lab. Anim., 29(5): 575-599. Doi: 10.1177/026119290102900513.

CHAPTER 3.4

RESPIRATORY OR SKIN SENSITIZATION

3.4.1 Definitions and general considerations

3.4.1.1 *Respiratory sensitization* refers to hypersensitivity of the airways occurring after inhalation of a substance or a mixture.

Skin sensitization refers to an allergic response occurring after skin contact with a substance or a mixture.

3.4.1.2 For the purpose of this chapter, sensitization includes two phases: the first phase is induction of specialized immunological memory in an individual by exposure to an allergen. The second phase is elicitation, i.e. production of a cell-mediated or antibody-mediated allergic response by exposure of a sensitized individual to an allergen.

3.4.1.3 For respiratory sensitization, the pattern of induction followed by elicitation phases is shared in common with skin sensitization. For skin sensitization, an induction phase is required in which the immune system learns to react; clinical symptoms can then arise when subsequent exposure is sufficient to elicit a visible skin reaction (elicitation phase). As a consequence, predictive tests usually follow this pattern in which there is an induction phase, the response to which is measured by a standardized elicitation phase, typically involving a patch test. The local lymph node assay is the exception, directly measuring the induction response. Evidence of skin sensitization in humans normally is assessed by a diagnostic patch test.

3.4.1.4 Usually, for both skin and respiratory sensitization, lower levels are necessary for elicitation than are required for induction. Provisions for alerting sensitized individuals to the presence of a particular sensitizer in a mixture can be found in 3.4.4.2.

3.4.1.5 The hazard class "respiratory or skin sensitization" is differentiated into:

    (a)   Respiratory sensitization; and

    (b)   Skin sensitization

3.4.2 Classification criteria for substances

3.4.2.1 *Respiratory sensitizers*

3.4.2.1.1 *Hazard categories*

3.4.2.1.1.1 Respiratory sensitizers shall be classified in Category 1 where subcategorization is not required by a competent authority or where data are not sufficient for subcategorization.

3.4.2.1.1.2 Where data are sufficient and where required by a competent authority, a refined evaluation according to 3.4.2.1.1.3 allows the allocation of respiratory sensitizers into sub-category 1A, strong sensitizers, or sub-category 1B for other respiratory sensitizers.

3.4.2.1.1.3 Effects seen in either humans or animals will normally justify classification in a weight of evidence assessment for respiratory sensitizers. Substances may be allocated to one of the two sub-categories 1A or 1B using a weight of evidence assessment in accordance with the criteria given in table 3.4.1 and on the basis of reliable and good quality evidence from human cases or epidemiological studies and/or observations from appropriate studies in experimental animals.

第 3.4 章

呼吸器感作性または皮膚感作性

3.4.1 定義および一般事項

3.4.1.1 *呼吸器感作性*とは、物質または混合物の吸入後に起こる、気道の過敏症をさす。

*皮膚感作性*とは、物質または混合物に皮膚接触した後に起こる、アレルギー性反応をさす。

3.4.1.2 本章では感作性に2つの段階を含んでいる。最初の段階はアレルゲンへのばく露による個人の特異的な免疫学的記憶の誘導 (induction) である。次の段階は惹起 (elicitation)、すなわち、感作された個人がアレルゲンにばく露することにより起こる細胞性あるいは抗体性のアレルギー反応である。

3.4.1.3 呼吸器感作性で、誘導から惹起段階へと続くパターンは一般に皮膚感作性でも同じである。皮膚感作性では、免疫システムが反応を学ぶ誘導段階を必要とする。続いて起こるばく露が視認できるような皮膚反応を惹起するのに十分であれば臨床症状となって現れる(惹起段階)。したがって、予見的試験は、まず誘導期があり、さらにそれへの反応が通常はパッチテストを含んだ標準化された惹起期によって測定されるパターンに従う。誘導反応を直接的に測定する局所のリンパ節試験は例外的である。ヒトでの皮膚感作性の証拠は普通診断学的パッチテストで評価される。

3.4.1.4 通常皮膚および呼吸器感作性では、惹起に必要なレベルは誘導に必要なレベルよりも低い。感作された人に混合物中の感作物質の存在を知らせるための対策を3.4.4.2に示した。

3.4.1.5 「呼吸器感作性または皮膚感作性」の有害性区分は次のように分かれる。

(a) 呼吸器感作性、および
(b) 皮膚感作性

3.4.2 物質の分類基準

3.4.2.1 *呼吸器感作性物質*

3.4.2.1.1 *有害性区分*

3.4.2.1.1.1 呼吸器感作性物質は、所管官庁によって細区分が要求されていない場合または細区分のためのデータが十分でない場合には、区分1に分類しなければならない。

3.4.2.1.1.2 データが十分にありまた所管官庁が要求している場合には、3.4.2.1.1.3にしたがって呼吸器感作性物質を細区分1A (強い感作性物質) または細区分1B (他の呼吸器感作性物質) に細かく評価する。

3.4.2.1.1.3 呼吸器感作性物質については、通常ヒトまたは動物で見られた影響は証拠の重み付け評価により分類の根拠となる。表3.4.1における判定基準にしたがいヒトの症例または疫学的研究および/または実験動物における適切な研究結果による信頼できる質の良い証拠に基づいて、証拠の重み付け評価により、物質は2つの細区分1Aまたは1Bのどちらかに割り当てられるであろう。

Table 3.4.1: Hazard category and sub-categories for respiratory sensitizers

CATEGORY 1:	Respiratory sensitizer
	A substance is classified as a respiratory sensitizer: (a) if there is evidence in humans that the substance can lead to specific respiratory hypersensitivity and/or (b) if there are positive results from an appropriate animal test[1].
Sub-category 1A:	Substances showing a high frequency of occurrence in humans; or a probability of occurrence of a high sensitization rate in humans based on animal or other tests[1]. Severity of reaction may also be considered.
Sub-category 1B:	Substances showing a low to moderate frequency of occurrence in humans; or a probability of occurrence of a low to moderate sensitization rate in humans based on animal or other tests[1]. Severity of reaction may also be considered.

3.4.2.1.2   *Human evidence*

3.4.2.1.2.1   Evidence that a substance can lead to specific respiratory hypersensitivity will normally be based on human experience. In this context, hypersensitivity is normally seen as asthma, but other hypersensitivity reactions such as rhinitis/conjunctivitis and alveolitis are also considered. The condition will have the clinical character of an allergic reaction. However, immunological mechanisms do not have to be demonstrated.

3.4.2.1.2.2   When considering the human evidence, it is necessary for a decision on classification to take into account, in addition to the evidence from the cases:

    (a)   the size of the population exposed;

    (b)   the extent of exposure.

3.4.2.1.2.3   The evidence referred to above could be:

    (a)   clinical history and data from appropriate lung function tests related to exposure to the substance, confirmed by other supportive evidence which may include:

        (i)   in vivo immunological test (e.g. skin prick test);

        (ii)   in vitro immunological test (e.g. serological analysis);

        (iii)   studies that may indicate other specific hypersensitivity reactions where immunological mechanisms of action have not been proven, e.g. repeated low-level irritation, pharmacologically mediated effects;

        (iv)   a chemical structure related to substances known to cause respiratory hypersensitivity;

    (b)   data from positive bronchial challenge tests with the substance conducted according to accepted guidelines for the determination of a specific hypersensitivity reaction.

3.4.2.1.2.4   Clinical history should include both medical and occupational history to determine a relationship between exposure to a specific substance and development of respiratory hypersensitivity. Relevant information includes aggravating factors both in the home and workplace, the onset and progress of the disease, family history and medical history of the patient in question. The medical history should also include a note of other allergic or airway disorders from childhood, and smoking history.

---

[1] *At present, recognized and validated animal models for the testing of respiratory hypersensitivity are not available. Under certain circumstances, data from animal studies may provide valuable information in a weight of evidence assessment.*

表 3.4.1：呼吸器感作性物質の有害性区分および細区分

区分1：	呼吸器感作性物質
	物質は呼吸器感作性物質として分類される： (a) ヒトに対し当該物質が特異的な呼吸器過敏症を引き起こす証拠がある場合、または (b) 適切な動物試験により陽性結果が得られている場合[1]。
細区分1A：	ヒトで高頻度に症例が見られる；または動物や他の試験[1]に基づいたヒトでの強い感作率の可能性がある。反応の重篤性についても考慮する。
細区分1B：	ヒトで低～中頻度に症例が見られる；または動物や他の試験[1]に基づいたヒトでの低～中程度の感作率の可能性がある。反応の重篤性についても考慮する。

3.4.2.1.2 *ヒトでの証拠*

3.4.2.1.2.1 物質が特異的な呼吸器過敏症を起こす可能性があるとする証拠は、通常はヒトでの経験をもとにして得られる。この場合、過敏症は通常喘息として観察されるが、例えば鼻炎/結膜炎および肺胞炎のようなその他の過敏症なども考えられる。アレルギー性反応の臨床的特徴を有することが条件となる。ただし、免疫学的メカニズムは示す必要はない。

3.4.2.1.2.2 ヒトでの証拠を考える場合、分類の決定には事例から得られる証拠に加えて、さらに下記のことに考慮する必要がある：

(a) ばく露された集団の大きさ；

(b) ばく露の程度。

3.4.2.1.2.3 上記に述べた証拠には下記のものが考えられる：

(a) 臨床履歴および当該物質へのばく露に関連する適切な肺機能検査より得られたデータで、下記の項目、およびその他の裏付け証拠により確認されたもの：

　(i) in vivo 免疫学的試験（例：皮膚プリック試験）；

　(ii) in vitro 免疫学的試験（例：血清学的分析）；

　(iii) 例えば反復低濃度刺激、薬理学的介在作用など、免疫学的メカニズムがまだ証明されていないその他の特異的過敏症反応の存在を示す試験

　(iv) 呼吸器過敏症の原因となることがわかっている物質に関連性のある化学構造；

(b) 特異的過敏症反応測定のために認められた指針に沿って実施された、当該物質についての気管支負荷試験の陽性結果。

3.4.2.1.2.4 臨床履歴には、特定の物質に対するばく露と呼吸器過敏症発生の間の関連性を決定するための、病歴および職歴の両方が記載されるべきである。該当する情報として、家庭および職場の両方での悪化要因、疾患の発症および経過、問題となっている患者の家族歴および病歴などが含まれる。この病歴にはさらに、子供時代からのその他のアレルギー性または気道障害についての記録および喫煙歴についても記載されるべきである。

---

[1] *現時点では、呼吸器過敏症の試験用として認められ、検証された動物モデルはない。ある場合には、動物実験によるデータは証拠の重み付け評価において貴重な情報を提供するであろう。*

3.4.2.1.2.5   The results of positive bronchial challenge tests are considered to provide sufficient evidence for classification on their own. It is however recognized that in practice many of the examinations listed above will already have been carried out.

3.4.2.1.3   *Animal studies*

Data from appropriate animal studies[1] which may be indicative of the potential of a substance to cause sensitization by inhalation in humans[2] may include:

(a)   measurements of Immunoglobulin E (IgE) and other specific immunological parameters, for example in mice;

(b)   specific pulmonary responses in guinea pigs.

3.4.2.2   *Skin sensitizers*

3.4.2.2.1   *Hazard categories*

3.4.2.2.1.1   Skin sensitizers shall be classified in Category 1 where subcategorization is not required by a competent authority or where data are not sufficient for subcategorization.

3.4.2.2.1.2   Where data are sufficient and where required by a competent authority, a refined evaluation according to 3.4.2.2.2 to 3.4.2.2.6 allows the allocation of skin sensitizers into sub-category 1A, strong sensitizers, or sub-category 1B for other skin sensitizers.

3.4.2.2.1.3   For classification of skin sensitizers, all available and relevant information is collected and its quality in terms of adequacy and reliability is assessed. Classification should be based on mutually acceptable data/results generated using methods and/or defined approaches that are validated according to international procedures. These include both OECD guidelines and equivalent methods/defined approaches (see 1.3.2.4.3). Sections 3.4.2.2.2 to 3.4.2.2.6 provide classification criteria for the different types of information that may be available.

3.4.2.2.1.4   A tiered approach (see 3.4.2.2.7) organizes the available information on skin sensitization into levels/tiers and provides for decision-making in a structured and sequential manner. Classification results directly when the information consistently satisfies the criteria. However, where the available information gives inconsistent and/or conflicting results within a tier, classification of a substance or a mixture is made on the basis of the weight of evidence within that tier. In some cases when information from different tiers gives inconsistent and/or conflicting results (see 3.4.2.2.7.7) or where data individually are insufficient to conclude on the classification, an overall weight of evidence assessment is used (see 1.3.2.4.9 and 3.4.2.2.7.6).

3.4.2.2.1.5   Guidance on the interpretation of criteria and references to relevant guidance documents are provided in 3.4.5.3.

3.4.2.2.2   *Classification based on human data (tier 1 in figure 3.4.1)*

3.4.2.2.2.1   A substance is classified as a skin sensitizer in category 1 if there is evidence in humans that the substance can lead to sensitization by skin contact in a substantial number of persons.

3.4.2.2.2.2   Substances showing a high frequency of occurrence in humans, can be presumed to have the potential to produce significant sensitization and are classified in sub-category 1A. Severity of reaction may also be considered. Human evidence for sub-category 1A can include:

(a)   positive responses at ≤ 500 μg/cm^2 (Human Repeated Insult Patch Test (HRIPT), (Human maximization test (HMT) – induction threshold);

---

[1]   *At present, recognized and validated animal models for the testing of respiratory hypersensitivity are not available. Under certain circumstances, data from animal studies may provide valuable information in a weight of evidence assessment.*
[2]   *The mechanisms by which substances induce symptoms of asthma are not yet fully known. For preventative measures, these substances are considered respiratory sensitizers. However, if on the basis of the evidence, it can be demonstrated that these substances induce symptoms of asthma by irritation only in people with bronchial hyperreactivity, they should not be considered as respiratory sensitizers.*

3.4.2.1.2.5　気管支負荷試験の陽性結果は、分類のための十分な証拠になると考えられる。しかし、臨床現場では、実際には上記の試験の多くはすでに実施されているであろう。

3.4.2.1.3　*動物試験*

ヒトに吸入された場合に過敏症[2]の原因となる可能性を示すような適切な動物試験[1]から得られるデータには、下記のようなものがある:

(a) 例えばマウスを用いた免疫グロブリン E (IgE) およびその他特異的免疫学的項目の測定;

(b) モルモットにおける特異的肺反応。

3.4.2.2　*皮膚感作性物質*

3.4.2.2.1　*有害性区分*

3.4.2.2.1.1　皮膚感作性物質は、所管官庁によって細区分が要求されていない場合または細区分のためのデータが十分でない場合には、区分1に分類しなければならない。

3.4.2.2.1.2　データが十分にありまた所管官庁が要求している場合には、3.4.2.2.2～3.4.2.2.6にしたがって皮膚感作性物質を細区分1A（強い感作性物質）または細区分1B（他の皮膚感作性物質）に細かく評価する。

3.4.2.2.1.3　皮膚感作性物質の分類については、利用可能なすべての関連情報を収集し、妥当性および信頼性の観点からその質を評価する。分類は、国際的な手順にしたがって検証された方法および／またはディファインドアプローチを用いて生成された、相互に受け入れ可能なデータ／結果に基づいて行われるべきである。これには、OECD ガイドラインと同等の方法／ディファインドアプローチの両方が含まれる（1.3.2.4.3 参照）。3.4.2.2.2～3.4.2.2.6 では、利用可能な異なるタイプの情報に対する分類基準を示している。

3.4.2.2.1.4　段階的アプローチ（3.4.2.2.7 参照）は、皮膚感作性に関する利用可能な情報をレベル／段階に整理し、構造的かつ連続的な方法で意思決定することを提供する。分類の結果は、情報が一貫して基準を満たす場合に直接得られる。しかし、利用可能な情報が段階内で一貫性のない、あるいは矛盾する結果を与える場合、物質または混合物の分類は、その段階内の証拠の重みに基づいて行われる。異なる段階からの情報が一貫性のない、あるいは矛盾する結果を与える場合（3.4.2.2.7.7 参照）、あるいは分類を結論づけるには個々のデータが不十分な場合、包括的な証拠の重み付け評価が行われる（1.3.2.4.9 および 3.4.2.2.7.6 参照）。

3.4.2.2.1.5　基準の解釈に関する手引きおよび関連の手引きの参照は、3.4.5.3 に記載されている。

3.4.2.2.2　*ヒトのデータに基づく分類（図3.4.1の段階1）*

3.4.2.2.2.1　皮膚接触によって物質がヒトにおいて相当数の感作を引き起こす可能性があるという証拠がある場合、その物質は区分1の皮膚感作性物質として分類される。

3.4.2.2.2.2　ヒトでの発生頻度が高い物質は、明らかな感作を引き起こす可能性があると推定され、細区分1Aに分類される。また、反応の重篤性も考慮される。細区分1Aのヒトでの証拠は、以下を含む;

(a) ≦500μg/cm² （HRIPT、HMT－誘導閾値）で陽性反応;

---

[1] 現時点では、呼吸器過敏症の試験用として認められ、検証された動物モデルはない。ある場合には、動物実験によるデータは証拠の重み付け評価において貴重な情報を提供するであろう。

[2] 物質が喘息の症状を誘発するメカニズムはまだ完全に解明されていない。予防のために、このような物質を呼吸器感作性物質であるとみなす。ただし、証拠をもとに、これらの物質が刺激作用により気管支過敏症の人にだけに喘息症状を誘発することが実証された場合、これらは呼吸器感作性物質であるとみなされるべきではない。

(b) diagnostic patch test data where there is a relatively high and substantial incidence of reactions in a defined population in relation to relatively low exposure;

(c) other epidemiological evidence where there is a relatively high and substantial incidence of allergic contact dermatitis in relation to relatively low exposure.

3.4.2.2.2.3 Substances showing a low to moderate frequency of occurrence in humans can be presumed to have the potential to produce sensitization and are classified in sub-category 1B. Severity of reaction may also be considered. Human evidence for sub-category 1B can include:

(a) positive responses at > 500 µg/cm^2 (HRIPT, HMT – induction threshold);

(b) diagnostic patch test data where there is a relatively low but substantial incidence of reactions in a defined population in relation to relatively high exposure;

(c) other epidemiological evidence where there is a relatively low but substantial incidence of allergic contact dermatitis in relation to relatively high exposure.

3.4.2.2.3 *Classification based on standard animal data (tier 1 in figure 3.4.1)*

3.4.2.2.3.1 A substance is classified as a skin sensitizer if there are positive results from an appropriate animal test. For Category 1, when an adjuvant type test method for skin sensitization is used, a response of at least 30 % of the animals is considered as positive. For a non-adjuvant Guinea pig test method a response of at least 15 % of the animals is considered positive. For Category 1, a stimulation index of three or more is considered a positive response in the radioisotopic local lymph node assay (LLNA). For the non-radioactive modifications to the LLNA, a stimulation index of 1.8 or more in the LLNA: DA, 1.6 or more in the LLNA: BrdU-ELISA, and 2.7 or more in the LLNA: BrdU-FCM are considered positive. Test methods for skin sensitization are described in OECD Guideline 406 (the Guinea Pig Maximisation test and the Buehler guinea pig test) and guidelines 429/442A/442B (Local Lymph Node Assays). Other methods may be used provided that they are well-validated and scientific justification is given. The Mouse Ear Swelling Test (MEST), appears to be a reliable screening test to detect moderate to strong sensitizers, and can be used as a first stage in the assessment of skin sensitization potential.

Table 3.4.2: Animal test results for Category 1

Assay	Criteria
Local lymph node assay	SI ≥ 3
Local lymph node assay: DA	SI ≥ 1.8
Local lymph node assay: BrdU-ELISA	SI ≥ 1.6
Local lymph node assay: BrdU-FCM	SI ≥ 2.7
Adjuvant Guinea pig test method	≥30% responding at any intradermal induction dose
Non-adjuvant Guinea pig test method	≥15% responding at any topical induction dose

3.4.2.2.3.2 Substances showing a high potency in animals, can be presumed to have the potential to produce significant sensitization in humans and are classified in sub-category 1A. Severity of reactions may also be considered. Animal test results for sub-category 1A can include data with values indicated in table 3.4.3 below:

Table 3.4.3: Animal test results for sub-category 1A

Assay	Criteria
Local lymph node assay	EC3 value ≤ 2 %
Guinea pig maximisation test	≥ 30 % responding at ≤ 0.1 % intradermal induction dose or ≥ 60 % responding at > 0.1 % to ≤ 1 % intradermal induction dose
Buehler assay	≥15 % responding at ≤ 0.2 % topical induction dose or ≥ 60 % responding at > 0.2 % to ≤ 20 % topical induction dose

Note: *For the LLNA: BrdU-ELISA, subcategorization criteria (1A: EC1.6 value ≤ 6%, 1B: EC1.6 value > 6%, Maeda and Takeyoshi, 2019; Kobayashi et al., 2020) have been proposed and validated by OECD, but no subcategorization criteria have yet been agreed internationally. Validated subcategorization criteria may still be accepted by some competent authorities. A competent authority may decide which subcategorization criteria, if any, should be applied for these test methods.*

(b) 比較的低レベルのばく露を受けた対象集団において、比較的高い率で相当程度の陽性反応を示すパッチテストのデータ；

(c) 比較的低レベルのばく露を受けた対象集団において、アレルギー性接触皮膚炎の比較的高い率で相当程度の陽性反応を示す他の疫学的な証拠。

3.4.2.2.2.3　ヒトでの発生頻度が低～中程度である物質は、感作を引き起こす可能性があると推定され、細区分 1B に分類される。また、反応の重症度も考慮することができる。細区分 1B のヒトでの証拠は以下を含むことができる：

(a) >500μg/cm² （HRIPT、HMT－誘導閾値）で陽性反応；

(b) 比較的高レベルのばく露を受けた対象集団において、比較的低い率ではあるが相当程度の陽性反応を示すパッチテストのデータ；

(c) 比較的高レベルのばく露を受けた対象集団において、アレルギー性接触皮膚炎の比較的低い率ではあるが相当程度の陽性反応を示す他の疫学的な証拠。

3.4.2.2.3　標準的な動物データに基づく分類（図 3.4.1 の段階 1）

3.4.2.2.3.1　適切な動物試験で陽性結果が得られた場合、その物質は皮膚感作性物質として分類される。皮膚感作性区分1について、アジュバント型の皮膚感作性試験法が用いられる場合、動物の≧30%で反応があれば陽性であると考えられる。アジュバントを用いないモルモット試験方法では、動物の少なくとも≧15%で反応があれば陽性であると考えられる。区分 1 に関して、放射性同位元素を用いた局所リンパ節試験（LLNA）において刺激指数（SI 値）が 3 以上であれば陽性反応と考えられる。LLNA の非放射性試験法については、LLNA：DA で 1.8 以上、LLNA：BrdU-ELISA で 1.6 以上、LLNA：BrdU-FCM で 2.7 以上の刺激指数を陽性とみなす。皮膚感作性に関する試験方法は、OECD ガイドライン 406（モルモット Maximisation 試験およびモルモット Buehler 試験）とガイドライン 429/442A/442B（局所リンパ節試験）に定められている。他の方法でも有効性が確認され科学的に妥当性が示されている場合には使用してもよい。マウス耳介腫脹試験（MEST）は、中程度から強い感作性物質を検出する信頼性の高いスクリーニング法であり、皮膚感作性評価の第一段階として用いることができる。

表 3.4.2：区分 1 の動物試験結果

試験方法	基準
局所リンパ節試験	SI ≧ 3
局所リンパ節試験: DA	SI ≧ 1.8
局所リンパ節試験: BrdU-ELISA	SI ≧ 1.6
局所リンパ節試験: BrdU-FCM	SI ≧ 2.7
アジュバント型のモルモット試験方法	いかなる皮内投与量でも ≧ 30% の反応
アジュバントを用いないモルモット試験方法	いかなる局所投与量でも ≧ 15% の反応

3.4.2.2.3.2　動物で高い感作能力を示す物質は、ヒトに明らかな感作を引き起こす可能性があると推定でき、細区分 1A に分類される。また、反応の重篤性も考慮する。細区分 1A の動物試験結果には、以下の表 3.4.3 に示す値を持つデータを含めることができる：

表 3.4.3：動物試験結果による細区分 1A

試験方法	判定基準
局所リンパ節試験	EC3 値 ≦2%
モルモット Maximisation 試験	皮内投与量 ≦0.1%で、≧30% の反応　または 皮内投与量 >0.1 %、≦1%で、≧60% の反応
モルモット Buehler 試験	局所投与量 ≦0.2%で、≧15% の反応　または 局所投与量 >0.2 %、≦20% で、≧60% の反応

注：LLNA：BrdU-ELISA については、細区分基準（1A：EC1.6値≦6%、1B：EC1.6値＞6%、Maeda and Takeyoshi, 2019; Kobayashi et al., 2020）が提案され、OECD により有効性が検証されているが、細区分基準は国際的にまだ合意されてはいない。検証された細区分基準は、一部の所管官庁に受け入れられているかもしれない。所管官庁は、これらの試験法について（もしあれば）どの細区分基準を適用すべきかを決定することができる。

As for the LLNA: DA and LLNA: BrdU-FCM, there are currently no validated and internationally agreed criteria for subcategorization of skin sensitizers. Therefore, these test methods can only be used to conclude on either classification in category 1 or no classification.

3.4.2.2.3.3 Substances showing a low to moderate potency in animals, can be presumed to have the potential to produce significant sensitization in humans and are classified in sub-category 1B. Severity of reactions may also be considered. Animal test results for sub-category 1B can include data with values indicated in table 3.4.4 below:

Table 3.4.4: Animal test results for sub-category 1B

Assay	Criteria
Local lymph node assay	EC3 value > 2 %
Guinea pig maximisation test	≥ 30 % to < 60 % responding at > 0.1 % to ≤ 1 % intradermal induction dose or ≥ 30 % responding at > 1 % intradermal induction dose
Buehler assay	≥ 15 % to < 60 % responding at > 0.2 % to ≤ 20 % topical induction dose or ≥ 15 % responding at > 20 % topical induction dose

Note: *For the LLNA: BrdU-ELISA, subcategorization criteria (1A: EC1.6 value ≤ 6%, 1B: EC 1.6 value > 6%, Maeda and Takeyoshi, 2019; Kobayashi et al., 2020) have been proposed and validated by OECD, but no subcategorization criteria have yet been agreed internationally. subcategorizationValidated subcategorization criteria may still be accepted by some competent authorities. A competent authority may decide which subcategorization criteria, if any, should be applied for these test methods.*

As for the LLNA: DA and LLNA: BrdU-FCM, there are currently no validated and internationally agreed criteria for subcategorization of skin sensitizers. Therefore, these test methods can only be used to conclude on either classification in category 1 or no classification.

3.4.2.2.4 *Classification based on defined approaches (tier 1 or tier 2 in figure 3.4.1)*

3.4.2.2.4.1 Defined approaches consist of a rule-based combination of data obtained from a predefined set of different information sources (e.g. *in chemico* methods, in vitro methods, physico-chemical properties, non-test methods). It is recognized that most single non animal methods are not able to replace in vivo methods fully for most regulatory endpoints. Thus, defined approaches can be useful strategies of combining data for classifying substances and mixtures. Results obtained with a defined approach validated according to international procedures, such as OECD Guideline 497 or an equivalent approach, are conclusive for classification for skin sensitization if the criteria of the defined approach are fulfilled (see table 3.4.7)[3]. Data from a defined approach can only be used for classification when the tested substance is within the applicability domain of the defined approach used. Additional limitations described in the published literature should also be taken into consideration.

3.4.2.2.4.2 Where the results from defined approaches are assigned a level of confidence as for example in OECD Guideline 497, a low confidence outcome of a defined approach in tier 1 is inconclusive and thus cannot be used on its own to classify but may be considered in combination with other data in tier 2.

3.4.2.2.4.3 Individual evidence used within a defined approach should not also be used outside of that defined approach.

3.4.2.2.5 *Classification based on in chemico/in vitro data (tier 1 or tier 2 in figure 3.4.1)*

3.4.2.2.5.1 The currently available *in chemico*/in vitro methods address specific biological mechanisms leading to the acquisition of skin sensitization as described, for example, in the OECD Adverse Outcome Pathway for Skin Sensitization (see OECD (2014)). Individual test methods that are validated according to international procedures and are accepted as stand-alone methods, can be used to conclude on the classification in tier 1. A competent authority may decide whether to use the method described in Appendix III to OECD Test Guideline 442C as a stand-alone method to discriminate between sub-category 1A and those not categorized as sub-category 1A (see 3.4.5.3.5).

3.4.2.2.5.2 Other non stand-alone *in chemico*/in vitro methods that are validated according to international procedures such as OECD test guidelines 442C (Annex I and II), 442D and 442E, are accepted as supportive evidence

---

[3] *Additional defined approaches have been proposed for skin sensitization (OECD 2017) but no classification criteria have yet been agreed internationally.*

LLNA: DA および *LLNA: BrdU-FCM* に関しては、現在、皮膚感作性物質の細区分のための有効で国際的に合意された基準は存在しない。したがって、これらの試験法は、区分 1 に分類されるか、分類されないかの結論を出すためにのみ使用することができる。

3.4.2.2.3.3 動物で低～中程度の感作能力を示す物質は、ヒトに明らかな感作を引き起こす可能性があると推定でき、細区分 1B に分類される。また、反応の重篤性も考慮する。細区分 1B の動物試験結果には、以下の表 3.4.4 に示す値を持つデータを含めることができる。:

表 3.4.4：動物試験結果による細区分 1B

試験方法	判定基準
局所リンパ節試験	EC3 値 >2%
モルモット Maximisation 試験	皮内投与量 >0.1%、≦1%で、≧30%、<60% の反応　　または 皮内投与量 >1% で、≧30% の反応
モルモット Buehler 試験	局所投与量 >0.2%、≦20%で、≧15%、<60% の反応　　または 局所投与量 >20% で、≧15% の反応

注：*LLNA: BrdU-ELISA* については、*細区分基準（1A: EC1.6 値≦6%、1B: EC1.6 値>6%、Maeda and Takeyoshi, 2019; Kobayashi et al., 2020）が提案され、OECD により有効性が検証されているが、細区分基準は国際的にまだ合意されてはいない。検証された細区分基準は、一部の所管官庁に受け入れられるかもしれない。所管官庁は、これらの試験法について（もしあれば）どの細区分基準を適用すべきかを決定することができる。*

LLNA: DA および *LLNA: BrdU-FCM* に関しては、現在、皮膚感作性物質の細区分のための有効で国際的に合意された基準は存在しない。したがって、これらの試験法は、区分 1 に分類されるか、分類されないかの結論を出すためにのみ使用することができる。

3.4.2.2.4　ディファインドアプローチに基づく分類（図 3.4.1 の段階 1 または段階 2）

3.4.2.2.4.1　ディファインドアプローチは、事前に定められた一連の異なる情報源（例：*in chemico* 法、*in vitro* 法、物理化学的特性、試験によらない方法）から得られたデータのルールに基づく組合せで構成される。ほとんどの規制エンドポイントにおいて、単一の動物試験によらない方法の多くは、*in vivo* 法に完全に置き換えることができないと認識されている。したがって、ディファインドアプローチは、物質や混合物を分類するためにデータを組み合わせることで有用な戦略となり得る。OECD ガイドライン 497 または同等のアプローチのような国際的な手順にしたがって検証されたディファインドアプローチで得られた結果は、ディファインドアプローチの基準が満たされる場合、皮膚感作性の分類において決定的となる（表 3.4.7 参照）[3]。ディファインドアプローチのデータは、試験物質が使用されたディファインドアプローチの適用範囲内にある場合にのみ、分類に用いることができる。公表文献に記載されている追加の制限も考慮されるべきである。

3.4.2.2.4.2　OECD ガイドライン 497 のように、ディファインドアプローチの結果に信頼性のレベルが付与されている場合、段階 1 のディファインドアプローチにおける信頼性の低い結果は決定的ではなく、単独で分類に用いることはできないが、段階 2 の他のデータと組み合わせて考慮することはできる。

3.4.2.2.4.3　ディファインドアプローチにおいて使用される個々の証拠は、そのディファインドアプローチ以外では使用されるべきではない。

3.4.2.2.5　*in chemico/in vitro* データに基づく分類（図 3.4.1 の段階 1 または段階 2）

3.4.2.2.5.1　現在利用可能な *in chemico/in vitro* 試験法は、例えば、OECD Adverse Outcome Pathway for Skin Sensitization（OECD (2014) 参照）に記載されているように、皮膚感作の獲得につながる具体的な生物学的メカニズムに対応している。国際的な手順にしたがって検証され、単独の方法として認められている個々の試験法は、段階 1 の分類について結論を出すために使用することができる。所管官庁は、細区分 1A と細区分 1A に分類されないものを識別するために、OECD テストガイドライン 442C の付録 III に記載された方法を単独で使用するかどうかを決定することができる（3.4.5.3.5 参照）。

3.4.2.2.5.2　OECD テストガイドライン 442C（補遺 I および II）、442D および 442E などの国際的な手順にしたがって検証された他の *in chemico/in vitro* 法は、裏付けとなる証拠として認められ、段階 1 で

---

[3] *皮膚感作性については、追加のディファインドアプローチが提案されているが（OECD 2017）、分類基準はまだ国際的に合意されていない。*

and should within tier 1 only be used in combination with other types of data in defined approaches. The use of these methods in tier 2 is described in 3.4.2.2.7.5.

3.4.2.2.5.3   Other validated *in chemico*/in vitro test methods accepted by some competent authorities are described in 3.4.5.3.6.2 [4]. A competent authority may decide which classification criteria, if any, should be applied for these test methods to conclude on classification.

3.4.2.2.5.4   *In chemico*/in vitro data can only be used for classification when the tested substance is within the applicability domain of the test method(s) used. Additional limitations described in the published literature should also be taken into consideration.

3.4.2.2.6   *Classification based on non-test methods (tier 2 in figure 3.4.1)*

3.4.2.2.6.1   Classification, including the conclusion not classified, can be based on non-test methods, with due consideration of reliability and applicability, on a case-by-case basis. Non-test methods include computer models predicting qualitative structure activity relationships (structural alerts, SAR) or quantitative structure-activity relationships (QSARs), computer expert systems, and read-across using analogue and category approaches.

3.4.2.2.6.2   Read-across using analogue or category approaches requires sufficiently reliable test data on similar substance(s) and justification of the similarity of the tested substance(s) with the substance to be classified. Where adequate justification of the read-across approach is provided, it has in general higher weight than (Q)SARs.

3.4.2.2.6.3   Classification based on (Q)SARs requires sufficient data and validation of the model. The validity of the computer models and the prediction should be assessed using internationally recognized principles for the validation of (Q)SARs. With respect to reliability, lack of alerts in a SAR or expert system is not sufficient evidence for no classification.

3.4.2.2.6.4   For conclusions on no classification from read-across and (Q)SARs the adequacy and robustness of the scientific reasoning and of the supporting evidence should be well substantiated and normally requires multiple negative substances with good structural and physical (related to toxicokinetics) similarity to the substance being classified, as well as a clear absence of positive substances with good structural and physical similarity to the substance being classified.

3.4.2.2.7   *Classification in a tiered approach (figure 3.4.1)*

3.4.2.2.7.1   A tiered approach to the evaluation of information should be considered, where applicable (figure 3.4.1), recognizing that not all tiers as well as information within a tier may be relevant. However, all available and relevant information of sufficient quality needs to be examined for consistency with respect to the resulting classification.

3.4.2.2.7.2   Tier 1 - Classification based on human data, standard animal data, defined approaches or stand-alone *in chemico*/in vitro methods

For classification of a substance, evidence in tier 1 may include data from any or all of the following lines of evidence. Where information from data within tier 1 is inconsistent and/or conflicting, the conclusion is determined in a weight of evidence assessment:

  (a) Experimental studies in humans (e.g. predictive patch testing, HRIPT, HMT (see paragraph 1.3.2.4.7, criteria in 3.4.2.2.2.2 (a) and 3.4.2.2.2.3 (a) and guidance in 3.4.5.3.2);

  (b) Epidemiological studies (e.g. case control studies, prospective studies) assessing allergic contact dermatitis (see paragraph 1.3.2.4.7, criteria in 3.4.2.2.2.2 (b) and (c) and 3.4.2.2.2.3 (b) and (c) and guidance in 3.4.5.3.2);

  (c) Well-documented cases of allergic contact dermatitis (see criteria in 3.4.2.2.2.2 (b) and 3.4.2.2.2.3 (b) and guidance in 3.4.5.3.2);

  (d) Appropriate animal studies (see criteria in 3.4.2.2.3 and guidance in 3.4.5.3.3);

---

[4] *Additional in chemico/in vitro methods have been proposed for skin sensitization (see 3.4.5.3.6.2) but no classification criteria have yet been agreed internationally.*

は、ディファインドアプローチで他の種類のデータと組み合わせてのみ使用されるべきである。段階2におけるこれらの方法の使用は、3.4.2.2.7.5 で説明されている。

3.4.2.2.5.3　一部の所管官庁が認めるその他の検証された in chemico/in vitro 法は、3.4.5.3.6.2 [4] に記載されている。所管官庁は、分類の結論を出すために、これらの試験法について分類基準がある場合には、どの基準を適用すべきかを決定することができる。

3.4.2.2.5.4　In chemico/in vitro データは、試験物質が使用された試験法の適用範囲内にある場合にのみ、分類に使用できる。公表文献に記載されている追加の制限も考慮されるべきである。

3.4.2.2.6　*試験によらない方法に基づく分類（図 3.4.1 の段階 2）*

3.4.2.2.6.1　ケースバイケースで、信頼性および適用性を十分に考慮した上で、区分に該当しないという結論も含めて、試験によらない方法に基づいて分類を行うことができる。試験によらない方法には、定性的構造活性相関（構造アラート、SAR）や定量的構造活性相関（QSARs）のコンピューターモデル、コンピューターエクスパートシステム、および、類似物質およびカテゴリーアプローチを用いたリード・アクロスがある。

3.4.2.2.6.2　類似物質やカテゴリーアプローチを用いたリード・アクロスでは、類似物質に関する十分に信頼できる試験データとそして、試験物質と分類対象物質の類似性の正当性が必要である。リード・アクロス手法に関して十分な正当性が示されている場合には、一般には(Q)SARs よりも高い重み付けがなされる。

3.4.2.2.6.3　(Q)SARs に基づく分類では十分なデータとモデルの妥当性検証が必要である。コンピュータモデルとその予測の妥当性は、国際的に認められた(Q)SAR の検証のための原則を用いて評価されるべきである。信頼性に関して、SAR における構造アラートまたはエキスパートシステムがないことは、区分に該当しない十分な証拠とはならない。

3.4.2.2.6.4　リード・アクロスおよび (Q)SAR から得られた区分に該当しない（分類されない）という結論については、科学的推論およびそれを裏付ける証拠の妥当性と頑健性が十分に実証されている必要があり、分類対象物質との構造的および物理的類似性が高い陽性物質が明らかに存在しないことを示すのと同様に、通常は分類対象物質との構造的および物理的(関連するトキシコキネティクスに) 類似性が高い複数の陰性物質が必要である。

3.4.2.2.7　*段階的アプローチによる分類（図 3.4.1）*

3.4.2.2.7.1　情報を評価するための段階的アプローチは、適用可能な場合、すべての段階および段階に含まれる情報が関連しているわけではないことを認識した上で検討されるべきである（図 3.4.1）。ただし、利用可能かつ質的にも十分な関連情報のすべてが、分類結果に関して一貫性があるかどうか精査される必要がある。

3.4.2.2.7.2　段階 1・ヒトデータ、標準的な動物データ、ディファインドアプローチまたは単独の *in chemico/in vitro* 法に基づく分類

物質の分類のために、段階 1 の証拠には、以下の一連の証拠（lines of evidence）のいずれかまたはすべてからのデータを含まれる。段階 1 のデータからの情報に一貫性がないおよび/または矛盾している場合、結論は証拠の重み付け評価で決定される：

(a) ヒトを対象とした実験的研究（例：予測パッチテスト、HRIPT、HMT（1.3.2.4.7、3.4.2.2.2.2 (a) および 3.4.2.2.2.3 (a) の基準ならびに 3.4.5.3.2 の手引きを参照のこと）；

(b) アレルギー性接触皮膚炎を評価する疫学研究（例：症例対照研究、前向き研究）（1.3.2.4.7、3.4.2.2.2.2 (b) および (c) ならびに 3.4.2.2.2.3 (b) および (c) の基準ならびに 3.4.5.3.2 の手引きを参照のこと）；

(c) アレルギー性接触皮膚炎の明確に文書化された症例（3.4.2.2.2.2 (b) および 3.4.2.2.2.3 (b) の基準および 3.4.5.3.2 の手引きを参照のこと）；

(d) 適切な動物実験（3.4.2.2.3 の基準、3.4.5.3.3 の手引きを参照）；

---

[4] *皮膚感作性については、追加のディファインドアプローチが提案されているが（OECD 2017）、分類基準はまだ国際的に合意されていない。*

(e) Defined approaches validated according to international procedures (see 3.4.2.2.4, guidance in 3.4.5.3.4 and table 3.4.7);

(f) Stand-alone *in chemico*/in vitro methods validated according to international procedures (see 3.4.2.2.5, guidance in 3.4.5.3.5 and table 3.4.8).

3.4.2.2.7.3 Tier 2 - Classification based on inconclusive data from tier 1, non stand-alone *in chemico*/in vitro methods or non-test methods.

In case a definitive conclusion on classification, including subcategorization where required by a competent authority, cannot be derived from tier 1, additional lines of evidence shall be considered in a weight of evidence assessment in tier 2. These may include:

(a) Data from non stand-alone *in chemico*/in vitro methods (see 3.4.2.2.5 and 3.4.5.3.5);

(b) Data from non-test methods (see 3.4.2.2.6).

3.4.2.2.7.4 Evidence from non stand-alone *in chemico*/in vitro methods and from non-test methods should not be considered at this stage if this data is already used in a defined approach under 3.4.2.2.7.2.

3.4.2.2.7.5 Individual non stand-alone *in chemico*/in vitro methods validated according to international procedures and non-test methods (including read-across) can be applied in a weight of evidence assessment together with inconclusive data from tier 1 and should be used in this second tier because they can usually not be used as stand-alone (with the exception of good quality read-across). However, a competent authority may decide that a positive result with one of these non stand-alone *in chemico*/in vitro methods, may be used on its own to classify in category 1 (see table 3.4.8).

3.4.2.2.7.6 Tier 3 - Classification based on overall weight of evidence assessment, including additional indicators

In case a definitive conclusion on classification including subcategorization where required by a competent authority, cannot be derived from the previous tiers, an overall weight of evidence assessment using expert judgment should be used that may include a combination of two or more indicators of skin sensitization as listed below:

(a) Isolated episodes of allergic contact dermatitis;

(b) Epidemiological studies of limited power, e.g. where chance, bias or confounders have not been ruled out fully with reasonable confidence;

(c) Data from animal tests, performed according to existing guidelines, which do not meet the criteria for a positive result described in 3.4.2.2.3, but which are sufficiently close to the limit to be considered significant;

(d) Data from non-standard methods.

3.4.2.2.7.7 Where information from the various tiers is inconsistent and/or conflicting with respect to the resulting classification, information of sufficient quality from a higher tier is generally given a higher weight than information from a lower tier. However, when information from a lower tier would result in a stricter classification than information from a higher tier and there is concern for misclassification, then classification is determined by an overall weight of evidence assessment (i.e. in tier 3). For example, having consulted the guidance in 3.4.5.3 as appropriate, classifiers concerned with a negative result for skin sensitization in a Buehler study when there is a clear positive result in humans for very similar substances (from read-across) would utilise an overall weight of evidence assessment.

3.4.2.2.8 *Immunological contact urticaria*

3.4.2.2.8.1 Substances meeting the criteria for classification as respiratory sensitizers may in addition cause immunological contact urticaria. Consideration should be given to classifying these substances also as skin sensitizers. Substances which cause immunological contact urticaria without meeting the criteria for respiratory sensitizers should also be considered for classification as skin sensitizers.

3.4.2.2.8.2 There is no recognized animal model available to identify substances which cause immunological contact urticaria. Therefore, classification will normally be based on human evidence which will be similar to that for skin sensitization.

(e) 国際的な手順にしたがって検証されたディファインドアプローチ（3.4.2.2.4、3.4.5.3.4 の手引きおよび表 3.4.7 を参照のこと）；

(f) 国際的な手順にしたがって検証された単独の *in chemico/in vitro* 法（3.4.2.2.5、3.4.5.3.5 の手引きおよび表 3.4.8 を参照のこと）。

3.4.2.2.7.3 段階 2- 段階 1 で決定的でないデータ、単独でない *in chemico/in vitro* 法または試験によらない方法に基づく分類

細区分を含め、所管官庁が要求する分類に関する決定的な結論が段階 1 で得られない場合、段階 2 の証拠の重み付け評価において、追加の一連の証拠（lines of evidence）を考慮するものとする。これらには以下のものが含まれる：

(a) 単独では決定的でない *in chemico/in vitro* 法からのデータ（3.4.2.2.5 および 3.4.5.3.5 を参照）；

(b) 試験によらない方法によるデータ（3.4.2.2.6 参照）；

3.4.2.2.7.4 単独では決定的でない *in chemico/in vitro* 試験法および試験によらない方法からの証拠は、このデータが既に 3.4.2.2.7.2 のディファインドアプローチで用いられている場合には、当段階では考慮しないものとする。

3.4.2.2.7.5 国際的な手順にしたがって検証された個々の単独では決定的でない *in chemico/in vitro* 法および試験によらない方法（リードアクロスを含む）は、段階 1 の決定的でないデータとともに証拠の重み付け評価に適用でき、通常単独で使用できないため（良質のリードアクロスを除く）、この段階 2 で用いるべきである。しかし、所管官庁は、これらの単独では決定的でない *in chemico/in vitro* 測定法の一つで陽性結果が得られた場合、それだけで区分 1 に分類することができると決定することができる（表 3.4.8 参照）。

3.4.2.2.7.6 段階 3- 追加の指標を含む包括的な証拠の重み付け評価に基づく分類

細区分を含め、所管官庁が必要とする分類に関する決定的な結論が前の段階から得られない場合、専門家の判断による包括的な証拠の重み付け評価を用いるべきであり、それには以下に示す皮膚感作性指標の 2 つ以上の組合せを含むことができる。

(a) アレルギー性接触皮膚炎の単独のエピソード；

(b) 偶然性、偏見、交絡因子が合理的な信頼性をもって完全に排除されていない場合など、検出力が限定的な疫学調査；

(c) 既存のガイドラインにしたがって実施された動物実験のデータで、3.4.2.2.3 に記載された陽性結果の基準を満たさないが、有意とみなされる限界値に十分近いもの；

(d) 非標準的な方法から得られたデータ；

3.4.2.2.7.7 様々な段階からの情報が、分類結果に対して一貫性がなく、かつ/または相反する場合、一般に、より上位の段階からの十分な質の情報は、より下位の段階からの情報よりも高い重み付けがなされる。しかし、下位の段階の情報が上位の段階の情報よりも厳しい分類となり、誤分類の懸念がある場合、分類は包括的な証拠の重み付け評価によって決定される（すなわち段階 3）。例えば、3.4.5.3 の手引きを適宜参照したうえで、分類者が Buehler 法試験における陰性の結果を、（リードアクロスによる）非常に類似した物質のヒトでの明確な陽性結果によって懸念を持つ場合、包括的な証拠の重み付け評価を利用することになるであろう。

3.4.2.2.8　免疫性接触じんましん

3.4.2.2.8.1　呼吸器感作性物質に分類するための判定基準に適合する物質は、さらに免疫性接触じんましんを引き起こすことがある。これらの物質を皮膚感作性物質としても分類することも検討するべきである。免疫性接触じんましんを誘発する物質で、呼吸器感作性物質の判定基準には適合しない物質もまた、皮膚感作性物質として分類することを検討すべきである。

3.4.2.2.8.2　免疫性接触じんましんを生じる物質を識別するのに利用可能な動物モデルは認められていない。したがって、分類は、通常、皮膚感作性物質と同様にヒトでの証拠に基づいて行われる。

Figure 3.4.1: Application of the tiered approach for skin sensitization[a]

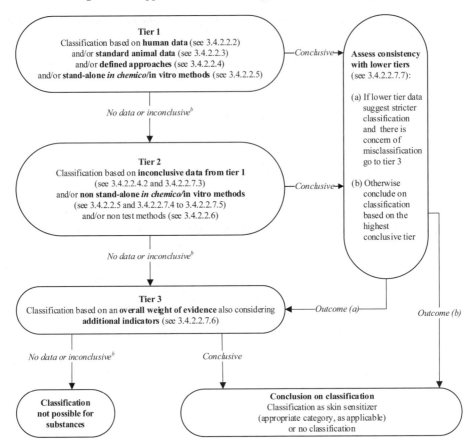

[a] Before applying the approach, the explanatory text in 3.4.2.2.7 as well as the guidance in 3.4.5.3 should be consulted. Only adequate and reliable data of sufficient quality should be included in applying the tiered approach.

[b] Information may be inconclusive for various reasons, e.g.:

- The available data may be of insufficient quality, or otherwise insufficient/inadequate for the purpose of classification, e.g. due to quality issues related to experimental design and/or reporting;

- Where competent authorities make use of the skin sensitization sub-categories 1A and 1B, the available data may not be capable of distinguishing between sub-category 1A and sub-category 1B.

3.4.3 Classification criteria for mixtures

3.4.3.1 *Classification of mixtures when data are available for the complete mixture*

When reliable and good quality evidence from human experience or appropriate studies in experimental animals, as described in the criteria for substances, is available for the mixture, then the mixture can be classified by weight of evidence assessment of these data. Care should be exercised in evaluating data on mixtures that the dose used does not render the results inconclusive. (For special labelling required by some competent authorities, see the note to table 3.4.5 and 3.4.4.2).

3.4.3.2 *Classification of mixtures when data are not available for the complete mixture: bridging principles*

3.4.3.2.1 Where the mixture itself has not been tested to determine its sensitizing properties, but there are sufficient data on both the individual ingredients and similar tested mixtures to adequately characterize the hazards of the

図 3.4.1 皮膚感作性 [a] に対する段階的アプローチの適用

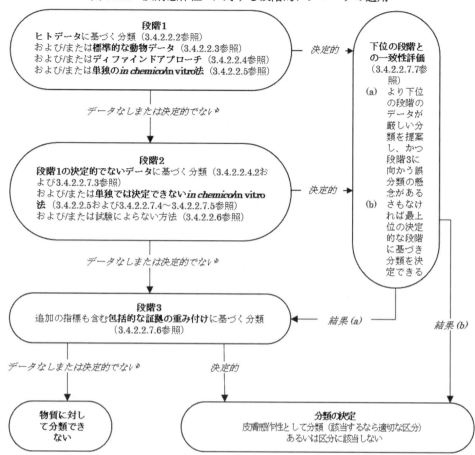

a 本アプローチを適用する前に、3.4.2.2.7の説明文及び3.4.5.3のガイダンスを参照すること。段階的アプローチの適用には、十分な質の信頼できるデータのみを含めるべきである。

b 情報は、以下のような様々な理由で決定的でない場合がある：

— 実験計画および/または報告に関連する品質上の問題により、利用可能なデータは質が十分ではない、または分類の目的には不十分/不適切である可能性がある；

— 所管官庁が皮膚感作性細区分1Aおよび1Bを使用する場合、利用可能なデータによっては細区分1Aと細区分1Bを区別することができない可能性がある。

3.4.3 混合物の分類基準

3.4.3.1 *混合物そのものについて試験データが利用可能な場合の混合物の分類*

混合物について、物質に関する分類判定基準で記述されている通り、ヒトの経験または適切な動物実験から信頼できる質の良い証拠が利用できる場合には、混合物はこのデータの証拠の重み付け評価によって分類できる。混合物に関するデータを評価する際には、使用された用量が、結論を不確かにさせていないかに注意を払うべきである。（一部の所管官庁による特別なラベル表示要件については、本章の表 3.4.5 の注記および3.4.4.2 参照）

3.4.3.2 *混合物そのものについて試験データが利用できない場合の混合物の分類：つなぎの原則（bridging principles）*

3.4.3.2.1 混合物そのものは感作性を決定する試験がなされていないが、当該混合物の有害性を適切に特定するための、個々の成分および類似の試験された混合物の両方に関して十分なデータがある場合、これ

mixture, these data will be used in accordance with the following agreed bridging principles. This ensures that the classification process uses the available data to the greatest extent possible in characterizing the hazards of the mixture without the necessity for additional testing in animals.

3.4.3.2.2   *Dilution*

If a tested mixture is diluted with a diluent which is not a sensitizer and which is not expected to affect the sensitization of other ingredients, then the new diluted mixture may be classified as equivalent to the original tested mixture.

3.4.3.2.3   *Batching*

The sensitizing properties of a tested production batch of a mixture can be assumed to be substantially equivalent to that of another untested production batch of the same commercial product when produced by or under the control of the same manufacturer, unless there is reason to believe there is significant variation such that the sensitization potential of the untested batch has changed. If the latter occurs, a new classification is necessary.

3.4.3.2.4   *Concentration of mixtures of the highest sensitizing category/sub-category*

If a tested mixture is classified in Category 1 or sub-category 1A, and the concentration of the ingredients of the tested mixture that are in Category 1 and sub-category 1A is increased, the resulting untested mixture should be classified in Category 1 or sub-category 1A without additional testing.

3.4.3.2.5   *Interpolation within one category/sub-category*

For three mixtures (A, B and C) with identical ingredients, where mixtures A and B have been tested and are in the same category/sub-category, and where untested mixture C has the same toxicologically active ingredients as mixtures A and B but has concentrations of toxicologically active ingredients intermediate to the concentrations in mixtures A and B, then mixture C is assumed to be in the same category/sub-category as A and B.

3.4.3.2.6   *Substantially similar mixtures*

Given the following:

(a)   Two mixtures: (i)   A + B;
                      (ii) C + B;

(b)   The concentration of ingredient B is essentially the same in both mixtures;

(c)   The concentration of ingredient A in mixture (i) equals that of ingredient C in mixture (ii);

(d)   Ingredient B is a sensitizer and ingredients A and C are not sensitizers;

(e)   A and C are not expected to affect the sensitizing properties of B.

If mixture (i) or (ii) is already classified based on test data, then the other mixture can be classified in the same hazard category.

3.4.3.2.7   *Aerosols*

An aerosolized form of the mixture may be classified in the same hazard category as the tested non-aerosolized form of the mixture provided that the added propellant does not affect the sensitizing properties of the mixture upon spraying.

3.4.3.3   *Classification of mixtures when data are available for all ingredients or only for some ingredients of the mixture*

The mixture should be classified as a respiratory or skin sensitizer when at least one ingredient has been classified as a respiratory or skin sensitizer and is present at or above the appropriate cut-off value/concentration limit for the specific endpoint as shown in table 3.4.5 for solid/liquid and gas respectively.

らのデータは以下の合意されたつなぎの原則にしたがって使用される。これによって、分類プロセスで動物試験を追加する必要もなく、混合物の有害性判定に利用可能なデータを可能な限り最大限に用いられるようになる。

3.4.3.2.2 *希釈*

試験された混合物が、感作性物質ではなく、また他の成分の感作に影響を与えないと予想される希釈剤で希釈される場合、新しい希釈された混合物は、元の試験された混合物と同等として分類してもよい。

3.4.3.2.3 *製造バッチ*

混合物の試験されていない製造バッチに感作特性があるかどうかは、同じ製造者によって、またはその管理下で生産された同じ商品の試験された別のバッチの毒性と本質的に同等とみなすことができる。ただし、試験されていないバッチで感作特性が変化するような有意な変動があると考えられる理由がある場合はこの限りではない。このような場合にはもし後者が起こるなら、新しい分類が必要である。

3.4.3.2.4 *毒性の強い混合物の濃縮*

試験された混合物が区分 1 または細区分 1A に分類され、区分 1 および細区分 1A にある試験された混合物の成分の濃度が増加する場合、試験されていない新しい混合物は、追加試験なしで区分 1 または細区分 1A に分類するべきである。

3.4.3.2.5 *1 つの毒性区分内での内挿*

3 つの混合物（A、B および C）は同じ成分を持ち、A と B は試験され同じ区分/細区分にある。試験されていない混合物 C は混合物 A および B と同じ毒性学的に活性な成分を持ち、毒性学的に活性な成分の濃度が混合物 A と B の中間である場合、混合物 C は A および B と同じ区分/細区分にあるとする。

3.4.3.2.6 *本質的に類似した混合物*

次を仮定する：

(a) 2 つの混合物： (i) A+B ;
 (ii) C+B ;
(b) 成分 B の濃度は、両方の混合物で本質的に同じである；
(c) 混合物(i)の成分 A の濃度は、混合物(ii)の成分 C の濃度に等しい；
(d) 成分 B は感作物質であり、成分 A と C は感作物質ではない；
(e) A と C は、B の感作性に影響しないと予想される。

混合物(i)または(ii)が既に試験データに基づいて分類されている場合には、他方の混合物は同じ有害性区分に分類することができる。

3.4.3.2.7 *エアゾール*

エアゾール形態の混合物は、添加された噴射剤が噴霧時に混合物の感作性に影響しないという条件下では、試験された非エアゾール形態の混合物と同じ有害性区分に分類してよい。

3.4.3.3 混合物の全成分について、または一部の成分だけについてデータが利用可能な場合の混合物の分類

混合物は、少なくとも 1 つの成分が呼吸器感作性物質または皮膚感作性物質として分類され、固体/液体と気体についてそれぞれ表 3.4.5 に示したように、それぞれの生体影響に示されたカットオフ値/濃度限界以上で存在する場合、呼吸器感作性物質または皮膚感作性物質として分類されるべきである。

Table 3.4.5: Cut-off values/concentration limits of ingredients of a mixture classified as either respiratory sensitizers or skin sensitizers that would trigger classification of the mixture

Ingredient classified as:	Cut-off values/concentration limits triggering classification of a mixture as:		
	respiratory sensitizer Category 1		skin sensitizer Category 1
	Solid/Liquid	Gas	All physical states
Respiratory sensitizer Category 1	≥ 0.1 % (see note)	≥ 0.1 % (see note)	--
	≥ 1.0 %	≥ 0.2 %	
Respiratory sensitizer sub-category 1A	≥ 0.1 %	≥ 0.1 %	--
Respiratory sensitizer sub-category 1B	≥ 1.0 %	≥ 0.2 %	
Skin sensitizer Category 1	--	--	≥ 0.1 % (see note)
	--	--	≥ 1.0 %
Skin sensitizer sub-category 1A	--	--	≥ 0.1 %
Skin sensitizer sub-category 1B	--	--	≥ 1.0 %

NOTE: *Some competent authorities may require SDS and/or supplemental labelling only, as described in 3.4.4.2 for mixtures containing a sensitizing ingredient at concentrations between 0.1 and 1.0 % (or between 0.1 and 0.2 % for a gaseous respiratory sensitizer). While the current cut-off values reflect existing systems, all recognize that special cases may require information to be conveyed below that level.*

3.4.4 Hazard communication

3.4.4.1 General and specific considerations concerning labelling requirements are provided in *Hazard communication: Labelling* (chapter 1.4). Annex 1 contains summary tables about classification and labelling. Annex 3 contains examples of precautionary statements and pictograms which can be used where allowed by the competent authority. Table 3.4.6 presents specific label elements for substances and mixtures classified into this hazard class based on the criteria in this chapter.

Table 3.4.6: Label elements for respiratory or skin sensitization

	Respiratory sensitization Category 1 and sub-categories 1A and 1B	Skin sensitization Category 1 and sub-categories 1A and 1B
Symbol	Health hazard	Exclamation mark
Signal word	Danger	Warning
Hazard statement	May cause allergy or asthma symptoms or breathing difficulties if inhaled	May cause an allergic skin reaction

3.4.4.2 Some chemicals that are classified as sensitizers may elicit a response, when present in a mixture in quantities below the cut-offs established in table 3.4.5, in individuals who are already sensitized to the chemicals. To protect these individuals, certain authorities may choose to require the name of the ingredients as a supplemental label element whether or not the mixture as a whole is classified as sensitizer.

3.4.5 Decision logic and guidance

The decision logics which follow are not part of the harmonized classification system but are provided here as additional guidance. It is strongly recommended that the person responsible for classification study the criteria before and during use of the decision logics.

表 3.4.5：混合物の分類基準となる呼吸器感作性物質または皮膚感作性物質として分類された混合物成分のカットオフ値/濃度限界

成分の分類：	混合物の分類基準となるカットオフ値/濃度限界		
	呼吸器感作性物質 区分 1		皮膚感作性物質 区分 1
	固体/液体	気体	すべての物理的状態
呼吸器感作性物質 区分 1	≧0.1%（注記）	≧0.1%（注記）	
	≧1.0%	≧0.2%	
呼吸器感作性物質 細区分 1A	≧0.1%	≧0.1%	
呼吸器感作性物質 細区分 1B	≧1.0%	≧0.2%	
皮膚感作性物質 区分 1			≧0.1%（注記）
			≧1.0%
皮膚感作性物質 細区分 1A			≧0.1%
皮膚感作性物質 細区分 1B			≧1.0%

注記：一部の所管官庁は、3.4.4.2に記載されているように0.1%～1.0%（または気体状の呼吸器感作性物質については0.1～0.2%）の間の濃度で感作性成分を含む混合物に対して、SDSおよび/または追加のラベル表示のみを要求してもよい。現行のカットオフ濃度は既存のシステムを反映したものであり、特別なケースでは、これ以下のレベルでも情報を伝えてもよいことは広く認められている。

3.4.4 危険有害性情報の伝達

3.4.4.1 表示要件に関する通則および細則は、*危険有害性に関する情報の伝達：表示（第1.4章）*に定める。附属書1に分類と表示に関する概要表を示す。附属書3には、注意書きおよび所管官庁が許可した場合に使用可能な注意絵表示の例を示す。下記の表3.4.6には、本章の判定基準に基づいて呼吸器感作性および皮膚感作性と分類される物質および混合物の具体的なラベル要素を示す。

表 3.4.6：呼吸器感作性および皮膚感作性のラベル要素

	呼吸器感作性 区分 1 細区分 1A および 1B	皮膚感作性 区分 1 細区分 1A および 1B
シンボル	健康有害性	感嘆符
注意喚起語	危険	警告
危険有害性情報	吸入するとアレルギー、喘息または、呼吸困難を起こすおそれ	アレルギー性皮膚反応を起こすおそれ

3.4.4.2 感作性ありと分類されている一部の化学品は、表3.4.5のカットオフ値よりも少ない量で混合物中に存在しても、すでに感作されている個人に反応を惹起することがあろう。これらの人々を保護するために、関係所管官庁は、混合物として感作性物質であるかないかにかかわらずラベルに補足的な情報として成分名の記載を要求することができる。

3.4.5 判定論理および手引き

以下の判定論理は、調和分類システムには含まれないが、追加的な手引きとしてここに示す。分類の責任者に対し、この判定論理を使用する前および使用する際に判定基準についてよく調べ理解することを強く勧める。

3.4.5.1 Decision logic 3.4.1 for respiratory sensitization

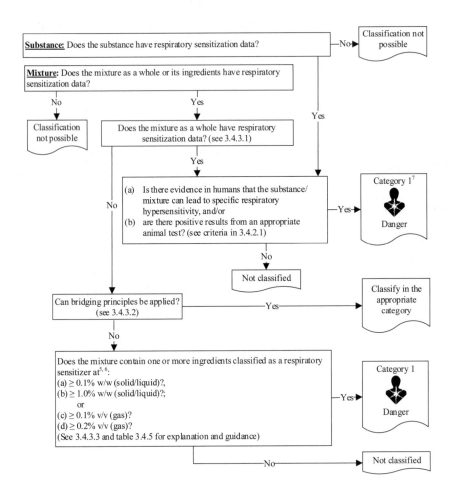

---

[5] *For specific concentration limits, see "The use of cut-off values/concentration limits" in chapter 1.3, paragraph 1.3.3.2.*
[6] *See 3.4.4.2.*
[7] *See 3.4.2.1.1 for details on use of Category 1 sub-categories.*

3.4.5.1 判定論理 3.4.1 呼吸器感作性

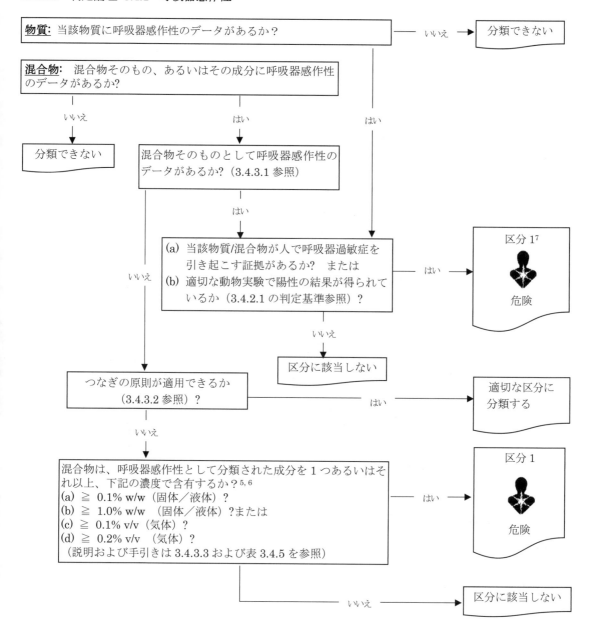

5　個々の濃度の限度については、第 1.3 章の 1.3.3.2 「カットオフ値/濃度限度の使用」を参照のこと。
6　3.4.4.2 を参照。
7　区分 1 の細区分の詳細については 3.4.2.1.1 を参照。

3.4.5.2 *Decision logic 3.4.2 for skin sensitization*

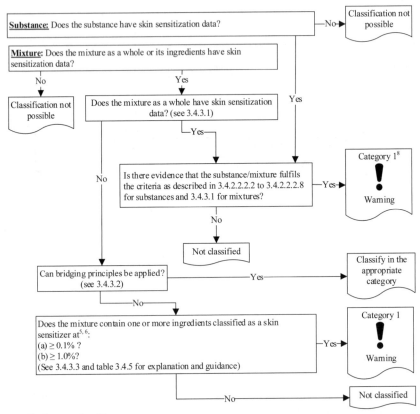

3.4.5.3 *Background guidance*

3.4.5.3.1 *Relevant guidance documents*

Mechanistic information on the process of skin sensitization is available in the OECD document on the Adverse Outcome Pathway for skin sensitization (see OECD (2014)). This information can be helpful in understanding the value of the individual *in chemico* and in vitro methods compared to the in vivo methods.

3.4.5.3.2 *Guidance on the use of human data*

3.4.5.3.2.1 The classification of a substance can be based on human evidence generated from a variety of sources. These sources include human predictive patch testing, epidemiological studies, case studies, case reports or histories, diagnostic patch testing and medical surveillance reports, and poison control centre information. This data may have been generated for consumers, workers, or the general population. When considering human evidence, consideration should be given to the size, exposure level, and exposure frequency of the exposed population. Guidance for evaluating human evidence and the criteria in 3.4.2.2.2 is provided by some competent authorities (e.g. ECHA Guidance on the Application of the CLP Criteria, 2017).

---

[5] *For specific concentration limits, see "The use of cut-off values/concentration limits" in chapter 1.3, paragraph 1.3.3.2.*

[6] *See 3.4.4.2.*

[8] *See 3.4.2.2.1 for details on use of Category 1 sub-categories.*

3.4.5.2 判定論理 3.4.2 皮膚感作性

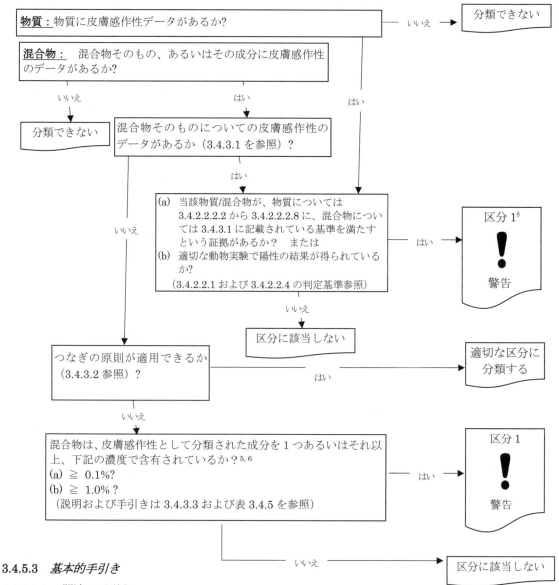

3.4.5.3 基本的手引き

3.4.5.3.1 関連の手引き

皮膚感作性のプロセスに関するメカニズム情報は、皮膚感作性の有害性発現経路(AOP)に関する OECD 文書(OECD (2014) 参照)で利用可能である。この情報は、in vivo 法と比較した個々の *in chemico* 法および in vitro 法の価値を理解する上で有用である。

3.4.5.3.2 ヒトのデータの使用に関する手引き

3.4.5.3.2.1 物質の分類は、様々な情報源から作成されたヒトの証拠に基づくことができる。 これらの情報源には、ヒトの予測パッチテスト、疫学的研究、症例研究、症例報告または履歴、診断用パッチテストおよび医学的監視報告、毒物管理センターの情報などがある。 このデータは、消費者、労働者、または一般集団を対象として作成されたものである可能性がある。 ヒトでの証拠を考慮する場合、ばく露された集団の規模、ばく露レベル、ばく露頻度を考慮する必要がある。ヒトでの証拠と 3.4.2.2.2 の基準を評価するための手引きは、いくつかの所管官庁によって提供されている(例:ECHA Guidance on the Application of the CLP Criteria, 2017)。

5 　個々の濃度の限度については、第 1.3 章の 1.3.3.2 「カットオフ値/濃度限度の使用」を参照のこと。
6 　3.4.4.2 を参照。
8 　区分 1 の細区分の使用に関する詳細は 3.4.2.2.1 を参照。

3.4.5.3.2.2   Positive data from predictive patch testing (HRIPT or HMT) conducted through human experimental and clinical studies, showing allergic contact dermatitis caused by the test substance can be used to classify substances for skin sensitization These studies are generally conducted in controlled clinical settings and in general the larger the population size, the more reliable the study outcome is. Criteria for evaluating this data are provided in paragraphs 3.4.2.2.2.1 and 3.4.2.2.2.

3.4.5.3.2.3   Positive data from well-run epidemiological studies (in accordance with WHO CIOMS guidelines, 2009) can be used for classifying substances for skin sensitization. Some examples of epidemiological studies may include case control studies, cohort studies, cross-sectional studies, or longitudinal studies. These studies should have large sample sizes with well-documented exposures to a substance.

3.4.5.3.2.4   A specific type of epidemiological study (such as randomized control studies or trials) may include information from diagnostic patch testing. Diagnostic patch testing is considered by some competent authorities to be the gold standard in diagnosing contact allergy in dermatitis patients (Johansen et al, 2015). Importantly, due consideration needs to be given to the appropriate selection of vehicle, substance and patch test concentrations for the purpose of not causing false negatives, false positives, irritant reactions or inducing contact allergy (skin sensitization). Positive data from experimental/clinical/diagnostic studies in humans and/or well-documented episodes of allergic contact dermatitis may be used to classify substances for skin sensitization, when it can be assumed with sufficient likelihood that the tested substance was indeed the most likely cause for induction of sensitization. Therefore, it should be established that there is at least a general likelihood that the respective patient(s) had been previously exposed to the substance. On the other hand, negative results from such tests are not sufficient to prove that the test substance should not be classified as a skin sensitizer.

3.4.5.3.2.5   Human data not generated in controlled experiments with volunteers for the purpose of hazard classification (e.g. case studies, case reports and case histories, and poison control centre information) can be used with caution. Consideration should be given to the frequency of cases, the inherent properties of the substances, as well as factors such as the exposure situation, bioavailability, individual predisposition, cross-reactivity and preventive measures taken.

3.4.5.3.2.6   Special consideration should be given to negative human data as full dose-response information is generally not available. For example, a negative result in an HRIPT or HMT at a low concentration may not allow for the conclusion that the substance does not have skin sensitizing properties as such effect at a higher concentration may not be excluded. In addition, negative human data should not necessarily be used to negate positive results from animal studies and/or defined approaches, but can be used as part of a weight of evidence assessment. For both animal and human data, consideration should be given to the impact of the vehicle (e.g. Wright et al, 2001 and Kligman, 1966).

3.4.5.3.2.7   For example, negative results from substances tested in a predictive patch test at a DSA (dose per skin area) < 500 µg/cm^2 imply that a classification for skin sensitization might not be needed at all, however, classification as sub-category 1A or 1B cannot be ruled out, because the concentration tested was not high enough to exclude these possibilities. The same holds for test results for which it is unknown whether the test concentration corresponded to a DSA < 500 µg/cm^2. Negative results from substances tested at a DSA ≥ 500 µg/cm^2 suggest that classification might not be needed. However, while classification as sub-category 1A can be ruled out, classification as sub-category 1B cannot, because a higher test concentration might have resulted in a positive test result. However, a negative test result at a concentration of 100% can justify no classification (based on this test). Nevertheless, negative results at low concentrations may be informative for mixtures containing the substance at similar or lower concentrations.

3.4.5.3.3   *Guidance on the use of standard animal data*

3.4.5.3.3.1   The most common assays used for dermal sensitization testing in animals are the Local Lymph Node Assay (LLNA, OECD test guidelines 429 and 442A and 442B), the Guinea Pig Maximization Test (GPMT, OECD Test Guideline 406) and the Buehler test (OECD Test Guideline 406). When evaluating the quality of the study, consideration should be given, as relevant, to the strain of the mouse and guinea pig used, the number, age, and sex of the animals, and the test conditions used (e.g. preparation of patch test site, dose level selection, chemical preparation, positive and negative test controls).

3.4.5.3.3.2   OECD test guidelines for the LLNA include the radioactive assay (OECD Test Guideline 429) and non-radioactive assays (OECD test guidelines 442A and 442B; LLNA:DA, LLNA:BrdU-ELISA, and LLNA:BrdU-FCM). In these tests, sensitizers are characterized by increasing the group mean stimulation index ("SI", a measure of lymph node proliferation) in treated groups versus concurrent vehicle controls by more than a predefined critical value which is different for each form of the LLNA (e.g. SI ≥ 3 for the radioactive LLNA, SI ≥ 1.6 for the LLNA:BrdU-ELISA). For sensitizers, subcategorization is performed based on the effective concentration (EC) causing an increase in SI of exactly

3.4.5.3.2.2　ヒトの実験的および臨床研究を通じて実施された予測的なパッチテスト（HRIPT または HMT）により、試験物質によるアレルギー性接触皮膚炎を示す陽性データは、皮膚感作性物質の分類に用いることができる。これらの試験は、一般に管理された臨床環境で実施され、一般に母集団の規模が大きいほど、研究結果の信頼性は高くなる。このデータの評価基準は、3.4.2.2.2.1 および 3.4.2.2.2 に記載している。

3.4.5.3.2.3　適切に実施された疫学研究の陽性データ（WHO CIOMS ガイドライン, 2009 に準拠）は、皮膚感作性物質の分類に使用することができる。疫学研究の例としては、症例対照研究、コホート研究、横断研究、縦断研究などがある。これらの研究は、サンプルサイズが大きく、物質へのばく露が十分に記録されている必要がある。

3.4.5.3.2.4　特定のタイプの疫学研究（ランダム化対照研究または試験など）には、診断用パッチテストからの情報を含めることができる。診断用パッチテストは、一部の所管官庁では、皮膚炎患者の接触アレルギーを診断する際のゴールドスタンダードと考えられている（Johansen et al, 2015）。重要なのは、偽陰性、偽陽性、刺激性反応、接触アレルギーの誘発（皮膚感作性）を引き起こさない目的で、媒体、物質、パッチテストの濃度を適切に選択することに十分配慮することである。ヒトにおける実験／臨床／診断研究からの陽性データおよび／または十分に記録されたアレルギー性接触皮膚炎の事例は、試験物質が感作性を誘発する最も可能性の高い原因であることが十分に仮定できる場合、皮膚感作性物質の分類に使用できる。したがって、それぞれの患者が過去にその物質にばく露された可能性が少なくとも一般的にあることが立証されなければならない。一方、このような試験で陰性の結果が得られても、試験物質が皮膚感作性物質として分類されるべきではないことを証明するのに十分ではない。

3.4.5.3.2.5　有害性の分類を目的としたボランティアによる管理された実験で作成されたものでないヒトデータ（例：ケーススタディ、症例報告および症例履歴、毒物管理センター情報）は、慎重に使用することができる。症例の頻度、物質の固有の特性、ならびにばく露状況、バイオアベイラビリティ、個人の素因、交差反応性および講じられた予防措置などの要因を考慮する必要がある。

3.4.5.3.2.6　一般に完全な用量反応情報が得られないため、ヒトの陰性データには特別な配慮が必要である。例えば、低濃度での HRIPT や HMT が陰性であっても、高濃度での影響が否定できないため、皮膚感作性を有しないという結論には至らないかもしれない。また、ヒトでの陰性データは、必ずしも動物実験および/またはディファインドアプローチによる陽性結果を否定するために使用されるべきではなく、証拠の重み付け評価の一部として使用することができる。動物およびヒトのデータの両方について、媒体の影響を考慮する必要がある（例：Wright et al, 2001 および Kligman, 1966）。

3.4.5.3.2.7　例えば、パッチテストで DSA（皮膚面積あたりの用量）＜500 μg/cm^2 で試験した物質が陰性だった場合、皮膚感作性の分類は全く必要ないかもしれないが、試験濃度が細区分 1A または 1B への分類の可能性を排除できるほど高くなかったため、細区分 1A または 1B への分類は除外できない。試験濃度が DSA ＜ 500 μg/cm^2 に相当するかどうか不明な試験結果についても同様である。DSA≧500 μg/cm^2 で試験した物質が陰性であった場合は、分類の必要がない可能性が示唆される。しかし、細区分 1A への分類は除外できるが、細区分 1B への分類は除外できない。なぜなら、より高い試験濃度で陽性となったかもしれないからである。しかし、100％の濃度で陰性であれば、（この試験に基づいた分類として）区分に該当しないことを正当化することができる。それでも、低濃度での陰性結果は、同程度またはそれ以下の濃度でその物質を含む混合物にとって有益である場合がある。

3.4.5.3.3　*標準的動物データの使用に関する手引き*

3.4.5.3.3.1　動物における経皮感作性試験に用いられる最も一般的な試験法は、局所リンパ節試験（LLNA、OECD テストガイドライン 429 および 442A、442B）、モルモット Maximization 試験（GPMT、OECD テストガイドライン 406）および Buehler 試験（OECD テストガイドライン 406）である。試験の質を評価する際には、使用したマウスやモルモットの系統、動物の数、年齢、性別、使用した試験条件（例えば、パッチテスト部位の準備、用量レベルの選択、化学物質の調製、陽性対照群と陰性対照群）を関連するものとして考慮すべきである。

3.4.5.3.3.2　LLNA に関する OECD テストガイドラインには、放射分析（OECD テストガイドライン 429）および非放射分析（OECD テストガイドライン 442A および 442B；LLNA:DA、LLNA:BrdU-ELISA、LLNA:BrdU-FCM）が含まれる。これらの試験において、感作性物質は、処理群対同時の媒体対照群における群平均刺激指数（「SI」、リンパ節増殖の指標）を、LLNA の形態ごとに異なる事前に定められた臨界値（例えば、LLNA-RI では SI≧3、LLNA:BrdU-ELISA では SI≧1.6）より増加させることにより特徴づけられる。感作性物質については、ちょうど臨界的な大きさの SI の増加を引き起こす有効濃度（EC）に

the critical magnitude (e.g. the EC3 under OECD Test Guideline 429 is the concentration leading to an exactly threefold increase in group mean SI versus control).

3.4.5.3.3.3 The respective OECD test guidelines for the different LLNA variants specify that a pre-screen test should be undertaken to determine the highest concentration to be tested. If such a test has not been performed and the LLNA was carried out with a test concentration < 100%, a rationale (e.g. based on solubility, local or systemic toxicity, see OECD test guidelines 429, and 442A and 442B) needs to be provided that the highest test concentration represents the maximum testable concentration. Otherwise, the reliability of a negative test result has to be considered compromised.

3.4.5.3.3.4 EC values are normally obtained by interpolation between adjacent test concentrations, i.e. between the highest test concentration causing an SI below, and the lowest test concentration causing an SI above the critical value. However, care must be taken when the EC value falls below the lowest concentration tested and can therefore only be estimated by extrapolation, which is associated with additional uncertainty. In some cases, the SI at the highest concentration tested falls only slightly below the critical SI value, which raises the question of upward extrapolation (unless the maximum testable concentration has been applied). These and other issues regarding the reliability of LLNA results are further discussed in Ryan et al. (2007) and Annex 3 of OECD Series on Testing and Assessment No. 336 (Supporting Document to OECD Guideline Document 497), which also provides a highly curated database of test guidelines 429 LLNA EC3 values.

3.4.5.3.3.5 Further limitations have been identified for the radioactive and non-radioactive LLNAs. For example, substances containing certain functional groups may interfere with the accuracy of the assay. These limitations as well as the possibility of borderline positive results are described in OECD test guidelines 429, and 442A and 442B. Variability in EC values for the same substance may also be the result of the vehicle used. For example, analysis has shown an underestimation of potency (i.e. higher EC3 values) with predominantly aqueous vehicles or propylene glycol (see Jowsey, 2008).

3.4.5.3.3.6 For OECD Test Guideline 406, the concentration of test chemical used for each induction exposure should be systemically well-tolerated using the highest dose to cause mild-to-moderate skin irritation. The concentration used for the challenge exposure should be the highest non-irritant dose. A positive result in a guinea pig test is defined as a grade above zero according to the applicable grading scale such as the Magnusson and Kligman grading scale for OECD Test Guideline 406 at one or more of the two observation time-points. A grade of 0.5, which is sometimes reported, is therefore also considered a positive result.

3.4.5.3.4 *Guidance on the use of defined approaches*

Defined approaches validated according to international procedures and described in OECD Guideline 497 have been characterized for the level of confidence that can be assigned to the predictions based on the applicability domain of the individual information sources used and the data interpretation procedure applied (see table 3.4.7). Other defined approaches under consideration but not yet validated according to international procedures and described in OECD Guidance Document 256 according to internationally agreed criteria for their reporting (OECD Guidance Document 255) may be accepted by some competent authorities.

3.4.5.3.5 *Guidance on the use of non-stand-alone in chemico/in vitro methods*

Individual in chemico/in vitro methods such as those reported in OECD test guidelines 442C, 442D and 442E, due to their limited mechanistic coverage, cannot be used on their own to conclude on Category 1 or no classification according to the criteria defined in table 3.4.8 and further data are necessary for classification in tier 2. In addition, although some of these methods provide quantitative information, these cannot be used for the purposes of subcategorization into sub-categories 1A and 1B since the criteria have not been validated according to international procedures. Nevertheless, such quantitative information may be accepted by a competent authority when used in a weight of evidence assessment under tier 2 for the purpose of subcategorization. This is also in line with the statement in these test guidelines that "*Depending on the regulatory framework, positive results generated with these methods may be used on their own to classify a chemical into UN GHS Category 1.*" Therefore, the GHS also allows a competent authority to decide that a positive result with one of these non stand-alone *in chemico*/in vitro methods, may be used on its own to classify in Category 1 and whether test guideline 442C (Appendix III) kinetic Direct Peptide Reactivity Assay (kDPRA) can be used to differentiate between sub-category 1A and no sub-category 1A.

3.4.5.3.6 *Guidance on the use of non-standard data*

3.4.5.3.6.1 Validated but not yet adopted *in chemico*/in vitro methods such as those reported under 3.4.5.3.6.2 as well as in vivo test methods which do not comply with internationally agreed guidelines for the identification of skin

基づいて細区分化が行われる（例えば、OECD テストガイドライン 429 の EC3 は、群平均 SI が対照に対してちょうど 3 倍増加する濃度である）。

3.4.5.3.3.3　LLNA の異なる改変法に関する OECD の各テストガイドラインでは、試験する最高濃度を決定するために予備スクリーニング試験を実施することが規定されている。このような試験が実施されておらず、LLNA が試験濃度＜100％で実施された場合、最高試験用量が試験可能な最大濃度を示す根拠（例えば、溶解性、局所毒性または全身毒性に基づく、OECD テストガイドライン 429、442A および 442B を参照）を提示する必要がある。そうでなければ、陰性試験結果の信頼性が損なわれると考えなければならない。

3.4.5.3.3.4　EC 値は通常、隣接する試験濃度間、すなわち臨界値を下回る SI を引き起こす最高試験濃度と臨界値を上回る SI を引き起こす最低試験濃度間の内挿法によって得られる。しかし、EC 値が試験した最低濃度を下回り、そのため外挿によってのみ推定できる場合には注意が必要であり、これにはさらなる不確実性が伴う。場合によっては、試験した最高濃度の SI 値が臨界 SI 値をわずかに下回るだけで、上方への外挿が問題となる（試験可能な最高濃度が適用されていない限り）。LLNA の結果の信頼性に関するこれらおよびその他の問題については、Ryan et al. (2007)、および OECD 試験および評価に関するシリーズ No.336（OECD ガイドライン文書 497 のサポート文書）の附属書 3 でさらに議論されており、同書ではテストガイドライン 429 で LLNA の EC3 値の高度に精選されたデータベースも提供している。

3.4.5.3.3.5　放射性 LLNA と非放射性 LLNA については、さらなる制限が確認されている。例えば、特定の官能基を含む物質は試験の精度を妨げる可能性がある。これらの限界や境界域陽性結果の可能性については、OECD テストガイドライン 429、442A および 442B に記載されている。同一物質の EC 値のばらつきは、使用した媒体の結果である可能性もある。例えば、主に水性媒体またはプロピレングリコールを用いた分析では、感作能の過小評価（すなわち、EC3 値が高い）が示されている（Jowsey, 2008 参照）。

3.4.5.3.3.6　OECD テストガイドライン 406 では、各誘導ばく露に用いる試験物質の濃度は、軽度から中等度の皮膚刺激を引き起こす最高用量を用いて、全身的に十分に許容されるべきである。惹起のばく露に使用する濃度は、非刺激性の最高用量とする。モルモット試験における陽性結果は、OECD テストガイドライン 406 の Magnusson and Kligman の等級付けなどの適用可能な等級付けに従い、2 つの観察時点のうち 1 つ以上において等級がゼロを超えるものと定義される。したがって、ときに報告される 0.5 の等級も陽性結果とみなされる。

3.4.5.3.4　*ディファインドアプローチの使用に関する手引き*

　国際的な手順にしたがって検証され、OECD ガイドライン 497 に記載されているディファインドアプローチは、使用される個々の情報源の適用可能領域と適用されるデータ解釈手順に基づいて、予測に割り当てられる信頼性のレベルについて特徴付けられている（表 3.4.7 参照）。国際的な手順にしたがって検証されておらず、かつ、それらの報告（OECD ガイダンス文書 255）に対して国際的に合意された基準にしたがっている OECD ガイダンス文書 256 にも記載されていない検討中のその他のディファインドアプローチは、一部の所管官庁に受け入れられる可能性がある。

3.4.5.3.5　*単独でない in chemico/in vitro 試験法の使用に関する手引き*

　OECD テストガイドライン 442C、442D、442E で報告されているような個別の in chemico/in vitro 試験法は、その機械論的な適用範囲が限定されているため、表 3.4.8 に定義された基準にしたがって区分 1 または区分に該当しないと結論づけるために単独で用いることはできず、段階 2 で分類するためにはさらなるデータが必要である。さらに、これらの方法の中には定量的な情報を提供するものもあるが、国際的な手順にしたがって基準が検証されていないため、細区分 1A および 1B への細区分分類の目的には使用できない。とはいえ、このような定量的情報は、細区分化を目的とした段階 2 の証拠の重み付け評価で使用される場合、所管官庁によって受け入れられる可能性がある。これは、当該テストガイドラインの「*規制の枠組みによっては、これらの方法で得られた陽性結果は、それ自体で化学物質を国連 GHS 区分 1 に分類するために使用することができる*」という記述とも一致する。したがって、GHS では、所管官庁が、これらの単独でない in chemico/in vitro 法の一つで陽性結果が出た場合、それだけで区分 1 に分類することができ、テストガイドライン 442C（付属書Ⅲ）の kinetic Direct Peptide Reactivity Assay（kDPRA）を細区分 1A と細区分 1A に該当しないを区別するために使用できるかどうかを決定することも可能である。

3.4.5.3.6　*非標準的なデータの使用に関する手引き*

3.4.5.3.6.1　3.4.5.3.6.2 で報告されているような、検証されているがまだ採用されていない *in chemico/in*

sensitizers or the assessment of skin sensitizing potency may provide supportive evidence when used in an overall weight of evidence assessment (i.e. tier 3).

3.4.5.3.6.2 A non-exhaustive list of other validated *in chemico*/in vitro test methods accepted by some competent authorities but not adopted as OECD test guidelines is provided below. A competent authority may decide which classification criteria, if any, should be applied for these test methods:

> (a)   The Genomic Allergen Rapid Detection (GARD) potency is a transcriptomics-based in vitro assay addressing the third key event of the skin sensitization Adverse Outcome Pathway (activation of dendritic cells) similar to the GARDskin but uses a different gene signature that provides subcategorization of skin sensitizers (Gradin et al., 2020; Zeller et al., 2017; Corsini et al. 2021);
>
> (b)   The SENS-IS assay is a genomic approach applied to a Reconstructed Human Epidermis (RhE) (Cottrez et al., 2015; Cottrez et al., 2016);
>
> (c)   The Epidermal Sensitization Assay (EpisensA) is based on the measurement of the upregulation of four genes in a reconstructed human epidermis (RhE) to discriminate between sensitizers and non-sensitizers (Saito et al., 2017).

3.4.5.3.7   *Guidance on the weight of evidence assessment for classifying substances and mixtures for skin sensitization*

3.4.5.3.7.1 There may be situations where results from tests and/or non-test methods are available but disagree with each other with respect to the classification. In these situations, the tiered approach to classification for skin sensitization requires a weight of evidence assessment consistent with the principles elaborated in sections 1.3.2.4.2 and 1.3.2.4.9 on test data quality and weight of evidence, respectively. In addition, some guidance on the weight of evidence assessment specific for skin sensitization is provided below which can be applied when the general principles do not result in a conclusion on the classification. It should be noted that human and animal results for a substance obtained at low concentrations may still be informative for classifying a mixture containing the substance at similar or lower concentrations.

3.4.5.3.7.2 Mutual compatibility of study results

3.4.5.3.7.2.1 In cases where results are in disagreement with each other (e.g. not classified versus Category 1, sub-category 1A or 1B; sub-category 1A versus 1B), a weight of evidence assessment becomes necessary. However, less obvious situations may also occur such as where certain studies may point to not classified or sub-category 1B, while it cannot be excluded that a stricter classification might have resulted under a different dosing regime. For example, a negative HMT result at a dose per skin area of 100 µg/cm^2 cannot exclude that a positive result might have been obtained at e.g. 300 µg/cm^2 (sub-category 1A) or 700 µg/cm^2 (sub-category 1B). The same holds for LLNA test results obtained from tests which have not been carried out using the highest possible test concentration (see OECD Test Guideline 429 for details).

3.4.5.3.7.2.2 In the following ambiguous cases, study results for substances and mixtures would not be in disagreement with another study result pointing at that stricter classification:

> (a)   A not classified result obtained at a lower test concentration does not exclude the possibility of a sub-category 1B outcome at a higher test concentration. Therefore, a not classified result obtained at a low concentration is compatible with other not classified outcomes, or with category 1 and sub-category 1B outcomes obtained at higher test concentrations.
>
> (b)   A not classified result at a very low-test concentration does not even exclude a possible outcome of sub-category 1A at a higher test concentration. Therefore, a not classified outcome obtained at a very low-test concentration is compatible with all possible classification outcomes (i.e. not classified, category 1, sub-category 1A or 1B) obtained at higher test concentrations.
>
> (c)   A sub-category 1B result at a higher test concentration does not exclude a sub-category 1A outcome at a lower test concentration. Therefore, a sub-category 1B classification tested at a high-test concentration is compatible with other outcomes of sub-category 1B, or even sub-category 1A, obtained at lower test concentrations.

vitro 試験法、および皮膚感作性物質の同定または皮膚感作性の評価のための国際的に合意されたガイドラインに準拠していない in vivo 試験法は、包括的な証拠の重み付け評価（すなわち、段階 3）において使用される場合、裏付けとなる証拠を提供する可能性がある。

3.4.5.3.6.2 　OECD テストガイドラインとしては採用されていないが、一部の所管官庁に認められている検証された他の in chemico/in vitro 試験法の非網羅的リストを以下に示す。所轄官庁は、これらの試験法について、分類基準がある場合、どの分類基準を適用すべきかを決定することができる：

 (a) Genomic Allergen Rapid Detection (GARD) potency はトランスクリプトミクスに基づく in vitro 試験法であり、GARDskin と同様に皮膚感作性 AOP（有害事象経路）の第 3 キーイベント（樹状細胞の活性化）を扱うが、皮膚感作性物質の細区分を提供する異なる遺伝子特徴（gene signature）を用いる（Gradin et al., 2020；Zeller et al., 2017；Corsini et al., 2021）。

 (b) SENS-IS アッセイは、再構築ヒト表皮（RhE）に適用されるゲノムアプローチである（Cottrez et al., 2015；Cottrez et al., 2016）。

 (c) Epidermal Sensitization Assay (EpisensA)は、感作性物質と非感作性物質を識別するために、再構築ヒト表皮(RhE)における 4 つの遺伝子のアップレギュレーションの測定に基づいている(Saito et al., 2017)。

3.4.5.3.7 　*物質および混合物の皮膚感作性分類のための証拠の重み付け評価に関する手引き*

3.4.5.3.7.1 　試験法および／または試験によらない方法の結果が利用可能であっても、分類に関して相互に一致しない状況が存在する可能性がある。このような状況では、皮膚感作性分類の段階的アプローチは、試験データの質および証拠の重みについて、それぞれ 1.3.2.4.2 項および 1.3.2.4.9 項で詳述した原則と一致する証拠の重み付け評価を必要とする。さらに、皮膚感作性に特有の証拠の重み付け評価に関する手引きを以下に示す。この手引きは、一般原則では分類に関する結論が得られない場合に適用できる。低濃度で得られた物質のヒトおよび動物の結果は、類似の濃度またはより低濃度でその物質を含む混合物を分類する上で、なお参考になる可能性があることに留意すべきである。

3.4.5.3.7.2 　試験結果の相互適合性

3.4.5.3.7.2.1 　結果が相互に不一致の場合（例えば、区分に該当しないと区分 1、細区分 1A または 1B、細区分 1A 対 1B）、証拠の重み付け評価が必要となる。しかし、例えば、ある研究が区分に該当しないまたは細区分 1B を指し示している一方で、異なる投与量設定の場合ではより厳しい分類がなされた可能性を排除できないなど、あまり明らかでない状況も起こり得る。例えば、皮膚面積当たりの投与量が 100 μg/cm² での HMT 結果が陰性であっても、例えば 300 μg/cm²（細区分 1A）または 700 μg/cm²（細区分 1B）で陽性結果が得られた可能性を排除することはできない。可能な限りの最高試験濃度で実施されていない試験で得られた LLNA 試験結果についても同じことが当てはまる（詳細は OECD テストガイドライン 429 を参照）。

3.4.5.3.7.2.2 　以下の曖昧なケースでは、物質および混合物の試験結果は、より厳しい分類を示す別の試験結果と矛盾しない：

 (a) 低い試験濃度で得られた区分に該当しないという分類結果は、より高い試験濃度で細区分 1B の結果が得られる可能性を排除するものではない。したがって、低い濃度で得られた区分に該当しないという結果は、他の区分に該当しないという結果、またはより高い濃度で得られた区分 1 および細区分 1B の結果と適合性がある。

 (b) 非常に低い試験濃度での区分に該当しないという結果は、より高い試験濃度での細区分 1A の結果の可能性を排除するものではない。したがって、非常に低い試験濃度で得られた区分に該当しないという結果は、より高い試験濃度で得られたすべての可能な分類結果（すなわち、区分に該当しない、区分 1、細区分 1A または 1B）と適合性がある。

 (c) より高い試験濃度における細区分 1B の結果は、より低い試験濃度における細区分 1A の結果を除外するものではない。したがって、高い試験濃度で試験された細区分 1B の分類は、より低い試験濃度で得られた細区分 1B、あるいは細区分 1A の他の結果と適合性がある。

3.4.5.3.7.2.3　　　　If at least one unambiguous study result allows for subcategorization of a substance or mixture and all other study results are not in disagreement (see above), then it can be classified into a sub-category. For example, if all study results are in the same sub-category (i.e. sub-category 1A or 1B), or with at least one study permitting subcategorization (i.e. either sub-category 1A or 1B) and all other studies classified into Category 1 without subcategorization, then the substance or mixture can be subcategorized.

3.4.5.3.7.3　　　　Weight of evidence considerations for giving one study result more weight than another

3.4.5.3.7.3.1　　　　Some classifiers or competent authorities may take various approaches to evaluate study results given the required level of expert judgement (see 1.3.2.4.8) required to perform a weight of evidence assessment. Competent authorities may specify their preferred approach in their own guidance. For example, through:

(a) Applying a precautionary approach, giving more weight to studies resulting in the stricter classification outcome;

(b) Giving human data higher weight than animal or non-test data;

(c) Giving certain animal data (e.g. LLNA data) more weight than other animal data (e.g. Buehler test data).

3.4.5.3.7.3.2　　　　Often, several results (of the same or different type) may have to be considered in the weight of evidence assessment. There are no generally recognized rules for this situation, however, possible solutions to integrating several results of the same type may include, for example:

(a) A precautionary approach where the strictest classification outcome from all studies of sufficient quality is assigned as the overall classification outcome.

(b) Averaging the obtained dose descriptors (e.g. LLNA EC3 values) or classification outcomes (no classification, Category, 1, 1A, 1B). A detailed discussion of such approaches can be found in Annex 3 (on LLNA data) and Annex 4 (on HMT/HRIPT data) of OECD Series on Testing and Assessment No. 336 (Supporting document to OECD Guideline Document 497).

3.4.5.3.7.2.3　物質または混合物の細区分を許容する明確な試験結果が少なくとも 1 つあり、他のすべての試験結果が矛盾していない場合（上記参照）、その物質は細区分に分類することができる。例えば、すべての試験結果が同じ細区分（すなわち、細区分 1A または 1B）である場合、または少なくとも 1 つの試験結果が細区分化を許容し（すなわち、細区分 1A または 1B のいずれか）、他のすべての試験結果が細区分化せずに区分 1 に分類される場合、物質または混合物を細区分化することができる。

3.4.5.3.7.3　ある試験結果を他の試験結果よりも重視するための証拠の重み付けの考慮事項

3.4.5.3.7.3.1　分類者または所管官庁によっては、証拠の重み付け評価を行うために必要な専門家の判断レベル（1.3.2.4.8 参照）を考慮した上で、試験結果を評価するために様々なアプローチをとることができる。所管官庁は、自らの手引きにおいて、望ましいアプローチを指定することができる。例えば；

(a) 予防的アプローチを適用し、より厳しい分類をもたらす試験結果をより重視する；

(b) ヒトのデータを動物試験データや試験によらないデータよりも重視する；

(c) 特定の動物データ（LLNA データなど）を他の動物データ（Buehler 試験データなど）より重視する。

3.4.5.3.7.3.2　多くの場合、証拠の重み付け評価において、複数の結果（同じ種類または異なる種類）を考慮しなければならない場合がある。このような状況に対して一般的に認識されたルールはないが、同じ種類の複数の結果を統合するために可能な解決策には、例えば、次のようなものがある：

(a) 十分な質を有する全ての研究から最も厳しい分類結果を総合的な分類結果とする予防的アプローチ。

(b) 得られた用量記述子（例えば、LLNA EC3 値）または分類結果（区分に該当しない、区分 1、1A、1B）の平均化。このようなアプローチの詳細な議論は、OECD Series on Testing and Assessment No.336（OECD ガイドライン文書 497 の付属書）の付属書 3（LLNA データに関する）および付属書 4（HMT/HRIPT データに関する）に記載されている。

Table 3.4.7: Criteria for defined approaches

Category	OECD Guideline 497 on Defined Approaches for Skin sensitization "2 out of 3" (2o3) defined approach	OECD Guideline 497 on Defined Approaches for Skin sensitization Integrated testing strategy (ITSv1) defined approach and Integrated testing strategy (ITSv2) defined approach)
	2o3 defined approach to skin sensitization hazard identification based on *in chemico* (key event 1 - Direct Peptide Reactivity Assay (KE1-DPRA)) and in vitro (key event 2-OECD 442D Appendix IA, key event 3 - human Cell Line Activation Test (KE3-h-CLAT)). Assays are run for two key events, and if these assays provide consistent results, then the chemical is predicted accordingly as sensitizer or non-sensitizer. If the first two assays provide discordant results, the assay for the remaining key event is run. The overall result is based on the two concordant findings taking into account the confidence on the obtained predictions as described in the guideline	**ITSv1** based on *in chemico* (KE1-DPRA) and in vitro (KE3-h-CLAT) data, and in silico (Derek Nexus) predictions. **ITSv2** based on *in chemico* (KE1 -DPRA) and in vitro (KE3 -h-CLAT) data, and in silico (OECD QSAR Toolbox) predictions. Quantitative results of h-CLAT and DPRA are converted into a score from 0 to 3. For the in silico prediction (Derek or OECD QSAR Toolbox), a positive outcome is assigned a score of 1; a negative outcome is assigned a score of 0. When these scores have been assessed, a total battery score ranging from 0 to 7, calculated by summing the individual scores, is used to predict the sensitizing potential (hazard identification; GHS Cat. 1 versus no classification) and potency (GHS sub-cat. 1A, sub-cat. 1B and no classification).
1	2 out of 3 or 3 out of 3 positive predictions	Total battery score ≥ 2
1A	Not applicable	Total battery score 6-7
1B	Not applicable	Total battery score 2-5
Not classified	2 out of 3 or 3 out of 3 negative predictions	Total battery score < 2

表 3.4.7: ディファインドアプローチの基準

カテゴリー	OECD ガイドライン 497 皮膚感作性のディファインドアプローチ "2 out of 3" (2o3) ディファインドアプローチ	OECD Guideline 497 皮膚感作性のディファインドアプローチ 統合的試験戦略 (ITSv1)ディファインドアプローチおよび 統合的試験戦略 (ITSv2 ディファインドアプローチ)
	in chemico (キーイベント 1 - Direct Peptide Reactivity Assay (KE1-DPRA)) および in vitro (キーイベント 2-OECD 442D 付録 IA, キーイベント 3 - human Cell Line Activation Test (KE3-h-CLAT)) に基づく皮膚感作性有害性特定の 2o3 ディファインドアプローチ アッセイは2つの主要な事象について実施され、これらのアッセイで一貫した結果が得られた場合、その化学物質は感作性または非感作性であると予測される。最初の2つのアッセイで不一致の結果が得られた場合、残りのキーイベントのアッセイが実行される。ガイドラインに記載されているように、得られた全体的な結果は、一致した2つの所見に基づく。予測の信頼性を考慮し、一致した2つの所見に基づく。	in chemico (KE1-DPRA) および in vitro (KE3-h-CLAT) データ、および in silico (Derek Nexus) 予測に基づく ITSv1 in chemico (KE1-DPRA) および in vitro (KE3-h-CLAT) データ、および in silico (OECD QSAR Toolbox) 予測に基づく ITSv2 h-CLAT および DPRA の定量的結果は 0～3 のスコアに変換される。in silico 予測 (Derek 又は OECD QSAR ToolBox) については、陽性結果には1のスコア、陰性結果には0のスコアが割り当てられる。これらのスコアが評価された場合、感作性 (危険有害性の同定；GHS 区分 1 対区分に該当しない) および感作性能 (GHS 細区分 1A, 1B および区分に該当しない) を予測するために、各スコアを合計して算出される 0～7のバッテリースコア合計値が用いられる。
1	3つのうち、2つもしくは3つの陽性予測	バッテリースコアの合計値 ≥ 2
1A	適用なし	バッテリースコアの合計値 6-7
1B	適用なし	バッテリースコアの合計値 2-5
区分に該当しない	3つのうち、2つもしくは3つの陰性予測	バッテリースコアの合計値 < 2

Table 3.4.8: Criteria for individual in chemico/in vitro methods

Category	OECD Test Guideline 442C Key event-based Test Guideline for in chemico skin sensitisation assays addressing the adverse outcome pathway (AOP) Key Event on covalent binding to proteins			OECD Test Guideline 442D Key event-based Test Guideline for in vitro skin sensitisation assays addressing the AOP Key Event on keratinocyte activation antioxidant response element-nuclear factor-erythroid 2-related factor 2 (ARE-Nrf2) luciferase methods		OECD Test Guideline 442E In vitro skin sensitisation assays addressing the AOP Key Event on activation of dendritic cells		
	Method described in Appendix I	Method described in Appendix II	Method described in Appendix III	Method described in Appendix 1A[a]	Method described in Appendix 1B	Method described in Annex I	Method described in Annex II	Method described in Annex III	Method described in Annex IV
	The Direct Peptide Reactivity Assay (DPRA)[a]	The Amino acid Derivative Reactivity Assay (ADRA)[a]	The kinetic Direct Peptide Reactivity Assay (kDPRA)[b]		Lusens[a]	human Cell Line Activation Assay (h-CLAT)[a]	U937 Cell Line Activation Test[a]	Interleukin-8 luciferase (IL-8 Luc) assay[a]	Genomic Allergen Rapid Detection for assessment of skin sensitizers[a]
	Methods: in chemico methods addressing the process of haptenation by quantifying the reactivity of test chemicals towards model synthetic peptides containing either lysine or cysteine (DPRA and kDPRA) or towards model synthetic amino acid derivatives containing either N-(2-(1-naphthyl) acetyl)-L-cysteine (NAC) or α-N-(2-(1-naphthyl) acetyl)-L-lysine (NAL) (ADRA). The criteria are based on the mean of cysteine and lysine peptides percent depletion (DPRA), kinetic rates of cysteine depletion (kDPRA) and mean NAC and NAL percent depletion value (ADRA). Predictions models based on the cysteine or NAC percent depletion value alone in case the unreacted lysine peptide or NAL cannot be reliably measured can be applied for the DPRA and ADRA.			Methods: cell-based methods addressing the process of keratinocytes activation, by assessing with the help of luciferase, the Nrf2-mediated activation of antioxidant response element (ARE)-dependent genes following exposure of the cells to the test chemical. Cell viability is quantitatively measured in parallel by enzymatic conversion of the dye 3-(4,5-Dimethylthiazol-2-yl)-2,5-diphenyltetrazolium bromide (MTT). The criteria are based on the induction of the luciferase gene above a given threshold, quantified at subtoxic concentrations. Criteria should be met in 2 of 2 or in 2 of 3 repetitions.		Methods: cell-based methods addressing the process of monocytes/dendritic cell activation by either quantifying the change in the expression of cell surface marker(s) (e.g. cluster of differentiation 54 (CD54), cluster of differentiation 86 (CD86)) or the change in IL-8 expression or the transcriptional patterns of an endpoint-specific genomic biomarker signature following exposure of the cells to the test chemical. Criteria should be met in 2 of 2 or in at least 2 of 3 repetitions for test methods described in annexes I, II and III or in three valid biological replicates for test method described in Annex IV.			

- 186 -

表 3.4.8: 個別 *in chemico*/*in vitro* 法の基準

カテゴリー	OECDテストガイドライン442C			OECDテストガイドライン442D			OECDテストガイドライン442E		
	タンパク質との共有結合に基づく有害性発現経路(AOP)キーイベントを扱う *in chemico* 皮膚感作性試験のテストガイドラインに基づくキーイベント			ケラチノサイト活性化抗酸化応答配列-赤血球系転写因子2-関連因子2(ARE-Nrf2)ルシフェラーゼ法のAOPキーイベントを扱う in vitro 皮膚感作性アッセイのテストガイドラインに基づくキーイベント			樹状細胞の活性化に関するAOPキーイベントを扱う In vitro 皮膚感作性試験		
	付録Ⅰで説明された方法	付録Ⅱで説明された方法	付録Ⅲで説明された方法	付録. 1A[a]で説明された方法	付録 1B で説明された方法		付属書Ⅰで説明された方法	付属書Ⅱで説明された方法	付属書Ⅲで説明された方法	付属書Ⅳで説明された方法
	ペプチド結合性試験(DPRA)[a]	アミノ酸誘導体結合性試験(ADRA)[a]	動的ペプチド結合試験(kDPRA)[b]	Lusens[a]			ヒト細胞株活性化試験 (h-CLAT)[a]	U937 細胞株活性化試験[a]	インターロイキン-8ルシフェラーゼ(IL-8 Luc)アッセイ[a]	皮膚感作性評価のためのゲノミックアレルゲン迅速検出[a]
	方法:リジンまたはシステインを含むモデル合成ペプチド(DPRAおよびkDPRA)、または N-(2-(1-ナフチル)アセチル)-L-システイン(NAC)または α-N-(2-(1-ナフチル)アセチル)-L-リジン(NAL)(ADRA)のいずれかを含むモデル合成アミノ酸誘導体に対する試験化学品の反応性を定量化することにより、ハプテン化のプロセスを扱う in chemico 法。基準は、システインとリジンへペプチド減少率の平均値(DPRA)、システイン減少率の速度論的速度(kDPRA)、NACとNALの減少率の平均値(ADRA)に基づいている。DPRAとADRAについては、システインペプチドまたはNACを確実に測定できない場合に、システインペプチドまたはNACの減少率のみに基づく予測モデルを適用することができる。			方法:細胞を試験薬液にばく露した後、抗酸化応答配列(ARE)依存性遺伝子のNrf2媒介活性化をルシフェラーゼの助けを借りて評価することにより、ケラチノサイトの活性化のプロセスを扱う細胞ベースの方法。細胞生存率は、色素 3-(4,5-Dimethylthiazol-2-yl)-2,5-diphenyltetrazolium bromide (MTT)の酵素変換によって、並行して定量的に測定される。基準は、毒性濃度で定量された所定の閾値以上のルシフェラーゼ遺伝子の誘導に基づいている。基準は、2回の繰り返し試行のうち 2回、または 3回の繰り返し試行のうち 2回で満たされる必要がある。			方法:試験物質を細胞にばく露した後のエンドポイント特異的ゲノムバイオマーカーシグネチャーの転写パターンまたは IL-8 の発現パターン、あるいは細胞表面マーカー(例:分化抗原群 54 (CD54)、分化抗原群 86 (CD86))の発現変化のいずれかを定量することによって単球/樹状細胞の活性化プロセスを扱う細胞ベースの方法。付属書Ⅰ、ⅡおよびⅢに記載された試験法については、2回の繰り返しの試行のうち 2 回または 3 回の繰り返し試行のうちすくなくとも 2 回で、付属書Ⅳに記載された試験法については、3 回の有効な生物学的複製で、基準を満たさなければならない。			

- 186 -

Table 3.4.8: Criteria for individual *in chemico/in vitro* methods (*cont'd*)

| Category | OECD Test Guideline 442C — Key event-based Test Guideline for *in chemico* skin sensitization assays addressing the adverse outcome pathway (AOP) Key Event on covalent binding to proteins ||| OECD Test Guideline 442D — Key event-based Test Guideline for in vitro skin sensitization assays addressing the AOP Key Event on keratinocyte activation antioxidant response element-nuclear factor-erythroid 2-related factor 2 (ARE-Nrf2) luciferase methods || OECD Test Guideline 442E — In vitro skin sensitization assays addressing the AOP Key Event on activation of dendritic cells |||
	Method described in Appendix I — **The Direct Peptide Reactivity Assay (DPRA)**[a]	Method described in Appendix II — **The Amino acid Derivative Reactivity Assay (ADRA)**[a]	Method described in Appendix III — **The kinetic Direct Peptide Reactivity Assay (kDPRA)**[b]	Method described in Appendix 1A[a]	Method described in Appendix 1B — **Lusens**[a]	Method described in Annex I — **human Cell Line Activation Assay (h-CLAT)**[a]	Method described in Annex II — **U937 Cell Line Activation Test**[a]	Method described in Annex III — **Interleukin-8 luciferase (IL-8 Luc) assay**[a]	Method described in Annex IV — **Genomic Allergen Rapid Detection for assessment of skin sensitizers**[a]
1	The mean cysteine/lysine % depletion > 6.38% Or the mean cysteine % depletion > 13.89 %	The mean NAC and NAL % depletion ≥ 4.9% Or NAC% depletion ≥5.6%	Not applicable	The following 4 conditions are all met in 2 of 2 or in the same 2 of 3 repetitions: 1. Imax equal or higher than (³) 1.5 fold and statistically significantly different to the solvent control 2. The cellular viability is higher than (>) 70% at the lowest concentration with induction of luciferase activity equal or above 1.5 fold 3. The EC₁.₅ value is less than (<) 1000 μM (or < 200 μg/mL for test chemicals with no defined molecular weight) 4. There is an apparent overall dose-dependent increase in luciferase induction	The following conditions are all met in 2 of 2 or in the same 2 of 3 repetitions: 1. A luciferase induction above or equal to (³) 1.5 fold as compared to the solvent control is observed in at least 2 consecutive non-cytotoxic tested concentrations (i.e. cellular viability is equal or higher than (³) 70%) 2. At least three tested concentrations should be non-cytotoxic (cellular viability equal or higher than (³) 70%).	At least one of the following conditions is met in 2 of 2 or in at least 2 of 3 independent runs: The Relative Fluorescence Intensity of CD86 is equal to or greater than 150% at any tested concentration (with cell viability ≥ 50%) or the Relative Fluorescence Intensity of CD54 is equal to or greater than 200% at any tested concentration (with cell viability ≥ 50%).	The following condition is met in 2 of 2 or in at least 2 of 3 independent runs: The stimulation index of CD86 is equal or higher than (³) than 150% and/or interference is observed	The induction of normalised interleukin-8 luciferase activity (Ind-IL8LA) is equal or higher than (³) 1.4 and the lower limit of the 95% confidence interval of Ind-IL8LA is equal or higher than (³) 1.0 in at least 2 out of a maximum of 4 independent runs	The mean Decision Value (DV) is ≥0
1A	Not applicable	Not applicable	log kmax ≥ -2.0	Not applicable	Not applicable	Not applicable	Not applicable	Not applicable	Not applicable
1B	Not applicable	Not applicable	Not applicable	Not applicable	Not applicable	Not applicable	Not applicable	Not applicable	Not applicable

表 3.4.8: 個別 in chemico/in vitro 法の基準（続き）

カテゴリー	OECDテストガイドライン442C タンパク質との共有結合に基づく有害性発現経路(AOP)キーイベントを扱う in chemico 皮膚感作性試験のテストガイドラインに基づくキーイベント		OECDテストガイドライン442D ケラチノサイト活性化酸化応答配列-抗血球系転写因子2(ARE-Nrf2)ルシフェラーゼ法のAOPキーイベントを扱う in vitro 皮膚感作性試験のテストガイドラインに基づくキーイベント		OECDテストガイドライン442E 樹状細胞の活性化に関するAOPキーイベントを扱う In vitro 皮膚感作性試験		
	付録Iで説明された方法	付録IIで説明された方法	付録1Aで説明された方法	付録1Bで説明された方法	付属書Iで説明された方法	付属書IIで説明された方法	付属書IIIで説明された方法	付属書IVで説明された方法
	ペプチド結合性試験(DPRA)[a]	アミノ酸誘導体結合性試験(ADRA)[a] 動的ペプチド結合性試験(kDPRA)[b]		Lusens[a]	ヒト細胞株活性化試験(h-CLAT)[a]	U937細胞株活性化試験[a]	インターロイキン-8ルシフェラーゼ(IL-8Luc)アッセイ[a]	皮膚感作性評価のためのダイナミックアレンジ試験[a]
1	システインおよびリジンの減少率の平均値 > 6.38% または システインの減少率 > 13.89%	NAC および NAL の減少率の平均値 ≥4.9% または NAC の減少率 ≥5.6% 適用なし	以下の4つの条件を2回中2回、または3回中同じ2回の繰り返しですべて満たす: 1. Imaxが1.5倍以上であり、溶媒対照と統計的有意差が認められる。 2. ルシフェラーゼ活性の誘導倍率が1.5倍以上となる最低濃度において細胞生存率が70%超(>)である。 3. EC1.5値が1000μM未満(<)である(分子量不明の試験化学品の場合は<200μg/mL)。 4. ルシフェラーゼ誘導が明らかに用量依存的に増加する。	適用なし	2回中2回、または3回中2回の独立した測定において以下の条件のうち少なくとも1つを満たす: CD86の相対蛍光強度がいずれかの試験濃度において150%以上である(細胞生存率は≧50%)、 または CD54の相対蛍光強度がいずれかの試験濃度において200%以上である(細胞生存率が≧50%)。	2回中2回、または3回中2回の独立した測定において以下の条件のうち少なくとも1つを満たす: CD86の刺激指数が150%以上、および/または干渉が観察される。	最大4回の独立した測定のうち少なくとも2回において、正規化インターロイキン-8ルシフェラーゼ活性(Ind-IL8LA)の誘導が1.4以上であり、かつInd-IL8LAの95%信頼区間の下限が1.0以上である。	平均決定値(DV)は≧0である。
1A	適用なし	適用なし log kmax ≥ -2.0	適用なし	適用なし	適用なし	適用なし	適用なし	適用なし
1B	適用なし	適用なし	適用なし	適用なし	適用なし	適用なし	適用なし	適用なし

Table 3.4.8: Criteria for individual *in chemico*/*in vitro* methods *(cont'd)*

Category	OECD Test Guideline 442C Key event-based Test Guideline for *in chemico* skin sensitization assays addressing the adverse outcome pathway (AOP) Key Event on covalent binding to proteins			OECD Test Guideline 442D Key event-based Test Guideline for *in vitro* skin sensitization assays addressing the AOP Key Event on keratinocyte activation antioxidant response element-nuclear factor-erythroid 2-related factor 2 (ARE-Nrf2) luciferase methods		OECD Test Guideline 442E In vitro skin sensitization assays addressing the AOP Key Event on activation of dendritic cells		
	Method described in Appendix I	Method described in Appendix II	Method described in Appendix III	Method described in Appendix 1A[a]	Method described in Appendix 1B	Method described in Annex I	Method described in Annex II	Method described in Annex III	Method described in Annex IV
	The Direct Peptide Reactivity Assay (DPRA)[a]	The Amino acid Derivative Reactivity Assay (ADRA)[a]	The kinetic Direct Peptide Reactivity Assay (kDPRA)[b]		Lusens[a]	human Cell Line Activation Assay (h-CLAT)[a]	U937 Cell Line Activation Test[a]	Interleukin-8 luciferase (IL-8 Luc) assay[a]	Genomic Allergen Rapid Detection for assessment of skin sensitizers[a]
Not classified	The mean cysteine/lysine % depletion ≤ 6.38% or the mean cysteine % depletion ≤ 13.89 %	The mean NAC and NAL % depletion < 4.9% or NAC% depletion < 5.6%	Not applicable	At least one of the conditions for Category 1 is not met	At least one of the conditions for Category 1 is not met	None of the conditions for Category 1 is met	The stimulation index of CD86 is < 150% at all non-cytotoxic concentrations (cell viability ≥ 70%) and if no interference is observed	The Ind-IL8LA is less than (<) 1.4 and/or the lower limit of the 95% confidence interval of Ind-IL8LA is less than (<) 1.0 in at least 3 out of a maximum of 4 independent runs	The mean Decision Value (DV) is < 0

[a] Data cannot be used as stand-alone to conclude on classification in Category 1 or on no classification in tier 1 but could be used to conclude on classification in category 1 in tier 2 depending on the decision of the competent authority for their regulatory framework.
[b] A competent authority may decide that data can be used as stand-alone to conclude on classification in sub-category 1A.

表 3.4.8: 個別 *in chemico*/*in vitro* 法の基準 (続き)

カテゴリー	OECD テストガイドライン 442C タンパク質との共有結合に基づく有害性発現経路 (AOP) キーイベントを扱う *in chemico* 皮膚感作性試験のテストガイドラインに基づくキーイベント			OECD テストガイドライン 442D ケラチノサイト活性抗酸化応答配列-赤血球系転写因子 2 関連の AOP キーイベントを扱う *in vitro* 皮膚感作性アッセイのテストガイドラインに基づくキーイベント		OECD テストガイドライン 442E 樹状細胞の活性化に関する AOP キーイベントを扱う In vitro 皮膚感作性試験		
	付録 I で説明された方法	付録 II で説明された方法	付録 III で説明された方法	付録 1A[a] で説明された方法	付録 1B で説明された方法	付属書 I で説明された方法	付属書 II で説明された方法	付属書 III で説明された方法	付属書 IV で説明された方法
	ペプチド結合性試験 (DPRA)[a]	アミノ酸誘導体結合性試験 (ADRA)[a]	動的ペプチド結合性試験 (kDPRA)[b]		Lusens[a]	ヒト細胞株活性化試験 (h-CLAT)[a]	U937 細胞株活性化試験[a]	インターロイキン-8 ルシフェラーゼ (IL-8 Luc) アッセイ[a]	皮膚感作性評価のためのゲノミックアレルゲン迅速検出[a]
区分に該当しない	システインおよびリジンの減少率の平均値 ≤ 6.38% または システインの減少率 ≤13.89 %	NAC および NAL の減少率の平均値 ≤ 4.9% または NAC の減少率 < 5.6%	適用なし	区分 1 の条件のうち少なくとも 1 つが満たされていない	区分 1B で説明された方法	区分 1 のいずれかの条件も満たさない。	CD86 の刺激指数は干渉が観察されない場合、細胞毒性濃度 (細胞生存率≥70%) で 150%である。	最大 4 回の独立した測定のうち少なくとも 3 回において、IL8LA が 1.4 未満 (<)、または IL8LA の 95%信頼区間の下限が 1.0 未満 (<) である。	平均決定値 (DV) は < 0 である。

[a] データを単独で使用して、区分 1 の分類を結論付けるため、区分 1 で区分に該当しないこと結論付けることはできないが、段階 1 または段階 1 で区分 1 の分類を結論付けるため、または区分 1 に該当しないことを結論付けるために使用することができる。

[b] 所管官庁は、細区分 1A の分類を結論付けるためにデータを単独で使用できると決定する場合がある。

- 188 -

* *References:*

Corsini, E., Clewell, R., Cotgreave, I., Eskes, C., Kopp-Schneider, A., Westmoreland, C., Alves, P.M., Navas, J.M. and Piersma, A., ESAC Opinion on the Scientific Validity of the GARDskin and GARDpotency Test Methods, Asturiol Bofill, D., Casati, S. and Viegas Barroso, J.F. editor(s), Publications Office of the European Union, Luxembourg, 2021, ISBN 978-92-76-40345-6, doi:10.2760/626728, JRC125963.

Cottrez F, Boitel E, Auriault C, Aeby P, Groux H. Genes specifically modulated in sensitized skins allow the detection of sensitizers in a reconstructed human skin model. Development of the SENS-IS assay. Toxicol In Vitro. 2015 Jun;29(4):787-802. Doi: 10.1016/j.tiv.2015.02.012.

Cottrez F, Boitel E, Ourlin JC, Peiffer JL, Fabre I, Henaoui IS, Mari B, Vallauri A, Paquet A, Barbry P, Auriault C, Aeby P, Groux H. SENS-IS, a 3D reconstituted epidermis based model for quantifying chemical sensitization potency: Reproducibility and predictivity results from an inter-laboratory study. Toxicol In Vitro 2016 Apr;32:248-60. Doi: 10.1016/j.tiv.2016.01.007.

ECHA Guidance on the Application of the CLP Criteria Guidance to Regulation (EC) No 1272/2008 on classification, labelling and packaging (CLP) of substances and mixtures Version 5.0 July 2017

Gradin R., Johansson A., Forreryd A., Aaltonen E., Jerre A., Larne O., Mattson U., Johansson H. (2020) The GARDpotency assay for potency-associated subclassification of chemical skin sensitizers – Rationale, method development, and ring trial results of predictive performance and reproducibility. Toxicol. Sci. 176(2):423-432. Doi: 10.1093/toxsci/kfaa068

Johansson H., Lindstedt M., Albrekt A.S., Borrebaeck C.A. (2011) A genomic biomarker signature can predict skin sensitizers using a cell-based in vitro alternative to animal tests. BMC Genomics 12:399. Doi: 10.1186/1471-2164-12-399.

Johansson H., Rydnert F., Kühnl J., Schepky A., Borrebaeck C., Lindstedt M. (2014) Genomic allergen rapid detection in-house validation – A proof of concept. Toxicol. Sci. 139(2):362- 370. Doi: 10.1093/toxsci/kfu046.

Johansson H., Gradin R., Forreryd A., Agemark M., Zeller K., Johansson A., Larne O., van Vliet E., Borrebaeck C., Lindstedt M. (2017) Evaluation of the GARD assay in a blind Cosmetics Europe study. ALTEX 34(4):515-523. Doi: 10.14573/altex.1701121

Jowsey IR, Clapp CJ, Safford B, Gibbons BT, Basketter DA. (2008). The impact of vehicle on the relative potency of skin-sensitizing chemicals in the local lymph node assay. Cutan Ocul Toxicol: 27 (2); 67-75. Doi: 10.1080/15569520801904655.

Kligman A.M. (1966): The identification of contact allergens by human assay: II. Factors influencing the induction and measurement of allergic contact dermatitis. Journal of Investigative Dermatology 47 (5), 375-392. Doi: 10.1038/jid.1966.159

Kobayashi T., Maeda Y., Kondo H., Takeyoshi M. (2020) Applicability of the proposed GHS subcategorization criterion for LLNA:BrdU-ELISA (OECD TG442B) to the CBA/J strain mouse. Journal of Applied Toxicology. 40(10):1435-1439

Maeda Y., Takeyoshi M. (2019) Proposal of GHS subcategorization criteria for LLNA: BrdU-ELISA (OECD TG442B). Regulatory Toxicology and Pharmacology. 107:104409.

OECD (2014). The Adverse Outcome Pathway for Skin Sensitisation Initiated by Covalent Binding to Proteins, OECD Series on Testing and Assessment, No. 168, OECD Publishing, Paris. Doi.org/10.1787/9789264221444-en

OECD (2017), Guidance Document on the Reporting of Defined Approaches and Individual Information Sources to be Used within Integrated Approaches to Testing and Assessment (IATA) for Skin Sensitisation, OECD Series on Testing and Assessment, No. 256, OECD Publishing, Paris. https//Doi.org/10.1787/9789264279285-en.

Ryan CA et al. (2007): Extrapolating local lymph node assay EC3 values to estimate relative sensitizing potency. Cutan Ocul Toxicol 26(2), 135-45.

Saito K, Takenouchi O, Nukada Y, Miyazawa M, Sakaguchi H. An in vitro skin sensitization assay termed EpiSensA for broad sets of chemicals including lipophilic chemicals and pre/pro-haptens.

Toxicol In Vitro. 2017 Apr;40:11-25. doi: 10.1016/j.tiv.2016.12.005.

Wright ZM, Basketter PA, Blaikie L, Cooper KJ, Warbrick EV, Dearman RJ, Kimber I. Vehicle effects on skin sensitizing potency of four chemicals: assessment using the local lymph node assay. Int J Cosmet Sci. 2001 Apr;23(2):75-83. doi: 10.1046/j.1467-2494.2001.00066.x.

Zeller K.S., Forreryd A., Lindberg T., Gradin R., Chawade A., Lindstedt M. (2017) The GARD platform for potency assessment of skin sensitizing chemicals. ALTEX 34(4):539-559. Doi: 10.14573/altex.1701101.

* *References:*

Corsini, E., Clewell, R., Cotgreave, I., Eskes, C., Kopp-Schneider, A., Westmoreland, C., Alves, P.M., Navas, J.M. and Piersma, A., ESAC Opinion on the Scientific Validity of the GARDskin and GARDpotency Test Methods, Asturiol Bofill, D., Casati, S. and Viegas Barroso, J.F. editor(s), Publications Office of the European Union, Luxembourg, 2021, ISBN 978-92-76-40345-6, doi:10.2760/626728, JRC125963.

Cottrez F, Boitel E, Auriault C, Aeby P, Groux H. Genes specifically modulated in sensitized skins allow the detection of sensitizers in a reconstructed human skin model. Development of the SENS-IS assay. Toxicol In vitro. 2015 Jun;29(4):787-802. Doi: 10.1016/j.tiv.2015.02.012.

Cottrez F, Boitel E, Ourlin JC, Peiffer JL, Fabre I, Henaoui IS, Mari B, Vallauri A, Paquet A, Barbry P, Auriault C, Aeby P, Groux H. SENS-IS, a 3D reconstituted epidermis based model for quantifying chemical sensitization potency: Reproducibility and predictivity results from an inter-laboratory study. Toxicol In vitro 2016 Apr;32:248-60. Doi: 10.1016/j.tiv.2016.01.007.

ECHA Guidance on the Application of the CLP Criteria Guidance to Regulation (EC) No 1272/2008 on classification, labelling and packaging (CLP) of substances and mixtures Version 5.0 July 2017

Gradin R., Johansson A., Forreryd A., Aaltonen E., Jerre A., Larne O., Mattson U., Johansson H. (2020) The GARDpotency assay for potency-associated subclassification of chemical skin sensitizers – Rationale, method development, and ring trial results of predictive performance and reproducibility. Toxicol. Sci. 176(2):423-432. Doi: 10.1093/toxsci/kfaa068

Johansson H., Lindstedt M., Albrekt A.S., Borrebaeck C.A. (2011) A genomic biomarker signature can predict skin sensitizers using a cell-based in vitro alternative to animal tests. BMC Genomics 12:399. Doi: 10.1186/1471-2164-12-399.

Johansson H., Rydnert F., Kühnl J., Schepky A., Borrebaeck C., Lindstedt M. (2014) Genomic allergen rapid detection in-house validation – A proof of concept. Toxicol. Sci. 139(2):362- 370. Doi: 10.1093/toxsci/kfu046.

Johansson H., Gradin R., Forreryd A., Agemark M., Zeller K., Johansson A., Larne O., van Vliet E., Borrebaeck C., Lindstedt M. (2017) Evaluation of the GARD assay in a blind Cosmetics Europe study. ALTEX 34(4):515-523. Doi: 10.14573/altex.1701121

Jowsey IR, Clapp CJ, Safford B, Gibbons BT, Basketter DA. (2008). The impact of vehicle on the relative potency of skin-sensitizing chemicals in the local lymph node assay. Cutan Ocul Toxicol: 27 (2); 67-75. Doi: 10.1080/15569520801904655.

Kligman A.M. (1966): The identification of contact allergens by human assay: II. Factors influencing the induction and measurement of allergic contact dermatitis. Journal of Investigative Dermatology 47 (5), 375-392. Doi: 10.1038/jid.1966.159

Kobayashi T., Maeda Y., Kondo H., Takeyoshi M. (2020) Applicability of the proposed GHS subcategorization criterion for LLNA:BrdU-ELISA (OECD TG442B) to the CBA/J strain mouse. Journal of Applied Toxicology. 40(10):1435-1439

Maeda Y., Takeyoshi M. (2019) Proposal of GHS subcategorization criteria for LLNA: BrdU-ELISA (OECD TG442B). Regulatory Toxicology and Pharmacology. 107:104409.

OECD (2014). The Adverse Outcome Pathway for Skin Sensitisation Initiated by Covalent Binding to Proteins, OECD Series on Testing and Assessment, No. 168, OECD Publishing, Paris. Doi.org/10.1787/9789264221444-en

OECD (2017), Guidance Document on the Reporting of Defined Approaches and Individual Information Sources to be Used within Integrated Approaches to Testing and Assessment (IATA) for Skin Sensitisation, OECD Series on Testing and Assessment, No. 256, OECD Publishing, Paris. https//Doi.org/10.1787/9789264279285-en.

Ryan CA et al. (2007): Extrapolating local lymph node assay EC3 values to estimate relative sensitizing potency. Cutan Ocul Toxicol 26(2), 135-45.

Saito K, Takenouchi O, Nukada Y, Miyazawa M, Sakaguchi H. An in vitro skin sensitization assay termed EpiSensA for broad sets of chemicals including lipophilic chemicals and pre/pro-haptens.

Toxicol In vitro. 2017 Apr;40:11-25. doi: 10.1016/j.tiv.2016.12.005.

Wright ZM, Basketter PA, Blaikie L, Cooper KJ, Warbrick EV, Dearman RJ, Kimber I. Vehicle effects on skin sensitizing potency of four chemicals: assessment using the local lymph node assay. Int J Cosmet Sci. 2001 Apr;23(2):75-83. doi: 10.1046/j.1467-2494.2001.00066.x.

Zeller K.S., Forreryd A., Lindberg T., Gradin R., Chawade A., Lindstedt M. (2017) The GARD platform for potency assessment of skin sensitizing chemicals. ALTEX 34(4):539-559. Doi: 10.14573/altex.1701101.

CHAPTER 3.5

GERM CELL MUTAGENICITY

3.5.1 Definitions and general considerations

3.5.1.1 *Germ cell mutagenicity* refers to heritable gene mutations, including heritable structural and numerical chromosome aberrations in germ cells occurring after exposure to a substance or mixture.

3.5.1.2 This hazard class is primarily concerned with chemicals that may cause mutations in the germ cells of humans that can be transmitted to the progeny. However, mutagenicity/genotoxicity tests in vitro and in mammalian somatic cells in vivo are also considered in classifying substances and mixtures within this hazard class.

3.5.1.3 In the present context, commonly found definitions of the terms "mutagenic", "mutagen", "mutations" and "genotoxic" are used. A *mutation* is defined as a permanent change in the amount or structure of the genetic material in a cell.

3.5.1.4 The term *mutation* applies both to heritable genetic changes that may be manifested at the phenotypic level and to the underlying DNA modifications when known (including, for example, specific base pair changes and chromosomal translocations). The term *mutagenic* and *mutagen* will be used for agents giving rise to an increased occurrence of mutations in populations of cells and/or organisms.

3.5.1.5 The more general terms *genotoxic* and *genotoxicity* apply to agents or processes which alter the structure, information content, or segregation of DNA, including those which cause DNA damage by interfering with normal replication processes, or which in a non-physiological manner (temporarily) alter its replication. Genotoxicity test results are usually taken as indicators for mutagenic effects.

3.5.2 Classification criteria for substances

3.5.2.1 The classification system provides for two different categories of germ cell mutagens to accommodate the weight of evidence available. The two-category system is described in the following.

3.5.2.2 To arrive at a classification, test results are considered from experiments determining mutagenic and/or genotoxic effects in germ and/or somatic cells of exposed animals. Mutagenic and/or genotoxic effects determined in in vitro tests may also be considered.

3.5.2.3 The system is hazard based, classifying substances on the basis of their intrinsic ability to induce mutations in germ cells. The scheme is, therefore, not meant for the (quantitative) risk assessment of substances.

3.5.2.4 Classification for heritable effects in human germ cells is made on the basis of well conducted, sufficiently validated tests, preferably as described in OECD test guidelines. Evaluation of the test results should be done using expert judgement and all the available evidence should be weighed for classification.

3.5.2.5 Examples of in vivo heritable germ cell mutagenicity tests are:

Rodent dominant lethal mutation test (OECD 478)
Mouse heritable translocation assay (OECD 485)
Mouse specific locus test

3.5.2.6 Examples of in vivo somatic cell mutagenicity tests are:

Mammalian bone marrow chromosome aberration test (OECD 475)
Mammalian erythrocyte micronucleus test (OECD 474)

第 3.5 章

生殖細胞変異原性

3.5.1 定義および一般事項

3.5.1.1. *生殖細胞変異原性*とは、物質または混合物へのばく露後に起こる、生殖細胞における構造的および数的な染色体の異常を含む、遺伝性の遺伝子変異をさす。

3.5.1.2 この有害性クラスは主として、ヒトにおいて次世代に受継がれる可能性のある突然変異を誘発すると思われる化学品に関するものである。一方、in vitro での変異原性/遺伝毒性試験、および *in vivo* での哺乳類体細胞を用いた試験も、この有害性クラスの中で分類する際に考慮される。

3.5.1.3 本文書では、変異原性、変異原性物質、突然変異および遺伝毒性についての一般的な定義が採用されている。ここで*突然変異*とは、細胞内遺伝物質の量または構造の恒久的変化として定義されている。

3.5.1.4 *突然変異*という用語は、表現型レベルで発現されるような経世代的な遺伝的変化と、その根拠となっている DNA の変化(例えば、特異的塩基対の変化および染色体転座など)の両方に適用される。*変異原性*および*変異原性物質*という用語は、細胞または生物の集団における突然変異の発生を増加させる物質について用いられる。

3.5.1.5 より一般的な用語である*遺伝毒性物質*および*遺伝毒性*とは、DNA の構造や含まれる遺伝情報、または DNA の分離を変化させる物質あるいはその作用に適用される。これには、正常な複製過程の妨害により DNA に損傷を与えるものや、非生理的な状況において(一時的に)DNA 複製を変化させるものもある。遺伝毒性試験結果は、一般的に変異原性作用の指標として採用される。

3.5.2 物質の分類基準

3.5.2.1 本分類システムは、利用可能な証拠の重みを取り入れられるように、生殖細胞に対する変異原性物質に2種類の区分を設けている。この2種類の区分によるシステムを以下に示す。

3.5.2.2 分類のためには、ばく露動物の生殖細胞または体細胞における変異原性または遺伝毒性作用を判定する実験より得られた試験結果が考慮される。 in vitro 試験で判定された変異原性または遺伝毒性作用もまた考慮されてよい。

3.5.2.3 本システムは有害性に基づき、生殖細胞に突然変異を誘発する性質を本来持っている物質を分類する。したがって本スキームは、物質の(定量的)リスク評価のためのものではない。

3.5.2.4 ヒト生殖細胞に対する経世代的な影響の分類は、適切に実施され、十分に有効性が確認された試験に基づいて行う。OECD テストガイドラインに定められた方法に従った試験を用いるのが望ましい。試験結果は専門家の判断により評価され、利用可能な証拠すべてを比較検討して分類すべきである。

3.5.2.5 *in vivo* 生殖細胞経世代変異原性試験の例

げっ歯類を用いる優性致死試験(OECD478)
マウスを用いる相互転座試験(OECD485)
マウスを用いる特定座位試験

3.5.2.6 *in vivo* 体細胞変異原性試験の例

哺乳類骨髄細胞を用いる染色体異常試験(OECD475)
哺乳類赤血球を用いる小核試験(OECD474)

Figure 3.5.1: Hazard categories for germ cell mutagens

CATEGORY 1:	**Substances known to induce heritable mutations or to be regarded as if they induce heritable mutations in the germ cells of humans**
Category 1A:	**Substances known to induce heritable mutations in germ cells of humans**
	Positive evidence from human epidemiological studies.
Category 1B:	**Substances which should be regarded as if they induce heritable mutations in the germ cells of humans**
	(a) Positive result(s) from in vivo heritable germ cell mutagenicity tests in mammals; or
	(b) Positive result(s) from in vivo somatic cell mutagenicity tests in mammals, in combination with some evidence that the substance has potential to cause mutations to germ cells. This supporting evidence may, for example, be derived from mutagenicity/genotoxic tests in germ cells in vivo, or by demonstrating the ability of the substance or its metabolite(s) to interact with the genetic material of germ cells; or
	(c) Positive results from tests showing mutagenic effects in the germ cells of humans, without demonstration of transmission to progeny; for example, an increase in the frequency of aneuploidy in sperm cells of exposed people.
CATEGORY 2:	**Substances which cause concern for humans owing to the possibility that they may induce heritable mutations in the germ cells of humans**
	Positive evidence obtained from experiments in mammals and/or in some cases from in vitro experiments, obtained from:
	(a) Somatic cell mutagenicity tests in vivo, in mammals; or
	(b) Other in vivo somatic cell genotoxicity tests which are supported by positive results from in vitro mutagenicity assays.
	NOTE: *Substances which are positive in in vitro mammalian mutagenicity assays, and which also show structure activity relationship to known germ cell mutagens, should be considered for classification as Category 2 mutagens.*

3.5.2.7 Examples of mutagenicity/genotoxicity tests in germ cells are:

    (a) Mutagenicity tests:

        Mammalian spermatogonial chromosome aberration test (OECD 483)
        Spermatid micronucleus assay
        Trasgenic Rodent Somatic and Germ Cell Gene Mutation Assays (OECD 488)

    (b) Genotoxicity tests:

        Sister chromatid exchange analysis in spermatogonia
        Unscheduled DNA synthesis test (UDS) in testicular cells

3.5.2.8 Examples of genotoxicity tests in somatic cells are:

    In vivo Mammalian Alkaline Comet Assay (OECD 489)
    Transgenic Rodent Somatic and Germ Cell Gene Mutation Assays (OECD 488)
    Liver Unscheduled DNA Synthesis (UDS) in vivo (OECD 486)
    Mammalian bone marrow Sister Chromatid Exchanges (SCE)

3.5.2.9 Examples of in vitro mutagenicity tests are:

    In vitro mammalian chromosome aberration test (OECD 473)
    In vitro mammalian cell gene mutation test (OECD 476 and 490)
    Bacterial reverse mutation tests (OECD 471)

3.5.2.10 The classification of individual substances should be based on the total weight of evidence available, using expert judgement. In those instances where a single well-conducted test is used for classification, it should provide clear and unambiguously positive results. If new, well validated, tests arise these may also be used in the total weight of evidence to be considered. The relevance of the route of exposure used in the study of the substance compared to the route of human exposure should also be taken into account.

図 3.5.1：生殖細胞変異原性物質の有害性区分

> **区分 1**：ヒト生殖細胞に経世代突然変異を誘発することが知られているかまたは経世代突然変異を誘発すると見なされている物質
>
> **区分 1A**：ヒト生殖細胞に経世代突然変異を誘発することが知られている物質
> ヒトの疫学的調査による陽性の証拠。
>
> **区分 1B**：ヒト生殖細胞に経世代突然変異を誘発すると見なされるべき物質
>
> (a) 哺乳類における *in vivo* 経世代生殖細胞変異原性試験による陽性結果、または
>
> (b) 哺乳類における *in vivo* 体細胞変異原性試験による陽性結果に加えて、当該物質が生殖細胞に突然変異を誘発する可能性についての何らかの証拠。この裏付け証拠は、例えば生殖細胞を用いる *in vivo* 変異原性/遺伝毒性試験より、あるいは、当該物質またはその代謝物が生殖細胞の遺伝物質と相互作用する機能があることの実証により導かれる。または
>
> (c) 次世代に受継がれる証拠はないがヒト生殖細胞に変異原性を示す陽性結果；例えば、ばく露されたヒトの精子中の異数性発生頻度の増加など。
>
> **区分 2**：ヒト生殖細胞に経世代突然変異を誘発する可能性がある物質
>
> 哺乳類を用いる試験、または場合によっては下記に示す in vitro 試験による陽性結果
>
> (a) 哺乳類を用いる *in vivo* 体細胞変異原性試験、または
> (b) in vitro 変異原性試験の陽性結果により裏付けられたその他の *in vivo* 体細胞遺伝毒性試験
>
> *注記：哺乳類を用いる in vitro 変異原性試験で陽性となり、さらに既知の生殖細胞変異原性物質と化学的構造活性相関を示す物質は、区分2 変異原性物質として分類されるとみなすべきである。*

3.5.2.7 生殖細胞を用いる *in vivo* 変異原性/遺伝毒性試験の例：

(a) 変異原性試験：

　哺乳類精原細胞を用いる染色体異常試験(OECD483)
　哺乳類精子細胞を用いる小核試験
　トランスジェニックげっ歯類の体細胞および生殖細胞を用いた遺伝子突然変異試験(OECD488)

(b) 遺伝毒性試験：

　哺乳類精原細胞を用いる姉妹染色分体交換(SCE)試験
　哺乳類精巣細胞を用いる不定期 DNA 合成(UDS)試験

3.5.2.8 体細胞を用いる *in vivo* 遺伝毒性試験の例：

　in vivo 哺乳類アルカリコメットアッセイ(OECD489)
　トランスジェニックげっ歯類の体細胞および生殖細胞を用いた遺伝子突然変異試験(OECD488)
　哺乳類肝臓を用いる不定期 DNA 合成(UDS)試験(OECD486)
　哺乳類骨髄細胞を用いる姉妹染色分体交換(SCE)試験

3.5.2.9 in vitro 変異原性試験の例：

　in vitro 哺乳類培養細胞を用いる染色体異常試験(OECD473)
　in vitro 哺乳類培養細胞を用いる遺伝子突然変異試験(OECD476 および 490)
　細菌を用いる復帰突然変異試験(OECD471)

3.5.2.10 個々の物質の分類は、専門家の判断を取り入れて、利用可能な証拠全体の重み付けに基づいて行うべきである。適切に実施された単一の試験を用いて分類する場合には、その試験から明確で疑いようのない陽性結果が得られているべきである。十分に有効性が確認された新しい試験法が開発されたならば、それらも考慮すべき包括的な証拠の重み付けのために採用することもできる。ヒトばく露経路と比較して、当該物質の試験に用いられたばく露経路が妥当であるかも考慮すべきである。

3.5.3 **Classification criteria for mixtures**

3.5.3.1 *Classification of mixtures when data are available for the mixture itself*

Classification of mixtures will be based on the available test data for the individual ingredients of the mixture using cut-off values/concentration limits for the ingredients classified as germ cell mutagens. The classification may be modified on a case-by-case basis based on the available test data for the mixture as a whole. In such cases, the test results for the mixture as a whole must be shown to be conclusive taking into account dose and other factors such as duration, observations and analysis (e.g. statistical analysis, test sensitivity) of germ cell mutagenicity test systems. Adequate documentation supporting the classification should be retained and made available for review upon request.

3.5.3.2 *Classification of mixtures when data are not available for the complete mixture: bridging principles*

3.5.3.2.1 Where the mixture itself has not been tested to determine its germ cell mutagenicity hazard, but there are sufficient data on both the individual ingredients and similar tested mixtures to adequately characterize the hazards of the mixture, these data will be used in accordance with the following agreed bridging principles. This ensures that the classification process uses the available data to the greatest extent possible in characterizing the hazards of the mixture without the necessity for additional testing in animals.

3.5.3.2.2 *Dilution*

If a tested mixture is diluted with a diluent which is not expected to affect the germ cell mutagenicity of other ingredients, then the new diluted mixture may be classified as equivalent to the original tested mixture.

3.5.3.2.3 *Batching*

The germ cell mutagenic potential of a tested production batch of a mixture can be assumed to be substantially equivalent to that of another untested production batch of the same commercial product, when produced by or under the control of the same manufacturer unless there is reason to believe there is significant variation in composition such that the germ cell mutagenic potential of the untested batch has changed. If the latter occurs, a new classification is necessary.

3.5.3.2.4 *Substantially similar mixtures*

Given the following:

(a) Two mixtures: (i) A + B;

(ii) C + B;

(b) The concentration of mutagen ingredient B is the same in both mixtures;

(c) The concentration of ingredient A in mixture (i) equals that of ingredient C in mixture (ii);

(d) Data on toxicity for A and C are available and substantially equivalent, i.e. they are in the same hazard category and are not expected to affect the germ cell mutagenicity of B.

If mixture (i) or (ii) is already classified based on test data, then the other mixture can be classified in the same hazard category.

3.5.3.3 *Classification of mixtures when data are available for all ingredients or only for some ingredients of the mixture*

The mixture will be classified as a mutagen when at least one ingredient has been classified as a Category 1 or Category 2 mutagen and is present at or above the appropriate cut-off value/concentration limit as shown in table 3.5.1 below for Category 1 and 2 respectively.

3.5.3 混合物の分類基準

3.5.3.1 混合物そのものについて試験データが入手できる場合の混合物の分類

混合物の分類は、当該混合物の個々の成分について入手できる試験データに基づき、生殖細胞変異原性物質として分類される成分のカットオフ値/濃度限界を使用して行われる。当該混合物そのものの試験データが入手できる場合には、分類はケースバイケースで修正されることがある。このような場合、混合物そのものの試験結果は、生殖細胞変異原性試験系の用量や、試験期間、観察、分析(例えば、統計学的解析、試験感度)などの他の要因を考慮した上で確実であることが示されなければならない。分類が適切であることの証拠書類を保持し、要請に応じて示すことができるようにするべきである。

3.5.3.2 混合物そのものについて試験データが入手できない場合の混合物の分類:つなぎの原則 (bridging principle)

3.5.3.2.1 混合物そのものは生殖細胞変異原性を決定する試験がなされていないが、当該混合物の有害性を適切に特定するための、個々の成分および類似の試験された混合物の両方に関して十分なデータがある場合、これらのデータは以下の合意されたつなぎの原則にしたがって使用される。これによって、分類プロセスで動物試験を追加する必要もなく、混合物の有害性判定に入手されたデータを可能な限り最大限に用いることができるようになる。

3.5.3.2.2 希釈

試験された混合物が、他の成分の生殖細胞変異原性に影響を与えないと予想される希釈剤で希釈される場合、新しい希釈された混合物は、試験された元の混合物と同等として分類してもよい。

3.5.3.2.3 製造バッチ

混合物の試験されていない製造バッチに生殖細胞変異原性があるかどうかは、同じ製造者によって、またはその管理下で生産された同じ商品の試験された別のバッチの毒性と本質的に同等とみなすことができる。ただし、試験されていないバッチの生殖細胞変異原性が変化するような有意な組成の変動があると考えられる理由がある場合はこの限りではない。このような場合には、新しい分類を行う必要がある。

3.5.3.2.4 本質的に類似した混合物

次を仮定する:

(a) 2つの混合物: (i) A+B ;
(ii) C+B ;

(b) 変異原性成分Bの濃度は、両方の混合物で同じである;

(c) 混合物(i)の成分Aの濃度は、混合物(ii)の成分Cの濃度に等しい;

(d) AとCの毒性に関するデータは利用でき、実質的に同等であり、すなわち、AとCは同じ有害性区分に属し、かつ、Bの生殖細胞変異原性に影響を与えることはないと判断される。

混合物(i)または(ii)が既に試験データに基づいて分類されている場合には、他方の混合物は同じ有害性区分に分類することができる。

3.5.3.3 混合物の全成分または一部の成分だけについてデータが入手できる場合の混合物の分類

混合物は、少なくとも1つの成分が区分1または区分2変異原性物質として分類され、区分1と2それぞれについて表3.5.1に示したような適切なカットオフ値/濃度限界以上で存在する場合、変異原性物質として分類される。

Table 3.5.1: Cut-off values/concentration limits of ingredients of a mixture classified as germ cell mutagens that would trigger classification of the mixture

Ingredient classified as:	Cut-off/concentration limits triggering classification of a mixture as:		
	Category 1 mutagen		Category 2 mutagen
	Category 1A	Category 1B	
Category 1A mutagen	≥ 0.1 %	--	--
Category 1B mutagen	--	≥ 0.1 %	
Category 2 mutagen	--	--	≥ 1.0 %

Note: *The cut-off values/concentration limits in the table above apply to solids and liquids (w/w units) as well as gases (v/v units).*

3.5.4 Hazard communication

General and specific considerations concerning labelling requirements are provided in *Hazard communication: Labelling* (chapter 1.4). Annex 1 contains summary tables about classification and labelling. Annex 3 contains examples of precautionary statements and pictograms which can be used where allowed by the competent authority. Table 3.5.2 presents specific label elements for substances and mixtures classified in this hazard class based on the criteria in this chapter.

Table 3.5.2: Label elements for germ cell mutagenicity

	Category 1 (Category 1A, 1B)	Category 2
Symbol	Health hazard	Health hazard
Signal word	Danger	Warning
Hazard statement	May cause genetic defects (state route of exposure if it is conclusively proven that no other routes of exposure cause the hazard)	Suspected of causing genetic defects (state route of exposure if it is conclusively proven that no other routes of exposure cause the hazard)

表 3.5.1：混合物の分類の基準となる混合物の生殖細胞変異原性物質として分類された
成分のカットオフ値/濃度限界

成分の分類：	混合物の分類基準となるカットオフ値/濃度限界		
	区分1 変異原性物質		区分2 変異原性物質
	区分1A	区分1B	
区分1A 変異原性物質	≧0.1%	--	--
区分1B 変異原性物質	--	≧0.1%	--
区分2 変異原性物質	--	--	≧1.0%

注記：*上の表のカットオフ値/濃度限界は、気体（体積/体積単位）および、固体と液体（重量/重量単位）にも適用される。*

3.5.4 危険有害性情報の伝達

表示要件に関する通則および細則は、*危険有害性に関する情報の伝達：表示*（第1.4章）に定める。附属書1に分類と表示に関する概要表を示す。附属書3には、注意書きおよび所管官庁が許可した場合に使用可能な注意絵表示の例を示す。表3.5.2には、本章の判定基準に基づいてこの危険有害性クラスに分類された物質および混合物の具体的なラベル要素を示す。

表 3.5.2：生殖細胞変異原性のラベル要素

	区分1 （区分1A、1B）	区分2
シンボル	健康有害性	健康有害性
注意喚起語	危険	警告
危険有害性情報	遺伝性疾患のおそれ （他の経路からのばく露が有害でないことが決定的に証明されている場合、有害なばく露経路を記載）	遺伝性疾患のおそれの疑い （他の経路からのばく露が有害でないことが決定的に証明されている場合、有害なばく露経路を記載）

3.5.5 Decision logic and guidance

3.5.5.1 *Decision logic for germ cell mutagenicity*

The decision logic which follows is not part of the harmonized classification system but is provided here as additional guidance. It is strongly recommended that the person responsible for classification study the criteria before and during use of the decision logic.

3.5.5.1.1 *Decision logic 3.5.1 for substances*

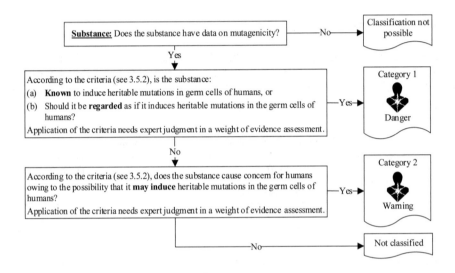

3.5.5 判定論理と手引き

3.5.5.1 *生殖細胞変異原性の判定論理*

以下の判定論理は、調和分類システムには含まれないが、追加的な手引きとしてここに示す。分類の責任者に対し、この判定論理を使用する前および使用する際に判定基準についてよく調べ理解することを強く勧める。

3.5.5.1.1. *物質の判定論理 3.5.1*

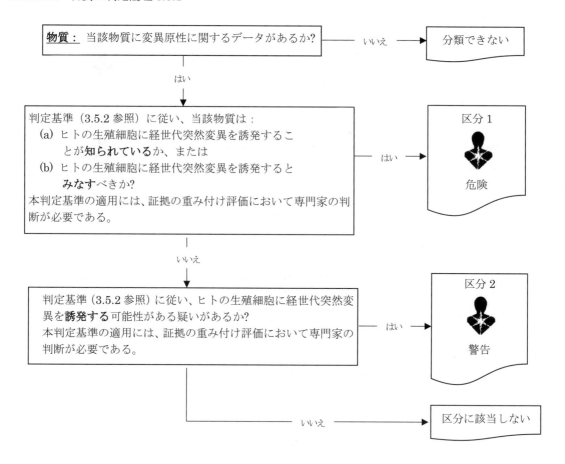

3.5.5.1.2   *Decision logic 3.5.2 for mixtures*

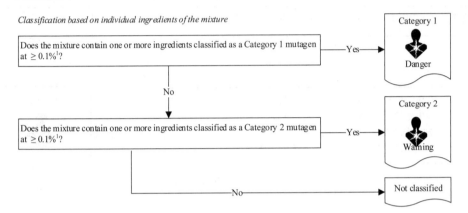

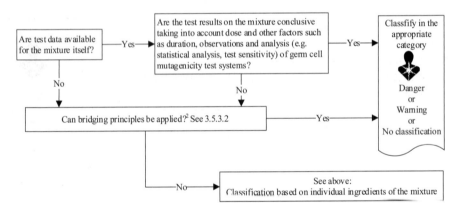

3.5.5.2   *Guidance*

It is increasingly accepted that the process of chemical-induced tumorigenesis in man and animals involves genetic changes in proto-oncogenes and/or tumour suppresser genes of somatic cells. Therefore, the demonstration of mutagenic properties of chemicals in somatic and/or germ cells of mammals in vivo may have implications for the potential classification of these chemicals as carcinogens (see also Carcinogenicity, chapter 3.6, paragraph 3.6.2.5.3).

---

[1] *For specific concentration limits, see "The use of cut-off values/concentration limits" in chapter 1.3, paragraph 1.3.3.2 and table 3.5.1 of this chapter.*

[2] *If data on another mixture are used in the application of bridging principles, the data on that mixture must be conclusive in accordance with 3.5.3.2.*

3.5.5.1.2 混合物の判定論理 3.5.2

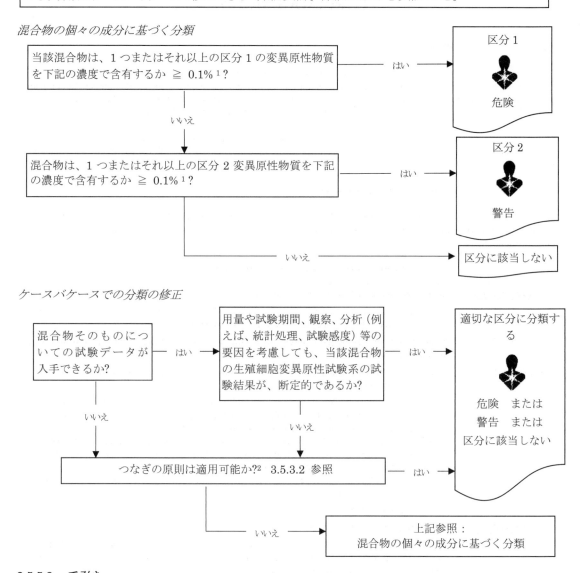

3.5.5.2 手引き

ヒトおよび動物において化学的に誘発される腫瘍形成の過程は、がん原遺伝子、または体細胞の腫瘍抑制遺伝子の遺伝的変化を伴うということはかなり認められるようになってきている。そのため、化学品が *in vivo* において哺乳動物の体細胞、または生殖細胞における変異原性を有することの証明は、その化学品は発がん性物質として分類され得るということの可能性を示すものである（第 3.6 章 3.6.2.5.3 発がん性、参照）。

1　個々の濃度限界については、第 1.3 章の 1.3.3.2 「カットオフ値/濃度限界の使用」および本章の表 3.5.1 を参照。
2　他の混合物のデータをつなぎの原則に用いた場合は、その混合物のデータは 3.5.3.2 に照らして決定的なものでなければならない。

CHAPTER 3.6

CARCINOGENICITY

3.6.1 Definitions

Carcinogenicity refers to the induction of cancer or an increase in the incidence of cancer occurring after exposure to a substance or mixture. Substances and mixtures which have induced benign and malignant tumours in well performed experimental studies on animals are considered also to be presumed or suspected human carcinogens unless there is strong evidence that the mechanism of tumour formation is not relevant for humans.

Classification of a substance or mixture as posing a carcinogenic hazard is based on its inherent properties and does not provide information on the level of the human cancer risk which the use of the substance or mixture may represent.

3.6.2 Classification criteria for substances

3.6.2.1 For the purpose of classification for carcinogenicity, substances are allocated to one of two categories based on strength of evidence and additional considerations (weight of evidence). In certain instances, route specific classification may be warranted.

Figure 3.6.1: Hazard categories for carcinogens

CATEGORY 1:	**Known or presumed human carcinogens**
	The placing of a substance in Category 1 is done on the basis of epidemiological and/or animal data. An individual substance may be further distinguished:
Category 1A:	**Known to have carcinogenic potential for humans; the placing of a substance is largely based on human evidence.**
Category 1B:	**Presumed to have carcinogenic potential for humans; the placing of a substance is largely based on animal evidence.**
	Based on strength of evidence together with additional considerations, such evidence may be derived from human studies that establish a causal relationship between human exposure to a substance and the development of cancer (known human carcinogen). Alternatively, evidence may be derived from animal experiments for which there is sufficient evidence to demonstrate animal carcinogenicity (presumed human carcinogen). In addition, on a case by case basis, scientific judgement may warrant a decision of presumed human carcinogenicity derived from studies showing limited evidence of carcinogenicity in humans together with limited evidence of carcinogenicity in experimental animals.
	Classification: Category 1 (A and B) Carcinogen
CATEGORY 2:	**Suspected human carcinogens**
	The placing of a substance in Category 2 is done on the basis of evidence obtained from human and/or animal studies, but which is not sufficiently convincing to place the substance in Category 1. Based on strength of evidence together with additional considerations, such evidence may be from either limited evidence of carcinogenicity in human studies or from limited evidence of carcinogenicity in animal studies.
	Classification: Category 2 Carcinogen

3.6.2.2 Classification as a carcinogen is made on the basis of evidence from reliable and acceptable methods and is intended to be used for substances which have an intrinsic property to produce such toxic effects. The evaluations should be based on all existing data, peer-reviewed published studies and additional data accepted by regulatory agencies.

3.6.2.3 *Carcinogen classification* is a one-step, criterion-based process that involves two interrelated determinations: evaluations of strength of evidence and consideration of all other relevant information to place substances with human cancer potential into hazard categories.

第 3.6 章

発がん性

3.6.1 定義

*発がん性*とは、物質または混合物へのばく露後に起こる、がんの誘発またはその発生率の増加をさす。動物を用いて適切に実施された実験研究で良性および悪性腫瘍を誘発した物質および混合物もまた、腫瘍形成のメカニズムがヒトには関係しないとする強力な証拠がない限りは、ヒトに対する発がん性物質として推定されるかまたはその疑いがあると考えられる。

物質または混合物の発がん有害性を有するものとしての分類は、それら固有の特性に基づきなされるものであり、このように分類されることによって、当該物質または混合物の使用により生ずる可能性のあるヒトのがんリスクの程度に関する情報を提供するものではない。

3.6.2 物質の分類基準

3.6.2.1 発がん性の分類では、物質は証拠の強さおよび追加検討事項(証拠の重み)をもとに2種類の区分のいずれかに割り当てられる。特殊な例では、経路に特化した分類を要すると判断される場合もある。

図 3.6.1：発がん性物質の有害性区分

区分 1：ヒトに対する発がん性が知られているあるいはおそらく発がん性がある

　　物質の区分 1 への分類は、疫学的データまたは動物データをもとに行う。個々の物質はさらに次のように区別されることもある：

区分 1A：ヒトに対する発がん性が知られている：主としてヒトでの**証拠**により**物質**をここに分類する。

区分 1B：ヒトに対しておそらく発がん性がある：主として**動物**での**証拠**により**物質**をここに分類する。

　　証拠の強さとその他の事項も考慮した上で、ヒトでの調査により物質に対するヒトのばく露と、がん発生の因果関係が確立された場合を、その証拠とする(ヒトに対する発がん性が知られている物質)。あるいは、動物に対する発がん性を実証する十分な証拠がある動物試験を、その証拠とすることもある(ヒトに対する発がん性があると考えられる物質)。さらに、試験からはヒトにおける発がん性の証拠が限られており、また実験動物での発がん性の証拠も限られている場合には、ヒトに対する発がん性があると考えられるかどうかは、ケースバイケースで科学的判定によって決定することもある。

分類：区分 1(A および B) 発がん性物質

区分 2：ヒトに対する発がん性が疑われる

　　物質の区分 2 への分類は、物質を確実に区分 1 に分類するには不十分な場合ではあるが、ヒトまたは動物での調査より得られた証拠をもとに行う。証拠の強さとその他の事項も考慮した上で、ヒトでの調査で発がん性の限られた証拠や、または動物試験で発がん性の限られた証拠が証拠とされる場合もある。

分類：区分 2 発がん性物質

3.6.2.2 発がん性物質としての分類は、信頼でき、かつ承認されている方法によって得られる証拠に基づいて行われるものである。また、この分類はこうした毒性を生じる固有の性質を有する物質を対象とすることを意図としている。評価は、すべての既存データ、ピアレビューされて発表された調査、および規制機関が承認した追加データに基づき行われるべきである。

3.6.2.3 *発がん性物質分類*は、一段階の1つの判定基準に基づくプロセスであるが、2種類の相互に関連した判断が関与する。すなわち、証拠の強さの評価と、他の関連情報の考慮(潜在的なヒトに対する発がん性を有する物質を有害性区分に分類することに関連する情報)である。

3.6.2.4   *Strength of evidence* involves the enumeration of tumours in human and animal studies and determination of their level of statistical significance. Sufficient human evidence demonstrates causality between human exposure and the development of cancer, whereas sufficient evidence in animals shows a causal relationship between the agent and an increased incidence of tumours. Limited evidence in humans is demonstrated by a positive association between exposure and cancer, but a causal relationship cannot be stated. Limited evidence in animals is provided when data suggest a carcinogenic effect but are less than sufficient. The terms "sufficient" and "limited" are used here as they have been defined by the International Agency for Research on Cancer (IARC) and are outlined in 3.6.5.3.1.

3.6.2.5   *Additional considerations (weight of evidence)*: Beyond the determination of the strength of evidence for carcinogenicity, a number of other factors should be considered that influence the overall likelihood that an agent may pose a carcinogenic hazard in humans. The full list of factors that influence this determination is very lengthy, but some of the important ones are considered here.

3.6.2.5.1   The factors can be viewed as either increasing or decreasing the level of concern for human carcinogenicity. The relative emphasis accorded to each factor depends upon the amount and coherence of evidence bearing on each. Generally, there is a requirement for more complete information to decrease than to increase the level of concern. Additional considerations should be used in evaluating the tumour findings and the other factors in a case-by-case manner.

3.6.2.5.2   Some important factors which may be taken into consideration, when assessing the overall level of concern are:

(a)   Tumour type and background incidence;

(b)   Multisite responses;

(c)   Progression of lesions to malignancy;

(d)   Reduced tumour latency;

Additional factors which may increase or decrease the level of concern include:

(e)   Whether responses are in single or both sexes;

(f)   Whether responses are in a single species or several species;

(g)   Structural similarity or not to a substance(s) for which there is good evidence of carcinogenicity;

(h)   Routes of exposure;

(i)   Comparison of absorption, distribution, metabolism and excretion between test animals and humans;

(j)   The possibility of a confounding effect of excessive toxicity at test doses;

(k)   Mode of action and its relevance for humans, such as mutagenicity, cytotoxicity with growth stimulation, mitogenesis, immunosuppression.

Guidance on how to consider important factors in classification of carcinogenicity is included in 3.6.5.3.

3.6.2.5.3   *Mutagenicity:* It is recognized that genetic events are central in the overall process of cancer development. Therefore, evidence of mutagenic activity in vivo may indicate that a substance has a potential for carcinogenic effects.

3.6.2.5.4   The following additional considerations apply to classification of substances into either Category 1 or Category 2. A substance that has not been tested for carcinogenicity may in certain instances be classified in Category 1 or Category 2 based on tumour data from a structural analogue together with substantial support from consideration of other important factors such as formation of common significant metabolites, e.g. for benzidine congener dyes.

3.6.2.5.5   The classification should also take into consideration whether or not the substance is absorbed by a given route(s); or whether there are only local tumours at the site of administration for the tested route(s), and adequate testing by other major route(s) show lack of carcinogenicity.

3.6.2.4 　証拠の強さには、ヒトおよび動物試験を用いた腫瘍数の計測およびその統計的有意性レベルの決定がかかわっている。ヒトで十分な証拠が得られたなら、ヒトのばく露とがん発生の間の因果関係が証明されるのに対し、動物で十分な証拠が得られたなら、その物質と腫瘍発生率の増加の因果関係が示される。ばく露とがんの間に陽性の関係があれば、ヒトでの限定された証拠が認められることになるが、因果関係を証明することはできない。データより発がん作用が示唆されれば、動物での限定された証拠となるが、それで十分とはならない。ここで用いた「十分」および「限定された」という言葉は、国際がん研究機関（IARC）により定義されていた通りに本書でも使われており、3.6.5.3.1 に概説した。

3.6.2.5 　*追加検討事項（証拠の重み）*：発がん性の証拠の強さの決定以外にも、その物質がヒトで発がん性を示すことについての全体的な可能性に影響するその他の多くの要因を考慮すべきである。この決定に影響する要因をすべて列挙すると非常に多くなるため、ここでは重要なものいくつかについて検討した。

3.6.2.5.1 　こうした要因は、ヒトに対する発がん性の懸念レベルを上昇または低下させるものと見なすことができる。各要因の相対的な重要度は、それぞれに付随している証拠の量および一貫性によって異なる。一般的に、懸念レベルを上げるより下げることの方により完全な情報が要求される。追加検討事項は、腫瘍の知見の評価等において、ケースバイケースで、活用されるべきである。

3.6.2.5.2 　総合的な懸念のレベルを評価する際に考慮される重要な要因をいくつか、下記に示した：

 (a) 　腫瘍の種類およびバックグランド発生率；
 (b) 　複数部位における反応；
 (c) 　病変から悪性腫瘍への進行；
 (d) 　腫瘍発生までの潜伏期間の短縮；

その他懸念レベルを上昇あるいは低下させる可能性のある要因には次のものが含まれる：

 (e) 　反応は雌雄いずれかであるか、または両方で認められるかどうか；
 (f) 　反応は単一種のみであるか、それともいくつかの生物種にも認められるかどうか；
 (g) 　発がん性の明確な証拠がある物質に構造的に類似しているかどうか；
 (h) 　ばく露経路；
 (i) 　試験動物とヒトの間の吸収、分布、代謝および排泄の比較；
 (j) 　試験用量での過剰な毒性作用が交絡要因となっている可能性；
 (k) 　変異原性、成長刺激を伴った細胞毒性、有糸分裂誘発性、免疫抑制などの作用様式およびヒトに対する関連性。

発がん性の分類における重要な因子に関する考え方の手引きは 3.6.5.3 に含まれている。

3.6.2.5.3 　*変異原性*：遺伝子レベルでの変化はがん発生の全体的な過程で中心的役割を占めることが認められている。したがって、*in vivo* での変異原性の証拠があれば、物質が発がん性を有する可能性が示唆される。

3.6.2.5.4 　下記の追加検討事項は、物質を区分1または区分2へ分類する際に適用する。発がん性について試験がなされていない物質は、構造的類似体の腫瘍データに加え、例えばベンジジン系の染料のように共通の重要な代謝物の生成などその他の重要な要因の検討より得られるしっかりした裏付けデータをもとに、区分1または区分2に分類される事例がある。

3.6.2.5.5 　分類に際しては、当該物質が投与経路で吸収されるかどうか、あるいは、試験経路では投与部位のみにしか局所腫瘍が認められないかどうか、更に、その他の主要経路による適切な試験から発がん性はないことが認められているかどうか等についても考慮すべきである。

3.6.2.5.6   It is important that whatever is known of the physico-chemical, toxicokinetic and toxicodynamic properties of the substances, as well as any available relevant information on chemical analogues, i.e. structure activity relationship, is taken into consideration when undertaking classification.

3.6.2.6   It is realized that some regulatory authorities may need flexibility beyond that developed in the hazard classification scheme. For inclusion into Safety Data Sheets, positive results in any carcinogenicity study performed according to good scientific principles with statistically significant results may be considered.

3.6.2.7   The relative hazard potential of a chemical is a function of its intrinsic potency. There is great variability in potency among chemicals, and it may be important to account for these potency differences. The work that remains to be done is to examine methods for potency estimation Carcinogenic potency as used here does not preclude risk assessment. The proceedings of a WHO/IPCS workshop on the *Harmonization of Risk Assessment for Carcinogenicity and Mutagenicity (Germ cells)-A Scoping Meeting (1995, Carshalton, UK)*, points to a number of scientific questions arising for classification of chemicals, e.g. mouse liver tumours, peroxisome proliferation, receptor-mediated reactions, chemicals which are carcinogenic only at toxic doses and which do not demonstrate mutagenicity. Accordingly, there is a need to articulate the principles necessary to resolve these scientific issues which have led to diverging classifications in the past. Once these issues are resolved, there would be a firm foundation for classification of a number of chemical carcinogens.

3.6.3   Classification criteria for mixtures

3.6.3.1   *Classification of mixtures when data are available for the complete mixture*

Classification of mixtures will be based on the available test data of the individual ingredients of the mixture using cut-off values/concentration limits for those ingredients. The classification may be modified on a case-by-case basis based on the available test data for the mixture as a whole. In such cases, the test results for the mixture as a whole must be shown to be conclusive taking into account dose and other factors such as duration, observations and analysis (e.g. statistical analysis, test sensitivity) of carcinogenicity test systems. Adequate documentation supporting the classification should be retained and made available for review upon request.

3.6.3.2   *Classification of mixtures when data are not available for the complete mixture: bridging principles*

3.6.3.2.1   Where the mixture itself has not been tested to determine its carcinogenic hazard, but there are sufficient data on both the individual ingredients and similar tested mixtures to adequately characterize the hazards of the mixture, these data will be used in accordance with the following agreed bridging principles. This ensures that the classification process uses the available data to the greatest extent possible in characterizing the hazards of the mixture without the necessity for additional testing in animals.

3.6.3.2.2   *Dilution*

If a tested mixture is diluted with a diluent that is not expected to affect the carcinogenicity of other ingredients, then the new diluted mixture may be classified as equivalent to the original tested mixture.

3.6.3.2.3   *Batching*

The carcinogenic potential of a tested production batch of a mixture can be assumed to be substantially equivalent to that of another untested production batch of the same commercial product, when produced by or under the control of the same manufacturer unless there is reason to believe there is significant variation in composition such that the carcinogenic potential of the untested batch has changed. If the latter occurs, a new classification is necessary.

3.6.3.2.4   *Substantially similar mixtures*

Given the following:

(a) Two mixtures: (i) A + B;
                   (ii) C + B;

(b) The concentration of carcinogen ingredient B is the same in both mixtures;

(c) The concentration of ingredient A in mixture (i) equals that of ingredient C in mixture (ii);

3.6.2.5.6　分類の際には、さらに、化学的構造類似体に関して利用可能な関連情報、すなわち構造活性相関と同様に、当該物質の物理化学的性質、トキシコキネティクス、トキシコダイナミクスがどの程度解明されているかについても、考慮することが重要である。

3.6.2.6　規制官庁によっては、有害性分類スキームにおいて策定されているものよりも広い柔軟性を要求する。優れた科学的な原則に則って実施された発がん性試験で、統計的に有意である陽性結果が得られたならば、安全データシートへの記載も考慮される場合がある。

3.6.2.7　化学品の相対的な有害性の強さは、その物質固有の特性である。化学品によって特性は大きく異なっており、こうした特性の違いを考慮することが重要な場合もある。こうした特性の推定方法の検討は残された課題である。ここで述べた発がん性特性は、リスク評価を排除するものではない。WHO/IPCS のワークショップ *発がん性と変異原性に関するリスク評価手法の調和——スコーピングのための会合*（1995,Carshalton,UK）において、化学品の分類に関して提起されている種々の科学的疑問、例えば、マウス肝腫瘍、ペルオキシソーム増殖、レセプター介在反応、毒性用量では発がん性であるが変異原性は示さない物質などが指摘されている。したがって、これまで一貫せず様々な分類を行う原因となったこれらの科学的課題を解決するために、必要な原則を明確に示す必要性がある。こうした課題が解決されれば、種々の発がん性化学物質の分類は確たるものとなるであろう。

3.6.3　混合物の分類基準

3.6.3.1　*混合物そのものについて試験データが入手できる場合の混合物の分類*

　混合物の分類は、当該混合物の個々の成分について入手できる試験データに基づき、各成分のカットオフ値/濃度限界を使用して行われる。当該混合物そのものの試験データが入手できる場合には、分類はケースバイケースで修正されることがある。このような場合、混合物そのものの試験結果は、発がん性試験系の用量や、試験期間、観察、分析（例えば、統計分析、試験感度）などの他の要因を考慮した上で確実であることが示されなければならない。分類が適切であることの証拠書類を保持し、要請に応じて示すことができるようにするべきである。

3.6.3.2　*混合物そのものについて試験データが入手できない場合の混合物の分類：つなぎの原則 (bridging principle)*

3.6.3.2.1　混合物そのものについては発がん性を決定する試験はなされていないが、当該混合物の有害性を適切に特定するための、個々の成分および類似の試験された混合物に関して十分なデータがある場合、これらのデータは以下の合意されたつなぎの原則にしたがって使用される。これによって、分類プロセスで動物試験を追加する必要もなく、混合物の有害性判定に入手されたデータを可能な限り最大限に用いることができるようになる。

3.6.3.2.2　*希釈*

　試験された混合物が、他の成分の発がん性に影響を与えないと予想される希釈剤で希釈される場合、新しい希釈された混合物は、試験された元の混合物と同等として分類してもよい。

3.6.3.2.3　*製造バッチ*

　混合物の試験されていない製造バッチに発がん性があるかどうかは、同じ製造者によって、またはその管理下で生産された同じ商品の試験された別のバッチの毒性と本質的に同等とみなすことができる。ただし、試験されていないバッチの発がん性が変化するような有意な組成の変動があると考えられる理由がある場合はこの限りではない。このような場合には、新しい分類を行う必要がある。

3.6.3.2.4　*本質的に類似した混合物*

　次を仮定する：

　　(a)　2 つの混合物：　(i)　　A+B；
　　　　　　　　　　　　(ii)　　C+B；

　　(b)　発がん性物質 B の濃度は、両方の混合物で同じである；

　　(c)　混合物(i)の成分 A の濃度は、混合物(ii)の成分 C の濃度に等しい；

(d) Data on toxicity for A and C are available and substantially equivalent, i.e. they are in the same hazard category and are not expected to affect the carcinogenicity of B.

If mixture (i) or (ii) is already classified based on test data, then the other mixture can be classified in the same hazard category.

3.6.3.3 *Classification of mixtures when data are available for all ingredients or only for some ingredients of the mixture*

The mixture will be classified as a carcinogen when at least one ingredient has been classified as a Category 1 or Category 2 carcinogen and is present at or above the appropriate cut-off value/concentration limit as shown in table 3.6.1 for Category 1 and 2 respectively.

Table 3.6.1: Cut-off values/concentration limits of ingredients of a mixture classified as carcinogen that would trigger classification of the mixture [a]

Ingredient classified as:	Cut-off/concentration limits triggering classification of a mixture as:		
	Category 1 carcinogen		Category 2 carcinogen
	Category 1A	Category 1B	
Category 1A carcinogen	≥ 0.1 %	--	--
Category 1B carcinogen	--	≥ 0.1 %	
Category 2 carcinogen	--	--	≥ 0.1 % (note 1)
			≥ 1.0 % (note 2)

[a] *This compromise classification scheme involves consideration of differences in hazard communication practices in existing systems. It is expected that the number of affected mixtures will be small; the differences will be limited to label warnings; and the situation will evolve over time to a more harmonized approach.*

NOTE 1: *If a Category 2 carcinogen ingredient is present in the mixture at a concentration between 0.1 % and 1 %, every regulatory authority would require information on the SDS for a product. However, a label warning would be optional. Some authorities will choose to label when the ingredient is present in the mixture between 0.1 % and 1 %, whereas others would normally not require a label in this case.*

NOTE 2: *If a Category 2 carcinogen ingredient is present in the mixture at a concentration of ≥ 1 %, both an SDS and a label would generally be expected.*

3.6.4 **Hazard communication**

General and specific considerations concerning labelling requirements are provided in *Hazard communication. Labelling* (chapter 1.4). Annex 1 contains summary tables about classification and labelling. Annex 3 contains examples of precautionary statements and pictograms which can be used where allowed by the competent authority. Table 3.6.2 presents specific label elements for substances and mixtures classified in this hazard class based on the criteria in this chapter.

(d) AとCの毒性に関するデータは利用でき、実質的に同等である、すなわち、AとCは同じ有害性区分に属し、かつ、Bの発がん性に影響を与えることはないと判断される。

混合物(i)または(ii)が既に試験データに基づいて分類されている場合には、他方の混合物は同じ有害性区分に分類することができる。

3.6.3.3 *混合物の全成分についてまたは一部の成分だけについてデータが入手できる場合の混合物の分類*

少なくとも1つの成分が区分1または区分2の発がん性物質として分類され、区分1と2それぞれについて表3.6.1に示したような適切なカットオフ値/濃度限界以上で存在する場合、混合物は、発がん性物質として分類される。

表3.6.1：混合物の分類基準となる発がん性成分のカットオフ値/濃度限界 [a]

成分の分類：	混合物の分類基準となるカットオフ値/濃度限界：		
	区分1 発がん性物質		区分2 発がん性物質
	区分1A	区分1B	
区分1A 発がん性物質	≧0.1%	--	--
区分1B 発がん性物質	--	≧0.1%	--
区分2 発がん性物質	--	--	≧0.1%（注記1）
			≧1.0%（注記2）

[a] この妥協案的分類体系は、既存システムの有害性に関する情報伝達の実施方法の相違を考慮したものである。影響を受ける混合物の数は少ないであろうし、そのシステム間の相違もラベル警告に限られるであろう。また、こうした状況は、時間と共に、より調和した手法に発展していくことが期待される。

注記1：区分2の発がん性物質成分が0.1%と1%の間の濃度で混合物中に存在する場合には、すべての規制官庁は、製品のSDSに情報の記載を要求することになろう。しかしながら、ラベルへの警告表示は任意となろう。一部官庁は成分が0.1%と1%の間で混合物中に存在する場合にラベル表示を選択するであろうが、他の官庁は、通常、このような場合にはラベル表示を要求しないであろう。

注記2：区分2発がん性物質成分が≧1%の濃度で混合物中に存在する場合、一般にSDSとラベルの両方が期待される。

3.6.4 危険有害性情報の伝達

表示要件に関する通則および細則は、*危険有害性に関する情報の伝達：表示*（第1.4章）に定める。附属書1に分類と表示に関する概要表を示す。附属書3には、注意書きおよび所管官庁が許可した場合に使用可能な注意絵表示の例を示す。表3.6.2には、本章の判定基準に基づいてこの危険有害性クラスに分類された物質および混合物の具体的なラベル要素を示す。

Table 3.6.2: Label elements for carcinogenicity

	Category 1 (Category 1A, 1B)	Category 2
Symbol	Health hazard	Health hazard
Signal word	Danger	Warning
Hazard statement	May cause cancer (state route of exposure if it is conclusively proven that no other routes of exposure cause the hazard)	Suspected of causing cancer (state route of exposure if it is conclusively proven that no other routes of exposure cause the hazard)

3.6.5 Decision logic and guidance

The decision logics which follow is not part of the harmonized classification system but is provided here as additional guidance. It is strongly recommended that the person responsible for classification study the criteria before and during use of the decision logic.

3.6.5.1 *Decision logic 3.6.1 for substances*

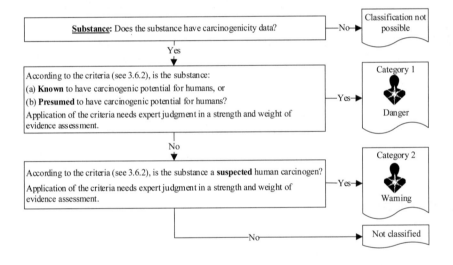

- 201 -

表 3.6.2：発がん性のラベル要素

	区分 1 （区分 1A、1B）	区分 2
シンボル	健康有害性	健康有害性
注意喚起語	危険	警告
危険有害性情報	発がんのおそれ （他の経路からのばく露が有害でないことが決定的に証明されている場合、有害なばく露経路を記載）	発がんのおそれの疑い （他の経路からのばく露が有害でないことが決定的に証明されている場合、有害なばく露経路を記載）

3.6.5 発がん性の判定論理と手引き

以下の判定論理は、調和分類システムには含まれないが、追加的な手引きとしてここに示す。分類の責任者に対し、この判定論理を使用する前および使用する際に、判定基準についてよく調べ理解することを強く勧める。

3.6.5.1 *物質の判定論理 3.6.1*

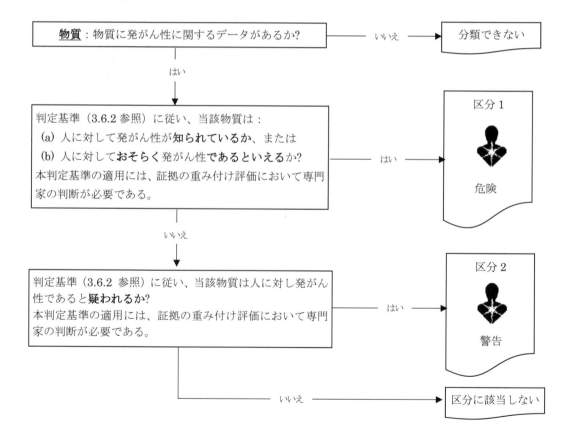

3.6.5.2   *Decision logic 3.6.2 for mixtures*

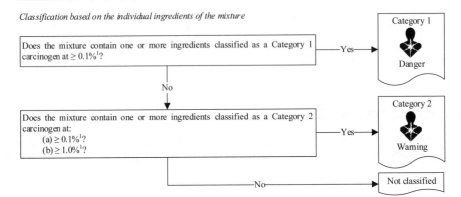

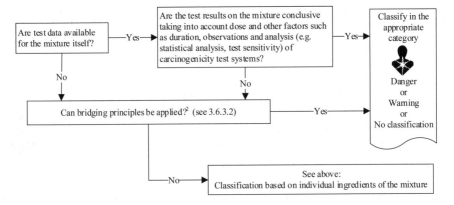

---

[1]   *For specific concentration limits, see "The use of cut-off values/concentration limits" in chapter 1.3, paragraph 1.3.3.2 and in table 3.6.1 of this chapter.*

[2]   *If data of another mixture are used in the application of bridging principles, the data on that mixture must be conclusive in accordance with 3.6.3.2.*

3.6.5.2 混合物の判定論理 3.6.2

> **混合物：**
> 混合物の分類は、成分のカットオフ値/濃度限界を用いて、当該混合物の**個々の成分**の利用可能な試験データに基づいて行われる。混合物そのものについての利用可能な試験データ、あるいはつなぎの原則に基づき、分類は**ケースバイケースで修正**できる。詳細は 3.6.2.7、 3.6.3.1 および 3.6.3.2 を参照のこと。

混合物の個々の成分に基づく分類

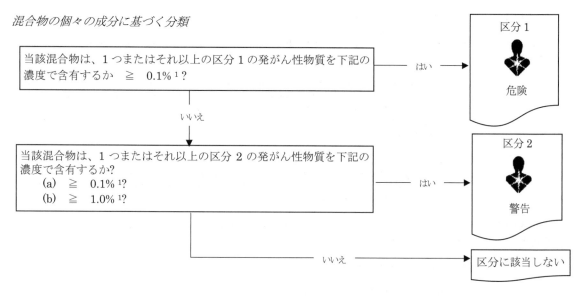

ケースバイケースでの分類の修正

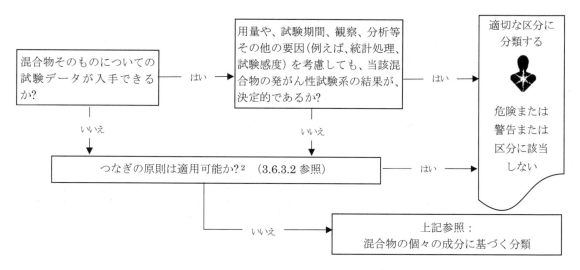

---

[1] 個々の濃度の限度については、第1.3章の1.3.3.2「カットオフ値/濃度限度の使用」および本章の表3.6.1を参照。
[2] 他の混合物のデータをつなぎの原則に用いた場合は、その混合物のデータは3.6.3.2に照らして決定的なものでなければならない。

- 202 -

3.6.5.3 *Background guidance*

3.6.5.3.1 Excerpts[3] from monographs of the International Agency for Research on Cancer (IARC) *Monographs programme on the evaluation of the strength of evidence of carcinogenic risks to humans* follow as in 3.6.5.3.1.1 and 3.6.5.3.1.2[4].

3.6.5.3.1.1 *Carcinogenicity in humans*

3.6.5.3.1.1.1 The evidence relevant to carcinogenicity from studies in humans is classified into one of the following categories:

(a) Sufficient evidence of carcinogenicity: the working group considers that a causal relationship has been established between exposure to the agent, mixture or exposure circumstance and human cancer. That is, a positive relationship has been observed between the exposure and cancer in studies in which chance, bias and confounding could be ruled out with reasonable confidence;

(b) Limited evidence of carcinogenicity: A positive association has been observed between exposure to the agent, mixture or exposure circumstance and cancer for which a causal interpretation is considered by the working group to be credible, but chance, bias or confounding could not be ruled out with reasonable confidence.

3.6.5.3.1.1.2 In some instances the above categories may be used to classify the degree of evidence related to carcinogenicity in specific organs or tissues.

3.6.5.3.1.2 *Carcinogenicity in experimental animals*

The evidence relevant to carcinogenicity in experimental animals is classified into one of the following categories:

(a) Sufficient evidence of carcinogenicity: The working group considers that a causal relationship has been established between the agent or mixture and an increased incidence of malignant neoplasms or of an appropriate combination of benign and malignant neoplasms in (i) two or more species of animals or (ii) in two or more independent studies in one species carried out at different times or in different laboratories or under different protocols;

(b) Exceptionally, a single study in one species might be considered to provide sufficient evidence of carcinogenicity when malignant neoplasms occur to an unusual degree with regard to incidence, site, type of tumour or age at onset;

(c) Limited evidence of carcinogenicity: the data suggest a carcinogenic effect but are limited for making a definitive evaluation because, e.g. (i) the evidence of carcinogenicity is restricted to a single experiment; or (ii) there are unresolved questions regarding the adequacy of the design, conduct or interpretation of the study; or (iii) the agent or mixture increases the incidence only of benign neoplasms or lesions of uncertain neoplastic potential, or of certain neoplasms which may occur spontaneously in high incidences in certain strains.

3.6.5.3.2 *Guidance on how to consider important factors in classification of carcinogenicity**

The guidance provides an approach to analysis rather than hard and fast rules. This section provides some considerations. The weight of evidence assessment called for in GHS is an integrative approach which considers important factors in determining carcinogenic potential along with the strength of evidence analysis. The IPCS "*Conceptual Framework for Evaluating a Mode of Action for Chemical carcinogenesis*" (2001), the International Life Sciences Institute (ILSI) "*Framework for Human Relevance Analysis of Information on Carcinogenic Modes of Action*" (Meek et al., 2003; Cohen et al., 2003, 2004) and the IARC (Preamble section 12(b)) provide a basis for systematic assessments which may be performed in a consistent fashion internationally; the IPCS also convened a panel in 2004 to

---

[3] The excerpts from IARC Monographs, which follow, are taken from the OECD Integrated Document on Harmonization of Classification and Labelling. They are not part of the agreed text on the harmonized classification system developed by the OECD Task Force-HCL but are provided here as additional guidance.

[4] See 3.6.2.4.

3.6.5.3 *基本的手引き*

3.6.5.3.1　国際がん研究機関（International Agency for Research on Cancer：IARC）[3]によるヒトの発がん性リスクの証拠の強さの評価についてのモノグラフからの抜粋を以下の 3.6.5.3.1.1 および 3.6.5.3.1.2 に示す[4]。

3.6.5.3.1.1　*ヒトに対する発がん性*

3.6.5.3.1.1.1　ヒトの調査から得られた発がん性に関する証拠は、次の区分のいずれかに分類される：

(a) 発がん性の十分な証拠：作業部会によれば、物質、混合物、ばく露環境におけるばく露とヒト発がんとの因果関係が確立されたもの。すなわち、偶然性、偏り、交絡要因が排除された研究において、ばく露とがんとの間に陽性の関係が観察されることを意味する；

(b) 発がん性の限定的証拠：物質、混合物に対するばく露、またはばく露環境におけるばく露と発がんとの間に陽性の関連性があると解釈され、その因果関係の信頼性を上記作業部会が認めているが、合理的信頼性を持って、偶然性、偏り、交絡要因が排除されていないものを意味する。

3.6.5.3.1.1.2　一部の事例においては、上記の区分は特定の臓器または組織における発がん性に関する証拠の重要度の分類に使用される場合もある。

3.6.5.3.1.2　*実験動物に対する発がん性*

実験動物に対する発がん性に関する証拠は、次の区分のうちいずれかに分類される：

(a) 発がん性の十分な証拠：作業部会によれば、(i) 複数種の動物または (ii) 1 種の動物に関して別個の時期、別個の研究機関、もしくは別個のプロトコールの下で行われた複数の独立した研究において、当該物質または混合物と悪性新生物、または良性および悪性新生物の特有な組み合わせの発生率の増加との間に因果関係が確立されていること；

(b) 例外的に、単一種の動物に関する単回の研究によっても、悪性新生物の発生率、発生箇所、腫瘍形態、発生時の年齢という観点から見て異常な程度の発生を示す場合は、発がん性の十分な証拠になると考えられる；

(c) 発がん性の限定的証拠：データは発がん作用を示しているが、断定的な評価を下すには限定的である場合、例えば、(i) 発がん性の証拠が単一の実験に限定される場合、(ii) 当該の研究の設計、実施、解釈の妥当性に関して未解決の疑問がある場合、または、(iii) 当該物質または混合物が良性新生物もしくは不特定の新生物性の病変、あるいは一部の系統に高い比率で自然発生することがある特定の新生物の発生数のみを増加させる場合である。

3.6.5.3.2　*発がん性分類における重要な因子についての考え方の手引き**

本手引きは厳密な規則というよりは分析方法を提供するものである。この節ではいくつかの考察について記述している。GHS で求めている証拠の重み付け評価は、証拠の強さの分析とともに発がん性の可能性を決定する重要な因子を検討する包括的な方法である。IPCS の "*Conceptual Framework for Evaluating a Mode of Action for Chemical carcinogenesis*" (2001)、国際生命科学研究所（ILSI）の "*Framework for Human Relevance Analysis of Information on Carcinogenic Modes of Action*" (Meek et al., 2003; Cohen et al., 2003, 2004) および IARC（前文 12(b)）が、国際的に統一された方法で実施可能な系統的な評価の基

---

[3] IARC モノグラフからの抜粋は OECD 分類と表示の調和に関する中間報告から来ているものである。これらは OECD 分類と表示の調和に関するタスクフォースで承認されたテキストではないが、ここで追加的手引きとして記載されている。

[4] 3.6.2.4 を参照

further develop and clarify the human relevance framework. However, the internationally available documents are not intended to dictate answers, nor provide lists of criteria to be checked off.

3.6.5.3.2.1 Mode of action

The various international documents on carcinogen assessment all note that mode of action in and of itself, or consideration of comparative metabolism, should be evaluated on a case-by-case basis and are part of an analytic evaluative approach. One must look closely at any mode of action in animal experiments taking into consideration comparative toxicokinetics/toxicodynamics between the animal test species and humans to determine the relevance of the results to humans. This may lead to the possibility of discounting very specific effects of certain types of chemicals. Life stage-dependent effects on cellular differentiation may also lead to qualitative differences between animals and humans. Only if a mode of action of tumour development is conclusively determined not to be operative in humans may the carcinogenic evidence for that tumour be discounted. However, a weight of evidence assessment for a substance calls for any other tumorigenic activity to be evaluated as well.

3.6.5.3.2.2 Responses in multiple animal experiments

Positive responses in several species add to the weight of evidence, that a chemical is a carcinogen. Taking into account all of the factors listed in 3.6.2.5.2 and more, such chemicals with positive outcomes in two or more species would be provisionally considered to be classified in GHS Category 1B until human relevance of animal results are assessed in their entirety. It should be noted, however, that positive results for one species in at least two independent studies, or a single positive study showing unusually strong evidence of malignancy may also lead to Category 1B.

3.6.5.3.2.3 Responses are in one sex or both sexes

Any case of gender-specific tumours should be evaluated in light of the total tumorigenic response to the substance observed at other sites (multi-site responses or incidence above background) in determining the carcinogenic potential of the substance.

If tumours are seen only in one sex of an animal species, the mode of action should be carefully evaluated to see if the response is consistent with the postulated mode of action. Effects seen only in one sex in a test species may be less convincing than effects seen in both sexes, unless there is a clear patho-physiological difference consistent with the mode of action to explain the single sex response.

3.6.5.3.2.4 Confounding effects of excessive toxicity or localized effects

Tumours occurring only at excessive doses associated with severe toxicity generally have doubtful potential for carcinogenicity in humans. In addition, tumours occurring only at sites of contact and/or only at excessive doses need to be carefully evaluated for human relevance for carcinogenic hazard. For example, forestomach tumours, following administration by gavage of an irritating or corrosive, non-mutagenic chemical, may be of questionable relevance. However, such determinations must be evaluated carefully in justifying the carcinogenic potential for humans; any occurrence of other tumours at distant sites must also be considered.

3.6.5.3.2.5 Tumour type, reduced tumour latency

Unusual tumour types or tumours occurring with reduced latency may add to the weight of evidence for the carcinogenic potential of a substance, even if the tumours are not statistically significant.

Toxicokinetic behaviour is normally assumed to be similar in animals and humans, at least from a qualitative perspective. On the other hand, certain tumour types in animals may be associated with toxicokinetics or toxicodynamics that are unique to the animal species tested and may not be predictive of carcinogenicity in humans. Very few such examples have been agreed internationally. However, one example is the lack of human relevance of kidney tumours in male rats associated with compounds causing $\alpha 2u$-globulin nephropathy (IARC, Scientific Publication N° 147). Even when a particular tumour type may be discounted, expert judgment must be used in assessing the total tumour profile in any animal experiment.

礎を提供している。IPCS はまた 2004 年にも、ヒトに関連した枠組みをさらに発展させ明確にするためのパネルを開催した。しかし国際的に利用可能な文書は、答えを与えることを意図したものでも、照合されるべき判断基準のリストを提供するものでもない。

3.6.5.3.2.1　作用様式

　発がん性評価に関する種々の国際的な文書はすべて、作用様式それ自体あるいは比較代謝の検討はケースバイケースで評価されるべきで、これらは分析評価方法の一部であるとしている。実験結果のヒトに対する妥当性を調べるために、実験動物種とヒトの間の比較トキシコキネティクス/トキシコダイナミクス（毒物動態/毒性動力学）を考慮しながら、動物実験におけるあらゆる作用様式も注意深く観察しなければならない。これによりある種の化学品が持つ非常に特異的な影響が重要ではないとみなされる可能性もあろう。細胞分化に与えるライフステージ依存的な影響はヒトと動物の質的な違いにつながる可能性もある。腫瘍発生のある作用様式がヒトでは機能しないと確定される場合に限り、その腫瘍に対する発がんの証拠は重要ではないとみなされるであろう。しかし化学品に対する証拠の重み付け評価は腫瘍形成に関与する他のいかなる作用も同様に評価することを求めている。

3.6.5.3.2.2　複数の動物実験での反応

　いくつかの動物種での陽性反応は、化学品が発がん性であるという、証拠の重み付けに加わる。3.6.2.5.2 のリストに含まれる全ての要因以上のものを考慮に入れて、2 つ以上の動物種で陽性結果を示す化学品は、動物試験の結果のヒトへの関連性が完全に評価されるまでは、暫定的に GHS 区分 1B に分類してよいであろう。一方、少なくとも 2 つの独立した研究における一種での陽性結果、あるいは悪性度の極めて強い証拠を示す 1 つの陽性結果もまた区分 1B になるであろう。

3.6.5.3.2.3　片方あるいは両性での反応

　性特異的腫瘍の場合、当該物質の発がん性を検討する際には、他の部位で観察された全ての腫瘍形成反応（多部位での反応やバックグラウンドを超えた発生率）をふまえて評価しなければならない。

　もし、ある動物種の片方の性でのみ腫瘍が見られた場合には、反応が想定される作用様式と一致しているかどうか、作用様式を注意深く評価しなければならない。1 つの実験種の片性でのみ見られた影響は、片性での反応を説明する作用様式と一致する明らかな病理－生理学的相違がない限り、両性で見られた影響よりも説得力は低いであろう。

3.6.5.3.2.4　過剰な毒性作用あるいは局所作用の交絡的影響

　重篤な毒性を伴う過剰用量でのみ生じる腫瘍は、通常、ヒトに対する発がん性の可能性は疑わしい。加えて、試験物質が直接ばく露する部位のみ、および/あるいは過剰用量でのみで生じる腫瘍も、ヒトでの発がん性の妥当性を注意深く評価する必要がある。例えば、刺激性あるいは腐食性を有するものの変異原性のない化学品の胃チューブによる経口投与によってできた前胃腫瘍は妥当性が疑わしいであろう。しかし、ヒトに対する発がん性の有無を明らかにするには、そのような決定は注意深く行わなければならない。投与遠位部位におけるいかなる他の腫瘍の発生も考慮されなければならない。

3.6.5.3.2.5　腫瘍のタイプ、腫瘍形成時間の減少

　通常見られないタイプの腫瘍あるいは形成までに要する時間が減少した腫瘍は、たとえ腫瘍の発生頻度が統計学的に有意でなかったとしても、化学品の発がん性に対する証拠の重み付けに加えられるであろう。

　通常、トキシコキネティクスは少なくとも質的な観点からは動物とヒトで同じであると仮定されている。一方、動物におけるある種の腫瘍は、試験に用いる動物種に特有なトキシコキネティクスやトキシコダイナミクスに関連している可能性があり、ヒトの発がん性の予測には使えないであろう。非常にわずかながらそのような例が国際的に認められている。1 つの例は、α2u-グロブリン腎障害を誘発する化合物による雄ラットでの腎腫瘍にはヒトでの妥当性がないというものである (IARC, Scientific Publication №147)。特殊な腫瘍タイプを割り引いて考えたとしても、動物実験における腫瘍形成の評価においては専門家の判断がなされなければならない。

* *References:*

Cohen, S.M., J. Klaunig, M.E. Meek, R.N. Hill, T. Pastoor, L. Lehman-McKeeman, J. Bucher, D.G. Longfellow, J. Seed, V. Dellarco, P. Fenner-Crisp, and D. Patton. 2004. Evaluating the human relevance of chemically induced animal tumors. Toxicol. Sci., 78(2): 181-186.

Cohen, S.M., M.E. Mkke, J.E. Klaunig, D.E. Patton, P.A. Fenner-Crisp. 2003. The human relevance of information on carcinogenic modes of action: overview. Crit. Rev. Toxicol. 33(6), 581-9.

Meek, M.E., J.R. Bucher, S.M. Cohen, V. Dellarco, R.N. Hill, L. Lehman-McKeeman, D.G. Longfellow, T. Pastoor, J. Seed, D.E. Patton. 2003. A framework for human relevance analysis of information on carcinogenic modes of action. Crit. Rev.Toxicol., 33(6), 591-653.

Sonich-Mullin, C., R. Fielder, J. Wiltse, K. Baetcke, J. Dempsey, P. Fenner-Crisp, D. Grant, M. Hartley, A. Knapp, D. Kroese, I. Mangelsdorf, E. Meek, J.M. Rice, and M. Younes. 2001. The Conceptual Framework for Evaluating a Mode of Action for Chemical Carcinogenesis. Reg. Tox. Pharm. 34, 146-152.

International Programme on Chemical Safety Harmonization Group. 2004 Report of the First Meeting of the Cancer Working Group. World Health Organization. Report IPCS/HSC-CWG-1/04. Geneva

International Agency for Research on Cancer. IARC Monographs on the Evaluation of Carcinogenic Risks to Human. Preambles to volumes. World Health Organization. Lyon, France.

S.M. Cohen, P.A.Fenner-Crisp, and D.E. Patton. 2003. Special Issue: Cancer Modes of Action and Human Relevance. Critical Reviews in Toxicology, R.O. McClellan, ed., Volume 33/Issue 6. CRC Press.

C.C. Capen, E. Dybing and J.D. Wilbourn. 1999. Species differences in Thyroid, Kidney and Urinary Bladder Carcinogenesis. International Agency for Research on Cancer, Scientific Publication N° 147.

* 参考文献

Cohen, S.M., J. Klaunig, M.E. Meek, R.N. Hill, T. Pastoor, L. Lehman-McKeeman, J.Bucher, D.G. Longfellow, J. Seed, V. Dellarco, P. Fenner-Crisp, and D. Patton. 2004. Evaluating the human relevance of chemically induced animal tumors. Toxicol. Sci., 78(2): 181-186.

Cohen, S.M., M.E. Mkke, J.E. Klaunig, D.E. Patton, P.A. Fenner-Crisp. 2003. The human1 relevance of information on carcinogenic modes of action: overview. Crit. Rev. Toxicol.33(6), 581-9.

Meek, M.E., J.R. Bucher, S.M. Cohen, V. Dellarco, R.N. Hill, L. Lehman-McKeeman, D.G. Longfellow, T. Pastoor, J. Seed, D.E. Patton. 2003. A framework for human relevance analysis of information on carcinogenic modes of action. Crit. Rev.Toxicol.,33(6), 591-653.

Sonich-Mullin, C., R. Fielder, J. Wiltse, K. Baetcke, J. Dempsey, P. Fenner-Crisp, D .Grant, M. Hartley, A. Knapp, D. Kroese, I. Mangelsdorf, E. Meek, J.M. Rice, and M.Younes. 2001. The Conceptual Framework for Evaluating a Mode of Action for Chemical Carcinogenesis. Reg. Tox. Pharm. 34, 146-152.

International Programme on Chemical Safety Harmonization Group. 2004 Report of the First Meeting of the Cancer Working Group. World Health Organization. Report IPCS/HSC-CWG-1/04. Geneva

International Agency for Research on Cancer. IARC Monographs on the Evaluation of Carcinogenic Risks to Human. Preambles to volumes. World Health Organization. Lyon, France.

S.M. Cohen, P.A.Fenner-Crisp, and D.E. Patton. 2003. Special Issue: Cancer Modes of Action and Human Relevance. Critical Reviews in Toxicology, R.O. McClellan, ed., Volume 33/Issue 6. CRC Press.

C.C. Capen, E. Dybing and J.D. Wilbourn. 1999. Species differences in Thyroid, Kidney and Urinary Bladder Carcinogenesis. International Agency for Research on Cancer, Scientific Publication N° 147.

CHAPTER 3.7

REPRODUCTIVE TOXICITY

3.7.1 Definitions and general considerations

3.7.1.1 *Reproductive toxicity*

Reproductive toxicity refers to adverse effects on sexual function and fertility in adult males and females, as well as developmental toxicity in the offspring, occurring after exposure to a substance or mixture. The definitions presented below are adapted from those agreed as working definitions in IPCS/EHC Document N°225 *Principles for evaluating health risks to reproduction associated with exposure to chemicals*. For classification purposes, the known induction of genetically based inheritable effects in the offspring is addressed in *Germ cell mutagenicity* (chapter 3.5), since in the present classification system it is considered more appropriate to address such effects under the separate hazard class of germ cell mutagenicity.

In this classification system, reproductive toxicity is subdivided under two main headings:

(a) Adverse effects on sexual function and fertility;

(b) Adverse effects on development of the offspring.

Some reproductive toxic effects cannot be clearly assigned to either impairment of sexual function and fertility or to developmental toxicity. Nonetheless, substances and mixtures with these effects would be classified as reproductive toxicants with a general hazard statement.

3.7.1.2 *Adverse effects on sexual function and fertility*

Any effect of chemicals that would interfere with sexual function and fertility. This may include, but not be limited to, alterations to the female and male reproductive system, adverse effects on onset of puberty, gamete production and transport, reproductive cycle normality, sexual behaviour, fertility, parturition, pregnancy outcomes, premature reproductive senescence, or modifications in other functions that are dependent on the integrity of the reproductive systems.

Adverse effects on or via lactation are also included in reproductive toxicity, but for classification purposes, such effects are treated separately (see 3.7.2.1). This is because it is desirable to be able to classify chemicals specifically for an adverse effect on lactation so that a specific hazard warning about this effect can be provided for lactating mothers.

3.7.1.3 *Adverse effects on development of the offspring*

Taken in its widest sense, developmental toxicity includes any effect which interferes with normal development of the conceptus, either before or after birth, and resulting from exposure of either parent prior to conception, or exposure of the developing offspring during prenatal development, or postnatally, to the time of sexual maturation. However, it is considered that classification under the heading of developmental toxicity is primarily intended to provide a hazard warning for pregnant women and men and women of reproductive capacity. Therefore, for pragmatic purposes of classification, developmental toxicity essentially means adverse effects induced during pregnancy, or as a result of parental exposure. These effects can be manifested at any point in the life span of the organism. The major manifestations of developmental toxicity include death of the developing organism, structural abnormality, altered growth and functional deficiency.

第 3.7 章

生殖毒性

3.7.1 定義および一般事項

3.7.1.1 *生殖毒性*

*生殖毒性*とは、物質または混合物へのばく露後におこる、雌雄の成体の性機能および生殖能に対する悪影響に加えて、子世代における発生毒性をさす。下記に示された定義は、IPCS/EHC の文書番号 225、化学品へのばく露と関連する生殖に対する健康リスクの評価原則における仮の定義にしたがって作成したものである。分類という目的から、遺伝子要因に基づく子への遺伝的影響の誘発については、生殖細胞に対する変異原性という別の有害性クラスの方がより適切であると思われるため、*生殖細胞変異原性*（第 3.5 章）に示してある。

本分類システムでは、生殖毒性は以下の 2 つの主項目に分けられている：

(a) 性機能および生殖能に対する悪影響；

(b) 子の発生に対する悪影響。

ある種類の生殖毒性の影響は、性機能および生殖能の損傷によるものであるか、または発生毒性によるものであるか明確に評価することはできない。それにもかかわらず、これらの影響を持つ物質および混合物は、一般的な危険有害性情報には生殖毒性物質と分類されるであろう。

3.7.1.2 *性機能および生殖能に対する悪影響*

化学品による性機能および生殖能を阻害するあらゆる影響。これには雌雄生殖器官の変化、生殖可能年齢の開始時期、配偶子の生成および移動、生殖周期の正常性、性的行動、受精能/受胎能、分娩、妊娠の予後に対する悪影響、生殖機能の早期老化、または正常な生殖系に依存する他の機能における変化などが含まれるが、必ずしもこれらに限られるわけではない。

授乳に対するまたは授乳を介した影響も生殖毒性に含められるが、この分類においては、別に扱っている（3.7.2.1 を参照）。なぜならば、特に授乳に対して悪影響を及ぼす化学品を分類することは、授乳中の母親に対して有害性情報を提供するためにも望ましいからである。

3.7.1.3 *子の発生に対する悪影響*

発生毒性を広義にとらえると、胎盤、胎児あるいは生後の子の正常な発生を妨害するあらゆる作用が含まれる。それは受胎の前のいずれかの親のばく露、胎児期における発生中の胎児のばく露、あるいは出生後の性的成熟期までのばく露によるものがある。ただし、発生毒性という分類においては、妊娠女性および生殖能のある男女に対して有害性警告を提供することを第一の目的としていると考えることができる。したがって、分類するという目的のために、発生毒性とは本質的に妊娠中または親のばく露によって誘発される悪影響をいう。このような影響は、その生体の生涯のいかなる時点においても発現され得る。発生毒性の発現には主として、発生中の生体の死亡、構造異常、生育異常、および機能不全が含まれる。

3.7.2 Classification criteria for substances

3.7.2.1 *Hazard categories*

For the purpose of classification for reproductive toxicity, substances are allocated to one of two categories. Effects on sexual function and fertility, and on development, are considered. In addition, effects on lactation are allocated to a separate hazard category.

Figure 3.7.1 (a): Hazard categories for reproductive toxicants

CATEGORY 1:	**Known or presumed human reproductive toxicant**
	This category includes substances which are known to have produced an adverse effect on sexual function and fertility or on development in humans or for which there is evidence from animal studies, possibly supplemented with other information, to provide a strong presumption that the substance has the capacity to interfere with reproduction in humans. For regulatory purposes, a substance can be further distinguished on the basis of whether the evidence for classification is primarily from human data (Category 1A) or from animal data (Category 1B).
CATEGORY 1A:	**Known human reproductive toxicant**
	The placing of the substance in this category is largely based on evidence from humans.
CATEGORY 1B:	**Presumed human reproductive toxicant**
	The placing of the substance in this category is largely based on evidence from experimental animals. Data from animal studies should provide clear evidence of an adverse effect on sexual function and fertility or on development in the absence of other toxic effects, or if occurring together with other toxic effects the adverse effect on reproduction is considered not to be a secondary non-specific consequence of other toxic effects. However, when there is mechanistic information that raises doubt about the relevance of the effect for humans, classification in Category 2 may be more appropriate.
CATEGORY 2:	**Suspected human reproductive toxicant**
	This category includes substances for which there is some evidence from humans or experimental animals, possibly supplemented with other information, of an adverse effect on sexual function and fertility, or on development, in the absence of other toxic effects, or if occurring together with other toxic effects the adverse effect on reproduction is considered not to be a secondary non-specific consequence of the other toxic effects, and where the evidence is not sufficiently convincing to place the substance in Category 1. For instance, deficiencies in the study may make the quality of evidence less convincing, and in view of this Category 2 could be the more appropriate classification.

Figure 3.7.1 (b): Hazard category for effects on or via lactation

EFFECTS ON OR VIA LACTATION

Effects on or via lactation are allocated to a separate single category. It is appreciated that for many substances there is no information on the potential to cause adverse effects on the offspring via lactation. However, substances which are absorbed by women and have been shown to interfere with lactation, or which may be present (including metabolites) in breast milk in amounts sufficient to cause concern for the health of a breastfed child, should be classified to indicate this property hazardous to breastfed babies. This classification can be assigned on the basis of:

(a) absorption, metabolism, distribution and excretion studies that would indicate the likelihood the substance would be present in potentially toxic levels in breast milk; and/or

(b) results of one or two generation studies in animals which provide clear evidence of adverse effect in the offspring due to transfer in the milk or adverse effect on the quality of the milk; and/or

(c) human evidence indicating a hazard to babies during the lactation period.

3.7.2 物質の分類基準

3.7.2.1 *有害性区分*

生殖毒性の分類目的に照らし、物質は2つの区分のうちの1つに割り当てられる。性機能および生殖能に対する作用に加えて、発生に対する作用も考慮の対象となる。更に、授乳に対する影響については、別の有害性区分が割り当てられている。

図 3.7.1(a)：生殖毒性物質の有害性区分

区分1：ヒトに対して生殖毒性があることが知られている、あるいはあると考えられる物質

この区分には、ヒトの性機能および生殖能あるいは発生に悪影響を及ぼすことが知られている物質、またはできれば他の補足情報もあることが望ましいが、動物試験によりその物質がヒトの生殖を阻害する可能性があることが強く推定される物質が含まれる。規制のためには、分類のための証拠が主としてヒトのデータによるものか（区分1A）、あるいは動物データによるものなのか（区分1B）によってさらに区別することもできる。

区分1A：ヒトに対して生殖毒性があることが知られている物質

この区分への物質の分類は、主にヒトにおける証拠をもとにして行われる。

区分1B：ヒトに対して生殖毒性があると考えられる物質

この区分への物質の分類は、主に実験動物による証拠をもとにして行われる。動物実験より得られたデータは、他の毒性作用のない状況で性機能および生殖能または発生に対する悪影響の明確な証拠を提供しているべきであるが、他の毒性作用も同時に生じている場合には、その生殖に対する悪影響が、他の毒性作用が原因となった二次的な非特異的影響ではないと考えられる。ただし、ヒトに対する影響の妥当性について疑いが生じるようなメカニズムに関する情報がある場合には、区分2に分類する方がより適切である。

区分2：ヒトに対する生殖毒性が疑われる物質

この区分に分類するのは次のような物質である。できれば他の補足情報もあることが望ましいが、ヒトまたは実験動物から、他の毒性作用のない状況で性機能および生殖能あるいは発生に対する悪影響についてある程度の証拠が得られている物質、または、他の毒性作用も同時に生じている場合には、他の毒性作用が原因となった二次的な非特異的影響ではないと考えられるものの、当該物質を区分1に分類するにはまだ証拠が十分でないような物質。例えば、試験に欠陥があり、証拠の信頼性が低いため、区分2とした方がより適切な分類であると思われる場合がある。

図 3.7.1(b)：授乳影響の有害性区分

授乳に対するまたは授乳を介した影響

授乳に対するまたは授乳を介した影響は別の区分に振り分けられる。多くの物質には、授乳によって子に悪影響を及ぼす可能性についての情報がないことが認められている。ただし、女性によって吸収され、母乳分泌に影響を与える、または授乳中の子供の健康に懸念をもたらすに十分な量で母乳中に存在すると思われる物質（代謝物も含めて）は、哺乳中の乳児に対するこの有害性に分類して示すべきである。この分類は下記の事項をもとに指定される：

(a) 吸収、代謝、分布および排泄に関する試験で、当該物質が母乳中で毒性を持ちうる濃度で存在する可能性が認められた場合；および/または
(b) 動物を用いた一世代または二世代試験の結果より、母乳中への移行による子への悪影響または母乳の質に対する悪影響の明らかな証拠が得られた場合；および/または
(c) 授乳期間中の乳児に対する有害性を示す証拠がヒトで得られた場合。

3.7.2.2 Basis of classification

3.7.2.2.1 Classification is made on the basis of the appropriate criteria, outlined above, and a total weight of evidence assessment. Classification as a reproductive toxicant is intended to be used for chemicals which have an intrinsic, specific property to produce an adverse effect on reproduction and chemicals should not be so classified if such an effect is produced solely as a non-specific secondary consequence of other toxic effects.

3.7.2.2.2 In the evaluation of toxic effects on the developing offspring, it is important to consider the possible influence of maternal toxicity.

3.7.2.2.3 For human evidence to provide the primary basis for a Category 1A classification there must be reliable evidence of an adverse effect on reproduction in humans. Evidence used for classification should ideally be from well conducted epidemiological studies which include the use of appropriate controls, balanced assessment, and due consideration of bias or confounding factors. Less rigorous data from studies in humans should be supplemented with adequate data from studies in experimental animals and classification in Category 1B should be considered.

3.7.2.3 Weight of evidence

3.7.2.3.1 Classification as a reproductive toxicant is made on the basis of a total weight of evidence assessment. This means that all available information that bears on the determination of reproductive toxicity is considered together. Included is information such as epidemiological studies and case reports in humans and specific reproduction studies along with sub-chronic, chronic and special study results in animals that provide relevant information regarding toxicity to reproductive and related endocrine organs. Evaluation of substances chemically related to the material under study may also be included, particularly when information on the material is scarce. The weight given to the available evidence will be influenced by factors such as the quality of the studies, consistency of results, nature and severity of effects, level of statistical significance for intergroup differences, number of endpoints affected, relevance of route of administration to humans and freedom from bias. Both positive and negative results are assembled together into a weight of evidence determination. However, a single, positive study performed according to good scientific principles and with statistically or biologically significant positive results may justify classification (see also 3.7.2.2.3).

3.7.2.3.2 Toxicokinetic studies in animals and humans, site of action and mechanism or mode of action study results may provide relevant information, which could reduce or increase concerns about the hazard to human health. If it can be conclusively demonstrated that the clearly identified mechanism or mode of action has no relevance for humans or when the toxicokinetic differences are so marked that it is certain that the hazardous property will not be expressed in humans then a substance which produces an adverse effect on reproduction in experimental animals should not be classified.

3.7.2.3.3 In some reproductive toxicity studies in experimental animals the only effects recorded may be considered of low or minimal toxicological significance and classification may not necessarily be the outcome. These include for example small changes in semen parameters or in the incidence of spontaneous defects in the foetus, small changes in the proportions of common foetal variants such as are observed in skeletal examinations, or in foetal weights, or small differences in postnatal developmental assessments.

3.7.2.3.4 Data from animal studies ideally should provide clear evidence of specific reproductive toxicity in the absence of other, systemic, toxic effects. However, if developmental toxicity occurs together with other toxic effects in the dam, the potential influence of the generalized adverse effects should be assessed to the extent possible. The preferred approach is to consider adverse effects in the embryo/foetus first, and then evaluate maternal toxicity, along with any other factors, which are likely to have influenced these effects, as part of the weight of evidence. In general, developmental effects that are observed at maternally toxic doses should not be automatically discounted. Discounting developmental effects that are observed at maternally toxic doses can only be done on a case-by-case basis when a causal relationship is established or refuted.

3.7.2.3.5 If appropriate information is available it is important to try to determine whether developmental toxicity is due to a specific maternally mediated mechanism or to a non-specific secondary mechanism, like maternal stress and the disruption of homeostasis. Generally, the presence of maternal toxicity should not be used to negate findings of embryo/foetal effects, unless it can be clearly demonstrated that the effects are secondary non-specific effects. This is especially the case when the effects in the offspring are significant, e.g. irreversible effects such as structural malformations. In some situations it is reasonable to assume that reproductive toxicity is due to a secondary consequence of maternal toxicity and discount the effects, for example if the chemical is so toxic that dams fail to thrive and there is severe inanition; they are incapable of nursing pups; or they are prostrate or dying.

3.7.2.2　分類の根拠

3.7.2.2.1　分類は、上記に概略を記した適切な判定基準、および包括的な証拠の重み付け評価をもとに行われる。生殖毒性物質としての分類は、生殖に対して、固有かつ特異的な性質の有害影響をもたらす化学品に適用されることを目的としており、もしそのような影響が単に他の毒性作用の非特異的な二次的影響として誘発されたにすぎないならば、化学品をそのように分類すべきではない。

3.7.2.2.2　発生中の子に対する毒性作用の評価では、母体に対する毒性が影響を及ぼしている可能性についても考慮することが重要である。

3.7.2.2.3　区分 1A 分類の重要な根拠となる、ヒトで得られた証拠は、ヒトの生殖に対する有害影響を示す信頼性のある証拠でなくてはならない。分類に用いる証拠は、理想的には、適切な対照群を設け、バランスのとれた評価が行われ、偏りまたは交絡要因について当然払うべき注意が払われているような、入念に実施された疫学的調査より得られたものにすべきである。ヒトから得られても厳密性を欠くデータは、実験動物を用いた試験により得られた十分なデータで補足すべきであり、区分 1B への分類も考えるべきである。

3.7.2.3　証拠の重み

3.7.2.3.1　生殖毒性物質としての分類は、包括的な証拠の重み付け評価をよりどころとして行われる。これはすなわち、生殖毒性の決定に関わるすべての利用可能な情報が一括して考慮されることを意味している。これには、ヒトでの疫学的調査や症例報告と共に、動物を用いた亜慢性、慢性および特定試験で生殖器官ならびに関連内分泌器官に対する毒性関連情報が得られる特異的生殖試験の結果も含まれる。当該物質自体に関する情報がわずかしかない場合には、試験対象である物質と化学的に関連性のある物質の評価も含まれることもある。利用可能な証拠に対する重みは、試験の質、結果の一貫性、作用の特徴および重篤度、群間差の統計的有意性のレベル、影響を受けるエンドポイントの数、投与経路がヒトとの関連性で妥当であるかどうか、および偏りが排除されているかによって異なってくる。陽性結果と陰性結果の両者を組み合わせて、証拠の重みが決定される。単一の陽性試験であっても、優れた科学的原則にしたがって実施され、また、統計的または生物学的に有意な陽性結果が得られたものならば、分類の正当性の判断理由となりうる（3.7.2.2.3 も参照）。

3.7.2.3.2　動物およびヒトでのトキシコキネティクスの試験、作用部位および作用機序または作用様式の試験結果からも関連情報が得られることがあり、これによってヒトの健康に対する有害性に関する懸念が増えることもあれば減ることもある。もし、作用機序または作用様式が明らかに特定され、それがヒトには関係ないことが最終的に実証されるならば、またはトキシコキネティクスの違いが著しく異なるためにヒトではこの有害性が発現されないことを明確に示すことができるならば、実験動物で生殖に有害影響を及ぼす物質であっても分類すべきでない。

3.7.2.3.3　実験動物を用いた生殖試験で、記録された作用が、毒性学的な重要性が低いかまたは最小限なものしかないと見なされるならば、必ずしも結果的に分類されるとは限らない。そうした作用の例として、例えば精液に関する測定項目のわずかな変化、または胎児の偶発的異常の発生率のわずかな変化、例えば骨格検査で測定されるような一般的な胎児奇形または胎児体重の比率のわずかな変化、または出生後の発生評価結果のわずかな違いなどがある。

3.7.2.3.4　動物試験より得られたデータは、理想的には、特異的な生殖毒性の明確な証拠を、その他の全身毒性を伴わない状況で示すべきである。ただし、発生毒性が母動物におけるその他の毒性影響と同時に起きる場合には、総合的な有害作用の潜在的影響について、できる限り評価すべきである。まず胚または胎児における有害影響を検討し、ついで母動物に対する毒性を評価し、こうした有害影響に影響していると思われるようなその他の要因も合わせて、証拠の重みの一部として評価することが望ましい方法である。一般的に、母動物に毒性を示す用量において認められる発生毒性を機械的に無視してしまうべきでない。母動物に毒性を示す用量で認められる発生毒性を割り引いて考えてよいのは、因果関係が確認されているまたは否定されている時にケースバイケースで判断する場合にのみ可能となる。

3.7.2.3.5　適切な情報が入手されたならば、発生毒性が、母動物の介在する特異的メカニズムによるものなのか、それとも例えば母動物のストレスやホメオスタシスのかく乱のような非特異的な二次的メカニズムによるものなのかを判断するよう試みることが重要である。一般的に、胚または胎児に対する影響が二次的な非特異的影響であることが明確に実証されない限り、母体に対する毒性があることを胚または胎児に対する影響の知見を否定するのに用いるべきではない。特に子における影響が明確である場合、例えば奇形のような非可逆的影響である場合にこれが当てはまる。また状況によっては、生殖毒性が母体に対する毒性の二次的結果であるとして、胚または胎児に対する作用を割り引いて考えることが合理的であることもある。例えば、その化学品の毒性が極めて強いために母動物が生長できず、重度の栄養障害があり子の哺育ができない、または衰弱したり瀕死の状態であったりする場合などである。

3.7.2.4 *Maternal toxicity*

3.7.2.4.1 Development of the offspring throughout gestation and during the early postnatal stages can be influenced by toxic effects in the mother either through non-specific mechanisms related to stress and the disruption of maternal homeostasis, or by specific maternally-mediated mechanisms. So, in the interpretation of the developmental outcome to decide classification for developmental effects it is important to consider the possible influence of maternal toxicity. This is a complex issue because of uncertainties surrounding the relationship between maternal toxicity and developmental outcome. Expert judgement and a weight of evidence assessment, using all available studies, should be used to determine the degree of influence that should be attributed to maternal toxicity when interpreting the criteria for classification for developmental effects. The adverse effects in the embryo/foetus should be first considered, and then maternal toxicity, along with any other factors which are likely to have influenced these effects, as weight of evidence, to help reach a conclusion about classification.

3.7.2.4.2 Based on pragmatic observation, it is believed that maternal toxicity may, depending on severity, influence development via non-specific secondary mechanisms, producing effects such as depressed foetal weight, retarded ossification, and possibly resorptions and certain malformations in some strains of certain species. However, the limited numbers of studies which have investigated the relationship between developmental effects and general maternal toxicity have failed to demonstrate a consistent, reproducible relationship across species. Developmental effects, which occur even in the presence of maternal toxicity are considered to be evidence of developmental toxicity, unless it can be unequivocally demonstrated on a case by case basis that the developmental effects are secondary to maternal toxicity. Moreover, classification should be considered where there is significant toxic effect in the offspring, e.g. irreversible effects such as structural malformations, embryo/foetal lethality, significant post-natal functional deficiencies.

3.7.2.4.3 Classification should not automatically be discounted for chemicals that produce developmental toxicity only in association with maternal toxicity, even if a specific maternally-mediated mechanism has been demonstrated. In such a case, classification in Category 2 may be considered more appropriate than Category 1. However, when a chemical is so toxic that maternal death or severe inanition results, or the dams are prostrate and incapable of nursing the pups, it may be reasonable to assume that developmental toxicity is produced solely as a secondary consequence of maternal toxicity and discount the developmental effects. Classification may not necessarily be the outcome in the case of minor developmental changes e.g. small reduction in foetal/pup body weight, retardation of ossification when seen in association with maternal toxicity.

3.7.2.4.4 Some of the end-points used to assess maternal toxicity are provided below. Data on these end points, if available, need to be evaluated in light of their statistical or biological significance and dose response relationship.

(a) Maternal mortality: an increased incidence of mortality among the treated dams over the controls should be considered evidence of maternal toxicity if the increase occurs in a dose-related manner and can be attributed to the systemic toxicity of the test material. Maternal mortality greater than 10 % is considered excessive and the data for that dose level should not normally be considered for further evaluation.

(b) Mating index (N° animals with seminal plugs or sperm/N° mated × 100)[1]

(c) Fertility index (N° animals with implants/N° of matings × 100)[1]

(d) Gestation length (if allowed to deliver)

(e) Body weight and body weight change: consideration of the maternal body weight change and/or adjusted (corrected) maternal body weight should be included in the evaluation of maternal toxicity whenever such data are available. The calculation of an adjusted (corrected) mean maternal body weight change, which is the difference between the initial and terminal body weight minus the gravid uterine weight (or alternatively, the sum of the weights of the foetuses), may indicate whether the effect is maternal or intrauterine. In rabbits, the body weight gain may not be useful indicators of maternal toxicity because of normal fluctuations in body weight during pregnancy.

(f) Food and water consumption (if relevant): the observation of a significant decrease in the average food or water consumption in treated dams compared to the control group may be useful in evaluating maternal toxicity, particularly when the test material is administered in the diet or drinking water. Changes in food or water consumption should be evaluated in conjunction with

---

[1] It is recognized that this index can also be affected by the male.

3.7.2.4　母体に対する毒性

3.7.2.4.1　妊娠期間中から出生後の早期段階に至るまでの子の発達は、ストレスおよび母体のホメオスタシスのかく乱に関係した非特異的メカニズム、または母体が介在する特異的メカニズムを通して、母体における毒性作用に影響されうる。そのため、発生毒性に関する分類決定のために発生の結果を解釈する際には、母体に対する毒性が影響している可能性を考慮することが重要である。このことは、母体に対する毒性と発生への影響の関係が明らかでないために、困難な問題である。発生毒性作用に関する分類のための判定基準を解釈する場合、母体の毒性に帰すべき影響の程度を決定するために、利用可能なあらゆるデータを用い、専門家の判断と証拠の重み付け評価を利用すべきである。まず胚または胎児に対する有害影響を検討し、次に母体に対する毒性に加え、こうした作用に影響する可能性があると思われるその他の要因があれば、証拠の重み付けとして、分類に関する結論に到達するのに役立てるべきである。

3.7.2.4.2　実際上の所見をもとに、母体に対する毒性は、その重篤度にもよるが、非特異的な二次的メカニズムによって発生に影響を及ぼし、胎児体重増加抑制、骨化遅延、ならびにある生物種の系統において組織吸収や奇形等の影響を誘発すると考えられている。しかしながら、発生に対する影響と母体に対する一般的な毒性の関連性を検討している限られた研究においても、種間における一貫した、再現性のある関連性を実証できていない。母体に対する毒性があったとしても発生に対する影響が認められた場合、その発生に対する作用がケースバイケースで母体に対する毒性の2次作用であると確実に実証されない限り、発生毒性の証拠であると見なされる。さらに、子に明確な毒性作用、例えば奇形、胚または胎児致死、出生後の明確な機能障害等の不可逆的作用などが認められる場合には、分類することを検討すべきである。

3.7.2.4.3　母体に対する毒性との関連性によってのみ発生毒性を生じるような化学品については、たとえ母体が介在する特異的メカニズムが示されているとしても、分類を機械的に否定すべきでない。そうした場合には、区分1に分類するより区分2に分類する方がふさわしいと考えられることもある。ただし、化学品の毒性がきわめて強いために母動物が死亡したり重度の栄養失調となるか、または母動物が衰弱して子の哺育ができない場合には、発生毒性は単に母体毒性に誘発された二次的結果にすぎないと推測して、発生影響を無視する方が合理的である。例えば、胎児または子の体重のわずかな低下や骨化の遅延などが母体に対する毒性との関連性で観察される場合には、必ずしも分類を行う必要はない。

3.7.2.4.4　母体に対する毒性評価に用いられる影響のいくつかを以下に示す。これらの影響に関するデータが利用可能であれば、その統計的または生物学的有意性ならびに用量反応関係に照らして評価する必要がある。

(a) 母体の死亡：対照群と比べて投与群母動物の死亡率が増加した場合、その増加に用量依存性があるならば、これは母体に対する毒性の証拠であると見なされる必要があり、被験物質の全身毒性を表すものとされる。母動物の死亡率が10％を超えているならば過度であると見なされ、その用量レベルで得られたデータは通常、それより先の評価に考慮されるべきではない。

(b) 交尾率（交尾栓または精子が認められた動物数/交配した動物数×100）[1]

(c) 受胎率（着床が認められた動物数/交尾動物数×100）[1]

(d) 妊娠期間（出産に至る場合）

(e) 体重および体重変化：母動物の体重変化または調整（補正）後の母体体重に関するデータが利用可能であるならば、これらは必ず評価に含めるべきである。試験開始時の母体体重より試験終了時の母体体重から妊娠子宮重量（または、胎児体重合計値）を除いた値を差し引いた差である調整（補正）後の母体平均体重の変化で、その作用が母体に対するものか、または子宮内に対する作用かがわかることもある。ウサギでは、妊娠期間中に体重変動があるのが普通であるため、体重増加率は母体に対する毒性の有効な指標とならない場合もある。

(f) 摂餌量および摂水量（該当する場合）：投与群母動物で対照群と比べて平均摂餌量または摂水量の有意な低下が認められれば、特にその被験物質を飼料中または飲料水中に混入して投与した場合に、母体に対する毒性評価に有用となる。観察された作用が母体に対する毒性を反映しているかどうか、

---

[1] この指標は雄によっても影響されることが認められている。

maternal body weights when determining if the effects noted are reflective of maternal toxicity or more simply, unpalatability of the test material in feed or water.

(g) <u>Clinical evaluations</u> (including clinical signs, markers, haematology and clinical chemistry studies): The observation of increased incidence of significant clinical signs of toxicity in treated dams relative to the control group may be useful in evaluating maternal toxicity. If this is to be used as the basis for the assessment of maternal toxicity, the types, incidence, degree and duration of clinical signs should be reported in the study. Examples of frank clinical signs of maternal intoxication include: coma, prostration, hyperactivity, loss of righting reflex, ataxia, or laboured breathing.

(h) <u>Post-mortem data</u>: increased incidence and/or severity of post-mortem findings may be indicative of maternal toxicity. This can include gross or microscopic pathological findings or organ weight data, e.g. absolute organ weight, organ-to-body weight ratio, or organ-to-brain weight ratio. When supported by findings of adverse histopathological effects in the affected organ(s), the observation of a significant change in the average weight of suspected target organ(s) of treated dams, compared to those in the control group, may be considered evidence of maternal toxicity.

3.7.2.5 *Animal and experimental data*

3.7.2.5.1 A number of internationally accepted test methods are available; these include methods for developmental toxicity testing (e.g. OECD Test Guideline 414, ICH Guideline S5A, 1993), methods for peri- and post-natal toxicity testing (e.g. ICH S5B, 1995) and methods for one or two-generation toxicity testing (e.g. OECD test guidelines 415, 416, 443).

3.7.2.5.2 Results obtained from Screening Tests (e.g. OECD Guidelines 421 - Reproduction/ Developmental Toxicity Screening Test, and 422 - Combined Repeated Dose Toxicity Study with Reproduction/Development Toxicity Screening Test) can also be used to justify classification, although it is recognized that the quality of this evidence is less reliable than that obtained through full studies.

3.7.2.5.3 Adverse effects or changes, seen in short- or long-term repeated dose toxicity studies, which are judged likely to impair reproductive function and which occur in the absence of significant generalized toxicity, may be used as a basis for classification, e.g. histopathological changes in the gonads.

3.7.2.5.4 Evidence from in vitro assays, or non-mammalian tests, and from analogous substances using structure-activity relationship (SAR), can contribute to the procedure for classification. In all cases of this nature, expert judgement must be used to assess the adequacy of the data. Inadequate data should not be used as a primary support for classification.

3.7.2.5.5 It is preferable that animal studies are conducted using appropriate routes of administration which relate to the potential route of human exposure. However, in practice, reproductive toxicity studies are commonly conducted using the oral route, and such studies will normally be suitable for evaluating the hazardous properties of the substance with respect to reproductive toxicity. However, if it can be conclusively demonstrated that the clearly identified mechanism or mode of action has no relevance for humans or when the toxicokinetic differences are so marked that it is certain that the hazardous property will not be expressed in humans then a substance which produces an adverse effect on reproduction in experimental animals should not be classified.

3.7.2.5.6 Studies involving routes of administration such as intravenous or intraperitoneal injection, which may result in exposure of the reproductive organs to unrealistically high levels of the test substance, or elicit local damage to the reproductive organs, e.g. by irritation, must be interpreted with extreme caution and on their own would not normally be the basis for classification.

3.7.2.5.7 There is general agreement about the concept of a limit dose, above which the production of an adverse effect may be considered to be outside the criteria which lead to classification. However, there was no agreement within the OECD Task Force regarding the inclusion within the criteria of a specified dose as a limit dose. Some test guidelines specify a limit dose, other test guidelines qualify the limit dose with a statement that higher doses may be necessary if anticipated human exposure is sufficiently high that an adequate margin of exposure would not be achieved. Also, due to species differences in toxicokinetics, establishing a specific limit dose may not be adequate for situations where humans are more sensitive than the animal model.

それとも、より単純に、飼料中または水中の被験物質の味が摂取に適していないためであるのかを決定する場合、摂餌量または摂水量の変化は、母体の体重と関連させて評価すべきである。

(g) 臨床評価（臨床症状、マーカー、血液学的検査および臨床化学検査等）：投与群母動物で対照群に比べて毒性の明確な臨床徴候の発生率の増加が認められれば、母体に対する毒性評価に有用となる。もしこれを母体に対する毒性評価の根拠として採用するならば、臨床徴候の種類、発生率、程度および継続期間の長さが試験で報告されているべきである。母体に対する毒性の臨床徴候として確実であるのは、昏睡、衰弱、自発運動亢進、立ち直り反射の消失、歩行失調または呼吸困難などである。

(h) 剖検データ：剖検所見の発生率または重篤度の上昇が、母体に対する毒性の指標となることもある。これには、肉眼または顕微鏡病理所見や、例えば臓器の絶対重量、体重に対する臓器重量比または脳に対する臓器重量比などの臓器重量データが含まれる。投与群母動物で対照群に比べて、標的臓器と推測される臓器平均重量に有意な変化が認められた場合、作用を受ける臓器に病理組織学的有害影響の所見が認められればそれが裏付けとなって、母動物に対する毒性の証拠であると見なしてもよい。

3.7.2.5 *動物データおよび実験データ*

3.7.2.5.1 国際的に容認されている試験方法として何種類かが利用可能である。例えば、発生毒性試験方法（例：OECD テストガイドライン 414、ICH ガイドライン S5A 1993）、周産期および出生後の毒性試験方法（例：ICH S5B 1995）および一世代または二世代生殖毒性試験方法（例：OECD テストガイドライン 415、416、443）がある。

3.7.2.5.2 スクリーニング試験（例：OECD テストガイドライン 421・生殖/発生毒性スクリーニング試験、および 422・反復投与毒性試験と生殖/発生毒性スクリーニング試験を組み合わせた試験）も分類の判断に用いることができるが、これより得られる証拠の質は、完全な試験より得られた証拠より信頼性に劣ることは認識されている。

3.7.2.5.3 例えば明確な一般的毒性を伴わずに生じる有害影響または変化が短期または長期反復投与毒性試験で認められ、生殖腺の組織病理学的変化など、生殖機能を損なう見込みがあると判断されたならば、分類の根拠として採用されることもある。

3.7.2.5.4 in vitro 試験または哺乳類以外の動物での試験より得られた証拠、および構造活性相関(SAR)を用いて類似物質より得られた証拠は、分類手順に役立てられる。その性格上、そのデータの妥当性の評価には専門家の判断が採用されなければならない。妥当性を欠くデータは分類の第一義的裏付けとして採用すべきでない。

3.7.2.5.5 動物試験は、ヒトでのばく露があり得る経路に関連した適切な投与経路により実施することが望ましい。ただし実際には、生殖毒性試験は一般的に経口経路により実施され、そうした試験ではその物質の生殖毒性に関する有害性評価に適切となる。ただし、明確な作用機序または作用様式が特定されたがヒトには該当しないこと、またはトキシコキネティクスの違いが著しいためにその有害性がヒトでは発現されないことが結論として実証できるならば、実験動物の生殖に有害影響を生じるような物質でも分類すべきでない。

3.7.2.5.6 静脈注射または腹腔内注射などの投与経路を用いる試験では、被験物質の生殖器官のばく露濃度が非現実的なほどに高濃度となってしまう場合、または、例えば刺激性などにより生殖器官に局所的損傷をもたらす場合には、細心の注意を払って解釈すべきであり、そうした試験だけでは通常分類の根拠とはならない。

3.7.2.5.7 それを超えると有害影響を誘発して分類の判定基準を外れるであろうと思われる限界用量の概念に関する一般的同意はなされている。しかし、OECD タスクフォース内部では、特定の用量を限界用量として判定基準に算入することは同意されていない。試験指針には限界用量を定めているものもあれば、またはヒトの予想ばく露濃度が高いために適切なばく露マージンが取れそうにない場合には、より高い用量が必要なこともあると述べた上で限界用量を認めているガイドラインもある。また、トキシコキネティクスには種差があるために、ヒトの感受性の方が動物モデルより高いような状況では、特定の限界用量を設定することは適切でない場合もある。

3.7.2.5.8　　In principle, adverse effects on reproduction seen only at very high dose levels in animal studies (for example doses that induce prostration, severe inappetence, excessive mortality) would not normally lead to classification, unless other information is available, e.g. toxicokinetics information indicating that humans may be more susceptible than animals, to suggest that classification is appropriate. Please also refer to the section on Maternal Toxicity for further guidance in this area.

3.7.2.5.9　　However, specification of the actual "limit dose" will depend upon the test method that has been employed to provide the test results, e.g. in the OECD Test Guideline for repeated dose toxicity studies by the oral route, an upper dose of 1000 mg/kg unless expected human response indicates the need for a higher dose level, has been recommended as a limit dose.

3.7.2.5.10　　Further discussions are needed on the inclusion within the criteria of a specified dose as a limit dose.

3.7.3　　Classification criteria for mixtures

3.7.3.1　　*Classification of mixtures when data are available for the complete mixture*

Classification of mixtures will be based on the available test data of the individual constituents of the mixture using cut-off values/concentration limits for the ingredients of the mixture. The classification may be modified on a case-by-case basis based on the available test data for the mixture as a whole. In such cases, the test results for the mixture as a whole must be shown to be conclusive taking into account dose and other factors such as duration, observations and analysis (e.g. statistical analysis, test sensitivity) of reproduction test systems. Adequate documentation supporting the classification should be retained and made available for review upon request.

3.7.3.2　　*Classification of mixtures when data are not available for the complete mixture: bridging principles*

3.7.3.2.1　　Where the mixture itself has not been tested to determine its reproductive toxicity, but there are sufficient data on both the individual ingredients and similar tested mixtures to adequately characterize the hazards of the mixture, these data will be used in accordance with the following agreed bridging rules. This ensures that the classification process uses the available data to the greatest extent possible in characterizing the hazards of the mixture without the necessity for additional testing in animals.

3.7.3.2.2　　*Dilution*

If a tested mixture is diluted with a diluent which is not expected to affect the reproductive toxicity of other ingredients, then the new diluted mixture may be classified as equivalent to the original tested mixture.

3.7.3.2.3　　*Batching*

The reproductive toxicity potential of a tested production batch of a mixture can be assumed to be substantially equivalent to that of another untested production batch of the same commercial product, when produced by or under the control of the same manufacturer unless there is reason to believe there is significant variation in composition such that the reproductive toxicity potential of the untested batch has changed. If the latter occurs, a new classification is necessary.

3.7.3.2.4　　*Substantially similar mixtures*

Given the following:

(a)　Two mixtures:　(i)　A + B;
　　　　　　　　　　(ii)　C + B;

(b)　The concentration of ingredient B, toxic to reproduction, is the same in both mixtures;

(c)　The concentration of ingredient A in mixture (i) equals that of ingredient C in mixture (ii);

(d)　Data on toxicity for A and C are available and substantially equivalent, i.e. they are in the same hazard category and are not expected to affect the reproductive toxicity of B.

If mixture (i) or (ii) is already classified based on test data, then the other mixture can be classified in the same hazard category.

3.7.2.5.8　原則として、動物試験できわめて高い用量段階（例えば、衰弱、重度の食欲不振、高い死亡率を生じるような用量）でのみ認められる生殖に対する有害影響は、例えばヒトの感受性の方が動物より高いことを示すトキシコキネティクスの情報のようなその他の情報が入手されて、その分類が適切であることを裏付けることがない限り、通常は分類の根拠とはならない。この分野の更なる手引きについては「母体に対する毒性」の項を参照されたい。

3.7.2.5.9　ただし、実際の「限界用量」の内容は、試験結果を得るために採用されている試験方法によって異なってくる。例えば経口経路による反復投与毒性に関するOECDテストガイドラインでは、ヒトで予想される反応から用量段階を高める必要性が示唆されない限りは、試験に採用する高い方の用量1000mg/kgが限界用量として推奨されている。

3.7.2.5.10　特定の用量を限界用量として判定基準に含めるには更なる議論が必要である。

3.7.3　混合物の分類基準

3.7.3.1　*混合物そのものについて試験データが入手できる場合の混合物の分類*

　混合物の分類は、当該混合物の個々の成分について入手できる試験データに基づき、成分のカットオフ値/濃度限界を使用して行われる。当該混合物そのものについて試験データが入手できる場合には、分類はケースバイケースで修正されることがある。このような場合、混合物そのものの試験結果は、生殖毒性試験系の用量や、試験期間、観察、分析（例えば、統計分析、試験感度）などの他の要因を考慮した上で確実であることが示されなければならない。分類が適切であることの証拠書類を保持し、要請に応じて示すことができるようにするべきである。

3.7.3.2　*混合物そのものについて試験データが入手できない場合の混合物の分類：つなぎの原則（bridging principle）*

3.7.3.2.1　混合物そのものは生殖毒性有害性を決定する試験がなされていないが、当該混合物の有害性を適切に特定するための、個々の成分および類似の試験された混合物の両方に関して十分なデータがある場合、これらのデータは以下の合意されたつなぎの原則にしたがって使用される。これによって、分類プロセスで動物試験を追加する必要もなく、混合物の有害性判定に入手されたデータを可能な限り最大限に用いることが可能になる。

3.7.3.2.2　*希釈*

　試験された混合物が、他の成分の生殖毒性に影響を与えないと予想される希釈剤で希釈される場合、新しい希釈された混合物は、試験された元の混合物と同等として分類してもよい。

3.7.3.2.3　*製造バッチ*

　混合物の試験されていない製造バッチに生殖毒性があるかどうかは、同じ製造者によって、またはその管理下で生産された同じ商品の試験された別のバッチの毒性と本質的に同等とみなすことができる。ただし、試験されていないバッチの生殖毒性能が変化するような有意な変動があると考えられる理由がある場合はこの限りではない。このような場合には、新しい分類が必要である。

3.7.3.2.4　*本質的に類似した混合物*

　次を仮定する：
　　(a)　2つの混合物：　(i)　　　A+B；
　　　　　　　　　　　　(ii)　　　C+B；
　(b)　生殖毒性をもつ成分Bの濃度は、両方の混合物で同じである；
　(c)　混合物(i)の成分Aの濃度は、混合物(ii)の成分Cの濃度に等しい；
　(d)　AとCの毒性に関するデータは利用でき、実質的に同等である、すなわち、AとCは同じ有害性区分に属し、かつ、Bの生殖毒性に影響を与えることはないと判断される。

　混合物(i)または(ii)が既に試験データに基づいて分類されている場合には、他方の混合物は同じ有害性区分に分類することができる。

3.7.3.3 *Classification of mixtures when data are available for all ingredients or only for some ingredients of the mixture*

3.7.3.3.1 The mixture will be classified as a reproductive toxicant when at least one ingredient has been classified as a Category 1 or Category 2 reproductive toxicant and is present at or above the appropriate cut-off value/concentration limit as shown in table 3.7.1 below for Category 1 and 2 respectively.

3.7.3.3.2 The mixture will be classified for effects on or via lactation when at least one ingredient has been classified for effects on or via lactation and is present at or above the appropriate cut-off value/concentration limit as shown in table 3.7.1 for the additional category for effects on or via lactation.

Table 3.7.1: Cut-off values/concentration limits of ingredients of a mixture classified as reproductive toxicants or for effects on or via lactation that would trigger classification of the mixtures[a]

Ingredient classified as:	Cut-off/concentration limits triggering classification of a mixture as:			
	Category 1 reproductive toxicant		Category 2 reproductive toxicant	Additional category for effects on or via lactation
	Category 1A	Category 1B		
Category 1A reproductive toxicant	≥ 0.1 % (note 1)	--	--	--
	≥ 0.3 % (note 2)			
Category 1B reproductive toxicant	--	≥ 0.1 % (note 1)	--	--
		≥ 0.3 % (note 2)		
Category 2 reproductive toxicant	--	--	≥ 0.1 % (note 3)	--
			≥ 3.0 % (note 4)	
Additional category for effects on or via lactation	--	--	--	≥ 0.1 % (note 1)
				≥ 0.3 % (note 2)

[a] *This compromise classification scheme involves consideration of differences in hazard communication practices in existing systems. It is expected that the number of affected mixtures will be small; the differences will be limited to label warnings; and the situation will evolve over time to a more harmonized approach.*

NOTE 1: *If a Category 1 reproductive toxicant or substance classified in the additional category for effects on or via lactation is present in the mixture as an ingredient at a concentration between 0.1 % and 0.3 %, every regulatory authority would require information on the SDS for a product. However, a label warning would be optional. Some authorities will choose to label when the ingredient is present in the mixture between 0.1 % and 0.3 %, whereas others would normally not require a label in this case.*

NOTE 2: *If a Category 1 reproductive toxicant or substance classified in the additional category for effects on or via lactation is present in the mixture as an ingredient at a concentration of ≥ 0.3 %, both an SDS and a label would generally be expected.*

NOTE 3: *If a Category 2 reproductive toxicant is present in the mixture as an ingredient at a concentration between 0.1 % and 3.0 %, every regulatory authority would require information on the SDS for a product. However, a label warning would be optional. Some authorities will choose to label when the ingredient is present in the mixture between 0.1 % and 3.0 %, whereas others would normally not require a label in this case.*

NOTE 4: *If a Category 2 reproductive toxicant is present in the mixture as an ingredient at a concentration of ≥ 3.0%, both an SDS and a label would generally be expected.*

3.7.3.3 混合物の全成分についてまたは一部の成分だけについてデータが入手できた場合の混合物の分類

3.7.3.3.1 混合物は、少なくとも 1 つの成分が区分 1 または区分 2 生殖毒性物質として分類され、区分 1 と 2 それぞれについて表 3.7.1 に示したような適切なカットオフ値/濃度限界以上で存在する場合、生殖毒性物質として分類される。

3.7.3.3.2 混合物は、少なくとも 1 つの成分が、授乳に対するまたは授乳を介した影響について分類され、授乳に対するまたは授乳を介した影響に関する追加区分のために表 3.7.1 に示したような適切なカットオフ値/濃度限界以上で存在する場合、授乳に対するまたは授乳を介した影響について分類される。

表 3.7.1：混合物の分類基準となる生殖毒性物質成分のカットオフ値/濃度限界 [a]

成分の分類：	混合物の分類基準となるカットオフ値/濃度限界：			
	区分 1 生殖毒性物質		区分 2 生殖毒性物質	授乳に対するまたは授乳を介した影響に関する追加区分
	区分 1A	区分 1B		
区分 1A 生殖毒性物質	≧0.1%（注記 1）	--	--	--
	≧0.3%（注記 2）	--	--	--
区分 1B 生殖毒性物質	--	≧0.1%（注記 1）	--	--
	--	≧0.3%（注記 2）	--	--
区分 2 生殖毒性物質	--	--	≧0.1%（注記 3）	--
	--	--	≧3.0%（注記 4）	--
授乳に対するまたは授乳を介した影響に関する追加区分	--	--	--	≧0.1%（注記 1）
	--	--	--	≧0.3%（注記 2）

[a] この妥協の産物である分類方法は現行の危険有害性の情報伝達における相違を考慮して作成された。影響を受ける混合物の数が少なく、相違はラベル表示に限られ、さらなる調和により状況が良くなることが期待される。

注記 1：区分 1 生殖毒性成分あるいは授乳に対するまたは授乳を介した影響のための追加区分に分類される物質が 0.1%と 0.3%の間の濃度で混合物に存在する場合には、すべての規制官庁は、製品の SDS に情報の記載を要求することになろう。しかし、ラベルへの警告表示は任意となろう。一部の規制官庁は、成分が 0.1%と 0.3%の間で混合物に存在する場合に表示を選択するであろうが、他の官庁は、通常、この場合に表示を要求しないことになろう。

注記 2：区分 1 生殖毒性成分あるいは授乳に対するまたは授乳を介した影響のための追加区分に分類される物質が ≧0.3%の濃度で混合物に存在する場合には、一般に SDS とラベル表示の両方に記載することになろう。

注記 3：区分 2 生殖毒性成分が 0.1%と 3.0%の間の濃度で混合物に存在する場合には、すべての規制官庁は、製品の SDS に情報の記載を要求することになろう。しかし、ラベルへの警告表示は任意となろう。一部の規制官庁は、成分が 0.1%と 3.0%の間で混合物に存在する場合に表示を選択するであろうが、他の官庁は、通常、この場合には表示を要求しないことになろう。

注記 4：区分 2 生殖毒性成分が ≧3.0%の濃度で混合物に存在する場合には、一般に SDS と表示の両方に記載することになろう。

3.7.4 Hazard communication

General and specific considerations concerning labelling requirements are provided in *Hazard communication: Labelling* (chapter 1.4). Annex 1 contains summary tables about classification and labelling. Annex 3 contains examples of precautionary statements and pictograms which can be used where allowed by the competent authority. Table 3.7.2 presents specific label elements for substances and mixtures classified into this hazard class based on the criteria in this chapter.

Table 3.7.2: Label elements for reproductive toxicity

	Category 1 (Category 1A, 1B)	Category 2	Additional category for effects on or via lactation
Symbol	Health hazard	Health hazard	*No symbol*
Signal word	Danger	Warning	*No signal word*
Hazard statement	May damage fertility or the unborn child (state specific effect if known) (state route of exposure if it is conclusively proven that no other routes of exposure cause the hazard)	Suspected of damaging fertility or the unborn child (state specific effect if known) (state route of exposure if it is conclusively proven that no other routes of exposure cause the hazard)	May cause harm to breast-fed children.

3.7.5 Decision logics for classification

3.7.5.1 *Decision logic for reproductive toxicity*

The decision logic which follows is not part of the harmonized classification system but is provided here as additional guidance. It is strongly recommended that the person responsible for classification study the criteria before and during use of the decision logic.

3.7.5.1.1 *Decision logic 3.7.1 for substances*

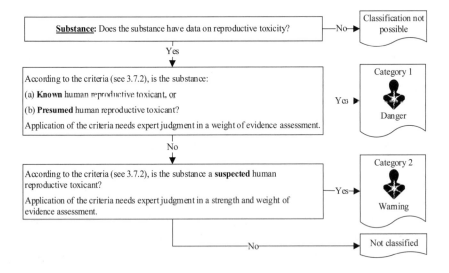

- 214 -

3.7.4 危険有害性情報の伝達

表示要件に関する通則および細則は、*危険有害性に関する情報の伝達：表示*（第 1.4 章）に定める。附属書 1 に分類と表示に関する概要表を示す。附属書 3 には、注意書きおよび所管官庁が許可した場合に使用可能な注意絵表示の例を示す。表 3.7.2 には、本章の判定基準に基づいてこの危険有害性クラスに分類された物質および混合物の具体的なラベル要素を示す。

表 3.7.2：生殖毒性のラベル要素

	区分 1 （区分 1A、1B）	区分 2	授乳に対するまたは授乳を介した影響に関する追加区分
シンボル	健康有害性	健康有害性	シンボルなし
注意喚起語	危険	警告	*注意喚起語なし*
危険有害性情報	生殖能または胎児への悪影響のおそれ （もし判れば影響の内容を記載する）（他の経路からのばく露が有害でないことが決定的に証明されている場合、有害なばく露経路を記載）	生殖能または胎児への悪影響のおそれの疑い （もし判れば影響の内容を記載する）（他の経路からのばく露が有害でないことが決定的に証明されている場合、有害なばく露経路を記載）	授乳中の子に害を及ぼすおそれ

3.7.5 分類判定論理

3.7.5.1 *生殖毒性の判定論理*

以下の判定論理は、調和分類システムには含まれないが、追加的な手引きとしてここに示す。分類の責任者に対し、この判定論理を使用する前および使用する際に判定基準についてよく調べ理解することを強く勧める。

3.7.5.1.1 *物質の判定論理 3.7.1*

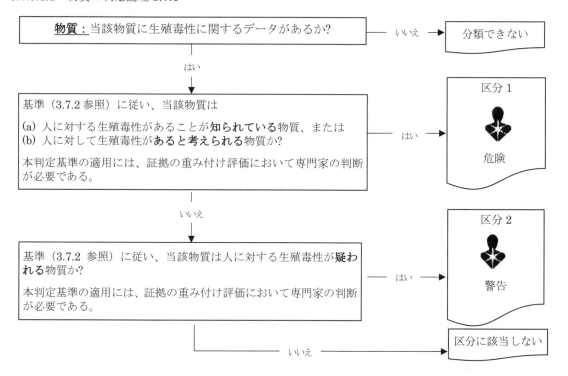

- 214 -

3.7.5.1.2   *Decision logic 3.7.2 for mixtures*

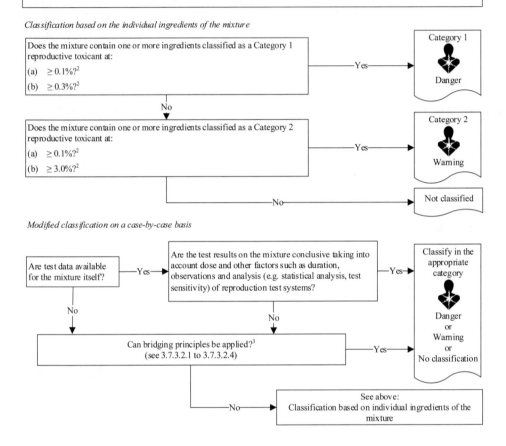

---

[2]   *For specific concentration limits, see "The use of cut-off values/concentration limits" in chapter 1.3, paragraph 1.3.3.2, and in table 3.7.1 of this chapter.*

[3]   *If data on another mixture are used in the application of bridging principles, the data on that mixture must be conclusive in accordance with 3.7.3.2.*

3.7.5.1.2　*混合物の判定論理 3.7.2*

<u>**混合物：**</u>　混合物の分類は、成分のカットオフ値/濃度限界を用いて、当該混合物の**個々の成分**の利用可能な試験データに基づいて行われる。混合物そのものについての利用可能な試験データ、あるいはつなぎの原則に基づき、分類は**ケースバイケースで修正できる**。以下のケースバイケースでの分類の修正を参照。詳細は 3.7.3.1, 3.7.3.2, および 3.7.3.3 を参照のこと。

混合物の個々の成分に基づく分類

ケースバイケースでの修正分類

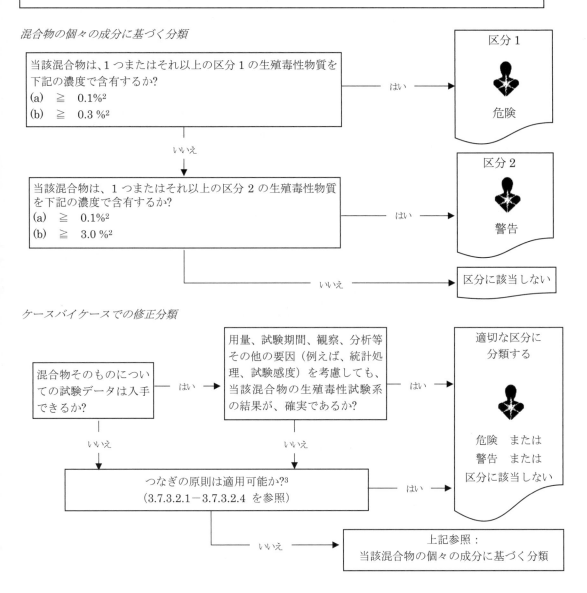

---

[2]　*個々の濃度の限度については、第 1.3 章の 1.3.3.2「カットオフ値/濃度限度の使用」および本章の表 3.7.1 を参照。*
[3]　*他の混合物のデータをつなぎの原則に用いた場合は、その混合物のデータは 3.7.3.2 に照らして断定的なものでなければならない。*

3.7.5.2 *Decision logic for effects on or via lactation*

3.7.5.2.1 *Decision logic 3.7.3 for substances*

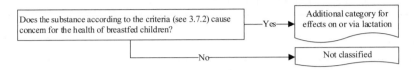

3.7.5.2.2 *Decision logic 3.7.4 for mixtures*

Mixture: Classification of mixtures will be based on the available test data for the **individual ingredients** of the mixture, using cut-off values/concentration limits for those ingredients. The classification may be **modified on a case-by-case basis** based on the available test data for the mixture as a whole or based on bridging principles. See modified classification on a case-by-case basis below. For further details see 3.7.3.1, 3.7.3.2 and 3.7.3.3.

Classification based on the individual ingredients of the mixture

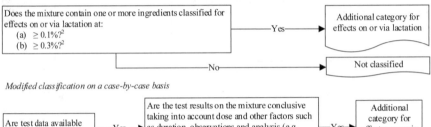

Modified classification on a case-by-case basis

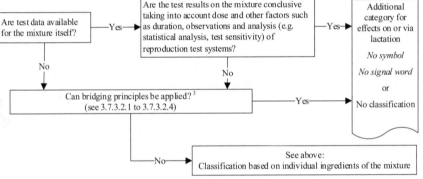

---

[2] For specific concentration limits, see "The use of cut-off values/concentration limits" in chapter 1.3, paragraph 1.3.3.2, and in table 3.7.1 of this chapter.

[3] If data on another mixture are used in the application of bridging principles, the data on that mixture must be conclusive in accordance with 3.7.3.2.

3.7.5.2 授乳に対するまたは授乳を介した影響の判定論理

3.7.5.2.1 物質の判定論理 3.7.3

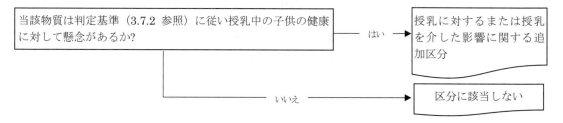

3.7.5.2.2 混合物の判定論理 3.7.4

> **混合物**: 混合物の分類は、成分のカットオフ値/濃度限界を用いて、当該混合物の**個々の成分**の利用可能な試験データに基づいて行われる。混合物そのものについての利用可能な試験データ、あるいはつなぎの原則に基づき、分類は**ケースバイケース**で修正できる。以下のケースバイケースでの分類の修正を参照。詳細は判定基準（3.7.3.1, 3.7.3.2, および 3.7.3.3）を参照。

混合物の個々の成分に基づく分類

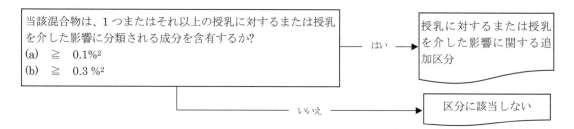

ケースバイケースでの分類の修正

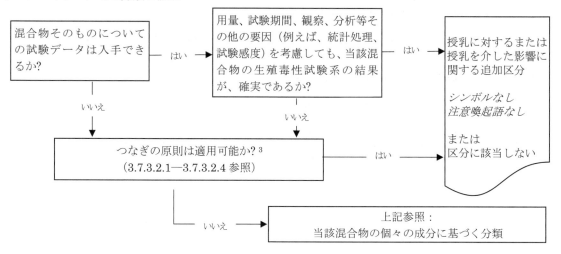

---

[2]　個々の濃度の限度については、第1.3章の1.3.3.2「カットオフ値/濃度限度の使用」および本章の表3.7.1を参照。
[3]　他の混合物のデータをつなぎの原則に用いた場合は、その混合物のデータは3.7.3.2に照らして決定的なものでなければならない。

- 216 -

CHAPTER 3.8

SPECIFIC TARGET ORGAN TOXICITY SINGLE EXPOSURE

3.8.1 Definitions and general considerations

3.8.1.1 *Specific target organ toxicity – single exposure* refers to specific, non-lethal toxic effects on target organs occurring after a single exposure to a substance or mixture. All significant health effects that can impair function, both reversible and irreversible, immediate and/or delayed and not specifically addressed in chapters 3.1 to 3.7 and 3.10 are included (see also 3.8.1.6).

3.8.1.2 Classification identifies the substance or mixture as being a specific target organ toxicant and, as such, it may present a potential for adverse health effects in people who are exposed to it.

3.8.1.3 Classification depends upon the availability of reliable evidence that a single exposure to the substance or mixture has produced a consistent and identifiable toxic effect in humans, or, in experimental animals, toxicologically significant changes which have affected the function or morphology of a tissue/organ, or has produced serious changes to the biochemistry or haematology of the organism and these changes are relevant for human health. It is recognized that human data will be the primary source of evidence for this hazard class.

3.8.1.4 Assessment should take into consideration not only significant changes in a single organ or biological system but also generalized changes of a less severe nature involving several organs.

3.8.1.5 Specific target organ toxicity can occur by any route that is relevant for humans, i.e. principally oral, dermal or inhalation.

3.8.1.6 Specific target organ toxicity following a repeated exposure is addressed in the GHS as described in chapter 3.9 and is therefore excluded from the present chapter. Substances and mixtures should be classified for single and repeated dose toxicity independently.

Other specific toxic effects, such as acute toxicity, skin corrosion/irritation, serious eye damage/eye irritation, respiratory or skin sensitization, germ cell mutagenicity, carcinogenicity, reproductive toxicity, and aspiration toxicity are assessed separately in the GHS and consequently are not included here.

3.8.1.7 The classification criteria in this chapter are organized as criteria for substances categories 1 and 2 (see 3.8.2.1), criteria for substances Category 3 (see 3.8.2.2) and criteria for mixtures (see 3.8.3). See also figure 3.8.1.

3.8.2 Classification criteria for substances

3.8.2.1 *Substances of Category 1 and Category 2*

3.8.2.1.1 Substances are classified for immediate or delayed effects separately, by the use of expert judgement on the basis of the weight of all evidence available, including the use of recommended guidance values (see 3.8.2.1.9). Then substances are placed in Category 1 or 2, depending upon the nature and severity of the effect(s) observed (figure 3.8.1).

第 3.8 章
特定標的臓器毒性
単回ばく露

3.8.1 定義および一般事項

3.8.1.1　*特定標的臓器毒性－単回ばく露*とは、物質または混合物への単回のばく露後に起こる、特異的な非致死性の標的臓器への影響をさす。可逆的と不可逆的、あるいは急性および遅発性両方の、かつ第 3.1 章から 3.7 章および 3.10 章において明確に扱われていない、機能を損ないうるすべての明確な健康への影響がこれに含まれる（3.8.1.6 も参照）。

3.8.1.2　この分類は、ある物質または混合物が、特定標的臓器毒性があるか、およびそれにばく露したヒトに対して健康に有害な影響を及ぼす可能性があるかどうかを確認する。

3.8.1.3　分類は、ある物質または混合物に対する単回ばく露がヒトにおける一貫性のある、かつ特定できる毒性影響を与えたこと、あるいは実験動物において、組織/臓器の機能または形態に影響する毒性学的に明確な変化が示されたか、または生物の生化学的項目または血液学的項目に重大な変化が示され、これらの変化がヒトの健康状態に関連性があるということについての信頼できる証拠が入手できるかに依存する。この有害性クラスに関しては、ヒトのデータを優先的な証拠とすることが確認されている。

3.8.1.4　評価においては、単一臓器または生物学的システムにおける明確な変化だけでなく、いくつかの臓器に対するそれほど重度でない一般的変化も考慮すべきである。

3.8.1.5　特定標的臓器毒性は、ヒトに関連するいずれの経路によっても、すなわち主として経口、経皮または吸入によって起こりうる。

3.8.1.6　反復ばく露により起きる特定標的臓器毒性は GHS 第 3.9 章で記述され、それゆえに本章からは除外されている。物質および混合物は、単回および反復投与による毒性に関して独立に分類されるべきである。

　急性毒性、皮膚腐食性/刺激性、眼に対する重篤な損傷性/眼刺激性、呼吸器または皮膚感作性、生殖細胞変異原性、発がん性、生殖毒性、および誤えん有害性のような、他の特定毒性影響は GHS の中で別に評価されるので、ここには含まれない。

3.8.1.7　この章における分類基準は、区分 1 および 2 の物質（3.8.2.1 参照）の基準、区分 3 の物質(3.8.2.2 参照)の基準および混合物の区分(3.8.3 参照)の基準として体系化されている。図 3.8.1.参照。

3.8.2 物質の分類基準

3.8.2.1 *区分 1 および区分 2 の物質*

3.8.2.1.1　物質は、勧告されたガイダンス値（3.8.2.1.9 参照）の使用を含む入手されたすべての証拠の重み付けに基づく専門家の判断によって、急性と遅発性の影響に分けて分類される。そして、観察された影響の性質および重度によって区分 1 または 2 のいずれかに分類される。（図 3.8.1）

Figure 3.8.1: Hazard categories for specific target organ toxicity following single exposure

CATEGORY 1:	**Substances that have produced significant toxicity in humans, or that, on the basis of evidence from studies in experimental animals can be presumed to have the potential to produce significant toxicity in humans following single exposure**
	Placing a substance in Category 1 is done on the basis of:
	(a) reliable and good quality evidence from human cases or epidemiological studies; or
	(b) observations from appropriate studies in experimental animals in which significant and/or severe toxic effects of relevance to human health were produced at generally low exposure concentrations. Guidance dose/concentration values are provided below (see 3.8.2.1.9) to be used as part of weight-of-evidence evaluation.
CATEGORY 2:	**Substances that, on the basis of evidence from studies in experimental animals can be presumed to have the potential to be harmful to human health following single exposure**
	Placing a substance in Category 2 is done on the basis of observations from appropriate studies in experimental animals in which significant toxic effects, of relevance to human health, were produced at generally moderate exposure concentrations. Guidance dose/concentration values are provided below (see 3.8.2.1.9) in order to help in classification.
	In exceptional cases, human evidence can also be used to place a substance in Category 2 (see 3.8.2.1.9).
CATEGORY 3:	**Transient target organ effects**
	There are target organ effects for which a substance/mixture may not meet the criteria to be classified in Categories 1 or 2 indicated above. These are effects which adversely alter human function for a short duration after exposure and from which humans may recover in a reasonable period without leaving significant alteration of structure or function. This category only includes narcotic effects and respiratory tract irritation. Substances/mixtures may be classified specifically for these effects as discussed in 3.8.2.2.

NOTE: *For these categories the specific target organ/system that has been primarily affected by the classified substance may be identified, or the substance may be identified as a general toxicant. Attempts should be made to determine the primary target organ/system of toxicity and classify for that purpose, e.g. hepatotoxicants, neurotoxicants. One should carefully evaluate the data and, where possible, not include secondary effects, e.g. a hepatotoxicant can produce secondary effects in the nervous or gastro-intestinal systems.*

3.8.2.1.2　　The relevant route of exposure by which the classified substance produces damage should be identified.

3.8.2.1.3　　Classification is determined by expert judgement, on the basis of the weight of all evidence available including the guidance presented below.

3.8.2.1.4　　Weight of evidence of all data, including human incidents, epidemiology, and studies conducted in experimental animals, is used to substantiate specific target organ toxic effects that merit classification.

3.8.2.1.5　　The information required to evaluate specific target organ toxicity comes either from single exposure in humans, e.g. exposure at home, in the workplace or environmentally, or from studies conducted in experimental animals. The standard animal studies in rats or mice that provide this information are acute toxicity studies which can include clinical observations and detailed macroscopic and microscopic examination to enable the toxic effects on target tissues/organs to be identified. Results of acute toxicity studies conducted in other species may also provide relevant information.

3.8.2.1.6　　In exceptional cases, based on expert judgement, it may be appropriate to place certain substances with human evidence of target organ toxicity in Category 2: (a) when the weight of human evidence is not sufficiently convincing to warrant Category 1 classification, and/or (b) based on the nature and severity of effects. Dose/concentration levels in humans should not be considered in the classification and any available evidence from animal studies should be consistent with the Category 2 classification. In other words, if there are also animal data available on the chemical that warrant Category 1 classification, the substance should be classified as Category 1.

図 3.8.1：特定標的臓器毒性 −単回ばく露のための区分

<u>区分 1</u>：ヒトに明確な毒性を示した物質、または実験動物での試験の証拠に基づいて単回ばく露によってヒトに明確な毒性を示す可能性があると考えられる物質

  区分 1 に物質を分類するには、次に基づいて行う：

  (a) ヒトの症例または疫学的研究からの信頼でき、かつ質の良い証拠；または、
  (b) 実験動物における適切な試験において、一般的に低濃度のばく露でヒトの健康に関連のある明確なおよび/または重篤な毒性影響を生じたという所見。証拠の重み付けによる評価の一環として使用すべき用量/濃度ガイダンス値は後述する (3.8.2.1.9 参照)。

<u>区分 2</u>：実験動物を用いた試験の証拠に基づき単回ばく露によってヒトの健康に有害である可能性があると考えられる物質

  物質を区分 2 に分類するには、実験動物での適切な試験において、一般的に中等度のばく露濃度でヒトの健康に関連のある明確な毒性影響を生じたという所見に基づいて行われる。ガイダンス用量/濃度値は分類を容易にするために後述する (3.8.2.1.9 参照)。

  例外的に、ヒトでの証拠も、物質を区分 2 に分類するために使用できる (3.8.2.1.9 参照)。

<u>区分 3</u>：一時的な特定臓器への影響

  物質または混合物が上記に示された区分 1 または 2 に分類される基準に合致しない特定臓器への影響がある。これらは、ばく露の後、短期間だけ、ヒトの機能に悪影響を及ぼし、構造または機能に明確な変化を残すことなく合理的な期間において回復する影響である。この区分は、麻酔の作用および気道刺激性のみを含む。物質/混合物は、3.8.2.2 において議論されているように、これらの影響に対して明確に分類できる。

注記：これらの区分においても、分類された物質によって一次的影響を受けた特定標的臓器/器官が明示されるか、または一般的な毒性物質であることが明示される。毒性の主標的臓器を決定し、その意義にそって分類する、例えば肝毒性物質、神経毒性物質のように分類するよう努力するべきである。そのデータを注意深く評価し、できる限り二次的影響を含めないようにすべきである。例えば、肝毒性物質は、神経または消化器官で二次的影響を起こすことがある。

3.8.2.1.2　分類した物質が障害を起こしたばく露経路を明示すべきである。

3.8.2.1.3　分類は、後述のガイダンス値を含む利用可能なすべての証拠の重み付けに基づいて、専門家の判断によって決定する。

3.8.2.1.4　ヒトでの疾患の発生、疫学および実験動物を用いて実施した試験を含むすべてのデータの証拠の重み付けは、分類を助ける特定標的臓器毒性影響を証明するために使用される。

3.8.2.1.5　特定標的臓器毒性を評価するために必要な情報は、ヒトにおける単回ばく露、例えば、家庭、職場あるいは環境中でのばく露か、または実験動物を用いて実施した試験のいずれからも得られる。この情報を提供するラットまたはマウスにおける標準的動物試験は急性毒性試験であり、標的組織/臓器に及ぼす毒性影響の確認をするための臨床所見および詳細な肉眼および顕微鏡による検査を含んでいる。他の動物種を用いて実施された急性毒性試験の結果も適切な情報となりうる。

3.8.2.1.6　例外的に、標的臓器毒性のヒトでの証拠を有するある種の物質を、専門家の判断に基づいて区分 2 に分類するのが適切な場合がある：それは(a)ヒトでの証拠の重み付けが区分 1 への分類を正当化することが十分には確信できない場合、または(b)影響の性質および重篤度に基づく場合である。ヒトにおける用量/濃度レベルは、分類において考慮すべきではなく、動物試験で入手されたいかなる証拠も、区分 2 への分類と矛盾しないことである。換言すれば、物質について区分 1 への分類を保証する動物試験データも入手されている場合、その物質は区分 1 として分類するべきである。

3.8.2.1.7   *Effects considered to support classification for Category 1 and 2*

3.8.2.1.7.1   Evidence associating single exposure to the substance with a consistent and identifiable toxic effect demonstrates support for classification.

3.8.2.1.7.2   It is recognized that evidence from human experience/incidents is usually restricted to reports of adverse health consequences, often with uncertainty about exposure conditions, and may not provide the scientific detail that can be obtained from well-conducted studies in experimental animals.

3.8.2.1.7.3   Evidence from appropriate studies in experimental animals can furnish much more detail, in the form of clinical observations, and macroscopic and microscopic pathological examination and this can often reveal hazards that may not be life-threatening but could indicate functional impairment. Consequently, all available evidence, and relevance to human health, must be taken into consideration in the classification process.

   Examples of relevant toxic effects in humans and/or animals are provided below:

   (a)   Morbidity resulting from single exposure;

   (b)   Significant functional changes, more than transient in nature, in the respiratory system, central or peripheral nervous systems, other organs or other organ systems, including signs of central nervous system depression and effects on special senses (e.g. sight, hearing and sense of smell);

   (c)   Any consistent and significant adverse change in clinical biochemistry, haematology, or urinalysis parameters;

   (d)   Significant organ damage that may be noted at necropsy and/or subsequently seen or confirmed at microscopic examination;

   (e)   Multifocal or diffuse necrosis, fibrosis or granuloma formation in vital organs with regenerative capacity;

   (f)   Morphological changes that are potentially reversible but provide clear evidence of marked organ dysfunction;

   (g)   Evidence of appreciable cell death (including cell degeneration and reduced cell number) in vital organs incapable of regeneration.

3.8.2.1.8   *Effects considered not to support classification for Category 1 and 2*

   It is recognized that effects may be seen that would not justify classification.

   Examples of such effects in humans and/or animals are provided below:

   (a)   Clinical observations or small changes in bodyweight gain, food consumption or water intake that may have some toxicological importance but that do not, by themselves, indicate "significant" toxicity;

   (b)   Small changes in clinical biochemistry, haematology or urinalysis parameters and/or transient effects, when such changes or effects are of doubtful or minimal toxicological importance;

   (c)   Changes in organ weights with no evidence of organ dysfunction;

   (d)   Adaptive responses that are not considered toxicologically relevant;

   (e)   Substance-induced species-specific mechanisms of toxicity, i.e. demonstrated with reasonable certainty to be not relevant for human health, should not justify classification.

3.8.2.1.7　区分1および2への分類を支持すると考えられる影響

3.8.2.1.7.1　物質への単回ばく露が、一貫した特定の毒性作用を示した場合には、分類への根拠となる。

3.8.2.1.7.2　ヒトでの経験/疾患の発生から得られる証拠は、通常、健康被害の報告に限定され、ばく露条件が不確実であることがしばしばあり、実験動物で適切に実施された試験から得られるような科学的な詳細情報が提供されないと理解されている。

3.8.2.1.7.3　実験動物における適切な試験の証拠は、臨床所見、肉眼および顕微鏡による病理組織学的検査の形をとって多くのより詳しい内容を供給することができ、そして、生命への危険に至らない機能障害を起こすかも知れない有害性を、しばしば明らかにすることができる。したがって、入手されたすべての証拠およびヒトの健康状態への関連性は、分類の過程において考慮を払う必要がある。

 ヒトまたは実験動物における関連性のある毒性影響の実例を以下に示す：

  (a) 単回ばく露に起因する罹患；

  (b) 中枢神経系抑制の徴候および特殊感覚器（例：視覚、聴覚および嗅覚）に及ぼす影響を含む本質的に一時的なものにとどまらない呼吸器系、中枢または末梢神経系、他の器官、あるいはその他の器官系における明確な機能変化；

  (c) 臨床生化学的検査、血液学的検査または尿検査の項目における一貫した明確で有害な変化；

  (d) 剖検時に観察され、またはその後の病理組織学的検査時に認められた、または確認された明確な臓器損傷；

  (e) 再生能力を有する生体臓器における多発性またはびまん性壊死、線維症または肉芽腫形成；

  (f) 潜在的に可逆的であるが、臓器の著しい機能障害の明確な証拠を提供する形態学的変化；

  (g) 再生が不可能な生体臓器における明白な細胞死（細胞の退化および細胞数の減少を含む）の証拠。

3.8.2.1.8　区分1および2への分類を支持しないと考えられる影響

分類を正当化しないと考えられる影響があることが認められている。

ヒトまたは実験動物におけるこのような影響の実例を以下に示す：

  (a) 毒性学的にはいくらかの重要性をもつかもしれないが、それだけでは「明確な」毒性を示すものではない臨床所見、または体重増加量、摂餌量または摂水量のわずかな変化；

  (b) 臨床生化学的検査、血液学的検査または尿検査の項目における軽度の変化または一時的な影響で、このような変化または影響に疑いがある場合、または毒性学的意義がほとんどない場合；

  (c) 臓器機能障害の証拠がない臓器重量の変化；

  (d) 毒性学的に重要と考えられない適応反応；

  (e) 物質が誘発する種特異的な毒性作用メカニズムで、合理的な確実性を持ってヒトの健康との関連性を持たないことが実証された場合は、分類を正当化すべきではない。

3.8.2.1.9   *Guidance values to assist with classification based on the results obtained from studies conducted in experimental animals for Category 1 and 2*

3.8.2.1.9.1   In order to help reach a decision about whether a substance should be classified or not, and to what degree it would be classified (Category 1 versus Category 2), dose/concentration "guidance values" are provided for consideration of the dose/concentration which has been shown to produce significant health effects. The principal argument for proposing such guidance values is that all chemicals are potentially toxic and there has to be a reasonable dose/concentration above which a degree of toxic effect is acknowledged.

3.8.2.1.9.2   Thus, in animal studies, when significant toxic effects are observed, that would indicate classification, consideration of the dose/concentration at which these effects were seen, in relation to the suggested guidance values, can provide useful information to help assess the need to classify (since the toxic effects are a consequence of the hazardous property(ies) and also the dose/concentration).

3.8.2.1.9.3   The guidance value ranges proposed for single-dose exposure which has produced a significant non-lethal toxic effect are those applicable to acute toxicity testing, as indicated in table 3.8.1.

Table 3.8.1: Guidance value ranges for single-dose exposures[a]

Route of exposure	Units	Guidance value ranges for:		
		Category 1	**Category 2**	**Category 3**
Oral (rat)	mg/kg body weight	$C \leq 300$	$2000 \geq C > 300$	Guidance values do not apply[b]
Dermal (rat or rabbit)	mg/kg body weight	$C \leq 1000$	$2000 \geq C > 1000$	
Inhalation (rat) gas	ppmV/4h	$C \leq 2500$	$20000 \geq C > 2500$	
Inhalation (rat) vapour	mg/l/4h	$C \leq 10$	$20 \geq C > 10$	
Inhalation (rat) dust/mist/fume	mg/l/4h	$C \leq 1.0$	$5.0 \geq C > 1.0$	

[a]   The guidance values and ranges mentioned in table 3.8.1. above are intended only for guidance purposes, i.e. to be used as part of the weight of evidence assessment, and to assist with decision about classification. They are not intended as strict demarcation values.

[b]   Guidance values are not provided since this classification is primarily based on human data. Animal data may be included in the weight of evidence assessment.

3.8.2.1.9.4   Thus it is feasible that a specific profile of toxicity is seen to occur at a dose/concentration below the guidance value, e.g. < 2000 mg/kg body weight by the oral route, however the nature of the effect may result in the decision not to classify. Conversely, a specific profile of toxicity may be seen in animal studies occurring at above a guidance value, e.g. ≥ 2000 mg/kg body weight by the oral route, and in addition there is supplementary information from other sources, e.g. other single dose studies, or human case experience, which supports a conclusion that, in view of the weight of evidence, classification would be the prudent action to take.

3.8.2.1.10   *Other considerations*

3.8.2.1.10.1   When a substance is characterized only by use of animal data (typical of new substances, but also true for many existing substances), the classification process would include reference to dose/concentration guidance values as one of the elements that contribute to the weight of evidence assessment.

3.8.2.1.10.2   When well-substantiated human data are available showing a specific target organ toxic effect that can be reliably attributed to single exposure to a substance, the substance may be classified. Positive human data, regardless of probable dose, predominates over animal data. Thus, if a substance is unclassified because specific target organ toxicity observed was considered not relevant or significant to humans, if subsequent human incident data become available showing a specific target organ toxic effect, the substance should be classified.

3.8.2.1.10.3   A substance that has not been tested for specific target organ toxicity may in certain instances, where appropriate, be classified on the basis of data from a validated structure activity relationship and expert judgement-based extrapolation from a structural analogue that has previously been classified together with substantial support from consideration of other important factors such as formation of common significant metabolites.

3.8.2.1.10.4   It is recognized that saturated vapour concentration may be used as an additional element by some regulatory systems to provide for specific health and safety protection.

3.8.2.1.9 　実験動物を用いて実施した試験で得られた結果に基づく区分1および2への分類を補助する
　　　　　 ガイダンス値

3.8.2.1.9.1 　物質を分類すべきであるか否か、また、どのランク（区分1か、区分2か）に分類するかについての決定を下すことを助ける目的で、明確な健康影響を生じることが認められた用量/濃度「ガイダンス値」を示した。そのようなガイダンス値を提案する主要な論拠は、すべての化学品は潜在的に有毒であり、それ以上ではある程度の毒性影響が認められる妥当な用量/濃度があるはずだからである。

3.8.2.1.9.2 　したがって、動物試験においては、分類を示す明確な毒性影響が認められた場合、提案されたガイダンス値に照らして、これらの影響の認められた用量/濃度の考察をすることは、分類の必要性を評価する有益な情報を提供する（毒性影響は、有害性と用量/濃度の結果であるから）。

3.8.2.1.9.3 　明確な非致死性の毒性影響を生じる単回投与ばく露について提案されたガイダンス値の範囲は、以下に示すように急性毒性試験に適用されるものである。

表 3.8.1：単回ばく露に関するガイダンス値の範囲 [a]

ばく露経路	単位	区分1	区分2	区分3
経口（ラット）	mg/kg 体重	$C \leq 300$	$2000 \geq C > 300$	ガイダンス値は、適用しない [b]
経皮（ラットまたはウサギ）	mg/kg 体重	$C \leq 1000$	$2000 \geq C > 1000$	
吸入（ラット）気体	ppmV/4時間	$C \leq 2500$	$20000 \geq C > 2500$	
吸入（ラット）蒸気	mg/l/4時間	$C \leq 10$	$20 \geq C > 10$	
吸入（ラット）粉塵/ミスト/ヒューム	mg/l/4時間	$C \leq 1.0$	$5.0 \geq C > 1.0$	

[a] 　上記の表3.8.1に記載したガイダンス値および範囲は、あくまでもガイダンスとしてのためのものである。すなわち、証拠の重み付け評価の一環として、分類の決定を助けるためのものであって、厳密な境界値として意図されたものではない。

[b] 　この分類は主としてヒトのデータに基づいているので、ガイダンス値は示されていない。動物のデータは、証拠の重み付け評価に含まれうる。

3.8.2.1.9.4 　特定の毒性プロフィールは、ガイダンス値未満の用量/濃度、例えば、＜2000 mg/kg 体重の経口投与で起こることがありうるが、影響の性質から分類をしない決定をする結果となる場合もある。逆に、特定の毒性プロフィールは、動物試験においてガイダンス値を超える用量/濃度、例えば、≧2000 mg/kg 体重の経口投与で認められ、そして、その他の情報源からの補足情報、例えば、他の単回投与試験またはヒトでの症例経験など結論を支持するものがある場合は、証拠の重み付けを考慮して分類することが賢明であろう。

3.8.2.1.10 　その他の考慮事項

3.8.2.1.10.1 　ある物質が動物データの使用だけによって特徴付けられている場合（新規物質では典型的な事例で、しかしまた、多くの既存物質にも当てはまる）、分類の過程では、証拠の重み付け評価への寄与要素の1つとして、用量/濃度ガイダンス値を参照することが含まれるであろう。

3.8.2.1.10.2 　物質に対する単回ばく露に確かに起因するとされる特定標的臓器毒性影響が明確に実証されたヒトのデータが入手できた場合、当該物質は分類できる。投与量が推定でしかなくても、ヒトの陽性データは、動物データに対して優先される。したがって、認められた特定標的臓器毒性がヒトとの関連性がない、または明らかでないと考えて物質を分類しなかった場合、もしその後に、特定標的臓器毒性影響を示すヒトでの疾患データが入手できれば、その物質を分類すべきである。

3.8.2.1.10.3 　特定標的臓器毒性について試験をされていない物質でも、場合によっては、検証された構造活性相関データ、および共通の重要な代謝物を生成することのような他の重要な要因の考慮からの実質的な支援も合わせて、すでに分類されている構造類似体から専門家の判断に基づいた外挿を用いて分類することも可能であろう。

3.8.2.1.10.4 　一部の規制システムでは、特別な健康および安全保護のために、飽和蒸気濃度を追加要因として利用してもよいと認められている。

3.8.2.2 *Substances of Category 3*

3.8.2.2.1 *Criteria for respiratory tract irritation*

The criteria for respiratory tract irritation as Category 3 are:

(a) Respiratory irritant effects (characterized by localized redness, edema, pruritis and/or pain) that impair function with symptoms such as cough, pain, choking, and breathing difficulties are included. It is recognized that this evaluation is based primarily on human data;

(b) Subjective human observations could be supported by objective measurements of clear respiratory tract irritation (RTI) (e.g. electrophysiological responses, biomarkers of inflammation in nasal or bronchoalveolar lavage fluids);

(c) The symptoms observed in humans should also be typical of those that would be produced in the exposed population rather than being an isolated idiosyncratic reaction or response triggered only in individuals with hypersensitive airways. Ambiguous reports simply of "irritation" should be excluded as this term is commonly used to describe a wide range of sensations including those such as smell, unpleasant taste, a tickling sensation, and dryness, which are outside the scope of this classification endpoint;

(d) There are currently no validated animal tests that deal specifically with RTI, however, useful information may be obtained from the single and repeated inhalation toxicity tests. For example, animal studies may provide useful information in terms of clinical signs of toxicity (dyspnoea, rhinitis etc) and histopathology (e.g. hyperemia, edema, minimal inflammation, thickened mucous layer) which are reversible and may be reflective of the characteristic clinical symptoms described above. Such animal studies can be used as part of weight of evidence assessment;

(e) This special classification would occur only when more severe organ effects including in the respiratory system are not observed.

3.8.2.2.2 *Criteria for narcotic effects*

The criteria for narcotic effects as Category 3 are:

(a) Central nervous system depression including narcotic effects in humans such as drowsiness, narcosis, reduced alertness, loss of reflexes, lack of coordination, and vertigo are included. These effects can also be manifested as severe headache or nausea, and can lead to reduced judgment, dizziness, irritability, fatigue, impaired memory function, deficits in perception and coordination, reaction time, or sleepiness;

(b) Narcotic effects observed in animal studies may include lethargy, lack of coordination righting reflex, narcosis, and ataxia. If these effects are not transient in nature, then they should be considered for classification as Category 1 or 2.

3.8.3 Classification criteria for mixtures

3.8.3.1 Mixtures are classified using the same criteria as for substances, or alternatively as described below. As with substances, mixtures should be classified for specific target organ toxicity for single and repeated exposure (chapter 3.9) independently.

3.8.3.2 *Classification of mixtures when data are available for the complete mixture*

When reliable and good quality evidence from human experience or appropriate studies in experimental animals, as described in the criteria for substances, is available for the mixture, then the mixture can be classified by weight of evidence assessment of this data. Care should be exercised in evaluating data on mixtures, that the dose, duration, observation or analysis, do not render the results inconclusive.

3.8.2.2　区分3の物質

3.8.2.2.1　気道刺激性の基準

区分3としての気道刺激性の基準は以下の通りである：

(a) 咳、痛み、息詰まり、呼吸困難等の症状で機能を阻害する（局所的な赤化、浮腫、かゆみあるいは痛みによって特徴付けられる）ものが気道刺激性に含まれる。この評価は、主としてヒトのデータに基づくと認められている；

(b) 主観的なヒトの観察は、明確な気道刺激性(RTI)の客観的な測定により支持されうる。（例：電気生理学的反応、鼻腔または気管支肺胞洗浄液での炎症に関する生物学的指標）；

(c) ヒトにおいて観察された症状は、他に見られない特有の反応または敏感な気道を持った個人においてのみ誘発された反応であることより、むしろばく露された個体群において生じる典型的な症状でもあるべきである。「刺激性」という単なる漠然とした報告については、この用語は、この分類のエンドポイントの範囲外にある臭い、不愉快な味、くすぐったい感じや乾燥といった感覚を含む広範な感覚を表現するために一般に使用されるので除外するべきである；

(d) 明確に気道刺激性を扱う検証された動物試験は現在存在しないが、有益な情報は、単回および反復吸入毒性試験から得ることができる。例えば、動物試験は、毒性の症候（呼吸困難、鼻炎等）および可逆的な組織病理（充血、浮腫、微少な炎症、肥厚した粘膜層）について有益な情報を提供することができ、上記で述べた特徴的な症候を反映しうる。このような動物実験は証拠の重み付け評価に使用できるであろう；

(e) この特別な分類は、呼吸器系を含むより重篤な臓器への影響は観察されない場合にのみ生じるであろう。

3.8.2.2.2　麻酔作用の判定基準

区分3としての麻酔作用の判定基準は以下の通りである；

(a) 眠気、うとうと感、敏捷性の低下、反射の消失、協調の欠如およびめまいといったヒトにおける麻酔作用を含む中枢神経系の抑制を含む。これらの影響は、ひどい頭痛または吐き気としても現れ、判断力低下、めまい、過敏症、倦怠感、記憶機能障害、知覚や協調の欠如、反応時間（の延長）や嗜眠に到ることもある；

(b) 動物試験において観察される麻酔作用は、嗜眠、協調・立ち直り反射の欠如、昏睡、運動失調を含む。これらの影響が本質的に一時的なものでないならば、区分1また2に分類されると考えるべきである。

3.8.3　混合物の分類基準

3.8.3.1　混合物は、物質に対するものと同じ判定基準、または以下に述べる判定基準を用いて分類される。物質と同じように、混合物は特定標的臓器毒性に関して、単回ばく露および反復ばく露（第3.9章）について独立に分類されるべきである。

3.8.3.2　混合物そのものについて試験データが入手できる場合の混合物の分類

物質に関する判定基準で述べたように、混合物についてヒトでの経験または適切な実験動物での試験から信頼できる質の良い証拠が入手された場合、当該混合物はこのデータの証拠の重み付け評価によって分類できる。混合物に関するデータを評価する際には、用量、ばく露期間、観察、または分析が、結論を不確かにさせることのないように注意を払うべきである。

3.8.3.3 *Classification of mixtures when data are not available for the complete mixture: bridging principles*

3.8.3.3.1 Where the mixture itself has not been tested to determine its specific target organ toxicity, but there are sufficient data on both the individual ingredients and similar tested mixtures to adequately characterize the hazards of the mixture, these data can be used in accordance with the following bridging principles. This ensures that the classification process uses the available data to the greatest extent possible in characterizing the hazards of the mixture without the necessity of additional testing in animals.

3.8.3.3.2 *Dilution*

If a tested mixture is diluted with a diluent which has the same or a lower toxicity classification as the least toxic original ingredient and which is not expected to affect the toxicity of other ingredients, then the new diluted mixture may be classified as equivalent to the original tested mixture.

3.8.3.3.3 *Batching*

The toxicity of a tested production batch of a mixture can be assumed to be substantially equivalent to that of another untested production batch of the same commercial product when produced by or under the control of the same manufacturer, unless there is reason to believe there is significant variation such that the toxicity of the untested batch has changed. If the latter occurs, a new classification is necessary.

3.8.3.3.4 *Concentration of highly toxic mixtures*

If in a tested mixture of Category 1, the concentration of a toxic ingredient is increased, the resulting concentrated mixture should be classified in Category 1 without additional testing.

3.8.3.3.5 *Interpolation within one hazard category*

For three mixtures (A, B and C) with identical ingredients, where mixtures A and B have been tested and are in the same hazard category, and where untested mixture C has the same toxicologically active ingredients as mixtures A and B but has concentrations of toxicologically active ingredients intermediate to the concentrations in mixtures A and B, then mixture C is assumed to be in the same hazard category as A and B.

3.8.3.3.6 *Substantially similar mixtures*

Given the following:

(a) Two mixtures: (i) A + B;
 (ii) C + B;
(b) The concentration of ingredient B is essentially the same in both mixtures;
(c) The concentration of ingredient A in mixture (i) equals that of ingredient C in mixture (ii);
(d) Data on toxicity for A and C are available and substantially equivalent, i.e. they are in the same hazard category and are not expected to affect the toxicity of B.

If mixture (i) or (ii) is already classified based on test data, then the other mixture can be classified in the same hazard category.

3.8.3.3.7 *Aerosols*

An aerosolized form of a mixture may be classified in the same hazard category as the tested, non-aerosolized form of the mixture for oral and dermal toxicity provided the added propellant does not affect the toxicity of the mixture on spraying. Classification of aerosolized mixtures for inhalation toxicity should be considered separately.

3.8.3.3 混合物そのものについてデータが入手できない場合の混合物の分類：つなぎの原則 (bridging principles)

3.8.3.3.1 混合物そのものは特定標的臓器毒性有害性を決定する試験がなされていないが、当該混合物の有害性を適切に特定するための、個々の成分および類似の試験された混合物の両方に関して十分なデータがある場合、これらのデータは以下の合意されたつなぎの原則にしたがって使用される。これによって、分類プロセスで動物試験を追加する必要もなく、混合物の有害性判定に入手されたデータを可能な限り最大限に利用できるようになる。

3.8.3.3.2 *希釈*

試験された混合物が、毒性の最も弱い成分と同等またはそれ以下の毒性分類に属する希釈剤で希釈され、希釈剤が他の成分の毒性に影響を与えないと予想されれば、新しい希釈された混合物は、試験された元の混合物と同等であると分類してもよい。

3.8.3.3.3 *製造バッチ*

混合物の試験されていない製造バッチに毒性があるかどうかは、同じ製造者によって、またはその管理下で生産された同じ商品の試験された別のバッチの毒性と本質的に同等とみなすことができる。ただし、試験されていないバッチの毒性が変化するような有意な変動があると考えられる理由がある場合はこの限りではない。このような場合、新しい分類が必要である。

3.8.3.3.4 *毒性の強い混合物の濃縮*

区分1の試験された混合物で、毒性成分の濃度が増加する場合には、結果として濃縮された混合物は追加試験なしで区分1に分類すべきである。

3.8.3.3.5 *1つの有害性区分内の内挿*

3つの混合物（A、BおよびC）は同じ成分を持ち、AとBは試験され同じ有害性区分にある。混合物Cは混合物AおよびBと同じ毒性学的に活性な成分を持ち、毒性学的に活性な成分の濃度が混合物AとBの中間である場合、混合物Cは、AおよびBと同じ有害性区分であると推定される。

3.8.3.3.6 *本質的に類似した混合物*

次を仮定する：

(a) 2つの混合物： (i)　　A+B；
(ii)　　C+B；

(b) 成分Bの濃度は、両方の混合物で本質的に同じである；

(c) 混合物(i)の成分Aの濃度は、混合物(ii)の成分Cの濃度に等しい；

(d) AとCの毒性に関するデータは利用でき、実質的に同等である、すなわち、AとCは同じ有害性区分に属し、かつ、Bの毒性に影響を与えることはないと判断される。

混合物(i)または(ii)が既に試験データに基づいて分類されている場合には、他方の混合物は同じ有害性区分に分類することができる。

3.8.3.3.7 *エアゾール*

エアゾール形態の混合物は、添加された噴射剤が噴霧の際、混合物の毒性に影響しないという条件下では、経口および経皮毒性について試験された非エアゾル形態の混合物の分類と同じ有害性区分に分類してよい。エアゾール化された混合物の吸入毒性に関する分類は、別個に考えるべきである。

3.8.3.4   *Classification of mixtures when data are available for all ingredients or only for some ingredients of the mixture*

3.8.3.4.1   Where there is no reliable evidence or test data for the specific mixture itself, and the bridging principles cannot be used to enable classification, then classification of the mixture is based on the classification of the ingredient substances. In this case, the mixture will be classified as a specific target organ toxicant (specific organ specified), following single exposure when at least one ingredient has been classified as a Category 1 or Category 2 specific target organ toxicant – single exposure and is present at or above the appropriate cut-off value/concentration limit as mentioned in table 3.8.2 below for Category 1 and 2 respectively.

Table 3.8.2: Cut-off values/concentration limits of ingredients of a mixture classified as a specific target organ toxicant that would trigger classification of the mixture as Category 1 or 2[a]

Ingredient classified as:	Cut-off/concentration limits triggering classification of a mixture as:	
	Category 1	**Category 2**
Category 1 target organ toxicant	≥ 1.0 % (note 1)	1.0 ≤ ingredient < 10 % (note 3)
	≥ 10 % (note 2)	
Category 2 target organ toxicant	--	≥ 1.0 % (note 4)
		≥ 10 % (note 5)

[a]   *This compromise classification scheme involves consideration of differences in hazard communication practices in existing systems. It is expected that the number of affected mixtures will be small; the differences will be limited to label warnings; and the situation will evolve over time to a more harmonized approach.*

NOTE 1:   *If a Category 1 specific target organ toxicant is present in the mixture as an ingredient at a concentration between 1.0 % and 10 %, every regulatory authority would require information on the SDS for a product. However, a label warning would be optional. Some authorities will choose to label when the ingredient is present in the mixture between 1.0 % and 10 %, whereas others would normally not require a label in this case.*

NOTE 2:   *If a Category 1 specific target organ toxicant is present in the mixture as an ingredient at a concentration of ≥ 10 %, both an SDS and a label would generally be expected.*

NOTE 3:   *If a Category 1 specific target organ toxicant is present in the mixture as an ingredient at a concentration between 1.0 % and 10 %, some authorities classify this mixture as a Category 2 specific target organ toxicant, whereas others would not.*

NOTE 4:   *If a Category 2 specific target organ toxicant is present in the mixture as an ingredient at a concentration between 1.0 % and 10 %, every regulatory authority would require information on the SDS for a product. However, a label warning would be optional. Some authorities will choose to label when the ingredient is present in the mixture between 1.0 % and 10 %, whereas others would normally not require a label in this case.*

NOTE 5:   *If a Category 2 specific target organ toxicant is present in the mixture as an ingredient at a concentration of ≥ 10 %, both an SDS and a label would generally be expected.*

3.8.3.4.2   Care should be exercised when toxicants affecting more than one organ system are combined that the potentiation or synergistic interactions are considered, because certain substances can cause target organ toxicity at < 1 % concentration when other ingredients in the mixture are known to potentiate its toxic effect.

3.8.3.4.3   Care should be exercised when extrapolating the toxicity of a mixture that contains Category 3 ingredient(s). A cut-off value/concentration limit of 20 % has been suggested; however, it should be recognized that this cut-off value concentration limit may be higher or less depending on the Category 3 ingredient(s) and that some effects such as respiratory tract irritation may not occur below a certain concentration while other effects such as narcotic effects may occur below this 20 % value. Expert judgment should be exercised. Respiratory tract irritation and narcotic effects are to be evaluated separately in accordance with the criteria given in 3.8.2.2. When conducting classifications for these hazards, the contribution of each ingredient should be considered additive, unless there is evidence that the effects are not additive.

3.8.3.4.4   In cases where the additivity approach is used for Category 3 ingredients, the "relevant ingredients" of a mixture are those which are present in concentrations ≥ 1 % (w/w for solids, liquids, dusts, mists, and vapours and v/v for gases), unless there is a reason to suspect that an ingredient present at a concentration < 1 % is still relevant when classifying the mixture for respiratory tract irritation or narcotic effects.

3.8.3.4 混合物の全成分について、または一部の成分だけについてデータが入手できる場合の混合物の分類

3.8.3.4.1 当該混合物自身について信頼できる証拠または試験データがなく、つなぎの原則を用いて分類できない場合には、混合物の分類は成分物質の分類に基づいて行われる。この場合、混合物の少なくとも1つの成分が区分1または区分2特定標的臓器毒性物質－単回ばく露として分類され、そして区分1または区分2それぞれについて以下の表 3.8.2 に示される適切なカットオフ値/濃度限界またはそれ以上で存在する場合、その混合物は、単回ばく露について特定標的臓器毒性物質（特定された当該臓器/器官の）として分類される。

表 3.8.2：混合物の分類の分類基準となる特定標的臓器毒性物質として
分類された混合物成分の区分1および2のカットオフ値/濃度限界[a]

成分の分類	混合物の分類基準となるカットオフ値/濃度限界：	
	区分1	区分2
区分1 標的臓器毒性物質	≧1.0%（注記1） ≧10%（注記2）	1.0%≦成分<10%（注記3）
区分2 標的臓器毒性物質	—	≧1.0%（注記4） ≧10%（注記5）

[a] この妥協の産物である分類方法は現行の危険有害性の情報伝達における相違を考慮して作成された。影響を受ける混合物の数が少なく、相違はラベル表示に限られ、さらなる調和により状況が良くなることが期待される。

注記1：区分1の特定標的臓器毒性物質が1.0%と10%の間の濃度で成分として混合物中に存在する場合は、すべての規制官庁は、製品のSDSに情報の記載を要求することになろう。しかし、ラベルへの警告表示は任意となろう。ある規制官庁は、成分が1.0%と10%の間で混合物中に存在する場合に表示を選択し、他の官庁は通常この場合に表示を要求しないことになろう。

注記2：区分1の特定標的臓器毒性物質が、≧10%の濃度で成分として混合物中に存在する場合には、一般にSDSと表示の両方が対象となろう。

注記3：区分1の特定標的臓器毒性物質が1.0%と10%の間の濃度で成分として混合物中に存在する場合には、ある規制官庁は、この混合物を区分2の標的臓器毒性物質として分類するのに対して、他の官庁はそうしないことになろう。

注記4：区分2の特定標的臓器毒性物質が1.0%と10%の間の濃度で成分として混合物中に存在する場合には、すべての規制官庁は、製品のSDSに情報の記載を要求することになろう。しかし、ラベルへの警告表示は任意となろう。ある規制官庁は、その成分が1.0%と10%の間で混合物中に存在する場合に表示を選択し、他の官庁は通常、この場合に表示を要求しないことになろう。

注記5：区分2の特定標的臓器毒性物質が、≧10%の濃度で成分として混合物中に存在する場合には、一般にSDSと表示の両方が対象となろう。

3.8.3.4.2 複数の臓器系に影響を与える毒性物質が組合せて使用される場合は、増強作用または相乗作用を考慮するように注意を払うべきである。なぜなら、一部の物質は、混合物中の他の成分がその毒性影響を増強することが知られている場合、<1%の濃度で標的臓器毒性を引き起こす可能性があるからである。

3.8.3.4.3 区分3の成分を含む混合物の毒性を外挿する際には、注意を払うべきである。20%のカットオフ値が提案されてきた。しかしながら、区分3の成分によっては、このカットオフ値がさらに大きくなったり小さくなったりすることがあること、気道刺激性の影響はある濃度以下では生じないが、麻酔作用等他の影響はこの20%の値以下でも生じうるということを認識するべきである。専門家の判断が行われるべきである。気道刺激性と麻酔作用は 3.8.2.2 に示された判定基準にしたがって別々に評価される。これらの有害性について分類するときは、影響が相加的でないという証拠がない限り、それぞれの成分の寄与について相加的に考えるべきである。

3.8.3.4.4 区分3に加成方式が使われる場合、混合物の「考慮すべき成分」とは、≥1%（固体、液体、粉塵、ミストおよび蒸気の場合の場合 w/w、ガスの場合 v/v）の濃度で存在するものである。ただし、気道刺激性または麻酔作用に関して混合物を分類するとき、<1%の濃度で存在する成分が関連していると疑われる理由がある場合を除く。

3.8.4 **Hazard communication**

3.8.4.1 General and specific considerations concerning labelling requirements are provided in *Hazard communication: Labelling* (chapter 1.4). Annex 1 contains summary tables about classification and labelling. Annex 3 contains examples of precautionary statements and pictograms which can be used where allowed by the competent authority. Table 3.8.3 presents specific label elements for substances and mixtures classified into this hazard class based on the criteria in this chapter.

Table 3.8.3: **Label elements for specific target organ toxicity after single exposure**

	Category 1	**Category 2**	**Category 3**
Symbol	Health hazard	Health hazard	Exclamation mark
Signal word	Danger	Warning	Warning
Hazard statement	Causes damage to organs (or state all organs affected, if known) (state route of exposure if it is conclusively proven that no other routes of exposure cause the hazard)	May cause damage to organs (or state all organs affected, if known) (state route of exposure if it is conclusively proven that no other routes of exposure cause the hazard)	May cause respiratory irritation; or May cause drowsiness or dizziness

3.8.5 **Decision logic for specific target organ toxicity following single exposure**

The decision logic which follows is not part of the harmonized classification system but is provided here as additional guidance. It is strongly recommended that the person responsible for classification study the criteria before and during use of the decision logic.

3.8.4 危険有害性情報の伝達

3.8.4.1　表示要件に関する通則および細則は、*危険有害性に関する情報の伝達：表示（第 1.4 章）*に定める。附属書 1 に分類と表示に関する概要表を示す。附属書 3 には、注意書きおよび所管官庁が許可した場合に使用可能な注意絵表示の例を示す。表 3.8.3 には、本章の判定基準に基づいてこの危険有害性クラスに分類された物質および混合物の具体的なラベル要素を示す。

表 3.8.3：単回ばく露による特定標的臓器毒性のラベル要素

	区分 1	区分 2	区分 3
シンボル	健康有害性	健康有害性	感嘆符
注意喚起語	危険	警告	警告
危険有害性情報	臓器の障害 （もし判れば影響を受けるすべての臓器を記載） （他の経路からのばく露が有害でないことが決定的に証明されている場合、有害なばく露経路を記載）	臓器の障害のおそれ （もし判れば影響を受けるすべての臓器を記載） （他の経路からのばく露が有害でないことが決定的に証明されている場合、有害なばく露経路を記載）	呼吸器への刺激のおそれ または 眠気またはめまいのおそれ

3.8.5 単回ばく露による特定標的臓器毒性の判定論理

以下の判定論理は、調和分類システムには含まれないが、追加的な手引きとしてここに示す。分類の責任者に対し、この判定論理を使用する前および使用する際に判定基準についてよく調べ理解することを強く勧める。

3.8.5.1   Decision logic 3.8.1

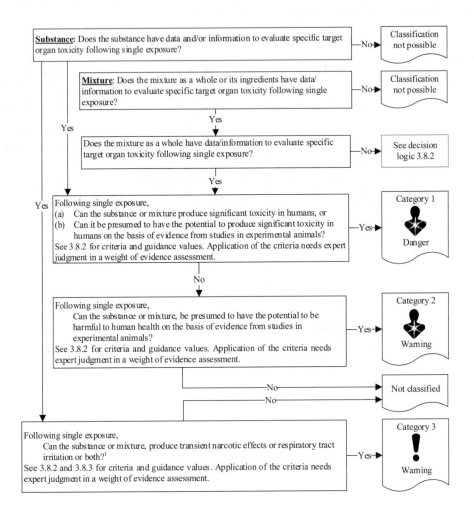

---

[1] *Classification in Category 3 would only occur when classification into Category 1 or Category 2 (based on more severe respiratory effects or narcotic effects that are not transient) is not warranted. See 3.8.2.2.1 (e) (respiratory effects) and 3.8.2.2.2 (b) (narcotic effects).*

3.8.5.1 判定論理 3.8.1

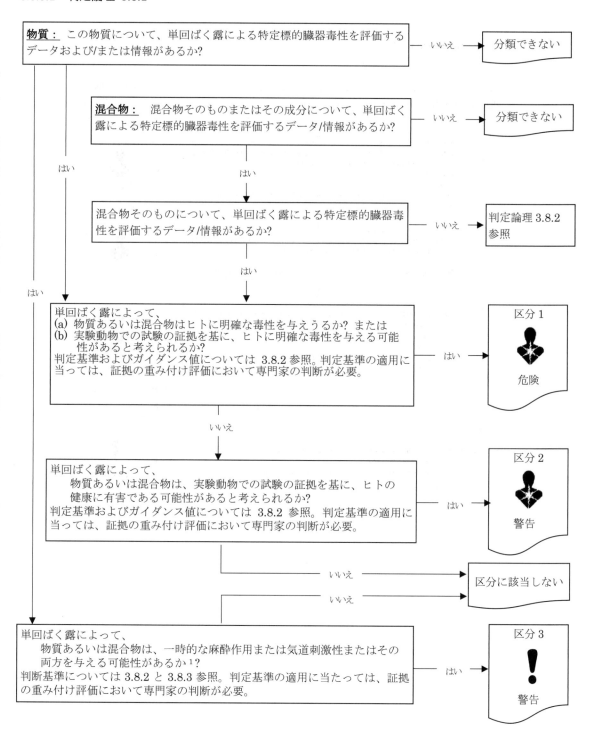

---

[1] 区分3への分類は、区分1または区分2（一時的でないさらに重篤な気道への影響または麻酔作用に基づく）への分類が確実にない場合のみ、行われる。3.8.2.2.1(e)（気道への影響）および3.8.2.2.2(b)（麻酔作用）を参照。

3.8.5.2 *Decision logic 3.8.2*

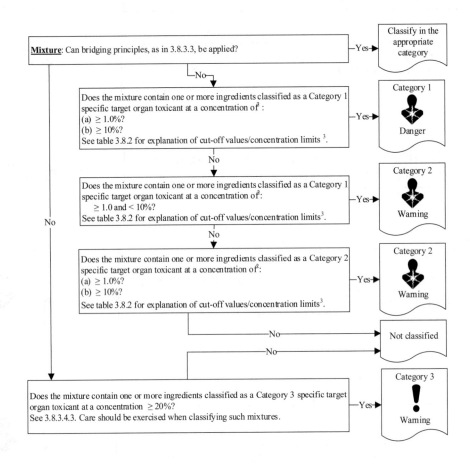

---

[2] *See 3.8.2 of this chapter and "The use of cut-off values/concentration limits" in chapter 1.3, paragraph 1.3.3.2.*
[3] *See 3.8.3.4 and table 3.8.2 for explanation and guidance.*

3.8.5.2 判定論理 3.8.2

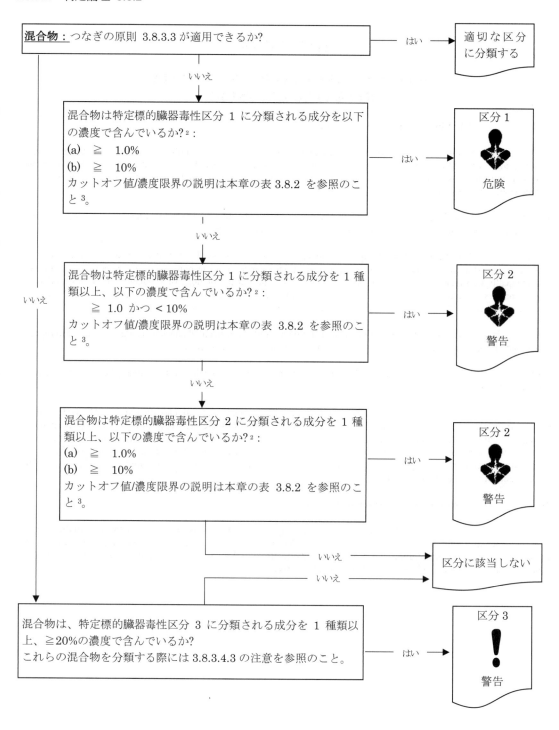

---

[2] 本章3.8.2および第1.3章1.3.3.2の「カットオフ値/濃度限界の使用」を参照のこと

[3] 説明ならびに手引きについては3.8.3.4ならびに表 3.8.2を参照のこと。

CHAPTER 3.9

SPECIFIC TARGET ORGAN TOXICITY REPEATED EXPOSURE

3.9.1 Definitions and general considerations

3.9.1.1 *Specific target organ toxicity-repeated exposure* refers to specific toxic effects on target organs occurring after repeated exposure to a substance or mixture. All significant health effects that can impair function, both reversible and irreversible, immediate and/or delayed and not specifically addressed in chapters 3.1 to 3.7 and 3.10 are included (see also 3.9.1.6).

3.9.1.2 Classification identifies the substance or mixture as being a specific target organ toxicant and, as such, it may present a potential for adverse health effects in people who are exposed to it.

3.9.1.3 Classification depends upon the availability of reliable evidence that a repeated exposure to the substance or mixture has produced a consistent and identifiable toxic effect in humans, or, in experimental animals, toxicologically significant changes which have affected the function or morphology of a tissue/organ, or has produced serious changes to the biochemistry or haematology of the organism and these changes are relevant for human health. It is recognized that human data will be the primary source of evidence for this hazard class.

3.9.1.4 Assessment should take into consideration not only significant changes in a single organ or biological system but also generalized changes of a less severe nature involving several organs.

3.9.1.5 Specific target organ toxicity can occur by any route that is relevant for humans, i.e. principally oral, dermal or inhalation.

3.9.1.6 Non-lethal toxic effects observed after a single-event exposure are classified in the GHS as described in *Specific target organ toxicity – Single exposure* (chapter 3.8) and are therefore excluded from the present chapter. Substances and mixtures should be classified for single and repeated dose toxicity independently. Other specific toxic effects, such as acute toxicity, skin corrosion/irritation, serious eye damage/eye irritation, respiratory or skin sensitization, germ cell mutagenicity, carcinogenicity, reproductive toxicity and aspiration toxicity are assessed separately in the GHS and consequently are not included here.

3.9.2 Classification criteria for substances

3.9.2.1 Substances are classified as specific target organ toxicant by expert judgement on the basis of the weight of all evidence available, including the use of recommended guidance values which take into account the duration of exposure and the dose/concentration which produced the effect(s) (see 3.9.2.9) and are placed in one of two categories, depending upon the nature and severity of the effect(s) observed.

第 3.9 章
特定標的臓器毒性
反復ばく露

3.9.1 定義および一般事項

3.9.1.1 *特定標的臓器毒性－反復ばく露*とは、物質または混合物への反復ばく露後に起こる、特異的な標的臓器への影響をさす。可逆的、不可逆的、あるいは急性または遅発性両方の機能を損ないうる、そして第 3.1 章から第 3.7 章および第 3.10 章では検討されていない、すべての明確な健康への影響がこれに含まれる（3.9.1.6 参照）。

3.9.1.2 この分類は、ある物質または混合物が、特定標的臓器毒性があるか、およびそれにばく露したヒトに対して健康に有害な影響を及ぼす可能性があるかどうかを確認する。

3.9.1.3 分類は、ある物質または混合物に対する反復ばく露がヒトにおける一貫性のある、かつ特定できる毒性影響を与えたこと、あるいは実験動物において組織/臓器の機能または形態に影響する毒性学的に明確な変化が示されたか、または生物の生化学的項目または血液学的項目に重大な変化が示され、これらの変化がヒトの健康状態に関連性があるということについて信頼できる証拠が入手できるかに依存する。この有害性クラスに関しては、ヒトのデータを優先的な証拠とすることが確認されている。

3.9.1.4 評価においては、単一の臓器または生物学的システムにおける明確な変化だけでなく、いくつかの臓器に対するそれほど重度でない一般的変化も考慮すべきである。

3.9.1.5 特定標的臓器毒性は、ヒトに関連するいずれの経路によっても、すなわち主として経口、経皮または吸入によって、起こり得る。

3.9.1.6 GHS における単回ばく露での非致死性毒性の分類については*特定標的臓器毒性－単回ばく露*（第 3.8 章）に述べられており、したがって本章からは除外されている。物質および混合物は、単回および反復投与による毒性に関して独立に分類されるべきである。急性毒性、皮膚腐食性/刺激性、眼に対する重篤な損傷性/眼刺激性、呼吸器または皮膚感作性、生殖細胞変異原性、発がん性、生殖毒性、および誤えん有害性のような、他の特定毒性影響は GHS の中で別に評価されるので、ここには含まれない。

3.9.2 物質の分類基準

3.9.2.1 物質は、影響を生ずるばく露期間および用量/濃度を考慮に入れて勧告されたガイダンス値（3.9.2.9 参照）の使用を含む、入手されたすべての証拠の重みに基づいて専門家の行った判断によって、特定標的臓器毒性物質として分類される。そして、観察された影響の性質および重度によって 2 つの区分のいずれかに分類される。

Figure 3.9.1: Hazard categories for specific target organ toxicity following repeated exposure

CATEGORY 1:	**Substances that have produced significant toxicity in humans, or that, on the basis of evidence from studies in experimental animals can be presumed to have the potential to produce significant toxicity in humans following repeated exposure**
	Placing a substance in Category 1 is done on the basis of:
	(a) reliable and good quality evidence from human cases or epidemiological studies; or,
	(b) observations from appropriate studies in experimental animals in which significant and/or severe toxic effects, of relevance to human health, were produced at generally low exposure concentrations. Guidance dose/concentration values are provided below (see 3.9.2.9) to be used as part of weight-of-evidence evaluation.
CATEGORY 2:	**Substances that, on the basis of evidence from studies in experimental animals can be presumed to have the potential to be harmful to human health following repeated exposure**
	Placing a substance in Category 2 is done on the basis of observations from appropriate studies in experimental animals in which significant toxic effects, of relevance to human health, were produced at generally moderate exposure concentrations. Guidance dose/concentration values are provided below (see 3.9.2.9) in order to help in classification.
	In exceptional cases human evidence can also be used to place a substance in Category 2 (see 3.9.2.6).

NOTE: For both categories the specific target organ/system that has been primarily affected by the classified substance may be identified, or the substance may be identified as a general toxicant. Attempts should be made to determine the primary target organ/system of toxicity and classify for that purpose, e.g. hepatotoxicants, neurotoxicants. One should carefully evaluate the data and, where possible, not include secondary effects, e.g. a hepatotoxicant can produce secondary effects in the nervous or gastro-intestinal systems.

3.9.2.2   The relevant route of exposure by which the classified substance produces damage should be identified.

3.9.2.3   Classification is determined by expert judgement, on the basis of the weight of all evidence available including the guidance presented below.

3.9.2.4   Weight of evidence of all data, including human incidents, epidemiology, and studies conducted in experimental animals, is used to substantiate specific target organ toxic effects that merit classification. This taps the considerable body of industrial toxicology data collected over the years. Evaluation should be based on all existing data, including peer-reviewed published studies and additional data acceptable to regulatory agencies.

3.9.2.5   The information required to evaluate specific target organ toxicity comes either from repeated exposure in humans, e.g. exposure at home, in the workplace or environmentally, or from studies conducted in experimental animals. The standard animal studies in rats or mice that provide this information are 28 day, 90 day or lifetime studies (up to 2 years) that include haematological, clinico-chemical and detailed macroscopic and microscopic examination to enable the toxic effects on target tissues/organs to be identified. Data from repeat dose studies performed in other species may also be used. Other long-term exposure studies, e.g. for carcinogenicity, neurotoxicity or reproductive toxicity, may also provide evidence of specific target organ toxicity that could be used in the assessment of classification.

3.9.2.6   In exceptional cases, based on expert judgement, it may be appropriate to place certain substances with human evidence of specific target organ toxicity in Category 2: (a) when the weight of human evidence is not sufficiently convincing to warrant Category 1 classification, and/or (b) based on the nature and severity of effects. Dose/concentration levels in humans should not be considered in the classification and any available evidence from animal studies should be consistent with the Category 2 classification. In other words, if there are also animal data available on the substance that warrant Category 1 classification, the substance should be classified as Category 1.

図 3.9.1：特定標的臓器毒性 −反復ばく露のための区分

<u>区分 1</u>：　ヒトに明確な毒性を示した物質、または実験動物での試験の証拠に基づいて反復ばく露によって<u>ヒトに明確な毒性を示す</u>可能性があると考えられる物質

   物質を区分 1 に分類するのは、次に基づいて行う：
   (a) ヒトの症例または疫学的研究からの信頼でき、かつ質の良い証拠；または、
   (b) 実験動物での適切な試験において、一般的に低いばく露濃度で、ヒトの健康に関連のある明確な、または重篤な毒性影響を生じたという所見。証拠の重み付けによる評価の一環として使用すべき用量/濃度のガイダンス値は後述する（3.9.2.9 参照）。

<u>区分 2</u>：　動物実験の証拠に基づき反復ばく露によって<u>ヒトの健康に有害である</u>可能性があると考えられる物質

   物質を区分 2 に分類するには、実験動物での適切な試験において、一般的に中等度のばく露濃度で、ヒトの健康に関連のある明確な毒性影響を生じたという所見に基づいて行う。分類に役立つ用量/濃度のガイダンス値は後述する（3.9.2.9 参照）。

   例外的なケースにおいてヒトでの証拠を、物質を区分 2 に分類するために使用できる（3.9.2.6 参照）。

   注記：いずれの区分においても、分類された物質によって最初に影響を受けた特定標的の臓器/器官が明示されるか、または一般的な毒性物質であることが明示される。毒性の主標的臓器を決定し（例えば肝毒性物質、神経毒性物質）、その目的にそって分類するよう努力すべきである。そのデータを注意深く評価し、できる限り二次的影響を含めないようにすべきである。例えば、肝毒性物質は、神経または消化器官に二次的影響を起こすことがある。

3.9.2.2　分類した物質が損傷を起こしたばく露経路を明示すべきである。

3.9.2.3　分類は、後述の手引きを含む、入手されたすべての証拠の重み付けに基づいて、専門家の判断によって決定する。

3.9.2.4　ヒトでの疾患の発生情報、疫学情報および実験動物を用いて実施した試験結果を含む、すべてのデータについての証拠の重み付けは、分類に役立つ特定標的臓器毒性影響を実証するために使用される。これは長年にわたって集められた大量の産業毒性学データを利用することになる。評価は、校閲され公表された研究論文および規制官庁が受理し得る追加データを含む、すべての既存データに基づくべきである。

3.9.2.5　特定標的臓器毒性を評価するために必要な情報は、ヒトにおける反復ばく露、例えば、家庭、作業場あるいは環境中でのばく露、または実験動物を用いて実施した試験のいずれからも得られる。この情報を提供するラットまたはマウスにおける標準的動物試験は 28 日間、90 日間または生涯試験（2 年間まで）であり、標的組織/臓器に対する毒性影響を確認するための血液学的検査、臨床化学的検査、詳細な肉眼的および病理組織学的検査を含んでいる。その他の動物種を用いて実施された反復投与試験のデータも利用し得る。また、その他の長期ばく露試験、例えば、発がん性試験、神経毒性試験または生殖毒性試験も、分類評価のために使用する特定標的臓器毒性の証拠を提供するかもしれない。

3.9.2.6　例外的に、特定標的臓器毒性のヒトでの証拠を有するある種の物質を、専門家の判断に基づいて、区分 2 に分類するのが適切な場合がある：それは(a)ヒトでの証拠の重み付けが区分 1 への分類を正当化することが十分には確信できない場合、または(b)影響の性質および重度に基づく場合である。ヒトにおける用量/濃度レベルは、分類において考慮すべきではなく、動物試験で入手された証拠が、区分 2 への分類と矛盾しないことである。換言すれば、物質について区分 1 への分類を保証する動物試験データが入手されている場合、その物質は区分 1 に分類するべきである。

3.9.2.7 *Effects considered to support classification*

3.9.2.7.1 Reliable evidence associating repeated exposure to the substance with a consistent and identifiable toxic effect demonstrates support for classification.

3.9.2.7.2 It is recognized that evidence from human experience/incidents is usually restricted to reports of adverse health consequences, often with uncertainty about exposure conditions, and may not provide the scientific detail that can be obtained from well-conducted studies in experimental animals.

3.9.2.7.3 Evidence from appropriate studies in experimental animals can furnish much more detail, in the form of clinical observations, haematology, clinical chemistry, macroscopic and microscopic pathological examination and this can often reveal hazards that may not be life-threatening but could indicate functional impairment. Consequently all available evidence, and relevance to human health, must be taken into consideration in the classification process. Examples of relevant toxic effects in humans and/or animals are provided below:

(a) Morbidity or death resulting from repeated or long-term exposure. Morbidity or death may result from repeated exposure, even to relatively low doses/concentrations, due to bioaccumulation of the substance or its metabolites, or due to the overwhelming of the de-toxification process by repeated exposure;

(b) Significant functional changes in the central or peripheral nervous systems or other organ systems, including signs of central nervous system depression and effects on special senses (e.g. sight, hearing and sense of smell);

(c) Any consistent and significant adverse change in clinical biochemistry, haematology, or urinalysis parameters;

(d) Significant organ damage that may be noted at necropsy and/or subsequently seen or confirmed at microscopic examination;

(e) Multifocal or diffuse necrosis, fibrosis or granuloma formation in vital organs with regenerative capacity;

(f) Morphological changes that are potentially reversible but provide clear evidence of marked organ dysfunction (e.g. severe fatty change in the liver);

(g) Evidence of appreciable cell death (including cell degeneration and reduced cell number) in vital organs incapable of regeneration.

3.9.2.8 *Effects considered not to support classification*

It is recognized that effects may be seen that would not justify classification. Examples of such effects in humans and/or animals are provided below:

(a) Clinical observations or small changes in bodyweight gain, food consumption or water intake that may have some toxicological importance but that do not, by themselves, indicate "significant" toxicity;

(b) Small changes in clinical biochemistry, haematology or urinalysis parameters and/or transient effects, when such changes or effects are of doubtful or minimal toxicological importance;

(c) Changes in organ weights with no evidence of organ dysfunction;

(d) Adaptive responses that are not considered toxicologically relevant;

(e) Substance-induced species-specific mechanisms of toxicity, i.e. demonstrated with reasonable certainty to be not relevant for human health, should not justify classification.

3.9.2.7 分類を支持すると考えられる影響

3.9.2.7.1 一貫して特定できる毒性作用を有する物質に反復ばく露したという証拠がある場合には、分類を支持する。

3.9.2.7.2 ヒトでの経験/疾患の発生から得られる証拠は、通常健康被害の報告に限定され、ばく露条件については不確実なことがしばしばであり、実験動物で適切に実施された試験から得られるような科学的な詳細情報は提供されないと理解されている。

3.9.2.7.3 実験動物での適切な試験からの証拠は、臨床所見、血液学検査、臨床化学検査、肉眼および顕微鏡による病理組織学的検査の形で、はるかに詳細な内容を提供することができ、そして、これは生命への危険には至らないが機能障害を起こすかもしれない有害性を、しばしば明らかにすることができる。したがって、入手されたすべての証拠およびヒトの健康との関連性は、分類の過程において考慮を払う必要がある。ヒトまたは実験動物における関連のある毒性影響の例を、以下に示す：

(a) 反復あるいは長期ばく露に起因する罹患または死亡。比較的低い用量/濃度においても、当該物質またはその代謝物の生物蓄積によって、あるいは反復ばく露によって解毒過程が機能しなくなることによって、反復ばく露で罹患または死亡に至る可能性がある；

(b) 中枢神経系抑制、および特定の感覚器（例えば視覚、聴覚および嗅覚）に及ぼす影響を含む、中枢または末梢神経系あるいはその他の器官系における明確な機能変化；

(c) 臨床生化学的検査、血液学的検査または尿検査の項目における、一貫した明確で有害な変化；

(d) 剖検時に観察され、またはその後の病理組織学的検査時に認められ、または確認された、明確な臓器損傷；

(e) 再生能力を有する生体臓器における多発性またはびまん性壊死、線維症または肉芽腫形成；

(f) 潜在的に可逆的であるが、臓器の著しい機能障害の明確な証拠を提供する形態学的変化（例えば、肝臓における重度の脂肪変化）；

(g) 再生が不可能な生体臓器における明白な細胞死の証拠（細胞の退化および細胞数の減少を含む）。

3.9.2.8 分類を支持しないと考えられる影響

分類を正当化しないと考えられている影響があることが認められている。ヒトまたは実験動物におけるこのような影響の例を、以下に示す：

(a) 毒性学的にはいくらか重要かもしれないが、それだけでは「明確な」毒性を示すものではない臨床所見、または体重増加量、摂餌量または摂水量のわずかな変化；

(b) 臨床生化学的検査、血液学的検査または尿検査の項目における軽度の変化または一時的な影響で、このような変化または影響に疑いがある場合、または毒性学的意義がほとんどない場合；

(c) 臓器機能障害の証拠のない臓器重量の変化；

(d) 毒性学的に重要と考えられない適応反応；

(e) 物質が誘発する種に特異な毒性メカニズムで、合理的確実性をもってヒトの健康との関係性を持たないことが実証されたものは分類を正当化すべきでない。

3.9.2.9 *Guidance values to assist with classification based on the results obtained from studies conducted in experimental animals*

3.9.2.9.1 In studies conducted in experimental animals, reliance on observation of effects alone, without reference to the duration of experimental exposure and dose/concentration, omits a fundamental concept of toxicology, i.e. all substances are potentially toxic, and what determines the toxicity is a function of the dose/concentration and the duration of exposure. In most studies conducted in experimental animals the test guidelines use an upper limit dose value.

3.9.2.9.2 In order to help reach a decision about whether a substance should be classified or not, and to what degree it would be classified (Category 1 versus Category 2), dose/concentration "guidance values" are provided in table 3.9.1 for consideration of the dose/concentration which has been shown to produce significant health effects. The principal argument for proposing such guidance values is that all chemicals are potentially toxic and there has to be a reasonable dose/concentration above which a degree of toxic effect is acknowledged. Also, repeated-dose studies conducted in experimental animals are designed to produce toxicity at the highest dose used in order to optimize the test objective and so most studies will reveal some toxic effect at least at this highest dose. What is therefore to be decided is not only what effects have been produced, but also at what dose/concentration they were produced and how relevant is that for humans.

3.9.2.9.3 Thus, in animal studies, when significant toxic effects are observed, that would indicate classification, consideration of the duration of experimental exposure and the dose/concentration at which these effects were seen, in relation to the suggested guidance values, can provide useful information to help assess the need to classify (since the toxic effects are a consequence of the hazardous property(ies) and also the duration of exposure and the dose/concentration).

3.9.2.9.4 The decision to classify at all can be influenced by reference to the dose/concentration guidance values at or below which a significant toxic effect has been observed.

3.9.2.9.5 The guidance values proposed refer basically to effects seen in a standard 90-day toxicity study conducted in rats. They can be used as a basis to extrapolate equivalent guidance values for toxicity studies of greater or lesser duration, using dose/exposure time extrapolation similar to Haber's rule for inhalation, which states essentially that the effective dose is directly proportional to the exposure concentration and the duration of exposure. The assessment should be done on a case-by-case basis; e.g. for a 28-day study the guidance values below would be increased by a factor of three.

3.9.2.9.6 Thus for Category 1 classification, significant toxic effects observed in a 90-day repeated-dose study conducted in experimental animals and seen to occur at or below the (suggested) guidance values as indicated in table 3.9.1 would justify classification:

Table 3.9.1: Guidance values to assist in Category 1 classification

Route of exposure	Units	Guidance values (dose/concentration)
Oral (rat)	mg/kg bw/d	≤ 10
Dermal (rat or rabbit)	mg/kg bw/d	≤ 20
Inhalation (rat) gas	ppmV/6h/d	≤ 50
Inhalation (rat) vapour	mg/litre/6h/d	≤ 0.2
Inhalation (rat) dust/mist/fume	mg/litre/6h/d	≤ 0.02

Note: "bw" is for "body weight", "h" for "hour" and "d" for "day".

3.9.2.9.7 For Category 2 classification, significant toxic effects observed in a 90-day repeated-dose study conducted in experimental animals and seen to occur within the (suggested) guidance value ranges as indicated in table 3.9.2 would justify classification:

3.9.2.9 実験動物を用いて実施した試験で得られた結果に基づいた分類を補助するガイダンス値

3.9.2.9.1 実験動物を使って行われた研究において、実験のばく露時間および用量/濃度を参照することなく影響の観察にのみ依存することは、「すべての物質は潜在的に毒性を有し、毒性は用量/濃度およびばく露時間の関数となる」という毒性学の基本概念の1つを無視していることになる。実験動物を使った研究の大半においては、試験指針には上限値の用量が使われている。

3.9.2.9.2 物質を分類すべきであるか否か、また、どのランク（区分1か、区分2か）に分類するかについての決定を下すことを助ける目的で、明確な健康影響を生じることが示されたことのある用量/濃度を考察するための用量/濃度「ガイダンス値」を表3.9.1に掲げる。そのようなガイダンス値を提案する主要な論拠は、すべての化学物質は潜在的に有毒であり、それ以上ではある程度の毒性影響が確認される妥当な用量/濃度が存在するに違いないからである。また、動物を用いて実施される反復投与試験は、試験目的を最も効果的にするために、使用した最高用量で毒性を生ずるよう設計され、ほとんどの試験では、少なくとも最高用量ではいくつかの毒性影響を示す。したがって、決定すべきことは、どのような作用が生ずるかだけでなく、どのような用量/濃度で作用が生じるか、そして、それをヒトに対してどのように関連づけるかである。

3.9.2.9.3 したがって、動物試験において、分類を示す明確な毒性影響が認められた場合、提案されたガイダンス値と比較して、試験したばく露期間およびこれらの影響が認められた用量/濃度を考察することは、分類の必要性を評価するのを助けるための有益な情報を提供する（毒性影響は有害性と、ばく露期間および用量/濃度との結果であるから）。

3.9.2.9.4 ガイダンス値またはそれ以下の用量/濃度で明確な毒性影響が観察されたかを参照することで、分類の決定が影響されることがある。

3.9.2.9.5 提案されたガイダンス値は、基本的にはラットを用いて実施した標準の90日間毒性試験で認められた影響に基づいている。これらのガイダンス値は、より長期の、またはより短期のばく露による毒性試験に対する等価ガイダンス値を外挿するための基礎として使用できる。これは、吸入毒性についてのハーバー則（有効な用量はばく露濃度とばく露期間に比例する）と同様な、用量やばく露時間に関する外挿をするものである。その評価はケースバイケースを原則に行うべきである。例えば、28日間の試験については、下記のガイダンス値を3倍して使用する。

3.9.2.9.6 したがって区分1への分類に当たっては、実験動物を使った90日間の反復投与試験において、表3.9.1に示すガイダンス値（案）またはこれを下回る値で観察された明確な毒性影響が、分類を正当化するものとなる。

表3.9.1：区分1への分類を助けるガイダンス値

ばく露経路	単位	ガイダンス値（用量/濃度）
経口（ラット）	mg/kg 体重/日	≦10
経皮（ラットまたはウサギ）	mg/kg 体重/日	≦20
吸入（ラット）気体	ppmV/6 時間/日	≦50
吸入（ラット）蒸気	mg/リットル/6 時間/日	≦0.2
吸入（ラット）粉塵/ミスト/ヒューム	mg/リットル/6 時間/日	≦0.02

注記：「bw」は「体重」、「h」は「時間」、「d」は「日」を表す。

3.9.2.9.7 区分2への分類については、実験動物を用いて実施した90日間反復投与試験で観察され、かつ表3.9.2に示すガイダンス値（案）の範囲内で起こることが認められた明確な毒性影響が、分類を正当化するものとなる。

Table 3.9.2: Guidance values to assist in Category 2 classification

Route of exposure	Units	Guidance value range (dose/concentration)
Oral (rat)	mg/kg bw/d	$10 < C \leq 100$
Dermal (rat or rabbit)	mg/kg bw/d	$20 < C \leq 200$
Inhalation (rat) gas	ppmV/6h/d	$50 < C \leq 250$
Inhalation (rat) vapour	mg/litre/6h/d	$0.2 < C \leq 1.0$
Inhalation (rat) dust/mist/fume	mg/litre/6h/d	$0.02 < C \leq 0.2$

Note: "bw" is for body weight, "h" for" hour" and "d" for "day".

3.9.2.9.8   The guidance values and ranges mentioned in 3.9.2.9.6 and 3.9.2.9.7 are intended only for guidance purposes, i.e. to be used as part of the weight of evidence assessment, and to assist with decisions about classification. They are not intended as strict demarcation values.

3.9.2.9.9   Thus it is feasible that a specific profile of toxicity is seen to occur in repeat-dose animal studies at a dose/concentration below the guidance value, eg. < 100 mg/kg bw/day by the oral route, however the nature of the effect, e.g. nephrotoxicity seen only in male rats of a particular strain known to be susceptible to this effect, may result in the decision not to classify. Conversely, a specific profile of toxicity may be seen in animal studies occurring at above a guidance value, eg. ≥ 100 mg/kg bw/day by the oral route, and in addition there is supplementary information from other sources, e.g. other long-term administration studies, or human case experience, which supports a conclusion that, in view of the weight of evidence, classification would be the prudent action to take.

3.9.2.10   *Other considerations*

3.9.2.10.1   When a substance is characterized only by use of animal data (typical of new substances, but also true for many existing substances), the classification process would include reference to dose/concentration guidance values as one of the elements that contribute to the weight of evidence assessment.

3.9.2.10.2   When well-substantiated human data are available showing a specific target organ toxic effect that can be reliably attributed to repeated or prolonged exposure to a substance, the substance may be classified. Positive human data, regardless of probable dose, predominates over animal data. Thus, if a substance is unclassified because no specific target organ toxicity was seen at or below the proposed dose/concentration guidance value for animal testing, if subsequent human incident data become available showing a specific target organ toxic effect, the substance should be classified.

3.9.2.10.3   A substance that has not been tested for specific target organ toxicity may in certain instances, where appropriate, be classified on the basis of data from a validated structure activity relationship and expert judgement-based extrapolation from a structural analogue that has previously been classified together with substantial support from consideration of other important factors such as formation of common significant metabolites.

3.9.2.10.4   It is recognized that saturated vapour concentration may be used as an additional element by some regulatory systems to provide for specific health and safety protection.

3.9.3   Classification criteria for mixtures

3.9.3.1   Mixtures are classified using the same criteria as for substances, or alternatively as described below. As with substances, mixtures should be classified for specific target organ toxicity for single exposure (see chapter 3.8) and repeated exposure independently.

3.9.3.2   *Classification of mixtures when data are available for the complete mixture*

When reliable and good quality evidence from human experience or appropriate studies in experimental animals, as described in the criteria for substances, is available for the mixture, then the mixture can be classified by weight of evidence assessment of this data. Care should be exercised in evaluating data on mixtures, that the dose, duration, observation or analysis, do not render the results inconclusive.

表 3.9.2：区分 2 への分類を助けるガイダンス値

ばく露経路	単位	ガイダンス値範囲（用量/濃度）
経口（ラット）	mg/kg 体重/日	$10 < C \leqq 100$
経皮（ラットまたはウサギ）	mg/kg 体重/日	$20 < C \leqq 200$
吸入（ラット）気体	ppmV/6 時間/日	$50 < C \leqq 250$
吸入（ラット）蒸気	mg/リットル/6 時間/日	$0.2 < C \leqq 1.0$
吸入（ラット）粉塵/ミスト/ヒューム	mg/リットル/6 時間/日	$0.02 < C \leqq 0.2$

注記：「bw」は「体重」、「h」は「時間」、「d」は「日」を表す。

3.9.2.9.8　3.9.2.9.6 および 3.9.2.9.7 に記載したガイダンス値および範囲は、あくまでもガイダンスとしてのためのものである。すなわち、証拠の重み付け評価の一環として、分類の決定を助けるためのものであって、厳密な境界値として意図されたものではない。

3.9.2.9.9　反復投与動物試験においてガイダンス値以下の用量/濃度、例えば< 100mg/kg 体重/日の経口投与で、ある毒性が観察されても、この影響を受けやすいことが知られている特定系統の雄ラットだけに認められた腎毒性のように、影響の性質によっては分類しないと決定することもありうる。逆に、特定の毒性プロフィールが、動物試験においてガイダンス値以上の用量/濃度、例えば≧100mg/kg 体重/日の経口投与で起こることがあり、そして他の情報源からの補足情報、例えば、他の長期投与試験またはヒトでの症例経験などその結論を支持するものがある場合は証拠の重み付けを考慮して、分類することが賢明であろう。

3.9.2.10　その他の考慮事項

3.9.2.10.1　物質が動物データのみによって特徴付けられる場合（新規物質に典型的な事例であるが、多くの既存物質も同様に）、分類プロセスには、証拠の重み付け評価への寄与要素の 1 つとして、用量/濃度ガイダンス値を参照することが含まれるであろう。

3.9.2.10.2　物質への反復または長期ばく露に確実に起因するとされる特定標的臓器毒性影響を示す、適正に実証されたヒトのデータが入手できた場合、その物質は分類できる。投与量が推定でしかなくても、ヒトの陽性データは動物データに優先する。したがって、ある物質が、動物試験のために提案された用量/濃度ガイダンス値、またはそれ以下の投与量で特定標的臓器毒性が認められず、分類されなかった場合、もしもその後に特定標的臓器毒性影響を示すヒトでの疾患の発生データが入手されれば、その物質を分類すべきである。

3.9.2.10.3　特定標的臓器毒性について試験をされていない物質でも、場合によっては、検証された構造活性相関データ、および共通の重要な代謝物を生成する等他の重要な要因の考慮からの実質的な支援も合わせて、すでに分類された構造類似体から専門家の判断に基づいて外挿して、分類することも可能であろう。

3.9.2.10.4　規制システムによっては、特別な健康および安全保護のために飽和蒸気濃度を追加要因として利用してもよいと認められている。

3.9.3　混合物の分類基準

3.9.3.1　混合物は、物質に対するものと同じ判定基準、または以下に述べる基準を用いて分類される。物質と同じように、混合物は特定標的臓器毒性に関して、単回ばく露（第 3.8 章参照）および反復ばく露について独立に分類されるべきである。

3.9.3.2　*混合物そのものについて試験データが入手できる場合の混合物の分類*

　物質に関する判定基準で述べたように、混合物についてヒトでの経験または適切な実験動物での試験から信頼できる質の良い証拠が入手された場合、当該混合物はこのデータの証拠の重み付け評価によって分類できる。混合物に関するデータを評価する際には、用量、ばく露期間、観察、または分析が、結論を不確かにさせていないかに注意を払うべきである。

3.9.3.3 *Classification of mixtures when data are not available for the complete mixture: bridging principles*

3.9.3.3.1 Where the mixture itself has not been tested to determine its specific target organ toxicity, but there are sufficient data on both the individual ingredients and similar tested mixtures to adequately characterize the hazards of the mixture, these data can be used in accordance with the following bridging principles. This ensures that the classification process uses the available data to the greatest extent possible in characterizing the hazards of the mixture without the necessity of additional testing in animals.

3.9.3.3.2 *Dilution*

If a tested mixture is diluted with a diluent which has the same or a lower toxicity classification as the least toxic original ingredient and which is not expected to affect the toxicity of other ingredients, then the new diluted mixture may be classified as equivalent to the original tested mixture.

3.9.3.3.3 *Batching*

The toxicity of a tested production batch of a mixture can be assumed to be substantially equivalent to that of another untested production batch of the same commercial product when produced by or under the control of the same manufacturer, unless there is reason to believe there is significant variation such that the toxicity of the untested batch has changed. If the latter occurs, a new classification is necessary.

3.9.3.3.4 *Concentration of highly toxic mixtures*

If in a tested mixture of Category 1, the concentration of a toxic ingredient is increased, the resulting concentrated mixture should be classified in Category 1 without additional testing.

3.9.3.3.5 *Interpolation within one hazard category*

For three mixtures (A, B and C) with identical ingredients, where mixtures A and B have been tested and are in the same hazard category, and where untested mixture C has the same toxicologically active ingredients as mixtures A and B but has concentrations of toxicologically active ingredients intermediate to the concentrations in mixtures A and B, then mixture C is assumed to be in the same hazard category as A and B.

3.9.3.3.6 *Substantially similar mixtures*

Given the following:

(a) Two mixtures: (i) A + B;
(ii) C + B;

(b) The concentration of ingredient B is essentially the same in both mixtures;

(c) The concentration of ingredient A in mixture (i) equals that of ingredient C in mixture (ii);

(d) Data on toxicity for A and C are available and substantially equivalent, i.e. they are in the same hazard category and are not expected to affect the toxicity of B.

If mixture (i) or (ii) is already classified based on test data, then the other mixture can be classified in the same hazard category.

3.9.3.3.7 *Aerosols*

An aerosolized form of a mixture may be classified in the same hazard category as the tested, non-aerosolized form of the mixture for oral and dermal toxicity provided the added propellant does not affect the toxicity of the mixture on spraying. Classification of aerosolized mixtures for inhalation toxicity should be considered separately.

3.9.3.3 混合物そのものについて試験データが入手できない場合の混合物の分類：つなぎの原則 (bridging principle)

3.9.3.3.1 混合物そのものは、特定標的臓器毒性を決定するために試験が行われていないが、当該混合物の有害性を適切に特定するための、個々の成分および類似の試験された混合物の両方に関して十分なデータがある場合、これらのデータは以下の合意されたつなぎの原則にしたがって使用される。これによって、分類プロセスに動物試験を追加する必要もなく、混合物の有害性判定に入手されたデータを可能な限り最大限に用いることができる。

3.9.3.3.2 *希釈*

試験された混合物が、毒性の最も弱い成分と同等以下の毒性分類に属する希釈剤で希釈され、希釈剤が他の成分の毒性に影響を与えないことが予想されれば、新しい希釈された混合物は、試験された元の混合物と同等であると分類してもよい。

3.9.3.3.3 *製造バッチ*

混合物の試験されていない製造バッチに毒性があるかどうかは、同じ製造者によって、またはその管理下で生産された同じ商品の試験された別のバッチの毒性と本質的に同等とみなすことができる。ただし、試験されていないバッチの毒性が変化するような有意な変動があると考えられる理由がある場合はこの限りではない。このような場合、新しい分類が必要である。

3.9.3.3.4 *毒性の強い混合物の濃縮*

試験された混合物が区分 1 に分類され、毒性成分の濃度が増加する場合には、結果として濃縮された混合物は追加試験なしで区分 1 に分類すべきである。

3.9.3.3.5 *1 つの有害性区分内の内挿*

3 つの混合物（A、B および C）は同じ成分を持ち、A と B は試験され同じ有害性区分にある。試験されていない混合物 C は混合物 A および B と同じ毒性学的に活性な成分を持ち、毒性学的に活性な成分の濃度が混合物 A と B の中間である場合、混合物 C は A および B と同じ有害性区分にあるとする。

3.9.3.3.6 *本質的に類似した混合物*

次を仮定する：

(a) 2 つの混合物： (i)　　A+B；
　　　　　　　　　(ii)　　C+B；

(b) 成分 B の濃度は、両方の混合物で本質的に同じである；

(c) 混合物(i)の成分 A の濃度は、混合物(ii)の成分 C の濃度に等しい；

(d) A と C の毒性に関するデータは利用でき、実質的に同等である、すなわち、A と C は同じ有害性区分に属し、かつ、B の毒性に影響を与えることはないと判断される。

混合物(i)または(ii)が既に試験データに基づいて分類されている場合には、他方の混合物は同じ有害性区分に分類することができる。

3.9.3.3.7 *エアゾール*

エアゾール形態の混合物は、添加された噴射剤が噴霧時に混合物の毒性に影響しないという条件下では、経口および経皮毒性について試験された非エアゾール形態の混合物と同じ有害性区分に分類してよい。エアゾール化された混合物の吸入毒性に関する分類は、個別に考慮されるべきである。

3.9.3.4 *Classification of mixtures when data are available for all ingredients or only for some ingredients of the mixture*

3.9.3.4.1 Where there is no reliable evidence or test data for the specific mixture itself, and the bridging principles cannot be used to enable classification, then classification of the mixture is based on the classification of the ingredient substances. In this case, the mixture will be classified as a specific target organ toxicant (specific organ specified), following repeated exposure, when at least one ingredient has been classified as a Category 1 or Category 2 specific target organ toxicant – repeated exposure and is present at or above the appropriate cut-off value/concentration limit as mentioned in table 3.9.3 for Category 1 and 2 respectively.

Table 3.9.3: Cut-off values/concentration limits of ingredients of a mixture classified as a specific target organ toxicant that would trigger classification of the mixture[a]

Ingredient classified as:	Cut-off/concentration limits triggering classification of a mixture as:	
	Category 1	**Category 2**
Category 1 target organ toxicant	≥ 1.0 % (note 1)	1.0 ≤ ingredient < 10 % (note 3)
	≥ 10 % (note 2)	1.0 ≤ ingredient < 10 % (note 3)
Category 2 target organ toxicant		≥ 1.0 % (note 4)
		≥ 10 % (note 5)

[a] *This compromise classification scheme involves consideration of differences in hazard communication practices in existing systems. It is expected that the number of affected mixtures will be small; the differences will be limited to label warnings; and the situation will evolve over time to a more harmonized approach.*

NOTE 1: *If a Category 1 specific target organ toxicant is present in the mixture as an ingredient at a concentration between 1.0 % and 10 %, every regulatory authority would require information on the SDS for a product. However, a label warning would be optional. Some authorities will choose to label when the ingredient is present in the mixture between 1.0 % and 10 %, whereas others would normally not require a label in this case.*

NOTE 2: *If a Category 1 specific target organ toxicant is present in the mixture as an ingredient at a concentration of ≥ 10 %, both an SDS and a label would generally be expected.*

NOTE 3: *If a Category 1 specific target organ toxicant is present in the mixture as an ingredient at a concentration between 1.0 % and 10 %, some authorities classify this mixture as a Category 2 target organ toxicant, whereas others would not.*

NOTE 4: *If a Category 2 specific target organ toxicant is present in the mixture as an ingredient at a concentration between 1.0 % and 10 %, every regulatory authority would require information on the SDS for a product. However, a label warning would be optional. Some authorities will choose to label when the ingredient is present in the mixture between 1.0 % and 10 %, whereas others would normally not require a label in this case.*

NOTE 5: *If a Category 2 specific target organ toxicant is present in the mixture as an ingredient at a concentration of ≥ 10 %, both an SDS and a label would generally be expected.*

3.9.3.4.2 These cut-off values and consequent classifications should be applied equally and appropriately to both single- and repeated-dose target organ toxicants.

3.9.3.4.3 Mixtures should be classified for either or both single- and repeated-dose toxicity independently.

3.9.3.4.4 Care should be exercised when toxicants affecting more than one organ system are combined that the potentiation or synergistic interactions are considered, because certain substances can cause specific target organ toxicity at < 1 % concentration when other ingredients in the mixture are known to potentiate its toxic effect.

3.9.3.4 混合物の全成分について、または一部の成分だけについてデータが入手できた場合の混合物の分類

3.9.3.4.1 当該混合物自身について信頼できる証拠または試験データがなく、つなぎの原則を用いて分類できない場合には、混合物の分類は成分物質の分類に基づいて行われる。この場合、混合物の少なくとも1つの成分が区分1または区分2特定標的臓器毒性物質－反復ばく露として分類され、そして区分1や区分2それぞれについて以下の表3.9.3に示される適切なカットオフ値/濃度限界またはそれ以上の濃度で存在する場合、その混合物は、反復ばく露について特定標的臓器毒性物質（特定された当該臓器/器官の）として分類される。

表 3.9.3：混合物の分類のための、特定標的臓器毒性物質として分類された混合物の成分のカットオフ値/濃度限界 [a]

成分の分類：	混合物の分類のためのカットオフ値/濃度限界：	
	区分1	区分2
区分1 標的臓器毒性物質	≧1.0%（注1） ≧10%（注2）	1.0%≦成分＜10%（注3） 1.0%≦成分＜10%（注3）
区分2 標的臓器毒性物質		≧1.0%（注4） ≧10%（注5）

[a] この妥協の産物である分類方法は現行の危険有害性の情報伝達における相違を考慮して作成された。影響を受ける混合物の数が少なく、相違はラベル表示に限られ、さらなる調和により状況が良くなることが期待される。

注記1：区分1の特定標的臓器毒性物質が1.0%と10%の間の濃度で成分として混合物中に存在する場合は、すべての規制官庁は、製品のSDSに情報の記載を要求することになろう。しかし、ラベルへの警告表示は任意となろう。ある規制官庁は、成分が1.0%と10%の間で混合物中に存在する場合に表示を選択し、他の官庁は通常この場合にラベル表示を要求しないことになろう。

注記2：区分1の特定標的臓器毒性物質が、≧10%の濃度で成分として混合物中に存在する場合には、一般にSDSと表示の両方が対象となろう。

注記3：区分1の特定標的臓器毒性物質が1.0%と10%の間の濃度で成分として混合物中に存在する場合には、ある規制官庁は、この混合物を区分2の標的臓器毒性物質として分類するのに対して、他の官庁はそうしないことになろう。

注記4：区分2の特定標的臓器毒性物質が1.0%と10%の間の濃度で成分として混合物中に存在する場合には、すべての規制官庁は、製品のSDSに情報の記載を要求することになろう。しかし、ラベルへの警告表示は任意となろう。ある規制官庁は、その成分が1.0%と10%の間で混合物中に存在する場合に表示を選択し、他の官庁は通常、この場合にラベル表示を要求しないことになろう。

注記5：区分2の特定標的臓器毒性物質が、≧10%の濃度で成分として混合物中に存在する場合には、一般にSDSと表示の両方が対象となろう。

3.9.3.4.2 これらのカットオフ値およびその結果として生じる分類は、単回および反復投与標的臓器毒性物質の両方に同等にそして適切に適用されるべきである。

3.9.3.4.3 混合物は、単回および反復投与毒性のいずれかまたは両方について、独立して分類されるべきである。

3.9.3.4.4 複数の臓器系に影響を与える毒性物質が組合せて使用される場合は、増強作用または相乗作用を考慮するように注意を払うべきである。なぜなら、一部の物質は、混合物中の他の成分がその毒性影響を増強することが知られている場合、＜1%の濃度で特定標的臓器毒性を引き起こす可能性があるからである。

3.9.4 **Hazard communication**

General and specific considerations concerning labelling requirements are provided in *Hazard communication: Labelling* (chapter 1.4). Annex 1 contains summary tables about classification and labelling. Annex 3 contains examples of precautionary statements and pictograms which can be used where allowed by the competent authority. Table 3.9.4 presents specific label elements for substances and mixtures classified into this hazard class based on the criteria in this chapter.

Table 3.9.4: Label elements for specific target organ toxicity following repeated exposure

	Category 1	Category 2
Symbol	Health hazard	Health hazard
Signal word	Danger	Warning
Hazard statement	Causes damage to organs (state all organs affected, if known) through prolonged or repeated exposure (state route of exposure if it is conclusively proven that no other routes of exposure cause the hazard)	May cause damage to organs (state all organs affected, if known) through prolonged or repeated exposure (state route of exposure if it is conclusively proven that no other routes of exposure cause the hazard)

3.9.5 **Decision logic for specific target organ toxicity following repeated exposure**

The decision logic which follows is not part of the harmonized classification system but is provided here as additional guidance. It is strongly recommended that the person responsible for classification studies the criteria before and during use of the decision logic.

3.9.5.1 *Decision logic 3.9.1*

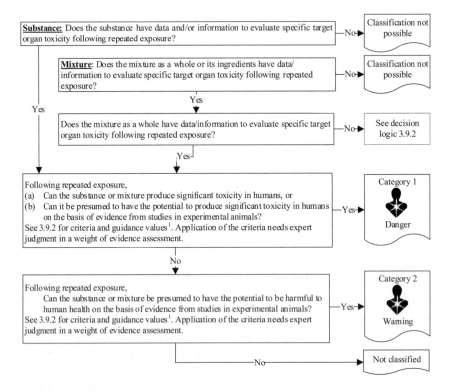

---

[1] *See 3.9.2, tables 3.9.1 and 3.9.2, and in chapter 1.3, paragraph 1.3.3.2 "The use of cut-off values/concentration limits".*

3.9.4 危険有害性情報の伝達

表示要件に関する通則および細則は、*危険有害性に関する情報の伝達：表示*（第1.4章）に定める。附属書1に分類と表示に関する概要表を示す。附属書3には、注意書きおよび所管官庁が許可した場合に使用可能な注意絵表示の例を示す。表3.9.4には、本章の判定基準に基づいてこの危険有害性クラスに分類された物質および混合物の具体的なラベル要素を示す。

表 3.9.4：反復ばく露による特定標的臓器毒性のラベル要素

	区分 1	区分 2
シンボル	健康有害性	健康有害性
注意喚起語	危険	警告
危険有害性情報	長期にわたる、または反復ばく露による臓器の障害（判っていれば影響を受けるすべての臓器名を記載）（他の経路からのばく露が有害でないことが決定的に証明されている場合、有害なばく露経路を記載）	長期にわたる、または反復ばく露による臓器の障害のおそれ（判っていれば影響を受けるすべての臓器名を記載）（他の経路からのばく露が有害でないことが決定的に証明されている場合、有害なばく露経路を記載）

3.9.5 反復ばく露による特定標的臓器毒性の判定論理

以下の判定論理は、調和分類システムには含まれないが、追加的な手引きとしてここに示す。分類の責任者に対し、この判定論理を使用する前および使用する際に判定基準についてよく調べ理解することを強く勧める。

3.9.5.1 判定論理 3.9.1

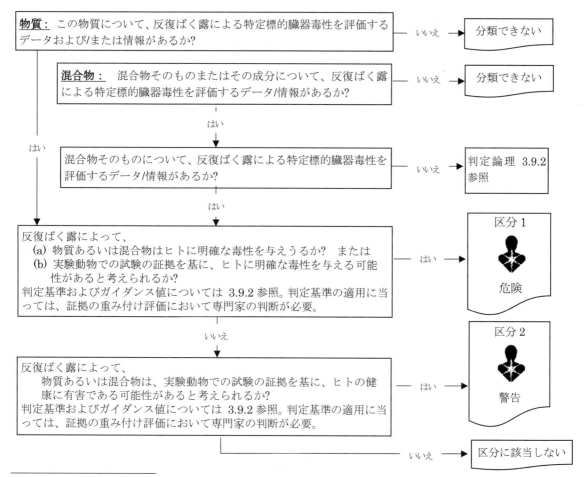

---

1 本章3.9.2、表3.9.1および3.9.2、および*第1.3章の1.3.3.2「カットオフ値/濃度限度の使用」*を参照のこと。

- 234 -

3.9.5.2 *Decision logic 3.9.2*

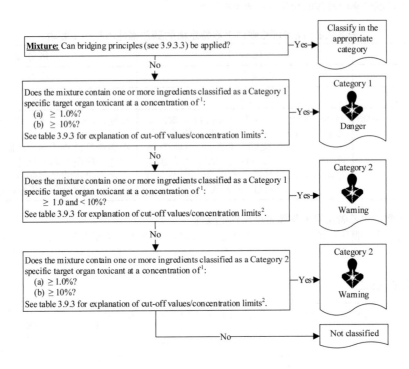

---

[1] *See 3.9.2, tables 3.9.1 and 3.9.2, and in chapter 1.3, paragraph 1.3.3.2 "The use of cut-off values/concentration limits".*

[2] *See 3.9.3.4 and 3.9.4 and table 3.9.3 for explanation and guidance.*

3.9.5.2 判定論理 3.9.2

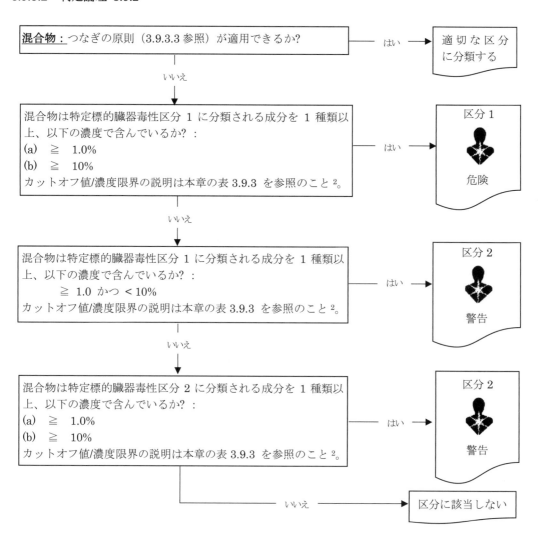

---
[1] 本章 3.9.2、表 3.9.1 および 3.9.2、および第 1.3 章の 1.3.3.2「カットオフ値/濃度限度の使用」を参照のこと。
[2] 説明ならびに手引きについては 3.9.3.4 および 3.9.4 ならびに表 3.9.3 を参照のこと。

CHAPTER 3.10

ASPIRATION HAZARD

3.10.1 Definitions and general considerations

3.10.1.1 *Aspiration* means the entry of a liquid or solid chemical directly through the oral or nasal cavity, or indirectly from vomiting, into the trachea and lower respiratory system.

3.10.1.2 *Aspiration hazard* refers to severe acute effects such as chemical pneumonia, pulmonary injury or death occurring after aspiration of a substance or mixture.

3.10.1.3 Aspiration is initiated at the moment of inspiration, in the time required to take one breath, as the causative material lodges at the crossroad of the upper respiratory and digestive tracts in the laryngopharyngeal region.

3.10.1.4 Aspiration of a substance or mixture can occur as it is vomited following ingestion. This may have consequences for labelling, particularly where, due to acute toxicity, a recommendation may be considered to induce vomiting after ingestion. However, if the substance/mixture also presents an aspiration toxicity hazard, the recommendation to induce vomiting may need to be modified.

3.10.1.5 *Specific considerations*

3.10.1.5.1 A review of the medical literature on chemical aspiration revealed that some hydrocarbons (petroleum distillates) and certain chlorinated hydrocarbons have been shown to pose an aspiration hazard in humans. Primary alcohols, and ketones have been shown to pose an aspiration hazard only in animal studies.

3.10.1.5.2 While a methodology for determination of aspiration hazard in animals has been utilized, it has not been standardized. Positive experimental evidence with animals can only serve as a guide to possible aspiration toxicity in humans. Particular care must be taken in evaluating animal data for aspiration hazards.

3.10.1.5.3 The classification criteria refer to kinematic viscosity. The following provides the conversion between dynamic and kinematic viscosity:

$$\frac{\text{Dynamic viscosity}(\text{mPa}\cdot\text{s})}{\text{Density}(\text{g/cm}^3)} = \text{Kinematic viscosity}(\text{mm}^2/\text{s})$$

3.10.1.5.4 Although the definition of aspiration in 3.10.1.2 includes the entry of solids into the respiratory system, classification according to (b) in table 3.10.1 for Category 1 or for Category 2 is intended to apply to liquid substances and mixtures only.

3.10.1.5.5 *Classification of aerosol/mist products*

Aerosol and mist products are usually dispensed in containers such as self-pressurized containers, trigger and pump sprayers. The key to classifying these products is whether a pool of product is formed in the mouth, which then may be aspirated. If the mist or aerosol from a pressurized container is fine, a pool may not be formed. On the other hand, if a pressurized container dispenses product in a stream, a pool may be formed that may then be aspirated. Usually, the mist produced by trigger and pump sprayers is coarse and therefore, a pool may be formed that then may be aspirated. When the pump mechanism may be removed and contents are available to be swallowed then the classification of the products should be considered.

第 3.10 章
誤えん有害性

3.10.1 定義および一般的事項

3.10.1.1 　*誤えん*とは、液体または固体の化学品が口または鼻腔から直接、または嘔吐によって間接的に、気管および下気道へ侵入することをいう。

3.10.1.2 　*誤えん有害性*とは、物質または混合物の誤えん後に起こる、化学肺炎、肺損傷あるいは死のような重篤な急性影響をさす。

3.10.1.3 　誤えんは、原因物質が喉頭咽頭部分の上気道と上部消化官の岐路部分に入り込むと同時になされる吸気により引き起こされる。

3.10.1.4 　物質または混合物の誤えんは、それを摂取した後に嘔吐した時も起こりうる。このことは、急性毒性を有するため摂取後吐かせることを推奨している場合、表示に影響を及ぼすかもしれない。物質/混合物が誤えんの危険性に分類される毒性も示す場合は、吐かせることについての推奨は修正する必要があるであろう。

3.10.1.5 *特別に留意すべき事項*

3.10.1.5.1 　化学品の誤えんに関する医学文献レビューでは、ある炭化水素（石油留分）およびある種の塩素化炭化水素は、ヒトに誤えん有害性をもつことを明らかにした。一級アルコール、およびケトンは動物実験でのみ誤えん有害性が示されている。

3.10.1.5.2 　動物における誤えん有害性を決定するための方法論は利用されているが、標準化されたものはない。動物実験で陽性であるという証拠は、ヒトに対して、誤えん有害性に分類される毒性があるかもしれないという指針として役立つ程度である。誤えん有害性に関する動物データを評価する際は、特別な注意をしなければならない。

3.10.1.5.3 　分類基準は動粘性率を参照している。以下に、粘性率と動粘性率の変換を示す。

$$粘性率（mPa·s） \div 密度（g/cm^3） = 動粘性率（mm^2/s）$$

3.10.1.5.4 　3.10.1.2 における誤えん有害性の定義には呼吸器系への固体の侵入を含んでいるが、区分 1 あるいは区分 2 に対する表 3.10.1 の(b)による分類は液体の物質および混合物のみへの適用を意図したものである。

3.10.1.5.5 　*エアゾール/ミスト製剤の分類*

　エアゾールおよびミスト製剤は、通常、自己加圧式容器や引き金ポンプ式噴霧器などの容器に入れられて供される。これらの製剤の分類の鍵は、製剤が噴霧後に誤えんされるほどに口内に溜まるかどうかである。加圧容器からのミストまたはエアゾールが微細であれば、口内には溜まらないかもしれないが、製剤が（霧状ではなく）流れのようになって供されれば、口内に溜まり誤えんされる可能性がある。通常、引き金ポンプ式噴霧器によって噴霧されるミストは粗い粒子であるため、口内に溜まり誤えんされる場合がある。ポンプ装置を取り外すことができ、直接内容物を飲み込むことが可能な場合には、分類を考慮すべきである。

3.10.2 Classification criteria for substances

Table 3.10.1: Hazard categories for aspiration toxicity

Categories	Criteria
Category 1: Chemicals known to cause human aspiration toxicity hazards or to be regarded as if they cause human aspiration toxicity hazard	A substance is classified in Category 1: (a) Based on reliable and good quality human evidence (see note 1); or (b) If it is a hydrocarbon and has a kinematic viscosity ≤ 20.5 mm^2/s, measured at 40° C.
Category 2: Chemicals which cause concern owing to the presumption that they cause human aspiration toxicity hazard	On the basis of existing animal studies and expert judgment that takes into account surface tension, water solubility, boiling point, and volatility, substances, other than those classified in Category 1, which have a kinematic viscosity ≤ 14 mm^2/s, measured at 40° C (see note 2).

NOTE 1: *Examples of substances included in Category 1 are certain hydrocarbons, turpentine and pine oil.*

NOTE 2: *Taking this into account, some authorities would consider the following to be included in this Category: n-primary alcohols with a composition of at least 3 carbon atoms but not more than 13; isobutyl alcohol, and ketones with a composition of no more than 13 carbon atoms.*

3.10.3 Classification criteria for mixtures

3.10.3.1 *Classification when data are available for the complete mixture*

A mixture is classified in Category 1 based on reliable and good quality human evidence.

3.10.3.2 *Classification of mixtures when data are not available for the complete mixture: bridging principles*

3.10.3.2.1 Where the mixture itself has not been tested to determine its aspiration toxicity, but there are sufficient data on both the individual ingredients and similar tested mixtures to adequately characterize the hazard of the mixture, these data will be used in accordance with the following bridging principles. This ensures that the classification process uses the available data to the greatest extent possible in characterizing the hazards of the mixture without the necessity of additional testing in animals.

3.10.3.2.2 *Dilution*

If a tested mixture is diluted with a diluent that does not pose an aspiration toxicity hazard, and which is not expected to affect the aspiration toxicity of other ingredients or the mixture, then the new diluted mixture may be classified as equivalent to the original tested mixture. However, the concentration of aspiration toxicant(s) should not drop below 10 %.

3.10.3.2.3 *Batching*

The aspiration toxicity of a tested production batch of a mixture can be assumed to be substantially equivalent to that of another untested production batch of the same commercial product, when produced by or under the control of the same manufacturer, unless there is reason to believe there is significant variation such that the aspiration toxicity, reflected by viscosity or concentration, of the untested batch has changed. If the latter occurs, a new classification is necessary.

3.10.3.2.4 *Concentration of Category 1 mixtures*

If a tested mixture is classified in Category 1, and the concentration of the ingredients of the tested mixture that are in Category 1 is increased, the resulting untested mixture should be classified in Category 1 without additional testing.

3.10.2 物質の分類基準

表 3.10.1：誤えん有害性の区分

区分	判定基準
区分1：ヒトへの誤えん有害性があると知られている、またはヒトへの誤えん有害性があるとみなされる化学品	区分1に分類される物質： (a) ヒトに関する信頼度が高く、かつ質の良い有効な証拠に基づく（注記1参照）；または (b) 40℃で測定した動粘性率が ≦20.5 mm²/s の炭化水素の場合。
区分2：ヒトへの誤えん有害性があると推測される化学品	40℃で測定した動粘性率が ≦14 mm²/s で区分1に分類されない物質であって、既存の動物実験、ならびに表面張力、水溶性、沸点および揮発性を考慮した専門家の判断に基づく（注記2参照）

注記1：区分1に含まれる物質の例はある種の炭化水素であるテレピン油およびパイン油である。

注記2：この点を考慮し、次の物質をこの区分に含める所管官庁もあると考えられる：3以上13を超えない炭素原子で構成された一級のノルマルアルコール；イソブチルアルコールおよび13を超えない炭素原子で構成されたケトン。

3.10.3 混合物の分類基準

3.10.3.1 *混合物そのものについてデータが利用できる場合の分類*

混合物は、ヒトに関する信頼度が高く、かつ質の良い有効な証拠に基づき区分1に分類される。

3.10.3.2 *混合物そのものについてデータが利用できない場合の混合物の分類：つなぎの原則 (bridging Principles)*

3.10.3.2.1 混合物そのものは誤えん有害性を決定するための試験がなされていないが、当該混合物の有害性を適切に特定するための、個々の成分および類似の試験された混合物の両方に関して十分なデータがある場合、これらのデータは以下のつなぎの原則にしたがって利用される。これによって、分類プロセスで動物試験を追加する必要もなく、混合物の有害性判定に利用可能なデータを可能な限り最大限に用いられるようになる。

3.10.3.2.2 *希釈*

試験された混合物が誤えん有害性をもたない物質で希釈され、その物質が他の成分または混合物の有害性に影響を与えないことが予想されれば、新しい希釈された混合物は、試験された元の混合物と同等として分類してもよい。しかし、誤えん有害性をもつ物質の濃度は10%の値以下に下げるべきではない。

3.10.3.2.3 *製造バッチ*

混合物の試験されていない製造バッチに誤えん有害性があるかどうかは、同じ製造者によって、またはその管理下で生産された同じ商品の試験された別のバッチの毒性と本質的に同等とみなすことができる。ただし、誤えん有害性が、粘性または濃度によりもたらされ、試験されていないバッチの有害性が変化するような有意な変動があると考えられる理由がある場合はこの限りではない。このような場合には、新しい分類が必要である。

3.10.3.2.4 *区分1の混合物の濃縮*

試験された混合物が区分1に分類され、区分1である試験された混合物の成分の濃度が増加すれば、結果として濃縮された混合物は、追加試験なしで区分1に分類するべきである。

3.10.3.2.5   *Interpolation within one hazard category*

For three mixtures (A, B and C) with identical ingredients, where mixtures A and B have been tested and are in the same hazard category, and where untested mixture C has the same toxicologically active ingredients as mixtures A and B but has concentrations of toxicologically active ingredients intermediate to the concentrations in mixtures A and B, then mixture C is assumed to be in the same hazard category as A and B.

3.10.3.2.6   *Substantially similar mixtures*

Given the following:

(a)   Two mixtures: (i)   A + B;
                         (ii)   C + B;

(b)   The concentration of ingredient B is essentially the same in both mixtures;

(c)   The concentration of ingredient A in mixture (i) equals that of ingredient C in mixture (ii);

(d)   Aspiration toxicity for A and C is substantially equivalent, i.e. they are in the same hazard category and are not expected to affect the aspiration toxicity of B.

If mixture (i) or (ii) is already classified based on the criteria in table 3.10.1, then the other mixture can be classified in the same hazard category.

3.10.3.3   *Classification of mixtures when data are available for all ingredients or only for some ingredients of the mixture*

3.10.3.3.1   The "relevant ingredients" of a mixture are those which are present in concentrations ≥ 1 %.

3.10.3.3.2   *Category 1*

3.10.3.3.2.1   A mixture is classified as Category 1 when the sum of the concentrations of Category 1 ingredients is ≥ 10 %, and the mixture has a kinematic viscosity ≤ 20.5 mm^2/s, measured at 40 °C.

3.10.3.3.2.2   In the case of a mixture which separates into two or more distinct layers, the entire mixture is classified as Category 1 if in any distinct layer the sum of the concentrations of Category 1 ingredients is ≥ 10 %, and it has a kinematic viscosity ≤ 20.5 mm^2/s, measured at 40 °C.

3.10.3.3.3   *Category 2*

3.10.3.3.3.1   A mixture is classified as Category 2 when the sum of the concentrations of Category 2 ingredients is ≥ 10 % and the mixture has a kinematic viscosity ≤ 14 mm^2/s, measured at 40 °C.

3.10.3.3.3.2   In classifying mixtures in this category, the use of expert judgment that considers surface tension, water solubility, boiling point, volatility is critical and especially when Category 2 substances are mixed with water.

3.10.3.3.3.3   In the case of classifying a mixture which separates into two or more distinct layers, the entire mixture is classified as Category 2 if in any distinct layer the sum of the concentrations of Category 2 ingredients is ≥ 10 %, and it has a kinematic viscosity ≤ 14 mm^2/s, measured at 40 °C.

3.10.4   Hazard communication

3.10.4.1   General and specific considerations concerning labelling requirements are provided in *Hazard communication: Labelling* (chapter 1.4). Annex 1 contains summary tables about classification and labelling. Annex 3 contains examples of precautionary statements and pictograms, which can be used where allowed by the competent authority. Table 3.10.2 presents specific label elements for substances and mixtures classified into this hazard class based on the criteria in this chapter.

3.10.3.2.5 　*1つの有害性区分内での内挿*

3つの混合物（A、BおよびC）は同じ成分を持ち、AとBは試験され同じ有害性区分にある。試験されていない混合物Cは混合物AおよびBと同じ毒性学的に活性な成分を持ち、毒性学的に活性な成分の濃度が混合物AとBの中間である場合、混合物CはAおよびBと同じ有害性区分にあるとする。

3.10.3.2.6 　*本質的に類似した混合物*

次を仮定する：

(a) 　2つの混合物： 　(i) 　　A + B ；
　　　　　　　　　　　(ii) 　　C + B ；

(b) 　成分Bの濃度は、両方の混合物で本質的に同じである；

(c) 　混合物(i)の成分Aの濃度は、混合物(ii)の成分Cの濃度に等しい；

(d) 　AとCの誤えん毒性に関するデータは実質的に同等であり、すなわちAとCは同じ有害性区分に属し、かつ、Bの誤えん毒性には影響を与えることはないと判断される。

混合物(i)または(ii)が既に表 3.10.1 の判定基準によって分類されている場合には、他方の混合物は同じ有害性区分に分類することができる。

3.10.3.3 　*混合物の全成分についてまたは一部の成分だけについてデータが利用できる場合の混合物の分類*

3.10.3.3.1 　混合物の「考慮すべき成分」は、≥1%の濃度で存在するものである。

3.10.3.3.2 　*区分1*

3.10.3.3.2.1 　区分1の成分の濃度の合計が ≥10%で、しかも40°Cにおける動粘性率が ≤20.5 mm²/s の時に混合物は区分1と分類される。

3.10.3.3.2.2 　混合物が2つ以上の相に明確に分離している場合、もしどの明確に分離している相においても、区分1の成分の濃度の合計が ≥ 10%で、しかも 40°C における動粘性率が ≤ 20.5 mm²/s の時には、混合物全体としては区分1と分類される。

3.10.3.3.3 　*区分2*

3.10.3.3.3.1 　区分2の成分の濃度の合計が ≥ 10%で、しかも 40°C における動粘性率が ≤ 14 mm²/s の時に混合物は区分2と分類される。

3.10.3.3.3.2 　混合物をこの区分に分類する場合、表面張力、水溶解度、沸点、揮発性を検討する専門家判断は不可欠であり、特に区分2の成分が水と混合されている場合にはそうである。

3.10.3.3.3.3 　混合物が2つ以上の層に明確に分離している場合、もしどの明確に分離している層においても、区分2の成分の濃度の合計が ≥10%で、しかも40°Cにおける動粘性率が ≤14 mm²/s の時には、混合物全体としては区分2と分類される。

3.10.4 　危険有害性情報の伝達

3.10.4.1 　表示要件に関する通則および細則は、*危険有害性に関する情報の伝達：表示*（第1.4章）に定める。附属書1に分類と表示に関する概要表を示す。附属書3には、注意書きおよび所管官庁が許可した場合に使用可能な注意絵表示の例を示す。表 3.10.2 には、本章の判定基準に基づいてこの危険有害性クラスに分類された物質および混合物の具体的なラベル要素を示す。

Table 3.10.2: Label elements for aspiration toxicity

	Category 1	Category 2
Symbol	Health hazard	Health hazard
Signal word	Danger	Warning
Hazard statement	May be fatal if swallowed and enters airways	May be harmful if swallowed and enters airways

3.10.5   Decision logic for aspiration toxicity

The decision logic which follows is not part of the harmonized classification system but is provided here as additional guidance. It is strongly recommended that the person responsible for classification study the criteria before and during use of the decision logic.

3.10.5.1   *Decision logic 3.10.1*

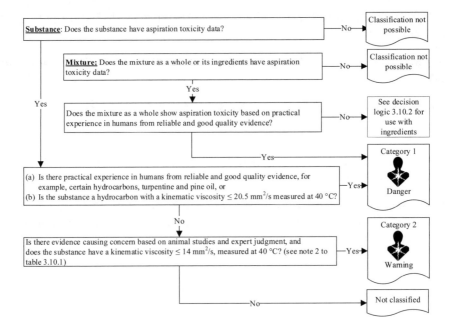

- 240 -

表 3.10.2：誤えん有害性のラベル要素

	区分1	区分2
シンボル	健康有害性	健康有害性
注意喚起語	危険	警告
危険有害性情報	飲み込んで気道に侵入すると生命に危険のおそれ	飲み込んで気道に侵入すると有害のおそれ

3.10.5 誤えん有害性の判定論理

以下の判定論理は、調和分類システムには含まれないが、追加的な手引きとしてここに示す。分類の責任者に対し、この判定論理を使用する前および使用する際に判定基準についてよく調べ理解することを強く勧める。

3.10.5.1 *判定論理 3.10.1*

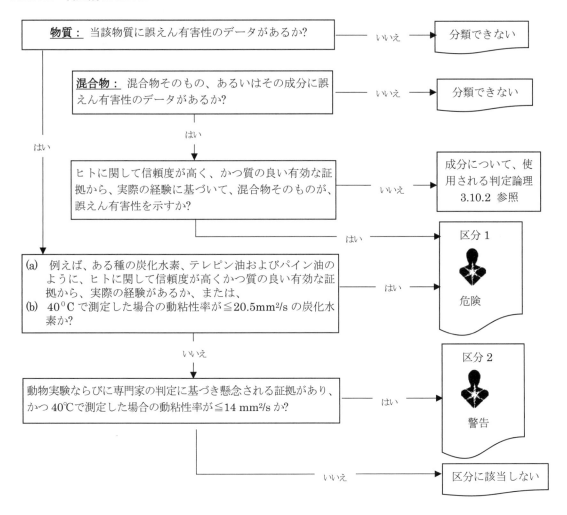

- 240 -

3.10.5.2 *Decision logic 3.10.2*

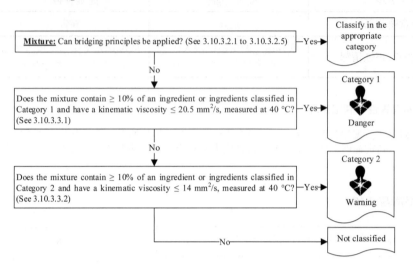

3.10.5.2　判定論理 3.10.2

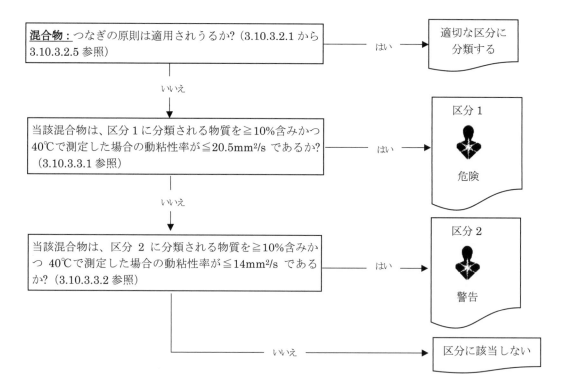

PART 4

ENVIRONMENTAL HAZARDS

PART 4

ENVIRONMENTAL HAZARDS

第4部

環境に対する有害性

CHAPTER 4.1

HAZARDOUS TO THE AQUATIC ENVIRONMENT

4.1.1 Definitions and general considerations

4.1.1.1 *Definitions*

Acute aquatic toxicity means the intrinsic property of a substance to be injurious to an organism in a short-term aquatic exposure to that substance.

Availability of a substance means the extent to which this substance becomes a soluble or disaggregate species. For metal availability, the extent to which the metal ion portion of a metal (M°) compound can disaggregate from the rest of the compound (molecule).

Bioavailability (or biological availability) means the extent to which a substance is taken up by an organism, and distributed to an area within the organism. It is dependent upon physico-chemical properties of the substance, anatomy and physiology of the organism, pharmacokinetics, and route of exposure. Availability is not a prerequisite for bioavailability.

Bioaccumulation means net result of uptake, transformation and elimination of a substance in an organism due to all routes of exposure (i.e. air, water, sediment/soil and food).

Bioconcentration means net result of uptake, transformation and elimination of a substance in an organism due to waterborne exposure.

Chronic aquatic toxicity means the intrinsic property of a substance to cause adverse effects to aquatic organisms during aquatic exposures which are determined in relation to the life-cycle of the organism.

Complex mixtures or multi-component substances or complex substances means mixtures comprising a complex mix of individual substances with different solubilities and physico-chemical properties. In most cases, they can be characterized as a homologous series of substances with a certain range of carbon chain length/number of degree of substitution.

Degradation means the decomposition of organic molecules to smaller molecules and eventually to carbon dioxide, water and salts.

ECx means the concentration associated with x % response.

Long-term (chronic) hazard, for classification purposes, means the hazard of a chemical caused by its chronic toxicity following long-term exposure in the aquatic environment.

NOEC (No Observed Effect Concentration) means the test concentration immediately below the lowest tested concentration with statistically significant adverse effect. The NOEC has no statistically significant adverse effect compared to the control.

Short-term (acute) hazard, for classification purposes, means the hazard of a chemical caused by its acute toxicity to an organism during short-term aquatic exposure to that chemical.

4.1.1.2 *Basic elements*

4.1.1.2.1 The basic elements for use within the harmonized system are:

    (a) acute aquatic toxicity;

    (b) chronic aquatic toxicity;

    (c) potential for or actual bioaccumulation; and

    (d) degradation (biotic or abiotic) for organic chemicals.

第 4.1 章

水生環境有害性

4.1.1 定義および一般事項

4.1.1.1 定義

*急性水生毒性*とは、物質への短期的な水生ばく露において、生物に対して有害な、当該物質の本質的な特性をいう。

物質の*利用性*とは、物質が溶解性ないし解離性を有するようになる程度を意味する。金属の利用性とは、金属化合物の金属イオン化した部分が同化合物の他の部分（分子）から解離する程度を意味する。

*生物学的利用性*とは、物質が生物に取り込まれ、生物内のある部位に分布する程度を意味する。これは物質の物理化学的特質、生物の体内組織および生理機能、ファーマコキネティクスならびにばく露の経路に依存する。単なる利用性は、生物学的利用性の必要条件とはならない。

*生物蓄積性*とは、あらゆるばく露経路（すなわち、空気、水、底質/土壌および食物）からの、生物体内への物質の取り込み、生物体内における物質の変化、および排泄からなる総体的な結果を意味する。

*生物濃縮*とは、水を媒体とするばく露による、生物体内への物質の取り込み・生物体内における物質の変化および排泄からなる総体的な結果を意味する。

*慢性水生毒性*とは、水生生物のライフサイクルに対応した水生ばく露期間に、水生生物に悪影響を及ぼすような、物質の本質的な特性を意味する。

複合混合物、または多成分物質もしくは複合物質とは、それぞれ異なる溶解性および物理化学的性質を有する個々の物質の複合体からなる混合物を意味する。多くの場合、これらはある範囲の炭素鎖の長さ/置換基の度数を持つ一連の類似物質として特徴付けられる。

*分解*とは、有機物分子がより小さな分子に、さらに最終的には二酸化炭素、水および塩類に分解することを意味する。

*EC_x*とは x%の反応を示す濃度をいう。

*長期（慢性）有害性*は、分類の目的では、水生環境における化学品への長期間のばく露を受けた後にその慢性毒性によって引き起こされる化学品の有害性を意味する。

*NOEC（無影響濃度）*とは、統計的に有意な悪影響を示す最低の試験濃度直下の試験濃度をいう。NOECではコントロール群と比べて有意な悪影響は見られない。

*短期（急性）有害性*は、分類の目的では、化学品への短期の水生ばく露の間にその急性毒性によって生物に引き起こされる化学品の有害性を意味する。

4.1.1.2 基本的要素

4.1.1.2.1 GHS において用いられる基本的要素は下記のとおり：

(a) 急性水生毒性；

(b) 慢性水生毒性；

(c) 潜在的な、または実際の生物蓄積性；

(d) 有機化学品の（生物的または非生物的）分解。

4.1.1.2.2   While data from internationally harmonized test methods are preferred, in practice, data from national methods may also be used where they are considered as equivalent. In general, it has been agreed that freshwater and marine species toxicity data can be considered as equivalent data and are preferably to be derived using OECD test guidelines or equivalent according to the principles of Good Laboratory Practices (GLP). Where such data are not available classification should be based on the best available data.

4.1.1.3   *Acute aquatic toxicity*

Acute aquatic toxicity would normally be determined using a fish 96 hour LC_{50} (OECD Test Guideline 203 or equivalent), a crustacea species 48 hour EC_{50} (OECD Test Guideline 202 or equivalent) and/or an algal species 72 or 96 hour EC_{50} (OECD Test Guideline 201 or equivalent). These species are considered as surrogate for all aquatic organisms and data on other species such as Lemna may also be considered if the test methodology is suitable.

4.1.1.4   *Chronic aquatic toxicity*

Chronic toxicity data are less available than acute data and the range of testing procedures less standardized. Data generated according to OECD test guidelines 210 (Fish Early Life Stage), or 211 (Daphnia Reproduction) and 201 (Algal Growth Inhibition) can be accepted (see also annex 9, pararaph A9.3.3.2). Other validated and internationally accepted tests could also be used. The NOECs or other equivalent ECx should be used.

4.1.1.5   *Bioaccumulation potential*

The potential for bioaccumulation would normally be determined by using the octanol/water partition coefficient, usually reported as a log K_{ow} determined by OECD Test Guideline 107, 117 or 123. While this represents a potential to bioaccumulate, an experimentally determined Bioconcentration Factor (BCF) provides a better measure and should be used in preference when available. A BCF should be determined according to OECD Test Guideline 305.

4.1.1.6   *Rapid degradability*

4.1.1.6.1   Environmental degradation may be biotic or abiotic (e.g. hydrolysis) and the criteria used reflect this fact (see 4.1.2.11.3). Ready biodegradation can most easily be defined using the biodegradability tests (A-F) of OECD Test Guideline 301. A pass level in these tests can be considered as indicative of rapid degradation in most environments. These are freshwater tests and thus the use of the results from OECD Test Guideline 306 which is more suitable for marine environments has also been included. Where such data are not available, a BOD (5 days)/COD ratio ≥ 0.5 is considered as indicative of rapid degradation.

4.1.1.6.2   Abiotic degradation such as hydrolysis, primary degradation, both abiotic and biotic, degradation in non-aquatic media and proven rapid degradation in the environment may all be considered in defining rapid degradability. Special guidance on data interpretation is provided in the guidance document (annex 9).

4.1.1.7   *Other considerations*

4.1.1.7.1   The harmonized system for classifying substances for the hazards they present to the aquatic environment is based on a consideration of existing systems listed in 4.1.1.7.3. The aquatic environment may be considered in terms of the aquatic organisms that live in the water, and the aquatic ecosystem of which they are part. To that extent, the proposal does not address aquatic pollutants for which there may be a need to consider effects beyond the aquatic environment such as the impacts on human health etc. The basis, therefore, of the identification of hazard is the aquatic toxicity of the substance, although this may be modified by further information on the degradation and bioaccumulation behaviour.

4.1.1.7.2   While the scheme is intended to apply to all substances and mixtures, it is recognized that for some substances, e.g. metals, poorly soluble substances, etc., special guidance will be necessary. Two guidance documents (see annexes 9 and 10) have been prepared to cover issues such as data interpretation and the application of the criteria defined below to such groups of substances. Considering the complexity of this endpoint and the breadth of the application of the system, the guidance documents are considered an important element in the operation of the harmonized scheme.

4.1.1.7.3   Consideration has been given to existing classification systems as currently in use, including the European Union supply and use scheme, the revised GESAMP hazard evaluation procedure, IMO scheme for marine pollutants, the European road and rail transport scheme (ADR/RID), the Canadian and United States of America pesticide systems and the United States of America land transport scheme. The harmonized scheme is considered suitable for use for packaged goods in both supply and use and multimodal transport schemes, and elements of it may be used for bulk land transport and bulk marine transport under MARPOL 73/78 Annex II insofar as this uses aquatic toxicity.

4.1.1.2.2 国際的に調和された試験方法によるデータが望ましいが、実際には各国独自の方法より得られたデータでも、それが同等であると判断されたならば、使用してよいであろう。一般に、淡水種および海水種での毒性データは同等であると合意されている。これらについては、OECD テストガイドラインまたは GLP 原則によって同等とみなせる方法でデータが導かれることが望ましい。こうしたデータが入手できない場合には、入手された最良のデータをもとに分類を行うべきである。

4.1.1.3 急性水生毒性

急性水生毒性は通常、魚類の 96 時間 LC_{50} (OECD テストガイドライン 203 またはこれに相当する試験)、甲殻類の 48 時間 EC_{50} (OECD テストガイドライン 202 またはこれに相当する試験) または藻類の 72 時間もしくは 96 時間 EC_{50} (OECD テストガイドライン 201 またはこれに相当する試験) により決定される。これらの生物種はすべての水生生物に代わるものとしてみなされるが、例えば Lemna (アオウキクサ) 等その他の生物種に関するデータも、試験方法が適切なものであれば、考慮されることもある。

4.1.1.4 慢性水生毒性

慢性毒性データは、急性毒性データほどは利用できるものがなく、一連の試験手順もそれほど標準化されていない。OECD テストガイドライン 210 (魚類の初期生活段階毒性試験) または 211 (ミジンコの繁殖試験) および 201 (藻類生長阻害試験) によって得られたデータは受け入れることができる (附属書 9 の A9.3.3.2 参照)。その他、有効性が確認され、国際的に容認された試験も採用できる。NOEC または相当する EC_x を採用するべきである。

4.1.1.5 生物蓄積性

生物蓄積性は通常、オクタノール/水分配係数を用いて決定され、一般的には OECD テストガイドライン 107、117 または 123 により決定された $\log K_{ow}$ として報告される。この値が生物蓄積性の潜在的な可能性を示しているのに対して、実験的に求められた生物濃縮係数(BCF)はより適切な尺度を与えるものであり、入手できれば BCF の方を採用すべきである。BCF は OECD テストガイドライン 305 にしたがって決定されるべきである。

4.1.1.6 急速分解性

4.1.1.6.1 環境中での分解は生物的分解と非生物的分解 (例えば加水分解) とがあり、採用される判定基準はこの事実を反映している (4.1.2.11.3 参照)。易生分解性は OECD テストガイドライン 301(A-F)にある生分解性試験により最も容易に定義づけできる。これらの試験で急速分解性とされるレベルは、ほとんどの環境中での急速分解性の指標とみなすことができる。これらは淡水系での試験であるため、海水環境により適合している OECD テストガイドライン 306 より得られる結果も取り入れることとされた。こうしたデータが利用できない場合には、BOD (5 日間) / COD 比が 0.5 より大きいことが急速分解性の指標と考えられている。

4.1.1.6.2 加水分解などの非生物的分解、生物的および非生物的の両方の一次分解、非水系媒体中での分解性および環境中で証明された急速分解性はいずれも、急速分解性を判定する際に考慮されてよい。データの解釈に関する特別な手引きは、附属書 9 に示される。

4.1.1.7 その他の考慮事項

4.1.1.7.1 水生環境有害性に関して物質を分類するための調和されたシステムは、4.1.1.7.3 にリストされる既存のシステムが考慮されている。水生環境とは、水中に生息する水生生物およびそれらが構成している水域生態系として考えることができる。その範囲では、本提案は、例えばヒトの健康に対する影響のような、水生環境の範囲を超える影響を考慮する必要があるような水質汚染物質には言及しない。したがって、その物質の水生毒性が有害性の特定の基礎となるが、分解性および生物蓄積性の挙動に関するさらなる情報によって変更されることもある。

4.1.1.7.2 このスキームはすべての物質と混合物に適用することを意図しているが、例えば金属や難溶性の物質など一部の物質については特別な指針の必要性が認識されている。このような物質群でのデータの解釈や下記に定める判定基準の適用などについての課題を対象とした 2 つの手引書 (附属書 9 および 10 参照) が作成された。このエンドポイントは複雑であり、システムが広範囲に適用されることを考慮すると、これらの手引書は調和されたスキームを活用する際には 1 つの重要な要素となると考えられる。

4.1.1.7.3 EU における供給および使用スキーム、改正された GESAMP 危険有害性評価手順、IMO 海洋汚染物質のためのスキーム、欧州道路鉄道輸送スキーム(ADR/RID)、カナダおよび米国の農薬システムや米国陸上輸送スキームなど、現在使用されている既存の分類システムについて検討を行った。調和されたスキームは、包装された物品の供給および使用、ならびに複合輸送スキームに使用するのが適切であると考えられており、水生毒性を扱う限りにおいて、その要素はばら積みの陸上輸送および MARPOL 条約 73/78 附属書 II によるばら積みの海上輸送にも用いることができよう。

4.1.2 **Classification criteria for substances**

4.1.2.1   Whilst the harmonized classification system consists of three short-term (acute) classification categories and four long-term (chronic) classification categories, the core part of the harmonized classification system for substances consists of three short-term (acute) classification categories and three long-term (chronic) classification categories (see table 4.1.1 (a) and (b)). The short-term (acute) and the long-term (chronic) classification categories are applied independently. The criteria for classification of a substance in Acute 1 to 3 are defined on the basis of the acute toxicity data only (EC_{50} or LC_{50}). The criteria for classification of a substance into Chronic 1 to 3 follow a tiered approach where the first step is to see if available information on chronic toxicity merits long-term hazard classification. In absence of adequate chronic toxicity data, the subsequent step is to combine two types of information, i.e. acute toxicity data and environmental fate data (degradability and bioaccumulation data) (see figure 4.1.1).

4.1.2.2   The system also introduces a "safety net" classification (Chronic 4) for use when the data available do not allow classification under the formal criteria but there are nevertheless some grounds for concern. The precise criteria are not defined with one exception: for poorly water soluble substances for which no toxicity has been demonstrated, classification can occur if the substance is both not rapidly degraded and has a potential to bioaccumulate. It is considered that for such poorly soluble substances, the toxicity may not have been adequately assessed in the short-term test due to the low exposure levels and potentially slow uptake into the organism. The need for this classification can be negated by demonstrating that the substance does not require classification for aquatic long-term (chronic) hazards.

4.1.2.3   Substances with acute toxicities well below 1 mg/l or chronic toxicities well below 0.1 mg/l (if non-rapidly degradable) and 0.01 mg/l (if rapidly degradable) contribute as ingredients of a mixture to the toxicity of the mixture even at a low concentration and should be given increased weight in applying the summation method (see note 2 to table 4.1.1 and paragraph 4.1.3.5.5.5).

4.1.2.4   Substances classified under the following criteria (table 4.1.1) will be categorized as "hazardous to the aquatic environment". These criteria describe in detail the classification categories. They are diagrammatically summarized in table 4.1.2.

Table 4.1.1: Categories for substances hazardous to the aquatic environment *(Note 1)*

(a)   Short-term (acute) aquatic hazard

Category Acute 1: *(Note 2)*	
96 hr LC_{50} (for fish)	≤ 1 mg/l and/or
48 hr EC_{50} (for crustacea)	≤ 1 mg/l and/or
72 or 96hr ErC_{50} (for algae or other aquatic plants)	≤ 1 mg/l *(Note 3)*
Category Acute 1 may be subdivided for some regulatory systems to include a lower band at $L(E)C_{50} ≤ 0.1$ mg/l	
Category Acute 2:	
96 hr LC_{50} (for fish)	> 1 but ≤ 10 mg/l and/or
48 hr EC_{50} (for crustacea)	> 1 but ≤ 10 mg/l and/or
72 or 96hr ErC_{50} (for algae or other aquatic plants)	> 1 but ≤ 10 mg/l *(Note 3)*
Category Acute 3:	
96 hr LC_{50} (for fish)	> 10 but ≤ 100 mg/l and/or
48 hr EC_{50} (for crustacea)	> 10 but ≤ 100 mg/l and/or
72 or 96hr ErC_{50} (for algae or other aquatic plants)	> 10 but ≤ 100 mg/l *(Note 3)*
Some regulatory systems may extend this range beyond an $L(E)C_{50}$ of 100 mg/l through the introduction of another category.	

(Cont'd on next page)

4.1.2 物質の分類基準

4.1.2.1 調和されたシステムは、3つの短期（急性）分類区分と4つの長期（慢性）分類区分で構成されているが、その主要部分を成すのは3つの短期（急性）分類区分と3つの長期（慢性）分類区分である（表4.1.1(a)および(b)を参照）。急性毒性および慢性毒性の分類区分は独立して適用される。急性区分1～3に分類するための判定基準は、急性毒性データ（EC_{50}またはLC_{50}）のみに基づいて定義される。慢性区分1～3に分類するための判定基準は段階的なアプローチに従う。すなわち、まず第一ステップで慢性毒性について得られた情報が長期（慢性）有害性の区分に役立つかどうかを調べ、そして慢性毒性の十分なデータがない場合には、次のステップで、2種類の情報すなわち急性毒性データと環境運命データ（分解性および生物蓄積性のデータ）を組み合わせることになる（図4.1.1を参照）。

4.1.2.2 調和されたシステムでは、利用できるデータからは正式の判定基準による分類ができないが、それにも関わらず何らかの懸念の余地がある場合に用いられるよう、分類の「セーフティネット」（区分：慢性4）を導入している。明確な判定基準が定められているわけではないが、例外が1つある。すなわち、水に難溶性の物質については、その毒性が証明されていなくてもその物質が速やかに分解せず、かつ生物蓄積性の可能性があるならば、分類されることがありうる。そのような難溶性物質に対しては、生物へのばく露レベルが低く、取込み速度も遅いため、短期試験では毒性を適切に評価できていない可能性がある。その物質が水生の長期（慢性）有害性について分類する必要がないことを実証することによって、このように分類する必要性を否定できる。

4.1.2.3 急性毒性が1mg/lを十分に下回るか、または慢性毒性が（急速分解性がない場合に）0.1mg/lを十分に下回り、（急速分解性がある場合は）0.01mg/lを十分に下回る物質は、濃度が低くても混合物の成分として混合物の毒性に関与する。加算法を適用する際にはその重み付けを増加させるべきである（表4.1.1の注記2と4.1.3.5.5.5を参照）。

4.1.2.4 次の判定基準（表4.1.1）にしたがって分類された物質は「水生環境有害性」の分類に入る。詳細な分類区分を表4.1.2に一覧表としてまとめた。

表 4.1.1：水生環境有害性物質の区分 *(注記 1)*

(a) 短期（急性）水生有害性

> **区分 急性1**（*注記 2*）
> 　　96時間 LC_{50}（魚類に対する） ≦1mg/l および/または
> 　　48時間 EC_{50}（甲殻類に対する） ≦1mg/l および/または
> 　　72または96時間 ErC_{50}（藻類または他の水生植物に対する） ≦1mg/l *(注記 3)*
>
> 規制体系によっては、急性1をさらに細分して、$L(E)C_{50}$≦0.1mg/l という、より低い濃度帯を含む場合もある。
>
> **区分 急性2**
> 　　96時間 LC_{50}（魚類に対する） >1mg/l だが ≦10mg/l および/または
> 　　48時間 EC_{50}（甲殻類に対する） >1mg/l だが ≦10mg/l および/または
> 　　72または96時間 ErC_{50}（藻類または他の水生植物に対する） >1mg/l だが ≦10mg/l *(注記 3)*
>
> **区分 急性3**
> 　　96時間 LC_{50}（魚類に対する） >10mg/l だが ≦100mg/l および/または
> 　　48時間 EC_{50}（甲殻類に対する） >10mg/l だが ≦100mg/l および/または
> 　　72または96時間 ErC_{50}（藻類または他の水生植物に対する） >10mg/l だが ≦100mg/l *(注記 3)*
>
> 規制体系によっては、$L(E)C_{50}$が100mg/lを超える、別の区分を設ける場合もある。

（次ページに続く）

Table 4.1.1: Categories for substances hazardous to the aquatic environment *(Note 1) (cont'd)*

(b)   Long-term (chronic) aquatic hazard *(see also figure 4.1.1)*

    (i)   Non-rapidly degradable substances (Note 4) for which there are adequate chronic toxicity data available

Category Chronic 1: *(Note 2)*	
Chronic NOEC or EC$_x$ (for fish)	≤ 0.1 mg/l and/or
Chronic NOEC or EC$_x$ (for crustacea)	≤ 0.1 mg/l and/or
Chronic NOEC or EC$_x$ (for algae or other aquatic plants)	≤ 0.1 mg/l
Category Chronic 2:	
Chronic NOEC or EC$_x$ (for fish)	≤ 1 mg/l and/or
Chronic NOEC or EC$_x$ (for crustacea)	≤ 1 mg/l and/or
Chronic NOEC or EC$_x$ (for algae or other aquatic plants)	≤ 1 mg/l

    (ii)   Rapidly degradable substances for which there are adequate chronic toxicity data available

Category Chronic 1: *(Note 2)*	
Chronic NOEC or EC$_x$ (for fish)	≤ 0.01 mg/l and/or
Chronic NOEC or EC$_x$ (for crustacea)	≤ 0.01 mg/l and/or
Chronic NOEC or EC$_x$ (for algae or other aquatic plants)	≤ 0.01 mg/l
Category Chronic 2:	
Chronic NOEC or EC$_x$ (for fish)	≤ 0.1 mg/l and/or
Chronic NOEC or EC$_x$ (for crustacea)	≤ 0.1 mg/l and/or
Chronic NOEC or EC$_x$ (for algae or other aquatic plants)	≤ 0.1 mg/l
Category Chronic 3:	
Chronic NOEC or EC$_x$ (for fish)	≤ 1 mg/l and/or
Chronic NOEC or EC$_x$ (for crustacea)	≤ 1 mg/l and/or
Chronic NOEC or EC$_x$ (for algae or other aquatic plants)	≤ 1 mg/l

    (iii)   Substances for which adequate chronic toxicity data are not available

Category Chronic 1: *(Note 2)*	
96 hr LC$_{50}$ (for fish)	≤ 1 mg/l and/or
48 hr EC$_{50}$ (for crustacea)	≤ 1 mg/l and/or
72 or 96hr ErC$_{50}$ (for algae or other aquatic plants)	≤ 1 mg/l *(Note 3)*
and the substance is not rapidly degradable and/or the experimentally determined BCF is ≥ 500 (or, if absent, the log K$_{ow}$ ≥ 4). *(Notes 4 and 5)*	
Category Chronic 2:	
96 hr LC$_{50}$ (for fish)	> 1 but ≤ 10 mg/l and/or
48 hr EC$_{50}$ (for crustacea)	> 1 but ≤ 10 mg/l and/or
72 or 96hr ErC$_{50}$ (for algae or other aquatic plants)	> 1 but ≤ 10 mg/l *(Note 3)*
and the substance is not rapidly degradable and/or the experimentally determined BCF is ≥ 500 (or, if absent, the log K$_{ow}$ ≥ 4). *(Notes 4 and 5)*	
Category Chronic 3:	
96 hr LC$_{50}$ (for fish)	> 10 but ≤ 100 mg/l and/or
48 hr EC$_{50}$ (for crustacea)	> 10 but ≤ 100 mg/l and/or
72 or 96hr ErC$_{50}$ (for algae or other aquatic plants)	> 10 but ≤ 100 mg/l *(Note 3)*
and the substance is not rapidly degradable and/or the experimentally determined BCF is ≥ 500 (or, if absent, the log K$_{ow}$ ≥ 4). *(Notes 4 and 5)*.	

(c)   "Safety net" classification

> **Category Chronic 4:**
>
> Poorly soluble substances for which no acute toxicity is recorded at levels up to the water solubility, and which are not rapidly degradable and have a log K$_{ow}$ ≥ 4, indicating a potential to bioaccumulate, will be classified in this category unless other scientific evidence exists showing classification to be unnecessary. Such evidence would include an experimentally determined BCF < 500, or a chronic toxicity NOECs > 1 mg/l, or evidence of rapid degradation in the environment.

表 4.1.1：水生環境有害性物質の区分 *(注記 1)（続き)*

(b) 長期（慢性）水生有害性 *(図 4.1.1 も参照)*

(i) 慢性毒性の十分なデータが得られる、急速分解性のない物質 *(注記 4)*

区分 慢性1 *(注記 2)*

慢性 NOEC または EC_x（魚類に対する）≦0.1mg/l および/または
慢性 NOEC または EC_x（甲殻類に対する）≦0.1mg/l および/または
慢性 NOEC または EC_x（藻類または他の水生植物に対する）≦0.1mg/l

区分 慢性2

慢性 NOEC または EC_x（魚類に対する）≦ 1mg/l および/または
慢性 NOEC または EC_x（甲殻類に対する）≦ 1mg/l および/または
慢性 NOEC または EC_x（藻類または他の水生植物に対する）≦ 1mg/l

(ii) 慢性毒性の十分なデータが得られる、急速分解性のある物質

区分 慢性1 *(注記 2)*

慢性 NOEC または EC_x（魚類に対する）≦0.01mg/l および/または
慢性 NOEC または EC_x（甲殻類に対する）≦0.01mg/l および/または
慢性 NOEC または EC_x（藻類または他の水生植物に対する）≦0.01mg/l

区分 慢性2

慢性 NOEC または EC_x（魚類に対する）≦0.1mg/l および/または
慢性 NOEC または EC_x（甲殻類に対する）≦0.1mg/l
および/または
慢性 NOEC または EC_x（藻類または他の水生植物に対する）≦0.1mg/l

区分 慢性3

慢性 NOEC または EC_x（魚類に対する）≦1mg/l および/または
慢性 NOEC または EC_x（甲殻類に対する）≦1mg/l および/または
慢性 NOEC または EC_x（藻類または他の水生植物に対する）≦1mg/l

(iii) 慢性毒性の十分なデータが得られない物質

区分 慢性1 *(注記 2)*

96 時間 LC_{50}（魚類に対する）≦1mg/l および/または
48 時間 EC_{50}（甲殻類に対する）≦1mg/l および/または
72 または 96 時間 ErC_{50}（藻類または他の水生植物に対する）≦1mg/l *(注記 3)*
であって急速分解性がないか、または実験的に求められた BCF≧500（またはデータがないときは $logK_{ow}$≧4）であること *(注記 4 および 5)*

区分 慢性2

96 時間 LC_{50}（魚類に対する）> 1mg/l だが ≦10mg/l および/または
48 時間 EC_{50}（甲殻類に対する）> 1mg/l だが ≦10mg/l および/または
72 または 96 時間 ErC_{50}（藻類または他の水生植物に対する）> 1mg/l だが ≦10mg/l *(注記 3)*
であって急速分解性がないか、または実験的に求められた BCF≧500（またはデータがないときは $logK_{ow}$≧4）であること *(注記 4 および 5)*

区分 慢性3

96 時間 LC_{50}（魚類に対する）> 10mg/l だが ≦100mg/l および/または
48 時間 EC_{50}（甲殻類に対する）> 10mg/l だが ≦100mg/l および/または
72 または 96 時間 ErC_{50}（藻類または他の水生植物に対する）>10mg/l だが≦100mg/l *(注記 3)*
であって急速分解性がないか、または実験的に求められた BCF≧500（またはデータがないときは $logK_{ow}$≧4）であること *(注記 4 および 5)*

(c) 「セーフティネット」分類

区分 慢性4

水溶性が低く水中溶解度までの濃度で急性毒性がみられないものであって、急速分解性ではなく、生物蓄積性を示す $logK_{ow}$≧4 であるもの。他に科学的証拠が存在して分類が必要でないことが判明している場合はこの限りでない。そのような証拠とは、実験的に求められた BCF＜500 であること、または慢性毒性 NOEC＞1mg/l であること、あるいは環境中において急速分解性であることの証拠などである。

NOTE 1: The organisms fish, crustacea and algae are tested as surrogate species covering a range of trophic levels and taxa, and the test methods are highly standardized. Data on other organisms may also be considered, however, provided they represent equivalent species and test endpoints.

NOTE 2: When classifying substances as Acute 1 and/or Chronic 1 it is necessary at the same time to indicate an appropriate M factor (see 4.1.3.5.5.5) to apply the summation method.

NOTE 3: Where the algal toxicity ErC_{50} [= EC_{50} (growth rate)] falls more than 100 times below the next most sensitive species and results in a classification based solely on this effect, consideration should be given to whether this toxicity is representative of the toxicity to aquatic plants. Where it can be shown that this is not the case, professional judgment should be used in deciding if classification should be applied. Classification should be based on the ErC_{50}. In circumstances where the basis of the EC_{50} is not specified and no ErC_{50} is recorded, classification should be based on the lowest EC_{50} available.

NOTE 4: Lack of rapid degradability is based on either a lack of ready biodegradability or other evidence of lack of rapid degradation. When no useful data on degradability are available, either experimentally determined or estimated data, the substance should be regarded as not rapidly degradable.

NOTE 5: Potential to bioaccumulate, based on an experimentally derived BCF ≥ 500 or, if absent, a log $K_{ow} \geq 4$, provided log K_{ow} is an appropriate descriptor for the bioaccumulation potential of the substance. Measured log K_{ow} values take precedence over estimated values and measured BCF values take precedence over log K_{ow} values.

Figure 4.1.1: Categories for substances long-term (chronic) hazardous to the aquatic environment

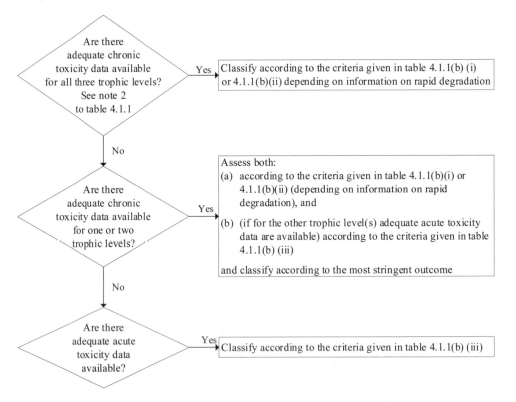

4.1.2.5 The system for classification recognizes that the core intrinsic hazard to aquatic organisms is represented by both the acute and chronic toxicity of a substance, the relative importance of which is determined by the specific regulatory system in operation. Distinction can be made between the short-term (acute) hazard and the long-term (chronic) hazard and therefore separate hazard categories are defined for both properties representing a gradation in the level of hazard identified. The lowest of the available toxicity values between and within the different trophic levels (fish, crustacean, algae) will normally be used to define the appropriate hazard category(ies). There may be circumstances, however, when a weight of evidence assessment may be used. Acute toxicity data are the most readily available and the tests used are the most standardized.

注記 1：魚類、甲殻類および藻類といった生物は、一連の栄養段階と分類群をカバーする代表種として試験されており、その試験方法は高度に標準化されている。その他の生物に関するデータも考慮されることもあるが、ただし同等の生物種およびエンドポイントによる試験であることが前提である。

注記 2：物質を急性1または慢性1と分類する場合は、同時に、加算法を適用するための適切な毒性乗率 M（4.1.3.5.5.5 参照）を示す必要がある。

注記 3：藻類に対する毒性値ErC_{50}〔＝EC_{50}（生長率）〕が、次に感受性の高い種より100倍以上小さく、この作用のみによって分類されることになる場合、この毒性が水生植物に対する毒性を代表しているかどうかについて考慮する必要がある。もし代表していないことが認められた場合には、分類すべきかどうかの決定には専門家の判断を用いる必要がある。分類はErC_{50}により行う必要がある。EC_{50} を得た根拠が特定されず、かつErC_{50}が記録されていないような状況では、入手されたEC_{50}最低値によって分類すべきである。

注記 4：急速分解性の欠如は、易生分解性の欠如、または急速分解性が欠如していることについてのその他の証拠より判断する。実験的に求められたデータ、または推定により求められたデータのいずれにせよ、分解性に関する有用なデータが得られない場合は、その物質は急速分解性がないものとみなすべきである。

注記 5：生物蓄積性は、実験により求められたBCFが500以上であるか、またはそのようなBCFが求められていない場合には$logK_{ow}≧4$が適切な指標である。実測により求められた$logK_{ow}$値の方が推定により求められた$logK_{ow}$値より優先され、また$logK_{ow}$値よりBCF実測値の方が優先される。

図4.1.1: 水生環境に対して長期（慢性）有害性のある物質の分類

3つの栄養段階すべてについて、慢性毒性の十分なデータが入手できるか？表 4.1.1 の注記2を参照 → はい → 急速分解性に関する情報に応じて、表 4.1.1 (b) (i)または表 4.1.1 (b) (ii)に示す基準にしたがって分類を行う

↓ いいえ

1つまたは2つの栄養段階について、慢性毒性の十分なデータが入手できるか？ → はい → 以下の両方、すなわち、
(a) （急速分解性に関する情報に応じて、表 4.1.1 (b) (i)または表 4.1.1 (b) (ii)に示す基準にしたがって、
(b) （他の栄養段階について急性毒性の十分なデータが得られる場合は）、表 4.1.1 (b) (iii)に示す基準にしたがって、
評価を行い、最も厳格な結果に合わせた分類を行う

↓ いいえ

急性毒性の十分なデータが入手できるか？ → はい → 表 4.1.1 (b) (iii)に示す基準にしたがって、分類を行う

4.1.2.5　GHS では、水生生物に対する固有の主要な有害性は、物質の急性および慢性両方の毒性によって代表されると認識されており、その相対的な重要性は、施行されている特定の規制システムによって決まる。短期（急性）有害性と長期（慢性）有害性を区別することが可能であるため、この双方の性質についてはそれぞれ有害性レベルの段階によって有害性区分が定められている。適切な有害性区分を決定するには、通常、異なる栄養段階（魚類、甲殻類、藻類）について入手された毒性値のうちの最低値が用いられる。しかし、証拠の重み付け評価が用いられるような場合もある。急性毒性データは最も容易に入手でき、試験も最も標準化されている。

4.1.2.6   Acute toxicity represents a key property in defining the hazard where transport of large quantities of a substance may give rise to short-term dangers arising from accidents or major spillages. Hazards categories up to $L(E)C_{50}$ values of 100 mg/l are thus defined although categories up to 1000 mg/l may be used in certain regulatory frameworks. The category Acute 1 may be further sub-divided to include an additional category for acute toxicity $L(E)C_{50} \leq 0.1$ mg/l in certain regulatory systems such as that defined by MARPOL 73/78 Annex II. It is anticipated that their use would be restricted to regulatory systems concerning bulk transport.

4.1.2.7   For packaged substances it is considered that the principal hazard is defined by chronic toxicity, although acute toxicity at $L(E)C_{50}$ levels ≤ 1 mg/l are also considered hazardous. Levels of substances up to 1 mg/l are considered as possible in the aquatic environment following normal use and disposal. At toxicity levels above this, it is considered that the acute toxicity itself does not describe the principal hazard, which arises from low concentrations causing effects over a longer time scale. Thus, a number of hazard categories are defined which are based on levels of chronic aquatic toxicity. Chronic toxicity data are not available for many substances, however, and in those cases it is necessary to use the available data on acute toxicity to estimate this property. The intrinsic properties of a lack of rapid degradability and/or a potential to bioconcentrate in combination with acute toxicity may be used to assign a substance to a long-term (chronic) hazard category. Where chronic toxicity is available showing NOECs greater than water solubility or greater than 1 mg/l, this would indicate that no classification in any of the long-term (chronic) hazard categories Chronic 1 to 3 would be necessary. Equally, for substances with an $L(E)C_{50} > 100$ mg/l, the toxicity is considered as insufficient to warrant classification in most regulatory systems.

4.1.2.8   Recognition is given to the classification goals of MARPOL 73/78 Annex II, which covers the transport of bulk quantities in ships tanks, which are aimed at regulating operational discharges from ships and assigning of suitable ship types. They go beyond that of protecting aquatic ecosystems, although that clearly is included. Additional hazard categories may thus be used which take account of factors such as physico-chemical properties and mammalian toxicity.

4.1.2.9   *Aquatic toxicity*

4.1.2.9.1   The organisms fish, crustacea and algae are tested as surrogate species covering a range of trophic levels and taxa, and the test methods are highly standardized. Data on other organisms may also be considered, however, provided they represent equivalent species and test endpoints. The algal growth inhibition test is a chronic test but the EC_{50} is treated as an acute value for classification purposes. This EC_{50} should normally be based on growth rate inhibition. If only the EC_{50} based on reduction in biomass is available, or it is not indicated which EC_{50} is reported, this value may be used in the same way.

4.1.2.9.2   Aquatic toxicity testing, by its nature, involves the dissolution of the substance under test in the water media used and the maintenance of a stable bioavailable exposure concentration over the course of the test. Some substances are difficult to test under standard procedures and thus special guidance will be developed on data interpretation for these substances and how the data should be used when applying the classification criteria.

4.1.2.10   *Bioaccumulation*

It is the bioaccumulation of substances within the aquatic organisms that can give rise to toxic effects over longer time scales even when actual water concentrations are low. The potential to bioaccumulate is determined by the partitioning between n-octanol and water. The relationship between the partition coefficient of an organic substance and its bioconcentration as measured by the BCF in fish has considerable scientific literature support. Using a cut-off value of log $K_{ow} \geq 4$ is intended to identify only those substances with a real potential to bioconcentrate. In recognition that the log K_{ow} is only an imperfect surrogate for a measured BCF, such a measured value would always take precedence. A BCF in fish of < 500 is considered as indicative of a low level of bioconcentration. Some relationships can be observed between chronic toxicity and bioaccumulation potential, as toxicity is related to the body burden.

4.1.2.11   *Rapid degradability*

4.1.2.11.1   Substances that rapidly degrade can be quickly removed from the environment. While effects can occur, particularly in the event of a spillage or accident, they will be localized and of short duration. The absence of rapid degradation in the environment can mean that a substance in the water has the potential to exert toxicity over a wide temporal and spatial scale. One way of demonstrating rapid degradation utilizes the biodegradation screening tests designed to determine whether a substance is "readily biodegradable". Thus, a substance which passes this screening test is one that is likely to biodegrade "rapidly" in the aquatic environment and is thus unlikely to be persistent. However, a fail in the screening test does not necessarily mean that the substance will not degrade rapidly in the environment. Thus, a further criterion was added which would allow the use of data to show that the substance did actually degrade biotically or abiotically in the aquatic environment by > 70 % in 28 days. Thus, if degradation could be demonstrated under environmentally realistic conditions, then the definition of "rapid degradability" would have been met. Many degradation

4.1.2.6　急性毒性は、ある物質の大量輸送の事故または大量漏出が原因となって、短期の危険が生じる場合の有害性を決定する重要な性質を表す。このために L(E)C$_{50}$ 値が 100mg/l に至る有害性区分が定められているが、特定の規制の枠組みにおいては 1000mg/l までの区分が用いられてもよい。急性区分 1 はさらに細分化して、例えば MARPOL 条約 73/78 附属書 II に定められているように、特定の規制システムにおいては、急性毒性 L(E)C$_{50}$≦0.1mg/l の区分を設けてもよい。その用途は、ばら積み輸送に関する規制システムに限られるであろうと予想される。

4.1.2.7　包装された物質の場合、主要な有害性は慢性毒性で決まると考えられているが、L(E)C$_{50}$ 値が ≦1mg/l の急性毒性もまた有害であると考えられる。通常の使用および廃棄後に、水生環境中の物質濃度は 1mg/l までになることもあり得ると考えられる。これより高い毒性レベルの場合は、急性毒性そのものでは、長い時間スケールで影響を及ぼすような低濃度によって生じる根本的な有害性を説明できないと考えられる。したがって、慢性水生毒性のレベルに基づいて多くの有害性区分が定められている。しかし、多くの物質では慢性毒性データを利用できず、こうした場合は、慢性毒性を評価するのに入手できる急性毒性のデータを用いなければならない。急速分解性の欠如または生物蓄積性の可能性といった本質的な特性と急性毒性とを組み合わせて、物質を長期（慢性）有害性区分に指定することもできよう。また、慢性毒性値が利用でき、NOEC が水溶解度よりも大きいか 1mg/l を超える場合、これは長期（慢性）有害性区分慢性 1～3 に分類する必要はないことを意味する。同様に、L(E)C$_{50}$>100mg/l の物質については、ほとんどの規制システムで、その毒性を分類する根拠になるほどではないと考えられている。

4.1.2.8　MARPOL 条約 73/78 附属書 II の分類目標にも考慮した。この規則は船舶タンクによるばら積み輸送を対象としたもので、船舶からの操業に伴う排出を規制すること、およびふさわしい船型要件を指定することを目標としている。水圏生態系の保護も明らかに対象に含まれているが、それにとどまらない目標を目指している。したがって、物理化学的性質や哺乳類に対する毒性等の要因を考慮に加えた追加の有害性区分が用いられるかもしれない。

4.1.2.9　水生毒性

4.1.2.9.1　魚類、甲殻類および藻類といった生物は、一連の栄養段階および分類群をカバーする代表種として試験されており、その試験方法は高度に標準化されている。その他の生物に関するデータも考慮されることもあるが、ただし同等の生物種およびエンドポイントによる試験であることが前提である。藻類生長阻害試験は慢性試験ではあるが、その EC$_{50}$ は分類の目的では急性値として扱われる。この EC$_{50}$ は通常、生長速度阻害をもとに得られるべきである。生物量の減少にもとづく EC$_{50}$（訳注：面積法による EC$_{50}$）しか得られない場合、またはどの EC$_{50}$ が報告されているか示されていない場合でも、これらの数値を同様に使用してもよいであろう。

4.1.2.9.2　水生毒性試験はその性格上、試験対象物質を、使用している水媒体に溶かし、生物学的利用性のあるばく露濃度を試験期間中に安定して維持することを必要とする。物質によっては標準手順で試験することが困難であり、したがってそうした物質に関するデータの解釈に関して、および分類基準に適用する際にどのようにデータを利用すべきかについて、特別の指針が策定されるであろう。

4.1.2.10　生物蓄積性

　実際の物質の水中濃度は低くても、長い時間スケールで毒性影響を発現しうるのが、水生生物への蓄積である。生物蓄積性は、n-オクタノール/水分配係数により測定される。有機物質の分配係数と、魚類を用いた BCF により測定された生物濃縮性との関連性は、多くの科学文献により支持されている。GHS においてカットオフ値として log K$_{ow}$≧4 を採用しているのは、現実的に生物濃縮性のあるような物質のみを識別するためである。log K$_{ow}$ は BCF 測定値の不完全な代替値にすぎないことから、BCF 実測値が常に優先されるべきである。魚類における BCF<500 という値は生物濃縮性が低レベルであることを意味すると考えられる。毒性が身体への負荷に関係があることから、慢性毒性と生物蓄積性との間には何らかの関係が認められる。

4.1.2.11　急速分解性

4.1.2.11.1　急速分解性を示す物質は、環境から速やかに除去される。特に漏出や事故などの際には影響が起こることもありうるが、それは局所的で短期間のものになろう。急速分解性を示さないということは、水中において物質が時間的にも空間的にも広い範囲で毒性を発現する可能性があることを意味する。急速分解性を示す 1 つの方法として、物質が「容易に生分解可能」かどうかを決定するよう設計された生分解性スクリーニングテストを採用している。このスクリーニングテストに合格する物質は、水中環境で「速やかに」生分解する可能性のある物質であり、したがって残留する見込みは小さい。しかし、このスクリーニング試験に不合格となったとしても、必ずしもその物質が環境中で速やかに分解しないことを意味するわけではない。そのため、その物質が水中環境において生物的または非生物的に 28 日間に 70%以上、実際に分解したことを示すデータを用いたさらなる基準が追加された。したがって、もし現実的な環境条

data are available in the form of degradation half-lives and these can also be used in defining rapid degradation. Details regarding the interpretation of these data are further elaborated in the guidance document of annex 9. Some tests measure the ultimate biodegradation of the substance, i.e. full mineralization is achieved. Primary biodegradation would not normally qualify in the assessment of rapid degradability unless it can be demonstrated that the degradation products do not fulfill the criteria for classification as hazardous to the aquatic environment.

4.1.2.11.2 It must be recognized that environmental degradation may be biotic or abiotic (e.g. hydrolysis) and the criteria used reflect this fact. Equally, it must be recognized that failing the ready biodegradability criteria in the OECD tests does not mean that the substance will not be degraded rapidly in the real environment. Thus, where such rapid degradation can be shown, the substance should be considered as rapidly degradable. Hydrolysis can be considered if the hydrolysis products do not fulfil the criteria for classification as hazardous to the aquatic environment. A specific definition of rapid degradability is shown below. Other evidence of rapid degradation in the environment may also be considered and may be of particular importance where the substances are inhibitory to microbial activity at the concentration levels used in standard testing. The range of available data and guidance on its interpretation are provided in the guidance document of annex 9.

4.1.2.11.3 Substances are considered rapidly degradable in the environment if the following criteria hold true:

(a) if in 28-day ready biodegradation studies, the following levels of degradation are achieved:

(i) tests based on dissolved organic carbon: 70 %;

(ii) tests based on oxygen depletion or carbon dioxide generation: 60 % of theoretical maxima;

These levels of biodegradation must be achieved within 10 days of the start of degradation which point is taken as the time when 10 % of the substance has been degraded, unless the substance is identified as a complex, multi-component substance with structurally similar constituents. In this case, and where there is sufficient justification, the 10-day window condition may be waived and the pass level applied at 28 days as explained in annex 9 (A9.4.2.2.3).

(b) if, in those cases where only BOD and COD data are available, when the ratio of BOD_5/COD is ≥ 0.5; or

(c) if other convincing scientific evidence is available to demonstrate that the substance can be degraded (biotically and/or abiotically) in the aquatic environment to a level > 70 % within a 28-day period.

4.1.2.12 *Inorganic compounds and metals*

4.1.2.12.1 For inorganic compounds and metals, the concept of degradability as applied to organic compounds has limited or no meaning. Rather the substance may be transformed by normal environmental processes to either increase or decrease the bioavailability of the toxic species. Equally the use of bioaccumulation data should be treated with care. Specific guidance will be provided on how these data for such materials may be used in meeting the requirements of the classification criteria.

4.1.2.12.2 Poorly soluble inorganic compounds and metals may be acutely or chronically toxic in the aquatic environment depending on the intrinsic toxicity of the bioavailable inorganic species and the rate and amount of this species which may enter solution. A protocol for testing these poorly soluble materials is included in annex 10. All evidence must be weighed in a classification decision. This would be especially true for metals showing borderline results in the Transformation/Dissolution Protocol.

4.1.2.13 *Use of QSARs*

While experimentally derived test data are preferred, where no experimental data are available, validated Quantitative Structure Activity Relationships (QSARs) for aquatic toxicity and log K_{ow} may be used in the classification process. Such validated QSARs may be used without modification to the agreed criteria, if restricted to chemicals for which their mode of action and applicability are well characterized. Reliable calculated toxicity and log K_{ow} values should be valuable in the safety net context. QSARs for predicting ready biodegradation are not yet sufficiently accurate to predict rapid degradation.

件下で分解が実証できた場合、「急速分解性」の定義に適合するであろう。多くの分解データは分解の半減期という形で入手されるが、これらもまた急速分解性を定義するのに用いることができる。これらデータの解釈の詳細に関しては附属書9の手引書に記述されている。いくつかの試験はその物質の究極の生分解性、すなわち完全な無機化の達成を測定するものである。分解生成物が水生環境有害性という分類判定基準を満足しない限り、急速分解性の評価において、通常は一次生分解性を用いないであろう。

4.1.2.11.2　環境中の分解は生物学的な場合もあれば非生物学的（例えば加水分解）な場合もあり、用いられる判定基準はこの事実を反映しているということが認識されなければならない。それと同様に、OECD試験で易生分解性の判定基準に適合しなくとも、その物質が現実の環境中で速やかに分解しないことを必ずしも意味するものではないことも認識されなければならない。したがって、こうした急速分解性が示されれば、その物質は急速分解性を示すと考えるべきである。加水分解による生成物が、水生環境有害性の分類基準を満たさないのであれば、加水分解性についても考慮に入れて良い。急速分解性の明確な定義を次項に示す。環境中の急速分解性についての別の証拠も考慮してよく、その物質が標準的試験で用いられる濃度レベルで微生物活性を阻害する場合には特に重要になろう。利用可能なデータ範囲とその解釈に関する指針は附属書9の手引きに示されている。

4.1.2.11.3　下記の判定基準にあてはまれば、物質は環境中で速やかに分解するとみなされる：

(a) 28日間の易生分解性試験で下記のいずれかの分解レベルが達成された場合：

　(i)　溶存有機炭素による試験：70%；

　(ii)　酸素消費量または二酸化炭素生成量による試験：理論的最高値の60%；

その物質が構造的に類似した構成要素を持つ複合的な多成分物質であると認められない場合、これらの生分解レベルは、分解開始後10日以内に達成されなければならず、分解開始点は物質の10%が分解された時点とする。多成分物質と認められる場合、附属書9(A9.4.2.2.3)で説明するように、十分な根拠があれば、10日間の時間ウィンドウ条件は免除され、28日間の合格レベルが適用される。

(b) BODまたはCODデータしか利用できないような場合には、BOD_5/CODが0.5以上となった場合；または

(c) 28日間以内に70%を超えるレベルで水生環境において分解（生物学的または非生物学的に）されることを証明するようなその他の有力な科学的証拠が入手された場合。

4.1.2.12　無機化合物および金属

4.1.2.12.1　無機化合物および金属については、有機化合物に適用される分解性の概念は限定された意味しか持たないか、または全く意味を持たない。これらの物質は分解というよりも、むしろ、通常の環境プロセスによって変換され、有毒な化学種の生物学的利用能を増加または減少させることがある。同様に、生物蓄積性データも注意して取扱わなければならない。これらの物質のデータを、分類基準の要求事項に適合させ、どのように使用するかに関しては特別な手引きが作成されることになろう。

4.1.2.12.2　難溶性の無機化合物と金属は、生物学的利用性のある無機化学種固有の毒性、およびこの無機化学種が溶液中に溶け込む速度と量に応じて、水生環境において急性毒性または慢性毒性をもつ可能性がある。これらの難溶性物質に関する試験手順は、附属書10に記載する。全ての証拠は分類判定の際に重み付けされなければならない。これは特に、変化/溶解プロトコルでボーダーラインの結果を示す金属にあてはまる。

4.1.2.13　QSARの利用

実験によって導かれた試験データの方が好ましいが、実験データが入手できない場合には、水生毒性と$\log K_{ow}$についての、有効性が確認されている定量的構造活性相関（QSAR）を分類プロセスに利用することもできる。このような有効性が確認されているQSARは、その作用様式および適用可能性がよく把握されている化学品に限定されるなら、合意された判定基準に適用できるであろう。信頼できる算定毒性値と$\log K_{ow}$の値は、上記のセーフティネットにおいて有効であろう。易生分解性を予測するためのQSARは、現在のところまだ急速分解性を予測するのに十分正確ではない。

4.1.2.14 *The classification criteria for substances diagrammatically summarized*

Table 4.1.2: Classification scheme for substances hazardous to the aquatic environment

Classification categories			
Short-term (acute) hazard (*Note 1*)	**Long-term (chronic) hazard** (*Note 2*)		
	Adequate chronic toxicity data available		Adequate chronic toxicity data not available (*Note 1*)
	Non-rapidly degradable substances (*Note 3*)	**Rapidly degradable substances** (*Note 3*)	
Category: Acute 1 $L(E)C_{50} \leq 1.00$	**Category: Chronic 1** NOEC or $EC_x \leq 0.1$	**Category: Chronic 1** NOEC or $EC_x \leq 0.01$	**Category: Chronic 1** $L(E)C_{50} \leq 1.00$ and lack of rapid degradability and/or BCF ≥ 500 or, if absent log $K_{ow} \geq 4$
Category: Acute 2 $1.00 < L(E)C_{50} \leq 10.0$	**Category: Chronic 2** $0.1 < $ NOEC or $EC_x \leq 1$	**Category: Chronic 2** $0.01 < $ NOEC or $EC_x \leq 0.1$	**Category: Chronic 2** $1.00 < L(E)C_{50} \leq 10.0$ and lack of rapid degradability and/or BCF ≥ 500 or, if absent log $K_{ow} \geq 4$
Category: Acute 3 $10.0 < L(E)C_{50} \leq 100$		**Category: Chronic 3** $0.1 < $ NOEC or $EC_x \leq 1$	**Category: Chronic 3** $10.0 < L(E)C_{50} \leq 100$ and lack of rapid degradability and/or BCF ≥ 500 or, if absent log $K_{ow} \geq 4$
	Category: Chronic 4 (*Note 4*) Example: (*Note 5*) No acute toxicity and lack of rapid degradability and BCF ≥ 500 or, if absent log Kow ≥ 4, unless NOECs > 1 mg/l		

NOTE 1: Acute toxicity band based on $L(E)C_{50}$ values in mg/l for fish, crustacea and/or algae or other aquatic plants (or QSAR estimation if no experimental data).

NOTE 2: Substances are classified in the various chronic categories unless there are adequate chronic toxicity data available for all three trophic levels above the water solubility or above 1 mg/l. ("Adequate" means that the data sufficiently cover the endpoint of concern. Generally, this would mean measured test data, but in order to avoid unnecessary testing it can, on a case-by-case basis, also be estimated data, e.g. (Q)SAR, or for obvious cases expert judgment).

NOTE 3: Chronic toxicity band based on NOEC or equivalent EC_x values in mg/l for fish or crustacea or other recognized measures for chronic toxicity.

NOTE 4: The system also introduces a "safety net" classification (referred to as category Chronic 4) for use when the data available do not allow classification under the formal criteria but there are nevertheless some grounds for concern.

NOTE 5: For poorly soluble substances for which no acute toxicity has been demonstrated at the solubility limit, and are both not rapidly degraded and have a potential to bioaccumulate, this category should apply unless it can be demonstrated that the substance does not require classification for aquatic long-term (chronic) hazards.

4.1.2.14 物質の分類基準の概要表

表4.1.2：水生環境有害性物質の分類スキーム

分類区分			
短期（急性）有害性 *(注記1)*	長期（慢性）有害性 *(注記2)*		
^	慢性毒性データが十分に入手できる場合		慢性毒性データが十分に入手できない場合 *(注記1)*
^	急速分解性のない物質 *(注記3)*	急速分解性のある物質 *(注記3)*	^
区分：急性1 $L(E)C_{50} \leqq 1.00$	区分：慢性1 NOECまたは$EC_x \leqq 0.1$	区分：慢性1 NOECまたは$EC_x \leqq 0.01$	区分：慢性1 $L(E)C_{50} \leqq 1.00$で急速分解性がないか、あるいはBCF$\geqq$500または、データがない場合$\log K_{ow} \geqq 4$
区分：急性2 $1.00 < L(E)C_{50} \leqq 10.0$	区分：慢性2 $0.1 <$ NOECまたは$EC_x \leqq 1$	区分：慢性2 $0.01 <$ NOECまたは$EC_x \leqq 0.1$	区分：慢性2 $1.00 < L(E)C_{50} \leqq 10.0$で急速分解性がないか、あるいはBCF$\geqq$500または、データがない場合$\log K_{ow} \geqq 4$
区分：急性3 $10.0 < L(E)C_{50} \leqq 100$		区分：慢性3 $0.1 <$ NOECまたは$EC_x \leqq 1$	区分：慢性3 $10.0 < L(E)C_{50} \leqq 100$で急速分解性がないか、あるいはBCF$\geqq$500または、データがない場合$\log K_{ow} \geqq 4$
	区分:慢性4 *(注記4)* 例： *(注記5)* NOECs＞1mg/lでない場合であって、急速毒性はなく、かつ急速分解性のデータもなく、さらにBCF$\geqq$500または、データがない時は$\log Kow \geqq 4$		

注記1：急性毒性値の帯域は、魚類、甲殻類または藻類あるいはその他の水生植物に対する$L(E)C_{50}$ (mg/l)（または実験データがない場合にはQSAR推定値）に基づく。

注記2：3つの栄養段階すべてで水溶解度または1mg/lを超える十分な慢性毒性データが存在する場合以外は、物質はさまざまな長期（慢性）区分に分類される。（「十分」というのは、データが対象のエンドポイントを十分にカバーしているという意味である。一般的にはこれは測定された試験データを意味するが、不必要な試験を回避するため、ケース・バイ・ケースで、推定値、例えば(Q)SAR推定値、もしくは明白な場合には専門家の判断ということもありうる）。

注記3：慢性毒性値の帯域は、魚類、甲殻類に対する$NOEC$(mg/l)または等価EC_x(mg/l)か、その他慢性毒性に関して公認されている手段に基づく。

注記4：このシステムは、利用できるデータからは正式な判定基準による分類ができないが、それにも関わらず何らかの懸念の余地がある場合に用いられるよう、分類の「セーフティネット」（区分 慢性4という。）を導入している。

注記5：溶解度の限界地点で急性毒性がないことが示されており、速やかに分解されず、生物蓄積性がある難溶性の物質については、その物質が水生の長期（慢性）有害性に区分する必要がないと立証されない場合は、この区分を適用すべきである。

4.1.3   Classification criteria for mixtures

4.1.3.1   The classification system for mixtures covers all classification categories which are used for substances, meaning categories Acute 1 to 3 and Chronic 1 to 4. In order to make use of all available data for purposes of classifying the aquatic environmental hazards of the mixture, the following assumption has been made and is applied where appropriate:

The "relevant ingredients" of a mixture are those which are present in a concentration equal to or greater than 0.1 % (w/w) for ingredients classified as Acute and/or Chronic 1 and equal to or greater than 1 % (w/w) for other ingredients, unless there is a presumption (e.g. in the case of highly toxic ingredients) that an ingredient present at a concentration less than 0.1 % can still be relevant for classifying the mixture for aquatic environmental hazards.

4.1.3.2   The approach for classification of aquatic environmental hazards is tiered and is dependent upon the type of information available for the mixture itself and for its ingredients. Elements of the tiered approach include classification based on tested mixtures, classification based on bridging principles, the use of "summation of classified ingredients" and/or an "additivity formula". Figure 4.1.2 outlines the process to be followed.

Figure 4.1.2: Tiered approach to classification of mixtures for short-term (acute) and long-term (chronic) aquatic environmental hazards

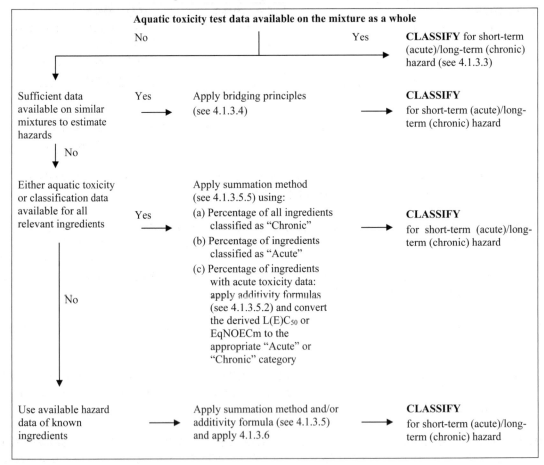

4.1.3 混合物の分類基準

4.1.3.1 混合物のための分類システムは、物質の分類のために用いるすべての分類区分、すなわち急性区分1～3および慢性区分1～4をカバーしている。混合物の水生環境有害性を分類するために入手できるすべてのデータを用いるために、以下の仮定が設定され、必要に応じて適用される。

混合物の「考慮すべき成分」とは、急性1または慢性1と分類される成分については濃度0.1%（w/w）以上で存在するもの、および他の成分については濃度1%（w/w）以上で存在するものをいう。ただし、0.1%未満の成分でも、その混合物の水生環境有害性を分類することに関連すると予想される場合（例えば毒性が強い成分の場合など）は、この限りではない。

4.1.3.2 水生環境有害性を分類するアプローチは段階的であり、混合物そのものおよびその各成分について入手できる情報の種類に依存する。この段階的アプローチの要素には、試験された混合物にもとづく分類、つなぎの原則にもとづく分類、「分類済み成分の加算」または「加算式」の使用、が含まれる。図4.1.2に従うべきプロセスの概略を示す。

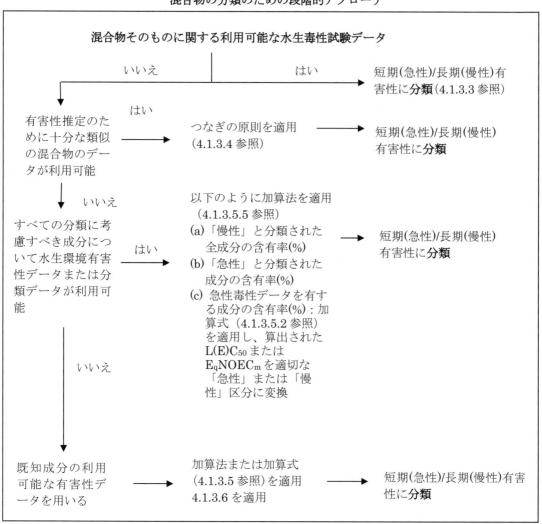

図4.1.2：短期（急性）および長期（慢性）水生環境有害性に関する混合物の分類のための段階的アプローチ

4.1.3.3 *Classification of mixtures when toxicity data are available for the complete mixture*

4.1.3.3.1 When the mixture as a whole has been tested to determine its aquatic toxicity, this information can be used for classifying the mixture according to the criteria that have been agreed for substances. The classification should normally be based on the data for fish, crustacea and algae/plants (see 4.1.1.3 and 4.1.1.4). When adequate acute or chronic data for the mixture as a whole are lacking, "bridging principles" or "summation method" should be applied (see paragraphs 4.1.3.4 and 4.1.3.5 and decision logic 4.1.5.2.2).

4.1.3.3.2 The long-term (chronic) hazard classification of mixtures requires additional information on degradability and in certain cases bioaccumulation. There are no degradability and bioaccumulation data for mixtures as a whole. Degradability and bioaccumulation tests for mixtures are not used as they are usually difficult to interpret, and such tests may be meaningful only for single substances.

4.1.3.3.3 *Classification for categories Acute 1, 2 and 3*

(a) When there are adequate acute toxicity test data (LC_{50} or EC_{50}) available for the mixture as a whole showing $L(E)C_{50} \leq 100$ mg/l:

Classify the mixture as Acute 1, 2 or 3 in accordance with table 4.1.1(a)

(b) When there are acute toxicity test data ($LC_{50}(s)$ or $EC_{50}(s)$) available for the mixture as a whole showing $L(E)C_{50}(s) > 100$ mg/l, or above the water solubility:

No need to classify for short-term (acute) hazard

4.1.3.3.4 *Classification for categories Chronic 1, 2 and 3*

(a) When there are adequate chronic toxicity data (EC_x or NOEC) available for the mixture as a whole showing EC_x or NOEC of the tested mixture ≤ 1 mg/l:

(i) Classify the mixture as Chronic 1, 2 or 3 in accordance with table 4.1.1 (b)(ii) (rapidly degradable) if the available information allows the conclusion that all relevant ingredients of the mixture are rapidly degradable;

(ii) Classify the mixture as Chronic 1 or 2 in all other cases in accordance with table 4.1.1 (b)(i) (non-rapidly degradable);

(b) When there are adequate chronic toxicity data (EC_x or NOEC) available for the mixture as a whole showing $EC_x(s)$ or NOEC(s) of the tested mixture > 1 mg/l or above the water solubility:

No need to classify for long-term (chronic) hazard, unless there are nevertheless reasons for concern

4.1.3.3.5 *Classification for category Chronic 4*

If there are nevertheless reasons for concern:

Classify the mixture as Chronic 4 (safety net classification) in accordance with table 4.1.1(c).

4.1.3.4 *Classification of mixtures when toxicity data are not available for the complete mixture: bridging principles*

4.1.3.4.1 Where the mixture itself has not been tested to determine its aquatic environmental hazard, but there are sufficient data on the individual ingredients and similar tested mixtures to adequately characterize the hazards of the mixture, this data will be used in accordance with the following agreed bridging principles. This ensures that the classification process uses the available data to the greatest extent possible in characterizing the hazards of the mixture without the necessity for additional testing in animals.

4.1.3.3 混合物そのものについて入手できるデータがある場合の混合物の分類

4.1.3.3.1 混合物そのものが水生毒性を判定するために試験されている場合には、物質に関して合意された判定基準にしたがって、その情報を混合物の分類に用いることができる。その場合、分類は通常、魚類、甲殻類、藻類/水生植物のデータに基づいて行うべきである（4.1.1.3 および 4.1.1.4 を参照）。混合物そのもの全体について急性または慢性の十分なデータがない場合は、「つなぎの原則」または「加算法」を適用すべきである（4.1.3.4 および 4.1.3.5 並びに判定論理 4.1.5.2.2 を参照）。

4.1.3.3.2 混合物の長期（慢性）有害性に係る分類を行うに当たっては、分解性や、一部のケースでは生物蓄積性に関する追加の情報が必要である。混合物そのものについては分解性や生物蓄積性に関するデータはない。混合物の分解性や生物蓄積性の試験のデータは、通常は解釈するのが難しいので用いられることがなく、そうした試験が有意義なのは単一の物質に対してだけである。

4.1.3.3.3 *急性1、2および3の区分の分類*

(a) 混合物そのもの全体について、$L(E)C_{50} \leq 100mg/l$ という急性毒性試験の十分なデータ（LC_{50} または EC_{50}）が得られる場合：

混合物を急性1、2または3に分類する（表4.1.1(a)を参照）。

(b) 混合物そのもの全体について、$L(E)C_{50}$ が $>100mg/l$ または水溶解度より大きいという急性毒性試験のデータ（$LC_{50}(s)$ または $EC_{50}(s)$）が得られる場合：

短期（急性）有害性についての分類は不要である。

4.1.3.3.4 *慢性1、2および3の区分の分類*

(a) 試験された混合物のEC_xまたは$NOEC$が$\leq 1mg/l$を示す混合物そのものについて、慢性毒性（EC_xまたは$NOEC$）の十分なデータが得られる場合：

(i) 入手した情報から混合物の関連成分すべてが急速分解性があるとの結論が認められた場合、表4.1.1(b)(ii)（急速分解性がある）にしたがって、その混合物を慢性1、2または3に分類する；

(ii) 他のすべてのケースでは、表4.1.1(b)(i)（急速分解性がない）にしたがって、その混合物を慢性1または2に分類する；

(b) 試験された混合物の$EC_x(s)$または$NOEC(s)$が$>1mg/l$または水溶解度より大きいことを示す混合物そのもの全体について、慢性毒性（EC_xまたは$NOEC$）の十分なデータが得られる場合：

それでも懸念の余地がある場合を除き、長期（慢性）有害性についての分類は不要である。

4.1.3.3.5 *慢性4の区分の分類*

それでも懸念の余地がある場合は：

表4.1.1(c)にしたがって、その混合物を慢性4（セーフティネット分類）に分類する。

4.1.3.4 混合物そのものについて水生試験毒性データが入手できない場合の混合物の分類：つなぎの原則（bridging Principles）

4.1.3.4.1 混合物そのものの水生環境有害性を決定する試験は行われていないが、当該混合物の有害性を適切に特定するための、個々の成分および類似の試験された混合物に関して十分なデータがある場合、以下のような合意されたつなぎの原則にしたがって、これらのデータが使用される。これによって、分類プロセスのために、追加の動物試験を行う必要なく入手できるデータを可能な限り最大限に用いて、混合物の有害性判定が可能になる。

4.1.3.4.2　　*Dilution*

Where a new mixture is formed by diluting a tested mixture or a substance with a diluent which has an equivalent or lower aquatic hazard classification than the least toxic original ingredient and which is not expected to affect the aquatic hazards of other ingredients, then the resulting mixture may be classified as equivalent to the original tested mixture or substance. Alternatively, the method explained in 4.1.3.5 could be applied.

4.1.3.4.3　　*Batching*

The aquatic hazard classification of a tested production batch of a mixture can be assumed to be substantially equivalent to that of another untested production batch of the same commercial product when produced by or under the control of the same manufacturer, unless there is reason to believe there is significant variation such that the aquatic hazard classification of the untested batch has changed. If the latter occurs, new classification is necessary.

4.1.3.4.4　　*Concentration of mixtures which are classified with the most severe classification categories (Chronic 1 and Acute 1)*

If a tested mixture is classified as Chronic 1 and/or Acute 1, and the ingredients of the mixture which are classified as Chronic 1 and/or Acute 1 are further concentrated, the more concentrated untested mixture should be classified with the same classification category as the original tested mixture without additional testing.

4.1.3.4.5　　*Interpolation within one hazard category*

For three mixtures (A, B and C) with identical ingredients, where mixtures A and B have been tested and are in the same hazard category and where untested mixture C has the same toxicologically active ingredients as mixtures A and B but has concentrations of toxicologically active ingredients intermediate to the concentrations in mixtures A and B, then mixture C is assumed to be in the same hazard category as A and B.

4.1.3.4.6　　*Substantially similar mixtures*

Given the following:

(a)　　Two mixtures:　　(i)　　A + B;
　　　　　　　　　　　　(ii)　　C + B;

(b)　　The concentration of ingredient B is essentially the same in both mixtures;

(c)　　The concentration of ingredient A in mixture (i) equals that of ingredient C in mixture (ii);

(d)　　Data on aquatic hazards for A and C are available and are substantially equivalent, i.e. they are in the same hazard category and are not expected to affect the aquatic toxicity of B.

If mixture (i) or (ii) is already classified based on test data, then the other mixture can be classified in the same hazard category.

4.1.3.5　　Classification of mixtures when toxicity data are available for all ingredients or only for some ingredients of the mixture

4.1.3.5.1　　The classification of a mixture is based on summation of the concentrations of its classified ingredients. The percentage of ingredients classified as "Acute" or "Chronic" will feed straight into the summation method. Details of the summation method are described in 4.1.3.5.5.

4.1.3.5.2　　Mixtures can be made of a combination of both ingredients that are classified (as Acute 1, 2, 3 and/or Chronic 1, 2, 3, 4) and those for which adequate toxicity test data is available. When adequate toxicity data are available for more than one ingredient in the mixture, the combined toxicity of those ingredients may be calculated using the following additivity formulas (a) or (b), depending on the nature of the toxicity data:

4.1.3.4.2 希釈

新しい混合物が、試験された混合物または物質を、毒性が最も弱い元の成分と比べて水生環境有害性分類が同等以下でありかつ他の成分の水生環境有害性に影響を与えることが予想されない希釈剤で希釈されて作られたものである場合、その結果生じる混合物は元の試験された混合物または物質と同等のものとして分類してもよい。また代わりに、4.1.3.5 で説明した方法を適用することもできる。

4.1.3.4.3 製造バッチ

混合物の試験されていない製造バッチに水生環境有害性があるかどうかは、同じ製造者によって、またはその管理下で生産された同じ商品の試験された別のバッチの毒性と本質的に同等とみなすことができる。ただし、その試験されていないバッチの水生環境有害性分類が変わってしまうような、有意な変動があると考えられる理由がある場合は、この限りではない。このような場合、新しい分類が必要である。

4.1.3.4.4 最も厳しい分類区分（慢性 1 および急性 1）に分類される混合物の濃縮

ある試験された混合物が慢性 1 または急性 1 に分類され、その混合物の慢性 1 または急性 1 に分類される成分がさらに濃縮される場合は、試験されていないより濃縮された混合物は、追加試験なしで、元の試験された混合物と同じ分類区分に分類すべきである。

4.1.3.4.5 有害性区分内での内挿

成分が同じ 3 つの混合物（A、B および C）については、混合物 A と混合物 B が試験されて同じ有害性区分に分類される場合および、試験されていない混合物 C が混合物 A および B と同じ毒性成分を持つが、その毒性成分の濃度が混合物 A と B の中間であるような場合、混合物 C は混合物 A および B と同じ有害性区分にあるとみなされる。この 3 種類の混合物において、成分内容は同じであることに注意すること。

4.1.3.4.6 本質的に類似した混合物

次を仮定する：

(a) 2 つの混合物： (i) A+B ;
(ii) C+B ;

(b) 成分 B の濃度は、両方の混合物で本質的に同じである；

(c) 混合物(i)の成分 A の濃度は、混合物(ii)の成分 C の濃度に等しい；

(d) A と C の水生有害性のデータが得られており、これらが実質的に同等である、すなわち、これらは同じ有害性区分に属し、かつ、B の水生毒性に影響を与えることはないと判断される。

混合物(i)または(ii)が既に試験データに基づいて分類されている場合は、他の混合物は同じ有害性区分に分類することができる。

4.1.3.5 混合物のすべての成分、または一部の成分についてのみ毒性データが入手できる場合の混合物の分類

4.1.3.5.1 混合物の分類は、その成分の分類の加算にもとづいて行われる。「急性」または「慢性」に分類された成分の含有率は、そのままで、この加算法に用いられることになる。この加算法の詳細については 4.1.3.5.5 で説明する。

4.1.3.5.2 混合物は、分類済みの成分（急性 1、2、3 または慢性 1、2、3、4）と十分な試験データが入手できる成分との組合せで構成されていることもある。混合物中の成分 2 種類以上について十分な毒性データが入手できる場合には、毒性データの性質に応じて下記の加算式(a)または(b)にしたがって、これらの成分の毒性加算値を算出できる：

(a) Based on acute aquatic toxicity:

$$\frac{\sum C_i}{L(E)C_{50_m}} = \sum_n \frac{C_i}{L(E)C_{50_i}}$$

where:

C_i	=	concentration of ingredient i (weight percentage);
$L(E)C\,50_i$	=	LC_{50} or EC_{50} for ingredient i, in (mg/l);
n	=	number of ingredients, and i is running from 1 to n;
$L(E)C\,50_m$	=	$L(E)\,C_{50}$ of the part of the mixture with test data;

The calculated toxicity may be used to assign that portion of the mixture a short-term (acute) hazard category which is then subsequently used in applying the summation method;

(b) Based on chronic aquatic toxicity:

$$\frac{\sum C_i + \sum C_j}{EqNOEC_m} = \sum_n \frac{C_i}{NOEC_i} + \sum_n \frac{C_j}{0.1 \times NOEC_j}$$

where:

C_i	=	concentration of ingredient i (weight percentage) covering the rapidly degradable ingredients;
C_j	=	concentration of ingredient j (weight percentage) covering the non-rapidly degradable ingredients;
$NOEC_i$	=	NOEC (or other recognized measures for chronic toxicity) for ingredient i covering the rapidly degradable ingredients, in mg/l;
$NOEC_j$	=	NOEC (or other recognized measures for chronic toxicity) for ingredient j covering the non-rapidly degradable ingredients, in mg/l;
n	=	number of ingredients, and i and j are running from 1 to n;
$EqNOEC_m$	=	Equivalent NOEC of the part of the mixture with test data;

The equivalent toxicity thus reflects the fact that non-rapidly degrading substances are classified one hazard category level more "severe" than rapidly degrading substances.

The calculated equivalent toxicity may be used to assign that portion of the mixture a long-term (chronic) hazard category, in accordance with the criteria for rapidly degradable substances (table 4.1.1(b)(ii)), which is then subsequently used in applying the summation method.

4.1.3.5.3　　When applying the additivity formula for part of the mixture, it is preferable to calculate the toxicity of this part of the mixture using for each ingredient toxicity values that relate to the same taxonomic group (i.e. fish, crustacean or algae) and then to use the highest toxicity (lowest value) obtained (i.e. use the most sensitive of the three groups). However, when toxicity data for each ingredient are not available in the same taxonomic group, the toxicity value of each ingredient should be selected in the same manner that toxicity values are selected for the classification of substances, i.e. the higher toxicity (from the most sensitive test organism) is used. The calculated acute and chronic toxicity may then be used to classify this part of the mixture as Acute 1, 2 or 3 and/or Chronic 1, 2 or 3 using the same criteria described for substances.

4.1.3.5.4　　If a mixture is classified in more than one way, the method yielding the more conservative result should be used.

(a) 急性水生毒性に基づく場合：

$$\frac{\sum C_i}{L(E)C_{50_m}} = \sum_n \frac{C_i}{L(E)C_{50_i}}$$

ここで、
- C_i = 成分iの濃度（重量パーセント）
- $L(E)C_{50i}$ = 成分iのLC_{50}またはEC_{50}（mg/l）
- n = 成分数（iは1からnまでの値をとる）
- $L(E)C_{50m}$ = 混合物の中で試験データが存在している部分の$L(E)C_{50}$

この毒性計算値を用いてその混合物の部分に短期（急性）有害性区分を割り振り、その後これを加算法に適用してもよい。

(b) 慢性水生毒性に基づく場合：

$$\frac{\sum C_i + \sum C_j}{EqNOEC_m} = \sum_n \frac{C_i}{NOEC_i} + \sum_n \frac{C_j}{0.1 \times NOEC_j}$$

ここで、
- C_i = 急速分解性のある成分iの濃度（重量パーセント）；
- C_j = 急速分解性のない成分を含む成分jの濃度（重量パーセント）；
- $NOEC_i$ = 急速分解性のある成分iのNOEC（あるいはその他慢性毒性に関して公認されている手段）（mg/l）；
- $NOEC_j$ = 急速分解性のない成分jのNOEC（あるいはその他慢性毒性に関して公認されている手段）（mg/l）；
- n = 成分数（iとjは1からnまでの値をとる）；
- $EqNOEC_m$ = 混合物のうち試験データが存在する部分の等価NOEC；

等価毒性は、急速分解性のない成分は急速分解性のある物質よりも1つ「厳しい」有害性区分レベルに分類されるという事実を反映している。

この等価毒性計算値を用いて、急速分解性物質の判定基準（表4.1.1(b)(ii)）に基づいて、その混合物の部分に長期（慢性）有害性区分を割り振り、その後これを加算法に適用してもよい。

4.1.3.5.3 混合物の一部にこの加算式を適用する場合、同一分類群（すなわち、魚類、甲殻類または藻類）について各物質の毒性値を用いて混合物のこの部分の毒性を計算し、得られた計算値の中の最も高い毒性値（最低毒性濃度、これら3つの分類群のうち感受性が最も高い群で得られた値）を採用することが望ましい。ただし、同一分類群での各成分の毒性データが入手できない場合には、物質の分類に毒性値を選択するのと同じやり方で各成分の毒性値を選択する。すなわち毒性の強い方の値（感受性が最も高い試験生物種で得られた値）を採用する。この計算された急性および慢性の毒性値を使い、物質の分類に関する判定基準と同じ基準を用いて、この混合物の一部を急性1、2または3あるいは慢性1、2または3と分類してもよい。

4.1.3.5.4 混合物の分類が1種類以上の方法で行われる場合、より保守的な（安全側の）結果となるような方法を採用すべきである。

4.1.3.5.5   *Summation method*

4.1.3.5.5.1   Rationale

4.1.3.5.5.1.1   In case of the ingredient classification categories Acute 1/Chronic 1 to Acute 3/Chronic 3, the underlying toxicity criteria differ by a factor of 10 in moving from one category to another. Ingredients with a classification in a high toxicity band may therefore contribute to the classification of a mixture in a lower band. The calculation of these classification categories therefore needs to consider the contribution of all ingredients classified Acute 1/Chronic 1 to Acute 3/Chronic 3 together.

4.1.3.5.5.1.2   When a mixture contains ingredients classified as Acute 1 or Chronic 1, attention should be paid to the fact that such ingredients, when their acute toxicity is well below 1 mg/l and/or chronic toxicity is well below 0.1 mg/l (if non rapidly degradable) and 0.01 mg/l (if rapidly degradable) contribute to the toxicity of the mixture even at a low concentration (see also *Classification of hazardous substances and mixtures* in chapter 1.3, paragraph 1.3.3.2.1). Active ingredients in pesticides often possess such high aquatic toxicity but also some other substances like organometallic compounds. Under these circumstances the application of the normal cut-off values/concentration limits may lead to an "under-classification" of the mixture. Therefore, multiplying factors should be applied to account for highly toxic ingredients, as described in 4.1.3.5.5.5.

4.1.3.5.5.2   Classification procedure

In general, a more severe classification for mixtures overrides a less severe classification, e.g. a classification with Chronic 1 overrides a classification with Chronic 2. As a consequence, the classification procedure is already completed if the result of the classification is Chronic 1. A more severe classification than Chronic 1 is not possible; therefore, it is not necessary to undergo the further classification procedure.

4.1.3.5.5.3   Classification for categories Acute 1, 2 and 3

4.1.3.5.5.3.1   First, all ingredients classified as Acute 1 are considered. If the sum of the concentrations (in %) of these ingredients multiplied by their corresponding M factor is $\geq$ 25 % the whole mixture is classified as Acute 1. If the result of the calculation is a classification of the mixture as Acute 1, the classification process is completed.

4.1.3.5.5.3.2   In cases where the mixture is not classified as Acute 1, classification of the mixture as Acute 2 is considered. A mixture is classified as Acute 2 if 10 times the sum of the concentrations (in %) of all ingredients classified as Acute 1 multiplied by their corresponding M factor plus the sum of the concentrations (in %) of all ingredients classified as Acute 2 is $\geq$ 25 %. If the result of the calculation is classification of the mixture as Acute 2, the classification process is completed.

4.1.3.5.5.3.3   In cases where the mixture is not classified either as Acute 1 or Acute 2, classification of the mixture as Acute 3 is considered. A mixture is classified as Acute 3 if 100 times the sum of the concentrations (in %) of all ingredients classified as Acute 1 multiplied by their corresponding M factor plus 10 times the sum of the concentrations (in %) of all ingredients classified as Acute 2 plus the sum of the concentrations (in %) of all ingredients classified as Acute 3 is $\geq$ 25%.

4.1.3.5.5.3.4   The classification of mixtures for short-term (acute) hazards based on this summation of the concentrations of classified ingredients is summarized in table 4.1.3.

Table 4.1.3: Classification of a mixture for short-term (acute) hazards based on summation of the concentrations of classified ingredients

Sum of the concentrations (in %) of ingredients classified as:		Mixture is classified as:
Acute 1 × M[a]	$\geq$ 25 %	Acute 1
(M × 10 × Acute 1) + Acute 2	$\geq$ 25 %	Acute 2
(M × 100 × Acute 1) + (10 × Acute 2) + Acute 3	$\geq$ 25 %	Acute 3

[a]   *For explanation of the M factor, see 4.1.3.5.5.5.*

4.1.3.5.5　加算法

4.1.3.5.5.1　原則の説明

4.1.3.5.5.1.1　急性1/慢性1から急性3/慢性3に至る、物質の分類区分では、ある区分から1つ区分を移ると、その根拠となっている毒性判定基準には10倍の差がある。このため、毒性の強い段階に分類されている物質が、より弱い段階にある混合物の分類に寄与することがある。したがって、これら分類区分の計算では、急性1/慢性1から急性3/慢性3の区分に分類される物質すべての関与を考慮する必要がある。

4.1.3.5.5.1.2　ある混合物に急性区分1または慢性区分1として分類される成分が含まれている場合、こうした成分では急性毒性濃度が1mg/lよりはるかに低い場合、または慢性毒性濃度が（急速分解性がない時に）0.1mg/lよりはるかに低いか（急速分解性がある時に）0.01mg/lよりはるかに低い場合、濃度が低くてもその混合物の毒性に関与するという事実に注意を払うべきである（1.3章1.3.3.2.1 *有害性物質および混合物の分類*も参照のこと）。農薬中の活性成分は、しばしば有機金属化合物のような強い水生毒性を有するが、同時に他の毒性も有する成分を含んでいる。そうした状況では、標準的なカットオフ値/濃度限界を適用すると、その混合物を「本来の毒性よりも弱い区分に分類（過小評価）」してしまうこともある。したがって、4.1.3.5.5.5で説明するように、強い毒性をもつ物質を考慮するには、毒性乗率を適用すべきである。

4.1.3.5.5.2　分類手順

一般的に、混合物に対するより厳しい分類は、厳しくない分類より優先して採用される。例えば、慢性1の分類は慢性2の分類より優先される。その結果、分類結果が慢性1であれば、それで分類手順はすでに完了している。慢性1よりも厳しい分類はありえないため、さらに分類手順を進める必要はない。

4.1.3.5.5.3　急性区分1、2および3への分類

4.1.3.5.5.3.1　まず急性1として分類されたすべての成分を検討する。これらの該当する毒性乗率Mをかけた成分の濃度(%)の合計が25%以上ならば、その混合物は全体として急性区分1として分類される。計算の結果、混合物の分類が急性1となった場合、分類プロセスはこれで完了である。

4.1.3.5.5.3.2　混合物が急性1に分類されない場合、その混合物が急性2として分類されないかを検討する。該当する毒性乗率Mをかけて急性1として分類されるすべての成分の濃度(%)の合計の10倍と急性2として分類されるすべての成分の濃度(%)の合計の総和が25%以上ならば、その混合物は急性2として分類される。計算の結果、混合物の分類が急性区分2となった場合、分類プロセスはこれで完了である。

4.1.3.5.5.3.3　混合物が急性1にも急性2にも分類されない場合、その混合物が急性3として分類されないかを検討する。該当する毒性乗率Mをかけて急性1として分類されるすべての成分の濃度(%)の合計の100倍と急性2として分類されるすべての成分の濃度(%)の合計の10倍および急性3として分類されるすべての成分の濃度(%)の合計の総和が25%以上ならば、その混合物は急性3として分類される。

4.1.3.5.5.3.4　分類された成分濃度(%)をこのように加算して行う混合物の短期（急性）有害性分類について、下記の表4.1.3に要約する。

表4.1.3：分類された成分の濃度の加算による混合物の短期（急性）有害性分類

分類される成分の濃度(%)の合計		混合物の分類
急性1×M [a]	≧25%	急性1
(M×10×急性1)+急性2	≧25%	急性2
(M×100×急性1)+(10×急性2)+急性3	≧25%	急性3

[a] *毒性乗率Mの説明は、4.1.3.5.5.5を参照*

4.1.3.5.5.4   Classification for categories Chronic 1, 2, 3 and 4

4.1.3.5.5.4.1   First, all ingredients classified as Chronic 1 are considered. If the sum of the concentrations (in %) of these ingredients multiplied by their corresponding M factor is ≥ 25 % the mixture is classified as Chronic 1. If the result of the calculation is a classification of the mixture as Chronic 1 the classification procedure is completed.

4.1.3.5.5.4.2   In cases where the mixture is not classified as Chronic 1, classification of the mixture as Chronic 2 is considered. A mixture is classified as Chronic 2 if 10 times the sum of the concentrations (in %) of all ingredients classified as Chronic 1 multiplied by their corresponding M factor plus the sum of the concentrations (in %) of all ingredients classified as Chronic 2 is ≥ 25 %. If the result of the calculation is classification of the mixture as Chronic 2, the classification process is completed.

4.1.3.5.5.4.3   In cases where the mixture is not classified either as Chronic 1 or Chronic 2, classification of the mixture as Chronic 3 is considered. A mixture is classified as Chronic 3 if 100 times the sum of the concentrations (in %) of all ingredients classified as Chronic 1 multiplied by their corresponding M factor plus 10 times the sum of the concentrations (in %) of all ingredients classified as Chronic 2 plus the sum of the concentrations (in %) of all ingredients classified as Chronic 3 is ≥ 25 %.

4.1.3.5.5.4.4   If the mixture is still not classified in either category Chronic 1, 2 or 3, classification of the mixture as Chronic 4 should be considered. A mixture is classified as Chronic 4 if the sum of the concentrations (in %) of ingredients classified as Chronic 1, 2, 3 and 4 is ≥ 25 %.

4.1.3.5.5.4.5   The classification of mixtures for long-term (chronic) hazards based on this summation of the concentrations of classified ingredients is summarized in table 4.1.4.

Table 4.1.4: Classification of a mixture for long-term (chronic) hazards based on summation of the concentrations of classified ingredients

Sum of the concentrations (in %) of ingredients classified as:		Mixture is classified as:
Chronic 1 × M[a]	≥ 25 %	Chronic 1
(M × 10 × Chronic 1) + Chronic 2	≥ 25 %	Chronic 2
(M × 100 × Chronic 1) + (10 × Chronic 2) + Chronic 3	≥ 25 %	Chronic 3
Chronic 1 + Chronic 2 + Chronic 3 + Chronic 4	≥ 25 %	Chronic 4

[a]   For explanation of the M factor, see 4.1.3.5.5.5.

4.1.3.5.5.5   Mixtures with highly toxic ingredients

Acute 1 or Chronic 1 ingredients with acute toxicities well below 1 mg/l and/or chronic toxicities well below 0.1 mg/l (if non-rapidly degradable) and 0.01 mg/l (if rapidly degradable) may influence the toxicity of the mixture and should be given increased weight in applying the summation method. When a mixture contains ingredients classified as Acute or Chronic 1, the tiered approach described in 4.1.3.5.5.3 and 4.1.3.5.5.4 should be applied using a weighted sum by multiplying the concentrations of Acute 1 and Chronic 1 ingredients by a factor, instead of merely adding up the percentages. This means that the concentration of "Acute 1" in the left column of table 4.1.3 and the concentration of "Chronic 1" in the left column of table 4.1.4 are multiplied by the appropriate multiplying factor. The multiplying factors to be applied to these ingredients are defined using the toxicity value, as summarized in table 4.1.5 below. Therefore, in order to classify a mixture containing Acute/Chronic 1 ingredients, the classifier needs to be informed of the value of the M factor in order to apply the summation method. Alternatively, the additivity formula (see 4.1.3.5.2) may be used when toxicity data are available for all highly toxic ingredients in the mixture and there is convincing evidence that all other ingredients, including those for which specific acute and/or chronic toxicity data are not available, are of low or no toxicity and do not significantly contribute to the environmental hazard of the mixture.

4.1.3.5.5.4　慢性区分 1、2、3 および 4 への分類

4.1.3.5.5.4.1　まず慢性 1 に分類されたすべての成分について考える。これらの該当する毒性乗率 M をかけた成分の濃度(%)の合計が 25%以上ならば、その混合物は慢性区分 1 に分類される。計算の結果、混合物の分類が慢性区分 1 となった場合、分類プロセスはこれで完了である。

4.1.3.5.5.4.2　混合物が慢性 1 に分類されない場合、その混合物が慢性 2 として分類されないかを検討する。該当する毒性乗率 M をかけて慢性 1 として分類されたすべての成分の濃度(%)の合計の 10 倍と慢性 2 として分類されたすべての成分の濃度(%)の合計の総和が 25%以上ならば、その混合物は慢性 2 として分類される。計算の結果、混合物の分類が慢性区分 2 となった場合、分類プロセスはこれで完了である。

4.1.3.5.5.4.3　混合物が慢性 1 にも慢性 2 にも分類されない場合、その混合物が慢性 3 として分類されないかを検討する。該当する毒性乗率 M をかけて慢性 1 として分類されたすべての成分の濃度(%)の合計の 100 倍と慢性 2 として分類されたすべての成分の濃度(%)の合計の 10 倍および慢性 3 として分類されたすべての成分の濃度(%)の合計の総和が 25%以上ならば、その混合物は慢性 3 として分類される。

4.1.3.5.5.4.4　その混合物が慢性 1、2 または 3 のいずれにも分類されない場合、その混合物が慢性 4 として分類されないかを検討するべきである。慢性 1、2、3 および 4 に分類された成分の濃度(%)の合計が 25%以上ならば、混合物は慢性 4 として分類される。

4.1.3.5.5.4.5　分類済み成分の濃度をこのように加算して行う混合物の長期（慢性）有害性分類について、下記の表 4.1.4 に要約する。

表 4.1.4：分類された成分の濃度の加算による混合物の長期（慢性）有害性分類

分類される成分の濃度(%)の合計		混合物の分類
慢性 1×M [a]	≧25%	慢性 1
(M×10×慢性 1)+慢性 2	≧25%	慢性 2
(M×100×慢性 1)+(10×慢性 2)+慢性 3	≧25%	慢性 3
慢性 1+慢性 2+慢性 3+慢性 4	≧25%	慢性 4

[a]　*毒性乗率 M の説明は、4.1.3.5.5.5 を参照*

4.1.3.5.5.5　強い毒性をもつ成分を含む混合物

急性毒性が 1mg/l よりはるかに低いか、または慢性毒性が（急速分解性がない時に）0.1mg/l よりはるかに低いか、（急速分解性がある時に）0.01mg/l よりはるかに低い場合の急性区分 1 または慢性区分 1 の成分は、混合物の毒性に影響する可能性があり、分類手法に加算法を適用する際にはその重み付けを増加させるべきである。急性 1 または慢性 1 として分類される成分が混合物に含まれている場合、4.1.3.5.5.3 および 4.1.3.5.5.4 に記載した段階的アプローチ、単に含有率を加算するのではなく、急性区分 1 または慢性 1 に分類される成分の濃度に毒性乗率をかけた、重み付け加算を用いるべきである。すなわち、表 4.1.3 の左側欄の「急性 1」の濃度および表 4.1.4 の左側欄の「慢性 1」の濃度に、適切な毒性乗率を掛けることを意味する。こうした成分に適用される毒性乗率は、下記の表 4.1.5 にまとめたように、毒性値を用いて定義される。したがって、急性/慢性 1 の成分を含む混合物を分類するには、分類担当者はこの加算法を適用するために毒性乗率 M の値を教えられておく必要がある。または、その混合物中の高毒性成分すべてについては毒性データが入手でき、かつその他の成分については、個々の急性または慢性毒性データが揃っていないような成分も含めて、毒性が弱いかまたはなく、その混合物の環境有害性に明確に影響しないという説得力のある証拠があれば、加算式（4.1.3.5.2）を用いてもよい。

Table 4.1.5: Multiplying factors for highly toxic ingredients of mixtures

Acute toxicity	M factor	Chronic toxicity	M factor	
L(E)C$_{50}$ value		NOEC value	NRD[a] ingredients	RD[b] ingredients
0.1 < L(E)C$_{50}$ ≤ 1	1	0.01 < NOEC ≤ 0.1	1	-
0.01 < L(E)C$_{50}$ ≤ 0.1	10	0.001 < NOEC ≤ 0.01	10	1
0.001 < L(E)C$_{50}$ ≤ 0.01	100	0.0001 < NOEC ≤ 0.001	100	10
0.0001 < L(E)C$_{50}$ ≤ 0.001	1000	0.00001 < NOEC ≤ 0.0001	1000	100
0.00001 < L(E)C$_{50}$ ≤ 0.0001	10000	0.000001 < NOEC ≤ 0.00001	10000	1000
(continue in factor 10 intervals)		(continue in factor 10 intervals)		

[a] *Non-rapidly degradable*
[b] *Rapidly degratdable*

4.1.3.6 *Classification of mixtures with ingredients without any useable information*

In the event that no useable information on acute and/or chronic aquatic toxicity is available for one or more relevant ingredients, it is concluded that the mixture cannot be attributed (a) definitive hazard category(ies). In this situation the mixture should be classified based on the known ingredients only, with the additional statement that: "× % of the mixture consists of ingredient(s) of unknown hazards to the aquatic environment". The competent authority can decide to specify that the additional statement is communicated on the label or on the SDS or both, or to leave the choice of where to place the statement to the manufacturer/supplier.

4.1.4 **Hazard communication**

General and specific considerations concerning labelling requirements are provided in *Hazard communication: Labelling* (chapter 1.4). Annex 1 contains summary tables about classification and labelling. Annex 3 contains examples of precautionary statements and pictograms which can be used where allowed by the competent authority. Table 4.1.6 presents specific label elements for substances and mixtures classified into this hazard class based on the criteria in this chapter.

Table 4.1.6: Label elements for hazardous to the aquatic environment

SHORT-TERM (ACUTE) AQUATIC HAZARD

	Category 1	Category 2	Category 3
Symbol	Environment	*No symbol*	*No symbol*
Signal word	Warning	*No signal word*	*No signal word*
Hazard statement	Very toxic to aquatic life	Toxic to aquatic life	Harmful to aquatic life

LONG-TERM (CHRONIC) AQUATIC HAZARD

	Category 1	Category 2	Category 3	Category 4
Symbol	Environment	Environment	*No symbol*	*No symbol*
Signal word	Warning	*No signal word*	*No signal word*	*No signal word*
Hazard statement	Very toxic to aquatic life with long lasting effects	Toxic to aquatic life with long lasting effects	Harmful to aquatic life with long lasting effects	May cause long lasting harmful effects to aquatic life

表 4.1.5：混合物中の高毒性成分に関する毒性乗率M

急性毒性 L(E)C$_{50}$値	毒性乗率 M	慢性毒性 NOEC値	毒性乗率 M NRD[a] 成分	RD[b] 成分
0.1＜L(E)C$_{50}$≦1	1	0.01＜NOEC≦0.1	1	-
0.01＜L(E)C$_{50}$≦0.1	10	0.001＜NOEC≦0.01	10	1
0.001＜L(E)C$_{50}$≦0.01	100	0.0001＜NOEC≦0.001	100	10
0.0001＜L(E)C$_{50}$≦0.001	1000	0.00001＜NOEC≦0.0001	1000	100
0.00001＜L(E)C$_{50}$≦0.0001	10000	0.000001＜NOEC≦0.00001	10000	1000
（以降10倍ずつ続く）		（以降10倍ずつ続く）		

[a] 急速分解性がない
[b] 急速分解性がある

4.1.3.6 利用可能な情報がない成分を含む混合物の分類

関連成分のうち1種類以上について急性または慢性水生毒性に関して利用可能な情報が揃っていない混合物については、決定的な有害性区分に帰属させることはできないと結論付けられる。そのような状況では、混合物は既知成分のみにもとづいて分類され、「本混合物の成分x%については水生環境有害性が不明である」という記述を追加しておくべきである。所管官庁はその追加的な記述をラベルまたはSDSあるいはその両方で伝達することを明記するかどうか、またその記述をどこにするかの選択を製造者/供給者に委ねるかどうかを決めることができる。

4.1.4 危険有害性情報の伝達

表示要件に関する通則および細則は、*危険有害性に関する情報の伝達：表示*（第1.4章）に定める。附属書1に分類と表示に関する概要表を示す。附属書3には、注意書きおよび所管官庁が許可した場合に使用可能な注意絵表示の例を示す。表4.1.6には、本章の判定基準に基づいてこの危険有害性クラスに分類された物質および混合物の具体的なラベル要素を示す。

表 4.1.6：水生環境有害性物質のラベル要素

短期（急性）水性有害性

	区分1	区分2	区分3
シンボル	環境	シンボルなし	シンボルなし
注意喚起語	警告	注意喚起語なし	注意喚起語なし
危険有害性情報	水生生物に非常に強い毒性	水生生物に毒性	水生生物に有害

長期（慢性）水性有害性

	区分1	区分2	区分3	区分4
シンボル	環境	環境	シンボルなし	シンボルなし
注意喚起語	警告	注意喚起語なし	注意喚起語なし	注意喚起語なし
危険有害性情報	長期継続的影響によって水生生物に非常に強い毒性	長期継続的影響によって水生生物に毒性	長期継続的影響によって水生生物に有害	長期継続的影響によって水生生物に有害のおそれ

4.1.5 Decision logic for substances and mixtures hazardous to the aquatic environment

The decision logics which follow are not part of the harmonized classification system but are provided here as additional guidance. It is strongly recommended that the person responsible for classification study the criteria before and during use of the decision logic.

4.1.5.1 *Short-term (acute) aquatic hazard classification*

4.1.5.1.1 *Decision logic 4.1.1 for substances and mixtures hazardous to the aquatic environment*

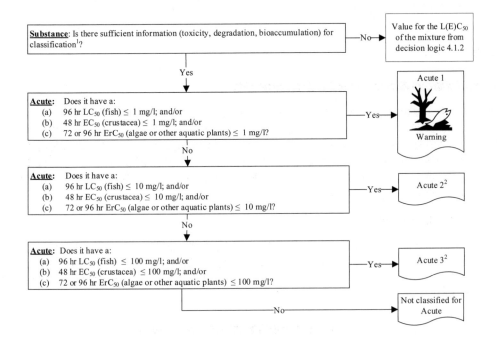

(cont'd on next page)

---

[1] *Classification can be based on either measured data and/or calculated data (see 4.1.2.13 and annex 9) and/or analogy decisions (see A9.6.4.5 in annex 9).*

[2] *Labelling requirements differ from one regulatory system to another, and certain classification categories may only be used in one or a few regulations.*

- 260 -

4.1.5 水生環境有害性のある物質および混合物の判定論理

以下の判定論理は、調和分類システムには含まれないが、追加的な手引きとしてここに示す。分類の責任者に対し、この判定論理を使用する前および使用する際に判定基準についてよく調べ理解することを強く勧める。

4.1.5.1 短期（急性）水生有害性の分類

4.1.5.1.1 水生環境有害性のある物質および混合物の判定論理4.1.1

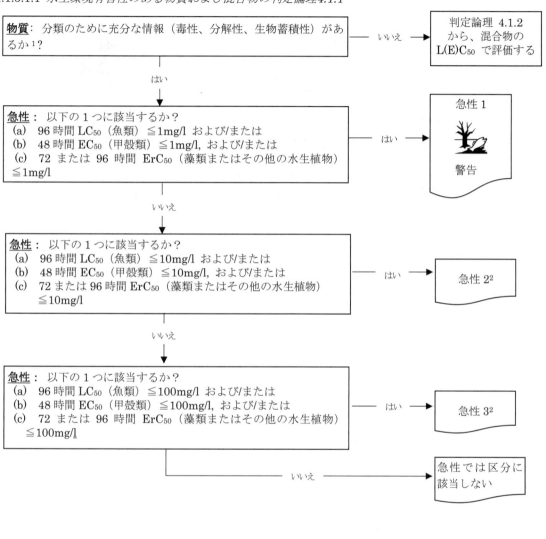

（次ページに続く）

---

[1] 分類は実測データまたは計算値（本章4.1.2.13 および附属書 9参照）または類似性判定（附属書9のA9.6.4.5 参照）に基づいてよい。

[2] 表示の要件は規制体系ごとに異なる。一部の分類区分は、1つまたは 少数の規則のみでしか使用されない場合もある。

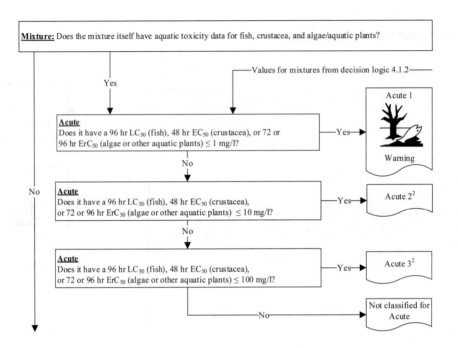

(cont'd on next page)

---

[2] *Labelling requirements differ from one regulatory system to another, and certain classification categories may only be used in one or a few regulations.*

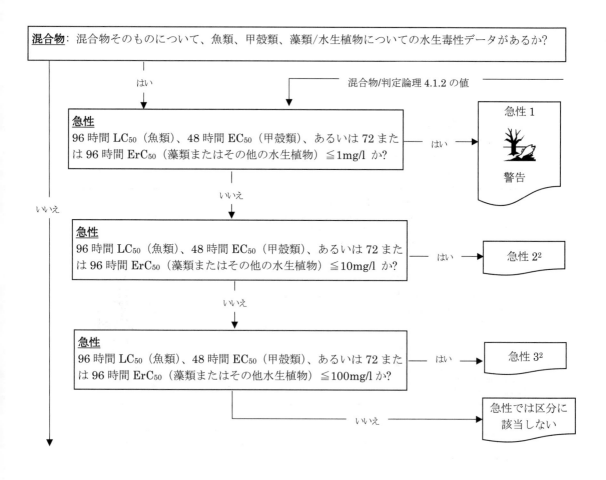

(次ページに続く)

---

[2] 表示の要件は規制体系ごとに異なる。一部の分類区分は、1 つまたは 少数の規則のみでしか使用されない場合もある。

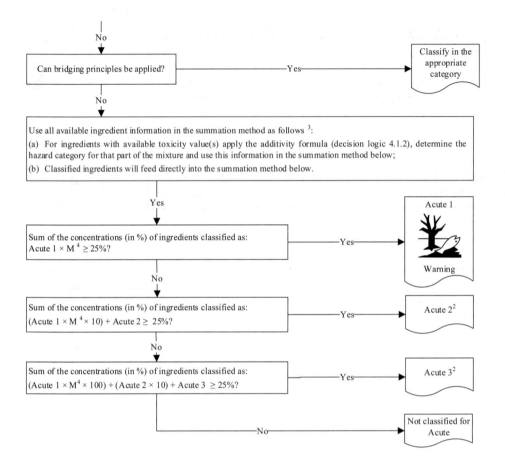

---

[2] Labelling requirements differ from one regulatory system to another, and certain classification categories may only be used in one or a few regulations.

[3] If not all ingredients have information, include the statement "x % of the mixture consists of ingredients(s) of unknown hazards to the aquatic environment" on the label. The competent authority can decide to specify that the additional statement be communicated on the label or on the SDS or both, or to leave the choice of where to place the statement to the manufacturer/supplier. Alternatively, in the case of a mixture with highly toxic ingredients, if toxicity values are available for these highly toxic ingredients and all other ingredients do not significantly contribute to the hazard of the mixture, then the additivity formula may be applied (see 4.1.3.5.5.5). In this case and other cases where toxicity values are available for all ingredients, the short-term (acute) classification may be made solely on the basis of the additivity formula.

[4] For explanation of M factor see 4.1.3.5.5.5.

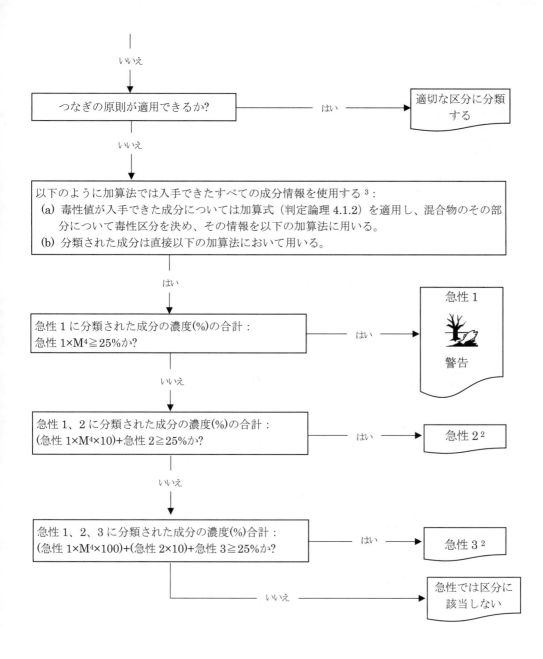

---

[2] 表示の要件は規制体系ごとに異なる。一部の分類区分は、1つまたは少数の規則のみでしか使用されない場合もある。
[3] すべての成分についての情報が揃っていない場合、ラベルに「混合物中 x %の成分は、水生環境有害性が未知のものである」という記述を入れる。所管官庁はその追加的な記述をラベルまたはSDSあるいはその両方で伝達することを明記するかどうか、またその記述をどこにするかの選択を製造者/供給者に委ねるかどうかを決めることができる。あるいは、非常に毒性の強い成分を含む混合物の場合、当該成分についての毒性データが入手でき、他の成分が混合物の有害性に明確な影響を及ぼさないものであれば、加算式を適用してもよい（4.1.3.5.5.5 参照）。この場合、およびすべての成分について毒性値が入手できた場合は、短期（急性）分類は加算式に基づいてのみ行うことができる。
[4] 毒性乗率Mの説明は4.1.3.5.5.5 を参照のこと。

4.1.5.1.2　　　*Decision logic 4.1.2 for mixtures (additivity formula)*

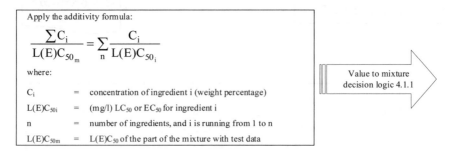

4.1.5.2　　*Long-term (chronic) aquatic hazard classification*

4.1.5.2.1　　　*Decision logic 4.1.3 (a) for substances*

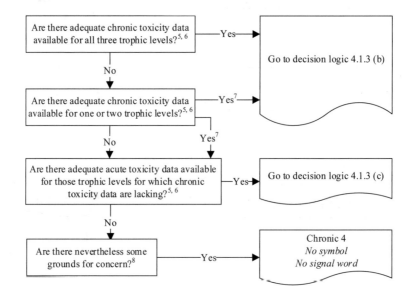

---

[5]　Data are preferably to be derived using internationally harmonized test methods (e.g. OECD test guidelines or equivalent) according to the principles of good laboratory practices (GLP), but data from other test methods such as national methods may also be used where they are considered as equivalent (see 4.1.1.2.2 and A9.3.2 of annex 9).

[6]　See figure 4.1.1.

[7]　Follow the flow chart in both ways and choose the most stringent classification outcome.

[8]　Note that the system also introduces a "safety net" classification (referred to as Category: Chronic 4) for use when the data available do not allow classification under the formal criteria but there are nevertheless some grounds for concern.

4.1.5.1.2　*判定論理 4.1.2 混合物（加算式）*

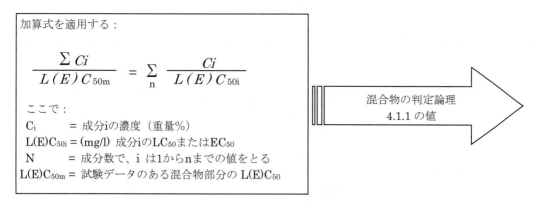

4.1.5.2　*長期（慢性）水生有害性の分類*

4.1.5.2.1　*物質の判定論理 4.1.3(a)*

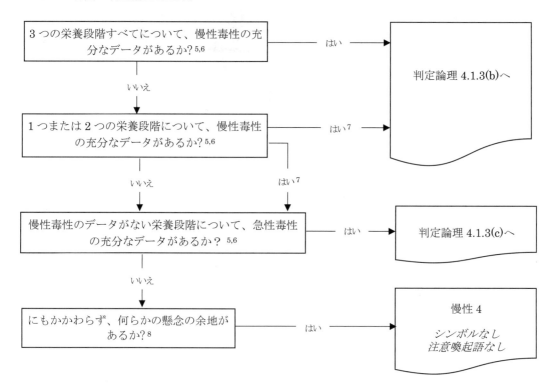

---

[5]　GLP原則に従った国際的に調和された試験方法（例えばOECDテストガイドラインまたはそれと同等なもの）を用いてデータを得るのが望ましいが、それらと同等なものとみなされれば、内国的な方法などの他の試験方法も用いても構わない（4.1.1.2.2および附属書9のA9.3.2を参照）。

[6]　図4.1.1を参照。

[7]　両方の方式でフローチャートをたどり、最も厳しい分類結果を選ぶ。

[8]　ただし、システムでは、利用できるデータからは正式の判定基準による分類ができないが、それにも関わらず何らかの懸念の余地がある場合に用いられるよう、分類の「セーフティネット」（区分：慢性4）を導入している。

4.1.5.2.2　　　　Decision logic 4.1.3 (b) for substances (when adequate chronic toxicity data are available for all three trophic levels)[5]

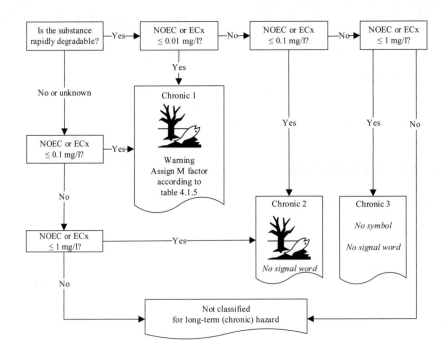

---

[5] Data are preferably to be derived using internationally harmonized test methods (e.g. OECD test guidelines or equivalent) according to the principles of good laboratory practices (GLP), but data from other test methods such as national methods may also be used where they are considered as equivalent (see 4.1.1.2.2 and A9.3.2 of annex 9).

4.1.5.2.2 （3つの栄養段階すべてについて、慢性毒性の十分なデータが得られた場合の）物質の判定論理 4.1.3(b)[5]

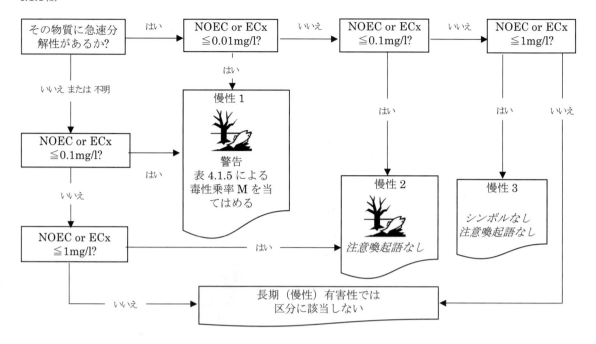

---

[5] GLP 原則に従った国際的に調和された試験方法（例えばOECDテストガイドラインまたはそれと同等なもの）を用いてデータを得るのが望ましいが、それらと同等なものとみなされれば、内国的な方法などの他の試験方法も用いても構わない（*4.1.1.2.2*および附属書9の*A9.3.2*を参照）。

4.1.5.2.3   Decision logic 4.1.3 (c) for substances (when adequate chronic toxicity data not are available for all three trophic levels)[5]

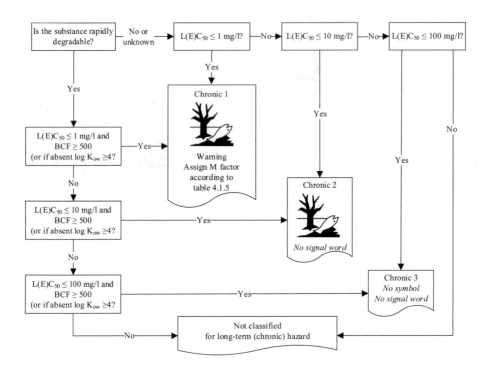

---

[5] Data are preferably to be derived using internationally harmonized test methods (e.g. OECD test guidelines or equivalent) according to the principles of good laboratory practices (GLP), but data from other test methods such as national methods may also be used where they are considered as equivalent (see 4.1.1.2.2 and A9.3.2 of annex 9).

4.1.5.2.3 （3つの栄養段階すべてについて、慢性毒性の十分なデータが得られない場合の）物質の判定論理4.1.3(c) [5]

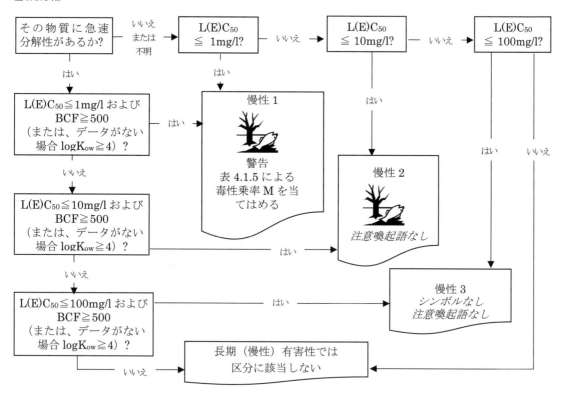

---

[5] GLP 原則に従った国際的に調和された試験方法（例えばOECD テストガイドラインまたはそれと同等なもの）を用いてデータを得るのが望ましいが、それらと同等なものとみなされれば、内国的な方法などの他の試験方法も用いても構わない（4.1.1.2.2および附属書9のA9.3.2を参照）。

4.1.5.2.4   *Decision logic 4.1.4 for mixtures*

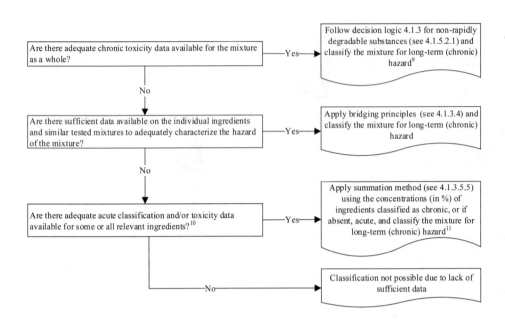

---

[9] Degradability and bioaccumulation tests for mixtures are not used as they are usually difficult to interpret, and such tests may be meaningful only for single substances. The mixture is therefore by default regarded as non-rapidly degradable. However, if the available information allows the conclusion that all relevant ingredients of the mixture are rapidly degradable the mixture can, for classification purposes, be regarded as rapidly degradable.

[10] In the event that no useable information on acute and/or chronic aquatic toxicity is available for one or more relevant ingredients, it is concluded that the mixture cannot be attributed (a) definitive hazard category(ies). In this situation the mixture should be classified based on the known ingredients only, with the additional statement that: "× % of the mixture consists of ingredient(s) of unknown hazards to the aquatic environment". The competent authority can decide to specify that the additional statement be communicated on the label or on the SDS or both, or to leave the choice of where to place the statement to the manufacturer/supplier.

[11] When adequate toxicity data are available for more than one ingredient in the mixture, the combined toxicity of those ingredients may be calculated using the additivity formulas (a) or (b) in 4.1.3.5.2, depending on the nature of the toxicity data. The calculated toxicity may be used to assign that portion of the mixture a short-term (acute) or long-term (chronic) hazard category which is then subsequently used in applying the summation method. (It is preferable to calculate the toxicity of this part of the mixture using for each ingredient a toxicity value that relate to the same taxonomic group (e.g. fish, crustacea or algae) and then to use the highest toxicity (lowest value) obtained (i.e. use the most sensitive of the three groups) (see 4.1.3.5.3)).

4.1.5.2.4 混合物についての判定論理4.1.4

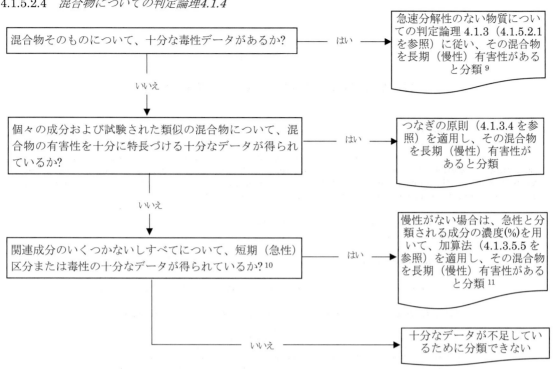

---

[9] 混合物の分解性や生物蓄積性の試験のデータは、通常は解釈するのが難しいので用いられることがなく、そうした試験が有意義なのは単一の物質に対してだけである。このため混合物は、当初の段階で急速分解性のないものとみなされる。とはいえ、入手した情報から混合物の関連成分すべてが急速分解性があるとの結論が認められた場合は、その混合物は、分類目的のために急速分解性があると分類することができる。

[10] 関連成分のうち1種類以上について急性または慢性水生毒性に関して利用可能な情報が揃っていない混合物については、決定的な有害性区分に帰属させることはできないと結論づけられる。そのような状況では、混合物は既知成分のみにもとづいて分類され、「本混合物の成分 x %については水生環境有害性が不明である」という記述を追加しておくべきである。所管官庁はその追加的な記述をラベルまたはSDSあるいはその両方で伝達するかどうかを明記するかどうか、またその記述をどこにするかの選択を製造者/供給者に委ねるかどうかを決めることができる。

[11] 混合物中の成分2種類以上について十分な毒性データが入手できる場合には、毒性データの性質に応じて、4.1.3.5.2の加算式(a)または(b)にしたがって、これらの成分の毒性加算値を算出できる。この毒性計算値を用いてその混合物の部分に急性または慢性の有害性区分を割り振り、その後これを加算法に適用してもよい。同一分類群（例えば魚類、甲殻類または藻類）について各成分の毒性値を用いて混合物のこの部分の毒性を計算し、得られた計算値の中の最も高い毒性値（最低毒性濃度、これら3つの分類群のうち感受性が最も高い群で得られた値）を採用することが望ましい（4.1.3.5.3を参照）。

CHAPTER 4.2

HAZARDOUS TO THE OZONE LAYER

4.2.1 Definitions

Ozone Depleting Potential (ODP) is an integrative quantity, distinct for each halocarbon source species, that represents the extent of ozone depletion in the stratosphere expected from the halocarbon on a mass-for-mass basis relative to CFC-11. The formal definition of ODP is the ratio of integrated perturbations to total ozone, for a differential mass emission of a particular compound relative to an equal emission of CFC-11.

Montreal Protocol is the Montreal Protocol on Substances that Deplete the Ozone Layer as either adjusted and/or amended by the Parties to the Protocol.

4.2.2 Classification criteria[1]

A substance or mixture shall be classified as Category 1 according to the following table:

Table 4.2.1: Criteria for substances and mixtures hazardous to the ozone layer

Category	Criteria
1	Any of the controlled substances listed in annexes to the Montreal Protocol; or Any mixture containing at least one ingredient listed in the annexes to the Montreal Protocol, at a concentration ≥ 0.1 %

4.2.3 Hazard communication

General and specific considerations concerning labelling requirements are provided in *Hazard Communication: Labelling* (chapter 1.4). Annex 1 contains summary tables about classification and labelling. Annex 3 contains examples of precautionary statements and pictograms which can be used where allowed by the competent authority. Table 4.2.2 presents specific label elements for substances and mixtures classified into this hazard class based on the criteria in this chapter.

Table 4.2.2: Label elements for substances and mixtures hazardous to the ozone layer

	Category 1
Symbol	Exclamation mark
Signal word	Warning
Hazard statement	Harms public health and the environment by destroying ozone in the upper atmosphere

---

[1] The criteria in this chapter are intended to be applied to substances and mixtures. Equipment, articles or appliances (such as refrigeration or air conditioning equipment) containing substances hazardous to the ozone layer are beyond the scope of these criteria. Consistent with 1.1.2.5 (a)(iii) regarding pharmaceutical products, GHS classification and labelling criteria do not apply to medical inhalers at the point of intentional intake.

第4.2 章

オゾン層への有害性

4.2.1 定義

*オゾン破壊係数（ODP）*とは、ハロカーボンによって見込まれる成層圏オゾンの破壊の程度を、CFC-11に対して質量ベースで相対的に表した積算量であり、ハロカーボンの種類ごとに異なるものである。ODPの正式な定義は、等量の CFC-11 排出量を基準にした、特定の化合物の排出に伴う総オゾンの擾乱量の積算値の比の値である。

*モントリオール議定書*とは、議定書の締約国によって調整および/または修正された、オゾン層破壊物質に関するモントリオール議定書をいう。

4.2.2 分類基準[1]

物質または混合物は次表にしたがって区分1に分類される。

表4.2.1：オゾン層への有害性のある物質および混合物の基準

区分	基準
1	モントリオール議定書の附属書に列記された、あらゆる規制物質；または モントリオール議定書の附属書に列記された成分を、濃度≧0.1%で少なくとも 1 つ含むあらゆる混合物

4.2.3 危険有害性に関する情報の伝達

表示要件に関する通則および細則は、*危険有害性に関する情報の伝達：表示*（第 1.4 章）に定める。附属書 1 に分類と表示に関する概要表を示す。附属書 3 には、注意書きおよび所管官庁が許可した場合に使用可能な注意絵表示の例を示す。表 4.2.2 には、本章の判定基準に基づいてこの危険有害性クラスに分類された物質および混合物の具体的なラベル要素を示す。

表4.2.2：オゾン層への有害性のある物質および混合物のラベル要素

	区分1
シンボル	感嘆符
注意喚起語	警告
危険有害性情報	オゾン層を破壊し、健康および環境に有害

---

[1] 本章の判定基準は、物質および混合物に適用されることを意図したものである。オゾン層有害性物質を含有する機器、品目または（冷蔵機器やエアコンなどの）電気器具は、この判定基準の適用範囲外である。医薬品に関する 1.1.2.5 (a)(iii)に合わせて、GHS 分類および表示基準は意図的な吸入という点で医療用吸入器には適用されない。

4.2.4 **Decision logic for substances and mixtures hazardous to the ozone layer**

The decision logic which follows is not part of the harmonized classification system but is provided here as additional guidance. It is strongly recommended that the person responsible for classification study the criteria before and during use of the decision logic.

Decision logic 4.2.1

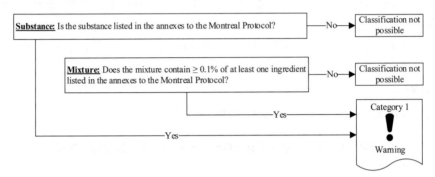

4.2.4 オゾン層有害性のある物質および混合物の判定論理

以下の判定論理は、調和分類システムには含まれないが、追加的な手引きとしてここに示す。分類の責任者に対し、この判定論理を使用する前および使用する際に判定基準についてよく調べ理解することを強く推奨する。

判定論理 4.2.1

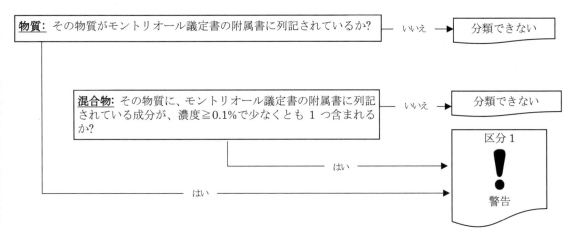

ANNEXES

附属書

ANNEX 1

CLASSIFICATION AND LABELLING SUMMARY TABLES

ANNEX 1

CLASSIFICATION AND LABELLING SUMMARY TABLES

附属書1

分類および表示のまとめ

ANNEX 1

CLASSIFICATION AND LABELLING SUMMARY TABLES

NOTE 1: *The codification of hazard statements is further explained in annex 3 (section 1). The hazard statement codes are intended to be used for reference purposes only. They are not part of the hazard statement text and should not be used to replace it.*

NOTE 2: *To provide clarity, assist labelling practitioners and enable comparison between equivalent classification and labelling systems under the GHS and the UN Model Regulations, transport hazard classes, divisions and pictograms are included in tables A1.1 to A1.30. However, it should be noted that in these tables the UN Model Regulations classification and labelling entries are provided for indicative purposes only. For transport purposes, the classification and labelling provisions prescribed by the UN Model Regulations shall be used (see also chapter 1.4, section 1.4.10 of the GHS).*

NOTE 3: *GHS hazard pictograms are displayed in the shape of a square set at a point with a black symbol on a white background with a red frame. The transport pictograms (commonly referred to as labels in the UN Model Regulations) shall be displayed on a background of contrasting colour or, where appropriate, shall have either a dotted or solid boundary line as provided in chapter 5.2, section 5.2.2.2 of the UN Model Regulations and in tables A1.1 to A1.30 below. For some hazard categories, the symbol, number and border line of the transport pictogram may be shown in white instead of black. Where such an alternative is available it is shown in the relevant tables below (see tables A1.2, A1.3, A1.5, A1.6, A1.12, A1.15 and A1.17).*

附属書1

分類および表示のまとめ

注記1：　危険有害性情報のコードについては附属書3（第1節）でさらに説明されている。危険有害性情報のコードは参照の目的だけに使用される。これらは危険有害性情報の一部ではないし、その代わりに用いるべきではない。

注記2：　明確さを提供し、表示を実際に行う人を支援し、さらにGHSおよびUNモデル規則の下での分類や表示システムの比較を可能にするために、輸送の危険有害性クラス、区分および絵表示を表A1.1からA1.30に含めた。ただし、これらの表ではUNモデル規則の分類や表示の記載は提示する目的にだけあることに注意すべきである。輸送の目的ではUNモデル規則に記載されている分類や表示の規定を使用しなければならない（GHSの第1.4章1.4.10も参照のこと）。

注記3：　GHSの危険有害性の絵表示は、赤枠に白地に黒のシンボルで、1つの頂点で正立させた正方形で示される。輸送の絵表示(通常UNモデル規則では標札とされている)は、UNモデル規則第5.2章5.2.2.2および下表A1.1からA1.30に示されているように、対照的な色の背景に表示されるか、必要に応じて点線または実線の境界線をもたなければならない。一部の危険有害性区分では、シンボル、番号および境界線が黒ではなく白で示されることがある。そのような代替が利用できる場合は、以下の関連する表に示されている（表A1.2、A1.3、A1.5、A1.6、A1.12、A1.15およびA1.17を参照のこと）。

A1.1 **Explosives** (see chapter 2.1 for classification criteria)

Classification			Labelling				GHS hazard statement codes
GHS hazard class	GHS hazard category	UN Model Regulations class or division	GHS pictogram	UN Model Regulations pictogram[a]	GHS signal word	GHS hazard statement	
Explosives	1	Not applicable	(explosive pictogram)	Not applicable	Danger	Explosive	H209 H210[b] H211[b]
	2A	1.1 1.2 1.3	(explosive pictogram)	1 (div)	Danger	Explosive	H209
		1.5		1.5			
		1.6		1.6			
	2B	1.4	(explosive pictogram)	1.4	Warning	Fire or projection hazard	H204
	2C		(exclamation mark)		Warning	Fire or projection hazard	H204

[a] Under the UN Model Regulations, (*) indicates the place for compatibility group and (**) indicates the place for division - to be left blank if explosive is the subsidiary hazard.

[b] Additional hazard statements for explosives that are sensitive to initiation or for which sufficient information on their sensitivity is not available (see chapter 2.1, section 2.1.3).

A1.1 爆発物（判定基準は第 2.1 章を参照のこと）

分類			表示					GHS 危険有害性情報コード
GHS 危険有害性クラス	GHS 危険有害性区分	UN モデル規則クラスまたは区分	GHS 絵表示	UN モデル規則絵表示[a]	GHS 注意喚起語	GHS 危険有害性情報		
爆発物	1	要求されない	(爆発絵表示)	要求されない	危険	爆発物			H209 H210[b] H211[b]
	2A	1.1	(爆発絵表示)	(1.1 絵表示)	危険	爆発物			H209
		1.2							
		1.3							
		1.5		1.5					
		1.6		1.6					
	2B	1.4	(爆発絵表示)	1.4	警告	火災または飛散危険性			H204
	2C		(!絵表示)		警告	火災または飛散危険性			H204

[a] UN モデル規則では、(*) は隔離区分番号の場所を、(**) は区分の場所を示す－爆発物が副次危険性の場合には空欄とする。

[b] 起爆に敏感な爆発物またはその敏感さに関する十分な情報が入手できない爆発物に関する追加の危険有害性情報（第 2.1 章 2.1.3 参照）。

A1.2 Flammable gases (see chapter 2.2 for classification criteria)

Classification			Labelling					GHS hazard statement codes
GHS hazard class	GHS hazard category		UN Model Regulations class or division	GHS pictogram	UN Model Regulations pictogram[a]	GHS signal word	GHS hazard statement	
Flammable gases	1A	Flammable gas	2.1	(flame pictogram)	(2.1 pictogram) or (2.1 pictogram)	Danger	Extremely flammable gas	H220
^	^	Pyrophoric gas	^	^	^	^	Extremely flammable gas	H220
^	^	^	^	^	^	^	May ignite spontaneously if exposed to air	H232
^	^	Chemically unstable gas — A	^	^	^	^	Extremely flammable gas	H220
^	^	^	^	^	^	^	May react explosively even in the absence of air	H230
^	^	Chemically unstable gas — B	^	^	^	^	Extremely flammable gas	H220
^	^	^	^	^	^	^	May react explosively even in the absence of air at elevated pressure and/or temperature	H231
^	1B							
^	2		Not applicable	No pictogram	Not applicable	Warning	Flammable gas	H221

[a] Under the UN Model Regulations, pyrophoric gases and chemically unstable gases (A and B) are classified based on their flammability in Class 2, Division 2.1.

A1.2　可燃性ガス（判定基準は第 2.2 章を参照のこと）

分類			表示					GHS 危険有害性情報コード
GHS 危険有害性クラス	GHS 危険有害性区分		UN モデル規則クラスまたは区分	GHS 絵表示	UN モデル規則絵表示 [a]	GHS 注意喚起語	GHS 危険有害性情報	
可燃性ガス	1A	可燃性ガス	2.1	(炎)	(クラス2.1 または 自然発火)	危険	極めて可燃性の高いガス	H220
^	^	自然発火性ガス	^	^	^	^	極めて可燃性の高いガス	H220
^	^	^	^	^	^	^	空気に触れると自然発火するおそれ	H232
^	^	化学的に不安定なガス A	^	^	^	^	極めて可燃性の高いガス	H220
^	^	^	^	^	^	^	空気が無くても爆発的に反応するおそれ	H230
^	^	化学的に不安定なガス B	^	^	^	^	極めて可燃性の高いガス	H220
^	^	^	^	^	^	^	圧力および/または温度が上昇した場合、空気が無くても爆発的に反応するおそれ	H231
^	1B			^	^	^	可燃性ガス	H221
^	2			要求されない	絵表示なし	要求されない	警告	^	^

[a] UN モデル規則では、自然発火性ガス、化学的に不安定なガス（A および B）はその燃焼性によってクラス 2、区分 2.1 に分類される。

A1.3   Aerosols and chemicals under pressure (see chapter 2.3 for classification criteria)

Classification			Labelling				GHS hazard statement codes
GHS hazard class	GHS hazard category	UN Model Regulations class or division	GHS pictogram	UN Model Regulations pictogram	GHS signal word	GHS hazard statement	
Aerosols (section 2.3.1)	1	2.1	(flame)	(flame 2) or (flame 2)	Danger	Extremely flammable aerosol	H222
						Pressurized container: may burst if heated	H229
	2	2.1	(flame)	(flame 2)	Warning	Flammable aerosol	H223
						Pressurized container: may burst if heated	H229
	3	2.2	*No pictogram*	(cylinder 2) or (cylinder 2)	Warning	Pressurized container: may burst if heated	H229
Chemicals under pressure (section 2.3.2)	1	2.1	(flame) and (cylinder)	(flame 2) or (flame 2)	Danger	Extremely flammable chemical under pressure: may explode if heated	H282
	2	2.1	(flame) and (cylinder)	(flame 2) or (flame 2)	Danger	Flammable chemical under pressure: may explode if heated	H283
	3	2.2	(cylinder)	(cylinder 2) or (cylinder 2)	Warning	Chemical under pressure: may explode if heated	H284

A1.3　エアゾールおよび加圧下化学品（判定基準は第 2.3 章を参照のこと）

分類			表示				GHS 危険有害性情報コード
GHS 危険有害性クラス	GHS 危険有害性区分	UN モデル規則クラスまたは区分	GHS 絵表示	UN モデル規則絵表示	GHS 注意喚起語	GHS 危険有害性情報	
エアゾール（2.3.1）	1	2.1	⬥炎⬥	⬥炎⬥ または ⬥炎⬥	危険	極めて可燃性の高いエアゾール	H222
						高圧容器：熱すると破裂のおそれ	H229
	2				警告	可燃性エアゾール	H223
						高圧容器：熱すると破裂のおそれ	H229
	3	2.2	絵表示なし	⬥ボンベ⬥ または ⬥ボンベ⬥	警告	高圧容器：熱すると破裂のおそれ	H229
加圧下化学品（2.3.2）	1	2.1	⬥炎⬥ および ⬥ボンベ⬥	⬥炎⬥ または ⬥炎⬥	危険	極めて可燃性の高い加圧下化学品：熱すると爆発のおそれ	H282
	2				警告	可燃性の加圧下化学品：熱すると爆発のおそれ	H283
	3	2.2	⬥ボンベ⬥	⬥ボンベ⬥ または ⬥ボンベ⬥	警告	加圧下化学品：熱すると爆発のおそれ	H284

A1.4 Oxidizing gases (see chapter 2.4 for classification criteria)

Classification			Labelling				GHS hazard statement code
GHS hazard class	GHS hazard category	UN Model Regulations class or division	GHS pictogram	UN Model Regulations pictogram[a]	GHS signal word	GHS hazard statement	
Oxidizing gases	1	2[a]	(flame over circle)	(Division 5.1 pictogram)	Danger	May cause or intensify fire; oxidizer	H270

[a] Under the UN Model Regulations, oxidizing gases are classified under the applicable Class 2 division according to their primary gas hazard and will display the applicable Class 2 transport pictogram. In addition, they are assigned a Division 5.1 (flame over circle) transport pictogram due to their oxidizing subsidiary hazard.

A1.5 Gases under pressure (see chapter 2.5 for classification criteria)

Classification			Labelling				Hazard statement codes
GHS hazard class	GHS hazard category	UN Model Regulations class or division	GHS pictogram	UN Model Regulations pictogram[a]	GHS signal word	GHS hazard statement	
Gases under pressure	Compressed gas	2.2	(gas cylinder)	(Division 2 pictogram) or (Division 2 refrigerated pictogram)	Warning	Contains gas under pressure; may explode if heated	H280
	Liquefied gas					Contains gas under pressure; may explode if heated	H280
	Refrigerated liquefied gas					Contains refrigerated gas; may cause cryogenic burns or injury	H281
	Dissolved gas					Contains gas under pressure; may explode if heated	H280

[a] Under the UN Model Regulations, this pictogram is not required for gases under pressure that are also toxic or flammable gases. In those cases, the applicable toxic or flammable gas hazard class pictogram is used instead.

A1.6 Flammable liquids (see chapter 2.6 for classification criteria)

Classification			Labelling				GHS hazard statement codes
GHS hazard class	GHS hazard category	UN Model Regulations class or division	GHS pictogram	UN Model Regulations pictogram	GHS signal word	GHS hazard statement	
Flammable liquids	1	3	(flame)	(Division 3 pictogram) or (Division 3 pictogram)	Danger	Extremely flammable liquid and vapour	H224
	2					Highly flammable liquid and vapour	H225
	3				Warning	Flammable liquid and vapour	H226
	4	Not applicable	No pictogram	Not required		Combustible liquid	H227

A1.4 酸化性ガス（判定基準は第2.4章を参照のこと）

分類			表示				GHS危険有害性情報コード
GHS危険有害性クラス	GHS危険有害性区分	UNモデル規則クラスまたは区分	GHS絵表示	UNモデル規則絵表示 [a]	GHS注意喚起語	GHS危険有害性情報	
酸化性ガス	1	2 [a]	（炎）	（5.1 円上の炎）	危険	発火または火炎助長のおそれ；酸化性物質	H270

[a] UNモデル規則では、酸化性ガスは主要なガスの危険性にしたがってクラス2に分類され、クラス2の輸送絵表示が示される。さらに酸化性の副次危険性にしたがって区分5.1（円上の炎）の輸送絵表示が割り当てられる。

A1.5 高圧ガス（判定基準は第2.5章を参照のこと）

分類			表示				GHS危険有害性情報コード
GHS危険有害性クラス	GHS危険有害性区分	UNモデル規則クラスまたは区分	GHS絵表示	UNモデル規則絵表示 [a]	GHS注意喚起語	GHS危険有害性情報	
高圧ガス	圧縮ガス	2.2	（ガスボンベ）	（ガスボンベ）または（ガスボンベ）	警告	高圧ガス；熱すると爆発のおそれ	H280
	液化ガス					高圧ガス；熱すると爆発のおそれ	H280
	深冷液化ガス					深冷液化ガス；凍傷または傷害のおそれ	H281
	溶解ガス					高圧ガス；熱すると爆発のおそれ	H280

[a] UNモデル規則では、この絵表示は毒性または可燃性ガスでもある高圧ガスには要求されていない。これらの場合、代わりに該当する毒性または可燃性ガスの絵表示が使用される。

A1.6 引火性液体（判定基準は第2.6章を参照のこと）

分類			表示				GHS危険有害性情報コード
GHS危険有害性クラス	GHS危険有害性区分	UNモデル規則クラスまたは区分	GHS絵表示	UNモデル規則絵表示	GHS注意喚起語	GHS危険有害性情報	
引火性液体	1	3	（炎）	（炎）または（炎）	危険	極めて引火性の高い液体および蒸気	H224
	2				危険	引火性の高い液体および蒸気	H225
	3				警告	引火性液体および蒸気	H226
	4	要求されない	絵表示なし	要求されない	警告	可燃性液体	H227

A1.7 Flammable solids (see chapter 2.7 for classification criteria)

Classification			Labelling				GHS hazard statement codes
GHS hazard class	GHS hazard category	UN Model Regulations class or division	GHS pictogram	UN Model Regulations pictogram	GHS signal word	GHS hazard statement	
Flammable solids	1	4.1	(flame pictogram)	(flame pictogram)	Danger	Flammable solid	H228
	2				Warning		

A1.8 Self-reactive substances and mixtures (see chapter 2.8 for classification criteria)

Classification			Labelling				GHS hazard statement codes
GHS hazard class	GHS hazard category	UN Model Regulations class or division	GHS pictogram	UN Model Regulations pictogram[a]	GHS signal word	GHS hazard statement	
Self-reactive substances and mixtures	Type A	4.1 Type A	(exploding bomb)	*(Transport may not be allowed)[b]*	Danger	Heating may cause an explosion	H240
	Type B	4.1 Type B	(exploding bomb) *and* (flame)	*and if applicable*	Danger	Heating may cause a fire or explosion	H241
	Types C and D	4.1 Types C and D	(flame)	(flame)	Danger	Heating may cause a fire	H242
	Types E and F	4.1 Types E and F			Warning		
	Type G	Type G	*No pictogram*	*Not applicable*	*No signal word*	*No hazard statement*	*None*

[a] Under the UN Model Regulations, where a Type B substance or mixture has an explosive subsidiary hazard, then the transport pictogram for Divisions 1.1, 1.2 or 1.3 shall also be used without the indication of the division number or the compatibility group. For a substance or mixture of hazard category Type B, special provision 181 may apply (Exemption of explosive label with competent authority approval. See chapter 3.3 of the UN Model Regulations for more details).

[b] May not be acceptable for transport in the packaging in which it is tested (See chapter 2.4, paragraph 2.4.2.3.2.1 of the UN Model Regulations).

A1.7 可燃性固体 （判定基準は第2.7章を参照のこと）

分類			表示					GHS 危険有害性情報コード
GHS 危険有害性クラス	GHS 危険有害性区分	UN モデル規則クラスまたは区分	GHS 絵表示	UN モデル規則絵表示	GHS 注意喚起語	GHS 危険有害性情報	
可燃性固体	1	4.1	炎	炎(縞)	危険	可燃性固体		H228
	2				警告			

A1.8 自己反応性物質および混合物 （判定基準は第2.8章を参照のこと）

分類			表示					GHS 危険有害性情報コード
GHS 危険有害性クラス	GHS 危険有害性区分	UN モデル規則クラスまたは区分	GHS 絵表示	UN モデル規則絵表示 a	GHS 注意喚起語	GHS 危険有害性情報	
自己反応性物質および混合物	タイプA	4.1 タイプA	爆発	(輸送が許可されないであろう) b	危険	熱すると爆発のおそれ || H240
	タイプB	4.1 タイプB	爆発 および 炎	炎(縞) および該当する場合 爆発	危険	熱すると火災または爆発のおそれ || H241
	タイプCとD	4.1 タイプCおよびD	炎	炎(縞)	危険	熱すると火災のおそれ || H242
	タイプEとF	4.1 タイプEおよびF			警告	熱すると火災のおそれ || H242
	タイプG	タイプG	絵表示なし	要求されない	注意喚起語なし	危険有害性情報なし || なし

a UNモデル規則では、タイプBの物質または混合物に副次危険性がある場合には、区分番号または隔離区分番号を示さずに、区分1.1、1.2または1.3に対する輸送絵表示も使用しなければならない。タイプBについては、UNモデル規則に基づく特別規則181条が適用される（所管官庁の許可による爆発物ラベル適用除外。詳細はUNモデル規則第3.3章を参照のこと）。

b 試験された包装容器での輸送は許可されないであろう(UNモデル規則第2.4章2.4.2.3.2.1参照のこと)。

A1.9 **Pyrophoric liquids** (see chapter 2.9 for classification criteria)

Classification			Labelling				GHS hazard statement code
GHS hazard class	GHS hazard category	UN Model Regulations class or division	GHS pictogram	UN Model Regulations pictogram	GHS signal word	GHS hazard statement	
Pyrophoric liquids	1	4.2	🔥	🔥	Danger	Catches fire spontaneously if exposed to air	H250

A1.10 **Pyrophoric solids** (see chapter 2.10 for classification criteria)

Classification			Labelling				GHS hazard statement code
GHS hazard class	GHS hazard category	UN Model Regulations class or division	GHS pictogram	UN Model Regulations pictogram	GHS signal word	GHS hazard statement	
Pyrophoric solids	1	4.2	🔥	🔥	Danger	Catches fire spontaneously if exposed to air	H250

A1.11 **Self-heating substances and mixtures** (see chapter 2.11 for classification criteria)

Classification			Labelling				GHS hazard statement codes
GHS hazard class	GHS hazard category	UN Model Regulations class or division	GHS pictogram	UN Model Regulations pictogram	GHS signal word	GHS hazard statement	
Self-heating substances and mixtures	1	4.2	🔥	🔥	Danger	Self-heating; may catch fire	H251
	2				Warning	Self-heating in large quantities; may catch fire	H252

A1.12 **Substances and mixtures, which in contact with water, emit flammable gases** (see chapter 2.12 for classification criteria)

Classification			Labelling				GHS hazard statement codes
GHS hazard class	GHS hazard category	UN Model Regulations class or division	GHS pictogram	UN Model Regulations pictogram	GHS signal word	GHS hazard statement	
Substances and mixtures, which in contact with water, emit flammable gases	1	4.3	🔥	🔥	Danger	In contact with water releases flammable gases which may ignite spontaneously	H260
	2			or	Danger	In contact with water releases flammable gases	H261
	3				Warning		

A1.9 自然発火性液体（判定基準は第2.9章を参照のこと）

分類			表示				GHS危険有害性情報コード
GHS危険有害性クラス	GHS危険有害性区分	UNモデル規則クラスまたは区分	GHS絵表示	UNモデル規則絵表示	GHS注意喚起語	GHS危険有害性情報	
自然発火性液体	1	4.2			危険	空気に触れると自然発火	H250

A1.10 自然発火性固体（判定基準は第2.10章を参照のこと）

分類			表示				GHS危険有害性情報コード
GHS危険有害性クラス	GHS危険有害性区分	UNモデル規則クラスまたは区分	GHS絵表示	UNモデル規則絵表示	GHS注意喚起語	GHS危険有害性情報	
自然発火性固体	1	4.2			危険	空気に触れると自然発火	H250

A1.11 自己発熱性物質および混合物（判定基準は第2.11章を参照のこと）

分類			表示				GHS危険有害性情報コード
GHS危険有害性クラス	GHS危険有害性区分	UNモデル規則クラスまたは区分	GHS絵表示	UNモデル規則絵表示	GHS注意喚起語	GHS危険有害性情報	
自己発熱性物質および混合物	1	4.2			危険	自己発熱；火災のおそれ	H251
	2				警告	大量の場合 自己発熱；火災のおそれ	H252

A1.12 水反応可燃性物質および混合物（判定基準は第2.12章を参照のこと）

分類			表示				GHS危険有害性情報コード
GHS危険有害性クラス	GHS危険有害性区分	UNモデル規則クラスまたは区分	GHS絵表示	UNモデル規則絵表示	GHS注意喚起語	GHS危険有害性情報	
水反応可燃性物質および混合物	1	4.3			危険	水に触れると自然発火するおそれのある可燃性ガスを発生	H260
	2			または	危険	水に触れると可燃性ガスを発生	H261
	3				警告		

A1.13 **Oxidizing liquids** (see chapter 2.13 for classification criteria)

Classification			Labelling					GHS hazard statement codes
GHS hazard class	GHS hazard category	UN Model Regulations class or division	GHS pictogram	UN Model Regulations pictogram	GHS signal word	GHS hazard statement	
Oxidizing liquids	1	5.1	(flame over circle)	(5.1 diamond)	Danger	May cause fire or explosion; strong oxidizer || H271
^	2	^	^	^	Danger	May intensify fire; oxidizer || H272
^	3	^	^	^	Warning	^ || ^

A1.14 **Oxidizing solids** (see chapter 2.14 for classification criteria)

Classification			Labelling					GHS hazard statement codes
GHS hazard class	GHS hazard category	UN Model Regulations class or division	GHS pictogram	UN Model Regulations pictogram	GHS signal word	GHS hazard statement	
Oxidizing solids	1	5.1	(flame over circle)	(5.1 diamond)	Danger	May cause fire or explosion; strong oxidizer || H271
^	2	^	^	^	Danger	May intensify fire; oxidizer || H272
^	3	^	^	^	Warning	^ || ^

A1.13 酸化性液体（判定基準は第2.13章を参照のこと）

分類			表示					GHS危険有害性情報コード
GHS危険有害性クラス	GHS危険有害性区分	UNモデル規則クラスまたは区分	GHS絵表示	UNモデル規則絵表示	GHS注意喚起語	GHS危険有害性情報		
酸化性液体	1	5.1			危険	火災または爆発のおそれ；強酸化性物質			H271
	2				危険	火災助長のおそれ；酸化性物質			H272
	3				警告				

A1.14 酸化性固体（判定基準は第2.14章を参照のこと）

分類			表示					GHS危険有害性情報コード
GHS危険有害性クラス	GHS危険有害性区分	UNモデル規則クラスまたは区分	GHS絵表示	UNモデル規則絵表示	GHS注意喚起語	GHS危険有害性情報		
酸化性固体	1	5.1			危険	火災または爆発のおそれ；強酸化性物質			H271
	2				危険	火災助長のおそれ；酸化性物質			H272
	3				警告				

A1.15 **Organic peroxides** (see chapter 2.15 for classification criteria)

Classification			Labelling					GHS hazard statement codes
GHS hazard class	GHS hazard category	UN Model Regulations class or division	GHS pictogram	UN Model Regulations pictogram[a]	GHS signal word	GHS hazard statement	
Organic peroxides	Type A	5.2 Type A	(exploding bomb)	*(Transport may not be allowed)*[b]	Danger	Heating may cause an explosion || H240
	Type B	5.2 Type B	(exploding bomb) and (flame)	(5.2) or (5.2) and if applicable[a]: (1)	Danger	Heating may cause a fire or explosion || H241
	Types C and D	5.2 Types C and D	(flame)	(5.2) or (5.2)	Danger	Heating may cause a fire || H242
	Types E and F	5.2 Types E and F			Warning	Heating may cause a fire || H242
	Type G	Type G	*No pictogram*	*Not applicable*	*No signal word*	*No hazard statement* || *None*

[a] Under the UN Model Regulations, where a Type B substance or mixture has an explosive subsidiary hazard, then the transport pictogram for Divisions 1.1, 1.2 or 1.3 shall also be used without the indication of the division number or the compatibility group. For a substance or mixture of hazard category Type B, special provision 181 may apply (Exemption of explosive label with competent authority approval. See chapter 3.3 of the UN Model Regulations for more details).

[b] May not be acceptable for transport in the packaging in which it is tested (See chapter 2.5, par. 2.5.3.2.2 of the UN Model Regulations).

A1.16 **Corrosive to metals** (see chapter 2.16 for classification criteria)

Classification			Labelling				GHS hazard statement code
GHS hazard class	GHS hazard category	UN Model Regulations class or division	GHS pictogram	UN Model Regulations pictogram	GHS signal word	GHS hazard statement	
Corrosive to metals	1	8	(corrosion)	(corrosion 8)	Warning	May be corrosive to metals	H290

A1.15 有機過酸化物 （判定基準は第2.15章を参照のこと）

分類			表示				GHS危険有害性情報コード
GHS危険有害性クラス	GHS危険有害性区分	UNモデル規則クラスまたは区分	GHS絵表示	UNモデル規則絵表示 [a]	GHS注意喚起語	GHS危険有害性情報	
有機過酸化物	タイプA	5.2 タイプA		（輸送が許可されないであろう）[b]	危険	熱すると爆発のおそれ	H240
	タイプB	5.2 タイプB	および	および該当する場合 [a]:	危険	熱すると火災または爆発のおそれ	H241
	タイプCおよびD	5.2 タイプCおよびD		または	危険	熱すると火災のおそれ	H242
	タイプEおよびF	5.2 タイプEおよびF		または	警告	熱すると火災のおそれ	H242
	タイプG	タイプG	絵表示なし	要求されない	注意喚起語なし	危険有害性情報なし	なし

[a] UNモデル規則では、タイプBの物質または混合物に副次危険性がある場合には、区分番号や隔離区分番号を示さずに、区分1.1、1.2または1.3に対する輸送絵表示も使用しなければならない。タイプBについては、UNモデル規則に基づく特別規則181条が適用される（所管官庁の許可による爆発物ラベルの適用除外。詳細はUNモデル規則第3.3章を参照のこと）。

[b] 試験された包装容器での輸送は許可されないであろう（UNモデル規則第2.5章2.5.3.2.2参照のこと）。

A1.16 金属腐食性 （判定基準は第2.16章を参照のこと）

分類			表示				GHS危険有害性情報コード
GHS危険有害性クラス	GHS危険有害性区分	UNモデル規則クラスまたは区分	GHS絵表示	UNモデル規則絵表示	GHS注意喚起語	GHS危険有害性情報	
金属腐食性	1	8			警告	金属腐食のおそれ	H290

A1.17 Desensitized explosives (see chapter 2.17 for classification criteria)

Classification			Labelling				GHS hazard statement codes
GHS hazard class	GHS hazard category	UN Model Regulations class or division[a]	GHS pictogram	UN Model Regulations pictogram[a]	GHS signal word	GHS hazard statement	
Desensitized explosives	1	3 or 4.1			Danger	Fire, blast projection hazard; increased risk of explosion if desensitizing agent is reduced	H206
	2				Danger	Fire or projection hazard; increased risk of explosion if desensitizing agent is reduced	H207
	3				Warning		
	4				Warning	Fire hazard; increased risk of explosion if desensitizing agent is reduced	H208

[a] Under the UN Model Regulations, liquid desensitized explosives are classified in Class 3 and solid desensitized explosives are classified in Division 4.1.

A1.18 Acute toxicity (see chapter 3.1 for classification criteria)

Classification			Labelling				GHS hazard statement codes
GHS hazard class	GHS hazard category	UN Model Regulations class or division[a]	GHS pictogram	UN Model Regulations pictogram[a]	GHS signal word	GHS hazard statement	
Acute toxicity	1, 2 Oral	2.3 or 6.1			Danger	Fatal if swallowed	H300
	1, 2 Dermal					Fatal in contact with skin	H310
	1, 2 Inhalation					Fatal if inhaled	H330
	3 Oral				Danger	Toxic if swallowed	H301
	3 Dermal					Toxic in contact with skin	H311
	3 Inhalation					Toxic if inhaled	H331
	4 Oral				Warning	Harmful if swallowed	H302
	4 Dermal					Harmful in contact with skin	H312
	4 Inhalation					Harmful if inhaled	H332
	5 Oral	Not applicable	No pictogram	Not applicable	Warning	May be harmful if swallowed	H303
	5 Dermal					May be harmful in contact with skin	H313
	5 Inhalation					May be harmful if inhaled	H333

[a] Under the UN Model Regulations, toxic gases are classified in Division 2.3 and toxic substances (as defined in the UN Model Regulations) are classified in Division 6.1.

A1.17 鈍性化爆発物 （判定基準は第2.17章を参照のこと）

分類			表示					GHS危険有害性情報コード
GHS危険有害性クラス	GHS危険有害性区分	UNモデル規則クラスまたは区分[a]	GHS絵表示	UNモデル規則絵表示[a]	GHS注意喚起語	GHS危険有害性情報	
鈍性化爆発物	1	3 または 4.1	(炎)	(炎) または (炎・縞)	危険	火災、爆風または飛散危険性；鈍性化剤が減少した場合には爆発の危険性の増加 || H206
	2				危険	火災または飛散危険性；鈍性化剤が減少した場合には爆発の危険性の増加		H207
	3				警告			
	4				警告	火災の危険性；鈍性化剤が減少した場合には爆発の危険性の増加		H208

[a] UNモデル規則では、液体の鈍性化爆発物はクラス3に分類され、固体の鈍性化爆発物は区分4.1に分類される。

A1.18 急性毒性 （判定基準は第3.1章を参照のこと）

分類			表示					GHS危険有害性情報コード
GHS危険有害性クラス	GHS危険有害性区分	UNモデル規則クラスまたは区分[a]	GHS絵表示	UNモデル規則絵表示[a]	GHS注意喚起語	GHS危険有害性情報	
急性毒性	1、2	経口	2.3 または 6.1	(どくろ)	(どくろ2) または (どくろ6)	危険	飲み込むと生命に危険	H300
		経皮					皮膚に接触すると生命に危険	H310
		吸入					吸入すると生命に危険	H330
	3	経口				危険	飲み込むと有毒	H301
		経皮					皮膚に接触すると有毒	H311
		吸入					吸入すると有毒	H331
	4	経口	要求されない	(!)	要求されない	警告	飲み込むと有害	H302
		経皮					皮膚に接触すると有害	H312
		吸入					吸入すると有害	H332
	5	経口		絵表示なし		警告	飲み込むと有害のおそれ	H303
		経皮					皮膚に接触すると有害のおそれ	H313
		吸入					吸入すると有害のおそれ	H333

[a] UNモデル規則では、毒性ガスは区分2.3に分類され、毒性物質は（UNモデル規則で定義されているように）区分6.1に分類される。

A1.19　　　**Skin corrosion/irritation** (see chapter 3.2 for classification criteria)

Classification			Labelling				GHS hazard statement codes
GHS hazard class	GHS hazard category	UN Model Regulations class or division	GHS pictogram	UN Model Regulations pictogram	GHS signal word	GHS hazard statement	
Skin corrosion/irritation	1 1A, 1B, 1C	8	(corrosion pictogram)	(class 8 pictogram)	Danger	Causes severe skin burns and eye damage	H314
	2	Not applicable	(exclamation mark)	Not applicable	Warning	Causes skin irritation	H315
	3		No pictogram		Warning	Causes mild skin irritation	H316

A1.20　　　**Serious eye damage/eye irritation** (see chapter 3.3 for classification criteria)

Classification			Labelling				GHS hazard statement codes
GHS hazard class	GHS hazard category	UN Model Regulations class or division	GHS pictogram	UN Model Regulations pictogram	GHS signal word	GHS hazard statement	
Serious eye damage/eye irritation	1	Not applicable	(corrosion pictogram)	Not applicable	Danger	Causes serious eye damage	H318
	2/2A		(exclamation mark)		Warning	Causes serious eye irritation	H319
	2B		No pictogram		Warning	Causes eye irritation	H320

A1.21　　　**Respiratory sensitization** (see chapter 3.4 for classification criteria)

Classification			Labelling				GHS hazard statement codes
GHS hazard class	GHS hazard category	UN Model Regulations class or division	GHS pictogram	UN Model Regulations pictogram	GHS signal word	GHS hazard statement	
Respiratory sensitization	1, 1A, 1B	Not applicable	(health hazard pictogram)	Not applicable	Danger	May cause allergy or asthma symptoms or breathing difficulties if inhaled	H334

A1.19 皮膚腐食性/刺激性（判定基準は第3.2章を参照のこと）

分類			表示					GHS危険有害性情報コード
GHS危険有害性クラス	GHS危険有害性区分	UNモデル規則クラスまたは区分	GHS絵表示	UNモデル規則絵表示	GHS注意喚起語	GHS危険有害性情報	
皮膚腐食性/刺激性	1 1A,1B,1C	8	◇	◇	危険	重篤な皮膚の薬傷および眼の損傷	H314
	2	要求されない	◇	要求されない	警告	皮膚刺激	H315
	3		絵表示なし		警告	軽度の皮膚刺激	H316

A1.20 眼に対する重篤な損傷性/眼刺激性（判定基準は第3.3章を参照のこと）

分類			表示					GHS危険有害性情報コード
GHS危険有害性クラス	GHS危険有害性区分	UNモデル規則クラスまたは区分	GHS絵表示	UNモデル規則絵表示	GHS注意喚起語	GHS危険有害性情報	
眼に対する重篤な損傷性/眼刺激性	1	要求されない	◇	要求されない	危険	重篤な眼の損傷	H318
	2/2A		◇		警告	強い眼刺激	H319
	2B		絵表示なし		警告	眼刺激	H320

A1.21 呼吸器感作性（判定基準は第3.4章を参照のこと）

分類			表示					GHS危険有害性情報コード
GHS危険有害性クラス	GHS危険有害性区分	UNモデル規則クラスまたは区分	GHS絵表示	UNモデル規則絵表示	GHS注意喚起語	GHS危険有害性情報	
呼吸器感作性	1、1A、1B	要求されない	◇	要求されない	危険	吸入するとアレルギー、喘息または呼吸困難を起こすおそれ	H334

A1.22 **Skin sensitization** (see chapter 3.4 for classification criteria)

Classification			Labelling				GHS hazard statement codes
GHS hazard class	GHS hazard category	UN Model Regulations class or division	GHS pictogram	UN Model Regulations pictogram	GHS signal word	GHS hazard statement	
Skin sensitization	1, 1A, 1B	Not applicable	⟨!⟩	Not required	Warning	May cause an allergic skin reaction	H317

A1.23 **Germ cell mutagenicity** (see chapter 3.5 for classification criteria)

Classification			Labelling				GHS hazard statement codes
GHS hazard class	GHS hazard category	UN Model Regulations class or division	GHS pictogram	UN Model Regulations pictogram	GHS signal word	GHS hazard statement	
Germ cell mutagenicity	1, 1A, 1B	Not applicable	⟨☠⟩	Not applicable	Danger	May cause genetic defects *(state route of exposure if it is conclusively proven that no other routes of exposure cause the hazard)*	H340
	2				Warning	Suspected of causing genetic defects *(state route of exposure if it is conclusively proven that no other routes of exposure cause the hazard)*	H341

A1.24 **Carcinogenicity** (see chapter 3.6 for classification criteria)

Classification			Labelling				GHS hazard statement codes
GHS hazard class	GHS hazard category	UN Model Regulations class or division	GHS pictogram	UN Model Regulations pictogram	GHS signal word	GHS hazard statement	
Carcinogenicity	1, 1A, 1B	Not applicable	⟨☠⟩	Not applicable	Danger	May cause cancer *(state route of exposure if it is conclusively proven that no other routes of exposure cause the hazard)*	H350
	2				Warning	Suspected of causing cancer *(state route of exposure if it is conclusively proven that no other routes of exposure cause the hazard)*	H351

A1.22 皮膚感作性（判定基準は第 3.4 章を参照のこと）

分類			表示				GHS危険有害性情報コード
GHS危険有害性クラス	GHS危険有害性区分	UNモデル規則クラスまたは区分	GHS絵表示	UNモデル規則絵表示	GHS注意喚起語	GHS危険有害性情報	
皮膚感作性	1、1A、1B	要求されない	！	要求されない	警告	アレルギー性皮膚反応を起こすおそれ	H317

A1.23 生殖細胞変異原性（判定基準は第 3.5 章を参照のこと）

分類			表示				GHS危険有害性情報コード
GHS危険有害性クラス	GHS危険有害性区分	UNモデル規則クラスまたは区分	GHS絵表示	UNモデル規則絵表示	GHS注意喚起語	GHS危険有害性情報	
生殖細胞変異原性	1、1A、1B	要求されない	(健康有害性)	要求されない	危険	遺伝性疾患のおそれ（他の経路からのばく露が有害でないことが決定的に証明されている場合、有害なばく露経路を記載）	H340
	2				警告	遺伝性疾患のおそれの疑い（他の経路からのばく露が有害でないことが決定的に証明されている場合、有害なばく露経路を記載）	H341

A1.24 発がん性（判定基準は第 3.6 章を参照のこと）

分類			表示				GHS危険有害性情報コード
GHS危険有害性クラス	GHS危険有害性区分	UNモデル規則クラスまたは区分	GHS絵表示	UNモデル規則絵表示	GHS注意喚起語	GHS危険有害性情報	
発がん性	1、1A、1B	要求されない	(健康有害性)	要求されない	危険	発がんのおそれ（他の経路からのばく露が有害でないことが決定的に証明されている場合、有害なばく露経路を記載）	H350
	2				警告	発がんのおそれの疑い（他の経路からのばく露が有害でないことが決定的に証明されている場合、有害なばく露経路を記載）	H351

A1.25 Reproductive toxicity (see chapter 3.7 for classification criteria)

Classification			Labelling					GHS hazard statement codes
GHS hazard class	GHS hazard category	UN Model Regulations class or division	GHS pictogram	UN Model Regulations pictogram	GHS signal word	GHS hazard statement	
Reproductive toxicity	1, 1A, 1B	*Not applicable*		*Not applicable*	Danger	May damage fertility or the unborn child *(state specific effect if known) (state route of exposure if it is conclusively proven that no other routes of exposure cause the hazard)*	H360
	2	*Not applicable*		*Not applicable*	Warning	Suspected of damaging fertility or the unborn child *(state specific effect if known) (state route of exposure if it is conclusively proven that no other routes of exposure cause the hazard)*	H361
	Additional category for effects on or via lactation		*No pictogram*		*No signal word*	May cause harm to breast-fed children	H362

A1.26 Specific target organ toxicity following single exposure (see chapter 3.8 for classification criteria)

Classification			Labelling					GHS hazard statement codes
GHS hazard class	GHS hazard category	UN Model Regulations class or division	GHS pictogram	UN Model Regulations pictograms	GHS signal word	GHS hazard statement	
Specific target organ toxicity – single exposure	1	*Not applicable*		*Not applicable*	Danger	Causes damage to organs *(or state all organs affected, if known) (state route of exposure if it is conclusively proven that no other routes of exposure cause the hazard)*	H370
	2	*Not applicable*		*Not applicable*	Warning	May cause damage to organs *(or state all organs affected, if known) (state route of exposure if it is conclusively proven that no other routes of exposure cause the hazard)*	H371
	3				Warning	May cause respiratory irritation *or* May cause drowsiness or dizziness	H335 H336

A1.25 生殖毒性 （判定基準は第 3.7 章を参照のこと）

分類			表示					GHS 危険有害性情報コード
GHS 危険有害性クラス	GHS 危険有害性区分	UN モデル規則クラスまたは区分	GHS 絵表示	UN モデル規則絵表示	GHS 注意喚起語	GHS 危険有害性情報	
生殖毒性	1、1A、1B	要求されない		要求されない	危険	生殖能または胎児への悪影響のおそれ（もし判れば影響の内容を記載する）（他の経路からのばく露が有害でないことが決定的に証明されている場合、有害なばく露経路を記載）	H360
	2				警告	生殖能または胎児への悪影響のおそれの疑い（もし判れば影響の内容を記載する）（他の経路からのばく露が有害でないことが決定的に証明されている場合、有害なばく露経路を記載）	H361
	追加区分 授乳に対するまたは授乳を介した影響		絵表示なし		注意喚起語なし	授乳中の子に害を及ぼすおそれ	H362

A1.26 特定標的臓器毒性－単回ばく露 （判定基準は第 3.8 章を参照のこと）

分類			表示					GHS 危険有害性情報コード
GHS 危険有害性クラス	GHS 危険有害性区分	UN モデル規則クラスまたは区分	GHS 絵表示	UN モデル規則絵表示	GHS 注意喚起語	GHS 危険有害性情報	
特定標的臓器毒性－単回ばく露	1	要求されない		要求されない	危険	臓器の障害（もし判れば影響を受ける全ての臓器を記載する）（他の経路からのばく露が有害でないことが決定的に証明されている場合、有害なばく露経路を記載）	H370
	2				警告	臓器の障害のおそれ（もし判れば影響を受ける全ての臓器を記載する）（他の経路からのばく露が有害でないことが決定的に証明されている場合、有害なばく露経路を記載）	H371
	3				警告	呼吸器への刺激のおそれ または 眠気またはめまいのおそれ	H335 H336

- 285 -

A1.27 Specific target organ toxicity following repeated exposure (see chapter 3.9 for classification criteria)

Classification			Labelling				GHS hazard statement codes
GHS hazard class	GHS hazard category	UN Model Regulations class or division	GHS pictogram	UN Model Regulations pictogram	GHS signal word	GHS hazard statement	
Specific target organ toxicity – repeated exposure	1	Not applicable		Not applicable	Danger	Causes damage to organs *(state all organs affected, if known)* through prolonged or repeated exposure *(state route of exposure if it is conclusively proven that no other routes of exposure cause the hazard)*	H372
	2				Warning	May cause damage to organs *(state all organs affected, if known)* through prolonged or repeated exposure *(state route of exposure if it is conclusively proven that no other routes of exposure cause the hazard)*	H373

A1.28 Aspiration hazard (see chapter 3.10 for classification criteria)

Classification			Labelling				GHS hazard statement codes
GHs hazard class	GHS hazard category	UN Model Regulations class or division	GHS pictogram	UN Model Regulations pictogram	GHS signal word	GHS hazard statement	
Aspiration hazard	1	Not applicable		Not applicable	Danger	May be fatal if swallowed and enters airways	H304
	2				Warning	May be harmful if swallowed and enters airways	H305

A1.29 (a) Hazardous to the aquatic environment, short-term (acute) (see chapter 4.1 for classification criteria)

Classification			Labelling				GHS hazard statement codes
GHS hazard class	GHS hazard category	UN Model Regulations class or division[a]	GHS pictogram	UN Model Regulations pictogram[a]	GHS signal word	GHS hazard statement	
Hazardous to the aquatic environment, short-term (Acute)	Acute 1	9		and	Warning	Very toxic to aquatic life	H400
	Acute 2	Not applicable	No pictogram	Not applicable	No signal word	Toxic to aquatic life	H401
	Acute 3					Harmful to aquatic life	H402

[a] Under the UN Model Regulations, for category Acute 1, environmentally hazardous substances are classified under Class 9 and shall bear both the Class 9 transport pictogram and the environmentally hazardous substance transport mark (see chapter 5.2, section 5.2.1.6 and chapter 5.3, section 5.3.2.3, of the UN Model Regulations). However, if the environmentally hazardous substance presents any other hazards covered by UN Model Regulations, the Class 9 transport pictogram shall be replaced by the transport pictogram(s) applicable to the hazard(s) present and the environmentally hazardous substance pictogram is not required.

A1.27 特定標的臓器毒性－反復ばく露（判定基準は第3.9章を参照のこと）

分類			表示				GHS危険有害性情報コード
GHS危険有害性クラス	GHS危険有害性区分	UNモデル規則クラスまたは区分	GHS絵表示	UNモデル規則絵表示	GHS注意喚起語	GHS危険有害性情報	
特定標的臓器毒性－反復ばく露	1	要求されない	☣	要求されない	危険	長期にわたる、または反復ばく露による臓器の障害（もし判れば影響を受ける全ての臓器を記載する）(他の経路からのばく露が有害でないことが決定的に証明されている場合、有害なばく露経路を記載)	H372
	2				警告	長期にわたる、または反復ばく露による臓器の障害のおそれ（もし判れば影響を受ける全ての臓器を記載する）(他の経路からのばく露が有害でないことが決定的に証明されている場合、有害なばく露経路を記載)	H373

A1.28 誤えん有害性（判定基準は第3.10章を参照のこと）

分類			表示				GHS危険有害性情報コード
GHS危険有害性クラス	GHS危険有害性区分	UNモデル規則クラスまたは区分	GHS絵表示	UNモデル規則絵表示	GHS注意喚起語	GHS危険有害性情報	
誤えん有害性	1	要求されない	☣	要求されない	危険	飲み込んで気道に侵入すると生命に危険のおそれ	H304
	2				警告	飲み込んで気道に侵入すると有害のおそれ	H305

A1.29（a） 水生環境有害性、短期（急性）（判定基準は第4.1章を参照のこと）

分類			表示				GHS危険有害性情報コード
GHS危険有害性クラス	GHS危険有害性区分	UNモデル規則クラスまたは区分	GHS絵表示	UNモデル規則絵表示 [a]	GHS注意喚起語	GHS危険有害性情報	
水生環境有害性、短期（急性）	急性1	9	🐟	9 および 🐟	警告	水生生物に非常に強い毒性	H400
	急性2	要求されない	絵表示なし	要求されない	注意喚起語なし	水生生物に毒性	H401
	急性3					水生生物に有害	H402

[a] UNモデル規則では、区分急性1に関して、環境有害物質は区分9に分類され、クラス9の輸送絵表示と環境有害物質輸送絵表示の両方を付けなければならない（UNモデル規則第5.2章5.2.1.6および第5.3章5.3.2.3参照のこと）。ただし当該物質がUNモデル規則でカバーする他の危険有害性がある場合には、クラス9の輸送絵表示は、該当する危険性に適用される輸送絵表示に置き換えられ、環境有害物質絵表示は要求されない。

A1.29 (b) Hazardous to the aquatic environment, long-term (chronic) (see chapter 4.1 for classification criteria)

Classification			Labelling					GHS hazard statement codes
GHS hazard class	GHS hazard category	UN Model Regulations class or division[a]	GHS pictogram	UN Model Regulations pictogram[a]	GHS signal word	GHS hazard statement	
Hazardous to the aquatic environment, long-term (Chronic)	Chronic 1	9			Warning	Very toxic to aquatic life with long lasting effects	H410
	Chronic 2				No signal word	Toxic to aquatic life with long lasting effects	H411
	Chronic 3	Not applicable	No pictogram	Not applicable	No signal word	Harmful to aquatic life with long lasting effects	H412
	Chronic 4				No signal word	May cause long lasting harmful effects to aquatic life	H413

[a] Under the UN Model Regulations, for categories Chronic 1 and 2, environmentally hazardous substances are classified under Class 9 and shall bear both the Class 9 transport pictogram and the environmentally hazardous substance transport mark (see chapter 5.2, section 5.2.1.6 and chapter 5.3, section 5.3.2.3, of the UN Model Regulations). However, if the environmentally hazardous substance presents any other hazards covered by UN Model Regulations, the Class 9 transport pictogram shall be replaced by the transport pictogram(s) applicable to the hazard(s) present and the environmentally hazardous substance pictogram is not required.

A1.30 Hazard to the ozone layer (see chapter 4.2 for classification criteria)

Classification			Labelling					GHS hazard statement code
GHS hazard class	GHS hazard category	UN Model Regulations class or division	GHS pictogram	UN Model Regulations pictograms	GHS signal word	GHS hazard statement	
Hazardous to the ozone layer	1	Not applicable		Not applicable	Warning	Harms public health and the environment by destroying ozone in the upper atmosphere	H420

A1.29（b） 水生環境有害性、長期（慢性）（判定基準は第 4.1 章を参照のこと）

分類			表示					GHS 危険有害性情報コード
GHS 危険有害性クラス	GHS 危険有害性区分	UN モデル規則クラスまたは区分 [a]	GHS 絵表示	UN モデル規則絵表示 [a]	GHS 注意喚起語	GHS 危険有害性情報	
水生環境有害性、長期（慢性）	慢性 1	9		および	警告	長期継続的影響によって水生生物に非常に強い毒性	H410
	慢性 2				注意喚起語なし	長期継続的影響によって水生生物に毒性	H411
	慢性 3	要求されない	絵表示なし	要求されない	注意喚起語なし	長期継続的影響によって水生生物に有害	H412
	慢性 4					長期継続的影響によって水生生物に有害のおそれ	H413

[a] *UN モデル規則では、区分慢性 1 および 2 に関して、環境有害物質は区分 9 に分類され、クラス 9 の輸送絵表示と環境有害物質輸送絵表示の両方を付けなければならない（UN モデル規則第 5.2 章 5.2.1.6 および第 5.3 章 5.3.2.3 参照のこと）。ただし当該物質が UN モデル規則でカバーする他の危険有害性がある場合には、クラス 9 の輸送絵表示は、該当する危険性に適用される輸送絵表示に置き換えられ、環境有害物質絵表示は要求されない。*

A1.30　オゾン層への有害性（判定基準は第 4.2 章を参照のこと）

分類			表示					GHS 危険有害性情報コード
GHS 危険有害性クラス	GHS 危険有害性区分	UN モデル規則クラスまたは区分	GHS 絵表示	UN モデル規則絵表示	GHS 注意喚起語	GHS 危険有害性情報	
オゾン層への有害性	1	要求されない		要求されない	警告	オゾン層を破壊し、健康および環境に有害	H420

ANNEX 7

(Memorial)

ANNEX 2

(Reserved)

附属書2

(保留)

附录 2

(略)

ANNEX 3

CODIFICATION OF HAZARD STATEMENTS, CODIFICATION AND USE OF PRECAUTIONARY STATEMENTS, CODIFICATION OF HAZARD PICTOGRAMS AND EXAMPLES OF PRECAUTIONARY PICTOGRAMS

ANNEX 3

CODIFICATION OF HAZARD STATEMENTS, CODIFICATION AND USE OF PRECAUTIONARY STATEMENTS, CODIFICATION OF HAZARD PICTOGRAMS AND EXAMPLES OF PRECAUTIONARY PICTOGRAMS

附属書3

危険有害性情報のコード
注意書きのコードと使用法
絵表示のコード
および注意絵表示の例

ANNEX 3

Section 1

CODIFICATION OF HAZARD STATEMENTS

A3.1.1 Introduction

A3.1.1.1 *Hazard statement* means a statement assigned to a hazard class and category that describes the nature of the hazards of a hazardous product, including, where appropriate, the degree of hazard.

A3.1.1.2 This section contains the recommended codes assigned to each of the hazard statements applicable to the hazard categories under the GHS.

A3.1.1.3 The hazard statement codes are intended to be used for reference purposes. They are not part of the hazard statement text and should not be used to replace it.

A3.1.2 Codification of hazard statements

A3.1.2.1 Hazard statements are assigned a unique alphanumerical code which consists of one letter and three numbers, as follows:

(a) the letter "H" (for "hazard statement");

(b) a number designating the type of hazard to which the hazard statement is assigned according to the numbering of the different parts of the GHS, as follows:

- "2" for physical hazards;
- "3" for health hazards;
- "4" for environmental hazards;

(c) two numbers corresponding to the sequential numbering of hazards arising from the intrinsic properties of the substance or mixture, such as explosivity (codes from 200 to 210), flammability (codes from 220 to 230), etc.

A3.1.2.2 The codes to be used for designating hazard statements are listed, in numerical order, in table A3.1.1 for physical hazards, table A3.1.2 for health hazards and table A3.1.3 for environmental hazards. Each table is divided into 4 columns containing the following information:

Column (1) The hazard statement code;

Column (2) The hazard statement text;

The text in bold should appear on the label, except as otherwise specified. The information in italics should also appear as part of the hazard statement when the information is known.

For example: "**causes damages to organs** (or state all organs affected, if known) **through prolonged or repeated exposure** (state route of exposure if it is conclusively proven that no other routes of exposure cause the hazard)".

Column (3) Hazard class, with a reference to the chapter of the GHS where information about the hazard class may be found.

Column (4) The hazard category or categories within a hazard class for which the use of a hazard statement is applicable.

A3.1.2.3 In addition to individual hazard statements, a number of combined hazard statements are given in table A3.1.2. The alphanumerical codes for the combined statements are constructed from the codes for the individual statements that are combined, conjoined with the plus ("+") sign. For example, H300 + H310 + H330 indicates that the text to appear on the label is "**Fatal if swallowed, in contact with skin or if inhaled**".

附属書 3

第 1 節

危険有害性情報のコード

A3.1.1 序文

A3.1.1.1 *危険有害性情報*とは、危険有害性クラスおよび危険有害性区分に割り当てられた文言であって、危険有害な製品の危険有害性の性質を、該当する程度も含めて記述する文言をいう。

A3.1.1.2 この節には GHS の危険有害性区分に適用される危険有害性情報にそれぞれ割り当てられた推奨コードを含む。

A3.1.1.3 危険有害性情報のコードは参照するためのものである。コードは危険有害性情報の文言の一部ではないので、文言の代わりに用いることはできない。

A3.1.2 危険有害性情報のコード

A3.1.2.1 危険有害性情報には 1 つの文字と 3 つの数字からなる英数字コードが、下記のように割り当てられている：

(a) 文字「H」（危険有害性情報"hazard statement"）；
(b) 危険有害性の種類を示す番号、割り当てられた危険有害性情報に対し下記のように番号が付けられている：
 －「2」物理化学的危険性；
 －「3」健康有害性；
 －「4」環境有害性；
(c) 連続した 2 つの番号、物質や混合物の性質に起因する危険有害性をあらわす、例えば、爆発性（コード 200 から 210）、可燃性（コード 220 から 230）他。

A3.1.2.2 割り当てられた危険有害性情報に使用するコードは、物理的化学危険性は表 A3.1.1 に、健康有害性は表 A3.1.2 に、環境有害性は表 A3.1.3 に番号順に記載してある：

欄（1） 危険有害性情報コード；

欄（2） 危険有害性情報；

 特別の指示がない限り、太字になっている文言がラベルに記載される。斜体の情報は、もし情報があれば、危険有害性情報の一部として記載する。

 例えば、「**長期にわたる、または反復ばく露**（他の経路からのばく露が有害でないことが決定的に証明されている場合、有害な経路を記載する）**による臓器の障害**（もしわかればすべての影響を受ける臓器を挙げる）」。

欄（3） 危険有害性クラスおよび該当する危険有害性クラスについての情報がある GHS 文書の章。

欄（4） 当該危険有害性情報が適用可能な危険有害性クラスでの区分。

A3.1.2.3 それぞれの危険有害性情報に加え、いくつかの結合された危険有害性情報を表 A3.1.2 に示した。結合された情報に対するアルファベットと数によるコードは、それぞれの情報コードがプラス「＋」によって結ばれている。例えば H300+H310+H330 は、ラベルに表示する文言が「**飲み込んだ場合や皮膚に接触した場合や吸入した場合は生命に危険**」であることを示す。

A3.1.2.4   All assigned hazard statements should appear on the label unless otherwise specified in 1.4.10.5.3.3. The competent authority may specify the order in which they appear. Also, where a combined hazard statement is permitted for two or more hazard statements (see A3.1.2.5), the competent authority may specify whether the combined hazard statement or the corresponding individual statements should appear on the label or may leave the choice to the manufacturer/supplier.

A3.1.2.5   In addition to the combinations found in table A3.1.2, it is also permitted to combine more than one health hazard statement of equivalent severity if, for example, there is insufficient space on the label. When hazard statements are combined, all hazards must be clearly conveyed and only the repetitive text may be deleted. Statements can be combined by using the word "and", additional punctuation, and changing the case of the initial letter of the word at the beginning of a statement. For example, H317 "**May cause an allergic skin reaction**" + H340 "**May cause genetic defects**" + H350 "**May cause cancer**" may all be combined because they are all for Category 1 health hazards (i.e. health hazard statements of equivalent severity) and have repetitive elements of the hazard statement (i.e. the statements begin with "may cause"). These statements may be combined to "**May cause an allergic skin reaction, genetic defects, and cancer**". The competent authority may limit the types of combinations permitted to ensure comprehensibility (e.g. limit the number of hazard statements that can be combined).

Table A3.1.1: Hazard statement codes for physical hazards

Code (1)	Physical hazard statements (2)	Hazard class (GHS chapter) (3)	Hazard category (4)
H200	*[Deleted]*		
H201	*[Deleted]*		
H202	*[Deleted]*		
H203	*[Deleted]*		
H204	**Fire or projection hazard**	Explosives (chapter 2.1)	2B, 2C
H205	*[Deleted]*		
H206	**Fire, blast or projection hazard; increased risk of explosion if desensitizing agent is reduced**	Desensitized explosives (chapter 2.17)	1
H207	**Fire or projection hazard; increased risk of explosion if desensitizing agent is reduced**	Desensitized explosives (chapter 2.17)	2, 3
H208	**Fire hazard; increased risk of explosion if desensitizing agent is reduced**	Desensitized explosives (chapter 2.17)	4
H209	**Explosive**	Explosives (chapter 2.1)	1, 2A
H210	**Very sensitive**	Explosives (chapter 2.1)	1
H211	**May be sensitive**	Explosives (chapter 2.1)	1
H220	**Extremely flammable gas**	Flammable gases (chapter 2.2)	1A
H221	**Flammable gas**	Flammable gases (chapter 2.2)	1B, 2
H222	**Extremely flammable aerosol**	Aerosols (chapter 2.3)	1
H223	**Flammable aerosol**	Aerosols (chapter 2.3)	2
H224	**Extremely flammable liquid and vapour**	Flammable liquids (chapter 2.6)	1
H225	**Highly flammable liquid and vapour**	Flammable liquids (chapter 2.6)	2
H226	**Flammable liquid and vapour**	Flammable liquids (chapter 2.6)	3
H227	**Combustible liquid**	Flammable liquids (chapter 2.6)	4
H228	**Flammable solid**	Flammable solids (chapter 2.7)	1, 2
H229	**Pressurized container: may burst if heated**	Aerosols (chapter 2.3)	1, 2, 3

A3.1.2.4　すべての割り当てられた危険有害性情報は、1.4.10.5.3.3での決まりに当てはまらない限り、ラベルに記載すべきである。所管官庁は記載の順序について決めてもよい。また、結合された危険有害性情報が2つ以上の危険有害性（A3.1.2.5を参照）を示すことを可能にする場合には、所管官庁は結合された危険有害性情報あるいはそれぞれの情報のどちらをラベルに記載するかを決めてもよいし、製造者/供給者に選択を委ねてもよい。

A3.1.2.5　表A3.1.2にある組み合わせに加え、ラベルに十分なスペースがない場合などには、同じ重篤度の健康有害性に関する記述を2つ以上組み合わせることも認められる。危険有害性情報を組み合わせる場合、すべての有害性を明確に伝えなければならず、繰り返される文言のみを削除することができる。危険有害性情報は、"and"（「および」）という単語を使用したり、句読点を追加したり、危険有害性情報の先頭の単語の大文字と小文字を変えたりすることで結合することができる。　例えば、H317「アレルギー性皮膚反応を起こすおそれ」＋H340「遺伝性疾患のおそれ」＋H350「発がんのおそれ」は、すべて区分1の健康有害性（すなわち、同等の重篤度の健康有害性に関する記述）であり、危険有害性情報の繰り返し要素（すなわち、"may cause"（「…のおそれ」）で始まる記述）があるため、組み合わせることができる。これらの記述は、"アレルギー性皮膚反応、遺伝性疾患および発がんを引き起こすおそれ"と組み合わせることができる。所管官庁は、わかりやすさを確保するために、許可される組み合わせの種類を制限することができる（例えば、組み合わせることができる危険有害性情報の数を制限する）。

表A3.1.1：物理化学的危険性の危険有害性情報コード

コード (1)	物理化学的危険性　危険有害性情報 (2)	危険有害性クラス（GHSの章）(3)	危険有害性区分 (4)
H200	(削除)		
H201	(削除)		
H202	(削除)		
H203	(削除)		
H204	火災または飛散危険性	爆発物（2.1章）	2B, 2C
H205	(削除)		
H206	火災、爆風または飛散危険性；鈍性化剤が減少した場合には爆発の危険性の増加	鈍性化爆発物（2.17章）	1
H207	火災または飛散危険性；鈍性化剤が減少した場合には爆発の危険性の増加	鈍性化爆発物（2.17章）	2, 3
H208	火災危険性；鈍性化剤が減少した場合には爆発の危険性の増加	鈍性化爆発物（2.17章）	4
H209	爆発物	爆発物（2.1章）	1, 2A
H210	非常に敏感	爆発物（2.1章）	1
H211	敏感である可能性	爆発物（2.1章）	1
H220	極めて可燃性の高いガス	可燃性ガス（2.2章）	1A
H221	可燃性ガス	可燃性ガス（2.2章）	1B, 2
H222	極めて可燃性の高いエアゾール	エアゾール（2.3章）	1
H223	可燃性エアゾール	エアゾール（2.3章）	2
H224	極めて引火性の高い液体および蒸気	引火性液体（2.6章）	1
H225	引火性の高い液体および蒸気	引火性液体（2.6章）	2
H226	引火性の液体および蒸気	引火性液体（2.6章）	3
H227	可燃性液体	引火性液体（2.6章）	4
H228	可燃性固体	可燃性固体（2.7章）	1, 2
H229	高圧容器：熱すると破裂のおそれ	エアゾール（2.3章）	1, 2, 3

Table A3.1.1: Hazard statement codes for physical hazards *(cont'd)*

Code (1)	Physical hazard statements (2)	Hazard class (GHS chapter) (3)	Hazard category (4)
H230	May react explosively even in the absence of air	Flammable gases (chapter 2.2)	1A, chemically unstable gas A
H231	May react explosively even in the absence of air at elevated pressure and/or temperature	Flammable gases (chapter 2.2)	1A, chemically unstable gas B
H232	May ignite spontaneously if exposed to air	Flammable gases (chapter 2.2)	1A, pyrophoric gas
H240	Heating may cause an explosion	Self-reactive substances and mixtures (chapter 2.8); and Organic peroxides (chapter 2.15)	Type A
H241	Heating may cause a fire or explosion	Self-reactive substances and mixtures (chapter 2.8); and Organic peroxides (chapter 2.15)	Type B
H242	Heating may cause a fire	Self-reactive substances and mixtures (chapter 2.8); and Organic peroxides (chapter 2.15)	Types C, D, E, F
H250	Catches fire spontaneously if exposed to air	Pyrophoric liquids (chapter 2.9); Pyrophoric solids (chapter 2.10)	1
H251	Self-heating; may catch fire	Self-heating substances and mixtures (chapter 2.11)	1
H252	Self-heating in large quantities; may catch fire	Self-heating substances and mixtures (chapter 2.11)	2
H260	In contact with water releases flammable gases which may ignite spontaneously	Substances and mixtures which, in contact with water, emit flammable gases (chapter 2.12)	1
H261	In contact with water releases flammable gas	Substances and mixtures which, in contact with water, emit flammable gases (chapter 2.12)	2, 3
H270	May cause or intensify fire; oxidizer	Oxidizing gases (chapter 2.4)	1
H271	May cause fire or explosion; strong oxidizer	Oxidizing liquids (chapter 2.13); Oxidizing solids (chapter 2.14)	1
H272	May intensify fire; oxidizer	Oxidizing liquids (chapter 2.13); Oxidizing solids (chapter 2.14)	2, 3
H280	Contains gas under pressure; may explode if heated	Gases under pressure (chapter 2.5)	Compressed gas Liquefied gas Dissolved gas
H281	Contains refrigerated gas; may cause cryogenic burns or injury	Gases under pressure (chapter 2.5)	Refrigerated liquefied gas
H282	**Extremely flammable chemical under pressure:** May explode if heated	Chemicals under pressure (chapter 2.3)	1
H283	**Flammable chemical under pressure:** May explode if heated	Chemicals under pressure (chapter 2.3)	2
H284	**Chemical under pressure:** May explode if heated	Chemicals under pressure (chapter 2.3)	3
H290	May be corrosive to metals	Corrosive to metals (chapter 2.16)	1

コード (1)	物理化学的危険性 危険有害性情報 (2)	危険有害性クラス（GHSの章） (3)	危険有害性区分 (4)
H230	空気が無くても爆発的に反応するおそれ	可燃性ガス（2.2章）	1A,化学的に不安定なガスA
H231	圧力および/または温度が上昇した場合、空気が無くても爆発的に反応するおそれ	可燃性ガス（2.2章）	1A,化学的に不安定なガスB
H232	空気に触れると自然発火のおそれ	可燃性ガス（2.2章）	1A,自然発火性ガス
H240	熱すると爆発のおそれ	自己反応性物質および混合物（2.8章） 有機過酸化物（2.15章）	タイプA
H241	熱すると火災または爆発のおそれ	自己反応性物質および混合物（2.8章） 有機過酸化物（2.15章）	タイプB
H242	熱すると火災のおそれ	自己反応性物質および混合物（2.8章） 有機過酸化物（2.15章）	タイプC,D,E,F
H250	空気に触れると自然発火	自然発火性液体（2.9章） 自然発火性固体（2.10章）	1
H251	自己発熱；火災のおそれ	自己発熱性物質および混合物（2.11章）	1
H252	大量の場合自己発熱；火災のおそれ	自己発熱性物質および混合物（2.11章）	2
H260	水に触れると自然発火するおそれのある可燃性ガスを発生	水反応可燃性物質および混合物（2.12章）	1
H261	水に触れると可燃性ガスを発生	水反応可燃性物質および混合物（2.12章）	2,3
H270	発火または火災助長のおそれ；酸化性物質	酸化性ガス（2.4章）	1
H271	火災または爆発のおそれ；強酸化性物質	酸化性液体（2.13章） 酸化性固体（2.14章）	1
H272	火災助長のおそれ；酸化性物質	酸化性液体（2.13章） 酸化性固体（2.14章）	2,3
H280	高圧ガス；熱すると爆発のおそれ	高圧ガス（2.5章）	高圧ガス 液化ガス 溶解ガス
H281	深冷液化ガス；凍傷または傷害のおそれ	高圧ガス（2.5章）	深冷液化ガス
H282	極めて可燃性の高い加圧下化学品：熱すると爆発のおそれ	加圧下化学品（第2.3章）	1
H283	可燃性の加圧下化学品：熱すると爆発のおそれ	加圧下化学品（第2.3章）	2
H284	加圧下化学品：熱すると爆発のおそれ	加圧下化学品（第2.3章）	3
H290	金属腐食のおそれ	金属腐食性（2.16章）	1

Table A3.1.2: Hazard statement codes for health hazards

Code (1)	Health hazard statements (2)	Hazard class (GHS chapter) (3)	Hazard category (4)
H300	**Fatal if swallowed**	Acute toxicity, oral (chapter 3.1)	1, 2
H301	**Toxic if swallowed**	Acute toxicity, oral (chapter 3.1)	3
H302	**Harmful if swallowed**	Acute toxicity, oral (chapter 3.1)	4
H303	**May be harmful if swallowed**	Acute toxicity, oral (chapter 3.1)	5
H304	**May be fatal if swallowed and enters airways**	Aspiration hazard (chapter 3.10)	1
H305	**May be harmful if swallowed and enters airways**	Aspiration hazard (chapter 3.10)	2
H310	**Fatal in contact with skin**	Acute toxicity, dermal (chapter 3.1)	1, 2
H311	**Toxic in contact with skin**	Acute toxicity, dermal (chapter 3.1)	3
H312	**Harmful in contact with skin**	Acute toxicity, dermal (chapter 3.1)	4
H313	**May be harmful in contact with skin**	Acute toxicity, dermal (chapter 3.1)	5
H314	**Causes severe skin burns and eye damage**	Skin corrosion/irritation (chapter 3.2)	1, 1A, 1B, 1C
H315	**Causes skin irritation**	Skin corrosion/irritation (chapter 3.2)	2
H316	**Causes mild skin irritation**	Skin corrosion/irritation (chapter 3.2)	3
H317	**May cause an allergic skin reaction**	Skin sensitization (chapter 3.4)	1, 1A, 1B
H318	**Causes serious eye damage**	Serious eye damage/eye irritation (chapter 3.3)	1
H319	**Causes serious eye irritation**	Serious eye damage/eye irritation (chapter 3.3)	2/2A
H320	**Causes eye irritation**	Serious eye damage/eye irritation (chapter 3.3)	2B
H330	**Fatal if inhaled**	Acute toxicity, inhalation (chapter 3.1)	1, 2
H331	**Toxic if inhaled**	Acute toxicity, inhalation (chapter 3.1)	3
H332	**Harmful if inhaled**	Acute toxicity, inhalation (chapter 3.1)	4
H333	**May be harmful if inhaled**	Acute toxicity, inhalation (chapter 3.1)	5
H334	**May cause allergy or asthma symptoms or breathing difficulties if inhaled**	Respiratory sensitization (chapter 3.4)	1, 1A, 1B
H335	**May cause respiratory irritation**	Specific target organ toxicity, single exposure; Respiratory tract irritation (chapter 3.8);	3
H336	**May cause drowsiness or dizziness**	Specific target organ toxicity, single exposure; Narcotic effects (chapter 3.8)	3
H340	**May cause genetic defects** (*state route of exposure if it is conclusively proven that no other routes of exposure cause the hazard*)	Germ cell mutagenicity (chapter 3.5)	1, 1A, 1B
H341	**Suspected of causing genetic defects** (*state route of exposure if it is conclusively proven that no other routes of exposure cause the hazard*)	Germ cell mutagenicity (chapter 3.5)	2

表 A3.1.2：健康有害性の危険有害性情報コード

コード (1)	健康有害性　危険有害性情報 (2)	危険有害性クラス（GHSの章） (3)	危険有害性区分 (4)
H300	飲み込むと生命に危険	急性毒性（経口）（3.1章）	1,2
H301	飲み込むと有毒	急性毒性（経口）（3.1章）	3
H302	飲み込むと有害	急性毒性（経口）（3.1章）	4
H303	飲み込むと有害のおそれ	急性毒性（経口）（3.1章）	5
H304	飲み込んで気道に侵入すると生命に危険のおそれ	誤えん有害性（3.10章）	1
H305	飲み込んで気道に侵入すると有害のおそれ	誤えん有害性（3.10章）	2
H310	皮膚に接触すると生命に危険	急性毒性（経皮）（3.1章）	1,2
H311	皮膚に接触すると有毒	急性毒性（経皮）（3.1章）	3
H312	皮膚に接触すると有害	急性毒性（経皮）（3.1章）	4
H313	皮膚に接触すると有害のおそれ	急性毒性（経皮）（3.1章）	5
H314	重篤な皮膚の薬傷および眼の損傷	皮膚腐食性/刺激性（3.2章）	1,1A,1B,1C
H315	皮膚刺激	皮膚腐食性/刺激性（3.2章）	2
H316	軽度の皮膚刺激	皮膚腐食性/刺激性（3.2章）	3
H317	アレルギー性皮膚反応を起こすおそれ	皮膚感作性（3.4章）	1, 1A,1B
H318	重篤な眼の損傷	眼に対する重篤な損傷性/眼刺激性（3.3章）	1
H319	強い眼刺激	眼に対する重篤な損傷性/眼刺激性（3.3章）	2/2A
H320	眼刺激	眼に対する重篤な損傷性/眼刺激性（3.3章）	2B
H330	吸入すると生命に危険	急性毒性（吸入）（3.1章）	1,2
H331	吸入すると有毒	急性毒性（吸入）（3.1章）	3
H332	吸入すると有害	急性毒性（吸入）（3.1章）	4
H333	吸入すると有害のおそれ	急性毒性（吸入）（3.1章）	5
H334	吸入するとアレルギー、喘息または呼吸困難を起こすおそれ	呼吸器感作性（3.4章）	1, 1A,1B
H335	呼吸器への刺激のおそれ	特定標的臓器毒性、単回ばく露、気道刺激性（3.8章）	3
H336	眠気またはめまいのおそれ	特定標的臓器毒性、単回ばく露、麻酔作用（3.8章）	3
H340	遺伝性疾患のおそれ *(他の経路からのばく露が有害でないことが決定的に証明されている場合、有害なばく露経路を記載する)*	生殖細胞変異原性（3.5章）	1,1A,1B
H341	遺伝性疾患のおそれの疑い *(他の経路からのばく露が有害でないことが決定的に証明されている場合、有害なばく露経路を記載する)*	生殖細胞変異原性（3.5章）	2

Table A3.1.2: Hazard statement codes for health hazards *(cont'd)*

Code (1)	Health hazard statements (2)	Hazard class (GHS chapter) (3)	Hazard category (4)
H350	**May cause cancer** *(state route of exposure if it is conclusively proven that no other routes of exposure cause the hazard)*	Carcinogenicity (chapter 3.6)	1, 1A, 1B
H351	**Suspected of causing cancer** *(state route of exposure if it is conclusively proven that no other routes of exposure cause the hazard)*	Carcinogenicity (chapter 3.6)	2
H360	**May damage fertility or the unborn child** *(state specific effect if known)(state route of exposure if it is conclusively proven that no other routes of exposure cause the hazard)*	Reproductive toxicity (chapter 3.7)	1, 1A, 1B
H361	**Suspected of damaging fertility or the unborn child** *(state specific effect if known)(state route of exposure if it is conclusively proven that no other routes of exposure cause the hazard)*	Reproductive toxicity (chapter 3.7)	2
H362	**May cause harm to breast-fed children**	Reproductive toxicity, effects on or via lactation (chapter 3.7)	Additional category
H370	**Causes damage to organs** *(or state all organs affected, if known) (state route of exposure if it is conclusively proven that no other routes of exposure cause the hazard)*	Specific target organ toxicity, single exposure (chapter 3.8)	1
H371	**May cause damage to organs** *(or state all organs affected, if known)(state route of exposure if it is conclusively proven that no other routes of exposure cause the hazard)*	Specific target organ toxicity, single exposure (chapter 3.8)	2
H372	**Causes damage to organs** *(state all organs affected, if known)* **through prolonged or repeated exposure** *(state route of exposure if it is conclusively proven that no other routes of exposure cause the hazard)*	Specific target organ toxicity, repeated exposure (chapter 3.9)	1
H373	**May cause damage to organs** *(state all organs affected, if known)* **through prolonged or repeated exposure** *(state route of exposure if it is conclusively proven that no other routes of exposure cause the hazard)*	Specific target organ toxicity, repeated exposure (chapter 3.9)	2
H300 + H310	**Fatal if swallowed or in contact with skin**	Acute toxicity, oral (chapter 3.1) and acute toxicity, dermal (chapter 3.1)	1, 2
H300 + H330	**Fatal if swallowed or if inhaled**	Acute toxicity, oral (chapter 3.1) and acute toxicity, inhalation (chapter 3.1)	1, 2
H310 + H330	**Fatal in contact with skin or if inhaled**	Acute toxicity, dermal (chapter 3.1) and acute toxicity, inhalation (chapter 3.1)	1, 2
H300 + H310 + H330	**Fatal if swallowed, in contact with skin or if inhaled**	Acute toxicity, oral (chapter 3.1), acute toxicity, dermal (chapter 3.1) and acute toxicity, inhalation (chapter 3.1)	1, 2

コード (1)	健康有害性　危険有害性情報 (2)	危険有害性クラス（GHSの章） (3)	危険有害性区分 (4)
H350	発がんのおそれ（他の経路からのばく露が有害でないことが決定的に証明されている場合、有害なばく露経路を記載する）	発がん性（3.6章）	1,1A,1B
H351	発がんのおそれの疑い（他の経路からのばく露が有害でないことが決定的に証明されている場合、有害なばく露経路を記載する）	発がん性（3.6章）	2
H360	生殖能または胎児への悪影響のおそれ（もしわかればすべての影響を受ける臓器を挙げる）（他の経路からのばく露が有害でないことが決定的に証明されている場合、有害なばく露経路を記載する）	生殖毒性（3.7章）	1,1A,1B
H361	生殖能または胎児への悪影響のおそれの疑い（もしわかればすべての影響を受ける臓器を挙げる）（他の経路からのばく露が有害でないことが決定的に証明されている場合、有害なばく露経路を記載する）	生殖毒性（3.7章）	2
H362	授乳中の子に害を及ぼすおそれ	生殖毒性、授乳に対するまたは授乳を介した影響（3.7章）	追加区分
H370	臓器の障害（もしわかればすべての影響を受ける臓器を挙げる）（他の経路からのばく露が有害でないことが決定的に証明されている場合、有害なばく露経路を記載する）	特定標的臓器毒性、単回ばく露（3.8章）	1
H371	臓器の障害のおそれ（もしわかればすべての影響を受ける臓器を挙げる）（他の経路からのばく露が有害でないことが決定的に証明されている場合、有害なばく露経路を記載する）	特定標的臓器毒性、単回ばく露（3.8章）	2
H372	長期にわたる、または反復ばく露（他の経路からのばく露が有害でないことが決定的に証明されている場合、有害なばく露経路を記載する）による臓器の障害（もしわかればすべての影響を受ける臓器を挙げる）	特定標的臓器毒性、反復ばく露（3.9章）	1
H373	長期にわたる、または反復ばく露（他の経路からのばく露が有害でないことが決定的に証明されている場合、有害なばく露経路を記載する）による臓器の障害のおそれ（もしわかればすべての影響を受ける臓器を挙げる）	特定標的臓器毒性、反復ばく露（3.9章）	2
H300 + H310	飲み込んだ場合や皮膚に接触した場合は生命に危険	急性毒性（経口）（3.1章）および急性毒性（経皮）（3.1章）	1, 2
H300 + H330	飲み込んだ場合や吸入した場合は生命に危険	急性毒性（経口）（3.1章）および急性毒性（吸入）（3.1章）	1, 2
H310 + H330	皮膚に接触した場合や吸入した場合は生命に危険	急性毒性（経皮）（3.1章）および急性毒性（吸入）（3.1章）	1, 2
H300 + H310 + H330	飲み込んだ場合や皮膚に接触した場合や吸入した場合は生命に危険	急性毒性（経口）（3.1章）および急性毒性（経皮）（3.1章）および急性毒性（吸入）（3.1章）	1, 2

Table A3.1.2: Hazard statement codes for health hazards *(cont'd)*

Code (1)	Health hazard statements (2)	Hazard class (GHS chapter) (3)	Hazard category (4)
H301 + H311	Toxic if swallowed or in contact with skin	Acute toxicity, oral (chapter 3.1) and acute toxicity, dermal (chapter 3.1)	3
H301 + H331	Toxic if swallowed or if inhaled	Acute toxicity, oral (chapter 3.1) and acute toxicity, inhalation (chapter 3.1)	3
H311 + H331	Toxic in contact with skin or if inhaled	Acute toxicity, dermal (chapter 3.1) and acute toxicity, inhalation (chapter 3.1)	3
H301 + H311 + H331	Toxic if swallowed, in contact with skin or if inhaled	Acute toxicity, oral (chapter 3.1), acute toxicity, dermal (chapter 3.1) and acute toxicity, inhalation (chapter 3.1)	3
H302 + H312	Harmful if swallowed or in contact with skin	Acute toxicity, oral (chapter 3.1) and acute toxicity, dermal (chapter 3.1)	4
H302 + H332	Harmful if swallowed or if inhaled	Acute toxicity, oral (chapter 3.1) and acute toxicity, inhalation (chapter 3.1)	4
H312 + H332	Harmful in contact with skin or if inhaled	Acute toxicity, dermal (chapter 3.1) and acute toxicity, inhalation (chapter 3.1)	4
H302 + H312 + H332	Harmful if swallowed, in contact with skin or if inhaled	Acute toxicity, oral (chapter 3.1), acute toxicity, dermal (chapter 3.1) and acute toxicity, inhalation (chapter 3.1)	4
H303 + H313	May be harmful if swallowed or in contact with skin	Acute toxicity, oral (chapter 3.1) and acute toxicity, dermal (chapter 3.1)	5
H303 + H333	May be harmful if swallowed or if inhaled	Acute toxicity, oral (chapter 3.1) and acute toxicity, inhalation (chapter 3.1)	5
H313 + H333	May be harmful in contact with skin or if inhaled	Acute toxicity, dermal (chapter 3.1) and acute toxicity, inhalation (chapter 3.1)	5
H303 + H313 + H333	May be harmful if swallowed, in contact with skin or if inhaled	Acute toxicity, oral (chapter 3.1), acute toxicity, dermal (chapter 3.1) and acute toxicity, inhalation (chapter 3.1)	5
H315 + H319	Causes skin irritation and serious eye irritation [a]	Skin corrosion/irritation (chapter 3.2) and serious eye damage/eye irritation (chapter 3.3)	2 (skin) + 2/2A (eye)
H315 + H320	Causes skin and eye irritation [a]	Skin corrosion/irritation (chapter 3.2) and serious eye damage/eye irritation (chapter 3.3)	2 (skin) + 2B (eye)

[a] *Competent authorities may select the applicable hazard statement(s) depending on the serious eye damage/eye irritation hazard categories implemented in their jurisdiction (2/2A or 2A/2B).*

コード (1)	健康有害性 危険有害性情報 (2)	危険有害性クラス（GHSの章）(3)	危険有害性区分 (4)
H301 + H311	飲み込んだ場合や皮膚に接触した場合は有毒	急性毒性（経口）（3.1章）および急性毒性（経皮）（3.1章）	3
H301 + H331	飲み込んだ場合や吸入した場合は有毒	急性毒性（経口）（3.1章）および急性毒性（吸入）（3.1章）	3
H311 + H331	皮膚に接触した場合や吸入した場合は有毒	急性毒性（経皮）（3.1章）および急性毒性（吸入）（3.1章）	3
H301 + H311 + H331	飲み込んだ場合や皮膚に接触した場合や吸入した場合は有毒	急性毒性（経口）（3.1章）および急性毒性（経皮）（3.1章）および急性毒性（吸入）（3.1章）	3
H302 + H312	飲み込んだ場合や皮膚に接触した場合は有害	急性毒性（経口）（3.1章）および急性毒性（経皮）（3.1章）	4
H302 + H332	飲み込んだ場合や吸入した場合は有害	急性毒性（経口）（3.1章）および急性毒性（吸入）（3.1章）	4
H312 + H332	皮膚に接触した場合や吸入した場合は有害	急性毒性（経皮）（3.1章）および急性毒性（吸入）（3.1章）	4
H302 + H312 + H332	飲み込んだ場合や皮膚に接触した場合や吸入した場合は有害	急性毒性（経口）（3.1章）および急性毒性（経皮）（3.1章）および急性毒性（吸入）（3.1章）	4
H303 + H313	飲み込んだ場合や皮膚に接触した場合は有害のおそれ	急性毒性（経口）（3.1章）および急性毒性（経皮）（3.1章）	5
H303 + H333	飲み込んだ場合や吸入した場合は有害のおそれ	急性毒性（経口）（3.1章）および急性毒性（吸入）（3.1章）	5
H313 + H333	皮膚に接触した場合や吸入した場合は有害のおそれ	急性毒性（経皮）（3.1章）および急性毒性（吸入）（3.1章）	5
H303 + H313 + H333	飲み込んだ場合や皮膚に接触した場合や吸入した場合は有害のおそれ	急性毒性（経口）（3.1章）および急性毒性（経皮）（3.1章）および急性毒性（吸入）（3.1章）	5
H315 + H319	皮膚刺激および強い眼刺激 [a]	皮膚腐食性/刺激性（3.2章）および眼に対する重篤な損傷性/眼刺激性（3.3章）	2（皮膚）+ 2/2A（眼）
H315 + H320	皮膚および眼刺激 [a]	皮膚腐食性/刺激性（3.2章）および眼に対する重篤な損傷性/眼刺激性（3.3章）	2（皮膚）+ 2B（眼）

[a] *所管官庁は、管轄区域で実施される眼に対する重篤な損傷／眼刺激性の有害性区分に応じて、該当する危険有害性情報を選択することができる（2/2A 又は 2A/2B）。*

Table A3.1.3: Hazard statement codes for environmental hazards

Code (1)	Environmental hazard statements (2)	Hazard class (GHS chapter) (3)	Hazard category (4)
H400	**Very toxic to aquatic life**	Hazardous to the aquatic environment, acute hazard (chapter 4.1)	1
H401	**Toxic to aquatic life**	Hazardous to the aquatic environment, acute hazard (chapter 4.1)	2
H402	**Harmful to aquatic life**	Hazardous to the aquatic environment, acute hazard (chapter 4.1)	3
H410	**Very toxic to aquatic life with long lasting effects**	Hazardous to the aquatic environment, long-term hazard (chapter 4.1)	1
H411	**Toxic to aquatic life with long lasting effects**	Hazardous to the aquatic environment, long-term hazard (chapter 4.1)	2
H412	**Harmful to aquatic life with long lasting effects**	Hazardous to the aquatic environment, long-term hazard (chapter 4.1)	3
H413	**May cause long lasting harmful effects to aquatic life**	Hazardous to the aquatic environment, long-term hazard (chapter 4.1)	4
H420	**Harms public health and the environment by destroying ozone in the upper atmosphere**	Hazardous to the ozone layer (chapter 4.2)	1

表 A3.1.3：環境有害性の危険有害性情報コード

コード (1)	環境有害性　危険有害性情報 (2)	危険有害性クラス（GHSの章）(3)	危険有害性区分 (4)
H400	水生生物に非常に強い毒性	水生環境有害性　短期（急性）（4.1章）	1
H401	水生生物に毒性	水生環境有害性　短期（急性）（4.1章）	2
H402	水生生物に有害	水生環境有害性　短期（急性）（4.1章）	3
H410	長期継続的影響によって水生生物に非常に強い毒性	水生環境有害性　長期（慢性）（4.1章）	1
H411	長期継続的影響によって水生生物に毒性	水生環境有害性　長期（慢性）（4.1章）	2
H412	長期継続的影響によって水生生物に有害	水生環境有害性　長期（慢性）（4.1章）	3
H413	長期継続的影響によって水生生物に有害のおそれ	水生環境有害性　長期（慢性）（4.1章）	4
H420	オゾン層を破壊し、健康および環境に有害	オゾン層への有害性（4.2章）	1

ANNEX 3

Section 2

CODIFICATION AND USE OF PRECAUTIONARY STATEMENTS

A3.2.1 Introduction

A3.2.1.1 A *precautionary statement* is a phrase (and/or pictogram) which describes recommended measures that should be taken to minimize or prevent adverse effects resulting from exposures to a hazardous product, or its improper storage or handling of a hazardous product (see 1.4.10.5.2 (c)).

A3.2.1.2 For the purposes of the GHS, there are five types of precautionary statements: **general, prevention, response** (in case of accidental spillage or exposure, emergency response and first-aid)**, storage** and **disposal**.

A3.2.1.3 This section provides guidance on the selection and use of precautionary statements for each GHS hazard class and category. It will be subject to further refinement and development over time though the overall approach set out below will remain.

A3.2.1.4 Precautionary statements should, as an important part of hazard communication, appear on GHS labels, generally along with the GHS hazard communication elements (pictograms, signal words and hazard statements). Additional supplemental information, such as directions for use, may also be provided at the discretion of the manufacturer/supplier and/or competent authority (see chapter 1.2 and chapter 1.4, paragraph 1.4.6.3). For some specific chemicals, supplementary first aid, treatment measures or specific antidotes or cleansing materials may be required. Poisons Centres and/or medical practitioners or specialist advice should be sought in such situations and included on labels.

A3.2.1.5 The starting point for assigning precautionary statements is the hazard classification of the substance or mixture. The system of classifying hazards in the GHS is based on the intrinsic properties of the substances or mixtures involved (see 1.3.2.2.1). In some systems, however, labelling may not be required for chronic hazards on consumer product if information shows that the respective risks can be excluded under conditions of normal handling, normal use or foreseeable misuse (see annex 5). If certain hazard statements are not required then the corresponding precautionary statements are also not necessary (see A5.1.1).

A3.2.1.6 The guidance for assigning the statements in this section has been developed to provide the essential minimum precautionary statements that are linked to relevant GHS hazard classification criteria and type of hazard.

A3.2.1.7 Precautionary statements from existing classification systems were used to the maximum extent as the basis for the development of this section. These existing systems have included the IPCS International Chemical Safety Card (ICSC) Compilers Guide, the American National Standards (ANSI Z129.1), the EU classification and labelling directives, the Emergency Response Guidebook (ERG 2004), and the Pesticide Label Review Manual of the Environmental Protection Agency (EPA) of the United States of America.

A3.2.1.8 The understanding and following of precautionary label information, specific safety guidelines, and the safety data sheet for each substance or mixture before use are part of occupational health and safety procedures. Consistent use of precautionary statements will reinforce safe handling procedures and will enable the key concepts and approaches to be emphasized in workplace training and education activities.

A3.2.1.9 In order to correctly implement precautionary measures concerning prevention, response, storage and disposal, it is also necessary to have information on the composition of the substance or mixture at hand, so that information shown on the container, label and safety data sheet can be taken into account when asking for further specialist advice.

A3.2.1.10 To protect people with different reading abilities, it might be useful to include both precautionary pictograms and precautionary statements in order to convey information in more than one way (see 1.4.4.1 (a)). It should be noted, however, that the protective effect of pictograms is limited and the examples in this annex do not cover all precautionary aspects to be addressed. While pictograms can be useful, they can be misinterpreted and are not a substitute for training.

附属書 3

第 2 節

注意書きのコードおよび使用

A3.2.1 序文

A3.2.1.1 *注意書き*は、危険有害性をもつ製品へのばく露、または、その不適切な保管や取扱いから生じる被害を防止し、または最小にするために取るべき推奨措置について記述した文言（および/または絵表示）である（1.4.10.5.2 (c) 参照）。

A3.2.1.2 GHS には 5 種類の注意書きがある、すなわち**一般**、**安全対策**、**応急措置**（事故的な漏出やばく露、応急措置および救急処置）、**保管**そして**廃棄**である。

A3.2.1.3 この節は、各 GHS 危険有害性クラスと区分に対する注意書きの選択と使用に関する手引きを提供する。以下に記載された全体的なアプローチは残るであろうが、この節は時の経過とともに更なる改良と開発を必要とするであろう。

A3.2.1.4 注意書きは、危険有害性情報伝達の重要な一部として GHS の危険有害性伝達要素（絵表示、注意喚起語および危険有害性情報）とともに、GHS ラベルに記載されるべきである。使用の指示のような追加の補助情報は、製造者/供給者、または所管官庁の指示により提供されてもよい（第 1.2 章 1.4.6.3 参照）。いくつかの特殊な化学品については、補足の応急処置、処置方法もしくは特別な解毒剤または洗浄剤が要求されるであろう。そのような状況では毒物センターもしくは臨床医または専門家のアドバイスが求められるべきであり、かつそれらはラベルに含まれるべきである。

A3.2.1.5 注意書きを割り当てる出発点は物質または混合物の危険有害性の分類である。GHS の中での危険有害性を分類するシステムは含まれる物質または混合物の固有の特性に基づく（1.3.2.2.1 参照）。しかしながら、通常の取扱い、通常使用または予見できる誤用の条件下では各々のリスクが除外されるという情報が提示されるのであれば、消費者製品に関する慢性の有害性に対してラベルが要求されないというシステムもある（附属書 5 参照）。ある種の危険有害性情報が要求されていなければ、対応する注意書きもまた不要である（A5.1.1 参照）。

A3.2.1.6 この節における注意書きを割り当てるための手引きは、関連する GHS 危険有害性分類基準および危険有害性の種類に関連している必須で最低限の注意書きを提供するために開発されてきた。

A3.2.1.7 既存の分類システムからの注意書きは、この節の開発のための基礎として最大限使用された。これらの既存のシステムには、IPCS 国際化学品安全性カード (ICSC) コンパイラーズガイド、米国規格 (ANSI Z129.1)、欧州分類と表示指令、緊急対応ガイドブック (ERG 2004) および米国環境保護庁表示検査マニュアルがある。

A3.2.1.8 使用前に、ラベルの注意書き情報、特定の安全手引きおよび各物質または混合物の安全データシートを理解しそれにしたがうことは、労働安全衛生手順の一部である。注意書きの継続的な使用は、安全取り扱い手順を強化し、職場での訓練や教育活動においてカギとなる概念や方法を明確にすることにもなる。

A3.2.1.9 安全対策、応急措置、保管および廃棄に関する注意処置を正確に実行するために、物質または混合物の組成に関する情報を手元に置くことも必要である。そうすることによって、更なる専門家の判断を必要とする際に、容器上に表示された情報、ラベルや安全データシートの情報を生かすことができる。

A3.2.1.10 さまざまな読解力の人々を保護するために、情報を一種類以上で伝えるように注意絵表示と注意書きの両方を含むことは有用であろう(1.4.4.1(a)参照)。しかしながら、絵表示の防護効果は限定されており、附属書 3 の例は記述されるすべての予防的観点をカバーしているわけではないことに注意するべきである。絵表示は有用であるが、誤解されることがあり、訓練に代替するものではない。

A3.2.2 **Codification of precautionary statements**

A3.2.2.1 Precautionary statements are assigned a unique alphanumerical code which consists of one letter and three numbers as follows:

(a) a letter "P" (for "precautionary statement")

(b) one number designating the type of precautionary statement as follows:

- "1" for general precautionary statements;
- "2" for prevention precautionary statements;
- "3" for response precautionary statements;
- "4" for storage precautionary statements;
- "5" for disposal precautionary statements;

(c) two numbers (corresponding to the sequential numbering of precautionary statements)

A3.2.2.2 The precautionary statement codes are intended to be used for reference purposes. They are not part of the precautionary statement text and should not be used to replace it.

A3.2.2.3 The codes to be used for designating precautionary statements are listed, in numerical order, in table A3.2.1 for general precautionary statements, table A3.2.2 for prevention precautionary statements, table A3.2.3 for response precautionary statements, table A3.2.4 for storage precautionary statements and table A3.2.5 for disposal precautionary statements.

A3.2.2.4 Where square brackets […] appear around a precautionary statement code, this indicates the precautionary statement is not appropriate in every case and should be used only in certain circumstances. In these cases, conditions for use explaining when the text should be used are given in column (5) of the tables.

A3.2.3 **Structure of the precautionary statement tables**

A3.2.3.1 Each precautionary statement table is divided into 5 columns containing the following information:

Column (1)   The precautionary statement code;

Column (2)   The precautionary statement text;

Column (3)   The hazard class and the route of exposure, where relevant, for which the use of a precautionary statement is recommended together with a reference to the chapter of the GHS where information about the hazard class may be found.

Column (4)   The hazard category or categories within a hazard class for which the use of a precautionary statement is applicable.

Column (5)   Where applicable, conditions relating to the use of a precautionary statement;

A3.2.4 **Use of precautionary statements**

A3.2.4.1 Tables A3.2.1 to A3.2.5 show the **core part of the precautionary statements in bold print**. This is the text, except as otherwise specified, that should appear on the label. Derogations from the recommended labelling statements are at the discretion of competent authorities (see A3.2.5).

A3.2.4.2 When a forward slash or diagonal mark "/" appears in a precautionary statement text, it indicates that a choice has to be made between the phrases they separate. In such cases, the manufacturer or supplier can choose, or the competent authorities may prescribe one or more appropriate phrase(s). For example P280 "**Wear protective gloves/protective clothing/eye protection/face protection/hearing protection/…**" could read "**wear eye protection**" or "**wear eye and face protection**".

A3.2.4.3 When three full stops "**…**" appears in a precautionary statement text, they indicate that all applicable conditions are not listed. For example in P241 "**Use explosion-proof [electrical/ventilating/lighting/...] equipment**", the use of "..." indicates that other equipment may need to be specified. Further details of the information to be provided may be found in column (5) of the tables. In such cases the manufacturer or supplier can choose, or the competent authorities may prescribe the other conditions to be specified.

A3.2.2　注意書きのコード化

A3.2.2.1　注意書きには１つの文字と３つの数字からなる英数字コードが、下記のように割り当てられている：
　　(a)　文字「P」（注意書き"precautionary statement"）；
　　(b)　注意書きの文言の種類により１つの番号が割り当てられる：
　　　　－「1」　一般的注意書き；
　　　　－「2」　安全対策の注意書き；
　　　　－「3」　応急措置の注意書き；
　　　　－「4」　保管の注意書き；
　　　　－「5」　廃棄の注意書き；
　　(c)　２つの数字（注意書きに対応した連続した数字）。

A3.2.2.2　注意書きのコードは参照するためのものである。コードは注意書きの文言の一部ではないので、文言の代わりに用いることはできない。

A3.2.2.3　割り当てられた注意書きに使用するコードは、一般的注意書きは表 A3.2.1 に、安全対策の注意書きは表 A3.2.2 に、応急措置の注意書きは表 A3.2.3 に、保管の注意書きは表 A3.2.4 に、廃棄の注意書きは表 A3.2.5 に番号順に記載してある。

A3.2.2.4　注意書きのコードの周囲に四角括弧【…】が表示されている場合、その注意書きが全ての場合において適切とは限らず、特定の条件のときのみ使用されるべきであることを示している。このような場合、文言が使用されるべき条件の説明は表の欄(5)に記載されている。

A3.2.3　注意書き表の構成

A3.2.3.1　それぞれの注意書き表は５つの欄に分割され下記の情報を含む：
　　欄(1)　注意書きコード；
　　欄(2)　注意書き；
　　欄(3)　危険有害性クラスおよび該当する場合のばく露経路、推奨される注意書きとともに危険有害性クラスに関する情報がある GHS 文書の章を記載；
　　欄(4)　当該危険有害性情報が適用可能な危険有害性クラスでの区分；
　　欄(5)　該当する場合、注意書きの使用に関する条件。

A3.2.4　注意書きの使用

A3.2.4.1　表 A3.2.1 から A3.2.5 には**注意書きの核となる部分を太字**で示している。特別の指示がない限り、この文言をラベルに使用すべきである。推奨されるラベル用文言の変更は所管官庁の選択による（A3.2.5 参照）。

A3.2.4.2　注意書きに斜線"/"がある時、これは分離された文言を選択しなければならないことを示す。このような場合、製造者や供給者は選択するか、あるいは所管官庁は１つ以上の文言を規定してもよいであろう。例えば、P280 の「**保護手袋/保護衣/保護眼鏡/保護面/聴覚保護具/…を着用すること**」を「保護眼鏡を着用すること」あるいは「保護眼鏡あるいは保護面を着用すること」としてもよい。

A3.2.4.3　注意書きに３つの句点"…"がある時、これらは全ての適用条件がそろっているわけではないことを示す。例えば、P241「**防爆型の【電気/換気/照明/…】機器を使用すること**」の"…"は他の機器が特定される必要があるかもしれないことを示している。さらなる記載すべき詳細な情報は表の欄(5)にあるかもしれない。このような場合、製造者や供給者は選択するか、あるいは所管官庁は最も適当な文言を規定してもよいであろう。

A3.2.4.4   Where square brackets [...] appear around some text in a precautionary statement, this indicates that the text in square brackets is not appropriate in every case and should be used only in certain circumstances. In these cases, conditions for use explaining when the text should be used are given in column (5) of the tables. For example, P264 states: "**Wash hands [and ...] thoroughly after handling.**". This statement is given with the condition for use: "*- text in square brackets to be used when the manufacturer/supplier or the competent authority specify other parts of the body to be washed after handling.*".The application of the condition for use should be interpreted as follows: If additional information is provided explaining what other part(s) of the body is to be washed following handling, then the text in brackets is to be used followed by the name of the relevant body part(s). However, if other part(s) of the body do not need to be specified, the text in square brackets should not be used, and the precautionary statement should read: "**Wash hands thoroughly after handling.**".

A3.2.4.5   In cases where additional information is required, or information either has to be or may be specified, this is indicated by a relevant entry in column (5) in plain text.

A3.2.4.6   When *text in italics* is used in the tables, this indicates specific conditions applying to the use or allocation of the precautionary statement. This may relate to conditions attaching to either the general use of a precautionary statement or its use for a particular hazard class and/or hazard category. For example, P241 "**Use explosion-proof [electrical/ventilating/lighting/...] equipment**", only applies for flammable solids "*if dust clouds can occur*". *Text in italics* that starts with "*– if*" or "*– specify*" is an explanatory conditional note for the application of the precautionary statements and is not intended to appear on the label.

A3.2.4.7   Where precautionary statements become obsolete, 'deleted' is inserted under the existing code in column 1 of the tables in this section to avoid potential confusion between codes used in different editions of the GHS.

A3.2.5   **Flexibility in the use of precautionary statements**

A3.2.5.1   *Omission of precautionary statements where the advice is not relevant*

A3.2.5.1.1   Subject to any requirements of competent authorities, those responsible for labelling may decide to omit other precautionary statements for a hazard class and category where the information is clearly not appropriate or is adequately addressed by other information on the label, taking into account the nature of the user (e.g. consumer, employers and workers), the quantity supplied, and the intended and foreseeable circumstances of use. Where a decision is made to omit a precautionary statement the manufacturer or supplier of the substance or mixture should be able to demonstrate that the precautionary statement is not appropriate for the intended and foreseeable use, including potential emergency situations.

A3.2.5.2   *Combination or consolidation of precautionary statements*

A3.2.5.2.1   To facilitate translation into the languages of users, precautionary statements have been broken down into individual sentences or parts of sentences in the tables in this section (see tables A3.2.1 to A3.2.5). In a number of instances, the text that appears on a GHS label requires that these be added back together. This is indicated in this annex by codes conjoined with a plus sign "+". For example, P305 + P351 + P338 indicates that the text to appear on the label is "**IF IN EYES: Rinse cautiously with water for several minutes. Remove contact lenses, if present and easy to do. Continue rinsing**". These combined precautionary statements can also be found at the end of each of the precautionary statement tables in this section. Translation of only the single precautionary statements is required, as this will enable the compilation of the combined precautionary statements.

A3.2.5.2.2   Flexibility in the application of individual, combinations or consolidations of precautionary statements is encouraged to save label space and improve readability. Precautionary statements can be combined by using the word "and", additional punctuation, and changing the case of the initial letter of the word at the beginning of a statement. For example, P302 + P335 + P334 "**IF ON SKIN: Brush off loose particles from skin and immerse in cool water [or wrap in wet bandages]**."The tables in this section and the matrix in section 3 of annex 3 include a number of combined precautionary statements. However, these are only examples and those responsible for labelling should further combine and consolidate phrases where this contributes to clarity and comprehensibility of label information.

A3.2.5.2.3   Combination of precautionary statements can also be useful for different types of hazard where the precautionary behaviour is similar. Examples are P370 + P372+ P380 + P373 "**In case of fire: Explosion risk. Evacuate area. DO NOT fight fire when fire reaches explosives**" and P210 + P403 "**Keep away from heat, hot surfaces, sparks, open flames and other ignition sources. No smoking. Store in a well-ventilated place**".

A3.2.4.4　注意書きの文の周囲に四角括弧【…】が表示されている場合、これは括弧の中の文言が全ての場合において適切とは限らず、特定の条件のときのみ使用されるべきであることを示している。このような場合、文言が使用されるべき条件の説明は表の欄（5）に記載されている。例えば、P264 では「取扱い後は手【および…】をよく洗うこと。」と記載されている。この記述は、「－ *製造者/供給者または所管官庁が、取扱い後に洗浄する身体の他の部分を指定した場合には括弧内の記述を用いる。*」という使用条件とともに示されている。その適用条件は次のように解釈される：取扱い後に洗浄すべき身体の他の部分を説明する追加情報が提供されている場合は、括弧内の文言に続いて該当する身体の部分の名称を使用する。ただし、身体のその他の部分を特定する必要がない場合は、括弧内の文言は使用せず、注意書きの記述は次のようにする：「取扱い後は手をよく洗うこと」。

A3.2.4.5　追加情報が必要または情報が特定されなければならないまたはされたほうがよい場合、これは関連項目として欄（5）に文言で示されている。

A3.2.4.6　表に*斜体字*が使用されている時は、注意書きの使用や割り当てに特別な条件が必要であることを示している。これは注意書きの一般的な使用や特別な危険有害性クラスおよび/または区分への使用条件に関わっている。例えば、P241「防爆型の【電気/換気/照明/…】機器を使用すること」は、可燃性固体では「*粉じん雲が発生する場合*」のみ適用される。「*―の場合（時）*」または「*―明示する*」の*斜体文字*は注意書きの適用に関する条件の注記であり、ラベルに記載されるものではない。

A3.2.4.7　不適切になった注意書きのところでは、GHS の他の版で使用されているコードとの混乱を防ぐために、本節の表の欄（1）における既存コードの下に「削除」が挿入されている。

A3.2.5　注意書きの使用における柔軟性

A3.2.5.1　*助言が適切でない場合の注意書きの省略*

A3.2.5.1.1　所管官庁の要求を満足したうえで、その情報が明らかに適切ではないあるいはラベル上の他の情報で十分に伝えられている場合には、使用者（例えば消費者、事業者及び労働者）の特性、供給量、および意図された予見可能な使用環境を考慮して、表示に責任のある者は、ある危険有害性クラスおよび区分に関する他の注意書きを省略してもよい。注意書きを省略すると決定した場合には、物質または混合物の製造者または供給者は、その注意書きが、潜在的な緊急事態も含めて、意図された予見可能な使用に対して適切でないことを示すことができなければならない。

A3.2.5.2　*注意書きの結合または統合*

A3.2.5.2.1　使用者の言語への翻訳を容易にするために、この節の表における注意書きは個々の文言あるいは部分的な文言に細分化されている（表 A3.2.1 から表 A3.2.5 参照）。多くの例で見られるように GHS ラベルで必要とされる文章はこれらを結合したものである。これは本附属書で加算マーク「+」用いたコードで示されている。例えば、P305+P351+P338 はラベルでは「眼に入った場合：水で数分注意深く洗うこと。次にコンタクトレンズを着用していて容易に外せる場合は外すこと。その後も洗浄を続けること。」となる。これらの結合注意書きは本節の注意書き表の最後に記載されている。まず、個別の注意書きの翻訳が必要であり、これが結合注意書きを可能にする。

A3.2.5.2.2　個々の、結合したあるいは統合した注意書きの適用に関する柔軟性は、ラベルのスペースを節約しさらに読みやすさの改善を促進する。注意書きは、「および」という単語を使用したり、句読点を追加したり、文頭の単語の最初の文字の大文字と小文字を変更したりすることで組み合わせることができる。例えば、P302 + P335 + P334 「皮膚に付着した場合：固着していない粒子を皮膚から払いのけ、冷たい水に浸すこと【または湿った包帯で覆うこと】。」本節の表および附属書 3、第 3 節のマトリクスには多くの結合された注意書きがある。しかしこれらは単なる例であり、そうすることがラベル情報を明快にわかりやすくするのであれば、表示に責任のある者はさらに文言の結合および統合をするべきである。

A3.2.5.2.3　注意書きの結合は、予防行動が同じであれば、別の種類の危険有害性に対しても有用である。例えば、P370+P372+P380+P373「火災の場合：爆発する危険性あり。区域より避難させること。炎が爆発物に届いたら消火活動をしないこと。」および P210+P403「熱、高温のもの、火花、裸火および他の着火源から遠ざけること。禁煙。換気の良い場所で保管すること。」があげられる。

A3.2.5.3 *Variations of text not affecting the safety message*

A3.2.5.3.1 Subject to any requirements of competent authorities, the precautionary statements that appear on labels or in safety data sheets may incorporate minor textual variations from those set out in the GHS where these variations assist in communicating safety information and the safety advice is not diluted or compromised. These may include spelling variations, synonyms or other equivalent terms appropriate to the region where the product is supplied and used.

A3.2.5.3.2 In all cases, clear plain language is essential to convey information on precautionary behaviour. Furthermore, to ensure clarity of safety messages any variations should be applied consistently on the label and in the safety data sheet.

A3.2.5.4 *Application of precautionary statements concerning medical response*

A3.2.5.4.1 Where a substance or mixture is classified for a number of health hazards, this may trigger multiple precautionary statements relating to medical response. In general, the following principles should be applied:

(a) Always combine medical response statements with at least one route of exposure or symptom ("IF" statement). However, this does not apply to P319 "**Get medical help if you feel unwell**" for specific target organ toxicity repeated exposure, or to P317 "**Get medical help**" for gases under pressure (refrigerated liquefied gas) which are not combined with a separate "IF" statement. Relevant "IF" statements describing symptoms (e.g. P332, P333, P337, P342) should be included in full;

(b) Where the same medical response statement is triggered for different routes of exposure, then the exposure routes should be combined. If the same response statement is triggered with three or more routes of exposure then P308, "**IF exposed or concerned:**", may be used instead. If a route of exposure is triggered multiple times it should only be included once;

(c) Where different medical response statements are triggered for the same route of exposure, then P316 "**Get emergency medical help immediately**" should have priority over P317 "**Get medical help**"; and P317 should have priority over P319 "**Get medical help if you feel unwell**". P318, "**If exposed or concerned get medical advice**", should always appear if triggered. To improve clarity and readability when more than one medical statement appears, supplemental text such as 'additionally' or 'also' should be inserted;

(d) Where different medical response statements are triggered for different routes of exposure, all the relevant precautionary statements for medical response should appear.

For example:

1. Where P301 and P304 "**IF SWALLOWED:**", "**IF INHALED:**" and P302 "**IF ON SKIN:**" (for acute toxicity oral 2, inhalation 1 and skin corrosion respectively) are triggered with P316 "**Get emergency medical help immediately**", then P301 + P304 + P302 + P316, "**IF SWALLOWED, INHALED OR ON SKIN: Get emergency medical help immediately**" should appear. Alternatively, "**IF SWALLOWED, INHALED OR ON SKIN:**" may be replaced by P308, "**IF exposed or concerned:**".

2. Where P301 "**IF SWALLOWED:**" (for both aspiration hazard and for acute toxicity oral 4) is triggered with P316 "**Get emergency medical help immediately**" and P317 "**Get medical help**" respectively, then P301 + P316 "**IF SWALLOWED: Get emergency medical help immediately**" should appear. Where the classification also includes germ cell mutagenicity, carcinogenicity or reproductive toxicity triggering P318 "**If exposed or concerned, get medical advice**", then "**IF SWALLOWED: Get emergency medical help immediately. Additionally, if exposed or concerned, get medical advice**" should appear.

3. Where P304, P301, P302 and P333 "**IF INHALED:**", "**IF SWALLOWED:**", "**IF ON SKIN:**" and "**If skin irritation or rash occurs:**" (for acute toxicity inhalation 2, acute toxicity oral 4 and skin sensitization respectively) are triggered with P316 "**Get emergency medical help immediately**", and P317 "**Get medical help**", then "**IF INHALED: Get emergency medical help immediately**" and "**IF SWALLOWED or if skin irritation or rash occurs: Get medical help**" should appear.

4. Where P302 and P305 "**IF ON SKIN:**" and "**IF IN EYES:**" (for acute toxicity dermal 2 and eye irritation respectively) are triggered with P316 "**Get emergency medical help immediately**", P317 "**Get medical help**" and P319 "**Get medical help if you feel unwell**" (for specific target

A3.2.5.3　安全警告に影響しない文節の多様性

A3.2.5.3.1　所管官庁の要求を満足したうえで、ラベルまたは安全データシートに使われる注意書きは、それらの変化が安全情報の伝達を助けまた安全助言が弱まったり損なわれたりしないかぎり、GHSにあるそれらと多少異なってもよい。これらには綴りの変化、同義語またはその製品が供給され使用される地域において適切な他の同等の言葉を含むであろう。

A3.2.5.3.2　すべての場合において、予防行動に関する情報を伝達するためには明確で平易な言葉が必須である。さらに安全情報を明快にするために、いかなる多様性もラベルおよび安全データシートに一貫して適用されるべきである。

A3.2.5.4　医療対応に関する注意書きの適用

A3.2.5.4.1　物質または混合物が多くの健康有害性に関して分類された場合、医療対応に関連した多くの注意書きが導出されるであろう。一般に以下のような原則が適用されるべきである：

(a) 医療対応の文言は、常に少なくとも一つのばく露経路または症状（「－の場合（時）」）と結合される。しかし「－の場合（時）」と結合していない、反復ばく露による標的臓器毒性に関するP319「気分が悪い時は、医療処置を受けること」または高圧ガス（深冷液化ガス）に関するP317「医療処置を受けること」にはこれは適用しない。直接関連した症状を記述した「－の場合（時）」（例えば、P332, P333, P337, P342）は略さずに記載されなければならない；

(b) 同じ医療対応の文言が異なるばく露経路で導出された場合には、ばく露経路は結合されるべきである。もし同じ対応の文言が三つ以上のばく露経路に用いられている場合には、代わりにP308「ばく露またはその懸念がある場合：」を使用してもよい。もし一つのばく露経路が何回も出てきた場合には、一回だけ記載されるべきである；

(c) 異なる医療対応の文言が同じばく露経路で導出された場合には、P316「すぐに救急の医療処置を受けること」がP317「医療処置を受けること」よりも優先されるべきであり；またP317「医療処置を受けること」はP319「気分が悪い時は、医療処置を受けること」よりも優先されるべきである。該当する場合にはP318「ばく露またはその懸念がある場合は、医学的助言を求めること」は常に示すべきである。二つ以上の医療対応が示された場合には、明確さおよび読み易さを改善するために「さらに」または「もまた」などの追加的な文言が挿入されるべきである；

(d) 異なる医療対応の文言が異なるばく露経路で導出された場合には、すべての医療対応に関連する注意書きの文言が示されるべきである。

例：

1. P301およびP304の「飲み込んだ場合：」、「吸入した場合：」およびP302「皮膚に付着した場合：」（それぞれ急性毒性経口2、吸入1および皮膚腐食性）がP316「すぐに救急の医療処置を受けること」と共に導出された場合、P301+P304+P302+P316「飲み込んだ場合、吸入した場合または皮膚に付着した場合：すぐに救急の医療処置を受けること」とするべきである。また「飲み込んだ場合、吸入した場合または皮膚に付着した場合：」はP308「ばく露またはその懸念がある場合：」に変えてもよい。

2. P301「飲み込んだ場合：」（誤えん有害性および急性毒性経口4）が、それぞれP316「すぐに救急の医療処置を受けること」およびP317「医療処置を受けること」と共に導出された場合、P301+P316「飲み込んだ場合：すぐに救急の医療処置を受けること」とするべきである。分類がさらにP318「ばく露またはその懸念がある場合は、医学的助言を求めること」を導出する生殖細胞変異原性、発がん性または生殖毒性を含む場合、「飲み込んだ場合：すぐに救急の医療処置を受けること。さらにばく露またはその懸念がある場合は、医学的助言を求めること」とするべきである。

3. P304、P301、P302およびP333の「吸入した場合：」、「飲み込んだ場合：」、「皮膚に付着した場合：」およびP333「皮膚刺激または発疹が生じた場合：」（それぞれ急性毒性吸入2、急性毒性経口4および皮膚感作性）が、P316「すぐに救急の医療処置を受けること」およびP317「医療処置を受けること」と共に導出された場合、「吸入した場合：すぐに救急の医療処置を受けること」および「飲み込んだ場合または皮膚刺激または発疹が生じた場合：医療処置を受けること」とするべきである。

4. P302およびP305の「皮膚に付着した場合：」および「眼に入った場合：」（それぞれ急性毒性経皮2および眼刺激）が、P316「すぐに救急の医療処置を受けること」、P317「医療処置を受けること」およびP319「気分が悪い時は、医療処置を受けること」（標的臓器毒性反復ばく露）

organ toxicity repeated exposure), then P302 + P316 "**IF ON SKIN: Get emergency medical help immediately**", P337 + P317 "**If eye irritation persists: Get medical help**" and P319 "**Get medical help if you feel unwell**" should all appear.

A3.2.5.4.2　　Precautionary statements setting out other relevant responses such as P330 to P336, P338, P340, P351 to P354 and P360 to P364 should also appear in full on the label as appropriate.

A3.2.6　　General precautionary measures

A3.2.6.1　　General precautionary measures should be adopted for all substances and mixtures which are classified as hazardous to human health or the environment. To this end, the needs of, and the information sources available to two groups of users should be taken into account: consumers and employers/workers.

A3.2.6.2　　In addition to the appropriate precautionary statements in the matrix, taking into account the guidance in this section, the general precautionary statements laid out in table A3.2.1 are appropriate for consumers and should also appear on GHS labels.

と共に導出された場合、P302+P316「皮膚に付着した場合：すぐに救急の医療処置を受けること」、P337+P317「眼の刺激が続く場合：医療処置を受けること」およびP319「気分が悪い時は、医療処置を受けること」のすべてを示すべきである。

A3.2.5.4.2　またP330からP336、P338、P340、P351からP354およびP360からP364のような、他の関連する行動に結び付く注意書きは、略さずに適切にラベルに示すべきである。

A3.2.6　一般的注意書き

A3.2.6.1　一般的注意書きは、ヒトの健康または環境に有害と分類されるすべての物質と混合物に適用されるべきである。この目的を達成するために、2つのグループに適用される注意書きの必要性と情報源が考慮されなければならない：消費者および雇用者/労働者。

A3.2.6.2　マトリクスにおける適切な注意書きに加えて、当節の手引きも考慮すると、表A3.2.1にある一般的注意書きは消費者に対して適切であり、GHSラベルにも使用されるべきである。

Table A3.2.1: General precautionary statements

Code (1)	General precautionary statements (2)	Hazard class (3)	Hazard category (4)	Conditions for use (5)
P101	**If medical advice is needed, have product container or label at hand.**	as appropriate		Consumer products
P102	**Keep out of reach of children.**	as appropriate		Consumer products
P103	**Read carefully and follow all instructions.**	as appropriate		Consumer products – *omit where P203 is used.*

Table A3.2.2: Prevention precautionary statements

Code (1)	Prevention precautionary statements (2)	Hazard class (3)	Hazard category (4)	Conditions for use (5)
P201	*[Deleted]*			
P202	*[Deleted]*			
P203	**Obtain, read and follow all safety instructions before use.**	Explosives (chapter 2.1)	1A, 1, 2A, 2B	
		Flammable gases (chapter 2.2)	Chemically unstable gas A	
			Chemically unstable gas B	
		Germ cell mutagenicity (chapter 3.5)	1, 1A, 1B, 2	
		Carcinogenicity (chapter 3.6)	1, 1A, 1B, 2	
		Reproductive toxicity (chapter 3.7)	1, 1A, 1B, 2	
		Reproductive toxicity, effects on or via lactation (chapter 3.7)	Additional category	
P210	**Keep away from heat, hot surfaces, sparks, open flames and other ignition sources. No smoking.**	Explosives (chapter 2.1)	1A, 1, 2A, 2B, 2C	
		Flammable gases (chapter 2.2)	Flammable gas	
			Pyrophoric gas	
			Chemically unstable gas A	
			Chemically unstable gas B	
		Aerosols (chapter 2.3)	1B, 2	
		Chemicals under pressure (chapter 2.3)	1, 2, 3	
		Flammable liquids (chapter 2.6)	1, 2, 3, 4	
		Flammable solids (chapter 2.7)	1, 2	
		Self-reactive substances and mixtures (chapter 2.8)	Types A, B, C, D, E, F	

- 306 -

表 A3.2.1 一般的注意書き

コード (1)	一般的注意書き (2)	危険有害性クラス (3)	危険有害性区分 (4)	使用の条件 (5)
P101	医学的助言が必要な時には、製品容器やラベルを手許に持っていくこと。	適宜		消費者製品
P102	子供の手の届かないところに置くこと。	適宜		消費者製品
P103	全ての指示をよく読み、従うこと。	適宜		消費者製品 —P203を使用するときは省略

表 A3.2.2 安全対策注意書き

コード (1)	安全対策注意書き (2)	危険有害性クラス (3)	危険有害性区分 (4)	使用の条件 (5)
P201	(削除)			
P202	(削除)			
P203	使用前にすべての安全説明書を入手し、読み、従うこと。	爆発物 (2.1章)	1, 2A, 2B	
		可燃性ガス (2.2章)	1A　化学的に不安定なガス A	
			化学的に不安定なガス B	
		生殖細胞変異原性 (3.5章)	1, 1A, 1B, 2	
		発がん性 (3.6章)	1, 1A, 1B, 2	
		生殖毒性 (3.7章)	1, 1A, 1B, 2	
		生殖毒性、授乳に対するまたは授乳を介した影響 (3.7章)	追加区分	
P210	熱、高温のもの、火花、裸火およびその他の着火源から遠ざけること。禁煙。	爆発物 (2.1章)	1, 2A, 2B, 2C	
		可燃性ガス (2.2章)	可燃性ガス　　　　　　　1A	
			自然発火性ガス	
			化学的に不安定なガス A	
			化学的に不安定なガス B	
		エアゾール (2.3章)	1B, 2	
		加圧下化学品 (2.3章)	1, 2, 3	
		引火性液体 (2.6章)	1, 2, 3	
		可燃性固体 (2.7章)	1, 2, 3, 4	
		自己反応性物質および混合物 (2.8章)	1, 2	
			タイプ A, B, C, D, E, F	

- 306 -

Table A3.2.2: Prevention precautionary statements *(cont'd)*

Code (1)	Prevention precautionary statements (2)	Hazard class (3)	Hazard category (4)	Conditions for use (5)
P210 *(cont'd)*	**Keep away from heat, hot surfaces, sparks, open flames and other ignition sources. No smoking.**	Pyrophoric liquids (chapter 2.9)	1	
		Pyrophoric solids (chapter 2.10)	1	
		Oxidizing liquids (chapter 2.13)	1, 2, 3	
		Oxidizing solids (chapter 2.14)	1, 2, 3	
		Organic peroxides (chapter 2.15)	Types A, B, C, D, E, F	
		Desensitized explosives (chapter 2.17)	1, 2, 3, 4	
P211	**Do not spray on an open flame or other ignition source.**	Aerosols (chapter 2.3)	1, 2	
		Chemicals under pressure (chapter 2.3)	1, 2	
P212	**Avoid heating under confinement or reduction of the desensitizing agent**	Desensitized explosives (chapter 2.17)	1, 2, 3, 4	
P220	**Keep away from clothing and other combustible materials.**	Oxidizing gases (chapter 2.4)	1	
		Oxidizing liquids (chapter 2.13)	1, 2, 3	
		Oxidizing solids (chapter 2.14)	1, 2, 3	
P222	**Do not allow contact with air.**	Flammable gases (chapter 2.2)	1A, Pyrophoric gas	*– if emphasis of the hazard statement is deemed necessary*.
		Pyrophoric liquids (chapter 2.9)	1	
		Pyrophoric solids (chapter 2.10)	1	
P223	**Do not allow contact with water.**	Substances and mixtures which, in contact with water, emit flammable gases (chapter 2.12)	1, 2	*– if emphasis of the hazard statement is deemed necessary*.
P230	**Keep diluted with …**	Explosives (chapter 2.1)	1, 2A, 2B, 2C	*– for explosive substances and mixtures that are diluted with solids or liquids, or wetted with, dissolved or suspended in water or other liquids to reduce their explosive properties.* …Manufacturer/supplier or competent authority to specify appropriate material.
		Desensitized explosives (chapter 2.17)	1, 2, 3, 4	…Manufacturer/supplier or the competent authority to specify appropriate material.

表 A3.2.2 安全対策注意書き (続き)

コード (1)	安全対策注意書き (2)	危険有害性クラス (3)	危険有害性区分 (4)	使用の条件 (5)
P210 (続き)	熱、高温のもの、火花、裸火および他の着火源から遠ざけること。禁煙。	自然発火性液体	1	
		自然発火性固体 (2.10 章)	1	
		酸化性液体 (2.13 章)	1, 2, 3	
		酸化性固体 (2.14 章)	1, 2, 3	
		有機過酸化物 (2.15 章)	タイプ A, B, C, D, E, F	
		鈍性化爆発物 (2.17 章)	1, 2, 3, 4	
P211	裸火または他の着火源に噴霧しないこと。	エアゾール (2.3 章)	1, 2	
P212	密閉状態での加熱または鈍化剤の減少を避ける	加圧下化学品 (2.3 章)	1, 2	
		鈍性化爆発物 (2.17 章)	1, 2, 3, 4	
P220	衣類および可燃物から遠ざけること。	酸化性ガス (2.4 章)	1	
		酸化性液体 (2.13 章)	1, 2, 3	
		酸化性固体 (2.14 章)	1, 2, 3	
P222	空気に接触させないこと。	可燃性ガス (2.2 章)	1A, 自然発火性ガス	
		自然発火性液体 (2.9 章)	1	
		自然発火性固体 (2.10 章)	1	
P223	水と接触させないこと。	水反応可燃性物質および混合物 (2.12 章)	1, 2	
P230	…にて希釈しておくこと。	爆発物 (2.1 章)	1,2A,2B,2C	一爆発物の性質を抑制するために、固体または液体で希釈された、または水または他の液体で湿らされた、または懸濁された物質及び混合物に対して …製造者/供給者または所管官庁が指定する適当な物質
		鈍性化爆発物 (2.17 章)	1,2,3,4	一危険有害性情報の強調が必要と考えられる場合 …製造者/供給者または所管官庁が指定する適当な物質

- 307 -

Table A3.2.2: Prevention precautionary statements (cont'd)

Code (1)	Prevention precautionary statements (2)	Hazard class (3)	Hazard category (4)	Conditions for use (5)
P231	Handle and store contents under inert gas/...	Pyrophoric liquids (chapter 2.9)	1	...Manufacturer/supplier or the competent authority to specify appropriate liquid or gas if "inert gas" is not appropriate.
		Pyrophoric solids (chapter 2.10)	1	...Manufacturer/supplier or the competent authority to specify appropriate liquid or gas if "inert gas" is not appropriate.
		Substances and mixtures which, in contact with water, emit flammable gases (chapter 2.12)	1, 2, 3	– if the substance or mixture reacts readily with moisture in air.
P232	Protect from moisture.	Substances and mixtures which, in contact with water, emit flammable gases (chapter 2.12)	1, 2, 3	
P233	Keep container tightly closed.	Flammable liquids (chapter 2.6)	1, 2, 3	– if the liquid is volatile and may generate an explosive atmosphere.
		Pyrophoric liquids (chapter 2.9)	1	
		Pyrophoric solids (chapter 2.10)	1	
		Desensitized explosives (chapter 2.17)	1, 2, 3, 4	
		Acute toxicity, inhalation (chapter 3.1)	1, 2, 3	– if the chemical is volatile and may generate a hazardous atmosphere.
		Respiratory sensitization (chapter 3.4)	1, 1A, 1B	
		Specific target organ toxicity, single exposure; respiratory tract irritation (chapter 3.8)	3	– if the chemical is volatile and may generate a hazardous atmosphere.
		Specific target organ toxicity, single exposure; narcotic effects (chapter 3.8)	3	
P234	Keep only in original packaging.	Explosives (chapter 2.1)	2A, 2B, 2C	- Omit where P236 is used.
		Self-reactive substances and mixtures (chapter 2.8)	Types A, B, C, D, E, F	
		Organic peroxides (chapter 2.15)	Types A, B, C, D, E, F	
		Corrosive to metals (chapter 2.16)	1	
P235	Keep cool.	Flammable liquids (chapter 2.6)	1, 2, 3	– for flammable liquids Category 1 and other flammable liquids that are volatile and may generate an explosive atmosphere.
		Self-reactive substances and mixtures (chapter 2.8)	Types A, B, C, D, E, F	– may be omitted if P411 is given on the label.
		Self-heating substances and mixtures (chapter 2.11)	1, 2	– may be omitted if P413 is given on the label.
		Organic peroxides (chapter 2.15)	Types A, B, C, D, E, F	– may be omitted if P411 is given on the label.

表 A3.2.2 安全対策注意書き（続き）

コード (1)	安全対策注意書き (2)	危険有害性クラス (3)	危険有害性区分 (4)	使用の条件 (5)
P231	不活性ガス…下で取扱い、保管すること。	自然発火性液体 (2.9 章)	1	…不活性ガスが適当でない場合、製造者供給者または所轄官庁が指定する適当な液体または… 一物質または混合物が空気中の湿気と速やかに反応する場合 …不活性ガスが適当でない場合、製造者供給者または所轄官庁が指定する適当な液体または…
		自然発火性固体 (2.10 章)	1	
		水反応可燃性物質および混合物 (2.12 章)	1, 2, 3	
P232	湿気を遮断すること。	水反応可燃性物質および混合物 (2.12 章)	1, 2, 3	
P233	容器を密閉しておくこと。	引火性液体 (2.6 章)	1, 2, 3	一液体が揮発性で爆発危険性を増す可能性がある場合
		自然発火性液体 (2.9 章)	1	
		自然発火性固体 (2.10 章)	1	
		鈍性化爆発物 (2.17 章)	1, 2, 3, 4	
		急性毒性-吸入 (3.1 章)	1, 2, 3	一化学品が揮発性で有害な環境となる可能性がある場合
		呼吸器感作性作用 (3.4 章)	1, 1A, 1B	
		特定標的臓器毒性、単回ばく露、気道刺激性 (3.8 章)	3	一化学品が揮発性で有害な環境となる可能性がある場合
		特定標的臓器毒性、単回ばく露、麻酔作用 (3.8 章)	3	
P234	他の容器に移し替えないこと。	爆発物 (2.1 章)	2A, 2B, 2C	一P236 が使用されている場合には省略
		自己反応性物質および混合物 (2.8 章)	タイプ A, B, C, D, E, F	
		有機過酸化物 (2.15 章)	タイプ A, B, C, D, E, F	
P235	涼しいところに置くこと。	金属腐食性 (2.16 章)	1	
		引火性液体 (2.6 章)	1, 2, 3	一引火性液体区分 1 および他の引火性液体で揮発性が高く爆発危険性を増す場合 一P411 がラベルに示されている場合には省略してもよい
		自己反応性物質および混合物 (2.8 章)	タイプ A, B, C, D, E, F	一P411 がラベルに示されている場合には省略してもよい
		有機過酸化物 (2.15 章)	タイプ A, B, C, D, E, F	
		自己発熱性物質および混合物 (2.11 章)	1, 2	一P413 がラベルに示されている場合には省略してもよい
		有機過酸化物 (2.15 章)	タイプ A, B, C, D, E, F	一P411 がラベルに示されている場合には省略してもよい

Table A3.2.2: Prevention precautionary statements *(cont'd)*

Code (1)	Prevention precautionary statements (2)	Hazard class (3)	Hazard category (4)	Conditions for use (5)
P236	**Keep only in original packaging; Division … in the transport configuration.**	Explosives (chapter 2.1)	2A, 2B, 2C	- To be applied for explosives assigned a division within Class 1 for transport. - May be omitted for single packaging where the transport pictogram displaying the division (within Class 1) appears. - May be omitted where the use of different outer packaging results in different divisions for transport. …Manufacturer/supplier or competent authority to specify the division for transport.
P240	**Ground and bond container and receiving equipment.**	Explosives (chapter 2.1)	1, 2A, 2B, 2C	– *if the explosive is electrostatically sensitive.*
		Flammable liquids (chapter 2.6)	1, 2, 3	– *if the liquid is volatile and may generate an explosive atmosphere.*
		Flammable solids (chapter 2.7)	1, 2	– *if the solid is electrostatically sensitive.*
		Self-reactive substances and mixtures (chapter 2.8)	Types A, B, C, D, E, F	– *if electrostatically sensitive and able to generate an explosive atmosphere.*
		Organic peroxides (chapter 2.15)	Types A, B, C, D, E, F	– *if electrostatically sensitive and able to generate an explosive atmosphere.*
P241	**Use explosion-proof [electrical/ventilating/lighting/…] equipment.**	Flammable liquids (chapter 2.6)	1, 2, 3	– *if the liquid is volatile and may generate an explosive atmosphere.* – *text in square brackets may be used to specify specific electrical, ventilating, lighting or other equipment if necessary and as appropriate.* – *precautionary statement may be omitted where local or national legislation introduces more specific provisions.*
		Flammable solids (chapter 2.7)	1, 2	– *if dust clouds can occur.* – *text in square brackets may be used to specify specific electrical, ventilating, lighting or other equipment if necessary and as appropriate.* – *precautionary statement may be omitted where local or national legislation introduces more specific provisions.*

表 A3.2.2 安全対策注意書き(続き)

コード(1)	安全対策注意書き(2)	危険有害性クラス(3)	危険有害性区分(4)	使用の条件(5)
P236	元の容器のままで保存すること：輸送の構成において区分…	爆発物 (2.1章)	2A, 2B, 2C	− 輸送のクラス1の区分が割り当てられた爆発物の区分に対して適用する − 単一包装で、区分(クラス1)を示す輸送絵表示がされていれば、省略してもよい − 輸送の異なる区分になっている他の外部包装が使用されていれば所管官庁が指定する輸送の区分 …製造者供給者または所管官庁が指定する輸送の区分
P240	容器を接地しアースを取ること。	爆発物 (2.1章)	1, 2A, 2B, 2C	− 爆発物が静電気的に敏感である場合
		引火性液体 (2.6章)	1, 2, 3	− 液体が揮発性であり、爆発する環境をつくる可能性があるとき
		可燃性固体 (2.7章)	1, 2	− 固体が静電気的に敏感である場合
		自己反応性物質および混合物 (2.8章)	タイプ A, B, C, D, E, F	− 静電気的に敏感であり、爆発する環境をつくる可能性があるとき
		有機過酸化物 (2.15章)	タイプ A, B, C, D, E, F	− 静電気的に敏感であり、爆発する環境をつくる可能性があるとき
P241	防爆型の[電気/換気/照明/...]機器を使用すること。	引火性液体 (2.6章)	1, 2, 3	− 液体が揮発性であり、爆発する環境をつくる可能性がある場合 − [ ] 内の文章は、電気機器、換気装置、照明機器あるいは他の適切な機器を特定するために、必要性がある場合に適切に使用される − 国内規制でより詳細な規定がある場合にはこの注意書きは省略してもよい
		可燃性固体 (2.7章)	1, 2	− 粉じん雲が発生する可能性がある場合 − [ ] 内の文章は、電気機器、換気装置、照明機器あるいは他の適切な機器を特定するために、必要性がある場合に適切に使用される − 国内規制でより詳細な規定がある場合にはこの注意書きは省略してもよい

Table A3.2.2: Prevention precautionary statements *(cont'd)*

Code (1)	Prevention precautionary statements (2)	Hazard class (3)	Hazard category (4)	Conditions for use (5)
P242	Use non-sparking tools.	Flammable liquids (chapter 2.6)	1, 2, 3	– *if the liquid is volatile and may generate an explosive atmosphere and if the minimum ignition energy is very low. (This applies to substances and mixtures where the minimum ignition energy is < 0.1 mJ, e.g. carbon disulphide).*
P243	Take action to prevent static discharges.	Flammable liquids (chapter 2.6)	1, 2, 3	– *if the liquid is volatile and may generate an explosive atmosphere.* – *may be omitted where local or national legislation introduces more specific provisions.*
P244	Keep valves and fittings free from oil and grease.	Oxidizing gases (chapter 2.4)	1	
P250	Do not subject to grinding/shock/friction/….	Explosives (chapter 2.1)	1, 2A, 2B, 2C	– *if the explosive is mechanically sensitive* …Manufacturer/supplier or the competent authority to specify applicable rough handling.
P251	Do not pierce or burn, even after use.	Aerosols (chapter 2.3)	1, 2, 3	
P260	Do not breathe dust/fume/gas/mist/vapours/spray.	Acute toxicity, inhalation (chapter 3.1)	1, 2	Manufacturer/supplier or the competent authority to specify applicable physical state(s).
		Respiratory sensitization (chapter 3.4)	1, 1A, 1B	
		Specific target organ toxicity, single exposure (chapter 3.8)	1, 2	
		Specific target organ toxicity, repeated exposure (chapter 3.9)	1, 2	
		Skin corrosion (chapter 3.2)	1, 1A, 1B, 1C	– *specify do not breathe dusts or mists*
		Reproductive toxicity, effects on or via lactation (chapter 3.7)	Additional category	– *if inhalable particles of dusts or mists may occur during use.*
P261	Avoid breathing dust/fume/gas/mist/vapours/spray.	Acute toxicity, inhalation (chapter 3.1)	3, 4	– *may be omitted if P260 is given on the label.* Manufacturer/supplier or the competent authority to specify applicable conditions.
		Skin sensitization (chapter 3.4)	1, 1A, 1B	
		Specific target organ toxicity, single exposure; respiratory tract irritation (chapter 3.8)	3	
		Specific target organ toxicity, single exposure; narcotic effects (chapter 3.8)	3	
P262	Do not get in eyes, on skin, or on clothing.	Acute toxicity, dermal (chapter 3.1)	1, 2, 3	
P263	Avoid contact during pregnancy and while nursing.	Reproductive toxicity – effects on or via lactation (chapter 3.7)	Additional category	

表 A3.2.2 安全対策注意書き (続き)

コード (1)	安全対策注意書き (2)	危険有害性クラス (3)	危険有害性区分 (4)	使用の条件 (5)
P242	火花を発生させない工具を使用すること。	引火性液体 (2.6章)	1, 2, 3	−液体が揮発性で爆発危険性を増す可能性がある場合および最少引火エネルギーが非常に低い場合（これは例えば二硫化炭素のように、最少引火エネルギーが0.1mJ 未満の物質や混合物に適用される。）のとき −国内規制でより詳細な規定がある場合にはこの注意書きは省略してもよい
P243	静電気放電に対する措置を講ずること。	引火性液体 (2.6章)	1, 2, 3	−液体が揮発性で爆発する可能性がある環境をつくる可能性があるとき −国内規制でより詳細な規定がある場合にはこの注意書きは省略してもよい
P244	バルブや付属品にはグリースおよび油を使用しないこと。	酸化性ガス (2.4章)	1	
P250	粉砕/衝撃/摩擦…のような取り扱いをしないこと。	爆発物 (2.1章)	1, 2A, 2B, 2C	−爆発物が力学的に敏感である場合…なお取り扱い
P251	使用後も、穴を開けたり燃やしたりしないこと。	エアゾール (2.3章)	1, 2, 3	
P260	粉じん/煙/ガス/ミスト/蒸気/スプレーを吸入しないこと。	急性毒性（吸入）(3.1章) 呼吸器感作性 (3.4章) 特定標的臓器毒性、単回ばく露 (3.8章) 特定標的臓器毒性、反復ばく露 (3.9章) 皮膚腐食性 (3.2章) 生殖毒性、授乳に対するまたは授乳を介した影響 (3.7章)	1, 2 1, 1A, 1B 1, 2 1, 2 1, 1A, 1B, 1C 追加区分	製造者/供給者または所管官庁が指定する該当の物理的状態 −粉じんやミストを吸入されうるほどりやミストの粒子が発生するかもしれない場合 −使用中に吸入されうるほどりやミストの粒子が発生するかもしれない場合
P261	粉じん/煙/ガス/ミスト/蒸気/スプレーの吸入を避けること。	急性毒性（吸入）(3.1章) 皮膚感作性 (3.4章) 特定標的臓器毒性、単回ばく露、気道刺激性 (3.8章) 特定標的臓器毒性、単回ばく露、麻酔作用 (3.8章)	3, 4 1, 1A, 1B 3 3	−P260 がラベルに記載される場合には省略してもよい 製造者/供給者または所管官庁が指定する適用条件
P262	眼、皮膚、または被服につけないこと。	急性毒性 (経皮) (3.1章)	1, 2, 3	
P263	妊娠中および授乳期中は接触を避けること。	生殖毒性、授乳に対するまたは授乳を介した影響 (3.7章)	追加区分	

Table A3.2.2: Prevention precautionary statements *(cont'd)*

Code (1)	Prevention precautionary statements (2)	Hazard class (3)	Hazard category (4)	Conditions for use (5)
P264	**Wash hands [and ...] thoroughly after handling.**	Acute toxicity, oral (chapter 3.1)	1, 2, 3, 4	- *text in square brackets to be used when the manufacturer/supplier or the competent authority specify other parts of the body to be washed after handling.*
		Acute toxicity, dermal (chapter 3.1)	1, 2, 3	
		Skin corrosion (chapter 3.2)	1, 1A, 1B, 1C	
		Skin irritation (chapter 3.2)	2	
		Serious eye damage (chapter 3.3)	1	
		Eye irritation (chapter 3.3)	2/2A, 2B	
		Reproductive toxicity, effects on or via lactation (chapter 3.7)	Additional category	
		Specific target organ toxicity, single exposure (chapter 3.8)	1, 2	
		Specific target organ toxicity, repeated exposure (chapter 3.9)	1	
P265	**Do not touch eyes.**	Serious eye damage (chapter 3.3)	1	
		Eye irritation (chapter 3.3)	2/2A, 2B	
P270	**Do not eat, drink or smoke when using this product.**	Acute toxicity, oral (chapter 3.1)	1, 2, 3, 4	
		Acute toxicity, dermal (chapter 3.1)	1, 2, 3	
		Reproductive toxicity, effects on or via lactation (chapter 3.7)	Additional category	
		Specific target organ toxicity, single exposure (chapter 3.8)	1, 2	
		Specific target organ toxicity, repeated exposure (chapter 3.9)	1	
P271	**Use only outdoors or with adequate ventilation.**	Acute toxicity, inhalation (chapter 3.1)	1, 2, 3, 4	Manufacturer/supplier to specify what type of ventilation would be adequate for safe use on the safety data sheet and in any supplemental safety instructions provided to consumers.
		Respiratory sensitization (chapter 3.4)	1, 1A, 1B	
		Specific target organ toxicity, single exposure; respiratory tract irritation (chapter 3.8)	3	
		Specific target organ toxicity, single exposure; narcotic effects (chapter 3.8)	3	
P272	**Contaminated work clothing should not be allowed out of the workplace.**	Skin sensitization (chapter 3.4)	1, 1A, 1B	

表 A3.2.2 安全対策注意書き（続き）

コード (1)	安全対策注意書き (2)	危険有害性クラス (3)	危険有害性区分 (4)	使用の条件 (5)
P264	取扱後は手【および…】をよく洗うこと。	急性毒性（経口）（3.1章） 急性毒性（経皮）（3.1章） 皮膚腐食性（3.2章） 皮膚刺激性（3.2章） 重篤な眼の損傷性（3.3章） 眼刺激性（3.3章） 生殖毒性、授乳に対するまたは授乳を介した影響（3.7章） 特定標的臓器毒性、単回ばく露（3.8章） 特定標的臓器毒性、反復ばく露（3.9章）	1, 2, 3, 4 1, 2, 3 1, 1A, 1B, 1C 2 1 2/2A, 2B 追加区分 1, 2 1	一製造者/供給者または所管官庁が、洗浄する体の他の部分を指定した場合には []内の文章を用いる
P265	眼には触らないこと。	重篤な眼の損傷性（3.3章） 眼刺激性（3.3章）	1 2/2A, 2B	
P270	この製品を使用する時に、飲食または喫煙をしないこと。	急性毒性（経口）（3.1章） 急性毒性（経皮）（3.1章） 生殖毒性、授乳に対するまたは授乳を介した影響（3.7章） 特定標的臓器毒性、単回ばく露（3.8章） 特定標的臓器毒性、反復ばく露（3.9章）	1, 2, 3, 4 1, 2, 3 追加区分 1, 2 1	
P271	屋外または十分な換気のある場所でのみ使用すること。	急性毒性（吸入）（3.1章） 呼吸器感作性（3.4章） 単回気道刺激性（3.8章） 特定標的臓器作用、単回ばく露、気道麻酔作用（3.8章）	1, 2, 3, 4 1, 1A, 1B 3 3	製造者/供給者は、安全データシートおよび消費者に提供する補足的な安全説明書に、安全な使用のためにどのような種類の換気が適切であるかを明記すること。
P272	汚染された作業衣は作業場から出さないこと。	皮膚感作性（3.4章）	1, 1A, 1B	

- 311 -

Table A3.2.2: Prevention precautionary statements *(cont'd)*

Code (1)	Prevention precautionary statements (2)	Hazard class (3)	Hazard category (4)	Conditions for use (5)
P273	**Avoid release to the environment.**	Hazardous to the aquatic environment, acute hazard (chapter 4.1)	1, 2, 3	*– if this is not the intended use.*
		Hazardous to the aquatic environment, long-term hazard (chapter 4.1)	1, 2, 3, 4	
P280	**Wear protective gloves/protective clothing/eye protection/face protection/hearing protection/…**	Explosives (chapter 2.1)	1, 2A, 2B, 2C	Manufacturer/supplier or the competent authority to specify the appropriate personal protective equipment.
		Flammable gases (chapter 2.2)	1A, Pyrophoric gas	
		Flammable liquids (chapter 2.6)	1, 2, 3, 4	
		Flammable solids (chapter 2.7)	1, 2	
		Self-reactive substances and mixtures (chapter 2.8)	Types A, B, C, D, E, F	
		Pyrophoric liquids (chapter 2.9)	1	
		Pyrophoric solids (chapter 2.10)	1	
		Self-heating substances and mixtures (chapter 2.11)	1, 2	
		Substances and mixtures which, in contact with water, emit flammable gases (chapter 2.12)	1, 2, 3	
		Oxidizing liquids (chapter 2.13)	1, 2, 3	
		Oxidizing solids (chapter 2.14)	1, 2, 3	
		Organic peroxides (chapter 2.15)	Types A, B, C, D, E, F	
		Desensitized explosives (chapter 2.17)	1, 2, 3, 4	
		Acute toxicity, dermal (chapter 3.1)	1, 2, 3, 4	*- Specify protective gloves/clothing.* Manufacturer/supplier or the competent authority may further specify type of equipment where appropriate.
		Skin corrosion (chapter 3.2)	1, 1A, 1B, 1C	*– Specify protective gloves/clothing and eye/face protection.* Manufacturer/supplier or the competent authority may further specify type of equipment where appropriate.
		Skin irritation (chapter 3.2)	2	*- Specify protective gloves.*
		Skin sensitization (chapter 3.4)	1, 1A, 1B	Manufacturer/supplier or the competent authority may further specify type of equipment where appropriate.
		Serious eye damage (chapter 3.3)	1	*- Specify protective gloves and eye/face protection.*

表 A3.2.2 安全対策注意書き (続き)

コード (1)	安全対策注意書き (2)	危険有害性クラス (3)	危険有害性区分 (4)	使用の条件 (5)
P273	環境への放出を避けること。	水生環境有害性（急性）(4.1章)	1, 2, 3	— 必要な時以外は
		水生環境有害性（慢性）(4.1章)	1, 2, 3, 4	
P280	保護手袋/保護衣/保護眼鏡/保護面/聴覚保護具/…を着用すること。	爆発物 (2.1章)	1, 2A, 2B, 2C	製造者/供給者または所管官庁が指定する適当な個人用保護具
		可燃性ガス (2.2章)	1A, 自然発火性ガス	
		引火性液体 (2.6章)	1, 2, 3, 4	
		可燃性固体 (2.7章)	1, 2	
		自己反応性物質および混合物 (2.8章)	タイプ A, B, C, D, E, F	
		自然発火性液体 (2.9章)	1	
		自然発火性固体 (2.10章)	1	
		自己発熱性物質および混合物 (2.11章)	1, 2	
		水反応可燃性物質および混合物 (2.12章)	1, 2, 3	
		酸化性液体 (2.13章)	1, 2, 3	
		酸化性固体 (2.14章)	1, 2, 3	
		有機過酸化物 (2.15章)	タイプ A, B, C, D, E, F	
		鈍性化爆発物 (2.17章)	1, 2, 3, 4	
		急性毒性（経皮）(3.1章)	1, 1A, 1B, 1C	— 保護手袋/保護衣を指定すること 製造者/供給者または所管官庁が装具の種類を適宜指定してもよい
		皮膚腐食性 (3.2章)	1, 1A, 1B, 1C	— 保護手袋/保護衣および保護面を指定すること 製造者/供給者または所管官庁が装具の種類を適宜指定してもよい
		皮膚刺激性 (3.2章)	2	— 保護手袋を指定すること 製造者/供給者または所管官庁が装具の種類を適宜指定してもよい
		皮膚感作性 (3.4章)	1, 1A, 1B	— 保護手袋を指定すること 製造者/供給者または所管官庁が装具の種類を適宜指定してもよい
		重篤な眼の損傷性 (3.3章)	1	— 保護手袋および眼鏡保護面を指定すること

Table A3.2.2: Prevention precautionary statements (*cont'd*)

Code (1)	Prevention precautionary statements (2)	Hazard class (3)	Hazard category (4)	Conditions for use (5)
P280 (*cont'd*)	**Wear protective gloves/protective clothing/eye protection/face protection/hearing protection/…**	Eye irritation (chapter 3.3)	2/2A	Manufacturer/supplier or the competent authority may further specify type of equipment where appropriate.
		Respiratory sensitization (chapter 3.4)	1, 1A, 1B	– Specify *protective gloves/clothing*. Manufacturer/supplier or the competent authority may further specify type of equipment where appropriate
		Germ cell mutagenicity (chapter 3.5)	1, 1A, 1B, 2	Manufacturer/supplier or the competent authority to specify the appropriate personal protective equipment.
		Carcinogenicity (chapter 3.6)	1, 1A, 1B, 2	
		Reproductive toxicity (chapter 3.7)	1, 1A, 1B, 2	
P282	**Wear cold insulating gloves and either face shield or eye protection.**	Gases under pressure (chapter 2.5)	Refrigerated liquefied gas	
P283	**Wear fire resistant or flame-retardant clothing.**	Oxidizing liquids (chapter 2.13)	1	
		Oxidizing solids (chapter 2.14)	1	
P284	**In case of inadequate ventilation wear respiratory protection.**	Acute toxicity, inhalation (chapter 3.1)	1, 2	Manufacturer/supplier to specify on the safety data sheet what type of ventilation would be adequate for safe use and provide additional information with the chemical at the point of use that explains what type of respiratory equipment may also be needed.
		Respiratory sensitization (chapter 3.4)	1, 1A, 1B	
P231 + P232	**Handle and store contents under inert gas/…. Protect from moisture.**	Substances and mixtures which, in contact with water, emit flammable gases (chapter 2.12)	1, 2, 3	– *if the substance or mixture reacts readily with moisture in air.* …Manufacturer/supplier or the competent authority to specify appropriate liquid or gas if "inert gas" is not appropriate.
P264	**Wash hands [and…] thoroughly after handling.**	Serious eye damage (chapter 3.3)	1	– *text in square brackets to be used when the manufacturer/supplier or the competent authority specify other parts of the hibody to be washed after handling.*
+ P265	**Do not touch eyes.**	Eye irritation (chapter 3.3)	2/2A, 2B	

表 A3.2.2 安全対策注意書き (続き)

コード (1)	安全対策注意書き (2)	危険有害性クラス (3)	危険有害性区分 (4)	使用の条件 (5)
P280 (続き)	保護手袋/保護衣/保護眼鏡/保護面/聴覚保護具…を着用すること。	眼刺激性 (3.3章)	2/2A	製造者/供給者または所管官庁が装具の種類を適宜指定してもよい
		呼吸器感作性 (3.4章)	1, 1A, 1B	一保護手袋/供給者または所管官庁が製造者/供給者または所管官庁指定する装具の種類を適宜指定してもよい
		生殖細胞変異原性 (3.5章)	1, 1A, 1B, 2	製造者/供給者または所管官庁が指定する適当な個人用保護具
		発がん性 (3.6章)	1, 1A, 1B, 2	
		生殖毒性 (3.7章)	1, 1A, 1B, 2	
		高圧ガス (2.5章)	深冷液化ガス	
P282	耐寒手袋および保護面または保護眼鏡を着用すること。			
P283	防火服または防炎服を着用すること。	酸化性液体 (2.13章)	1	
		酸化性固体 (2.14章)	1	
P284	換気が不十分な場合は呼吸用保護具を着用すること。	急性毒性 (吸入) (3.1章)	1, 2	製造者/供給者は、安全データシートに、安全な使用のためにどのような種類の換気が適切であるかを明記し、使用時にどのような種類の呼吸装置も必要となる可能性があるかを説明する追加情報を、化学品とともに提供すること。
		呼吸器感作性 (3.4章)	1, 1A, 1B	
P231 + P232	湿気を遮断し、不活性ガス/…下で取り扱い保管すること。	水反応可燃性物質および混合物 (2.12章)	1, 2, 3	一物質あるいは混合物が空気中の湿度と速やかに反応する場合…もし「不活性ガス」が適当でない場合には、製造者/供給者または所管官庁が指定する適当な液体または気体を指定した場合には、取扱後に洗浄する体の他の部分を指定には [ ] 内の文章を用いる
P264 + P265	取扱後は手 [および…] をよく洗うこと。眼には触らないこと。	重篤な眼の損傷性 (3.3章)	1	一製造者/供給者または所管官庁が、取扱後に洗浄する体の他の部分を指定した場合には [ ] 内の文章を用いる
		眼刺激性 (3.3章)	2/2A, 2B	

Table A3.2.3: Response precautionary statements

Code (1)	Response precautionary statements (2)	Hazard class (3)	Hazard category (4)	Conditions for use (5)
P301	**IF SWALLOWED:**	Acute toxicity, oral (chapter 3.1)	1, 2, 3, 4, 5	
		Skin corrosion (chapter 3.2)	1, 1A, 1B, 1C	
		Aspiration hazard (chapter 3.10)	1, 2	
P302	**IF ON SKIN:**	Pyrophoric liquids (chapter 2.9)	1	
		Pyrophoric solids (chapter 2.10)	1	
		Substances and mixtures which, in contact with water, emit flammable gases (chapter 2.12)	1, 2	
		Acute toxicity, dermal (chapter 3.1)	1, 2, 3, 4, 5	
		Skin corrosion (chapter 3.2)	1, 1A, 1B, 1C	
		Skin irritation (chapter 3.2)	2	
		Skin sensitization (chapter 3.4)	1, 1A, 1B	
P303	**IF ON SKIN (or hair):**	Flammable liquids (chapter 2.6)	1, 2, 3	
P304	**IF INHALED:**	Acute toxicity, inhalation (chapter 3.1)	1, 2, 3, 4, 5	
		Skin corrosion (chapter 3.2)	1, 1A, 1B, 1C	
		Respiratory sensitization (chapter 3.4)	1, 1A, 1B	
		Specific target organ toxicity, single exposure; respiratory tract irritation (chapter 3.8)	3	
		Specific target organ toxicity, single exposure; narcotic effects (chapter 3.8)	3	
P305	**IF IN EYES:**	Skin corrosion (chapter 3.2)	1, 1A, 1B, 1C	
		Serious eye damage (chapter 3.3)	1	
		Eye irritation (chapter 3.3)	2/2A, 2B	
P306	**IF ON CLOTHING:**	Oxidizing liquids (chapter 2.13)	1	
		Oxidizing solids (chapter 2.14)	1	
P308	**IF exposed or concerned:**	Specific target organ toxicity, single exposure (chapter 3.8)	1, 2	
P310	*[Deleted]*			
P311	*[Deleted]*			
P312	*[Deleted]*			
P313	*[Deleted]*			
P314	*[Deleted]*			
P315	*[Deleted]*			

表 A3.2.3 応急措置注意書き

コード (1)	応急措置注意書き (2)	危険有害性クラス (3)	危険有害性区分 (4)	使用の条件 (5)
P301	飲み込んだ場合：	急性毒性（経口）(3.1章) 皮膚腐食性 (3.2章) 誤えん有害性 (3.10章)	1, 2, 3, 4, 5 1, 1A, 1B, 1C 1, 2	
P302	皮膚に付着した場合：	自然発火性液体 (2.9章) 自然発火性固体 (2.10章) 水反応可燃性物質および混合物 (2.12章) 急性毒性（経皮）(3.1章) 皮膚腐食性 (3.2章) 皮膚刺激性 (3.2章) 皮膚感作性 (3.4章)	1 1 1,2 1, 2, 3, 4, 5 1, 1A, 1B, 1C 2 1, 1A, 1B	
P303	皮膚（または髪）に付着した場合：	引火性液体 (2.6章)	1, 2, 3	
P304	吸入した場合：	急性毒性（吸入）(3.1章) 皮膚腐食性 (3.2章) 呼吸器感作性 (3.4章) 特定標的臓器毒性、単回ばく露、気道刺激性 (3.8章) 特定標的臓器毒性、単回ばく露、麻酔作用 (3.8章)	1,2, 3, 4, 5 1, 1A, 1B, 1C 1, 1A, 1B 3 3	
P305	眼に入った場合：	皮膚腐食性 (3.2章) 眼に対する重篤な損傷性 (3.3章) 眼刺激性 (3.3章)	1, 1A, 1B, 1C 1 2/2A, 2B	
P306	衣類にかかった場合：	酸化性液体 (2.13章) 酸化性固体 (2.14章)	1 1	
P308	ばく露またはその懸念がある場合：	特定標的臓器毒性、単回ばく露 (3.8章)	1, 2	
P310	(削除)			
P311	(削除)			
P312	(削除)			
P313	(削除)			
P314	(削除)			
P315	(削除)			

Table A3.2.3: Response precautionary statements *(cont'd)*

Code (1)	Response precautionary statements (2)	Hazard class (3)	Hazard category (4)	Conditions for use (5)
P316	**Get emergency medical help immediately.**	Acute toxicity, oral (chapter 3.1)	1, 2, 3	Competent authority or manufacturer/supplier may add, "Call" followed by the appropriate emergency telephone number, or the appropriate emergency medical help provider, for example, a poison centre, emergency centre or doctor.
		Acute toxicity, dermal (chapter 3.1)	1, 2, 3	
		Acute toxicity, inhalation (chapter 3.1)	1, 2, 3	
		Skin corrosion (chapter 3.2)	1, 1A, 1B, 1C	
		Respiratory sensitization (chapter 3.4)	1, 1A, 1B	
		Specific target organ toxicity, single exposure; (chapter 3.8)	1, 2	
		Aspiration hazard (chapter 3.10)	1, 2	
P317	**Get medical help.**	Gases under pressure (chapter 2.5)	Refrigerated liquefied gas	
		Acute toxicity, oral (chapter 3.1)	4, 5	
		Acute toxicity, dermal (chapter 3.1)	4, 5	
		Acute toxicity, inhalation (chapter 3.1)	4, 5	
		Skin irritation (chapter 3.2)	2, 3	
		Serious eye damage (chapter 3.3)	1	
		Eye irritation (chapter 3.3)	2/2A, 2B	
		Skin sensitization (chapter 3.4)	1, 1A, 1B	
P318	**IF exposed or concerned, get medical advice.**	Germ cell mutagenicity (chapter 3.5)	1, 1A, 1B, 2	
		Carcinogenicity (chapter 3.6)	1, 1A, 1B, 2	
		Reproductive toxicity (chapter 3.7)	1, 1A, 1B, 2	
		Reproductive toxicity, effects on or via lactation (chapter 3.7)	Additional category	
P319	**Get medical help if you feel unwell.**	Specific target organ toxicity, single exposure; respiratory tract irritation (chapter 3.8)	3	
		Specific target organ toxicity, single exposure; narcotic effects (chapter 3.8)	3	
		Specific target organ toxicity, repeated exposure (chapter 3.9)	1, 2	
P320	**Specific treatment is urgent (see ... on this label).**	Acute toxicity, inhalation (chapter 3.1)	1, 2, 3	*– if immediate administration of antidote is required.* ...Reference to supplemental first aid instruction.

表 A3.2.3 応急措置注意書き(続き)

コード (1)	応急措置注意書き (2)	危険有害性クラス (3)	危険有害性区分 (4)	使用の条件 (5)
P316	すぐに救急の医療処置を受けること。	急性毒性(経口)(3.1章) 急性毒性(経皮)(3.1章) 急性毒性(吸入)(3.1章) 皮膚腐食性(3.2章) 呼吸器感作性(3.4章) 特定標的臓器毒性、単回ばく露(3.8章) 誤えん有害性(3.10章)	1, 2, 3 **1, 2, 3** **1, 2, 3** 1, 1A, 1B, 1C 1, 1A, 1B 1, 2 1, 2	所管官庁または製造者/供給者は「電話」に続けて、適当な時電話番号すなわち救急時医療提供者、例えば中毒センター、救急センターまたは医師などを追加してもよい。
P317	医療処置を受けること。	高圧ガス(2.5章) 急性毒性(経口)(3.1章) 急性毒性(経皮)(3.1章) 急性毒性(吸入)(3.1章) 皮膚刺激性(3.2章) 眼に対する重篤な損傷性(3.3章) 皮膚感作性(3.4章)	深冷液化ガス 4, 5 4, 5 4, 5 2, 3 1 2/2A, 2B	
P318	ばく露またはその懸念がある場合は、医学的助言を求めること。	生殖細胞変異原性(3.5章) 発がん性(3.6章) 生殖毒性(3.7章) 生殖毒性、授乳に対するまたは授乳を介した影響(3.7章)	1, 1A, 1B 1, 1A, 1B, 2 1, 1A, 1B, 2 1, 1A, 1B, 2 追加区分	
P319	気分が悪い時は、医療処置を受けること。	特定標的臓器毒性、単回ばく露、気道刺激性(3.8章) 特定標的臓器毒性、単回ばく露、麻酔作用(3.8章) 特定標的臓器毒性、反復ばく露(3.9章)	3 3 1, 2	
P320	特別な処置が緊急に必要である(このラベルの…を見よ)。	急性毒性(吸入)(3.1章)	1, 2	一緊急の解毒剤の投与が必要な場合 …補足的な応急措置の説明

Table A3.2.3: Response precautionary statements *(cont'd)*

Code (1)	Response precautionary statements (2)	Hazard class (3)	Hazard category (4)	Conditions for use (5)
P321	**Specific treatment (see ... on this label).**	Acute toxicity, oral (chapter 3.1)	1, 2, 3	– *if immediate administration of antidote is required.* ...Reference to supplemental first aid instruction.
		Acute toxicity, dermal (chapter 3.1)	1, 2, 3, 4	– *if immediate measures such as specific cleansing agent is advised.* ...Reference to supplemental first aid instruction.
		Acute toxicity, inhalation (chapter 3.1)	3	– *if immediate specific measures are required.* ...Reference to supplemental first aid instruction.
		Skin corrosion (chapter 3.2)	1, 1A, 1B, 1C	...Reference to supplemental first aid instruction.
		Skin irritation (chapter 3.2)	2	Manufacturer/supplier or the competent authority may specify a cleansing agent if appropriate.
		Skin sensitization (chapter 3.4)	1, 1A, 1B	
		Specific target organ toxicity, single exposure (chapter 3.8)	1	– *if immediate measures are required.* ...Reference to supplemental first aid instruction.
P330	**Rinse mouth.**	Acute toxicity, oral (chapter 3.1)	1, 2, 3, 4	
		Skin corrosion (chapter 3.2)	1, 1A, 1B, 1C	
		Skin corrosion (chapter 3.2)	1, 1A, 1B, 1C	
P331	**Do NOT induce vomiting.**	Aspiration hazard (chapter 3.10)	1, 2	
P332	**If skin irritation occurs:**	Skin irritation (chapter 3.2)	2, 3	– *may be omitted if P333 is given on the label.*
P333	**If skin irritation or rash occurs:**	Skin sensitization (chapter 3.4)	1, 1A, 1B	
P334	**Immerse in cool water [or wrap in wet bandages].**	Pyrophoric liquids (chapter 2.9)	1	– *text in square brackets to be used for pyrophoric liquids and solids.*
		Pyrophoric solids (chapter 2.10)	1	
		Substances and mixtures which, in contact with water, emit flammable gases (chapter 2.12)	1, 2	– *use only "Immerse in cool water." Text in square brackets should not be used.*
P335	**Brush off loose particles from skin.**	Pyrophoric solids (chapter 2.10)	1	
		Substances and mixtures which, in contact with water, emit flammable gases (chapter 2.12)	1, 2	
P336	**Immediately thaw frosted parts with lukewarm water. Do not rub affected area.**	Gases under pressure (chapter 2.5)	Refrigerated liquefied gas	

- 316 -

表 A3.2.3 応急措置注意書き（続き）

コード (1)	応急措置注意書き (2)	危険有害性クラス (3)	危険有害性区分 (4)	使用の条件 (5)
P321	特別な処置が必要である（このラベルの…を見よ）。	急性毒性（経口）（3.1章）	1, 2, 3	一緊急の解毒剤の投与が必要な場合 …補足的な応急措置の説明
		急性毒性（経皮）（3.1章）	1, 2, 3, 4	一緊急の洗浄剤などを推薦する場合 …補足的な応急措置の説明
		急性毒性（吸入）（3.1章）	3	一補足的な応急措置の説明
		皮膚腐食性（3.2章）	1, 1A, 1B, 1C	一補足的な応急措置の説明
		皮膚刺激性（3.2章）	2	一製造者/供給者または所管官庁が指定する適切な洗浄剤がある場合
		皮膚感作性（3.4章）	1, 1A, 1B	一緊急の特別な処置が必要な場合 …補足的な応急措置の説明
		特定標的臓器毒性、単回ばく露（3.8章）		
P330	口をすすぐこと。	急性毒性（経口）（3.1章）	1, 2, 3, 4	
		皮膚腐食性（3.2章）	1, 1A, 1B, 1C	
P331	無理に吐かせないこと。	皮膚腐食性（3.2章）	1, 1A, 1B, 1C	
		誤えん有害性（3.10章）	1, 2	
P332	皮膚刺激が生じた場合：	皮膚刺激性（3.2章）	2, 3	
P333	皮膚刺激または発疹が生じた場合：	皮膚感作性（3.4章）	1, 1A, 1B	ーP333 がラベルに記載されている場合には省略してもよい
P334	冷たい水に浸すこと【または湿った包帯で覆うこと】。	自然発火性液体（2.9章）	1	ー[ ] 内の文章は自然発火性液体及び固体にのみ使用する
		自然発火性固体（2.10章）	1	
		水反応可燃性物質および混合物（2.12章）	1, 2	ー「冷たい水に浸すこと」のみ使用する。[ ] 内の文章は使用しない
P335	固着していない粒子を皮膚から払いのけること。	自然発火性固体（2.10章）	1	
		水反応可燃性物質および混合物（2.12章）	1, 2	
P336	凍った部分をすぐにぬるま湯でとかすこと。受傷部はこすらないこと。	高圧ガス（2.5章）	深冷液化ガス	

- 316 -

Table A3.2.3: Response precautionary statements *(cont'd)*

Code (1)	Response precautionary statements (2)	Hazard class (3)	Hazard category (4)	Conditions for use (5)
P337	**If eye irritation persists:**	Eye irritation (chapter 3.3)	2/2A, 2B	
P338	**Remove contact lenses, if present and easy to do. Continue rinsing.**	Skin corrosion (chapter 3.2)	1, 1A, 1B, 1C	
		Serious eye damage (chapter 3.3)	1	
		Eye irritation (chapter 3.3)	2/2A, 2B	
P340	**Remove person to fresh air and keep comfortable for breathing.**	Acute toxicity, inhalation (chapter 3.1)	1, 2, 3, 4	
		Skin corrosion (chapter 3.2)	1, 1A, 1B, 1C	
		Respiratory sensitization (chapter 3.4)	1, 1A, 1B	
		Specific target organ toxicity, single exposure; respiratory tract irritation (chapter 3.8)	3	
		Specific target organ toxicity, single exposure; narcotic effects (chapter 3.8)	3	
P342	**If experiencing respiratory symptoms:**	Respiratory sensitization (chapter 3.4)	1, 1A, 1B	
P351	**Rinse cautiously with water for several minutes.**	Eye irritation (chapter 3.3)	2/2A, 2B	
P352	**Wash with plenty of water/…**	Acute toxicity, dermal (chapter 3.1)	1, 2, 3, 4	…Manufacturer/supplier or the competent authority may specify a cleansing agent if appropriate, or may recommend an alternative agent in exceptional cases if water is clearly inappropriate.
		Skin irritation (chapter 3.2)	2	
		Skin sensitization (chapter 3.4)	1, 1A, 1B	
P353	**Rinse affected areas with water [or shower].**	Flammable liquids (chapter 2.6)	1, 2, 3	– *text in square brackets to be included where the manufacturer/supplier or the competent authority considers it appropriate for the specific chemical.*
P354	**Immediately rinse with water for several minutes.**	Skin corrosion (chapter 3.2)	1, 1A, 1B, 1C	
		Serious eye damage (chapter 3.3)	1	
P360	**Rinse immediately contaminated clothing and skin with plenty of water before removing clothes.**	Oxidizing liquids (chapter 2.13)	1	
		Oxidizing solids (chapter 2.14)	1	
P361	**Take off immediately all contaminated clothing.**	Flammable liquids (chapter 2.6)	1, 2, 3	
		Acute toxicity, dermal (chapter 3.1)	1, 2, 3	
		Skin corrosion (chapter 3.2)	1, 1A, 1B, 1C	
P362	**Take off contaminated clothing.**	Acute toxicity, dermal (chapter 3.1)	4	
		Skin irritation (chapter 3.2)	2	
		Skin sensitization (chapter 3.4)	1, 1A, 1B	
P363	**Wash contaminated clothing before reuse.**	Skin corrosion (chapter 3.2)	1, 1A, 1B, 1C	

表 A3.2.3 応急措置注意書き(続き)

コード (1)	応急措置注意書き (2)	危険有害性クラス (3)	危険有害性区分 (4)	使用の条件 (5)
P337	眼の刺激が続く場合:	眼刺激性 (3.3章)	2/2A, 2B	
P338	コンタクトレンズを着用していて容易に外せる場合は外すこと。その後も洗浄を続けること。	皮膚腐食性 (3.2章) 眼に対する重篤な損傷性 (3.3章) 眼刺激性 (3.3章)	1, 1A, 1B, 1C 1 2/2A, 2B	
P340	空気の新鮮な場所に移し、呼吸しやすい姿勢で休息させること。	急性毒性 (吸入) (3.1章) 皮膚腐食性 (3.2章) 呼吸器感作性 (3.4章) 特定標的臓器毒性、単回ばく露、気道刺激性 (3.8章) 特定標的臓器毒性、単回ばく露、麻酔作用 (3.8章)	1, 2, 3, 4 1, 1A, 1B, 1C 1, 1A, 1B 3 3	
P342	呼吸に関する症状が出た場合:	呼吸器感作性 (3.4章)	1, 1A, 1B	
P351	水で数分間注意深く洗うこと。	眼刺激性 (3.3章)	2/2A, 2B	
P352	多量の水/…で洗うこと	急性毒性 (経皮) (3.1章) 皮膚刺激性 (3.2章) 皮膚感作性 (3.4章)	1, 2, 3, 4 2 1, 1A, 1B	…製造者/供給者または所管官庁が指定する適切な洗浄剤がある場合、または明らかに水が不適切で他の薬剤を推薦する場合
P353	接触部位を水 [またはシャワー] で洗うこと。	引火性液体 (2.6章)	1, 2, 3	
P354	すぐに水で数分間洗うこと。	皮膚腐食性 (3.2章) 重篤な眼の損傷性 (3.3章)	1, 1A, 1B, 1C 1	
P360	服を脱ぐ前に、直ちに汚染された衣類及び皮膚を多量の水で洗うこと。	酸化性液体 (2.13章) 酸化性固体 (2.14章)	1 1	
P361	汚染された衣類を直ちにすべて脱ぐこと。	引火性液体 (2.6章) 急性毒性 (経皮) (3.1章) 皮膚腐食性 (3.2章)	1, 2, 3 1, 1A, 1B, 1C 1, 1A, 1B, 1C	一製造者/供給者または所管官庁が特定の化学品に対してこれが適当だとした場合には [ ] 内の文章を含める
P362	汚染された衣類を脱ぐこと。	急性毒性 (経皮) (3.1章) 皮膚刺激性 (3.2章) 皮膚感作性 (3.4章)	4 2 1, 1A, 1B	
P363	汚染された衣類を再使用する場合には洗濯をすること。	皮膚腐食性 (3.2章)	1, 1A, 1B, 1C	

- 317 -

Table A3.2.3: Response precautionary statements *(cont'd)*

Code (1)	Response precautionary statements (2)	Hazard class (3)	Hazard category (4)	Conditions for use (5)
P364	**And wash it before reuse.**	Acute toxicity, dermal (chapter 3.1)	1, 2, 3, 4	
		Skin irritation (chapter 3.2)	2	
		Skin sensitization (chapter 3.4)	1, 1A, 1B	
P370	**In case of fire:**	Explosives (chapter 2.1)	1, 2A, 2B, 2C	
		Chemicals under pressure (chapter 2.3)	1, 2	
		Oxidizing gases (chapter 2.4)	1	
		Flammable liquids (chapter 2.6)	1, 2, 3, 4	
		Flammable solids (chapter 2.7)	1, 2	
		Self-reactive substances and mixtures (chapter 2.8)	Types A, B, C, D, E, F	
		Pyrophoric liquids (chapter 2.9)	1	
		Pyrophoric solids (chapter 2.10)	1	
		Substances and mixtures which, in contact with water, emit flammable gases (chapter 2.12)	1, 2, 3	
		Oxidizing liquids (chapter 2.13)	1, 2, 3	
		Oxidizing solids (chapter 2.14)	1, 2, 3	
		Organic peroxides (chapter 2.15)	Types A, B, C, D, E, F	
		Desensitized explosives (chapter 2.17)	1, 2, 3	
P371	**In case of major fire and large quantities:**	Oxidizing liquids (chapter 2.13)	1	
		Oxidizing solids (chapter 2.14)	1	
		Desensitized explosives (chapter 2.17)	4	
P372	**Explosion risk.**	Explosives (chapter 2.1)	1, 2A, 2B	
		Self-reactive substances and mixtures (chapter 2.8)	Type A	
		Organic peroxides (chapter 2.15)	Type A	
P373	**DO NOT fight fire when fire reaches explosives.**	Explosives (chapter 2.1)	1, 2A, 2B	
		Self-reactive substances and mixtures (chapter 2.8)	Type A	
		Organic peroxides (chapter 2.15)	Type A	

表 A3.2.3 応急措置注意書き(続き)

コード (1)	応急措置注意書き (2)	危険有害性クラス (3)	危険有害性区分 (4)	使用の条件 (5)
P364	そして再使用する場合には洗濯をすること。	急性毒性 (経皮) (3.1章)	1, 2, 3, 4	
		皮膚刺激性 (3.2章)	2	
		皮膚感作性 (3.4章)	1, 1A, 1B	
P370	火災の場合：	爆発物 (2.1章)	1, 2A, 2B, 2C	
		加圧下化学品 (2.3章)	1, 2	
		酸化性ガス (2.4章)	1	
		引火性液体 (2.6章)	1, 2, 3, 4	
		可燃性固体 (2.7章)	1, 2	
		自己反応性物質および混合物 (2.8章)	タイプ A, B, C, D, E, F	
		自然発火性液体 (2.9章)	1	
		自然発火性固体 (2.10章)	1	
		水反応可燃性物質おおび混合物 (2.12章)	1, 2, 3	
		酸化性液体 (2.13章)	1, 2, 3	
		酸化性固体 (2.14章)	1, 2, 3	
		有機過酸化物 (2.15章)	タイプ A, B, C, D, E, F	
		鈍性化爆発物 (2.17章)	1, 2, 3	
P371	大火災の場合で大量にある場合：	酸化性液体 (2.13章)	1	
		酸化性固体 (2.14章)	1	
		鈍性化爆発物 (2.17章)	4	
P372	爆発する危険性あり。	爆発物 (2.1章)	1, 2A, 2B	
		自己反応性物質および混合物 (2.8章)	タイプ A	
		有機過酸化物 (2.15章)	タイプ A	
P373	炎が爆発物に届いたら消火活動をしないこと。	爆発物 (2.1章)	1, 2A, 2B	
		自己反応性物質および混合物 (2.8章)	タイプ A	
		有機過酸化物 (2.15章)	タイプ A	

Table A3.2.3: Response precautionary statements *(cont'd)*

Code (1)	Response precautionary statements (2)	Hazard class (3)	Hazard category (4)	Conditions for use (5)
P375	**Fight fire remotely due to the risk of explosion.**	Explosives (chapter 2.1)	2C	
		Self-reactive substances and mixtures (chapter 2.8)	Type B	
		Oxidizing liquids (chapter 2.13)	1	
		Oxidizing solids (chapter 2.14)	1	
		Organic peroxides (chapter 2.15)	Type B	
		Desensitized explosives (chapter 2.17)	1, 2, 3, 4	
P376	**Stop leak if safe to do so.**	Chemicals under pressure (chapter 2.3)	1, 2, 3	
		Oxidizing gases (chapter 2.4)	1	
P377	**Leaking gas fire: Do not extinguish, unless leak can be stopped safely.**	Flammable gases (chapter 2.2)	Flammable gas 1A	
			Pyrophoric gas	
			Chemically unstable gas A	
			Chemically unstable gas B	
P378	**Use ... to extinguish.**	Chemicals under pressure (chapter 2.3)	1B, 2	– *if water increases risk.* ... Manufacturer/supplier or the competent authority to specify appropriate media.
		Flammable liquids (chapter 2.6)	1, 2	
		Flammable solids (chapter 2.7)	1, 2, 3, 4	
		Self-reactive substances and mixtures (chapter 2.8)	1, 2	
		Pyrophoric liquids (chapter 2.9)	Types B, C, D, E, F	
		Pyrophoric solids (chapter 2.10)	1	
		Substances and mixtures which, in contact with water, emit flammable gases (chapter 2.12)	1	
		Oxidizing liquids (chapter 2.13)	1, 2, 3	
		Oxidizing solids (chapter 2.14)	1, 2, 3	
		Organic peroxides (chapter 2.15)	Types B, C, D, E, F	
P380	**Evacuate area.**	Explosives (chapter 2.1)	1, 2A, 2B, 2C	
		Self-reactive substances and mixtures (chapter 2.8)	Types A, B	
		Oxidizing liquids (chapter 2.13)	1	
		Oxidizing solids (chapter 2.14)	1	
		Organic peroxides (chapter 2.15)	Types A, B	
		Desensitized explosives (chapter 2.17)	1, 2, 3, 4	

表 A3.2.3 応急措置注意書き(続き)

コード (1)	応急措置注意書き (2)	危険有害性クラス (3)	危険有害性区分 (4)	使用の条件 (5)
P375	爆発の危険性があるため、離れた距離から消火すること。	爆発物 (2.1 章)	2C	
		自己反応性物質および混合物 (2.8 章)	タイプ B	
		酸化性液体 (2.13 章)	1	
		酸化性固体 (2.14 章)	1	
		有機過酸化物 (2.15 章)	タイプ B	
		鈍性化爆発物 (2.17 章)	1, 2, 3, 4	
P376	安全に対処できるならば漏洩を止めること。	加圧下化学品 (2.3 章)	1, 2, 3	
		酸化性ガス (2.4 章)	1	
P377	漏洩ガス火災の場合：漏えいを安全に停止されない限り消火しないこと。	可燃性ガス (2.2 章)	可燃性ガス 1A 1B, 2 自然発火性ガス 化学的に不安定なガス A 化学的に不安定なガス B	
P378	消火するために...を使用すること。	加圧下化学品 (2.3 章)	1, 2	
		引火性液体 (2.6 章)	1, 2, 3, 4	
		可燃性固体 (2.7 章)	1, 2	
		自己反応性物質および混合物 (2.8 章)	タイプ B, C, D, E, F	
		自然発火性液体 (2.9 章)	1	
		自然発火性固体 (2.10 章)	1	
		水反応可燃性物質および混合物 (2.12 章)	1, 2, 3	
		酸化性液体 (2.13 章)	1, 2, 3	
		酸化性固体 (2.14 章)	1, 2, 3	
		有機過酸化物 (2.15 章)	タイプ B, C, D, E, F	
P380	区域より退避させること。	爆発物 (2.1 章)	1, 2A, 2B, 2C	一水がリスクを増大させる場合 ...製造者/供給者または所管官庁が指定する適当な手段
		自己反応性物質および混合物 (2.8 章)	タイプ A, B	
		酸化性液体 (2.13 章)	1	
		酸化性固体 (2.14 章)	1	
		有機過酸化物 (2.15 章)	タイプ A, B	
		鈍性化爆発物 (2.17 章)	1, 2, 3, 4	

Table A3.2.3: Response precautionary statements (cont'd)

Code (1)	Response precautionary statements (2)	Hazard class (3)	Hazard category (4)	Conditions for use (5)
P381	In case of leakage, eliminate all ignition sources.	Flammable gases (chapter 2.2)	Flammable gas: 1A Pyrophoric gas Chemically unstable gas A Chemically unstable gas B: 1B, 2	
		Chemicals under pressure (chapter 2.3)	1, 2	
P390	Absorb spillage to prevent material-damage.	Corrosive to metals (chapter 2.16)	1	
P391	Collect spillage.	Hazardous to the aquatic environment, acute hazard (chapter 4.1)	1	
		Hazardous to the aquatic environment, long-term hazard (chapter 4.1)	1, 2	
P301 + P316	IF SWALLOWED: Get emergency medical help immediately.	Acute toxicity, oral (chapter 3.1)	1, 2, 3	Competent authority or manufacturer / supplier may add, "Call" followed by the appropriate emergency telephone number, or the appropriate emergency medical help provider, for example, a poison centre, emergency centre or doctor.
		Aspiration hazard (chapter 3.10)	1, 2	
P301 + P317	IF SWALLOWED: Get medical help.	Acute toxicity, oral (chapter 3.1)	4, 5	
P302 + P317	IF ON SKIN: Get medical help.	Acute toxicity, dermal (chapter 3.1)	5	
P302 + P334	IF ON SKIN: Immerse in cool water or wrap in wet bandages.	Pyrophoric liquids (chapter 2.9)	1	
P302 + P352	IF ON SKIN: Wash with plenty of water/...	Acute toxicity, dermal (chapter 3.1)	1, 2, 3, 4	...Manufacturer/supplier or the competent authority may specify a cleansing agent if appropriate or may recommend an alternative agent in exceptional cases if water is clearly inappropriate.
		Skin irritation (chapter 3.2)	2	
		Skin sensitization (chapter 3.4)	1, 1A, 1B	
P304 + P317	IF INHALED: Get medical help.	Acute toxicity, inhalation (chapter 3.1)	5	

表 A3.2.3 応急措置注意書き（続き）

コード (1)	応急措置注意書き (2)	危険有害性クラス (3)	危険有害性区分 (4)		使用の条件 (5)
P381	漏えいした場合、着火源を除去すること。	可燃性ガス (2.2章)	可燃性ガス	1A	
			自然発火性ガス		
			化学的に不安定なガス A	1B, 2	
			化学的に不安定なガス B		
		加圧下化学品 (2.3章)		1, 2	
P390	物的被害を防止するために流出したものを吸収すること	金属腐食性 (2.16章)	1		
P391	漏出物を回収すること。	水生環境有害性（急性）(4.1章)	1		
		水生環境有害性（慢性）(4.1章)	1, 2		
P301 + P316	飲み込んだ場合：すぐに救急の医療処置を受けること。	急性毒性（経口）(3.1章)	1, 2, 3		所管官庁または製造者/供給者が電話に続けて、適当な救急時電話番号など、適当な救急時医療提供者、例えば中毒センター、救急センターまたは医師などを追加してもよい
		誤えん有害性 (3.10章)	1, 2		
P301 + P317	飲み込んだ場合：医療処置を受けること。	急性毒性（経口）(3.1章)	4, 5		
P302 + P317	皮膚に付着した場合：冷たい水に浸すこと（または湿った包帯で覆うこと）。	急性毒性（経皮）(3.1章)	5		
P302 + P334	皮膚に付着した場合：冷たい水に浸すこと（または湿った包帯で覆うこと）。	自然発火性液体 (2.9章)	1		
P302 + P352	皮膚に付着した場合：多量の水の水／…で洗う	急性毒性（経皮）(3.1章)	1, 2, 3, 4		製造者/供給者または所管官庁が指定する適切な洗浄剤がある場合、または明らかに水が不適切な場合は他の薬剤を推薦すること
		皮膚刺激性 (3.2章)	2		
		皮膚感作性 (3.4章)	1, 1A, 1B		
P304 + P317	吸入した場合：医療処置を受けること。	急性毒性（吸入）(3.1章)	5		

- 320 -

Table A3.2.3: Response precautionary statements *(cont'd)*

Code (1)	Response precautionary statements (2)	Hazard class (3)	Hazard category (4)	Conditions for use (5)
P304 + P340	IF INHALED: Remove person to fresh air and keep comfortable for breathing.	Acute toxicity, inhalation (chapter 3.1)	1, 2, 3, 4	
		Skin corrosion (chapter 3.2)	1, 1A, 1B, 1C	
		Respiratory sensitization (chapter 3.4)	1, 1A, 1B	
		Specific target organ toxicity, single exposure; respiratory tract irritation (chapter 3.8)	3	
		Specific target organ toxicity, single exposure; narcotic effects (chapter 3.8)	3	
P306 + P360	IF ON CLOTHING: Rinse immediately contaminated clothing and skin with plenty of water before removing clothes.	Oxidizing liquids (chapter 2.13)	1	
		Oxidizing solids (chapter 2.14)	1	
P308 + P316	IF exposed or concerned: Get emergency medical help immediately.	Specific target organ toxicity, single exposure (chapter 3.8)	1, 2	Competent authority or manufacturer / supplier may add, "Call" followed by the appropriate emergency telephone number, or the appropriate emergency medical help provider, for example, a poison centre, emergency centre or doctor.
P332 + P317	If skin irritation occurs: Get medical help.	Skin irritation (chapter 3.2)	2, 3	– *may be omitted when P333+P317 is given on the label.*
P333 + P317	If skin irritation or rash occurs: Get medical help.	Skin sensitization (chapter 3.4)	1, 1A, 1B	
P336 + P317	Immediately thaw frosted parts with lukewarm water. Do not rub affected area. Get medical help.	Gases under pressure (chapter 2.5)	Refrigerated liquefied gas	
P337 + P317	If eye irritation persists: Get medical help.	Eye irritation (chapter 3.3)	2/2A, 2B	
P342 + P316	If experiencing respiratory symptoms: Get emergency medical help immediately.	Respiratory sensitization (chapter 3.4)	1, 1A, 1B	Competent authority or manufacturer/supplier may add, "Call" followed by the appropriate emergency telephone number, or the appropriate emergency medical help provider, for example, a poison centre, emergency centre or doctor.
P361 + P364	Take off immediately all contaminated clothing and wash it before reuse.	Acute toxicity, dermal (chapter 3.1)	1, 2, 3	

表 A3.2.3 応急措置注意書き(続き)

コード (1)	応急措置注意書き (2)	危険有害性クラス (3)	危険有害性区分 (4)	使用の条件 (5)
P304 + P340	吸入した場合：空気の新鮮な場所に移し、呼吸しやすい姿勢で休息させること。	急性毒性（吸入）(3.1章)	1, 2, 3, 4	
		皮膚腐食性 (3.2章)	1, 1A, 1B, 1C	
		呼吸器感作性 (3.4章)	1, 1A, 1B	
		特定標的臓器毒性、単回ばく露、気道刺激性 (3.8章)	3	
		特定標的臓器毒性、単回ばく露、麻酔作用 (3.8章)	3	
P306 + P360	衣類にかかった場合：服を脱ぐ前に、直ちに汚染された衣類及び皮膚を多量の水で洗うこと。	酸化性液体 (2.13章)	1	
		酸化性固体 (2.14章)	1	
P308 + P316	ばく露またはその懸念がある場合：すぐに救急の医療処置を受けること。	特定標的臓器毒性、単回ばく露、麻酔作用 (3.8章)	1, 2	所管官庁または製造者/供給者は、「電話」に続けて、適当な救急時電話番号すなわち、例えば中毒医療提供者、救急センター、救急センターまたは医師などを追加してもよい。
P332 + P317	皮膚刺激が生じた場合：医療処置を受けること。	皮膚刺激性 (3.2章)	2, 3	−P333+P317 がラベル上にあるときは省略してもよい
P333 + P317	皮膚刺激または発疹が生じた場合：医療処置を受けること。	皮膚感作性 (3.4章)	1, 1A, 1B	
P336 + P317	凍った部分をすぐにぬるま湯でとかすこと。受傷部はこすらないこと。医療処置を受けること。	高圧ガス (2.5章)	深冷液化ガス	
P337 + P317	眼の刺激が続く場合：医療処置を受けること。	眼刺激性 (3.3章)	2/2A, 2B	
P342 + P316	呼吸に関する症状が出た場合：すぐに救急の医療処置を受けること。	呼吸器感作性 (3.4章)	1, 1A, 1B	所管官庁または製造者/供給者は、「電話」に続けて、適当な救急時電話番号すなわち、例えば中毒医療提供者、救急センター、救急センターまたは医師などを追加してもよい
P361 + P364	汚染された衣類を直ちにすべて脱ぎ、再使用する場合には洗濯をすること。	急性毒性（経皮）(3.1章)	1, 2, 3	

Table A3.2.3: Response precautionary statements *(cont'd)*

Code (1)	Response precautionary statements (2)	Hazard class (3)	Hazard category (4)	Conditions for use (5)
P362 + P364	**Take off contaminated clothing and wash it before reuse.**	Acute toxicity, dermal (chapter 3.1)	4	
		Skin irritation (chapter 3.2)	2	
		Skin sensitization (chapter 3.4)	1, 1A, 1B	
P370 + P376	**In case of fire: Stop leak if safe to do so.**	Oxidizing gases (chapter 2.4)	1	
P370 + P378	**In case of fire: Use ... to extinguish.**	Chemicals under pressure (chapter 2.3)	1, 2	– *if water increases risk.*
		Flammable liquids (chapter 2.6)	1, 2, 3, 4	... Manufacturer/supplier or the competent authority to specify appropriate media.
		Flammable solids (chapter 2.7)	1, 2	
		Self-reactive substances and mixtures (chapter 2.8)	Types C, D, E, F	
		Pyrophoric liquids (chapter 2.9)	1	
		Pyrophoric solids (chapter 2.10)	1	
		Substances and mixtures which, in contact with water, emit flammable gases (chapter 2.12)	1, 2, 3	
		Oxidizing liquids (chapter 2.13)	1, 2, 3	
		Oxidizing solids (chapter 2.14)	1, 2, 3	
		Organic peroxides (chapter 2.15)	Types C, D, E, F	
P301 + P330 + P331	**IF SWALLOWED: Rinse mouth. Do NOT induce vomiting.**	Skin corrosion (chapter 3.2)	1, 1A, 1B, 1C	
P302 + P335 + P334	**IF ON SKIN: Brush off loose particles from skin and immerse in cool water [or wrap in wet bandages].**	Pyrophoric solids (chapter 2.10)	1	– *text in square brackets to be used for pyrophoric solids*
		Substances and mixtures which, in contact with water, emit flammable gases (chapter 2.12)	1, 2	– *use only "Immerse in cool water." Text in square brackets should not be used.*
P302 + P361 + P354	**IF ON SKIN: Take off immediately all contaminated clothing. Immediately rinse with water for several minutes.**	Skin corrosion (chapter 3.2)	1, 1A, 1B, 1C	

表 A3.2.3 応急措置注意書き(続き)

コード (1)	応急措置注意書き (2)	危険有害性クラス (3)	危険有害性区分 (4)	使用の条件 (5)
P362 + P364	汚染された衣類を脱ぎ、再使用する場合には洗濯をすること。	急性毒性（経皮）(3.1章)	4	
		皮膚刺激性 (3.2章)	2	
		皮膚感作性 (3.4章)	1, 1A, 1B	
P370 + P376	火災の場合：安全に対処できるならば漏洩を止めること。	酸化性ガス (2.4章)	1	
P370 + P378	火災の場合：消火するために…を使用すること。	加圧下化学品 (2.3章)	1, 2	―水がリスクを増大させる場合
		引火性液体 (2.6章)	1, 2, 3, 4	…製造者／供給者または所管官庁が指定する
		可燃性固体 (2.7章)	1, 2	適当な手段
		自己反応性物質および混合物 (2.8章)	タイプC, D, E, F	
		自然発火性液体 (2.9章)	1	
		自然発火性固体 (2.10章)	1	
		水反応可燃性物質および混合物 (2.12章)	1, 2, 3	
		酸化性液体 (2.13章)	1, 2, 3	
		酸化性固体 (2.14章)	1, 2, 3	
		有機過酸化物 (2.15章)	タイプC, D, E, F	
P301 + P330 + P331	飲み込んだ場合：口をすすぐこと。無理に吐かせないこと。	皮膚腐食性 (3.2章)	1, 1A, 1B, 1C	
P302 + P335 + P334	皮膚に付着した場合：固着していない粒子を皮膚から払いのけ、冷たい水に浸すこと［または湿った包帯で覆うこと］。	自然発火性固体 (2.10章)	1	― [ ] 内の文章は自然発火性固体に使用する
		水反応可燃性物質および混合物 (2.12章)	1, 2	― [ ] 内の文章は水に浸すことのみ使用する。 [ ] 内の文章は使用しない
P302 + P361 + P354	皮膚に付着した場合：直ちに汚染された衣類をすべて脱ぐこと。すぐに水で数分間洗うこと。	皮膚腐食性 (3.2章)	1, 1A, 1B, 1C	

- 322 -

Table A3.2.3: Response precautionary statements (cont'd)

Code (1)	Response precautionary statements (2)	Hazard class (3)	Hazard category (4)	Conditions for use (5)
P303 + P361 + P353	IF ON SKIN (or hair): Take off immediately all contaminated clothing. Rinse affected areas with water [or shower].	Flammable liquids (chapter 2.6)	1, 2, 3	– text in square brackets to be included where the manufacturer/supplier or the competent authority considers it appropriate for the specific chemical.
P305 + P351 + P338	IF IN EYES: Rinse cautiously with water for several minutes. Remove contact lenses, if present and easy to do. Continue rinsing.	Eye irritation (chapter 3.3)	2/2A, 2B	
P305 + P354 + P338	IF IN EYES: Immediately rinse with water for several minutes. Remove contact lenses, if present and easy to do. Continue rinsing.	Skin corrosion (chapter 3.2)	1, 1A, 1B, 1C	
		Serious eye damage (chapter 3.3)	1	
P370 + P380 + P375	In case of fire: Evacuate area. Fight fire remotely due to the risk of explosion.	Explosives (chapter 2.1)	2C	
		Desensitized explosives (chapter 2.17)	1, 2, 3	
P371 + P380 + P375	In case of major fire and large quantities: Evacuate area. Fight fire remotely due to the risk of explosion.	Oxidizing liquids (chapter 2.13)	1	
		Oxidizing solids (chapter 2.14)	1	
		Desensitized explosives (chapter 2.17)	4	
P370 + P372 + P380 + P373	In case of fire: Explosion risk. Evacuate area. DO NOT fight fire when fire reaches explosives.	Explosives (chapter 2.1)	1, 2A, 2B	
		Self-reactive substances and mixtures (chapter 2.8)	Type A	
		Organic peroxides (chapter 2.15)	Type A	
P370 + P380 + P375 [+ P378]	In case of fire: Evacuate area. Fight fire remotely due to the risk of explosion. [Use…to extinguish].	Self-reactive substances and mixtures (chapter 2.8)	Type B	– text in square brackets to be used if water increases risk.
		Organic peroxides (chapter 2.15)	Type B	… Manufacturer/supplier or the competent authority to specify appropriate media.

- 323 -

表 A3.2.3 応急措置注意書き(続き)

コード (1)	応急措置注意書き (2)	危険有害性クラス (3)	危険有害性区分 (4)	使用の条件 (5)
P303 + P361 + P353	皮膚(または髪)に付着した場合：直ちに汚染された衣類をすべて脱ぐこと。接触部位を水【またはシャワー】で洗うこと。	引火性液体 (2.6 章)	1, 2, 3	一製造者供給者または所管官庁が特定の化学品に対してそれが適当だとした場合には{ }内の文章を含める
P305 + P351 + P338	眼に入った場合：水で数分間注意深く洗うこと。コンタクトレンズを着用していて容易に外せる場合は外すこと。その後も洗浄を続けること。	眼刺激性 (3.3 章)	2/2A, 2B	
P305 + P354 + P338	眼に入った場合：すぐに水で数分間洗うこと。コンタクトレンズを着用していて容易に外せる場合は外すこと。	皮膚腐食性 (3.2 章) 重篤な眼の損傷性 (3.3 章)	1, 1A, 1B, 1C 1	
P370 + P380 + P375	火災の場合：区域にある爆発の危険性があるため、離れた距離から消火すること。	爆発物 (2.1 章) 鈍化爆発物 (2.17 章)	2C 1, 2, 3	
P371 + P380 + P375	大火災の場合で大量にある場合：区域より退避させ、爆発の危険性があるため、離れた距離から消火すること。	酸化性液体 (2.13 章) 酸化性固体 (2.14 章) 鈍化爆発物 (2.17 章)	1 1 4	
P370 + P372 + P380 + P373	火災の場合：爆発する危険性あり、区域より退避させること。炎が爆発物に届いたら消火活動をしないこと。	爆発物 (2.1 章) 自己反応性物質および混合物 (2.8 章) 有機過酸化物 (2.15 章)	1, 2A, 2B タイプ A タイプ A	
P370 + P380 + P375 [+ P378]	火災の場合：区域より退避させ、爆発の危険性があるため、離れた距離から消火すること。【消火するために…を使用すること。】	自己反応性物質および混合物 (2.8 章) 有機過酸化物 (2.15 章)	タイプ B タイプ B	一{ }内の文章は水がリスクを大きくする場合に使用する。 一{ }内の使用者供給者または所管官庁が指定する ……適当な手段

- 323 -

Table A3.2.4: Storage precautionary statements

Code (1)	Storage precautionary statements (2)	Hazard class (3)	Hazard category (4)	Conditions for use (5)
P401	**Store in accordance with...**	Explosives (chapter 2.1)	1, 2A, 2B, 2C	... Manufacturer/supplier or the competent authority to specify local/regional/ national/international regulations as applicable.
		Desensitized explosives (chapter 2.17)	1, 2, 3, 4	
P402	**Store in a dry place.**	Substances and mixtures which, in contact with water, emit flammable gases (chapter 2.12)	1, 2, 3	
P403	**Store in a well-ventilated place.**	Flammable gases (chapter 2.2)	Flammable gas — 1A	
			Pyrophoric gas — 1A	
			Chemically unstable gas A — 1A	
			Chemically unstable gas B — 1A	
			1B, 2	
		Chemicals under pressure (chapter 2.3)	1, 2, 3	
		Oxidizing gases (chapter 2.4)	1	
		Gases under pressure (chapter 2.5)	Compressed gas	
			Liquefied gas	
			Refrigerated liquefied gas	
			Dissolved gas	
		Flammable liquids (chapter 2.6)	1, 2, 3, 4	*– for flammable liquids Category 1 and other flammable liquids that are volatile and may generate an explosive atmosphere.*
		Self-reactive substances and mixtures (chapter 2.8)	Types A, B, C, D, E, F	*– except for temperature controlled self-reactive substances and mixtures or organic peroxides because condensation and consequent freezing may take place.*
		Organic peroxides (chapter 2.15)	Types A, B, C, D, E, F	
		Acute toxicity, inhalation (chapter 3.1)	1, 2, 3	*– if the chemical is volatile and may generate a hazardous atmosphere.*
		Respiratory sensitization (chapter 3.4)	1, 1A, 1B	
		Specific target organ toxicity, single exposure; respiratory tract irritation (chapter 3.8)	3	*– if the chemical is volatile and may generate a hazardous atmosphere.*
		Specific target organ toxicity, single exposure; narcotic effects (chapter 3.8)	3	

表 A3.2.4 保管注意書き

コード (1)	保管注意書き (2)	危険有害性クラス (3)	危険有害性区分 (4)	使用の条件 (5)
P401	…にしたがって保管すること。	爆発物 (2.1 章)	1, 2A, 2B, 2C	…製造者/供給者または所管官庁が特定する国の規則で適用できる国際/都道府県/市町村の規則
		鈍感化爆発物 (2.17 章)	1, 2, 3, 4	
P402	乾燥した場所に保管すること。	水反応可燃性物質および混合物 (2.12 章)	1, 2, 3	
P403	換気の良い場所で保管すること。	可燃性ガス (2.2 章) 1A 可燃性ガス		
		自然発火性ガス		
		化学的に不安定なガス A		
		化学的に不安定なガス B		
			1B, 2	
		加圧下化学品 (2.3 章)	1, 2, 3	
		酸化性ガス (2.4 章)	1	
		高圧ガス (2.5 章) 圧縮ガス		
		液化ガス		
		深冷液化ガス		
		溶解ガス		
		引火性液体 (2.6 章)	1, 2, 3, 4	一引火性液体区分 1 および他の引火性液体で揮発性が高く爆発する環境をつくる可能性があるとき
		自己反応性物質および混合物 (2.8 章)	タイプ A, B, C, D, E, F	一温度が管理されている自己反応性物質および混合物または有機過酸化物は、濃縮およびそれに伴う凍結が起きるので、除外する
		有機過酸化物 (2.15 章)	タイプ A, B, C, D, E, F	
		急性毒性 (吸入) (3.1 章)	1, 2, 3	一化学品が揮発性で有害な環境をつくりだす場合
		呼吸器感作性 (3.4 章)	1, 1A, 1B	
		特定標的臓器毒性、単回ばく露、気道刺激性 (3.8 章)	3	一化学品が揮発性で有害な環境をつくりだす場合
		特定標的臓器毒性、単回ばく露、麻酔作用 (3.8 章)	3	一化学品が揮発性で有害な環境をつくりだす場合

Table A3.2.4: Storage precautionary statements *(cont'd)*

Code (1)	Storage precautionary statements (2)	Hazard class (3)	Hazard category (4)	Conditions for use (5)
P404	**Store in a closed container.**	Substances and mixtures which, in contact with water, emit flammable gases (chapter 2.12)	1, 2, 3	
P405	**Store locked up.**	Acute toxicity, oral (chapter 3.1)	1, 2, 3	
		Acute toxicity, dermal (chapter 3.1)	1, 2, 3	
		Acute toxicity, inhalation (chapter 3.1)	1, 2, 3	
		Skin corrosion (chapter 3.2)	1, 1A, 1B, 1C	
		Germ cell mutagenicity (chapter 3.5)	1, 1A, 1B, 2	
		Carcinogenicity (chapter 3.6)	1, 1A, 1B, 2	
		Reproductive toxicity (chapter 3.7)	1, 1A, 1B, 2	
		Specific target organ toxicity, single exposure (chapter 3.8)	1, 2	
		Specific target organ toxicity, single exposure; respiratory tract irritation (chapter 3.8)	3	
		Specific target organ toxicity, single exposure; narcotic effects (chapter 3.8)	3	
		Aspiration hazard (chapter 3.10)	1, 2	
P406	**Store in a corrosion resistant/...container with a resistant inner liner.**	Corrosive to metals (chapter 2.16)	1	*– may be omitted if P234 is given on the label* *... Manufacturer/supplier or the competent authority to specify other compatible materials.*
P407	**Maintain air gap between stacks or pallets.**	Self-heating substances and mixtures (chapter 2.11)	1, 2	
P410	**Protect from sunlight.**	Aerosols (chapter 2.3)	1, 2, 3	
		Chemicals under pressure (chapter 2.3)	1, 2, 3	*– may be omitted for chemicals under pressure filled in transportable cylinders in accordance with packing instruction P200 or P206 of the UN Model Regulations, unless those chemicals under pressure are subject to (slow) decomposition or polymerization, or the competent authority provides otherwise.*

表 A3.2.4 保管注意書き (続き)

コード (1)	保管注意書き (2)	危険有害性クラス (3)	危険有害性区分 (4)	使用の条件 (5)
P404	密閉容器に保管すること。	水反応可燃性物質および混合物 (2.12章)	1, 2, 3	
P405	施錠して保管すること。	急性毒性（経口）(3.1章)	1, 2, 3	
		急性毒性（経皮）(3.1章)	1, 2, 3	
		急性毒性（吸入）(3.1章)	1, 2, 3	
		皮膚腐食性 (3.2章)	1, 1A, 1B, 1C	
		生殖細胞変異原性 (3.5章)	1, 1A, 1B, 2	
		発がん性 (3.6章)	1, 1A, 1B, 2	
		生殖毒性 (3.7章)	1, 1A, 1B, 2	
		特定標的臓器毒性、単回ばく露、(3.8章)	1, 2	
		特定標的臓器毒性、単回ばく露、気道刺激性 (3.8章)	3	
		特定標的臓器毒性、単回ばく露、麻酔作用 (3.8章)	3	
		誤えん有害性 (3.10章)	1, 2	
P406	耐腐食性/耐腐食性内張りのある…容器に保管すること。	金属腐食性 (2.16章)	1	－P234 がラベルに記載されている場合には省略してもよい。 …製造者/供給者または所管官庁が指定する他の互換性がある材料
P407	積荷またはパレット間にすきまをあけること。	自己発熱性物質および混合物 (2.11章)	1, 2	
P410	日光から遮断すること。	エアゾール (2.3章)	1, 2, 3	
		加圧下化学品 (2.3章)	1, 2, 3	－加圧下化学品が UN モデル規則の包装指示 P200 または P206 にしたがっている場合は輸送用シリンダーに充填されている場合には削除してもよい、ただしこれらの加圧下化学品が (遅い) 分解や高分子化を起こさず、または所管官庁が他の方法を示さない場合に限る

Table A3.2.4: Storage precautionary statements (cont'd)

Code (1)	Storage precautionary statements (2)	Hazard class (3)	Hazard category (4)	Conditions for use (5)
P410 (con't)	**Protect from sunlight.**	Gases under pressure (chapter 2.5)	Compressed gas Liquefied gas Dissolved gas	– *may be omitted for gases filled in transportable gas cylinders in accordance with packing instruction P200 of the UN Recommendations on the Transport of Dangerous Goods, Model Regulations, unless those gases are subject to (slow) decomposition or polymerisation, or the competent authority provides otherwise.*
		Self-heating substances and mixtures (chapter 2.11)	1, 2	
		Organic peroxides (chapter 2.15)	Types A, B, C, D, E, F	
P411	**Store at temperatures not exceeding …°C/…°F.**	Self-reactive substances and mixtures (chapter 2.8)	Types A, B, C, D, E, F	– *if temperature control is required (according to section 2.8.2.3 or 2.15.2.3 of the GHS) or if otherwise deemed necessary.*
		Organic peroxides (chapter 2.15)	Types A, B, C, D, E, F	…Manufacturer/supplier or the competent authority to specify temperature using applicable temperature scale.
P412	**Do not expose to temperatures exceeding 50°C/ 122 °F.**	Aerosols (chapter 2.3)	1, 2, 3	Manufacturer/supplier or the competent authority to use applicable temperature scale.
P413	**Store bulk masses greater than … kg/…lbs at temperatures not exceeding …°C/…°F.**	Self-heating substances and mixtures (chapter 2.11)	1, 2	…Manufacturer/supplier or the competent authority to specify mass and temperature using applicable scale.
P420	**Store separately.**	Self-reactive substances and mixtures (chapter 2.8)	Types A, B, C, D, E, F	
		Self-heating substances and mixtures (chapter 2.11)	1, 2	
		Oxidizing liquids (chapter 2.13)	1	
		Oxidizing solids (chapter 2.14)	1	
		Organic peroxides (chapter 2.15)	Types A, B, C, D, E, F	
P402 + P404	**Store in a dry place. Store in a closed container.**	Substances and mixtures which, in contact with water, emit flammable gases (chapter 2.12)	1, 2, 3	

表 A3.2.4 保管注意書き（続き）

コード (1)	保管注意書き (2)	危険有害性クラス (3)	危険有害性区分 (4)	使用の条件 (5)
P410 (継続)	日光から遮断すること。	高圧ガス (2.5 章)	圧縮ガス 液化ガス 溶解ガス	―ガスが UN モデル規則の包装指示 P200 にしたがっている場合はガスシリンダーに充填されている場合には削除して（遅い）もよい、ただしこれらのガスが分解や高分子化を起こさず、また所管官庁が他の方法を示さない場合に限る
		自己発熱性物質および混合物 (2.11 章)	1, 2	
		有機過酸化物 (2.15 章)	タイプ A, B, C, D, E, F	
P411	…°C 以下の温度で保管すること。	自己反応性物質および混合物 (2.8 章)	タイプ A, B, C, D, E, F	―温度管理が必要な場合 (GHS2.8.2.3 または 2.15.2.3 により) あるいは他の方法が必要と考えられる場合 …製造者/供給者または所管官庁が指定する適用可能な温度計を用いて指定した温度
		有機過酸化物 (2.15 章)	タイプ A, B, C, D, E, F	
P412	50°C 以上の温度にばく露しないこと。	エアゾール (2.3 章)	1, 2, 3	
P413	…kg 以上の大量品は、…°C 以下の温度で保管すること。	自己発熱性物質および混合物 (2.11 章)	1, 2	―製造者/供給者または所管官庁が指定する適用可能な計測器を用いた量と温度
P420	隔離して保管すること。	自己反応性物質および混合物 (2.8 章)	タイプ A, B, C, D, E, F	
		自己発熱性物質および混合物 (2.11 章)	1, 2	
		酸化性液体 (2.13 章)	1	
		酸化性固体 (2.14 章)	1	
		有機過酸化物 (2.15 章)	タイプ A, B, C, D, E, F	
P402 + P404	乾燥した場所または密閉容器に保管すること。	水反応可燃性物質および混合物 (2.12 章)	1, 2, 3	

Table A3.2.4: Storage precautionary statements *(cont'd)*

Code (1)	Storage precautionary statements (2)	Hazard class (3)	Hazard category (4)	Conditions for use (5)
P403 + P233	**Store in a well-ventilated place. Keep container tightly closed.**	Acute toxicity, inhalation (chapter 3.1)	1, 2, 3	– if the chemical is volatile and may generate a hazardous atmosphere.
		Specific target organ toxicity, single exposure; respiratory tract irritation (chapter 3.8)	3	
		Specific target organ toxicity, single exposure; narcotic effects (chapter 3.8)	3	
P403 + P235	**Store in a well-ventilated place. Keep cool.**	Flammable liquids (chapter 2.6)	1, 2, 3	– for flammable liquids Category 1 and other flammable liquids that are volatile and may generate an explosive atmosphere.
P410 + P403	**Protect from sunlight. Store in a well-ventilated place.**	Chemicals under pressure (chapter 2.3)	1, 2, 3	– P410 may be omitted for chemicals under pressure filled in transportable cylinders in accordance with packing instruction P200 or P206 of the UN Model Regulations, unless those chemicals under pressure are subject to (slow) decomposition or polymerization, or the competent authority provides otherwise.
		Gases under pressure (chapter 2.5)	Compressed gas Liquefied gas Dissolved gas	– P410 may be omitted for gases filled in transportable gas cylinders in accordance with packing instruction P200 of the UN Model Regulations, unless those gases are subject to (slow) decomposition or polymerisation, or the competent authority provides otherwise.
P410 + P412	**Protect from sunlight. Do not expose to temperatures exceeding 50°C/122°F.**	Aerosols (chapter 2.3)	1, 2, 3	Manufacturer/supplier or the competent authority to use applicable temperature scale.

表 A3.2.4 保管注意書き(続き)

コード (1)	保管注意書き (2)	危険有害性クラス (3)	危険有害性区分 (4)	使用の条件 (5)
P403 + P233	換気の良い場所で保管すること。容器を密閉しておくこと。	急性毒性（吸入）(3.1章) 特定標的臓器毒性、単回ばく露、気道刺激性 (3.8章) 特定標的臓器毒性、単回ばく露、麻酔作用 (3.8章)	1, 2, 3 3 3	―化学品が揮発性で有害な環境を作る可能性があるとき
P403 + P235	換気の良い場所で保管すること。涼しいところに置くこと。	引火性液体 (2.6章)	1, 2, 3	―引火性液体区分 1 および揮発性があり爆発する環境を作る可能性のある液体
P410 + P403	日光から遮断し、換気の良い場所で保管すること。	加圧下化学品 (2.3章)	1, 2, 3	―P410 は、加圧下化学品が UN モデル規則の包装用指示 P200 または P206 にしたがっている輸送用シリンダーに充填されている場合には削除してもよい、ただしこれからの加圧下化学品が（遅い）分解や高分子化を起こさず、または所管官庁が他の方法を示さない場合に限る
		高圧ガス (2.5章)	圧縮ガス 液化ガス 溶解ガス	―P410 は、ガスが UN モデル規則の包装用指示 P200 にしたがっている運送用ガスシリンダーに充填されている場合には削除してもよい、ただしこれらのガスが（遅い）分解や高分子化を起こさず、または所管官庁が他の方法を示さない場合に限る。
P410 + P412	日光から遮断し、50℃以上の温度にばく露しないこと。	エアゾール (2.3章)	1, 2, 3	―製造者/供給者または所管官庁が指定する適用可能な温度計を用いる

Table A3.2.5: Disposal precautionary statements

Code (1)	Disposal precautionary statements (2)	Hazard class (3)	Hazard category (4)	Conditions for use (5)
P501	**Dispose of contents/container to ...**	Flammable liquids (chapter 2.6)	1, 2, 3, 4	... in accordance with local/regional/national/international regulation (to be specified). Manufacturer/supplier or the competent authority to specify whether disposal requirements apply to contents, container or both.
		Self-reactive substances and mixtures (chapter 2.8)	Types A, B, C, D, E, F	
		Substances and mixtures which, in contact with water, emit flammable gases (chapter 2.12)	1, 2, 3	
		Oxidizing liquids (chapter 2.13)	1, 2, 3	
		Oxidizing solids (chapter 2.14)	1, 2, 3	
		Organic peroxides (chapter 2.15)	Types A, B, C, D, E, F	
		Desensitized explosives (chapter 2.17)	1, 2, 3, 4	
		Acute toxicity, oral (chapter 3.1)	1, 2, 3, 4	
		Acute toxicity, dermal (chapter 3.1)	1, 2, 3, 4	
		Acute toxicity, inhalation (chapter 3.1)	1, 2, 3	
		Skin corrosion (chapter 3.2)	1, 1A, 1B, 1C	... in accordance with local/regional/national/international regulation (to be specified). Manufacturer/supplier or the competent authority to specify whether disposal requirements apply to contents, container or both.
		Respiratory sensitization (chapter 3.4)	1, 1A, 1B	
		Skin sensitization (chapter 3.4)	1, 1A, 1B	
		Germ cell mutagenicity (chapter 3.5)	1, 1A, 1B, 2	
		Carcinogenicity (chapter 3.6)	1, 1A, 1B, 2	
		Reproductive toxicity (chapter 3.7)	1, 1A, 1B, 2	
		Specific target organ toxicity, single exposure (chapter 3.8)	1, 2	
		Specific target organ toxicity, single exposure; respiratory tract irritation (chapter 3.8)	3	
		Specific target organ toxicity, single exposure; narcotic effects (chapter 3.8)	3	
		Specific target organ toxicity, repeated exposure (chapter 3.9)	1, 2	
		Aspiration hazard (chapter 3.10)	1, 2	
		Hazardous to the aquatic environment, acute hazard (chapter 4.1)	1, 2, 3	
		Hazardous to the aquatic environment, long-term hazard (chapter 4.1)	1, 2, 3, 4	

表 A3.2.5 廃棄注意書き

コード (1)	廃棄注意書き (2)	危険有害性クラス (3)	危険有害性区分 (4)	使用の条件 (5)
P501	内容物/容器を…に廃棄すること。	引火性液体 (2.6章)	1, 2, 3, 4	…国際/国/都道府県/市町村の規則(明示する)に従って製造者/供給者または所管官庁が指定する内容物、容器またはその両者に適用する廃棄物要件
		自己反応性物質および混合物 (2.8章)	タイプ A, B, C, D, E, F	
		水反応可燃性物質および混合物 (2.12章)	1, 2, 3	
		酸化性液体 (2.13章)	1, 2, 3	
		酸化性固体 (2.14章)	1, 2, 3	
		有機過酸化物 (2.15章)	タイプ A, B, C, D, E, F	
		鈍性化爆発物 (2.17章)	1, 2, 3, 4	
		急性毒性 (経口) (3.1章)	1, 2, 3, 4	
		急性毒性 (経皮) (3.1章)	1, 2, 3, 4	
		急性毒性 (吸入) (3.1章)	1, 2, 3	
		皮膚腐食性 (3.2章)	1, 1A, 1B, 1C	
		呼吸器感作性 (3.4章)	1, 1A, 1B	
		皮膚感作性 (3.4章)	1, 1A, 1B	
		生殖細胞変異原性 (3.5章)	1, 1A, 1B, 2	
		発がん性 (3.6章)	1, 1A, 1B, 2	…国際/国/都道府県/市町村の規則(明示する)に従って製造者/供給者または所管官庁が指定する内容物、容器またはその両者に適用する廃棄物要件
		生殖毒性 (3.7章)	1, 1A, 1B, 2	
		特定標的臓器毒性、単回ばく露、(3.8章)	1, 2	
		特定標的臓器毒性、単回ばく露、気道刺激性 (3.8章)	3	
		特定標的臓器有害性、単回ばく露、麻酔作用 (3.8章)	3	
		特定標的臓器毒性、反復ばく露、(3.9章)	1, 2	
		誤えん有害性 (3.10章)	1, 2	
		水生環境有害性 (急性) (4.1章)	1, 2, 3	
		水生環境有害性 (慢性) (4.1章)	1, 2, 3, 4	

Table A3.2.5: Disposal precautionary statements *(cont'd)*

Code (1)	Disposal precautionary statements (2)	Hazard class (3)	Hazard category (4)	Conditions for use (5)
P502	**Refer to manufacturer or supplier for information on recovery or recycling.**	Hazardous to the ozone layer (chapter 4.2)	1	
P503	**Refer to manufacturer/supplier… for information on disposal/recovery/recycling.**	Explosives (chapter 2.1)	1A, 2A, 2B, 2C	… Manufacturer/supplier or the competent authority to specify appropriate source of information in accordance with local/regional/national/international regulations as applicable

表 A3.2.5 廃棄注意書き(続き)

コード (1)	廃棄注意書き (2)	危険有害性クラス (3)	危険有害性区分 (4)	使用の条件 (5)
P502	回収またはリサイクルに関する情報について製造者または供給者に問い合わせる。	オゾン層への有害性 (4.2章)	1	
P503	廃棄/回収/リサイクルに関する情報について製造者/供給者/…に問い合わせる。	爆発物 (2.1章)	1, 2A, 2B, 2C	…製造者/供給者または所管官庁が特定する適用可能な国際/国/都道府県/市町村の規則に従った適当な情報源

- 329 -

ANNEX 3

Section 3

MATRIX OF PRECAUTIONARY STATEMENTS BY HAZARD CLASS/CATEGORY

A3.3.1   **Introduction**

A3.3.1.1   This section sets out a matrix listing the recommended precautionary statements for each hazard class and hazard category of the GHS by type of precautionary statement (see A3.2.1.2 and A3.2.2.1) except for general precautionary statements that do not have specific hazard class or categories. The matrix guides the selection of appropriate precautionary statements and includes elements for all categories of precautionary action. All specific elements relating to particular hazard classes should be used. In addition, general precautionary statements not linked to a certain hazard class or category should also be used where relevant (see A3.2.6).

A3.3.1.2   The tables making up the matrix show the core part of the precautionary statements in bold print. This is the text, except as otherwise specified, that should appear on the label. However, it is not necessary to insist on identical sets of words in all situations.

A3.3.1.3   In the majority of cases, the recommended precautionary statements are independent, e.g. the phrases for explosive hazard do not modify those related to certain health hazards and products that are classified for both hazard classes should bear appropriate precautionary statements for both.

附属書 3

第 3 節

危険有害性クラス/区分にしたがった注意書きのマトリクス

A3.3.1　序文

A3.3.1.1　当節では、GHSの各危険有害性クラスおよび危険有害性区分に対して推奨される注意書きを、特定の危険有害性クラスまたは区分を持たない一般的な注意書きを除いて、注意書きのタイプ（A3.2.1.2 および A3.2.2.1 参照）によってマトリクスに配列した。マトリクスは適切な注意書きの選択を示し、予防行動のすべての種類の要素を含んでいる。特定の危険有害性クラスに関係するすべての特異的な項目が使用されなければならない。加えて、ある危険有害性または区分に関係していない一般的な注意書きも、適切な場合には（A3.2.6 参照）、使用されなければならない。

A3.3.1.2　マトリクスを構成している表には注意書きの核心部分を太字で示す。特別の指示がない限りこの文言がラベルに記載される。しかしながら、すべての場合にまったく同一の言葉の組合せを強制することは不要である。

A3.3.1.3　多くの場合推奨される注意書きは独立している。例えば、爆発危険性の警句は健康有害性に関するものを制限しない、また危険性と有害性の両方に分類されるものは、どちらに対しても注意書きを持つべきである。

EXPLOSIVES
(CHAPTER 2.1)

Hazard category	Symbol	Signal word	Hazard statement	
1	Exploding bomb	Danger	H209	Explosive
			H210	Very sensitive
			H211	May be sensitive

Precautionary statements

Prevention	Response	Storage	Disposal
P203 **Obtain, read and follow all safety instructions before use.** **P210** **Keep away from heat, hot surfaces, sparks, open flames and other ignition sources. No smoking.** **P230** **Keep diluted with…** – *for explosive substances and mixtures that are diluted with solids or liquids, or wetted with, dissolved or suspended in water or other liquids to reduce their explosives properties* …Manufacturer/supplier or the competent authority to specify appropriate material. **P240** **Ground and bond container and receiving equipment.** – *if the explosive is electrostatically sensitive.* **P250** **Do not subject to grinding/shock/friction/…** – *if the explosive is mechanically sensitive.* …Manufacturer/supplier or the competent authority to specify applicable rough handling. **P280** **Wear protective gloves/protective clothing/eye protection/face protection/hearing protection/…** Manufacturer/supplier or the competent authority to specify the appropriate personal protective equipment.	**P370 + P372 + P380 + P373** **In case of fire: Explosion risk. Evacuate area. DO NOT fight fire when fire reaches explosives.**	**P401** **Store in accordance with…** …Manufacturer/supplier or the competent authority to specify local/regional/national/international regulations as applicable.	**P503** **Refer to manufacturer/ supplier/… for information on disposal/recovery/ recycling** …Manufacturer/supplier or the competent authority to specify appropriate source of information in accordance with local/regional/ national/international regulations as applicable.

- 332 -

爆発物
(第 2.1 章)

危険有害性区分	シンボル	注意喚起語	危険有害性情報
1	爆弾の爆発	危険	H209 爆発物 H210 非常に敏感 H211 敏感である可能性

注意書き

安全対策	応急措置	保管	廃棄
P203 使用前にすべての安全説明書を入手し、読み、従うこと。 P210 熱、高温のもの、火花、裸火および他の着火源から遠ざけること。禁煙。 P230 …にて希釈しておくこと。 一爆発物の性質を抑制するために、固体または液体で希釈された、又は水または他の液体で湿らされた、懸濁された物質及び混合物に対して …製造者/供給者または所管官庁が指定する適当な物質 P240 容器を接地しアースを取ること 一爆発物が静電気的に敏感である場合 P250 粉砕/衝撃/摩擦/…のような取り扱いをしないこと 一爆発物が力学的に敏感である場合 …製造者/供給者または所管官庁が指定する乱暴な取扱い P280 保護手袋/保護衣/保護眼鏡/保護面/聴覚保護具/…を着用すること。 製造者/供給者または所管官庁が指定する適切な個人用保護具の種類	P370+P372+P380+P373 火災の場合：爆発する危険性あり。区域より退避させること。炎が爆発物に届いたら消火活動をしないこと。	P401 …にしたがって保管すること。 …製造者/供給者または所管官庁が特定する適用できる国際/国/都道府県/市町村の規則	P503 廃棄/回収/リサイクルに関する情報について製造者/供給者/…に問い合わせる。 …製造者/供給者または所管官庁が特定する適用できる国際/国/都道府県/市町村の規則

EXPLOSIVES
(CHAPTER 2.1)

Hazard category	Symbol	Signal word	Hazard statement
2A	Exploding bomb	Danger	H209 Explosive
2B	Exploding bomb	Warning	H204 Fire or projection hazard

Precautionary statements

Prevention	Response	Storage	Disposal
P203 **Obtain, read and follow all safety instructions before use.** P210 **Keep away from heat, hot surfaces, sparks, open flames and other ignition sources. No smoking.** P230 **Keep diluted with...** - *for explosive substances and mixtures that are diluted with solids or liquids, or wetted with, dissolved or suspended in water or other liquids to reduce their explosives properties* … Manufacturer/supplier or the competent authority to specify appropriate material. P234 **Keep only in original packaging.** - *Omit where P236 is used.* P236 **Keep only in original packaging; Division ... in the transport configuration.** - *to be applied for explosives assigned a division within Class 1 for transport.* - *may be omitted for single packaging where the transport pictogram displaying the division (within Class 1) appears.* - *may be omitted where the use of different outer packaging results in different divisions for transport.* … Manufacturer/supplier or the competent authority to specify the division for transport.	P370 + P372 + P380 + P373 **In case of fire: Explosion risk. Evacuate area. DO NOT fight fire when fire reaches explosives.**	P401 **Store in accordance with...** … Manufacturer/supplier or the competent authority to specify local/regional/ national/international regulations as applicable.	P503 **Refer to manufacturer/ supplier/... for information on disposal/recovery/ recycling** … Manufacturer/supplier or the competent authority to specify appropriate source of information in accordance with local/regional/national/ international regulations as applicable.

(Cont'd on next page)

爆発物
(第 2.1 章)

危険有害性区分	シンボル	注意喚起語	危険有害性情報
2A 2B	爆弾の爆発 爆弾の爆発	危険 警告	H209 爆発物 H204 火災または飛散危険性

注意書き

安全対策	応急措置	保管	廃棄
P203 使用前にすべての安全説明書を入手し、読み、従うこと。 P210 熱、高温のもの、火花、裸火および他の着火源から遠ざけること。禁煙。 P230 …にて希釈しておくこと。 一爆発性の性質を抑制するために、固体または液体で希釈された、又は水または他の液体で湿らされた、懸濁された物質及び混合物に対して …製造者/供給者または所管官庁が指定する適当な物質 P234 他の容器に移し替えないこと。 一P236 が使用されている場合には省略 P236 元の容器のまま保存すること：輸送の構成において区分… 一輸送のクラス1の区分が割り当てられた爆容器に対して適用する 一単一包装で、区分 (クラス1) を示す輸送絵表示が示されていれば、省略してもよい 一輸送の異なる区分になっている容器の外部包装が使用されていれば省略してもよい …製造者/供給者または所管官庁が指定する輸送の区分	P370+P372+P380+P373 火災の場合：爆発する危険性あり。区域より退避させること。炎が爆発物に届いたら消火活動をしないこと。	P401 …にしたがって保管すること。 …製造者/供給者または所管官庁が特定する適用できる国際/国/都道府県/市町村の規則	P503 廃棄/回収/リサイクルに関する情報について製造者/供給者/…に問い合わせること。 …製造者/供給者または所管官庁が特定する適用できる国際/国/都道府県/市町村の規則

(次ページに続く)

EXPLOSIVES
(CHAPTER 2.1) *(cont'd)*

Hazard category	Symbol	Signal word	Hazard statement
2A	Exploding bomb	Danger	H209 Explosive
2B	Exploding bomb	Warning	H204 Fire or projection hazard

Precautionary statements

Prevention	Response	Storage	Disposal
P240 **Ground and bond container and receiving equipment.** – *if the explosive is electrostatically sensitive.* P250 **Do not subject to grinding/shock/friction/…** – *if the explosive is mechanically sensitive.* …Manufacturer/supplier or the competent authority to specify applicable rough handling. P280 **Wear protective gloves/protective clothing/eye protection/face protection/hearing protection/…** Manufacturer/supplier or the competent authority to specify the appropriate personal protective equipment.			

- 334 -

爆発物
(第 2.1 章) (続き)

危険有害性区分	シンボル	注意喚起語	危険有害性情報
2A	爆弾の爆発	危険	H209 爆発物
2B	爆弾の爆発	警告	H204 火災または飛散危険性

注意書き			
安全対策	応急措置	保管	廃棄
P240 容器を接地してアースを取ること。 ―爆発物が静電気的に敏感である場合 P250 粉砕/衝撃/摩擦...のような取り扱いをしないこと。 ―爆発物が力学的に敏感である場合 P280 保護手袋/保護衣/保護眼鏡/保護面/聴覚保護具...を着用すること。 製造者/供給者または所管官庁が指定する適切な個人用保護具の種類			

- 334 -

EXPLOSIVES
(CHAPTER 2.1)

Hazard category	Symbol	Signal word	Hazard statement
2C	Exclamation mark	Warning	H204 Fire or projection hazard

Precautionary statements

Prevention	Response	Storage	Disposal
P210 **Keep away from heat, hot surfaces, sparks, open flames and other ignition sources. No smoking.** **P230** **Keep diluted with…** *- for explosive substances and mixtures that are diluted with solids or liquids, or wetted with, dissolved or suspended in water or other liquids to reduce their explosives properties* *… Manufacturer/supplier or the competent authority to specify appropriate material.* **P234** **Keep only in original packaging.** *- Omit where P236 is used.* **P236** **Keep only in original packaging; Division … in the transport configuration.** *- to be applied for explosives assigned a division within Class 1 for transport.* *- may be omitted for single packaging where the transport pictogram displaying the division (within Class 1) appears.* *- may be omitted where the use of different outer packaging results in different divisions for transport.* *… Manufacturer/supplier or the competent authority to specify the division for transport.*	**P370 + P380 + P375** **In case of fire: Evacuate area. Fight fire remotely due to the risk of explosion**	**P401** **Store in accordance with…** *… Manufacturer/supplier or the competent authority to specify local/regional/ national/international regulations as applicable.*	**P503** **Refer to manufacturer/ supplier/… for information on disposal/recovery/ recycling** *… Manufacturer/supplier or the competent authority to specify appropriate source of information in accordance with local/regional/national/ international regulations as applicable.*

(Cont'd on next page)

爆発物
(第 2.1 章)

危険有害性区分	シンボル	注意喚起語	危険有害性情報
2C	感嘆符	警告	H204 火災または飛散危険性

注意書き

安全対策	応急措置	保管	廃棄
P210 熱、高温のもの、火花、裸火および他の着火源から遠ざけること。禁煙。 P230 …で湿らせておくこと。 ―爆発性を減少あるいは抑制するために鈍感剤を含む鈍性化剤で湿らせ、希釈、溶解あるいは懸濁させた物質および混合物の場合（鈍性化爆発物） …製造者/供給者または所管官庁が指定する適切な物質 P234 他の容器に移し替えないこと。 ―P236 が使用されている場合には省略 P236 元の容器のまま保存すること：輸送の構成において区分… ―輸送のクラス 1 の区分が割り当てられた爆発物に対して適用する ―単一包装で、区分（クラス 1）を示す輸送絵表示が示されていれば、省略してもよい ―製造者異なる区分になっている包装の外部包装が使用されていれば省略してもよい。 …製造者/供給者または所管官庁が指定する輸送の区分	P370+P380+P375 火災の場合：区域より退避させ、爆発の危険性があるため、離れた距離から消火すること。	P401 …にしたがって保管すること。 …製造者/供給者または所管官庁が特定する国際国/都道府県/市町村の規則	P503 廃棄/回収/リサイクルに関する情報について製造者/供給者…に問い合わせる。 …製造者/供給者または所管官庁が指定する適用できる国際国/都道府県/市町村の規則

（次ページに続く）

- 335 -

EXPLOSIVES
(CHAPTER 2.1) *(cont'd)*

Hazard category	Symbol	Signal word	Hazard statement
2C	Exclamation mark	Warning	H204 Fire or projection hazard

Precautionary statements

Prevention	Response	Storage	Disposal
P240 **Ground and bond container and receiving equipment.** – *if the explosive is electrostatically sensitive.* P250 **Do not subject to grinding/shock/friction/…** – *if the explosive is mechanically sensitive.* …Manufacturer/supplier or the competent authority to specify applicable rough handling. P280 **Wear protective gloves/protective clothing/eye protection/face protection/hearing protection/…** Manufacturer/supplier or the competent authority to specify the appropriate personal protective equipment.			

- 336 -

爆発物
(第 2.1 章)（続き）

危険有害性区分	シンボル	注意喚起語	危険有害性情報
2C	!	警告	H204 火災または飛散危険性

注意書き

安全対策	応急措置	保管	廃棄
P240 容器を接地しアースを取ること。 ―爆発物が静電気的に敏感である場合 P250 粉砕/衝撃/摩擦...のような取り扱いをしないこと。 ―爆発物が力学的に敏感である場合 ...製造者/供給者または所管官庁が指定する乱暴な取扱い P280 保護手袋/保護衣/保護眼鏡/保護面/聴覚保護具...を着用すること。 製造者/供給者または所管官庁が指定する適切な個人用保護具の種類			

- 336 -

FLAMMABLE GASES
(CHAPTER 2.2)

Hazard category	Symbol	Signal word	Hazard statement	
1A	Flame	Danger	H220	Extremely flammable gas
1B	Flame	Danger	H221	Flammable gas
2	*No symbol*	Warning	H221	Flammable gas

Precautionary statements

Prevention	Response	Storage	Disposal
P210 Keep away from heat, hot surfaces, sparks, open flames and other ignition sources. No smoking.	P377 Leaking gas fire: Do not extinguish, unless leak can be stopped safely. P381 In case of leakage, eliminate all ignition sources.	P403 Store in a well-ventilated place.	

可燃性ガス
(第 2.2 章)

危険有害性区分	シンボル	注意喚起語	危険有害性情報
1A	炎	危険	H220 極めて可燃性の高いガス
1B	炎	危険	H221 可燃性ガス
2	なし	警告	H221 可燃性ガス

注意書き

安全対策	応急措置	保管	廃棄
P210 熱、高温のもの、火花、裸火および他の着火源から遠ざけること。禁煙。	P377 漏洩ガス火災の場合：漏洩が安全に停止されない限り消火しないこと。 P381 漏えいした場合、着火源を除去すること。	P403 換気の良い場所で保管すること。	

FLAMMABLE GASES
(CHAPTER 2.2)
(Pyrophoric gases)

Hazard category	Symbol	Signal word	Hazard statement
1A, Pyrophoric gas	Flame	Danger	H220 Extremely flammable gas H232 May ignite spontaneously if exposed to air

Precautionary statements

Prevention	Response	Storage	Disposal
P210 **Keep away from heat, hot surfaces, sparks, open flames and other ignition sources. No smoking.** P222 **Do not allow contact with air.** *– if emphasis of the hazard statement is deemed necessary.* P280 **Wear protective gloves/protective clothing/eye protection/face protection/hearing protection/…** Manufacturer/supplier or the competent authority to specify the appropriate personal protective equipment.	P377 **Leaking gas fire: Do not extinguish, unless leak can be stopped safely.** P381 **In case of leakage, eliminate all ignition sources.**	P403 **Store in a well-ventilated place.**	

可燃性ガス
(第 2.2 章)
(自然発火性ガス)

危険有害性区分
1A、自然発火性ガス

シンボル
炎

注意喚起語
危険

危険有害性情報
H220 極めて可燃性の高いガス
H232 空気に触れると自然発火のおそれ

注意書き

安全対策	応急措置	保管	廃棄
P210 熱、高温のもの、火花、裸火および他の着火源から遠ざけること。禁煙。 P222 空気に接触させないこと。 P280 保護手袋/保護衣/保護眼鏡/保護面/聴覚保護具…… 一危険有害性情報の強調が必要と考えられる場合 製造者/供給者または所管官庁が指定する適切な個人用保護具の種類	P377 漏洩ガス火災の場合：漏洩が安全に停止されない限り消火しないこと。 P381 漏えいした場合、着火源を除去すること。	P403 換気の良い場所で保管すること。	

FLAMMABLE GASES
(CHAPTER 2.2)
(Chemically unstable gases)

Hazard category	Symbol	Signal word	Hazard statement	
1A, chemically unstable gas A	Flame	Danger	H220	Extremely flammable gas
			H230	May react explosively even in the absence of air
1A, chemically unstable gas B	Flame	Danger	H220	Extremely flammable gas
			H231	May react explosively even in the absence of air at elevated pressure and/or temperature

Precautionary statements

Prevention	Response	Storage	Disposal
P203 **Obtain, read and follow all safety instructions before use.** P210 **Keep away from heat, hot surfaces, sparks, open flames and other ignition sources. No smoking.**	P377 **Leaking gas fire: Do not extinguish, unless leak can be stopped safely.** P381 **In case of leakage, eliminate all ignition sources.**	P403 **Store in a well-ventilated place.**	

- 339 -

可燃性ガス
(第 2.2 章)
(化学的に不安定なガス)

危険有害性区分	シンボル	注意喚起語
1A, 化学的に不安定なガス A	炎	危険
1A, 化学的に不安定なガス B	炎	危険

危険有害性情報
H220 極めて可燃性の高いガス
H230 空気が無くても爆発的に反応するおそれ

H220 極めて可燃性の高いガス
H231 圧力および/または温度が上昇した場合、空気が無くても爆発的に反応するおそれ

注意書き

安全対策	応急措置	保管	廃棄
P203 使用前にすべての安全説明書を入手し、読み、従うこと。 P210 熱、高温のもの、火花、裸火および他の着火源から遠ざけること。禁煙。	P377 漏洩ガス火災の場合：漏洩が安全に停止されない限り消火しないこと。 P381 漏えいした場合、着火源を除去すること。	P403 換気の良い場所で保管すること。	

AEROSOLS
(CHAPTER 2.3, SECTION 2.3.1)

Hazard category	Symbol	Signal word	Hazard statement
1	Flame	Danger	H222 Extremely flammable aerosol H229 Pressurized container: may burst if heated
2	Flame	Warning	H223 Flammable aerosol H229 Pressurized container: may burst if heated

Precautionary statements

Prevention	Response	Storage	Disposal
P210 **Keep away from heat, hot surfaces, sparks, open flames and other ignition sources. No smoking.** P211 **Do not spray on an open flame or other ignition source.** P251 **Do not pierce or burn, even after use.**		P410 + P412 **Protect from sunlight. Do not expose to temperatures exceeding 50 °C/122 °F.** Manufacturer/supplier or the competent authority to use applicable temperature scale.	

エアゾール
(第 2.3 章、2.3.1)

危険有害性区分　　シンボル　　注意喚起語　　危険有害性情報
1　　　　　　　　炎　　　　　危険　　　　　H222　極めて可燃性の高いエアゾール
　　　　　　　　　　　　　　　　　　　　　　H229　高圧容器：熱すると破裂のおそれ
2　　　　　　　　炎　　　　　警告　　　　　H223　可燃性エアゾール
　　　　　　　　　　　　　　　　　　　　　　H229　高圧容器：熱すると破裂のおそれ

注意書き

安全対策	応急措置	保管	廃棄
P210 熱、高温のもの、火花、裸火およびその他の着火源から遠ざけること。禁煙。 P211 裸火または他の着火源に噴霧しないこと。 P251 使用後も含め、穴をあけたり燃やしたりしないこと。		P410 + P412 日光から遮断し、50 ℃以上の温度にばく露しないこと。 製造者/供給者または所管官庁が指定する適用可能な温度計を用いる	

- 340 -

AEROSOLS
(CHAPTER 2.3, SECTION 2.3.1)

Hazard category	Symbol	Signal word	Hazard statement
3	*No symbol*	Warning	H229 Pressurized container: may burst if heated

Precautionary statements

Prevention	Response	Storage	Disposal
P210 **Keep away from heat, hot surfaces, sparks, open flames and other ignition sources. No smoking.** P251 **Do not pierce or burn, even after use.**		P410 + P412 **Protect from sunlight. Do not expose to temperatures exceeding 50 °C/122 °F.** Manufacturer/supplier or the competent authority to use applicable temperature scale.	

エアゾール
(第 2.3 章、2.3.1)

危険有害性区分	シンボル	注意喚起語	危険有害性情報
3	なし	警告	H229 高圧容器：熱すると破裂のおそれ

注意書き

安全対策	応急措置	保管	廃棄
P210 熱、高温のもの、火花、裸火および他の着火源から遠ざけること。禁煙。 P251 使用後も含め、穴をあけたり燃やしたりしないこと。		P410 + P412 日光から遮断し、50 ℃以上の温度にばく露しないこと。 製造者/供給者または所管官庁が指定する適用可能な温度計を用いる	

CHEMICALS UNDER PRESSURE
(CHAPTER 2.3, SECTION 2.3.2)

Hazard category	Symbol	Signal word	Hazard statement	
1	Flame and gas cylinder	Danger	H282	Extremely flammable chemical under pressure: May explode if heated
2	Flame and gas cylinder	Warning	H283	Flammable chemical under pressure: May explode if heated

Precautionary statements

Prevention	Response	Storage	Disposal
P210 **Keep away from heat, hot surfaces, sparks, open flames and other ignition sources. No smoking.** P211 Do not spray on an open flame or other ignition source.	P381 **In case of leakage, eliminate all ignition sources.** P376 **Stop leak if safe to do so.** P370 + P378 **In case of fire, use …. to extinguish.** – *if water increases risk.* … Manufacturer/supplier or the competent authority to specify appropriate media.	P410 + P403 **Protect from sunlight. Store in a well-ventilated place.** *P410 may be omitted for chemicals under pressure filled in transportable cylinders in accordance with packing instruction P200 or P206 of the UN Model Regulations, unless those chemicals under pressure are subject to (slow)decomposition or polymerisation, or the competent authority provides otherwise.*	

加圧下化学品
(第 2.3 章、2.3.2)

危険有害性区分	シンボル	注意喚起語	危険有害性情報
1	炎 およびガスボンベ	危険	H282 極めて可燃性の高い加圧下化学品： 熱すると爆発のおそれ
2	炎 およびガスボンベ	警告	H283 可燃性の加圧下化学品： 熱すると爆発のおそれ

注意書き

安全対策	応急措置	保管	廃棄
P210 熱、高温のもの、火花、裸火および他の着火源から遠ざけること。禁煙。 P211 裸火または他の着火源に噴霧しないこと。	P381 漏洩した場合、着火源を除去すること。 P376 安全に対処できるなら漏洩を止めること。 P370+P378 火災の場合：消火するために…を使用すること。 一水がリスクを増大させる場合…製造者供給者または所管官庁が指定する適当な手段	P410＋P403 日光から遮断し、換気の良い場所で保管すること。 P410は、加圧下化学品がUNモデル規則の包装指示P200またはP206にしたがっている場合の運送用シリンダーに充填されている加圧下化学品には削除してもよい。ただしこれらの加圧下化学品が（遅い）分解や高分子化を起こさず、または所管官庁が他の方法を示さない場合に限る。	

CHEMICALS UNDER PRESSURE
(CHAPTER 2.3, SECTION 2.3.2)

Hazard category	Symbol	Signal word	Hazard statement
3	Gas cylinder	Warning	H284 Chemical under pressure: May explode if heated

Precautionary statements

Prevention	Response	Storage	Disposal
P210 **Keep away from heat, hot surfaces, sparks, open flames and other ignition sources. No smoking.**	P376 **Stop leak if safe to do so.**	P410 + P403 **Protect from sunlight. Store in a well-ventilated place.** *P410 may be omitted for chemicals under pressure filled in transportable cylinders in accordance with packing instruction P200 or P206 of the UN Model Regulations, unless those chemicals under pressure are subject to (slow) decomposition or polymerisation, or the competent authority provides otherwise.*	

- 343 -

加圧下化学品
(第 2.3 章、2.3.2)

危険有害性区分	シンボル	注意喚起語	危険有害性情報
3	ガスボンベ	警告	H284 加圧下化学品：熱すると爆発のおそれ

注意書き

安全対策	応急措置	保管	廃棄
P210 熱、高温のもの、火花、裸火および他の着火源から遠ざけること。禁煙。	P376 安全に対処できるなら漏洩を止めること。	P410 + P403 日光から遮断し、換起の良い場所で保管すること。 P410は、加圧下化学品がUNモデル規則の包装指示P200またはP206にしたがっている場合には削除してもよい。ただしこれらの加圧下化学品が（遅い）分解や高分子化を起こさず、また所管官庁が他の方法を示さない場合に限る。	

- 343 -

OXIDIZING GASES
(CHAPTER 2.4)

Hazard category	Symbol	Signal word	Hazard statement
1	Flame over circle	Danger	H270 May cause or intensify fire; oxidizer

Precautionary statements

Prevention	Response	Storage	Disposal
P220 **Keep away from clothing and other combustible materials.** P244 **Keep valves and fittings free from oil and grease.**	P370 + P376 **In case of fire: Stop leak if safe to do so.**	P403 **Store in a well-ventilated place.**	

- 344 -

酸化性ガス
(第 2.4 章)

危険有害性区分	シンボル	注意喚起語	危険有害性情報
1	円状の炎	危険	H270 発火または火災助長のおそれ；酸化性物質

注意書き

安全対策	応急措置	保管	廃棄
P220 衣類および可燃物から遠ざけること。 P244 バルブや付属品にはグリースおよび油を使用しないこと。	P370 + P376 火災の場合：安全に対処できるならば漏洩を止めること。	P403 換気の良い場所で保管すること。	

GASES UNDER PRESSURE
(CHAPTER 2.5)

Hazard category	Symbol	Signal word	Hazard statement
Compressed gas	Gas cylinder	Warning	H280 Contains gas under pressure; may explode if heated
Liquefied gas	Gas cylinder	Warning	H280 Contains gas under pressure; may explode if heated
Dissolved gas	Gas cylinder	Warning	H280 Contains gas under pressure; may explode if heated

Precautionary statements

Prevention	Response	Storage	Disposal
		P410 + P403 **Protect from sunlight. Store in a well-ventilated place.** – *P410 may be omitted for gases filled in transportable gas cylinders in accordance with packing instruction P200 of the UN Model Regulations, unless those gases are subject to (slow) decomposition or polymerisation, or the competent authority provides otherwise.*	

高圧ガス
(第 2.5 章)

危険有害性区分		シンボル	注意喚起語	危険有害性情報	
圧縮ガス		ガスボンベ	警告	H280	高圧ガス；熱すると爆発のおそれ
液化ガス		ガスボンベ	警告	H280	高圧ガス；熱すると爆発のおそれ
溶解ガス		ガスボンベ	警告	H280	高圧ガス；熱すると爆発のおそれ

注意書き

安全対策	応急措置	保管	廃棄
		P410 + P403 日光から遮断し、換気の良い場所で保管すること。 P410 はガスが UN モデル規則の包装指示 P200 にしたがっている運送用ガスシリンダーに充填されている場合には削除してもよい。ただしこれらのガスが（遅い）分解や高分子化を起こさず、また所管官庁が他の方法を示さない場合に限る	

- 345 -

GASES UNDER PRESSURE
(CHAPTER 2.5)

Hazard category	Symbol	Signal word	Hazard statement
Refrigerated liquefied gas	Gas cylinder	Warning	H281 Contains refrigerated gas; may cause cryogenic burns or injury

Precautionary statements

Prevention	Response	Storage	Disposal
P282 **Wear cold insulating gloves and either face shield or eye protection.**	P336 + P317 **Immediately thaw frosted parts with lukewarm water. Do not rub affected area. Get medical help.**	P403 **Store in a well-ventilated place.**	

高圧ガス
(第 2.5 章)

危険有害性区分　　　　シンボル　　　　注意喚起語　　　　危険有害性情報
深冷液化ガス　　　　　ガスボンベ　　　　警告　　　　　　　H281 深冷液化ガス；凍傷または傷害のおそれ

注意書き

安全対策	応急措置	保管	廃棄
P282 耐寒手袋および保護面または保護眼鏡を着用すること。	P336+P317 凍った部分をすぐにぬるま湯でとかすこと。受傷部はこすらないこと。医療処置を受けること。	P403 換気の良い場所で保管すること。	

- 346 -

FLAMMABLE LIQUIDS
(CHAPTER 2.6)

Hazard category	Symbol	Signal word	Hazard statement	
1	Flame	Danger	H224	Extremely flammable liquid and vapour
2	Flame	Danger	H225	Highly flammable liquid and vapour
3	Flame	Warning	H226	Flammable liquid and vapour

Precautionary statements

Prevention	Response	Storage	Disposal
P210 **Keep away from heat, hot surfaces, sparks, open flames and other ignition sources. No smoking.** **P233** **Keep container tightly closed.** – *if the liquid is volatile and may generate an explosive atmosphere.* **P240** **Ground and bond container and receiving equipment.** – *if the liquid is volatile and may generate an explosive atmosphere.* **P241** **Use explosion-proof [electrical/ventilating/lighting/...] equipment.** – *if the liquid is volatile and may generate an explosive atmosphere.* – *text in square brackets may be used to specify specific electrical, ventilating, lighting or other equipment if necessary and as appropriate.* – *precautionary statement may be omitted where local or national legislation introduces more specific provisions.* **P242** **Use non-sparking tools.** – *if the liquid is volatile and may generate an explosive atmosphere and if the minimum ignition energy is very low. (This applies to substances and mixtures where the minimum ignition energy is < 0.1 mJ, e.g. carbon disulphide).* **P243** **Take action to prevent static discharges.** – *if the liquid is volatile and may generate an explosive atmosphere.* – *may be omitted where local or national legislation introduces more specific provisions.* **P280** **Wear protective gloves/protective clothing/eye protection/face protection/hearing protection/...** Manufacturer/supplier or the competent authority to specify the appropriate personal protective equipment.	P303 + P361 + P353 **IF ON SKIN (or hair): Take off immediately all contaminated clothing. Rinse affected areas with water [or shower].** – *text in square brackets to be included where the manufacturer/supplier or the competent authority considers it appropriate for the specific chemical.* P370 + P378 **In case of fire: Use ... to extinguish.** – *if water increases risk.* ... *Manufacturer/supplier or the competent authority to specify appropriate media.*	P403 + P235 **Store in a well-ventilated place. Keep cool.** – *for flammable liquids Category 1 and other flammable liquids that are volatile and may generate an explosive atmosphere.*	P501 **Dispose of contents/container to...** ... in accordance with local/regional/national/international regulations (to be specified). Manufacturer/supplier or the competent authority to specify whether disposal requirements apply to contents, container or both.

引火性液体
(第 2.6 章)

危険有害性区分	シンボル	注意喚起語	危険有害性情報
1	炎	危険	H224 極めて引火性の高い液体および蒸気
2	炎	危険	H225 引火性の高い液体および蒸気
3	炎	警告	H226 引火性液体および蒸気

注意書き

安全対策	応急措置	保管	廃棄
P210 熱、高温のもの、火花、裸火および他の着火源から遠ざけること。禁煙。 P233 容器を密閉しておくこと。 ―液体が爆発性で爆発する環境をつくる可能性がある環境 P240 容器を接地してアースを取ること。 ―液体が爆発性で爆発する環境を作る可能性があるとき P241 防爆型の【電気換気照明...】機器を使用すること。 ―液体が爆発性で爆発する環境をつくる可能性があるとき ―［］内の文章は、電気機器、換気装置、照明機器あるいは他の機器を特定するために、必要性がある場合に適切に使用される ―国内規制でより詳細な規定がある場合にはこの注意書きは省略してもよい P242 火花を発生させない工具を使用すること。 ―液体が爆発性で爆発する環境をつくる可能性がある（これは例えば二硫化炭素のように、最少引火エネルギーが非常に低い場合 0.1mJ 未満の物質や混合物に適用される。） P243 静電気放電に対する予防措置を講ずること。 ―液体が爆発性で爆発する環境をつくる可能性があるとき ―国内規制でより詳細な規定がある場合にはこの注意書きは省略してもよい P280 保護手袋／保護衣／保護眼鏡／保護面／聴覚保護具／...を着用すること。 製造者／供給者または所管官庁が指定する適切な個人用保護具の種類	P303 + P361 + P353 皮膚（または髪）に付着した場合：直ちに汚染された衣類をすべて脱ぐこと。接触部位を流水【またはシャワー】で洗うこと。 ―製造者／供給者または所管官庁が特定の化学品に対してそれが適切だとした場合には［］内の文章を含める P370 + P378 火災の場合：消火するために...を使用すること。 ―水がリスクを増大させる場合 ―製造者／供給者または所管官庁が指定する適切な手段	P403 + P235 換気の良い場所で保管すること。涼しいところに置くこと。 ―引火性液体区分１および爆発性があり爆発する環境をつくる可能性がある他の液体	P501 内容物／容器を...に廃棄すること。 ...国際／国都道府県／市町村の規則（明示する）に従って 製造者／供給者または所管官庁が指定する内容物、容器またはその両者に適用する廃棄要件

FLAMMABLE LIQUIDS
(CHAPTER 2.6)

Hazard category	Symbol	Signal word	Hazard statement
4	*No symbol*	Warning	H227 Combustible liquid

Precautionary statements

Prevention	Response	Storage	Disposal
P210 **Keep away from heat, hot surfaces, sparks, open flames and other ignition sources. No smoking.** P280 **Wear protective gloves/protective clothing/eye protection/face protection/hearing protection/…** Manufacturer/supplier or the competent authority to specify the appropriate personal protective equipment.	P370 + P378 **In case of fire: Use … to extinguish.** – *if water increases risk.* … Manufacturer/supplier or the competent authority to specify appropriate media.	P403 **Store in a well-ventilated place.** – *for flammable liquids Category 1 and other flammable liquids that are volatile and may generate an explosive atmosphere.*	P501 **Dispose of contents/container to…** …in accordance with local/regional/ national/international regulations (to be specified). Manufacturer/supplier or the competent authority to specify whether disposal requirements apply to contents, container or both.

引火性液体
(第 2.6 章)

危険有害性区分	シンボル	注意喚起語	危険有害性情報
4	なし	警告	H227 可燃性液体

注意書き

安全対策	応急措置	保管	廃棄
P210 熱、高温のもの、火花、裸火および他の着火源から遠ざけること。禁煙。 P280 保護手袋/保護衣/保護眼鏡/保護面/聴覚保護具…を着用すること。 製造者/供給者または所管官庁が指定する適切な個人用保護具の種類	P370 + P378 火災の場合：消火するために…を使用すること。 ―水がリスクを増大させる場合 製造者/供給者または所管官庁が指定する適切な手段	P403 換気の良い場所で保管すること。 ―引火性液体区分1および他の引火性液体で揮発性が高く爆発する環境をつくる可能性があるとき	P501 内容物/容器を…に廃棄すること。 …国際/国/都道府県/市町村の規則（明示する）に従って 製造者/供給者または所管官庁が指定する内容物、容器またはその両者に適用する廃棄物要件

- 348 -

FLAMMABLE SOLIDS
(CHAPTER 2.7)

Hazard category	Symbol	Signal word	Hazard statement
1	Flame	Danger	H228 Flammable solid
2	Flame	Warning	H228 Flammable solid

Precautionary statements

Prevention	Response	Storage	Disposal
P210 **Keep away from heat, hot surfaces, sparks, open flames and other ignition sources. No smoking.** P240 **Ground and bond container and receiving equipment.** – *if the solid is electrostatically sensitive.* P241 **Use explosion-proof [electrical/ventilating/ lighting/…]equipment.** – *if dust clouds can occur.* – *text in square brackets may be used to specify specific electrical, ventilating, lighting or other equipment if necessary and as appropriate.* – *precautionary statement may be omitted where local or national legislation introduces more specific provisions.* P280 **Wear protective gloves/ protective clothing/eye protection/face protection/hearing protection/…** Manufacturer/supplier or the competent authority to specify the appropriate personal protective equipment.	P370 + P378 **In case of fire: Use … to extinguish** – *if water increases risk.* … Manufacturer/supplier or the competent authority to specify appropriate media.		

- 349 -

可燃性固体
(第 2.7 章)

危険有害性区分	シンボル	注意喚起語	危険有害性情報
1	炎	危険	H228 可燃性固体
2	炎	警告	H228 可燃性固体

注意書き

安全対策	応急措置	保管	廃棄
P210 熱、高温のもの、火花、裸火および他の着火源から遠ざけること。－禁煙。 P240 容器を接地しアースを取ること。 ―固体が静電気的に敏感である場合 P241 防爆型の【電気/換気/照明/…】機器を使用すること。 ―粉じん雲が発生する可能性のある場合 ―[] 内の文責は、電気機器、換気装置、照明機器あるいは他の機器を特定するために、必要性がある場合に適切に使用される ―国内規制によりより詳細な規定がある場合にはこの注意書きは省略してもよい P280 保護手袋/保護衣/保護眼鏡/保護面/聴覚保護具…を着用すること。 製造者/供給者または所管官庁が指定する適切な個人用保護具の種類	P370 + P378 火災の場合：消火するために…を使用すること。 ―水がリスクを増大させる場合 …製造者/供給者または所管官庁が指定する適切な手段。		

- 349 -

SELF-REACTIVE SUBSTANCES AND MIXTURES
(CHAPTER 2.8)

Hazard category	Symbol	Signal word	Hazard statement
Type A	Exploding bomb	Danger	H240 Heating may cause an explosion

Precautionary statements

Prevention	Response	Storage	Disposal
P210 **Keep away from heat, hot surfaces, sparks, open flames and other ignition sources. No smoking.** P234 **Keep only in original packaging.** P235 **Keep cool.** *– may be omitted if P411 is given on the label.* P240 **Ground and bond container and receiving equipment.** *– if electrostatically sensitive and able to generate an explosive atmosphere.* P280 **Wear protective gloves/protective clothing/eye protection/face protection/hearing protection/...** Manufacturer/supplier or the competent authority to specify the appropriate personal protective equipment.	P370 + P372 + P380 + P373 **In case of fire: Explosion risk. Evacuate area. DO NOT fight fire when fire reaches explosives.**	P403 **Store in a well-ventilated place.** *– except for temperature controlled self-reactive substances and mixtures or organic peroxides because condensation and consequent freezing may take place.* P411 **Store at temperatures not exceeding ...°C/...°F.** *– if temperature control is required (according to section 2.8.2.3 or 2.15.2.3 of the GHS) or if otherwise deemed necessary.* ... Manufacturer/supplier or the competent authority to specify temperature using applicable temperature scale. P420 **Store separately.**	P501 **Dispose of contents/container to...** ... in accordance with local/regional/national/international regulations (to be specified). Manufacturer/supplier or the competent authority to specify whether disposal requirements apply to contents, container or both.

自己反応性物質および化学品
（第 2.8 章）

危険有害性区分	シンボル	注意喚起語	危険有害性情報
タイプ A	爆弾の爆発	危険	H240 熱すると爆発のおそれ

注意書き

安全対策	応急措置	保管	廃棄
P210 熱、高温のもの、火花、裸火および他の着火源から遠ざけること。禁煙。 P234 他の容器に移し替えないこと。 P235 涼しいところに置くこと。 ―P411がラベルに示されている場合には省略してもよい P240 容器を接地レアースを取ること。 ―静電気的に敏感で、爆発する環境をつくる可能性があるとき P280 保護手袋/保護衣/保護眼鏡/保護面/聴覚保護具…を着用すること。 ―製造者供給者または所管官庁が指定する適切な個人用保護具の種類	P370 + P372 + P380 + P373 火災の場合：爆発する危険性あり。区域より退避させること。炎が爆発物に届いたら消火活動をしないこと。	P403 換気の良い場所で保管すること。 ―温度が管理されている自己反応性物質およびそれに混合物または有機過酸化物は、濃縮およびそれに伴う凍結が起きるので、除外する P411 …℃以下の温度で保管すること。 ―温度管理が必要な場合（GHS2.8.2.3 または 2.15.2.3 により）あるいは他の方法が必要と考えられる場合 ―製造者供給者または所管官庁が適用可能な温度計を用いて指定した温度 P420 隔離して保管すること。	P501 内容物容器を…に廃棄すること。 ―国際/国/都道府県/市町村の規則（明示する）に従って 製造者供給者または容器または廃棄の両者に適用する内容物、容器または廃棄の両者に適用する廃棄物要件

- 350 -

SELF-REACTIVE SUBSTANCES AND MIXTURES
(CHAPTER 2.8)

Hazard category	Symbol	Signal word	Hazard statement
Type B	Exploding bomb and flame	Danger	H241 Heating may cause a fire or explosion

Precautionary statements

Prevention	Response	Storage	Disposal
P210 **Keep away from heat, hot surfaces, sparks, open flames and other ignition sources. No smoking.** P234 **Keep only in original packaging.** P235 **Keep cool** – *may be omitted if P411 is given on the label.* P240 **Ground and bond container and receiving equipment.** – *if electrostatically sensitive and able to generate an explosive atmosphere.* P280 **Wear protective gloves/protective clothing/eye protection/face protection/hearing protection/...** Manufacturer/supplier or the competent authority to specify the appropriate personal protective equipment.	P370 + P380 + P375 [+ P378] **In case of fire: Evacuate area. Fight fire remotely due to the risk of explosion. [Use...to extinguish]** – *text in square brackets to be included if water increases risk.* ... Manufacturer/supplier or the competent authority to specify appropriate media.	P403 **Store in a well-ventilated place.** – *except for temperature controlled self-reactive substances and mixtures or organic peroxides because condensation and consequent freezing may take place.* P411 **Store at temperatures not exceeding ...°C/...°F.** – *if temperature control is required (according to section 2.8.2.3 or 2.15.2.3 of the GHS) or if otherwise deemed necessary.* ... Manufacturer/supplier or the competent authority to specify temperature using applicable temperature scale. P420 **Store separately.**	P501 **Dispose of contents/container to...** ...in accordance with local/regional/national/international regulations (to be specified). Manufacturer/supplier or the competent authority to specify whether disposal requirements apply to contents, container or both.

自己反応性物質および化学品
(第 2.8 章)

危険有害性区分	シンボル	注意喚起語	危険有害性情報
タイプ B	爆弾の爆発および炎	危険	H241 熱すると火災または爆発のおそれ

注意書き

安全対策	応急措置	保管	廃棄
P210 熱、高温のもの、火花、裸火および他の着火源から遠ざけること。禁煙。 P234 他の容器に移し替えないこと。 P235 涼しいところに置くこと。 —P411 がラベルに示されている場合には省略してもよい —静電気を接地し接続すること。 —静電気的に敏感で、爆発危険性を増す可能性がある場合 P280 保護手袋/保護衣/保護眼鏡/保護面/聴覚保護具…を着用すること。 製造者/供給者または所管官庁が指定する適切な個人用保護具の種類	P370 + P380 + P375 [+P378] 火災の場合：区域より退避させ、爆発の危険性があるため、離れた距離から消火すること。[消火するために…を使用すること。] —[]内の文章は水はりスクを大きくする場合に使用する。 —製造者/供給者または所管官庁が指定する適切な手段	P403 換気の良い場所で保管すること。 —温度管理が必要とされている有機過酸化物は、濃縮および/またはその混合物または凍結が起きるので、除外する P411 …℃以下の温度で保管すること。 —温度管理が必要な場合（GHS2.8.2.3 または2.15.2.3 により）あるいは他の方法が必要と考えられる場合 —製造者/供給者または所管官庁が適用可能な温度計を用いて指定した温度 P420 隔離して保管すること。	P501 内容物/容器を…に廃棄すること。 …国際/国/都道府県/市町村の規則（明示する）に従って 製造者/供給者または所管官庁が指定する内容物、容器またはその両者に適用する廃棄物要件

- 351 -

SELF-REACTIVE SUBSTANCES AND MIXTURES
(CHAPTER 2.8)

Hazard category	Symbol	Signal word	Hazard statement	
Type C	Flame	Danger	H242	Heating may cause a fire
Type D	Flame	Danger	H242	Heating may cause a fire
Type E	Flame	Warning	H242	Heating may cause a fire
Type F	Flame	Warning	H242	Heating may cause a fire

Precautionary statements

Prevention	Response	Storage	Disposal
P210 **Keep away from heat, hot surfaces, sparks, open flames and other ignition sources. No smoking.** P234 **Keep only in original packaging.** P235 **Keep cool** – *may be omitted if P411 is given on the label.* P240 **Ground and bond container and receiving equipment.** – *if electrostatically sensitive and able to generate an explosive atmosphere.* P280 **Wear protective gloves/protective clothing/eye protection/face protection/hearing protection/...** Manufacturer/supplier or the competent authority to specify the appropriate personal protective equipment.	P370 + P378 **In case of fire: Use ... to extinguish** – *if water increases risk.* ... Manufacturer/supplier or the competent authority to specify appropriate media.	P403 **Store in a well-ventilated place.** – *except for temperature controlled self-reactive substances and mixtures or organic peroxides because condensation and consequent freezing may take place.* P411 **Store at temperatures not exceeding ...°C/...°F.** – *if temperature control is required (according to section 2.8.2.3 or 2.15.2.3 of the GHS) or if otherwise deemed necessary.* ...Manufacturer/supplier or the competent authority to specify temperature using applicable temperature scale. P420 **Store separately.**	P501 **Dispose of contents/container to...** ...in accordance with local/regional/national/international regulations (to be specified). Manufacturer/supplier or the competent authority to specify whether disposal requirements apply to contents, container or both.

自己反応性物質および化学品
(第 2.8 章)

危険有害性区分	シンボル	注意喚起語	危険有害性情報
タイプ C	炎	危険	H242 熱すると火災のおそれ
タイプ D	炎	危険	H242 熱すると火災のおそれ
タイプ E	炎	警告	H242 熱すると火災のおそれ
タイプ F	炎	警告	H242 熱すると火災のおそれ

注意書き

安全対策	応急措置	保管	廃棄
P210 熱、高温のもの、火花、裸火および他の着火源から遠ざけること。禁煙。 P234 他の容器に移し替えないこと。 P235 涼しいところに置くこと。 ―P411がラベルに示されている場合には省略してもよい。 P240 容器を接地してアースを取ること。 ―静電気的に敏感で、爆発危険性を増す可能性がある場合 P280 保護手袋/保護衣/保護眼鏡/保護面/聴覚保護具…を着用すること。 製造者/供給者または所管官庁が指定する適切な個人用保護具の種類	P370 + P378 火災の場合：消火するために…を使用すること。 ―水がリスクを増大させる場合 …製造者/供給者または所管官庁が指定する適切な手段	P403 換気の良い場所で保管すること。 ―温度が管理されている自己反応性物質および混合物または有機過酸化物は、濃縮およびそれに伴う凝結が起きるので、除外する P411 …℃以下の温度で保管すること。 ―温度管理が必要な場合 (GHS2.8.2.3 または 2.15.2.3により) あるいは他の方法が適用可能と考えられる場合 …製造者/供給者または所管官庁が適用可能な温度を用いて指定した温度 P420 隔離して保管すること。	P501 内容物/容器を…に廃棄すること。 …国際国/都道府県/市町村の規則 (明示する) に従って 製造者/供給者または所管官庁が指定する内容物、容器またはその両者に適用する廃棄物要件

PYROPHORIC LIQUIDS
(CHAPTER 2.9)

Hazard category	Symbol	Signal word	Hazard statement
1	Flame	Danger	H250 Catches fire spontaneously if exposed to air

Precautionary statements

Prevention	Response	Storage	Disposal
P210 **Keep away from heat, hot surfaces, sparks, open flames and other ignition sources. No smoking.** P222 **Do not allow contact with air.** *– if emphasis of the hazard statement is deemed necessary.* P231 **Handle and store contents under inert gas/…** …Manufacturer/supplier or the competent authority to specify appropriate liquid or gas if "inert gas" is not appropriate. P233 **Keep container tightly closed.** P280 **Wear protective gloves/protective clothing/eye protection/face protection/hearing protection/…** Manufacturer/supplier or the competent authority to specify the appropriate personal protective equipment.	P302 + P334 **IF ON SKIN: Immerse in cool water or wrap in wet bandages.** P370 + P378 **In case of fire: Use … to extinguish** *– if water increases risk.* …Manufacturer/supplier or the competent authority to specify appropriate media.		

自然発火性液体
(第 2.9 章)

危険有害性区分	シンボル	注意喚起語	危険有害性情報
1	炎	危険	H250 空気に触れると自然発火

注意書き

安全対策	応急措置	保管	廃棄
P210 熱、高温のもの、火花、裸火および他の着火源から遠ざけること。禁煙。 P222 空気に接触させないこと。 一危険有害性情報の強調が必要と考えられる場合 P231 不活性ガス...下で取扱い、保管すること。 ...「不活性ガス」が適当でない場合、製造者供給者または所管官庁が指定する適当な液体または...ガス P233 容器を密閉しておくこと。 P280 保護手袋/保護衣/保護眼鏡/保護面/聴覚保護具...を着用すること。 製造者供給者または所管官庁が指定する適切な個人用保護具の種類	P302 + P334 皮膚に付着した場合：冷たい水に浸すことまたは湿った包帯で覆うこと。 P370 + P378 火災の場合：消火するために...を使用すること。 一水がリスクを増大させる場合 ...製造者供給者または所管官庁が指定する適切な手段。		

- 353 -

PYROPHORIC SOLIDS
(CHAPTER 2.10)

Hazard category	Symbol	Signal word	Hazard statement
1	Flame	Danger	H250 Catches fire1 spontaneously if exposed to air

Precautionary statements

Prevention	Response	Storage	Disposal
P210 **Keep away from heat, hot surfaces, sparks, open flames and other ignition sources. No smoking.** P222 **Do not allow contact with air.** – *if emphasis of the hazard statement is deemed necessary.* P231 **Handle and store contents under inert gas/…** …Manufacturer/supplier or the competent authority to specify appropriate liquid or gas if "inert gas" is not appropriate. P233 **Keep container tightly closed.** P280 **Wear protective gloves/protective clothing/eye protection/face protection/hearing protection/…** Manufacturer/supplier or the competent authority to specify the appropriate personal protective equipment.	P302 + P335 + P334 **IF ON SKIN: Brush off loose particles from skin and immerse in cool water or wrap in wet bandages.** P370 + P378 **In case of fire: Use … to extinguish** – *if water increases risk.* …Manufacturer/supplier or the competent authority to specify appropriate media.		

- 354 -

自然発火性固体
（第 2.10 章）

注意喚起語
危険

危険有害性情報
H250 空気に触れると自然発火

危険有害性区分　　シンボル
1　　　　　　　　炎

注意書き

安全対策	応急措置	保管	廃棄
P210 熱、高温のもの、火花、裸火および他の着火源から遠ざけること。禁煙。 P222 空気に接触させないこと。 一危険有害性情報の強調が必要と考えられる場合 P231 不活性ガス/...下で取扱い、保管すること。 「不活性ガス」が適当でない場合、製造者/供給者または所管官庁が指定する適当な液体または はガス P233 容器を密閉しておくこと。 P280 保護手袋/保護衣/保護眼鏡/保護面/聴覚保護具...を着用すること。 製造者/供給者または所管官庁が指定する適切な個人用保護具の種類	P302+P335+P334 皮膚に付着した場合：固着していない粒子を皮膚から払いのけ、冷たい水に浸すことまたは湿った包帯で覆うこと。 P370 + P378 火災の場合：消火するために...を使用すること。 一水がリスクを増大させる場合 ...製造者/供給者または所管官庁が指定する適切な手段		

- 354 -

SELF-HEATING SUBSTANCES AND MIXTURES
(CHAPTER 2.11)

Hazard category	Symbol	Signal word	Hazard statement
1	Flame	Danger	H251 Self-heating; may catch fire
2	Flame	Warning	H252 Self-heating in large quantities; may catch fire

Precautionary statements

Prevention	Response	Storage	Disposal
P235 **Keep cool.** — *may be omitted if P413 is given on the label.* P280 **Wear protective gloves/protective clothing/eye protection/face protection/hearing protection/...** Manufacturer/supplier or the competent authority to specify the appropriate personal protective equipment.		P407 **Maintain air gap between stacks or pallets.** P410 **Protect from sunlight** P413 **Store bulk masses greater than ... kg/...lbs at temperatures not exceeding ...°C/...°F.** ... Manufacturer/supplier or the competent authority to specify mass and temperature using applicable scale. P420 **Store separately.**	

自己発熱性物質および化学品
（第 2.11 章）

危険有害性区分	シンボル	注意喚起語	危険有害性情報
1	炎	危険	H251 自己発熱；発火のおそれ
2	炎	警告	H252 大量の場合自己発熱；火災のおそれ

注意書き

安全対策	応急措置	保管	廃棄
P235 涼しい所に置くこと。 ―P413がラベルに示されている場合には省略してもよい P280 保護手袋/保護衣/保護眼鏡/保護面/聴覚保護具…を着用すること。 製造者/供給者または所管官庁が指定する適切な個人用保護具の種類		P407 積荷またはパレット間にすきまをあけること。 P410 日光から遮断すること。 P413 …kg 以上の大量品は、…℃ 以下の温度で保管すること。 …製造者/供給者または所管官庁が指定する適用可能な計測器を用いた量と温度 P420 隔離して保管すること。	

- 355 -

SUBSTANCES AND MIXTURES WHICH, IN CONTACT WITH WATER, EMIT FLAMMABLE GASES
(CHAPTER 2.12)

Hazard category	Symbol	Signal word	Hazard statement
1	Flame	Danger	H260 In contact with water releases flammable gases, which may ignite spontaneously
2	Flame	Danger	H261 In contact with water releases flammable gas

Precautionary statements

Prevention	Response	Storage	Disposal
P223 **Do not allow contact with water.** – *if emphasis of the hazard statement is deemed necessary.* P231 + P232 **Handle and store contents under inert gas/… Protect from moisture.** – *if the substance or mixture reacts readily with moisture in air.* …Manufacturer/supplier or the competent authority to specify appropriate liquid or gas if "inert gas" is not appropriate. P280 **Wear protective gloves/protective clothing/eye protection/face protection/hearing protection/…** Manufacturer/supplier or the competent authority to specify the appropriate personal protective equipment.	P302 + P335 + P334 **IF ON SKIN: Brush off loose particles from skin and immerse in cool water.** P370 + P378 **In case of fire: Use … to extinguish** – *if water increases risk.* …Manufacturer/supplier or the competent authority to specify appropriate media.	P402 + P404 **Store in a dry place. Store in a closed container.**	P501 **Dispose of contents/container to…** …in accordance with local/regional/national/international regulations (to be specified). Manufacturer/supplier or the competent authority to specify whether disposal requirements apply to contents, container or both.

水反応可燃性物質および化学品
(第 2.12 章)

危険有害性区分	シンボル	注意喚起語	危険有害性情報
1	炎	危険	H260 水に触れると自然発火するおそれのある可燃性ガスを発生
2	炎	危険	H261 水に触れると可燃性ガスを発生

注意書き

安全対策	応急措置	保管	廃棄
P223 水と接触させないこと。 P231+P232 湿気を遮断し、不活性ガス…下で取り扱い保管すること。 一物質あるいは混合物が空気中の水分と速やかに反応する場合 …もし「不活性ガス」が適切でない場合に、製造者/供給者または所管官庁が指定する適切な液体又はガス P280 保護手袋/保護衣/保護眼鏡/保護面/聴覚保護具…を着用すること。 製造者/供給者または所管官庁が指定する適切な個人用保護具の種類	P302+P335+P334 皮膚に付着した場合：固着していない粒子を皮膚から払いのけ、冷たい水に浸すこと。 一危険有害性情報の強調が必要と考えられる場合 P370 + P378 火災の場合：消火するために…を使用すること。 一水がリスクを増大させる場合 …製造者/供給者または所管官庁が指定する適切な手段	P402 + P404 乾燥した場所または密閉容器に保管すること。	P501 内容物/容器を…に廃棄すること。 …国際（国/都道府県/市町村）の規則（明示する）に従って 製造者/供給者または所管官庁が指定する内容物、容器またはその両者に適用する廃棄物要件

SUBSTANCES AND MIXTURES WHICH, IN CONTACT WITH WATER, EMIT FLAMMABLE GASES (CHAPTER 2.12)

Hazard category	Symbol	Signal word	Hazard statement
3	Flame	Warning	H261 In contact with water releases flammable gas

Precautionary statements

Prevention	Response	Storage	Disposal
P231 + P232 **Handle and store contents under inert gas/… Protect from moisture.** – *if the substance or mixture reacts readily with moisture in air.* …Manufacturer/supplier or the competent authority to specify appropriate liquid or gas if "inert gas" is not appropriate. P280 **Wear protective gloves/protective clothing/eye protection/face protection/hearing protection/…** Manufacturer/supplier or the competent authority to specify the appropriate personal protective equipment.	P370 + P378 **In case of fire: Use … to extinguish.** – *if water increases risk.* …Manufacturer/supplier or the competent authority to specify appropriate media.	P402 + P404 **Store in a dry place. Store in a closed container.**	P501 **Dispose of contents/container to…** … in accordance with local/regional/national/international regulations (to be specified). Manufacturer/supplier or the competent authority to specify whether disposal requirements apply to contents, container or both.

水反応可燃性物質および化学品
（第 2.12 章）

危険有害性区分	シンボル	注意喚起語	危険有害性情報
3	炎	警告	H261 水に触れると可燃性ガスを発生

注意書き

安全対策	応急措置	保管	廃棄
P231＋P232 湿気を遮断し、不活性ガス…下で取り扱い保管すること。 ―物質あるいは混合物が空気中の水分と速やかに反応する場合 …もし不活性ガスが適切でない場合には、製造者/供給者または所管官庁が指定する適切な波体又はガス P280 保護手袋/保護衣/保護眼鏡/保護面/聴覚保護具…を着用すること。 製造者/供給者または所管官庁が指定する適切な個人用保護具の種類	P370＋P378 火災の場合：消火するために…を使用すること。 ―水がリスクを増大させる場合 …製造者/供給者または所管官庁が指定する適切な手段	P402＋P404 乾燥した場所または密閉された容器中で保管すること。	P501 内容物/容器を…に廃棄すること。 …国際/国/都道府県/市町村の規則（明示する）に従って、製造者/供給者または所管官庁が指定する内容物、容器またはその両者に適用する廃棄物要件

OXIDIZING LIQUIDS
(CHAPTER 2.13)

Hazard category	Symbol	Signal word	Hazard statement
1	Flame over circle	Danger	H271 May cause fire or explosion; strong oxidizer

Precautionary statements

Prevention	Response	Storage	Disposal
P210 **Keep away from heat, hot surfaces, sparks, open flames and other ignition sources. No smoking.** P220 **Keep away from clothing and other combustible materials.** P280 **Wear protective gloves/protective clothing/eye protection/face protection/hearing protection/...** Manufacturer/supplier or the competent authority to specify the appropriate personal protective equipment. P283 **Wear fire resistant or flame retardant clothing.**	P306 + P360 **IF ON CLOTHING: Rinse immediately contaminated clothing and skin with plenty of water before removing clothes.** P371 + P380 + P375 **In case of major fire and large quantities: Evacuate area. Fight fire remotely due to the risk of explosion.** P370 + P378 **In case of fire: Use ... to extinguish.** – *if water increases risk.…* Manufacturer/supplier or the competent authority to specify appropriate media.	P420 **Store separately.**	P501 **Dispose of contents/container to…** …in accordance with local/regional/national/international regulations (to be specified). Manufacturer/supplier or the competent authority to specify whether disposal requirements apply to contents, container or both.

酸化性液体
(第 2.13 章)

危険有害性区分	シンボル	注意喚起語	危険有害性情報
1	円上の炎	危険	H271 火災または爆発のおそれ；強酸化性物質

注意書き

安全対策	応急措置	保管	廃棄
P210 熱、高温のもの、火花、裸火および他の着火源から遠ざけること。禁煙。 P220 衣類および可燃物から遠ざけること。 P280 保護手袋/保護衣/保護眼鏡/保護面/聴覚保護具…を着用すること。 製造者供給者または所管官庁が指定する適切な個人用保護具の種類 P283 防火服または防炎服を着用すること。	P306 + P360 衣類にかかった場合：服を脱ぐ前に、直ちに汚染された衣類および皮膚を多量の水で洗うこと。 P371 + P380 + P375 大火災の場合で大量にある場合：区域より退避させ、爆発の危険性があるため、離れた距離から消火すること。 P370 + P378 火災の場合：消火するために…を使用すること。 一水がリスクを増大させる場合 …製造者供給者または所管官庁が指定する適切な手段	P420 隔離して保管すること。	P501 内容物容器を…に廃棄すること。 …国際/国/都道府県/市町村の規則（明示する）に従って 製造者供給者または所管官庁が指定する内容物、容器またはその両者に適用する廃棄物要件

- 358 -

OXIDIZING LIQUIDS
(CHAPTER 2.13)

Hazard category	Symbol	Signal word	Hazard statement
2	Flame over circle	Danger	H272 May intensify fire; oxidizer
3	Flame over circle	Warning	H272 May intensify fire; oxidizer

Precautionary statements

Prevention	Response	Storage	Disposal
P210 **Keep away from heat, hot surfaces, sparks, open flames and other ignition sources. No smoking.** P220 **Keep away from clothing and other combustible materials.** P280 **Wear protective gloves/protective clothing/eye protection/face protection/hearing protection/...** Manufacturer/supplier or the competent authority to specify the appropriate personal protective equipment.	P370 + P378 **In case of fire: Use ... to extinguish.** – *if water increases risk.* ... Manufacturer/supplier or the competent authority to specify appropriate media.		P501 **Dispose of contents/container to...** ...in accordance with local/regional/ national/international regulations (to be specified). Manufacturer/supplier or the competent authority to specify whether disposal requirements apply to contents, container or both.

- 359 -

酸化性液体
(第 2.13 章)

危険有害性区分	シンボル	注意喚起語	危険有害性情報
2	円上の炎	危険	H272 火災助長のおそれ；酸化性物質
3	円上の炎	警告	H272 火災助長のおそれ；酸化性物質

注意書き

安全対策	応急措置	保管	廃棄
P210 熱、高温のもの、火花、裸火および他の着火源から遠ざけること。禁煙。 P220 衣類および可燃物から遠ざけること。 P280 保護手袋/保護衣/保護眼鏡/保護面/聴覚保護具…を着用すること。 …製造者/供給者または所管官庁が指定する適切な個人用保護具の種類	P370 + P378 火災の場合：消火するために…を使用すること。 一水がリスクを増大させる場合 …製造者/供給者または所管官庁が指定する適切な手段		P501 内容物/容器を…に廃棄すること。 …国際/国/都道府県/市町村の規則（明示する）に従って 製造者/供給者または所管官庁が指定する内容物、容器またはその両者に適用する廃棄物要件

OXIDIZING SOLIDS
(CHAPTER 2.14)

Hazard category	Symbol	Signal word	Hazard statement
1	Flame over circle	Danger	H271 May cause fire or explosion; strong oxidizer

Precautionary statements

Prevention	Response	Storage	Disposal
P210 **Keep away from heat, hot surfaces, sparks, open flames and other ignition sources. No smoking.** P220 **Keep away from clothing and other combustible materials.** P280 **Wear protective gloves/protective clothing/eye protection/face protection/hearing protection/...** Manufacturer/supplier or the competent authority to specify the appropriate personal protective equipment. P283 **Wear fire resistant or flame-retardant clothing.**	P306 + P360 **IF ON CLOTHING: Rinse immediately contaminated clothing and skin with plenty of water before removing clothes.** P371 + P380 + P375 **In case of major fire and large quantities: Evacuate area. Fight fire remotely due to the risk of explosion.** P370 + P378 **In case of fire: Use ... to extinguish.** – *if water increases risk.* ... Manufacturer/supplier or the competent authority to specify appropriate media.	P420 **Store separately.**	P501 **Dispose of contents/container to...** ...in accordance with local/regional/national/international regulations (to be specified). Manufacturer/supplier or the competent authority to specify whether disposal requirements apply to contents, container or both.

酸化性固体
(第 2.14 章)

危険有害性区分	シンボル	注意喚起語	危険有害性情報
1	円上の炎	危険	H271 火災または爆発のおそれ；強酸化性物質

注意書き

安全対策	応急措置	保管	廃棄
P210 熱、高温のもの、火花、裸火および他の着火源から遠ざけること。禁煙。 P220 衣類および可燃物から遠ざけること。 P280 保護手袋／保護衣／保護眼鏡／保護面／聴覚保護具…を着用すること。 製造者／供給者または所管官庁が指定する適切な個人用保護具の種類 P283 防火服または防炎服を着用すること。	P306 + P360 衣類にかかった場合：服を脱ぐ前に、直ちに汚染された衣類および皮膚を多量の水で洗うこと。 P371 + P380 + P375 大火災の場合で大量にある場合：区域より退避させ、爆発の危険性があるため、離れた距離から消火すること。 P370 + P378 火災の場合：消火するために…を使用すること。 ―水がリスクを増大させる場合 …製造者／供給者または所管官庁が指定する適切な手段	P420 隔離して保管すること。	P501 内容物／容器を…に廃棄すること。 …国際／国／都道府県／市町村の規則（明示する）に従って 製造者／供給者または所管官庁が指定する内容物、容器またはその両者に適用する廃棄物要件

OXIDIZING SOLIDS
(CHAPTER 2.14)

Hazard category	Symbol	Signal word	Hazard statement
2	Flame over circle	Danger	H272 May intensify fire; oxidizer
3	Flame over circle	Warning	H272 May intensify fire; oxidizer

Precautionary statements

Prevention	Response	Storage	Disposal
P210 **Keep away from heat, hot surfaces, sparks, open flames and other ignition sources. No smoking.** P220 **Keep away from clothing and other combustible materials.** P280 **Wear protective gloves/protective clothing/eye protection/face protection/hearing protection/...** Manufacturer/supplier or the competent authority to specify the appropriate personal protective equipment.	P370 + P378 **In case of fire: Use ... to extinguish.** – *if water increases risk.* ... Manufacturer/supplier or the competent authority to specify appropriate media.		P501 **Dispose of contents/container to...** ... in accordance with local/regional/national/international regulations (to be specified). Manufacturer/supplier or the competent authority to specify whether disposal requirements apply to contents, container or both.

- 361 -

酸化性固体
(第 2.14 章)

危険有害性区分	シンボル	注意喚起語	危険有害性情報
2	円上の炎	危険	H272 火災助長のおそれ；酸化性物質
3	円上の炎	警告	H272 火災助長のおそれ；酸化性物質

注意書き

安全対策	応急措置	保管	廃棄
P210 熱、高温のもの、火花、裸火および他の着火源から遠ざけること。禁煙。 P220 衣類および可燃物から遠ざけること。 P280 保護手袋/保護衣/保護眼鏡/保護面/聴覚保護具…を着用すること。 製造者/供給者または所管官庁が指定する適切な個人用保護具の種類	P370 + P378 火災の場合：消火するために…を使用すること。 -水がリスクを増大させる場合 製造者/供給者または所管官庁が指定する適切な手段		P501 内容物/容器を…に廃棄すること。 …国際物/国都道府県/市町村の規則（明示する）に従って 製造者/供給者または所管官庁が指定する内容物、容器またはその両者に適用する廃棄物要件

- 361 -

ORGANIC PEROXIDES
(CHAPTER 2.15)

Hazard category	Symbol	Signal word	Hazard statement
Type A	Exploding bomb	Danger	H240 Heating may cause an explosion

Precautionary statements

Prevention	Response	Storage	Disposal
P210 **Keep away from heat, hot surfaces, sparks, open flames and other ignition sources. No smoking.** P234 **Keep only in original packaging.** P235 **Keep cool.** – *may be omitted if P411 is given on the label.* P240 **Ground and bond container and receiving equipment.** – *if electrostatically sensitive and able to generate an explosive atmosphere.* P280 **Wear protective gloves/ protective clothing/eye protection/face protection/hearing protection/…** Manufacturer/supplier or the competent authority to specify the appropriate personal protective equipment.	P370 + P372 + P380 + P373 **In case of fire: Explosion risk. Evacuate area. DO NOT fight fire when fire reaches explosives.**	P403 **Store in a well-ventilated place.** – *except for temperature controlled self-reactive substances and mixtures or organic peroxides because condensation and consequent freezing may take place.* P410 **Protect from sunlight.** P411 **Store at temperatures not exceeding …°C/…°F.** – *if temperature control is required (according to section 2.8.2.3 or 2.15.2.3 of the GHS) or if otherwise deemed necessary.* … Manufacturer/supplier or the competent authority to specify temperature using applicable temperature scale P420 **Store separately.**	P501 **Dispose of contents/container to…** … in accordance with local/regional/national/international regulations (to be specified). Manufacturer/supplier or the competent authority to specify whether disposal requirements apply to contents, container or both.

有機過酸化物
(第 2.15 章)

危険有害性区分	シンボル	注意喚起語	危険有害性情報
タイプ A	爆弾の爆発	危険	H240 熱すると爆発のおそれ

注意書き

安全対策	応急措置	保管	廃棄
P210 熱、高温のもの、火花、裸火および他の着火源から遠ざけること。禁煙。 P234 他の容器に移し替えないこと。 P235 涼しいところに置くこと。 ーP411 がラベルに示されている場合には省略してもよい P240 容器を接地しアースを取ること。 ー静電気的に敏感で、爆発する環境をつくる可能性があるとき P280 保護手袋/保護衣/保護眼鏡/保護面/聴覚保護具…を着用すること。 ー製造者供給者または所管官庁が指定する適切な個人用保護具の種類	P370+P372+P380+P373 火災の場合：爆発する危険性あり。区域より退避させること。炎が爆発物に届いたら消火活動をしないこと。	P403 換気の良い場所で保管すること。 ー温度が管理されている自己反応性物質おおよび混合物または有機過酸化物は、濃縮おおよびそれに伴う凍結が起きるので、除外する P410 日光から遮断すること。 P411 …℃以下の温度で保管すること。 ー温度管理が必要な場合（GHS2.8.2.3 または2.15.2.3 により）あるいは他の方法が適用可能と考えられる場合 …製造者供給者または所管官庁が指定した温度計を用いて指定した温度 P420 隔離して保管すること。	P501 内容物/容器を…に廃棄すること。 …国際/国/都道府県/市町村の規則（明示する）に従って 製造者供給者または所管官庁が指定する内容物、容器またはその両者に適用する廃棄物要件

ORGANIC PEROXIDES
(CHAPTER 2.15)

Hazard category	Symbol	Signal word	Hazard statement
Type B	Exploding bomb and flame	Danger	H241 Heating may cause a fire or explosion

Precautionary statements

Prevention	Response	Storage	Disposal
P210 **Keep away from heat, hot surfaces, sparks, open flames and other ignition sources. No smoking.** P234 **Keep only in original packaging.** P235 **Keep cool.** *– may be omitted if P411 is given on the label.* P240 **Ground and bond container and receiving equipment.** *– if electrostatically sensitive and able to generate an explosive atmosphere.* P280 **Wear protective gloves/ protective clothing/eye protection/face protection/hearing protection/…** Manufacturer/supplier or the competent authority to specify the appropriate personal protective equipment.	P370 + P380 + P375 [+ P378] **In case of fire: Evacuate area. Fight fire remotely due to the risk of explosion. [Use…to extinguish]** *– text in square brackets to be used if water increases risk.* …Manufacturer/supplier or the competent authority to specify appropriate media.	P403 **Store in a well-ventilated place.** *– except for temperature controlled self-reactive substances and mixtures or organic peroxides because condensation and consequent freezing may take place.* P410 **Protect from sunlight.** P411 **Store at temperatures not exceeding …°C/…°F.** *– if temperature control is required (according to section 2.8.2.3 or 2.15.2.3 of the GHS) or if otherwise deemed necessary.* …Manufacturer/supplier or the competent authority to specify temperature using applicable temperature scale. P420 **Store separately.**	P501 **Dispose of contents/container to…** … in accordance with local/regional/national/international regulations (to be specified). Manufacturer/supplier or the competent authority to specify whether disposal requirements apply to contents, container or both.

- 363 -

有機過酸化物
(第 2.15 章)

危険有害性区分	シンボル	注意喚起語	危険有害性情報
タイプ B	爆弾の爆発 および炎	危険	H241 熱すると火災または爆発のおそれ

注意書き

安全対策	応急措置	保管	廃棄
P210 熱、高温のもの、火花、裸火および他の着火源から遠ざけること。禁煙。 P234 他の容器に移し替えないこと。 P235 涼しいところに置くこと。 ―P411がラベルに示されている場合には省略してもよい P240 容器を接地しアースを取ること。 ―静電気的に敏感で、爆発する環境をつくる可能性があるとき P280 保護手袋/保護衣/保護眼鏡/保護面/聴覚保護具…を着用すること。 ―製造者供給者または所管官庁が指定する適切な個人用保護具の種類	P370＋P380＋P375 [＋P378] 火災の場合：区域より退避させ、爆発の危険性があるため、離れた距離から消火すること。[消火するために…を使用すること。] ―[ ]内の文章は水がリスクを大きくする場合に使用する。 ―製造者供給者または所管官庁が指定する適切な手段	P403 換気の良い場所で保管すること。 ―温度管理が必要とされている自己反応性物質および混合物または有機過酸化物は、濃縮および凍結に伴う凍結が起きるので、除外する P410 日光から遮断すること。 P411 …℃以下の温度で保管すること。 ―温度管理が必要な場合（GHS2.8.2.3または2.15.2.3により）あるいは他の方法が適用可能と考えられる場合 ―製造者供給者または所管官庁が適用可能温度計を用いて指定した温度 P420 隔離して保管すること。	P501 内容物/容器を…に廃棄すること。 ―国際/国都道府県市町村の規則（明示する）に従って ―製造者供給者または所管官庁が指定する内容物、容器またはその両者に適用する廃棄物要件

- 363 -

ORGANIC PEROXIDES
(CHAPTER 2.15)

Hazard category	Symbol	Signal word	Hazard statement
Type C	Flame	Danger	H242 Heating may cause a fire
Type D	Flame	Danger	H242 Heating may cause a fire
Type E	Flame	Warning	H242 Heating may cause a fire
Type F	Flame	Warning	H242 Heating may cause a fire

Precautionary statements

Prevention	Response	Storage	Disposal
P210 **Keep away from heat, hot surfaces, sparks, open flames and other ignition sources. No smoking.** P234 **Keep only in original packaging.** P235 **Keep cool.** — *may be omitted if P411 is given on the label.* P240 **Ground and bond container and receiving equipment.** — *if electrostatically sensitive and able to generate an explosive atmosphere.* P280 **Wear protective gloves/ protective clothing/eye protection/face protection/hearing protection/...** Manufacturer/supplier or the competent authority to specify the appropriate personal protective equipment.	P370 + P378 **In case of fire: Use ... to extinguish.** — *if water increases risk.* ... Manufacturer/supplier or the competent authority to specify appropriate media.	P403 **Store in a well-ventilated place.** — *except for temperature controlled self-reactive substances and mixtures or organic peroxides because condensation and consequent freezing may take place.* P410 **Protect from sunlight.** P411 **Store at temperatures not exceeding ...°C/...°F.** — *if temperature control is required (according to section 2.8.2.3 or 2.15.2.3 of the GHS) or if otherwise deemed necessary....* Manufacturer/supplier or the competent authority to specify temperature using applicable temperature scale. P420 **Store separately.**	P501 **Dispose of contents/container to...** ... in accordance with local/regional/national/international regulations (to be specified). Manufacturer/supplier or the competent authority to specify whether disposal requirements apply to contents, container or both.

有機過酸化物
(第 2.15 章)

危険有害性区分	シンボル	注意喚起語	危険有害性情報
タイプ C	炎	危険	H242 熱すると火災のおそれ
タイプ D	炎	危険	H242 熱すると火災のおそれ
タイプ E	炎	警告	H242 熱すると火災のおそれ
タイプ F	炎	警告	H242 熱すると火災のおそれ

注意書き

安全対策	応急措置	保管	廃棄
P210 熱、高温のもの、火花、裸火および他の着火源から遠ざけること。禁煙。 P234 他の容器に移し替えないこと。 P235 涼しいところに置くこと。 ―P411がラベルに示されている場合には省略してもよい P240 容器を接地しアースを取ること。 ―静電気的に敏感で、爆発する環境をつくる可能性があるとき P280 保護手袋/保護衣/保護眼鏡/保護面/聴覚保護具/…を着用すること。 製造者/供給者または所管官庁が指定する適切な個人用保護具の種類	P370+P378 火災の場合：消火するために…を使用すること。 ―水が火災リスクを増大させる場合 製造者/供給者または所管官庁が指定する適切な手段	P403 換気の良い場所で保管すること。 ―温度管理されている有機過酸化物、混合物または自己反応性物質および濃縮およびそれに伴う凍結が起こるもの、除外する P410 日光から遮断すること。 P411 …℃以下の温度で保管すること。 ―温度管理が必要な場合（GHS2.8.2.3 または2.15.2.3により）あるいは他の方法が適用可能な場合 製造者/供給者または所管官庁が指定する温度計を用いて指定した温度 P420 隔離して保管すること。	P501 内容物/容器を…に廃棄すること。 …国際/国都道府県/市町村の規則（明示する）に従って 製造者/供給者または所管官庁が指定する内容物、容器またはその両者に適用する廃棄物要件

CORROSIVE TO METALS
(CHAPTER 2.16)

Hazard category	Symbol	Signal word	Hazard statement
1	Corrosion	Warning	H290 May be corrosive to metals

Precautionary statements

Prevention	Response	Storage	Disposal
P234 **Keep only in original packaging.**	P390 **Absorb spillage to prevent material-damage.**	P406 **Store in a corrosion resistant/... container with a resistant inner liner.** – *may be omitted if P234 is given on the label* ... Manufacturer/supplier or the competent authority to specify other compatible materials.	

金属腐食性
(第 2.16 章)

危険有害性区分	シンボル	注意喚起語	危険有害性情報
1	腐食性	警告	H290 金属腐食のおそれ

注意書き

安全対策	応急措置	保管	廃棄
P234 他の容器に移し替えないこと。	P390 物的被害を防止するためにも流出したものを吸収すること。	P406 耐腐食性/耐腐食性内張りのある...容器に保管すること。 -P234がラベルにある場合省略してもよい ...製造者/供給者または所管官庁が指定する他の互換性がある材料	

- 365 -

DESENSITIZED EXPLOSIVES
(CHAPTER 2.17)

Hazard category	Symbol	Signal word	Hazard statement	
1	Flame	Danger	H206	Fire, blast or projection hazard; increased risk of explosion if desensitizing agent is reduced
2	Flame	Danger	H207	Fire or projection hazard; increased risk of explosion if desensitizing agent is reduced
3	Flame	Warning	H207	Fire or projection hazard; increased risk of explosion if desensitizing agent is reduced

Precautionary statements

Prevention	Response	Storage	Disposal
P210 **Keep away from heat, hot surfaces, sparks, open flames and other ignition sources. No smoking.** P212 **Avoid heating under confinement or reduction of the desensitizing agent.** P230 **Keep diluted with…** …Manufacturer/supplier or the competent authority to specify appropriate material. P233 **Keep container tightly closed.** P280 **Wear protective gloves/protective clothing/eye protection/face protection/hearing protection/…** Manufacturer/supplier or the competent authority to specify the appropriate personal protective equipment.	P370+P380+P375 **In case of fire: Evacuate area. Fight fire remotely due to the risk of explosion.**	P401 **Store in accordance with…** …Manufacturer/supplier or the competent authority to specify local/regional/ national/international regulations as applicable.	P501 **Dispose of contents/containers to…** …in accordance with local/regional/national/international regulations (to be specified). Manufacturer/supplier or the competent authority to specify whether disposal requirements apply to contents, container or both.

鈍性化爆発物
(第 2.17 章)

危険有害性区分	注意喚起語	シンボル
1	危険	炎
2	危険	炎
3	警告	炎

危険有害性情報
H206 火災、爆風または飛散危険性；鈍性化剤が減少した場合には爆発の危険性の増加
H207 火災または飛散危険性；鈍性化剤が減少した場合には爆発の危険性の増加
H207 火災または飛散危険性；鈍性化剤が減少した場合には爆発の危険性の増加

注意書き

安全対策	応急措置	保管	廃棄
P210 熱、高温のもの、火花、裸火および他の着火源から遠ざけること。禁煙。 P212 密閉状態での加熱または鈍性化剤の減少を避ける P230 …にて希釈しておくこと。 …製造者/供給者または所管官庁が指定する適切な物質 P233 容器を密閉しておくこと。 P280 保護手袋/保護衣/保護眼鏡/保護面/聴覚保護具…を着用すること。 製造者/供給者または所管官庁が指定する適切な個人用保護具の種類	P370+P380+P375 火災の場合：区域より退避させ、爆発の危険性があるため、離れた距離から消火すること。	P401 …にしたがって保管すること。 …製造者/供給者または所管官庁が指定する適用できる国際国/国/都道府県/市町村の規則	P501 内容物/容器を…に廃棄すること。 …国際国/都道府県/市町村の規則（明示する）に従って製造者/供給者または所管官庁が指定する内容物、容器または容器に適用するその両者に適用する廃棄物要件

DESENSITIZED EXPLOSIVES
(CHAPTER 2.17)

Hazard category	Symbol	Signal word	Hazard statement
4	Flame	Warning	H208 Fire hazard; increased risk of explosion if desensitizing agent is reduced

Precautionary statements

Prevention	Response	Storage	Disposal
P210 **Keep away from heat, hot surfaces, sparks, open flames and other ignition sources. No smoking.** P212 **Avoid heating under confinement or reduction of the desensitizing agent.** P230 **Keep diluted with…** …Manufacturer/supplier or the competent authority to specify appropriate material. P233 **Keep container tightly closed.** P280 **Wear protective gloves/protective clothing/eye protection/face protection/hearing protection/…** Manufacturer/supplier or the competent authority to specify the appropriate personal protective equipment.	P371+P380+P375 **In case of major fire and large quantities: Evacuate area. Fight fire remotely due to the risk of explosion.**	P401 **Store in accordance with…** …Manufacturer/supplier or the competent authority to specify local/regional/ national/international regulations as applicable.	P501 **Dispose of contents/containers to…** …in accordance with local/regional/national /international regulations (to be specified). Manufacturer/supplier or the competent authority to specify whether disposal requirements apply to contents, container or both.

危険有害性区分	シンボル	注意喚起語	危険有害性情報
4	炎	警告	H208 火災危険性；鈍性化剤が減少した場合には爆発の危険性の増加

鈍性化爆発物
(第 2.17 章)

注意書き

安全対策	応急措置	保管	廃棄
P210 熱、高温のもの、火花、裸火および他の着火源から遠ざけること。禁煙。 P212 密閉状態での加熱または鈍性化剤の減少を避ける P230 …にて希釈しておくこと。 …製造者/供給者または所管官庁が指定する適切な物質 P233 容器を密閉しておくこと。 P280 保護手袋/保護衣/保護眼鏡/保護面/聴覚保護具…を着用すること。 製造者/供給者または所管官庁が指定する適切な個人用保護具の種類	P371+P380+P375 大火災の場合で大量で区域にある場合：区域より退避させ、爆発の危険性があるため、離れた距離から消火すること。	P401 …にしたがって保管すること。 …製造者/供給者または所管官庁が特定する適用できる国際/国/都道府県/市町村の規則	P501 内容物/容器を…に廃棄すること。 …国際/国/都道府県/市町村の規則（明示する）に従って 製造者/供給者または所管官庁が指定する内容物、容器またはその両者に適用する廃棄物要件

- 367 -

ACUTE TOXICITY – ORAL
(CHAPTER 3.1)

Hazard category	Symbol	Signal word	Hazard statement	
1	Skull and crossbones	Danger	H300	Fatal if swallowed
2	Skull and crossbones	Danger	H300	Fatal if swallowed
3	Skull and crossbones	Danger	H301	Toxic if swallowed

Precautionary statements

Prevention	Response	Storage	Disposal
P264 **Wash hands [and…] thoroughly after handling.** – *text in square brackets to be used when the manufacturer/supplier or the competent authority specify other parts of the body to be washed after handling.* P270 **Do not eat, drink or smoke when using this product.**	P301 + P316 **IF SWALLOWED: Get emergency medical help immediately.** Competent authority or manufacturer / supplier may add, 'Call' followed by the appropriate emergency telephone number, or the appropriate emergency medical help provider, for example, a Poison Centre, Emergency Centre or Doctor. P321 **Specific treatment (see … on this label)** – *if immediate administration of antidote is required.* … Reference to supplemental first aid instruction. P330 **Rinse mouth.**	P405 **Store locked up.**	P501 **Dispose of contents/container to…** … in accordance with local/regional/national/international regulations (to be specified). Manufacturer/supplier or the competent authority to specify whether disposal requirements apply to contents, container or both.

- 368 -

急性毒性（経口）
（第 3.1 章）

危険有害性区分	シンボル	注意喚起語	危険有害性情報
1	どくろ	危険	H300 飲み込むと生命に危険
2	どくろ	危険	H300 飲み込むと生命に危険
3	どくろ	危険	H301 飲み込むと有毒

注意書き

安全対策	応急措置	保管	廃棄
P264 取扱後は手【および…】をよく洗うこと。 ―製造者/供給者又は所管官庁が、取扱後に洗浄する体の他の部分を指定した場合には[ ]内の文章を用いる P270 この製品を使用する時に、飲食または喫煙をしないこと。	P301 + P316 飲み込んだ場合：すぐに救急の医療処置を受けること。 ―製造者/供給者又は所管官庁が、「電話」に続けて、適当な救急時電話番号などから適当な救急時医療提供者、例えば中毒センター、救急センターまたは医師などを追加してもよい。 P321 特別な処置が必要である。 （このラベルの…を参照） ―緊急の解毒剤の投与が必要な場合 …補足的な応急措置の説明 P330 口をすすぐこと	P405 施錠して保管すること。	P501 内容物/容器を…に廃棄すること。 …国際/国都道府県/市町村の規則（明示する）に従って ―製造者/供給者または所管官庁が指定する内容物、容器または/その両者に適用する廃棄物要件

- 368 -

ACUTE TOXICITY – ORAL
(CHAPTER 3.1)

Hazard category	Symbol	Signal word	Hazard statement
4	Exclamation mark	Warning	H302 Harmful if swallowed

Precautionary statements

Prevention	Response	Storage	Disposal
P264 **Wash hands [and…] thoroughly after handling.** *- text in square brackets to be used when the manufacturer/supplier or the competent authority specify other parts of the body to be washed after handling.* P270 **Do not eat, drink or smoke when using this product.**	P301 + P317 **IF SWALLOWED: Get medical help.** P330 **Rinse mouth.**		P501 **Dispose of contents/container to…** … in accordance with local/regional/national/international regulations (to be specified). Manufacturer/supplier or the competent authority to specify whether disposal requirements apply to contents, container or both.

危険有害性区分	シンボル	注意喚起語	危険有害性情報
急性毒性（経口）(第 3.1 章)	感嘆符	警告	H302 飲み込むと有害
4			

注意書き

安全対策	応急措置	保管	廃棄
P264 取扱後は手［および…］をよく洗うこと。 ―製造者/供給者または所管官庁が、取扱後に洗浄する体の他の部分を指定した場合には[]内の文章を用いる P270 この製品を使用する時に、飲食または喫煙をしないこと。	P301 + P317 飲み込んだ場合：医療処置を受けること。 P330 口をすすぐこと。		P501 内容物/容器を…に廃棄すること。 …国際/国/都道府県/市町村の規則（明示する）に従って 製造者/供給者または所管官庁が指定する内容物、容器またはその両者に適用する廃棄物要件

- 369 -

ACUTE TOXICITY – ORAL
(CHAPTER 3.1)

Hazard category	Symbol	Signal word	Hazard statement
5	*No symbol*	Warning	H303 May be harmful if swallowed

Precautionary statements

Prevention	Response	Storage	Disposal
	P301 + P317 **IF SWALLOWED: Get medical help.**		

急性毒性（経口）
（第 3.1 章）

危険有害性区分	シンボル	注意喚起語	危険有害性情報
5	なし	警告	H303 飲み込むと有害のおそれ

注意書き

安全対策	応急措置	保管	廃棄
	P301 + P317 飲み込んだ場合：医療処置を受けること。		

ACUTE TOXICITY – DERMAL
(CHAPTER 3.1)

Hazard category	Symbol	Signal word	Hazard statement
1	Skull and crossbones	Danger	H310 Fatal in contact with skin
2	Skull and crossbones	Danger	H310 Fatal in contact with skin

Precautionary statements

Prevention	Response	Storage	Disposal
P262 **Do not get in eyes, on skin, or on clothing.** P264 **Wash hands [and...] thoroughly after handling.** – *text in square brackets to be used when the manufacturer/supplier or the competent authority specify other parts of the body to be washed after handling.* P270 **Do not eat, drink or smoke when using this product.** P280 **Wear protective gloves/protective clothing.** Manufacturer/supplier or the competent authority may further specify type of equipment where appropriate.	P302 + P352 **IF ON SKIN: Wash with plenty of water/...** ...Manufacturer/supplier or the competent authority may specify a cleansing agent if appropriate, or may recommend an alternative agent in exceptional cases if water is clearly inappropriate. P316 **Get emergency medical help immediately.** Competent authority or manufacturer / supplier may add, 'Call' followed by the appropriate emergency telephone number, or the appropriate emergency medical help provider, for example, a Poison Centre, Emergency Centre or Doctor. P321 **Specific treatment (see ... on this label)** – *if immediate measures such as specific cleansing agent is advised.* ...Reference to supplemental first aid instruction. P361+ P364 **Take off immediately all contaminated clothing and wash it before reuse.**	P405 **Store locked up.**	P501 **Dispose of contents/container to...** ... in accordance with local/regional/national/international regulations (to be specified). Manufacturer/supplier or the competent authority to specify whether disposal requirements apply to contents, container or both.

急性毒性（経皮）
（第 3.1 章）

危険有害性区分	シンボル	注意喚起語	危険有害性情報
1	どくろ	危険	H310 皮膚に接触すると生命に危険
2	どくろ	危険	H310 皮膚に接触すると生命に危険

注意書き

安全対策	応急措置	保管	廃棄
P262 眼、皮膚、または衣類につけないこと。 P264 取扱後は手［および…］をよく洗うこと。 ―製造者供給者または所管官庁が、取扱後に洗浄する体の他の部分を指定した場合には[]内の文章を用いる P270 この製品を使用する時に、飲食または喫煙をしないこと。 P280 保護手袋/保護衣を着用すること。 ―製造者供給者または所管官庁が装具の種類を指定してもよい。	P302 + P352 皮膚に付着した場合：多量の水/…で洗うこと。 ―製造者供給者または所管官庁が指定する適切な洗浄剤がある場合、または明らかに水が不適切である場合、他の薬剤を推薦する場合 P316 すぐに緊急の医療処置を受けること。 P321 特別な処置が必要である（このラベルの…を参照）。 ―緊急の洗浄剤などを推薦する場合 …補足的な応急措置の説明 P361+ P364 汚染された衣類を直ちにすべて脱ぎ、再使用する場合には洗濯をすること。	P405 施錠して保管すること。	P501 内容物/容器を…に廃棄すること。 …製造者供給者または所管官庁が指定する内容に従って ―国際/国/都道府県/市町村の規則（明示する）に 製造者供給者または所管官庁が適用する廃棄物要件物、容器または／およびその両者に適用する廃棄物要件

- 371 -

ACUTE TOXICITY – DERMAL
(CHAPTER 3.1)

Hazard category	Symbol	Signal word	Hazard statement
3	Skull and crossbones	Danger	H311 Toxic in contact with skin

Precautionary statements

Prevention	Response	Storage	Disposal
P262 **Do not get in eyes, on skin, or on clothing.** P264 **Wash hands [and …] thoroughly after handling.** – *text in square brackets to be used when the manufacturer/supplier or competent authority specify other parts of the body to be washed after handling.* P270 **Do not eat, drink or smoke when using this product.** P280 **Wear protective gloves/protective clothing.** Manufacturer/supplier or the competent authority may further specify type of equipment where appropriate.	P302 + P352 **IF ON SKIN: Wash with plenty of water/…** …Manufacturer/supplier or the competent authority may specify a cleansing agent if appropriate, or may recommend an alternative agent in exceptional cases if water is clearly inappropriate. P316 **Get emergency medical help immediately.** Competent authority or manufacturer / supplier may add, 'Call' followed by the appropriate emergency telephone number, or the appropriate emergency medical help provider, for example, a Poison Centre, Emergency Centre or Doctor. P321 **Specific treatment (see … on this label)** – *if immediate measures such as specific cleansing agent is advised.* … Reference to supplemental first aid instruction. P361 + P364 **Take off immediately all contaminated clothing and wash it before reuse.**	P405 Store locked up.	P501 **Dispose of contents/container to…** … in accordance with local/regional/national/international regulations (to be specified). Manufacturer/supplier or the competent authority to specify whether disposal requirements apply to contents, container or both.

急性毒性（経皮）
(第 3.1 章)

危険有害性区分	シンボル	注意喚起語	危険有害性情報
3	どくろ	危険	H311 皮膚に接触すると有毒

注意書き

安全対策	応急措置	保管	廃棄
P262 眼、皮膚、または衣類につけないこと。 P264 取扱後は手【および…】をよく洗うこと。 ―製造者供給者または所管官庁が、取扱後に洗浄する体の他の部分を指定した場合には [ ] 内の文章を用いる P270 この製品を使用する時に、飲食または喫煙をしないこと。 P280 保護手袋／保護衣を着用すること。 ―製造者供給者または所管官庁が装具の種類を指定してもよい。	P302＋P352 皮膚に付着した場合：多量の水／…で洗うこと。 ―製造者供給者または所管官庁が指定する適切な洗浄剤がある場合、または明らかに水が不適切で他の薬剤を推薦する場合 P316 すぐに救急の医療処置を受けること。 ―所管官庁または製造者供給者は、「電話」に続けて、適当な救急時電話番号／または適当な救急時医療提供者、例えば中毒センター、救急センターまたは医師などを追加してもよい。 P321 特別な処置が必要である（このラベルの…を参照）。 ―緊急の洗浄剤などを推薦する場合 …補足的な応急措置の説明 P361＋P364 汚染された衣類を直ちにすべて脱ぎ、再使用する場合には洗濯をすること。	P405 施錠して保管すること。	P501 内容物／容器を…に廃棄すること。 …国際（国／都道府県／市町村）の規則（明示する）に従って 製造者供給者または所管官庁が指定する内容物、容器またはその両者に適用する廃棄物要件

ACUTE TOXICITY – DERMAL
(CHAPTER 3.1)

Hazard category	Symbol	Signal word	Hazard statement
4	Exclamation mark	Warning	H312 Harmful in contact with skin

Precautionary statements

Prevention	Response	Storage	Disposal
P280 **Wear protective gloves/protective clothing** Manufacturer/supplier or the competent authority may further specify type of equipment where appropriate.	P302 + P352 **IF ON SKIN: Wash with plenty of water/…** …Manufacturer/supplier or the competent authority may specify a cleansing agent if appropriate, or may recommend an alternative agent in exceptional cases if water is clearly inappropriate. P317 **Get medical help.** P321 **Specific treatment (see … on this label)** – *if immediate measures such as specific cleansing agent is advised.* …Reference to supplemental first aid instruction. P362 + P364 **Take off contaminated clothing and wash it before reuse.**		P501 **Dispose of contents/container to…** … in accordance with local/regional/national/international regulations (to be specified). Manufacturer/supplier or the competent authority to specify whether disposal requirements apply to contents, container or both.

危険有害性区分	シンボル	注意喚起語	危険有害性情報
急性毒性（経皮） （第 3.1 章） 4	感嘆符	警告	H312 皮膚に接触すると有害

注意書き

安全対策	応急措置	保管	廃棄
P280 保護手袋/保護衣を着用すること。 製造者/供給者または所管官庁が指定する装具の種類を指定してもよい。	P302 + P352 皮膚に付着した場合：多量の水/…で洗うこと。 …製造者/供給者または所管官庁が指定する場合、または明らかに水が不適切な洗浄剤がある場合、または明らかに水が不適切で他の薬剤を推薦する場合 P317 医療処置を受けること。 P321 特別な処置が必要である（このラベルの…を見よ）。 ―緊急の洗浄剤など…を推薦する場合 …補足的な応急措置の説明 P362+ P364 汚染された衣類を脱ぎ、再使用する場合には洗濯をすること。		P501 内容物/容器を…に廃棄すること。 …国際/国/都道府県/市町村の規則（明示する）に従って 製造者/供給者または所管官庁が指定する内容物、容器またはその両者に適用する廃棄物要件

ACUTE TOXICITY – DERMAL
(CHAPTER 3.1)

Symbol	Signal word	Hazard statement
No symbol	Warning	H313 May be harmful in contact with skin

Hazard category			
5			

Precautionary statements

Prevention	Response	Storage	Disposal
	P302 + P317 **IF ON SKIN: Get medical help.**		

急性毒性（経皮）
(第 3.1 章)

注意喚起語
警告

危険有害性情報
H313 皮膚に接触すると有害のおそれ

危険有害性区分
5

シンボル
なし

注意書き

安全対策	応急措置	保管	廃棄
	P302+P317 皮膚に付着した場合：医療処置を受けること。		

- 374 -

ACUTE TOXICITY – INHALATION
(CHAPTER 3.1)

Hazard category	Symbol	Signal word	Hazard statement
1	Skull and crossbones	Danger	H330 Fatal if inhaled
2	Skull and crossbones	Danger	H330 Fatal if inhaled

Precautionary statements

Prevention	Response	Storage	Disposal
P260 **Do not breathe dust/fume/gas/mist/vapours/spray.** Manufacturer/supplier or the competent authority to specify applicable physical state(s). P271 **Use only outdoors or with adequate ventilation.** Manufacturer/supplier to specify what type of ventilation would be adequate for safe use on the safety data sheet and in any supplemental safety instructions provided to consumers. P284 **In case of inadequate ventilation wear respiratory protection.** Manufacturer/supplier to specify on the safety data sheet what type of ventilation would be adequate for safe use and provide additional information with the chemical at the point of use that explains what type of respiratory equipment may also be needed.	P304 + P340 **IF INHALED: Remove person to fresh air and keep comfortable for breathing.** P316 **Get emergency medical help immediately.** Competent authority or manufacturer / supplier may add, 'Call' followed by the appropriate emergency telephone number, or the appropriate emergency medical help provider, for example, a Poison Centre, Emergency Centre or Doctor. P320 **Specific treatment is urgent (see ... on this label)** – *if immediate administration of antidote is required.* ... Reference to supplemental first aid instruction.	P403 + P233 **Store in a well-ventilated place. Keep container tightly closed.** – *if the chemical is volatile and may generate a hazardous atmosphere.* P405 **Store locked up.**	P501 **Dispose of contents/container to...** ... in accordance with local/regional/national/international regulations (to be specified). Manufacturer/supplier or the competent authority to specify whether disposal requirements apply to contents, container or both.

急性毒性（吸入）
(第 3.1 章)

危険有害性区分	シンボル	注意喚起語	危険有害性情報
1	どくろ	危険	H330 吸入すると生命に危険
2	どくろ	危険	H330 吸入すると生命に危険

注意書き

安全対策	応急措置	保管	廃棄
P260 粉じん/煙/ガス/ミスト/蒸気/スプレーを吸入しないこと。 製造者/供給者または所管官庁が指定する物理的状態 P271 屋外または十分な換気のある場所でのみ使用すること。 製造者/供給者は、安全データシートおよび消費者に提供する安全説明書に、安全な使用のためにどのような種類の換気が適切であるかを明記すること。 P284 換気が不十分な場合は呼吸用保護具を着用すること。 製造者/供給者は、安全データシートに、安全な使用のためにどのような種類の換気が適切であるかを明記し、使用時にどのような種類の呼吸保護具を必要となる可能性があるかを説明する追加情報を、化学品とともに提供すること。	P304 + P340 吸入した場合：空気の新鮮な場所に移し、呼吸しやすい姿勢で休息させること。 P316 すぐに救急の医療処置を受けること。「電話」に続けて、適切な救急時電話番号すなわち適当な救急時医療提供者、例えば中毒センター、救急センターまたは医師などを追加してもよい。 P320 特別な処置が緊急に必要である（このラベルの場合 ...を参照） 一緊急の解毒剤の投与が必要な場合 一補足的な応急措置の説明	P403 + P233 換気の良いところで保管すること。容器を密閉しておくこと。 一化学品が揮発性で有害な環境を作る可能性があるとき P405 施錠して保管すること。	P501 内容物/容器を...に廃棄すること。 ...国際/国都道府県/市町村の規則（明示すること）に従って 製造者/供給者または所管官庁が指定する内容物、容器またはその両者に適用する廃棄物要件

- 375 -

ACUTE TOXICITY – INHALATION
(CHAPTER 3.1)

Hazard category	Symbol	Signal word	Hazard statement
3	Skull and crossbones	Danger	H331 Toxic if inhaled

Precautionary statements

Prevention	Response	Storage	Disposal
P261 **Avoid breathing dust/fume/gas/mist/vapours/spray.** – *may be omitted if P260 is given on the label* Manufacturer/supplier or the competent authority to specify applicable conditions. P271 **Use only outdoors or with adequate ventilation.** Manufacturer/supplier to specify what type of ventilation would be adequate for safe use on the safety data sheet and in any supplemental safety instructions provided to consumers.	P304 + P340 **IF INHALED: Remove person to fresh air and keep comfortable for breathing.** P316 **Get emergency medical help immediately.** Competent authority or manufacturer / supplier may add, 'Call' followed by the appropriate emergency telephone number, or the appropriate emergency medical help provider, for example, a Poison Centre, Emergency Centre or Doctor. P321 **Specific treatment (see ... on this label)** – *if immediate specific measures are required.* ... Reference to supplemental first aid instruction.	P403 + P233 **Store in a well-ventilated place. Keep container tightly closed.** – *if the chemical is volatile and may generate a hazardous atmosphere.* P405 **Store locked up.**	P501 **Dispose of content/container to...** ... in accordance with local/regional/national/international regulations (to be specified). Manufacturer/supplier or the competent authority to specify whether disposal requirements apply to contents, container or both.

急性毒性（吸入）
（第 3.1 章）

危険有害性区分	シンボル	注意喚起語	危険有害性情報
3	どくろ	危険	H331 吸入すると有毒

注意書き

安全対策	応急措置	保管	廃棄
P261 粉じん/煙/ガス/ミスト/蒸気/スプレーの吸入を避けること。 －P260 がラベルに記載される場合には省略してもよい 製造者/供給者または所管官庁が指定する適用条件 P271 屋外または十分な換気のある場所でのみ使用すること。 製造者/供給者は、安全データシートおよび安全説明書に、補足する補足的な安全説明書に、安全な使用のためにどのような種類の換気が適切であるかを明記すること。	P304 + P340 吸入した場合：空気の新鮮な場所に移し、呼吸しやすい姿勢で休息させること。 P316 すぐに救急の医療処置を受けること。 所管官庁または製造者/供給者は救急時電話番号すなわち適当な毒物センター、救急センターを医師などに追加してもよい。時医療提供者、例えば中毒センター、救急センターまたは医師などに追加してもよい。 P321 特別な処置が必要である。（このラベルの...を参照） －緊急の特別な処置が必要な場合 ...補足的な応急措置の説明	P403 + P233 換気の良いところで保管すること。容器を密閉しておくこと。 －化学品が揮発性で有害な環境を作る可能性があるとき P405 施錠して保管すること。	P501 内容物/容器を...に廃棄すること。 ...国際/国/都道府県/市町村の規則（明示する）に従って 製造者/供給者または所管官庁が指定する内容物、容器またはその両者に適用する廃棄物要件

- 376 -

ACUTE TOXICITY – INHALATION
(CHAPTER 3.1)

Hazard category	Symbol	Signal word	Hazard statement
4	Exclamation mark	Warning	H332 Harmful if inhaled

Precautionary statements

Prevention	Response	Storage	Disposal
P261 **Avoid breathing dust/fume/gas/mist/vapours/spray.** – *may be omitted if P260 is given on the label* Manufacturer/supplier or the competent authority to specify applicable conditions. P271 **Use only outdoors or with adequate ventilation.** Manufacturer/supplier to specify what type of ventilation would be adequate for safe use on the safety data sheet and in any supplemental safety instructions provided to consumers.	P304 + P340 **IF INHALED: Remove person to fresh air and keep comfortable for breathing.** P317 **Get medical help.**		

- 377 -

危険有害性区分	シンボル	注意喚起語	危険有害性情報
急性毒性（吸入）(第 3.1 章)			
4	感嘆符	警告	H332 吸入すると有害

注意書き

安全対策	応急措置	保管	廃棄
P261 粉じん/煙/ガス/ミスト/蒸気/スプレーの吸入を避けること。 ―P260 がラベルに記載される場合には省略してもよい 製造者供給者または所管官庁が指定する適用条件 P271 屋外または十分な換気のある場所でのみ使用すること。 製造者供給者は、安全データシートおよび消費者に提供する補足的な安全説明書に、安全な使用のためにどのような種類の換気が適切であるかを明記すること。	P304 + P340 吸入した場合：空気の新鮮な場所に移し、呼吸しやすい姿勢で休息させること。 P317 医療処置を受けること。		

- 377 -

ACUTE TOXICITY – INHALATION
(CHAPTER 3.1)

Hazard category	Symbol	Signal word	Hazard statement
5	*No symbol*	Warning	H333 May be harmful if inhaled

Precautionary statements

Prevention	Response	Storage	Disposal
	P304 + P317 **IF INHALED:** **Get medical help.**		

急性毒性（吸入）
(第 3.1 章)

危険有害性区分	シンボル	注意喚起語	危険有害性情報
5	なし	警告	H333 吸入すると有害のおそれ

注意書き

安全対策	応急措置	保管	廃棄
	P304 + P317 吸入した場合：医療処置を受けること。		

SKIN CORROSION/IRRITATION
(CHAPTER 3.2)

Hazard category	Symbol	Signal word	Hazard statement
1, 1A, 1B, 1C	Corrosion	Danger	H314 Causes severe skin burns and eye damage

Precautionary statements

Prevention	Response	Storage	Disposal
P260 **Do not breathe dusts or mists.** – *if inhalable particles of dusts or mists may occur during use.* P264 **Wash hands [and…] thoroughly after handling.** - *text in square brackets to be used when the manufacturer/supplier or the competent authority specify other parts of the body to be washed after handling.* P280 **Wear protective gloves/protective clothing/eye protection/face protection.** Manufacturer/supplier or the competent authority may further specify type of equipment where appropriate.	P301 + P330 + P331 **IF SWALLOWED: Rinse mouth. Do NOT induce vomiting.** P302 + P361 + P354 **IF ON SKIN: Take off immediately all contaminated clothing. Immediately rinse with water for several minutes.** P363 **Wash contaminated clothing before reuse.** P304 + P340 **IF INHALED: Remove person to fresh air and keep comfortable for breathing.** P316 **Get emergency medical help immediately.** Competent authority or manufacturer / supplier may add, 'Call' followed by the appropriate emergency telephone number, or the appropriate emergency medical help provider, for example, a Poison Centre, Emergency Centre or Doctor. P321 **Specific treatment (see … on this label)** … Reference to supplemental first aid instruction. Manufacturer/supplier or the competent authority may specify a cleansing agent if appropriate. P305 + P354 + P338 **IF IN EYES: Immediately rinse with water for several minutes. Remove contact lenses, if present and easy to do. Continue rinsing.**	P405 **Store locked up.**	P501 **Dispose of contents/container to…** … in accordance with local/regional/national/international regulations (to be specified). Manufacturer/supplier or the competent authority to specify whether disposal requirements apply to contents, container or both.

- 379 -

皮膚腐食性/刺激性
(第 3.2 章)

危険有害性区分	シンボル	注意喚起語	危険有害性情報
1, 1A から 1C	腐食性	危険	H314 重篤な皮膚の薬傷および眼の損傷

注意書き

安全対策	応急措置	保管	廃棄
P260 粉じんまたはミストを吸入しないこと。 ―使用中に吸い込みうるほこりやミストの粒子が発生するかもしれない場合 P264 取扱後は手【および...】をよく洗うこと。 ―製造者供給者または所管官庁が、取扱後に洗浄する体の他の部分を指定した場合には [ ] 内の文章を用いる P280 保護手袋/保護衣/保護眼鏡/保護面を着用すること。 製造者供給者または所管官庁が指定した装具の種類を指定してもよい	P301+ P330 + P331 飲み込んだ場合：口をすすぐこと。無理に吐かせないこと。 P302+ P361 + P354 皮膚に付着した場合：直ちに汚染された衣類をすべて脱ぐこと。すぐに水で数分間洗うこと。 P363 汚染した衣類を再使用する場合には洗濯すること。 P304 + P340 吸入した場合：空気の新鮮な場所に移し、呼吸しやすい姿勢で休息させること。 P316 すぐに緊急の医療処置を受けること。 所管官庁または製造者供給者は、「電話」に続けて、適当な救急時電話番号すなわちおよび救急時医療提供者、例えば中毒センター、救急センターまたは医師などを追加してもよい。 P321 特別な処置が必要である（このラベルの...を参照）。 ...補足的な応急措置 製造者供給者または所管官庁が指定する洗浄剤がある場合 P305 + P354 + P338 眼に入った場合：すぐに水で数分間洗うこと。コンタクトレンズを着用していて容易に外せる場合は外すこと。その後も洗浄を続けること。	P405 施錠して保管すること。	P501 内容物容器を...に廃棄すること。 ...国際/国/都道府県/市町村の規則（明示する）に従って 製造者供給者または所管官庁が指定する内容物、容器またはその両者に適用する廃棄物要件

SKIN CORROSION/IRRITATION
(CHAPTER 3.2)

Hazard category	Symbol	Signal word	Hazard statement
2	Exclamation mark	Warning	H315 Causes skin irritation

Precautionary statements

Prevention	Response	Storage	Disposal
P264 **Wash hands [and…] thoroughly after handling.** *- text in square brackets to be used when the manufacturer/supplier or the competent authority specify other parts of the body to be washed after handling.* P280 **Wear protective gloves.** Manufacturer/supplier or the competent authority may further specify type of equipment where appropriate.	P302 + P352 **IF ON SKIN: Wash with plenty of water/…** *…Manufacturer/supplier or the competent authority may specify a cleansing agent if appropriate, or may recommend an alternative agent in exceptional cases if water is clearly inappropriate.* P321 **Specific treatment (see … on this label)** *… Reference to supplemental first aid instruction.* Manufacturer/supplier or the competent authority may specify a cleansing agent if appropriate. P332 + P317 **If skin irritation occurs: Get medical help.** *– may be omitted when P333+P317 appears on the label.* P362 + P364 **Take off contaminated clothing and wash it before reuse.**		

皮膚腐食性刺激性
（第 3.2 章）

危険有害性区分	シンボル	注意喚起語	危険有害性情報
2	感嘆符	警告	H315 皮膚刺激

注意書き

安全対策	応急措置	保管	廃棄
P264 取扱後は手【および…】をよく洗うこと。 ―製造者/供給者または所管官庁が、取扱後に洗浄する身体の他の部分を指定した場合または洗浄する場合には口内の文章を用いる P280 保護手袋を着用すること。 製造者/供給者または所管官庁が装具の種類を指定してもよい	P302 + P352 皮膚に付着した場合：多量の水/…で洗うこと。 ―製造者/供給者または所管官庁が指定する適切な洗浄剤がある場合、または明らかに水が不適切で他の薬剤を推薦する場合 P321 特別な処置が必要である（このラベルの…を参照）。 …補足的な応急措置の説明 製造者/供給者または所管官庁が指定する適切な洗浄剤がある場合 P332 + P317 皮膚刺激が生じた場合：医療処置を受けること。 ―P333+P317がラベル上にあるときは省略してもよい P362+P364 汚染された衣類を脱ぎ、再使用する場合には洗濯をすること。		

- 380 -

SKIN CORROSION/IRRITATION
(CHAPTER 3.2)

Hazard category	Symbol	Signal word	Hazard statement
3	*No symbol*	Warning	H316 Causes mild skin irritation

Precautionary statements

Prevention	Response	Storage	Disposal
	P332 + P317 **If skin irritation occurs: Get medical help.** *– may be omitted when P333+P317 appears on the label.*		

皮膚腐食性/刺激性
(第 3.2 章)

危険有害性区分	シンボル	注意喚起語	危険有害性情報
3	なし	警告	H316 軽度の皮膚刺激

注意書き

安全対策	応急措置	保管	廃棄
	P332 + P317 皮膚刺激が生じた場合：医療処置を受けること。 －P333+P317がラベル上にあるときは省略してもよい		

- 381 -

SERIOUS EYE DAMAGE/EYE IRRITATION
(CHAPTER 3.3)

Hazard category	Symbol	Signal word	Hazard statement
1	Corrosion	Danger	H318 Causes serious eye damage

Precautionary statements

Prevention	Response	Storage	Disposal
P264+P265 **Wash hands [and…] thoroughly after handling. Do not touch eyes.** - *text in square brackets to be used when the manufacturer/supplier or the competent authority specify other parts of the body to be washed after handling.* P280 **Wear eye protection/face protection.** - *Specify protective gloves and eye/face protection. Manufacturer/supplier or the competent authority may further specify type of equipment where appropriate.*	P305 + P354 + P338 **IF IN EYES: Immediately rinse with water for several minutes. Remove contact lenses, if present and easy to do. Continue rinsing.** P317 **Get medical help.**		

眼に対する重篤な損傷性/眼刺激性
（第 3.3 章）

危険有害性区分	シンボル	注意喚起語	危険有害性情報
1	腐食性	危険	H318 重篤な眼の損傷

注意書き

安全対策	応急措置	保管	廃棄
P264 + P265 取扱後は手 [および…] をよく洗うこと。眼には触らないこと。 ―製造者供給者または所管官庁が、取扱後に洗浄する体の他の部分を指定した場合には [ ] 内の文章を用いる P280 保護眼鏡/保護面を着用すること。 ―保護手袋および眼鏡/保護面を指定すること ―製造者供給者または所管官庁が装具の種類を指定してもよい	P305 + P354 + P338 眼に入った場合：すぐに水で数分間洗うこと。コンタクトレンズを着用していて容易に外せる場合は外すこと。その後も洗浄を続けること。 P317 医療処置を受けること。		

- 382 -

SERIOUS EYE DAMAGE/EYE IRRITATION
(CHAPTER 3.3)

Hazard category	Symbol	Signal word	Hazard statement
2/2A	Exclamation mark	Warning	H319 Causes serious eye irritation

Precautionary statements

Prevention	Response	Storage	Disposal
P264+P265 **Wash hands [and…] thoroughly after handling. Do not touch eyes.** - *text in square brackets to be used when the manufacturer/supplier or the competent authority specify other parts of the body to be washed after handling.* P280 **Wear eye protection/face protection.** - *Specify protective gloves and eye/face protection. Manufacturer/supplier or the competent authority may further specify type of equipment where appropriate.*	P305 + P351 + P338 **IF IN EYES: Rinse cautiously with water for several minutes. Remove contact lenses, if present and easy to do. Continue rinsing.** P337 + P317 **If eye irritation persists: Get medical help.**		

眼に対する重篤な損傷性/眼刺激性
（第 3.3 章）

危険有害性区分	シンボル	注意喚起語	危険有害性情報
2/2A	感嘆符	警告	H319 強い眼刺激

注意書き

安全対策	応急措置	保管	廃棄
P264＋P265 取扱後は手［および…］をよく洗うこと。眼に触らないこと。 ―製造者/供給者または所管官庁が、取扱後に洗浄する体の他の部分を指定した場合には口内の文章を用いる P280 保護眼鏡/保護面を着用すること。 ―製造者/供給者または所管官庁が装具の種類を指定してもよい	P305＋P351＋P338 眼に入った場合：水で数分間注意深く洗うこと。次に、コンタクトレンズを着用していて容易に外せる場合は外すこと。その後も洗浄を続けること。 P337＋P317 眼の刺激が続く場合：医療処置を受けること。		

- 383 -

SERIOUS EYE DAMAGE/EYE IRRITATION
(CHAPTER 3.3)

Hazard category	Symbol	Signal word	Hazard statement
2B	*No symbol*	Warning	H320 Causes eye irritation

Precautionary statements

Prevention	Response	Storage	Disposal
P264+P265 **Wash hands [and…] thoroughly after handling. Do not touch eyes.** *- text in square brackets to be used when the manufacturer/supplier or the competent authority specify other parts of the body to be washed after handling.*	P305 + P351 + P338 **IF IN EYES: Rinse cautiously with water for several minutes. Remove contact lenses, if present and easy to do. Continue rinsing.** P337 + P317 **If eye irritation persists: Get medical help.**		

眼に対する重篤な損傷性/眼刺激性
(第 3.3 章)

危険有害性区分	シンボル	注意喚起語	危険有害性情報
2B	なし	警告	H320 眼刺激

注意書き

安全対策	応急措置	保管	廃棄
P264 + P265 取扱後は手［および…］をよく洗うこと。眼には触らないこと。取扱後に洗浄する体の他の部分を指定した場合は 11 内の文章を用いる ―製造者供給者または所管官庁が、	P305 + P351 + P338 眼に入った場合：水で数分間注意深く洗うこと。次に、コンタクトレンズを着用していて容易に外せる場合は外すこと。その後も洗浄を続けること。 P337 + P317 眼の刺激が続く場合：医療処置を受けること。		

- 384 -

RESPIRATORY SENSITIZATION
(CHAPTER 3.4)

Hazard category	Symbol	Signal word	Hazard statement
1, 1A, 1B	Health hazard	Danger	H334 May cause allergy or asthma symptoms or breathing difficulties if inhaled

Precautionary statements

Prevention	Response	Storage	Disposal
P233 **Keep container tightly closed.** P260 **Do not breathe dust/fume/gas/mist/vapours/spray.** Manufacturer/supplier or the competent authority to specify applicable physical state(s). P271 **Use only outdoors or with adequate ventilation.** Manufacturer/supplier to specify what type of ventilation would be adequate for safe use on the safety data sheet and in any supplemental safety instructions provided to consumers. P280 **Wear protective gloves/protective clothing.** Manufacturer/supplier or the competent authority may further specify type of equipment where appropriate. P284 **In case of inadequate ventilation wear respiratory protection.** Manufacturer/supplier to specify on the safety data sheet what type of ventilation would be adequate for safe use and provide additional information with the chemical at the point of use that explains what type of respiratory equipment may also be needed.	P304 + P340 **IF INHALED: remove person to fresh air and keep comfortable for breathing.** P342 + P316 **If experiencing respiratory symptoms: Get emergency medical help immediately.** Competent authority or manufacturer / supplier may add, 'Call' followed by the appropriate emergency telephone number, or the appropriate emergency medical help provider, for example, a Poison Centre, Emergency Centre or Doctor.	P403 **Store in a well-ventilated place.**	P501 **Dispose of contents/container to...** ... in accordance with local/regional/national/international regulations (to be specified). Manufacturer/supplier or the competent authority to specify whether disposal requirements apply to contents, container or both.

呼吸器感作性
(第 3.4 章)

危険有害性区分	シンボル	注意喚起語	危険有害性情報
1, 1A, 1B	健康有害性	危険	H334 吸入するとアレルギー、喘息または呼吸困難を起こすおそれ

注意書き

安全対策	応急措置	保管	廃棄
P233 容器を密閉しておくこと。 P260 粉じん/煙/ガス/ミスト/蒸気/スプレーを吸入しないこと。 製造者/供給者または所管官庁が指定する物理的状態。 P271 屋外または十分な換気のある場所でのみ使用すること。 製造者/供給者、安全データシートおよび消費者に提供する補足的な安全説明書に、安全な使用のためにどのような種類の換気が適切であるかを明記すること。 P280 保護手袋/保護衣を着用すること。 製造者/供給者または所管官庁が、必要に応じて装具の種類を指定してもよい。 P284 換気が不十分な場合は呼吸用保護具を着用すること。 製造者/供給者は、安全データシートに、安全な使用のためにどのような種類の換気が適切であるかを明記し、使用時にどのような種類の呼吸用装置も必要となる可能性があるかを説明する追加情報を、化学品とともに提供すること。	P304 + P340 吸入した場合：空気の新鮮な場所に移し、呼吸しやすい姿勢で休息させること。 P342 + P316 呼吸に関する症状が出た場合：すぐに救急の医療処置を受けること。 製造者/供給者または所管官庁が指定する内容。 所管官庁または製造者/供給者は、適切な救急時電話番号すなわち適当な救急時医療提供者、例えば中毒センター、救急センターまたは医師などを追加してもよい。	P403 換気の良いところで保管すること。	P501 内容物/容器を…に廃棄すること。 …国際的/国/都道府県/市町村の規則（明示する）に従って 製造者/供給者または所管官庁が指定する内容物、容器またはその両者に適用する廃棄物要件

- 385 -

SKIN SENSITIZATION
(CHAPTER 3.4)

Hazard category	Symbol	Signal word	Hazard statement
1, 1A, 1B	Exclamation mark	Warning	H317 May cause an allergic skin reaction

Precautionary statements

Prevention	Response	Storage	Disposal
P261 **Avoid breathing dust/fume/gas/mist/vapours/spray.** – *may be omitted if P260 is given on the label* Manufacturer/supplier or the competent authority to specify applicable conditions. P272 **Contaminated work clothing should not be allowed out of the workplace.** P280 **Wear protective gloves.** Manufacturer/supplier or the competent authority may further specify type of equipment where appropriate.	P302 + P352 **IF ON SKIN: Wash with plenty of water/…** …Manufacturer/supplier or the competent authority may specify a cleansing agent if appropriate, or may recommend an alternative agent in exceptional cases if water is clearly inappropriate. P333 + P317 **If skin irritation or rash occurs: Get medical help.** P321 **Specific treatment (see … on this label)** … Reference to supplemental first aid instruction. Manufacturer/supplier or the competent authority may specify a cleansing agent if appropriate. P362 + P364 **Take off contaminated clothing and wash it before reuse.**		P501 **Dispose of contents/container to…** … in accordance with local/regional/national/international regulations (to be specified). Manufacturer/supplier or the competent authority to specify whether disposal requirements apply to contents, container or both.

危険有害性区分	シンボル	注意喚起語	危険有害性情報
1, 1A, 1B	感嘆符	警告	H317 アレルギー性皮膚反応を起こすおそれ

皮膚感作性
(第 3.4 章)

注意書き

安全対策	応急措置	保管	廃棄
P261 粉じん/煙/ガス/ミスト/蒸気/スプレーの吸入を避けること。 —P260 がラベルに記載される場合には省略してもよい 製造者/供給者または所管官庁が指定する適用条件 P272 汚染された作業衣は作業場から出さないこと。 P280 保護手袋を着用すること。 製造者/供給者または所管官庁が装具の種類を指定してもよい	P302 + P352 皮膚に付着した場合：多量の水/…で洗うこと。 …製造者/供給者または所管官庁が指定する適切な洗浄剤が記載される場合には省略して、または明らかに水が不適切で他の薬剤を推奨する場合 P333 + P317 皮膚刺激または発疹が生じた場合：医療処置を受けること。 P321 特別な処置が必要である（このラベルの…を参照）。 …補足的な応急措置の説明 製造者/供給者または所管官庁が指定する適切な洗浄剤がある場合 P362+P364 汚染された衣類を脱ぎ、再使用する場合には洗濯をすること。		P501 内容物/容器を…に廃棄すること。 …国際/国he道府県/市町村の規則（明示する）に従って 製造者/供給者または所管官庁が適用する内容物、容器または両者に適用する廃棄物要件

GERM CELL MUTAGENICITY
(CHAPTER 3.5)

Hazard category	Symbol	Signal word	Hazard statement	
1, 1A, 1B	Health hazard	Danger	H340	May cause genetic defects <...>
2	Health hazard	Warning	H341	Suspected of causing genetic defects <...>
			<...>	(state route of exposure if it is conclusively proven that no other routes of exposure cause the hazard)

Precautionary statements

Prevention	Response	Storage	Disposal
P203 **Obtain, read and follow all safety instructions before use.** P280 **Wear protective gloves/protective clothing/eye protection/face protection/hearing protection/...** Manufacturer/supplier or the competent authority to specify the appropriate personal protective equipment.	P318 **IF exposed or concerned, get medical advice.**	P405 **Store locked up.**	P501 **Dispose of contents/container to...** ... in accordance with local/regional/national/international regulations (to be specified). Manufacturer/supplier or the competent authority to specify whether disposal requirements apply to contents, container or both.

生殖細胞変異原性
（第 3.5 章）

危険有害性区分	シンボル	注意喚起語	危険有害性情報
1,1A,1B	健康有害性	危険	H340 遺伝性疾患のおそれ <...>
2	健康有害性	警告	H341 遺伝性疾患のおそれの疑い <...>

<...>には、他の経路からのばく露が有害でないことが決定的に証明されている場合、有害な経路を記載する

注意書き

安全対策	応急措置	保管	廃棄
P203 使用前にすべての安全説明書を入手し、読み、従うこと。 P280 保護手袋/保護衣/保護眼鏡/保護面/聴覚保護具…を着用すること。 製造者/供給者または所管官庁が保護具の種類を指定してもよい	P318 ばく露またはその懸念がある場合は、医学的助言を求めること。	P405 施錠して保管すること。	P501 内容物/容器を…に廃棄すること。 …国際/国/都道府県/市町村の規則（明示する）に従って 製造者/供給者または所管官庁が指定する内容物、容器または両者に適用する廃棄物要件を定してもよい

CARCINOGENICITY
(CHAPTER 3.6)

Hazard category	Symbol	Signal word	Hazard statement
1, 1A, 1B	Health hazard	Danger	H350 May cause cancer <...>
2	Health hazard	Warning	H351 Suspected of causing cancer <...>
			<...> (state route of exposure if it is conclusively proven that no other routes of exposure cause the hazard)

Precautionary statements

Prevention	Response	Storage	Disposal
P203 **Obtain, read and follow all safety instructions before use.** P280 **Wear protective gloves/protective clothing/eye protection/face protection/hearing protection/...** Manufacturer/supplier or the competent authority to specify the appropriate personal protective equipment.	P318 **IF exposed or concerned, get medical advice.**	P405 **Store locked up.**	P501 **Dispose of contents/container to...** ... in accordance with local/regional/national/international regulations (to be specified). Manufacturer/supplier or the competent authority to specify whether disposal requirements apply to contents, container or both.

発がん性
(第 3.6 章)

危険有害性区分
1, 1A, 1B
2

シンボル
健康有害性
健康有害性

注意喚起語
危険
警告

危険有害性情報
H350 発がんのおそれ <...>
H351 発がんのおそれの疑い <...>

<...> には、他の経路からのばく露が有害でないことが決定的に証明されている場合、有害な経路を記載する

注意書き

安全対策	応急措置	保管	廃棄
P203 使用前にすべての安全説明書を入手し、読み、従うこと。 P280 保護手袋/保護衣/保護眼鏡/保護面鏡/聴覚保護具…を着用すること。 製造者/供給者または所管官庁が装具の種類を指定してもよい	P318 ばく露またはその懸念がある場合は、医学的助言を求めること。	P405 施錠して保管すること。	P501 内容物/容器を…に廃棄すること。 …国際(国/都道府県/市町村)の規則(明示する)に従って 製造者/供給者または所管官庁が指定する内容物、容器またはその両者に適用する廃棄物要件を記載する

REPRODUCTIVE TOXICITY
(CHAPTER 3.7)

Hazard category	Symbol	Signal word	Hazard statement		
1, 1A, 1B	Health hazard	Danger	H360	May damage fertility or the unborn child <...> <<...>>	
2	Health hazard	Warning	H361	Suspected of damaging fertility or the unborn child <...> <<...>>	
			<...>	*(state specific effect if known)*	
			<<...>>	*(state route of exposure if it is conclusively proven that no other routes of exposure cause the hazard)*	

Precautionary statements

Prevention	Response	Storage	Disposal
P203 **Obtain, read and follow all safety instructions before use.** P280 **Wear protective gloves/protective clothing/eye protection/face protection/hearing protection/...** Manufacturer/supplier or the competent authority to specify the appropriate personal protective equipment.	P318 **IF exposed or concerned, get medical advice.**	P405 **Store locked up.**	P501 **Dispose of contents/container to...** ... in accordance with local/regional/national/international regulations (to be specified). Manufacturer/supplier or the competent authority to specify whether disposal requirements apply to contents, container or both.

生殖毒性
(第 3.7 章)

危険有害性区分	シンボル	注意喚起語	危険有害性情報
1, 1A, 1B	健康有害性	危険	H360 生殖能または胎児への悪影響のおそれ <...> <<...>>
2	健康有害性	警告	H361 生殖能または胎児への悪影響のおそれの疑い <...> <<...>>

<...>には影響の内容を記載する
<<...>>には、もし判れば影響の内容を記載する
<...>には、他の経路からのばく露が有害でないこと が決定的に証明されている場合、有害な経路を記載する

注意書き

安全対策	応急措置	保管	廃棄
P203 使用前にすべての安全説明書を入手し、読み、従うこと。 P280 保護手袋/保護衣/保護眼鏡/保護面/聴覚保護具… を着用すること。製造者/供給者または所管官庁が装具の種類を指定してもよい	P318 ばく露またはその懸念がある場合は、医学的助言を求めること。	P405 施錠して保管すること。	P501 内容物/容器を…に廃棄すること。 …国際/国/都道府県/市町村の規則(明示する)に 従って 製造者/供給者または所管官庁が指定する内容物、容器またはその両者に適用する廃棄物要件 が有害な経路を記載する

- 389 -

REPRODUCTIVE TOXICITY
(CHAPTER 3.7)
(effects on or via lactation)

Hazard category	Symbol	Signal word	Hazard statement
(additional)	*No symbol*	*No signal word*	H362 May cause harm to breast-fed children

Precautionary statements

Prevention	Response	Storage	Disposal
P203 **Obtain, read and follow all safety instructions before use.** P260 **Do not breathe dusts or mists.** *– if inhalable particles of dusts or mists may occur during use.* P263 **Avoid contact during pregnancy and while nursing.** P264 **Wash hands [and…] thoroughly after handling.** *- text in square brackets to be used when the manufacturer/supplier or the competent authority specify other parts of the body to be washed after handling.* P270 **Do not eat, drink or smoke when using this product.**	P318 **IF exposed or concerned, get medical advice.**		

生殖毒性
(第 3.7 章)
(授乳に対するまたは授乳を介した影響)

危険有害性区分	シンボル	注意喚起語	危険有害性情報
(追加)	なし	なし	H362 授乳中の子に害を及ぼすおそれ

注意書き

安全対策	応急措置	保管	廃棄
P203 使用前にすべての安全説明書を入手し、読み、従うこと。 P260 粉じん/ミストを吸入しないこと。 ―使用中に吸入されるほこりやミストの粒子が発生するかもしれない場合 P263 妊娠中および授乳期中は接触を避けること。 P264 取扱後は手［および...］をよく洗うこと。 ―製造者供給者または所管官庁が、取扱後に洗浄する体の他の部分を指定した場合には［］内の文章を用いる P270 この製品を使用する時に、飲食または喫煙をしないこと。	P318 ばく露またはその懸念がある場合は、医学的助言を求めること。		

- 390 -

SPECIFIC TARGET ORGAN TOXICITY (SINGLE EXPOSURE)
(CHAPTER 3.8)

Hazard category	Symbol	Signal word	Hazard statement		
1	Health hazard	Danger	H370	Causes damage to organs <...> <<...>>	
			<...>	(or state all organs affected if known)	
			<<...>>	(state route of exposure if it is conclusively proven that no other routes of exposure cause the hazard)	

Precautionary statements

Prevention	Response	Storage	Disposal
P260 **Do not breathe dust/fume/gas/mist/vapours/spray.** Manufacturer/supplier or the competent authority to specify applicable physical state(s). P264 **Wash hands [and…] thoroughly after handling.** - *text in square brackets to be used when the manufacturer/supplier or the competent authority specify other parts of the body to be washed after handling.* P270 **Do not eat, drink or smoke when using this product.**	P308 + P316 **IF exposed or concerned: Get emergency medical help immediately.** Competent authority or manufacturer / supplier may add, 'Call' followed by the appropriate emergency telephone number, or the appropriate emergency medical help provider, for example, a Poison Centre, Emergency Centre or Doctor. P321 **Specific treatment (see … on this label)** – *if immediate measures are required.* … Reference to supplemental first aid instruction.	P405 **Store locked up.**	P501 **Dispose of contents/container to…** … in accordance with local/regional/national/international regulations (to be specified). Manufacturer/supplier or the competent authority to specify whether disposal requirements apply to contents, container or both.

特定標的臓器毒性（単回ばく露）
（第 3.8 章）

危険有害性区分	シンボル	注意喚起語	危険有害性情報
1	健康有害性	危険	H370 臓器の障害 <...> <<...>> <...> （もしわかればすべての影響を受ける臓器を挙げる） <<...>> （他の経路からのばく露が有害でないことが決定的に証明されている場合、有害な経路を記載する）

注意書き

安全対策	応急措置	保管	廃棄
P260 粉じん／煙／ガス／ミスト／蒸気／スプレーを吸入しないこと。 …製造者／供給者または所管官庁が指定する物理的状態 P264 取扱後は手【および…】をよく洗うこと。 …製造者／供給者または所管官庁が、取扱後に洗浄する身体の他の部分を指定した場合には【】内の文章を用いる P270 この製品を使用する時に、飲食または喫煙をしないこと。	P308 + P316 ばく露またはその懸念がある場合：すぐに救急の医療処置を受けること。 …製造者または供給者または所管官庁が、適当な救急時電話番号ならびは製造者供給者、例えば中毒センター、救急センターまたは医師などを追加してもよい。 P321 特別な処置が必要である（このラベルの…）を参照。 …緊急の処置が必要な場合 …補足的な応急措置の説明	P405 施錠して保管すること。	P501 内容物／容器を…に廃棄すること。 …国際／国／都道府県／市町村の規則（明示する）に従って 製造者／供給者または所管官庁が指定する内容物、容器またはその両者に適用する廃棄物要件

- 391 -

SPECIFIC TARGET ORGAN TOXICITY (SINGLE EXPOSURE)
(CHAPTER 3.8)

Hazard category	Symbol	Signal word	Hazard statement
2	Health hazard	Warning	H371 May cause damage to organs <...> <<...>> <...> *(or state all organs affected, if known)* <<...>> *(state route of exposure if it is conclusively proven that no other routes of exposure cause the hazard)*

Precautionary statements

Prevention	Response	Storage	Disposal
P260 **Do not breathe dust/fume/gas/mist/vapours/spray.** *Manufacturer/supplier or the competent authority to specify applicable physical state(s).* P264 **Wash hands [and…] thoroughly after handling.** *- text in square brackets to be used when the manufacturer/supplier or the competent authority specify other parts of the body to be washed after handling.* P270 **Do not eat, drink or smoke when using this product.**	P308 + P316 **IF exposed or concerned: Get emergency medical help immediately.** *Competent authority or manufacturer / supplier may add, 'Call' followed by the appropriate emergency telephone number, or the appropriate emergency medical help provider, for example, a Poison Centre, Emergency Centre or Doctor.*	P405 **Store locked up.**	P501 **Dispose of contents/container to…** *… in accordance with local/regional/national/international regulations (to be specified). Manufacturer/supplier or the competent authority to specify whether disposal requirements apply to contents, container or both.*

- 392 -

特定標的臓器毒性（単回ばく露）
（第 3.8 章）

危険有害性区分	シンボル	注意喚起語
2	健康有害性	警告

危険有害性情報
H371 臓器の障害のおそれ ＜…＞ ＜＜…＞＞

＜…＞ （もしよくわかればすべての影響を受ける臓器を挙げる）
＜＜…＞＞ （他の経路からのばく露が有害でないことが決定的に証明されている場合、有害な経路を記載する）

注意書き

安全対策	応急措置	保管	廃棄
P260 粉じん/煙/ガス/ミスト/蒸気/スプレーを吸入しないこと。 製造者/供給者または所管官庁が指定する物理的状態 P264 取扱後は手【および…】をよく洗うこと。 一製造者/供給者または所管官庁が、取扱後に洗浄する体の他の部分を指定した場合には [ ] 内の文章を用いる P270 この製品を使用する時に、飲食または喫煙をしないこと。	P308 + P316 ばく露またはその懸念がある場合：すぐに救急の医療処置を受けること。 所管官庁または製造者/供給者は緊急時電話番号があれば「電話」に続けて、適当な救急電話番号、例えば中毒センター、救急センター、または医療提供者などを追加してもよい。	P405 施錠して保管すること。	P501 内容物/容器を…に廃棄すること。 …国際/国/都道府県/市町村の規則（明示する）に従って製造者/供給者または所管官庁が指定する内容物、容器またはその両者に適用する廃棄物要件

- 392 -

SPECIFIC TARGET ORGAN TOXICITY (SINGLE EXPOSURE)
(CHAPTER 3.8)

Hazard category	Symbol	Signal word	Hazard statement
3	Exclamation mark	Warning	H335 May cause respiratory irritation H336 May cause drowsiness or dizziness

Precautionary statements

Prevention	Response	Storage	Disposal
P261 **Avoid breathing dust/fume/gas/mist/vapours/spray.** – *may be omitted if P260 is given on the label.* Manufacturer/supplier or the competent authority to specify applicable conditions. P271 **Use only outdoors or with adequate ventilation.** Manufacturer/supplier to specify what type of ventilation would be adequate for safe use on the safety data sheet and in any supplemental safety instructions provided to consumers.	P304 + P340 **IF INHALED: Remove person to fresh air and keep comfortable for breathing.** P319 **Get medical help if you feel unwell.**	P403 + P233 **Store in a well-ventilated place. Keep container tightly closed.** – *if the chemical is volatile and may generate a hazardous atmosphere.* P405 **Store locked up.**	P501 **Dispose of contents/container to…** … in accordance with local/regional/national/international regulations (to be specified). Manufacturer/supplier or the competent authority to specify whether disposal requirements apply to contents, container or both.

特定標的臓器毒性（単回ばく露）
(第 3.8 章)

危険有害性区分　　シンボル　　注意喚起語　　危険有害性情報
3　　　　　　　　感嘆符　　　警告　　　　　H335 呼吸器への刺激のおそれ、または、
　　　　　　　　　　　　　　　　　　　　　H336 眠気またはめまいのおそれ

注意書き

安全対策	応急措置	保管	廃棄
P261 粉じん／煙／ガス／ミスト／蒸気／スプレーの吸入を避けること。 ―P260がラベルに記載される場合には省略してもよい 製造者／供給者または所管官庁が指定する適用条件 P271 屋外または十分な換気のある場所でのみ使用すること。 製造者／供給者は、安全データシートおよび消費者に提供する安全な説明書に、安全な使用のためにどのような種類の換気が適切であるかを明記すること。	P304 + P340 吸入した場合：空気の新鮮な場所に移し、呼吸しやすい姿勢で休息させること。 P319 気分が悪い時は、診察を受けること。	P403 + P233 換気の良いところで保管すること。容器を密閉しておくこと。 ―化学品が揮発性で有害な環境を作る可能性があるとき P405 施錠して保管すること。	P501 内容物／容器を...に廃棄すること。 ...国際／国／都道府県／市町村の規則（明示する）に従って 製造者／供給者または所管官庁が指定する内容物、容器またはその両者に適用する廃棄物要件

- 393 -

SPECIFIC TARGET ORGAN TOXICITY (REPEATED EXPOSURE)
(CHAPTER 3.9)

Hazard category	Symbol	Signal word	Hazard statement		
1	Health hazard	Danger	H372	Causes damage to organs <...> through prolonged or repeated exposure <<...>>	
			<...>	*(state all organs affected, if known)*	
			<<...>>	*(state route of exposure if it is conclusively proven that no other routes of exposure cause the hazard)*	

Precautionary statements

Prevention	Response	Storage	Disposal
P260 **Do not breathe dust/fume/gas/mist/vapours/spray.** *- Manufacturer/supplier or the competent authority to specify applicable physical state(s).* P264 **Wash hands [and...] thoroughly after handling.** *- text in square brackets to be used when the manufacturer/supplier or the competent authority specify other parts of the body to be washed after handling.* P270 **Do not eat, drink or smoke when using this product.**	P319 **Get medical help if you feel unwell.**		P501 **Dispose of contents/container to...** *... in accordance with local/regional/national/international regulations (to be specified).* Manufacturer/supplier or the competent authority to specify whether disposal requirements apply to contents, container or both.

特定標的臓器毒性（反復ばく露）
（第 3.9 章）

危険有害性区分	シンボル	注意喚起語	危険有害性情報
1	健康有害性	危険	H372 長期にわたる、または反復ばく露 <<...>> による臓器 <<...>> の障害 <...> （もしかかわればすべての影響を受ける臓器を挙げる） <<...>> （他の経路からのばく露が有害でないことが決定的に証明されている場合、有害な経路を記載する）

注意書き

安全対策	応急措置	保管	廃棄
P260 粉じん／煙／ガス／ミスト／蒸気／スプレーを吸入しないこと。 －製造者／供給者または所管官庁が指定する物理的状態 P264 取扱後は手 [および...] をよく洗うこと。 －製造者／供給者または所管官庁が、取扱後に洗浄する身体の他の部分を指定した場合には[]内の文章を用いる P270 この製品を使用する時に、飲食または喫煙をしないこと。	P319 気分が悪い時は、診察を受けること。		P501 内容物／容器を…に廃棄すること。 …国際／国／都道府県／市町村の規則（明示する）に従って 製造者／供給者または所管官庁が指定する内容物、容器またはその両者に適用する廃棄物要件を記載する

- 394 -

SPECIFIC TARGET ORGAN TOXICITY (REPEATED EXPOSURE)
(CHAPTER 3.9)

Hazard category	Symbol	Signal word	Hazard statement
2	Health hazard	Warning	H373 May cause damage to organs <...> through prolonged or repeated exposure <<...>> <...> *(state all organs affected, if known)* <<...>> *(state route of exposure if it is conclusively proven that no other routes of exposure cause the hazard)*

Precautionary statements

Prevention	Response	Storage	Disposal
P260 **Do not breathe dust/fume/gas/mist/vapours/spray.** Manufacturer/supplier or the competent authority to specify applicable physical state(s).	P319 **Get medical help if you feel unwell.**		P501 **Dispose of contents/container to...** ... in accordance with local/regional/national/international regulations (to be specified). Manufacturer/supplier or the competent authority to specify whether disposal requirements apply to contents, container or both.

特定標的臓器毒性（反復ばく露）
(第 3.9 章)

危険有害性区分　シンボル　注意喚起語　危険有害性情報
2　健康有害性　警告　H373 長期にわたる、または反復ばく露<<...>>による
　　　　　　　　　　　　臓器<...>の障害のおそれ

　　　　　　　　　　　　<...>　（もしかかればすべての影響を受ける臓器を
　　　　　　　　　　　　　　　　挙げる）
　　　　　　　　　　　　<<...>>（他の経路からのばく露が有害でないことが
　　　　　　　　　　　　　　　　決定的に証明されている場合、有害な経路を
　　　　　　　　　　　　　　　　記載する）

注意書き

安全対策	応急措置	保管	廃棄
P260 粉じん/煙/ガス/ミスト/蒸気/スプレーを吸入しないこと。 製造者/供給者または所管官庁が指定する物理的状態	P319 気分が悪い時は、診察を受けること。		P501 内容物/容器を…に廃棄すること。 国際/国/都道府県/市町村の規則（明示する）に従って 製造者/供給者または所管官庁が指定する内容物、容器または両者に適用する廃棄物要件

- 395 -

ASPIRATION HAZARD
(CHAPTER 3.10)

Hazard category	Symbol	Signal word	Hazard statement	
1	Health hazard	Danger	H304	May be fatal if swallowed and enters airways
2	Health hazard	Warning	H305	May be harmful if swallowed and enters airways

Precautionary statements

Prevention	Response	Storage	Disposal
	P301 + P316 **IF SWALLOWED: Get emergency medical help immediately.** Competent authority or manufacturer / supplier may add, 'Call' followed by the appropriate emergency telephone number, or the appropriate emergency medical help provider, for example, a Poison Centre, Emergency Centre or Doctor. P331 **Do NOT induce vomiting.**	P405 **Store locked up.**	P501 **Dispose of contents/container to…** … in accordance with local/regional/national/international regulations (to be specified). Manufacturer/supplier or the competent authority to specify whether disposal requirements apply to contents, container or both.

誤えん有害性
（第 3.10 章）

危険有害性区分	シンボル	注意喚起語	危険有害性情報
1	健康有害性	危険	H304 飲み込んで気道に侵入すると生命に危険のおそれ
2	健康有害性	警告	H305 飲み込んで気道に侵入すると有害のおそれ

注意書き

安全対策	応急措置	保管	廃棄
	P301 + P316 飲み込んだ場合：すぐに救急の医療処置を受けること。 所管官庁または製造者/供給者は、「電話」に続けて、適当な救急時電話番号すなわち適当な救急時医療提供者、例えば毒物センター、救急センターまたは医師などを追加してもよい。 P331 無理に吐かせないこと。	P405 施錠して保管すること。	P501 内容物/容器を…に廃棄すること。 …国際/国/都道府県/市町村の規則（明示する）に従って製造者/供給者または所管官庁が指定する内容物、容器またはその両者に適用する廃棄物要件

HAZARDOUS TO THE AQUATIC ENVIRONMENT – SHORT-TERM (ACUTE) HAZARD
(CHAPTER 4.1)

Hazard category	Symbol	Signal word	Hazard statement
1	Environment	Warning	H400 Very toxic to aquatic life

Precautionary statements

Prevention	Response	Storage	Disposal
P273 **Avoid release to the environment.** *– if this is not the intended use.*	P391 **Collect spillage.**		P501 **Dispose of contents/container to...** ... in accordance with local/regional/national/international regulations (to be specified). Manufacturer/supplier or the competent authority to specify whether disposal requirements apply to contents, container or both.

水生環境有害性 短期（急性）
（第 4.1 章）

危険有害性区分	シンボル	注意喚起語	危険有害性情報
1	環境	警告	H400 水生生物に非常に強い毒性

注意書き

安全対策	応急措置	保管	廃棄
P273 環境への放出を避けること。 ―必要な時以外は	P391 漏出物を回収すること。		P501 内容物/容器を…に廃棄すること。 …国際/国都道府県/市町村の規則（明示する）に従って 製造者/供給者または所管官庁が指定する内容物、容器またはその両者に適用する廃棄物要件

- 397 -

HAZARDOUS TO THE AQUATIC ENVIRONMENT – SHORT-TERM (ACUTE) HAZARD
(CHAPTER 4.1)

Hazard category	Symbol	Signal word	Hazard statement	
2	No symbol	No signal word	H401	Toxic to aquatic life
3	No symbol	No signal word	H402	Harmful to aquatic life

Precautionary statements

Prevention	Response	Storage	Disposal
P273 **Avoid release to the environment.** *– if this is not the intended use.*			P501 **Dispose of contents/container to…** … in accordance with local/regional/national/international regulations (to be specified). Manufacturer/supplier or the competent authority to specify whether disposal requirements apply to contents, container or both.

水生環境有害性 短期(急性)
(第 4.1 章)

危険有害性区分	シンボル	注意喚起語	危険有害性情報
2	なし	なし	H401 水生生物に毒性
3	なし	なし	H402 水生生物に有害

注意書き

安全対策	応急措置	保管	廃棄
P273 環境への放出を避けること。 ―必要な時以外は			P501 内容物/容器を…に廃棄すること。 …国際/国/都道府県/市町村の規則(明示する)に従って 製造者/供給者または所管官庁が指定する内容物、容器またはその両者に適用する廃棄物要件

- 398 -

HAZARDOUS TO THE AQUATIC ENVIRONMENT – LONG-TERM (CHRONIC) HAZARD
(CHAPTER 4.1)

Hazard category	Symbol	Signal word	Hazard statement
1	Environment	Warning	H410 Very toxic to aquatic life with long lasting effects
2	Environment	*No signal word*	H411 Toxic to aquatic life with long lasting effects

Precautionary statements

Prevention	Response	Storage	Disposal
P273 **Avoid release to the environment.** *– if this is not the intended use.*	P391 **Collect spillage.**		P501 **Dispose of contents/container to...** ... in accordance with local/regional/national/international regulations (to be specified). Manufacturer/supplier or the competent authority to specify whether disposal requirements apply to contents, container or both.

水生環境有害性 長期（慢性）
（第 4.1 章）

危険有害性区分	シンボル	注意喚起語	危険有害性情報
1	環境	警告	H410 長期継続的影響によって水生生物に非常に強い毒性
2	環境	なし	H411 長期継続的影響により水生生物に毒性

注意書き

安全対策	応急措置	保管	廃棄
P273 環境への放出を避けること。 一必要な時以外は	P391 漏出物を回収すること。		P501 内容物/容器を…に廃棄すること。 …国際/国/都道府県/市町村の規則（明示する）に従って 製造者/供給者または所管官庁が指定する内容物、容器またはその両者に適用する廃棄物要件

- 399 -

HAZARDOUS TO THE AQUATIC ENVIRONMENT – LONG-TERM (CHRONIC) HAZARD
(CHAPTER 4.1)

Hazard category	Symbol	Signal word	Hazard statement	
3	No symbol	No signal word	H412	Harmful to aquatic life with long lasting effects
4	No symbol	No signal word	H413	May cause long lasting harmful effects to aquatic life

Precautionary statements

Prevention	Response	Storage	Disposal
P273 **Avoid release to the environment.** *– if this is not the intended use.*			P501 **Dispose of contents/container to…** … in accordance with local/regional/national/international regulations (to be specified). Manufacturer/supplier or the competent authority to specify whether disposal requirements apply to contents, container or both.

- 400 -

水生環境有害性 長期（慢性）
(第 4.1 章)

危険有害性区分	シンボル	注意喚起語	危険有害性情報
3	なし	なし	H412 長期継続的影響によって水生生物に有害
4	なし	なし	H413 長期継続的影響によって水生生物に有害のおそれ

注意書き

安全対策	応急措置	保管	廃棄
P273 環境への放出を避けること。 ―必要な時以外は			P501 内容物/容器を…に廃棄すること。 …国際/国/都道府県/市町村の規則（明示する）に従って 製造者/供給者または所管官庁が指定する内容物、容器またはその両者に適用する廃棄物要件

- 400 -

HAZARDOUS TO THE OZONE LAYER
(CHAPTER 4.2)

Hazard category	Symbol	Signal word	Hazard statement
1	Exclamation mark	Warning	H420　Harms public health and the environment by destroying ozone in the upper atmosphere

Precautionary statements

Prevention	Response	Storage	Disposal
			P502 **Refer to manufacturer or supplier for information on recovery or recycling.**

オゾン層への有害性
(第 4.2 章)

危険有害性区分	シンボル	注意喚起語	危険有害性情報
1	感嘆符	警告	H420 オゾン層の破壊により健康および環境に有害

注意書き

安全対策	応急措置	保管	廃棄
			P502 回収またはリサイクルに関する情報について製造者または供給者に問い合わせる

- 401 -

ANNEX 3

Section 4

CODIFICATION OF HAZARD PICTOGRAMS

A3.4.1 Introduction

A3.4.1.1 *Pictogram* means a graphical composition that may include a symbol plus other graphic elements, such as a border, background pattern or colour that is intended to convey specific information.

A3.4.1.2 This section contains the recommended code assigned to each of the pictograms prescribed by the GHS for sectors other than transport.

A3.4.1.3 The pictogram code is intended to be used for references purposes. It is not part of the pictogram and should not appear on labels or in section 2 of the safety data sheet.

A3.4.2 Codification of pictograms

A3.4.2.1 GHS pictograms for sectors other than transport are assigned a unique alphanumerical code as follows:

(a) the letters "GHS"; and

(b) a sequential number "01", "02", "03" etc. assigned in accordance with table A3.4.1 below.

Table A3.4.1

Code	Hazard pictogram	Symbol
GHS01		Exploding bomb
GHS02		Flame
GHS03		Flame over circle
GHS04		Gas cylinder
GHS05		Corrosion
GHS06		Skull and crossbones
GHS07		Exclamation mark
GHS08		Health hazard
GHS09		Environment

附属書3

第4節

危険有害性絵表示のコード

A3.4.1　序文

A3.4.1.1　絵表示とは、特定の情報を伝達することを意図したシンボルと境界線、背景のパターンまたは色のような図的要素から構成されるものをいう。

A3.4.1.2　本節では、輸送以外の分野に対してGHSで規定される絵表示に割り当てられた推奨されるコードについて述べる。

A3.4.1.3　絵表示のコードは参照の目的で使用されるように意図されている。これは絵表示の一部ではなく、ラベルまたは安全データシートの第2節に記載するべきではない。

A3.4.2　絵表示のコード化

A3.4.2.1　輸送分野以外の分野に対するGHS絵表示には次のようなアルファベットと数字を組み合わせたコードが割り当てられる：

(a)　文字「GHS」；および
(b)　連続した番号「01」、「02」、「03」などを以下の表A3.4.1のように割り当てる。

表 A3.4.1

コード	危険有害性絵表示	シンボル
GHS01		爆弾の爆発
GHS02		炎
GHS03		円上の炎
GHS04		ガスボンベ
GHS05		腐食性
GHS06		どくろ
GHS07		感嘆符
GHS08		健康有害性
GHS09		環境

ANNEX 3

Section 5

EXAMPLES OF PRECAUTIONARY PICTOGRAMS

A3.5.1   **Precautionary pictograms**

From European Union (Council Directive 92/58/EEC of 24 June 1992)

A3.5.2   **Precautionary pictograms "Keep out of reach of children"**

The following examples convey the meaning of precautionary statement P102 "Keep out of reach of children" and may be used to convey information in more than one way in accordance with sections 1.4.4.1 (a) and A3.2.1.10.

A3.5.2.1   *International Association for Soaps, Detergents and Maintenance Products (AISE) precautionary pictogram "Keep out of reach of children"*

The pictogram was developed by AISE and has been in use since 2004 in Europe and other jurisdictions for household care products.

Comprehensibility tests were carried out on the AISE Precautionary pictogram "Keep out of reach of children". The studies conducted in several countries in accordance with annex 6 of the GHS demonstrated that this icon was adequately understood by 88.6 % of the respondents, and that critical confusion about the icon was only sporadic (< 1 %).

附属書 3

第 5 節

注意絵表示の例

A3.5.1 注意絵表示（ピクトグラム）

欧州連合（1992 年 6 月 24 日付け理事会指令 92/58/EEC）から

A3.5.2 注意絵表示「子供の手の届かないところに置くこと。」

下記の例は注意書き P102「子供の手の届かないところに置くこと。」の意味を伝達し、1.4.4.1(a)および A3.2.1.10 にしたがった 1 つ以上の方法で情報を伝えるために使用される。

A3.5.2.1 欧州石鹸洗剤工業連合会 (AISE) による注意絵表示「子供の手の届かないところに置くこと。」

この絵表示は AISE によって開発され、家庭用製品に対して 2004 年以来欧州および他の法令で使用されてきた。

AISE の注意絵表示「子供の手の届かないところに置くこと。」について理解度試験が行われていた。その調査は GHS 附属書 6 にしたがっていくつかの国で行われ、この絵が回答者の 88.6%に的確に理解され,しかも絵についての決定的な誤解はほんの少し（＜1%）であったことが示されている。

A3.5.2.2 *Japan Soap and Detergent Association (JSDA) precautionary pictogram "Keep out of reach of children"*

The pictogram was developed by JSDA for voluntarily use on the label/packaging of consumer detergents products in Japan.

The JSDA safe use icon was tested in accordance with the Japanese standard JIS S 0102: "Testing procedure for graphical warning symbols for consumers". The icon passed the JIS S 0102 success criteria (comprehension > 85 %) with 96 % correct understanding and only 1.7 % of critical confusion.

A3.5.2.2　*日本石鹸洗剤工業会（JSDA）による注意絵表示「子供の手の届かないところに置くこと。」*

　この絵表示は、日本での消費者用洗剤のラベル/包装に対して、自主的な使用を目的として JSDA が開発したものである。

　この JSDA の安全表示は、日本産業規格 JIS S 0102：「消費者用警告図記号－試験の手順」にしたがって試験された。この絵は、96 点の正解とわずか 1.7％の極端に誤った理解で、JIS S 0102 の合格基準（理解度＞85 点）を満たした。

ANNEX 4

GUIDANCE ON THE PREPARATION OF SAFETY DATA SHEETS (SDS)

附属書4

安全データシート（SDS）
作成指針

ANNEX 4

GUIDANCE ON THE PREPARATION OF SAFETY DATA SHEETS (SDS)

A4.1 Introduction

A4.1.1 This annex provides guidance on the preparation of an SDS under the requirements of the Globally Harmonized System of Classification and Labelling of Chemicals (GHS). SDS's are an important element of hazard communication in the GHS, as explained in chapter 1.5. Use of this guidance document should support compliance with competent authority requirements and should allow the SDS to be prepared in accordance with the GHS.

A4.1.2 The use of this guidance document is dependent on importing countries requirements for SDS. It is hoped that the application of the GHS worldwide will eventually lead to a fully harmonized situation.

A4.1.3 Unless otherwise stated, all chapters, sections and tables referred to in this annex can be found in the main text of the GHS.

A4.2 General guidance for compiling an SDS

A4.2.1 *Scope and application*

Safety Data Sheets (SDS) should be produced for all substances and mixtures which meet the harmonized criteria for physical, health or environmental hazards under the GHS and for all mixtures which contain ingredients that meet the criteria for carcinogenic, toxic to reproduction or target organ toxicity in concentrations exceeding the cut-off limits for SDS specified by the criteria for mixtures (see table 1.5.1 in chapter 1.5). The competent authority may also require SDS for mixtures not meeting the criteria for classification as hazardous, but which contain hazardous ingredients in certain concentrations (see chapter 3.2). The competent authority may also require SDS for substances or mixtures that meet the criteria for classification as hazardous for non-GHS classes/end-points. An SDS is a well-accepted and effective method for the provision of information and may be used to convey information for substances or mixtures that do not meet or are not included in the GHS classification criteria.

A4.2.2 *General guidance*

A4.2.2.1 The writer of the SDS needs to keep in mind that an SDS must inform its audience of the hazards of a substance or a mixture and provide information on the safe storage, handling and disposal of the substance or a mixture. An SDS contains information on the potential health effects of exposure and how to work safely with the substance or mixture. It also contains hazard information derived from physicochemical properties or environmental effects, on the use, storage, handling and emergency response measures related to that substance or mixture. The purpose of this guidance is to ensure consistency and accuracy in the content of each of the mandatory headings required under GHS, so that the resulting safety data sheets will enable users to take the necessary measures relating to protection of health and safety at the workplace, and the protection of the environment. The information in the SDS shall be written in a clear and concise manner. The SDS shall be prepared by a competent person who shall take into account the specific needs of the user audience, as far as it is known. Persons placing substances and mixtures on the market shall ensure that refresher courses and training on the preparation of SDS be regularly attended by the competent persons.

A4.2.2.2 When writing the SDS, information should be presented in a consistent and complete form, with the workplace audience firmly in mind. However, it should be considered that all or part of the SDS can be used to inform workers, employers, health and safety professionals, emergency personnel, relevant government agencies, as well as members of the community.

A4.2.2.3 Language used in the SDS should be simple, clear and precise, avoiding jargon, acronyms and abbreviations. Vague and misleading expressions should not be used. Phrases such as "may be dangerous", "no health effects", "safe under most conditions of use", or "harmless" are also not recommended. It may be that information on certain properties is of no significance or that it is technically impossible to provide; if so, the reasons for this must be clearly stated under each heading. If it is stated that a particular hazard does not exist, the safety data sheet should clearly differentiate between cases where no information is available to the classifier, and cases where negative test results are available.

附属書 4

安全データシート（SDS）作成指針

A4.1　序文

A4.1.1　本文書は化学品の分類および表示に関する世界調和システム(GHS)の要求に基づく手引きを提供するものである。SDS は、第 1.5 章に説明があるように GHS における危険有害性情報の伝達の重要な要素の 1 つである。本手引き文書の使用により、所管官庁の要求事項の遵守を支援し、GHS に従った SDS の作成を可能にする。

A4.1.2　この手引き文書が使用されるか否かは SDS に対する輸入国の要求事項にかかっている。GHS が世界中に適用されれば、いずれは完全に調和された状況となることが期待されている。

A4.1.3　特に言及しない場合は、本附属書で言及しているすべての章、節および表は、GHS の本文に含まれている。

A4.2　SDS 作成のための一般的な手引き

A4.2.1　*適用範囲と実施*

　安全データシート(SDS)は、GHS の物理化学的危険性、健康または環境に対する有害性の調和した区分に適合するすべての物質とその混合物に対し、また、混合物の基準（第 1.5 章の表 1.5.1 参照）に規定されている SDS のカットオフ値を超える濃度で、発がん性、生殖毒性、または特定標的臓器毒性の基準を満たす物質を含有するすべての混合物に対して作成される。所管官庁は、危険有害性としての分類基準は満たさないが一定の濃度（第 3.2 章参照）で危険有害性物質を含有する混合物についても SDS を要求することができる。また所管官庁は GHS のクラス/エンドポイントの危険有害性物質として分類されない物質または混合物に対して SDS を要求することができる。SDS は情報提供に関して広く受け入れられている効果的な手法であり、GHS 分類基準に合致しないか、または含まれない物質または混合物の情報提供に用いられることができる。

A4.2.2　*一般的手引き*

A4.2.2.1　SDS の作成にあたっては、SDS は、対象とする人に物質または混合物の危険有害性情報を提供するものであり、また物質または混合物の安全な保管、取扱いと廃棄についての情報を提供するものでなくてはならないことを念頭におく必要がある。SDS にはばく露による潜在的健康影響と、物質または混合物を扱う際の作業方法に関する情報が含められている。また SDS にはその物質または混合物についての使用、保管、取扱いと緊急事態対策に関する物理化学的特性または環境影響に由来する危険有害性情報が含まれている。この手引きの目的とするところは GHS の下で要求される各必須項目の内容の整合性と正確さにあり、このために作成された安全データシートによって、使用者は作業場における健康保護と安全、および環境保全に関連した必要措置を採ることができるようになる。SDS に盛り込まれる情報は明確かつ簡潔に書かれていなければならない。この SDS は対象とする使用者の特定の必要性をできるだけ深く考慮に入れながら、適格者が作成しなければならない。物質と混合物を上市する者は、適格者が SDS の再教育講座及び研修に定期的に参加することを確実にしなければならない。

A4.2.2.2　SDS の作成にあたっては、作業場の対象者をはっきりと念頭において一貫した完全な形で情報が提供されるべきである。しかしながら、SDS の全体または一部が、地域社会の構成員に対すると同様に、労働者、雇用者、健康と安全の専門家、救急隊員、関係行政機関に対する情報提供のために使用されることを考慮すべきである。

A4.2.2.3　SDS で用いられる言葉は専門用語、頭文字語と略語の使用を避けて易しく明確かつ正確であるべきである。あいまいで紛らわしい表現を使用すべきではない。「危険かもしれない」、「健康への影響なし」、「ほとんどすべての条件下で使用しても安全」、または「無害」などを使うことも推奨できない。ある特性についての情報が重要ではないか、または技術的理由から情報の提供ができないことがある。その場合には各項目にその理由が明確に記載されなければならない。特定の危険が存在しない旨を記載する場合には、安全データシートは、分類するにあたって、情報がない場合と否定的な試験結果がある場合とを区別すべきである。

A4.2.2.4   The date of issue of the SDS should be stated and be very apparent. The date of issue is the date the SDS version was made public. This generally occurs shortly after the SDS authoring and publishing process is completed. Revised SDS's should clearly state the date of issue as well as a version number, revision number, supersedes date or some other indication of what version is replaced.

A4.2.3   *SDS format*

A4.2.3.1   The information in the SDS should be presented using the following 16 headings in the order given below (see also 1.5.3.2.1):

1. Identification;
2. Hazard identification;
3. Composition/information on ingredients;
4. First-aid measures;
5. Fire-fighting measures;
6. Accidental release measures;
7. Handling and storage;
8. Exposure controls/personal protection;
9. Physical and chemical properties;
10. Stability and reactivity;
11. Toxicological information;
12. Ecological information;
13. Disposal considerations;
14. Transport information;
15. Regulatory information;
16. Other information

A4.2.3.2   An SDS is not a fixed length document. The length of the SDS should be commensurate with the hazard of the material and the information available.

A4.2.3.3   All pages of an SDS should be numbered and some indication of the end of the SDS should be given (for example: "page 1 of 3"). Alternatively, number each page and indicate whether there is a page following (e.g. "Continued on next page" or "End of SDS").

A4.2.4   *SDS content*

A4.2.4.1   General information on SDS content can be found in 1.5.3.3. More practical information is given below.

A4.2.4.2   The minimum information outlined in section A4.3 of this annex should be included on the SDS, where applicable and available[1] under the relevant headings. When information is not available or lacking this should be clearly stated. The SDS should not contain any blanks.

A4.2.4.3   In addition, the SDS should contain a brief summary/conclusion of the data given, making it easy even for non-experts in the field to identify all the hazards for the hazardous substance/mixture.

A4.2.4.4   Use of abbreviations is not recommended because they may lead to confusion or decreased understanding.

A4.2.5   *Other information requirements*

A4.2.5.1   There are information requirements for the preparation of an SDS. The minimum information requirements are outlined in A4.3.

A4.2.5.2   In addition to the minimum information requirements (see A4.2.4.2), the SDS may also contain "additional information". Where a material has additional relevant and available information about its nature and/or use, that information should be included in the SDS (see A4.3.16 for further advice on additional information requirements).

---

[1]   Where "applicable" means where the information is applicable to the specific product covered by the SDS. Where "available" means where the information is available to the supplier or other entity that is preparing the SDS.

A4.2.2.4　安全データシートの発行日は明確に記載すべきである。発行日は、SDS が公開された日である。これは SDS の認定および公表まで完了した後すぐである。改訂された SDS は、バージョン番号、改訂番号、差し替えた日またはどのバージョンを差し替えたかの表示と同様に、発行日を明確に記載すべきである。

A4.2.3　*SDS の様式*

A4.2.3.1　SDS の情報は、次の 16 項目を使用し、下に示す順序で記載するべきである（1.5.3.2.1.参照）:
1. 物質または混合物および会社情報；
2. 危険有害性の要約；
3. 組成および成分情報；
4. 応急措置；
5. 火災時の措置；
6. 漏出時の措置；
7. 取扱いおよび保管上の注意；
8. ばく露防止および保護措置；
9. 物理的および化学的性質；
10. 安定性および反応性；
11. 有害性情報；
12. 環境影響情報；
13. 廃棄上の注意；
14. 輸送上の注意；
15. 適用法令；
16. その他の情報

A4.2.3.2　SDS は長さの決まった文書ではない。SDS の長さはその物質の危険有害性と入手可能な情報に相応の長さにすべきである。

A4.2.3.3　SDS のページにはすべてページ数を付け、SDS の終わりを示す何らかの表示（例えば、1/3 のように）をすべきである。他の方法として、各ページにページ数を付けるとともに次ページの有無を示してもよい。（例えば「次ページに続く」あるいは「SDS 終わり」）

A4.2.4　*SDS の内容*

A4.2.4.1　SDS の内容に関する一般情報は 1.5.3.3 で見ることができる。さらに実際的な情報は以下に述べる。

A4.2.4.2　本附属書の第 4.3 節で概説されている最小限の情報は、該当しまた入手できる場合には SDS の関連する項目に記載すべきである[1]。この情報が入手できない場合または欠けている場合、このことをはっきりと記述すべきである。SDS はいかなる空欄も残すべきではない。

A4.2.4.3　また SDS には取得されたデータの概要/結論を含むべきである。これによってこの分野の専門家でなくても危険有害な物質/混合物の危険有害性のすべてを認識するのが容易になる。

A4.2.4.4　略語が混乱や理解不足を招くおそれがあるため、略語の使用は推奨できない。

A4.2.5　*他の情報の要求事項*

A4.2.5.1　SDS の作成のための情報の要求事項が存在する。最小限の情報の要求事項は A4.3 で概説されている。

A4.2.5.2　最小限の情報の要求事項（A4.2.4.2 参照）に加えて、SDS には追加情報を含めることができる。ひとつの物質の性質および/または使用法についての関連追加情報がある場合にはその情報は SDS に含めるべきである。追加情報要求事項に関するさらなる助言については、A4.3.16 を参照すること。

---

[1]　「該当する」とは、*SDS を添付する製品について、情報が該当するということをいう。*
　　「入手できる」とは、*SDS を作成する供給者等が、情報を入手できるということをいう。*

A4.2.6   *Units*

Numbers and quantities should be expressed in units appropriate to the region into which the product is being supplied. In general, the International System of Units (SI) should be used.

A4.3   **Information requirements for the preparation of the SDS**

This section describes the GHS information requirements for SDS's. Additional information may be required by competent authorities.

A4.3.1   *SECTION 1: Identification*

Identify the substance or mixture and provide the name of the supplier, recommended uses and the contact detail information of the supplier including an emergency contact in this section.

A4.3.1.1   *GHS product identifier*

The identity of the substance or mixture (GHS product identifier) should be exactly as found on the label. If one generic SDS is used to cover several minor variants of a substance or mixture, all names and variants should be listed on the SDS or the SDS should clearly delineate the range of substances included.

A4.3.1.2   *Other means of identification*

In addition, or as an alternative, to the GHS product identifier, the substance or mixture may be identified by alternative names, numbers, company product codes, or other unique identifiers. Provide other names or synonyms by which the substance or mixture is labelled or commonly known, if applicable.

A4.3.1.3   *Recommended use of the chemical and restrictions on use*

Provide the recommended or intended use of the substance or mixture, including a brief description of what it actually does, e.g. flame retardant, anti-oxidant, etc. Restrictions on use should, as far as possible, be stated including non-statutory recommendations by the supplier.

A4.3.1.4   *Supplier's details*

The name, full address and phone number(s) of the supplier should be included on the SDS.

A4.3.1.5   *Emergency phone number*

References to emergency information services should be included in all SDS. If any restrictions apply, such as hours of operation (e.g. Monday - Friday, 8:00 a.m. - 6:00 p.m., or 24 hours) or limits on specific types of information (e.g. medical emergencies, or transportation emergencies), this should be clearly stated.

A4.3.2   *SECTION 2: Hazard identification*

This section describes the hazards of the substance or mixture and the appropriate warning information (signal word, hazard statement(s) and precautionary statement(s)) associated with those hazards. The section should include a brief summary/conclusion of the data given as described in A4.2.4.3.

A4.3.2.1   *Classification of the substance or mixture*

A4.3.2.1.1   This subsection indicates the hazard classification of the substance or mixture.

A4.3.2.1.2   If the substance or mixture is classified in accordance with Parts 2, 3 and/or 4 of the GHS, generally the classification is communicated by providing the appropriate hazard class and category/subcategory to indicate the hazard (for example, flammable liquid Category 1 and skin corrosive, Category 1A). However, when classification is differentiated within a hazard class and results in unique hazard statements, then the classification should also reflect that differentiation. For example, the route of exposure differentiates the acute toxicity classification as follows: acute oral toxicity Category 1, acute dermal toxicity Category 1 and acute inhalation toxicity Category 1. If a substance or mixture is classified into more than one category in a hazard class that is differentiated, then all classifications should be communicated.

A4.2.6　単位

数と量は製品が供給される地域において適切な単位で記述すべきである。一般的には、国際単位系(SI)を用いるべきである。

A4.3　SDSの作成のための必要情報

本章では SDS に必要な GHS の最小限の情報要求事項について説明する。追加情報が所管官庁によって要求されることがある。

A4.3.1　*第1節：物質または混合物および会社情報*

物質または混合物を特定し、供給者名、推奨される使用法、本節にある緊急連絡を含む供給者の詳細な連絡先の情報を提供すること。

A4.3.1.1　*GHSの製品特定名*

物質または混合物の特定（GHS の製品特定名）はラベル表示と完全に一致すべきである。物質または混合物の若干性質の異なるものを単一の共通した SDS でカバーする場合には、すべての名称および若干性質の異なるものが SDS に一覧されるかまたは含まれる物質の範囲が SDS に明確に示されるべきである。

A4.3.1.2　*他の特定手段*

物質または混合物は別称、製品番号、会社の製品コード、または他の独自の識別方法によって特定することができる。該当する場合には、物質または混合物がラベル表示されているか一般に知られている他の名称または別称を使用すること。

A4.3.1.3　*化学品の推奨用途と使用上の制限*

例えば難燃化剤、抗酸化剤などの物質または混合物の実際の働きについて簡単な説明を含む推奨もしくは意図された用途を示すこと。使用上の制限は、供給者は法令に定めのない推奨も含めてできるだけ多く記載すべきである。

A4.3.1.4　*供給者の詳細*

供給者の名称、省略のない住所、電話番号を SDS に記載すべきである。

A4.3.1.5　*緊急電話番号*

すべての SDS には緊急時情報提供を記載すべきである。作業時間（例えば月曜日から金曜日、午前8時から午後6時、あるいは24時間営業）、または特定情報の限界（例えば医学的緊急事態、または輸送緊急事態）など何らかの制約がある場合、明確に記述すべきである。

A4.3.2　*第2節：危険有害性の要約*

本節では、物質または混合物の危険有害性とこれらの危険有害性に関連する適切な警告情報（注意喚起語、危険有害性情報および注意書き）を記載する。本節では、A4.2.4.3 に記述されているデータの簡単な要約/結論を含むべきである。

A4.3.2.1　*物質または混合物の分類*

A4.3.2.1.1　本小節では物質または混合物の危険有害性分類を示す。

A4.3.2.1.2　物質または混合物が GHS の第2部、第3部または第4部に従って分類される場合には、一般にその分類は危険有害性を示す適切な危険有害性クラスおよび区分/細区分（例えば、引火性液体 区分1および皮膚腐食性 区分1A）によって伝えられる。しかし、分類が同じクラス内で危険有害性情報が異なる場合には、分類はその相違を反映していなければならない。例えば、ばく露の経路は以下のように急性毒性分類の違いとなる：急性経口毒性区分1、急性経皮毒性区分1および急性吸入毒性区分1。物質あるいは混合物が1つの危険有害性クラスで複数の区分に分類される場合には、全ての分類は伝達されなければならない。

A4.3.2.2   *GHS label elements, including precautionary statements*

A4.3.2.2.1   Based on the classification, provide the appropriate label elements: signal word(s), hazard statement(s) and precautionary statement(s).

A4.3.2.2.2   Pictograms (or hazard symbols) may be provided as a graphical reproduction of the symbols in black and white or the name of the symbol, e.g. "flame", "skull and crossbones".

A4.3.2.3   *Other hazards which do not result in classification*

Provide information on other hazards which do not result in classification but may contribute to the overall hazards of the material, for example, formation of air contaminants during hardening or processing, dust explosion hazards, suffocation, freezing or environmental effects such as hazards to soil-dwelling organisms. To communicate combustible dust hazards, and thus a potential risk of dust explosions under the approach described in annex 11 in a standardized manner, competent authorities may allow the use of the phrases identified in A11.2.7.3 on labels, SDSs and/or in operating instructions or may leave the choice to the manufacturer or supplier.

A4.3.3   SECTION 3: Composition/information on ingredients

Identify the ingredient(s) of the product in this section. This includes identifying impurities and stabilizing additives which are themselves classified and which contribute to the classification of the substance. This section may also be used to provide information on complex substances.

NOTE: *For information on ingredients, the competent authority rules for confidential business information take priority over the rules for product identification. When applicable, indicate that confidential information about the composition was omitted.*

A4.3.3.1   *Substances*

A4.3.3.1.1   Chemical identity of the substance

The identity of a substance is provided by its common chemical name. The chemical name can be identical to the GHS product identifier.

NOTE: *The "common chemical name" may, for example, be the CAS name or IUPAC name, as applicable.*

A4.3.3.1.2   Common name(s), synonym(s) of the substance

Common names and synonyms should be provided where appropriate.

A4.3.3.1.3   CAS number and other unique identifiers for the substance

The Chemical Abstract Service (CAS) registry number provides a unique chemical identification and should be provided when available. Other unique identifiers specific to a country or region, such as the European Community (EC) number could be added.

A4.3.3.1.4   Impurities and stabilizing additives which are themselves classified and which contribute to the classification of the substance

Identify any impurities and/or stabilizing additives, which are themselves classified and which contribute to the classification of the substance.

A4.3.3.2   *Mixtures*

A4.3.3.2.1   For a mixture, provide the chemical identity, identification number (within the meaning of A4.3.3.1.3) and concentration or concentration ranges of all hazardous ingredients, which are hazardous to health or the environment within the meaning of the GHS, and are present above their cut-off levels. Manufacturers or suppliers may choose to list all ingredients, including non-hazardous ingredients.

A4.3.2.2　注意書きを含む GHS ラベル要素

A4.3.2.2.1.　分類に基づいて適切なラベル要素を示す：注意喚起語、危険有害性情報および注意書き。

A4.3.2.2.2.　絵表示（または危険有害性シンボル）は、白黒のシンボルにより記載するか、または例えば、「炎」、「どくろ」のようなシンボルの名称を用いてもよい。

A4.3.2.3　分類に結び付かない他の危険有害性

分類に結び付かないが材料の全体としての危険有害性に寄与するかもしれない他の危険有害性に関する情報、例えば、硬化または処理中の空気汚染物の形成、粉じん爆発危険、窒息、凍結、または土壌生息生物に対する有害性のような環境上の影響などを提供すること。附属書 11 に記載されている標準化された方法によって、可燃性粉じんの危険性および粉じん爆発の潜在的なリスクを伝えるために、所管官庁は A11.2.7.3 に記載されている文言をラベル、SDS および/または操作指示書で使用することを許可しても、あるいはその選択を製造者または供給者に委ねてもよい。

A4.3.3　第3節：組成および成分情報

この節では、製品の成分を示すこと。それ自体は分類されており、なおかつ物質の分類に寄与する不純物と分解防止添加物の成分を示すことが含まれる。本節では錯化合物に関する情報を提供してもよい。

注記：成分に関する情報については、営業秘密情報についての所管官庁の規則が製品特定の規則に優先される。該当する場合には成分に関する営業秘密情報が省略されていることを示すこと。

A4.3.3.1　物質

A4.3.3.1.1　物質の化学的特定名

物質の特定には一般的な化学名が用いられる。化学名は GHS の製品特定名と同一であることがある。

注記：「一般的な化学名」は、必要に応じて、例えば CAS 名または IUPAC 名でよい。

A4.3.3.1.2.　物質の慣用名と別名

必要に応じて物質の慣用名と別名を記載すべきである。

A4.3.3.1.3　物質の CAS 番号とその他の特定名

ケミカル・アブストラクツ・サービス（CAS）の登録番号は、唯一の化学的特定名を与える。そして利用可能である場合には示すべきである。例えば、欧州委員会(EC)番号といった国または地域に特有の他の特定名を追加することができる。

A4.3.3.1.4　それ自体分類されており、なおかつ物質の分類に資する不純物と分解防止添加物

それ自体分類されており、なおかつ物質の分類に寄与するすべての不純物または分解防止添加物を特定すること。

A4.3.3.2　混合物

A4.3.3.2.1　混合物については、GHS の基準において健康または環境に有害で、かつカットオフ値を超えて含有されている、すべての危険有害性成分の物質の特定名と、(A4.3.3.1.3 の意味の範囲内の) 特定番号、濃度または濃度範囲を示すこと。製造者または供給者は、危険有害性のない成分も含めて、すべての成分を示してもよい。

A4.3.3.2.2 The concentrations of the ingredients of a mixture should be described as:

(a) exact percentages in descending order by mass or volume; or

(b) ranges of percentages in descending order by mass or volume if such ranges are acceptable to the appropriate competent national authority.

A4.3.3.2.3 When using a proportion range, the health and environmental hazard effects should describe the effects of the highest concentration of each ingredient, provided that the effects of the mixture as a whole are not available.

NOTE: *The "proportion range" refers to the concentration or percentage range of the ingredient in the mixture.*

A4.3.4 SECTION 4: First-aid measures

This section describes the initial care that can be given by an untrained responder without the use of sophisticated equipment and without a wide selection of medications available. If medical attention is required, the instructions should state this, including its urgency. It may be useful to provide information on the immediate effects, by route of exposure, and indicate the immediate treatment, followed by possible delayed effects with specific medical surveillance required.

A4.3.4.1 *Description of necessary first-aid measures*

A4.3.4.1.1 Provide first-aid instructions by relevant routes of exposure. Use sub-headings to indicate the procedure for each route (e.g. inhalation, skin, eye and ingestion). Describe expected immediate and delayed symptoms.

A4.3.4.1.2 Provide advice whether:

(a) immediate medical attention is required and if delayed effects can be expected after exposure;

(b) movement of the exposed individual from the area to fresh air is recommended;

(c) removal and handling of clothing and shoes from the individual is recommended; and

(d) personal protective equipment (PPE) for first-aid responders is recommended.

A4.3.4.2 *Most important symptoms/effects, acute and delayed*

Provide information on the most important symptoms/effects, acute and delayed, from exposure.

A4.3.4.3 *Indication of immediate medical attention and special treatment needed, if necessary*

Where appropriate, provide information on clinical testing and medical monitoring for delayed effects, specific details on antidotes (where they are known) and contraindications.

A4.3.5 SECTION 5: Fire-fighting measures

This section covers the requirements for fighting a fire caused by the substance or mixture or arising in its vicinity.

A4.3.5.1 *Suitable extinguishing media*

Provide information on the appropriate extinguishing media. In addition, indicate whether any extinguishing media are inappropriate for a particular situation involving the substance or mixture (e.g. avoid high pressure media which could cause the formation of a potentially explosive dust-air mixture).

A4.3.5.2 *Specific hazards arising from the chemical*

Provide advice on specific hazards that may arise from the chemical, such as hazardous combustion products that form when the substance or mixture burns. For example:

(a) "may produce toxic fumes of carbon monoxide if burning"; or

A4.3.3.2.2 　混合物の成分の濃度範囲に関して以下のように記述すべきである：

(a) 正確な百分率が降順により重量または体積で表示；または
(b) 適切な国の所管官庁によって受け入れられる場合には降順により重量または体積を百分率の範囲で表示。

A4.3.3.2.3 　成分割合比率を用いる場合、健康および環境に対する有害性については、混合物全体の影響を示すことができないならば、その成分の最も高濃度のものの影響を記載すべきである。

注記：「成分割合比率」は、混合物における成分の濃度または百分率範囲を意味する。

A4.3.4 　第4節：応急措置

この節では、訓練を受けていない対応者が、高度な装置を用いずに、かつ使用できる医薬品の選択肢が少ない中で行う初期手当について記載する。医療が必要ならばその緊急度も含めてその指示を記載するべきである。ばく露経路による急性影響に関する情報、救急治療法、特別な医学的監視を必要とする遅発影響に対する指示が有用であろう。

A4.3.4.1 　*必要な応急措置の説明*

A4.3.4.1.1 　それぞれのばく露経路ごとの応急処置を指示すること。小項目を用いて各経路（例えば、吸入、皮膚、眼および経口摂取）を示すこと。予想される急性の症状と発症が遅い症状について記載すること。

A4.3.4.1.2 　以下の場合に助言を行うこと：

(a) 速やかな治療が必要でありまたばく露後に遅発影響のおそれがある場合；
(b) ばく露した人を新鮮な空気のあるところへの搬出が推奨される場合；
(c) ばく露した人から衣服と靴を脱がせることとその処理が推奨される場合；
(d) 応急処置に対処するための個人用保護具（PPE）が推奨される場合。

A4.3.4.2 　*最も重要な急性および遅発症状/影響*

　必要に応じて、ばく露に由来する最も重要な急性および遅発症状/影響についての情報を提供すること。

A4.3.4.3 　*必要に応じた速やかな治療と必要とされる特別な治療の指示*

　必要に応じて遅発効果に対する臨床検査と医学的な監視、特定の解毒剤の詳細（知られている場合）および禁忌についての情報を提供すること。

A4.3.5 　第5節：火災時の措置

　本節は、物質または混合物によって、もしくはその近傍から発生した火災消火に当たる際の要求事項を示す。

A4.3.5.1 　*適切な消火剤*

　適切なタイプの消火用機器についての情報を提供すること。さらに消火用機器が物質または混合物にかかわる特定の状況において不適であるかどうかを示すこと（例えば、爆発可能性のある粉じん－空気混合物の形成を起こしうる高圧媒体を避ける）。

A4.3.5.2 　*化学品から生じる特定の危険性*

　物質または混合物が燃える際に有害な燃焼副産物が発生するなど、化学品から生じるおそれのある特定の危険有害性について助言すること。例えば：

(a) 「燃焼する際に一酸化炭素の毒性ガス発生のおそれがある」；または

(b) "produces oxides of sulphur and nitrogen on combustion".

A4.3.5.3 *Special protective actions for fire-fighters*

A4.3.5.3.1 Provide advice on any protective actions to be taken during fire fighting. For example, "keep containers cool with water spray".

A4.3.6 **SECTION 6: Accidental release measures**

This section recommends the appropriate response to spills, leaks, or releases in order to prevent or minimize the adverse effects on persons, property and the environment in this section. Distinguish between responses for large and small spills where the spill volume has a significant impact on the hazard. The procedures for containment and recovery may indicate that different practices are required.

A4.3.6.1 *Personal precautions, protective equipment and emergency procedures*

A4.3.6.1.1 For non-emergency personnel

Provide advice related to accidental spills and release of the substance or mixture such as:

(a) the wearing of suitable protective equipment (including personal protective equipment, see section 8 of the SDS) to prevent any contamination of skin, eyes and personal clothing;

(b) removal of ignition sources and provision of sufficient ventilation; and

(c) emergency procedures such as the necessity to evacuate the danger area or to consult an expert.

A4.3.6.1.2 For emergency responders

Provide advice related to suitable fabric for personal protective clothing (e.g.: "appropriate: Butylene; not appropriate: PVC).

A4.3.6.2 *Environmental precautions*

Provide advice on any environmental precautions related to accidental spills and release of the substance or mixture, such as keeping away from drains, surface and ground water.

A4.3.6.3 *Methods and materials for containment and cleaning up*

A4.3.6.3.1 Provide appropriate advice on how to contain and clean up a spill. Appropriate containment techniques may include:

(a) bunding[2], covering of drains; and

(b) capping procedures[3].

A4.3.6.3.2 Appropriate clean up procedures may include:

(a) neutralization techniques;

(b) decontamination techniques;

(c) adsorbent materials;

(d) cleaning techniques;

---

[2] A **bund** is a provision of liquid collection facilities which, in the event of any leak or spillage from tanks or pipe work, will capture well in excess of the volume of liquids held, e.g. an embankment. Bunded areas should drain to a capture tank which should have facilities for water/oil separation.

[3] i.e. providing a cover or protection (e.g. to prevent damage or spillage).

(b) 「燃焼する際に硫黄と窒素の酸化物が発生する」。

A4.3.5.3　*消防士用の特別な防具と予防措置*

A4.3.5.3.1　消火活動において遵守すべきすべての予防措置について助言すること。例えば、「格納容器は水噴霧によって低温に保つ」など。

A4.3.6　*第6節：漏出時の措置*

本節では、この節にある、人、施設および環境に与える有害影響の予防または最小限に抑えるための流出、漏れ、放出に対する適切な対応について勧告する。漏出量が危険有害性に大きく影響する場合、多量あるいは少量での対処の違いを示すこと。囲い込んで回収する方法には異なった処置が求められることを示してもよい。

A4.3.6.1　*人への予防措置、防具、および応急処置法*

A4.3.6.1.1　非緊急事態要員に対して

以下のような物質または混合物の不測の流出および放出に関する助言を提供すること：

(a) 皮膚、眼および個人の衣服の汚染を防止するため、適切な保護具の着用（個人の保護具を含む、SDS の第8節参照）；
(b) 着火源の除去および充分な換気；および
(c) 危険区域から避難または専門家に助言を求める必要性などの応急処置。

A4.3.6.1.2　緊急事態要員に対して

防護服の適切な素材（例えば、「適切」ブチレン、「不適切」PVC）に関する情報を提供する。

A4.3.6.2　*環境上の予防措置*

下水溝、地表水と地下水から離して置くなど物質または混合物の不測の流出と放出に関連する環境上の予防措置について助言すること。

A4.3.6.3　*封じ込めと流出物洗浄の方法および用具*

A4.3.6.3.1　流出を封じ込めて浄化する方法について適切な助言をすること。適切な封じ込め技術には以下のものがある：

(a) 土手を作る[2]、下水溝を覆う；および
(b) 被覆措置[3]。

A4.3.6.3.2　適切な洗浄方法には以下のものを含むこと：

(a) 中和方法；
(b) 汚染除去方法；
(c) 吸着材；
(d) 洗浄方法；

---

[2] "*土手 (bund)*" とは、タンクまたはパイプ作業から漏れや流出の際に、貯めておける液体の量を超えた際に回収する液体回収設備の提供をいう。土手が築かれる範囲は、水/油の分離の設備を持つべき回収タンクに排出されるべきである。

[3] *覆いまたは防護を用意（例えば、損壊または流出を防止）*

(e) vacuuming techniques; and

(f) equipment required for containment/clean up (include the use of non-sparking tools and equipment where applicable).

A4.3.6.3.3　Provide any other issues relating to spills and releases. For example, including advice on inappropriate containment or clean up techniques.

A4.3.7　***SECTION 7: Handling and storage***

This section provides guidance on safe handling practices that minimize the potential hazards to people, property and the environment from the substance or mixture. Emphasize precautions that are appropriate to the intended use and to the unique properties of the substance or mixture.

A4.3.7.1　*Precautions for safe handling*

A4.3.7.1.1　Provide advice that:

(a) allows safe handling of the substance or mixture;

(b) prevents handling of incompatible substances or mixtures;

(c) draws attention to operations and conditions which create new risks by altering the properties of the substance or mixture, and to appropriate countermeasures; and

(d) minimizes the release of the substance or mixture to the environment.

A4.3.7.1.2　It is good practice to provide advice on general hygiene. For example:

(a) "eating, drinking and smoking in work areas is prohibited";

(b) "wash hands after use"; and

(c) "remove contaminated clothing and protective equipment before entering eating areas".

A4.3.7.2　*Conditions for safe storage, including any incompatibilities*

Ensure that the advice provided is consistent with the physical and chemical properties in section 9 (Physical and chemical properties) of the SDS. If relevant, provide advice on specific storage requirements including:

(a) How to avoid:

(i) explosive atmospheres;

(ii) corrosive conditions;

(iii) flammability hazards;

(iv) incompatible substances or mixtures;

(v) evaporative conditions; and

(vi) potential ignition sources (including electrical equipment).

(b) How to control the effects of:

(i) weather conditions;

(ii) ambient pressure;

(iii) temperature;

(e) 真空装置による吸い取り方法；および

(f) 封じ込め/浄化に必要な装置（適切な場合、防爆器具や装置の使用も含める）。

A4.3.6.3.3 流出と放出などについての他のすべての事柄を提供する。例えば不適切な封じ込めまたは洗浄方法を含む。

A4.3.7 *第7節：取扱いおよび保管上の注意*

本節では、物質または混合物による、人、施設、環境に対する潜在的な危険有害性を最小限にするための安全な取扱いに関する手引きを提供する。物質または混合物の意図された使用と特性に適切な予防措置に重点を置くこと。

A4.3.7.1 *安全な取扱のための予防措置*

A4.3.7.1.1 以下の助言を行うこと：

(a) 物質または混合物の安全な取扱いを可能にすること；

(b) 混触危険性物質または混合物の取扱いの防止；

(c) 物質や混合物の性質を変えることによって新たなリスクを生む操作および条件、さらに適切な対策に注意をはらう；および

(d) 物質または混合物の環境への放出の最少化。

A4.3.7.1.2 一般的な衛生についての助言を提示することは望ましい。例えば：

(a) 「作業域内での飲食と喫煙の禁止」；

(b) 「使用後の手洗い」；および

(c) 「食事する場所に入る前の、汚染された衣類と防具の取り外し」。

A4.3.7.2 *混触危険性を含む、安全な保管条件*

物理化学的特性に基づいて提供する助言が、SDSの第9節（物理化学的特性）と矛盾していないことを確認すること。もし関連すれば、以下を含めた特定の保管要求事項について助言すること：

(a) 以下を回避する方法：
  (i) 爆発性；
  (ii) 腐食性条件；
  (iii) 燃焼危険性；
  (iv) 混触危険性物質または混合物；
  (v) 揮発性条件；および
  (vi) 潜在的発火源（電気設備を含む）。

(b) 以下の影響の制御方法：
  (i) 気象条件；
  (ii) 大気圧；
  (iii) 温度；

(iv) sunlight;

(v) humidity; and

(vi) vibration.

(c) How to maintain the integrity of the substance or mixture by the use of:

(i) stabilizers; and

(ii) anti-oxidants.

(d) Other advice including:

(i) ventilation requirements;

(ii) specific designs for storage rooms/vessels;

(iii) quantity limits under storage conditions (if relevant); and

(iv) packaging compatibilities.

A4.3.8 *SECTION 8: Exposure controls/personal protection*

Within this guidance the term "occupational exposure limit(s)" refers to limits in the air of the workplace or biological limit values. In addition, for the purposes of this document "exposure control" means the full range of specific protection and prevention measures to be taken during use in order to minimize worker and environmental exposure. Engineering control measures that are needed to minimize exposure to, and risks associated with the hazards of, the substance or mixture should be included in this section.

A4.3.8.1   *Control parameters*

A4.3.8.1.1   Where available, list the occupational exposure limits (limits in the air of the workplace or biological limit values), including notations, for a substance and for each of the ingredients of a mixture. If air contaminants are formed when using the substance or mixture as intended available occupational exposure limits for these should also be listed. If an occupational exposure limit exists for the country or region in which the SDS is being supplied, this should be listed. The source of the occupational exposure limit should be stated on the SDS. When listing occupational exposure limits, use the chemical identity as specified in section 3 (Composition/Information on ingredients) of the SDS.

A4.3.8.1.2   Where available, list the biological limit values, including notations, for a substance and for each of the ingredients of a mixture. Where possible, the biological limit value should be relevant to the countries or regions in which the SDS is being supplied. The source of the biological limit value should be stated on the SDS. When listing biological limit values, use the chemical identity as specified in section 3 of the SDS.

A4.3.8.1.3   Where a control banding approach is recommended for providing protection in relation to specific uses then sufficient detail should be given to enable effective management of the risk. The context and limitations of the specific control banding recommendation should be made clear.

A4.3.8.2   *Appropriate engineering controls*

The description of appropriate exposure control measures should relate to the intended modes of use of the substance or mixture. Sufficient information should be provided to enable a proper risk assessment to be carried out. Indicate when special engineering controls are necessary and specify which type. Examples include:

(a) "maintain air concentrations below occupational exposure standards", using engineering controls if necessary;

(b) "use local exhaust ventilation when…";

(c) "use only in an enclosed system";

(d) "use only in spray paint booth or enclosure";

(iv) 直射日光；

(v) 湿度；および

(vi) 振動。

(c) 以下を用いた物質または混合物の品質維持方法：

(i) 安定化剤；および

(ii) 抗酸化剤。

(d) 以下を含めたその他の情報提供：

(i) 換気要求事項；

(ii) 保管室/容器のための特別な設計；

(iii) 保管条件下での数量制限（関連がある場合）；および

(iv) 輸送容器の適合性。

A4.3.8　*第8節：ばく露防止および保護措置*

本手引きにおいては、「職業ばく露限界」という用語は作業場の空気中の限界値または生物学的限界値を意味する。また、この文書の目的上、「ばく露の管理」は、使用中に労働者と環境へのばく露を最小限にするために講じるべきすべての特別な防護と予防策をいう。物質または混合物へのばく露と物質または混合物の危険有害性に関連するリスクを最小限にするために必要な工学的制御方法は本節で詳細に記載されるべきである。

A4.3.8.1　*管理パラメーター*

A4.3.8.1.1　入手できる場合には、物質と混合物の各成分についての注釈を含めて職業性ばく露限界値（作業場の空気中の限界値または生物学的限界値）を示す。物質または混合物を意図して使用するときに、空気の汚染が生ずる場合は、これらの入手できる職業性ばく露限界値もまた示すべきである。SDSが供給されている国または地域における職業性ばく露限界値が存在する場合には、これを示すべきである。職業性ばく露限界値の出所をSDSにおいて記載すべきである。職業性ばく露限界値を示す場合には、SDSの第3節（組成/成分情報）に記載されている物質の特性を使用すべきである。

A4.3.8.1.2　入手できる場合には、物質および混合物の成分ごとの生物学的限界値を注釈つきで示す。可能であれば生物学的限界値はそのSDSが供給されている国や地域に関連づけるべきである。生物学的限界値の出所をSDSにおいて記載するべきである。生物学的限界値を示す場合、SDSの第3節で指定されている物質の特性を使用すべきである。

A4.3.8.1.3　特定の使用に関連して安全を確保するために、コントロールバンディングを推奨する場合には、効果的なリスク管理を可能とするために充分な詳細情報が提供されるべきである。特定のコントロールバンディングを行うにあたっての状況及び限界が明確にされるべきである。

A4.3.8.2　*適切な工学的管理方法*

適切なばく露管理対策の説明は、物質または混合物の使用状態に関連づけるべきである。適切なリスク評価を実施するために十分な情報が提供されるべきである。特別な工学的管理方法が必要である場合を示し、特定の型を明記する。その例には以下のものがある：

(a) 必要ならば工学的管理方法を用いて「職業的ばく露標準を下回る空気濃度を維持する」；

(b) 「～する場合、局所排気装置を用いる」；

(c) 「密閉系のみで使用」；

(d) 「スプレー塗装ブースまたは密閉系のみで使用」；

(e) "use mechanical handling to reduce human contact with materials"; or

(f) "use explosive dust handling controls".

The information provided here should complement that provided under section 7 (Handling and storage) of the SDS.

A4.3.8.3   *Individual protection measures, such as personal protective equipment (PPE)*

A4.3.8.3.1   Consistent with good occupational hygiene practices, personal protective equipment (PPE) should be used in conjunction with other control measures, including engineering controls, ventilation and isolation. See also section 5 (Fire- fighting measures) of the SDS for specific fire/chemical PPE advice.

A4.3.8.3.2   Identify the PPE needed to minimize the potential for illness or injury due to exposure from the substance or mixture, including:

(a) Eye/face protection: specify the type of eye protection and/or face shield required, based on the hazard of the substance or mixture and potential for contact;

(b) Skin protection: specify the protective equipment to be worn (e.g. type of gloves, boots, bodysuit) based on the hazards associated with the substance or mixture and the potential for contact;

(c) Respiratory protection: specify appropriate types of respiratory protection based on the hazard and potential for exposure, including air-purifying respirators and the proper purifying element (cartridge or canister) or breathing apparatus; and

(d) Thermal hazards: when specifying protective equipment to be worn for materials that represent a thermal hazard, special consideration should be given to the construction of the PPE.

A4.3.8.3.3   Special requirements may exist for gloves or other protective clothing to prevent skin, eye or lung exposure. Where relevant, this type of PPE should be clearly stated. For example, "PVC gloves" or "nitrile rubber gloves", and thickness and breakthrough time of the glove material. Special requirements may exist for respirators.

A4.3.9   ***SECTION 9: Physical and chemical properties***

A4.3.9.1   This section of annex 4 provides guidance for SDS preparers and is provided for information purposes. This guidance does not prescribe how this information should be presented on the SDS. The guidance is divided into three tables as described below.

A4.3.9.2   Table A4.3.9.1 provides guidance on the physical and chemical properties specified by chapter 1.5, table 1.5.2. The SDS preparer should clearly describe/identify the physical and chemical properties specified in table 1.5.2. In cases where the specific physical and chemical properties required by table 1.5.2 do not apply or are not available under a particular subheading, this should be clearly indicated.

A4.3.9.3   Table A4.3.9.2 lists properties/safety characteristics and test results that are not required on the SDS but may be useful to communicate when a substance or mixture is classified in the respective physical hazard class. Data which is deemed relevant with regard to a specific physical hazard but not resulting in classification (e.g. negative test results close to the criterion) may also be useful to communicate.

A4.3.9.4   Table A4.3.9.3 lists further properties/safety characteristics and test results that are not required on the SDS but may be useful to communicate for a substance or mixture. Other physical properties/safety characteristics of the substance or mixture not identified in this table may also be useful to communicate.

NOTE: *The properties in tables A4.3.9.1, A4.3.9.2 and A4.3.9.3 may be presented with or without any division (that is, as a list). Also the order of the properties may be adjusted if deemed appropriate.*

A4.3.9.5   Generally, the information given in this section of the SDS should relate to standard conditions for temperature and pressure (temperature of 20 °C and absolute pressure of 101.3 kPa). If other conditions apply, these should be indicated together with the respective property.

A4.3.9.6   Data on the SDS should be provided in appropriate units. Where the data relate to a hazard class, the units of measure should be as specified in the criteria for that hazard class.

    (e) 「人が材料に接触しないように機械的な操作にする」；または
    (f) 「爆発性粉塵の操作管理を行う」。

ここで提供される情報は、SDS の第 7 節（取扱いおよび保管）で提供される情報を補足するものであるべきである。

A4.3.8.3　*個人用保護具（PPE）などの個人保護措置*

A4.3.8.3.1　個人用保護具は良好な労働衛生の手順と矛盾しない工学的管理方法、換気、隔離を含めた他の管理手法と併用されるべきである。特定の火災/化学用 PPE についての情報提供は、SDS の第 5 節（火災時の措置）も参照のこと。

A4.3.8.3.2　以下を含め、物質または混合物へのばく露による疾病または傷害の起こる可能性を最小限にするために必要な個人用保護具（PPE）を特定すること：

    (a) 眼/顔面の保護：物質または混合物による危険有害性と接触の可能性に基づいて、必要な眼の保護または顔面保護具を特定する；
    (b) 皮膚の保護：物質または混合物に関する危険有害性と接触の可能性に基づいて、着用する保護具を指定する（例えば手袋、長靴、防護服の型）；
    (c) 呼吸器の保護：危険有害性とばく露の可能性に基づいて、空気浄化装置と適切な空気浄化部品（カートリッジまたは吸収缶）または呼吸装置を含めて、適切な呼吸器の種類を特定する；および
    (d) 高熱の危険性：高熱の危険性を有する材料に対して、着用する保護具を特定する。これにあたっては PPE の材質に特別の配慮をはらうべきである。

A4.3.8.3.3　皮膚、眼または肺のばく露防止のための手袋または他の保護衣に対して、特別な要求事項が挙げられることがある。該当する場合には PPE の種類を明確に記載すべきである。例えば、「PVC 手袋」または「ニトリルゴム手袋」、加えて、手袋の材料の厚さおよび透過時間など。人工呼吸器には、特別な要求事項があるであろう。

A4.3.9　*第 9 節：物理的および化学的性質*

A4.3.9.1　附属書 4 の本節は SDS 作成者へ手引きを提供し、また情報を提供するためにある。この手引きは情報を SDS にどのように表すかについては規定していない。この手引きは下記のように 3 つの表に分かれている。

A4.3.9.2　表 A4.3.9.1 は、第 1.5 章、表 1.5.2 に明記されている物理的および化学的性質に関するガイダンスである。SDS 作成者ははっきりと表 1.5.2 に明記されている物理的および化学的性質について記載/確認しなければならない。表 1.5.2 で要求されている特定の物理的および化学的性質が該当しないあるいは特定の小項目のなかのこれらが入手できない場合には、そのことを明示するべきである。

A4.3.9.3　表 A4.3.9.2 では SDS には要求されていない性質/安全特性および試験結果を列挙したが、物質または混合物が各物理化学的危険性クラスに分類されたときには情報伝達することが有用であろう。特定の物理化学的危険性に関係しているとみなされながら分類に至らなかったデータ（例えば判定基準に近い陰性の試験結果）もまた情報伝達することが有用であろう。

A4.3.9.4　表 A4.3.9.3 ではさらに SDS には要求されていない性質/安全特性および試験結果を列挙したが、物質または混合物に関して情報伝達することが有用であろう。本表には記載されていない物質または混合物の他の物理化学的性質/安全特性もまた情報伝達することが有用であろう。

注記：　表 A4.3.9.1、A4.3.9.2 および A4.3.9.3 の性質は分割してもしなくてもよい（すなわちリストとして）。性質の順序も適切と見なされれば変更してもよい。

A4.3.9.5　一般に、SDS の本節で与えられる情報は温度及び気圧が標準状態（温度 20°C、絶対気圧 101.3 kPa）のものとするべきである。他の条件を適用する場合には、それらを各性質とともに示すべきである。

A4.3.9.6　SDS のデータは適切な単位で示されなければならない。データが危険性クラスと関連する場合には、測定単位は当該危険性クラスの判定基準に特定されているものにするべきである。

A4.3.9.7   If relevant for the interpretation of the information or numeric value given, indicate the determination method (e.g. open-cup/closed-cup for flash point) or state whether the value was calculated.

A4.3.9.8   In the case of a mixture, where valid data is available for the mixture as a whole, it should be provided. When data for the mixture as a whole cannot be provided, data for the most relevant ingredient(s) may be provided, and this data should clearly indicate to which ingredient(s) the data apply.

A4.3.9.9   Other appropriate physical or chemical parameters or safety characteristics, in addition to those listed below, may also be included in this section of the SDS.

Table A4.3.9.1: Basic physical and chemical properties

This table lists basic physical and chemical properties and safety characteristics. Relevant information as required should be indicated for every property listed in this table, such as a short description, value(s), unit, conditions (e.g. temperature, pressure), method, each as appropriate.

If specific properties or safety characteristics do not apply (based on the respective information about applicability in the column "Remarks/Guidance") they should still be listed in the SDS with the statement "not applicable".

If information on specific properties or safety characteristics is not available, they should still be listed in the SDS with the statement "not available". It is recommended that, where appropriate, a short explanation is included as to why the data is not available, e.g. "melts", "decomposes", "dissolves".

Property	Remarks/Guidance
Physical state	− generally at standard conditions − for definitions for gas, liquid and solid see chapter 1.2
Colour	− indicate the colour of the substance or mixture as supplied − in cases where one SDS is used to cover variants of a mixture which may have different colours the term 'various' can be used to describe the colour (see A4.3.1.1 for an SDS for variants of a mixture)
Odour	− give a qualitative description of the odour if it is well-known or described in the literature − if available, indicate the odour threshold (qualitatively or quantitatively)
Melting point/ freezing point	− not applicable to gases − at standard pressure − indicate up to which temperature no melting point was observed in case the melting point is above the measuring range of the method − indicate if decomposition or sublimation occurs prior to or during melting − for waxes and pastes the softening point/range may be indicated instead − for mixtures indicate if it is technically not possible to determine the melting point/freezing point
Boiling point or initial boiling point and boiling range	− generally at standard pressure (a boiling point at lower pressure might be indicated in case the boiling point is very high or decomposition occurs before boiling) − indicate up to which temperature no boiling point was observed in case the boiling point is above the measuring range of the method − indicate if decomposition occurs prior to or during boiling − for mixtures indicate if it is technically not possible to determine the boiling point or range; in that case indicate also the boiling point of the lowest boiling ingredient

Flammability	− applicable to gases, liquids and solids
	− indicate whether the substance or mixture is ignitable (capable of catching on fire or being set on fire, even if not classified for flammability)
	− if available and appropriate, further information may be indicated in addition, e.g.
	   • whether the effect of ignition is other than a normal combustion (e.g. an explosion)
	   • ignitability under non-standard conditions
	− more specific information on the flammability may be indicated based on the respective hazard classification in accordance with table A4.3.9.2

A4.3.9.7 情報または与えられた数値の解釈が関係する場合には、測定方法（例えば引火点での開放/密閉カップ）を示すか、または値は計算されたものかどうかを記述する。

A4.3.9.8 混合物の場合、混合物全体として有効なデータがあるときには、それを示すべきである。混合物全体としてのデータを入手できない場合には、最も関連する成分のデータを示してもよく、さらにこのデータはどの成分に該当するものであるかを明示するべきである。

A4.3.9.9 他の適切な物理的または化学的性質もしくは安全特性は、下記のリストされたものに加えて、SDSの本節に含めてもよい。

表 A4.3.9.1：基本的な物理的および化学的性質

この表には基本的な物理的および化学的性質と安全特性を示した。要求されている関連情報は本表に示されているすべての性質に関して示されるべきである。例えば短い記述、値、単位、条件（例えば温度、圧力）、方法などである。

もし特定の性質または安全特性に該当しない場合にも（「注釈/手引き」欄における適用性についての各情報に基づく）、それらは「該当しない」との記述とともにSDSに記載されるべきである。

特定の性質または安全特性に関する情報が無い場合にも、それらは「入手できない」との記述とともに SDS に記載されるべきである。データが入手できない理由に関して、例えば「融解した」、「分解した」、「溶解した」など、短い説明が適切になされたほうがよい。

性質	注釈/手引き
物理状態	− 一般に標準状態下 − ガス、液体および固体の定義は第 2.1 章を参照
色	− 供給された物質または混合物の色を示す − 異なる色を持つような混合物のわずかな色の差異を 1 つの SDS でカバーする場合には、色を記述するために「さまざまな」という言葉を使うことができる（混合物のわずかな違いに対する SDS に関しては A4.3.1.1 を参照）
臭い	− よく知られているまたは文献に記述がある臭いならば、その性質を記載する − 入手可能であれば臭いの閾値を示す（定性的または定量的に）
融解点/凝固点	− ガスは該当しない − 標準圧力下 − 融点が測定方法の範囲を超えていた場合には、何度まで融点が観察できなかったかを示す − 融解前または融解中に分解や昇華が起きたかどうか示す − ワックスやペーストでは、代わりに軟化温度/範囲を示してもよい − 混合物で融点/凝固点を測定するのが技術的に可能ではない場合にはそれを示す
沸点または初留点および沸点範囲	− 一般的に標準圧力下（沸点が非常に高いまたは沸騰前に分解が起きる場合には、より低い圧力下での沸点が示されてもよい） − 沸点が測定方法の範囲を超えていた場合には、何度まで沸点が観察できなかったかを示す − 沸騰前または沸騰中に分解が起きたかどうか示す − 混合物で沸点または沸点範囲を測定するのが技術的に可能ではない場合にはそれを示す；その場合沸点が最も低い成分の沸点も示す
可燃性	− ガス、液体および固体が該当する − 物質または混合物が着火するかどうか示す（たとえ可燃性に分類されない場合でも、周囲に火炎があるまたは火炎を近づけた状態で着火するか） − もし入手可能で適切であれば、さらなる情報を追加的に示してもよい、例えば、 　• 着火後の状態が通常の燃焼以上（例えば爆発）か 　• 非標準状態下での着火の可能性　など − 表 A4.3.9.2 に従い各危険性分類に基づいた可燃性に関してのより具体的な情報を示してもよい

Table A4.3.9.1: Basic physical and chemical properties *(cont'd)*

Property	Remarks/Guidance

Lower and upper explosion limit/ flammability limit	– not applicable to solids
	– for flammable liquids indicate at least the lower explosion limit:
	  • if the flash point is approximately > -25°C, it might be not possible to determine the upper explosion limit at standard temperature; in that case it is recommended to indicate the upper explosion limit at elevated temperature
	  • if the flash point is > +20 °C the same holds for both the lower and upper explosion limit
	Note: *Depending on the region of the world the term "explosion limit" or "flammability limit" is used but is supposed to mean the same.*

Flash point	– not applicable to gases, aerosols and solids – for information on test methods etc., see chapter 2.6, paragraph 2.6.4.2 for mixtures: – indicate a value for the mixture itself if available, otherwise indicate the flash point(s) of those substances with the lowest flash point(s) as these are generally the main contributing ones
Auto-ignition temperature	– applicable to gases and liquids only for mixtures: – indicate a value for the mixture itself if available, otherwise indicate the auto-ignition temperature(s) of those ingredients with the lowest auto-ignition temperature(s)

Decomposition temperature	– applicable to self-reactive substance and mixtures and organic peroxides and other substances and mixtures which may decompose
	– indicate
	  • the SADT (self-accelerating decomposition temperature), together with the volume to which it applies or
	  • the decomposition onset temperature (see also section 20.3.3.3 of the *Manual of Tests and Criteria*)
	– indicate whether the temperature given is the SADT or the decomposition onset temperature
	– if no decomposition was observed, indicate up to which temperature no decomposition was observed, e.g. as "no decomposition observed up to x °C/°F"

pH	– not applicable to gases – applicable to aqueous liquids and solutions (the pH is linked to aqueous media by definition; measurements carried out in other media do not give the pH) – indicate the concentration of the test substance in water – where the pH is ≤ 2 or ≥ 11.5, see table A4.3.9.3 for information on acid/alkaline reserve
Kinematic viscosity	– applicable to liquids only – use preferably mm^2/s as unit (as the classification criteria for the hazard class aspiration hazard are based on this unit) – the dynamic viscosity may be indicated in addition. The kinematic viscosity is linked to the dynamic viscosity by the density: $$\text{Kinematic viscosity}(mm^2/s) = \frac{\text{Dynamic viscosity}(mPa \cdot s)}{\text{Density }(g/cm^3)}$$ – for non-Newtonian liquids, indicate thixotropic or rheopexic behaviour
Solubility	– generally at standard temperature – indicate the solubility in water – the solubility in other (non-polar) solvents may also be included – for mixtures, indicate if it is fully or only partially soluble in or miscible with water or other solvent
Partition coefficient n-octanol/water (log value)	– not applicable to inorganic and ionic liquids – generally not applicable to mixtures – may be calculated (using QSAR – Quantitative structure-activity relationship) – indicate whether the value is based on testing or on calculation

表 A4.3.9.1：基本的な物理的および化学的性質 (続き)

性質	注釈/手引き
爆発下限界及び爆発上限界/可燃限界	− 固体は該当しない − 引火性液体には少なくとも爆発下限を示す： 　• もし引火点が約 > -25 °C の場合、標準温度では 爆発上限を測定するのは可能ではないであろう；そのような場合上昇させた温度での爆発上限を示すことを勧める 　• もし引火点が > +20 °C の場合、同様のことが爆発下限及び爆発上限にもいえる *注記：世界の地域により「爆発限界」または「可燃限界」という用語が使用されているが、同じ意味と推測される*
引火点	− ガス、エアゾールおよび固体は該当しない − 試験方法に関する情報など、第 2.6 章 2.6.4.2 を参照 混合物： − 入手可能であれば混合物自体の値を示し、それがない場合には最も低い引火点を持つ物質の引火点を示す、通常、それらが主として引火点に寄与するため
自然発火点	− ガスおよび液体のみが該当する 混合物： − 入手可能であれば混合物自体の値を示す、それがない場合には最も低い自然発火温度をもつ成分の自然発火温度を示す
分解温度	− 自己反応性物質及び混合物、有機過酸化物および分解可能性のある物質及び混合物が該当する − 下記事項を示す 　• 適用される容量とともに SADT（自己加速分解温度）、または 　• 分解開始温度（*試験方法及び判定基準のマニュアル*の 20.3.3.3 参照） − 示された温度が SADT かまたは自己分解開始温度かを示す − もし分解が観察されない場合には、どの温度まで自己分解が観察されなかったかを示す、例えば「x °C/°F まで自己分解は観察されず」
pH	− ガスは該当しない − 水性液体および溶液が該当する（pH は定義により水媒体と関連している；他の媒体での測定では pH は得られない） − 水中での試験物質の濃度を示す − pH が ≤ 2 または ≥ 11.5 の時は、酸/アルカリ予備について表 A4.3.9.3 を参照
動粘性率	− 液体のみが該当する − 単位として mm^2/s が望ましい（誤えん有害性の分類基準はこの単位に基づいているため） − 追加的に粘性率を示してもよい。動粘性率は密度によって粘性率と関連している： $$動粘性率\ (mm^2/s) = \frac{粘性率\ (mPa \cdot s)}{密度\ (g/cm^3)}$$ − 非ニュートン液体については、チキソトロピーまたはレオペクシー動態について示す
溶解度	− 一般に標準温度下 − 水への溶解を示す − 他の（非極性）溶媒への溶解度も含まれてよい − 混合物の場合、それの全部または一部が水または他の溶媒に溶解または混和するのかを示す
n-オクタノール/水分配係数(log値)	− 無機およびイオン性液体は該当しない − 一般に混合物は該当しない − 計算できるかもしれない（QSAR 使用－定量的構造活性相関） − 値は試験によるものかまたは計算によるものかを示す

Table A4.3.9.1: Basic physical and chemical properties *(cont'd)*

Property	Remarks/Guidance
Vapour pressure	− generally at standard temperature − indicate the vapour pressure at 50°C for volatile fluids in addition (in order to enable distinction between gases and liquids based on the definitions in chapter 1.2) − in cases where one SDS is used to cover variants of a liquid mixture or liquefied gas mixture indicate a range for the vapour pressure − for liquid mixtures or liquefied gas mixtures, indicate a range for the vapour pressure or at least the vapour pressure of the most volatile ingredient(s) where the vapour pressure of the mixture is predominantly determined by this/these ingredient(s) − for liquid mixtures or liquefied gas mixtures, the vapour pressure may be calculated using the activity coefficients of the ingredients − the saturated vapour concentration (SVC) in ml/m^3 or in g/m^3 (=mg/l) may be indicated in addition. The saturated vapour concentration can be estimated as follows: $$\text{SVC in } ml/m^3: SVC = VP \cdot c_1$$ $$\text{SVC in } g/m^3: SVC = VP \cdot MW \cdot c_2$$ where • VP is the vapour pressure in hPa (=mbar) • MW is the molecular weight in g/mol and • c_1 and c_2 are conversion factors where $c_1 = 987.2$ $ml/(m^3 \cdot hPa)$ and $c_2 = 0.0412$ $mol/(m^3 \cdot hPa)$
Density and/or relative density	− applicable to liquids and solids only − generally at standard conditions − indicate as appropriate • the absolute density and/or • the relative density based on water at 4°C as reference (sometimes also called the specific gravity) − a range may be indicated in cases where variations in density are possible, e.g. due to batch manufacture, or where one SDS is used to cover several variants of a substance or mixture ***NOTE***: *For clarity, the SDS should indicate if absolute density (indicate units) and/or relative density (no units) is being reported.*
Relative vapour density	− applicable to gases and liquids only − for gases, indicate the relative density of the gas based on air at 20 °C as reference (=MW/29) − for liquids, indicate the relative vapour density based on air at 20 °C as reference (=MW/29) − for liquids, the relative density (D_m) of the vapour/air-mixture at 20 °C (air = 1) may be indicated in addition. It can be calculated as follows: $$D_m = 1 + (VP_{20} \cdot (MW - MW_{air}) \cdot c_3$$ where • VP_{20} is the vapour pressure at 20 °C in hPa (=mbar) • MW is the molecular weight in g/mol • MW_{air} is the molecular weight of air, $MW_{air} = 29$ g/mol • c_3 is a conversion factor, $c_3 = 34 \cdot 10^{-6} \frac{mol}{g \cdot hPa}$
Particle characteristics	− applicable to solids only − indicate the particle size (median and range) − if available and appropriate, further properties may be indicated in addition, e.g. • size distribution (range) • shape and aspect ratio • specific surface area

表 A4.3.9.1：基本的な物理的および化学的性質 *(続き)*

性質	注釈/手引き
蒸気圧	- 一般に標準温度下 - 追加的に 50 °C における揮発性液体の蒸気圧を示す（第 1.2 章の定義に基づきガスと液体の境界を明確にするため） - 若干組成の異なる液体混合物または液化ガス混合物を 1 つの SDS でカバーする場合には、蒸気圧の範囲を示す - 液体混合物または液化ガス混合物では、蒸気圧の範囲または混合物の蒸気圧が主としてもっとも揮発性の大きな成分によって決まる場合には少なくともこれらの蒸気圧を示す - 液体混合物または液化ガス混合物では、蒸気圧は成分の活性係数を用いて計算してもよい - 飽和蒸気濃度（SVC）を ml/m^3 または g/m^3 $(=mg/l)$ で追加的に示してもよい。飽和蒸気濃度は次のように推算できる： SVC in ml/m^3: $\quad SVC = VP \cdot c_1$ SVC in g/m^3: $\quad SVC = VP \cdot MW \cdot c_2$ ここで • VP は蒸気圧 hPa (=mbar) • MW は分子量 g/mol • c_1 および c_2 は換算係数、ここで $c_1 = 987.2$ $ml/(m^3 \cdot hPa)$ および $c_2 = 0.0412$ $mol/(m^3 \cdot hPa)$

密度および/または相対密度	- 液体と固体のみが該当する
	- 一般に標準状態下
	- 適切に示す
	  • 絶対密度および/または
	  • 参照として 4 °C の水を基準とした相対密度（ときに比重と呼ばれる）
	- 密度の変動が起きる場合には範囲で示してもよい、例えばバッチ製造による、または若干組成の異なる物質または混合物を 1 つの SDS でカバーする場合
	注記：明確にするために、絶対密度（単位を示す）および/または相対密度（単位なし）の報告がある場合には SDS に記載するべきである。

相対ガス密度	- ガスと液体のみが該当する - ガスでは参照として 20 °C の空気（=MW/29）を基準とした相対密度を示す - 液体では参照として 20 °C の空気（=MW/29）を基準とした相対蒸気密度を示す - 液体では追加的に 20 °C の蒸気/空気-混合物の相対密度（D_m）（空気＝1）を示してもよい。これは次のように計算する： $$D_m = 1 + (VP_{20} \cdot (MW - MW_{air}) \cdot c_3)$$ ここで • VP_{20} は 20 °C の蒸気圧 hPa (=mbar) • MW は分子量 g/mol • MWair は空気の分子量 MWair＝29g/mol • c_3 は換算係数 $\quad c_3 = 34 \cdot 10^{-6} \frac{mol}{g \cdot hPa}$

粒子特性	- 固体のみが該当する
	- 粒子サイズを示す（中央値及び範囲）
	- 入手可能で適切であれば、他の性質を追加的に示してもよい、例えば
	  • 粒径分布（範囲）
	  • 形およびアスペクト比
	  • 比表面積

Table A4.3.9.2: Data relevant with regard to physical hazard classes (supplemental)

This table lists properties/safety characteristics and test results that are not required on the SDS but may be useful to communicate when a substance or mixture is classified in the respective physical hazard class. Data which is deemed relevant with regard to a specific physical hazard but not resulting in classification (e.g. negative test results close to the criterion) may also be useful to communicate. Include any relevant information, such as a short description, value(s), unit, conditions (e.g. temperature, pressure), method, each as appropriate.

The name of the hazard class the data relates to may be indicated together with the data but it is not necessary to do so because the resulting classification is already indicated in section 2 of the SDS. Thus, the data may be listed in the same way as the data according to table A4.3.9.1.

Unless otherwise specified, the test methods referred to in this table are described in the *Manual of Tests and Criteria*.

Chapter	Hazard class	Property/Safety characteristic/Test result and Remarks/Guidance
2.1	Explosives	− indicate the sensitivity to shock, generally determined by the UN gap test: test 1 (a) and/or test 2 (a) (section 11.4 or 12.4 of the *Manual of Tests and Criteria*) (indicate at least + or −) − indicate the effect of heating under confinement, generally determined by the Koenen test: test 1 (b) and/or test 2 (b) (section 11.5 or 12.5 of the *Manual of Tests and Criteria*) (indicate preferably the limiting diameter) − indicate the effect of ignition under confinement, generally determined by test 1 (c) and/or test 2 (c) (section 11.6 or 12.6 of the *Manual of Tests and Criteria*) (indicate at least + or −) − indicate the sensitiveness to impact, generally determined by test 3 (a) (section 13.4 of the *Manual of Tests and Criteria*) (indicate preferably the limiting impact energy) − indicate the sensitiveness to friction, generally determined by test 3 (b) (section 13.5 of the *Manual of Tests and Criteria*) (indicate preferably the limiting load) − indicate the thermal stability, generally determined by test 3 (c) (section 13.6 of the *Manual of Tests and Criteria*) (indicate at least + or −) − in addition, this entry is also applicable to substances and mixtures which are exempted based on note 2 in chapter 2.1, section 2.1.3 and to other substances and mixtures which show a positive effect if heated under confinement − indicate the package (type, size, net mass of substance or mixture) based on which the division was assigned or based on which the substance or mixture was exempted
2.2	Flammable gases	for pure flammable gases: − no data on the explosion/flammability limits is needed because these are indicated based on table A4.3.9.1 − indicate the T_{Ci} (maximum content of flammable gas which, when mixed with nitrogen, is not flammable in air, in %) as per ISO 10156 − indicate the fundamental burning velocity if the gas is classified as Category 1B based on fundamental burning velocity, generally determined by ISO 817:2014, Annex C for flammable gas mixtures: − indicate the explosion/flammability limits, if tested, or indicate whether the classification and category assignment is based on the calculation as per ISO 10156 − indicate the fundamental burning velocity if the gas mixture is classified as Category 1B based on fundamental burning velocity, generally determined by ISO 817:2014, Annex C
2.3, section 2.3.1	Aerosols	− indicate the total percentage (by mass) of flammable components unless the Aerosol is classified as Aerosol cat. 1 because it contains more than 1 % flammable components or has a heat of combustion of at least 20 kJ/g and is not submitted to the flammability classification procedures (see note 2 in chapter 2.3, 2.3.1.2.1)
2.3, section 2.3.2	Chemicals under pressure	− indicate the total percentage (by mass) of flammable components − indicate the specific heat of combustion (generally in kJ/g)

表 A4.3.9.2：物理化学的危険性クラスに関連するデータ（補遺）

この表では、SDS には要求されていないものの、物質または混合物を各物理化学的危険性クラスに分類する際に情報伝達することが有用であろう性質/安全特性および試験結果を記載している。特定の物理化学的危険性に関連すると思われるが分類結果には至らない（例えば判定基準に近い否定的な結果）データも情報伝達することが有用であろう。短い説明、値、単位、条件（例えば温度、圧力）、方法のような、どのような関連情報でも含める。

データが関連する危険性クラスの名称をデータとともに示してもよいが、これは SDS の第 2 節において分類結果はすでに示されているので必須ではない。データは表 A4.3.9.1 のデータと同様の方法で示してもよい。

他に特記されていなければ、表に参照されている試験方法は、*試験方法及び判定基準のマニュアル*に記載されている。

章	危険性クラス	性質/安全特性/試験結果および注釈/手引き
2.1	爆発物	− 通常、国連ギャップ試験で測定される衝撃に対する感度を示す：試験 1 (a) および/または試験 2 (a)（*試験方法及び判定基準のマニュアルの 11.4 節または 12.4 節*）（少なくとも+または−で示す） − 密閉状態での熱の影響を示す、通常、ケーネン試験で測定する：試験 1 (b) および/または試験 2 (b)（*試験方法及び判定基準のマニュアルの 11.5 節または 12.5 節*）（限界径を示すことが望ましい） − 密閉状態での点火の影響を示す、通常、試験 1 (c)および/または試験 2 (c)（*試験方法及び判定基準のマニュアルの 11.6 節または 12.6 節*）で測定する（少なくとも+または−で示す） − 打撃に対する感度を示す、通常、試験 3 (a)で測定する（*試験方法及び判定基準のマニュアルの 13.4 節*）（限界衝撃エネルギーを示すことが望ましい） − 摩擦に対する感度を示す、通常、試験 3 (b)で測定する（*試験方法及び判定基準のマニュアルの 13.5 節*）（限界負荷を示すことが望ましい） − 熱安定性を示す、通常、試験 3 (c)で測定する（*試験方法及び判定基準のマニュアルの 13.6 節*）（少なくとも+または−で示す） − 追記、この項目は第 2.1 章 2.1.3 の注記に基づき除外されている物質および混合物、さらに密閉状態で加熱された場合にプラスの影響を示す物質および混合物にも該当する − 割り当てられている区分に基づいた、または除外されている物質または混合物に基づいた包装（タイプ、サイズ、物質または混合物の正味量）を示す
2.2	可燃性ガス	単一の可燃性ガス： − 表 A4.3.9.1 に基づいて示されているように、爆発/可燃限界に関するデータがないことを示す必要がある − ISO 10156 により T_{Ci}（窒素と混合した時の、空気中で燃えない可燃性ガスの最大%濃度）を示す − 基本的な燃焼速度に基づいてガスを区分 1B と分類する場合には、基本的な燃焼速度を示す、通常は ISO 817:2014、附属書 C で測定される 可燃性ガス混合物： − 試験されている場合には爆発/可燃限界を示す、または ISO 10156 の計算に基づいて分類、区分されたかどうかを示す − 基本的な燃焼速度に基づいてガス混合物を区分 1B と分類する場合には、基本的な燃焼速度を示す、通常は ISO 817:2014、附属書 C で測定される
2.3 2.3.1	エアゾール	− 1%を超える可燃性成分を含むまたは燃焼熱が少なくとも 20 kJ/g であり、さらに可燃性に関する分類手順を踏まないということで（第 2.3 章 2.3.1.2.1 の注記 2 を参照）、エアゾールがエアゾール区分 1 に分類されていない場合には、可燃性成分の合計%（質量）を示す
2.3 2.3.2	加圧下化学品	− 可燃性成分の全割合（質量）を示す − 燃焼熱を示す（一般に kJ/g）

Table A4.3.9.2: Data relevant with regard to physical hazard classes (supplemental) *(cont'd)*

Chapter	Hazard class	Property/Safety characteristic/Test result and Remarks/Guidance
2.4	Oxidizing gases	for pure oxidizing gases: − indicate the C_i (coefficient of oxygen equivalency) as per ISO 10156 for oxidizing gas mixtures: − indicate "Oxidizing gas Category 1 (tested as per ISO 10156)" for tested mixtures or indicate the calculated oxidizing power (OP) as per ISO 10156
2.5	Gases under pressure	for pure gases: − indicate the critical temperature for gas mixtures: − indicate the pseudo-critical temperature; it is estimated as the mole weighted average of the critical temperatures of the components as follows: $$\sum_{i=1}^{n} x_i \cdot T_{\text{Crit } i}$$ where • x_i is molar fraction of component i • $T_{\text{Crit } i}$ is the critical temperature of component i
2.6	Flammable liquids	− no additional data is needed because the boiling point and the flash point are indicated based on table A4.3.9.1 − indicate information on sustained combustibility if exemption based on Test L.2 (section 32.5.2 of the *Manual of Tests and Criteria)*, in accordance with note 2 in chapter 2.6, section 2.6.2, is considered
2.7	Flammable solids	− indicate the burning rate (or burning time for metal powders), generally determined by Test N.1 (section 33.2.1 of the *Manual of Tests and Criteria*) − indicate whether the wetted zone has been passed or not
2.8	Self-reactive substances and mixtures	− for the SADT (self-accelerating decomposition temperature), see the entry for the decomposition temperature in table A4.3.9.1 − indicate the decomposition energy (value and method of determination) − indicate detonation properties (Yes/Partial/No), also in packaging where relevant − indicate deflagration properties (Yes rapidly/Yes slowly/No), also in packaging where relevant − indicate the effect of heating under confinement (Violent/Medium/Low/No), also in packaging where relevant − indicate the explosive power if applicable (Not low/Low/None)
2.9	Pyrophoric liquids	− indicate whether spontaneous ignition or charring of the filter paper occurs, generally determined by Test N.3 (section 33.3.1.5 of the *Manual of Tests and Criteria*) (indicate e.g. "the liquid ignites spontaneously in air" or "a filter paper with the liquid chars in air")
2.10	Pyrophoric solids	− indicate whether spontaneous ignition occurs when poured or within five minutes thereafter, generally determined by Test N.2 (section 33.3.1.4 of the *Manual of Tests and Criteria*) (e.g. "the solid ignites spontaneously in air") − indicate whether pyrophoric properties could be altered over time, e.g. by formation of a protective surface layer through slow oxidation
2.11	Self-heating substances and mixtures	− indicate whether spontaneous ignition occurs, include possible screening data and/or method used (generally Test N.4, section 33.3.1.6 of the *Manual of Tests and Criteria*) and note the maximum temperature rise obtained − indicate the results of screening tests according to chapter 2.11, paragraph 2.11.4.2, if relevant and available
2.12	Substances and mixtures which, in contact with water, emit flammable gases	− indicate the identity of the emitted gas, if known − indicate whether the emitted gas ignites spontaneously − indicate the gas evolution rate, generally determined by Test N.5 (section 33.4.1.4 of the *Manual of Tests and Criteria*), unless the test has not been completed e.g. because the gas ignites spontaneously

表 A4.3.9.2：物理化学的危険性クラスに関連するデータ（補遺）*(続き)*

章	危険性クラス	性質/安全特性/試験結果および注釈/手引き
2.4	酸化性ガス	単一の酸化性ガス： − ISO 10156 による C_i（酸素当量係数）を示す 酸化性混合ガス − 試験された混合ガスには「酸化性ガス区分1（ISO 10156 による試験）」または ISO 10156 により計算された酸化力（OP）を示す
2.5	高圧ガス	単一のガス： − 臨界温度を示す 混合ガス：： − 擬臨界温度を示す；これは成分の臨界温度の分子加重平均として以下のように求められる： $\sum_{i=1}^{n} X_i \cdot T_{Crit\ i}$ ここで ・x_i は成分 i の分子比率 ・$T_{Crit\ i}$ は成分 i の臨界温度
2.6	引火性液体	− 沸点および引火点は表 A4.3.9.1 に示されているので、追加的なデータは必要ない − 第 2.6 章 2.6.2 の注記 2 にしたがい試験 L.2（*試験方法及び判定基準のマニュアルの 32.5.2 節*）に基づいた除外を考慮する場合には、持続的な燃焼性についての情報を示す
2.7	可燃性固体	− 燃焼速度（または金属粉の燃焼時間）を示す、通常、試験 N.1 で測定する（*試験方法及び判定基準のマニュアルの 33.2.1 節*） − 湿潤部分を超えたかどうかを示す
2.8	自己反応性物質および混合物	− SADT（自己加速分解温度）について、表 A4.3.9.1 の分解温度の項目を参照 − 分解エネルギーを示す（値および測定方法） − 爆発の性質を示す（はい/部分的/いいえ）、関連する包装も − 爆燃の性質を示す（はい急速に/はいゆっくりと/いいえ）、関連する包装も − 密閉状態での熱の影響を示す（激しく/中くらい/低い/いいえ）、関連する包装も − 該当すれば爆発力を示す（低くない/低い/ない）
2.9	自然発火性液体	− 自然発火またはろ紙を黒く焦がすかどうか示す、通常、試験 N.3 で測定する（*試験方法及び判定基準のマニュアルの 33.3.1.5 節*）（例えば「空気中で液体が自然発火する」または「空気中で液体がろ紙を黒く焦がす」）
2.10	自然発火性固体	− 注いでいる時またはそれから 5 分以内に自然発火が起きるかどうか示す、通常、試験 N.2 で測定する（*試験方法及び判定基準のマニュアルの 33.3.1.4 節*）（例えば「固体は空気中で自然発火する」） − 時間の経過とともに自然発火性が変わるかどうかを示す、例えばゆっくりした酸化によって保護面が形成される
2.11	自己発熱性物質および混合物	− 自然発火が起きるかどうか示す、スクリーニングデータおよび/または使用された方法（通常、試験 N.4、*試験方法及び判定基準のマニュアルの 33.3.1.6 節*）を含め、得られた最大温度上昇を記す − もし適切で、入手可能であれば、第 2.11 章、2.11.4.2 に従ったスクリーニングテストの結果を示す
2.12	水反応可燃性物質および混合物	− もし知られていれば、発生するガスを特定する − 発生したガスが自然に発火するかどうかを示す − 例えばガスが自然発火するなど試験が完結しない場合を除いて、ガス発生速度を示す、通常、試験 N.5（*試験方法及び判定基準のマニュアルの 33.4.1.4 節*）で測定する

Table A4.3.9.2: Data relevant with regard to physical hazard classes (supplemental) *(cont'd)*

Chapter	Hazard class	Property/Safety characteristic/Test result and Remarks/Guidance
2.13	Oxidizing liquids	– indicate whether spontaneous ignition occurs when mixed with cellulose, generally determined by Test O.2 (section 34.4.2 of the *Manual of Tests and Criteria*) (e.g. "the mixture with cellulose (prepared for Test O.2) ignites spontaneously")
2.14	Oxidizing solids	– indicate whether spontaneous ignition occurs when mixed with cellulose, generally determined by Test O.1 or Test O.3 (sections 34.4.1 or 34.4.3 of the *Manual of Tests and Criteria*) (e.g. "the mixture with cellulose (prepared for Test O.1 or O.3) ignites spontaneously")
2.15	Organic peroxides	– for the SADT (self-accelerating decomposition temperature) see the entry for the decomposition energy in table A4.3.9.1 – indicate the decomposition energy (value and method of determination), if available – indicate detonation properties (Yes/Partial/No), also in packaging where relevant – indicate deflagration properties (Yes rapidly/Yes slowly/No), also in packaging where relevant – indicate the effect of heating under confinement (Violent/Medium/Low/No), also in packaging where relevant – indicate the explosive power if applicable (Not low/Low/None)
2.16	Corrosive to metals	– indicate which metals are corroded by the substance or mixture (e.g. "corrosive to aluminium" or "corrosive to steel" etc.), if available – indicate the corrosion rate and whether it refers to steel or aluminium, generally determined by Test C.1 (section 37.4 of the *Manual of Tests and Criteria*), if available – include a reference to other sections of the SDS with regard to compatible or incompatible materials (e.g. to packaging compatibilities in section 7 or to incompatible materials in section 10), as appropriate
2.17	Desensitized explosives	– indicate what desensitizing agent is used – indicate the exothermic decomposition energy – indicate the corrected burning rate A_c

表 A4.3.9.2：物理化学的危険性クラスに関連するデータ（補遺）*(続き)*

章	危険性クラス	性質/安全特性/試験結果および注釈/手引き
2.13	酸化性液体	− セルロースと混ぜて自然発火が起きるかどうか示す、通常、試験 O.2（*試験方法及び判定基準のマニュアルの 34.4.2 節*）で測定する（例えば「セルロースとの混合物（試験 O.2 用に調整）は自然発火する」）
2.14	酸化性固体	− セルロースと混ぜて自然発火が起きるかどうか示す、通常、試験 O.1 または試験 O.3（*試験方法及び判定基準のマニュアルの 34.4.1 節または 34.4.3 節*）で測定する（例えば「セルロースとの混合物（試験 O.1 または試験 O.3 用に調整）は自然発火する」）
2.15	有機過酸化物	− SADT（自己加速分解温度）について、表 A4.3.9.1 の分解エネルギーの項目を参照 − 入手可能であれば分解エネルギーを示す（値および測定方法） − 爆発の性質を示す（はい/部分的/いいえ）、関連すれば包装も − 爆燃の性質を示す（はい急速に/はいゆっくりと/いいえ）、関連すれば包装も − 密閉状態での熱の影響を示す（激しく/中くらい/いいえ）、関連すれば包装も − 可能であれば爆発力を示す（低くない/低い/ない）
2.16	金属腐食性	− 入手可能であれば、物質または混合物でどの金属が腐食したかを示す（例えば「アルミニウムに腐食性」または「鋼に腐食性」など） − 入手可能であれば、腐食速度およびそれが鋼またはアルミニウムのどちらに言及しているかを示す、通常、試験 C.1 で測定する（*試験方法及び判定基準のマニュアルの 37.4 節*） − 適切に、同梱可能または混触危険に関する SDS の他節の参照を含める（例えば 7 節の同梱可能または 10 節の混触危険物質）
2.17	鈍性化爆発物	− どのような鈍性化剤が使われているか示す − 発熱分解エネルギーを示す − 補正燃焼速度 A_c を示す

Table A4.3.9.3: Further safety characteristics (supplemental)

This table lists further properties/safety characteristics and test results that are not required on the SDS but may be useful to communicate for a substance or mixture. Other physical properties/safety characteristics of the substance or mixture not identified in this table may also be useful to communicate. Include all relevant information, such as a short description, value(s), unit, conditions (e.g. temperature, pressure), method, each as appropriate.

Safety characteristic and/or test result	Remarks/Guidance
Mechanical sensitivity	− applicable to energetic substances and mixtures with an exothermic decomposition energy ≥ 500 J/g in accordance with the *Manual of Tests and Criteria*, appendix 6, section 3.3 (c) − indicate the sensitiveness to impact, generally determined by test 3 (a) (section 13.4 of the *Manual of Tests and Criteria*) (indicate preferably the limiting impact energy) − indicate the sensitiveness to friction, generally determined by test 3 (b) (section 13.5 of the *Manual of Tests and Criteria*) (indicate preferably the limiting load)
SAPT (self-accelerating polymerization temperature)	− applicable to substances and mixtures which may self-polymerize thereby generating dangerous amounts of heat and gas or vapour − indicate the volume for which the SAPT is given

Formation of explosible dust/air mixtures	− not applicable to gases and liquids
	− not applicable to solids containing only substances which are fully oxidized (e.g. silicon dioxide)
	− in case formation of explosible dust/air mixtures might be possible based on section 2 of the SDS, relevant safety characteristics may be indicated in addition, such as
	  • lower explosion limit / minimum explosible concentration
	  • minimum ignition energy
	  • deflagration index (K_{st})
	  • maximum explosion pressure
	− indicate the particle characteristics to which the data apply if different from the particle characteristics as indicated based on table A4.3.9.1
	
	NOTE 1: *The ability to form explosible dust/air mixtures may be determined e.g. by VDI 2263-1 "Dust Fires and Dust Explosions; Hazards - Assessment - Protective Measures; Test Methods for the Determination of the Safety Characteristics of Dusts" or by ISO/IEC 80079-20-2 "Explosive atmospheres - Part 20-2: Material characteristics - Combustible dusts test methods" (in preparation).*
	
	NOTE 2: *Explosion characteristics are specific for the tested dust. Normally they cannot be transferred to other dusts even if these are comparable. Fine-sized dusts of a particular substance tend to react stronger than coarser dusts.*

Acid/alkaline reserve	− applicable to substances and mixtures which have an extreme pH (pH ≤ 2 or ≥ 11.5) − indicate acid/alkaline reserve when used for evaluating skin and eye hazards

表 A4.3.9.3：さらなる安全特性（補足）

　この表では、物質または混合物に関して SDS では要求されていないものの情報伝達が有用であろう、さらなる性質/安全特性および試験結果を示している。この表では特定されていない他の物質または混合物の物理化学的性質/安全特性も情報伝達が有用であろう。短い記述、値、単位（例えば温度、圧力）、方法などそれぞれ適切にすべての関連情報を含める。

安全特性および/または試験結果	注釈/手引き
機械的感度	− （試験方法及び判定基準のマニュアル、付録 6、3.3 節 (c)）に従って≥ 500 J/g の発熱分解エネルギーを持った物質または混合物が該当する − 衝撃に対する感度を示す、通常、試験 3 (a)によって測定される（*試験方法及び判定基準のマニュアルの 13.4 節*）（限界衝撃エネルギーを示すことがのぞましい） − 摩擦に対する感度を示す、通常、試験 3 (b)によって測定される（*試験方法及び判定基準のマニュアルの 13.5 節*）（限界負荷を示すことがのぞましい）
SAPT（自己加速重合温度）	− 自己重合し、それによって危険な量の熱およびガスまたは蒸気を発生する物質や混合物が該当する − SAPT が与えられている容量を示す
爆発性粉じん/空気混合の形成	− ガスおよび液体は該当しない − 完全に酸化された物質（例えば二酸化ケイ素）のみ含む固体は該当しない − SDS の 2 節に基づき、爆発性の粉じん/空気混合物が形成される可能性があるとき、関連する安全特性は追加的に以下のようなものであろう 　• 爆発下限/爆発最少濃度 　• 最少発火エネルギー 　• 爆燃指数（K_{st}） 　• 最大爆発圧力 − もし表 A4.3.9.1 に基づいて示された粒子特性と異なる場合には、データを適用する粒子特性を示す *注記 1：爆発性粉じん/空気混合物を形成する能力は、例えば VDI* 2263-1「粉じん火災および粉じん爆発；危険−評価−保護対策；粉じんの安全特性の測定に関する試験方法」または ISO/IEC80079-20-2「爆発雰囲気−第 20.2 部：物質の特性−燃焼粉じん試験法」（準備中）* *注記 2：爆発特性は試験された粉じんに特有のものである。通常それらは、たとえ類似のものであっても、他の粉じんに拡大することはできない。特別な物質の微細な粉じんは粗い粉じんよりも強く反応する傾向がある。*
酸/アルカリ予備	− 極端な pH（pH ≤2 または≥11.5）をもつ物質や混合物が該当する − 皮膚および眼への有害性の評価を行う場合には、酸/アルカリ予備を示す

A4.3.10 **SECTION 10: Stability and reactivity**

A4.3.10.1   *Reactivity*

A4.3.10.1.1   Describe the reactivity hazards of the substance or mixture in this section. Provide specific test data for the substance or mixture as a whole, where available. However, the information may also be based on general data for the class or family of chemical if such data adequately represent the anticipated hazard of the substance or mixture.

A4.3.10.1.2   If data for mixtures are not available, ingredient data should be provided. In determining incompatibility, consider the substances, containers and contaminants that the substance or mixture might be exposed to during transportation, storage and use.

A4.3.10.2   *Chemical stability*

Indicate if the substance or mixture is stable or unstable under normal ambient and anticipated storage and handling conditions of temperature and pressure. Describe any stabilizers which are, or may need to be, used to maintain the product. Indicate the safety significance of any change in the physical appearance of the product.

A4.3.10.3   *Possibility of hazardous reactions*

If relevant, state if the substance or mixture will react or polymerize, releasing excess pressure or heat, or creating other hazardous conditions. Describe under what conditions the hazardous reactions may occur.

A4.3.10.4   *Conditions to avoid*

List conditions such as heat, pressure, shock, static discharge, vibrations or other physical stresses that might result in a hazardous situation.

A4.3.10.5   *Incompatible materials*

List classes of chemicals or specific substances with which the substance or mixture could react to produce a hazardous situation (e.g. explosion, release of toxic or flammable materials, liberation of excessive heat).

A4.3.10.6   *Hazardous decomposition products*

List known and reasonably anticipated hazardous decomposition products produced as a result of use, storage and heating. Hazardous combustion products should be included in section 5 (Fire-fighting measures) of the SDS.

A4.3.11 **SECTION 11: Toxicological information**

A4.3.11.1   This section is used primarily by medical professionals, occupational health and safety professionals and toxicologists. A concise but complete and comprehensible description of the various toxicological (health) effects, and the available data used to identify those effects, should be provided. Under GHS classification, the relevant hazards, for which data should be provided, are:

(a) acute toxicity;

(b) skin corrosion/irritation;

(c) serious eye damage/irritation;

(d) respiratory or skin sensitization;

(e) germ cell mutagenicity;

(f) carcinogenicity;

(g) reproductive toxicity;

(h) STOT-single exposure;

A4.3.10　第10節：安定性および反応性

A4.3.10.1　反応性

A4.3.10.1.1 本節では物質または混合物の反応性に関する危険性について記載する。可能な場合、全体として、物質または混合物についての特定の試験データを明記する。しかしながらそのデータが物質または混合物の予想される危険を適切に示す場合には、その情報もまた化学品のクラスまたはグループに関する一般データに基づいてもよい。

A4.3.10.1.2　混合物についてのデータが入手できない場合には、成分のデータが提供されるべきである。混触禁止の判定にあたっては物質、格納容器および物質または混合物が輸送、保管、使用の途中のばく露可能性がある不純物を考慮に入れること。

A4.3.10.2　化学的安定性

物質または混合物が標準大気および予測される保管および取扱いの温度と圧力条件下で安定か不安定かを示すこと。その製品を維持するために使用される、またはその必要がある安定剤を記述する。その製品の物理化学的外観におけるあらゆる変化に関する安全性の重要性を示す。

A4.3.10.3　危険有害反応可能性

該当する場合には物質または混合物が反応または重合して、過剰な圧力または熱を放出する、または危険有害な状態になるかを記載すること。いかなる条件下でその危険有害反応が起こりうるかを記載すること。

A4.3.10.4　避けるべき条件

危険有害な状況を招く可能性のある熱、圧力、衝撃、静電放電、振動または他の物理的応力などの諸条件を示すこと。

A4.3.10.5　混触禁止物質

物質または混合物と一緒に反応を起こして有害な状況（例えば爆発、有毒ガスまたは可燃性物質の放出、極度な放熱）を起こす物質または特定の物質の種類を示すこと。

A4.3.10.6　有害な分解生成物

使用、保管、加熱の結果生じる既知の合理的に予測可能な有害な分解生成物を示すこと。有害な分解生成物は、SDSの第5節（火災時の措置）に含まれるべきである。

A4.3.11　第11節：有害性情報

A4.3.11.1　本節は主として医学の専門家、産業衛生・安全の専門家、および毒物研究者によって使用される。さまざまな毒物学的（健康）影響についての簡潔で完結した分かりやすい説明とその影響を特定するために利用したデータが提供されるべきである。GHS分類においてデータを提供するべき関係する有害性は以下のとおりである：

　　(a)　急性毒性；
　　(b)　皮膚腐食性/刺激性；
　　(c)　眼に対する重篤な損傷/刺激性；
　　(d)　呼吸器または皮膚感作性；
　　(e)　生殖細胞変異原性；
　　(f)　発がん性；
　　(g)　生殖毒性；
　　(h)　特定標的臓器毒性－単回ばく露；

(i) STOT-repeated exposure; and

(j) aspiration hazard.

These hazards should always be listed on the SDS.

A4.3.11.2 The health effects included in the SDS should be consistent with those described in the studies used for the classification of the substance or mixture.

A4.3.11.3 Where there is a substantial amount of test data on the substance or mixture, it may be desirable to summarize results e.g. by route of exposure (see A4.3.11.1).

A4.3.11.4 The data included in this subsection should apply to the substance or mixture as used. The toxicological data should describe the mixture. If that information is not available, the classification under GHS and the toxicological properties of the hazardous ingredients should be provided.

A4.3.11.5 General statements such as "Toxic" with no supporting data or "Safe if properly used" are not acceptable as they may be misleading and do not provide a description of health effects. Phrases such as "not applicable", "not relevant", or leaving blank spaces in the health effects section can lead to confusion and misunderstanding and should not be used. For health effects where information is not available, this should be clearly stated. Health effects should be described accurately and relevant distinctions made. For example, allergic contact dermatitis and irritant contact dermatitis should be distinguished from each other.

A4.3.11.6 If data for any of these hazards are not available, they should still be listed on the SDS with a statement that data are not available. Also provide information on the relevant negative data (see A4.2.2.3). If data are available showing that the substance or mixture does not meet the criteria for classification, it should be stated on the SDS that the substance or mixture has been evaluated and based on available data, does not meet the classification criteria. Additionally, if a substance or mixture is found to be not classified for other reasons, for example, due to technical impossibility to obtain data, or inconclusive data, this should be clearly stated on the SDS.

A4.3.11.7 *Information on the likely routes of exposure*

Provide information on the likely routes of exposure and the effects of the substance or mixture via each possible route of exposure, that is, through ingestion (swallowing), inhalation or skin/eye exposure. A statement should be made if health effects are not known.

A4.3.11.8 *Symptoms related to the physical, chemical and toxicological characteristics*

Describe the potential adverse health effects and symptoms associated with exposure to the substance or mixture and its ingredients or known by-products. Provide information on the symptoms related to the physical, chemical and toxicological characteristics of the substance or mixture following exposure related to the intended uses. Describe the first symptoms at the lowest exposures through to the consequences of severe exposure; for example, "headaches and dizziness may occur, proceeding to fainting or unconsciousness; large doses may result in coma and death".

A4.3.11.9 *Delayed and immediate effects and also chronic effects from short- and long-term exposure*

Provide information on whether delayed or immediate effects can be expected after short- or long-term exposure. Also provide information on acute and chronic health effects relating to human exposure to the substance or mixture. Where human data are not available, animal data should be summarised, and the species clearly identified. It should be indicated in the SDS whether toxicological data is based on human or animal data.

A4.3.11.10 *Numerical measures of toxicity (such as acute toxicity estimates)*

Provide information on the dose, concentration or conditions of exposure that may cause adverse health effects. Where appropriate, doses should be linked to symptoms and effects, including the period of exposure likely to cause harm.

A4.3.11.11 *Interactive effects*

Information on interactions should be included if relevant and readily available.

(i)　特定標的臓器毒性－反復ばく露；および

   (j)　誤えん有害性。

　これらの危険有害性は常にSDSに記載するべきである。

A4.3.11.2　SDSの中に記載されている健康影響は物質または混合物の分類について使用された諸研究で記述されたものと整合しているべきである。

A4.3.11.3　物質または混合物に関してかなりの量の試験データがある場合には、例えばばく露経路ごとに結果をまとめることが望ましい。(A4.3.11.1.参照)。

A4.3.11.4　本節のデータは使用される物質または混合物に適用されるべきである。毒性データは混合物としての毒性記載とするべきである。その情報が利用できない場合には、有害な成分の毒性やGHS分類が提供されるべきである。

A4.3.11.5　支持するデータなしに「毒性がある」とか「正しく使用すれば安全である」などという一般的な説明は、誤解を招き、健康影響の説明をしていないため、適切でない。「あてはまらない」、「関係がない」といった表現または健康影響の記入欄を空欄にしておくと混乱と誤解を招くので使用するべきでない。情報がない場合、健康影響にはその旨を明確に記載するべきである。健康影響は正確に説明されかつ関連する事項との違いを説明するべきである。例えばアレルギー性接触性皮膚炎と刺激性接触皮膚炎はお互いに区別されるべきである。

A4.3.11.6　これらの危険有害性でデータが入手できないものがある場合、データが入手できない旨の説明を付してSDSに示されるべきである。関連する否定的データもまた情報提供すること（A4.2.2.3参照）。物質あるいは混合物が分類判定基準に合致しないことを示すデータがある場合には、入手できるデータに基づいて評価された物質あるいは混合物は判定基準に合致しないということをSDSに記載するべきである。さらに物質あるいは混合物が他の理由により分類されなかった場合には、例えばデータ採取が技術的に不可能である、あるいはデータが確定的でないなど、このこともSDSにはっきりと記載するべきである。

A4.3.11.7　*可能性のあるばく露経路の情報*

　可能性のあるそれぞれのばく露経路、すなわち経口摂取（飲み込み）、吸入または皮膚/眼のばく露を通じた物質または混合物のばく露と影響についての情報を提供する。この健康影響が知られていない場合にはその旨を記載するべきである。

A4.3.11.8　*物理的、化学的および毒物学的特性に関連する症状*

　物質または混合物とその成分または既知の副生物に対するばく露に関連する潜在的な健康への悪影響と症状を記載すること。意図する用途に関連したばく露による、物質または混合物の物理的、化学的および毒物学的な特性に関連する症状についての情報提供を行うこと。最低レベルの初期症状から重度のばく露結果までを記載すべきである。例えば、「失神または意識不明への進行に至る前に頭痛とめまいが起こることがある；重度のばく露により昏睡または死に至ることがある」とする。

A4.3.11.9　*短期および長期ばく露による遅発的・即時的影響ならびに慢性的影響*

　短期および長期ばく露の後に遅発または即時的影響が予測できるかどうかについての情報を提供すること。物質または混合物への人へのばく露に関連する急性および慢性の健康影響についても情報提供すること。人のデータが入手できない場合には、動物のデータを要約すべきで、その際には動物種を明示するべきである。SDSには毒性学的データが人によるものか動物によるものかを示すべきである。

A4.3.11.10　*毒性の数値化(急性毒性の推定など)*

　傷害を及ぼすおそれのある用量、濃度およびばく露条件について情報を提供すること。可能であれば、悪影響を及ぼすおそれのあるばく露期間も含め、用量と症状・影響との関連づけを行うべきである。

A4.3.11.11　*相互作用*

　関連性が認められかつ速やかに入手できる場合には、相互作用についての情報を含めるべきである。

A4.3.11.12   *Where specific chemical data are not available*

It may not always be possible to obtain information on the hazards of a substance or mixture. In cases where data on the specific substance or mixture are not available, data on the chemical class, if appropriate, may be used. Where generic data are used or where data are not available, this should be stated clearly in the SDS.

A4.3.11.13   *Mixtures*

If a mixture has not been tested for its health effects as a whole then information on each ingredient listed under A4.3.3.2.1 should be provided and the mixture should be classified using the processes that are described in the GHS (section 1.3.2.3 and subsequent chapters).

A4.3.11.14   *Mixture versus ingredient information*

A4.3.11.14.1   Ingredients may interact with each other in the body resulting in different rates of absorption, metabolism and excretion. As a result, the toxic actions may be altered, and the overall toxicity of the mixture may be different from its ingredients.

A4.3.11.14.2   It is necessary to consider whether the concentration of each ingredient is sufficient to contribute to the overall health effects of the mixture. The information on toxic effects should be presented for each ingredient, except:

    (a)   if the information is duplicated, it is not necessary to list this more than once. For example, if two ingredients both cause vomiting and diarrhoea, it is not necessary to list this twice. Overall, the mixture is described as causing vomiting and diarrhoea;

    (b)   if it is unlikely that these effects will occur at the concentrations present. For example, when a mild irritant is diluted in a non-irritating solution, there comes a point where the overall mixture would be unlikely to cause irritation;

    (c)   Predicting the interactions between ingredients is extremely difficult, and where information on interactions is not available, assumptions should not be made and instead the health effects of each ingredient should be listed separately.

A4.3.11.15   *Other information*

Other relevant information on adverse health effects should be included even when not required by the GHS classification criteria.

A4.3.12   **SECTION 12: Ecological information**

A4.3.12.1   The information that shall be provided in this section is to enable evaluation of the environmental impact of the substance or mixture if it were released to the environment. This information can assist in handling spills, and evaluating waste treatment practices, control of release, accidental release measures, and transport.

A4.3.12.2   A concise but complete and comprehensible description of the various ecotoxicological (environment) properties, and the available data used to identify those properties, should be provided. The basic properties, for which data should be provided, are:

    (a)   Toxicity;

    (b)   Persistence and degradability;

    (c)   Bioaccumulative potential;

    (d)   Mobility in soil;

    (e)   Other adverse effects.

These properties should always be listed on the SDS. Species, media, units, test duration and test conditions should be clearly indicated. (If data for any of these properties are not available, they should still be listed on the SDS with a statement that data are not available).

A4.3.11.12 *特定の化学的データがない場合*
　物質または混合物の危険有害性についての情報は必ずしも入手できるとは限らない。特定の物質または混合物についてのデータが入手できない場合には必要に応じてその物質の同類のデータを用いてもよい。一般的なデータが使用されるかデータが利用できない場合には、その旨をSDSに明記すべきである。

A4.3.11.13 *混合物*
　混合物全体として健康影響について試験されていない場合には、A4.3.3.2.1において示されている各々の成分についての情報を提供すべきであり、混合物は、GHSにおいて記述されている方法を使用し分類されるべきである。（第1.3.2.3節およびその後の章）

A4.3.11.14 *混合物対成分情報*

A4.3.11.14.　各成分は体内において互いに反応を起こし吸収、代謝および排泄の速度を変えることがある。その結果毒性作用に変化が生じ混合物の総合的な毒性がその成分と異なる可能性がある。

A4.3.11.14.2　各成分の濃度がその混合物に由来する総合的健康影響に対して充分な原因となっているかどうかを考察する必要がある。以下の場合を除いて、毒性影響の情報を、各成分について示すべきである。

　　(a)　情報に重複がある場合には、繰り返し記載する必要はない。例えば2つの成分がともに嘔吐と下痢を引き起こす場合には二度記載する必要はない。総合的に見てその混合物が嘔吐と下痢を起こすとして記載される。

　　(b)　考えている濃度でこれら影響が起こりそうにない場合。例えば、弱い刺激性物質が非刺激性溶液中に希釈される場合にその混合物全体が刺激を起こさないことがある。

　　(c)　成分間で生じる相互作用を予測するのは極めて難しく、相互作用についての情報が利用できない場合には仮定をすべきではなく、それに代えて各成分の健康影響を個別に示すべきである。

A4.3.11.15 *その他の情報*
　GHSの分類基準により要求されない場合でも有害な健康影響についての他の関連情報を含めるべきである。

A4.3.12　*第12節：環境影響情報*

A4.3.12.1　この節で提供すべき情報は、物質または混合物が、環境に放出される場合に環境に及ぼす影響の評価を可能にするものである。この情報は漏洩時の取扱い、廃棄物処理方法、放出の管理、事故流出時対策および輸送に役立つ。

A4.3.12.2　多様な環境毒物学的（環境）性質に関するまとまった包括的な記述およびそれらの性質の同定に使用されたデータが提供されるべきである。データを示すべき基本的な性質には以下のものがある：

　　(a)　毒性；
　　(b)　残留性と分解性；
　　(c)　生物蓄積性；
　　(d)　土壌中の移動度；
　　(e)　他の有害影響。

　これらの性質は常にSDSに記載するべきである。生物種、媒体、装置、試験継続期間及び試験条件を明記するべきである。（これら危険有害性のすべてのデータが入手できない場合、データが入手できない旨の説明を付してSDSに示されるべきである。）

A4.3.12.3   Some ecotoxicological properties are substance specific, i.e. bioaccumulation, persistence and degradability. The information should therefore be given, where available and appropriate, for each relevant ingredient of the mixture (i.e. those which are required to be listed in section 3 of the SDS).

A4.3.12.4   Provide also a short summary of the data given under A4.3.12.5 to A4.3.12.9 in relation to the hazard classification criteria. Where data are not available for classification, this should be clearly stated on the SDS for each basic property concerned. Additionally, if data are available showing that the substance or mixture does not meet the criteria for classification, it should be stated on the SDS that the substance or mixture has been evaluated and, based on available data, does not meet the classification criteria. Additionally, if a substance or mixture is found to be not classified for other reasons, for example, due to technical impossibility to obtain the data, or inconclusive data, this should be clearly stated on the SDS.

A4.3.12.5   *Toxicity*

Information on toxicity can be provided using data from tests performed on aquatic and/or terrestrial organisms. This should include relevant available data on both acute and chronic aquatic toxicity for fish, crustaceans, algae and other aquatic plants. In addition, toxicity data on other organisms (including soil micro-and macro-organisms) such as birds, bees and plants, should be included when available. Where the substance or mixture has inhibitory effects on the activity of micro-organisms, the possible impact on sewage treatment plants should be mentioned.

A4.3.12.6   *Persistence and degradability*

Persistence and degradability is the potential for the substance or the appropriate constituents of a mixture to degrade in the environment, either through biodegradation or other processes, such as oxidation or hydrolysis. Test results relevant to assess persistence and degradability should be given where available. If degradation half-lives are quoted it must be indicated whether these half-lives refer to mineralization or to primary degradation. The potential of the substance or certain constituents (see also A4.3.12.8) of a mixture to degrade in sewage treatment plants should also be mentioned.

A4.3.12.7   *Bioaccumulative potential*

Bioaccumulation is the potential for the substance or certain constituents of a mixture to accumulate in biota and, possibly, pass through the food chain. Test results relevant to assess the bioaccumulative potential should be given. This should include reference to the octanol-water partition coefficient (K_{ow}) and bioconcentration factor (BCF), if available.

A4.3.12.8   *Mobility in soil*

Mobility in soil is the potential of a substance or the constituents of a mixture, if released to the environment, to move under natural forces to the groundwater or to a distance from the site of release. The potential for mobility in soil should be given where available. Information on mobility can be determined from relevant mobility data such as adsorption studies or leaching studies. For example, K_{oc} values can be predicted from octanol/water partition coefficients (K_{ow}). Leaching and mobility can be predicted from models.

NOTE: *Where real data on the substance or mixture is available this data will take precedence over models and predictions.*

A4.3.12.9   *Other adverse effects*

Information on any other adverse effects to the environment should be included where available, such as environmental fate (exposure), ozone depletion potential, photochemical ozone creation potential, endocrine disrupting potential and/or global warming potential.

A4.3.13   **SECTION 13: Disposal considerations**

A4.3.13.1   *Disposal methods*

A4.3.13.1.1   Provide information for proper disposal, recycling or reclamation of the substance or mixture and/or its container to assist in the determination of safe and environmentally preferred waste management options, consistent with the requirements of the national competent authority. For the safety of persons conducting disposal, recycling or reclamation activities, please refer to the information in section 8 (exposure controls and personal protection) of the SDS.

A4.3.12.3　生物蓄積性、残留性および分解性など、いくつかの生態毒性をあらわす特性は物質に特異的である。入手可能で適切である場合には、混合物に含まれる該当する各物質（すなわち SDS の第 3 節に記載が要求されているもの）について情報を提供するべきである。

A4.3.12.4　危険有害性分類判定基準に関連して A4.3.12.5 から A4.3.12.9 にあるデータの概要も記載するべきである。これら分類のためのデータが入手できない時は、それぞれ該当する性質についてその旨を SDS に記載するべきである。物質あるいは混合物が分類判定基準に合致しないことを示すデータがある場合には、入手できるデータに基づいて評価された物質あるいは混合物は判定基準に合致しないということを SDS に記載するべきである。さらに物質あるいは混合物が他の理由により分類されなかった場合には、例えばデータ採取が技術的に不可能である、あるいはデータが確定的でないなど、このことも SDS にはっきりと記載するべきである。

A4.3.12.5　*毒性*
　毒性情報は水中または陸上の生物の試験データを用いて提供できる。これには魚類、甲殻類、藻類および他の水生植物についての急性および慢性の両者の関連する利用可能なデータを含むべきである。その上に入手可能であれば、鳥類、ハチ類、植物種などその他の生物（土壌中に生息する微小・大型生物）の毒性データを含むべきである。物質または混合物が微生物の生命に対して阻害作用がある場合には下水処理場に及ぼす影響の可能性に言及すべきである。

A4.3.12.6　*残留性と分解性*
　残留性と分解性は、物質または混合物の特有の成分が、例えば酸化または加水分解といった生分解の過程または他の過程のいずれかを経て、環境中において分解する性質である。分解半減期が引用される場合、これらの半減期が無機化または一次分解についてのものかどうか示されなければならない。物質または混合物の特有の成分が下水処理場で分解する可能性（A4.3.12.8 参照）にも言及すべきである。

A4.3.12.7　*生物蓄積性*
　生物蓄積性は、物質または混合物の特定の成分が生物相中に濃縮し最終的に食物連鎖を通り抜ける性質である。生物蓄積性を評価する適切な試験結果を示すべきである。利用可能である場合には、これにはオクタノール/水分配係数（Kow）と生物濃縮係数（BCF）についての参照を含めるべきである。

A4.3.12.8　*土壌中の移動性*
　土壌中の移動性は、環境に放出された場合に、物質または混合物の成分が、自然力により地下水に、または放出場所から離れた場所に移動する性質である。入手可能な場合は、土壌中の移動性について示すべきである。移動性の情報は、吸着試験や浸出試験のような適正な移動性データで決定できる。例えば Koc 値はオクタノール/水分配係数（Kow）から予測できる。浸出および移動性はモデルで予測できる。

注記：物質または混合物の真のデータが入手できる場合は、これがモデルと予測に優先する。

A4.3.12.9　*他の有害影響*
　環境に対するその他の有害影響についての情報が利用可能な場合には含めるべきである。これに該当するものには環境運命（ばく露）、オゾン層破壊の可能性、光化学的オゾン発生の可能性、内分泌かく乱の可能性または地球温暖化の可能性などがある。

A4.3.13　*第 13 節：廃棄上の注意*

A4.3.13.1　*廃棄方法*

A4.3.13.1.1　国の所管官庁の要求事項と整合性を保ちながら安全で環境的に望ましい廃棄物管理の選択肢を決定するために、物質または混合物もしくはその廃棄物用容器の適切な廃棄、リサイクルまたは埋立てについての情報の提供を行うこと。廃棄、リサイクルまたは埋立てに関わる人の安全については SDS の第 8 節（ばく露制御および保護措置）の情報を参照すること。

A4.3.13.1.2   Specify disposal containers and methods.

A4.3.13.1.3   Discuss physical/chemical properties that may affect disposal options.

A4.3.13.1.4   Discourage sewage disposal.

A4.3.13.1.5   Where appropriate, identify any special precautions for incineration or landfill.

A4.3.14   *SECTION 14: Transport information*

This section provides basic classification information for the transporting/shipment of a hazardous substance or mixture by road, rail, sea or air. Where information is not available or relevant this should be stated.

A4.3.14.1   *UN Number*

Provide the UN Number (i.e. four-figure identification number of the substance or article) from the *UN Model Regulations* [4].

A4.3.14.2   *UN Proper Shipping Name*

Provide the UN proper shipping name from the *UN Model Regulations*. For substances or mixtures the UN proper shipping name should be provided in this subsection if it has not appeared as the GHS product identifier or national or regional identifiers.

A4.3.14.3   *Transport hazard class(es)*

Provide the transport class (and subsidiary risks) assigned to the substances or mixtures according to the most predominant hazard that they present in accordance with the *UN Model Regulations*.

A4.3.14.4   *Packing group, if applicable*

Provide the packing group number from the UN Model Regulations, if applicable. The packing group number is assigned to certain substances in accordance with their degree of hazard.

A4.3.14.5   *Environmental hazards*

Indicate whether the substance or mixture is a known marine pollutant according to the IMDG Code, and if so, whether it is a "marine pollutant" or a "severe marine pollutant". Also indicate whether the substance or mixture is environmentally hazardous according to the *UN Model Regulations*, ADR, RID and ADN.

A4.3.14.6   *Special precautions for user*

Provide information on any special precautions, which a user needs to be aware of, or needs to comply with in connection with transport.

A4.3.14.7   *Transport in bulk according to IMO instruments*

This subsection only applies when cargoes are intended to be carried in bulk according to IMO instruments: e.g. chapter VI or VII of SOLAS Annex II or Annex V of MARPOL, the IBC Code, the IMSBC Code and the IGC Code (or earlier versions, i.e.: EGC Code or GC Code).

For liquid bulk cargoes, provide the product name (if name is different to that given in A4.3.1.1) as required by the shipment document and in accordance with the name used in the lists of product names given in chapters 17 or 18 of the IBC Code or the latest edition of the IMO's MEPC.2/Circular. Indicate ship type required and pollution category.

For solid bulk cargoes, provide the bulk cargo shipping name, whether or not the cargo is considered harmful to the marine environment (HME) according to MARPOL Annex V, whether it is a material hazardous only in bulk (MHB) according to the IMSBC Code, and which group it should be shipped according to the IMSBC.

For liquefied gas cargoes in bulk provide the product name and ship type according to the IGC Code (or earlier versions, i.e.: EGC Code or GC Code).

A4.3.13.1.2 廃棄物用容器と廃棄方法を特定すること。

A4.3.13.1.3 廃棄方法に影響を及ぼす可能性のある物理的/化学的特性について議論すること。

A4.3.13.1.4 下水への廃棄は推奨しないこと。

A4.3.13.1.5 該当する場合には、焼却または埋立てに関する特別な注意事項を示すこと。

A4.3.14　第14節：輸送上の注意

本節では、物質または混合物の陸上、鉄道、海上および航空による輸送/出荷ための基本的な分類情報を提供する。情報が入手できないか該当しない場合にはその旨を記載すべきである。

A4.3.14.1　*国連番号*
UN モデル規則による国連番号（すなわち、物質または物品の 4 桁の番号）を提供すること。

A4.3.14.2　*国連出荷名*
UN モデル規則による国連出荷正式名を提供すること。物質または混合物について、国連による適切な出荷名は GHS の製品特定名または国または地域の特定名として表されない場合、この節で示すべきである。

A4.3.14.3　*輸送時の危険性クラス*
UN モデル規則に従って物質が示す最も顕著な危険性に従って物質または混合物に割り当てられる輸送クラス（及び付随的なリスク）を記載すること。

A4.3.14.4　*該当する場合、容器等級*
該当する場合には *UN* モデル規則による容器等級番号を示すこと。容器等級番号は危険の程度に従って特定の物質に割り当てられる。

A4.3.14.5　*環境有害性*
物質または混合物が IMDG-code による海洋汚染物質として知られているか否か、もし知られている場合には「海洋汚染物質」または「深刻な海洋汚染物質」であるかを示すこと。また、物質または混合物が、*UN* モデル規則、ADR、RID、ADN に従って、環境有害性があるか否かも示すこと。

A4.3.14.6　*使用者のための特別予防措置*
使用者が認識しておく必要のある、または輸送に関連して守るべき特別予防措置のすべてに関する情報を提供すること。

A4.3.14.7　*IMO の方法によるばら積み輸送*

この小節は、荷を IMO 文書に従ってばら積み輸送を意図した場合にのみ適用される：例えば SOLAS 第 VI 又は VII 章、MARPOL 附属書 II または附属書 V、IBC Code、IMSBC Code および IGC Code（または前版 EGC Code または GC Code）。

液体のばら積み貨物に関しては（もし名称が A4.3.1.1 にあるものと異なる場合には）、IBC Code 第 17 または 18 章あるいは IMO's MEPC.2/Circular 最新版で付与されている製品名リストに記載されている名称に従って、船積書類で要求されているように製品名を示す。要求されている船型および汚染区分を示す。

固体ばら積み貨物に関しては、ばら積み発送名、MARPOL 附属書 V に従って荷が海洋環境に有害（HME）あるいは無害と考えられているか、IMSBC Code に従ってばら積み時のみ化学的危険性を有する物質（MHB）かどうかおよび IMSBC に従いどのグループで発送すべきかを示す。

ばら積みの液化ガスに関しては、製品名および IGC Code（または前版 EGC Code または GC Code）に従った船型を示す。

A4.3.15　　*SECTION 15: Regulatory information*

Describe any other regulatory information on the substance or mixture that is not provided elsewhere in the SDS (e.g. whether the substance or mixture is subject to the Montreal Protocol, the Stockholm Convention or the Rotterdam Convention).

A4.3.15.1　　*Safety, health and environmental regulations specific for the product in question*

Provide relevant national and/or regional information on the regulatory status of the substance or mixture (including its ingredients) under relevant safety, health and environmental regulations. This should include whether the substance is subject to any prohibitions or restrictions in the country or region into which it is being supplied.

A4.3.16　　*SECTION 16: Other information*

Provide information relevant to the preparation of the SDS in this section. This should incorporate other information that does not belong in sections 1 to 15 of the SDS, including information on preparation and revision of the SDS such as:

(a) the date of preparation of the latest revision of the SDS. When revisions are made to an SDS, unless it has been indicated elsewhere, clearly indicate where the changes have been made to the previous version of the SDS. Suppliers should maintain an explanation of the changes and be willing to provide it upon request;

(b) a key/legend to abbreviations and acronyms used in the SDS; and

(c) key literature references and sources for data used to compile the SDS.

NOTE:　　*While references are not necessary in SDS's, references may be included in this section if desired.*

A4.3.15　第15節：適用法令

物質または混合物について SDS のどこにも示されていない他の規制情報をすべて記載すること（例えば、物質または混合物が、モントリオール議定書、ストックホルム条約またはロッテルダム条約の対象であるかどうか）。

A4.3.15.1　*該当製品に特有な安全、健康および環境に関する規制*

該当する安全、健康および環境規則の下における物質または混合物（その成分を含める）の規制状況について国または地域に関連する情報を提供すること。物質が供給される国または地域におけるあらゆる禁止または制限であるかどうかを含むべきである。

A4.3.16　第16節：その他の情報

本節では、SDS の作成に関連する情報を提供すること。これには、以下のような SDS の作成と改訂に関する情報を含め、SDS の第 1 節から第 15 節にない他の情報を盛り込むべきである：

(a) SDS の最新改訂版作成の日付。SDS に改訂が加えられる時には、それが他で示されていない場合には、SDS の旧版で変更された箇所を明確すること。供給者は変更の説明を保管し、要求に応じて提供すべきである；

(b) SDS で用いられている略語と頭字語の意味/凡例；および

(c) SDS 作成に用いられたデータの主要な文献参照と出典。

注記：SDS に文献参照が必要ない場合でも、要求があれば、文献一覧を本節に含めることができる。

ANNEX 5

CONSUMER PRODUCT LABELLING BASED ON THE LIKELIHOOD OF INJURY

附属書5

危害の可能性に基づく消費者製品の表示

ANNEX 5

CONSUMER PRODUCT LABELLING BASED ON THE LIKELIHOOD OF INJURY

A5.1 Introduction

A5.1.1　　The Globally Harmonized System of Classification and Labelling of Chemicals is based on an assessment of the intrinsic hazardous properties of the chemicals involved. However, it has been recognized that some systems provide information about chronic health hazards in consumer products only after considering additional data regarding potential exposures to consumers under normal conditions of use or foreseeable misuse. These systems thus provide information based on an assessment of risk, or the likelihood of injury occurring from exposure to these products. Where this exposure assessment and determination of likelihood of injury reveal that the potential for harm to occur as a result of the expected exposures is insignificant, chronic health hazards may not be included on the product label for consumer use. This type of system was recognized in a paper clarifying the scope of the GHS work in 1998[1]:

"The application of the components of the system may vary by type of product or stage of the life cycle. Once a chemical is classified, the likelihood of adverse effects may be considered in deciding what informational or other steps should be taken for a given product or use setting".

A5.1.2　　The work on the GHS has not addressed harmonization of this type of approach. Therefore, specific procedures to apply this approach would have to be developed and applied by the competent authority. However, in recognition that it is an approach that has been used, and will continue to be used in the future, this annex is being provided to give additional guidance on how such an approach may work in practice.

A5.1.3　　Exposure assessments for some consumer products are used to determine what information is included on a label in this type of approach. Regulators and manufacturers obtain exposure data or generate hypothetical exposure data based on customary use or foreseeable misuse. These assumptions are then used to determine whether a chronic health hazard is included on a consumer product label, and what precautions are to be followed, under a risk-based approach. These decisions are thus made on the basis of considerations regarding the likelihood of harm occurring in the consumer exposure situations that have been identified.

A5.1.4　　Consumer product labels in some systems are based on a combination of hazard and risk. However, acute and physical hazards may be indicated on the label, while chronic health effects labelling based on risk is not indicated. This may be due in part to the expectation that exposures to some consumer products are of short duration, and thus may not be sufficient to lead to the development of chronic health effects as a result of those exposures. These expectations may not be accurate where consumer products are used in a workplace, e.g. paints or adhesives used by construction workers on a regular basis

A5.1.5　　While intrinsic hazards of a chemical can be determined for all sectors, information about exposure, and thus risk, varies significantly among the sectors covered by the GHS. The vehicle by which this information is then transmitted to the user also varies. In some cases, particularly in the consumer setting, the label is the sole source of information, while in others, especially the workplace, it is one piece of a comprehensive system, supplemented by SDS's and worker training. In transport, a label transmits the primary information, but additional information is provided by the transport documentation.

A5.2 General principles

A5.2.1　　While the specific risk assessment approach has not been addressed or harmonized in the GHS, certain general principles are as follows:

　　(a)　All chemicals should be classified based on GHS classification criteria

　　　　The first step in the process of classifying hazards and communicating information should always be classification of intrinsic hazards based on the GHS criteria for substances and mixtures;

---

[1] IOMC *Description and Further Clarification of the Anticipated Application of the Globally Harmonized System (GHS), IFCS/ISG3/98.32B.*

附属書5
危害の可能性に基づく消費者製品の表示

A5.1 序

A5.1.1　化学品の分類および表示に関する世界調和システムは、対象化学品に固有な危険有害性の評価に基づいている。しかし、消費者製品の慢性的な健康有害性についての情報を提供するのに先だって、通常の使用条件または予見される誤使用における消費者のばく露可能性に関する追加データを考慮に入れるシステムもあることが認められている。したがって、こうしたシステムは、製品へのばく露に由来するリスク評価、すなわち危害の可能性に関する評価に基づいて情報を提供するシステムといえる。ばく露評価および危害の可能性を判断した結果、予想されるばく露によって危害が生じる可能性がある程度以下であることが明らかになった場合は、慢性的な健康有害性に関する情報を消費者製品の表示に含めなくてもよい。このような種類のシステムは、1998年のGHSの作業範囲の設定文書[1]においても認められている。

「GHSの構成要素の適用は、製品の種類またはライフサイクルの段階によって異なってもよい。いったんある化学品を分類すれば、起こりうる影響を考慮して特定の製品または利用状況において必要な情報やその他の対策を決定することが可能になる。」

A5.1.2　GHSの作業では、この種の手法についてはまだ調和が図られていない。したがって所管官庁によって、この手法を採用するための特別な手続きが開発され、適用されなければならないであろう。しかし、こうした手法が現在まで用いられてきて、今後も用いられ続けていくことを認識した上で、本附属書は、そうした手法が実際上どのように機能するかについて追加的ガイダンスを示すものである。

A5.1.3　ある種の消費者製品に対しては、この種の手法においてラベルに含めるべき情報を決定するためにばく露評価が用いられる。規制機関および製造者は、ばく露データを入手し、または日常的な使用や予見される誤使用に基づいて仮定的なばく露データを取得する。次にこれらの仮定に基づいて、慢性的な健康有害性を消費者製品のラベルに含めるか、そしてどのような予防措置を取るべきかをリスクに基づく手法の下で決定する。したがって、こうした決定は、特定の消費者のばく露状況下で起こる危害の可能性に関する検討に基づいて行われるものである。

A5.1.4　あるシステムでは、消費者製品のラベルは危険有害性とリスクの組合せに基づいている。しかし、急性の有害性と物理的危険性はラベルに記載されていても、リスクに基づく慢性的な健康影響の表示はなされていないことがある。この理由の1つは、ある消費者製品へのばく露は短時間のものと想定され、したがって、そうしたばく露が慢性的な健康影響を生じるには十分でないと考えられるからである。こうした想定は、例えば建設作業員が定常的に用いる塗料や接着剤など、消費者製品が作業場で使用される場合にはあてはまらないであろう。

A5.1.5　化学品固有の危険有害性はすべての部門に対して特定できるが、ばく露、それゆえリスクに関する情報はGHSが対象とする部門間で著しく異なっている。この情報を利用者に伝達する媒体もまた多様である。ある場合、特に消費者の利用状況では、ラベルが唯一の情報源であるが、また他の場合、特に作業場においては、ラベルはSDSおよび作業者の訓練などにより補完される包括的なシステムの一部でしかない。輸送においては、ラベルが基本的な情報を伝達するが、追加情報が輸送関連文書によって提供される。

A5.2 一般原則

A5.2.1　GHSでは、個々のリスク評価の手法については扱われておらず、また調和も図られていないが、次のような一般原則が存在する：

(a) すべての化学品はGHSの分類判定基準に基づいて分類されるべきである。

危険有害性を分類し、情報を提供する第1段階は、常に物質および混合物に関するGHSの判定基準に基づく固有の危険有害性の分類であるべきである；

---

[1]　*IOMCによる世界調和システム（GHS）の予想される適用とその明確化(IFCS/ISG3/98.32B)*

(b) Risk-based labelling can only be applied by the competent authorities to the chronic health hazards of chemicals in the consumer product setting. All acute health, environmental and physical hazards should be labelled based on intrinsic hazards

The hazard classification should lead directly to labelling of acute health effects, environmental and physical hazards. The labelling approach that involves a risk assessment should only be applied to chronic health hazards, e.g. carcinogenicity, reproductive toxicity, or target organ toxicity based on repeated exposure. The only chemicals it may be applied to are those in the consumer product setting where consumer exposures are generally limited in quantity and duration;

(c) Estimates of possible exposures and risks to consumers should be based on conservative, protective assumptions to minimise the possibility of underestimating exposure or risk

Exposure assessments or estimates should be based on data and/or conservative assumptions.

Assessment of the risk and the approach to extrapolating animal data to humans should also involve a conservative margin of safety through establishment of uncertainty factors.

A5.2.2 *An example of risk-based labelling used in the United States Consumer Product Safety Commission*

A5.2.2.1 In general, consumers rely on product labels for information about the effects of a chemical. Whereas other sectors have additional sources of information (e.g. safety data sheets, transport documents) to expand upon or refine product information and relate risk to the hazard information provided, the consumer sector generally does not.

A5.2.2.2 As noted above, the general rule for the GHS is that the label information will be based on intrinsic properties (hazards) of the chemical in all sectors. The rationale for hazard based labelling in the GHS has been described earlier in this document, and may be applied to consumer products as well as products in other sectors.

A5.2.2.3 In particular, the principle of the user's "right-to-know" about the intrinsic hazards of the chemical is important and widely supported by many stakeholders. Hazard information is an incentive to choose less hazardous chemicals for use. It may not be possible to accurately predict the exposures when the products are used, and consumer protective measures are less certain than those in other more structured sectors.

A5.2.2.4 On the other hand some research has indicated [2-7] that a consumer's attention can be diverted by too much information on a label regarding all potential hazards. It appears there is some evidence that warnings focused on specific hazards that are likely to cause injury enhance consumer protection.

A5.2.2.5 To ensure that consumers have the information needed to take appropriate protective measures, a risk-based labelling approach examines likely or possible exposures and communicates information related to the actual risks of exposure. Consumer exposures from use, foreseeable use and accidents can be estimated since products are designed for specific use(s).

A5.2.2.6 The following process has not been harmonized in the GHS. It is consistent with US Consumer Product Safety Commission Guidelines[8] and with other national and international guidelines on conducting risk assessments[9-11]. A substance or product under evaluation for chronic hazard labelling for consumer use in the US must satisfy a two-part test. First, it must present one of the chronic hazards covered, i.e. be classified as a chronic hazard based on specific criteria. Second, a risk assessment must be carried out to establish whether it has the potential to cause substantial illness or injury during or as a result of "reasonably foreseeable handling or use or from ingestion by children". If the result of the risk assessment indicates the risk is very low, the substance or product need not be labelled for chronic hazard. In other words, whether a given substance is labelled for a chronic effect depends not only on whether it is hazardous, but also on exposure and risk.

A5.2.2.7 The extent of the exposure assessment would depend on the hazard. For example, for non-cancer chronic endpoints, an "acceptable daily intake" (ADI) would be calculated from the "no observed adverse effect level" (NOAEL). For a conservative estimate of exposure, one can assume that the consumer will use the entire consumer product in a day and/or assume that all of the hazardous substance/mixture that the consumer is exposed to will be absorbed. If the resulting exposure is lower than the "acceptable daily intake" no hazard communication would be required. If the exposure level is higher than the ADI, then a more refined quantitative assessment could be performed before making a final labelling decision. If refined data are not available, or a refined analysis is not done, the hazard would be communicated on the label.

(b) リスクに基づく表示は、所管官庁によって、消費者の製品使用状況における化学品の慢性的な健康有害性に関してのみ適用されてもよい。急性の健康有害性、環境有害性、物理的危険性は、固有の危険有害性に基づいて表示を行うべきである

危険有害性の分類は、急性の健康影響および環境影響、物理的危険性の表示に直結すべきである。リスク評価を含む表示方法は、慢性的な健康有害性、例えば発がん性、生殖毒性、反復ばく露による特定標的臓器毒性などにのみ適用すべきである。これを適用できる唯一の化学品は、消費者のばく露量とばく露期間が一般に限定されている消費者製品のみであろう；

(c) 消費者のばく露およびリスクの可能性の推定は、ばく露またはリスクを過小評価する可能性を最小限に抑えるため、慎重かつ保護的な仮定に基づくべきである

ばく露の評価または推定は、データまたは慎重な仮定に基づくべきである。

リスク評価と動物データのヒトへの外挿においても、不確実性係数を設定して、慎重に安全側の余裕を見込むべきである。

A5.2.2 米国消費者製品安全委員会による、リスクに基づく表示の例

A5.2.2.1 一般に、消費者は化学品の影響に関する情報を製品ラベルに頼っている。他の部門では、製品情報を拡充または詳細化し、提供された危険有害性情報にリスクを関連付けるための他の情報源（例えば安全データシートや輸送文書）を持っているのに対し、消費者部門は一般にそうではない。

A5.2.2.2 上述のように、GHSの一般原則は、ラベルの情報はすべての部門において化学品固有の性質（危険有害性）に基づくというものである。GHSの危険有害性に基づく表示の根拠については本文書で先に述べたが、これは他部門の製品だけでなく消費者製品にも適用されよう。

A5.2.2.3 特に、化学品の固有の危険有害性に関する消費者の「知る権利」の原則は重要であり、多くの利害関係者によって広く支持されている。危険有害性情報は、より危険有害性の低い化学品の使用を選択する動機になる。製品が使用される際のばく露を正確に予測することは不可能であろうし、また消費者保護措置は他のより体系化された部門に比べて確実性が低い。

A5.2.2.4 他方、ある研究[2-7]によると、すべての潜在的な危険有害性についてあまりにも多くの情報がラベルに表示されると、消費者の注意が散漫になることもあると指摘されている。傷害を起こしやすい特定の危険有害性に重点を置いた警告が消費者保護を高めるという証拠もありそうである。

A5.2.2.5 消費者が適切な保護措置を講じるのに必要な情報を確実に得られるようにするため、リスクに基づく表示方法は、可能性または蓋然性のあるばく露を分析し、ばく露による実際のリスクに関する情報を伝達する。製品は特定の使用のために設計されているので、その使用、ならびに予見される使用および事故における消費者のばく露を推定することができる。

A5.2.2.6 以下に述べる手順はGHSにおいて調和が図られたものではなく、米国消費者製品安全委員会指針[8]およびリスク評価を実施するための他の国際的、国内的指針[9-11]に従ったものである。米国では、消費者向けの慢性有害性表示のために評価される物質または製品は、2段階のテストを満たさなければならない。第一に、該当する物質または製品は、適用対象となる慢性有害性のいずれかを示さなければならない。すなわち、特定の判定基準に基づいて慢性有害性に分類されなければならない。第二に、該当する物質または製品が、「合理的に予見される取り扱いまたは使用、もしくは幼児による摂取」の最中あるいはその結果として、重大な疾病もしくは危害を引き起こす可能性を持つかどうかを判断するためのリスク評価を行わなければならない。リスク評価の結果、リスクが非常に低いことが示されれば、該当する物質または製品には慢性有害性についての表示をする必要はない。言い換えれば、ある物質にその慢性影響に関して表示が付されるか否かは、その物質が有害性をもつかだけでなく、ばく露およびリスクにも依存する。

A5.2.2.7 ばく露評価の範囲は、危険有害性によって異なる。例えば、発がん性以外の慢性影響については、「無毒性量（NOAEL）」から「1日許容摂取量（ADI）」が算定されるであろう。ばく露を慎重に評価するためには、消費者がその消費者製品全部を1日で使い切ること、または消費者がばく露を受ける有害な物質/混合物のすべてが吸収されることを仮定する。その結果生じるばく露が「1日許容摂取量」よりも低いレベルであれば、有害性に関する情報の伝達は必要ないであろう。ばく露レベルがADIより高い場合は、表示に関する最終的な決定を行う前に、より精度の高い定量的評価を行うこともできる。より精度の高いデータが入手できない場合、または精度の高い分析を行わない場合は、当該有害性をラベルで伝達することになろう。

A5.2.2.8 For carcinogens, a unit risk from exposure to the carcinogen would be calculated based on linear extrapolation with the multistage model as a default model. Life time exposures can be calculated either by assuming worst case scenarios (such as all of the substance in a product is reaching the target tissue at each use, exposure is daily/weekly/monthly), or by determining actual exposures during use, or some combination of these approaches.

A5.2.2.9 The competent authority will need to establish what level of risk is acceptable to implement such an approach to consumer product labelling for chronic effects. For example, CPSC recommends labelling for a cancer hazard if the lifetime excess risk exceeds one-in-a-million from exposure during "reasonably foreseeable handling and use."

References

1. ILO. 1999. *Current Sector Uses of Risk Communication*, IOMC/ILO/HC3/99.7.

2. A. Venema, M. Trommelen, and S. Akerboom. 1997. *Effectiveness of labelling of household chemicals*, Consumer Safety Institute, Amsterdam.

3. Leen Petre. 1994. *Safety information on dangerous products: consumer assessment*, COFACE, Brussels, Belgium.

4. European Commission. 1999. *DGIII Study on Comprehensibility of labels based on Directive 88/379/EEC on Dangerous Preparations.*

5. Magat, W.A., W.K. Viscusi, and J. Huber, 1988. *Consumer processing of hazard warning information*, Journal of Risk and Uncertainty, 1, 201-232.

6. Abt Associates, Inc. 1999. *Consumer Labelling Initiative: Phase II Report*, Cambridge, Massachusetts, Prepared for US EPA.

7. Viscusi, W.K. 1991. *Toward a proper role for hazard warnings in products liability cases*, Journal of Products Liability, 13, 139-163.

8. US Consumer Product Safety Commission. 2001. *Code of Federal Regulations, Subchapter C – Federal Hazardous Substances Act Regulations*, 16, Part 1500.

9. Saouter, E., G. Van Hoof, C. Pittinger, and T. Feijtel. 2000. *A retrospective analysis of the environmental profile of laundry detergents*, submitted to: International Journal of life cycle analysis, October 2000.

10. IPCS. 2001. *Principles for evaluating health risks to reproduction associated with exposure to chemicals*, Environmental Health Criteria No. 225.

11. IPCS. 2000. *Human exposure assessment*, Environmental Health Criteria No. 214.

12. IPCS. 1999. *Principles for assessment of risks to human health from exposure to chemicals*, Environmental Health Criteria No. 210.

A5.2.2.8　発がん性物質に関しては、既定モデルとして多段階モデルを用いた線形外挿法に基づいて、発がん性物質へのばく露から発がんに至るまでのユニットリスクを算定できる。生涯を通じてのばく露量は、最悪の場合のシナリオ（製品に含まれる物質のすべてが日/週/月単位の使用毎に標的組織に達するというようなシナリオ）を仮定するか、または使用中の実際のばく露量を定量するか、もしくはこれらの手法を組み合わせて算定することができる。

A5.2.2.9　所管官庁は、こうした手法を消費者製品の慢性影響に関する表示に適用する際、どのレベルのリスクなら許容できるかを決める必要があるだろう。例えば CPSC は、「合理的に予見できる取り扱いと使用」によるばく露からの生涯の超過リスクが 100 万分の 1 を超える場合に、発がん有害性の表示を勧告している。

参考文献

1. ILO. 1999. *Current Sector Uses of Risk Communication*, IOMC/ILO/HC3/99.7.
2. A. Venema, M. Trommelen, and S. Akerboom. 1997. *Effectiveness of labelling of household chemicals*, Consumer Safety Institute, Amsterdam.
3. Leen Petre. 1994. *Safety information on dangerous products: consumer assessment*, COFACE, Brussels, Belgium.
4. European Commission. 1999. *DGIII Study on Comprehensibility of labels based on Directive 88/379/EEC on Dangerous Preparations*.
5. Magat, W.A., W.K. Viscusi, and J. Huber, 1988. *Consumer processing of hazard warning information*, Journal of Risk and Uncertainty, 1, 201-232.
6. Abt Associates, Inc. 1999. *Consumer Labelling Initiative: Phase II Report*, Cambridge, Massachusetts, Prepared for US EPA.
7. Viscusi, W.K. 1991. *Toward a proper role for hazard warnings in products liability cases*, Journal of Products Liability, 13, 139-163.
8. US Consumer Product safety Commission. 2001. *Code of Federal Regulations, Subchapter C – Federal Hazardous Substances Act Regulations*, 16, Part 1500.
9. Saouter, E., G. Van Hoof, C. Pittinger, and T. Feijtel. 2000. *A retrospective analysis of the environmental profile of laundry detergents*, submitted to: International Journal of life cycle analysis, October 2000.
10. IPCS. 2001. *Principles for evaluating health risks to reproduction associated with exposure to chemicals*, Environmental Health Criteria No.225.
11. IPCS. 2000. *Human exposure assessment*, Environmental Health Criteria No.214.
12. IPCS. 1999. *Principles for assessment of risks to human health from exposure to chemicals*, Environmental Health Criteria No.210.

ANNEX 6

COMPREHENSIBILITY TESTING METHODOLOGY

附属書6

理解度に関する試験方法

ANNEX 6

COMPREHENSIBILITY TESTING METHODOLOGY[1]

A6.1   This instrument aims to provide a methodology for the assessment of the comprehensibility of labels and Safety Data Sheets (SDS's) for chemical hazards. The tool has been developed with a particular focus on addressing the needs of workers and consumers in developing countries. The emphasis of instrument development has been to provide a tool that is, as far as possible, globally applicable taking into account varied levels of literacy and differences in cultural experience.

A6.2   Overview of the instrument

A6.2.1   The instrument is organized into a number of modules, directions for each of which is covered in this annex. Broadly speaking, the instrument consists of four parts:

(a)   Module 1: This is a focus group, whose main purpose is to ensure that the instruments used in Modules 2 to 11 are sensible across diverse cultures and settings. Its use is recommended in all categories of target populations (see table A6.2 below) but it should be mandatory to commence with this module in groups of workers and community members from cultures different to the settings in which labels and SDS's have been produced;

(b)   Modules 2 to 8: These include a general questionnaire (module 2) and a set of label and Safety Data Sheet questions and exercises (modules 3 to 8). Depending on whether the subject is a worker and makes use of a Safety Data Sheet, some elements of these modules may not apply;

(c)   Module 9: This is a simulation exercise. One version is intended for workers and is applicable to most people involved in production, while the other version (module 9a) is adapted for a consumer setting;

(d)   Module 10: Module 10 contains a final post-test questionnaire. It is applicable to all participants in the questionnaires (modules 2 to 8) and the simulations (module 9). It is also administered to participants in the group exercise (module 11). The questionnaire is focused on training, and past experience, and offers an opportunity for open-ended feedback and comment on the testing process;

(e)   Module 11: This is a group exercise for workers that draws on all elements contained in previous modules and is intended to test comprehensibility in the context of group learning. It is designed to complement modules 2 to 10 but is carried out on different subjects to those in modules 1, 2 to 8, and 9.

A6.2.3   It is further proposed that follow-up testing be conducted at one and twelve months after comprehensibility testing. This testing should be repeated on the same subjects who underwent initial testing. Depending on resources and logistics, it may be possible to avoid re-testing on all the modules completed at baseline. Repeat testing would be important to gain insight into retention and real benefits of exposure to hazard messages.

A6.2.4   Table A6.1 summarizes the modules in the instrument, the main activities in the modules, and the objectives and outcomes to be derived from each module.

A6.2.5   Although the testing instrument has been designed as a self-contained package, it may be possible to make use of selective modules from the battery where there are local priorities and needs. Moreover, it is recognized that as global harmonization of hazard communication evolves, new needs for testing may arise. The instrument may be adapted to take account of new testing priorities over time by using adapted testing materials (labels and SDS's) in the same testing formats. For example, if new icons for hazard symbols are under consideration, module 4 can be amended to include new symbols.

---

[1] *Developed by a multidisciplinary team at the University of Cape Town, for the International Labour Office (ILO) Working Group on Hazard Communication as part of international efforts to promote a Global Harmonised System (GHS) for hazard communication.*

附属書6

理解度に関する試験方法[1]

A6.1 本試験方法は、化学品の危険有害性に関するラベルおよび安全データシート（SDS）に関する理解度の評価方法の提供を目的としている。また、開発途上国における労働者および消費者の要求に対応することに特に焦点を当てて、手法を開発した。本試験方法の開発にあたっては、知識のレベルの違いや文化的経験における相違を考慮に入れ、できるだけ世界的規模で適用できる手法を提供することに重点を置いた。

A6.2 試験方法の概要

A6.2.1 本試験方法は本附属書に記載されている多数の作業と指示書で編成されているが、おおむね、4つの部分から構成されている：

(a) 作業1：フォーカスグループの主な目的は、作業2〜11で使用される試験方法が多様な文化や背景によらず理にかなったものであることを確認することである。対象集団のすべての区分（表A6.2を参照）にその使用が推奨されているが、ラベルおよびSDSが作成された状況とは異なる文化的背景を持つ労働者および地域社会構成員のグループでこの作業を開始することが必須である；

(b) 作業2〜8：一般的な質問（作業2）、ラベルおよび安全データシートに関する質問、実習（作業3〜8）から構成されている。被験者が労働者であるかどうか、安全データシートを利用しているかどうかにより、これらの作業のうちいくつかの要素は適用されない場合がある；

(c) 作業9：模擬実習である。2種類あり、ひとつは作業者向けを目的としていて製造に関わるほとんどの人々に適用できる。もうひとつの種類（作業9a）は消費者に適用される；

(d) 作業10：最終試験後の質問であり、質問（作業2〜8）と模擬実習（作業9）への参加者全員に適用される。また、グループ実習（作業11）への参加者にも適用される。質問は、訓練と過去の経験に焦点が置かれ、自由形式の評価と試験手順についてのコメントの機会が与えられる；

(e) 作業11：それ以前の作業に含まれているすべての要素を含んだ、労働者向けのグループ実習であり、グループ学習という状況下における理解度に関する試験が目的である。作業2〜10までを補完することを意図しているが、作業1、2〜8、9までとは異なった被験者で実行される。

A6.2.3 理解度に関する試験の1ヶ月後と12ヶ月後にさらに追跡試験を行うことが、提案されている。この試験は、最初の試験と同じ被験者で繰り返すべきである。人的資源や実務の状況に応じて、最初に完了した全ての作業についての再試験は省くことができる場合がある。繰り返し試験は、記憶力を計る上で重要であり、また危険有害性情報に接するという実際的な利点もある。

A6.2.4 表A6.1では、本試験方法での作業、作業内の主な活動、各作業の目標および各作業から得られる成果が要約されている。

A6.2.5 本試験方法は自己充足型パッケージとして作成されているが、地域的な優先事項および必要性から作業を選択して使用することも可能な場合がある。さらに、世界的に危険有害性情報の伝達が調和することに伴い、試験に対する新たな必要性も生じうることが認識されている。本試験方法は、同じ試験様式で適用された試験材料（ラベルおよびSDS）を用いて、時とともに新しい試験優先事項を考慮し、改訂されうる。例えば、新しい危険有害性シンボルの図が考慮される場合、新しいシンボルを含むように作業4を改訂することもできる。

---

[1] *GHSを促進する国際的な取り組みのひとつである国際労働機関（ILO）の危険有害性情報伝達に関する作業班のために、ケープタウン大学の学際的チームによって開発された。*

A6.3 **Use of annex 6 and of the testing instrument**

A6.3.1   Each module is the actual test questionnaire for a specific set of comprehensibility testing objectives. The layout of the modules is such that instructions are clearly marked in the questionnaires for those administering the comprehensibility tests. Accompanying each module, but presented separately, is a set of detailed guidance notes comprising the manual for the particular module. The manuals also outline the different labels and/or SDS's to be used in each module and the outputs and time requirements of each module.

A6.3.2   To avoid rendering the modules to lengthy, instructions on the modules have been kept to a minimum in the text of the modules, reserving the elaboration on instructions for the manual sections. Where key instructions are present in modules 3 to 11, they are listed in bold text within shaded boxes to improve ease of administration. Italic font is used throughout the modules for all text to be read out to the subject.

A6.3.3   Some modules (modules 3, 4, 6, 7, 8 and 9) require random selection of labels and/or SDS's. A box of cards is provided to the interviewer to expedite the selection of a random label/SDS or set of labels/SDS's. The interviewer will have a specific box of such cards marked for every relevant module.

A6.3.4   Labels and Safety Data Sheets are provided but should be to conform to the normative styles and presentations existing in the countries in which the tool is to be applied. The GHS will bring a certain degree of standardization in the content and layout of hazard communication methods but a great deal of variation will still arise in relation to local traditions, styles, size and preferences. Labels and SDS's used in testing must as far as possible reflect the typical local usage patterns. Therefore, although sample labels and SDS's are provided with this manual, users are encouraged to adapt the test materials within the limits of the experimental design requirements so that the materials appear as authentic as possible to local subjects.

A6.3.5   Notwithstanding attempts to simplify the relatively complex testing procedures required to measure hazard communication comprehensibility, the test instrument require careful administration and quality control. Training of interviewers is therefore critical. This is dealt with in more detail in the manuals for modules 1 and 2.

A6.3　附属書6および試験方法の利用

A6.3.1　各作業は、特定の理解度に関する試験目的についての実際の試験の質問表である。作業の割り付けは、理解度に関する試験の管理者用に指示事項を質問表に明記して作成されている。特定の作業に関する説明書となる詳細な手引きが、各作業に附随して、ただし別の形式で、示されている。説明書には、各作業に使用されている各種ラベルやSDSおよび各作業の成果と必要時間数の概要が説明されている。

A6.3.2　作業が冗長になるのを避けるため、説明書の項に対する指示の詳細さは失わないようにして、作業の本文では作業の指示は最小にとどめてある。作業 3～11 で主な指示が示されている箇所では、管理を容易にするために、影をつけたボックス内に太字で表になっている。被験者に対して読み上げる本文すべてには、イタリック書体が使用されている。

A6.3.3　幾つかの作業（作業 3、4、6、7、8 および 9）では、ラベルやSDSの無作為な選択が必要とされる。面接者には、無作為なラベルまたは SDS の選択を効率良く行うために、カードの入った箱が提供される。面接者は、各関連作業用に印が付けられた特定のカードの入った箱を持つこととなる。

A6.3.4　ラベルと安全データシートが提供されるが、これらは適用国における規範的な形式および既存の表現に適合させるべきである。GHS は、危険有害性の伝達の内容と割り付けに関してある程度の標準化をもたらすが、その国の伝統、形式、サイズおよび好みとの関係で、まだかなり多様なものが出現するだろう。試験に使用されたラベルおよび SDS は、可能な限りその国の典型的な使用パターンを反映しなければならない。それゆえ、ラベルおよび SDS のサンプルがこの説明書とともに提供されるが、使用者は、実験上デザインに求められている範囲内で、被験者に対してできるだけ典型的なものに見えるような材料を選択することが推奨される。

A6.3.5　危険有害性情報の伝達の理解度を測るために必要な比較的複雑な試験手順を単純化する試みの一方で、試験方法は注意深い管理および品質管理が要求される。面接者の訓練はそれゆえ非常に重要である。これに関する詳細は作業 1 と 2 用の説明書に記載してある。

Table A6.1: Comprehensibility testing: Objectives and outcomes by module

Module	Contents	Objectives	Outcome
Module 1	Focus groups	To shape research tool to the context, language, and cultural interpretations of the specific target group. To identify cultural specific definitions of words. To test whether ranking, the use of colour for attributing hazard, and the quantitative estimation of ambiguous variables are culturally transferable. Testing strategies used in subsequent modules are piloted for face validity and identify alternatives. To identify potential biases in the testing situation arising from cultural use of items.	Culturally consistent explanations for difficult words. Appropriate use of colour in local context. Account of cultural factors that would bias comprehensibility tests. Validation of colour blindness test methods. Interpretability of psychometric scales for non-Western populations. Contextual testing. Instruments to capture workers' experience. "Dummy" symbols.
Module 2	General interview	To ascertain demographic and other data as a basis for analysis of comprehensibility. To clarify competence in colour and visual acuity necessary for some of the subsequent tests. To collect data on work experience, critical to interpretation of comprehensibility assessments.	Relevant demographic and other data for linking to study results and analysis. Colour and visual acuity assessed. Role work experience plays in comprehensibility.
Module 3	Recall, reading, and comprehensibility of labels and SDS's	To evaluate subjects' familiarity with a label and an SDS. To test subjects' recall of label elements. To evaluate the sequence used to look at label elements. To test the comprehensibility of signal words, colours, symbols and hazard statements. To assess the impact of the label on the subjects': - Ranking of hazard, both to self and to spouse or child, - Intention to use, store and dispose of the chemical. Whether ranking and reporting change after questions on comprehensibility. Can subjects correctly identify the appropriate SDS? Can subjects correctly identify information on chemical name, health hazard, physical hazard and use of protective clothing?	Identify a priori familiarity with labels and SDS's. An assessment of the impact of different label fonts. Identification of poorly understood elements terms. Identify statements with highest comprehensibility. Hazard ranking, and intention to behave as a result of the label. The effect of detailed questions on comprehensibility on subjects' perceptions of hazard as a proxy for training. The impact of the Hawthorne effect will be gauged. Comparison of ranking of hazard to self differs from ranking of hazard to a close relative. Identifying whether subjects can link data from a label to an appropriate SDS in a meaningful way.
Module 4	Rating and understanding of hazards: Signal words, colours and symbols	To test subjects' relative ranking for severity of hazard for: - signal words, colours and symbols; - combinations of symbols and multiple symbols; - selected combinations of symbols, colour and signal words. To test understanding of signal words, colours, and symbols. To test opinion on the ability of signal words, colours, and symbols to attract attention. To test whether subjects' perception of the label will influence their reported intention to use, store or dispose of the chemical. To explore subjects' views as to why hazard elements are present on a label.	Signal words, Colours and Symbols will be rated for ability to denote level of hazard, and for comprehension both separately, and for selected combinations of elements. Quality control assessment of face validity of ranking. Ability of label elements to attract attention. Label rated highest for attracting attention will be explored for its ability to: Prompt the subject to identify further information, particularly health hazard information. Influence reported intention to behave in safe ways.

表 A6.1：理解度に関する試験：作業ごとの課題と結果

作業	内容	目標	成果
作業 1	フォーカスグループ	特定の対象集団に関する状況、言語および文化的解釈に合わせて調査手法を構築する。言葉の文化的な固有の定義を特定する。順位付け、危険有害性に応じた色の使用および試験に使用可能かどうかを試験する。一連の作業に使用された試験方法は、見た目の妥当性と代替方法を特定することを指向している。	難解な言葉に対する文化的に一貫性のある使用。理解の状況下での適切な色の使用。理解度に関する試験方法の検証。色音試験方法に基づく説明。非西洋人に対する心理測定尺度の解釈。前後関係に基づく試験。作業者の経験を知る手段。「ダミー」記号。
作業 2	一般面接	理解度の分析の基礎としての人口統計及び他のデータを把握する。一連の試験のためにつなげて必要な色と視覚的鋭敏度の適性を明らかにする。理解度に関する評価に重要な労働経験のデータ収集を行う。	結果と分析の研究を関連させるための適切な人口統計及び他のデータ。色覚と視力の評価。理解度に関する役割演技演習。
作業 3	ラベルと SDS に関する記憶、読み取り、理解	ラベルや SDS に関する被験者の熟知度を評価する。各種ラベルの要素を見るのに使う順序を評価する。注意喚起語、色、シンボルおよびラベルの影響を評価する。被験者へのラベルの影響を評価する： — 自分自身、配偶者または子供にとっての危険有害性の理解の目的。 — 化学品の使用、貯蔵および廃棄の目的。 理解度に関する質問の後、貯蔵付けや報告が変わるかどうか。被験者が適切な SDS を正しく識別できるか。理解度が化学品、健康有害性、物理化学的危険性および保護衣の使用に関する情報を正しく識別できるか。	ラベルと SDS への先験的な熟知度の確認。各種ラベル書体の影響の評価。要素の分かりにくい用語の確認。非常に分かりやすい危険有害性情報の理解の確認。危険有害性の順位付けとラベルを見た結果としての被験者の危険有害性の認識についての理解度に関する詳細な質問の効果。ホーソン効果の影響の測定。自分自身への危険有害性の順位付けと親族・配偶者への危険有害性の順位付けの違いの比較。被験者が、意味のある方法でラベルのデータを適切な SDS に結びつけることができるかの確認。
作業 4	危険有害性に関する評価と理解 注意喚起語、色およびシンボル	被験者に関連した危険有害性の程度の順位付けについて試験する： — 注意喚起語、色、およびシンボル； — シンボルおよび複数のシンボルの組み合わせ； — 注意喚起語、色、およびシンボルの選択組み合わせ。注意喚起語、色、およびシンボルの機能に関する意見を調査する。被験者のラベルに対する認識により化学品の使用、貯蔵または廃棄に関して影響があるかどうか試験する。危険有害性要素のラベル表示の理由について、被験者の見解を調査する。	注意喚起語、色、およびシンボルは、危険有害性の程度を表示する能力、要素の単独および選択された組み合わせの両方で評価される。順位付けの見た目の妥当性に関する品質管理評価。注意を喚起するための項目のラベル性能。ラベルは注意を喚起する点において高位置に格付けされ、そのラベルの機能が下記事項を促進させる：安全な方法で行動することに対する報告概念の影響。被験者に健康有害性の情報の確認を促す。

Table A6.1: Comprehensibility testing: Objectives and outcomes by module

Module	Contents	Objectives	Outcome
Module 5	Comprehension of hazard symbols with and without text	To test subjects' understanding of symbols representing hazard classes. To test subjects' understanding of concepts of hazard classes. To identify whether adding text words improves understanding of selected symbols representing hazard classes: reproductive, carcinogenic, and mutagenic. To identify whether adding signal words improve understanding of symbols representing classes.	Ability to identify the correct symbol for a hazard class. Identification of hazard classes for which symbols perform poorly; and of symbols which perform poorly as indicators of a hazard class. Identify symbols with ambiguous interpretations. The effectiveness of adding text to symbols for reproductive, carcinogenic and mutagenic hazards. The effectiveness of adding signal words to symbols denoting hazard class.
Module 6	Size, placement background colour and border of symbols/pictograms	To test the impact of varying symbol size, border and placement. To test the impact of varying background colour and varying icon size in a pictogram relative to border.	Impact of the symbol size, border and placement: - ability to identify chemical name; - perception of risk; - recall of symbol as proxy for attention to symbol; - recall of hazard statement as proxy for attention to hazard statement; - reported intention to behave; - sequence of reading; Comparison of whether ranking of hazard to self differs from ranking of hazard to a close relative.
Module 7	Pictogram comprehension – additional testing (Pesticides)	To test subjects' ability to identify information: - chemical name; - health hazards; To assess subjects' rating of hazard. To test subjects' understanding of pictograms. To assess subjects' sequence of reading.	Comprehensibility of pictograms: understanding, ranking of hazard, attention, access to key information. Comparison of whether ranking of hazard to self differs from ranking of hazard to a close relative.
Module 8	Comprehensibility of safety data sheets (SDS's) by organization of data	To test subjects' ability to identify safety information from an SDS. To test the understanding of hazard information on an SDS. To evaluate what the subject reads on an SDS and the sequence in which subjects report reading the elements of the SDS. To assess what information is useful, appropriate and understandable. To assess whether SDS information is related to intention to behave in safe ways. To evaluate the impact of different organisation of SDS information on the above.	Comprehension of SDS hazard information assessed from different aspects: 1) Interpretation of health hazard information; 2) Self-assessment of understandability to others; 3) Scoring of how the subject explains a hazard statement to a third party; 4) Reported intention to behave Agreement between these four measures of understanding will be estimated. The impact of different ways to organize SDS information will be estimated. Subjective assessment of the usefulness and appropriateness of sub-elements to identify areas for further review of SDS development.

表 A6.1：理解度に関する試験：作業ごとの課題と結果

作業	内容	目標	成果
作業 5	文字表記がある場合とない場合の危険有害性シンボルの理解	危険有害性クラスを表すシンボルに関する概念についての被験者の理解を試験する。危険有害性クラスに関する被験者の理解を試験する。生殖、発がん性、突然変異といった文字表記の追加により危険有害性クラスを表す指定シンボルの追加により、クラスを表すシンボルの理解が向上するかどうかを確認する。注意喚起語のシンボルへの追加により、クラスを表すシンボルの理解が向上するかどうかを確認する。	危険有害性クラスの正しいシンボルを識別する能力。シンボルがよく機能しない危険有害性クラスおよび危険有害性クラスのラベルとしてよく機能しないシンボルの識別。不明瞭な解釈を与えるシンボルの識別。生殖、発がん性、突然変異有害性のシンボルへの文字表記の追加の有効性。危険有害性クラスを表すシンボルへの注意喚起語追加の有効性。
作業 6	シンボルまたは絵表示のサイズ、配置、背景色および境界線	シンボルのサイズ、境界線、配置、色の変更による影響に関して試験する。背景色の変更および枠に関連した絵表示中の図のサイズの変更による影響に関して試験する。	シンボルのサイズおよび配置の影響： — 化学名を識別できるか； — リスク認識； — シンボルへの注意の代用としての危険有害性の想起； — 危険有害性情報の代用としての危険有害性の想起； — 説明された取扱い方法； — 読み上げの順序。 自分自身への危険有害性の順位付けと親族・配偶者への危険有害性の順位付けに違いがあるかどうかの比較。
作業 7	絵表示の理解-追加試験（農薬）	下記情報を識別できる被験者の能力を試験する： — 化学名； — 健康有害性。 被験者の有害性の順位付けを評価する。 被験者の絵表示の理解について試験する。 被験者の一連の絵表示の読み取りを評価する。	絵表示の理解度：理解、有害性の順位付け、注意、主要情報の入手。 自分自身への有害性ランキングと親族・配偶者への有害性ランキングの比較。
作業 8	データ構成による安全データシート (SDS) の理解度	被験者の、SDS からの安全情報の理解を試験する。SDS に関する被験者の理解について試験する。被験者の SDS に関する読み取りおよび SDS の要素の読み取りに関する一連の報告を評価する。SDS 情報がどの情報が役立ち、適切でかつ理解できるかを評価する。SDS 情報が、安全な取扱い方法に関係しているかどうか評価する。上記に関し、異なった構成の SDS 情報の影響について評価する。	異なった観点から評価した SDS 危険有害性情報の理解： 1) 健康有害性情報 2) 他者に対する理解度の自己評価； 3) 被験者が第三者にどのように危険有害性情報を説明するかについての探求。 4) これら 4 つの理解度測定する種々の方法の影響が評価される。 SDS 情報からなる方法についての合意が評価される。SDS 開発のための展望を特定化するための副要素の有効性および適切さに関する主観評価。

- 442 -

Table A6.1: Comprehensibility testing: Objectives and outcomes by module

Module	Contents	Objectives	Outcome
Module 9	**Simulation exercise: impact of the use of labels and SDS's, and of symbols and signal words on labels on safe chemical practices**	To assess safety practices in relation to a simulated exercise in which a chemical is handled. To evaluate whether safety practices are improved by the presence of the signal word "Danger" and/or by the size of the hazard symbol "Skull and Crossbones". To identify whether past experience in relation to chemicals plays a significant role in both safety practices, and in the impact of signal words and symbols on safety practices.	Measures of actual behaviour observed and related to use of labels, SDS's prior to, and during the task. Safety behaviours include use of PPE and other preventive hygiene practices. The impact of varying label elements (with or without "Danger"; with different size hazard symbol) and SDS layout (explicit heath hazard heading versus health hazard data under regulatory information). Relationship between understanding, practice and experimental conditions to be explored.
Module 10	Post interview/post simulation interview	To ascertain past history of contact with chemicals and training. To test the effect of a brief explanation of symbols, signal words, colours and hazard statements on ranking for severity of hazard, and comprehension. To identify chemical information needs from subjects.	Variables derived from training and past experience for stratified analysis of responses to modules 3 to 9. Results will help to indicate whether training should be the subject of more detailed evaluation in the long term. Responses to questions on needs for chemical information can be useful to GHS efforts on chemical safety.
Module 11	Group exercise - comprehension	To test whether learning about hazard communication happens differently in a group context than with individuals. To test whether subjects working as a group come up with significantly different answers than when individual subjects are asked a questionnaire.	A quality control assessment on the affect of group versus individual learning. Groups coming up with significantly different responses from individuals indicate that the testing model needs to be revised. Implications for how training should be addressed in future as an element of hazard communication.

表 A6.1：理解度に関する試験：作業ごとの課題と結果

作業	内容	目標	成果
作業 9	模擬実習：ラベルと SDS の使用の影響、安全な化学品の措置に関するシンボルおよび注意喚起語の影響	化学品が取り扱われる模擬実習に関連した安全措置を評価する。 安全措置が、危険有害性シンボル、注意喚起語、「危険」や「注意」の掲示により向上するかどうか評価する。 化学品に対する過去の経験が、安全措置、シンボルおよび注意喚起語の影響に関して重要な役割をしているかどうかを識別する。	課題実行前および実行中のラベル、SDS の使用に関する実際の行動の観察。 安全行動は、PPE（個人保護具）の使用および他の予防衛生実習を含む。 ラベル要素（「危険」を使うまたは使わない；異なったサイズの危険有害性シンボル）、および SDS の割り付け（規制情報の下での健康有害性データ）に対する、明示的な健康有害性の表題の変更による影響。 理解度と調査対象となる実習、実験条件の関係。
作業 10	試験後の面接/模擬実習後の面接	化学品との過去の関連経験と訓練について確認する。 シンボル、注意喚起語、色、危険有害性の重大さの程度の順位付けについての試験。 危険有害性情報の簡単な説明の効果と理解に関するニーズを認識する。	作業 3～9 に対する反応を層別しての分析して得られる、訓練と過去の経験に起因した変化。 結果は、訓練が長期的にもっと詳細な評価の対象となるべきかどうかを指摘する際の一助となる。 化学品情報に関するニーズについての質問に対する反応は、化学品安全面での GHS の取組に有効である。
作業 11	グループ学習と理解	危険有害性の伝達に関する学習が、個人よりもグループ環境によって違うかどうか試験する。 質問を受けた時、グループ活動の被験者が、個人被験者と明らかに違った回答を提示するかどうか試験する。	グループ学習対個人学習の効果に関する精度管理の評価。 個人と明らかに違った反応を示したグループは、試験様式の訂正の必要性を提示する。 危険有害性の伝達の一要素として、将来どのように訓練に取り組むべきかの示唆。

A6.3.6   *Consent*: Before conducting any of the modules in this instrument, participants should first give informed consent. To do so, the purpose of the exercises should be explained to them as well as the procedures that will be asked of them. Participants should not be coerced into participating and should know that they have the right to withdraw their participation at any time. The nature of the information provided in the consent procedure is sufficiently generic so as not to give away the explicit hypotheses being tested.

A6.3.7   *Consent procedures* are outlined in the opening sections of modules 1 (focus group), 2 (commencement of interviews) and 10 (simulation exercises). Irrespective of whether the same subjects complete all modules or not, all three consent procedure should be applied when required. The consent procedure for the simulation is by necessity more of an explanation to obviate the obvious bias to be introduced by alerting the subject to the purpose of the exercise.

A6.3.8   Policy on rewards or compensation to participants: Each participating respondent in this study is to be given some form of compensation or incentive for participating in the study. Participating respondents should be told in consenting to the testing that at the end of the study some form of compensation will be presented to them. Compensation may vary from country to country depending on what is culturally appropriate and locally available. Some suggestions (based on other studies) are food (lunch), hats/caps, mugs, food (sugar, rice, mealie, meal), certificates, etc. It is up to the countries applying the tool to develop an appropriate policy on compensation for participants.

A6.4   Sampling

A6.4.1   *Target populations*

A6.4.1.1   Target populations are outlined in table A6.2 below. These are largely adult working populations, typical of groups who use, distribute or manage chemicals, either directly or indirectly. Children are another important potential audience. However, although the ability to provide understandable safety messages to children is recognized as critically important, it has not been possible to address this area in this manual because of the specialised methods required for evaluation. Further development at some future point may be able to extend the comprehensibility testing to methods suitable for children.

A6.4.1.2   Proposed methods for attaining representative samples are outlined in the Manual sections for modules 1 and 2. University students should not be used as they have been extensively used in previous hazard communication studies and are not considered representative of the target populations identified in this study.

A6.4.2   *Focus groups*

A6.4.2.1   Given the aim of the focus groups to ensure that the instruments used in modules 2 to 11 are sensible across diverse cultures and settings, participants for focus groups should be as far as possible typical of the target groups to be evaluated. Emphasis should be placed on targeting groups of workers and community members from cultures different to the settings in which labels and SDS's have been produced. This will mainly apply to farm workers, non-agricultural workers and community/residents/consumer groups, both literate and non-literate, groups whose cultural and linguistic backgrounds may make hazard communication complex. Categories for focus groups are recommended in the table A6.2 below.

A6.4.2.2   At least 2 focus groups are recommended per category. However, where results from a focus group in one category (e.g. non-literate farm workers) appear highly similar to an analogous group (e.g. non-literate non-agricultural workers), it may be possible to dispense with further groups. This should only be done if the testers are confident that no different results would be anticipated from additional testing. In general, once findings from different focus groups are consistent, it is recommended to proceed directly to the main evaluation (modules 2 onward). Where findings appear vastly discrepant, or where inadequate information to inform the rest of the instrument has been obtained, it is recommended to continue assembling focus groups until such information is obtained. Under such circumstances, testing until results are consistent or clarity is achieved may require more groups than the 2 per category recommended.

A6.4.2.3   Focus group participants should preferably not be the same workers included in the testing under modules 2 to 11 as some learning will take place through the focus group itself. Groups should aim, wherever possible, to be homogenous for language, inasmuch as all participants should be able to communicate in at least one common language.

A6.5   Questionnaire and experimental design

A6.5.1   Different sub-populations of working and non-working people will have different experiences that influence their comprehension of hazard communication messages. Modules 2 to 8, and module 10 will test comprehension under different experimental conditions. Sample size calculations combined with considerations of logistical ease suggest that the minimum numbers of subjects to be tested are those contained in the table A6.2 below.

A6.3.6　*同意*：本方法の作業を実施する前に、参加者はまず十分に説明を受けた上で同意するべきである。手続を説明するとともに、実習の目的を参加者に説明することになる。参加者は、参加を強制されるべきではなく、いつでも参加を取止める権利を有することを知らされるべきである。試験内容の仮定が明示されることはないように、同意手続において提供される情報は十分に一般的なものである。

A6.3.7　*同意手続*に関しては、作業1（フォーカスグループ）、2（一般面接）および10（模擬実習）の最初の項に概要が説明されている。同一被験者が作業のすべてを完了するかどうかは別として、要求時に3つの同意手続すべてが適用される。実習目的を注意することによって被験者に明らかな先入観を与えることを避けるために、模擬に関する同意手続はむしろ説明の必要によるものである。

A6.3.8　*参加者への報酬または補償に関する方針*：この研究への参加者には、何らかの形で、研究参加に対する補償または奨励金が与えられる。研究参加者には試験への同意時に何らかの形での補償が研究終了時に提供される旨を伝えられるべきである。補償は、文化的妥当性やその地で入手できるものにより異なる。食事（昼食）、帽子/キャップ、マグカップ、食料（砂糖、米、コーンミール）、証明書等が示唆されている（他の研究例より）。参加者への補償に関する適切な方針に関しては、本試験方法の利用国にその開発を委ねられる。

A6.4　集団の抽出

A6.4.1　*対象集団*

A6.4.1.1　対象集団に関して、下記表A6.2に概要が説明されている。主に大人の労働者で、直接または間接的に化学品を使用、配送または管理している典型的な集団である。子供もまた重要な潜在的対象者である。子供が理解しやすい安全に関するメッセージを提供する機能が非常に重要なことは認識されているが、評価のために専門的な方法が要求されるため、本説明書で取り組むことは不可能である。将来のある時点で、理解度に関する試験が、子供向けに拡大される可能性はある。

A6.4.1.2　代表例となる人々を獲得する方法についての提案の概略が説明書の項の作業1および2に説明されている。大学生は以前の危険有害性の伝達の研究において頻繁に利用されていること及び今回の研究で特定された対象集団の代表とも考えられていないことから利用すべきではない。

A6.4.2　*フォーカスグループ*

A6.4.2.1　作業2～11で使用される試験方法が多様な文化や状況を超越して識別可能であることを確認するというフォーカスグループの目的を考慮すると、フォーカスグループの参加者は、評価される対象集団としてできるだけ典型的であるべきである。特に、ラベルやSDSが作成された環境とは異なる文化からの労働者と構成員から成る対象集団は重要視するべきである。これは主に、知識のあるなしを問わず、農作業者、非農業労働者および地域社会/住民/消費者に適用され、その文化的および言語的背景が危険有害性の伝達の複雑さを形成する可能性のあるグループである。フォーカスグループに関する区分が下記表A6.2に推奨されている。

A6.4.2.2　少なくとも2つのフォーカスグループが、区分ごとに推奨される。しかし、ある区分におけるフォーカスグループ（例：知識がない農作業者）の結果が類似したグループ（例：知識がない非農業労働者）に非常に似ている場合には、それ以上のグループの試験は省略することができる。これは、試験者が、追加試験から期待される結果に相違が見られないという確信がある場合のみに行われるべきである。一般的には、複数のフォーカスグループからの所見に一貫性があれば、いったん、主評価に直接進行することを推奨する（作業2より先へ進む）。所見に大いに相違が見られる場合または試験方法の残りの部分を伝えるのに情報が不十分な場合、そのような情報が得られるまでフォーカスグループを集め続けることを推奨する。そのような状況下では、結果が一貫性を持つまたは明瞭となるまで、試験には区分ごとに2つ以上のグループが必要となろう。

A6.4.2.3　フォーカスグループの参加者は、作業2～11での試験時とは全く違う労働者であるべきである。というのも、フォーカスグループそのものを通してある種の学習が行われるためである。グループは可能な限り、すべての参加者が少なくともひとつの共通語で意思疎通できる程度には言語に関して均一であることを目標とするべきである。

A6.5　質問表および実験計画法

A6.5.1　労働者と非労働者の互いに異なる部分母集団は、試験の経験も異なり、それが危険有害性の伝達内容の理解力に影響を与える。作業2～8および10では、異なった実験条件下での理解が試験される。実務的な容

Modules 6 (effect of label font and layout on comprehensibility) and 9 (simulation test) include comparisons of different label types (8 and 11 strata respectively). Thus, larger numbers are needed for these modules to generate sufficient cases within each stratum. The other interview modules (3, 4, 5, 7 and 8) have fewer strata (vary from one to four maximum) and thus can be managed with fewer subjects. Users of this instrument may choose to apply all the modules to all participants, in which case the minimum number of participants recommended is as for modules 6 and 9 in table A6.2. Modules 2 and 10 must be completed by all participants as indicated.

A6.5.2 In view of the length of the full battery of tests (see table A6.3), it may be necessary for logistic reasons to break up the instrument by having different subjects complete only some of the modules. In this way, more participants are recruiting to the study but they complete only some parts of the evaluation. If this is the case, remember that all subjects must complete modules 2 and 10, irrespective of how many of the other modules they complete. For example, the battery of modules could be sub-divided into sets consisting of:

(a) Modules, 2, 3, 8 and 10;

(b) Modules 2, 4 and 10;

(c) Modules 2, 5, 6, 7 and 10;

(d) Modules 2 and 11;

(e) Modules 9, 2 and 10.

However, it is preferable that, if possible, participants are given the full battery of tests contained in the instrument and are adequately compensated for their effort.

Table A6.2: Sample size - recommended numbers

Category	Sub-category		Focus group: MODULE 1	Interviews: Modules 2, 6 and 10; Simulation: Module 9	Interview: Modules 3, 4, 5, 7 and 8
Target Group 1: Workplace a) Management	**Population 1**: Production Managers, engineering, technical		Optional	30-50[a]	25
	Population 2: Supervisory Managers in industry, agriculture		Optional	30-50[a]	25
b) Workers	Population: Farm workers	**3.** Literate	At least one group	100	50
		4. Non-literate	At least one group	100[a]	50
	Population: Workers other than in agriculture	**5.** Literate	At least one group	100	50
		6. Non-literate	At least one group	100[a]	50
Target Group 2: Transport	**Population 7**. Transport workers		Optional	30-50	25
Target Group 3: Community Residents/ Consumers/general public	**Population 8**: Literate		At least one group	100	50
	Population 9: Non-literate		At least one group	100[a]	50
	Population 10: Retailers and distributors		Optional	30-50[a]	25
Target Group 4: Emergency Responders	**Population 11**: Health Professionals, Technical Extension staff and Emergency Responders		Optional	30-50[a]	25
Target Group 5: Other	**Population 12**: Enforcement / Regulatory		Optional	30-50[a]	25

[a] *Recognizing the practical difficulties in organizing a simulation test, it is suggested that in these groups simulation testing only be carried out where resources are available and where practically feasible.*

易さを合わせて考えた参加集団サイズの計算によれば、被試験被験者の最小人数は下記表 A6.2 に示された構成人数である。作業 6（ラベルの書体および割付の理解度への影響）および 9（模擬試験）は、種々のラベルの種類の比較（各 8～11 層）を含んでいる。ゆえに、各層内で十分な事例を形成するために、これらの作業に対して多人数が必要とされる。他の面接用作業（3、4、5、7 および 8）は、ほとんど階層を持たず（1 ないし最大 4）、被験者も少なくてすむ。本方法の利用者は、参加者全員に対してすべての作業を適用することを選択することもできるが、その場合、推奨される最少参加人数は、表 A6.2 の作業 6 および 9 に対しての人数となる。作業 2 および 10 は、指示されている通り、参加者全員により完成されなければならない。

A6.5.2 試験の全期間から見ると（表 A6.3 参照）、異なった被験者にモジュールの幾つかのみを完成させることにより、方法を細分化することが実務的な理由から必要かもしれない。この方法だと、より多くの参加者が研究を補強することになるが、評価に関してある特定部分のみを完成することとなる。この場合、被験者全員が、他にいくつ作業を完成したかに関係なく作業 2 および 10 を完成しなければならない。例えば、作業集団は以下のように細分化され、構成されてもよい：

(a) 作業 2、3、8 および 10；
(b) 作業 2、4 および 10；
(c) 作業 2、5、6、7 および 10；
(d) 作業 2 および 11；
(e) 作業 9、2 および 10。

しかし、可能なら、参加者は本試験方法に含まれる試験項目のすべてを与えられ、その取組に対して適切に補償されるのが望ましい。

表 A6.2：参加集団 サイズ－推奨数

区分	副区分		フォーカスグループ：作業 1	面接：作業 2, 6 & 10；模擬実習作業 9	面接作業：3, 4, 5, 7, 8.
対象集団 1：作業場 a) 経営管理者	集団 1：製造管理者、エンジニアリング、技術		任意	30-50[a]	25
	集団 2：工業、農業の監督管理者		任意	30-50[a]	25
b) 労働者	集団：農作業者	3. 知識あり	最低 1 グループ	100	50
		4. 知識なし	最低 1 グループ	100[a]	50
	集団：農業以外の労働者	5. 知識あり	最低 1 グループ	100	50
		6. 知識なし	最低 1 グループ	100[a]	50
対象集団 2：輸送	集団 7：輸送関係者		任意	30-50	25
対象集団 3：地域住人/消費者/一般市民	集団 8：知識あり		最低 1 グループ	100	50
	集団 9：知識なし		最低 1 グループ	100[a]	50
	集団 10：小売業者および卸業者		任意	30-50[a]	25
対象集団 4：緊急時対応者	集団 11：保健専門員、技術扶助員および緊急時対応者		任意	30-50[a]	25
対象集団 5：その他	集団 12：施行/規制		任意	30-50[a]	25

[a] 模擬試験の実施が現実的には困難な場合、人的資源が得られかつ実際に実行可能な場所でのみ模擬試験を行うことを提案する。

A6.5.3   As far as possible, the selection of sub-groups should be done an as representative a sample as possible, using random selection of the population for participation. This is critical for generalizability of the results. Even where different participants are chosen from the same sub-group to complete different parts of the instrument, for reasons of length of the battery, selection of participants should emphasize representativeness. However, it is recognized that random selection may be very difficult to achieve in practice. Nonetheless, it should be borne in mind that whatever, selection is used, it should seek to generate a sample as representative as possible.

A6.5.4   Note that within the modules, randomization of subjects within the groups is essential and cannot be compromised on. Randomization is necessary for internal validity of the comparisons and is not the same as random selection of the sample, which is needed for generalizability of the study results.

A6.5.5   *Simulation studies:* Because simulations studies are relatively resource intensive exercises, it is proposed that the simulations only be conducted with limited target populations - workers, both agricultural and non-agricultural, transporters, and consumers. However, where resources permit, these simulations can easily be applied to other strata as desired.

A6.5.6   Contamination and co-intervention

A6.5.6.1   The testing design requires control circumstances. For this reason, the situation should be avoided where a participant is able to see or be told of the experimental materials of another participant. This will invalidate the comparisons being made where manipulation of the independent variable is key to the evaluation. Such events occurring in an experimental set up are called contamination.

A6.5.6.2   To avoid contamination, participants should avoid contact with each other whilst testing is being conducted. This may require considerable effort on the part of the testing team to ensure that chance meetings of subjects does not occur. Although difficult, every effort should be made to minimise the probability of contamination.

A6.5.6.3   A distinct but related problem is co-intervention, where both experimental groups are subjected to an intervention occurring independent of the experimental situation. This would occur when, for example, every worker in factory received detailed hazard safety training in the week before the testing was done. It may result in a masking of the effect of the different hazard communication elements and may lead to an under-estimation of the effect of different formulations of the label and SDS. Where this is not preventable, note should be taken of the possibility that co-intervention took place.

A6.5.7   Group learning

Module 11 is included to test comprehensibility in the context of group learning. It is applied only to workers (populations 3 to 6 in table A6.2 above) and will need a sample separate from workers completing modules 2 to 8. Ten groups should be tested in total including 5 groups of factory workers and 5 groups of farm workers. Groups should aim to be homogeneous for literacy level and approximately equal numbers literate and non-literate groups. Each group should not be larger than 10 and not smaller than 6.

A6.5.8   Context

A6.5.8.1   The context under which comprehensibility testing is carried out is crucial to the accurate evaluation of meaning and understanding. This is particularly so amongst workers with little formal education who use contextual cues to improve their understanding of hazard messages. For this reason, the bulk of testing in this instrument makes use of complete labels rather than elements of a label or SDS. While well-educated subjects may find it conceptually easier to respond to the isolated elements, the interpretation of such elements may have little bearing to real world learning situations. For this reason, all testing is to be conducted using realistic labels and SDS's.

A6.5.8.2   To maximize realism, an in-site label attached to a container will be used. To attach a different label to each container may pose an unnecessary burden on the tester, so it is proposed that the label be attached to a standard container and removed after testing. This procedure may require an assistant to the interviewer if overly burdensome for the interviewer. It is important that every visual cue be offered to subjects to maximise their possibilities of comprehension, particularly for workers with low levels of formal education who rely on contextual information to a greater degree. Therefore, the labels should be presented attached to container at all times. A Velcro strip attached to the container may make the procedure relatively simple.

A6.5.8.3   To standardize opportunities for comprehension, the actual chemicals identified in the labels will be spurious chemicals, although made to look as if they could be genuine agents. This aims to retain context, while not disadvantaging those unfamiliar with a particular chemical.

A6.5.3 可能な限り、副次的なグループの選択は無作為に行い、可能な限り参加集団の代表となるようにするべきである。これは結果を一般化するために非常に重要である。試験の連続期間の理由から、試験方法の別の部分を完成するために同じ副次的なグループ内から異なる参加者が選ばれたとしても、参加者の選択に関しては代表性に重点が置かれるべきである。しかし、無作為選択は実際には、達成が非常に難しいことが確認されている。それにもかかわらず、選択を行う場合、可能な限り参加集団の一般的代表となるようにするよう努めるべきである。

A6.5.4 作業を進める上で、グループ内で被験者を無作為に選択することは必須であり、それに関して妥協するべきではない。無作為化は、比較の内部有効性に必要で、これは研究結果の一般化に必要な参加集団の無作為選択とは異なる。

A6.5.5 *模擬研究*：模擬研究は、比較的人的資源要素の強い実習であるので、模擬実習は、農業/非農業労働者、輸送関係者および消費者といった限られた対象集団でのみ行うことが提案されている。しかし、人的資源が許すならば、これら模擬実習は、要望に応じて他の社会層にも簡単に適用することができる。

A6.5.6 *混成および相互干渉*

A6.5.6.1 試験計画には環境の制御が必要である。この理由から、参加者が他の参加者の実験材料を見聞きできる場所設定は避けるべきである。このような設定は、個々の相違の扱いが評価の鍵である比較を無効化するからである。実験的な設定でのそのような事例は、混成と呼ばれる。

A6.5.6.2 混成を回避するために、試験が行われている間、参加者は互いに接触を避けるべきである。試験チームの側にとって、被験者同士が接触しないようにすることは相当な努力を要するが、困難といえども、混成の可能性を最小限にするためにあらゆる努力をはらうべきである。

A6.5.6.3 混成とは別ではあるが関連性のある問題が相互干渉である。実験グループが実験状況に関係なく干渉を起こす傾向が見られる。例えば、各工場労働者が、試験が行われる1週間前に詳細な危険有害性安全訓練を受けた時に起こる可能性がある。これは、種々の危険有害性の伝達要素の効果を遮蔽する結果となり、異なるラベルやSDSの様式の効果に対する過小評価につながる可能性がある。これを防ぎようがない場合、相互干渉が起きる可能性を知っておくべきである。

A6.5.7 *グループ学習*

作業11は、グループ学習状況下での理解度に関する試験に含まれている。これは（上記表A6.2の集団3～6の）労働者のみに適用され、作業2～8を完成する労働者とは違った参加集団を必要とする。合計10グループが試験されるべきであり、そのうち5グループが工場作業者で、5グループが農作業者であるべきである。グループは、知識レベルに関しては均一で、知識のあるグループとないグループの数がほぼ同じ状態を目指さなければならない。各グループは6人以上10人以下でなければならない。

A6.5.8 *状況*

A6.5.8.1 意味と理解に関する正確な評価において、理解度に関する試験が実行される状況は非常に重要である。これは、正式な教育をほとんど受けておらず、危険有害性情報の理解を向上させるために状況を手がかりとして使う労働者の間では特に重要である。この理由から、この方法における試験の大部分では、ラベルやSDSの要素よりも完全なラベルが使用される。一方、高等教育を受けた被験者は、むしろ個々の要素に反応する方が概念的により易しいと感じるかもしれないが、そのような要素の解釈は、現実社会の学習状況とほとんど関連がない。この理由から、すべての試験は実際のラベルとSDSを用いて行われる。

A6.5.8.2 できるだけ現実感を出すために、容器に貼られたラベルを使用する。各容器に異なったラベルを貼付するのは、試験者に不必要な負担を負わせることとなるので、標準的な容器に貼付し、試験の後に剥がすことを提案する。この手順が面接者にとって過度な負担となる場合、助手が必要となろう。理解の可能性を最大に引き出すために、被験者、特に低レベルの正式教育のみを受け、かなり状況情報に頼っている労働者にあらゆる視覚的な手がかりを提供することが重要である。ゆえに、いかなる時もラベルを容器に貼付して提供するべきである。ベルクロ帯（マジックテープ）の容器への貼付により、手順はかなり単純化されよう。

A6.5.8.3 理解の機会を標準化するために、実際にラベルに名称が記されている化学品は、擬似物質が本物の物質のように見えるように作られている。これは、特定の化学品に慣れていない参加者に不利にならないような状況を保つことを目的としている。

A6.5.8.4   As indicated above, users are encouraged to adapt the test materials within the limits of the experimental design requirements so that the materials appear as authentic as possible to local subjects so as to maximise context.

A6.5.9   *Sample sizes for sub-studies*

Sample sizes for the sub-studies have been calculated based on a two-sided alpha error of 0.1 and a power of 0.8 but have also been tempered by considerations of logistical feasibility. Preliminary piloting of the instrument confirms these estimates. In particular, the simulation exercise has been considered relatively selectively for a smaller number of subjects and target groups, largely because of anticipated logistical constraints.

A6.5.10   *Translations*

A6.5.10.1   Language is key to much hazard communication. Although the instrument seeks to take account as far as possible of language differences, poor and unstandardized translation may introduce considerable error into the testing. For this reason, careful attention needs to be paid to accurate translation. The following procedure should be followed:

(a) Two persons fluent in English (the language of the current instrument) independently translate the questionnaire into the index language (the language of the target group);

(b) Both translations are then translated back into English by a further pair of translators independent of each other and of the original translators.

A6.5.10.2   Back-translations should aim to achieve less than 5 % errors on first round. Clarification of the errors in the translation should be conducted to correct ambiguities. Where possible, a combined translation should try to include all elements correctly translated and back translated from either questionnaire.

A6.5.10.3   If the latter is not possible, the translation with the lower rate of errors should be taken as the translation of preference. A second round of back translation will be necessary if errors exceed 5 %.

A6.5.11   *Timing of interviews and focus groups*

A6.5.11.1   Interviews and focus groups must be set up at a convenient time for both the interviewee and their employer (when this applies). Farm workers should not be requested to attend an interview during a crucial and busy period for farmers (e.g. planting, ploughing, spraying, or harvest). Workers should be interviewed during working time and should not suffer financial loss for their participation. It is not recommended that workers participate in their own time (lunch or after hours) without adequate compensation. If workers agree to participate during lunch break, the time must be adequate and suitable recompense provided (time back, lunch provided, etc).

A6.5.11.2   Table A6.3 gives the estimated time needed for completion of individual modules based on preliminary piloting with two South African factories. Depending on the module and how skilled the administrators of the modules are, total testing time could vary from 20 minutes to 2 hours. Testing times will be prolonged with non-literate workforces.

Table A6.3: Approximate testing times for hazard communication comprehensibility testing

Module	Time (minutes)
1	60 – 120
2	30 – 45
3	45 – 75
4	75 – 105
5	20 – 30
6	20 – 30
7	20 – 30
8	45 – 75
9	30
10	30 - 45
11	120 – 180

A6.5.8.4　既に述べたように、使用者は実験計画のできる範囲内で、被験者にとって試験材料が本物らしく見えるようにした方がよい。

A6.5.9　副研究の標本数

副研究の標本数は、二面アルファエラー値 0.1 およびべき乗数 0.8 として計算されるが、実務的実行可能性を考慮して調節される。本試験方法の初期実験的導入段階でこれらの概算が確認される。特に、模擬実習が想定される場合、主に予測される実務的制約のために、相対的に少人数の被験者および少数の対象集団が選択される。

A6.5.10　翻訳

A6.5.10.1　言語は危険有害性の伝達の重要なポイントである。本試験方法はできるだけ言語の違いを考慮に入れるよう努めているが、粗末で標準に満たない翻訳は試験に大きな誤りをもたらす可能性がある。この理由から、正確な翻訳への細心の注意が必要であり、下記手順にしたがって実施されるべきである：

(a) 英語（本文書の言語）の流暢な人員2名が別々に質問表を目標言語（対象集団の言語）に翻訳する；

(b) この2つの訳文を別のもう一組の翻訳者が各々、英語へ再翻訳する。

A6.5.10.2　再翻訳は、初回で5%以下の誤訳を目標とするべきである。誤訳を解明し、それにより不明瞭さを直すべきである。可能なら、2つを合わせた翻訳は、各質問表から正しく翻訳および再翻訳したすべての要素を含むよう努力すべきである。

A6.5.10.3　後者が不可能な場合、誤訳の割合の低い翻訳を優先させるべきである。誤訳が5%を超える場合、2回目の再翻訳が必要である。

A6.5.11　面接のタイミングおよびフォーカスグループ

A6.5.11.1　面接およびフォーカスグループに関して、面接を受ける者と雇用者（これが当てはまる場合）双方にとり、都合の良い時間を設定しなければならない。農業従事者にとって大切な繁忙期（例：植え付け、耕作、農薬散布、収穫期）には、面接に出席する要請をすべきではない。労働者は勤務時間内に面接を受けるべきで、参加のために経済的損失を被るべきではない。労働者が、自分の自由時間（昼食、または労働時間後）に、適切な補償なしに参加することは奨励されない。労働者が昼食休憩時間に参加することに同意した場合、その時間は適切で、しかも妥当な報酬が与えられなければならない（別に自由時間を与えられる、昼食が供給される等）。

A6.5.11.2　表 A6.3 は、南アフリカの 2 工場での初期実験的導入に基づく各作業の完成に要する時間の概算を表したものである。作業と作業管理者の技量によって、総試験時間は 20 分から 2 時間と差が出る。知識がない労働者の場合、試験時間は長くなる。

表 A6.3：危険有害性の伝達の理解度に関する試験に要するおおよその時間

作業	時間（分）
1	60－120
2	30－45
3	45－75
4	75－105
5	20－30
6	20－30
7	20－30
8	45－75
9	30
10	30－45
11	120－180

A6.5.12 *Rating and coding of responses*

A6.5.12.1　Rating of responses to comprehensibility testing requires expert judgement as to the correctness of the response. Previous experience in Zimbabwe has shown that content analysis of open-ended responses may be feasible where observers are carefully standardized in their approach.

A6.5.12.2　This instrument requires the presence of a set of experts to conduct the rating required for comprehension. The panel of experts should be identified before commencing the study in a process outlined below:

 (a) Select a panel with a range of experience, including (one or more) employees, employers and practitioners, as well as researchers skilled in the field of coding and rating;

 (b) Convene a workshop with the panel to review the nature of potential responses to questions in each of the modules listed. Review the documentation of the GHS process and aim to arrive at consensus as to what responses would constitute the following categories:

  (i) Correct: Meaning is identical, or fully consistent with intention of the GHS construct. This includes responses which are not 100 % the same as the GHS meaning but would suffice as the basis for a safety action or precaution;

  (ii) Partly correct: Some element of the meaning is correct but it would be insufficient to ensure adequate safety action or precaution;

  (iii) Incorrect: Meaning given is either completely wrong, or has very poor relation to the GHS intended meaning;

  (iv) Opposite meaning (critical confusions): Meaning given is not only incorrect but indicates an understanding opposite of the intention of the GHS system. Such a critical confusion may result in a dangerous behaviour or action;

  (v) Cannot answer/does not know;

 (c) Pilot the questionnaire amongst 5 or 10 subjects. Review the results in relation to the criteria selected;

 (d) If the results show significant discrepancy, iterate the process above until agreement reached about criteria.

A6.5.12.3　Further coding of responses to questions in the different modules is discussed under each module, where appropriate.

A6.5.13 *Analyses*

Analyses proposed for these modules are simple computations of proportions and means in relation to different strata. More complex analyses may be undertaken and are indicated in the different modules. An overall estimate for comprehensibility may be attempted by combining results from subjects in the different strata but should be adjusted for weightings by stratum and by other demographic factors known to affect comprehensibility.

A6.5.14 *Feedback and follow up*

All subjects should be offered the opportunity of seeing the results of the comprehensibility evaluations, and to give feedback on the interview and testing procedures.

A6.5.15 *Follow up evaluation*

Subjects participating in these evaluations should be re-interviewed after 1 month and 1 year to assess retention and the medium and long-term benefits of exposure to the GHS hazard messages. Depending on resources and logistics, it may be possible to avoid re-testing on all the modules completed at baseline.

A6.5.12 *反応の評価とコード化*

A6.5.12.1　理解度に関する試験に対する反応の評価は、反応の正しさに関しての専門的な判断を必要とする。ジンバブエでの過去の経験では、自由形式の反応の内容分析は、観察者の手法が十分に標準化されたところでは実行可能であることを示している。

A6.5.12.2　本試験方法には、理解に関して要求される評価を行う一連の専門家が必要である。下記に概要を説明した手順に沿って研究を始める前に、専門家委員会が特定されるべきである：

(a) 経験の範囲を考慮して委員を選定する。それには（1人またはそれ以上の）被雇用者、雇用者、従業者、およびコード化と評価の分野での熟練した研究者が含まれるべきである；

(b) 表にある各作業の質問に対する潜在的な反応特性を点検するために、委員とともにワークショップを開催する。GHS 手続書類を点検して、どんな反応が下記の区分に相当するかに関して合意に到達することを目指す；

　　(i) 正：意味が同一である、または GHS の構成概念の意図に完全に一致する。これは GHS の意味することと 100%同じではないが安全行動や予防措置の基礎として満足できる反応を含む；

　　(ii) 部分的に正：意味するところの要素は部分的には正しいが、適切な安全行動や予防措置を保証するに不十分である；

　　(iii) 誤：提示された意味が完全に間違っているかまたは GHS の意図した意味との関連が非常に薄い；

　　(iv) 反意（重大な混乱）：提示された意味が間違っているばかりか GHS システムの意図に対して反対の理解を示している。そのような重大な混乱は、結果として危険な態度又は行動となる可能性がある；

　　(v) 答えられない/知らない；

(c) 5〜10人の被験者を相手に質問を行う。選択した基準に関連した結果を点検する；

(d) 結果が重大な不一致を示す場合、基準について同意に達するまで上記の手順を繰り返す。

A6.5.12.3　必要に応じて、各作業の質問に対する反応に関し、さらなるコード化が各作業について討論される。

A6.5.13 *分析*

　　これら作業に関して提出される分析は、種々の社会層に関連した簡単な比率計算および平均値である。もっと複雑な分析が計画され、違う作業で示されるかもしれない。理解度に関する総合評価は、種々の社会層の被験者からまとめた結果により行われるが、階層や他の人口統計学的要因など理解度に作用することが知られている重み付けで調整されなければならない。

A6.5.14 *フィードバックおよび追跡*

　　被験者全員に、理解度の評価の結果を見る機会と面接および試験手順に関してのフィードバックを提供すべきである。

A6.5.15 *追跡評価*

　　これらの評価への参加被験者は、GHS 危険有害性情報に関する記憶力と中期間および長期間での同情報にふれた場合の利点を評価するために、1ヶ月後と1年後に再面接を受けるべきである。人的資源や実務に応じて、最初に完了したすべての作業についての再試験は省くことができる場合がある。

ANNEX 7

EXAMPLES OF ARRANGEMENTS OF THE GHS LABEL ELEMENTS

附属書7

ＧＨＳラベル要素の配置例

ANNEX 7

EXAMPLES OF ARRANGEMENTS OF THE GHS LABEL ELEMENTS

The following examples are provided for illustrative purposes and are subject to further discussion and consideration by the GHS Sub-Committee.

Example 1:   **Combination packaging for a Category 2 flammable liquid**

        Outer packaging: Box with a flammable liquid transport label*
        Inner packaging: Plastic bottle with GHS hazard warning label**

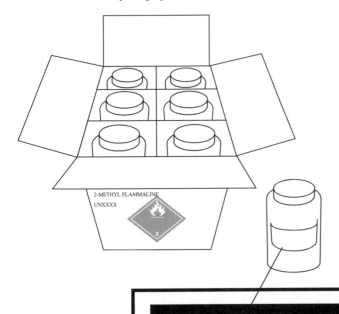

---

*   *Only the UN transport markings and labels are required for outer packagings.*
**   *A flammable liquid pictogram as specified in the "UN Model Regulations" may be used in place of the GHS pictogram shown on the inner packaging label.*

附属書7

GHSラベル要素の配置例

以下は図解のために示された例で、GHS小委員会でさらに議論と検討が行われる。

例1： 引火性液体区分2の組合せ容器包装

　　　外装容器：引火性液体輸送標札を付した箱＊
　　　内装容器：GHS危険有害性警告ラベルを附したプラスチック・びん＊＊

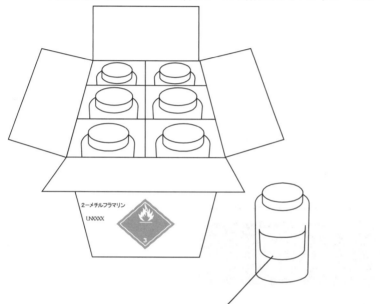

＊　　外装容器には国連輸送標札のみが必要。
＊＊　「UNモデル規則」に定める引火性液体の絵表示は、内装容器に表示するGHS絵表示に替えて使用することができる。

- 451 -

Example 2: Combination packaging for a Category 1 specific target organ toxicant and Category 2 flammable liquid

Outer packaging: Box with a flammable liquid transport label*
Inner packaging: Plastic bottle with GHS hazard warning label**

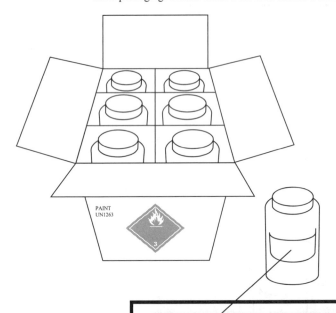

PAINT (FLAMMALINE, LEAD CHROMOMIUM)

Product identifier

(see 1.4.10.5.2 (d))

SIGNAL WORD (see 1.4.10.5.2 (a))

**

Hazard statements (see 1.4.10.5.2 (b))

Precautionary statements (see 1.4.10.5.2 (c))

Additional information as required by the competent authority as appropriate.

Supplier identification (see 1.4.10.5.2 (e))

---

*   *Only the UN transport markings and labels are required for outer packagings.*
** *A flammable liquid pictogram as specified in the UN Model Regulations may be used in place of the GHS pictogram shown on the inner packaging label.*

例2： 特定標的臓器毒性物質区分1および引火性液体区分2の組合せ容器包装

外装容器：引火性液体輸送標札を付した箱*
内装容器：GHS危険有害性警告ラベルを附したプラスチック・びん**

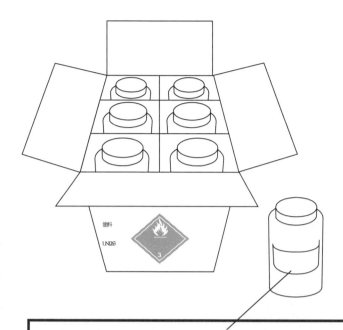

塗料（フラマリン、レッドクロモミウム）　　　製品特定名
　　　　　　　　　　　　　　　　　　　　　　（1.4.10.5.2(d) 参照）

　　　　　　　注意喚起語（1.4.10.5.2(a) 参照）

**

　　　　　　　危険有害性情報（1.4.10.5.2(b) 参照）

注意書き（1.4.10.5.2(c) 参照）
　所管官庁が指定する追加情報があればここに記載する。
供給者名称（1.4.10.5.2(e) 参照）

* 　外装容器には国連輸送標札のみが必要。
** 　「UNモデル規則」に定める引火性液体の絵表示は、内装容器に表示するGHS絵表示に替えて使用することができる。

Example 3: Combination packaging for a Category 2 skin irritant and Category 2A eye irritant

Outer packaging: Box with no label required for transport*
Inner packaging: Plastic bottle with GHS hazard warning label

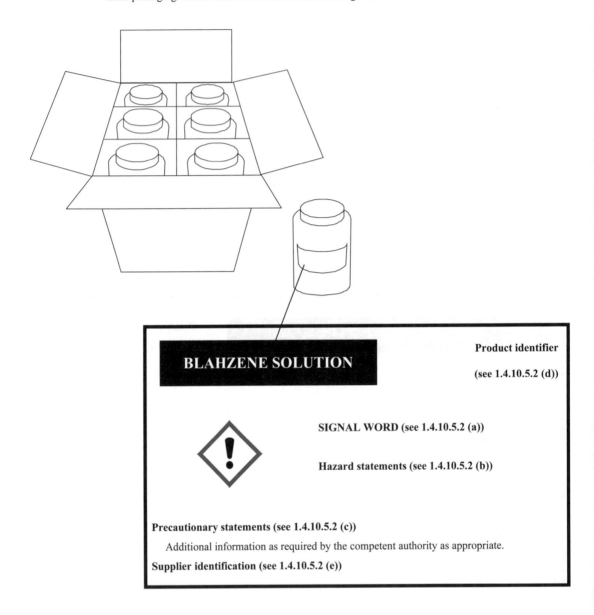

* *Some competent authorities may require a GHS label on the outer packaging in the absence of a transport label.*

例3： 皮膚刺激性物質区分2および眼刺激性物質区分2Aの組合せ容器包装

外装容器：輸送標札が不要な箱*
内装容器：GHS危険有害性警告ラベルを付したプラスチック・びん

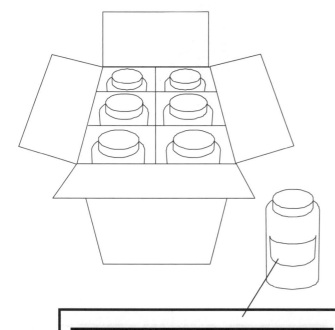

ブラーゼン溶液	製品特定名 (1.4.10.5.2(d) 参照)

注意喚起語 (1.4.10.5.2(a) 参照)

危険有害性情報 (1.4.10.5.2(b) 参照)

注意書き (1.4.10.5.2(c) 参照)
　所管官庁が指定する追加情報があればここに記載する。
供給者名称 (1.4.10.5.2(e) 参照)

* 輸送標札がない場合に、所管官庁によっては外装容器にGHSラベルを要求することがある。

- 453 -

Example 4: Single packaging (200 *l* drum) for a Category 2 flammable liquid

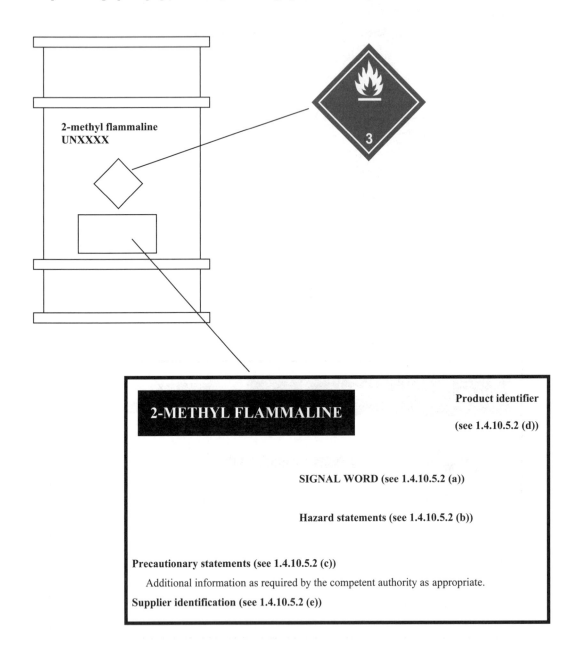

Note: *The GHS label and the flammable liquid pictogram and markings required by the "UN Model Regulations" may also be presented in a combined format.*

例4: 引火性液体区分2の単一容器包装（200リットルドラム）

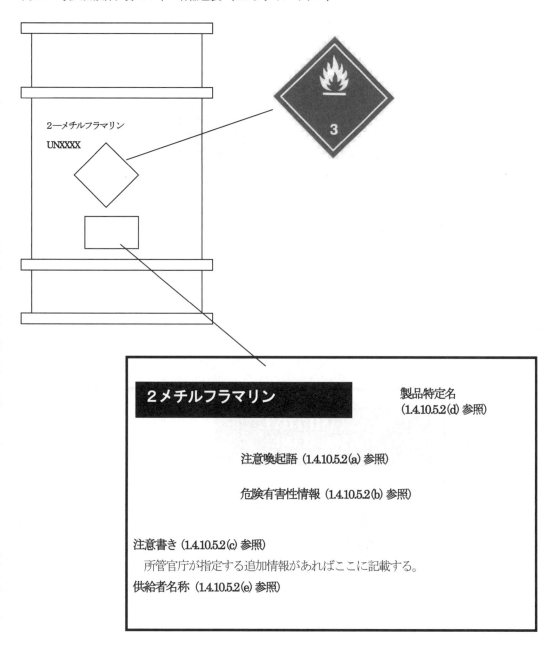

注記: GHSラベルと「UNモデル規則」に定める引火性液体の標札および表示は組み合わせた形式で表示することができる。

Example 5: Single packaging for a Category 1 specific target organ toxicant and Category 2 flammable liquid

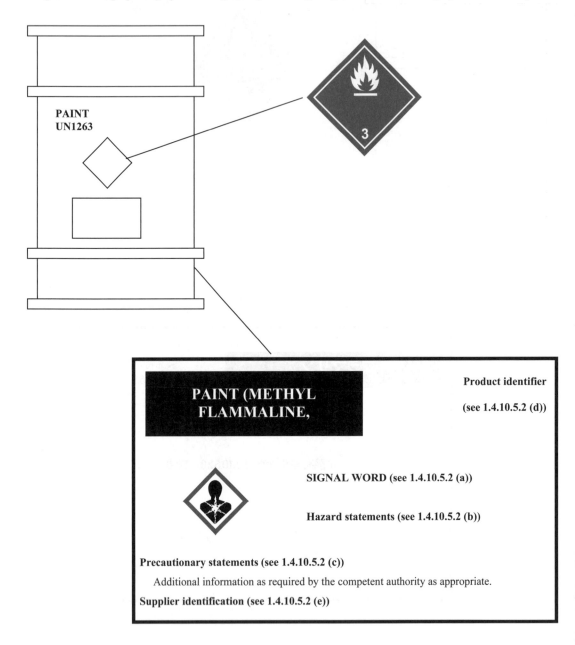

Note: *The GHS label and the flammable liquid pictogram and markings required by the "UN Model Regulations" may also be presented in a combined format.*

例5： 特定標的臓器毒性物質区分1および引火性液体区分2の単一容器包装

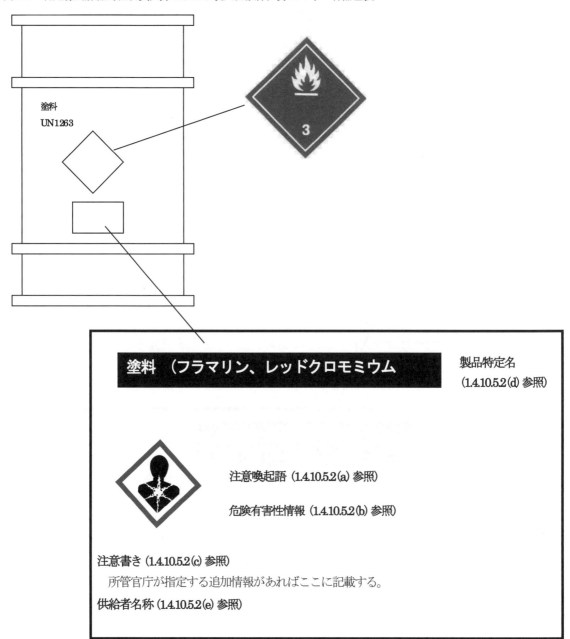

注記：GHSラベルと「UNモデル規則」に定める引火性液体の標札および表示は組み合わせた形式で表示することができる。

Example 6: Single packaging for a Category 2 skin irritant and Category 2A eye irritant

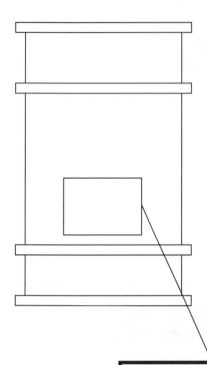

BLAHZENE SOLUTION

Product identifier
(see 1.4.10.5.2 (d))

SIGNAL WORD (see 1.4.10.5.2 (a))

Hazard statements (see 1.4.10.5.2 (b))

Precautionary statements (see 1.4.10.5.2 (c))

Additional information as required by the competent authority as appropriate.

Supplier identification (see 1.4.10.5.2 (e))

例6: 皮膚刺激性物質区分2および眼刺激性物質区分2Aの単一容器包装

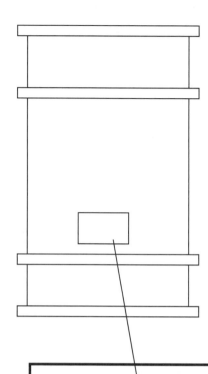

ブラーゼン溶液　　　　　　　　　製品特定名
　　　　　　　　　　　　　　　　(1.4.10.5.2(d) 参照)

　　　　　　注意喚起語 (1.4.10.5.2(a) 参照)

　　　　　　危険有害性情報 (1.4.10.5.2(b) 参照)

注意書き (1.4.10.5.2(c) 参照)
　所管官庁が指定する追加情報があればここに記載する。
供給者名称 (1.4.10.5.2(e) 参照)

Example 7: **Additional guidance when transport and other GHS information appear on single packagings**

(a) Where transport and other GHS information appear on a single packaging (e.g. a 200 l drum), consideration must be given to ensure that the label elements are placed in a manner that addresses the needs of the different sectors;

(b) Transport pictograms must convey information immediately in an emergency situation. They must be able to be seen from a distance, as well as in conditions that are smoky or otherwise partially obscure the package;

(c) The transport-related pictograms are distinct in appearance from pictograms intended solely for non-transport purposes which helps to distinguish them;

(d) The transport pictograms may be placed on a separate panel of a GHS label to distinguish them from the other information or may be placed adjacent to the other GHS information on the packaging;

(e) The pictograms may be distinguished by adjusting their size. Generally speaking, the size of the non-transport pictograms should be proportional to the size of the text of the other label elements. This would generally be smaller than the transport-related pictograms, but such size adjustments should not affect the clarity or comprehensibility of the non-transport pictograms;

Following is an example of how such a label may appear:

例7： 輸送および他のGHS情報が単一容器包装上に付される時の追加手引き

(a) 輸送および他のGHS情報が単一容器包装（例えば、200リットルドラム）上に付される場合には、異なった分野の必要性に対応する方法で配置されるよう考慮されなければならない；

(b) 輸送絵表示は緊急事態において直ちに情報を伝えなければならない。くすんでいる、または部分的に不鮮明な条件でも、また離れた所からでも見えなければならない；

(c) 輸送関連絵表示は、非輸送目的にのみ意図された絵表示と比べて、外見において明瞭である；

(d) 輸送絵表示は他の情報から識別するために、GHSラベルから分離した表示板に配置されてもよい、または、包装上の他のGHS情報と隣接して配置されてもよい；

(e) 絵表示はそれらの大きさを調整することにより識別されてもよい。一般的に、非輸送絵表示の大きさは他のラベル要素の文言の大きさとつりあっているべきである。これは一般的に輸送関連絵表示より小さいが、そのような大きさの調整は非輸送絵表示の明瞭さとわかりやすさに影響を与えるべきではない；

以下は、そのようなラベルをどのように付すのかの例である：

Single packaging using 3 adjacent panels to convey multiple hazards.

Product classified as: (a) Category 2 Flammable liquid; (b) Category Acute 4 (by inhalation); and (c) Category 2 Specific target organ toxicant following repeated exposure.

UN Number
Proper shipping name

[Universal Product Code (UPC)]

Danger
Keep out of the reach of children.
Read label before use.

Highly flammable liquid and vapour.
Harmful if inhaled.
May cause liver and kidney damage through prolonged or repeated exposure.

Keep container tightly closed.
Keep away from heat, hot surfaces, sparks, open flames and other ignition sources. No smoking.
Use only outdoors or in a well-ventilated area.
Do not breathe dust/fume/gas/mist/vapours/spray.
Wear protective gloves/protective clothing/eye protection/face protection/hearing protection/…
Ground and bond container and receiving equipment.

In case of fire: Use [as specified] to extinguish.

FIRST AID
IF INHALED: Remove person to fresh air and keep comfortable for breathing.
Call a POISON CENTER/doctor if you feel unwell.
Store in a well-ventilated place. Keep cool

CODE
PRODUCT NAME

COMPANY NAME

Street Address
City, State, Postal Code, Country
Phone Number
Emergency Phone Number

DIRECTIONS FOR USE:
XXXXXXXXXXXXXXXXXX
XXXXXXXXXXXXXXXXXX
XXXXXXXXXXXXXXXXXX

Fill weight: XXXX
Gross weight: XXXX
Expiration date: XXXX
Fill date: XXXX
Lot number: XXXX

- 458 -

複数の危険有害性を表示するために隣接する3つの表示板を用いた単一容器包装

製品の分類：(a) 引火性液体・区分2；(b) 急性毒性・区分4（吸入）；および (c) 特定標的臓器毒性・区分2（反復ばく露）

危険

子供の手の届かないように保管すること
使用前にラベルを読むこと

国連番号
正式輸送品名

[ユニバーサルプロダクトコード
（UPC）]

引火性の高い液体および蒸気。
吸入すると有害。
長期にわたる、または、反復ばく露による肝臓及び腎臓の障害のおそれ。

容器を密閉しておくこと。
熱火花/裸火から遠ざけること—禁煙。
屋外または換気のよい場所でのみ使用すること。
粉じん/煙/ガス/ミスト/蒸気/スプレーを吸入しないこと。
保護手袋および保護眼鏡/保護面/聴覚保護具/...を着用すること。
容器を接地/アースを取ること。

火災の場合には、[指定された]消火剤を用いること。

救急処置

吸入した場合：空気の新鮮な場所に移動し、呼吸しやすい姿勢で休息させること。
気分が悪いときは、医師に連絡すること。

換気の良い場所で保管すること。涼しいところに置くこと。

コード

製品名

会社名

町名
国名、県名、市名、郵便番号、
電話番号
緊急連絡先電話番号

使用法：
XXXXXXXXXXXXXXXXXXX
XXXXXXXXXXXXXXXXXXX
XXXXXXXXXXXXXXXXXXX

充填重量：XXXX　　ロット番号：XX
総重量：XXXX　　充填日：　　XXXX
有効期限：XXXXXX

- 458 -

Example 8:   **Labelling of small packagings**

Small immediate container that cannot be labelled based on shape/size and restrictions relating to the method of use, contained in an outside packaging which can display the entire information required on the GHS label

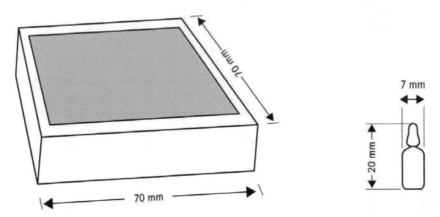

Cardboard box containing glass ampoules of a product used as laboratory reagent.
Each ampoule contains 0.5 g.

The working solution of the reagent is prepared by removing the top of the ampoule and placing the bottom half (containing the product) in the required amount of deionized water. Consequently, labels cannot be applied to the actual ampoules as they may contaminate the working solution, which may affect subsequent reactions. It is impossible to put all applicable GHS label elements on the immediate container (i.e. the glass ampoule) due to its size and shape.

The area available on the outer cardboard box is large enough to carry a legible version of the required GHS label elements.

The unlabelled glass ampoule is sealed in a polythene sleeve with an end tag for a label – the ampoule is not removed from the polythene sleeve until the point of intended use, i.e. preparation of the working solution. The area available for a label on the end tag is not sufficient to include all the required label elements. The labelling includes at least:

- the product identifier, signal word and name plus telephone number of the supplier on one side of the end tag;
- the hazard pictograms on the other side of the end tag.

This ensures that the user is aware of the product identity (enables identification of the associated safety data sheet), its hazards (indicates that the product is hazardous and needs to be handled/stored appropriately) and the name/contact details of the supplier (if needed in an emergency situation). The signal word and the pictogram are not on the same side in order to ensure the presence of safety information on both sides of the end tag.

Inner packaging
sleeve with minimum required GHS label elements

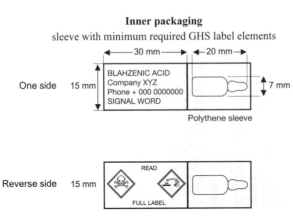

- 459 -

例 8 :　　小さな容器包装のラベル

形/サイズおよび使用方法による制限からラベルを貼ることができない小さい直接容器が、GHS ラベルで要求されているすべての情報が表示できる外容器に容れられている

試験用試薬として使用される製品のガラスアンプルを容れた段ボール箱、それぞれのアンプルは 0.5g 含む

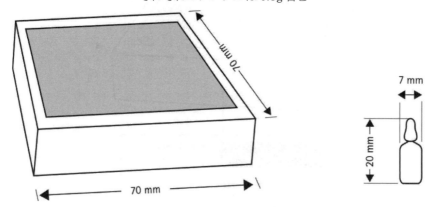

　試薬の作業用液体は、アンプルの頭部を除去し、下半分（製品を含む）を必要量の脱イオン水の中に入れることで作成する。したがって、作業用液体を汚染する可能性のあるラベルをアンプルに貼り付けることができない。これが続いて起きる反応に影響を与える可能性がある。サイズおよび形状により、直接容器（ガラスアンプル）に GHS の該当表示要素をすべて記載するのは不可能である。

　外側の段ボール箱の面積は大きく、要求される GHS ラベル要素を読みやすい大きさで記載できる。

　ラベルのないガラスアンプルはラベル用のタグを端につけたポリエチレンスリーブで密閉される－アンプルは使用直前、すなわち作業用液体を調整するまでポリエチレンスリーブから外さない。タグの端にあるラベル用の面積はすべての要求されるラベル要素を含むには十分ではない。表示には最小限以下を含む：

- タグの端の片面に製品特定名、注意喚起語および供給者の名称および電話番号；
- タグの端のもう一方の片面に危険有害性絵表示。

　これは、使用者が製品の本体（関連した SDS の確認を可能にする）、その危険有害性（製品は危険有害であり取扱い/保管が適当に行われる必要があることを示す）および供給者の名称/連絡先（緊急時に必要ならば）に気づくことを確実にする。端のタグの両面で安全情報を確実に示すために、注意喚起語および絵表示は同じ面にはしない。

内装容器：

最小限の GHS ラベル要素が記載されているスリーブ

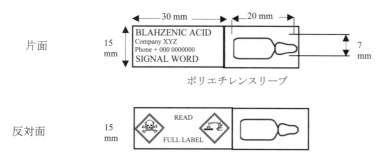

- 459 -

Outer packaging
All required GHS label elements (including hazard and precautionary statements) appear on the outside packaging

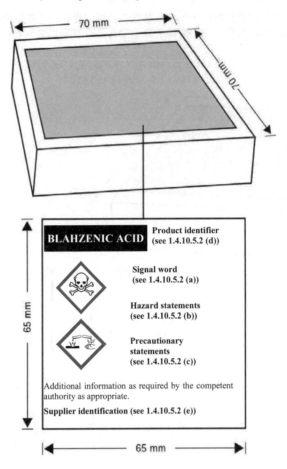

外装容器：
外包装にはすべての要求される GHS ラベル要素（危険有害性情報および注意喚起語）が表示される

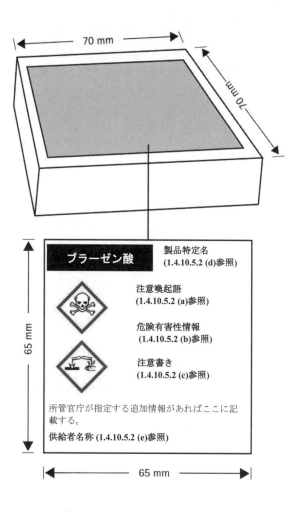

Example 9: Labelling of small packagings: Fold-out labels

This example illustrates one way to label containers where the manufacturer/supplier or competent authority has determined there is insufficient space to place the GHS pictogram(s), signal word, and hazard statement(s) together, as provided in 1.4.10.5.4.1, on the surface of the container. This might occur, for example, when the container is small, there are a large number of hazard statements assigned to the chemical, or the information needs to be displayed in multiple languages, so that the information may not be printed on the label in a size that is easily legible.

Metal container

A fold-out label is securely affixed to the immediate container (i.e. the fold-out label is attached so that it remains affixed during the foreseeable conditions and period of use). The fold-out label is produced in such a way that the front part cannot be detached from the remainder of the label and the label can repeatedly be closed again so it is not hanging loose.

The information is structured as follows and is provided, if applicable, in all the languages used for the label:

Front page

Information to be provided on the front page of the multilayer/fold-out label should contain at least:

GHS information:

- Product identifier[1]
- Hazard pictogram(s)
- Signal word
- Supplier identification (name, address and telephone number of the company)

Additional information:

- A symbol to inform the user that the label can be opened to illustrate that additional information is available on inside pages
- If more than one language is used on the fold-out label: the country codes or language codes

---

[1] *The product identifier on the front and back page does not include hazardous components. If hazardous components are required on the label they are displayed in the appropriate languages on the text pages.*

例9： 小さな容器包装のラベル：折りたたみラベル

この例は、製造者/供給者または所管官庁が、1.4.10.5.4.1に示されているように、容器の表面にGHS絵表示、注意喚起語、および危険有害性情報を一緒に示す場所が不十分であると決定した場合のラベルの1つの方法を示している。これは、例えば容器が小さい、化学品に多くの危険有害性が割り当てられている、あるいは情報を多言語で示す必要がある場合などに起こり、結果として情報を容易に読めるサイズでラベルに印刷できないであろう。

折りたたみラベルは直接容器にしっかり固定される（すなわち折りたたみラベルは見ることができる状態で使用期間中固定されて添付される）。折りたたみラベルは、表面部分がラベルの他と分離しないように、さらに折りたたみラベルは繰り返し閉じられてもだらしなく垂れさがらないように作られる。

もし可能であれば、ラベルに使用されるすべての言語において、情報は以下のように構成され示される。

表紙

多層/折りたたみラベルの表紙頁に示される情報には最低限以下のものが含まれる：

GHS情報：
- 製品特定名 [1]
- 危険有害性絵表示
- 注意喚起語
- 供給者情報（会社名、会社の住所及び電話番号）

追加的情報：
- ラベルは開くことができ内部頁に追加的な情報があることを使用者に示すためのシンボル
- 折りたたみラベルに1つ以上の言語を使用する場合：国コードまたは言語コード

---

[1] *表紙および裏表紙の製品特定名には危険有害成分は含まない。危険有害成分がラベルに要求されている場合には、それらは適当な言語で文章に示す。*

- 461 -

Text pages/Pages inside

GHS information:

- Product identifier including, as applicable, hazardous components contributing to the classification
- Signal word
- Hazard statements
- Precautionary statements
- Additional information (e.g. directions for use, information required by other regulations, etc.)

Additional information:

- If more than one language is used on the fold-out label: the country codes or language codes

Back page (affixed to the immediate container):

- Product identifier*
- Hazard pictogram(s)
- Signal word
- Supplier identification (name, address and telephone number of the company)

The product identifier (if applicable) and the signal words on the front page and the back page are in all languages used on the label.

If there is enough space on the front or on the back page, these pages can also be used to display text.

The text on the inside pages (text pages) can also be distributed on more than one page, if the available space is not sufficient. In general, it is better to spread the text across more than one page than to have smaller letters that make the text difficult to read. In all cases, the visibility and easy legibility of the label elements should be ensured without the aid of any device other than corrective lenses and contrasted with any other information on the hazardous product or the container.

It is recognized that some regulatory systems (e.g. pesticides) may have specific requirements for the application of labels using a multilayer or booklet style format. Where this is the case, labelling would be undertaken in accordance with the competent authority's requirements.

The size of the fold-out label and the number of folds should be in a rational relationship to the size of the container. This may limit the number of languages, which can be displayed on the fold-out label.

文章/内部頁：

GHS 情報：

- 製品特定名、可能であれば分類に寄与した危険有害成分も含む
- 注意喚起語
- 危険有害性情報
- 注意書き
- 追加的情報（例えば使用の指示、他の法令で要求されている情報など）

追加的情報：

- 折りたたみラベルに１つ以上の言語を使用する場合：国コードまたは言語コード

裏表紙（直接容器に固定）：

- 製品特定名*
- 危険有害性絵表示
- 注意喚起語
- 供給者情報（会社名、会社の住所及び電話番号）

表紙および裏表紙に製品特定名（可能であれば）と注意喚起語を、使用するすべての言語でラベルに記載する。

表紙または裏表紙に十分なスペースがあれば、ここに文言を記載することもできる。

スペースが十分でない場合には、内部頁（文章）の文言を 1 頁以上にわたり配分してもよい。一般に、文言を読むことが困難な小さい文字よりも、1 頁以上にわたり文言を展開したほうがよい。すべての場合において、ラベル要素の視認性および読み取り性が、矯正レンズ以外のいかなる道具の助けなしに確保され、有害な製品または容器上の他の情報より際立っていなければならない。

いくつかの法システム（例えば農薬）では多層または小冊子スタイルを使用したラベルの適用に関して特別な要求事項があることが知られている。この場合、表示は所管官庁の要求事項にしたがって行われるであろう。

折りたたみラベルのサイズおよび折りたたみの回数は、容器のサイズと合理的な関係でなければならない。これは折りたたみラベルに示すことができる言語の数を制限するかもしれない。

Examples:

Application of the labelling principles discussed in this example are illustrated for a multilingual label in the accordion style below:

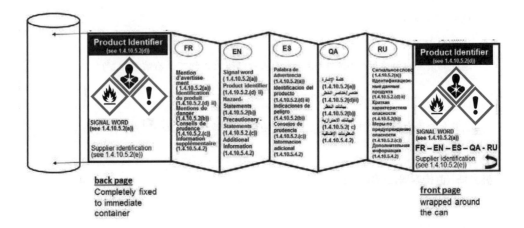

Additionally, the labelling principles discussed in this example could also be applied to any other fold-out label styles such as e.g. book style, order book style and window style.

Book style

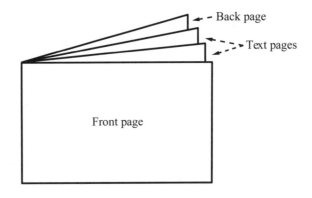

例：

この例で検討される表示の原則の適用は、下記のアコーディオンスタイルの多言語ラベルで示した。

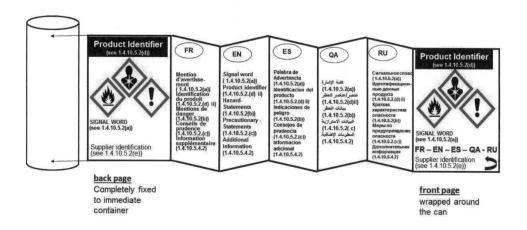

さらに、この例で検討された表示の原則は、例えば本型、注文帳型および扉型等どのような他の折りたたみラベルにも適用できるであろう。

本型

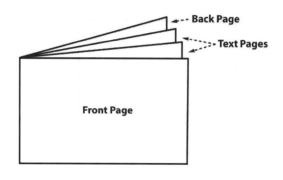

- 463 -

Order book style

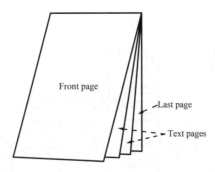

Window style

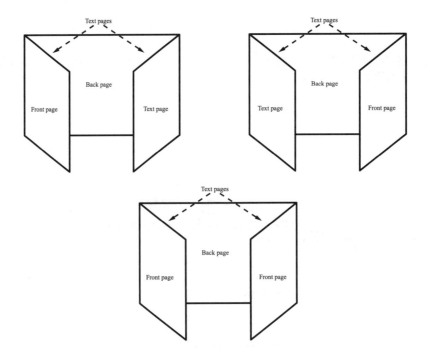

注文帳型

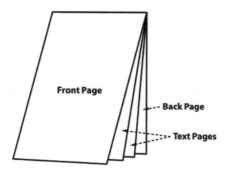

扉型

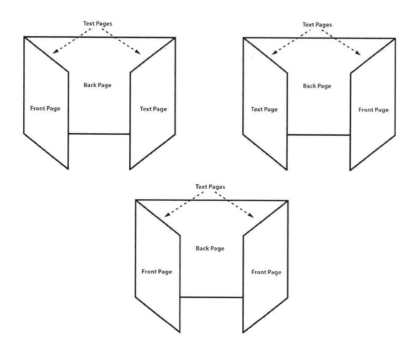

Example 10: Labelling of - sets or kits

A set or kit is a combination packaging intended for defined applications. Generally, a set or kit contains two or more small removable inner containers. Each inner container contains different products which can be hazardous or not hazardous substances or mixtures.

This example illustrates ways to label sets or kits where the manufacturer/supplier or competent authority has determined there is insufficient space to place together on each inner container within the kit, the GHS pictogram(s), signal word and hazard statement(s) in accordance with 1.4.10.5.4.1. This may occur when, for example, the inner containers are small, or there are a large number of hazard statements assigned to the chemical, or the information needs to be presented in multiple languages, so all the information cannot be printed on the label in a size that is easily legible. Two different scenarios where this may arise are illustrated, together with ways to provide the necessary GHS information.

Scenario A

The set or kit comprises an outer packaging containing the following inner containers: four cuvettes, each filled with the same substance or mixture (reagent 1) and two larger containers each filled with another substance or mixture (reagent 2).

The approach is to provide minimum information on each of the inner containers containing hazardous substances or mixtures, and to provide the full GHS label information for each hazardous substance or mixture on the outer packaging. For clarity, the full label information for each hazardous substance or mixture is grouped together on the outer packaging.

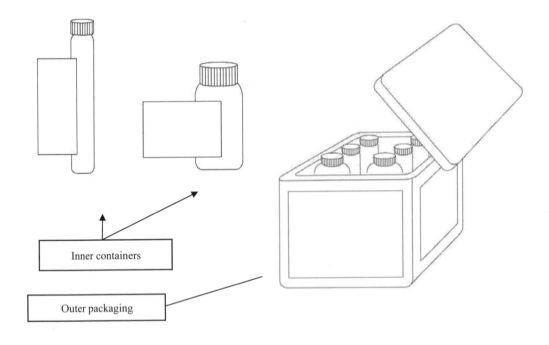

例 10： セットまたはキットのラベル

セットまたはキットは、特定の用途を意図した組み合わせ容器包装である。一般にセットまたはキットには 2 つ以上の取り出せる小さな内容器が入っている。それぞれの内容器は、危険有害なまたは危険有害ではない物質または混合物の異なる製品を含んでいる。

この例は、製造者/供給者または所管官庁が 1.4.10.5.4.1 にしたがった GHS の絵表示、注意喚起語および危険有害性情報をキット内のそれぞれの内容器にまとめて表示するスペースが不十分であると決定した場合に、セットまたはキットに表示する方法を示している。これは例えば、内容器が小さい、またはたくさんの危険有害性情報が化学品に割り当てられている、または情報を多くの言語で示す必要があるなど、つまりすべての情報が簡単に読めるサイズではラベルに印刷できない場合に起こりうる。このことが起こりうる二通りの異なる例を、必要な GHS 情報を提供する方法と共に示した。

例 A

セットまたはキットは以下の内容器を含んだ外装容器からなる：それぞれが同じ物質または混合物（試薬 1）で満たされている 4 つのキュベット、またそれぞれが別の同じ物質または混合物（試薬 2）で満たされている 2 つのより大きな容器。

解決方法は、危険有害な物質または混合物を含む内容器のそれぞれに最小限の情報を記載すること、そして外装容器にそれぞれの危険有害な物質または混合物に対するすべての GHS ラベル情報を記載することである。明確にするために、それぞれの危険有害な物質および混合物のすべてのラベル情報は外装容器にまとめられている。

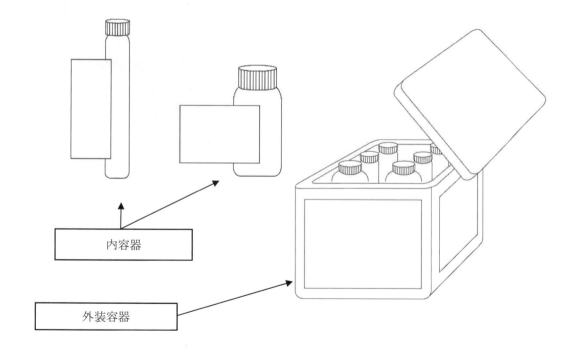

Inner container label

As the area available for a label on the inner containers is not sufficient to include all the required GHS label elements, the following minimum information is included on the label of each hazardous substance or mixture:

- Product identifier[2], and an identifier for each substance or mixture matching the identifier used on the outer packaging label and SDS for that substance or mixture, e.g. "Reagent 1" and "Reagent 2"

- Pictogram(s)

- Signal word

- The statement "Read full label"

- Supplier identification (i.e. name and telephone number)

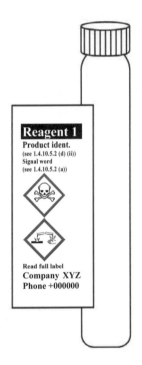
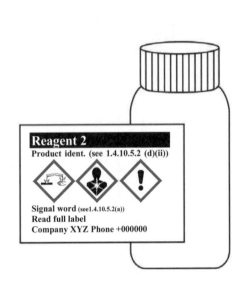

Outer packaging label

In addition to the set or kit identifier, in this case Reagent Kit for water analysis (see below), all the required GHS label elements appear on the outer packaging for each hazardous substance or mixture.

The label elements for each substance or mixture on the outer packaging are grouped together in order to distinguish clearly which label elements are assigned to which substance or mixture.

However, the supplier identification need appear only once on the outer packaging. Where possible any supplemental information may also be included on the outer packaging.

When a large number of precautionary statements are required, the precautionary statements may be located separately from the rest of the label elements, though general precautionary statements (table A3.2.1) and

---

[2] Where hazardous components are required to be identified on the label, they are displayed in the appropriate languages on the outer packaging label.

内容器ラベル

内容器のラベルに関してその利用できる面積が、すべての要求されている GHS ラベル要素を含むためには不十分なので、下記の最小限の情報をそれぞれの危険有害な物質または混合物のラベルに記載する：

- 製品特定名 [2]、それぞれの物質または混合物の特定名は、外装容器および SDS で使用されている物質または混合物の特定名と一致させる、例えば「試薬 1」および「試薬 2」
- 絵表示
- 注意喚起語
- 「すべてのラベル情報を読むこと」という文言
- 供給者名（すなわち名前および電話番号）

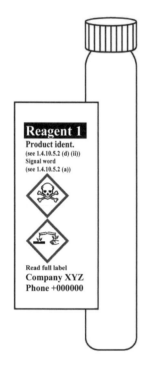
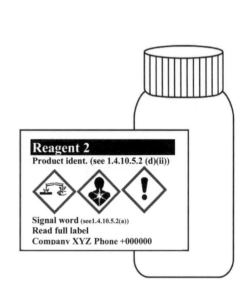

外装容器ラベル

この水分析の薬品キット（下記参照）の場合には、セットまたはキットの特定名に加えて、すべての要求されている GHS ラベル要素がそれぞれの危険有害物質または混合物に対して外装容器に示される。

外装容器におけるそれぞれの危険有害な物質または混合物に対するラベル要素は、どのラベル要素がどの物質または混合物に割り当てられたかを明確に区別するために、グループ化される。

しかし供給者の特定名は外装容器の一か所だけに示される必要がある。可能であれば補足的な情報が外装容器に含まれていてもよい。

不適切な記載を避けるために、一般的注意書き（表 3.2.1）および保管に関する注意書きは一か所だけ記載される必要がある（附属書 A3.2.5 注意書き使用における柔軟性参照）ものの、多くの注意書きが要

---

[2] 危険有害な成分をラベルに記載することが要求されている場合には、それらを適当な言語で外装容器のラベルに示す。

precautionary statements for storage need only appear once (see also A3.2.5 in annex 3 on flexibility in the use of precautionary statements) to avoid inappropriate statements, taking into account the nature of the user (e.g. consumers, employers and workers) the quantities supplied, and the intended and foreseeable circumstances of use. In these circumstances, the precautionary statements for each substance or mixture should be grouped together on the same side of the outer packaging and on a surface that is visible under normal conditions of use.

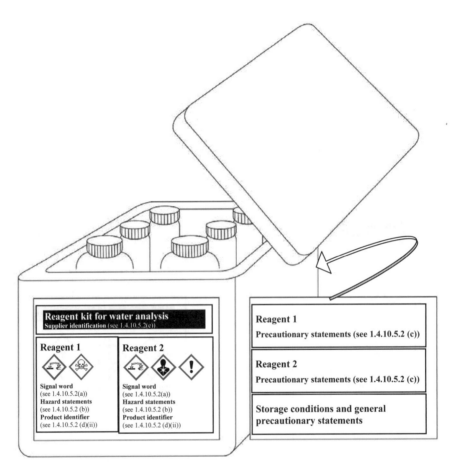

Scenario B

This scenario addresses the situation where it is not possible to affix all appropriate GHS labelling elements for each hazardous substance or mixture in the kit directly on the outer packaging label (due to technical reasons such as the size and shape of this packaging).

This scenario presents a sample kit used for marketing purposes which consist of a large number of different substances or mixtures in individual containers (sample bottles) presented in an outer packaging (e.g. a box). Depending upon the contents of each bottle, some or all of the different substances or mixtures may be classified as hazardous. The individual inner containers (e.g. bottles) are stored in the outer packaging throughout the lifecycle of the sample kit. Customers may select individual bottles and remove them from the box to check clarity, colour or odour and then replace them into the open slot within the outer packaging.

求されている場合、使用者（例えば消費者、事業者及び労働者）の特性、供給量および意図され予見可能な使用環境を考慮して、注意書きは他のラベル要素とは別に記載されてもよい。これらの状況では、それぞれの物質または混合物に対する注意書きは外装容器の同じ面でしかも通常の使用下で見える面にグループ化されているべきである。

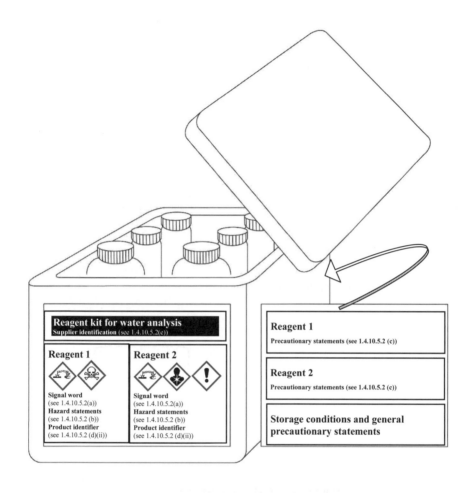

例 B

　この例は、キットのそれぞれの危険有害な物質または混合物に対するすべての適当な GHS ラベル要素を外装容器ラベルに貼ることができない（容器包装のサイズや形状などの技術的理由によって）場合について検討する。

　この例は、外装容器（例えば箱）に並べられている、それぞれの容器（サンプル瓶）に入ったたくさんの異なる物質または混合物からなるマーケティング目的に使用されるサンプルキットである。それぞれの瓶の内容物によるが、異なる物質または混合物のいくつかまたはすべては危険有害であると分類されるかもしれない。個々の内容器（例えば瓶）はサンプルキットのライフサイクルを通じて外装容器内に収められている。顧客は個々の瓶を選択し、透明度、色または臭いをチェックするために箱から取り出し、外装容器内にある小さい穴に戻すであろう。

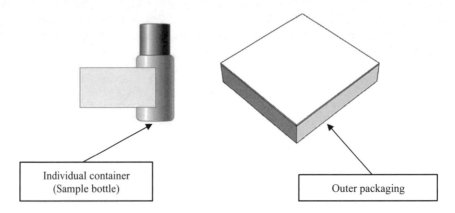

Individual container label

As the area available for a label on the different individual containers is not sufficient to include all required GHS label elements, the following minimum required information should be required:

- supplier identification (i.e. name and telephone number);
- product identifier[2];
- pictogram(s);
- signal word;
- the statement "Read full label enclosed".

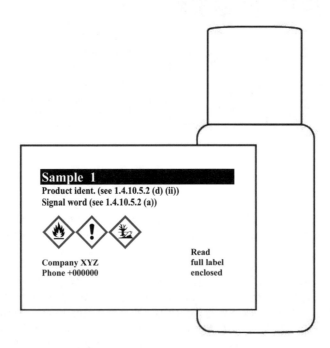

Example of individual container label

---

[2] *Where hazardous components are required to be identified on the label they are displayed in the appropriate languages as part of the full label information attached to the inside of the kit.*

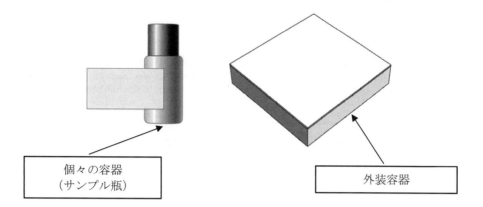

個々の容器
（サンプル瓶）

外装容器

個々の容器ラベル

異なるそれぞれの容器の面積がすべての必要な GHS ラベル要素を記載するには十分ではないので、以下の最小限の必要な情報が記載されるべきである：

- 供給者特定名（すなわち名前および電話番号）；
- 製品特定名[2]；
- 絵表示；
- 注意喚起語；
- 「すべてのラベル情報を読むこと」という文言。

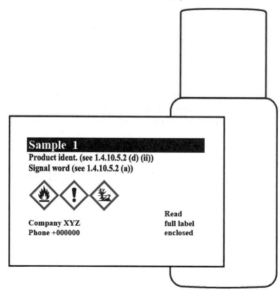

個々の内装容器のラベル例

---

[2] 危険有害な成分をラベルに記載することが要求されている場合には、それらはキットの中に添付してあるすべてのラベル情報の一部として適当な言語で示される。

Full label information

Attached to the inside of the outer packaging is the full GHS label information for each individual container containing a hazardous substance or mixture. The individual product identifiers on the label align with the product identifier on the individual container label. An example of the content of the full label information is shown below.

Product identifier (see 1.4.10.5.2 (d) (ii))	Pictogram(s) (see 1.4.10.4)	Signal word (see 1.4.10.5.2 (a))	Hazard statement(s) (see 1.4.10.5.2 (b))	Precautionary statement(s) (see 1.4.10.5.2 (c))	Supplemental information (see 1.4.10.5.4.2)
123	(flame) (exclamation mark) (environment)	Warning	Flammable liquid and vapour. Causes skin irritation. Toxic to aquatic life with long lasting effects.	Keep away from heat, hot surfaces, sparks, open flames and other ignition sources. No smoking. Keep container tightly closed. Use explosion-proof equipment. Use non-sparking tools. Take action to prevent static discharge. Avoid release to the environment. Wear protective gloves. IF ON SKIN (or hair): Take off immediately all contaminated clothing. Rinse affected areas with water. In case of fire: Use dry sand, dry chemical or alcohol-resistant foam for extinction. Store in a well-ventilated place. Keep cool.	

Although the contents of each inner container may not be classified as hazardous in accordance with the GHS, and thus would not need to be identified, it may be identified with a statement such as "Not meeting classification criteria" or "Not classified as hazardous" so as to eliminate confusion on the part of the user if the contents of an inner container is omitted from the full label information.

The document containing the full GHS label information should be organized and printed in a format that allows the user to readily identify the information for each individual container. The visibility of the label elements should be ensured without the aid of any device other than corrective lenses. The approach of this scenario may become infeasible if, given the number of samples, required languages, and precautionary statements, the document grows so large it becomes difficult to locate quickly the label information for a particular inner container.

As shown to the right, full label information regarding each inner container is contained within the outer packaging.
The sheets of full label information are permanently connected to the inside of the combination packaging using a secure method of attachment (e.g. fold out label adhered to box tie or tag as shown)

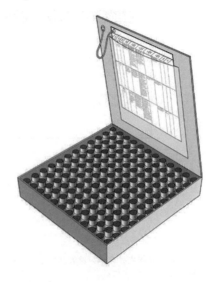

すべてのラベル情報

　危険有害な物質または混合物を含むそれぞれ個々の容器に対するすべての GHS ラベル情報は、外装容器の内側に貼付されている。このラベルにある個々の製品の特定名は、個々の容器のラベルにある製品の特定名と一致している。すべてのラベル情報の内容の例を以下に示した。

製品特定名 (1.4.10.5.2 (d)(ii)参照)	絵表示 (1.4.10.4 参照)	注意喚起語 (1.4.10.5.2(a)参照)	危険有害性情報 (1.4.10.5.2(b)参照)	注意書き (1.4.10.5.2(c)参照)	補足情報 (1.4.10.5.4.2参照)
123	(炎)(!)(環境)	警告	引火性の液体および蒸気 皮膚刺激 長期継続的影響により水生生物に毒性	熱、高温のもの、火花、裸火および他の着火源から遠ざけること。 禁煙。 容器を密閉しておくこと。 防爆型の機器を使用すること。 火花を発生させない工具を使用すること。 静電気放電に対する措置を講ずること。 環境への放出を避けること。 保護手袋を着用すること。 皮膚（または髪）に付着した場合：直ちに汚染された衣類をすべて脱ぐこと。接触部位を水で洗うこと。 火災の場合：消火するために乾燥砂、粉末薬品または耐アルコール泡を使用すること。 換気の良い場所で保管すること	

　たとえそれぞれの内容器の内容物が GHS により危険有害とは分類されない、つまり記載する必要がないとしても、「分類判定基準に合致しない」または「危険有害性の区分に該当しない」のような記載は必要であろう。もし内容器の内容が完全なラベル情報から省略されていても使用者の混乱を取り除くことができる。

　すべての GHS ラベル情報を含んだ文書は、使用者が即座にそれぞれ個々の容器に対する情報を認識することができるように、構成され印刷されるべきである。ラベル要素の視認性は、矯正レンズ以外のどのような道具を使用することもなく、確保されるべきである。この例の方法は、たくさんのサンプルがあり、複数言語が要求されさらに注意書きがあると、文書は多くなり、特定の内容器に対するラベル情報を素早く見つけ出すことが困難になり、実行不可能になるであろう。

　右に示すように、それぞれの内容器に関するすべてのラベル情報は外装容器内にある。
　すべてのラベル情報のシートは添付の確実な方法で組み合わせ包装の内部に恒久的に連結されている。（例えば図のように、折りたたまれたラベルがひもやタグなどで結ばれている）

Outer packaging label

The outer box, given the limited area for labelling, will display:

- kit identifier (name of kit);
- supplier identification (see 1.4.10.5.2(e));
- storage and general precautionary statements for the kit as a whole;
- pictograms for each single hazardous substance or mixture, without duplication;
- signal word (the most stringent assigned to any component);
- the statement "Read full label enclosed".

MARKET KIT

Product ident. (see 1.4.10.5.2 (d) (ii))

Signal word (see 1.4.10.5.2 (a))

Precautionary storage statements (see 1.4.10.5.2 (c))

Read full label enclosed

Supplier identification (see 1.4.10.5.2 (e))

外装容器ラベル

ラベル用の面積が限られている外箱について示す：

- キット特定名（キットの名称）；
- 供給者名（1.4.10.5.2(e)参照）；
- キット全体に対する保管および一般注意書き；
- それぞれの危険有害な物質または混合物に対する絵表示、重複なし；
- 注意喚起語（どれかの成分に割り当てられた中で最も厳しいもの）；
- 「すべてのラベル情報を読むこと」という文言。

ANNEX 8

AN EXAMPLE OF CLASSIFICATION IN THE GLOBALLY HARMONIZED SYSTEM

附属書8
世界調和システムにおける
分類例

ANNEX 8

AN EXAMPLE OF CLASSIFICATION IN THE GLOBALLY HARMONIZED SYSTEM

A8.1　Classification proposal

The following classification proposal draws on the GHS criteria. The document includes both brief statements about the proposal for each health hazard class and details of all the available scientific evidence.

Classification is proposed for both the acute toxicity and the corrosivity of this substance based on standard and non-standard animal studies.

Proposed classification	GHS:	Acute oral toxicity Category 4
		Acute dermal (skin) toxicity Category 3
		Skin irritation/corrosion Category 1C
		Eye irritation/serious eye damage Category 1
		Flammable liquid Category 4

A8.2　Identification of the substance

1.1	EINECS Name If not in EINECS IUPAC Name	Globalene Hazexyl Systemol
		CAS No. 999-99-9 EINECS No. 222-222-2
1.2	Synonyms (state ISO name if available)	2-Hazanol Globalethylene
1.3	Molecular formula	$C_xH_yO_z$
1.4	Structural formula	
1.5	Purity (w/w)	
1.6	Significant impurities or additives	
1.7	Known uses	*Industrial:* Solvent for surface coatings and cleaning solutions. Chemical intermediate for Globalexyl UNoxy ILOate. *General public:* Toilet cleaner

附属書 8

世界調和システムにおける分類例

A8.1 分類に関する提案

下記の分類に関する提案は GHS 判定基準によるものである。本文書には、各健康有害性クラスの提案および入手可能なすべての科学的証拠の詳細に関する簡単な記述が含まれる。

標準および標準外動物研究に基づき、この物質には急性毒性および腐食性の両方の分類が提案される。

分類案	GHS： 急性経口毒性　区分 4
	急性経皮（皮膚）毒性　区分 3
	皮膚腐食性/刺激性　区分 1C
	眼に対する重篤な損傷性/眼刺激性　区分 1
	引火性液体　区分 4

A8.2 物質の特定

1.1 EINECS 名称 　　EINECS にない場合は 　　IUPAC 名	グロバリンハーゼキシルシステモール
	CAS No. 999-99-9 EINECS No.　222-222-2
1.2 同義語 　　（ISO 名がある場合記入する）	2-ハザノール グローバルエチレン
1.3 分子式	$C_XH_YO_Z$
1.4 構造式	
1.5 純度 (w/w)	
1.6 重要な不純物または添加物	
1.7 既知の使用法	*工業用:* 表面コーティング溶剤および洗浄液。Globalexyl UNoxy ILOate の中間体 *一般用:* トイレクリーナー

- 473 -

A8.3 Physico-chemical characteristics

Classification as a Category 4 flammable liquid is proposed for the physico-chemical endpoints.

2.1	Physical form	Liquid
2.2	Molecular weight	146.2
2.3	Melting point/range (°C)	-45
2.4	Initial boiling point/ boiling range (°C)	208.3
2.5	Decomposition temperature	
2.6	Vapour pressure (Pa(°C))	7
2.7	Relative density (g/cm^3)	0.887 - 0.890
2.8	Vapour density (air = 1)	5.04
2.9	Fat solubility (mg/kg, °C)	
2.10	Water solubility (mg/kg, °C)	Slightly soluble (0.99 % w/w)
2.11	Partition coefficient (log Pow)	
2.12	Flammability flash point (°C) explosivity limits (%,v/v) auto-flammability temp. (°C)	closed cup: 81.7 open cup: 90.6 lower limit: 1.2 upper limit: 8.4
2.13	Explosivity	No data available
2.14	Oxidizing properties	
2.15	Other physico-chemical properties	

A8.4 Health and environmental characteristics

A8.4.1 *Acute toxicity*

A8.4.1.1 *Oral*

Classification under GHS Category 4 (300-2000 mg/kg) is justified.

Species	LD$_{50}$ (mg/kg)	Observations and remarks	Ref.
Rat	1480	No further details were available.	2
Rat	1500 (males) 740 (females)	The LD$_{50}$ values in mg/kg were calculated from ml/kg using the known density for EGHE of 0.89 g/cm^3.	8

A8.4.1.2 *Inhalation*

There were no deaths or signs of overt toxicity in animals exposed to the saturated vapour concentration of approximately 0.5 mg/l and therefore, the available data do not support classification.

Species	LC$_{50}$ (mg/l)	Exposure time (h)	Observations and remarks	Ref.
Rat	> 83 ppm. (approx = 0.5 mg/l).	4	No deaths, clinical signs or gross lesions occurred at 83 ppm (85 ppm is stated to be the saturated vapour concentration at room temperature).	3
Rat	Not stated	6	The animals were exposed to the saturated vapour concentration at room temperature (assumed to be 85 ppm). No deaths occurred and no signs of gross pathology were observed.	8
Rat	Not stated	8	No deaths occurred with exposure to the "saturated vapour concentration" at room temperature (assumed to be 85 ppm).	2

A8.3 物理化学的特性

物理化学危険性として区分 4 引火性液体の分類が提案される。

2.1	物理的形状	液体
2.2	分子量	146.2
2.3	融点/範囲 (℃)	-45
2.4	初留点/沸点範囲 (℃)	208.3
2.5	分解温度	
2.6	蒸気圧 (Pa(℃))	7
2.7	比重 (g/cm3)	0.887〜0.890
2.8	蒸気密度 (空気=1)	5.04
2.9	脂溶性 (mg/kg, ℃)	
2.10	水溶性 (mg/kg, ℃)	わずかに水溶性がある (0.99% w/w)
2.11	分配係数 (log Pow)	
2.12	可燃性 引火点 (℃) 爆発限界 (%,v/v) 自然発火温度 (℃)	クローズドカップ法：81.7　　オープンカップ法：90.6 下限：　　　　　1.2　　上限：　　　　　8.4
2.13	爆発性	データがない
2.14	酸化特性	
2.15	他の物理化学的特性	

A8.4 健康および環境特性

A8.4.1 急性毒性

A8.4.1.1 経口

GHS 区分 4（300-2000 mg/kg）の分類とする証拠が示されている。

種類	LD$_{50}$ (mg/kg)	観察および備考	文献
ラット	1480	詳細な情報がない。	2
ラット	1500（雄） 740（雌）	mg/kg 単位の LD$_{50}$ 値は、EGHE の既知の比重 0.89 g/cm³ を用いて ml/kg 値から計算した。	8

A8.4.1.2 吸入

約 0.5mg/l の飽和蒸気濃度に対してばく露された実験動物に、死亡または明白な毒性の徴候はなかった。ゆえに、入手可能なデータからは分類するのに十分な根拠が示されない。

種類	LC$_{50}$ (mg/l)	ばく露時間 (時)	観察および備考	文献
ラット	> 83 ppm （ほぼ 0.5 mg/l に相当）	4	83 ppm では、死亡、臨床的症状または肉眼的病変は起きない。（85 ppm が室温で飽和蒸気濃度であると記載されている）	3
ラット	記述なし	6	実験動物は、室温で（85 ppm と想定される）飽和蒸気濃度でばく露された。死亡は起きず、臨床病理学的症状は観察されなかった。	8
ラット	記述なし	8	室温では、（85 ppm と想定される）"飽和蒸気濃度" に対するばく露で死亡が起きなかった。	2

A8.4.1.3   *Skin*

Classification under GHS Category 3 (200-1000 mg/kg) is justified.

Species	LD$_{50}$ (mg/kg)	Observations and remarks	Ref.
Rat	790	No further details were available.	2
Rabbit (5/sex/ group)	720 (males) 830 (females)	Animals were exposed to up to 3560 mg/kg for 24 hours. All but 2 of the animals that died did so during the application period. Following the exposure period, local toxicity (erythema, oedema, necrosis and ecchymoses) was reported in an unstated number of animals, and persisted throughout the 14 day post-application observation period. Ulceration was also noted in an unstated number of animals at the end of the observation period.	8

A8.4.2   *Skin irritation/corrosion*

There are conflicting reports concerning the irritant nature of this substance. In a dedicated skin irritation study reported in the same paper as the acute dermal study, the author states that "necrosis" was observed in 3 of 6 treated rabbits which was still present on the last day of observation (day 7), along with mild to moderate erythema. Mild to marked oedema was also observed during the course of the study but had resolved within the 7-day observation period. Given that one animal showed no evidence of any skin response in this study and that only slight to moderate skin irritation was observed in the other animals the observation of "necrosis" in three of the animals is somewhat surprising. An acute dermal (skin) toxicity study in rabbits also reported signs of skin irritation including the description "necrosis" and ulceration but did not quantify the number of animals affected. In contrast to these findings, an old and briefly reported study indicated that there was little or no indication of skin irritation in rabbits.

Similarly, mixed skin irritation findings have been observed with a closely related substance, for which both necrosis and no skin irritation has been reported. In addition, a secondary source indicates that some other similar substances cause "moderate" skin irritation, and that prolonged exposure to these group of substances may cause burns. However, much shorter chain similar substances are not considered to be skin irritants.

It was considered that the reported necrosis in both the acute dermal and skin irritation studies cannot be dismissed and, taken together with the findings seen with structurally similar substances, this justifies classification. There are three Categories under the GHS for classification as corrosive. The data do not match the criteria readily, but Category 1C would be appropriate since the necrotic lesions observed occurred after an exposure period of 4 hours. There is no evidence to suggest that significantly shorter exposures would produce skin corrosion.

Species	No. of animals	Exposure time (h)	Conc. (w/w)	Dressing: (occlusive, semi-occlusive, open)	Observations and remarks (specify degree and nature of irritation and reversibility)	Ref.
Rabbit	6	4	0.5 ml of 100 %	Occlusive	No signs of irritation were observed in one animal, and only slight erythema (grade 1) in another on day 1, which had resolved by day 7. Four animals showed a mild to moderate erythema (grade 1-2) and a mild to marked oedema (grade 1-3) after removal of the dressing. The oedema had resolved by day 7 post-exposure. "Necrosis" at the application site was reported in 3/6 rabbits from day 1 until the end of the observation period on day 7. Desquamation was observed in 4/6 rabbits on day 7.	8
Rabbit (albino)	5	24	100 % (volume not stated)	Not stated	Little or no signs of skin irritation were found in this poorly reported study.	2

A8.4.1.3 経皮

GHS区分3 (200-1000 mg/kg)の分類とすることの根拠が示されている。

種類	LD$_{50}$ (mg/kg)	観察および備考	文献
ラット	790	詳細な情報がない。	2
ウサギ (5/性/グループ)	720 (雄) 830 (雌)	実験動物は、24時間3560 mg/kgまでばく露された。2匹以外全ては適用期間中に死亡した。ばく露後に、局所毒性(紅斑、浮腫、壊死および斑状出血)が、動物の数は記述されてないが報告されており、14日間の適用後観察期間中ずっと続いた。観察期間の終わりに、動物の数は記述されてないが潰瘍形成についても言及されていた。	8

A8.4.2 皮膚腐食性/刺激性

この物質の刺激性に関し、矛盾した報告がある。皮膚刺激性試験を目的とした急性経皮研究と同じ報告書の中で、報告者は、観察の最終日(7日目)にまだ生存していた6匹の処置済みのウサギ中、3匹に軽い～中程度の紅斑に加えて「壊死」が観察されたと記述している。軽い浮腫も研究過程の間中観察されたが、7日間の観察期間内で消退した。この研究で、動物のうち1匹は全く皮膚反応の証拠を見せず、他の動物に微かな程度から中程度の皮膚刺激性が観察された記述があるので、動物のうち3匹の「壊死」の観察は若干の驚きである。ウサギを用いた急性経皮(皮膚)毒性の研究でも「壊死」および潰瘍形成の記述を含む皮膚の徴候について報告されたが、病気に冒された動物の数については示されていない。これらの所見と対照的に、古くて簡単な研究報告には、ウサギには皮膚刺激性の特性がほとんどないか皆無であると明示されている。

同様に矛盾した皮膚刺激性の所見が、近い関係の物質で観察されており、その物質に対して壊死と皮膚刺激性がないとの両方の報告がなされている。加えて、二次文献には、他の幾つかの類似した物質が「中程度」の皮膚刺激性を起こし、これらの物質グループへの長期のばく露は熱傷の原因となることが示されている。しかし、より短い連鎖の類似物質は、皮膚刺激性物質になるとは考えられていない。

急性経皮・皮膚刺激性の研究において報告された壊死は無視できないと考えられ、さらに構造的に類似した物質での所見もあり、分類が正当化される。GHS分類では、腐食性として3つの区分がある。これらのデータは、区分に容易に整合しないが、4時間のばく露時間の後壊死病変が起きたことが観察されているので、区分1Cが適切である。著しく短いばく露が皮膚腐食性を引き起こすことを示す証拠はない。

種類	動物の数	ばく露時間(時)	濃度(w/w)	包帯:(閉塞性、半閉塞性、解放性)	観察および備考(刺激性・可逆性の程度と性質の特定)	文献
ウサギ	6	4	0.5 mlで100%	閉塞性	動物の1匹には刺激性の徴候が見られず、1日目に他の1匹に微かな紅斑(悪性度1)が見られ、7日目に消退した。包帯をはずした後、4匹に軽い～中程度の紅斑(悪性度1～2)が、軽い～顕著な程度の浮腫(悪性度1～3)が見られた。ばく露後7日目に浮腫は消退した。適用場所での「壊死」が、1日目から観察期間最終日の7日目までに6匹中3匹に起きたと報告されている。7日目に6匹中4匹に皮膚の剥離が観察されている。	8
ウサギ(アルビノ)	5	24	100%(容量は記述なし)	記述なし	この不十分な研究報告では、皮膚刺激性の徴候は、ほとんどまたは全く見られない。	2

A8.4.3 *Serious damage to eyes/eye irritation*

The only available study involved exposure of rabbits to considerably lower amounts of the test substance than the standard protocols for this endpoint recommend. Relatively severe (e.g. conjunctival redness grade 3) but reversible effects were seen. It is predictable that under standard test conditions, the effects on the eye would be very severe and consequently GHS Category 1 (irreversible effects on the eye) would be justified.

Species	No. of animals	Conc. (w/w)	Observations and remarks (specify degree and nature if irritation, any serious lesions, reversibility)	Ref.
Rabbit	6	0.005 ml of 100 %	One hour post-instillation conjunctival redness (grade 3) and discharge (grade 2.8) observed. The mean scores for the 24, 48 and 72 hour readings for corneal opacity, iris, conjunctival redness, chemosis and discharge were all approx 0.5. All lesions had resolved by day 7.	8
Rabbit	60	1 and 5 %	A report in the secondary literature of severe eye injury observed in rabbits associated with instillation of an unstated amount of 5 %, could not be substantiated as the information was not found in the reference stated.	1

A8.4.4 *Skin and respiratory sensitization*

No data are available. There are no additional grounds for concern (e.g. structure activity relationships) and no classification proposed.

A8.4.5 *Specific target organ toxicity following single or repeated exposure*

A8.4.5.1 *Toxicity following single exposure*

There is no reliable information available about the potential of this substance to produce specific, non-lethal target organ toxicity arising from a single exposure. Therefore, under GHS, no classification for specific target organ toxicity (STOT) single exposure is proposed.

A8.4.5.2 *Toxicity following repeated exposure*

A8.4.5.2.1 Oral

No oral repeat dose studies or human evidence are available and therefore no classification is proposed.

A8.4.5.2.2 Inhalation

There was no evidence of adverse toxicity in a 13-week rat inhalation study at 0.43 mg/l (approx. 72 ppm), an exposure level close to the saturated vapour concentration. No classification is justified according to GHS criteria.

Species	conc. mg/l	Exposure time (h)	Duration of treatment	Observations and remarks (specify group size, NOEL, effects of major toxicological significance)	Ref.
Rat (F344) 20/sex / group (plus 10/ sex/group - 4 week recovery groups)	0.12, 0.24 and 0.425	6	5 d/wk for 13 weeks	No deaths occurred. Decreased weight gain was observed in high dose animals of both sexes and medium dose females. There were no toxicologically significant changes in haematological or urinalysis parameters. High dose females showed an increase in alkaline phosphatase. High and medium dose males showed a statistically significant increase in absolute and relative kidney weight. A small increase in absolute liver weight (12 %) was observed in high dose females. However, there were no gross or histopathological changes in any organs examined.	3

A8.4.3 眼に対する重篤な損傷性/眼刺激性

この有害性に推奨される標準規定よりかなり低用量の試験物質にばく露されたウサギでの研究のみ入手可能であった。比較的重篤（例えば結膜発赤悪性度 3）であるが、可逆的影響が見られた。標準的な試験状態では、眼に対する影響が非常に重篤で、よって GHS 区分 1（眼に対する不可逆的影響）と評価されてもよいであろう。

種類	動物の数	濃度(w/w)	観察および備考（刺激性・重篤な病変・可逆性の程度と性質の特定）	文献
ウサギ	6	0.005 ml で 100%	滴下から 1 時間後に結膜の発赤（悪性度 3）および分泌物（悪性度 2.8）が観察された。角膜混濁、虹彩、結膜発赤、結膜浮腫および分泌物に関する 24 時間、48 時間および 72 時間の平均結果は全て約 0.5 だった。すべての病変は 7 日目までに消退した。	8
ウサギ	60	1 および 5%	量は記載されてないが 5%の液を滴下したウサギの眼において重篤な損傷が観察されたとする二次文献での報告があるが、その報告で挙げられている文献にそのような情報がないので、裏付けることができない。	1

A8.4.4 皮膚感作性および呼吸器感作性

データがない。関連した付加的な証拠（例えば構造活性相関）もなく、分類は提案されない。

A8.4.5 単回または反復ばく露による特定標的臓器毒性

A8.4.5.1 単回ばく露による毒性

この物質が単回ばく露によって特異的、非特異的な特定標的臓器毒性を起こす可能性があるかどうかについて信頼できる情報がない。ゆえに、単回ばく露特定標的臓器毒性の GHS 分類は提案されない。

A8.4.5.2 反復ばく露による毒性

A8.4.5.2.1 経口

反復経口投与の研究あるいは人での証拠が得られず、ゆえに分類も提案されない。

A8.4.5.2.2 吸入

0.43mg/l（約 72 ppm）という飽和蒸気濃度に近いばく露レベルでの 13 週間に渡るラットの吸入に関する研究において、有害毒性の証拠は無かった。GHS の基準に基づく分類に合致する十分な証拠が示されなかった。

種類	濃度 mg/l	ばく露時間(時)	処置期間	観察および備考（グループサイズ・NOEL・主な毒物学的影響の有意性）	文献
ラット (F344) 20/性/グループ（プラス 10/性/グループ - 4 週間リカバリーグループ）	0.12, 0.24 および 0.425	6	13 週間 5 日/週	死亡は起きなかった。体重増加抑制が高用量の雄・雌及び中用量の雌に観察された。血液または尿検査のパラメータには毒物学的に有意な変化は無かった。高用量の雌には、アルカリ・フォスファターゼの増加が見られた。高用量と中用量の雄には、絶対的および相対的な腎臓重量に、統計的に有意な増加が見られた。高用量の雌に、絶対的な肝臓重量に少量の増加（12%）が観察された。しかし、観察した臓器に、肉眼的または組織病理的な変化はなかった。	3

A8.4.5.2.3   Dermal

Unquantified haematological changes were reported in rabbits exposed to 444 mg/kg dermally for 11 days. However, due to the limited information provided, no conclusions can be drawn from this study and no classification is proposed.

Species	Dose mg/kg	Exposure time (h)	Duration of treatment	Observations and remarks (specify group size, NOEL, effects of major toxicological significance)	Ref.
Rabbit	0, 44, 222 and 444	6	9 doses applied over 11 days	This is an unpublished study reported in the secondary literature. Unquantified decreases in haematological parameters were noted in top dose animals. No description of local effects was provided.	1

A8.4.6   *Carcinogenicity (including chronic toxicity studies)*

No data available – no classification proposed.

A8.4.7   *Germ cell mutagenicity*

Negative results have been reported in vitro from Ames, cytogenetics, and gene mutation tests reported in the secondary literature. There are no in vivo data available. These data do not support classification.

In vitro studies

Test	Cell type	Conc. range	Observations and remarks	Ref.
Ames	Salmonella (strains unstated)	0.3-15 mg/plate	**Negative**, in the presence and absence of metabolic activation. This is an unpublished study described in a secondary source and no further information is available.	5
IVC	CHO	0.1-0.8 mg/ml (-S9), 0.08-0.4 mg/ml (+S9)	**Negative**, in the presence and absence of metabolic activation. This is an unpublished study described in a secondary source and no further information is available.	6
Gene mutation	CHO	Not stated	**Negative**. This is an unpublished study described in a secondary source and no further information is available.	7
SCE	CHO	Not stated	**Negative**. This is an unpublished study described in a secondary source and no further information is available.	7

A8.4.8   *Reproductive toxicity-Fertility*

No data available – no classification proposed.

A8.4.9   *Reproductive toxicity*

There was no evidence of reproductive toxicity in rats or rabbits following inhalation exposure to levels inducing slight maternal toxicity. It is noted that although shorter chain related substances are classified for reproductive toxicity, this toxicity decreases with increasing chain length such that there is no evidence of this hazard. No classification is proposed.

Species	Route	Dose	Exposure	Observations and remarks	Ref.
Rat	Inhalation	21, 41 and 80 ppm (0.12, 0.24 and 0.48 mg/l)	days 6-15 of gestation	The substance was tested up to approximately the saturated vapour concentration. Decreases in dam body weight gain, associated with decreases in food consumption, were observed in the medium and high dose groups during the exposure period. There was no evidence of reproductive toxicity.	4
Rabbit	Inhalation	21, 41 and 80 ppm (0.12, 0.24 and 0.48 mg/l)	days 6-18 of gestation	The substance was tested up to approximately the saturated vapour concentration. Decrease in absolute body weight during the exposure period was observed in the high dose animals. There was no evidence of reproductive toxicity.	4

A8.4.5.2.3　経皮

11 日間、444 mg/kg を経皮ばく露したウサギに関して、数量化表示されていないが、血液変化が報告された。しかし提供された情報が限られているため、この研究からは結果が引き出せず、分類も提案されなかった。

種類	投与 mg/kg	ばく露時間（時）	処置期間	観察および備考（グループサイズ・NOEL・主な毒物学的重要性の特定化）	参照
ウサギ	0, 44, 222 および 444	6	11 日間に 9 回投与がなされた。	これは、二次文献で報告された未刊行の研究である。数量表示がないが、血液のパラメータにおける減少が最高用量のウサギに見られた。局所的作用についての記述はない。	1

A8.4.6　発がん性（慢性毒性研究を含む）

データがない — 分類提案はない。

A8.4.7　生殖細胞変異原性

エームス（Ames）、細胞遺伝学、遺伝子突然変異に関する in vitro の試験で陰性の結果が二次文献で報告されている。in vivo のデータは得られていない。これらのデータからは分類はできない。

in vitro 研究

試験	細胞の型	濃度 範囲	観察および備考	文献
Ames	サルモネラ（菌株に関する記述はない。）	0.3-15 mg/プレート	代謝活性化の有無において**陰性**。これは二次文献に記述された未刊行の研究であり、これより詳細な情報が得られない。	5
IVC	CHO	0.1-0.8 mg/ml (-S9), 0.08-0.4 mg/ml (+S9)	代謝活性化の有無において**陰性**。これは二次文献に記述された未刊行の研究であり、これより詳細な情報が得られない。	6
遺伝子突然変異	CHO	記述なし	**陰性**。これは二次文献に記述された未刊行の研究であり、これより詳細な情報が得られない。	7
SCE	CHO	記述なし	**陰性**。これは二次文献に記述された未刊行の研究であり、これより詳細な情報が得られない。	7

A8.4.8　生殖毒性－受精

データがない — 分類提案はない。

A8.4.9　生殖毒性

わずかに母体毒性を誘引するレベルでの吸引ばく露を受けたラットまたはウサギにおける発生毒性の証拠はない。より短い鎖の関連物質が発生毒性に分類されることが知られているが、鎖の長さが増加するに従い、この有害性は減少し、この有害性に関する証拠はない。分類の提案はない。

種類	経路	投与	ばく露	観察および備考	文献
ラット	吸入	21, 41 および 80 ppm（0.12, 0.24 および 0.48 mg/l）	妊娠期間 6～15 日	物質は、ほぼ飽和蒸気濃度まで試験された。ばく露期間中、摂餌量の減少に関連する母体の体重増加の減少が、中用量および高用量のグループに見られた。発生毒性の証拠はなかった。	4
ウサギ	吸入	21, 41 および 80 ppm（0.12, 0.24 および 0.48 mg/l）	妊娠期間 6～18 日	物質は、ほぼ飽和蒸気濃度まで試験された。ばく露期間中、高用量のウサギに、絶対体重の減少が観察された。発生毒性の証拠はなかった。	4

A8.5 **References**

1. Patty, F. (Ed.) (1994). Industrial Hygiene and Toxicology. 4th Ed. pxxxx-xx New York: Wiley-Interscience.

2. Smyth, H.F., Carpenter, C.P., Weil, C.S. and Pozzani, U.S. (1954). Range finding toxicity data. *Arch. Ind. Hyg. Occup. Med*.

3. Fasey, Headrick, Silk and Sundquist (1987). Acute, 9-day, and 13-week vapour inhalation studies on Globalene Hazexyl Systemol. *Fundamental and Applied Toxicology*.

4. Wyeth, Gregor, Pratt and Obadia (1989). Evaluation of the developmental toxicity of Globalene Hazexyl Systemol in Fischer 344 rats and New Zealand White rabbits. *Fundamental and Applied Toxicology*.

5. Etc.

A8.5 文献

1. Patty, F. (Ed.) (1994). Industrial Hygiene and Toxicology. 4th Ed. pxxxx-xx New York: Wiley-Interscience.

2. Smyth, H.F., Carpenter, C.P., Weil, C.S. and Pozzani, U.S. (1954). Range finding toxicity data. *Arch. Ind. Hyg. Occup. Med.*

3. Fasey, Headrick, Silk and Sundquist (1987). Acute, 9-day, and 13-week vapour inhalation studies on Globalene Hazexyl Systemol. *Fundamental and Applied Toxicology.*

4. Wyeth, Gregor, Pratt and Obadia (1989). Evaluation of the developmental toxicity of Globalene Hazexyl Systemol in Fischer 344 rats and New Zealand White rabbits. *Fundamental and Applied Toxicology.*

5. その他

ANNEX 9

GUIDANCE ON HAZARDS TO THE AQUATIC ENVIRONMENT

附属書 9

水生環境有害性に関する手引き

Annex 9

GUIDANCE ON HAZARDS TO THE AQUATIC ENVIRONMENT

Contents

			Page
A9.1	Introduction		**483**
A9.2	**The harmonized classification scheme**		486
	A9.2.1	Scope	486
	A9.2.2	Classification categories and criteria	486
	A9.2.3	Rationale	486
	A9.2.4	Application	487
	A9.2.5	Data availability	488
	A9.2.6	Data quality	488
A9.3	**Aquatic toxicity**		489
	A9.3.1	Introduction	489
	A9.3.2	Description of tests	463
	A9.3.3	Aquatic toxicity concepts	491
	A9.3.4	Weight of evidence	493
	A9.3.5	Difficult to test substances	493
	A9.3.6	Interpreting data quality	498
A9.4	**Degradation**		498
	A9.4.1	Introduction	498
	A9.4.2	Interpretation of degradability data	499
	A9.4.3	General interpretation problems	503
	A9.4.4	Decision scheme	505
A9.5	**Bioaccumulation**		506
	A9.5.1	Introduction	506
	A9.5.2	Interpretation of bioconcentration data	506
	A9.5.3	Chemical classes that need special attention with respect to BCF and K_{ow} values	509
	A9.5.4	Conflicting data and lack of data	511
	A9.5.5	Decision scheme	511
A9.6	**Use of QSAR**		512
	A9.6.1	History	512
	A9.6.2	Experimental artifacts causing underestimation of hazard	512
	A9.6.3	QSAR modelling issues	513
	A9.6.4	Use of QSARs in aquatic classification	514
A9.7	**Classification of metals and metal compounds**		516
	A9.7.1	Introduction	516
	A9.7.2	Application of aquatic toxicity data and solubility data for classification	518
	A9.7.3	Assessment of environmental transformation	521
	A9.7.4	Bioaccumulation	521
	A9.7.5	Application of classification criteria to metals and metal compounds	522
Appendix I	Determination of degradability of organic substances		533
Appendix II	Factors influencing degradability in the aquatic environment		539
Appendix III	Basic principles of the experimental and estimation methods for determination of BCF and K_{ow} of organic substances		543
Appendix IV	Influence of external and internal factors on the bioconcentration potential of organic substances		547
Appendix V	Test guidelines		549
Appendix VI	References		553

附属書9

水生環境有害性に関する手引き

目次

		頁
A9.1	序	483
A9.2	調和された分類スキーム	486
	A9.2.1 適用範囲	486
	A9.2.2 分類区分および分類基準	486
	A9.2.3 根拠	486
	A9.2.4 適用	487
	A9.2.5 データの利用可能性	488
	A9.2.6 データの質	488
A9.3	水生毒性	489
	A9.3.1 序	489
	A9.3.2 試験の説明	490
	A9.3.3 水生毒性の概念	491
	A9.3.4 証拠の重み	493
	A9.3.5 試験困難な物質	493
	A9.3.6 データの質の解釈	498
A9.4	分解性	498
	A9.4.1 序	498
	A9.4.2 分解性データの解釈	499
	A9.4.3 解釈についての一般的な問題	503
	A9.4.4 判定スキーム	505
A9.5	生物蓄積性	506
	A9.5.1 序	506
	A9.5.2 生物濃縮性データの解釈	506
	A9.5.3 BCF および K_{ow} 値に関して特別な注意が必要な化学品クラス	509
	A9.5.4 矛盾するデータおよびデータの欠如	511
	A9.5.5 判定スキーム	511
A9.6	QSAR の使用	512
	A9.6.1 経緯	512
	A9.6.2 有害性の過小評価を起こす実験技術上の誤差	512
	A9.6.3 QSAR モデル化の課題	513
	A9.6.4 水生環境有害性分類への QSAR の使用	514
A9.7	金属および金属化合物の分類	516
	A9.7.1 序	516
	A9.7.2 分類への水生毒性データおよび溶解度データの適用	518
	A9.7.3 環境における変化に関する評価	521
	A9.7.4 生物蓄積性	521
	A9.7.5 金属および金属化合物に関する分類基準の適用	522
付録 I	有機物質の分解性の測定	533
付録 II	水生環境中の分解性に影響する因子	539
付録 III	有機物質の BCF および K_{ow} 測定のための実験法および推定法の基本原理	543
付録 IV	有機物質の生物濃縮性に対する体外および体内要因の影響	547
付録 V	テストガイドライン	549
付録 VI	参考文献	553

ANNEX 9

GUIDANCE ON HAZARDS TO THE AQUATIC ENVIRONMENT

NOTE: *The text of annex 9 is largely based on the "Guidance Document on the use of the harmonised system for the classification of chemicals which are hazardous for the aquatic environment" published by OECD in 2001, as Series on Testing and Assessment No.27 (ENV/JM/MONO(2001)8). The guidance document has remained unchanged since its publication in 2001, but since then, new OECD test guidelines or guidance documents have been adopted which are an additional source of information. For a list of updated references, refer to appendices V and VI to annex 9.*

A9.1 Introduction

A9.1.1 In developing the set of criteria for identifying substances hazardous to the aquatic environment, it was agreed that the detail needed to properly define the hazard to the environment resulted in a complex system for which some suitable guidance would be necessary. Therefore, the purpose of this document is twofold:

(a) to provide a description of and guidance to how the system will work;

(b) to provide a guidance to the interpretation of data for use in applying the classification criteria.

A9.1.2 The hazard classification scheme has been developed with the object of identifying those substances that present, through the intrinsic properties they possess, a danger to the aquatic environment. In this context, the aquatic environment is taken as the aquatic ecosystem in freshwater and marine, and the organisms that live in it. For most substances, the majority of data available addresses this environmental compartment. The definition is limited in scope in that it does not, as yet, include aquatic sediments, nor higher organisms at the top end of the aquatic food-chain, although these may to some extent be covered by the criteria selected.

A9.1.3 Although limited in scope, it is widely accepted that this compartment is both vulnerable, in that it is the final receiving environment for many harmful substances, and the organisms that live there are sensitive. It is also complex since any system that seeks to identify hazards to the environment must seek to define those effects in terms of wider effects on ecosystems rather than on individuals within a species or population. As will be described in detail in the subsequent sections, a limited set of specific properties of substances have been selected through which the hazard can be best described: acute aquatic toxicity; chronic aquatic toxicity; lack of degradability; and potential or actual bioaccumulation. The rationale for the selection of these data as the means to define the aquatic hazard will be described in more detail in section A9.2.

A9.1.4 This annex is limited at this stage, to the application of the criteria to substances. The term substances covers a wide range of chemicals, many of which pose difficult challenges to a classification system based on rigid criteria. The following sections will thus provide some guidance as to how these challenges can be dealt with based both on experience in use and clear scientific rationale. While the harmonized criteria apply most easily to the classification of individual substances of defined structure (see definition in chapter 1.2), some materials that fall under this category are frequently referred to as "complex mixtures". In most cases they can be characterized as a homologous series of substances with a certain range of carbon chain length/number or degree of substitution. Special methodologies have been developed for testing which provides data for evaluating the intrinsic hazard to aquatic organisms, bioaccumulation and degradation. More specific guidance is provided in the separate sections on these properties. For the purpose of this guidance document, these materials will be referred to as "complex substances" or "multi-component substances".

A9.1.5 Each of these properties (i.e. acute aquatic toxicity, chronic aquatic toxicity, degradability, bioaccumulation) can present a complex interpretational problem, even for experts. While internationally agreed testing guidelines exist and should be used for any and all new data produced, many data usable in classification will not have been generated according to such standard tests. Even where standard tests have been used, some substances, such as complex substances, hydrolytically unstable substances, polymers etc, present difficult interpretational problems when the results have to be used within the classification scheme. Thus, data are available for a wide variety of both standard and non-standard test organisms, both marine and freshwater, of varying duration and utilizing a variety of endpoints. Degradation data may be biotic or abiotic and can vary in environmental relevance. The potential to bioaccumulate can, for many organic chemicals, be indicated by the octanol-water partition coefficient. It can however be affected by many other factors and these will also need to be taken into account.

附属書　9

水生環境有害性に関する手引き

注記：附属書9の文章は主にOECDが2001年に試験評価シリーズNo.27（*ENV/JM/MONO(2001)8*）として発行した「水生環境に有害な化学品の分類のための調和システムの使用に関するガイダンス文書」に基づいている。ガイダンス文書は2001年の発行以来変更されていないが、それ以降、追加の情報源である新しいOECDテストガイドラインまたはガイダンス文書が採択されている。更新された参考文献のリストについては附属書9の付録Vおよび付録VIを参照のこと。

A9.1　序

A9.1.1　水生環境に有害な物質を特定するための判定基準のセットを開発するにあたり、環境に対する有害性を正しく定義するのに必要となる詳細な事項は、何らかの適切な手引きを必要とするであろう複合システムとなることが合意された。したがって、本文書の目的は以下の2つである：

(a)　このシステムがどのように機能するかについての説明と手引きを提供すること；

(b)　分類基準を適用する際に用いられるデータの解釈に対する手引きを提供すること。

A9.1.2　有害性分類スキームは、本来有する特性を介した水生環境への危険性を有するような物質を特定することを目的として開発された。ここでは水生環境とは、淡水および海洋中の生態系、ならびにそこに生息する生物体と解釈される。ほとんどの物質について、利用可能なデータの大部分はこの環境コンパートメントを扱っている。この定義は、適用範囲に水中底質、あるいは水中食物連鎖の頂点にある高位の生物がまだ含まれていないため、選択された判定基準である程度はカバーされているとはいえ、限界がある。

A9.1.3　適用範囲の点で限界はあるものの、このコンパートメントは多くの有害物質を最終的に受け入れる環境である点で無防備なものであり、そこに住む生物は高感受性であると広く理解されている。さらに、環境有害性の特定を追求するどのようなシステムでも、種または個体群中のそれぞれの個体に対するよりは、むしろ生態系に対するより広範な影響という意味での影響を特定することを追求せねばならないことから、このコンパートメントは複雑でもある。以下の各節に詳しく説明されるように、物質の有害性について最も適切に説明できる特定の性質の限定されたデータセットが選択されている。すなわち、急性水生毒性、慢性水生毒性、分解性の欠如、および潜在的な、または実際の生物蓄積性である。これらのデータを水生有害性の判定の手段として選択する根拠については、A9.2節に詳述する。

A9.1.4　この段階では、本附属書は物質への判定基準の適用に限定される。物質という用語は広範囲な化学品に対して用いられ、その化学品の多くは、柔軟性のない判定基準にもとづいた分類システムに対して困難な課題をもたらすものである。このため、以降の各節は、こうした課題に対して、使用経験と明確な科学的根拠に基づいてどのように対処できるかについての、いくつかの手引きを与えるものである。調和された判定基準は決まった構造をもつ物質を分類するのに最も容易に適用できる（1.2節の定義を参照のこと）が、この区分の対象となる物質が「複合混合物」とされる場合もしばしばある。ほとんどの例では、こうした物質は炭素鎖の長さ/置換基の数または置換度が一定の範囲内にある一連の同族物質として特徴づけることができる。水生生物に対する本質的有害性、生物蓄積性および分解性を評価するためのデータが得られるように、試験のための特別な方法論が開発されている。こうした特性については別々の節で、より具体的な手引きが示されている。本手引きの目的にそって、こうした物質は「複合物」または「多成分物質」と呼ばれる。

A9.1.5　これらの各性質（すなわち、急性水生毒性、慢性水生毒性、分解性、生物蓄積性）は専門家にとっても、解釈上複雑な問題を提起することがある。国際的に合意されたテストガイドラインが存在し、いかなる新規データにも採用されることになっているが、分類に使用できるデータの多くはこうした標準試験にしたがって作成されていないことがある。たとえ標準試験法が採用されていても、例えば複合物、加水分解性不安定物質、ポリマー等のように、結果を分類スキームの範囲内で使用しなければならない場合に、解釈上複雑な問題を提起する物質もある。したがって、標準の、ならびに標準でない試験生物種、海水および淡水、様々な試験期間と多様なエンドポイントを採用したデータが利用可能である。分解データには生物的なものもあれば非生物的なものもあり、環境との関連性も違うことがある。多くの有機化合物では、生物蓄積性はオクタノール/水分配係数で示すことができる。しかし、この生物蓄積性はその他多くの要因に影響されるので、こうした要因についても考慮することが必要となる。

A9.1.6   It is clearly the objective of a globally harmonized system that, having agreed on a common set of criteria, a common dataset should also be used so that once classified, the classification is globally accepted. For this to occur, there must first be a common understanding of the type of data that can be used in applying the criteria, both in type and quality, and subsequently a common interpretation of the data when measured against the criteria. For that reason, it has been felt necessary to develop a transparent guidance document that would seek to expand and explain the criteria in such a way that a common understanding of their rationale and a common approach to data interpretation may be achieved. This is of particular importance since any harmonized system applied to the "universe of chemicals" will rely heavily on self-classification by manufacturers and suppliers, classifications that must be accepted across national boundaries without always receiving regulatory scrutiny. This guidance document, therefore, seeks to inform the reader, in a number of key areas, and as a result lead to classification in a consistent manner, thus ensuring a truly harmonized and self-operating system.

A9.1.7   Firstly, it will provide a detailed description of the criteria, a rationale for the criteria selected, and an overview of how the scheme will work in practice (section A9.2). This section will address the common sources of data, the need to apply quality criteria, how to classify when the dataset is incomplete or when a large data-set leads to an ambiguous classification, and other commonly encountered classification problems.

A9.1.8   Secondly, the guidance will provide detailed expert advice on the interpretation of data derived from the available databases, including how to use non-standard data, and specific quality criteria that may apply for individual properties. The problems of data interpretation for "difficult substances", those substances for which standard testing methods either do not apply or give difficult interpretational problems, will be described and advice provided on suitable solutions. The emphasis will be on data interpretation rather than testing since the system will, as far as possible, rely on the best available existing data and data required for regulatory purposes. The four core properties, acute and chronic aquatic toxicity (section A9.3), degradability (section A9.4) and bioaccumulation (section A9.5) are treated separately.

A9.1.9   The range of interpretational problems can be extensive and as a result such interpretation will always rely on the ability and expertise of the individuals responsible for classification. However, it is possible to identify some commonly occurring difficulties and provide guidance that distils accepted expert judgement that can act as an aid to achieving a reliable and consistent result. Such difficulties can fall into a number of overlapping issues:

        (a)   The difficulty in applying the current test procedures to a number of types of substance;

        (b)   The difficulty in interpreting the data derived both from these "difficult to test" substances and from other substances;

        (c)   The difficulty in interpretation of diverse datasets derived from a wide variety of sources.

A9.1.10   For many organic substances, the testing and interpretation of data present no problems when applying both the relevant OECD Guideline and the classification criteria. There are a number of typical interpretational problems, however, that can be characterized by the type of substance being studied. These are commonly called "difficult substances":

        (a)   <u>poorly soluble substances</u>: these substances are difficult to test because they present problems in solution preparation, and in concentration maintenance and verification during aquatic toxicity testing. In addition, many available data for such substances have been produced using "solutions" in excess of the water solubility resulting in major interpretational problems in defining the true $L(E)C_{50}$ or NOEC for the purposes of classification. Interpretation of the partitioning behaviour can also be problematic where the poor solubility in water and octanol may be compounded by insufficient sensitivity in the analytical method. Water solubility may be difficult to determine and is frequently recorded as simply being less than the detection limit, creating problems in interpreting both aquatic toxicity and bioaccumulation studies. In biodegradation studies, poor solubility may result in low bioavailability and thus lower than expected biodegradation rates. The specific test method or the choice of procedures used can thus be of key importance;

        (b)   <u>unstable substances</u>: such substances that degrade (or react) rapidly in the test system present both testing and interpretational problems. It will be necessary to determine whether the correct methodology has been used, whether it is the substance or the degradation/reaction product that has been tested, and whether the data produced is relevant to the classification of the parent substance;

        (c)   <u>volatile substances</u>: such substances that can clearly present testing problems when used in open systems should be evaluated to ensure adequate maintenance of exposure concentrations. Loss of

A9.1.6　世界調和システムの明確な目標は、判定基準の共通セットに合意した上で、いったん分類されたら、その分類が世界的に受け入れられるように共通のデータセットも使用されるべきであるということである。これを実現させるためには、まず判定基準を適用するのに使用できるデータのタイプおよび質についての共通の理解、それゆえデータを判定基準に対して評価する場合の共通の解釈がなければならない。この理由により、判定基準の論理的基礎についての共通の理解およびデータ解釈への共通のアプローチが達成できるようなやり方で、判定基準を展開し説明することを追求する、透明性のある手引きの作成が必要であると感じられるようになった。「化学品の全領域」に調和システムを適用する場合、製造者や供給者自身による分類、すなわち必ずしも常に行政の監視を受けるとは限らないが、国境を越えて適用される分類に依存することが大きいので、こうした手引きは特に重要である。したがって本手引きは、読者に多くの重要な領域について情報を提供し、その結果として一貫性のあるやり方での分類に導き、真に調和された自律的なシステムを確立することを求めている。

A9.1.7　第一に、本手引きは判定基準に関する詳細な説明、選択された判定基準の論理的根拠、およびこのスキームが実際どのように機能するかについての概観を示す（A9.2 節）。この節では、データの一般的な入手源、信頼性の判定基準を適用する必要性、データセットが不完全な場合、またはデータセットが大きくて分類が多義的になる場合の分類の方法、およびその他一般的に遭遇する分類上の問題について述べる。

A9.1.8　第二に、本手引きは、非標準的なデータの使用を含めて、利用可能なデータベースから得られるデータの解釈、および個々の性質に適用される特別な信頼性の判定基準について、詳細な専門的助言を示す。「試験困難物質」、すなわち標準的な試験法が適用されないか、または解釈上の困難を生じる物質におけるデータ解釈の問題について説明がなされ、適切な解決法についての助言が示される。本システムは、可能な限り、利用可能な最も良い既存データおよび規制目的のために必要なデータに依存するので、試験よりむしろデータ解釈の方に重点がおかれる。核となる4つの特性、すなわち急性および慢性水生毒性（A 9.3節）、分解性（A9.4 節）および生物蓄積性（A9.5 節）は個別に取り扱う。

A9.1.9　解釈上の問題の範囲は広大となる可能性があり、その結果、つねに解釈は分類を担当する者の能力と専門知識に依存することになる。しかし、共通に起こる困難な問題をいくつか特定して、信頼できる一貫性のある結果を達成する手助けとなる、受け入れられる専門的判断を選び出す手引きを示すことは可能である。こうした困難な問題は以下に述べる、部分的に重複したいくつかの課題に分類できる：

　(a) 現行の試験手順をいくつかのタイプの物質に適用する際の困難；

　(b) こうした「試験困難な」物質とその他の物質から得られたデータを解釈する際の困難；

　(c) 広範囲な情報源から得られた多様なデータセットの解釈における困難。

A9.1.10　有機物の多くについては、適切な OECD ガイドラインと分類基準を適用すれば、試験とデータ解釈に問題は何も起こらない。しかし、試験される物質のタイプによって性格づけられる、いくつかの典型的な解釈上の問題が存在する。これらは一般に「試験困難物質」と呼ばれている：

　(a) <u>難溶性の物質</u>：これらの物質は、溶液の調製、濃度の維持および水生毒性試験中の濃度の維持と確認に問題を生じるため試験が困難である。その上、こうした物質について利用可能なデータの多くは、水に対する溶解度を超える濃度の「溶液」を用いて作成されてきており、その結果、分類のための真の $L(E)C_{50}$ または NOEC を決定する際に解釈上の大きな問題となる。水およびオクタノール中の溶解度が低く、分析方法の感度が十分でないことが加わる場合には、分配挙動の解釈もまた問題である。水に対する溶解度は測定が難しく、また単に検出限界より小さいと記録されていることも多く、水生毒性と生物蓄積性のいずれの試験でも解釈上の問題を起こす。生分解性の試験では、溶解度が低いと生物学的利用性が低くなり、予測される生分解速度より低くなることがある。したがって特別な試験方法または採用する手順の選択が極めて重要となり得る；

　(b) <u>不安定な物質</u>：試験系内で速やかに分解（または反応）する物質は、試験上および解釈上の問題を生じる。正しい方法論が使われているか、試験されたのはその物質なのかまたは分解/反応生成物であるか、および得られたデータは親物質の分類に適しているかを判定する必要がある；

　(c) <u>揮発性物質</u>：開放系で用いた場合に試験上の問題を生じることが明らかな物質は、ばく露濃度を適

test material during biodegradation testing is inevitable in certain methods and will lead to misinterpretation of the results;

(d) complex or multi-component substances: such substances, for example, hydrocarbon mixtures, frequently cannot be dissolved into a homogeneous solution, and the multiple components make monitoring impossible. Consideration therefore needs to be given to using the data derived from the testing of water accommodated fractions (WAFs) for aquatic toxicity, and the utilization of such data in the classification scheme. Biodegradation, bioaccumulation, partitioning behaviour and water solubility all present problems of interpretation, where each component of the mixture may behave differently;

(e) polymers: such substances frequently have a wide range of molecular masses, with only a fraction being water soluble. Special methods are available to determine the water-soluble fraction and these data will need to be used in interpreting the test data against the classification criteria;

(f) inorganic compounds and metals: such substances, which can interact with the media, can produce a range of aquatic toxicities dependant on such factors as pH, water hardness etc. Difficult interpretational problems also arise from the testing of essential elements that are beneficial at certain levels. For metals and inorganic metal compounds, the concept of degradability as applied to organic compounds has limited or no meaning. Equally the use of bioaccumulation data should be treated with care;

(g) surface active substances: such substances can form emulsions in which the bioavailablity is difficult to ascertain, even with careful solution preparation. Micelle formation can result in an overestimation of the bioavailable fraction even when "solutions" are apparently formed. This presents significant problems of interpretation in each of the water solubility, partition coefficient, bioaccumulation and aquatic toxicity studies;

(h) ionizable substances: such substances can change the extent of ionization according to the level of counter ions in the media. Acids and bases, for example, will show radically different partitioning behaviour depending on the pH;

(i) coloured substances: such substance can cause problems in the algal/aquatic plant testing because of the blocking of incident light;

(j) impurities: some substances can contain impurities that can change in % and in chemical nature between production batches. Interpretational problems can arise where either or both the toxicity and water solubility of the impurities are greater than the parent substance, thus potentially influencing the toxicity data in a significant way.

A9.1.11 These represent some of the problems encountered in establishing the adequacy of data, interpreting the data and applying that data to the classification scheme. Detailed guidance on how to deal with these problems, as well as other issues related will be presented in the following sections. The interpretation of data on acute and on chronic aquatic toxicity will be covered in section A9.3. This section will deal with the specific interpretational problems encountered for the above "difficult substances", including providing some advice on when and how such data can be used within the classification scheme. Also covered will be a general description of the test data used and the testing methodologies suitable for producing such data.

A9.1.12 A wide range of degradation data are available that must be interpreted according to the criteria for rapid degradability. Guidance is thus needed on how to use these data obtained by employing non-standard test methods, including the use of half-lives where these are available, of primary degradation, of soil degradation rates and their suitability for extrapolation to aquatic degradation and of environmental degradation rates. A short description of estimation techniques for evaluating degradability in relation to the classification criteria is also included. This guidance will be provided in section A9.4.

A9.1.13 Methods by which the potential to bioaccumulate can be determined will be described in section A9.5. This section will describe the relationship between the partition coefficient criteria and the bioconcentration factor (BCF), provide guidance on the interpretation of existing data, how to estimate the partition coefficient by the use of QSARs when no experimental data are available and in particular deal with the specific problems identified above for difficult substances. The problems encountered when dealing with substances of high molecular mass will also be covered.

切に維持して評価すべきである。方法によっては生分解性試験中の被験物質の濃度低下が避けられず、結果が誤って解釈されてしまうこともある；

(d) <u>複合的または多成分物質</u>：例えば炭化水素混合物等の物質は、溶解して均一な溶液にできないことが多く、また多成分であるために濃度測定が不可能となる。したがって、水和性の分画（WAFS）の水生毒性試験から得られたデータを使用すること、およびそうしたデータを分類スキームに利用することを検討する必要がある。混合物の各成分の挙動が異なる場合、生分解性、生物蓄積性、分配特性、および水に対する溶解性はいずれも解釈上の問題を生じる；

(e) <u>ポリマー</u>：これらの物質は広範囲な分子量域を有し、その一部分しか水に溶けないことが多い。水に可溶な分画を測定する特殊な方法が利用可能で、試験データを分類基準と対応させて解釈する際には、これらのデータを使用する必要がある；

(f) <u>無機化合物および金属</u>：媒体と相互作用しうる物質は、pHや水の硬度等の要因によって、一定範囲の水生毒性を生じうる。あるレベルでは有益となるような必須元素の試験からも解釈上困難な問題が生じる。金属および無機金属化合物では、有機化合物に適用されるような分解性の概念は限定された意味、または無意味である。同様に、生物蓄積性データも注意して扱う必要がある；

(g) <u>界面活性物質</u>：これらの物質はエマルジョンを形成することがあり、注意を払って溶液を調製しても、その生物学的利用性を確認することは困難である。ミセルが形成されると、外見上「溶液」が調製された時にも、生物学的に利用できる分画を過大に推定する結果になる。これは、水溶性、分配係数、生物蓄積性および水生毒性試験のそれぞれで解釈上重大な問題となる；

(h) <u>解離性物質</u>：これらは、媒体中の対イオンのレベルによって、イオン化の程度が変化しうる。例えば酸および塩基は、pHによって著しく異なる分配特性を示す；

(i) <u>着色物質</u>：これらは入射光を遮断するので、藻類/水生植物の試験では問題を生じる；

(j) <u>不純物</u>：製造バッチ間で、含有率（％）や化学的特性が異なる不純物を含む物質もある。こうした不純物の毒性と水に対する溶解度のいずれかまたは両方が親物質のそれより大きい場合、解釈上の問題が生じ、毒性データに無視できない影響を及ぼす可能性がある。

A9.1.11 これらは、データの妥当性の確立、データの解釈、およびデータを分類スキームに適用する際に遭遇するいくつかの問題の代表例となる。これらの問題、および関連する事項をどう扱うかについての詳しい手引きを、以下の各節で述べる。急性および慢性の水生毒性に関するデータの解釈についてはA9.3節で扱う。この節は、上述の「試験困難性物質」で遭遇する特別な解釈上の問題を扱っており、分類スキームにおいて、いつどのようにデータを使用できるかについての助言も与えている。さらに、用いる試験データの全般的な説明およびそのようなデータの作成に適した試験の方法論も扱っている。

A9.1.12 広範囲に及ぶ分解性データが利用可能であり、これらは急速分解性の判定基準にしたがって解釈されなければならない。このため、標準的でない試験法で得られたデータをどのように用いるかについての手引きが必要とされる。例えば半減期が利用可能な場合、あるいは一次分解、土壌中での分解速度などのデータの扱い方、ならびにそれらを水中での分解に外挿することの適切さ、および環境中での分解速度についての手引きがある。分類基準と関連して分解性を評価するための推定方法についての簡単な説明も含まれている。その手引きはA9.4節に示す。

A9.1.13 生物蓄積のポテンシャルを測定できる方法はA9.5節に記述される。この節では、分配係数の判定基準と生物濃縮係数（BCF）の関係を説明し、既存データの解釈に関する手引き、すなわち実験データが利用できない場合にQSARを利用して分配係数を推定する方法を示し、そして特に試験困難な物質について上記で特定された問題を扱う。高分子量物質の取扱いで遭遇する問題についても述べる。

A9.1.14   A section is also included which covers general issues concerning the use of QSARs within the system, when and how they may be used, for each of the three properties of concern. As a general approach, it is widely accepted that experimental data should be used rather than QSAR data when such data are available. The use of QSARs will thus be limited to such times when no reliable data are available. Not all substances are suitable for the application of QSAR estimations, however, and the guidance in section A9.6 will address this issue.

A9.1.15   Finally, a section is devoted to the special problems associated with the classification of metals and their compounds. Clearly, for these compounds, a number of the specific criteria such as biodegradability and octanol-water partition coefficient cannot be applied although the principle of lack of destruction via degradation, and bioaccumulation remain important concepts. Thus, it is necessary to adopt a different approach. Metals and metal compounds can undergo interactions with the media which affect the solubility of the metal ion, partitioning from the water column, and the species of metal ion that exists in the water column. In the water column, it is generally the dissolved metal ions which are of concern for toxicity. The interaction of the substance with the media may either increase or decrease the level of ions and hence toxicity. It is thus necessary to consider whether metal ions are likely to be formed from the substance and dissolve in the water, and if so whether they are formed rapidly enough to cause concern. A scheme for interpreting the results from this type of study is presented in section A9.7.

A9.1.16   While the guidance document provides useful advice on how to apply the criteria to a wide variety of situations, it remains a guidance only. It cannot hope to cover all situations that arise in classification. It should therefore be seen as a living document that in part describes the fundamental principles of the system, e.g. hazard based rather than risk based, and the fixed criteria. It must also, in part, be a repository for the accumulated experience in using the scheme to include the interpretations which allow the apparently fixed criteria to be applied in a wide variety of non-standard situations.

A9.2   The harmonized classification scheme

A9.2.1   *Scope*

The criteria were developed taking into account existing systems for hazard classification, such as EU-Supply and Use System, the Canadian and US Pesticide systems, GESAMP hazard evaluation procedure, IMO Scheme for Marine Pollutant, the European Road and Rail Transport Scheme (RID/ADR), and the US Land Transport. These systems include supply and subsequent use of chemicals, the sea transport of chemicals as well as transport of chemicals by road and rail. The harmonized criteria are therefore intended to identify hazardous chemicals in a common way for use throughout all these systems. To address the needs for all different sectors (transport, supply and use) it was necessary to create two different sub-classes, one sub-class for short-term (acute) aquatic hazards, consisting of three categories and one sub-class for lont-term (chronic) aquatic hazards, consisting of 4 categories. The short-term (acute) classification sub-class makes provision for two short-term (acute) hazard categories (Acute 2 and 3) not normally used when considering packaged goods. For chemicals transported in bulk, there are a number of regulatory decisions that can uniquely arise because of the bulk quantities being considered. For these situations, for example where decisions are required on the ship type to be used, consideration of all short-term (acute) hazard categories as well as the long-term (chronic) hazard categories are considered important. The following paragraphs describe in detail the criteria to be used in defining each of these hazard categories.

A9.2.2   *Classification categories and criteria*

The hazard categories for acute and chronic aquatic toxicity and their related criteria are set out in chapter 4.1, paragraph 4.1.2.4 and table 4.1.1.

A9.2.3   *Rationale*

A9.2.3.1   The harmonized system for classification recognizes that the intrinsic hazard to aquatic organisms is represented by both the acute and chronic or long-term toxicity of a substance, the relative importance of which is determined by the specific regulatory regimes in operation. Distinction can be made between the short-term (acute) hazard and the long-term (chronic) hazard and therefore hazard classes are defined for both properties representing a gradation in the level of hazard identified. Clearly the hazard identified by Chronic 1 is more severe than Chronic 2. Since the acute (short-term) hazard and long-term (chronic) hazard represent distinct types of hazard, they are not comparable in terms of their relative severity. Both hazard sub-classes should be applied independently for the classification of substances to establish a basis for all regulatory systems.

A9.2.3.2   The principal hazard classes defined by the criteria relate largely to the potential for long-term (chronic) hazard. This reflects the overriding concern with respect to chemicals in the environment, namely that the effects caused are usually sub-lethal, e.g. effects on reproduction, and caused by longer-term exposure. While recognizing that the long-

A9.1.14　QSAR を使用することについての一般的な問題、すなわち関心のある 3 つの特性それぞれについて、QSAR がいつどのように利用できるかを述べた節も含まれている。一般的なアプローチとして、実験データが利用可能な場合には、QSAR データよりも実験データの方を使用すべきであることは広く受け入れられている。このため、QSAR データの利用は信頼できるデータが得られない場合のみに限られる。しかしすべての物質が QSAR による推定の利用に適しているというわけではなく、本手引きの A9.6 節でこの問題について取りあげる。

A9.1.15　最後に、金属および金属化合物の分類に関連する特殊な問題に一節が割かれている。明らかにこのような化合物については、生分解性およびオクタノール/水分配係数といった多くの特定の判定基準は適用できないが、分解による崩壊がないという原則や、生物蓄積は重要なコンセプトとして残される。したがって別のアプローチを適用する必要がある。金属および金属化合物は媒体との相互作用を受け、その媒体は金属イオンの溶解度や、水相との分配、および水相に存在する金属イオンの種類に影響する。水相で毒性が問題となるのは、一般に溶解している金属イオンである。物質と媒体との相互作用はイオンレベルを、ひいては毒性を増すこともあれば減ることもある。このため、金属イオンがその物質から生成されて水に溶け出すかどうか、もしそうなら、金属イオンは問題を生じるほど速やかに形成されるかどうかを検討することが必要である。このタイプの試験での結果を解釈するためのスキームは A9.7 節に示されている。

A9.1.16　本手引きは、広く多様な状況において判定基準をどのように適用するかに関する有用な助言を与えてはいるが、それでも単なる手引きに過ぎない。分類の際に生じるあらゆる状況を網羅することは望めない。したがって、これはシステムの基本的な原則、例えばリスクベースよりも有害性ベースをまた確定された判定基準の一部について説明しているにすぎない、変更の可能性のある文書とみなさなければならない。さらに本手引きは、見かけでは確定している判定基準でも広く多様な非標準的状況に、適用できるようにする解釈を含んだ、スキームを使用する上で蓄積された経験の貯蔵所でもある。

A9.2　調和された分類スキーム

A9.2.1　*適用範囲*

　EU の供給および使用システム、カナダおよび米国の殺虫剤システム、GESAMP 有害性評価手順、IMO 海洋汚染物質に関するスキーム、欧州道路鉄道輸送スキーム（RID/ADR）および米国陸上輸送等、既存の有害性分類システムを考慮して判定基準が策定された。これらのシステムには、化学品の供給およびそれに続く使用、化学品の海上輸送、ならびに道路および鉄道による化学品輸送が含まれている。したがって、調和された判定基準はこれらすべてのシステムに共通のやり方で、危険有害性をもつ化学品を特定することを意図している。すべての異なった分野（輸送、供給および使用）についてのニーズを取り扱うために、2 種類の異なったクラス、すなわち 3 つの区分から構成される短期（急性）水性有害性クラス、および 4 つの区分から構成される長期（慢性）水性有害性クラスを作成する必要があった。急性分類区分クラスは、包装された物品を考える場合には通常は使用しない 2 種類の短期（急性）有害性区分（急性 2 および 3）についての規定を定めている。ばら積みで輸送される物質の場合、想定される量が大量であるために独自に生じてくる規制上の多くの決定がある。そうした状況では、例えば使用する船舶のタイプについて決定が求められる場合、すべての短期（急性）有害性区分および長期（慢性）有害性区分を考慮することが重要であると考えられる。以下の各節でこうした有害性区分を定義するのに用いられる判定基準について詳しく説明する。

A9.2.2　*分類区分および分類基準*

　急性および慢性水生毒性の有害性区分とそれに関連する判定基準は、4.1 章の 4.1.2.4、および表 4.1.1 に規定されている。

A9.2.3　*根拠*

A9.2.3.1　分類のための調和システムは、水生生物に対する本質的な有害性は、物質の急性毒性および慢性または長期間毒性の両方によって表されることを認めており、その相対的な重要性は、関係する特定の規制制度によって決まる。短期（急性）有害性と長期（慢性）有害性とは区別することができるので、有害性区分は、特定された有害性のレベルにおいて段階を示している両方の特性について定義される。慢性 1 と特定された有害性が慢性 2 によるものより重大であるのは明らかである。短期（急性）有害性と長期（慢性）有害性は明らかに異なったタイプの有害性を代表しているため、その相対的な重大さについて比較することはできない。物質の分類のためには両方の有害性クラスを独立に適用して、すべての規制システムの根拠を確立すべきである。

A9.2.3.2　判定基準によって定義された主要な有害性クラスは、長期（慢性）有害可能性に大きく関係している。これは環境中での化学品に関する最優先の関心事を反映している。すなわち生じた影響は通常は致死レベル以下で、例えば繁殖に対する影響、および長期ばく露により生じる影響なのである。特に包装

term (chronic) hazard represents the principal concern, particularly for packaged goods where environmental release would be limited in scope, it must also be recognized that chronic toxicity data are expensive to generate and generally not readily available for most substances. On the other hand, acute toxicity data are frequently readily available, or can be generated to highly standardised protocols. It is this acute toxicity which has therefore been used as the core property in defining both the acute and the long-term (chronic) hazard if no adequate chronic test data are available. Nevertheless, it has been recognized that chronic toxicity data, if available should be preferred in defining the long-term (chronic) hazard category.

A9.2.3.3   The combination of chronic toxicity and intrinsic fate properties reflects the potential hazard of a substance. Substances that do not rapidly degrade have a higher potential for longer term exposures and therefore should be classified in a more severe category than substances which are rapidly degradable (see A9.3.3.2.2).

A9.2.3.4   While recognizing that acute toxicity itself is not a sufficiently accurate predictor of chronic toxicity to be used solely and directly for establishing hazard, it is considered that, in combination with either a potential to bioaccumulate (i.e. a log $K_{ow} \geq 4$ unless BCF < 500) or potential longer-term exposure (i.e. lack of rapid degradation) it can be used as a suitable surrogate for classification purposes. Substances rapidly biodegrading that show acute toxicity with a significant degree of bioaccumulation will normally show chronic toxicity at a significantly lower concentration. Equally substances that do not rapidly degrade have a higher potential for giving rise to longer term exposures which again may result in long-term toxicity being realized. Thus, for example, in absence of adequate chronic test data, category Chronic 1 should be assigned if either of the following criteria are met:

(a) $L(E)C_{50}$ for any appropriate aquatic species ≤1 mg/l <u>and</u> a potential to bioaccumulate (log $K_{ow} \geq 4$ unless BCF < 500);

(b) $L(E)C_{50}$ for any appropriate aquatic species ≤1 mg/l <u>and</u> a lack of rapid degradation.

A9.2.3.5 and A9.5.   The precise definitions of the core elements of this system are described in detail in sections A9.3, A9.4

A9.2.3.6   For some poorly soluble substances, which are normally considered as those having a water solubility < 1 mg/l, no acute toxicity is expressed in toxicity tests performed at the solubility limit. If for such a substance, however, the BCF ≥ 500, or if absent, the log $K_{ow} \geq 4$ (indicating a bioaccumulating potential) and the substance is also not rapidly degradable, a safety net classification is applied, Chronic 4. For these types of substance the exposure duration in short term tests may well be too short for a steady state concentration of the substance to be reached in the test organisms. Thus, even though no acute toxicity has been measured in a short term (acute) test, it remains a real possibility that such non-rapidly degradable and bioaccumulative substances may exert chronic effects, particularly since such low degradability may lead to an extended exposure period in the aquatic environment.

A9.2.3.7   In defining aquatic toxicity, it is not possible to test all species present in an aquatic ecosystem. Representative species are therefore chosen which cover a range of trophic levels and taxonomic groupings. The taxa chosen, fish, crustacea and aquatic plants that represent the "base-set" in most hazard profiles, represent a minimum dataset for a fully valid description of hazard. The lowest of the available toxicity values will normally be used to define the hazard category. Given the wide range of species in the environment, the three tested can only be a poor surrogate and the lowest value is therefore taken for cautious reasons to define the hazard category. In doing so, it is recognized that the distribution of species sensitivity can be several orders of magnitude wide and that there will thus be both more and less sensitive species in the environment. Thus, when data are limited, the use of the most sensitive species tested gives a cautious but acceptable definition of the hazard. There are some circumstances where it may not be appropriate to use the lowest toxicity value as the basis for classification. This will usually only arise where it is possible to define the sensitivity distribution with more accuracy than would normally be possible, such as when large datasets are available. Such large datasets should be evaluated with due caution.

A9.2.4   *Application*

A9.2.4.1   Generally speaking, in deciding whether a substance should be classified, a search of appropriate databases and other sources of data should be made for the following data elements:

(a)   water solubility;

(b)   acute aquatic toxicity ($L(E)C_{50}$s);

(c)   chronic aquatic toxicity (NOECs and/or equivalent ECx);

(d)   available degradation (and specifically evidence of ready biodegradability);

(e)   stability data, in water;

された物品については環境への放出が量的に限られるので、長期（慢性）有害性が主要な関心事であると認識されている一方、慢性毒性データは作成に経費がかかり、一般にほとんどの物質で容易には利用できないことも認めなければならない。他方、急性毒性データは容易に入手できることが多く、または高度に標準化されたプロトコールによって作成することができる。したがって、十分な慢性試験データが得られない場合は、急性有害性、長期（慢性）有害性のどちらも定義する上で、中心的特性として利用されているのは急性毒性なのである。それでも、慢性毒性データが利用可能な場合には、長期（慢性）有害性区分を定義するには、慢性毒性データの方を優先すべきであると認識されている。

A9.2.3.3　慢性毒性と物質固有の運命特性を組み合わせが物質の潜在的な有害性に反映する。速やかに分解しない物質は、より長期ばく露が生じる可能性が高いため、速やかに分解する物質に比べて厳しい区分に分類されるべきである(A9.3.3.2.2)。

A9.2.3.4　急性毒性値それ自体は、有害性を立証するために単独かつ直接に使用しては、十分に正確な慢性毒性予測とはならないと認識されているが、生物蓄積性（すなわち $\log K_{ow} \geq 4$、ただし BCF<500 でない場合）または長期ばく露の可能性（すなわち急速分解性がない）のいずれかと組み合わせれば、分類の目的では適切な代用として使用できると考えられている。急性毒性および有意な程度で生物蓄積性を有する物質は、急速分解性を有していても通常はかなりの低濃度で慢性毒性を示す。同様に、速やかに分解しない物質には、長期ばく露が生じる可能性が高くなり、この場合もまた、長期毒性が十分に考えられることになる。したがって、例えば、十分な慢性試験のデータがない場合は、以下の判定基準に適合するなら、慢性1が指定されるべきである：

(a) 適切な水生生物種に対する $L(E)C_{50}$ が 1mg/l 以下、かつ生物蓄積の可能性がある（$\log K_{ow} \geq 4$ ただし BCF<500 でない）：

(b) 適切な水生生物種に対する $L(E)C_{50}$ が 1mg/l 以下、かつ急速分解性がないこと。

A9.2.3.5　このシステムの中心的要素についての正確な定義はそれぞれ A9.3 節、A9.4 節、ならびに A9.5 節で説明する。

A9.2.3.6　難溶性の物質は通常、水に対する溶解度が 1mg/l 未満の物質であるとみなされているが、これらの物質のなかに、溶解する限界濃度で実施された毒性試験で、急性毒性が発現しないものがある。しかし、こうした物質が BCF≧500、または BCF データがない場合に $\log K_{ow} \geq 4$（生物蓄積性の可能性を示唆している）であり、さらにその物質が急速分解性ではない場合、慢性4という安全ネットの分類が適用される。このようなタイプの物質では、短期試験におけるばく露期間では短すぎて、試験生物種の体内で物質が定常状態濃度に到達しないことがある。このため、短期（急性）試験で急性毒性が認められなかったとしても、特に分解性が低いと水生環境中ではばく露期間が延長されることになるので、急速分解性がなく生物蓄積性のある物質が慢性作用を生じる可能性を考える必要がある。

A9.2.3.7　水生毒性を決定する際に、水系生態系に存在しているすべての生物種を試験することは不可能である。したがって、ある範囲の栄養段階と分類群ごとに、代表する生物種が選ばれる。選択された分類群、すなわち魚類、甲殻類、および水生植物はほとんどの有害性プロフィールで「基本セット」になっており、有害性を十分に有効に表現するための最少のデータセットとなっている。利用可能な毒性値のうち最低の値が通常、有害性区分の決定に用いられる。環境中には広範囲な生物種が存在していることを考えれば、これら3種類の試験では不十分な代用データにしかなり得ず、したがって慎重を期すという意味で、最低値を有害性区分の決定に採用している。これを実施するにあたり、生物種の感受性の範囲は数桁のオーダーにわたる可能性があること、したがってこれらの生物種より感受性が高い種も低い種も、環境中に存在していることは認識されている。このため、データが少ない場合には、試験した中で最も感受性の高い生物種を用いることで、慎重ではあるが受け入れられる有害性の定義が得られる。毒性の最低値を分類の根拠に用いることが適切でないかもしれない状況もある。これが該当するのは、例えば大量のデータセットが利用可能な場合等、通常よりも正確に感受性分布を決定できる場合である。このような大量のデータセットは相応の注意を払って評価すべきである。

A9.2.4　*適用*

A9.2.4.1　一般論として、ある物質を分類すべきかどうかを決定するには、以下のデータ項目について適切なデータベース、およびその他のデータ源を検索しなければならない：

(a) 水に対する溶解性；
(b) 急性水生毒性（$L(E)C_{50}$）；
(c) 慢性水生毒性（NOECまたは同等のEC_x）；
(d) 利用可能な分解性データ（特に易生分解性の証拠）；
(e) 水中での安定性データ；

(f) fish bioconcentration factor (BCF);

(g) octanol/water partition coefficient (log K_{ow});

The water solubility and stability data, although not used directly in the criteria, are nevertheless important since they are a valuable help in the data interpretation of the other properties (see A9.1.10).

A9.2.4.2 To classify, a review should first be made of the available aquatic toxicity data. It will be necessary to consider all the available data and select those which meet the necessary quality criteria for classification. If there are no data available that meet the quality criteria required by the internationally standardized methods, it will be necessary to examine any available data to determine whether a classification can be made. If the data indicate that the acute aquatic toxicity L(E)C$_{50}$ is greater than 100 mg/l for soluble substances and the chronic aquatic toxicity is greater than 1 mg/l, then the substance is not classified as hazardous. There are a number of cases where no effects are observed in the test and the aquatic toxicity is thus recorded as a > water solubility value, i.e. there is no acute toxicity within the range of the water solubility in the test media. Where this is the case, and the water solubility in the test media is ≥ 1 mg/l, again, no classification need be applied.

A9.2.4.3 If chronic aquatic toxicity data are available, cut-off values will depend on whether the substance is rapidly degradable or not. Therefore, for non-rapidly degradable substances and those for which no information on degradation is available, the cut-off levels are higher than for those substances where rapid degradability can be confirmed (see chapter 4.1, tables 4.1.1 and 4.1.2).

A9.2.4.4 Where the lowest acute aquatic toxicity data are below 100 mg/l and no adequate chronic toxicity data are available, it is necessary to first decide which hazard category the toxicity falls in, and then to determine whether the chronic and/or the acute sub-class should be applied. This can simply be achieved by examining the available data on the partition coefficient, log K_{ow} and the available data on degradation. If either the log K_{ow} ≥ 4 or the substance cannot be considered as rapidly degradable, then the appropriate long-term (chronic) hazard category and the corresponding acute (short-term) hazard category are applied independently. It should be noted that, although the log K_{ow} is the most readily available indication of a potential to bioaccumulate, an experimentally derived BCF is preferred. Where this is available, this should be used rather than the partition coefficient. In these circumstances, a BCF ≥ 500 would indicate bioaccumulation sufficient to classify in the appropriate long-term (chronic) hazard category. If the substance is both rapidly degradable and has a low potential to bioaccumulate (BCF < 500 or, if absent, log K_{ow} < 4) then it should not be assigned to a long-term (chronic) hazard category, unless the chronic toxicity data indicate otherwise (see A9.2.4.3).

A9.2.4.5 For poorly soluble substances, generally speaking, those with a water solubility in the test media of < 1 mg/l, for which no aquatic toxicity has been found, should be further examined to determine whether Chronic Category 4 needs to be applied. Thus, if the substance is both not rapidly degradable and has a potential to bioaccumulate (BCF ≥ 500 or, if absent log K_{ow} ≥ 4), the Chronic 4 should be applied.

A9.2.5 *Data availability*

The data used to classify a substance can be drawn from data required for regulatory purposes as well as the relevant literature, although a number of internationally recognized databases exist which can act as a good starting point. Such databases vary widely in quality and comprehensiveness and it is unlikely that any one database will hold all he information necessary for classification to be made. Some databases specialize in aquatic toxicity and others in environmental fate. There is an obligation on the chemical supplier to make the necessary searches and checks to determine the extent and quality of the data available and to use it in assigning the appropriate hazard category.

A9.2.6 *Data quality*

A9.2.6.1 The precise use of the available data will be described in the relevant section but, as a general rule, data generated to standard international guidelines and to GLP is to be preferred over other types of data. Equally, however, it is important to appreciate that classification can be made based on the best available data. Thus, if no data is available which conforms to the quality standard detailed above, classification can still be made provided the data used is not considered invalid. To assist this process, a quality scoring guide has been developed and used extensively in a number of fora and generally conforms to the following categories:

(a) Data derived from official data sources that have been validated by regulatoryauthorities, such as EU Water Quality Monographs, US-EPA Water Quality Criteria. These data can be considered as valid for classification purposes. No assumption should be made that these are the only data available, however, and due regard should be given to the date of the relevant report. Newly available data may not have been considered;

(f) 魚類における生物濃縮係数（BCF）；
(g) オクタノール/水分配係数（log K_{ow}）；

水に対する溶解度および水中での安定性のデータは、判定基準では直接には用いられていないが、これらは他の特性のデータ解釈において貴重な参考データとなるので重要である（A9.1.10節参照）。

A9.2.4.2　分類のためには、まず始めに利用可能な水生毒性データをレビューすべきである。利用可能なすべてのデータを考慮して、分類に必要な品質基準に適合するデータを選択することが必要となる。国際的に標準化された方法において要求される品質基準に適合するデータが利用できない場合には、分類が可能かどうかを決定するために、利用可能なすべてのデータを検証する必要がある。データから溶解度の高い物質について急性水生毒性が $L(E)C_{50}>100mg/l$ で、かつ慢性水生毒性が 1mg/l を上回ることが示されたならば、その物質は有害であるとは分類されない。試験では影響が認められず、このため水生毒性が水に対する溶解値より大きい値と記録される、すなわちその試験媒体中での水に対する溶解度の範囲内では急性毒性はないとされるケースも多い。そのような場合に、試験媒体中での水に対する溶解度が 1mg/l 以上であれば、分類を適用する必要はない。

A9.2.4.3　慢性水生毒性データが得られる場合、カットオフ値は物質が速やかに分解するかどうかによって決まる。したがって、速やかに分解しない物質や、分解性についての情報がない物質については、カットオフ・レベルは急速分解性を確認できる物質よりも高くなる（4.1章、表4.1.1および表4.1.2参照）。

A9.2.4.4　水生毒性データの最低値が 100mg/l 以下で、かつ慢性毒性の十分なデータが利用できない場合、第一にその毒性が該当するのはどの有害性区分であるかを決定し、次に慢性クラスまたは急性クラスを適用すべきかどうかについて判定する必要がある。これは単に、分配係数 log K_{ow} および分解に関して利用可能なデータを検討することで達成できる。もし、log K_{ow} ≧4 であるか、またはその物質は急速分解性がないと考えられるならば、適切な長期（慢性）有害性および対応する短期（急性）有害性区分が別々に適用される。ただし、log K_{ow} は生物蓄積の可能性について最も容易に入手できる指標ではあるものの、実験的に求められた BCF の方が望ましいことに注意すべきである。BCF が利用可能であるなら、分配係数でなくこちらを用いるべきである。そのような状況では、BCF ≧500 ということは適切な長期（慢性）有害性クラスに分類するのに十分な程度の生物蓄積性であることを示す。もしその物質が急速分解性を有し、かつ生物蓄積性が低い（BCF<500、または BCF がなければ log K_{ow}<4）ならば、慢性毒性のデータが慢性毒性を有すると示さない限り、長期（慢性）有害性区分に指定されるべきではない。（A9.2.4.3 参照）。

A9.2.4.5　難溶性の物質、すなわち一般に試験媒体中での水に対する溶解度が 1mg/l 未満である物質で、水生毒性が認められていない物質については、慢性 4 を適用する必要があるかを判定するためにさらに検討を加えるべきである。例えば、もしその物質が急速分解性でなく、かつ生物蓄積の可能性がある（BCF ≧500、または BCF がなければ log K_{ow} ≧4）ならば、慢性 4 を適用すべきである。

A9.2.5　データの利用可能性

ある物質の分類に使用されるデータは、規制の目的で必要なデータ、および関連文献から導き出すことができるが、適切な出発点として利用できる国際的に認められた数々のデータベースも存在している。こうしたデータベースの質や包括性には大きな差があり、どのデータベースでも単独では分類を行うのに必要な情報がすべて揃うわけではない。水生毒性を専門としているデータベースもあれば、環境運命に詳しいデータベースもある。化学品供給者には、利用可能なデータの規模と信頼性を判定するために必要な調査や確認をし、適切な有害性区分の指定にそのデータを用いる義務がある。

A9.2.6　データの質

A9.2.6.1　利用可能なデータの正確な使い方については、関連する節で説明するが、一般的な規則として、標準的な国際ガイドラインおよび GLP に準拠して作成されたデータは、他の種類のデータよりも望ましいとされている。しかし同様に、利用可能な最良のデータに基づいても分類できることを十分に認識することも重要である。したがって、上に述べた品質基準を満たすデータが利用できない時であっても、用いるデータが無効であると考えられない限り、分類をすることが可能である。このプロセスを支援するために、質の得点付け手引きが策定されており、多くの場で広く用いられている。得点付け手引きは一般に以下の分類を充足している：

(a) 公的データ源から得られたデータで、EU 水質モノグラフ、USEPA クライテリアなど、規制官庁により有用性を確認済みのもの。これらのデータは、分類の目的には有効であるとみなすことができる。しかし、これらが唯一の利用できるデータであると考えるべきではなく、さらに関連報告書の日付に十分注意を払う必要がある。新たに入手できるデータがまだ考慮されていないこともある；

(b) Data derived from recognized international guidelines (e.g. OECD Guidelines) or national guidelines of equivalent quality. Subject to the data interpretation issues raised in the following sections, these data can be used for classification;

(c) Data derived from testing which, while not strictly according to a guideline detailed above, follows accepted scientific principles and procedures and/or has been peer reviewed prior to publication. For such data, where all the experimental detail is not recorded, some judgement may be required to determine validity. Normally, such data may be used within the classification scheme;

(d) Data derived from testing procedures which deviate significantly from standard guidelines and are considered as unreliable, should not be used in classification;

(e) QSAR data. The circumstances of use and validity of QSAR data are discussed in the relevant sections;

(f) Data derived from secondary sources such as handbooks, reviews, citation, etc. where the data quality cannot be directly evaluated. Such data should be examined where data from quality 1, 2 and 3 are not available, to determine whether it can be used. Such data should have sufficient detail to allow quality to be assessed. In determining the acceptability of these data for the purposes of classification, due regard should be given to the difficulties in testing that may have affected data quality and the significance of the reported result in terms of the level of hazard identified (see A9.3.6.2.3).

A9.2.6.2 Classification may also be made on incomplete toxicity datasets, e.g. where data are not available on all three trophic levels. In these cases, the classification may be considered as "provisional" and subject to further information becoming available. In general, all the data available will need to be considered prior to assigning a classification. Where good quality data are not available, lower quality data will need to be considered. In these circumstances, a judgement will need to be made regarding the true level of hazard. For example, where good quality data are available for a particular species or taxa, this should be used in preference to any lower quality data which might also be available for that species or taxa. However, good quality data may not always be available for all the basic data set trophic levels. It will be necessary to consider data of lower quality for those trophic levels for which good quality data are not available. Consideration of such data, however, will also need to consider the difficulties that may have affected the likelihood of achieving a valid result. For example, the test details and experimental design may be critical to the assessment of the usability of some data, such as that from hydrolytically unstable chemicals, while less so for other chemicals. Such difficulties are described further in section A9.3.

A9.2.6.3 Normally, the identification of hazard, and hence the classification will be based on information directly obtained from testing of the substance being considered. There are occasions, however, where this can create difficulties in the testing, or the outcomes do not conform to common sense. For example, some chemicals, although stable in the bottle, will react rapidly (or slowly) in water giving rise to degradation products that may have different properties. Where such degradation is rapid, the available test data will frequently define the hazard of the degradation products since it will be these that have been tested. These data may be used to classify the parent substance in the normal way. However, where degradation is slower, it may be possible to test the parent substance and thus generate hazard data in the normal manner. The subsequent degradation may then be considered in determining whether a short-term (acute) or long-term (chronic) hazard category should apply. There may be occasions, however, when a substance so tested may degrade to give rise to a more hazardous product. In these circumstances, the classification of the parent should take due account of the hazard of the degradation product, and the rate at which it can be formed under normal environmental conditions.

A9.3   Aquatic toxicity

A9.3.1   *Introduction*

The basis for the identification of hazard to the aquatic environment for a substance is the aquatic toxicity of that substance. Classification is predicated on having toxicity data for fish, crustacea, and algae/aquatic plant available. These taxa are generally accepted as representative of aquatic fauna and flora for hazard identification. Data on these particular taxa are more likely to be found because of this general acceptance by regulatory authorities and the chemical industry. Other information on the degradation and bioaccumulation behaviour is used to better delineate the aquatic hazard. This section describes the appropriate tests for ecotoxicity, provides some basic concepts in evaluating the data and using combinations of testing results for classification, summarizes approaches for dealing with difficulty substances, and includes a brief discussion on interpretation of data quality.

(b) 国際的に承認されたガイドライン（例：OECD テストガイドライン）またはそれと同等の品質の国内ガイドラインから得られたデータ。以降の各節に掲げられたデータ解釈上の問題があることを留意事項として、こうしたデータを分類に用いることができる；

(c) 上に述べたガイドラインに厳密にはしたがってはいないが、一般に容認された科学的な原則および手順にしたがっているデータ、または発表前に校閲されているデータ。こうしたデータについて、実験の詳細がすべて記録されていない場合、有効性の判定には何らかの判断が必要かもしれない。通常、こうしたデータは、分類スキームの中で用いられてもよい；

(d) 標準的なガイドラインから著しく逸脱した、または信頼性がないと思われる試験手順により求めたデータは、分類に用いるべきではない；

(e) QSAR データ。QSAR の使用状況および有効性については、関連の各節で議論する；

(f) ハンドブック、総説、引用等のように、データの質を直接には評価できない二次的な情報源から求めたデータ。データの質 1、2 および 3 からのデータが利用できない場合に、こうしたデータが使えるかどうかを判断するため精査すべきである。こうしたデータは、その質が評価できる程度に詳細である必要がある。これらのデータを分類目的に受け入れることが可能かどうかを判断する際には、データの質に影響しているかも知れない試験中の問題および報告された結果の有意性について、特定された有害性のレベルに関し、十分な考慮を払うべきである（A9.3.6.2.3 を参照）。

A9.2.6.2　分類は不完全な毒性データセットに基づいてなされることもある。例えば 3 つの栄養段階すべてについてのデータが利用できない場合である。このようなケースでは、分類は「暫定的」であるとみなされ、利用可能な追加情報の入手が必要になる。一般に、利用可能なすべてのデータが、分類を決定する前に考慮される必要がある。質の良いデータが利用できない場合には、より質の劣るデータでも考慮する必要がある。そのような状況では、真の有害性レベルについての判断がなされることが必要となる。例えば、特定の生物種または分類群に関して良質のデータが利用可能な場合、同じ種または分類群について得られるかもしれない、他の質の劣るデータに優先して、そうしたデータを採用すべきである。しかし、栄養段階すべてについての基礎的なデータセットに、必ずしも良質なデータが揃っているとは限らない。良質なデータが利用できない栄養段階については、質が劣るデータを検討することが必要となろう。しかし、こうしたデータを考慮するには、有効な結果に達する可能性に影響したと思われる問題についても考える必要がある。例えば、加水分解性の不安定な化学品のデータなど、試験の詳細および実験の設計が、その有用性の評価に決定的になるデータもある。そうした問題については、A9.3 節でさらに説明する。

A9.2.6.3　通常、有害性の特定、ひいては分類は、対象の物質を試験して直接得られた情報をもとになされるであろう。しかし、このことが試験に困難を生じ、または結果が常識にそぐわないこともある。例えば、保存びんの中では安定であっても、水と混合すると速やかに（またはゆっくりと）反応して、もとの化学品とは異なる性質の分解生成物を生じさせる物質もある。この場合、分解が速やかであるなら、実際に試験されたのは分解生成物であるため、利用可能なデータは分解生成物の有害性を特定することもしばしばある。こうしたデータが通常の方法で親物質を分類するのに用いられることもありうる。しかし、分解がより遅い場合、親物質を試験することが可能であり、有害性データが正常なやり方で作成される。遅れて起こる分解は、短期（急性）ないし長期（慢性）有害性区分を適用すべきかを決める際に考慮される。しかし、試験された物質が分解して、より有害な生成物を生じることも起こり得る。こうした場合には、親物質の分類では分解生成物の有害性、および通常の環境条件で分解生成物が形成される速度を十分に考慮しなければならない。

A9.3　水生毒性

A9.3.1　*序*

　物質の水生環境に対する有害性を特定する基礎は、その物質の水生毒性である。分類は、魚類、甲殻類、および藻類/水生植物の毒性データを入手することに基づいている。こうした分類群は、有害性を特定するための水中動物相および植物相の代表として、広く受け入れられている。規制官庁や化学企業がこれを受け入れているので、こうした特定の分類群に関するデータは見つけられる可能性が高い。分解性および生物蓄積性についての挙動に関するその他の情報は、水生有害性をよりよく描写するのに利用される。この節では、生態毒性に関する適切な試験について説明し、データを評価し試験結果の組み合わせを分類に用いるいくつかの基本概念を示し、分類の困難な物質を扱う手法をまとめ、データの質の解釈について簡単な考察を加える。

A9.3.2 *Description of tests*

A9.3.2.1 For classifying substances in the harmonized system, freshwater and marine species toxicity data can be considered as equivalent data. It should be noted that some types of substances, e.g. ionizable organic chemicals or organometallic substances may express different toxicities in freshwater and marine environments. Since the purpose of classification is to characterize hazard in the aquatic environment, the result showing the highest toxicity should be chosen.

A9.3.2.2 The GHS criteria for determining health and environmental hazards should be test method neutral, allowing different approaches as long as they are scientifically sound and validated according to international procedures and criteria already referred to in existing systems for the endpoints of concern and produce mutually acceptable data. According to the proposed system (OECD 1998):

> "Acute toxicity would normally be determined using a fish 96 hour LC_{50} (OECD Test Guideline 203 or equivalent), a crustacea species 48 hour EC_{50} (OECD Test Guideline 202 or equivalent) and/or an algal species 72 or 96 hour EC_{50} (OECD Test Guideline 201 or equivalent). These species are considered as surrogate for all aquatic organisms and data on other species such as the duckweed Lemna may also be considered if the test methodology is suitable."

Chronic testing generally involves an exposure that is lingering or continues for a longer time; the term can signify periods from days to a year, or more depending on the reproductive cycle of the aquatic organism. Chronic tests can be done to assess certain endpoints relating to growth, survival, reproduction and development.

> "Chronic toxicity data are less available than acute data and the range of testing procedures less standardised. Data generated according to OECD test guidelines 210 (Fish Early Life Stage), 202 Part 2 or 211 (Daphnia Reproduction) and 201 (Algal Growth Inhibition) can be accepted. Other validated and internationally accepted tests could also be used. The NOECs or other equivalent L(E)Cx should be used."

An OECD document describes the main statistical methods for the analysis of data of standardized ecotoxicity tests (OECD 2006).

A9.3.2.3 It should be noted that several of the OECD guidelines cited as examples for classification are being revised or are being planned for updating. Such revisions may lead to minor modifications of test conditions. Therefore, the expert group that developed the harmonized criteria for classification intended some flexibility in test duration or even species used.

A9.3.2.4 Guidelines for conducting acceptable tests with fish, crustacea, and algae can be found in many sources (OECD, 1999; EPA, 1996; ASTM, 1999; ISO EU). The OECD monograph No.11, Detailed Review Paper on Aquatic Toxicity Testing for Industrial Chemicals and Pesticides, is a good compilation of pelagic test methods and sources of testing guidance. This document is also a source of appropriate test methodologies.

A9.3.2.5 *Fish tests*

A9.3.2.5.1 Acute testing

Acute tests are generally performed with young juveniles 0.1 - 5 g in size for a period of 96 hours. The observational endpoint in these tests is mortality. Fish larger than this range and/or durations shorter than 96 hours are generally less sensitive. However, for classification, they could be used if no acceptable data with the smaller fish for 96 hours are available or the results of these tests with different size fish or test durations would influence classification in a more hazardous category. Tests consistent with OECD Test Guideline 203 (Fish 96 hour LC_{50}) or equivalent should be used for classification.

A9.3.2.5.2 Chronic testing

Chronic or long-term tests with fish can be initiated with fertilized eggs, embryos, juveniles, or reproductively active adults. Tests consistent with OECD Test Guideline 210 (Fish Early Life Stage), the fish life-cycle test (US EPA 850.1500), or equivalent can be used in the classification scheme. Durations can vary widely depending on the test purpose (anywhere from 7 days to over 200 days). Observational endpoints can include hatching success, growth (length and weight changes), spawning success, and survival. Technically, the OECD 210 Guideline (Fish Early Life Stage) is not a "chronic" test, but a sub-chronic test on sensitive life stages. It is widely accepted as a predictor of chronic toxicity and is used as such for purposes of classification in the harmonized system. Fish early life stage toxicity data are much more available than fish life cycle or reproduction studies.

A9.3.2 試験の説明

A9.3.2.1 本調和システムで物質を分類する目的では、淡水生物種と海水生物種の毒性データは同等とみなすことができる。イオン化する有機化合物や有機金属化合物等、ある種の物質では淡水環境と海洋環境では現れる毒性が異なることも指摘しておかなければならない。分類の目的は水生環境の有害性を記述することなので、最も高い毒性が示された結果が選ばれなければならない。

A9.3.2.2 健康および環境に対する有害性を判定するためのGHS判定基準は、中立の試験方法であるが、他方、科学的に正当であり、対象のエンドポイントについての既存システムについて先に述べた国際的な手順および判定基準にしたがって正当性が確認され、かつ相互に受け入れられるデータを作成できるなら、異なる手法も許容されるべきである。システム案（OECD 1998）では以下のようにされている：

「急性毒性は通常、魚類1種での96時間LC_{50}値（OECD テストガイドライン203 または同等のもの）、甲殻類1種での48時間EC_{50}値（OECD テストガイドライン202 または同等のもの）または藻類1種での72時間または96時間EC_{50}値（OECD テストガイドライン201 または同等のもの）により決定される。こうした生物種はすべての水生生物種の代表であるとみなされるが、ウキクサLemna等、その他の種に関するデータも、試験方法が妥当なものであれば、考慮されることもある。」

慢性試験では一般的に、長時間続くばく露が行われる。その期間は水生生物種の繁殖周期に応じて、数日間から1年間、またはそれ以上に至る。慢性試験は、成長、生存、繁殖および発育に関する特定のエンドポイントを評価するように実施してもよい。

「慢性毒性データは、急性データに比べて入手し難く、試験手順の範囲もそれほど標準化されていない。OECD テストガイドライン210（魚類の初期生活段階毒性試験）、202 Part 2 または211（ミジンコの繁殖試験）および201（藻類生長阻害試験）にしたがって得られたデータは受け入れられる。他の、正当性が確認され、国際的に受け入れられている試験も使用できよう。NOECまたは同等の$L(E)C_x$を用いるべきである。」

OECDの文書には標準的な生態毒性試験のデータ分析のための主な統計手法について記載したものがある（OECD 2006）。

A9.3.2.3 分類の例として引用したOECDガイドラインには、改正されているものや更新が予定されているものもあることを指摘しておかなければならない。こうした改正により試験条件がわずかながら変更されることもある。したがって、分類のための調和された判定基準を策定した専門家グループは、試験期間について、あるいは使用生物種についてさえも、ある程度の柔軟性を持たせるよう試みた。

A9.3.2.4 魚類、甲殻類、および藻類を用いた、受け入れられる試験を実施するためのガイドラインは多くの資料に見出される（OECD, 1999; EPA, 1996; ASTM, 1999; ISO EU）。OECDモノグラフNo.11工業用化学品および農薬の水生毒性試験に関する詳細レビュー文書は、広範囲の試験法の優れた集大成であり、また試験指針の資料である。この文書はまた、適切な試験方法論の情報源でもある。

A9.3.2.5 魚類を用いた試験

A9.3.2.5.1 急性試験

急性試験は一般に、体重0.1－5gの大きさの幼稚仔を用いて96時間の試験期間で実施される。こうした試験で観察するエンドポイントは死亡率である。このサイズより大きい魚または96時間より短い試験期間では、一般的に感度が落ちる。しかし、幼稚仔で96時間の受け入れられるデータが利用できない場合、またはこれら魚のサイズまたは試験期間が異なる試験結果が、より有害性の高い区分における分類に影響する際には、分類のためにこれらの結果を採用することもできよう。分類には、OECD テストガイドライン203（魚類96時間LC_{50}）またはこれと同等のガイドラインに従った試験を採用すべきである。

A9.3.2.5.2 慢性試験

魚類を用いた慢性試験または長期試験は、受精卵、胚、幼稚仔または繁殖行動の認められる成魚で開始できる。OECD テストガイドライン210（魚類の初期生活段階毒性試験）、魚類ライフサイクル試験（US EPA 850.1500）またはこれらと同等の試験法が分類スキームで用いられる。試験期間は試験の目的により大きく異なる（7日程度から200日以上に至るまで）。観察するエンドポイントとしては、孵化率、成長（体長および体重変化）、産卵数、および生存率がある。技術的にはOECD ガイドライン210（魚類の初期生活段階毒性試験）は「慢性」試験ではなく、感受性の高いライフステージにおける亜慢性試験である。この試験は慢性毒性の予測指標として広く受け入れられており、本調和システムでは分類の目的のためにこのようなものとして採用されている。魚類初期生活段階毒性データは、魚類ライフサイクル試験や魚類繁殖試験よりはるかに多く利用できる。

A9.3.2.6   *Crustacea tests*

A9.3.2.6.1   Acute testing

Acute tests with crustacea generally begin with first instar juveniles. For daphnids, a test duration of 48 hours is used. For other crustacea, such as mysids or others, a duration of 96 hours is typical. The observational endpoint is mortality or immobilization as a surrogate to mortality. Immobilization is defined as unresponsive to gentle prodding. Tests consistent with OECD Test Guideline 202 Part 1 (Daphnia acute) or USA-EPA OPPTS 850.1035 (Mysid acute toxicity) or their equivalents should be used for classification.

A9.3.2.6.2   Chronic testing

Chronic tests with crustacea also generally begin with first instar juveniles and continue through maturation and reproduction. For daphnids, 21 days is sufficient for maturation and the production of 3 broods. For mysids, 28 days is necessary. Observational endpoints include time to first brood, number of offspring produced per female, growth, and survival. It is recommended that tests consistent with OECD Test Guideline 202 Part 2 (Daphnia reproduction) or US-EPA 850.1350 (Mysid chronic) or their equivalents be used in the classification scheme.

A9.3.2.7   *Algae/Plant tests*

A9.3.2.7.1   *Tests in algae*

Algae are cultured and exposed to the test substance in a nutrient-enriched medium. Tests consistent with OECD Test Guideline 201 (Algal growth inhibition) should be used. Standard test methods employ a cell density in the inoculum in order to ensure exponential growth through the test, usually 3 to 4 days duration.

The algal test is a short-term test that provides both acute and chronic endpoints. The preferred observational endpoint in this study is algal growth rate inhibition because it is not dependent on the test design, whereas biomass depends both on growth rate of the test species as well as test duration and other elements of test design. If the endpoint is reported only as reduction in biomass or is not specified, then this value may be interpreted as an equivalent endpoint.

A9.3.2.7.2   *Tests in aquatic macrophytes*

The most commonly used vascular plants for aquatic toxicity tests are duckweeds (Lemna gibba and Lemna minor). The Lemna test is a short-term test and, although it provides both acute and sub-chronic endpoints, only the acute EC_{50} is used for classification in the harmonized system. The tests last for up to 14 days and are performed in nutrient enriched media similar to that used for algae but may be increased in strength. The observational endpoint is based on change in the number of fronds produced. Tests consistent with OECD Test Guideline on Lemna (in preparation)[1] and US-EPA 850.4400 (aquatic plant toxicity, Lemna) should be used.

A9.3.3   *Aquatic toxicity concepts*

This section addresses the use of acute and chronic toxicity data in classification, and special considerations for exposure regimes, algal toxicity testing, and use of QSARs. For a more detailed discussion of aquatic toxicity concepts, one can refer to Rand (1996).

A9.3.3.1   *Acute toxicity*

A9.3.3.1.1   Acute toxicity for purposes of classification refers to the intrinsic property of a substance to be injurious to an organism in a short-term exposure to that substance. Acute toxicity is generally expressed in terms of a concentration which is lethal to 50 % of the test organisms (LC_{50}), causes a measurable adverse effect to 50 % of the test organisms (e.g. immobilization of daphnids), or leads to a 50 % reduction in test (treated) organism responses from control (untreated) organism responses (e.g. growth rate in algae).

A9.3.3.1.2   Substances with an acute toxicity determined to be less than one part per million (1 mg/l) are generally recognized as being very toxic. The handling, use, or discharge into the environment of these substances poses a high degree of hazard and they are classified in Chronic 1 and/or Acute 1. Decimal bands are accepted for categorizing acute toxicity above this category. Substances with an acute toxicity measured from one to ten parts per million (1 - 10 mg/l) are classified in

---

[1]   *Published. OECD Test Guideline No. 221: Lemna sp. Growth Inhibition Test.*

A9.3.2.6 甲殻類を用いた試験

A9.3.2.6.1 急性試験

甲殻類を用いた試験は、一般に第一齢幼体（訳者注：孵化後24時間以内のもの）から開始する。ミジンコ類の場合には、48時間の試験期間が採用されている。その他の甲殻類、例えばアミやその他の種類では、96時間が基準である。観察するエンドポイントは死亡率、または死亡率の代用としての遊泳阻害である。遊泳阻害とは、軽い刺激に対する無応答として定義されている。OECDテストガイドライン202 Part 1（ミジンコ急性試験）またはUS EPA OPPTS 850.1035（アミ類急性毒性試験）あるいはこれらと同等のガイドラインを分類に用いるべきである。

A9.3.2.6.2 慢性試験

甲殻類を用いた慢性試験もまた、一般に第一齢幼体から開始し、成熟期および繁殖期まで継続される。ミジンコ類の場合、成熟および3回の産仔には21日間で十分である。アミでは28日必要である。観察するエンドポイントとして、最初の産仔までの期間、雌1頭あたりの産仔数、成長および生存率がある。OECDテストガイドライン202 Part 2（訳者注：現行ではテストガイドライン211）（ミジンコの繁殖試験）またはUS EPA OPPTS 850.1350（アミ類慢性毒性試験）あるいはこれらと同等のガイドラインを分類スキームに用いるべきである。

A9.3.2.7 藻類/植物を用いた試験

A9.3.2.7.1 藻類を用いた試験

藻類を栄養添加培地中で培養して被験物質にばく露する。OECDテストガイドライン201（藻類生長阻害試験）と同等の試験を用いるべきである。標準的な試験方法では、試験期間中（通常は3-4日間）の指数増殖を確認するために、植種源の細胞濃度を採用している。

藻類試験は、急性・慢性の両方のエンドポイントが得られる短期試験である。この試験で観察するエンドポイントとして望ましいのは藻類生長速度阻害である。この理由は、生長速度阻害は試験のデザインに依存しないのに対し、バイオマス法は試験生物種の生長速度や、試験期間やその他の試験のデザインの内容にも依存するためである。エンドポイントがバイオマスの減少のみ、または特定されないで報告されている場合は、この値は同等のエンドポイントとして解釈される場合がある。

A9.3.2.7.2 大型水生植物を用いた試験

水生毒性試験に最も多く用いられる維管束植物はウキクサ（*Lemna gibba*および*Lemna minor*）である。ウキクサ試験は短期試験であり、急性および亜慢性のエンドポイントが与えられるが、本調和システムでの分類には急性EC_{50}のみが使用される。この試験は14日間以内であり、藻類に用いるのと同様な栄養添加培地中で実施されるが、栄養強度が増加されることもある。観察するエンドポイントは生じた葉状体の数の変化に基づいている。Lemnaに関するOECDテストガイドライン（作成中）[1]およびUS EPA 850.4400（水生植物毒性試験、Lemna）を用いるべきである。

A9.3.3 水生毒性の概念

この節では、分類の際の急性毒性および慢性毒性の使用について、またばく露方式、藻類毒性試験およびQSARの利用についての特別な考察について取りあげる。水生毒性の概念についてさらに詳しい議論としては、Rand (1996)を参照できる。

A9.3.3.1 急性毒性

A9.3.3.1.1 分類の目的において急性毒性は、短期ばく露で生物種に有害性のある物質の本来の性質である。急性毒性は一般に、試験生物種の50%に対して致死的である濃度（LC_{50}）として、試験生物種の50%に測定可能な有害作用を及ぼす濃度として（例えば、ミジンコの遊泳阻害）、または対照（未処置）生物の反応と比べて試験（処置群）生物の反応（例：藻類の生長速度）が50%低下する濃度として表現される。

A9.3.3.1.2 急性毒性が1ppm（1mg/l）以下であると判定された物質は、一般的に非常に強い毒性であると認められている。こうした物質の取扱、使用、または環境中への放出は高度の有害性をもたらし、またこうした物質は慢性1および/または急性1に分類されている。十進法による毒性区分の帯域が、急性1より上の急性毒性区分用として受け入れられている。すなわち、急性毒性が1-10ppm（1-10mg/l）として

---

[1] 発行済。OECDテストガイドライン221（ウキクサ成長阻害試験：*Lemna sp. Growth Inhibition Test*）

Acute 2, from ten to one hundred parts per million (10 - 100 mg/l) are classified in Acute 3, and those over one hundred parts per million (> 100 mg/l) are regarded as practically non-toxic.

A9.3.3.2   *Chronic toxicity*

A9.3.3.2.1   Chronic toxicity, for purposes of classification, refers to the intrinsic property of a substance to cause adverse effects to aquatic organisms during exposures which are determined in relation to the life cycle of the organism. Such chronic effects usually include a range of sublethal endpoints and are generally expressed in terms of a No Observable Effect Concentration (NOEC), or an equivalent ECx. Observable endpoints typically include survival, growth and/or reproduction. Chronic toxicity exposure durations can vary widely depending on test endpoint measured and test species used.

A9.3.3.2.2   For the classification based on chronic toxicity a differentiation is made between rapidly degradable and non-rapidly degradable substances. Substances that do rapidly degrade are classified in category Chronic 1 when a chronic toxicity determined to be ≤ 0.01 mg/l. Decimal bands are accepted for categorizing chronic toxicity above this category. Substances with a chronic toxicity measured from 0.01 to 0.1 mg/l are classified in category Chronic 2 for chronic toxicity, from 0.1 to 1.0 mg/l are classified in category Chronic 3 for chronic toxicity, and those over 1.0 mg/l are regarded as practically non-toxic. For substances that do not rapidly degrade or where no information on rapid degradation is available two chronic categories are used: Chronic 1 when a chronic toxicity determined to be ≤ 0.1 mg/l and Chronic 2 when chronic toxicity is measured from 0.1 to 1.0 mg/l.

A9.3.3.2.3   Since chronic toxicity data are less common in certain sectors than acute data, for classification schemes, the potential for chronic toxicity is, in absence of adequate chronic toxicity data, identified by appropriate combinations of acute toxicity, lack of degradability and/or the potential or actual bioaccumulation. However, where adequate chronic toxicity data exist, this shall be used in preference over the classification based on the combination of acute toxicity with degradability and/or bioaccumulation. In this context, the following general approach should be used:

(a)   If adequate chronic toxicity data are available for all three trophic levels this can be used directly to determine an appropriate long-term (chronic) hazard category;

(b)   If adequate chronic toxicity data are available for one or two trophic levels, it should be examined if acute toxicity data are available for the other trophic level(s). A potential classification is made for the trophic level(s) with chronic data and compared with that made using the acute toxicity data for the other trophic level(s). The final classification shall be made according to the most stringent outcome;

(c)   In order to remove or lower a chronic classification, using chronic toxicity data, it must be demonstrated that the NOEC(s) (or equivalent ECx) used would be suitable to remove or lower the concern for all taxa which resulted in classification based on acute data in combination with degradability, and/or bioaccumulation. This can often be achieved by using a long-term NOEC for the most sensitive species identified by the acute toxicity. Thus, if a classification has been based on a fish acute LC_{50}, it would generally not be possible to remove or lower this classification using a long-term NOEC from an invertebrate toxicity test. In this case, the NOEC would normally need to be derived from a long-term fish test of the same species or one of equivalent or greater sensitivity. Equally, if classification has resulted from the acute toxicity to more than one taxa, it is likely that NOECs from each taxa will be needed. In case of classification of a substance as Chronic 4, sufficient evidence should be provided that the NOEC or equivalent ECx for each taxa is greater than 1 mg/l or greater than the water solubility of the substances under consideration.

A9.3.3.2.4   Testing with algae/Lemna cannot be used for removing or lowering a classification because:

(a)   the algae and Lemna tests are not long-term studies;

(b)   the acute to chronic ratio is generally narrow; and

(c)   the endpoints are more consistent with the acute endpoints for other organisms.

However where classification is applied solely due to the acute toxicity ($L(E)C_{50}$) observed in single algae/aquatic plant tests, but there is evidence from a range of other algae tests that the chronic toxicity (NOECs) for this taxonomic group is in the toxicity band corresponding to a less stringent classification category or above 1mg/l, this evidence could be used to consider removing or lowering a classification. At present this approach cannot be applied to aquatic plants since no standardized chronic toxicity tests have been developed.

測定された物質は急性 2 に、10－100ppm（10－100mg/l）として測定された物質は急性 3 に、および100ppm（100mg/l）を超える物質は実質的に毒性はないとみなされる。

A9.3.3.2　*慢性毒性*

A9.3.3.2.1　分類の目的において慢性毒性は、生物種のライフサイクルに関連して決定されたばく露期間中に、物質がその生物種に有害作用を及ぼす物質固有の特性をいう。こうした慢性作用には通常、一連の亜致死的なエンドポイントが含まれており、一般に無影響濃度（NOEC）または同等の影響濃度（ECx）として表される。観察するエンドポイントとして典型的なものには、生存、成長、または繁殖がある。慢性毒性のばく露期間は、測定するエンドポイントおよび用いる生物種によって広く異なる。

A9.3.3.2.2　慢性毒性に基づく分類については、速やかに分解する物質と速やかに分解しない物質とでは区別がなされている。慢性毒性値が≦0.01mg/lである場合、速やかに分解する物質は慢性1の区分に分類される。十進法による毒性区分の帯域が、このカテゴリーより上の慢性毒性区分用として受け入れられている。慢性毒性値が0.01mg/lから0.1mg/lである物質は慢性毒性区分の慢性2に、0.1mg/lから1.0mg/lである物質は慢性毒性区分の慢性3に分類され、1.0mg/l以上の物質では実質的に毒性がないとみなされる。速やかに分解しない物質、または急速分解性についての情報がない場合は、次の2つの慢性区分が用いられる：慢性毒性値が≦0.1mg/lである場合は慢性1、慢性毒性値が0.1mg/lから1.0mg/lである場合は慢性2とする。

A9.3.3.2.3　特定の分野では慢性毒性データは急性毒性データよりも一般的ではないため、慢性毒性データが十分に存在しないときは、分類のスキームにおいては、慢性毒性の可能性は、急性毒性、分解性の無さおよび/または、潜在的ないしは実際の生物蓄積性を適切に組み合わせることによって判断される。しかし、十分な慢性毒性データが存在する場合は、急性毒性と分解性および/または生物蓄積性との組み合わせに基づく分類よりも、慢性毒性データを優先的に利用すべきである。このことから、以下に述べる一般的なアプローチを採用すべきである：

(a)　3 つのすべての栄養段階について十分な慢性毒性データが得られる場合は、適切な長期（慢性）有害性区分の決定に直接このデータを利用できる；

(b)　1 つまたは 2 つの栄養段階について十分な慢性毒性データが得られる場合は、他の栄養段階について急性毒性データが得られるどうかを検証するべきである。慢性毒性データのある栄養段階について潜在的な分類がなされ、それが急性毒性データを用いてなされた他の栄養段階について潜在的な分類と比較される。最終的な分類は最も厳しい結果にしたがって実施されるべきである；

(c)　慢性毒性データを用いて慢性区分を解除あるいは格下げするには、使用されたNOEC（または同等のECx）が、急性毒性データと分解性および/または生物蓄積性との組み合わせに基づくすべての分類群について、慢性毒性区分から除外し、あるいは格下げするのに適切であることを証明しなければならない。これは多くの場合、急性毒性で最も感受性が高いとされた種に対する長期NOECを用いて達成される。このため、分類が魚の急性 LC_{50} に基づいている場合は、一般的に、無脊椎動物の毒性試験から得た長期 NOEC を使って、この区分を解除あるいは格下げすることはできない。この場合、NOECは通常、同一種の魚あるいは同等かより大きな感受性をもった魚の長期試験から導出される必要がある。同様に、複数の分類群について急性毒性を用いて分類された場合、それぞれの分類群から得た NOEC が必要となる。慢性 4 に分類された物質の場合は、それぞれの分類群について、NOEC または同等の ECx が 1mg/l より大きいかもしくは物質の水溶解度を上回るという十分な証拠が提供されるべきである。

A9.3.3.2.4　藻類/Lemnaを用いる試験は、分類の解除あるいは格下げに用いてはならない、なぜなら：

(a)　藻類およびウキクサを用いた試験は長期試験ではない；

(b)　一般的に慢性毒性と急性毒性の比が小さい；および

(c)　エンドポイントが他の生物における急性のエンドポイントよりも一貫性があるからである。

ただし、単一の藻類/水生植物試験で認められた急性毒性（L(E)C₅₀）のみを用いて分類された場合で、他の藻類を用いた一連の試験からこの分類群での慢性毒性（NOEC）が、より厳しくない分類区分に相当するか、もしくは1mg/l を超えるという証拠が得られている場合には、これらのデータを分類区分の解除もしくは格下げの検討に用いてもよい。現時点では、水生植物に対する標準化された慢性毒性試験法がまだ開発されていないため、このアプローチを水生植物に適用することはできない。

A9.3.3.3   *Exposure regimes*

Four types of exposure conditions are employed in both acute and chronic tests and in both freshwater and saltwater media: static, static-renewal (semi-static), recirculation, and flow-through. The choice for which test type to use usually depends on test substance characteristics, test duration, test species, and regulatory requirements.

A9.3.3.4   *Test media for algae*

Algal tests are performed in nutrient-enriched media and the use of one common constituent, EDTA, or other chelators, should be considered carefully. When testing the toxicity of organic chemicals, trace amounts of a chelator like EDTA are needed to complex micronutrients in the culture medium; if omitted, algal growth can be significantly reduced and compromise test utility. However, chelators can reduce the observed toxicity of metal test substances. Therefore, for metal compounds, it is desirable that data from tests with high concentration of chelators and/or tests with stoichiometrical excess of chelator relative to iron should be critically evaluated. Free chelator may mask heavy metal toxicity considerably, in particular with strong chelators like EDTA. However, in the absence of available iron in the medium the growth of algae can become iron limited, and consequently data from tests with no or with reduced iron and EDTA should be treated with caution.

A9.3.3.5   *Use of QSARs*

For purpose of classification, and in the absence of experimental data, QSARs can be relied upon to provide predictions of acute toxicity for fish, daphnia, and algae for non-electrolyte, non-electrophilic, and otherwise non-reactive substances (See section A9.6 on *Use of QSAR*). Problems remain for substances such as organophosphates which operate by means of special mechanisms such as functional groups which interact with biological receptors, or which can form sulfhydryl bonds with cellular proteins. Reliable QSARs have been derived for chemicals acting by a basic narcosis mechanism. These chemicals are nonelectrolytes of low reactivity such as hydrocarbons, alcohols, ketones and certain aliphatic chlorinated hydrocarbons which produce their biological effects as a function of their partition coefficients. Every organic chemical can produce narcosis. However, if the chemical is an electrolyte or contains specific functional groups leading to non-narcotic mechanisms as well, any calculations of toxicity based on partition coefficient alone would severely underestimate the toxicity. QSARs for acute aquatic toxicity of parent compounds cannot be used to predict the effects of toxic metabolites or degradates, when these arise after a longer time period than the duration of acute tests.

A9.3.4   Weight of evidence

A9.3.4.1   The best quality data should be used as the fundamental basis for classification. Classification should preferably be based on primary data sources. It is essential that test conditions are clearly and completely articulated.

A9.3.4.2   Where multiple studies for a taxonomic group are available, a decision on what is the most sensitive and highest quality must be made. A judgement has to be made on a case by case basis whether a non-GLP study with a more sensitive observation is used in lieu of a GLP study. It would appear that results that indicate high toxicity from tests performed according to non-standard or non-GLP guidelines should be able to be used for classification, whereas studies, which demonstrate negligible toxicity, would require more careful consideration. Substances, which are difficult to test, may yield apparent results that are more or less severe than the true toxicity. Expert judgement would also be needed for classification in these cases.

A9.3.4.3   Where more than one acceptable test is available for the same taxonomic group, the most sensitive (the one with the lowest $L(E)C_{50}$ or NOEC) is generally used for classification. However, this must be dealt with on a case-by-case basis. When larger data sets (4 or more values) are available for the same species, the geometric mean of toxicity values may be used as the representative toxicity value for that species. In estimating a mean value, it is not advisable to combine tests of different species within a taxa group or in different life stages or tested under different conditions or duration.

A9.3.5   Difficult to test substances

A9.3.5.1   Valid aquatic toxicity tests require the dissolution of the test substance in the water media under the test conditions recommended by the guideline. In addition, a bioavailable exposure concentration should be maintained for the duration of the test. Some substances are difficult to test in aquatic systems and guidance has been developed to assist in testing these materials (DoE 1996; ECETOC 1996; and US EPA 1996). OECD Guidance Document on aquatic toxicity testing of difficult substances and mixtures (OECD, 2000) is a good source of information on the types of substances that are difficult to test and the steps needed to ensure valid conclusions from tests with these materials.

A9.3.5.2   Nevertheless, much test data exist that may have used testing methodologies which, while not in conformity with what might be considered best practice today, can still yield information suitable for application of the classification

A9.3.3.3　ばく露方式

　急性および慢性の両方の試験、および淡水と塩水の両方の媒体に、4 種類のばく露条件が採用されている。すなわち、止水、止水-交換（半止水）、再循環、および流水である。どの種類の試験を採用するかの選択は、被験物質の性質、試験期間、試験生物種、および規制官庁による要求項目による。

A9.3.3.4　藻類試験用の培地

　藻類を用いる試験は、栄養添加培地中で実施されるが、ある種の一般的成分、すなわち EDTA またはその他のキレート剤の使用について注意を払う必要がある。有機化合物の毒性を試験する場合、微量の EDTA 等のキレート剤が培地中の複合微量成分として必要である。これを加えないと、藻類の生長は著しく阻害され、試験の有用性が損なわれる。しかしキレート剤は、金属被験物質の見かけの毒性を低下させることがある。したがって金属化合物の場合には、高濃度のキレート剤を加えた試験結果、または鉄に対して化学量論的に過剰なキレート剤を加えた試験結果を厳密に評価しなければならない。フリーのキレート剤、特に EDTA のような強力なキレート剤は重金属の毒性を著しく低下させるかもしれない。しかしながら、培地に鉄が不足している場合には、藻類の生長は鉄で制限されることになり、したがって、鉄および EDTA のない、あるいは減少させた試験の結果は注意して扱わなければならない。

A9.3.3.5　QSAR の利用

　分類を目的とし、なおかつ実験データがない場合には、非電解性、非親電子性、また非反応性の物質であるならば、魚類、ミジンコ、および藻類に対する急性毒性を予測するのに、QSAR に頼ることが可能である（QSAR の利用に関する A9.6 節参照）。有機リン化合物等の物質の場合には問題が残る。これらは、体内レセプターと相互作用をし、または細胞内タンパク質とスルフヒドリル結合を形成できる官能基等、特殊なメカニズムにより作用する。基本的な麻酔メカニズムで作用する化学品については、信頼できる QSAR が導かれている。こうした化学品とは、例えば炭化水素、アルコール、ケトン、およびある種の脂肪族塩素化炭化水素など、低反応性の非電解質であり、これらはその生物学的作用をその分配係数の関数として生じる。有機化学品ならば何でも、麻酔作用を生じることができる。しかし、化学品が電解質であったり、または麻酔以外のメカニズムが導かれる特殊な官能基を含んでいるならば、分配係数のみによる毒性の算定をすると、毒性を著しく過小評価することになろう。親化合物の急性水生毒性に関する QSAR は、毒性のある代謝物または分解物が急性試験の期間よりもずっと後になって生成する場合には、これら代謝物または分解物の作用を予測するのには利用できない。

A9.3.4　証拠の重み

A9.3.4.1　分類の基本的な基盤として最高品質のデータを用いるべきである。一次のデータ源に基づいて分類するのが望ましい。試験条件が明瞭かつ完全に表現されていることが不可欠である。

A9.3.4.2　1 つの分類群について複数の試験結果が入手されたならば、どれが最も高感度であり高品質であるかの決定をせねばならない。GLP 試験の代わりに非 GLP でもより高感度の所見の得られている試験を採用することについては、ケースバイケースで判断しなければならない。標準的でない、または非 GLP のガイドラインにしたがって実施された試験から得られた結果が強毒性を示唆している場合には、その結果を分類に使用できるべきであるが、こうした試験が、無視できる毒性を示した場合には、より注意深い検討が必要となろう。試験が困難な物質から得られる結果は、見かけ上、真の毒性より弱い毒性を示すこともあれば、強い毒性を示すこともある。こうしたケースでの分類には専門家の判断が必要であろう。

A9.3.4.3　同一の分類群について複数の受け入れられる試験結果が利用可能な場合、一般に、最も高感度の結果（L(E)C$_{50}$ 値または NOEC 値が最低値の結果）を分類に採用する。しかし、これはケースバイケースで扱わなければならない。同一生物種についてより大きなデータセット（4 個以上の数）が利用可能なときは、毒性値の幾何平均値をその生物種を代表する毒性値として用いてもよい。平均値を推定する際、分類群が同一でも異なった種の試験や、または異なったライフステージで行われた試験、あるいは条件または試験期間が異なる試験を組み合わせることは望ましくない。

A9.3.5　試験困難な物質

A9.3.5.1　妥当な水生毒性試験は、ガイドラインで勧告されている試験条件下において被験物質が水媒体中で溶解することを必要とする。さらに、試験期間を通じて、生物学的に利用できるばく露濃度が維持されなければならない。ある種の物質は水系での試験が困難であり、そうした物質の試験を支援するための手引きが作成されている（DoE 1996; ECETOC 1996; US EPA 1996）。試験困難物質および混合物の水生毒性試験に関する OECD ガイダンス文書（OECD 2000）は試験困難な物質の種類と、これらの物質を用いた試験から有効な結論を確実に得るために必要な手順に関するよい情報源である。

A9.3.5.2　それにもかかわらず、今日において最良と考えられている方法論にはしたがっていないが、分類基準に適用するのに適した情報を得ることのできる試験方法論を用いたと思われる多くの試験データが

criteria. Such data require special guidance on interpretation, although ultimately, expert judgement must be used in determining data validity. Such difficult to test substances may be poorly soluble, volatile, or subject to rapid degradation due to such processes as phototransformation, hydrolysis, oxidation, or biotic degradation. When testing algae, coloured materials may interfere with the test endpoint by attenuating the light needed for cell growth. In a similar manner, substances tested as cloudy dispersions above solubility may give rise to false toxicity measurements. Loading of the water column with test material can be an issue for particulates or solids such as metals. Petroleum distillate fractions can also pose loading problems, as well as difficult interpretational problems when deciding on the appropriate concentrations for determining $L(E)C_{50}$ values. The draft Guidance Document on Aquatic Toxicity Testing of Difficult Substances and Mixtures describes the more common properties of many types of substances which are likely to pose testing difficulties.

(a) Stability: If test chemical concentrations are expected to fall below 80 % of nominal, testing, in order to be valid, may require exposure regimes which provide for renewal of the test material. Semi-static or flow-through conditions are preferred. Special problems arise, therefore, with respect to testing on algae, where the standard guidelines generally include static tests to be conducted. While alternative exposure regimes are possible for crustacea and fish, these tests are frequently conducted on static conditions as included in the internationally agreed guidelines. In these tests, a certain level of degradation as well as other relevant factors have to be tolerated and appropriate account must be taken in calculations of toxic concentrations. Some approaches on how this can be dealt with are covered in A9.3.5.6. Where degradation occurs, it is also important to consider the influence of the toxicity of the degradation products on the recorded toxicity in the test. Expert judgement will need to be exercised when deciding if the data can be used for classification;

(b) Degradation: When a compound breaks down or degrades under test condition, expert judgement should be used in calculating toxicity for classification, including consideration of known or likely breakdown products. Concentrations of the parent material and all significant toxic degradates are desirable. If degradates are expected to be relatively non-toxic, renewable exposure regimes are desirable in order to ensure that levels of the parent compounds are maintained;

(c) Saturation: For single component substances, classification should be based only on toxic responses observed in the soluble range, and not on total chemical loading above solubility. Frequently, data are available which indicate toxicity at levels in excess of water solubility and, while these data will often be regarded as not valid, some interpretation may be possible. These problems generally apply when testing poorly soluble substances, and guidance on how to interpret such data is included in A9.3.5.7 (see also the Guidance Document on aquatic toxicity testing of difficult substances and mixtures);

(d) Perturbation of test media: Special provisions may be needed to ensure dissolution of difficult to test substances. Such measures should not lead to significant changes in the test media when such changes are likely to lead to an increase or decrease in the apparent toxicity and hence the classification level of the test substance;

(e) Complex substances: Many substances covered by the classification scheme are in fact mixtures, for which measurement of exposure concentrations is difficult, and in some cases impossible. Substances such as petroleum distillate fractions, polymers, substances with significant levels of impurities, etc can pose special problems since the toxic concentration is difficult to define and impossible to verify. Typical testing procedures often rely on the formation of a Water-Soluble Fraction (WSF) or Water Accommodated Fraction (WAF) and data are reported in terms of loading rates. These data may be used in applying the classification criteria.

A9.3.5.3　For classification of organic compounds, it is desirable to have stabilized and analytically measured test concentrations. Although measured concentrations are preferred, classification may be based on nominal concentration studies when these are the only valid data available under certain circumstances. If the material is likely to substantially degrade or otherwise be lost from the water column, care must be taken in data interpretation and classification should be done taking the loss of the toxicant during the test into account, if relevant and possible. Additionally, metals present their own set of difficulties and are discussed separately. table A9.3.1 lists several properties of difficult to test substances and their relevance for classification.

A9.3.5.4　In most difficult to test conditions, the actual test concentration is likely to be less than the nominal or expected test concentration. Where acute toxicities ($L(E)C_{50}$s) are estimated to be < 1 mg/l for a difficult to test substance, one can be fairly confident the classification in the Acute 1 (and Chronic 1 if appropriate) is warranted. However, if the estimated acute toxicity is greater than 1 mg/l, the estimated toxicity is likely to under-represent the toxicity. In these circumstances, expert judgement is needed to determine the acceptability of a test with a difficult to test substance for use in classification.

存在している。そうしたデータを解釈するには特別の手引きが必要であるが、最終的にはデータの妥当性判定に専門家の判断を用いなければならない。こうした試験困難な物質とは、溶解度が低い、揮発性である、または光変換、加水分解、酸化、あるいは生物学的分解等のプロセスにより速やかな分解を受ける物質である。藻類を試験する場合、着色物質は細胞成長に必要な光を弱めて、試験のエンドポイントに干渉することもある。同様に、溶解度以上の濁った分散状態で試験された物質は、誤った毒性の測定を起こすことがあろう。水相に被験物質を加える場合、粒子状物質または金属等の固体では問題が生じることがある。石油の蒸留分画も添加時に問題を生じることがあり、また L(E)C$_{50}$ 値を決定するための適切な濃度を設定する際に解釈上困難な問題を生じる。「試験困難物質および混合物の水生毒性試験に関するガイダンス文書」原案は、試験上困難を生じる可能性の高い多くのタイプの物質について、より一般的な特性を説明している。

(a) **安定性**：被験物質の濃度が設定濃度の 80%より低くなると予測されるときは、試験を有効とするためには、被験物質を液交換するようなばく露方式が求められる。半止水または流水条件が望ましい。このため藻類の試験については、標準的ガイドラインが一般に止水条件で実施されるので、特有の問題が生じる。甲殻類および魚類では別のばく露方式も可能であるが、国際的に了承されたガイドラインに記載されている止水条件での試験が実施されることが多い。これらの試験において、あるレベルの分解、その他の関連因子も許容しなければならず、毒性濃度の計算には適切な考慮を払わねばならない。こうしたことの取扱いについてのアプローチを A9.3.5.6 に示す。分解が起こる場合には、試験で記録された毒性に対する分解生成物の毒性の影響も考慮することが重要である。そのデータを分類に使用できるかどうかについては、専門家の判断が必要であろう；

(b) **分解**：化合物が試験条件下で壊れる、または分解する場合は、既知の、あるいは想定される分解生成物の考察も含めて、分類のための毒性計算に専門家の判断が必要である。親物質およびすべての有毒分解物の濃度測定が望ましい。分解物に比較的毒性がないことが予測されるならば、親物質の濃度を維持するため、試験液を交換するばく露方式が望ましい；

(c) **飽和**：単一成分の物質に関しては、溶解度以下の濃度範囲で測定された毒性反応のみから分類すべきであり、溶解度を超える物質負荷に基づくべきでない。水への溶解度を超えた濃度において毒性を示すデータが入手されることがしばしばあり、こうしたデータは妥当でないと考えられることが多いが、何らかの解釈も可能である。これらの問題は一般に、溶解度の低い物質の試験にあてはまり、こうしたデータをどう解釈するかについての手引きは A9.3.5.7 に記載する（「試験困難な物質および混合物の水生毒性試験に関する手引き」も参照のこと）；

(d) **試験媒体の変動**：試験困難な物質が確実に溶解するよう特別な手段が必要であろう。試験媒体を変えると見かけの毒性が増加または減少し、これによって被験物質の分類レベルも変わる可能性があるので、このような手段は、試験媒体を著しく変更しないようなものとするべきである；

(e) **複合物質**：本分類スキームの対象となる物質の多くは、現実的には混合物であるが、この場合ばく露濃度の測定が困難であり、また測定が不可能な場合もある。石油蒸留画分、ポリマー、不純物を多く含む物質等は、毒性濃度の測定が困難であり、検証も不可能なので、特有の問題が生じる。典型的な試験手順はしばしば「水可溶性画分（WSF）」または「水和画分（WAF）」の生成に依存しており、データは添加率で報告されている。こうしたデータを分類基準に用いてもよいであろう。

A9.3.5.3　有機化合物の分類では、試験濃度を安定にし、分析して測定してあることが望ましい。濃度が測定されている方が望ましいが、ある状況において設定濃度だけが妥当なデータとして利用可能な場合は、設定濃度に基づいて分類する場合もある。物質が実質的に分解し、または水相から消失する可能性がある場合は、データ解釈には注意が必要であり、もし妥当かつ可能であるならば、試験中の毒物の消失分を考慮して分類すべきである。その他に、金属が特有の問題を提起するが、これについては別に述べる。表 A9.3.1 に、試験困難物質のいくつかの性質、およびその分類との関連性についてまとめた。

A9.3.5.4　ほとんどの試験困難な状況において、実際の試験濃度は、設定上の、または期待される試験濃度より低くなる傾向がある。試験困難な物質の急性毒性値（L(E)C$_{50}$）が 1mg/l より低いと見積もられた場合、急性 1（および適合する場合には慢性 1）の分類が保証されたと確信してよい。しかし、急性毒性推定値が 1mg/l を超える場合には、その毒性推定値は毒性を過小に示している傾向がある。そのような状況で、試験困難な物質での試験結果を分類に用いることができるかを決定するには、専門家の判断が必要で

Where the nature of the testing difficulty is believed to have a significant influence on the actual test concentration when acute toxicity is estimated to be greater than 1 mg/l and the test concentration is not measured, then the test should be used with due caution in classification.

A9.3.5.5 The following paragraphs provide some detailed guidance on some of these interpretational problems. In doing so it should be remembered that this is guidance and hard and fast rules cannot be applied. The nature of many of the difficulties mean that expert judgement must always be applied both in determining whether there is sufficient information in a test for a judgement to be made on its validity, and also whether a toxicity level can be determined suitable for use in applying the classification criteria.

A9.3.5.6 *Unstable substances*

A9.3.5.6.1 While testing procedures should ideally have been adopted which minimized the impacts of instability in the test media, in practice, in certain tests, it can be almost impossible to maintain a concentration throughout the test. Common causes of such instability are oxidation, hydrolysis, photodegradation and biodegradation. While the latter forms of degradation can more readily be controlled, such controls are frequently absent in much existing testing. Nevertheless, for some testing, particularly acute and chronic fish toxicity testing, a choice of exposure regimes is available to help minimize losses due to instability, and this should be taken into account in deciding on the test data validity.

A9.3.5.6.2 Where instability is a factor in determining the level of exposure during the test, an essential prerequisite for data interpretation is the existence of measured exposure concentrations at suitable time points throughout the test. In the absence of analytically measured concentrations at least at the start and end of test, no valid interpretation can be made and the test should be considered as invalid for classification purposes. Where measured data are available, a number of practical rules can be considered by way of guidance in interpretation:

(a) where measured data are available for the start and end of test (as is normal for the acute Daphnia and algal tests), the $L(E)C_{50}$, for classification purposes, may be calculated based on the geometric mean of the start and end of test concentrations. Where the end of test concentrations are below the analytical detection limit, such concentrations shall be considered to be half that detection limit;

(b) where measured data are available at the start and end of media renewal periods (as may be available for the semi-static tests), the geometric mean for each renewal period should be calculated, and the mean exposure over the whole exposure period calculated from these data;

(c) where the toxicity can be attributed to a degradation breakdown product, and the concentrations of this are known, the $L(E)C_{50}$ for classification purposes, may be calculated based on the geometric mean of the degradation product concentration, back calculated to the parent substance;

(d) similar principles may be applied to measured data in chronic toxicity testing.

A9.3.5.7 *Poorly soluble substances*

A9.3.5.7.1 These substances, usually taken to be those with a solubility in water < 1 mg/l, are frequently difficult to dissolve in the test media, and the dissolved concentrations will often prove difficult to measure at the low concentrations anticipated. For many substances, the true solubility in the test media will be unknown and will often be recorded as < detection limit in purified water. Nevertheless, such substances can show toxicity, and where no toxicity is found, judgement must be applied to whether the result can be considered valid for classification. Judgement should err on the side of caution and should not underestimate the hazard.

A9.3.5.7.2 Ideally, tests using appropriate dissolution techniques and with accurately measured concentrations within the range of water solubility should be used. Where such test data are available, they should be used in preference to other data. It is normal, however, particularly when considering older data, to find such substances with toxicity levels recorded in excess of the water solubility, or where the dissolved levels are below the detection limit of the analytical method. Thus, in both circumstances, it is not possible to verify the actual exposure concentrations using measured data. Where these are the only data available on which to classify, some practical rules can be considered by way of general guidance:

(a) where the acute toxicity is recorded at levels in excess of the water solubility, the $L(E)C_{50}$ for classification purposes, may be considered to be equal to or below the measured water solubility. In such circumstances it is likely that Chronic 1 and/or Acute 1 should be applied. In making this decision, due attention should be paid to the possibility that the excess undissolved substance may have given rise to physical effects on the test organisms. Where this is considered the likely cause of the effects observed, the test should be considered as invalid for classification purposes;

ある。急性毒性の推定値が 1mg/l より大きく、かつ試験濃度が測定されていない場合で、試験の困難さの性格が実際の試験濃度に大きく影響していると考えられる場合には、相当の注意を払ってその試験を分類に用いるべきである。

A9.3.5.5　以下の各項では、こうした解釈上の問題について詳しい手引きを示している。解釈を行うにあたっては、これはあくまでも手引きであり、厳密かつ固定的な法則を適用できないことを念頭に置くべきである。試験の困難さの性質の多くは、試験にその妥当性を判定するに十分な情報があるか、また分類基準を適用するのに用いられるように毒性レベルを決定できるかどうか、その両方の判定に専門家の判断が必ず必要であることを意味している。

A9.3.5.6　*不安定な物質*

A9.3.5.6.1　試験手順は、被験物質が試験媒体中で不安定であることの影響を最小限に抑えるよう適用するのが理想であるが、現実には、ある種の試験では、試験期間を通じて濃度を維持することはほとんど不可能なことである。一般にそのような不安定さの原因は、酸化、加水分解、光分解および生分解である。光分解や生分解は比較的コントロールが容易であるが、多くの既存の試験法ではこうしたコントロールがなされていないことがしばしばである。それにもかかわらず、ある種の試験、特に魚類を用いた急性および慢性毒性試験では、不安定による損失を最小限に抑えるようばく露方式が選べるようになっており、またこのことは試験データの妥当性を決定する際に考慮されるべきである。

A9.3.5.6.2　試験中のばく露レベルの決定に不安定さが 1 つの要因となっている場合、データの解釈に不可欠な前提条件は、試験期間を通じて適切な複数の時点で測定されたばく露濃度があることである。少なくとも試験開始時と終了時に濃度測定の分析値が得られていないならば、正当な解釈ができず、その試験は分類目的には妥当でないと考えるべきである。分析データが利用可能な場合には、以下のようないくつかの実用規則が解釈上の手引きとして考えられる：

(a) 試験開始時および終了時の測定値がある場合（ミジンコ急性試験および藻類急性試験では標準の要件）、分類の目的では、$L(E)C_{50}$ 値を開始時および終了時の各濃度の幾何平均に基づいて計算してよい。試験終了時の濃度が分析上の検出限界より低い場合、濃度は検出限界の半分の値であると考える；

(b) 試験媒体交換期間の開始時と終了時の測定値がある場合（半止水条件での試験が利用可能）、更新期間ごとの幾何平均を計算し、そのデータから全ばく露期間にわたる平均ばく露量を算出すべきである；

(c) 毒性が分解生成物に帰属でき、その生成物の濃度が既知である場合、分類のための $L(E)C_{50}$ は、分解生成物濃度の幾何平均値に基づいて計算してから、親物質に逆算してもよい；

(d) 同様の原則が慢性毒性試験での測定値にも適用できよう。

A9.3.5.7　*難溶性の物質*

A9.3.5.7.1　難溶性の物質は、一般に水に対する溶解度が 1mg/l 未満の物質であるとされているが、これらの物質は試験媒体に難溶であることが多く、予測される低い濃度では、しばしば溶解濃度は測定困難であることが示される。多くの物質では試験媒体中の真の溶解度は知られず、純水中で検出限界以下であるとして記録されることが多い。にもかかわらず、そうした物質でも毒性を示すことがあり、毒性が認められなかった場合には、その結果が分類のために妥当であるとみなせるかの判断がなされなければならない。判断は慎重な側に寄せるべきで、有害性を過小評価することがあってはならない。

A9.3.5.7.2　理想的には、適切な溶解技術を採用し、水に対する溶解度の範囲内で濃度を正確に測定した試験を採用すべきである。そのような試験データが利用可能であれば、他のデータより優先して用いるべきである。しかし、以前のデータを考慮する場合などは特に、毒性レベルが水に対する溶解度より上で記録されている物質や、溶解度が分析方法の検出限界より下で記録されている場合等が普通に見られる。したがって、この両者の状況では、測定されたデータから実際のばく露濃度を検証することはできない。分類するのに、そのようなデータしか利用できない場合には、一般的な手引きとして以下のような実用的な規則が考えられる：

(a) 急性毒性が水に対する溶解度を超過する濃度で記録されている場合、分類目的での $L(E)C_{50}$ は水に対する溶解度の測定値以下であると考えてよい。その場合、慢性１および/または急性１の各区分を適用すべきであると思われる。この決定をする際、溶けなかった過剰の物質が試験生物体に物理的な影響を及ぼした可能性に十分注意を払うべきである。そのことが観察された影響の原因であると思われた場合には、その試験は分類のためには妥当でないとみなすべきである；

(b) where no acute toxicity is recorded at levels in excess of the water solubility, the $L(E)C_{50}$ for classification purposes may be considered to be greater than the measured water solubility. In such circumstances, consideration should be given to whether the Chronic 4 should apply. In making a decision that the substance shows no acute toxicity, due account should be taken of the techniques used to achieve the maximum dissolved concentrations. Where these are not considered as adequate, the test should be considered as invalid for classification purposes;

(c) where the water solubility is below the detection limit of the analytical method for a substance, and acute toxicity is recorded, the $L(E)C_{50}$ for classification purposes, may be considered to be less than the analytical detection limit. Where no toxicity is observed, the $L(E)C_{50}$ for classification purposes, may be considered to be greater than the water solubility. Due consideration should also be given to the quality criteria mentioned above;

(d) where chronic toxicity data are available, the same general rules should apply. Again, where these data cannot be validated by consideration of measured concentrations, the techniques used to achieve the maximum dissolved concentrations must be considered as appropriate.

A9.3.5.8 *Other factors contributing to concentration loss*

A number of other factors can also contribute to losses of concentration and, while some can be avoided by correct study design, interpretation of data where these factors have contributed may, from time to time, be necessary.

(a) sedimentation: this can occur during a test for a number of reasons. A common explanation is that the substance has not truly dissolved despite the apparent absence of particulates, and agglomeration occurs during the test leading to precipitation. In these circumstances, the $L(E)C_{50}$ or NOEC for classification purposes, may be considered to be based on the end of test concentrations. Equally, precipitation can occur through reaction with the media. This is considered under instability above;

(b) adsorption: this can occur for substances of high adsorption characteristics such as high log Kow substances. Where this occurs, the loss of concentration is usually rapid and exposure may best be characterized by the end of test concentrations;

(c) bioaccumulation: losses may occur through the bioaccumulation of a substance into the test organisms. This may be particularly important where the water solubility is low and log K_{ow} correspondingly high. The $L(E)C_{50}$ or NOEC for classification purposes, may be calculated based on the geometric mean of the start and end of test concentrations.

A9.3.5.9 *Perturbation of the test media*

A9.3.5.9.1 Strong acids and bases may appear toxic because they may alter pH. Generally, however changes of the pH in aquatic systems are normally prevented by buffer systems in the test medium. If no data are available on a salt, the salt should generally be classified in the same way as the anion or cation, i.e. as the ion that receives the most stringent classification. If the effect concentration is related to only one of the ions, the classification of the salt should take the molecular weight difference into consideration by correcting the effect concentration by multiplying with the ratio: MW_{salt}/MW_{ion}.

A9.3.5.9.2 Polymers are typically not available in aquatic systems. Dispersible polymers and other high molecular mass materials can perturb the test system and interfere with uptake of oxygen and give rise to mechanical or secondary effects. These factors need to be taken into account when considering data from these substances. Many polymers behave like complex substances, however, having a significant low molecular mass fraction which can leach from the bulk polymer. This is considered further below.

A9.3.5.10 *Complex substances*

A9.3.5.10.1 Complex substances are characterized by a range of chemical structures, frequently in a homologous series, but covering a wide range of water solubilities and other physico-chemical characteristics. On addition to water, an equilibrium will be reached between the dissolved and undissolved fractions which will be characteristic of the loading of the substance. For this reason, such complex substances are usually tested as a WSF or WAF, and the $L(E)C_{50}$ recorded based on the loading or nominal concentrations. Analytical support data are not normally available since the dissolved fraction will itself be a complex mixture of components. The toxicity parameter is sometimes referred to as LL_{50}, related to the lethal loading level. This loading level from the WSF or WAF may be used directly in the classification criteria.

(b) 水に対する溶解度を超過する濃度で急性毒性の記録がない場合、分類のための L(E)C$_{50}$ は水に対する溶解度測定値より大きいとみなしてよい。その場合、慢性 4 を指定すべきかどうかについて検討すべきである。その物質は急性毒性を示さないという判定をする際には、最大の溶解濃度を達成するのに用いた手段について十分な配慮がなされるべきである。その手段が適切であるとみなせない場合には、その試験は分類目的には妥当でないと考えるべきである；

(c) 水に対する溶解度がその分析法の検出限界より低く、かつ急性毒性が記録されている場合には、分類のための L(E)C$_{50}$ は分析の検出限界より小さいとみなしてよい。毒性が記録されていない場合には、分類のための L(E)C$_{50}$ は水に対する溶解度より大きいとみなしてよい。上述の品質判定基準には十分な考慮を払うべきである；

(d) 慢性毒性データが利用可能な場合には、同じ一般則を適用すべきである。原則として、水に対する溶解度で、または 1mg/l より上の濃度で、影響が認められないデータのみを考慮する必要がある。この場合にも、濃度測定の考察によってこうしたデータの妥当性を確認できないならば、最高溶解濃度の達成に用いた手段が適切なものであるか検討しなければならない。

A9.3.5.8 濃度低下に関与するその他の要因

その他にも多くの要因が、濃度低下に関与している可能性があり、正しい試験の設計により回避できる要因もあるが、こうした要因が関与したデータを解釈しなければならない場合がしばしばある。

(a) 沈殿：これはさまざまな理由によって、試験中に生じることがある。見かけ上粒子が見えなくとも、物質が実際には溶解しておらず、試験中に凝集して沈殿する、というのが一般的な説明である。この場合、分類のための L(E)C$_{50}$ または NOEC は試験終了時の濃度に基づくと考えてよい。同様に、媒体との反応により沈殿が生じることもある。このことについては、上述の不安定さのところで考察した；

(b) 吸着：これは例えば log K$_{ow}$ 値の大きい物質等、吸着性の高い物質で起こる。これが起こるときは濃度低下が通常速やかであり、ばく露は試験終了時の濃度で最もよく記述できる；

(c) 生物蓄積：物質が試験生物体内に蓄積することで濃度低下が起こることがある。水に対する溶解度が低く、それに関連して log K$_{ow}$ が高い場合には特に重要であろう。分類のための L(E)C$_{50}$ または NOEC は、試験の開始時濃度および終了時濃度の幾何平均値から計算してもよいであろう。

A9.3.5.9 試験媒体の変動

A9.3.5.9.1 強酸および強塩基は pH を変化させるので、有毒であるように思われるかもしれない。しかし一般に、水系での pH 変化は試験媒体の緩衝システムによって阻止されるのが普通である。塩についてのデータが得られていないならば、その塩は一般に陽イオンまたは陰イオンと同じ様に、すなわち最も厳しい分類がなされたイオンと同じに、分類される。作用濃度がイオン種のうち 1 種類だけに関係しているならば、その塩の分類では、作用濃度に分子量の比（MW$_{salt}$/MW$_{ion}$）を掛けて補正することで、分子量の差を考慮に加えるべきである。

A9.3.5.9.2 ポリマーは一般に水系では利用性はない。分散性ポリマーその他の高分子量物質は、試験系を攪乱して酸素取り込みを妨害し、機械的または二次的な影響を生じる。これらの物質のデータを検討する際には、このような要因を考慮する必要がある。しかし、ポリマーの多くは複合物質のように挙動し、原体ポリマーから浸出するかなり分子量の低い画分を有する。これについては、以下に考察する。

A9.3.5.10 複合物質

A9.3.5.10.1 複合物質は、しばしば同族の系列である一定範囲の化学構造で特徴づけられるが、水に対する溶解度、その他の物理化学的性質は広範囲にわたる。水に添加すると、溶解した部分と溶解していない部分の間で平衡に達するが、これは物質の負荷量によって決定される。この理由により、このような複合物質は通常 WSF または WAF として試験され、L(E)C$_{50}$ は添加濃度または設定濃度をもとに記録される。溶解部分はそれ自体が各成分の複雑な混合物なので、分析上に補助的に用いるデータは利用できないことが多い。この毒性パラメータは時に致死添加濃度に関連した LL$_{50}$ で表される。WSF または WAF から得られるこの添加レベルは、直接分類基準に用いてもよい。

A9.3.5.10.2   Polymers represent a special kind of complex substance, requiring consideration of the polymer type and their dissolution/dispersal behaviour. Polymers may dissolve as such without change, (true solubility related to particle size), be dispersible, or portions consisting of low molecular weight fractions may go into solution. In the latter case, in effect, the testing of a polymer is a test of the ability of low molecular mass material to leach from the bulk polymer, and whether this leachate is toxic. It can thus be considered in the same way as a complex mixture in that a loading of polymer can best characterize the resultant leachate, and hence the toxicity can be related to this loading.

Table A9.3.1: Classification of difficult test substances

Property	Nature of difficulty	Relevance for classification
Poorly water soluble	Achieving/maintaining required exposure concentration. Analysing exposure	When toxic responses are observed above apparent solubility, expert judgement is required to confirm whether effects are due to chemical toxicity or a physical effect; if no effects are observed, it should be demonstrated that full, saturated dissolution has been achieved
Toxic at low concentrations	Achieving/maintaining required exposure concentration. Analysing exposure	Classified based on toxicity < 1 mg/l
Volatile	Maintaining and measuring exposure concentration	Classification should be based on reliable measurement of concentrations
Photo-degradable	Maintaining exposure concentrations. Toxicity of breakdown products	Classification requires expert judgement and should be based on measured concentrations. Toxicity of significant breakdown products should be characterized
Hydrolytically unstable	Maintaining exposure concentrations. Toxicity of breakdown products. Comparison of degradation half-lives to the exposure regimen used in testing	Classification requires expert judgement, should be based on measured concentrations, and needs to address the toxicity of significant breakdown products
Oxidizable	Achieving, maintaining and measuring exposure concentration. Toxicity of modified chemical structures or breakdown products. Comparison of degradation half-lives to the exposure regimen used in testing	Classification requires expert judgement, should be based on measured concentrations, and needs to address the toxicity of significant breakdown products
Subject to corrosion/ transformation (this refers to metals /metal compounds)	Achieving, maintaining and measuring exposure concentration. Comparison of partitioning from the water column half-lives to the exposure regimen used in testing	Classification requires expert judgement, should be based on measured concentrations, and needs to address the toxicity of significant breakdown products
Biodegradable	Maintaining exposure concentrations. Toxicity of breakdown products. Comparison of degradation half-lives to the exposure regimen used in testing	Classification requires expert judgement, should be based on measured concentrations, and needs to address the toxicity of significant breakdown products
Adsorbing	Maintaining exposure concentrations. Analysing exposure. Toxicity mitigation due to reduced availability of test substance	Classification should use measured concentration of available material
Chelating	Distinguishing chelated and non-chelated fractions in media	Classification should use measurement of concentration of bioavailable material
Coloured	Light attenuation (an algal problem)	Classification must distinguish toxic effects from reduced growth due to light attenuation
Hydrophobic	Maintaining constant exposure concentrations	Classification should use measured concentration
Ionized	Maintaining exposure concentrations. Toxicity of breakdown products. Comparison of degradation half-lives to the exposure regime used in testing	Classification requires expert judgement, should be based on measured concentrations, and needs to address the toxicity of significant breakdown products
Multi-component	Preparing representative test batches	Considered same as complex mixture

A9.3.5.10.2　ポリマーは特別な種類の複合物質の代表であり、ポリマーの種類および溶解/分散挙動についての考察が必要である。ポリマーは、変化せずそのままの形で溶解することもあれば（真の溶解度は粒子サイズに関係している）、分散することや、または低分子量画分で構成される部分が溶液となることもある。最後のケースでは、ポリマーの試験は、低分子量物質が原体ポリマーから浸出する可能性、およびこの浸出物が毒性かどうかの試験となる。したがって、ポリマー添加量がそれによって生じる浸出物の特性を最も適切に決定し、これによって毒性はこの添加量に関連づけられる点で、複合混合物の場合と同じように考えることができる。

表 A9.3.1：試験困難な物質の分類

特性	困難さの内容	分類のための妥当性
水に対する溶解度が低い	求められるばく露濃度の達成/維持。ばく露濃度の分析。	見かけ上の溶解度より上の濃度で毒性反応が認められている場合には、影響が化学的な毒性によるものか、または物理的影響によるものかを確認するのに専門家の判断が求められる。影響は認められない場合には、完全な飽和溶解度が達成されていることを示す必要がある。
低濃度で毒性	求められるばく露濃度の達成/維持。ばく露濃度の分析。	毒性値が 1mg/l 未満である場合にこれに分類される
揮発性	ばく露濃度の維持および測定。	信頼のおける濃度測定をもとに分類すること。
光分解性	ばく露濃度の維持。分解生成物の毒性。	分類には専門家の判断が必要であり、測定濃度をもとに分類すること。主要な分解生成物の毒性を判定すること。
加水分解性で不安定	ばく露濃度の維持。分解生成物の毒性。分解半減期を試験で採用しているばく露方式と比較すること。	分類には専門家の判断が必要であり、測定濃度をもとに分類すること。また主要な分解生成物の毒性を取りあげる必要性あり。
酸化性	ばく露濃度の達成、維持および測定。化学構造の変化した物質、または分解生成物の毒性。分解半減期を試験で採用しているばく露方式と比較すること。	分類には専門家の判断が必要であり、測定濃度をもとに分類すること。また主要な分解生成物の毒性を取りあげる必要性あり。
腐食/変換を受ける（金属/金属化合物について）	ばく露濃度の達成、維持および測定。水相からの分配の半減期と試験で採用しているばく露方式の比較。	分類には専門家の判断が必要であり、測定濃度をもとに分類すること。また主要な分解生成物の毒性を取りあげる必要性あり。
生分解性	ばく露濃度の維持。分解生成物の毒性。分解の半減期と試験で採用しているばく露方式の比較。	分類には専門家の判断が必要であり、測定濃度をもとに分類すること。また主要な分解生成物の毒性を取りあげる必要性あり。
吸着性	ばく露濃度の維持。ばく露濃度の分析。被験物質の利用可能性の減少による毒性の低下。	分類は利用性のある物質の測定濃度をもとに行うこと。
キレート化	媒体中のキレート化画分およびキレート化していない部分の区別。	分類は生物学的利用性のある物質の生長阻害をもとに行うこと。
着色	光の減衰（藻類では問題）	分類では、毒性作用を、光量の減衰による生長阻害と区別しなければならない。
疎水性	ばく露濃度を一定に維持すること。	分類では測定濃度を用いること。
イオン化	ばく露濃度の維持。分解生成物の毒性。分解半減期と試験に用いたばく露方式の比較。	分類には専門家の判断が必要であり、測定濃度をもとに分類すること。また主要な分解生成物の毒性を取りあげる必要性あり。
多成分物質	代表的な試験バッチの調製。	複合混合物と同一として考慮すべきである。

A9.3.6   *Interpreting data quality*

A9.3.6.1   *Standardization*

Many factors can influence the results of toxicity tests with aquatic organisms. These factors include characteristics of the test water, experimental design, chemical characteristics of the test material, and biological characteristics of the test organisms. Therefore, it is important in conducting aquatic toxicity tests to use standardized test procedures to reduce the influence of these sources of extraneous variability. The goal of test standardization and international harmonization of these standards is to reduce test variability and improve precision, reproducibility, and consistency of test results.

A9.3.6.2   *Data hierarchies*

A9.3.6.2.1   Classification should be based on primary data of good quality. Preference is given to data conforming to OECD test guidelines or equivalent and Good Laboratory Practices (GLP). While data from internationally harmonized test methods performed on standard test species are preferred, results of tests performed using widely recognized international or national methods or their equivalent may also be used, e.g. ISO or ASTM methods. Data from tests that appear to conform to accepted guidelines, but which lacks provisions for GLP can be used in the absence of pertinent GLP data.

A9.3.6.2.2   Pedersen et al (1995) provides a data quality-scoring system, which is compatible with many others in current use, including that, used by the US-EPA for its AQUIRE database. See also Mensink et al (1995) for discussions of data quality. The data quality scoring system described in Pedersen *et al.* includes a reliability ranking scheme, which can be a model for use with in classifying under the harmonized scheme. The first three levels of data described by Pedersen are for preferred data.

A9.3.6.2.3   Data for classification under the harmonized scheme should come from primary sources. However, since many nations and regulatory authorities will perform classification using the globally harmonized scheme, classification should allow for use of reviews from national authorities and expert panels as long as the reviews are based on primary sources. Such reviews should include summaries of test conditions, which are sufficiently detailed for weight of evidence and classification decisions to be made. It may be possible to use the reviews, which were made by a well-recognized group such as GESAMP for which the primary data are accessible.

A9.3.6.2.4   In the absence of empirical test data, validated Quantitative Structure Activity Relationships (QSARs) for aquatic toxicity may be used. Test data always take precedence over QSAR predictions, providing the test data are valid.

A9.4   **Degradation**

A9.4.1   *Introduction*

A9.4.1.1   Degradability is one of the important intrinsic properties of substances that determine their potential environmental hazard. Non-degradable substances will persist in the environment and may consequently have a potential for causing long-term adverse effects on biota. In contrast, degradable substances may be removed in the sewers, in sewage treatment plants or in the environment.

Classification of substances is primarily based on their intrinsic properties. However, the degree of degradation depends not only on the intrinsic recalcitrance of the molecule, but also on the actual conditions in the receiving environmental compartment as e.g. redox potential, pH, presence of suitable micro-organisms, concentration of the substances and occurrence and concentration of other substrates. The interpretation of the degradation properties in an aquatic hazard classification context therefore requires detailed criteria that balance the intrinsic properties of the substance and the prevailing environmental conditions into a concluding statement on the potential for long-term adverse effects. The purpose of the present section is to present guidance for interpretation of data on degradability of organic substances. The guidance is based on an analysis of the above-mentioned aspects regarding degradation in the aquatic environment. Based on the guidance a detailed decision scheme for use of existing degradation data for classification purposes is proposed. The types of degradation data included in this guidance document are ready biodegradability data, simulation data for transformation in water, aquatic sediment and soil, BOD_5/COD-data and techniques for estimation of rapid degradability in the aquatic environment. Also considered are anaerobic degradability, inherent biodegradability, sewage treatment plant simulation test data, abiotic transformation data such as hydrolysis and photolysis, removal process such as volatilization and finally, data obtained from field investigations and monitoring studies.

A9.4.1.2   The term degradation is defined in chapter 4.1 as the decomposition of organic molecules to smaller molecules and eventually to carbon dioxide, water and salts. For inorganic compounds and metals, the concept of degradability as applied to organic compounds has limited or no meaning. Rather the substance may be transformed by normal environmental processes to either increase or decrease the bioavailability of the toxic species. Therefore, the

A9.3.6　データの質の解釈

A9.3.6.1　標準化

　水生生物を用いた毒性試験の結果に影響を及ぼす要因は多い。例えば、試験水の性質、実験の設計、被験物質の化学的性質、および試験生物種の生物学的性質等がある。したがって、水生毒性試験を実施する際には標準化された試験手順を採用して、こうした非本質的な変動の影響を減らすことが重要である。試験を標準化する、またこれらの標準を国際的に調和させる目的は、試験の変動性を減少させ、試験結果の正確性、再現性および一貫性を改善することである。

A9.3.6.2　データの序列

A9.3.6.2.1　分類は、高品質の一次データに基づくべきである。OECD テストガイドラインまたはそれと同等のもの、および優良試験所実施基準（GLP）に従ったデータが最も望ましい。標準的な試験生物種を用いて実施された国際的に調和された試験法から得られるデータの方が望ましいが、広く承認されている国際的または国内の試験方法またはそれと同等の方法、例えば ISO または ASTM で規定された方法で実施された試験の結果もまた使用してもよい。承認されたガイドラインにしたがっているようでも、GLP への対応に欠けている試験から得られたデータも、適切な GLP データがない場合には使用してよい。

A9.3.6.2.2　Pedersen ら（1995）はデータの質を得点付けするシステムを提案しているが、これは例えば US-EPA の AQUIRE データベースに使用されているような、その他多くの現用システムと互換性がある。データの質についての考察は Mensink ら（1995）も参照のこと。Pedersen らが報告しているデータの質得点付けシステムは信頼性ランキングスキームを含んでいるが、これは本調和スキームにしたがって分類する際に用いるモデルとなり得る。Pedersen の述べている最初の三段階のデータは、優先データに関するものである。

A9.3.6.2.3　本調和スキームのもとでの分類のためのデータは、一次情報源から得られているべきである。しかし、多くの国家や規制官庁は、世界的に調和されたスキームにより分類を実施するので、分類では国家当局および専門家パネルのレビューを採用する余地を持たせるべきである。ただしこのレビューは一次データ源にもとづいている必要がある。こうしたレビューには、証拠の重みおよびなされる分類決定のために、十分に詳述された試験条件の要約を含むべきである。例えば一次データの入手が可能な GESAMP 等の、実績の認められたグループによって作成されたレビューを用いることもできよう。

A9.3.6.2.4　実験に基づいた試験データがない場合には、水生毒性に関する定量的構造活性相関（QSAR）を用いてもよい。試験データが有効である場合には、常に QSAR より優先される。

A9.4　分解性

A9.4.1　序

A9.4.1.1　分解性は、物質の環境に対する潜在的有害性を決定する、重要な本質的特性の 1 つである。非分解性物質は環境中に残留し、結果として生物相に対する長期間に及ぶ有害作用の可能性を有している。反対に、分解性の物質は、下水、汚水処理施設、または環境中で除去されるであろう。

　物質の分類は主に物質の本来の性質に基づいて行われる。しかし分解の程度は、分子本来の抵抗性だけでなく、物質を受け入れる環境コンパートメントの実際の条件、例えば酸化還元電位、pH、適切な微生物の存在、物質の濃度、および他の基質の存在と濃度に依存する。したがって、水生有害性を分類する観点から分解性について解釈するには、その物質本来の特性と環境中で優先する状態とを比較考慮して、長期間の有害影響の可能性に関する結論を導く、詳細な判定基準が必要である。この節の目的は、有機物質の分解性に関するデータを解釈するための手引きを提示することである。この手引きは、水生環境における分解に関する上述の見地の分析に基づいている。この手引きをもとに、既存の分解性データを物質の分類目的に使用するための、詳細な判定スキームが提案されている。本手引きに含まれる分解性データのタイプは、易生分解性データ、水中、底質中および土壌中での物質変換に関するシミュレーションデータ、BOD$_5$/COD データ、ならびに水生環境における急速分解性を評価する技術である。さらに、嫌気的分解性、本質的生分解性、下水処理施設シミュレーション試験データ、加水分解や光分解等の非生物的物質変換データ、揮発等の除去プロセス、また最後には野外研究やモニタリング調査から得られたデータについても考察している。

A9.4.1.2　分解という用語は 4.1 章に、有機分子がより小さい分子に、最終的には二酸化炭素、水および塩類に分解すること、と定義されている。無機化合物および金属については、有機化合物に適用される分解性の概念は、その意味が限られるか、または意味をもたない。むしろ物質は、通常の環境プロセスで変換されて、毒性分子種の生物学的利用性を増加させることもあれば、低下させることもある。したがってこの節で

present section deals only with organic substances and organo-metals. Environmental partitioning from the water column is discussed in section A9.7.

A9.4.1.3   Data on degradation properties of a substance may be available from standardized tests or from other types of investigations, or they may be estimated from the structure of the molecules. The interpretation of such degradation data for classification purposes often requires detailed evaluation of the test data. Guidance is given in the present section and more details can be found in two paragraphs describing available methods (appendix A9.I) and factors influencing degradation in aquatic environments (appendix A9.II).

A9.4.2   *Interpretation of degradability data*

A9.4.2.1   *Rapid degradability*

Aquatic hazard classification of substances is normally based on existing data on their environmental properties. Only seldom will test data be produced with the main purpose of facilitating a classification. Often a diverse range of test data is available that does not necessarily fits directly with the classification criteria. Consequently, guidance is needed on interpretation of existing test data in the context of the aquatic hazard classification. Based on the harmonized criteria, guidance for interpretation of degradation data is prepared below for the three types of data comprised by the expression "rapid degradation" in the aquatic environment (see A9.1.8, A9.1.9, A9.1.2.3.1 to A9.2.3.4 and the definition in chapter 4.1, paragraph 4.1.2.11.3).

A9.4.2.2   *Ready biodegradability*

A9.4.2.2.1   Ready biodegradability is defined in OECD test guidelines No. 301 (OECD 1992). All organic substances that degrade to a level higher than the pass level in a standard OECD ready biodegradability test or in a similar test should be considered readily biodegradable and consequently also rapidly degradable. Many literature test data, however, do not specify all of the conditions that should be evaluated to demonstrate whether or not the test fulfils the requirements of a ready biodegradability test. Expert judgement is therefore needed as regards the validity of the data before use for classification purposes. Before concluding on the ready biodegradability of a test substance, however, at least the following parameters should be considered.

A9.4.2.2.2   Concentration of test substance

Relatively high concentrations of test substance are used in the OECD ready biodegradability tests (2-100 mg/l). Many substances may, however, be toxic to the inocula at such high concentrations causing a low degradation in the tests although the substances might be rapidly degradable at lower non-toxic concentrations. A toxicity test with micro-organisms (as e.g. the OECD Test Guideline 209 "Activated Sludge, Respiration Inhibition Test", the ISO 9509 nitrification inhibition test, or the ISO 11348 luminescent bacteria inhibition test) may demonstrate the toxicity of the test substance. When it is likely that inhibition is the reason for a substance being not readily degradable, results from a test employing lower non-toxic concentrations of the test substance should be used when available. Such test results could on a case by case basis be considered in relation to the classification criteria for rapid degradation, even though surface water degradation test data with environmentally realistic microbial biomass and non toxic realistic low concentration of the test substance in general are preferred, if available.

A9.4.2.2.3   Time window

The harmonized criteria (see 4.1.2.11.3) include a general requirement for all of the ready biodegradability tests on achievement of the pass level within 10 days. This is not in line with OECD Test Guideline 301 in which the 10-days time window applies to the OECD ready biodegradability tests except to the MITI I test (OECD Test Guideline 301C). In the Closed Bottle test (OECD Test Guideline 301D), a 14-days window may be used instead when measurements have not been made after 10 days. Moreover, often only limited information is available in references of biodegradation tests. Thus, as a pragmatic approach the percentage of degradation reached after 28 days may be used directly for assessment of ready biodegradability when no information on the 10-days time window is available. This should, however, only be accepted for existing test data and data from tests where the 10-days window does not apply.

Where there is sufficient justification, the 10-day window condition may be waived for complex, multi-component substances and the pass level applied at 28 days. The constituents of such substances may have different chain-lengths, degree and/or site of branching or stereo-isomers, even in their most purified commercial forms. Testing of each individual component may be costly and impractical. If a test on the complex, multi-component substance is performed and it is anticipated that a sequential biodegradation of the individual structures is taking place, then the 10-day window should not be applied to interpret the results of the test. A case by case evaluation should however take place on whether a biodegradability test on such a substance would give valuable information regarding its biodegradability as such (i.e.

は、有機物質および有機金属のみを扱う。水相からの環境中への分配についてはA9.7節で述べる。

A9.4.1.3　物質の分解性に関するデータは、標準化された試験によって、またはその他の研究から入手されるし、分子構造から推定できることもある。こうした分解性データを分類の目的で解釈するには、試験データの詳しい評価が必要とされることが多い。手引きについてはこの節に示し、さらに詳しいことは、利用できる方法について（付録A9.I）、および水生環境での分解に影響する要因について（付録A9.II）説明している2つの付録に見出すことができる。

A9.4.2　*分解性データの解釈*

A9.4.2.1　*急速分解性*

　物質の水生有害性の分類は通常、それらの物質の環境特性に関する既存データをもとに行われる。分類を促進することを主な目的として試験データが提供されることはまれにしかない。広範囲に及ぶデータが揃っていても、分類基準に必ずしも直接適合するわけではないことがほとんどである。結果的に、既存の試験データを水生有害性の分類と関係付けて解釈することに関する手引きが必要とされる。調和された判定基準に基づいて、分解データを解釈するための手引きが、下記のように水生環境における「急速分解性」という表現によりあらわされる3種類のデータについて作成された（A9.1.8, A9.1.9, A9.1.2.3.1 からA9.2.3.4 ならびに4.1 章パラグラフ 4.1.2.11.3 における定義を参照）。

A9.4.2.2　*易生分解性*

A9.4.2.2.1　易生分解性については、OECDテストガイドライン No.301 (OECD, 1992)に定義されている。標準的な OECD の易生分解性試験または同様な試験での易生分解性とされるレベルよりも高いレベルで分解する有機化合物はすべて易生分解性であり、したがって急速分解性でもあるとみなされるべきである。しかし文献にある試験データの多くは、その試験が易生分解性試験の要件を満たしていることを証明するために評価されるべきすべての条件を、特定しているとは限らない。したがって、データを分類の目的で使用する前に、その有意性に関して専門家による判断が必要とされる。しかし、被験物質の易生分解性について結論づける前に、少なくとも下記のパラメータについて考慮すべきである。

A9.4.2.2.2　被験物質の濃度

　OECD の易生分解性試験では比較的高い被験物質濃度が用いられている（2－100mg/l）。しかし多くの物質は、このように高い試験濃度では植種源に対して毒性となり、より低い非毒性濃度条件では速やかに分解するのに、この試験では分解性が低くなることがある。微生物を用いる毒性試験（例えば OECD テストガイドライン 209 活性汚泥呼吸阻害試験、ISO 9509 硝化阻害試験、または ISO 11348 発光バクテリア阻害試験）で被験物質の毒性を実証してもよい。ある物質について、阻害が易分解性でないことの理由となっていると思われる場合には、可能なら被験物質について、より低い無毒性濃度を採用した試験の結果を採用するべきである。環境的に現実に近い微生物相と被験物質を現実的な非毒性低濃度で用いた、河川水分解試験ができるなら一般に望ましいとされるが、上に述べたような低濃度試験の結果を、ケースバイケースで、急速分解性の分類基準に関して考慮することもできよう。

A9.4.2.2.3　試験期間（時間ウィンドウ）

　調和された判定基準（4.1.2.11.3 参照）には、易生分解性試験すべてについて、10 日間以内に易生分解性とされるレベルを達成するという一般的要件が含まれている。これは、MITI I 試験（OECD テストガイドライン 301C）以外の OECD 易生分解性試験（OECD テストガイドライン 301C）に 10 日間の時間ウィンドウが適用されるということと一致していない。クローズドボトル試験（OECD テストガイドライン 301D）では、10 日後に測定されなかった時には、かわりに 14 日間の時間ウィンドウを採用してもよい。さらに、生分解性試験に関して限られた情報しか利用できないことが多い。したがって、実用的なアプローチとして、10 日間の試験で情報が得られない場合には、28 日後に達成された分解率（%）を易生分解性評価に直接採用することもある。ただし、このことは既存の試験データ、および 10 日間のウィンドウが適用されない試験で得られたデータに対してのみ、受け入れられるべきである。

　十分な根拠がある場合は、複合的で多成分な物質については10日間の時間ウィンドウ条件は免除され、28 日以内の易生分解性試験が適用される。このような物質の成分は、その最も高純度の商用の試料でも、異なる鎖の長さ、枝分かれの数や位置もしくは立体異性体を含んでいる。それぞれの個別成分に対して試験を実施するのはコストがかさみ、実用的でない可能性がある。複合的で多成分な物質に対して試験を実施し、個々の構造体について連続的な生分解が起きると予想される場合は、試験結果の解釈に 10 日間の

regarding the degradability of all the constituents) or whether instead an investigation of the degradability of carefully selected individual components of the complex, multi-component substance is required.

A9.4.2.3 BOD_5/COD

Information on the 5-day biochemical oxygen demand (BOD_5) will be used for classification purposes only when no other measured degradability data are available. Thus, priority is given to data from ready biodegradability tests and from simulation studies regarding degradability in the aquatic environment. The BOD_5 test is a traditional biodegradation test that is now replaced by the ready biodegradability tests. Therefore, this test should not be performed today for assessment of the ready biodegradability of substances. Older test data may, however, be used when no other degradability data are available. For substances where the chemical structure is known, the theoretical oxygen demand (ThOD) can be calculated and this value should be used instead of the chemical oxygen demand (COD).

A9.4.2.4   *Other convincing scientific evidence*

A9.4.2.4.1   Rapid degradation in the aquatic environment may be demonstrated by other data than referred to in chapter 4.1, paragraph 4.1.2.11.3 (a) and (b). These may be data on biotic and/or abiotic degradation. Data on primary degradation can only be used where it is demonstrated that the degradation products shall not be classified as hazardous to the aquatic environment, i.e. that they do not fulfil the classification criteria.

A9.4.2.4.2   The fulfilment of 4.1.2.11.3 (c) requires that the substance is degraded in the aquatic environment to a level of > 70 % within a 28-day period. If first-order kinetics are assumed, which is reasonable at the low substance concentrations prevailing in most aquatic environments, the degradation rate will be relatively constant for the 28-day period. Thus, the degradation requirement will be fulfilled with an average degradation rate constant, $k > -(\ln 0.3 - \ln 1)/28 = 0.043$ day^{-1}. This corresponds to a degradation half-life, $t_{½} < \ln 2/0.043 = 16$ days.

A9.4.2.4.3   Moreover, as degradation processes are temperature dependent, this parameter should also be taken into account when assessing degradation in the environment. Data from studies employing environmentally realistic temperatures should be used for the evaluation. When data from studies performed at different temperatures need to be compared, the traditional Q10 approach could be used, i.e. that the degradation rate is halved when the temperature decreases by 10°C.

A9.4.2.4.4   The evaluation of data on fulfilment of this criterion should be conducted on a case-by-case basis by expert judgement. However, guidance on the interpretation of various types of data that may be used for demonstrating a rapid degradation in the aquatic environment is given below. In general, only data from aquatic biodegradation simulation tests are considered directly applicable. However, simulation test data from other environmental compartments could be considered as well, but such data require in general more scientific judgement before use.

A9.4.2.4.5   Aquatic simulation tests

Aquatic simulation tests are tests conducted in laboratory but simulating environmental conditions and employing natural samples as inoculum. Results of aquatic simulation tests may be used directly for classification purposes, when realistic environmental conditions in surface waters are simulated, i.e.:

(a)   substance concentration that is realistic for the general aquatic environment (often in the low µg/l range);

(b)   inoculum from a relevant aquatic environment;

(c)   realistic concentration of inoculum (10^3-10^6 cells/ml);

(d)   realistic temperature (e.g. 5°C to 25°C); and

(e)   ultimate degradation is determined (i.e. determination of the mineralization rate or the individual degradation rates of the total biodegradation pathway).

Substances that under these conditions are degraded at least 70 % within 28 days, i.e. with a half-life < 16 days, are considered rapidly degradable.

時間ウィンドウを適用すべきではない。しかしながら、個別の物質に対する生分解性試験が全体の生分解性（すなわち、すべての成分についての分解性）に関する価値ある情報を提供するものかどうか、あるいはその代わりに、複合的で多成分な物質の構成物質の中から慎重に選択された個別の成分の分解性の調査が必要となるかどうかについては、個別に評価されるべきである。

A9.4.2.3 *BOD₅/COD*

5日間生化学的酸素要求量（BOD_5）に関する情報は、他に分解性に関する測定データが得られていない場合にのみ、分類の目的で使用されよう。このように易生分解性試験や、水生環境中での分解性に関するシミュレーション試験から得られるデータの方が優先される。BOD_5試験は伝統的な生分解性試験であり、現在では易生分解性試験によって、とって替わられている。したがって、BOD_5試験は今日、物質の易生分解性を評価するために実施するべきではない。しかし、他に分解性データが利用できない場合には、古い試験データが用いられることもある。化学構造がわかっている物質については理論的酸素要求量（ThOD）を計算でき、この数値の方を化学的酸素要求量（COD）の代わりに用いるべきである。

A9.4.2.4 その他説得力ある科学的証拠

A9.4.2.4.1 水生環境での急速分解性は、4.1章 4.1.2.11.3（a）および（b）で参照される以外のデータによって証明されることもある。そのデータは、生分解性および/または非生分解性のデータでありうる。一次分解に関するデータは、分解生成物が水生環境に対して有害であると分類されない、すなわちこれらが分類基準を満たさないことを、証明できる場合にのみ使用できる。

A9.4.2.4.2 4.1章 4.1.2.11.3 の判定基準（c）を満たすには、その物質が水生環境において28日以内 70%より高いレベルで分解する必要がある。ほとんどの水生環境では物質濃度が低いことが多いので合理的なことであるが、一次の反応速度を推定すると、分解速度は28日間の間、比較的一定になる。これは、ほとんどの水生環境では物質の濃度が低いことを考えれば合理的である。したがって、分解に関する要件は、平均の分解速度定数 $k > -(\ln 0.3 - \ln 1)/28 = 0.043/day^{-1}$ で満たされる。これは分解の半減期 $t_{1/2} < \ln 2/0.043 = 16$ 日に相当する。

A9.4.2.4.3 さらに、分解プロセスは温度に依存するため、環境中での分解を推定する際には、このパラメータも考慮するべきである。環境の面から見て現実に即した温度を採用した試験から得られたデータを評価に用いるべきである。さまざまな温度で実施された試験から得られたデータを比較する必要がある場合、伝統的な$Q10$アプローチを用いてもよい。すなわち、温度が10℃低下する毎に分解速度を半分にするのである。

A9.4.2.4.4 データがこの判定基準を満たしているかの評価は、ケースバイケースで、専門家の判断で行うべきである。水生環境における急速分解性を証明するのに用いられる可能性のある多様なタイプのデータを解釈するための手引きを下記に示す。一般に、水中での生分解シミュレーション試験から得られたデータだけが直接適用できると考えられている。しかし、他の環境コンパートメントから得られたシミュレーション試験データも同様に考慮してもよいが、そのようなデータは一般に、使用する前により多くの科学的な判断が必要となる。

A9.4.2.4.5 水中シミュレーション試験

水中シミュレーション試験は実験室内で実施される試験であるが、環境条件をシミュレートし、自然サンプルを植種源として用いる。水中シミュレーション試験の結果は、河川での現実に近い環境条件をシミュレートしている場合、すなわち以下のような場合には直接、分類のために採用してよい：

(a) 物質濃度が一般的な水生環境に対して現実的である（数 µg/l の範囲にあることが多い）；
(b) 関連した水生環境からの植種源を用いている；
(c) 植種源の濃度が現実的である（1ml あたり生菌数 $10^3 - 10^6$ 個）；
(d) 温度が現実的である（例えば、5℃ - 25℃）；および
(e) 究極の分解まで測定されている（すなわち、無機化度、または生分解経路全体での個々の分解度の測定）。

こうした条件下で28日間に少なくとも70%分解される物質、すなわち半減期が16日間より短い物質は急速分解性であると考えられる。

A9.4.2.4.6 Field investigations

Parallels to laboratory simulation tests are field investigations or mesocosm experiments. In such studies, fate and/or effects of chemicals in environments or environmental enclosures may be investigated. Fate data from such experiments might be used for assessing the potential for a rapid degradation. This may, however, often be difficult, as it requires that an ultimate degradation can be demonstrated. This may be documented by preparing mass balances showing that no non-degradable intermediates are formed, and which take the fractions into account that are removed from the aqueous system due to other processes such as sorption to sediment or volatilization from the aquatic environment.

A9.4.2.4.7 Monitoring data

Monitoring data may demonstrate the removal of contaminants from the aquatic environment. Such data are, however, very difficult to use for classification purposes. The following aspects should be considered before use:

(a) Is the removal a result of degradation, or is it a result of other processes such as dilution or distribution between compartments (sorption, volatilization)?

(b) Is formation of non-degradable intermediates excluded?

Only when it can be demonstrated that removal as a result of ultimate degradation fulfils the criteria for rapid degradability, such data be considered for use for classification purposes. In general, monitoring data should only be used as supporting evidence for demonstration of either persistence in the aquatic environment or a rapid degradation.

A9.4.2.4.8 Inherent biodegradability tests

Substances that are degraded more than 70 % in tests for inherent biodegradability (OECD Test Guidelines 302) have the potential for ultimate biodegradation. However, because of the optimum conditions in these tests, the rapid biodegradability of inherently biodegradable substances in the environment cannot be assumed. The optimum conditions in inherent biodegradability tests stimulate adaptation of the micro-organisms thus increasing the biodegradation potential, compared to natural environments. Therefore, positive results in general should not be interpreted as evidence for rapid degradation in the environment[2].

A9.4.2.4.9 Sewage treatment plant simulation tests

Results from tests simulating the conditions in a sewage treatment plant (STP) (e.g. OECD Test Guideline 303)[3] cannot be used for assessing the degradation in the aquatic environment. The main reasons for this are that the microbial biomass in a STP is significantly different from the biomass in the environment, that there is a considerably different composition of substrates, and that the presence of rapidly mineralized organic matter in waste water facilitates degradation of the test substance by co-metabolism.

A9.4.2.4.10 Soil and sediment degradation data

It has been argued that for many non-sorptive (non-lipophilic) substances more or less the same degradation rates are found in soil and in surface water. For lipophilic substances, a lower degradation rate may generally be expected in soil than in water due to partial immobilization caused by sorption. Thus, when a substance has been shown to be degraded rapidly in a soil simulation study, it is most likely also rapidly degradable in the aquatic environment. It

---

[2] *In relation to interpretation of degradation data equivalent with the harmonised OECD criteria for Chronic 4, the standing EU working group for environmental hazard classification of substances is discussing whether certain types of data from inherent biodegradability tests may be used in a case by case evaluation as a basis for not classifying substances otherwise fulfilling this classification criterion.*

The inherent biodegradability tests concerned are the Zahn Wellens test (OECD TG 302 B) and the MITI II test (OECD TG 302 C). The conditions for use in this regard are:

(a) *The methods must not employ pre-exposed (pre-adapted) micro-organisms;*

(b) *The time for adaptation within each test should be limited, the test endpoint should refer to the mineralization only and the pass level and time for reaching these should be, respectively:*

  (i) *MITI II pass level > 60 % within 14 days*

  (ii) *Zahn Wellens Test > 70 % within 7 days.*

[3] *OECD test guidelines 311 and 314 are also available).*

A9.4.2.4.6　野外調査

　実験室でのシミュレーション試験に匹敵するものは、野外調査、あるいはメソコスムでの実験である。そのような研究では、環境中または環境閉鎖系における化学品の運命または影響が調査されることもある。こうした実験から得られた運命データが、急速分解性の可能性を評価するのに用いられることもある。しかしそれは、究極分解することが実証できる必要があるため、困難なことが多い。このことは、非分解性中間体は生じないことを示す物質収支を作成し、他のプロセス、例えば底質への吸着や水生環境からの揮発によって、水系システムから除去される分を考慮すれば証拠付けできるかもしれない。

A9.4.2.4.7　モニタリングデータ

　モニタリングデータは、水生環境からの汚染物の除去を示すかもしれない。しかし、こうしたデータを分類の目的で使用することは非常に難しい。使用する前に、以下の観点で検討するべきである：

　(a)　この除去は分解の結果なのか、あるいは、希釈、コンパートメント間の分配（吸着、揮発）等、のプロセスの結果であるのか？

　(b)　非分解性中間体が生成することはないか？

　究極分解の結果である除去が、急速分解性の判断基準を満たすことを証明できる場合のみ、こうしたデータを分類の目的で使用することを考慮すべきである。一般に、モニタリングデータは、水生環境中での難分解性または急速分解性を証明する裏付け証拠としてのみ用いるべきである。

A9.4.2.4.8　本質的生分解性試験

　本質的生分解性に関する試験（OECD テストガイドライン 302）で 70％以上の分解をする物質は、究極生分解性の可能性がある。しかし、こうした試験では最適条件を採用しているので、本質的に生分解性のある物質でも、環境中で急速分解性があるとは推論できない。本質的生分解性試験で最適条件を使用すれば、微生物の馴化が誘発され、自然環境に比べて生分解性が向上する。したがって一般に、陽性結果が得られても、環境中での急速分解性の証拠と解釈すべきでない[2]。

A9.4.2.4.9　下水処理施設シミュレーション試験

　下水処理施設（STP）内の条件をシミュレートする試験（例えば OECD テストガイドライン 303）[3]から得られた結果は、水生環境中での分解を評価するのには使用できない。その主な理由は、STP における微生物相は環境中の生物相とは著しく異なっていること、基質組成にかなりの違いがあること、および廃水中に存在する速やかに無機化される有機物が共代謝によって被験物質の分解を促進すること、である。

A9.4.2.4.10　土壌および底質中の分解データ

　非吸着性の物質（すなわち非親油性物質）は、程度に多少差はあるものの、土壌中と表層水中とで同様の分解度が見られると主張されてきた。親油性物質では、吸着による部分的な不動化のために、水中より土壌中の方が一般に分解度が低いと予想される。このため、土壌中シミュレーション試験で、ある物質が急速分解性であることが認められた場合、水生環境でも急速分解性である可能性がきわめて高い。し

---

[2]　慢性4に関する OECD 判定基準にそった分解性の解釈について、環境有害性分類に関する EU の作業班が検討しているが、それは本質的生分解性試験によるある種のデータを、分類基準を満たしていても、ケースバイケースで分類しないという根拠に使用できるかどうかということである。
　　関連した本質的生分解性試験には Zahn Wellen 試験（OECD TG 302 B）と MITI II 試験（OECD TG 302 C）がある。使用条件は以下のとおりである。

(a)　事前にばく露した（馴化した）微生物を用いてはならない。

(b)　それぞれの試験において、馴化時間は限られるべきであり、試験の指標は無機化であり、合格レベルおよびそれに到達する時間はそれぞれ以下のようである：

　　(i)　MITI II 合格レベル＞60％（14 日以内）

　　(ii)　Zahn Wellens 試験＞70％（7 日以内）

[3]　OECD テストガイドライン 311 および 314 も利用可能である。

is therefore proposed that an experimentally determined rapid degradation in soil is sufficient documentation for a rapid degradation in surface waters when:

(a) no pre-exposure (pre-adaptation) of the soil micro-organisms has taken place; and

(b) an environmentally realistic concentration of substance is tested; and

(c) the substance is ultimately degraded within 28 days with a half-life < 16 days corresponding to a degradation rate > 0.043 day^{-1}.

The same argumentation is considered valid for data on degradation in sediment under aerobic conditions.

A9.4.2.4.11   Anaerobic degradation data

Data regarding anaerobic degradation cannot be used in relation to deciding whether a substance should be regarded as rapidly degradable, because the aquatic environment is generally regarded as the aerobic compartment where the aquatic organisms, such as those employed for aquatic hazard classification, live.

A9.4.2.4.12   Hydrolysis

Data on hydrolysis (e.g. OECD Test Guideline 111) might be considered for classification purposes only when the longest half-life $t_{½}$ determined within the pH range 4-9 is shorter than 16 days. However, hydrolysis is not an ultimate degradation and various intermediate degradation products may be formed, some of which may be only slowly degradable. Only when it can be satisfactorily demonstrated that the hydrolysis products formed do not fulfil the criteria for classification as hazardous for the aquatic environment, data from hydrolysis studies could be considered.

When a substance is quickly hydrolysed (e.g. with $t_{½}$ < a few days), this process is a part of the degradation determined in biodegradation tests. Hydrolysis may be the initial transformation process in biodegradation.

A9.4.2.4.13   Photochemical degradation

Information on photochemical degradation (e.g. OECD, 1997) is difficult to use for classification purposes. The actual degree of photochemical degradation in the aquatic environment depends on local conditions (e.g. water depth, suspended solids, turbidity) and the hazard of the degradation products is usually not known. Probably only seldom will enough information be available for a thorough evaluation based on photochemical degradation.

A9.4.2.4.14   Estimation of degradation

A9.4.2.4.14.1   Certain QSARs have been developed for prediction of an approximate hydrolysis half-life, which should only be considered when no experimental data are available. However, a hydrolysis half-life can only be used in relation to classification with great care, because hydrolysis does not concern ultimate degradability (see "Hydrolysis" of this section). Furthermore, the QSARs developed until now have a rather limited applicability and are only able to predict the potential for hydrolysis on a limited number of chemical classes. The QSAR program HYDROWIN (version 1.67, Syracuse Research Corporation) is for example only able to predict the potential for hydrolysis on less than 1/5th of the existing EU substances which have a defined (precise) molecular structure (Niemelä, 2000).

A9.4.2.4.14.2   In general, no quantitative estimation method (QSAR) for estimating the degree of biodegradability of organic substances is yet sufficiently accurate to predict rapid degradation. However, results from such methods may be used to predict that a substance is not rapidly degradable. For example, when in the Biodegradation Probability Program (e.g. BIOWIN version 3.67, Syracuse Research Corporation) the probability is < 0.5 estimated by the linear or non-linear methods, the substances should be regarded as not rapidly degradable (OECD, 1994; Pedersen et al., 1995 & Langenberg et al., 1996). Also, other (Q)SAR methods may be used as well as expert judgement, for example, when degradation data for structurally analogue compounds are available, but such judgement should be conducted with great care. In general, a QSAR prediction that a substance is not rapidly degradable is considered a better documentation for a classification than application of a default classification, when no useful degradation data are available.

A9.4.2.4.15   Volatilization

Chemicals may be removed from some aquatic environments by volatilization. The intrinsic potential for volatilization is determined by the Henry's Law constant (H) of the substance. Volatilization from the aquatic environment is highly dependent on the environmental conditions of the specific water body in question, such as the water depth, the gas exchange coefficients (depending on wind speed and water flow) and stratification of the water body. Because volatilization only represents removal of a chemical from water phase, the Henry's Law constant cannot be used

たがって、土壌中での急速分解性が実験的に測定されることは、以下の場合には、表層水中での急速分解性の十分な証拠となると提案されている：

(a) 土壌微生物の事前のばく露（事前の馴化）を行わなかった；

(b) 試験した物質の濃度は環境的に現実に即したものである；

(c) その物質は 28 日間以内に半減期 16 日未満（分解速度定数＞0.043 day⁻¹ に相当）で究極分解に至る。

底質中における好気的条件下での分解に関するデータについても、同様な論拠が有効であると考えられている。

A9.4.2.4.11 嫌気的分解性データ

嫌気的分解に関するデータは、ある物質を急速分解性であるとみなすべきかを決定するのには使用できない。一般に水生環境は、その中に水生有害性分類に採用されているような水生生物種が生活している、好気的コンパートメントであるとみなされているからである。

A9.4.2.4.12 加水分解

加水分解に関するデータ（例：OECD テストガイドライン 111）は、pH4－pH9 の範囲内で測定された半減期 $t_{1/2}$ の最高値が 16 日より短い場合にのみ、分類において考慮できよう。しかし、加水分解は究極分解ではなく、様々な分解中間生成物が形成され、その中には分解が遅いものもあろう。生成された加水分解生成物が、水生環境に有害であるとする分類基準を満たさないと満足に証明できる場合にのみ、加水分解試験のデータを考慮できる。

ある物質が急速に加水分解されるならば（例えば $t_{1/2}$＜2,3 日）、このプロセスは生分解性試験で測定される分解の一部となる。加水分解は生分解における物質変換の初期プロセスでありうる。

A9.4.2.4.13 光化学的分解

光化学的分解に関する情報（例えば OECD, 1997）を分類の目的で使用するのは難しい。水生環境での実際の光化学的分解度は、局所条件（例えば水深、懸濁固体、濁度）に依存し、分解生成物の有害性は通常わからない。光化学的分解をもとに徹底した評価のための十分な情報が得られることは、おそらく極めて稀である。

A9.4.2.4.14 分解の推定

A9.4.2.4.14.1　加水分解の半減期を概算で予測するために、ある種の QSAR が開発されているが、これは実験データが全く利用できない場合にのみ考慮されるべきである。しかし、加水分解は究極分解を考慮していないので（この節の「加水分解」参照）、加水分解の半減期は、細心の注意を払ってのみ、分類に関して使用できる。さらに、これまでに考案された QSAR はその適用性にはやや限界があり、限られた数の化合物群についてしか加水分解ポテンシャルを予測できない。例えば、HYDROWIN（バージョン 1.67、Syracuse Research Corporation）という QSAR プログラムは、分子構造が（正確に）決定された既存 EU 物質の 1/5 未満の物質しか加水分解ポテンシャルを推測できない。（Niemelä 2000）

A9.4.2.4.14.2　一般に、有機物質の生分解度を推定する定量的評価法（QSAR）はまだ、急速分解性を予測するには精度が不十分である。しかし、こうした方法により得られた結果は、ある物質が急速分解性でないことを予測するのに使ってもよい。例えば、生分解確率プログラム（例：BIOWIN　バージョン 3.67、Syracuse Research Corporation）で、線形あるいは非線形の方法で推定された確率が 0.5 未満である場合、その物質は急速分解性ではないとみなされるべきである（OECD, 1994; Pedersen ら、1995 および Langenberg ら、1996）。さらに、例えば構造が類似している物質の分解データが利用可能な場合は、専門家の判断に加えてその他の(Q)SAR 法を用いることもできるが、その判断は細心の注意を払って行うべきである。一般に、有用な分解データが利用できない場合に、デフォルトの分類を適用するよりも、物質が急速分解性ではないという QSAR による予測の方が、分類のためにはよりよい考証であるとみなされている。

A9.4.2.4.15 揮発

化学品が水生環境から揮発によって除去されることもある。揮発の本質的なポテンシャルは、その物質のヘンリー定数（H）により決定される。水生環境からの揮発は、問題となっている特定の水系の環境条

for assessment of degradation in relation to aquatic hazard classification of substances. Substances that are gases at ambient temperature may however for example be considered further in this regard (see also Pedersen et al., 1995).

A9.4.2.5 *No degradation data available*

When no useful data on degradability are available - either experimentally determined or estimated data - the substance should be regarded as not rapidly degradable.

A9.4.3 ***General interpretation problems***

A9.4.3.1 *Complex substances*

The harmonized criteria for classification of chemicals as hazardous for the aquatic environment focus on single substances. A certain type of intrinsically complex substance are multi-component substances. They are typically of natural origin and need occasionally to be considered. This may be the case for chemicals that are produced or extracted from mineral oil or plant material. Such complex chemicals are normally considered as single substances in a regulatory context. In most cases they are defined as a homologous series of substances within a certain range of carbon chain length and/or degree of substitution. When this is the case, no major difference in degradability is foreseen and the degree of degradability can be established from tests of the complex chemical. One exception would be when a borderline degradation is found because in this case some of the individual substances may be rapidly degradable and other may be not rapidly degradable. This requires a more detailed assessment of the degradability of the individual components in the complex substance. When not-rapidly-degradable components constitute a significant part of the complex substance (e.g. more than 20 %, or for a hazardous component, an even lower content), the substance should be regarded as not rapidly degradable.

A9.4.3.2 *Availability of the substance*

A9.4.3.2.1 Degradation of organic substances in the environment takes place mostly in the aquatic compartments or in aquatic phases in soil or sediment. Hydrolysis, of course, requires the presence of water. The activity of micro-organisms depends on the presence of water. Moreover, biodegradation requires that the micro-organisms are directly in contact with the substance. Dissolution of the substance in the water phase that surrounds the micro-organisms is therefore the most direct way for contact between the bacteria and fungi and the substrate.

A9.4.3.2.2 The present standard methods for investigating degradability of substances are developed for readily soluble test compounds. However, many organic substances are only slightly soluble in water. As the standard tests require 2-100 mg/l of the test substance, sufficient availability may not be reached for substances with a low water solubility. Tests with continuous mixing and/or an increased exposure time, or tests with a special design where concentrations of the test substance lower than the water solubility have been employed, may be available on slightly soluble compounds.

A9.4.3.3 *Test duration less than 28 days*

A9.4.3.3.1 Sometimes degradation is reported for tests terminated before the 28-day period specified in the standards (e.g. the MITI, 1992). These data are of course directly applicable when a degradation greater than or equal to the pass level is obtained. When a lower degradation level is reached, the results need to be interpreted with caution. One possibility is that the duration of the test was too short and that the chemical structure would probably have been degraded in a 28-day biodegradability test. If substantial degradation occurs within a short time period, the situation may be compared with the criterion $BOD_5/COD \geq 0.5$ or with the requirements on degradation within the 10-days time window. In these cases, a substance may be considered readily degradable (and hence rapidly degradable), if:

(a) the ultimate biodegradability > 50 % within 5 days; or

(b) the ultimate degradation rate constant in this period is > 0.1 day^{-1} corresponding to a half-life of 7 days.

A9.4.3.3.2 These criteria are proposed in order to ensure that rapid mineralization did occur, although the test was ended before 28 days and before the pass level was attained. Interpretation of test data that do not comply with the prescribed pass levels must be made with great caution. It is mandatory to consider whether a biodegradability below the pass level was due to a partial degradation of the substance and not a complete mineralization. If partial degradation is the probable explanation for the observed biodegradability, the substance should be considered not readily biodegradable.

件、例えば水深、(風速および水流量に依存する)ガス交換係数、および水本体の層構造などに高度に依存している。揮発は水相からの物質除去の一例でしかないので、物質の水生有害性の分類に関する分解評価に、ヘンリー定数を使用することはできない。しかし、例えば環境温度で気体であるような物質については、この観点でさらに検討されてもよい(Pedersenら、1995も参照のこと)。

A9.4.2.5 *分解データが利用できない場合*

分解性に関する有用なデータが利用されていない場合—実験による測定データか、推定データかにかかわらず—その物質は急速分解性ではないとみなすべきである。

A9.4.3 *解釈についての一般的な問題*

A9.4.3.1 *複雑な物質*

化学品を水生環境に対して有害であると分類するための、調和された判定基準は単一の物質に焦点を合わせている。ある種の本質的な複合物質は、多成分物質である。これらの典型的なものは自然起源であるが、時には考慮に加える必要がある。鉱物油または植物材料から生成または抽出された物質がその例である。こうした複雑な物質は、規制の場合には通常、単一物質とみなされる。ほとんどの場合、これらは炭素鎖の長さまたは置換えが一定範囲内にある類似物質群として定義されている。そのような場合、分解性が大きく異なることは予想されず、分解度は、その複合物質についての諸試験で確定される。ただ1つ例外は分解に境界線が見出される場合で、なぜならこの場合、個々の物質のあるものは急速に分解し、他のものは急速には分解しないからである。その場合、複合物質に含まれる個々の成分の分解性をより詳しく評価する必要がある。それほどの急速分解性をもたない成分がその複合物質の相当部分を占める場合(例えば20%以上、または有害成分の場合はより低い含量)には、その物質は急速分解性ではないとみなされるべきである。

A9.4.3.2 *物質の適用性*

A9.4.3.2.1 環境中で有機物質の分解が起こるのは、水系コンパートメント中か、または土壌あるいは底質の水相中においてである。もちろん加水分解には水の存在が必要である。微生物の活動性は、水の存在に依存する。さらに、生分解には微生物が物質と直接接触する必要がある。したがって、微生物を取り巻いている水相中での物質の溶解は、バクテリアや菌類および基質と接触する、最も直接的な方法である。

A9.4.3.2.2 物質の分解性を調べるための現行の標準的な方法は、易溶解性の被験物質のために開発されている。しかし、有機物質の多くは水に僅かしか溶解しない。標準的な試験では被験物質濃度として 2—100mg/l が要求されるので、水への溶解度が低い物質については十分な利用性が満たされていないことがある。連続的に混合を続ける、またはばく露時間を延長する試験、あるいは水に対する溶解度よりも低い物質濃度が採用されている特殊な設計の試験法ならば、僅かにしか溶解しない化合物にも適用できるかもしれない。

A9.4.3.3 *28日より短い試験期間*

A9.4.3.3.1 基準(例えば MITI, 1992)に規定された 28 日間以前に終了した試験で分解が報告されることもある。もちろん、こうしたデータも、易生分解性とされるレベル以上の分解が達成されているならば、直接、適用できる。分解レベルがそれより低い場合には、その結果は注意して解釈しなければならない。1つの可能性は、試験期間が短すぎたということ、そうして 28 日の生分解性試験でなら、おそらく化学品の構造が分解されていたであろう、ということである。短期間で本質的な分解が起こるならば、そうした状況は $BOD_5/COD \geq 0.5$ という判定基準、または 10 日間の時間ウィンドウ内での分解に関する要件と比較できよう。そのような場合には、以下のいずれかの条件が合えば、物質は易分解性(したがって、急速分解性)であるとみなしてもよい:

(a) 究極生分解性が 5 日間で 50%を超える;または

(b) この期間中の究極分解の速度定数が 0.1 day^{-1} (半減期 7 日間に相当) より大きい。

A9.4.3.3.2 こうした判定基準は、試験が 28 日以前に、かつ易生分解性とされるレベルに達する前に終了されても、速やかな無機化が起こったのだということを確認するために提案されている。易生分解性とされる規定されたレベルにしたがっていない試験データの解釈は、細心の注意を払ってなさなければならない。易生分解性とされるレベルより低い生分解性となったのは、その物質の部分分解によるものであり、完全な無機化によるものでないのかどうかを検討することが不可欠である。観察された生分解性について、部分分解が可能性のある解釈であるなら、その物質は易生分解性であるとみなすべきではない。

A9.4.3.4   *Primary biodegradation*

In some tests, only the disappearance of the parent compound (i.e. primary degradation) is determined for example by following the degradation by specific or group specific chemical analyses of the test substance. Data on primary biodegradability may be used for demonstrating rapid degradability only when it can be satisfactorily demonstrated that the degradation products formed do not fulfil the criteria for classification as hazardous to the aquatic environment.

A9.4.3.5   *Conflicting results from screening tests*

A9.4.3.5.1   The situation where more degradation data are available for the same substance introduces the possibility of conflicting results. In general, conflicting results for a substance which has been tested several times with an appropriate biodegradability test could be interpreted by a weight of evidence assessment. This implies that if both positive (i.e. higher degradation than the pass level) and negative results have been obtained for a substance in ready biodegradability tests, then the data of the highest quality and the best documentation should be used for determining the ready biodegradability of the substance. However, positive results in ready biodegradability tests could be considered valid, irrespective of negative results, when the scientific quality is good and the test conditions are well documented, i.e. guideline criteria are fulfilled, including the use of non-pre-exposed (non-adapted) inoculum. None of the various screening tests are suitable for the testing of all types of substances, and results obtained by the use of a test procedure which is not suitable for the specific substance should be evaluated carefully before a decision on the use is taken.

A9.4.3.5.2   Thus, there are a number of factors that may explain conflicting biodegradability data from screening tests:

   (a)   inoculum;
   (b)   toxicity of test substance;
   (c)   test conditions;
   (d)   solubility of the test substance; and
   (e)   volatilization of the test substance.

A9.4.3.5.3   The suitability of the inoculum for degrading the test substance depends on the presence and amount of competent degraders. When the inoculum is obtained from an environment that has previously been exposed to the test substance, the inoculum may be adapted as evidenced by a degradation capacity, which is greater than that of an inoculum from a non-exposed environment. As far as possible the inoculum must be sampled from an unexposed environment, but for substances that are used ubiquitously in high volumes and released widespread or more or less continuously, this may be difficult or impossible. When conflicting results are obtained, the origin of the inoculum should be checked in order to clarify whether or not differences in the adaptation of the microbial community may be the reason.

A9.4.3.5.4   As mentioned above, many substances may be toxic or inhibitory to the inoculum at the relatively high concentrations tested in ready biodegradability tests. Especially in the Modified MITI (I) test (OECD Test Guideline 301C) and the Manometric Respirometry test (OECD Test Guideline 301F) high concentrations (100 mg/l) are prescribed. The lowest test substance concentrations are prescribed in the Closed Bottle test (OECD Test Guideline 301D) where 2-10 mg/l is used. The possibility of toxic effects may be evaluated by including a toxicity control in the ready biodegradability test or by comparing the test concentration with toxicity test data on micro-organisms, e.g. the respiration inhibition tests (OECD Test Guideline 209), the nitrification inhibition test (ISO 9509) or, if other microbial toxicity tests are not available, the bioluminescence inhibition test (ISO 11348). When conflicting results are found, this may becaused by toxicity of the test substance. If the substance is not inhibitory at environmentally realistic concentrations, the greatest degradation measured in screening tests may be used as a basis for classification. If simulation test data are available in such cases, consideration of these data may be especially important, because a low non inhibitory concentration of the substance may have been employed, thus giving a more reliable indication of the biodegradation half-life of the substance under environmentally realistic conditions.

A9.4.3.5.5   When the solubility of the test substance is lower than the concentrations employed in a test, this parameter may be the limiting factor for the actual degradation measured. In these cases, results from tests employing the lowest concentrations of test substance should prevail, i.e. often the Closed Bottle test (OECD Test Guideline 301D). In general, the DOC Die-Away test (OECD Test Guideline 301A) and the Modified OECD Screening test (OECD Test Guideline 301E) are not suitable for testing the biodegradability of poorly soluble substances (e.g. OECD Test Guideline 301).

A9.4.3.4 一次生分解

一部の試験においては、例えば被験物質に特異的な、または被験物質を含む物質集団に特異的な化学分析によって分解を追跡するような、親化合物の消失(すなわち、一次分解)のみを測定する。一次生分解性に関するデータは、生じた分解生成物が水生環境に対して有害であるとする分類基準に適合しないことが十分に証明できる場合にのみ、急速分解性の証拠に使用してよい。

A9.4.3.5 スクリーニング試験での矛盾する結果

A9.4.3.5.1 同一の物質に関して多くの分解データが利用可能な状況では、結果が矛盾する可能性が生じる。一般に、ある物質について1種類の適切な生分解性試験を何回か実施して得られた結果が矛盾していた場合、証拠の重み付け評価による解釈ができる。これはすなわち、ある物質に対して易生分解性試験で陽性(すなわち、易生分解性とされるレベルよりも高い分解性)と陰性の両方の結果が得られたならば、その物質の易生分解性を判定するのに、質が最も高く、記録証拠が最も適切なデータを使用すべきであることを意味している。しかし、たとえ陰性結果も得られていたとしても、科学的品質が良好で試験条件がよく記録されている、すなわち事前にばく露されていない(非馴化の)植種源を使用することも含めて、ガイドラインの判定基準が充足されているならば、易生分解性試験における陽性結果の方が有意であると考えることもできる。様々なスクリーニング方法のどれ1つとして、すべてのタイプの物質を試験するのには適してない。ある特定の物質に適していない試験手順を用いて得られた結果は、採用を判定する前に、注意して評価すべきである。

A9.4.3.5.2 このように、スクリーニング試験から得られた矛盾した生分解性データを説明する、以下のような多くの要因がある：

  (a) 植種源；
  (b) 被験物質の毒性；
  (c) 試験条件；
  (d) 被験物質の溶解度；および
  (e) 被験物質の揮発。

A9.4.3.5.3 被験物質の分解に植種源がどの程度適しているかは、能力のある分解者の存在と量に依存する。植種源がその被験物質にばく露されたことのある環境から採取されている場合、その植種源が馴化しているかもしれず、そのことは、ばく露されていない環境から採取した植種源より分解能力が大きいことで証拠立てられる。植種源は可能な限り非ばく露環境から採取しなければならないが、物質が普遍的・大量に使用されていて、広くまたはある程度継続的に放出されている場合には、非ばく露環境からの採取は困難であるか、または不可能である。結果が矛盾しているならば、その微生物集団の馴化度の違いが原因となっているかどうかを明らかにするために、植種源の起源を確認すべきである。

A9.4.3.5.4 先に述べたように、多くの物質は、易生分解性試験での比較的高い試験濃度で、植種源に対して毒性または阻害作用を示す。特に修正MITI (I) 試験(OECDテストガイドライン 301C)およびマノメータ呼吸測定試験(OECDテストガイドライン 301F)では高い濃度(100mg/l)が規定されている。最も低い試験濃度はクローズドボトル試験(OECDテストガイドライン 301D)で規定されており、ここでは2－10mg/lが用いられる。毒性作用の影響の可能性については、易生分解性試験に毒性対照を加えること、または試験濃度を微生物に対する毒性試験データと比較することで評価できよう。例えば呼吸阻害試験(OECDテストガイドライン 209)、硝化阻害試験(ISO 9509)、またはその他の微生物毒性試験が使用できなければ、生物発光阻害試験(ISO 11348)がある。矛盾する結果が見出された場合、その原因が被験物質の毒性であるかもしれない。その物質が環境的に現実性のある濃度で阻害しないならば、スクリーニング試験で測定した分解の最高値を分類の根拠として用いてよい。その場合にシミュレーション試験データが利用可能ならば、こうしたデータを考慮することは特に重要である。なぜなら、そうしたデータならば、その物質が阻害を示さない低濃度が採用されているため、環境的に現実性のある条件下での、その物質の生分解半減期をより高い信頼性をもって示すことができるからである。

A9.4.3.5.5 被験物質の溶解度が試験で用いる濃度より低い場合、このパラメータは測定された実際の分解性に対して限定要因であるかもしれない。そのような場合、被験物質の濃度が最も低い試験、すなわち、クローズドボトル 試験(OECDテストガイドライン 301D)であることが多いが、それの結果が優先されるべきである。一般に、溶解度が低い物質の生分解性を試験するのには、DOC ダイアウェイ試験(OECDテストガイドライン 301A)および修正 OECDスクリーニング 試験(OECDテストガイドライン 301E)は適切でない(例えばOECDテストガイドライン 301)。

A9.4.3.5.6   Volatile substances should only be tested in closed systems as the Closed Bottle test (OECD Test Guideline 301D), the MITI I test (OECD Test Guideline 301C) and the Manometric Respirometry test (OECD Test Guideline 301F). Results from other tests should be evaluated carefully and only considered if it can be demonstrated, e.g. by mass balance estimates, that the removal of the test substance is not a result of volatilization.

A9.4.3.6   *Variation in simulation test data*

A number of simulation test data may be available for certain high priority chemicals. Often such data provide a range of half lives in environmental media such as soil, sediment and/or surface water. The observed differences in half-lives from simulation tests performed on the same substance may reflect differences in test conditions, all of which may be environmentally relevant. A suitable half life in the higher end of the observed range of half lives from such investigations should be selected for classification by employing a weight of evidence assessment and taking the realism and relevance of the employed tests into account in relation to environmental conditions. In general, simulation test data of surface water are preferred relative to aquatic sediment or soil simulation test data in relation to the evaluation of rapid degradability in the aquatic environment.

A9.4.4   *Decision scheme*

The following decision scheme may be used as a general guidance to facilitate decisions in relation to rapid degradability in the aquatic environment and classification of chemicals hazardous to the aquatic environment.

A substance is considered to be not rapidly degradable unless at least one of the following is fulfilled:

(a)   the substance is demonstrated to be readily biodegradable in a 28-day test for ready biodegradability. The pass level of the test (70 % DOC removal or 60 % theoretical oxygen demand) must be achieved within 10 days from the onset of biodegradation, if it is possible to evaluate this according to the available test data. If this is not possible, then the pass level should be evaluated within a 14 days time window if possible, or after the end of the test; or

(b)   the substance is demonstrated to be ultimately degraded in a surface water simulation test[4] with a half-life of < 16 days (corresponding to a degradation of > 70 % within 28 days); or

(c)   the substance is demonstrated to be primarily degraded (biotically or abiotically) in the aquatic environment with a half-life < 16 days (corresponding to a degradation of > 70 % within 28 days) and it can be demonstrated that the degradation products do not fulfil the criteria for classification as hazardous to the aquatic environment.

When these data are not available rapid degradation may be demonstrated if either of the following criteria are justified:

(d)   the substance is demonstrated to be ultimately degraded in an aquatic sediment or soil simulation test[4] with a half-life of < 16 days (corresponding to a degradation of > 70 % within 28 days); or

(e)   in those cases where only BOD_5 and COD data are available, the ratio of BOD_5/COD is ≥ 0.5. The same criterion applies to ready biodegradability tests of a shorter duration than 28 days, if the half-life furthermore is < 7 days.

If none of the above types of data are available, then the substance is considered as not rapidly degradable. This decision may be supported by fulfilment of at least one of the following criteria:

(i)   the substance is not inherently degradable in an inherent biodegradability test; or

(ii)   the substances is predicted to be slowly biodegradable by scientifically valid QSARs, e.g. for the Biodegradation Probability Program, the score for rapid degradation (linear or non-linear model) < 0.5; or

(iii)   the substance is considered to be not rapidly degradable based on indirect evidence, as e.g. knowledge from structurally similar substances; or

(iv)   no other data regarding degradability are available.

---

[4]   *Simulations tests should reflect realistic environmental conditions such as low concentration of the chemical, realistic temperature and employment of ambient microbial biomass not pre-exposed to the chemical.*

A9.4.3.5.6　揮発性物質は、クローズドボトル試験（OECD テストガイドライン 301D）、MITI（I）試験（OECD テストガイドライン 301C）およびマノメータ呼吸測定試験（OECD テストガイドライン 301F）など、閉鎖系の試験系でのみ試験すべきである。その他の試験法による結果の評価には注意が必要であり、かつ、そうした結果は、例えば物質収支推定値により、被験物質の移動が揮発によるものでないことを実証できる場合にのみ考慮すべきである。

A9.4.3.6　*シミュレーション試験データにおける変動*

　ある種の優先度の高い化学品については、多くのシミュレーション試験データが入手できよう。そうしたデータは、土壌、底質または表層水のような環境媒体中における一連の半減期を与えていることがしばしばある。同一の物質で実施されたシミュレーション試験で求められた半減期が異なっていることが観察された場合には、試験条件の違いが反映されていると考えられるが、そうした条件はいずれも環境的に適切なものである。分類には、証拠の重み付け評価を採用し、かつ採用した試験法がどの程度環境条件に関して現実的でありかつ関連性があるかを考慮して、そのような研究から得られた一連の半減期測定値のうち、大きい数値の方から適切な半減期を選択すべきである。一般に、水生環境中における急速分解性の評価に関しては、河川水のシミュレーション試験データの方が、水中の底質または土壌中でのシミュレーション試験データよりも望ましいとされる。

A9.4.4　*判定スキーム*

　水生環境における急速分解性に関する判定、および水生環境に対して有害な化学品の分類を促進するための一般的な手引きとして、下記の判定スキームが使用できよう。

　以下の項目のうち、少なくとも1つを満たさなければ、物質は急速分解性ではないと考えられる：

(a) その物質が、28日間の易生分解性試験において、易生分解性であると証明される。利用可能な試験データから評価できる場合には、生分解開始から10日以内に試験の易生分解性とされるレベル（DOC 除去率70％または理論的酸素要求量60％）が達成される。それが可能でなければ、この易生分解性とされるレベルをできれば14日間の時間ウィンドウ以内、または試験終了後に評価すべきである；または

(b) 物質が表層水中のシミュレーション試験[3]で、半減期＜16日（28日以内に70％より高い分解に相当する）で究極分解されることが証明される；または

(c) 物質が水生環境中において半減期＜16日（28日以内に70％より高い分解に相当する）で初期段階の分解を受け（生分解ないし非生物分解によって）、かつ分解生成物は水生環境に対して有害であるという分類基準を充足しないことが証明できる。

　上述のデータが利用できない場合、以下の判定基準のいずれかが立証されれば、急速分解性と認めてよい：

(d) 底質中または土壌中のシミュレーション試験[4]で、物質が半減期＜16日（28日以内に70％より高い分解に相当する）で究極分解されることが証明できる；

(e) BOD_5 および COD データしか利用できない場合、BOD_5/COD 比が0.5以上である。さらに半減期が7日間未満であれば、28日間より短い期間で行う易生分解性試験にも同様な基準が適用される；

　上述のどのタイプのデータも利用できないならば、その物質は急速分解性ではないとみなすべきである。この決定は、以下の判定基準のいずれかを満たすことで支持される；

(i) 本質的生分解性試験で、その物質が本質的に分解性でないと認められる；または

(ii) 科学的に有意な QSAR、例えば Biodegradation Probability Program、によって、その物質がゆっくり生分解されると予測され、急速分解性のスコアが（線形ないし非線形のモデルで）0.5未満である；または

(iii) 間接的証拠、例えば構造的に類似した物質による知見から、その物質が急速分解性ではないと思われる；または

(iv) 分解性に関するその他のデータが利用できない。

---

[4] シミュレーション試験は、低濃度の化学品、現実的な温度さらに事前にその化学品にばく露されていない環境中の微生物群を採用するなど、現実の環境条件を反映していなくてはならない。

A9.5 **Bioaccumulation**

A9.5.1 *Introduction*

A9.5.1.1 Bioaccumulation is one of the important intrinsic properties of substances that determine the potential environmental hazard. Bioaccumulation of a substance into an organism is not a hazard in itself, but bioconcentration and bioaccumulation will result in a body burden, which may or may not lead to toxic effects. In the harmonized integrated hazard classification system for human health and environmental effects of chemical substances (OECD, 1998), the wording "potential for bioaccumulation" is given. A distinction should, however, be drawn between bioconcentration and bioaccumulation. Here bioconcentration is defined as the net result of uptake, transformation, and elimination of a substance in an organism due to waterborne exposure, whereas bioaccumulation includes all routes of exposure (i.e. via air, water, sediment/soil, and food). Finally, biomagnification is defined as accumulation and transfer of substances via the food chain, resulting in an increase of internal concentrations in organisms on higher levels of the trophic chain (European Commission, 1996). For most organic chemicals uptake from water (bioconcentration) is believed to be the predominant route of uptake. Only for very hydrophobic substances does uptake from food becomes important. Also, the harmonized classification criteria use the bioconcentration factor (or the octanol/water partition coefficient) as the measure of the potential for bioaccumulation. For these reasons, the present guidance document only considers bioconcentration and does not discuss uptake via food or other routes.

A9.5.1.2 Classification of a substance is primarily based on its intrinsic properties. However, the degree of bioconcentration also depends on factors such as the degree of bioavailability, the physiology of test organism, maintenance of constant exposure concentration, exposure duration, metabolism inside the body of the target organism and excretion from the body. The interpretation of the bioconcentration potential in a chemical classification context therefore requires an evaluation of the intrinsic properties of the substance, as well as of the experimental conditions under which bioconcentration factor (BCF) has been determined. Based on the guide, a decision scheme for application of bioconcentration data or log K_{ow} data for classification purposes has been developed. The emphasis of the present section is organic substances and organo-metals. Bioaccumulation of metals is also discussed in section A9.7.

A9.5.1.3 Data on bioconcentration properties of a substance may be available from standardized tests or may be estimated from the structure of the molecule. The interpretation of such bioconcentration data for classification purposes often requires detailed evaluation of test data. In order to facilitate this evaluation two additional appendixes are enclosed. These appendixes describe available methods (appendix III of annex 9) and factors influencing the bioconcentration potential (appendix IV of annex 9). Finally, a list of standardized experimental methods for determination of bioconcentration and K_{ow} are attached (appendix V of annex 9) together with a list of references (appendix VI of annex 9).

A9.5.2 *Interpretation of bioconcentration data*

A9.5.2.1 Environmental hazard classification of a substance is normally based on existing data on its environmental properties. Test data will only seldom be produced with the main purpose of facilitating a classification. Often a diverse range of test data is available which does not necessarily match the classification criteria. Consequently, guidance is needed on interpretation of existing test data in the context of hazard classification.

A9.5.2.2 Bioconcentration of an organic substance can be experimentally determined in bioconcentration experiments, during which BCF is measured as the concentration in the organism relative to the concentration in water under steady-state conditions and/or estimated from the uptake rate constant (k_1) and the elimination rate constant (k_2) (OECD 305, 1996). In general, the potential of an organic substance to bioconcentrate is primarily related to the lipophilicity of the substance. A measure of lipophilicity is the n-octanol-water partition coefficient (K_{ow}) which, for lipophilic non-ionic organic substances, undergoing minimal metabolism or biotransformation within the organism, is correlated with the bioconcentration factor. Therefore, K_{ow} is often used for estimating the bioconcentration of organic substances, based on the empirical relationship between log BCF and log K_{ow}. For most organic substances, estimation methods are available for calculating the K_{ow}. Data on the bioconcentration properties of a substance may thus be (i) experimentally determined, (ii) estimated from experimentally determined K_{ow}, or (iii) estimated from K_{ow} values derived by use of Quantitative Structure Activity Relationships (QSARs). Guidance for interpretation of such data is given below together with guidance on assessment of chemical classes, which need special attention.

A9.5.2.3 *Bioconcentration factor (BCF)*

A9.5.2.3.1 The bioconcentration factor is defined as the ratio on a weight basis between the concentration of the chemical in biota and the concentration in the surrounding medium, here water, at steady state. BCF can thus be experimentally derived under steady-state conditions, on the basis of measured concentrations. However, BCF can also be calculated as the ratio between the first-order uptake and elimination rate constants; a method which does not require equilibrium conditions.

A9.5 生物蓄積性

A9.5.1 序

A9.5.1.1　生物蓄積は、環境への潜在的有害性を決定する、物質本来の重要な特性の1つである。物質の生体内への生物蓄積それ自体は有害ではないが、生物濃縮および生物蓄積は、身体への負荷をもたらし、その結果、毒性影響が導かれたり、あるいは導かれなかったりする。化学物質の人への健康および環境への影響のための調和された統合有害性分類システム（OECD, 1998）では、「生物蓄積の可能性」という用語が与えられている。しかし生物濃縮と生物蓄積とは区別すべきである。ここでは、生物濃縮とは、水系でのばく露による生物体内での物質の取り込み、変換および排泄のネットと定義されているのに対し、生物蓄積はすべての経路（すなわち、空気、水、堆積物/土壌、および食物）のばく露を包含している。最後に、食物連鎖による生物濃縮（biomagnification）は、栄養連鎖が高位であるほど生物の体内濃度が高くなるような、食物連鎖による物質の蓄積および移動と定義されている（European Commission, 1996）。ほとんどの有機物質では、水系からの取り込み（生物濃縮）が、支配的な取り込み経路であると信じられている。また、調和された分類基準は、生物濃縮係数（またはオクタノール/水分配係数）を生物蓄積性の可能性の尺度として用いている。こうした理由により、今回の手引きは、生物濃縮性のみについて考察し、食物またはその他の経路による取り込みについては議論しない。

A9.5.1.2　物質の分類は、主にその本来の性質に基づいている。しかし生物濃縮の程度は、生物学的利用可能な程度、被験生物の生理学的状態、ばく露濃度の定常的維持、ばく露期間の長さ、被験生物の体内における代謝、および体内からの排泄等の要因によっても影響される。したがって化学品を分類する上で、生物濃縮性について解釈するには、物質本来の特性の評価と共に、生物濃縮係数（BCF）を測定した実験条件の評価も必要である。指針に基づいて、分類のために生物濃縮データまたは log K_{ow} を適用するための判定スキームが開発されている。本節では有機化合物および有機金属を中心として論じる。金属の生物蓄積についても、A9.7節で検討する。

A9.5.1.3　物質の生物濃縮性に関するデータは、標準的な試験から得られることもあれば、その分子構造から推定されることもある。分類を目的としたこれら生物濃縮性データの解釈には、試験データの詳細な評価がしばしば必要となる。この評価を容易にするために、2つの付録を加える。これらの付録は、採用できる試験方法について（附属書9付録III）、および生物濃縮性に影響する要因について（附属書9付録IV）述べる。最後に、生物濃縮性および K_{ow} を測定するための標準的な実験方法のリスト（附属書9付録V）および参考文献リスト（附属書9付録VI）を添付する。

A9.5.2 生物濃縮性データの解釈

A9.5.2.1　物質の環境有害性分類は通常、その物質の環境特性に関する既存のデータに基づいている。分類を速やかに行うことを主な目的とした試験データは、稀にしか作成されない。広範な試験データが入手されることがしばしばあるが、それらは必ずしも分類判断基準に適合しない。したがって、有害性分類との関連で既存の試験データを解釈することについて手引きが必要になる。

A9.5.2.2　有機物質の生物濃縮性は、生物濃縮性実験によって実験的に測定できる。実験では、BCFは定常状態における水中濃度に対する生物体内濃度として測定されるか、または取り込み速度定数（k_1）および排泄速度定数（k_2）から推定される（OECD 305, 1996）。一般に、有機物質が生物濃縮する可能性は、主にその物質の親油性に関係している。親油性の尺度はn-オクタノール/水分配係数（K_{ow}）であり、親油性の非イオン性有機物質において、生物体内で代謝または生物変換がわずかしか進行しない物質では、K_{ow} は生物濃縮係数と関連づけられている。したがって K_{ow} 値は、log BCF と log K_{ow} の間の経験的な関係に基づいて、有機物質の生物濃縮の推定にしばしば用いられる。ほとんどの有機物質に対しては、K_{ow} 算出のための推定方法が利用できる。このため、ある物質の生物濃縮性に関するデータは、(1) 実験的に測定する、(2) 実験的に測定された K_{ow} 値から推定する、または(3) 定量的構造活性相関（QSAR）を用いて導かれた K_{ow} 値から推定する、ことによって求められる。こうしたデータの解釈に関する手引きについて、各化学品族の評価に関する手引きと共に以下に示すが、これには特別の注意が必要である。

A9.5.2.3 生物濃縮係数（BCF）

A9.5.2.3.1　生物濃縮係数は生物体内と、定常状態における周囲の媒体、この場合は水中との化学品の濃度比（重量ベース）として定義される。したがってBCFは定常状態における濃度測定値として、実験的に導くことができる。しかし BCF は取り込みと排泄の一次反応速度定数の比としても計算でき、この方法でならば平衡状態を必要としない。

A9.5.2.3.2   Different test guidelines for the experimental determination of bioconcentration in fish have been documented and adopted, the most generally applied being the OECD test guideline (OECD 305, 1996).

A9.5.2.3.3   Experimentally derived BCF values of high quality are ultimately preferred for classification purposes as such data override surrogate data, e.g. K_{ow}.

A9.5.2.3.4   High quality data are defined as data where the validity criteria for the test method applied are fulfilled and described, e.g. maintenance of constant exposure concentration; oxygen and temperature variations, and documentation that steady-state conditions have been reached, etc. The experiment will be regarded as a high-quality study, if a proper description is provided (e.g. by Good Laboratory Practice (GLP)) allowing verification that validity criteria are fulfilled. In addition, an appropriate analytical method must be used to quantify the chemical and its toxic metabolites in the water and fish tissue (see section 1, appendix III for further details).

A9.5.2.3.5   BCF values of low or uncertain quality may give a false and too low BCF value; e.g. application of measured concentrations of the test substance in fish and water, but measured after a too short exposure period in which steady-state conditions have not been reached (OECD 306, 1992, regarding estimation of time to equilibrium). Therefore, such data should be carefully evaluated before use and consideration should be given to using K_{ow} instead.

A9.5.2.3.6   If there is no BCF value for fish species, high-quality data on the BCF value for other species may be used (e.g. BCF determined on blue mussel, oyster, scallop (ASTM E 1022-94)). Reported BCFs for microalgae should be used with caution.

A9.5.2.3.7   For highly lipophilic substances, e.g. with log K_{ow} above 6, experimentally derived BCF values tend to decrease with increasing log K_{ow}. Conceptual explanations of this non-linearity mainly refer to either reduced membrane permeation kinetics or reduced biotic lipid solubility for large molecules. A low bioavailability and uptake of these substances in the organism will thus occur. Other factors comprise experimental artefacts, such as equilibrium not being reached, reduced bioavailability due to sorption to organic matter in the aqueous phase, and analytical errors. Special care should thus be taken when evaluating experimental data on BCF for highly lipophilic substances as these data will have a much higher level of uncertainty than BCF values determined for less lipophilic substances.

A9.5.2.3.8   BCF in different test species

A9.5.2.3.8.1   BCF values used for classification are based on whole body measurements. As stated previously, the optimal data for classification are BCF values derived using the OECD 305 test method or internationally equivalent methods, which uses small fish. Due to the higher gill surface to weight ratio for smaller organisms than larger organisms, steady-state conditions will be reached sooner in smaller organisms than in larger ones. The size of the organisms (fish) used in bioconcentration studies is thus of considerable importance in relation to the time used in the uptake phase, when the reported BCF value is based solely on measured concentrations in fish and water at steady state. Thus, if large fish, e.g. adult salmon, have been used in bioconcentration studies, it should be evaluated whether the uptake period was sufficiently long for steady state to be reached or to allow for a kinetic uptake rate constant to be determined precisely.

A9.5.2.3.8.2   Furthermore, when using existing data for classification, it is possible that the BCF values could be derived from several different fish or other aquatic species (e.g. clams) and for different organs in the fish. Thus, to compare these data to each other and to the criteria, some common basis or normalization will be required. It has been noted that there is a close relationship between the lipid content of a fish or an aquatic organism and the observed BCF value. Therefore, when comparing BCF values across different fish species or when converting BCF values for specific organs to whole body BCFs, the common approach is to express the BCF values on a common lipid content. If e.g. whole body BCF values or BCF values for specific organs are found in the literature, the first step is to calculate the BCF on a % lipid basis using the relative content of fat in the fish (cf. literature/test guideline for typical fat content of the test species) or the organ. In the second step the BCF for the whole body for a typical aquatic organism (i.e. small fish) is calculated assuming a common default lipid content. A default value of 5 % is most commonly used (Pedersen et al., 1995) as this represents the average lipid content of the small fish used in OECD 305 (1996).

A9.5.2.3.8.3   Generally, the highest valid BCF value expressed on this common lipid basis is used to determine the wet weight based BCF-value in relation to the cut off value for BCF of 500 of the harmonized classification criteria (see chapter 4.1, table 4.1.1).

A9.5.2.3.2　魚類での生物濃縮を実験的に測定する各種のテストガイドラインが作成され採用されているが、最も一般的に適用されているのは OECD テストガイドライン（OECD 305, 1996）である。

A9.5.2.3.3　実験から導かれた高品質の BCF 値は、分類目的には最も望ましいとされ、その他の代替データ、例えば K_{ow} 値等より優先される。

A9.5.2.3.4　高品質データとは、適用された試験法に関する有意性判定基準が満たされており、そしてそのことが記述されている、例えばばく露濃度、酸素および温度に関する変数が一定に維持されていること、および定常状態に到達したことの記載がある、等として定義されている。的確な説明が記載され（例えば優良試験所規準（GLP）による）、有意性判定基準が満たされていると確認できるなら、その実験は高品質の試験であるとみなされる。さらに、水中および魚組織中の化学品とその代謝物を定量するために、適切な分析法を用いなくてはならない（詳細は付録 III 第 1 節参照）。

A9.5.2.3.5　低品質の、または品質が不明な BCF 値は、誤った、低すぎる値を与えるかもしれない。例えば、魚および水中の被験物質の濃度測定値を適用しても、測定までのばく露時間が短すぎ、定常状態にまだ達してない場合がある（OECD 306 (1996) 平衡達成時間の推定について、を参照）。したがって、このようなデータは使用前に注意して評価し、代わりに K_{ow} を採用することも検討すべきである。

A9.5.2.3.6　魚類についての BCF 値がない場合には、他の種の BCF 値に関する高品質データを用いてもよい（例えばムラサキイガイ、カキ、ホタテガイについて測定された BCF 値（ASTM E 1022-94））。微細藻類の BCF 報告値は注意して使用すべきである。

A9.5.2.3.7　例えば $\log K_{ow}$ 値が 6 より大きいような、高度に親油性の物質では、実験的に求めた BCF 値は $\log K_{ow}$ が大きいほど低くなる傾向がある。このような非線形性の概念的な説明は、主として、膜の浸透速度の減少、または体内脂質の大きな分子に対する溶解性の低下と関連づけられている。このようにして、生物体内ではこれらの物質の生物学的利用可能性および取り込みの低下が起こるであろう。別の原因として、平衡に達しなかった、水相中の有機物に吸着したために生物学的利用可能性が低下した、および分析誤差等、実験技術上の要因が考えられる。このため、親油性が高い物質の BCF 値に関する実験データを評価する際は、親油性が低い方の物質より不確実性レベルがはるかに高いので、特に注意が必要である。

A9.5.2.3.8　異なった試験生物種における BCF

A9.5.2.3.8.1　分類に用いられる BCF 値は、全身についての測定に基づいている。すでに述べたように、分類に最適なデータとは、小型魚類を用いる OECD 305 試験法、またはこれと同等の国際的な方法によって導かれた BCF 値である。小型種では大型種よりも体重に対する鰓表面積の比が大きいため、大型種より小型種の方が早く定常状態に到達する。このため、BCF 報告値が定常状態における魚および水中の測定濃度のみをもとに決定されているときには、生物濃縮研究に用いられる生物（魚類）のサイズは、取り込み段階に要する時間との関係で非常に重要である。したがって、例えばサケの成魚等、大型魚類を生物濃縮試験に使っているときには、取り込み期間が十分に長く、定常状態が達成されているかどうか、または動力学的な取り込み速度定数が正確に決定できるような一定の状態にあるかを評価すべきである。

A9.5.2.3.8.2　さらに、分類に既存データを用いる場合には、何か他の魚種または他の水生生物種（例えばアサリ）から、および魚のいろいろな臓器について、BCF 値が求められていることがある。そこで、こうしたデータを相互に、また判定基準と比較するための、何らかの共通の基盤または標準化が必要となるであろう。魚あるいは他の水生生物の体内脂質含量と、BCF 測定値の間には密接な関係があることが認められている。したがって、異なった魚種の間で BCF 値を比較する場合、または特定臓器についての BCF 値を全身の BCF 値に換算する場合、BCF 値を共通の脂質含量に対して表現することが一般的なアプローチである。もし例えば、文献中に全身の BCF 値または特定臓器の BCF が示されているならば、その魚類中または臓器の相対的な脂質含量（試験動物種の代表的な脂質含量に関する文献/テストガイドライン参照のこと）を用いて、脂質含量%に対する BCF 値を算出することが第一段階である。第二段階では、共通のデフォルト脂質含量を想定して、典型的な水生動物種（すなわち小型魚種）の全身の BCF を計算する。デフォルト値 5%は、OECD 305（1996）で採用されている小型魚種の平均脂質含量を代表していることから、この数値が最も多く用いられている（Pedersen ら、1995）。

A9.5.2.3.8.3　一般に、この共通の脂質含量に対して表記された BCF の最高有効値を用いて、調和された分類基準の BCF500 というカットオフ値と関連させて、湿体重量あたりの BCF 値を決定する。（4.1 章、表 4.1.1 を参照のこと）

A9.5.2.3.9   Use of radiolabelled substances

A9.5.2.3.9.1   The use of radiolabelled test substances can facilitate the analysis of water and fish samples. However, unless combined with a specific analytical method, the total radioactivity measurements potentially reflect the presence of the parent substance as well as possible metabolite(s) and possible metabolized carbon, which have been incorporated in the fish tissue in organic molecules. BCF values determined by use of radiolabelled test substances are therefore normally overestimated.

A9.5.2.3.9.2   When using radiolabelled substances, the labelling is most often placed in the stable part of the molecule, for which reason the measured BCF value includes the BCF of the metabolites. For some substances it is the metabolite which is the most toxic and which has the highest bioconcentration potential. Measurements of the parent substance as well as the metabolites may thus be important for the interpretation of the aquatic hazard (including the bioconcentration potential) of such substances.

A9.5.2.3.9.3   In experiments where radiolabelled substances have been used, high radiolabel concentrations are often found in the gall bladder of fish. This is interpreted to be caused by biotransformation in the liver and subsequently by excretion of metabolites in the gall bladder (Comotto et al., 1979; Wakabayashi et al., 1987; Goodrich et al., 1991; Toshima et al., 1992). When fish do not eat, the content of the gall bladder is not emptied into the gut, and high concentrations of metabolites may build up in the gall bladder. The feeding regime may thus have a pronounced effect on the measured BCF. In the literature many studies are found where radiolabelled compounds are used, and where the fish are not fed. As a result, high concentrations of radioactive material are found in the gall bladder. In these studies, the bioconcentration may in most cases have been overestimated. Thus, when evaluating experiments, in which radiolabelled compounds are used, it is essential to evaluate the feeding regime as well.

A9.5.2.3.9.4   If the BCF in terms of radiolabelled residues is documented to be $\geq$ 1000, identification and quantification of degradation products, representing $\geq$ 10 % of total residues in fish tissues at steady-state, are for e.g. pesticides strongly recommended in OECD guideline No. 305 (1996). If no identification and quantification of metabolites are available, the assessment of bioconcentration should be based on the measured radiolabelled BCF value. If, for highly bioaccumulative substances (BCF $\geq$ 500), only BCFs based on the parent compound and on radiolabelled measurements are available, the latter should thus be used in relation to classification.

A9.5.2.4   *Octanol-water-partitioning coefficient (K_{ow})*

A9.5.2.4.1   For organic substances experimentally derived high-quality K_{ow} values, or values which are evaluated in reviews and assigned as the "recommended values", are preferred over other determinations of Kow. When no experimental data of high quality are available, validated Quantitative Structure Activity Relationships (QSARs) for log K_{ow} may be used in the classification process. Such validated QSARs may be used without modification to the agreed criteria if they are restricted to chemicals for which their applicability is well characterized. For substances like strong acids and bases, substances which react with the eluent, or surface-active substances, a QSAR estimated value of K_{ow} or an estimate based on individual *n*-octanol and water solubilities should be provided instead of an analytical determination of K_{ow} (EEC A.8., 1992; OECD 117, 1989). Measurements should be taken on ionizable substances in their non-ionized form (free acid or free base) only by using an appropriate buffer with pH below pK for free acid or above the pK for free base.

A9.5.2.4.2   Experimental determination of K_{ow}

For experimental determination of K_{ow} values, several different methods, Shake-flask, and HPLC, are described in standard guidelines, e.g. OECD Test Guideline 107 (1995); OECD Test Guideline 117 (1989); EEC A.8. (1992); EPA-OTS (1982); EPA-FIFRA (1982); ASTM (1993). The shake-flask method is recommended when the log K_{ow} value falls within the range from –2 to 4. The shake-flask method applies only to essential pure substances soluble in water and *n*-octanol. For highly lipophilic substances, which slowly dissolve in water, data obtained by employing a slow-stirring method are generally more reliable. Furthermore, the experimental difficulties, associated with the formation of microdroplets during the shake-flask experiment, can to some degree be overcome by a slow-stirring method where water, octanol, and test compound are equilibrated in a gently stirred reactor. With the slow-stirring method (OECD Test Guideline 123) a precise and accurate determination of K_{ow} of compounds with log K_{ow} of up to 8.2 is allowed. As for the shake-flask method, the slow-stirring method applies only to essentially pure substances soluble in water and *n*-octanol. The HPLC method, which is performed on analytical columns, is recommended when the log K_{ow} value falls within the range 0 to 6. The HPLC method is less sensitive to the presence of impurities in the test compound compared to the shake-flask method. Another technique for measuring log K_{ow} is the generator column method (USEPA 1985).

As an experimental determination of the K_{ow} is not always possible, e.g. for very water-soluble substances, very lipophilic substances, and surfactants, a QSAR-derived K_{ow} may be used.

A9.5.2.3.9　放射性標識物質の使用

A9.5.2.3.9.1　放射性標識された被験物質を使用すれば、水および魚サンプルの分析が容易になる。しかし、特有の分析法を組み合わせて用いなければ、総放射能測定値は親化合物だけでなく、考えられる代謝生成物、および有機分子として魚組織中に取り込まれた、考えられる代謝をうけた炭素の存在を反映している可能性がある。したがって、放射性標識された被験物質を用いて決定されたBCF値は、通常、過大評価されている。

A9.5.2.3.9.2　放射性標識物質を用いる場合、分子の安定な部分が標識されることが最も多いが、このことがBCF測定値に代謝物のBCFが含まれる理由である。物質によっては、代謝物が最も毒性が強く、生物濃縮性が最も高いこともある。したがって、水生有害性（生物濃縮性も含めて）を解釈するには、親物質だけでなく、代謝物の測定も重要になることがある。

A9.5.2.3.9.3　放射性標識物質を用いる実験では、しばしば、魚の胆嚢に高い放射性標識濃度が見出される。これは、肝臓内での生物学的変換、およびその後の胆嚢中への代謝物の排泄によって起こると解釈されている（Comottoら, 1979; Wakabayashiら, 1987; Goodrichら, 1991; Toshimaら, 1992）。魚が摂餌しないと、胆嚢内容物は腸内に排出されず、胆嚢内に高濃度の代謝物が蓄積されることがある。したがって、BCF測定値に給餌方法が大きく影響することがある。文献には、放射性標識化合物を用いており、そして魚に給餌していない多くの試験が見られる。結果的に高濃度の放射性物質が胆嚢内に検出される。こうした試験では、ほとんどの例で生物濃縮が過大に推定されている。したがって、放射性標識化合物が用いた実験を評価する際は、給餌方式についても評価することが不可欠である。

A9.5.2.3.9.4　残留放射能の点からBCFが1000以上と記録されているならば、定常状態で魚組織中の合計残留放射能の10%以上を占めている分解生成物を同定および定量することが、例えば殺虫剤などについて、OECDテストガイドライン305（1996）では強く勧告されている。代謝物の同定も定量も可能でないならば、生物濃縮性の評価は放射性標識BCF測定値に基づいて行うべきである。生物蓄積性の高い物質（BCF≧500）について、親化合物についてのBCF、および放射性測定値によるBCFだけが利用可能な場合は、分類に関しては後者を採用すべきである。

A9.5.2.4　オクタノール/水分配係数（K_{ow}）

A9.5.2.4.1　有機物質では、実験から求められた高品質のK_{ow}値、またはレビューにおいて評価され「推奨値」として指定された数値が、その他のK_{ow}測定よりも望ましい。高品質の実験データが利用できない場合には、log K_{ow}のための有効性評価済みの構造活性相関（QSAR）を分類プロセスに使用してもよい。こうした有効性評価済みQSARは、その適用性がよく確かめられている化学品にのみ限定されているならば、合意済みの判定基準に対して変更することなく使用してよい。例えば強酸や強塩基のような物質、溶出液と反応する物質や、界面活性作用のある物質では、QSARにより推定されたK_{ow}値、またはn-オクタノールと水とに対する個別の溶解性をもとにした推定値が、分析によるK_{ow}の測定のかわりに与えられるべきである（EEC A.8., 1992; OECD 117, 1989）。解離性物質ではイオン化していない形態（遊離酸または遊離塩基）について、遊離酸についてはpKより低い、および遊離塩基についてはpKより高いpHの適切な緩衝液を用いてのみ、測定を行うべきである。

A9.5.2.4.2　K_{ow}の実験的測定

　K_{ow}値を実験的に測定するには、例えばフラスコ振盪法やHPLC等、いくつかの異なった試験法が、標準のガイドラインに記載されている。例えばOECDテストガイドライン107（1995）、OECDテストガイドライン117（1989）、EEC A.8（1992）、EPA-OTS（1982）、EPA-FIFRA（1982）、ASTM（1993）等がある。フラスコ振盪法はlog K_{ow}が−2〜4の範囲内になる場合に推奨される。フラスコ振盪法は、水およびn−オクタノールに可溶な、本質的に純粋な物質にのみ適用される。水中溶解速度が遅い、高度に親油性の物質には、低速撹拌法を用いて得られたデータの方が、一般により信頼性が高い。さらに、フラスコ振盪法の実験では、微小滴の生成に伴う実験的な困難があるが、これは、オクタノールおよび被験物質が低速で撹拌された反応容器内で平衡に達する低速撹拌法によって、ある程度まで克服できる。低速撹拌法（作成中のOECDテストガイドライン123）を用いることで、log K_{ow}が8.2までの化合物のK_{ow}を、正確かつ精密に測定できる。フラスコ振盪法と同様に、低速撹拌法は水およびn−オクタノールに可溶な本質的に純粋な物質にのみ適用される。HPLC法は分析カラムを用いる方法であり、log K_{ow}値が0から6の範囲内となる場合に推奨される。HPLC法の方がフラスコ振盪法に比べて、被験物質中の不純物の存在による影響を受けにくい。log K_{ow}を測定する方法には他にジェネレータ・カラム法（USEPA, 1985）がある。

　実験的なK_{ow}の測定は、例えば極めて水溶性の高い物質や極めて親油性の物質および界面活性剤など、必ずしも可能というわけではないため、QSARにより導いたK_{ow}を採用してもよい。

A9.5.2.4.3   Use of QSARs for determination of log K_{ow}

When an estimated K_{ow} value is found, the estimation method has to be taken into account. Numerous QSARs have been and continue to be developed for the estimation of K_{ow}. Four commercially available PC programmes (CLOGP, LOGKOW (KOWWIN), AUTOLOGP, SPARC) are frequently used for risk assessment if no experimentally derived data are available. CLOGP, LOGKOW and AUTOLOGP are based upon the addition of group contributions, while SPARC is based upon a more fundamental chemical structure algorithm. Only SPARC can be employed in a general way for inorganic or organometallic compounds. Special methods are needed for estimating log K_{ow} for surface-active compounds, chelating compounds and mixtures. CLOGP is recommended in the US EPA/EC joint project on validation of QSAR estimation methods (US EPA/EC 1993). Pedersen *et al.* (1995) recommended the CLOGP and the LOGKOW programmes for classification purposes because of their reliability, commercial availability, and convenience of use. The following estimation methods are recommended for classification purposes (table A9.5.1).

Table A9.5.1:   Recommended QSARs for estimation of K_{ow}

Model	log K_{ow} range	Substance utility
CLOGP	0 < log K_{ow} < 9 [a]	The program calculates log K_{ow} for organic compounds containing C, H, N, O, Hal, P, and/or S.
LOGKOW (KOWWIN)	-4 < log K_{ow} < 8 [b]	The program calculates log K_{ow} for organic compounds containing C, H, N, O, Hal, Si, P, Se, Li, Na, K, and/or Hg. Some surfactants (e.g. alcohol ethoxylates, dyestuffs, and dissociated substances may be predicted by the program as well.
AUTOLOGP	log K_{ow} > 5	The programme calculates log K_{ow} for organic compounds containing C, H, N, O, Hal, P and S. Improvements are in progress in order to extend the applicability of AUTOLOGP.
SPARC	Provides improved results over KOWWIN and CLOGP for compounds with log K_{ow} > 5	SPARC is a mechanistic model based on chemical thermodynamic principles rather than a deterministic model rooted in knowledge obtained from observational data. Therefore, SPARC differs from models that use QSARs (i.e. KOWWIN, CLOGP, AUTOLOGP) in that no measured log K_{ow} data are needed for a training set of chemicals. Only SPARC can be employed in a general way for inorganic or organometallic compounds.

[a]   *A validation study performed by Niemelä, who compared experimental determined log K_{ow} values with estimated values, showed that the program precisely predicts the log K_{ow} for a great number of organic chemicals in the log K_{ow} range from below 0 to above 9 (n = 501, r^2 = 0.967) (TemaNord 1995: 581).*

[b]   *Based on a scatter plot of estimatedversus experimental log K_{ow} (Syracuse Research Corporation, 1999), where 13058 compound have been tested, the LOGKOW is evaluated being valid for compounds with a log K_{ow} in the interval -4 - 8.*

A9.5.3   *Chemical classes that need special attention with respect to BCF and K_{ow} values*

A9.5.3.1   There are certain physico-chemical properties, which can make the determination of BCF or its measurement difficult. These may be substances, which do not bioconcentrate in a manner consistent with their other physico-chemical properties, e.g. steric hindrance or substances which make the use of descriptors inappropriate, e.g. surface activity, which makes both the measurement and use of log K_{ow} inappropriate.

A9.5.3.2   *Difficult substances*

A9.5.3.2.1   Some substances are difficult to test in aquatic systems and guidance has been developed to assist in testing these materials (DoE, 1996; ECETOC 1996; US EPA 1996; OECD 2000). The OECD Guidance Document on aquatic toxicity testing of difficult substances and mixtures (OECD, 2000) is also a good source of information for bioconcentration studies, on the types of substances that are difficult to test and the steps needed to ensure valid conclusions from tests with these substances. Difficult to test substances may be poorly soluble, volatile, or subject to rapid degradation due to such processes as phototransformation, hydrolysis, oxidation, or biotic degradation.

A9.5.3.2.2   To bioconcentrate organic compounds, a substance needs to be soluble in lipids, present in the water, and available for transfer across the fish gills. Properties which alter this availability will thus change the actual bioconcentration of a substance, when compared with the prediction. For example, readily biodegradable substances may only be present in the aquatic compartment for short periods of time. Similarly, volatility, and hydrolysis will reduce the concentration and the time during which a substance is available for bioconcentration. A further important parameter,

A9.5.2.4.3 QSARを用いたlog K$_{ow}$の決定

K$_{ow}$の推定値が見出された場合には、それを推定した方法を考慮する必要がある。K$_{ow}$の推定のために、数々のQSARが考案され、また現在でも開発されている。実験的に導いたデータが利用できないならば、市販されている4種のPCプログラム（CLOGP, LOGKOW (KOWWIN), AUTOLOGP, SPARC）がリスク評価に多く用いられている。CLOGP, LOGKOWおよびAUTOLOGPは、官能基の関与の加算に基づいているのに対し、SPARCはより基本的な化学構造のアルゴリズムに基づいている。SPARCのみが一般的な方法で、無機化合物または有機金属化合物に採用できる。界面活性物質、錯体形成化合物および混合物のlog K$_{ow}$推定には、特別な方法が必要である。QSAR推定法の有意性評価に関するUS EPA/EC合同プロジェクトではCLOGPが推奨されている（US EPA/EC 1993）。Pedersenら(1995)は、CLOGPおよびLOGKOWプログラムを、その信頼性、市販されていること、および使用上の簡便さの理由から、分類目的に推奨している。分類の目的には以下の推定法が推奨されている（表A9.5.1）。

表A9.5.1：K$_{ow}$推定に推奨されているQSAR

モデル	log Kow の範囲	使用できる物質
CLOGP	log K$_{ow}$<0 、 log K$_{ow}$>9[a]	このプログラムは、C, H, N, O, ハロゲン, P または S を含む有機化合物の log K$_{ow}$ を計算する。
LOGKOW (KOWWIN)	-4 <log K$_{ow}$ <8[b]	このプログラムはC, H, N, O, ハロゲン, Si, P, Se, Li, Na, K または Hg を含む有機化合物の log K$_{ow}$ を計算する。界面活性物質（例：アルコールエトキシレート）、染料および解離物質にもこのプログラムで予測できるものもある。
AUTOLOGP	log K$_{ow}$>5	このプログラムは C, H, N, O, ハロゲン, P および S を含む有機化合物の log K$_{ow}$ を計算する。AUTOLOGP の適用性を拡張するために改良が行われている。
SPARC	log K$_{ow}$>5 の物質では、KOWWIN および CLOGP よりも優れた結果が得られる。	SPARC は観察データから得られた知識に由来する決定論的モデルというよりむしろ化学的な熱力学原理にもとづいたメカニズムモデルである。このため、SPARC は QSAR を用いる他のモデル（すなわち KOWWIN, CLOGP, AUTOLOGP）とは、トレーニング用の化学品セットについては log K$_{ow}$ 測定データが必要でないという点で異なっている。無機化合物や有機金属化合物に一般的な方法で適用できるのは、SPARC だけである。

[a] *Niemelä は実験的に求められた log K$_{ow}$ 値と推定値を比較する有意性評価研究を行い、log K$_{ow}$ の範囲が 0 以下から 9 以上までの多数の有機化合物の log K$_{ow}$ について、このプログラムが正確に予測することを確認した (n=501, r²=0.967) (TemaNord 1995: 581)*

[b] *log K$_{ow}$ 推定値を実験値に対して分散プロットし、13058 物質について検討したところ (Syracuse Research Corporation, 1999)、LOGKOW は log K$_{ow}$ が−4 から 8 の範囲にある化合物について有効であると評価されている。*

A9.5.3 BCFおよびK$_{ow}$値に関して特別な注意が必要な化学品クラス

A9.5.3.1 BCFの決定または測定を困難にする可能性のある特定の物理化学的性質がある。こうした物質には、その生物濃縮性が当該物質の他の物理化学的性質とは一貫しないようなものもある。例えば立体障害や物理化学的なパラメータの記述を不適切にするもの、または例えば界面活性等のように log K$_{ow}$ の測定と使用の両方を不適切にしてしまうようなものがある。

A9.5.3.2 試験困難な物質

A9.5.3.2.1 物質には、水系システムでの試験が困難なものもあり、こうした物質の試験を支援するために手引きが作成されている（DoE, 1996; ECETOC 1996; US EPA 1996; OECD, 2000）。試験困難な物質および混合物の水生毒性試験に関するOECDガイダンス文書（OECD, 2000）はまた生物濃縮試験の情報源として、またこれらの物質の試験から有意な結論が得られるようにするために必要なステップについても、適切な情報源である。試験困難な物質は、溶解性が低いか、揮発性であるか、または光変換や加水分解、酸化または生分解等のプロセスのために急速分解性をもつ。

A9.5.3.2.2 有機化合物の生物濃縮には、その物質が脂肪に可溶であり、水中に存在し、そして魚の鰓を通過して移動可能であることが要件となる。したがって、こうした存在および移動可能性を変化させるような物質の性質は、予測値と比べて、物質の実際の生物濃縮を変化させる。例えば、易生分解性の物質は、

which may reduce the actual exposure concentration of a substance, is adsorption, either to particulate matter or to surfaces in general. There are a number of substances, which have shown to be rapidly transformed in the organism, thus leading to a lower BCF value than expected. Substances that form micelles or aggregates may bioconcentrate to a lower extent than would be predicted from simple physico-chemical properties. This is also the case for hydrophobic substances that are contained in micelles formed as a consequence of the use of dispersants. Therefore, the use of dispersants in bioaccumulation tests is discouraged.

A9.5.3.2.3 In general, for difficult to test substances, measured BCF and K_{ow} values – based on the parent substance – are a prerequisite for the determination of the bioconcentration potential. Furthermore, proper documentation of the test concentration is a prerequisite for the validation of the given BCF value.

A9.5.3.3 *Poorly soluble and complex substances*

Special attention should be paid to poorly soluble substances. Frequently the solubility of these substances is recorded as less than the detection limit, which creates problems in interpreting the bioconcentration potential. For such substances the bioconcentration potential should be based on experimental determination of log K_{ow} or QSAR estimations of log K_{ow}.

When a multi-component substance is not fully soluble in water, it is important to attempt to identify the components of the mixture as far as practically possible and to examine the possibility of determining its bioaccumulation potential using available information on its components. When bioaccumulating components constitute a significant part of the complex substance (e.g. more than 20 % or for hazardous components an even lower content), the complex substance should be regarded as being bioaccumulating.

A9.5.3.4 *High molecular weight substances*

Above certain molecular dimensions, the potential of a substance to bioconcentrate decreases. This is possibly due to steric hindrance of the passage of the substance through gill membranes. It has been proposed that a cut-off limit of 700 for the molecular weight could be applied (e.g. European Commission, 1996). However, this cut-off has been subject to criticism and an alternative cut-off of 1000 has been proposed in relation to exclusion of consideration of substances with possible indirect aquatic effects (CSTEE, 1999). In general, bioconcentration of possible metabolites or environmental degradation products of large molecules should be considered. Data on bioconcentration of molecules with a high molecular weight should therefore be carefully evaluated and only be used if such data are considered to be fully valid in respect to both the parent compound and its possible metabolites and environmental degradation products.

A9.5.3.5 *Surface-active agents*

A9.5.3.5.1 Surfactants consist of a lipophilic (most often an alkyl chain) and a hydrophilic part (the polar headgroup). According to the charge of the headgroup, surfactants are subdivided into classes of anionic, cationic, non-ionic, or amphoteric surfactants. Due to the variety of different headgroups, surfactants are a structurally diverse class of compounds, which is defined by surface activity rather than by chemical structure. The bioaccumulation potential of surfactants should thus be considered in relation to the different subclasses (anionic, cationic, non-ionic, or amphoteric) instead of to the group as a whole. Surface-active substances may form emulsions, in which the bioavailability is difficult to ascertain. Micelle formation can result in a change of the bioavailable fraction even when the solutions are apparently formed, thus giving problems in interpretation of the bioaccumulation potential.

A9.5.3.5.2 Experimentally derived bioconcentration factors

Measured BCF values on surfactants show that BCF may increase with increasing alkyl chain length and be dependant of the site of attachment of the head group, and other structural features.

A9.5.3.5.3 Octanol-water-partition coefficient (K_{ow})

The octanol-water partition coefficient for surfactants can not be determined using the shake-flask or slow stirring method because of the formation of emulsions. In addition, the surfactant molecules will exist in the water phase almost exclusively as ions, whereas they will have to pair with a counter-ion in order to be dissolved in octanol. Therefore, experimental determination of K_{ow} does not characterize the partition of ionic surfactants (Tolls, 1998). On the other hand, it has been shown that the bioconcentration of anionic and non-ionic surfactants increases with increasing lipophilicity (Tolls, 1998). Tolls (1998) showed that for some surfactants, an estimated log K_{ow} value using LOGKOW could represent the bioaccumulation potential; however, for other surfactants some 'correction' to the estimated log K_{ow} value using the method of Roberts (1989) was required. These results illustrate that the quality of the relationship between

水環境中には短時間しか存在しない。同様に、揮発および加水分解は濃度を低下させ、また物質が生物濃縮に利用される時間を短縮させる。物質の実際のばく露濃度を低下させる可能性のあるさらに重要なパラメータは、一般に粒子状物質ないし表面への吸着である。生物体内で速やかに変換され、このために予測されるよりも低い BCF 値を導くことが示された物質もたくさんある。ミセルまたは凝集体を形成する物質は、単純な物理化学的特性から予測されるよりも生物濃縮の程度が低くなるであろう。分散剤を使用することによって形成されたミセル内に含まれている疎水性物質にも、このことがあてはまる。したがって、生物蓄積の試験に分散剤を用いることは望ましくない。

A9.5.3.2.3　一般に、試験困難な物質には、親物質に基づいた BCF および K_{ow} の測定値が、生物濃縮性を決定するために不可欠である。さらに、求められた BCF 値の有意性評価には、被験物質濃度の正しい記録が不可欠である。

A9.5.3.3　*溶解性の低い物質および複合物質*

溶解性の低い物質には特別の注意が必要である。こうした物質の溶解性は検出限界より低いと記録されていることが多く、生物濃縮性を解釈する際に問題となる。こうした物質では、生物濃縮性は log K_{ow} の実験による測定または log K_{ow} の QSAR 推定に基づくべきである。

多成分物質が水に完全には溶解しない場合、混合物の成分を実際的に可能な限り同定し、その成分について利用可能な情報を用いて、生物濃縮性が決定できる可能性を検討するよう試みることが重要である。生物蓄積性のある成分がその複合物質のかなりの部分（例えば 20%以上、有害成分ではより低い含量）を占めるときには、その複合物質は生物蓄積性であるとみなされるべきである。

A9.5.3.4　*分子量の大きい物質*

ある分子量を超えると、物質が生物濃縮する可能性は減少する。これはおそらく、物質が鰓膜を通過する際の立体障害によるものと思われる。分子量のカットオフ限界として 700 が適用できるのではないか、と提案されている（例えば European Commission, 1996）。しかし、このカットオフ値は批判の対象になっており、間接的な水の影響の可能性がある物質を考慮から除外して、別のカットオフ値 1000 が提案されている（CSTEE, 1999）。一般に巨大分子については、その想定される代謝物または環境中での分解生成物の生物濃縮性を考慮すべきである。したがって、巨大分子量分子の生物濃縮性に関するデータは、評価に注意が必要であり、親化合物、および想定される代謝物ならびに環境中での分解生成物の両方について完全に有効であると考えられる場合にのみ、そのデータを用いるべきである。

A9.5.3.5　*界面活性剤*

A9.5.3.5.1　界面活性剤は、親油性部分（最も多いのはアルキル鎖）と親水性部分（極性の頭部基）から構成されている。頭部基の電荷により、界面活性剤は陰イオン、陽イオン、非イオン、および両性の界面活性剤に細分化される。頭部基は多様であるため、界面活性剤は構造的に多様な群に属する物質であり、化学構造よりむしろ界面活性によって定義される。したがって、界面活性剤の生物蓄積性は、界面活性剤全体としてでなく、異なった小分類（陰イオン性、陽イオン性、非イオン性、または両性）に関して考慮すべきである。界面活性物質はエマルジョンを形成することもあり、その場合には生物学的利用可能性を確かめることは困難である。ミセルが形成されると、外見上溶液となっていても、生物学的に利用可能な部分が変化する可能性があり、生物蓄積性の解釈上の問題が生じる。

A9.5.3.5.2　実験的に求めた生物濃縮係数

界面活性剤に対する BCF 測定値から、BCF はアルキル鎖が長いほど増加し、また頭部基のついている部位、ならびに他の構造的特徴に依存することが示されている。

A9.5.3.5.3　オクタノール/水分配係数（K_{ow}）

界面活性剤のオクタノール/水分配係数は、エマルジョンが形成されるため、フラスコ振騰法や低速攪拌法では測定できない。さらに、界面活性剤分子は水相中で殆ど例外なくイオンとしてのみ存在するが、それに対して、オクタノール中に溶解するには対イオンとペアを作らねばならない。したがって、K_{ow} を実験的に測定しても、イオン性界面活性剤の分配の特性を記述するわけではない (Tolls, 1998)。その一方で、陰イオン性および非イオン性界面活性物質は、親油性が高いほど生物濃縮性も高いことが示されている (Tolls, 1998)。Tolls (1998) は、ある界面活性剤については、LOGKOW を用いて推定した log K_{ow} 値で生物蓄積性を表すことができるが、他の界面活性剤では、この log K_{ow} 推定値に Roberts (1989) の方法を用いて「補正」が必要であることを示した。これらの結果は、log K_{ow} 推定値と生物濃縮の関係について

log K_{ow} estimates and bioconcentration depends on the class and specific type of surfactants involved. Therefore, the classification of the bioconcentration potential based on log K_{ow} values should be used with caution.

A9.5.4 *Conflicting data and lack of data*

A9.5.4.1 *Conflicting BCF data*

In situations where multiple BCF data are available for the same substance, the possibility of conflicting results might arise. In general, conflicting results for a substance, which has been tested several times with an appropriate bioconcentration test, should be interpreted by a weight of evidence assessment. This implies that if experimental determined BCF data, both ≥ and < 500, have been obtained for a substance the data of the highest quality and with the best documentation should be used for determining the bioconcentration potential of the substance. If differences still remain, if e.g. high-quality BCF values for different fish species are available, generally the highest valid value should be used as the basis for classification.

When larger data sets (4 or more values) are available for the same species and life stage, the geometric mean of the BCF values may be used as the representative BCF value for that species.

A9.5.4.2 *Conflicting log K_{ow} data*

The situations, where multiple log K_{ow} data are available for the same substance, the possibility of conflicting results might arise. If log K_{ow} data both ≥ and < 4 have been obtained for a substance, then the data of the highest quality and the best documentation should be used for determining the bioconcentration potential of the substance. If differences still exist, generally the highest valid value should take precedence. In such situation, QSAR estimated log K_{ow} could be used as a guidance.

A9.5.4.3 *Expert judgement*

If no experimental BCF or log K_{ow} data or no predicted log K_{ow} data are available, the potential for bioconcentration in the aquatic environment may be assessed by expert judgement. This may be based on a comparison of the structure of the molecule with the structure of other substances for which experimental bioconcentration or log K_{ow} data or predicted K_{ow} are available.

A9.5.5 *Decision scheme*

A9.5.5.1 Based on the above discussions and conclusions, a decision scheme has been elaborated which may facilitate decisions as to whether or not a substance has the potential for bioconcentration in aquatic species.

A9.5.5.2 Experimentally derived BCF values of high quality are ultimately preferred for classification purposes. BCF values of low or uncertain quality should not be used for classification purposes if data on log K_{ow} are available because they may give a false and too low BCF value, e.g. due to a too short exposure period in which steady-state conditions have not been reached. If no BCF is available for fish species, high quality data on the BCF for other species (e.g. mussels) may be used.

A9.5.5.3 For organic substances, experimentally derived high quality K_{ow} values, or values which are evaluated in reviews and assigned as the "recommended values", are preferred. If no experimentally data of high quality are available validated Quantitative Structure Activity Relationships (QSARs) for log K_{ow} may be used in the classification process. Such validated QSARs may be used without modification in relation to the classification criteria, if restricted to chemicals for which their applicability is well characterized. For substances like strong acids and bases, metal complexes, and surface-active substances a QSAR estimated value of K_{ow} or an estimate based on individual *n*-octanol and water solubilities should be provided instead of an analytical determination of K_{ow}.

A9.5.5.4 If data are available but not validated, expert judgement should be used.

A9.5.5.5 Whether or not a substance has a potential for bioconcentration in aquatic organisms could thus be decided in accordance with the following scheme:

    (a)   Valid/high quality experimentally determined BCF value = YES:

          (i)   BCF ≥ 500: *The substance has a potential for bioconcentration*
          (ii)   BCF < 500: *The substance does not have a potential for bioconcentration.*

のデータの質は、関係している界面活性剤の種類や特定のタイプに依存することを示している。したがって、log K_{ow} にもとづいた生物濃縮性の分類は注意して用いるべきである。

A9.5.4 矛盾するデータおよびデータの欠如

A9.5.4.1 *矛盾する BCF データ*

同一の物質に対して多数の BCF データが利用可能な状況では、矛盾する結果が生じる可能性がある。一般に、ある適切な生物濃縮性試験で何回か試験されて、同一物質について、適切な生物濃縮性試験で何回か試験が行われ、矛盾する結果が得られた場合は証拠の重み付け評価によって解釈すべきである。これはすなわち、ある物質について、500 以上と 500 未満の両方の BCF データが、実験による測定で得られている場合には、最高の品質かつ記録資料が揃っているデータを、その物質の生物濃縮性判定に用いるべきであることを意味している。それでも差がある場合、例えば異なった魚種について高品質の BCF 値が利用可能な場合には、一般には分類の根拠として有効な値のうち最も高いものを用いるべきである。

同一生物種の同一ライフステージに関して、より大きなデータセット（4 件以上の数値）が利用可能な場合には、その種を代表する BCF 値として、BCF 値の幾何平均を使用してもよい。

A9.5.4.2 *矛盾する log K_{ow} データ*

同一の物質に対して複数の log K_{ow} データが利用可能な状況では、矛盾する結果が生じる可能性がある。ある物質に対して 4 以上と 4 未満の両方の log K_{ow} 値が得られたならば、その物質の生物濃縮性の決定には最高品質かつ記録が最適であるデータを採用すべきである。それでも違いがあるならば、一般に有効な値のうち最高のものを優先すべきである。このような状況では、QSAR で推定された log K_{ow} 値を、ガイダンスとして用いることもできよう。

A9.5.4.3 *専門家の判断*

BCF や log K_{ow} の実験データも、log K_{ow} の予測データも利用できないならば、水系環境中での生物濃縮性は専門家の判断で評価されることもある。これは、その分子の構造を、生物濃縮または log K_{ow} の実験データまたは K_{ow} 予測値が利用可能な別の物質の構造と比較することに基づくであろう。

A9.5.5 *判定スキーム*

A9.5.5.1 上述の考察と結論をもとに、物質が水生生物中に生物濃縮性があるかどうかの判定を容易にする、1 つの判定スキームが考案された。

A9.5.5.2 分類目的のためには、高品質の実験による BCF 値が究極的に望ましいものである。低品質の、または不確かな品質の BCF 値は、例えばばく露期間が短すぎて定常状態に達しなかったなどの理由で、誤った、あるいは低すぎる値を与える可能性があるので、log K_{ow} に関するデータが利用可能であれば、このような BCF データを使用すべきでない。魚類の BCF が利用できないならば、別の種（例えばイガイ）に関する高品質の BCF データを採用してもよい。

A9.5.5.3 有機物質では、実験から導かれた高品質の K_{ow} 値、あるいは排泄によって評価され「推奨値」として指定された数値が望ましい。高品質の実験データが利用できないならば、log K_{ow} として、有意性評価された構造活性相関（QSAR）を分類のために用いることもできる。こうした有意性評価された QSAR は、その適用性が十分に判定されている化学品に限るなら、分類基準に関して修正することなく使用できよう。強酸、強塩基、金属錯体、および界面活性物質等の物質では、K_{ow} の分析測定のかわりに、QSAR による K_{ow} の推定値、または n−オクタノールならびに水に対する個々の溶解度に基づいた推定値を求めるべきである。

A9.5.5.4 データが利用可能であるが、有意性評価がなされていないならば、専門家の判断を採用すべきである。

A9.5.5.5 したがって、物質に水生生物における生物濃縮性があるかどうかは、以下のスキームにしたがって判定できる:

(a) 有効/高品質な実験から得られた BCF 値→ あり：

　(i) BCF≧500：*その物質には生物濃縮の可能性がある*
　(ii) BCF＜500：*その物質には生物濃縮の可能性はない。*

(b) Valid/high quality experimentally determined BCF value = NO:

Valid/high quality experimentally determined log K_{ow} value =YES:

(i) log $K_{ow} \geq 4$: *The substance has a potential for bioconcentration*
(ii) log $K_{ow} < 4$: *The substance does not have a potential for bioconcentration.*

(c) Valid/high quality experimentally determined BCF value = NO:

Valid/high quality experimentally determined log K_{ow} value =NO:

Use of validated QSAR for estimating a log K_{ow} value = YES:

(i) log $K_{ow} \geq 4$: *The substance has a potential for bioconcentration*
(ii) log $K_{ow} < 4$: *The substance does not have a potential for bioconcentration.*

A9.6   Use of QSAR

A9.6.1   *History*

A9.6.1.1   Quantitative Structure-Activity Relationships (QSAR) in aquatic toxicology can be traced to the work of Overton in Zürich (Lipnick, 1986) and Meyer in Marburg (Lipnick, 1989a). They demonstrated that the potency of substances producing narcosis in tadpoles and small fish is in direct proportion to their partition coefficients measured between olive oil and water. Overton postulated in his 1901 monograph "Studien über die Narkose," that this correlation reflects toxicity taking place at a standard molar concentration or molar volume within some molecular site within the organism (Lipnick, 1991a). In addition, he concluded that this corresponds to the same concentration or volume for a various organisms, regardless of whether uptake is from water or via gaseous inhalation. This correlation became known in anaesthesia as the Meyer-Overton theory.

A9.6.1.2   Corwin Hansch and co-workers at Pomona College proposed the use of n-octanol/water as a standard partitioning system, and found that these partition coefficients were an additive, constitutive property that can be directly estimated from chemical structure. In addition, they found that regression analysis could be used to derive QSAR models, providing a statistical analysis of the findings. Using this approach, in 1972 these workers reported 137 QSAR models in the form log $(1/C) = A$ log $K_{ow} + B$, where K_{ow} is the n-octanol/water partition coefficient, and C is the molar concentration of a chemical yielding a standard biological response for the effect of simple non-electrolyte non-reactive organic compounds on whole animals, organs, cells, or even pure enzymes. Five of these equations, which relate to the toxicity of five simple monohydric alcohols to five species of fish, have almost identical slopes and intercepts that are in fact virtually the same as those found by Könemann in 1981, who appears to have been unaware of Hansch's earlier work. Könemann and others have demonstrated that such simple non-reactive non-electrolytes all act by a narcosis mechanism in an acute fish toxicity test, giving rise to minimum or baseline toxicity (Lipnick, 1989b).

A9.6.2   *Experimental artifacts causing underestimation of hazard*

A9.6.2.1   Other non-electrolytes can be more toxic than predicted by such a QSAR, but not less toxic, except as a result of a testing artefact. Such testing artefacts include data obtained for compounds such as hydrocarbons which tend to volatilize during the experiment, as well as very hydrophobic compounds for which the acute testing duration may be inadequate to achieve steady state equilibrium partitioning between the concentration in the aquatic phase (aquarium test solution), and the internal hydrophobic site of narcosis action. A QSAR plot of log K_{ow} *vs* log C for such simple non-reactive non-electrolytes exhibits a linear relationship so long as such equilibrium is established within the test duration. Beyond this point, a bilinear relationship is observed, with the most toxic chemical being the one with the highest log K_{ow} value for which such equilibrium is established (Lipnick, 1995).

A9.6.2.2   Another testing problem is posed by water solubility cut-off. If the toxic concentration required to produce the effect is above the compound's water solubility, no effect will be observed even at water saturation. Compounds for which the predicted toxic concentration is close to water solubility will also show no effect if the test duration is insufficient to achieve equilibrium partitioning. A similar cut-off is observed for surfactants if toxicity is predicted at a concentration beyond the critical micelle concentration. Although such compounds may show no toxicity under these conditions when tested alone, their toxic contributions to mixtures are still present. For compounds with the same log K_{ow} value, differences in water solubility reflect differences in enthalpy of fusion related to melting point. Melting point is a reflection of the degree of stability of the crystal lattice and is controlled by intermolecular hydrogen bonding, lack of conformational flexibility, and symmetry. The more highly symmetric a compound, the higher the melting point (Lipnick, 1990).

(b)　有効/高品質な実験から得られた BCF 値→　なし：

有効/高品質な実験から得られた log K$_{ow}$ 値→　あり：

(i)　log K$_{ow}$ ≧4：その物質には*生物濃縮の可能性がある*
(ii)　log K$_{ow}$ <4：その物質には*生物濃縮の可能性はない*

(c)　有効/高品質な実験から得られた BCF 値→　なし：

有効/高品質な実験から得られた log K$_{ow}$ 値→　なし：

log K$_{ow}$ 値の推定に有意性の評価された QSAR を使用→　使用可：

(i)　log K$_{ow}$ ≧4：その物質には*生物濃縮の可能性がある*
(ii)　log K$_{ow}$ <4：その物質には*生物濃縮の可能性はない*。

A9.6　QSAR の使用

A9.6.1　*経緯*

A9.6.1.1　水生毒性分野における定量的構造活性相関（QSAR）は、チューリッヒの Overton（Lipnick, 1986）およびマールブルグの Meyer（Lipnick, 1989a）の研究にまで遡ることができる。彼らは、物質がオタマジャクシおよび小型魚類に麻酔作用を及ぼす効力が、オリーブオイルと水の間で測定した分配係数と直接比例していることを示した。Overton は 1901 年に著したモノグラフ"Studien über die Narkose（麻酔に関する研究）"で、このような相関性は、生物体内の何らかの分子的な部位において標準モル濃度またはモル容積で生じている毒性を反映していると主張した（Lipnick, 1991a）。さらに、取り込みが水からなのか気体吸入によるのかに関係なく、このことは多様な生物に対して同一濃度または容積に相当すると結論づけた。この相関性は麻酔学分野では Meyer－Overton 理論として知られるようになった。

A9.6.1.2　ポモナ大学の Corwin Hansch および共同研究者らは、n－オクタノール/水を標準的な分配システムとして使用することを提案し、そしてこの分配係数は化学構造から直接推定できる、加算的かつ構成的な性質であることを発見した。さらに、試験結果の統計解析を前提として、QSAR を導くために回帰分析が利用できることを見出した。このアプローチを用いて、これらの研究者らは 1972 年に、log (1/C) = A log K$_{ow}$ + B の形をとった 137 の QSAR モデルを報告した。この場合 K$_{ow}$ は n－オクタノール/水分配係数、C は非電解質かつ非反応性の単純な有機化合物が、動物の全身、臓器、細胞、場合によっては純粋な酵素に及ぼす影響に対する、標準的な生物反応を生じる化学品のモル濃度である。これらの公式のうち 5 つは 5 種類の単純な一価アルコールの毒性を 5 種類の魚類と関連づけたもので、ほぼ同一の勾配および切片を有し、Könemann によって 1981 年に発見されたものと実質上、一致している。後者は Hansch の先行研究を知らなかったように思われる。Könemann らは、このような単純な非反応性かつ非電解性の物質はすべて、魚類を用いた急性毒性試験では麻酔薬のメカニズムで作用して、最小のまたはベースラインの毒性となることを示した（Lipnick, 1989b）。

A9.6.2　*有害性の過小評価を起こす実験技術上の誤差*

A9.6.2.1　それ以外の非電解性の物質は、こうした QSAR で予測されるよりも毒性が高いことはあるが、試験の技術的な誤差を除けば、毒性が低くなることはない。こうした試験の技術的な誤差としては、実験中に揮発しがちな炭化水素等の化合物、あるいは急性の試験期間では、水相（水槽中の試験溶液）中の濃度と、麻酔作用を起こす体内の疎水性部位との間に定常状態の平衡的な分配を達成するのに至っていない、非常に親水性の高い化合物、などから得られたデータがある。このような単純な非反応性かつ非電解性の物質の log K$_{ow}$ と log C の QSAR プロットは、試験期間中にこうした平衡に達している限りは線形の関係を示す。この点を超えると、双線形の関係が観察され、最も毒性の高い物質は平衡が達成され log K$_{ow}$ が最高となる物質である（Lipnick, 1995）。

A9.6.2.2　試験の際のもう 1 つの問題とは、水溶解性によるカットオフによって起こるものである。影響を生じるのに必要な毒性濃度が、その化合物の水溶解度より高いと、水に飽和していたとしても、何の影響も観察されないであろう。予測される毒性濃度が水溶解度に近い化合物もまた、試験期間が平衡的な分配が達成するのに十分でないときは、影響は見られないであろう。界面活性剤についても、臨界ミセル濃度より高い濃度で毒性が予測されるなら、同様なカットオフが観察される。こうした化合物は、個別に試験した場合にはこれらの条件下で毒性が認められないこともあるが、混合物の毒性に対する寄与はなお存在している。log K$_{ow}$ 値が同じ化合物でも、水に対する溶解性の差は、融点に関連する融合エンタルピーの違いを反映する。融点は、結晶格子の安定度を反映するものであり、分子間水素結合、コンホメーションの柔軟性の欠如、および対称性によって左右される。化合物は対称的であるほど、融点は高くなる（Lipnick, 1990）。

A9.6.3   QSAR modelling issues

A9.6.3.1   Choosing an appropriate QSAR implies that the model will yield a reliable prediction for the toxicity or biological activity of an untested chemical. Generally speaking, reliability decreases with increasing complexity of chemical structure, unless a QSAR has been derived for a narrowly defined set of chemicals similar in structure to the candidate substance. QSAR models derived from narrowly defined classes of chemicals are commonly employed in the development of pharmaceuticals once a new lead compound is identified and there is a need to make minor structural modifications to optimize activity (and decrease toxicity). Overall, the objective is to make estimates by interpolation rather than extrapolation.

A9.6.3.2   For example, if 96-h LC_{50} test data for fathead minnow are available for ethanol, n-butanol, n-hexanol, and n-nonanol, there is some confidence in making a prediction for this endpoint for n-propanol and n-pentanol. In contrast, there is would have less confidence in making such a prediction for methanol, which is an extrapolation, with fewer carbon atoms than any of the tested chemicals. In fact, the behaviour of the first member of such a homologous is typically the most anomalous and should not be predicted using data from remaining members of the series. Even the toxicity of branched chain alcohols may be an unreasonable extrapolation, depending upon the endpoint in question. Such extrapolation becomes more unreliable to the extent that toxicity is related to production of metabolites for a particular endpoint, as opposed to the properties of the parent compound. Also, if toxicity is mediated by a specific receptor binding mechanism, dramatic effects may be observed with small changes in chemical structure.

A9.6.3.3   What ultimately governs the validity of such predictions is the degree to which the compounds used to derive the QSAR for a specific biological endpoint, are acting by a common molecular mechanism. In many and perhaps most cases, a QSAR does not represent such a mechanistic model, but merely a correlative one. A truly valid mechanistic model must be derived from a series of chemicals all acting by a common molecular mechanism and fit to an equation using one or more parameters that relate directly to one or more steps of the mechanism in question. Such parameters or properties are more generally known as molecular descriptors. It is also important to keep in mind that many such molecular descriptors in common use may not have a direct physical interpretation. For a correlative model, the statistical fit of the data are likely to be poorer than a mechanistic one given these limitations. Mechanisms are not necessarily completely understood, but enough information may be known to provide confidence in this approach. For correlative models, the predictive reliability increases with the narrowness with which each is defined, e.g. categories of electrophiles, such as acrylates, in which the degree of reactivity may be similar and toxicity can be estimated for a "new" chemical using a model based solely on the log K_{ow} parameter.

A9.6.3.4   As an example, primary and secondary alcohols containing a double or triple bond that is conjugated with the hydroxyl function (i.e. allylic or propargylic) are more toxic than would be predicted for a QSAR for the corresponding saturated compounds. This behaviour has been ascribed to a proelectrophile mechanism involving metabolic activation by the ubiquitous enzyme alcohol dehydrogenase to the corresponding α,β-unsaturated aldehydes and ketones which can act as electrophiles via a Michael-type acceptor mechanism (Veith et al., 1989). In the presence of an alcohol dehydrogenase inhibitor, these compounds behave like other alcohols and do not show excess toxicity, consistent with the mechanistic hypothesis.

A9.6.3.5   The situation quickly becomes more complex once one goes beyond such a homologous series of compounds. Consider, for example, simple benzene derivatives. A series of chlorobenzenes may be viewed as similar to a homologous series. Not much difference is likely in the toxicities of the three isomeric dichlorobenzenes, so that a QSAR for chlorobenzenes based upon test data for one of these isomers is likely to be adequate. What about the substitution of other functional groups on benzene ring? Unlike an aliphatic alcohol, addition of a hydroxyl functionality to a benzene ring produces a phenol which is no longer neutral, but an ionizable acidic compound, due to the resonance stabilization of the resulting negative charge. For this reason, phenol does not act as a true narcotic agent. With the addition of electron withdrawing substituents to phenol (e.g. chlorine atoms), there is a shift to these compounds acting as uncouplers of oxidative phosphorylation (e.g. the herbicide dinoseb). Substitution of an aldehyde group leads to increased toxicity via an electrophile mechanism for such compounds react with amino groups, such as the lysine ε-amino group to produce a Schiff Base adduct. Similarly, a benzylic chloride acts as an electrophile to form covalent abducts with sulfhydryl groups. In tackling a prediction for an untested compound, the chemical reactivity of these and many other functional groups and their interaction with one another should be carefully studied, and attempts made to document these from the chemical literature (Lipnick, 1991b).

A9.6.3.6   Given these limitations in using QSARs for making predictions, it is best employed as a means of establishing testing priorities, rather than as a means of substituting for testing, unless some mechanistic information is available on the untested compound itself. In fact, the inability to make a prediction along with known environmental release and exposure may in itself be adequate to trigger testing or the development of a new QSAR for a class of chemicals for which such decisions are needed. A QSAR model can be derived by statistical analysis, e.g. regression analysis, from such a data set. The most commonly employed molecular descriptor, log K_{ow}, may be tried as a first attempt.

A9.6.3　QSAR モデル化の課題

A9.6.3.1　適切な QSAR を選択するということは、そのモデルが、未試験の化学品の毒性または生物活性について、信頼できる予測値を与えるということを意味している。一般的に言えば、対象物質に構造が類似している化学品の厳密に定義されたセットから QSAR が導かれているのでなければ、化学構造の複雑さが増すほど信頼性は低下する。厳密に定義された化学品群から導かれた QSAR モデルは、医薬品の開発に一般的に採用されており、例えば、新しいリード化合物が見つかっており、活性の最適化（および毒性の低下）をさせるため、僅かな構造の修正を行いたい場合に用いられる。全体的に、その目標は外挿よりむしろ内挿による推定を行うことである。

A9.6.3.2　例えば、エタノール、n－ブタノール、n－ヘキサノール、および n－ノナノールについて、ファットヘッドミノーを用いた 96 時間の LC50 試験データが揃っているならば、n－プロパノールおよび n－ペンタノールについて、このエンドポイントの毒性値を、ある程度の信頼度をもって予測できる。反対にメタノールでは、試験されたどの物質よりも炭素原子数が少なく、外挿となるので、予測の信頼性は劣ることになろう。実際、このような同族体グループにおける最初のメンバー物質の挙動は、一般的に最も変則的であり、その系列の他のメンバー物質からのデータを用いて予測すべきではない。分岐型アルコールの毒性でさえ、問題とされるエンドポイントによっては、不合理な外挿になる。こうした外挿結果は、親化合物の性質に反して、特定のエンドポイントに対する毒性が代謝物の生成に関与する程度に応じて、さらに信頼性が低くなる。また、毒性が特定の受容体結合メカニズムに介在されている場合、化学構造がわずかに変化しただけで、劇的な作用が観察されることもある。

A9.6.3.3　こうした予測結果の有意性を究極的に支配するのは、ある特定の生物学的なエンドポイントについて QSAR を導き出すのに用いられた化合物群が、共通の分子機構によってどの程度まで作用しているか、である。多くの、そして恐らくほとんどの例で、QSAR はこうした機構モデルを示すことはなく、単に相関的なモデルを示すだけである。真に有意な機構モデルは、共通の分子機構により作用する、一連の化学品群から導かれるはずであり、問題となっている機構の一段階または複数段階に直接関連している 1 個または複数のパラメータを用いた公式に適合するはずである。このようなパラメータまたは特性はより一般的には分子的ディスクリプターとして知られている。また、こうした共通に使用される分子的ディスクリプターの多くは、直接の物理的な解釈がない場合もあることを留意しておくことも重要である。相関モデルについては、こうした限界を考えると、データの統計的適合性が機構モデルよりも劣るように思われる。機構は完全には把握されていなくてもよいが、このアプローチにおいて信頼性をもてるだけの十分な情報がわかっていることもある。相関モデルの場合、各モデルが定義される厳密さが増すほど、その予測の信頼性も増す。例えばアクリレート等の求電子物質のカテゴリーを指定すれば、この範囲内では反応性の程度が同様であり、「新規」化学品について、log K_{ow} パラメータのみにもとづいたモデルを用いて毒性の推定が可能である。

A9.6.3.4　一例として、水酸基と共役している二重結合または三重結合を含む一級および二級アルコール（すなわちアリルまたはプロパギルアルコール）は、対応する飽和化合物に関する QSAR に対して予測されるよりも毒性が高い。このような挙動は、至る所に存在する酵素であるアルコールデヒドロゲナーゼによる代謝活性化によって、対応する α, β －不飽和アルデヒドおよびケトンに代謝され、これがミカエル型のアクセプター機構により求電子物質として作用するという求電子物質前駆体機構によるものとされている（Veith ら、1989）。アルコールデヒドロゲナーゼ阻害物質が存在すると、これらの化合物は他のアルコール類と同様に挙動し、機構による仮説と一致して、過剰な毒性を示さない。

A9.6.3.5　こうした一連の類似化合物から外れると、状況は一気に、さらに複雑化する。例えば、単純なベンゼン誘導体を考えてみる。クロロベンゼン類は一連の類似化合物と同様であるとみなしてよい。3 種類のジクロロベンゼン異性体の毒性にはそれほど大きな違いがあるとは思われないため、これら異性体のうち 1 つに関する試験データをもとにしたクロロベンゼン類に関する QSAR は妥当であるように思われる。ベンゼン環に他の官能基が置換された場合はどうか？脂肪族アルコールとは異なり、水酸基をベンゼン環に付加するとフェノールが生成されて、もはや中性ではなく、結果的に生じる負の電荷が共鳴により安定化することにより、イオン化性のある酸性化合物となる。この理由で、フェノールは真の麻酔薬として作用しない。フェノールに電子吸引性の置換基（例えば塩素原子）を付加すると、こうした化合物は酸化的リン酸化の脱共役剤として作用する化合物となる（例えば除草剤のジノセブ）。アルデヒド基を置換すると、こうした化合物は、リジンの ε －アミノ基のような、アミノ基と反応してシッフ塩基付加物を生成するので、求電子機構により毒性が増加する。同様にベンジルクロライドは、求電子物質としてスルフヒドリル基と共有結合付加物を生成する。試験していない化合物の予測を試みる際には、これらの、ならびに他の多くの官能基の化学的反応性と、それらの間の相互作用について注意深く検討すべきであり、これらを化学文献から証拠揃えする努力を払わねばならない（Lipnick, 1991b）。

A9.6.3.6　予測を行うのに QSAR を用いるには限界があることから、まだ試験していない化合物自体について何らかの機構についての情報が入手されていない限り、試験の代替手段としてよりも、試験優先順位設定の手段として採用するのが最善である。実際、環境放出および環境ばく露が分かっていても予測が不可能であるということ自体が、試験を実施する、あるいは判定が必要な化学品クラスについて新しい QSAR を開発するきっかけとなるであろう。こうしたデータセットからの統計解析、例えば回帰分析等によって、QSAR モデルを導くことができる。最も一般的に採用されている分子的指標である log K_{ow} を、最初の企てとして試みてもよい。

A9.6.3.7 By contrast, derivation of a mechanism based QSAR model requires an understanding or working hypothesis of molecular mechanism and what parameter or parameters would appropriately model these actions. It is important to keep in mind that this is different from a hypothesis regarding mode of action, which relates to biological/physiological response, but not molecular mechanism.

A9.6.4 *Use of QSARs in aquatic classification*

A9.6.4.1 The following inherent properties of substances are relevant for classification purposes concerning the aquatic environment:

(a) partition coefficient n-octanol-water log K_{ow};

(b) bioconcentration factor BCF;

(c) degradability - abiotic and biodegradation;

(d) acute aquatic toxicity for fish, daphnia and algae;

(e) prolonged toxicity for fish and daphnia.

A9.6.4.2 Test data always take precedence over QSAR predications, providing the test data are valid, with QSARs used for filling data gaps for purposes of classification. Since the available QSARs are of varying reliability and application range, different restrictions apply for the prediction of each of these endpoints. Nevertheless, if a tested compound belongs to a chemical class or structure type (see above) for which there is some confidence in the predictive utility of the QSAR model, it is worthwhile to compare this prediction with the experimental data, as it is not unusual to use this approach to detect some of the experimental artefacts (volatilization, insufficient test duration to achieve equilibrium, and water solubility cut-off) in the measured data, which would mostly result in classifying substances as lower than actual toxicity.

A9.6.4.3 When two or more QSARs are applicable or appear to be applicable, it is useful to compare the predictions of these various models in the same way that predicted data should be compared with measured (as discussed above). If there is no discrepancy between these models, the result provides encouragement of the validity of the predictions. Of course, it may also mean that the models were all developed using data on similar compounds and statistical methods. On the other hand, if the predictions are quite different, this result needs to be examined further. There is always the possibility that none of the models used provides a valid prediction. As a first step, the structures and properties of the chemicals used to derive each of the predictive models should be examined to determine if any models are based upon chemicals similar in both of these respects to the one for which a prediction is needed. If one data set contains such an appropriate analogue used to derive the model, the measured value in the database for that compound vs model prediction should be tested. If the results fit well with the overall model, it is likely the most reliable one to use. Likewise, if none of the models contain test data for such an analogue, testing of the chemical in question is recommended.

A9.6.4.4 The U.S. EPA has recently posted a draft document on its website "Development of Chemical Categories in the HPV Challenge Program," that proposes the use of chemical categories to "... voluntarily compile a Screening Information Data Set (SIDS) on all chemicals on the US HPV list ... [to provide] basic screening data needed for an initial assessment of the physicochemical properties, environmental fate, and human and environmental effects of chemicals" (US EPA, 1999). This list consists of "...about 2,800 HPV chemicals which were reported for the Toxic Substances Control Act's 1990 Inventory Update Rule (IUR)".

A9.6.4.5 One approach being proposed "...where this is scientifically justifiable ... is to consider closely related chemicals as a group, or category, rather than test them as individual chemicals. In the category approach, not every chemical needs to be tested for every SIDS endpoint". Such limited testing could be justified providing that the "...final data set must allow one to assess the untested endpoints, ideally by interpolation between and among the category members." The process for defining such categories and in the development of such data are described in the proposal.

A9.6.4.6 A second potentially less data intensive approach being considered (US EPA, 2000a) is "... applying SAR principles to a single chemical that is closely related to one or more better characterized chemicals ("analogs")." A third approach proposed consists of using "... a combination of the analogue and category approaches ... [for] individual chemicals ... [similar to that] used in ECOSAR (US EPA, 2000b), a SAR-based computer program that generates ecotoxicity values. ". The document also details the history of the use of SARs within the U.S. EPA new chemicals program, and how to go about collecting and analysing data for the sake of such SAR approaches.

A9.6.3.7　反対に、メカニズムに基づいたQSARモデルの導出には、分子機構、およびどのようなパラメータがこうした作用を適切にモデル化できるか、の理解または作業仮説が必要である。これは、分子機構ではなく、生物学的/生理学的反応と関連づけられた、作用様式に関する仮説とは違うことを留意しておくことが重要である。

A9.6.4　水生環境有害性分類へのQSARの使用

A9.6.4.1　水生環境に関する分類の目的には、以下のような物質本来の特性が関係する：

(a) n－オクタノール/水分配係数 log K_{ow} ；

(b) 生物濃縮係数 BCF ；

(c) 分解性－非生物的および生物的分解；

(d) 魚類、ミジンコおよび藻類に対する急性水生毒性；

(e) 魚類およびミジンコに対する長期毒性。

A9.6.4.2　試験データが有効であり、QSARは分類のためのデータの欠損を補うのに用いられるという前提で、常に試験データの方がQSARによる予測より優先される。利用できるQSARの信頼性および適用範囲は多様であるので、これらエンドポイントそれぞれの予測にさまざまな制限が適用される。それにもかかわらず、試験された化合物が、あるQSARモデルの予測のための有用性にある程度信頼できるような化学品のクラスまたは構造タイプ（上記参照）に属しているならば、この予測結果を実験データと比較してみる価値がある。これは、測定データ中の実験技術上の誤差（揮発、平衡に達するのに不十分な試験期間、水溶解度によるカットオフ）を検出するのに、このようなアプローチを用いることは珍しいことではないからである。実験技術上の誤差の多くは物質を実際の毒性より低く分類してしまう原因となる。

A9.6.4.3　2種類以上のQSARが適用できるか、またはそう思われる場合には、予測データを実測データと（上述のように）比較するのと同じように、さまざまなモデルの予測を比較することは有用である。それらのモデル間に相違がなければ、その結果は予測の有意性を高めることになる。もちろんこれは、すべてのモデルが類似した化合物のデータと統計的方法を用いて開発されたことも意味している。他方、予測値が全く違っていたならば、結果についてさらに検証する必要がある。用いたモデルのいずれも有効な予測値を与えなかった可能性も常にある。第一段階として、各予測モデルを導いた化学品の構造および特性を検証して、こうしたモデルは、予測が必要とされる構造および特性の両方に関して類似している化学品をもとに作成されているかどうかを判定しなければならない。あるデータセットに、そのモデルの導出に用いられた適切な類似物質が含まれていれば、モデルの予測値に対してその化合物のデータベース中の測定値を検証すべきである。結果がモデル全体に十分適合するのであれば、それが最も信頼できるモデルであると思われる。また、そのような類似物質に関する試験データが含まれているモデルがないならば、問題の化学品を試験することが推奨される。

A9.6.4.4　米国EPAは最近ウェブサイトに「HPVチャレンジプログラムにおける化学品カテゴリーの策定」というドラフト文書を提示し、「…米国HPVリストにあるすべての化学品のスクリーニング情報データセット（SIDS）を自主的に編纂し…化学品の物理化学的性質、環境運命、および人と環境に対する影響の初期評価に必要な基礎的スクリーニングデータを［提供する］」ために、化学品区分を用いることを提案している（US EPA 1999）。このリストは「毒性物質管理法の1990年インベントリー更新規則（IUR）のために報告された約2,800種のHPV化学品」からなっている。

A9.6.4.5　提案されている1つのアプローチは「…科学的に正当であるならば…密接な関連性のある化学品を、個々の化学品として試験するよりも、グループとして、または区分として考えることである。この区分によるアプローチでは、SIDSのエンドポイントごとにあらゆる化学品を試験する必要はない。」というものである。こうした限定的な試験が適正であると判断されるのは「…最終データセットは、理想的には区分に含まれる物質間の内挿「ここに強調を付言」により、まだ試験していないエンドポイントの評価を可能にするものでなければならない」ことが前提となっている。こうした区分を定義し、データを作成するプロセスが提案のなかで説明されている。

A9.6.4.6　検討されている第二の、それほどデータにたよらないアプローチ（US EPA, 2000a）は「…より詳しく特性化されている化学品（「類似物質」）に密接に関連した単一化学品にSAR原則を適用する」ことである。提案されている第三のアプローチは、「…生態毒性値を作成するSARにもとづくコンピュータプログラムである、ECOSAR（US EPA, 2000b）に用いられている［のと類似している］個々の化学品［に対する］…類似物質によるアプローチと区分によるアプローチの組み合わせ…」を用いる方法である。この資料ではまた、米国EPA新規化学品プログラム内でのSAR使用の経緯、およびこうしたSARアプローチのためにどのようにデータを収集し解析するかについて詳述している。

A9.6.4.7   The Nordic Council of Ministers issued a report (Pederson et al., 1995) entitled "Environmental Hazard Classification," that includes information on data collection and interpretation, as well as a section (5.2.8) entitled "QSAR estimates of water solubility and acute aquatic toxicity". This section also discusses the estimation of physicochemical properties, including log K_{ow}. For the sake of classification purposes, estimation methods are recommended for prediction of "minimum acute aquatic toxicity," for "...neutral, organic, non-reactive and non-ionizable compounds such as alcohols, ketones, ethers, alkyl, and aryl halides, and can also be used for aromatic hydrocarbons, halogenated aromatic and aliphatic hydrocarbons as well as sulphides and disulphides," as cited in an earlier OECD Guidance Document (OECD, 1995). The Nordic document also includes diskettes for a computerized application of some of these methods.

A9.6.4.8   The European Centre for Ecotoxicology and Toxicology of Chemicals (ECETOC) has published a report entitled "QSARs in the Assessment of the Environmental Fate and Effects of Chemicals," which describes the use of QSARs to "...check the validity of data or to fill data gaps for priority setting, risk assessment and classification" (ECETOC, 1998). QSARs are described for predicting environmental fate and aquatic toxicity. The report notes that "a consistent dataset for [an endpoint] covered ... for a well-defined scope of chemical structures ("domain") [is needed] ... from which a training set is developed. The document also discusses the advantage of mechanism-based models, the use of statistical analysis in the development of QSARs, and how to assess "outliers".

A9.6.4.9   *Octanol-water-partition coefficient (K_{ow})*

A9.6.4.9.1   Computerized methods such as CLOGP (US EPA, 1999), LOGKOW (US EPA, 2000a) and SPARC (US EPA, 2000b) are available to calculate log Kow directly from chemical structure. CLOGP and LOGKOW are based upon the addition of group contributions, while SPARC is based upon a more fundamental chemical structure algorithm. Caution should be used in using calculated values for compounds that can undergo hydrolysis in water or some other reaction, since these transformations need to be considered in the interpretation of aquatic toxicity test data for such reactive chemicals. Only SPARC can be employed in a general way for inorganic or organometallic compounds. Special methods are needed in making estimates of log K_{ow} or aquatic toxicity for surface-active compounds, chelating compounds, and mixtures.

A9.6.4.9.2   Values of log K_{ow} can be calculated for pentachlorophenol and similar compounds, both for the ionized and unionized (neutral) forms. These values can potentially be calculated for certain reactive molecules (e.g. benzotrichloride), but the reactivity and subsequent hydrolysis also need to be considered. Also, for such ionizable phenols, pKa is a second parameter. Specific models can be used to calculate log K_{ow} values for organometallic compounds, but they need to be applied with caution since some of these compounds really exist in the form of ion pairs in water.

A9.6.4.9.3   For compounds of extremely high lipophilicity, measurements up to about 6 to 6.5 can be made by shake flask and can be extended up to about log K_{ow} of 8 using the slow stirring approach (Bruijn et al., 1989). Calculations are considered useful even in extrapolating beyond what can be measured by either of these methods. Of course, it should be kept in mind that if the QSAR models for toxicity, etc. are based on chemicals with lower log K_{ow} values, the prediction itself will also be an extrapolation; in fact, it is known that in the case of bioconcentration, the relationship with log K_{ow} becomes non-linear at higher values. For compounds with low log K_{ow} values, the group contribution can also be applied, but this is not very useful for hazard purposes since for such substances, particularly with negative log K_{ow} values, little if any partitioning can take place into lipophilic sites and as Overton reported, these substances produce toxicity through osmotic effects (Lipnick, 1986).

A9.6.4.10   *Bioconcentration factor BCF*

A9.6.4.10.1   If experimentally determined BCF values are available, these values should be used for classification. Bioconcentration measurements must be performed using pure samples at test concentrations within water solubility, and for an adequate test duration to achieve steady state equilibrium between the aqueous concentration and that in the fish tissue. Moreover, with bioconcentration tests of extended duration, the correlation with log K_{ow} levels off and ultimately decreases. Under environmental conditions, bioconcentration of highly lipophilic chemicals takes place by a combination of uptake from food and water, with the switch to food taking place at log $K_{ow} \approx 6$. Otherwise log K_{ow} values can be used with a QSAR model as a predictor of the bioaccumulation potential of organic compounds. Deviations from these QSARs tend to reflect differences in the extent to which the chemicals undergo metabolism in the fish. Thus, some chemicals, such as phthalate, can bioconcentrate significantly less than predicted for this reason. Also, caution should be applied in comparing predicted BCF values with those using radiolabeled compounds, where the tissue concentration thus detected may represent a mix of parent compound and metabolites or even covalently bound parent or metabolite.

A9.6.4.7　北欧閣僚会議は「環境有害性分類」と題する報告書を発行した（Pederson ら, 1995）。これはデータ収集および解釈に関する情報と共に、「水に対する溶解性および急性水生毒性の QSAR 推定」という題の節（5.2.8）も記載されている。この節では、log K$_{ow}$ も含めた物理化学的性質の推定についても論じている。分類目的のためには、先の OECD ガイドライン（OECD, 1995）に引用されているように、「最少の急性水生毒性」予測のための推定方法が推奨されており、これは、「アルコール、ケトン、エーテル、ハロゲン化アルキルおよびハロゲン化アリール等、中性の、非反応性・非イオン化性の有機化合物…、および芳香族炭化水素、ハロゲン化された芳香族および脂肪族炭化水素、ならびにスルフィドやジスルフィドにも用いることが可能である」。この北欧の文書には、これらの方法のいくつかをコンピュータで利用するためのディスクも含まれている。

A9.6.4.8　欧州化学品生態毒性および毒性センター（ECETOC）は「化学品の環境運命および影響の評価における QSAR」と題する報告書を発行している。これは QSAR を「…データの有意性の確認、または優先順位の設定、リスク評価および分類のためにデータの欠落を補充する」ために用いることについて述べている（ECETOC, 1998）。環境運命および水生毒性予測のための QSAR の説明がされている。この報告書は「…対象となっている［エンドポイント］のための一貫性のあるデータセットが…十分に定義された化学構造の範囲（ドメイン）［が必要であり］…それから訓練用セットを開発する」と述べている。この文書はまた、機構に基づいたモデルの長所、QSAR 開発における統計解析の使用、および「かけ離れたデータ」をどう評価するかに付いて論じている。

A9.6.4.9　*n−オクタノール/水分配係数（K$_{ow}$）*

A9.6.4.9.1　化学構造から直接 log K$_{ow}$ を計算するために、CLOGP（US EPA, 1999）、LOGKOW（US EPA, 2000a）、および SPARC（US EPA, 2000b）等のコンピュータ化された方法が利用できる。CLOGP および LOGKOW は官能基の関与の加算に基づいているのに対し、SPARC はより基礎的な化学構造アルゴリズムに基づいている。水中で加水分解またはその他の反応を受ける可能性のある化合物について計算値を使用する際には注意が必要である。こうした変化は、このような反応性化学品に関する水生毒性試験データを解釈する際に考慮されなければならない。無機化合物および有機金属化合物には、SPARC だけが一般的な方法で採用できる。界面活性物質、キレート形成化合物、および混合物の log K$_{ow}$ または水生毒性を推定するには、特別な方法が必要である。

A9.6.4.9.2　ペンタクロロフェノールおよび類似化合物については、イオン化および非イオン化（中性）の両方の形態について log K$_{ow}$ 値を計算できる。こうした数値は特定の反応性分子（例えばベンゾトリクロリド）について計算できる可能性もあるが、反応性およびその後の加水分解についても考慮する必要がある。また、こうしたイオン化しうるフェノールについては、pKa が第二のパラメータである。有機金属化合物の log K$_{ow}$ 値を計算するのに特定のモデルを用いることができるが、こうした化合物のあるものは、実際には水中でイオン対として存在しているものもあるので、注意して使用する必要がある。

A9.6.4.9.3　極めて親油性の高い化合物では、log K$_{ow}$ 6 から 6.5 までの測定はフラスコ振騰法で行えるが、低速撹拌法を用いれば log K$_{ow}$ を約 8 まで拡大できる（Bruijn ら, 1989）。これらの方法で測定できる範囲を越えて外挿する際にも、計算が有用であると考えられる。もちろん、毒性等に関する QSAR モデルが log K$_{ow}$ 値の低い化学品に基づいているならば、予測それ自体も外挿となることに留意しておく必要がある。実際、生物濃縮性については、数値が高いときには log K$_{ow}$ との関係が非線形になることが知られている。log K$_{ow}$ 値の低い化合物では、官能基の関与の考え方も適用できるが、そうした物質、特に log K$_{ow}$ 値がマイナスの物質では、親油性部位への分配が起ったとしても、僅かであり、Overton が述べているように、これらの物質は浸透圧作用によって毒性を生じるため、有害性分類目的にはそれほど有用ではない（Lipnick, 1986）。

A9.6.4.10　*生物濃縮係数 BCF*

A9.6.4.10.1　実験的に測定された BCF が利用可能なら、これらの数値を分類に用いるべきである。生物濃縮試験の測定は、純粋サンプルを用いて、水に対する溶解度の範囲内の試験濃度で、また水中濃度と魚組織中濃度の平衡が定常状態を達成するのに十分な試験期間で実施しなければならない。さらに、期間を延長して生物濃縮試験をすると、log K$_{ow}$ との相関性は水平となり、最終的には減少する。環境条件下では、親油性の高い化学品の生物濃縮は食物ならびに水からの取り込みの組み合せによって起こり、log K$_{ow}$≈6 で食物からの取り込みに切り替わる。そうでなければ、log K$_{ow}$ 値は QSAR モデルと共に、有機化合物の生物濃縮性を予測するパラメータとして用いることができる。こうした QSAR からのずれは、化学品が魚体内で代謝される程度の差を反映している傾向がある。したがって、フタレート等の化学品は、この理由によって生物濃縮が予測されるより著しく低くなることがある。さらに、BCF 予測値を、放射性化合物を用いた BCF 値と比較するには注意が必要である。なぜなら、こうして検出された組織中濃度は、親化合物と代謝物の混合、また共有結合した親化合物または代謝物の分も含んでいることあるからである。

A9.6.4.10.2   Experimental log K_{ow} values are to be used preferentially. However, older shake flask values above 5.5 are not reliable and in many cases, it is better to use some average of calculated values or to have these remeasured using the slow stirring method (Bruijn et al., 1989). If there is reasonable doubt about the accuracy of the measured data, calculated log K_{ow} values shall be used.

A9.6.4.11   *Degradability - abiotic and biodegradation*

QSARs for abiotic degradation in water phases are narrowly defined linear free energy relationships (LFERs) for specific classes of chemicals and mechanisms. For example, such LFERs are available for hydrolysis of benzylic chlorides with various substituents on the aromatic ring. Such narrowly defined LFER models tend to be very reliable if the needed parameters are available for the Substituent(s) in question. Photo degradation, i.e. reaction with UV produced reactive species, may be extrapolated from estimates for the air compartment. While these abiotic processes do not usually result in complete degradation of organic compounds, they are frequently significant starting points, and may be rate limiting. QSARs for calculating biodegradability are either compound specific (OECD, 1995) or group contribution models like the BIODEG program (Hansch and Leo, 1995; Meylan and Howard 1995; Hilal et al., 1994; Howard et al., 1992; Boethling et al., 1994; Howard and Meylan 1992; Loonen et al., 1999). While validated compound class specific models are very limited in their application range, the application range of group contribution models is potentially much broader but limited to compounds containing the model substructures. Validation studies have suggested that the biodegradability predictions by currently available group contribution models may be used for prediction of "not ready biodegradability" (Pedersen et al., 1995; Langenberg et al., 1996; USEPA, 1993) – and thus in relation to aquatic hazard classification "not rapid degradability."

A9.6.4.12   *Acute aquatic toxicity for fish, daphnia and algae*

The acute aquatic toxicity of non-reactive, non-electrolyte organic chemicals (baseline toxicity) can be predicted from their log K_{ow} value with a quite high level of confidence, provided the presence of electrophile, proelectrophile, or special mechanism functional groups (see above) were not detected. Problems remain for such specific toxicants, for which the appropriate QSAR has to be selected in a prospective manner. Since straightforward criteria for the identification of the relevant modes of action are still lacking, empirical expert judgement needs to be applied for selecting a suitable model. Thus, if an inappropriate QSAR is employed, the predictions may be in error by several orders of magnitude, and in the case of baseline toxicity, will be predicted less toxic, rather than more.

A9.6.4.13   *Prolonged toxicity for fish and Daphnia*

Calculated values for chronic toxicity to fish and Daphnia should not be used to overrule classification based on experimental acute toxicity data. Only a few validated models are available for calculating prolonged toxicity for fish and Daphnia. These models are based solely on log K_{ow} correlations and are limited in their application to non-reactive, non-electrolyte organic compounds, and are not suitable for chemicals with specific modes of action under prolonged exposure conditions. The reliable estimation of chronic toxicity values depends on the correct discrimination between non-specific and specific chronic toxicity mechanisms; otherwise, the predicted toxicity can be wrong by orders of magnitude. It should be noted that although for many compounds, excess toxicity[5] in a chronic test correlates with excess toxicity in an acute test, this is not always the case.

A9.7   Classification of metals and metal compounds

A9.7.1   *Introduction*

A9.7.1.1   The harmonized system for classifying substances is a hazard-based system, and the basis of the identification of hazard is the aquatic toxicity of the substances, and information on the degradation and bioaccumulation behaviour (OECD 1998). Since this document deals only with the hazards associated with a given substance when the substance is dissolved in the water column, exposure from this source is limited by the solubility of the substance in water and bioavailability of the substance in species in the aquatic environment. Thus, the hazard classification schemes for metals and metal compounds are limited to the hazards posed by metals and metal compounds when they are available (i.e. exist as dissolved metal ions, for example, as M^+ when present as $M\text{-}NO_3$), and do not take into account exposures to metals and metal compounds that are not dissolved in the water column but may still be bioavailable, such as metals in foods. This section does not take into account the non-metallic ion (e.g. CN-) of metal compounds which may be toxic. For such metal compounds the hazards of the non-metallic ions must also be considered.

---

[5]   *Excess toxicity, T_e = (Predicted baseline toxicity)/Observed toxicity.*

A9.6.4.10.2 　実験による log K$_{ow}$ を最も優先して使用すべきである。しかし、旧来のフラスコ振騰法では 5.5 以上の値は信頼できず、多くの場合、計算値の平均値を使用するか、または低速攪拌法で測定し直した方がよい（Bruijn ら, 1989）。測定データの精度に、理にかなった疑いがあるなら、log K$_{ow}$ 計算値を使用すべきである。

A9.6.4.11 　*分解性 － 非生物的分解および生分解*

　水相中の非生物的分解についての QSAR は、特定の化学品群およびメカニズムについての厳密に定義された線形自由エネルギー関係（LFER）である。例えば、こうした LFER は、芳香環にいろいろに置換基を有する塩化ベンジルの加水分解に利用できる。こうした厳密に定義された LFER モデルは、必要なパラメータが問題の置換基について利用できるなら、非常に信頼性が高くなる。光分解、すなわち UV により生じた反応性分子種との反応は、大気コンパートメントについて推定値から外挿できることもある。こうした非生物的プロセスは通常は有機化合物の完全分解までには至らないが、しばしば重要な開始点であり、また律速でありうる。生分解性を計算するための QSAR は、化合物に固有のもの（OECD, 1995）であるか、または BIODEG 等の官能基関与モデルのいずれかである（Hansch と Leo, 1995; Meylan と Howard, 1995; Hilal ら, 1994; Howard ら, 1992; Boethling ら, 1994; Howard と Meylan 1992; Loonen ら, 1999）。有意性の評価された化合物クラス固有のモデルは適用範囲が極めて限られているのに対し、官能基関与モデルの適用範囲はより広い可能性がある。しかし、モデルとなる下部構造を含む化合物に限定されている。有意性評価研究から、現在利用できる官能基関与モデルによる生物分解性予測は「易生分解性でない」ことの予測に用いられること（Pederson ら, 1995; Langenberg ら, 1996; US EPA, 1993）、したがって、水生有害性分類上「急速分解性でない」ことと関連づけられることが示唆されている。

A9.6.4.12 　*魚、ミジンコおよび藻類に対する急性水生毒性*

　非反応性かつ非電解質の有機化合物の急性水生毒性（ベースライン毒性）は、その log K$_{ow}$ 値から極めて高いレベルの信頼性をもって予測できるが、ただし求電子、前求電子、または特殊な機能をもつ官能基（上記参照）の存在が検出されなかったことが前提である。こうした特殊な毒性物質については、適切な QSAR を予測的な方法で選択しなければならないという問題が残る。これは、関連性のある作用機序を特定するための直接で簡単な判定基準がまだないので、適切なモデルを選択するために、専門家の経験的判断を用いる必要があるからである。したがって、もし適切でない QSAR が採用されると、予測値に何桁もの誤差が生じ、ベースライン毒性の場合には、毒性が高い方よりむしろ低い方に予測されることになろう。

A9.6.4.13 　*魚およびミジンコに対する長期毒性*

　魚およびミジンコに対する慢性毒性の計算値を、急性毒性実験データに基づいた分類を否定するのに用いてはならない。魚およびミジンコに対する長期毒性を計算するのに利用できる有意性判定済みのモデルは少ししかない。これらのモデルは、log K$_{ow}$ 相関関係のみに基づいており、その適用性は非反応性、非電解質の有機化合物に限られ、また、長期ばく露条件下で特殊な作用機序をもつ化学品には適していない。信頼できる慢性毒性値の推定は、慢性毒性機構を非特異的なものと特異的なものの間で正しく区別することに依存している。さもないと、予測された毒性は何桁も誤る可能性がある。多くの化合物では慢性試験で得られた過剰毒性[5]が急性試験での過剰毒性と相関している場合があるものの、必ずしもそれが該当しない場合もあることに、注意しなければならない。

A9.7 　金属および金属化合物の分類

A9.7.1 　*序*

A9.7.1.1 　物質を分類するための調和システムは有害性に基づくシステムであり、有害性を特定する根拠は物質の水生毒性、ならびに分解性および生物蓄積作用についての情報である（OECD 1998）。本文書は、ある物質が水相中に溶解している際の、その物質に伴う有害性のみを扱っているので、これに由来するばく露は、当該物質の水への溶解度と、水生環境中の生物種における当該物質の生物学的利用能によって制限される。したがって、金属および金属化合物に関する有害性分類スキームも、金属および金属化合物が利用されうる（すなわち例えば M-NO$_3$ として存在する場合の M$^+$ のように、溶存金属イオンとして存在する）場合に示される有害性に限定され、食物中の金属のように、水相中には溶解していないが、なお生物学的に利用されうるであろう金属および金属化合物に対するばく露を考慮に入れるものではない。本節では、毒性を持つ可能性のある金属化合物の非金属イオン（例えば CN-）は考慮しない。このような金属化合物については、その非金属イオンの有害性についても検討しなければならない。

---

[5] 　*過剰毒性、T$_e$＝（ベースライン毒性予測値）/毒性実測値。*

A9.7.1.1.1   Organometallic compounds (e.g. methyl mercury or tributyltin,…) and organometallic salts may also be of concern given that they may pose bioaccumulation or persistence hazards in case they do not quickly dissociate or dissolve in water. Unless they act as a significant source of the metal ion (as a result of the dissociation and/or degradation processes), the organic moieties and the inorganic components should be assessed individually (OECD 2015). They are therefore excluded from the guidance of this section and should be classified according to the general guidance provided in section 4. Alternatively, those metal compounds that contain an organic component but that dissociate or dissolve easily in water as the metal ion should be treated in the same way as metal compounds and classified according to this annex (e.g. Zinc acetate, …).

A9.7.1.2   The level of the metal ion which may be present in solution following the addition of the metal and/or its compounds, will largely be determined by two processes: the extent to which it can be dissolved, i.e. its water solubility, and the extent to which it can react with the media to transform to water soluble forms. The rate and extent at which this latter process, known as "transformation" for the purposes of this guidance, takes place can vary extensively between different compounds and the metal itself, and is an important factor in determining the appropriate hazard class. Where data on transformation are available, they should be taken into account in determining the classification. The protocol for determining this rate is available in annex 10.

A9.7.1.3   Generally speaking, the rate at which a substance dissolves is not considered relevant to the determination of its intrinsic toxicity. However, for metals and many poorly soluble inorganic metal compounds, the difficulties in achieving dissolution through normal solubilization techniques is so severe that the two processes of solubilization and transformation become indistinguishable. Thus, where the compound is sufficiently poorly soluble that the levels dissolved following normal attempts at solubilization do not exceed the available $L(E)C_{50}$, it is the rate and extent of transformation, which must be considered. The transformation will be affected by a number of factors, not least of which will be the properties of the media with respect to pH, water hardness, temperature etc. In addition to these properties, other factors such as the size and specific surface area of the particles which have been tested, the length of time over which exposure to the media takes place and, of course the mass or surface area loading of the substance in the media will all play a part in determining the level of dissolved metal ions in the water. Transformation data can generally, therefore, only be considered as reliable for the purposes of classification if conducted according to the standard protocol in annex 10.

A9.7.1.4   This protocol aims at standardizing the principal variables such that the level of dissolved ion can be directly related to the loading of the substance added. It is this loading level which yields the level of metal ion equivalent to the available $L(E)C_{50}$ that can then be used to determine the hazard category appropriate for classification. The testing methodology is detailed in annex 10. The strategy to be adopted in using the data from the testing protocol, and the data requirements needed to make that strategy work, will be described.

A9.7.1.5   In considering the classification of metals and metal compounds, both readily and poorly soluble, recognition has to be paid to a number of factors. As defined in chapter 4.1, the term "degradation" refers to the decomposition of organic molecules. For inorganic compounds and metals, clearly the concept of degradability, as it has been considered and used for organic substances, has limited or no meaning. Rather, the substance may be transformed by normal environmental processes to either increase or decrease the bioavailability of the toxic species. Equally, the log K_{ow} cannot be considered as a measure of the potential to accumulate. Nevertheless, the concepts that a substance, or a toxic metabolite/reaction product may not be rapidly lost from the environment and/or may bioaccumulate are as applicable to metals and metal compounds as they are to organic substances.

A9.7.1.6   Speciation of the soluble form can be affected by pH, water hardness and other variables, and may yield particular forms of the metal ion which are more or less toxic. In addition, metal ions could be made non-available from the water column by a number of processes (e.g. partitioning or chemical speciation to a non-soluble and hence not-bioavailable form). Sometimes these processes can be sufficiently rapid to be analogous to degradation in assessing long-term (chronic) classification. However, partitioning of the metal ion from the water column to other environmental media does not necessarily mean that it is no longer bioavailable, nor does it mean that the metal has been made permanently unavailable.

A9.7.1.7   Information pertaining to the extent of the partitioning of a metal ion from the water column, or the extent to which a metal has been or can be converted to a form that is less toxic or non-toxic is frequently not available over a sufficiently wide range of environmentally relevant conditions, and thus, a number of assumptions will need to be made as an aid in classification. These assumptions may be modified if available data show otherwise. In the first instance it should be assumed that the metal ions, once in the water, are not rapidly partitioned from the water column and thus these compounds do not meet the criteria. Underlying this is the assumption that, although speciation can occur, the species will remain available under environmentally relevant conditions. This may not always be the case, as described above, and any evidence available that would suggest changes to the bioavailability over the course of 28 days, should be carefully examined. The bioaccumulation of metals and inorganic metal compounds is a complex process and

A9.7.1.1.1　有機金属化合物（例：メチル水銀やトリブチルスズなど）および有機金属塩も、水中で速やかに解離または溶解しない場合、生物蓄積性または残留性の危険を引き起こす可能性があり、懸念される。（解離および/または分解プロセスの結果として）金属イオンの重要な供給源として機能しない限り、有機部分と無機成分は個別に評価されるべきである（OECD 2015）。 したがって、これらはこのセクションのガイダンスから除外され、セクション 4 で提供される一般的なガイダンスに従って分類されるべきである。あるいは、有機成分を含むが、金属イオンとして水に容易に解離または溶解する金属化合物は、金属化合物と同じ方法で処理し、この附属書にしたがって分類すべきである（例：酢酸亜鉛など）。

A9.7.1.2　金属またはその化合物を添加した後、溶液中に存在する金属イオンのレベルは、その溶出の程度すなわち水への溶解度と、媒体との相互作用によって水に溶解しうる形態への変化を起こす程度という主に 2 種類の過程によって決定される。本手引きの目的に照らして「変化」と呼ぶ、この後者の過程が起こる速度とその程度は、様々な化合物および金属自体の間で大きく異なることがあり、適切な有害性クラスを決定するにあたって重要な要素となる。変換に関するデータが利用可能な場合、分類決定においてはそれを考慮に入れるべきである。この速度を測定するためのプロトコールは附属書10に記載されている。

A9.7.1.3　一般的に言って、物質が溶解する速度が、その本質的な毒性の決定に関係するとは考えられていない。しかし、金属および多くの難溶性の無機金属化合物に関しては、通常の溶解技術によって溶解を達成させることが非常に難しいので、溶解と変化の 2 つの過程は区別しにくくなる。したがって、化合物が十分に難溶性であって、可溶化させる通常の試みによって溶解するレベルが、利用可能な L(E)C$_{50}$ を超えない場合に考慮しなければならないのは、変化の速度および程度である。この変化は多くの要因によって影響されるが、その中で無視できないものは媒体のpH 値、水硬度、温度などの性質であろう。これらの性質に加えて、試験を行った微粒子の粒径や比表面積、媒体に対してばく露された時間、そして言うまでもなく媒体内における当該物質の容積または表面積負荷など、他の要素もまたすべて、水中に溶解された金属イオンのレベルを決めるのに役割を果たす。したがって一般に、附属書 10 にある標準プロトコールにしたがって行われた変化のデータのみが、分類の目的に対して信頼できると考えられる。

A9.7.1.4　このプロトコールは、溶解したイオンのレベルが、添加される物質の負荷に直接関係付けられるように、主要な変動要因を標準化することを目的にしている。分類に適した有害性クラスを決定するのに使用できるのは、得られたL(E)C$_{50}$に相当する金属イオンのレベルを達成させる、この負荷レベルである。この試験方法については附属書 10 に詳述されている。この試験プロトコールによるデータを使用するにあたって適用される戦略、およびこの戦略を機能させる上で必要となるデータの要件について後に述べる。

A9.7.1.5　易溶性および難溶性の金属および金属化合物の分類を検討するにあたっては、数多くの要素について理解しなければならない。第 4.1 章で定義したように、「分解」という用語は有機分子の分解を意味するものである。無機化合物および金属については、有機化合物について考慮され、用いられている分解性の概念が意味をなさず、あるいは限定的な意味しか持たないことは明らかである。むしろ、物質は通常の環境過程によって、有毒物質種の生物学的利用可能性を増加または減少させるように変化させられる可能性がある。同様に、log K$_{ow}$ もまた蓄積性の尺度と考えることはできない。それにもかかわらず、物質または有毒な代謝物/反応生成物が速やかに環境から消失せず、および/または生物蓄積するという考え方は、有機物質に対してと同様に、金属および金属化合物にも当てはまることである。

A9.7.1.6　可溶性の形態への分化は、pH 値、水硬度、およびその他の変動要因によって影響を受け、多少とも毒性を持つ特別な形態の金属イオンを生成することがある。それに加えて、金属イオンは数多くの作用（例えば不溶性でゆえに生物学的に利用可能でない形態への分配又は化学種同定）によって水相において利用不可能な形態に変換される可能性もある。時にこれらの過程が長期（慢性）毒性の分類評価における分解と同じ位、速やかに起こることもある。しかし、水相から他の環境媒体への金属イオンの分配は、必ずしも生物学的に利用できなくなったことを意味するわけではなく、またこの金属が恒久的に利用できなくなったことを意味するものでもない。

A9.7.1.7　水相からの金属イオンの分配の程度、または金属がより毒性の低い、もしくは無毒性の形態に転換された、あるいは転換されうる程度に関する情報は、十分に広範囲な、環境上ありうる条件全体に対しては、利用できないことがしばしばあるので、したがって分類の助けとするために数多くの仮定を置くことが必要となろう。こうした仮定は、他の方法を示したデータがある場合には、変更してよい。第一に、金属イオンは、一度水に導入されると、水相から速やかには分配されず、したがってこれらの化合物は判定基準に適合しない、と仮定すべきである。この背景にあるのは、金属の分化は起こりうるが、その分化種は環境上ありうる条件下で利用可能である、という前提である。前述のように、このことは常に成り立つわけではないので、28 日の試験期間において生物学的利用可能性の変化を示唆する証拠が得られた場合

bioaccumulation data should be used with care. The application of bioaccumulation criteria will need to be considered on a case-by-case basis taking due account of all the available data.

A9.7.1.8   A further assumption that can be made, which represents a cautious approach, is that, in the absence of any solubility data for a particular metal compound, either measured or calculated, the substance will be sufficiently soluble to cause toxicity at the level of the ecotoxicity reference value (ERV), being the acute ERV (expressed as $L(E)C_{50}$), and/or the chronic ERV (expressed as the NOEC/ECx), and thus may be classified in the same way as other soluble salts. Again, this is clearly not always the case, and it may be wise to generate appropriate solubility data.

A9.7.1.9   This section deals with metals and metal compounds. For how this guidance applies to organometallic compounds and organometallic salts, see A9.7.1.1.1. Within the context of this guidance document, metals and metal compounds are characterized as follows:

  (a) metals, M^0, in their elemental state are not soluble in water but may transform to yield the available form. This means that a metal in the elemental state may react with water or a dilute aqueous electrolyte to form soluble cationic or anionic products, and in the process the metal will oxidize, or transform, from the neutral or zero oxidation state to a higher one;

  (b) in a simple metal compound, such as an oxide or sulphide, the metal already exists in the oxidized state, so that further metal oxidation is unlikely to occur when the compound is introduced into an aqueous medium.

However, while oxidization may not change, interaction with the media may yield more soluble forms. A sparingly soluble metal compound can be considered as one for which a solubility product can be calculated, and which will yield a small amount of the available form by dissolution. However, it should be recognized that the final solution concentration may be influenced by a number of factors, including the solubility product of some metal compounds precipitated during the Transformation/Dissolution test, e.g. aluminium hydroxide.

A9.7.2 *Application of aquatic toxicity data and solubility data for classification*

A9.7.2.1 *Interpretation of aquatic toxicity data*

A9.7.2.1.1   Aquatic toxicity studies carried out according to a recognized protocol should normally be acceptable as valid for the purposes of classification. Section A9.3 should also be consulted for generic issues that are common to assessing any aquatic toxicity data point for the purposes of classification.

A9.7.2.1.1.1   Ecotoxicity data of soluble inorganic compounds are used and combined to derive the acute and chronic ecotoxicity reference value of the dissolved metal ion (ERV or ERV_{ion}). The ecotoxicity of soluble inorganic metal compounds is dependent on the physico-chemistry of the medium, irrespective of the original metal species released in the environment.

A9.7.2.1.1.2   When evaluating ecotoxicity data and deriving ERVs, the general "weight of evidence" principle is also applicable to metals (see section A9.3.4).

A9.7.2.1.1.3   The ecotoxicity data selected should be evaluated for their adequacy. Adequacy covers here both the reliability (inherent quality of a test relating to test methodology and the way that the performance and results of a test are described) and the relevance (extent to which a test is appropriate to be used for the derivation of an ecotoxicity reference value) of the available ecotoxicity data (see sections A9.2.6 and A9.3.6):

  (a) Under the reliability criteria, metal specific considerations include the description of some abiotic parameters in the test conditions for enabling the consideration of the bioavailable metal concentration and free metal ion concentration:

    (i) Description of the physical test conditions: in addition to the general parameters (O_2, T°, pH, ...), measurements of abiotic parameters such as dissolved organic carbon (DOC), hardness, alkalinity of the water that govern the speciation and hence the metal bioavailability are recommended;

    (ii) Description of test materials and methods: to calculate the free metal ion concentration with speciation models the concentrations of dissolved major ions and cations (e.g. aluminium, iron, magnesium, and calcium) are recommended;

は、慎重に検討すべきである。金属および無機金属化合物の生物蓄積は複雑な過程であり、生物蓄積性データは注意して使用すべきである。生物蓄積性の判定基準を適用する際は、利用可能なすべてのデータを十分に考慮して、ケースバイケースで判断する必要がある。

A9.7.1.8 慎重なアプローチの例となるもう 1 つの前提は、特定の金属化合物について、測定されたあるいは計算された溶解度データが存在しない際、当該物質が急性 ERV（$L(E)C_{50}$ で表される）および／または慢性 ERV（NOEC/ECx で表される）の生態毒性参照値（ERV）のレベルで毒性を引き起こす十分な溶解度を有し、したがって他の可溶性塩類と同じ様に分類することができる、という仮定である。これもまた明らかに必ずしも当てはまらない場合があるので、適切な溶解度データを得ることが賢明であろう。

A9.7.1.9 本節では、金属および金属化合物を扱う。本手引きが有機金属化合物及び有機金属塩にどのように適用されるかについては、A9.7.1.1.1 を参照のこと。本手引きの中では、金属および金属化合物は次のように特徴付けられている：

(a) 元素の状態 M^0 では水に溶解しないが、利用能のある形態に変化しうる金属。これはすなわち、元素状態の金属が水または希薄な水性電解質と反応して、溶解性の陽イオン性もしくは陰イオン性の生成物を形成し、その過程で金属が中性ないしゼロ酸化の状態から、より酸化数の高い状態へ酸化もしくは変化されることを意味する；

(b) 酸化物や硫化物などの単純な金属化合物においては、この金属は既に酸化された状態で存在し、したがってそうした金属が水性媒体に導入されても、さらに金属の酸化が起こることは考えられない。

しかし、酸化状態に変化は起こらなくとも、媒体との相互作用によってより可溶性の形態が得られる可能性はある。難溶性の金属化合物は、その溶解度積を計算でき、かつ溶解によって少量の利用できる形態を生じる化合物と考えることができる。しかし、例えば水酸化アルミニウムのように、変化/溶解試験の過程で析出する金属化合物の溶解度積など、数多くの要因によって最終的な溶解濃度が影響されることを認識すべきである。

A9.7.2 分類への水生毒性データおよび溶解度データの適用

A9.7.2.1 水生毒性データの解釈

A9.7.2.1.1 承認されたプロトコールにしたがって実施された水生毒性試験は、通常、分類の目的に対して有意なデータとして受入れられるべきである。分類を目的とした水生毒性データポイントの評価に共通する一般的な問題については、A9.3 章もまた参照すべきである。

A9.7.2.1.1.1 溶解性無機化合物の生態毒性データは、溶存金属イオンの急性および慢性生態毒性基準値（ERV または ERV_{ion}）を導出するために使用され、組み合わせられる。溶解性無機金属化合物の生態毒性は、環境中に放出された元の金属種に関係なく、媒体の物理化学的性質に依存する。

A9.7.2.1.1.2 生態毒性データを評価し、ERV を導出する場合、一般的な「証拠の重み付け」原則は金属にも適用される（A9.3.4 節参照）。

A9.7.2.1.1.3 選択された生態毒性データは、その妥当性について評価されなければならない。ここでいう妥当性には、利用可能な生態毒性データの信頼性（試験の方法論、試験の性能と結果の記述方法に関する試験固有の質）と妥当性（生態毒性基準値の導出に用いることが適切な試験の範囲）の両方を含む。(A9.2.6 節および A9.3.6 節を参照)。

(a) 信頼性基準では、金属特有の考慮事項として、生物学的に利用可能な金属濃度と遊離金属イオン濃度を考慮できるように試験条件におけるいくつかの非生物学的パラメータを記述する：

(i) 物理的試験条件の説明：水の一般的なパラメータ（O_2、$T°$、pH、…）に加え、金属の化学種の決定、ひいては金属の生物学的利用能を支配する水の非生物学的パラメータである溶存有機炭素（DOC）、硬度、アルカリ度などの測定が推奨される；

(ii) 試験材料と試験方法の説明：化学種分化モデルを用いて遊離金属イオン濃度を算出するために、溶存主要イオンと陽イオン（アルミニウム、鉄、マグネシウム、カルシウムなど）の濃度を推奨する；

(iii) Concentration-effect relationship; hormesis: sometimes an increased performance in growth or reproduction is seen at low metal doses that exceed the control values, referred to as hormesis. Such effects can occur especially with major trace nutrients such as iron, zinc and copper but can also occur with a wide variety of non-essential substances. In such cases, positive effects should not be considered in the derivation of acute ERVs and especially chronic ERVs. Other models than the conventional log-logistic dose-response model should be used to fit the dose-response curve and consideration should be given to the adequacy of the control diet/exposure. Due to the essential nutritional needs, caution is needed with regards to extrapolation of the dose-response curve (e.g. to derive an acute or chronic ERV) below the lowest tested concentration.

(b) Under the relevancy criteria, certain considerations need to be made, related to the relevancy of the test substance and to acclimatisation/adaptation:

(i) Relevance of the test substance: tests conducted with soluble metal salts should be used for the purpose of deriving acute and chronic ERVs. The ecotoxicity adapted from organic metal compounds exposure should not be used;

(ii) Acclimatisation/adaptation: For essential metals, the culture medium should contain a minimal concentration not causing deficiency for the test species used. This is especially relevant for organisms used for chronic toxicity tests where the margin between essentiality and toxicity may become small. For this reason a proper description of culture conditions related to the level of essential metals is required.

A9.7.2.1.2   Metal complexation and speciation

A9.7.2.1.2.1   The toxicity of a particular metal in solution, appears to depend primarily on (but is not strictly limited to) the level of dissolved free metal ions. Abiotic factors including alkalinity, ionic strength and pH can influence the toxicity of metals in two ways:

(a) by influencing the chemical speciation of the metal in water (and hence affecting the availability); and

(b) by influencing the uptake and binding of available metal by biological tissues.

For the classification of metals and metal compounds, transformation/dissolution testing is carried out over a pH range (see A10.2.3.2). If evidence is available that the aquatic toxicity of the dissolved metal depends on pH, then transformation/dissolution data and aquatic toxicity are compared at a similar pH. If such evidence is not available, then the aquatic toxicity cannot be grouped according to pH. The highest aquatic toxicity observed is then compared to the transformation/dissolution data obtained at the pH which causes maximum transformation and dissolution.

A9.7.2.1.2.2   Where speciation is important, it may be possible to model the concentrations of the different forms of the metal, including those that are likely to cause toxicity. Analysis methods for quantifying exposure concentrations, which are capable of distinguishing between the complexed and uncomplexed fractions of a test substance, may not always be available.

A9.7.2.1.2.3   Complexation of metals to organic and inorganic ligands in test media and natural environments can be estimated from metal speciation models. Speciation models for metals, including pH, hardness, DOC, and inorganic substances such as MINTEQ (Brown and Allison, 1987), WHAM (Tipping, 1994; Tipping et al., 2011) and CHESS (Santore and Driscoll, 1995) can be used to calculate the uncomplexed and complexed fractions of the metal ions. Alternatively, the Biotic Ligand Model (BLM), allows for the calculation of the concentration of metal ion responsible for the toxic effect at the level of the organism, which may be affected by the DOC concentration, the pH, and the concentrations of competing ions such as calcium and magnesium. Such models may be investigated to better understand the impact of test medium composition on metal toxicity. The BLM model has at present only been validated for specific metals, organisms, and end-points (Santore and Di Toro, 1999; Garman et al., 2020). The models and formula used for the characterization of metal complexation in the media should always be clearly reported, allowing for their translation back to natural environments (OECD, 2000). In case a metal-specific BLM is available covering an appropriate pH range, a comparison of aquatic toxicity data can be made using the entire effects database for different reference pH values, relevant to the transformation/dissolution data.

(iii) 濃度と影響の関係；恒常性（ホルミシス）：恒常性と呼ばれるように、管理値を超える低用量の金属で、成長や繁殖における性能の向上が見られることがある。このような効果は、特に鉄、亜鉛、銅のような主要微量栄養素で起こりうるが、多種多様な非必須物質でも起こりうる。このような場合、急性 ERV や特に慢性 ERV の導出において、正の効果を考慮すべきではない。用量反応曲線のあてはめには、従来の対数論理学的用量反応モデル以外のモデルを使用し、対照餌／ばく露の適切性を考慮すべきである。必要不可欠な栄養のため、試験された最低濃度以下の用量反応曲線の外挿（急性または慢性 ERV の導出など）には注意が必要である。

(b) 関連性基準では、被験物質の関連性及び順応／馴化に関連して、一定の検討を行う必要がある：

(i) 被験物質の関連性：可溶性金属塩を用いて実施される試験は、急性および慢性の ERV を導き出す目的で使用されるべきである。有機金属化合物のばく露による生態毒性を利用すべきではない。

(ii) 順応/馴化：必須金属の場合、培地には、使用する試験種に対して欠乏を引き起こさない最小濃度が含まれている必要がある。これは、必須性と毒性の間のマージンが小さくなる可能性がある慢性毒性試験に使用される生物に特に関係する。このため、必須金属のレベルに関連する培養条件の適切な説明が必要である。

A9.7.2.1.2　金属の錯体形成と分化

A9.7.2.1.2.1　溶液中における特定金属の毒性は、主に（しかし、厳密に限定されずに）溶存する自由金属イオンのレベルによって決定されるように思われる。アルカリ度、イオン強度、pH 値を含む非生物的な要素は 2 つの方法で、金属の毒性に影響しうる：

(a) 水中における金属の化学種に影響する（したがってその利用可能性に影響する）こと；および

(b) 利用される金属の生物組織による取り込みと結合に影響すること

金属および金属化合物の分類のために、変化/溶解試験は pH の範囲にわたって実施される（A10.2.3.2 参照）。溶存金属の水生毒性が pH に依存するという証拠が利用可能な場合は、変化/溶解データと水生毒性は同じ pH で比較される。そのような証拠が利用できない場合、水生毒性を pH に従ってグループ化することはできない。すると、観察された最も高い水生毒性が、最大の変化と溶解を引き起こす pH で得られた変化/溶解データと比較される。

A9.7.2.1.2.2　金属の分化が重要な場合には、毒性を引き起こしやすいものも含め、金属の様々な形態の濃度をモデル化することが可能であろう。試験物質の錯化および非錯化フラクションを区別してばく露濃度を定量するための分析法は、常に利用できるわけではないであろう。

A9.7.2.1.2.3　試験媒体中および自然環境中における、金属の有機および無機配位子（リガンド）への錯体化は、金属の種分化モデルから見積もることができる。MINTEQ（Brown と Allison, 1987）、WHAM（Tipping, 1994; Tipping et al., 2011）並びに CHESS（Santore と Driscoll, 1995）などの pH 値、硬度、DOC、および無機物質を含む、金属の種分化モデルは、金属イオンの非錯化および錯化フラクションを算定するのに使用できる。またその代わりに、生物リガンドモデル（BLM）を用いて、生物レベルにおける毒性影響の原因となる金属イオン濃度を計算することもでき、この濃度は、DOC 濃度、pH、カルシウムやマグネシウムなどの競合イオンの濃度に影響される可能性がある。このようなモデルは、試験培地組成が金属の毒性に及ぼす影響をよりよく理解するために研究される可能性がある。この BLM モデルは現在のところ、ある限られた金属、生物、およびエンドポイントについて、検証されている（Santore と Di Toro, 1999; Garman et al., 2020）。媒体内における金属錯体形成の特性分析に用いられたモデルおよび算定式に関しては、自然環境に戻して検討できるような形で、常に明確に記録、報告を行うべきである（OECD, 2000）。適切な pH の範囲をカバーする金属特異的な BLM が利用可能な場合、変化/溶解データに関連する異なる参照 pH 値の影響データベース全体を用いて、水生毒性データの比較を行うことができる。

A9.7.2.2 *Interpretation of solubility data*

A9.7.2.2.1 When considering the available data on solubility, their validity and applicability to the identification of the hazard of metal compounds should be assessed. Where these are the only information available and the solubility data cannot provide a clear answer on the solubility rate and equilibrium, it is highly recommended that solubility data be generated using the Transformation/Dissolution Protocol (annex 10).

A9.7.2.2.2 Assessment of existing data

Existing data will be in one of three forms. For some well-studied metals, there will be solubility products and/or solubility data for the various inorganic metal compounds. It is also possible that the pH relationship of the solubility will be known. However, for many metals or metal compounds, it is probable that the available information will be descriptive only, e.g. poorly soluble. Unfortunately, there appears to be very little (consistent) guidance about the solubility ranges for such descriptive terms. Where these are the only information available it is probable that solubility data will need to be generated using the Transformation/Dissolution Protocol (annex 10).

A9.7.2.2.3 Screening test for assessing solubility of metal compounds

In the absence of solubility data for metal compounds, a screening test for assessing solubility should be performed as described in the Transformation/Dissolution Protocol (annex 10). The screening test is conducted at the high loading rate (100 mg/l) and under rapid and vigorous agitation for 24 h. The function of the screening test is:

(a) To identify those metal compounds which undergo either dissolution or rapid transformation such that their ecotoxicity potential is indistinguishable from soluble forms in that they may be classified based on the dissolved ion concentration.

(b) To verify the pH dependency of the dissolution, in preparation of the full transformation/dissolution test. Where data at different pH are available from the screening test, then the full test should at least be conducted at the pH which maximises the solubility. Where data are not available over the full pH range, a check should be made that this maximum solubility has been achieved by reference to suitable thermodynamic speciation models or other suitable methods (see A9.7.2.1.2.3). In the absence of suitable data or models, it is highly recommended that solubility data are generated to cover the full pH range. It should be noted that this screening test is only intended to be used for metal compounds. Metals should be assessed at the level of the full test (see A9.7.2.2.4).

A9.7.2.2.4 Full test for assessing solubility of metals and metal compounds

A9.7.2.2.4.1 The full test should at least be carried out at the pH[6] that maximizes the concentration of dissolved metal ions in solution. The pH may be chosen following the same guidance as given for the screening test.

A9.7.2.2.4.2 Based on the data from the full Test, it is possible to generate a concentration of the metal ions in solution after 7 days for each of the three loadings (i.e. 1 mg/l as "low", 10 mg/l as "medium" and 100 mg/l as "high") used in the test. If the purpose of the test is to assess the long-term (chronic) hazard of the substance, then the loadings[7] should be 0.01 mg/l, 0.1 mg/l or 1 mg/l depending on the transformation rate, and the duration of the test should be extended to 28 days.

---

[6] *The Transformation/Dissolution Protocol specifies a pH range of 6 to 8.5 for the 7-day test and 5.5 to 8.5 for the 28-day test. Considering the difficulty in carrying out transformation/dissolution tests at pH 5.5, OECD only validated the test in the pH range of 6 to 8.*

[7] *Lower loading rates than 1 mg/l may not be practically feasible for each case. While transformation/dissolution testing at lower loading rates is in principle the best way forward it is technically often not feasible. Extensive experience with the Transformation/Dissolution Protocol demonstrated that reliable predictions can be made for other loading rates. In order to make maximal use of existing transformation/dissolution data, the 28-day results for the lower loading rates (0.1 and 0.01 mg/l) can therefore often be derived by extrapolation from evidence at other loading rates. This approach should be justified on a case-by-case basis and supported by reliable information on the transformation/dissolution at different loading rates. It should be further noted that the relationship between loading rate and dissolved metal concentration may not be linear. Therefore, extrapolating transformation/dissolution data to lower loadings should preferably be made by using the equations of section A10.6.1 or alternatively by extrapolating in a precautionary way.*

A9.7.2.2　溶解度データの解釈

A9.7.2.2.1　溶解度に関して利用可能なデータについて考察する場合、その正当性、および金属化合物の有害性の特定に向けた適用性を評価すべきである。特に、データが作られた時のpH値は判っていなければならない。

A9.7.2.2.2　既存データの評価

既存データは次の3種類のうちいずれかの形になる。すなわち、一部の十分に研究された金属に関しては、種々の無機金属化合物について溶解度積または溶解度データが存在するであろう。また、溶解度とpH値との関係が知られている可能性もある。しかし、多くの金属または金属化合物に関しては、利用可能な情報が、例えば、難溶性である、などの記述的なものでしかないことがありうる。不幸にして、このような記述的な表現に対する溶解度の範囲についての(一貫性のある)手引きは非常に少ないように思われる。これらが唯一の利用可能な情報であり、溶解度データから溶解速度および平衡に関する明確な回答が得られない場合、変化/溶解プロトコール(附属書10)を用いて溶解度データをとることが強く推奨される。

A9.7.2.2.3　金属化合物の溶解度を評価するためのスクリーニング試験

金属化合物の溶解度データがない場合、溶解度を評価するためのスクリーニング試験を変化/溶解プロトコール(附属書10)に記載されている方法で実施する。スクリーニング試験は、高負荷率(100 mg/l)で、急速かつ激しい撹拌下で24時間実施される。スクリーニング検査の機能は次のとおりである：

(a) 溶解イオン濃度に基づいて分類できるという点で、溶解または急速な変化を起こすためにその生態毒性ポテンシャルが可溶性の形態と区別できないような金属化合物を特定する。

(b) 変化/溶解詳細試験の準備として、溶解のpH依存性を検証する。スクリーニング試験から異なるpHでのデータが得られる場合、詳細試験は少なくとも溶解度が最大となるpHで実施する。すべてのpHの範囲にわたってデータが得られない場合は、適切な熱力学的種分化モデルまたはその他の適切な方法を参照し、その最大溶解度が達成されていることを確認する(A9.7.2.1.2.3参照)。適切なデータまたはモデルがない場合は、すべてのpHの範囲をカバーする溶解度データをとることが強く推奨される。このスクリーニング試験は金属化合物にのみ使用されることを目的としていることに注意すべきである。金属は詳細試験のレベルで評価されるべきである(A9.7.2.2.4参照)。

A9.7.2.2.4　金属および金属化合物の溶解度を評価するための詳細試験

A9.7.2.2.4.1　詳細試験は少なくとも溶液中における溶解した金属イオンの濃度が最大になるpH値[6]で実施すべきである。その場合、pH値はスクリーニング試験と同様の手引きにしたがって選ばれる。

A9.7.2.2.4.2　詳細試験のデータに基づいて、試験で使用した3種類の負荷量(すなわち1 mg/lの「低負荷」、10 mg/lの「中負荷」、および100mg/lの「高負荷」)それぞれについて、7日後の溶液中の金属イオン濃度を計算することができる。試験の目的が物質の長期(慢性)有害性の評価である場合、負荷量[7]は変化速度に応じて0.01 mg/l、0.1 mg/l、または1 mg/lとし、試験期間を28日間まで延長すべきである。

---

[6] 変化/溶解プロトコールは、7日間試験ではpH6〜8.5、28日間試験ではpH5.5〜8.5を指定している。pH5.5での変化/溶解試験の実施が困難であることを考慮し、OECDはpH6〜8の範囲でのみ試験を検証した。
[7] 1mg/lより低い負荷率は、それぞれの場合において現実的に実行できない可能性がある。より低い負荷率での変化/溶解試験は、原理的には最善の方法であるが、技術的には実行不可能な場合が多い。変化/溶解プロトコールに関する広範な経験から、他の負荷率でも信頼できる予測が可能であることが実証された。既存の変化/溶解データを最大限に活用するため、低負荷率(0.1および0.01 mg/l)の28日間の結果は、他の負荷率におけるエビデンスから外挿することによって導き出すことができる。この方法は、ケースバイケースで正当化されるべきであり、異なる負荷率における変化/溶解に関する信頼できる情報によって裏付けられるべきである。さらに、溶出速度と溶出金属濃度の関係は直線的でない可能性があることに注意すべきである。従って、より低い負荷率への変化/溶解データの外挿は、A10.6.1節の式を用いるか、予防的な方法で外挿することが望ましい。

A9.7.2.3　　　*Comparison of aquatic toxicity data and solubility data*

A decision on how to classify the substance will be made by comparing aquatic toxicity data and solubility data. Depending on the available data, two approaches can be followed:

(a) If limited information on the transformation/dissolution at different pH levels is available, or if the aquatic toxicity of the dissolved metal does not depend on pH, then the lowest ERV and the highest transformation/dissolution result, each potentially derived at different pH levels, should provide the basis for classification (this should be used as the default approach);

(b) If evidence is available that the aquatic toxicity of the dissolved metal depends on pH, and sufficient toxicity data are available at varying pH levels, then a split of the acute and chronic ERVs can be performed according to the pH band. If in addition transformation/dissolution data at different pH levels are available, then the classification may be derived by comparing transformation/dissolution data with the ERV at corresponding pH levels, meaning that toxicity data and transformation/dissolution data are in this case always compared at the same pH band. This split of the effects data into pH bands would apply in an equal way to the acute and the chronic effects data sets. The most stringent classification outcome across all pH bands should be used.

A9.7.3　　　*Assessment of environmental transformation*

A9.7.3.1　　　Environmental transformation of one species of a metal to another species of the same does not constitute degradation as applied to organic compounds and may increase or decrease the availability and bioavailability of the toxic species. However, as a result of naturally occurring geochemical processes metal ions can partition from the water column. Data on water column residence time, the processes involved at the water – sediment interface (i.e. deposition and re-mobilization) are fairly extensive but have not been integrated into a meaningful database. Nevertheless, using the principles and assumptions discussed above in A9.7.1, it may be possible to incorporate this approach into classification.

A9.7.3.2　　　Such assessments are very difficult to give guidance for and will normally be addressed on a case by case approach. However, the following may be taken into account:

(a) Changes in speciation if they are to non-available forms, however, the potential for the reverse change to occur must also be considered;

(b) Changes to a metal compound which is considerably less soluble than that of the metal compound being considered.

Some caution is recommended, see A9.7.1.5 and A9.7.1.6.

A9.7.4　　　*Bioaccumulation*

A9.7.4.1　　　While log K_{ow} is a good predictor of BCF for certain types of organic compounds e.g. non-polar organic substances, it is irrelevant for inorganic substances such as inorganic metal compounds because metals, in contrast to organic substances, are not lipophilic and are generally not transported through cellular membranes by passive processes. Uptake of metal ions typically occurs through active processes.

A9.7.4.2　　　The mechanisms for uptake and depuration rates of metals are very complex and variable and there is at present no general model to describe this. Instead the bioaccumulation of metals according to the classification criteria should be evaluated on a case-by-case basis using expert judgement.

A9.7.4.3　　　While BCFs are indicative of the potential for bioaccumulation there may be a number of complications in interpreting measured BCF values for metals and inorganic metal compounds. For some metals and inorganic metal compounds the relationship between water concentration and BCF in some aquatic organisms is inverse, and bioconcentration data should be used with care. This is particularly relevant for metals that are biologically essential. Metals that are biologically essential are actively regulated in organisms in which the metal is essential (homeostasis). Removal and sequestration processes that minimize toxicity are complemented by an ability to up-regulate concentrations for essentiality. Since nutritional requirement of the organisms can be higher than the environmental concentration, this active regulation can result in high BCFs and an inverse relationship between BCFs and the concentration of the metal in water. When environmental concentrations are low, high BCFs may be expected as a natural consequence of metal uptake to meet nutritional requirements and in these instances can be viewed as a normal phenomenon. Additionally, if internal

A9.7.2.3 水生毒性データと溶解度データの比較

物質をどのように分類するかは、水生毒性データと溶解度データを比較して決定する。利用可能なデータに応じて次の2つのアプローチをとることができる：

(a) 異なる pH レベルでの変化/溶解に関する情報が限られている場合、あるいは溶存金属の水生毒性が pH に依存しない場合、異なる pH レベルで得られる可能性のある、最も低い ERV と最も高い変化/溶解の結果をそれぞれ分類の基礎とすべきである（これは既定の方法として使用されるべきである）。

(b) 溶存金属の水生毒性が pH に依存するという証拠が利用可能で、様々な pH 水準で十分な毒性データが利用可能な場合、急性及び慢性 ERV を pH 値域に従って分割することができる。さらに、異なる pH 水準での変化/溶解データが利用可能であれば、変化/溶解データと対応する pH 水準での ERV を比較することによって分類を導き出すことができる。この pH 値域への影響データの分割は、急性および慢性影響データセットに同様に適用される。すべての pH 値域で最も厳しい分類結果が使用されるべきである。

A9.7.3 環境における変化に関する評価

A9.7.3.1 環境における、ある金属物質から同じ金属の別の化学種への変化は、有機化合物に適用されるような分解に基づくものではなく、毒性のある化学種の利用可能性および生物学的利用可能性を増減させる可能性があるものである。しかし、自然に起こる地球化学的な過程の結果として、金属イオンは水相から分配しうるものである。水相滞留時間や、水—底質界面でのプロセス（沈積および再可動化）についてのデータはかなりあるが、まだ意味のあるデータベースに統合されてはいない。しかしながら、上記 A9.7.1 で述べた原則および前提を用いることで、この手法を分類に取り入れることは可能であろう。

A9.7.3.2 このような評価の手引きを示すことは非常に困難であり、通常はケースバイケースのアプローチにおいて対処すべきものであろう。しかし、以下の事項は考慮できる：

(a) 利用可能性を持たない形態への種分化。ただし、逆方向の変化が起こる可能性も考慮しなければならない；

(b) 対象の金属化合物の溶解性よりも大幅に溶解性が低い金属化合物への変化。

ある程度の慎重さが求められる。A9.7.1.5 および A9.7.1.6 を参照。

A9.7.4 生物蓄積性

A9.7.4.1 log K_{ow} は、非極性有機物などあるタイプの有機化合物については、BCF の良好な予測因子であるが、これはもちろん、無機金属化合物などの無機物質には当てはまらない。なぜなら、有機物質とは対照的に、金属は親油性ではなく、一般に受動的なプロセスでは細胞膜を通って輸送されないからである。金属イオンの取り込みは、通常、能動的なプロセスを通じて起こる。

A9.7.4.2 金属の取り込みおよび排泄速度のメカニズムは非常に複雑かつ多様であり、現在のところこれを記述する一般的なモデルはない。かわりに、分類基準に従った金属の生物蓄積性を、専門家の判断に基づいて、ケースバイケースで評価すべきである。

A9.7.4.3 生物濃縮係数 BCF は生物蓄積性の尺度であるが、金属および非有機金属化合物について測定された BCF の値を解釈するには、多くの複雑な要素がある。ある金属および無機金属化合物に関しては、水中濃度と、ある水生生物における BCF との関係が逆相関なので、生物濃縮性データは慎重に使用しなければならない。このことは、特に生物学的に必須の金属に当てはまることである。生物学的に必須の金属は、その金属を必須としている生体内で能動的に制御される（ホメオスタシス）。毒性を最小化する除去と隔離のプロセスは、必須性のために濃度を上昇させる能力によって補完される。生体の栄養的な要求度が環境濃度より高い場合もあるので、この能動的な制御の結果として BCF の値は高くなり、BCF と水中における当該金属濃度には逆相関の関係になる。環境における濃度が低い場合には、栄養面での必要性を満たすために金属を取り込む自然な結果として BCF の値が高くなることが予想され、この事例においては正常な現象とみなすことができる。加えて、生体内の濃度が生体によって制御されていれば、測定された BCF の値は外部の濃度が上昇するにつれて低下することになろう。外部の濃度が非常に高くなって一定の限界値を超えるか、または制御メカニズムを圧倒するようになると、これは当該の生体に危害を及ぼすものとなる。また、金属が、ある生物にとっては必須であっても、他の生物には必須でない場合がある。

concentration is regulated by the organism, then measured BCFs may decline as external concentration increases. When external concentrations are so high that they exceed a threshold level or overwhelm the regulatory mechanism, this can cause harm to the organism. Also, while a metal may be essential in a particular organism, it may not be essential in other organisms. Therefore, where the metal is not essential or when the bioconcentration of an essential metal is above nutritional levels special consideration should be given to the potential for bioconcentration and environmental concern.

A9.7.4.4   For essential elements, measured BCFs decline as external concentrations increase because the internal concentrations are regulated by the organism. Non-essential metals are also actively regulated to some extent and therefore also for non-essential metals, an inverse relationship between the metal concentration and the external concentration may be observed (McGeer et al., 2003). When external concentrations are so high that they exceed a threshold level, or overwhelm the regulatory mechanism, this can cause harm to the organism. BCF and BAF may be used to estimate metal accumulation by:

    (a)   Considering information on essentiality and homeostasis of metals/metal compounds. As a result of such regulation, the "bioaccumulative" criterion is not applicable to metals;

    (b)   Assessing bioconcentration factors for non-essential metals, should preferably be done from BCF studies using environmentally relevant concentrations in the test media.

A9.7.5   *Application of classification criteria to metals and metal compounds*

A9.7.5.1   *Introduction to the classification strategy for metals and metal compounds*

A9.7.5.1.1   Short-term (acute) and long-term (chronic) hazards are assessed individually for metals and metal compounds. For long-term hazards preference should be given in applying the approach based on chronic toxicity data. Such evidence is often available for the readily soluble metal salts. The schemes for the determination of short-term and long-term aquatic hazards of metals and metal compounds are described below and summarized in the figures:

    (a)   A9.7.1 (short-term hazard classification of metals);

    (b)   A9.7.2 and A9.7.3 (long-term hazard classification of metals);

    (c)   A9.7.4 (short-term hazard classification of metal compounds);

    (d)   A9.7.5 (long-term hazard classification of metal compounds).

A9.7.5.1.1.1   There are several stages in these schemes where data are used for decision purposes. It is not the intention of the classification schemes to generate new data. In the absence of valid data, it will be necessary to use all available data and expert judgement.

A9.7.5.1.1.2   In the following sections, the reference to the acute and chronic ERVs refer to the data point(s) that will be used to select the hazard categories for the metal or metal compound.

A9.7.5.1.2   When considering acute and chronic ERVs for metal compounds (ERV$_{compound}$), it is important to ensure that the data point to be used as the justification for the classification is expressed in the weight of the molecule of the metal compound to be classified. This is known as correcting for molecular weight. Thus while most metal data are expressed in, for example, mg/l of the dissolved metal ion (abbreviated ERV$_{ion}$), this value will need to be adjusted to the corresponding weight of the metal compound. Thus:

$$\text{ERV}_{compound} = \text{ERV}_{ion} \times \left(\frac{\text{molecular weight of metal compound}}{\sum \text{atomic weight of the atom(s) of the metal in the compound}} \right)$$

where:

    ERV$_{compound}$ = ERV of the metal compound

    ERV$_{ion}$ = ERV of the dissolved metal ion

したがって、金属が必須でない場合、または必須金属の生物蓄積度が栄養レベルを超える場合には、生物蓄積性および環境上の問題について特に配慮すべきである。

A9.7.4.4　必須元素については、内部濃度が生物によって調節されるため、BCFの測定値は外部濃度が増加するにつれて減少する。非必須金属もある程度能動的に調節されるため、非必須金属についても、金属濃度と外部濃度との間に逆相関が観察されることがある（McGeer et al., 2003）。外部濃度が閾値レベルを超えるほど高い場合、あるいは調節機構を圧倒する場合、生物に害を及ぼす可能性がある。BCF と BAF は、次のような方法で金属蓄積量を推定することができる：

>　(a) 金属／金属化合物の本質性とホメオスタシスに関する情報を考慮する。このような調節の結果、「生物蓄積性」基準は金属には適用されない。
>
>　(b) 非必須金属の生物濃縮係数の評価は、試験媒体中の環境的に適切な濃度を用いた BCF 研究から行うことが望ましい。

A9.7.5　金属および金属化合物に関する分類基準の適用

A9.7.5.1　金属および金属化合物に関する分類戦略の概要

A9.7.5.1.1　金属および金属化合物については、短期（急性）および長期（慢性）の有害性を個別に評価する。長期有害性については、慢性毒性データに基づくアプローチを優先的に適用する。このような証拠は、易溶性金属塩ではしばしば利用可能である。金属および金属化合物の短期および長期の水生有害性を決定するためのスキームを以下に記述し、図に要約する：

>　(a) A9.7.1（金属の短期有害性分類）；
>
>　(b) A9.7.2 及び A9.7.3（金属の長期有害性分類）；
>
>　(c) A9.7.4（金属化合物の短期有害性分類）；
>
>　(d) A9.7.5（金属化合物の長期有害性分類）；

A9.7.5.1.1.1　これらのスキームには、決定目的でデータが使用されるいくつかの段階がある。分類スキームは、新たなデータを生成することを意図していない。有効なデータがない場合は、利用可能なすべてのデータと専門家の判断を利用する必要がある。

A9.7.5.1.1.2　以下の節では、急性および慢性 ERV への言及は、金属または金属化合物の危険有害性区分を選定するために使用されるデータポイントを指す。

A9.7.5.1.2　金属化合物の急性および慢性 ERV（$ERV_{compound}$）を考慮する場合、分類の正当性の根拠として用いられるデータポイントが、分類すべき金属化合物分子の重量で表されていることを確認することが重要である。これは分子量補正として知られる作業である。したがって、大半の金属データは例えば溶存金属イオン（ERV_{ion} と略す）mg/l のように表されるが、この値は対応する金属化合物の重量に調整を行う必要がある。したがって：

$$ERV_{compound} = ERV_{ion} \times (金属化合物の分子量／化合物中の金属原子の原子量の総和)$$

ここで

$ERV_{compound}$ ＝ 金属化合物の ERV

ERV_{ion} ＝ 溶存金属イオンの ERV

A9.7.5.2   *Classification strategy for metals*

A9.7.5.2.1   Short-term (acute) aquatic hazard of metals

A9.7.5.2.1.1   The scheme for determining the short-term (acute) aquatic hazard of metals is described in this section and summarized in figure A9.7.1.

A9.7.5.2.1.2   Where the acute ERV of the dissolved metal ion is greater than 100 mg/l, the metals need not be considered further in the classification scheme.

A9.7.5.2.1.3   Where the acute ERV of the dissolved metal ions is less than or equal to 100 mg/l, consideration must be given to the data available on the rate and extent to which these ions can be generated from the metal. Such data, to be valid and useable, should have been generated using the Transformation/Dissolution Protocol (annex 10).

A9.7.5.2.1.4   Where 7-day data from the Transformation/Dissolution Protocol are available, then, the results should be used to aid classification according to the following rules. Classify the metal as:

    (a)   Category Acute 1 if the dissolved metal ion concentration at the low loading rate is greater than or equal to the acute ERV. Assign an Acute M factor according to table A9.7.1;

    (b)   Category Acute 2 if the dissolved metal ion concentration at the low loading rate is less than the acute ERV, but at the medium loading rate it is greater than or equal to the acute ERV;

    (c)   Category Acute 3 if the dissolved metal ion concentration at the low and the medium loading rates is less than the acute ERV, but at the high loading rate it is greater than or equal to the acute ERV.

Do not classify the metal for short-term aquatic hazard if the dissolved metal concentration at all loading rates is below the acute ERV.

Figure A9.7.1: Classification strategy for determining the short-term (acute) aquatic hazard of metals

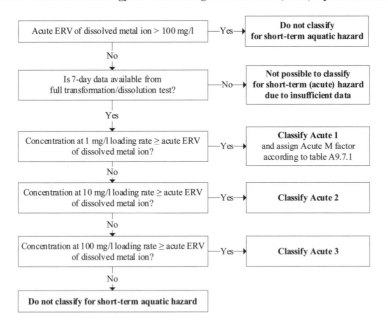

A9.7.5.2 金属の分類戦略

A9.7.5.2.1 金属の短期（急性）水生有害性

A9.7.5.2.1.1 金属の短期（急性）水生有害性を決定するスキームは本節に記載され、図 A9.7.1 に要約される。

A9.7.5.2.1.2 溶存金属イオンの急性 ERV が 100 mg/l より大きい場合、その金属は分類スキームにおいてそれ以上考慮する必要はない。

A9.7.5.2.1.3 溶存金属イオンの急性 ERV が 100mg/l 以下である場合、これらのイオンが当該金属から生成される速度と程度に関する利用可能なデータを考慮しなければならない。そのようなデータが有効で利用可能であるためには、変化/溶解プロトコール（附属書10）を用いて生成されるべきである。

A9.7.5.2.1.4 変化/溶解プロトコールの7日間のデータが利用可能な場合、その結果は次のルールに従って分類を支援するために使用されるものとする。金属を次のように分類する：

(a) 低負荷率における溶存金属イオン濃度が急性 ERV 以上の場合、区分急性 1 とする。表 A9.7.1 に従って急性 M 因子を割り当てる；

(b) 低負荷率での溶存金属イオン濃度が急性 ERV 未満であるが、中負荷率では急性 ERV 以上である場合、区分急性 2 とする；

(c) 低負荷率および中負荷率における溶存金属イオン濃度が急性 ERV 未満であるが、高負荷率では急性 ERV 以上である場合、区分急性 3 とする。

すべての負荷率における溶存金属濃度が急性 ERV 未満である場合、その金属を短期水生有害性として分類しない。

図 A9.7.1：金属の短期（急性）水生有害性を決定するための分類戦略

A9.7.5.2.2   Long-term (chronic) aquatic hazard of metals

The scheme for determining the long-term (chronic) aquatic hazard of metals is described in this section and summarized in figures A9.7.2 and A9.7.3. Metals can be classified for long-term aquatic hazard using chronic toxicity data when available or using the surrogate approach in absence of appropriate chronic toxicity data.

A9.7.5.2.2.1   Approach based on available chronic toxicity data

A9.7.5.2.2.1.1   Where the chronic ERV of the dissolved metal ion is greater than 1 mg/l, the metal need not be considered further in the classification scheme.

A9.7.5.2.2.1.2   Where the chronic ERV of the dissolved metal ion is less than or equal to 1 mg/l, consideration must be given to the available data on the rate and extent to which these ions can be generated from the metal. To be valid and useable, such data should have been generated or calculated using the Transformation/Dissolution Protocol (annex 10) for 28 days (see A9.7.2.2.4). If such data are unavailable, the surrogate approach should be used (see A9.7.5.2.2.2). Where 28-day transformation/dissolution data are available, then classify the metal as:

(a)   Category Chronic 1 if the dissolved metal ion concentration obtained at a loading rate of 0.1 mg/l (0.01 mg/l if there is evidence of rapid environmental transformation) is greater than or equal to the chronic ERV. Assign a Chronic M factor according to table A9.7.1;

(b)   Category Chronic 2 if the dissolved metal ion concentration obtained at a loading rate of 1 mg/l (0.1 mg/l if there is evidence of rapid environmental transformation) is greater than or equal to the chronic ERV;

(c)   Category Chronic 3 if the dissolved metal ion concentration obtained at a loading rate of 1 mg/l is greater than the chronic ERV and there is evidence of rapid environmental transformation.

A9.7.5.2.2.1.3   Classify the metal as category Chronic 4 if the data available do not allow classification under the formal criteria but there are nevertheless some grounds for concern (see 4.1.2.2).

A9.7.5.2.2.1.4   Do not classify the metal for long-term aquatic hazard if the dissolved metal ion concentration obtained from the 28-day transformation/dissolution test at a loading rate of 1 mg/l is less than the chronic ERV of the dissolved metal ion.

A9.7.5.2.2　金属の長期（慢性）水生有害性

　　　金属の長期（慢性）水生有害性を決定するスキームは本節に記載され、図 A9.7.2 および A9.7.3 に要約される。金属は利用可能な場合には慢性毒性データを用いて、また適切な慢性毒性データがない場合には代替手法を用いて、長期水生有害性について分類することができる。

A9.7.5.2.2.1　　利用可能な慢性毒性データに基づくアプローチ

A9.7.5.2.2.1.1　溶存金属イオンの慢性 ERV が 1mg/l より大きい場合、その金属は分類スキーム においてそれ以上考慮する必要はない。

A9.7.5.2.2.1.2　溶存金属イオンの慢性 ERV が 1mg/l 以下の場合、これらのイオンが当該金属から生成される速度と程度に関する利用可能なデータを考慮しなければならない。そのようなデータが有効で利用可能であるためには、28 日間の変化/溶解プロトコール（附属書 10）を用いて生成又は計算されたものでなければならない（A9.7.2.2.4 参照）。そのようなデータが利用できない場合は代替法を用いる（A9.7.5.2.2.2 参照）。28 日間の変化/溶解データが利用可能な場合、その金属を次のように分類する：

(a) 0.1 mg/l（急速な環境変化の証拠がある場合は 0.01mg/l）の負荷率で得られた溶存金属イオン濃度が、慢性 ERV 以上の場合は、区分慢性 1 とする。表 A9.7.1 に従って慢性毒性乗率 M を割り当てる；

(b) 1 mg/l（急速な環境変化の証拠がある場合は 0.1mg/l）の負荷率で得られた溶存金属イオン濃度が、慢性 ERV 以上の場合は、区分慢性 2 とする；

(c) 1 mg/l の負荷率で得られた溶存金属イオン濃度が、慢性 ERV よりも大きく、急速な環境変化の証拠がある場合は、区分慢性 3 とする。

A9.7.5.2.2.1.3　利用可能なデータでは正式な基準に基づく分類はできないが、それでも懸念すべき根拠がある場合は、その金属を区分慢性 4 に分類する（4.1.2.2 を参照）。

A9.7.5.2.2.1.4　負荷率 1 mg/l での 28 日間の変化/溶解試験で得られた溶存金属イオン濃度が、溶存金属イオン濃度の慢性 ERV 未満である場合、その金属を長期水生有害性には分類しない。

Figure A9.7.2: Classification strategy for determining long-term aquatic hazard of metals on the basis of chronic data

```
                    ┌─────────────────────────────┐
        ┌─── No ────│  Is chronic ERV available?  │
        │           └─────────────────────────────┘
        │                        │ Yes
        │                        ▼
        │           ┌─────────────────────────────┐          ┌──────────────────────┐
        │           │  Is the chronic ERV ≤ 1 mg/l?│── No ──▶│ Do not classify for  │
        │           └─────────────────────────────┘          │ long-term aquatic    │
        │                        │ Yes                        │ hazard               │
        │                        ▼                            └──────────────────────┘
┌───────────────┐       ┌─────────────────────────────────┐
│ Go to figure  │◀─No───│ Is 28-day data available from   │
│ A9.7.3        │       │ full transformation/dissolution │
│ (surrogate    │       │ test?                           │
│ approach)     │       └─────────────────────────────────┘
└───────────────┘                   │ Yes
                                    ▼
                        ┌───────────────────────────────┐
                        │ Is there evidence of rapid    │
                        │ environmental transformation? │
                        └───────────────────────────────┘
                          Yes                      No
```

| Classify Chronic 1 and assign Chronic M factor according to Table A9.7.1 | ◀─Yes─ | Concentration at 0.01 mg/l loading rate ≥ chronic ERV of dissolved metal ion? | | Concentration at 0.1 mg/l loading rate ≥ chronic ERV of dissolved metal ion? | ─Yes─▶ | Classify Chronic 1 and assign Chronic M factor according to table A9.7.1 |

| Classify Chronic 2 | ◀─Yes─ | Concentration at 0.1 mg/l loading rate ≥ chronic ERV of dissolved metal ion? | | Concentration at 1 mg/l loading rate ≥ chronic ERV of dissolved metal ion? | ─Yes─▶ | Classify Chronic 2 |

| Classify Chronic 3 | ◀─Yes─ | Concentration at 1 mg/l loading rate ≥ chronic ERV of dissolved metal ion? | ─No─▶ | Are there grounds for concern (see 4.1.2.2)? | ─Yes─▶ | Classify Chronic 4 |

                                         │ No
                                         ▼
                    ┌──────────────────────────────────────┐
                    │ Do not classify for long-term   │
                    │ aquatic hazard   │
                    └──────────────────────────────────────┘

A9.7.5.2.2.2 The surrogate approach

A9.7.5.2.2.2.1 Where appropriate chronic toxicity data and/or transformation/dissolution data are not available, but the metal is classified for short-term (acute) aquatic hazard, then classify the metal as (unless there is evidence of rapid environmental transformation and no bioaccumulation):

(a) Category Chronic 1 if the metal is classified for short-term (acute) aquatic hazard as category Acute 1. Assign the same M factor as for category Acute 1;

(b) Category Chronic 2 if the metal is classified for short-term (acute) aquatic hazard as category Acute 2;

(c) Category Chronic 3 if the metal is classified for short-term (acute) aquatic hazard as category Acute 3.

A9.7.5.2.2.2.2 In the lack of a short-term aquatic hazard classification due to missing transformation/dissolution data, and there is no clear data of sufficient validity to show that the transformation of metal ions will not occur, the safety net classification (Chronic 4) should be applied when the known classifiable toxicity of these soluble forms is considered to produce sufficient concern. For example, this is the case when the acute ERV_{ion} is equal to or below 100 mg/l, and/or if the chronic ERV_{ion} is equal to or below 1mg/l. In these cases, testing according to the Transformation/Dissolution Protocol may be considered.

A9.7.5.2.2.2.3 Do not classify the metal for long-term aquatic hazard if the metal is not classified for short-term aquatic hazard and if there are no grounds for concern.

図A9.7.2 慢性データに基づいて金属の長期（慢性）水生有害性を決定するための分類戦略

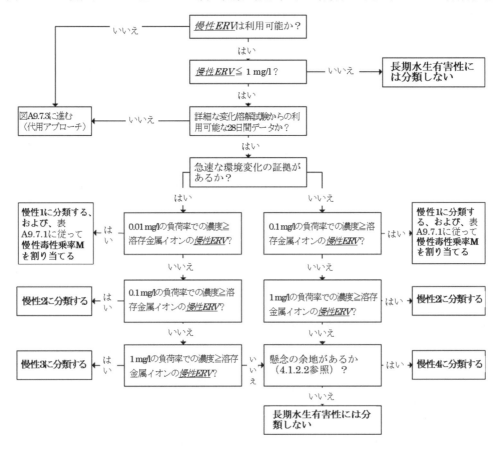

A9.7.5.2.2.2 代用アプローチ

A9.7.5.2.2.2.1 適切な慢性毒性データおよび/または変化/溶解データが利用できないが、金属が短期（急性）水生有害性として分類されている場合、その金属を次のように分類する（急速な環境変化の証拠があり、生物蓄積性がない場合を除く）：

(a) その金属が短期（急性）水生有害性の区分急性1に分類されている場合は、区分慢性1に分類する。区分急性1と同じ毒性乗率Mを割り当てる；

(b) その金属が短期（急性）水生有害性の区分急性2に分類される場合は、区分慢性2に分類する；

(c) その金属が短期（急性）水生有害性で区分急性3に分類される場合は、区分慢性3に分類する。

A9.7.5.2.2.2.2 変化/溶解データが欠落しているために短期水生有害性の分類が存在せず、金属イオンの変化が起こらないことを示す十分な有効性を示す明確なデータがない場合、これらの可溶性形態の既知の分類可能な毒性が十分な懸念を引き起こすと考えられる場合には、セーフティネット分類（慢性4）を適用するものとする。例えば、急性ERV_{ion}が100 mg/l以下、および/または慢性ERV_{ion}が1 mg/l以下の場合である。このような場合、変化/溶解プロトコールに従った試験が考慮される。

A9.7.5.2.2.2.3 金属が短期水生有害性に分類されず、懸念の根拠がない場合、その金属は長期水生有害性に分類しない。

Figure A9.7.3: Classification strategy for determining long-term aquatic hazard of metals in absence of appropriate chronic toxicity reference data and/or 28-day transformation/dissolution data

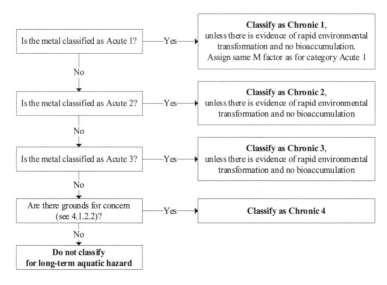

A9.7.5.3   *Classification strategy for metal compounds*

Metal compounds will be considered as readily soluble if the water solubility (measured e.g. through a screening test according to the Transformation/Dissolution Protocol, or estimated e.g. from the solubility product), expressed as the concentration of dissolved metal ion, is greater than or equal to the acute ERV_{ion}. In the context of the classification criteria, metal compounds will also be considered as readily soluble if such data are unavailable, i.e. there are no clear data of sufficient validity to show that the transformation to metal ions will not occur. Care should be exercised for compounds whose solubility is close to the acute ERV as the conditions under which solubility is measured could differ significantly from those of the acute toxicity test. In these cases the results of the screening test are preferred. Metal compounds will be considered as poorly soluble if the water solubility (measured e.g. through a screening test, or estimated e.g. from the solubility product), expressed as the concentration of dissolved metal ion, is less than the acute ERV_{ion}.

A9.7.5.3.1   *Short-term (acute) aquatic hazard of metal compounds*

A9.7.5.3.1.1   Readily soluble metal compounds are classified on the basis of the acute $ERV_{compound}$. Classify the readily soluble metal compound as:

(a) Category Acute 1 if the acute $ERV_{compound}$ is equal to or less than 1 mg/l column. Assign an Acute M factor according to table A9.7.1;

(b) Category Acute 2 if the acute $ERV_{compound}$ is greater than 1 mg/l but less than or equal to 10 mg/l;

(c) Category Acute 3 if the acute $ERV_{compound}$ is greater than 10 mg/l but less than or equal to 100 mg/l.

Do not classify the readily soluble metal compound for short-term aquatic hazard if the acute $ERV_{compound}$ is greater than 100 mg/l.

A9.7.5.3.1.2   Poorly soluble metal compounds are classified on the basis of the acute ERV of the dissolved metal ion and 7-day transformation/dissolution data. Classify the poorly soluble metal compound as:

(a) Category Acute 1 if the dissolved metal ion concentration at the low loading rate is equal to or greater than the acute ERV_{ion}, and assign Acute M factor according to table A9.7.1;

(b) Category Acute 2 if the dissolved metal ion concentration at the medium loading rate is equal to or greater than the acute ERV_{ion};

図 A9.7.3　適切な慢性毒性参照データおよび/または 28 日間の変化/溶解データがない場合の金属の長期水生有害性を決定するための分類戦略

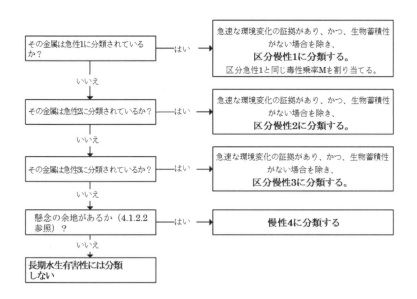

A9.7.5.3　*金属化合物の分類戦略*

　　金属化合物は、溶解した金属イオンの濃度として表される水溶解度（例えば変化/溶解プロトコルに従ったスクリーニング試験により測定されるか、例えば溶解度積から推定される）が急性 ERV_{ion} 以上の場合、易溶性とみなされる。分類基準の文脈において、金属化合物は、そのようなデータが利用できない場合、すなわち、金属イオンへの変換が起こらないことを示す十分な有効性を有する明確なデータがない場合にも、易溶性とみなされる。溶解度の測定条件が急性毒性試験の条件と大きく異なる可能性があるため、溶解度が急性 ERV に近い化合物については注意が必要である。このような場合は、スクリーニング試験の結果を優先する。金属化合物は、溶解した金属イオンの濃度として表される水溶解度（例えばスクリーニング試験により測定されたもの、又は溶解度積から推定されたもの）が急性 ERV_{ion} より低い場合、難溶性とみなされる。

A9.7.5.3.1　*金属化合物の短期（急性）水生有害性*

A9.7.5.3.1.1　易溶性金属化合物は急性 $ERV_{compound}$ に基づいて分類する。易溶性金属化合物を次のように分類する：

　　(a)　急性 $ERV_{compound}$ が 1 mg/l 以下の場合、分類は区分急性 1 とし、表 A9.7.1 に従って急性毒性乗率 M を割り当てる；

　　(b)　急性 $ERV_{compound}$ が 1 mg/l より大きく 10 mg/l 以下の場合、区分急性 2 とする；

　　(c)　急性 $ERV_{compound}$ が 10 mg/l より大きく 100 mg/l 以下の場合、区分急性 3 とする。

　　急性 $ERV_{compound}$ が 100 mg/l より大きい場合、易溶性金属化合物を短期水生有害性に分類しない。

A9.7.5.3.1.2　難溶性金属化合物は溶存金属イオン急性 ERV と 7 日間変化/溶解データに基づいて分類する。難溶性金属化合物を次のように分類する：

　　(a)　低負荷率での溶存金属イオン濃度が急性 ERV_{ion} 以上の場合、区分急性 1 とし、表 A9.7.1 に従って急性毒性乗率 M を割り当てる；

　　(b)　中負荷率での溶存金属イオン濃度が急性 ERV_{ion} 以上の場合、区分急性 2 とする；

(c) Category Acute 3 if the dissolved metal ion concentration at the high loading rate is equal to or greater than the acute ERV_{ion}.

Do not classify the poorly soluble metal compound for short-term (acute) aquatic hazard if the dissolved metal ion concentration is below the acute ERV of the dissolved metal ion at all loading rates.

Figure A9.7.4: Classification strategy for determining the short-term (acute) aquatic hazard of metal compounds

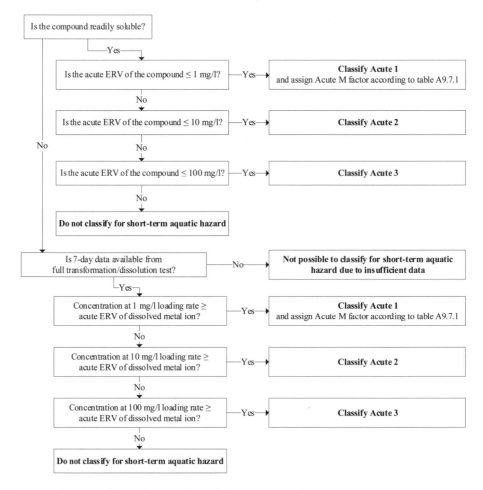

A9.7.5.3.2 *Long-term (chronic) aquatic hazard of metal compounds*

The scheme for determining the long-term (chronic) aquatic hazard of metal compounds is described in this section and summarised in figure A9.7.5. Metal compounds can be classified for long-term aquatic hazard using chronic toxicity data when available, or using the surrogate approach in absence of appropriate chronic toxicity data.

A9.7.5.3.2.1 Approach based on available chronic toxicity data

A9.7.5.3.2.1.1 Where the chronic $ERV_{compound}$ is greater than 1 mg/l, the metal compound need not to be considered further in the classification scheme for long-term hazard.

A9.7.5.3.2.1.2 Readily soluble metal compounds are classified on the basis of the chronic $ERV_{compound}$. If there is no evidence of rapid environmental transformation, then classify the readily soluble metal compound as:

(a) Category Chronic 1 if the chronic $ERV_{compound}$ is equal to or less than 0.1 mg/l (0.01 mg/l if there is evidence of rapid environmental transformation). Assign a chronic M factor according to table A9.7.1;

(c) 高負荷率での溶存金属イオン濃度が急性 ERV$_{ion}$ 以上の場合、区分急性 3 とする。

すべての負荷率で溶存金属イオン濃度がその溶存金属イオンの急性 ERV 未満である場合、難溶性金属化合物を短期(急性)水生有害性に分類しない。

図 A9.7.4　金属化合物の短期(急性)水生有害性を決定するための分類戦略

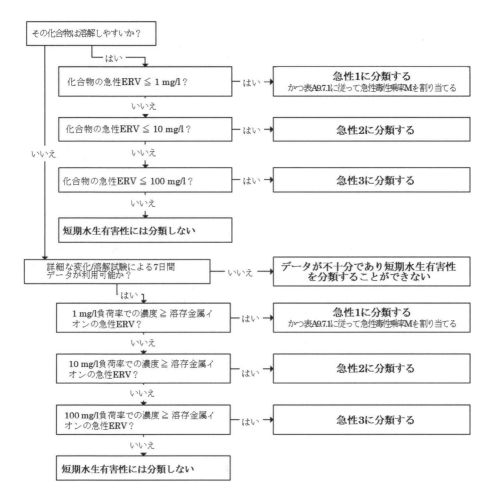

A9.7.5.3.2　*金属化合物の長期(慢性)水生有害性*

金属化合物の長期(慢性)水生有害性を決定するスキームを本節で説明し、図 A9.7.5 に要約する。金属化合物の長期水生有害性は、慢性毒性データがある場合はそれを用いて、適切な慢性毒性データがない場合は代用アプローチを用いて分類することができる。

A9.7.5.3.2.1　利用可能な慢性毒性データに基づくアプローチ

A9.7.5.3.2.1.1　慢性 ERV$_{compound}$ が 1 mg/l より大きい場合、金属化合物は長期有害性の分類スキームでさらに考慮する必要はない。

A9.7.5.3.2.1.2　易溶性金属化合物は慢性 ERV$_{compound}$ に基づいて分類される。急速な環境変化の証拠がない場合は、易溶性金属化合物を次のように分類する:

(a) 慢性 ERV$_{compound}$ が 0.1 mg/l (急速な環境変化の証拠がある場合は 0.01 mg/l) 以下の場合、分類は区分慢性 1 とし、表 A9.7.1 に従って慢性毒性乗率 M を割り当てる;

(b) Category Chronic 2 if the chronic ERV$_{compound}$ is equal to or less than 1 mg/l (0.1 mg/l if there is evidence of rapid environmental transformation);

(c) Category Chronic 3 if the chronic ERV$_{compound}$ is equal to or less than 1 mg/l and there is evidence of rapid environmental transformation;

(d) Category Chronic 4 if the data available do not allow classification under the formal criteria but there are nevertheless some grounds for concern (see 4.1.2.2).

A9.7.5.3.2.1.3   Poorly soluble metal compounds: Consideration must be given to the data available on the rate and extent to which these ions can be generated from the metal compound. For such rate and extent data, to be valid and useable, they should have been generated using the Transformation/Dissolution Protocol for a 28-day period. Where such 28-day transformation/dissolution data are unavailable, the surrogate approach should be used (see A9.7.5.3.2.2). Where 28-day transformation/dissolution data are available, then classify the poorly soluble metal compound as:

(a) Category Chronic 1 if the dissolved metal ion concentration obtained at a loading rate of 0.1 mg/l (0.01 mg/l if there is evidence of rapid environmental transformation) is greater than or equal to the chronic ERV of the dissolved metal ion. Assign a chronic M factor according to table A9.7.1;

(b) Category Chronic 2 if the dissolved metal ion concentration obtained at a loading rate of 1 mg/l (0.1 mg/l if there is evidence of rapid environmental transformation) is greater than or equal to the chronic ERV of the dissolved metal ion;

(c) Category Chronic 3 if the dissolved metal ion concentration obtained at a loading rate of 1 mg/l is greater than or equal to the chronic ERV of the dissolved metal ion and there is evidence of rapid environmental transformation;

(d) Category Chronic 4 if the data available do not allow classification under the formal criteria but there are nevertheless some grounds for concern (see 4.1.2.2).

Do not classify the poorly soluble metal compound for long-term (chronic) aquatic hazard if the dissolved metal ion concentration obtained from the 28-day transformation/dissolution test at a loading rate of 1 mg/l is less than the chronic ERV of the dissolved metal ion.

(b) 慢性 $ERV_{compound}$ が 1 mg/l（急速な環境変化の証拠がある場合は 0.1 mg/l）以下の場合、区分慢性 2 とする；

(c) 慢性 $ERV_{compound}$ が 1 mg/l 以下であり、急速な環境変化の証拠がある場合は、区分慢性 3 とする；

(d) 利用可能なデータでは正式な基準に基づいて分類できないが、それでも懸念の根拠がある場合は、区分慢性 4 に分類する（4.1.2.2 参照）。

A9.7.5.3.2.1.3　難溶性金属化合物：金属化合物からこれらのイオンが生成される速度と程度に関する利用可能なデータを考慮する必要がある。このような速度と程度のデータが有効で利用可能であるためには、28 日間変化/溶解プロトコールを用いて生成されたものでなければならない。そのような 28 日間変化/溶解データが利用できない場合は、代用アプローチ（A9.7.5.3.2.2 参照）を用いる。28 日間変化/溶解データが利用可能な場合、難溶性金属化合物を次のように分類する：

(a) 0.1 mg/l（急速な環境変化の証拠がある場合は 0.01 mg/l）の負荷率で得られた溶存金属イオン濃度が、溶存金属イオンの慢性 ERV_{ion} 以上である場合、分類は区分慢性 1 とし、表 A9.7.1 に従って慢性毒性乗率 M を割り当てる；

(b) 1 mg/l（急速な環境変化の証拠がある場合は 0.1 mg/l）の負荷率で得られた溶存金属イオン濃度が、溶存金属イオンの慢性 ERV_{ion} 以上である場合、分類は区分慢性 2 とする；

(c) 1 mg/l の負荷率で得られた溶存金属イオン濃度が、溶存金属イオンの慢性 ERV_{ion} 以上であり、急速な環境変化の証拠がある場合、分類は区分慢性 3 とする；

(d) 利用可能なデータでは正式な基準に基づいて分類できないが、それでも懸念の根拠がある場合は、区分慢性 4 に分類する(4.1.2.2 参照)。

1 mg/l の負荷率での 28 日間変化/溶解試験で得られた溶存金属イオン濃度が、溶存金属イオンの慢性 ERV 未満である場合、難水溶性金属化合物を長期（慢性）水生有害物質に分類しない。

Figure A9.7.5: Classification strategy for determining long-term aquatic hazard of metal compounds on the basis of chronic data

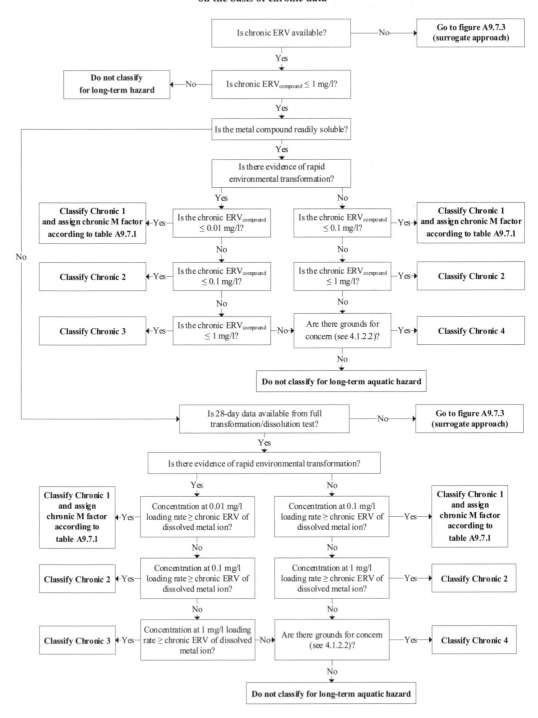

図 A9.7.5：慢性データに基づいて金属化合物の長期水生有害性を決定するための分類戦略

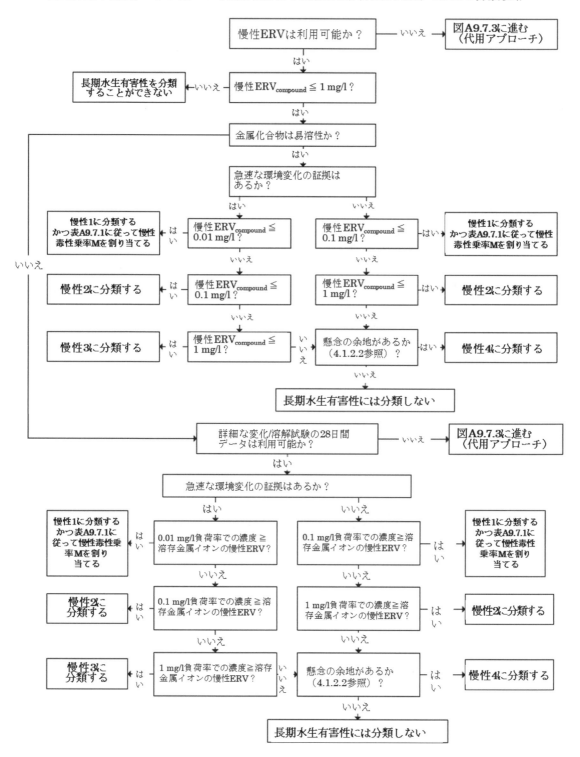

A9.7.5.3.3.2   The surrogate approach

Where appropriate chronic toxicity data and/or transformation/dissolution data are not available, but the metal compound is classified for short-term (acute) aquatic hazard, then the metal compound is classified according to the surrogate approach. The surrogate approach for metal compounds is identical to that for metals (see A9.7.5.2.2.2).

A9.7.5.4   *Particle size and surface area*

A9.7.5.4.1   Particle size, or moreover surface area, is a crucial parameter in that any variation in the size or surface area tested may cause a significant change in the levels of metals ions released in a given time-window. Thus, this particle size or surface area is fixed for the purposes of the transformation test, allowing the comparative classifications to be based solely on the loading level. Normally, the classification data generated would have used the smallest particle size marketed to determine the extent of transformation. There may be cases where data generated for a particular metal powder is not considered as suitable for classification of the massive forms. For example, where it can be shown that the tested powder is structurally a different material (e.g. different crystallographic structure) and/or it has been produced by a special process and cannot be generated from the massive metal, classification of the massive can be based on testing of a more representative particle size or surface area, if such data are available. The powder may be classified separately based on the data generated on the powder. However, in normal circumstances it is not anticipated that more than two classification proposals would be made for the same metal.

A9.7.5.4.2   Metals with a particle size smaller than the default diameter value of 1 mm can be tested on a case-by-case basis. One example of this is where metal powders are produced by a different production technique or where the powders give rise to a higher dissolution (or reaction) rate than the massive form leading to a more stringent classification.

A9.7.5.4.3   The particle sizes tested depend on the substance being assessed and are shown in the table below:

Type	Particle size	Comments
Metal compounds	Smallest representative size sold	Never larger than 1 mm
Metals – powders	Smallest representative size sold	May need to consider different sources if yielding different crystallographic/morphologic properties
Metals – massive	1 mm	Default value may be altered if sufficient justification

Massive forms will usually be tested as 1 mm particles. Alternatively, the transformation/dissolution testing of materials with different surface areas may result in highly reliable dissolution kinetic equations that allow to define the "Critical Particle Diameter" (CPD) for appropriate loadings for the acute and long-term hazard assessment.

A9.7.5.4.4   For some forms of metals, it may be possible, using the Transformation/Dissolution Protocol (OECD 2001), to obtain a correlation between the concentration of the metal ion after a specified time interval as a function of the surface area loadings of the forms tested. Such correlations should be established for the relevant pH ranges as specified in the Transformation/Dissolution Protocol. In such cases, it could then be possible to estimate the level of dissolved metal ion concentration of the metal with different particles, using the critical surface area approach as proposed by Skeaff *et. al.* (2000) (see reference in appendix VI, part 5, Metals and metal compounds). From this correlation and a linkage to the appropriate toxicity data at corresponding pH level, it is possible to determine a critical surface area of the substance that delivers the L(E)C$_{50}$ to the dissolution medium and then to convert the critical surface area to a critical particle diameter (see example). This critical particle diameter at appropriate mass loadings for acute and long-term hazard assessment can then be used to:

(a)   determine the classification category of powders based on the finest representative powder on the market; and

(b)   determine an accurate classification of the massive metal by applying a 1 mm (default) diameter.

A9.7.5.4.4.1   Within the critical surface area approach an equation is developed to predict metal ion release (based on previously measured metal ion release from different loadings of the metal), which is correlated to measured surface area, and a corresponding calculated equivalent particle diameter. The basis of the critical surface area approach is that the release of metal ions is dependent on the surface area of the substance, with this release being predictable once the relationship has been established. The critical surface area as the surface area loading (mm^2/l) to a medium that delivers a selected ecotoxicity reference value to that medium. The term SA is the measured specific surface area (m^2/g) of the metal sample. The measured specific critical surface area (SA$_{crit}$) (m^2/g) is the measured specific surface area for the corresponding low, medium and high loadings which are associated with the respective acute and long-term aquatic

A9.7.5.3.2.2　代用アプローチ

適切な慢性毒性データおよび/または変化/溶解データが利用できないが、金属化合物が短期（急性）水生有害性に分類される場合、その金属化合物は代用アプローチに従って分類される。金属化合物の代用アプローチは、金属の代用アプローチと同じである（A9.7.5.2.2.2 参照）。

A9.7.5.4　*粒子径と表面積*

A9.7.5.4.1　試験物質の粒子径または表面積における変動が、一定の時間ウィンドウ内に放出される金属イオンのレベルに著しい変化を引き起こす可能性があることから、粒子径、なかんずく表面積は決定的なパラメータである。したがって、負荷レベルのみに基づく相対的な分類が行えるように、変化試験のために粒子径または表面積は一定とされる。一般に、生成された分類データは、変化の程度を判定するのに、市販されている最も粒子径の小さな試料を用いている。特別な金属粉について生成されたデータは塊状形態の分類に適さないとみなされる場合がある。例えば、試験粉末が構造的に別の物質（例えば別の結晶構造）であることが明らかな場合、または試験粉末が特別な工程によって得られたものであって、塊状の金属からは生成できないような場合、この塊状物質の分類は、データが利用可能なら、より代表的な粒子径もしくは表面積を持つ物質の試験データに基づくことができる。この物質の粉末は、粉末に関して生成されたデータに基づいて別個に分類されることになろう。しかし、通常の環境においては、同一の金属について 2 つを超える分類案が示されることは予想されない。

A9.7.5.4.2　粒子径がデフォルトの直径 1mm を下回る金属は、ケースバイケースで試験されてもよい。例えば、別個の生産技術によって生産された金属粉や、金属粉が塊状形態よりも高い溶解度（または反応性）を示し、その結果より厳しい区分への分類になる場合である。

A9.7.5.4.3　次の表に示すように、試験を行う粒子径は評価対象となる物質によって異なる：

種別	粒子径	備考
金属化合物	市販通常製品のうち最小	最大 1 mm
金属 － 粉末	市販通常製品のうち最小	様々な結晶/形態上の性質がある場合には、別個のデータを考慮する必要も考えられる。
金属 － 塊状	1 mm	十分な正当性があれば、デフォルト値を変更することができる。

塊状形態は通常 1 mm の粒子として試験されるだろう。あるいは、異なる表面積を有する物質の変化/溶解試験により、急性および長期の有害性評価のための適切な負荷量に対する「臨界粒子径」（CPD）を定義することを可能にするような信頼性の高い溶解速度式が得られる可能性がある。

A9.7.5.4.4　一部の形態の金属に関しては、変化/溶解プロトコール（OECD 2001）を用いて、試験した形態の表面積負荷量の関数として、指定された時間間隔後の金属イオン濃度の相関を得ることが可能であろう。そのような場合、Skeaff ら（2000）が提案している臨界表面積の考え方（付録 VI 第 5 部「金属および金属化合物」参考文献参照）を用いて、様々な粒子径を持つ金属の溶存金属イオン濃度レベルを見積ることが可能となるであろう。すなわち、この相関性および対応する pH レベルにおける適切な毒性データとの関連から、媒体に L(E)C$_{50}$ を与える物質の臨界表面積を特定し、次に臨界表面積を臨海粒子径に変換することが可能である（例参照）。この臨界粒子径は、急性および長期の有害性評価のための適切な質量負荷量において、次の目的に使用することができる：

(a) 市販されている最も細かい代表的な粉末に基づいて粉末の分類カテゴリーを決定する；および

(b) 直径 1 mm（既定値）を適用して、塊状金属の正確な分類を決定する。

A9.7.5.4.4.1　臨界表面積アプローチでは、測定された表面積および対応する計算された同等の粒子径と相関関係にある金属イオンの放出を予測する式が作成される（過去に測定された異なる負荷量の金属からの金属イオン放出に基づいて）。臨界表面積アプローチの基本は、金属イオンの放出が物質の表面積に依存することであり、その関係が確立されればこの放出は予測可能である。臨界表面積は、選択された生態毒性基準値をその媒体に与える媒体に負荷される表面積（mm²/l）として表される。SA という用語は、金属試料の測定された比表面積（m²/g）である。測定された比臨界表面積（SA$_{crit}$）（m²/g）は、金属および金

toxicity classification categories in the classification scheme for metals and metal compounds. A typical equation for this relationship for a given substance, aquatic medium, pH and retention time is:

$$\log(C_{Me(aq)}, mg/l) = a + b \log(A_{meas})$$

$C_{Me(aq)}$ = total dissolved concentration of metal ion (mg/l) at a particular length of test time (i.e. 168 hours for short-term hazard assessment) under certain conditions (i.e. pH, specified medium, etc.), as determined by transformation/dissolution testing of different surface area loadings

a, b = regression coefficients

A_{meas} = initial surface area loading (in mm²/l), calculated as follows:

$$A_{meas} = SA \times \text{substance mass loading} \times 10^6$$

where:

SA = specific surface area (in m²/g) measured with the Brunauer-Emmet Teller (BET) nitrogen adsorption-desorption technique

Substance mass loading in g/l.

A9.7.5.5   *Setting M factors for metals and inorganic metal compounds*

A9.7.5.5.1   For the hazard class "Hazardous to the aquatic environment", where the application of the normal cut-off values or concentrations limits may lead to an "under-classification" of the mixture, the M factor concept is used. The M factors are used in application of the summation method for the classification of mixtures containing substances that are classified as very ecotoxic. The concept of M factors has been established to give an increased weight to very toxic substances when classifying mixtures. This ensures that the magnitude of their toxicity is not lost in the derivation of the mixtures classification. M factors are only applicable to the concentration of a substance classified as hazardous to the aquatic environment (Categories Acute 1 and Chronic 1) and are used to derive by the summation method the classification of a mixture in which the substance is present. They are, however, substance specific and it is important that they are established when classifying substances. It is important to note that separate Acute and Chronic M factors should be derived and these may not necessarily be of the same value, depending how each was determined (e.g. the basis of the separate acute and chronic ERV values).

A9.7.5.5.2   For readily soluble metal compounds M factors are applied as for organic substances (see table A9.7.1).

A9.7.5.5.3   For poorly soluble metal compounds and metals M factors are applied based on the ratio of the dissolved metal ion concentration (obtained from transformation/dissolution testing after respectively 7 and 28 days for the loading that was used to establish the classification of Category Acute 1 or Category Chronic 1) and the ERV of the dissolved metal ion. If that ratio is below 10 then an M factor of 1 is applied; if that ratio is ≥ 10 and < 100 then an M factor of 10 is applied; if that ratio is ≥ 100 and < 1000 then an M factor of 100 is applied… (continue this approach in factor 10 intervals).

Table A9.7.1: M factors for readily soluble metal compounds

Acute ERV$_{compound}$ (mg/l)	Acute multiplication factors (M)
0.1 < Acute ERV ≤ 1	1
0.01 < Acute ERV ≤ 0.1	10
0.001 < Acute ERV ≤ 0.01	100
0.0001 < Acute ERV ≤ 0.001	1000
Continue in factor 10 intervals	

属化合物の分類スキームにおけるそれぞれの急性および長期水生毒性分類カテゴリーに関連する、対応する低負荷、中負荷、高負荷の測定比表面積である。与えられた物質、水性媒体、pH、および保持時間に対するこの関係の一般的な式は以下のとおりである：

$$\log(C_{Me(aq)}, mg/l) = a + b \log(A_{meas})$$

$C_{Me(aq)}$ = さまざまな表面積負荷量の変化/溶解試験によって決定される、特定の条件（pH、特定媒体など）下での特定の試験時間（つまり、短期有害性評価の場合は 168 時間）における金属イオンの総溶解濃度（mg/l）

a, b = 回帰係数

A_{meas} = 初期表面積負荷量（単位：mm²/l）は、次のように計算される：

$$A_{meas} = SA \times 物質の質量負荷量 \times 10^6$$

ここで：
SA = Brunauer-Emmet Teller (BET)窒素吸脱着法で測定した比表面積（単位：m²/g）
物質の質量負荷量（g/l）

A9.7.5.5 金属および無機金属化合物の毒性乗率 M の設定

A9.7.5.5.1 危険有害性クラス「水生環境有害性」の場合、通常のカットオフ値又は濃度限界値を適用すると、混合物の「過小分類」につながる可能性があるため、毒性乗率 M の概念が使用される。毒性乗率 M は、非常に生態毒性が強いと分類された物質を含む混合物の分類に、加算法を適用する際に使用される。毒性乗率 M の概念は、混合物を分類する際に非常に毒性の高い物質の大きな重みを与えるために確立された。これにより、混合物の分類を導き出す際に、その毒性の大きさが失われないようにしている。毒性乗率 M は、水生環境に有害と分類された物質（急性 1 および慢性 1）の濃度にのみ適用され、その物質が含まれる混合物の分類を加算法によって導き出すために使用される。しかし、これらは物質固有のものであり、物質を分類する際に設定することが重要である。急性および慢性の毒性乗率 M は別々に導出されるべきであり、それぞれの決定方法（例えば、急性および慢性の個別の ERV 値の根拠）によっては、これらは必ずしも同じ値ではないことに注意することが重要である。

A9.7.5.5.2 易溶性金属化合物には、有機物質と同様に毒性乗率 M が適用される（表 A9.7.1 参照）。

A9.7.5.5.3 難溶性金属化合物および金属の場合、毒性乗率 M は、（急性区分 1 または慢性区分 1 の分類を確立するために使用された負荷のそれぞれ 7 日および 28 日後の変化/溶解試験から得られる）溶解金属イオン濃度の比、及び溶解金属イオンの ERV に基づいて適用される。その比率が 10 よりも小さい場合、毒性乗率 M は 1 が適用され、≧10 かつ < 100 の場合、毒性乗率 M は 10 が適用される。その比率が ≧ 100 かつ < 1000 の場合、毒性乗率 M は 100 が適用される（このアプローチは、乗率 10 の間隔で続く）。

表 A9.7.1: 易溶性金属化合物の毒性乗率 M

急性 ERV$_{compound}$ (mg/l)	急性毒性乗率 (M)
0.1 < 急性 ERV ≦ 1	1
0.01 < 急性 ERV ≦ 0.1	10
0.001 < 急性 ERV ≦ 0.01	100
0.0001 < 急性 ERV ≦ 0.001	1000
以降ファクターが 10 倍ごとに続く	

Chronic ERV$_{compound}$ (mg/l)	Chronic multiplication factors (M)	
	No rapid environmental transformation	**Rapid environmental transformation**
0.01 < Chronic ERV ≤ 0.1	1	-
0.001 < Chronic ERV ≤ 0.01	10	1
0.0001 < Chronic ERV ≤ 0.001	100	10
0.00001 < Chronic ERV ≤ 0.0001	1000	100
Continue in factor 10 intervals		

慢性 ERV$_{compound}$ (mg/l)	慢性毒性乗率 (M)	
	急速な環境変化なし	急速な環境変化あり
0.01 ＜ 慢性 ERV ≦ 0.1	1	-
0.001 ＜ 慢性 ERV ≦ 0.01	10	1
0.0001 ＜ 慢性 ERV ≦ 0.001	100	10
0.00001 ＜ 慢性 ERV ≦ 0.0001	1000	100
以降ファクターが 10 倍ごとに続く		

Annex 9

APPENDIX I

Determination of degradability of organic substances

1. Organic substances may be degraded by abiotic or biotic processes or by a combination of these. A number of standard procedures or tests for determination of the degradability are available. The general principles of some of these are described below. It is by no way the intention to present a comprehensive review of degradability test methods, but only to place the methods in the context of aquatic hazard classification.

2. **Abiotic degradability**

2.1 Abiotic degradation comprises chemical transformation and photochemical transformation. Usually abiotic transformations will yield other organic compounds but will not cause a full mineralization (Schwarzenbach *et al.*, 1993). Chemical transformation is defined as transformation that happens without light and without the mediation of organisms whereas photochemical transformations require light.

2.2 Examples of relevant chemical transformation processes in aqueous environment are hydrolysis, nucleophilic substitution, elimination, oxidation and reduction reactions (Schwarzenbach et al., 1993). Of these, hydrolysis is often considered the most important and it is the only chemical transformation process for which international test guidelines are generally available. The tests for abiotic degradation of chemicals are generally in the form of determination of transformation rates under standardized conditions.

2.3 *Hydrolysis*

2.3.1 Hydrolysis is the reaction of the nucleophiles H_2O or OH^- with a chemical where a (leaving) group of the chemical is exchanged with an OH group. Many compounds, especially acid derivatives, are susceptible to hydrolysis. Hydrolysis can both be abiotic and biotic, but in regard to testing only abiotic hydrolysis is considered. Hydrolysis can take place by different mechanisms at different pHs, neutral, acid- or base-catalysed hydrolysis, and hydrolysis rates may be very dependent on pH.

2.3.2 Currently two guidelines for evaluating abiotic hydrolysis are generally available, the OECD Test Guideline 111 Hydrolysis as a function of pH (corresponding to OPPTS 835.2110) and OPPTS 835.2130 Hydrolysis as a function of pH and temperature. In OECD Test Guideline 111, the overall hydrolysis rate at different pHs in pure buffered water is determined. The test is divided in two, a preliminary test that is performed for chemicals with unknown hydrolysis rates and a more detailed test that is performed for chemicals that are known to be hydrolytically unstable and for chemicals for which the preliminary test shows fast hydrolysis. In the preliminary test the concentration of the chemical in buffered solutions at pHs in the range normally found in the environment (pHs of 4, 7 and 9) at 50ºC is measured after 5 days. If the concentration of the chemical has decreased less than 10 % it is considered hydrolytically stable, otherwise the detailed test may be performed. In the detailed test, the overall hydrolysis rate is determined at three pHs (4, 7 and 9) by measuring the concentration of the chemical as a function of time. The hydrolysis rate is determined at different temperatures so that interpolations or extrapolations to environmentally relevant temperatures can be made. The OPPTS 835.2130 test is almost identical in design to the OECD Test Guideline 111, the difference mainly being in the treatment of data.

2.3.3 It should be noted that apart from hydrolysis the hydrolysis rate constants determined by the tests include all other abiotic transformations that may occur without light under the given test conditions. Good agreement has been found between hydrolysis rates in natural and in pure waters (OPPTS 835.2110).

2.4 *Photolysis*

2.4.1 According to the definitions set out in the OECD Guidance Document concerning aquatic direct photolysis (OECD, 1997), phototransformation of compounds in water can be in the form of primary or secondary phototransformation, where the primary phototransformation (photolysis) can be divided further into direct and indirect photolysis. Direct phototransformation (photolysis) is the case where the chemical absorbs light and as a direct result hereof undergoes transformation. Indirect phototransformation is the case where other excited species transfer energy, electrons or H-atoms to the chemical and thereby induces a transformation (sensitized photolysis). Secondary phototransformation is the case where chemical reactions occur between the chemical and reactive short lived species like hydroxy radicals, peroxy radicals or singlet oxygen that are formed in the presence of light by reactions of excited species like excited humic or fulvic acids or nitrate.

附属書 9

付録 I

有機物質の分解性の測定

1. 有機物質は、非生物的または生物的プロセスのいずれか、あるいはその組み合わせによって分解される。分解性を測定するための数多くの標準的手順、または試験が使用できる。これら試験法のうちいくつかの一般原則について下記に説明する。分解性試験法についての包括的なレビューを提示する意図ではなく、水生有害性分類に関連して、手法を並べたものに過ぎない。

2. 非生物分解

2.1 非生物分解は、化学的変換と光化学変換を含んでいる。非生物的変換からは通常、別の有機化合物が生成されるが、完全な無機化を起こすことはない（Schwarzenbach ら, 1993）。化学変換は、光および生物体の介在なしに起こる変換であると定義されているが、光化学変換には光が必要である。

2.2 水生環境に関連した化学変換プロセスの例として、加水分解、求核置換反応、脱離反応、酸化および還元反応がある（Schwarzenbach ら, 1993）。これらのうちで加水分解は、最も重要であると見なされる場合が多く、また国際的なテストガイドラインが一般的に利用できる唯一の化学変換である。化学品の非生物分解試験は一般的に、標準化された条件下での変換速度の測定という形で行われる。

2.3 加水分解

2.3.1 加水分解は、求核物質 H_2O または OH^- と化学品との反応であり、化学品の（離脱）グループが OH 基と交換される。化合物の多く、特に酸誘導体は加水分解を受けやすい。加水分解は生物的にも非生物的にもおこなわれるが、試験に関しては非生物加水分解のみが考慮される。加水分解は異なった pH、すなわち中性、酸性または塩基触媒の加水分解で、異なったメカニズムが起こり、加水分解速度は pH にきわめて依存している。

2.3.2 現時点では、非生物加水分解を評価するのに、一般に利用できるガイドラインは、OECD テストガイドライン 111「pH の関数としての加水分解」（OPPTS 835.2110 に該当する）および OPPTS 835.2130「pH および温度の関数としての加水分解」の 2 種類である。OECD テストガイドライン 111 では、純粋な緩衝液中において pH を変えた場合の全体的な加水分解速度を測定する。この試験は二段階、すなわち加水分解速度が未知である化学品について実施する予備試験、および加水分解的に不安定なことがわかっている化学品および予備試験で急速な加水分解が認められた化学品について実施する、より詳しい試験とに分けられる。予備試験では、環境中で通常見られる pH（pH4, 7 および 9）範囲にした、温度 50℃ の緩衝液中で、化学品の濃度を 5 日間後に測定する。その化学品の濃度が 10%未満であれば、加水分解的に安定であると見なすが、そうでない場合には詳しい試験を実施する。詳しい試験段階では、3 種類の pH（4, 7 および 9）条件において、その化学品濃度を時間の関数として測定し、全体的な加水分解速度を測定する。加水分解速度は各種温度条件で測定し、環境的に関連性のある温度に内挿または外挿できるようにする。OPPTS 835.2130 試験は実験のデザインでは OECD テストガイドライン 111 にほぼ同一であるが、データ処理に主な違いがある。

2.3.3 試験で測定される加水分解定数は、加水分解以外にも、所定の試験条件下で生じる光を伴わないその他すべての非生物的変換を含んでいることに注意しなければならない。天然水と純水の間の加水分解速度には良好な一致が認められている（OPPTS 835.2110）。

2.4 光分解

2.4.1 水中での光分解に関する OECD ガイダンス文書にある定義によると、水中での化合物の光変換は、一次または二次の光変換の形をとっており、一次の光変換（光分解）はさらに、直接光分解および間接光分解に分けられる。直接光変換（光分解）は、化学品が光を吸収し、その直接の結果として変換を受ける場合をいう。間接光変換とは、別の励起された分子種がエネルギーや電子または水素原子をその化学品に移動させ、これによって変換を誘発する場合である（増感光分解）。二次の光変換とは、化学品と反応性の短命な分子種、例えば励起されたフミン酸、フルボ酸または硝酸塩等の励起分子種の反応により、光の存在下で生成されたヒドロキシラジカルや過酸化ラジカルまたは一重項酸素等の分子種との間に化学反応が生じる場合である。

2.4.2 The only currently available guidelines on phototransformation of chemicals in water are therefore OPPTS 835.2210 *Direct photolysis rate in water by sunlight,* OECD Test Guideline 316 *Phototransformation of chemicals in water-direct photolysis,* and OPPTS 835.5270 *Indirect photolysis screening test*. The OPPTS 835.2210 test as well as OECD Test Guideline 316 uses a tiered approach. In tier 1 the maximum direct photolysis rate constant (minimum half-life) is calculated from a measured molar absorptivity. In tier 2 there are two phases. In Phase 1 the chemical is photolysed with sunlight and an approximate rate constant is obtained. In Phase 2, a more accurate rate constant is determined by using an actinometer that quantifies the intensity of the light that the chemical has actually been exposed to. From the parameters measured, the actual direct photodegradation rate at different temperatures and for different latitudes can be calculated. This degradation rate will only apply to the uppermost layer of a water body, e.g. the first 50 cm or less and only when the water is pure and air saturated which may clearly not be the case in environment. However, the results can be extended over other environmental conditions by the use of a computer programme incorporating attenuation in natural waters and other relevant factors.

2.4.3 The OPPTS 835.5270 screening test concerns indirect photolysis of chemicals in waters that contain humic substances. The principle of the test is that in natural waters exposed to natural sunlight a measured phototransformation rate will include both direct and indirect phototransformation, whereas only direct phototransformation will take place in pure water. Therefore, the difference between the direct photodegradation rate in pure water and the total photodegradation in natural water is the sum of indirect photolysis and secondary photodegradation according to the definitions set out in the annex 9 guidance document. In the practical application of the test, commercial humic substances are used to make up a synthetic humic water, which mimics a natural water. It should be noted that the indirect phototransformation rate determined is only valid for the season and latitude for which it is determined and it is not possible to transfer the results to other latitudes and seasons.

3. Biotic degradability

3.1 Only a brief overview of the test methods is given below. For more information, the comprehensive OECD Detailed Review Paper on Biodegradability Testing (OECD, 1995) should be consulted.

3.2 *Ready biodegradability*

3.2.1 Standard tests for determination of the ready biodegradability of organic substances are developed by a number of organisations including OECD (OECD test guidelines 301A-F), EU (C.4 tests), OPPTS (835.3110) and ISO (9408, 9439, 10707).

3.2.2 The ready biodegradability tests are stringent tests, which provide limited opportunity for biodegradation and acclimatization to occur. The basic test conditions ensuring these specifications are:

(a) high concentration of test substance (2-100 mg/l);

(b) the test substance is the sole carbon and energy source;

(c) low to medium concentration of inoculum (10^4-10^8 cells/ml);

(d) no pre-adaptation of inoculum is allowed;

(e) 28 days test period with a 10-days time window (except for the MITI I method (OECD Test Guideline 301C)) for degradation to take place;

(f) test temperature < 25°C; and

(g) pass levels of 70 % (DOC removal) or 60 % (O_2 demand or CO_2 evolution) demonstrating complete mineralization (as the remaining carbon of the test substance is assumed to be built into the growing biomass).

3.2.3 It is assumed that a positive result in one of the ready biodegradability tests demonstrates that the substance will degrade rapidly in the environment (OECD test guidelines).

3.2.4 Also the traditional BOD_5 tests (e.g. the EU C.5 test) may demonstrate whether a substance is readily biodegradable. In this test, the relative biochemical oxygen demand in a period of 5 days is compared to the theoretical oxygen demand (ThOD) or, when this is not available, the chemical oxygen demand (COD). The test is completed within five days and consequently, the pass level defined in the proposed hazard classification criteria at 50 % is lower than in the ready biodegradability tests.

2.4.2　水中の化学品の光変換に関して、現在利用できるガイドラインは日光による、OECD テストガイドライン 316 *水中での直接光分解による化学品の光変換*および OPPTS 835.5270 *間接光分解スクリーニング試験*だけである。OECD テストガイドライン 316 と同様 OPPTS 835.2210 試験は段階的アプローチを採用している。第 1 段階では、モル吸収率測定値から最大の直接光分解速度定数（最小半減期）を算出する。第 2 段階には 2 つのフェーズがある。フェーズ 1 では、化学品を日光で光分解し、およその速度定数を得る。フェーズ 2 では、その化学品が実際にばく露された光の強度を測定するアクチノメータを用いて、より正確な速度定数を測定する。測定したパラメータから、温度および緯度が異なる場合の実際の直接光分解速度が計算できる。この分解速度が適用できるのは、水の最上部の層、例えば一番上の 50cm 以内で、水が純水かつ空気が飽和している場合のみであるが、このような状態は環境中では実現しないことは明らかである。しかし、自然界の水系およびその他の関連要因を組み入れたコンピュータプログラムを用いれば、この結果を他の環境条件にも拡張できる。

2.4.3　OPPTS 835.5270 スクリーニング試験は、フミン系物質を含む水の中での化学品の間接光分解に関するものである。この試験の原理とは、自然の日光にさらされた自然水系では、光変換速度の測定値には直接および間接両方の光変換が含まれるが、純水中では直接の光変換しか起こらない、ということである。したがって、純水中の直接光分解速度と、自然水系中の総合的な光分解の差は、附属書 9 の手引きに定められた定義によれば、間接光分解と二次光分解の合計である。この試験法を実際に応用するには、市販されているフミン系物質を用いて、自然水系を模した合成腐植水を作成する。間接光変換速度の測定値は、それが測定された季節と緯度にのみ有意であること、およびその結果を他の緯度や季節に換算することは不可能であることに注意が必要である。

3.　生分解性

3.1　試験法の簡単な説明だけを下記に示す。詳しい情報は、包括的な「生分解性試験に関する詳細レビュー文書」（OECD, 1995)を参照されたい。

3.2　*易生分解性*

3.2.1　有機物質の易生分解性を測定する標準的試験法が、OECD（OECD テストガイドライン 301A-F）、EU（C.4 テスト）、 OPPTS（835.3110)および ISO（9408, 9439, 10707)等、多くの機関により開発されている。

3.2.2　易生分解性試験は厳格な試験であり、生分解および馴化が生じる機会が限定されている。このような仕様を確実なものにしている、基礎的な試験条件は次のものである：

(a)　被験物質濃度が高い（2－100mg/l）こと；

(b)　被験物質だけが炭素およびエネルギーの供給源であること；

(c)　植種源の濃度は低いか中程度である（生菌数 10^4-10^8 個/mL）こと；

(d)　植種源の事前の馴化を許さないこと；

(e)　分解が生じる時間ウィンドウは 10 日間であり、試験期間は 28 日（MITI I 法（OECD テストガイドライン 301C）を除いて）とすること；

(f)　試験温度は＜25℃のこと；および

(g)　合格レベルは 70％（DOC 除去）または 60％（酸素要求量または CO_2 発生量）で、完全な無機化が認められること（被験物質の残存炭素は、成長しているバイオマスに取り込まれたと考えられる）。

3.2.3　こうした易生分解性試験の 1 つでの陽性結果は、その物質が環境中で急速分解性であることを示す（OECD テストガイドライン）。

3.2.4　従来からの BOD_5 試験（例えば EU C.5 テスト）によって、物質が易生分解性であるかどうかが示されることもある。この試験では、5 日間の相対的生化学的酸素要求量を理論的酸素要求量（ThOD）と比較するか、または ThOD が利用できない場合には化学的酸素要求量（COD）と比較する。この試験は 5 日間で完了するので、提案された有害性分類基準で定められている 50％という合格レベルは、易生分解性試験の合格レベルより低い。

3.2.5 The screening test for biodegradability in seawater (OECD Test Guideline 306) may be seen as seawater parallel to the ready biodegradability tests. Substances that reach the pass level in OECD Test Guideline 306 (i.e. > 70 % DOC removal or > 60 theoretical oxygen demand) may be regarded as readily biodegradable, since the degradation potential is normally lower in seawater than in the freshwater degradation tests.

3.3 *Inherent biodegradability*

3.3.1 Tests for inherent biodegradability are designed to assess whether a substance has any potential for biodegradation. Examples of such tests are the OECD test guidelines 302A-C tests, the EU C.9 and C.12 tests, and the ASTM E 1625-94 test.

3.3.2 The basic test conditions favouring an assessment of the inherent biodegradation potential are:

(a) a prolonged exposure of the test substance to the inoculum allowing adaptation within the test period;

(b) a high concentration of micro-organisms;

(c) a favourable substance/biomass ratio.

3.3.3 A positive result in an inherent test indicates that the test substance will not persist indefinitely in the environment, however a rapid and complete biodegradation can not be assumed. A result demonstrating more than 70 % mineralization indicates a potential for ultimate biodegradation, a degradation of more than 20 % indicates inherent, primary biodegradation, and a result of less than 20 % indicates that the substance is persistent. Thus, a negative result means that non-biodegradability (persistence) should be assumed (OECD test guidelines).

3.3.4 In many inherent biodegradability tests only the disappearance of the test substance is measured. Such a result only demonstrates a primary biodegradability and not a total mineralization. Thus, more or less persistent degradation products may have been formed. Primary biodegradation of a substance is no indication of ultimate degradability in the environment.

3.3.5 The OECD inherent biodegradation tests are very different in their approach and especially, the MITI II test (OECD Test Guideline 302C) employs a concentration of inoculum that is only three times higher than in the corresponding MITI I ready biodegradability test (OECD Test Guideline 301C). Also, the Zahn-Wellens test (OECD Test Guideline 302B) is a relatively "weak" inherent test. However, although the degradation potential in these tests is not very much stronger than in the ready biodegradability tests, the results can not be extrapolated to conditions in the ready biodegradability tests and in the aquatic environment.

3.4 *Aquatic simulation tests*

3.4.1 A simulation test attempts to simulate biodegradation in a specific aquatic environment. As examples of a standard test for simulation of degradation in the aquatic environment may be mentioned the ISO/DS14592 Shake flask batch test with surface water or surface water/sediment suspensions (Nyholm and Toräng, 1999), the ASTM E1279-89(95) test on biodegradation by a shake-flask die-away method and the similar OPPTS 835.3170 test. Such test methods are often referred to as river die-away tests.

3.4.2 The features of the tests that ensure simulation of the conditions in the aquatic environment are:

(a) use of a natural water (and sediment) sample as inoculum; and

(b) low concentration of test substance (1-100 µg/l) ensuring first-order degradation kinetics.

3.4.3 The use of a radiolabelled test compound is recommended as this facilitates the determination of the ultimate degradation. If only the removal of the test substance by chemical analysis is determined, only the primary degradability is determined. From observation of the degradation kinetics, the rate constant for the degradation can be derived. Due to the low concentration of the test substance, first-order degradation kinetics are assumed to prevail.

3.4.4 The test may also be conducted with natural sediment simulating the conditions in the sediment compartment. Moreover, by sterilizing the samples, the abiotic degradation under the test conditions can be determined.

3.2.5　海水中の生分解性スクリーニング試験（OECD テストガイドライン 306）は、易生分解性試験の海水条件に対応するとみなしてよい。OECD テストガイドライン 306 の合格レベル（すなわち DOC 除去が＞70％または理論的酸素要求量が＞60％）に達する物質は、易生分解性であるとみなしてよい。なぜなら分解性は通常、海水中では、淡水での分解性試験より低くなるからである。

3.3　本質的生分解性

3.3.1　本質的生分解性試験は、ある物質に生分解の可能性があるかどうかを評価するよう設計されている。こうした試験の例として、OECD テストガイドライン 302A-C の各試験、EU C.9 および C.12 の各試験、および ASTM E 1625-94 試験等がある。

3.3.2　本質的生分解性の評価を目的とした基本的な試験の条件は以下のものである：

(a)　試験期間中に馴化させるよう、被験物質の植種源に対する長いばく露時間；

(b)　高い微生物濃度；

(c)　好適な物質/バイオマスの比率。

3.3.3　本質的生分解性試験での陽性結果は、その物質が環境中で無限には存続しないことを意味するが、速かで完全な生分解を推論することはできない。結果から 70％を超える無機化が示された場合、究極の生分解の可能性を意味し、20％を超える分解は本質的な一次生分解を示す。また 20％以下の分解は、その物質は難分解性であることを意味している。したがって、陰性の結果は、非生分解性（難分解性）と考えるべきであることを意味している（OECD テストガイドライン）。

3.3.4　本質的生分解性試験の多くは、被験物質の消失のみを測定する。このような結果は、一次の生分解だけで、総合的な無機化は示されない。したがって、多少にかかわらず、難分解性の分解生成物が生成している可能性もある。物質の一次生分解は、環境中における本質的生分解性を示すものではない。

3.3.5　OECD の各本質的生分解性試験は、そのアプローチの点で極めて違いが大きく、特に MITI II 試験（OECD テストガイドライン 302C）は、相当する MITI I 易生分解性試験（OECD テストガイドライン 301C）より 3 倍程度高いだけの植種源濃度を採用している。また、Zahn-Wellens 試験（OECD テストガイドライン 302B）は、比較的「弱い」本質的生分解性試験である。しかし、こうした試験で認められる分解性は易生分解性試験で認められる分解性よりそれほど強くはないにもかかわらず、その結果は易生分解性試験および水生環境における条件に外挿することはできない。

3.4　水系シミュレーション試験

3.4.1　シミュレーション試験は、ある水生環境における分解をシミュレートしようと試みる。水生環境における分解のシミュレーションのための標準的な試験法の例としては、ISO/DS14592「表層水または表層水/底質懸濁物のフラスコ振騰バッチテスト」（Nyholm と Toräng, 1999）、フラスコ振騰ダイアウェイ試験法による生分解性の ASTM E 1279-89(95)試験、および同様な OPPTS 835.3170 試験が挙げられる。これらの試験法は河川ダイアウェイ試験として参照されることが多い。

3.4.2　水生環境の条件をシミュレートできるようにする試験の特徴は以下のものである：

(a)　自然水（および底質）サンプルを植種源として使用；および

(b)　一次反応の分解速度になるような低い被験物質濃度（1−100μg/l）。

3.4.3　放射性標識された被験物質の使用は、本質的分解の測定を容易にするので、推奨されている。化学分析によって、被験物質の除去だけを測定すると、初期の分解だけしか測定されない。分解の動力学の観察から、分解の速度定数を導き出すことができる。被験物質濃度が低いので、一次反応の分解速度が優先すると推定できる。

3.4.4　試験はまた、底質コンパートメント中の条件をシミュレートして、自然界の底質を用いて実施してもよい。さらに、サンプルを滅菌することによって、試験条件下における非生物分解も測定できる。

3.5　　*STP simulation tests*

Tests are also available for simulating the degradability in a sewage treatment plant (STP), e.g. the OECD Test Guideline 303A Coupled Unit test, ISO 11733 Activated sludge simulation test, and the EU C.10 test. Recently, a new simulation test employing low concentrations of organic pollutants has been proposed (Nyholm et. al., 1996).

3.6　　*Anaerobic degradability*

3.6.1　　Test methods for anaerobic biodegradability determine the intrinsic potential of the test substance to undergo biodegradation under anaerobic conditions. Examples of such tests are the ISO 11734:1995(E) test, the ASTM E 1196-92 test and the OPPTS 835.3400 test.

3.6.2　　The potential for anaerobic degradation is determined during a period of up to eight weeks and with the test conditions indicated below:

　　(a)　performance of the test in sealed vessels in the absence of O_2 (initially in a pure N_2 atmosphere);

　　(b)　use of digested sludge;

　　(c)　a test temperature of 35 °C; and

　　(d)　determination of head-space gas pressure (CO_2 and CH_4 formation).

3.6.3　　The ultimate degradation is determined by determining the gas production. However, also primary degradation may be determined by measuring the remaining parent substance.

3.7　　*Degradation in soil and sediment*

3.7.1　　Many substances end up in the soil or sediment compartments and an assessment of their degradability in these environments may therefore be of importance. Among standard methods may be mentioned the OECD Test Guideline 304A test on inherent biodegradability in soil, which corresponds to the OPPTS 835.3300 test.

3.7.2　　The special test characteristics ensuring the determination of the inherent degradability in soil are:

　　(a)　natural soil samples are used without additional inoculation;

　　(b)　radiolabelled test substance is used; and

　　(c)　evolution of radiolabelled CO_2 is determined.

3.7.3　　A standard method for determining the biodegradation in sediment is the OPPTS 835.3180 Sediment/water microcosm biodegradation test. Microcosms containing sediment and water are collected from test sites and test compounds are introduced into the system. Disappearance of the parent compound (i.e. primary biodegradation) and, if feasible, appearance of metabolites or measurements of ultimate biodegradation may be made.

3.7.4　　Two OECD test guidelines address aerobic and anaerobic transformation in soil and in aquatic sediments (OECD test guidelines 307 and 308 respectively). The experiments are performed to determine the rate of transformation of the test substance and the nature and rates of formation and decline of transformation products under environmentally realistic conditions including a realistic concentration of the test substance. Either complete mineralization or primary degradability may be determined depending on the analytical method employed for determining the transformation of the test substance.

3.8　　*Methods for estimating biodegradability*

3.8.1　　In recent years, possibilities for estimating environmental properties of substances have been developed and, among these, also methods for predicting the biodegradability potential of organic substances (e.g. the Syracuse Research Corporation's Biodegradability Probability Program, BIOWIN). Reviews of methods have been performed by OECD (1993) and by Langenberg et al. (1996). They show that group contribution methods seem to be the most successful methods. Of these, the Biodegradation Probability Program (BIOWIN) seems to have the broadest application. It gives a qualitative estimate of the probability of slow or fast biodegradation in the presence of a mixed population of environmental micro-organisms. The applicability of this program has been evaluated by the US EPA/EC Joint Project on the Evaluation of (Q)SARs (OECD, 1994), and by Pedersen et al. (1995). The latter is briefly referred below.

3.5 STPシミュレーション試験

下水処理施設（STP）における分解性をシミュレートする試験もある。例えば OECD テストガイドライン 303A "Coupled Unit" 試験、ISO 11733「活性汚泥シミュレーション試験」、EU C.10 試験等がある。最近になって、低濃度の有機汚染物質を用いる、新しいシミュレーション試験が提案された（Nyholmら，1996）。

3.6 嫌気的分解性

3.6.1 嫌気的生分解性のための試験法は、被験物質が嫌気的条件化で生分解を受ける本来の可能性を測定する。こうした試験法の例は、ISO 11734:1995(E)試験、ASTM E 1196-92 試験、および OPPTS 835.3400 試験等である。

3.6.2 嫌気的分解性は、8 週間までの試験期間で、下に示すような試験条件で測定される：
 (a) 酸素の存在しない状態（初期は純粋な窒素雰囲気）で密閉容器内での試験実施；
 (b) 消化された汚泥の使用；
 (c) 試験温度 35℃；および
 (d) ヘッドスペースのガス圧を測定（CO_2 および CH_4 の生成）。

3.6.3 本質的生分解はガスの生成によって判定される。しかし、親化合物の残存量を測定して初期段階の分解も測定できる。

3.7 土壌および底質中の分解

3.7.1 物質の多くは最終的に土壌または底質コンパートメントに行き着くので、こうした環境中における物質の分解性評価が重要かもしれない。標準的な方法として、土壌中の本質的生分解性に関する OECD テストガイドライン 304A 試験があり、これは OPPTS 835.3300 試験に相当する。

3.7.2 土壌中の本質的な分解性が測定できるようにする特殊な試験の特徴は以下のものである：

 (a) 自然界から得た土壌サンプルを、追加の植種源なしで使用する；
 (b) 放射性標識された被験物質を用いる；および
 (c) 放射性標識された二酸化炭素の生成量を測定する。

3.7.3 底質中の生分解を測定するための標準的な方法は、OPPTS 835.3180「底質/水ミクロコズム生分解性試験」である。底質および水を含むミクロコズムを試験地点から採取し、その系に被験物質を加える。親化合物の消失（すなわち初期段階の分解）、そして可能であれば代謝物の出現、あるいは本質的生分解の測定を行ってもよい。

3.7.4 2 つの OECD テストガイドラインが土および水堆積物中の好気性および嫌気性変換を扱っている（それぞれテストガイドライン 307 および 308）。実験は、被験物質の現実的な濃度を含む、環境的に現実に近い条件下での、被験物質変換速度、および変換生成物の物質種、ならびに生成と減少の各速度を測定するために行なわれる。被験物質の変換の測定に採用する分析方法に依存して、完全な無機化、あるいは初期段階の分解のいずれかが測定される。

3.8 生分解性推定の方法

3.8.1 近年になって、物質の環境特性を推定する可能性が発展してきており、その中で、有機物質の生分解性を予測する方法も開発された（例えば Syracuse Research Corporation の Biodegradability Probability Program, BIOWIN）。方法についてのレビューは OECD（1993)および Langenberg ら(1996)によって行われた。そのレビューによれば、官能基の寄与を見る方法が最も成功しているように思われる。そのうちでも、Biodegradability Probability Program （BIOWIN)は適用範囲が広いように思われる。このプログラムでは、環境微生物の混合集団の存在下で、生分解が遅いか速いかを定性的に推定する。このプログラムの応用範囲については、US EPA/EC による(Q)SAR 評価に関する合同プロジェクト（OECD, 1994）および Pedersen ら(1995)が評価している。この後者について下記に簡略に述べる。

3.8.2	A validation set of experimentally determined biodegradation data was selected among the data from MITI (1992) but excluding substances for which no precise degradation data were available and substances already used for development of the programme. The validation set then consisted of 304 substances. The biodegradability of these substances was estimated by use of the programme's non-linear estimation module (the most reliable) and the results compared with the measured data. 162 substances were predicted to degrade "fast", but only 41 (25 %) were actually readily degradable in the MITI I test. 142 substances were predicted to degrade "slowly", which was confirmed by 138 (97 %) substances being not readily degradable in the MITI I test. Thus, it was concluded that the programme may be used for classification purposes only when no experimental degradation data can be obtained, and when the programme predicts a substance to be degraded "slowly". In this case, the substance can be regarded as not rapidly degradable.

3.8.3	The same conclusion was reached in the US EPA/EC Joint Project on the Evaluation of (Q)SARs by use of experimental and QSAR data on new substances notified in the EU. The evaluation was based on an analysis of QSAR predictions on 115 new substances also tested experimentally in ready biodegradability tests. Only 9 of the substances included in this analysis were readily biodegradable. The employed QSAR methodology is not fully specified in the final report of the Joint US EPA/EC project (OECD, 1994), but it is likely that the majority of predictions were made by using methods which later have been integrated in the Biodegradation Probability Program.

3.8.4	Also in the EU TGD (EC, 1996) it is recommended that estimated biodegradability by use of the Biodegradation Probability Program is used only in a conservative way, i.e. when the programme predicts fast biodegradation, this result should not be taken into consideration, whereas predictions of slow biodegradation may be considered (EC, 1996).

3.8.5	Thus, the use of results of the Biodegradability Probability Program in a conservative way may fulfil the needs for evaluating biodegradability of some of the large number of substances for which no experimental degradation data are available.

3.8.2 実験的に測定された生分解性データの有意性確認のためのセットを MITI (1992) のデータから選び出したが、正確な分解データが入手されていない物質や、上述のプログラムの開発にすでに使用されている物質は除外した。このようにして有意性確認用セットは 304 物質から構成された。これらの物質の生分解性について、プログラムの（最も信頼性が高い）非線形推定モジュールを用いて推定し、結果を測定データと比較した。162 種の物質が「速やかに」分解すると予測されたが、MITI I 試験では、41 種（25%）だけが実際に易分解性であった。142 種は「ゆっくり」分解すると予測されたが、MITI I 試験で実際に急速分解性でないとされたのは 138 種（97%）であった。したがって、このプログラムは、分解実験データが得られず、かつプログラムによる推定が、その物質は「ゆっくりと」分解されるとなる場合にのみ、有害性分類に使用できると結論づけられた。そうした場合、物質は速かな分解性ではないとみなすことができる。

3.8.3 EU に届け出された新規物質についての、実験データおよび QSAR データを使用した、(Q)SAR の評価に関する US EPA/EC 合同プロジェクトでも、同様な結論が得られた。易生分解性試験で実験的に試験された 115 の新規物質について、QSAR 予測を解析したものに基づいて評価した。この解析に含まれた物質のうち、9 種だけが易生分解性であった。採用された QSAR の方法については、US EPA/EC 合同プロジェクトの最終報告書（OECD, 1994）でも十分には記述されていないが、予測のほとんどは、後になって Biodegradation Probability Program (BIOWIN) に組み込まれた方法によって行われた可能性が高い。

3.8.4 EU TGD (EC, 1996)においても、Biodegradation Probability Program を用いた、生分解性の推定は控えめに使うのみとすることが推奨されている。すなわち、このプログラムで速かな生分解性が予測される場合、その結果を考慮に加えるべきではなく、反対に生分解性が遅いと予測されたならば、考慮してもよい（EC, 1996）。

3.8.5 したがって、Biodegradation Probability Program の結果を控えめに使うことは、分解性に関する実験データが利用できない多数の物質のいくつかについて生分解性を評価する必要を満たすかもしれない。

Annex 9

APPENDIX II

Factors influencing degradability in the aquatic environment

1. Introduction

1.1 The OECD classification criteria are considering the hazards to the aquatic environment only. However, the hazard classification is primarily based on data prepared by conduction of tests under laboratory conditions that only seldom are similar to the conditions in the environment. Thus, the interpretation of laboratory test data for prediction of the hazards in the aquatic environment should be considered.

1.2 Interpretation of test results on biodegradability of organic substances has been considered in the OECD Detailed Review Paper on Biodegradability Testing (OECD, 1995).

1.3 The conditions in the environment are typically very different from the conditions in the standardized test systems, which make the extrapolation of degradation data from laboratory tests to the environment difficult. Among the differences, the following have significant influence on the degradability:

(a) Organism related factors (presence of competent micro-organisms);

(b) Substrate related factors (concentration of the substance and presence of other substrates); and

(c) Environment related factors (physico-chemical conditions, presence of nutrients, bioavailability of the substance).

These aspects will be discussed further below.

2. Presence of competent micro-organisms

2.1 Biodegradation in the aquatic environment is dependent on the presence of competent micro-organisms in sufficient numbers. The natural microbial communities consist of a very diverse biomass and when a 'new' substance is introduced in a sufficiently high concentration, the biomass may be adapted to degrade this substance. Frequently, the adaptation of the microbial population is caused by the growth of specific degraders that by nature are competent to degrade the substance. However, also other processes as enzyme induction, exchange of genetic material and development of tolerance to toxicity may be involved.

2.2 Adaptation takes place during a "lag" phase, which is the time period from the onset of the exposure until a significant degradation begins. It seems obvious that the length of the lag phase will depend on the initial presence of competent degraders. This will again depend on the history of the microbial community, i.e. whether the community formerly has been exposed to the substance. This means that when a xenobiotic substance has been used and emitted ubiquitously in a number of years, the likelihood of finding competent degraders will increase. This will especially be the case in environments receiving emissions as e.g. biological wastewater treatment plants. Often more consistent degradation results are found in tests where inocula from polluted waters are used compared to tests with inocula from unpolluted water (OECD, 1995; Nyholm and Ingerslev, 1997).

2.3 A number of factors determine whether the potential for adaptation in the aquatic environment is comparable with the potential in laboratory tests. Among other things adaptation depends on:

(a) initial number of competent degraders in the biomass (fraction and number);

(b) presence of surfaces for attachment;

(c) concentration and availability of substrate; and

(d) presence of other substrates.

2.4 The length of the lag phase depends on the initial number of competent degraders and, for toxic substances, the survival and recovery of these. In standard ready biodegradability tests, the inoculum is sampled in sewage treatment plants. As the load with pollutants is normally higher than in the environment, both the fraction and the number

附属書　9

付録Ⅱ

水生環境中の分解性に影響する因子

1. 序

1.1　OECDの分類基準は、水生環境に対する有害性のみを考慮している。しかし、有害性分類は主に、環境中の条件に類似していることは極めて稀なような実験室条件で、試験を実施して作成されたデータに基づいている。したがって、水生環境での有害性の予測のために、実験室での試験データの解釈を考えるべきである。

1.2　有機物質の生分解性に関する試験結果の解釈は、OECD生分解性試験に関する詳細レビュー文書(OECD, 1995)で検討されている。

1.3　環境中の状態は、標準化された試験系における条件とは特徴的に非常に異なっているので、実験室での試験から得られた分解性データを環境に外挿することを困難にしている。差異の中でも、以下の要因は分解性に著しい影響を及ぼす：

(a)　生物体に関連した要因（分解能力をもつ微生物の存在）；

(b)　物質に関連した要因（物質濃度および他の基質の存在）；および

(c)　環境関連要因（物理化学的条件、栄養分の存在、物質の生物学的利用性）

これらの各点について、以下にさらに議論する。

2. 分解能力をもつ微生物の存在

2.1　水生環境における生分解は、分解能力のある微生物が十分な数で存在することに依存している。自然界の微生物集団はきわめて多様なバイオマスで構成されており、「新規」物質が十分高濃度で導入されると、バイオマスはその物質を分解するよう馴化することもある。多くの場合、微生物集団の馴化は、本来その物質を分解する能力を有する特定の分解者の増殖によって起こる。しかし、酵素の誘導、遺伝物質の交換および毒性に対する耐性の獲得等、その他のプロセスもかかわることがある。

2.2　馴化は「ラグ（遅れ）」相で起こる。これはばく露開始から著しい分解が開始されるまでの時間である。ラグ相の長さが、分解能力の高い分解者が最初から存在しているかどうかに依存することは明らかなように思われる。このことはまた微生物集団の経緯、すなわち集団が以前にその物質にばく露されたかどうか、に依存する。これはすなわち、人工物質が使用され、何年間も至るところで放出されていると、分解能力の高い分解者が見つかる可能性が高くなることを意味している。これが特にあてはまるのは、例えば生物学的な下水処理施設等からの排出を受けている環境である。汚染されていない水系から得た植種源を用いる試験に比べて、汚染された水系から植種源を得ている試験の方が、分解結果により一貫性が見られることが多い（OECD, 1995; NyholmとIngerslev, 1997）。

2.3　水生環境での馴化性が、実験室における試験での馴化性と比較できるかを決定するいくつかの要因がある。そうした要因のうち、馴化性が依存するのは下記のような要因である：

(a)　バイオマス中の分解能力の高い分解者の初期の数（％および数）；

(b)　固着するための表面の存在；

(c)　基質の濃度および利用性；および

(d)　他の基質の存在。

2.4　ラグ相の長さは、分解能力の高い分解者の初期の数および有毒な物質の場合には、これら分解者の生存および回復に依存する。標準的な易生分解性試験では、植種源は下水処理施設から採取されている。こ

of competent degraders may be higher than in the less polluted aquatic environment. It is, however, difficult to estimate how much longer the lag phase will be in the aquatic environment than in a laboratory test due to the likely lower initial number of competent degraders.

2.5 Over long periods of time, the initial concentration of competent degraders is not important as they will grow up when a suitable substrate is present in sufficient concentrations. However, if the degradability in a short period of time is of concern, the initial concentration of competent degrading micro-organisms should be considered (Scow, 1982).

2.6 The presence of flocs, aggregates and attached micro-organisms may also enhance adaptation by e.g. development of microbial niches with consortia of micro-organisms. This is of importance when considering the capability of adaptation in the diverse environments in sewage treatment plants or in sediment or soil. However, the total number of micro-organisms in ready biodegradability tests and in the aquatic environment are of the same orders of magnitude (10^4-10^8 cells/ml in ready biodegradability tests and 10^3-10^6 cells/ml or more in surface water (Scow, 1982). Thus, this factor is probably of minor importance.

2.7 When discussing the extrapolation to environmental conditions it may be valuable to discriminate between oligotrophic and eutrophic environments. Micro-organisms thriving under oligotrophic conditions are able to mineralize organic substrates at low concentrations (fractions of mg C/L), and they normally have a greater affinity for the substrate but lower growth rates and higher generation times than eutrophic organisms (OECD, 1995). Moreover, oligotrophs are unable to degrade chemicals in concentrations higher than 1 mg/l and may even be inhibited at high concentrations. Opposite to that, eutrophs require higher substrate concentrations before mineralization begins and they thrive at higher concentrations than oligotrophs. Thus, the lower threshold limit for degradation in the aquatic environment will depend on whether the microbial population is an oligotroph or an eutroph population. It is, however, not clear whether oligotrophs and eutrophs are different species or whether there is only an oligotrophic and an eutrophic way of life (OECD, 1995). Most pollutants reach the aquatic environment directly through discharge of wastewater and consequently, these recipients are mostly eutrophic.

2.8 From the above discussion it may thus be concluded that the chance of presence of competent degraders is greatest in highly exposed environments, i.e. in environments continuously receiving substances (which more frequently occurs for high production volume chemicals than for low production volume chemicals). These environments are often eutrophic and therefore, the degradation may require relatively high concentrations of substances before onset. On the other hand, in pristine waters competent species may be lacking, especially species capable of degradation of chemicals only occasionally released as low production volume chemicals.

3. Substrate related factors

3.1 *Concentration of test substance*

3.1.1 In most laboratory tests, the test substance is applied in very high concentrations (2-100 mg/l) compared to the concentrations in the lower µg/l range that may be expected in the aquatic environment. In general, growth of micro-organisms is not supported when a substrate is present in concentrations below a threshold level of around 10 µg/l and at lower concentrations, even the energy requirement for maintenance is not met (OECD, 1995). The reason for this lower threshold level is possibly a lack of sufficient stimulus to initiate an enzymatic response (Scow, 1982). This means in general that the concentrations of many substances in the aquatic environment are at a level where they can only hardly be the primary substrate for degrading micro-organisms.

3.1.2 Moreover, the degradation kinetics depends on substance concentration (S_0) compared with the saturation constant (K_s) as described in the Monod equation. The saturation constant is the concentration of the substrate resulting in a specific growth rate of 50 % of the maximum specific growth rate. At substrate concentrations much lower than the saturation constant, which is the normal situation in most of the aquatic environment, the degradation can be described by first order or logistic kinetics (OECD, 1995). When a low density of micro-organisms (lower than 10^3-10^5 cells/ml) prevails (e.g. in oligotrophic waters), the population grows at ever decreasing rates which is typical of logistic kinetics. At a higher density of micro-organisms (e.g. in eutrophic waters), the substrate concentration is not high enough to support growth of the cells and first order kinetics apply, i.e. the degradation rate is proportional with the substance concentration. In practice, it may be impossible to distinguish between the two types of degradation kinetics due to uncertainty of the data (OECD, 1995).

3.1.3 In conclusion, substances in low concentrations (i.e. below 10 µg/l) are probably not degraded as primary substrates in the aquatic environment. At higher concentrations, readily degradable substances will probably be degraded as primary substrates in the environment at a degradation rate more or less proportional with the concentration of the substance. The degradation of substances as secondary substrates is discussed below.

こでは汚染物質の負荷量が一般に環境中より多いので、分解能力の高い分解者の比率および数は、より汚染されていない水生環境中に比べて大きいであろう。しかし、水生環境では分解能力の高い分解者の初期の数が、実験室での試験より小さいので、ラグ相が水生環境でどの程度長くなるかを推定するのは困難である。

2.5 　長期間にわたる場合には、分解能力の高い分解者は、適切な基質が十分な濃度で存在していれば増殖するので、その初期濃度は重要ではない。しかし、短期間での分解性を問題にする場合は、分解能力のある微生物の初期濃度を考慮する必要がある（Scow, 1982）。

2.6 　フロック、凝集物および付着した微生物が存在することによっても、例えば微生物共同体による微生物ニッチの形成等により、馴化が増強される。このことは、下水処理施設、底質あるいは土壌中など多様な環境における馴化能力を考える場合には重要である。しかし、易生分解性試験および水生環境中の微生物の総数は、ほぼ同じ桁数である（易生分解性試験では生菌数は 10^4-10^8 個/ml、表層水中では 10^3-10^6 個/ml またはそれ以上）（Scow, 1982）。

2.7 　環境条件への外挿を考える場合、貧栄養環境と富栄養環境を区別することは有益であろう。貧栄養条件で成育している微生物は、低濃度（mgC/L 程度）の有機基質を無機化することができ、通常は富栄養条件にある生物体より基質に対する親和性は大きいが、成長速度は低く発生回数は多い（OECD, 1995）。さらに貧栄養微生物は濃度が 1mg/l を超える化学品を分解することができず、高濃度では抑制されることさえある。これとは反対に富栄養微生物は、無機化の開始前に高濃度の基質を必要とし、貧栄養微生物よりも高濃度で成育している。このように、水生環境における分解の低い閾値は、その微生物集団が貧栄養集団か富栄養集団かに依存する。しかし、貧栄養微生物と富栄養微生物とが異なった種であるか、またそれぞれに貧栄養的な生活方法と富栄養的な生活方法しかないのかは不明である（OECD, 1995）。ほとんどの汚染物質は廃水の放出によって直接水生環境に到達し、したがって受け入れる環境はほとんどが富栄養となる。

2.8 　上述の議論から、高度にばく露されている環境、すなわち連続的に物質を受け入れている環境（生産量の低い化学品よりも生産量の大きい化学品で、より多く起こる）において、分解能力の高い分解者の存在する機会が最高となると結論づけてよいだろう。こうした環境は富栄養となることが多く、したがって、分解が始まる前に比較的高濃度の物質が必要となることもある。その一方で、清潔な水系では分解能力の高い微生物種、特に生産量の低い化学品として、まれにしか放出されない化学品を分解する能力のある微生物種が欠乏しているであろう。

3. 基質関連要因

3.1 *被験物質の濃度*

3.1.1 　ほとんどの実験室での試験で、被験物質は、水生環境で予測される数 µg/l 域の濃度に比べて、極めて高濃度（2－100mg/l）で添加される。一般に、基質が約 10µg/l という閾値より低い濃度で存在するときは、微生物の増殖が支持されず、維持のためのエネルギー要求量さえ満たされない（OECD, 1995）。このように低い閾値レベルの理由は、酵素的反応を開始する十分な刺激が足りないためであると思われる（Scow, 1982）。これは一般に水生環境での多くの物質の濃度が、分解性微生物の初期の基質にかろうじてなりえるレベルでしか存在していないことを意味している。

3.1.2 　さらに、分解の反応速度は、Monod の式における飽和定数（K_s）と物質濃度（S_o）に依存する。飽和定数は、最大比増殖速度の 50％の比増殖速度となる基質濃度である。飽和定数よりはるかに低い基質濃度は、ほとんどの水生環境では普通の状況であるが、この状態では分解は一次反応またはロジスティック速度論で説明できる（OECD, 1995）。微生物密度が低い（10^3-10^5 個/ml）状態が優先的である場合（例えば貧栄養水系）、集団はさらに低い速度で増殖するが、これはロジスティック速度論で典型的なものである。微生物密度がそれより高い場合（例えば富栄養水系）には、細胞増殖を支えられるほど基質濃度は高くなく、一次反応速度論が適用される、すなわち、分解速度は物質濃度に比例する。実際には、データの不確実性のために、この 2 つの種類の分解速度論を区別することは不可能であろう（OECD, 1995）。

3.1.3 　まとめとして、低濃度（すなわち 10µg/l 以下）の物質は、おそらく水生環境中で主要な基質として分解されることはないと思われる。濃度がそれより高ければ、易生分解性物質はほぼ物質濃度に比例した分解速度で、環境中の主要物質として分解されると思われる。二次基質としての物質の分解については以下で述べる。

3.2 *Presence of other substrates*

3.2.1 In the standard tests, the test substance is applied as the sole substrate for the micro-organisms while in the environment, a large number of other substrates are present. In natural waters, concentrations of dissolved organic carbon are often found in the range 1-10 mg C/l, i.e. up to a factor 1000 higher than a pollutant. However, much of this organic carbon is relatively persistent with an increasing fraction of persistent matter the longer the distance from the shore.

3.2.2 Bacteria in natural waters are primarily nourishing on exudates from algae. These exudates are mineralized very quickly (within minutes) demonstrating that there is a high degradation potential in the natural micro-organism communities. Thus, as micro-organisms compete for the variety of substrates in natural waters, there is a selection pressure among micro-organisms resulting in growth of opportunistic species capable of nourishing on quickly mineralized substrates, while growth of more specialized species is suppressed. Experiences from isolation of bacteria capable of degrading various xenobiotics have demonstrated that these organisms are often growing relatively slowly and survive on complex carbon sources in competition with more rapidly growing bacteria. When competent micro-organisms are present in the environment, their numbers may increase if the specific xenobiotic substrate is continuously released and reach a concentration in the environment sufficient to support growth. However, most of the organic pollutants in the aquatic environment are present in low concentrations and will only be degraded as secondary substrates not supporting growth.

3.2.3 On the other hand, the presence of quickly mineralized substrates in higher concentrations may facilitate an initial transformation of the xenobiotic molecule by co-metabolism. The co-metabolized substance may then be available for further degradation and mineralization. Thus, the presence of other substrates may increase the possibilities for a substance to be degraded.

3.2.4 It may then be concluded that the presence of a variety of substrates in natural waters and among them quickly mineralized substrates, may on the one hand cause a selection pressure suppressing growth of micro-organisms competent of degrading micro-pollutants. On the other hand, it may facilitate an increased degradation by an initial co-metabolism followed by a further mineralization. The relative importance of these processes under natural conditions may vary depending on both the environmental conditions and the substance and no generalization can yet be established.

4. Environment related factors

4.1 The environmental variables control the general microbial activity rather than specific degradation processes. However, the significance of the influence varies between different ecosystems and microbial species (Scow, 1982).

4.2 *Redox potential*

One of the most important environment related factors influencing the degradability is probably the presence of oxygen. The oxygen content and the related redox potential determines the presence of different types of micro-organisms in aquatic environments with aerobic organisms present in the water phase, in the upper layer of sediments and in parts of sewage treatment plants, and anaerobic organisms present in sediments and parts of sewage treatment plants. In most parts of the water phase, aerobic conditions are prevailing, and the prediction of the biodegradability should be based on results from aerobic tests. However, in some aquatic environments the oxygen content may be very low in periods of the year due to eutrophication and the following decay of produced organic matter. In these periods, aerobic organisms will not be able to degrade the chemical, but anaerobic processes may take over if the chemical is degradable under anaerobic conditions.

4.3 *Temperature*

Another important parameter is the temperature. Most laboratory tests are performed at 20-25°C (standard aerobic ready biodegradability tests), but anaerobic tests may be performed at 35 °C as this better mimics the conditions in a sludge reactor. Microbial activity is found in the environment at temperatures ranging from below 0°C to 100°C. However, optimum temperatures are probably in the range from 10°C to 30°C and roughly, the degradation rate doubles for every 10°C increase of temperature in this range (de Henau, 1993). Outside this optimum range the activity of the degraders is reduced drastically although some specialized species (termo- and psycrophilic bacteria) may thrive. When extrapolating from laboratory conditions, it should be considered that some aquatic environments are covered by ice in substantial periods of the year and that only minor or even no degradation can be expected during the winter season.

3.2 その他の基質の存在

3.2.1 標準的な試験では、被験物質はその微生物に対して単一の基質として添加されるが、環境中では、他の基質が多数存在している。自然水系では、溶存有機炭素濃度は、しばしば 1－10mgC/l の範囲で、すなわち汚染物質よりも 1000 倍高い濃度で検出される。しかし、こうした有機炭素の多くは比較的難分解性であり、岸辺から遠いほど難分解性物質の比率が高くなる。

3.2.2 自然水系のバクテリアは、藻類の浸出液を主な栄養源としている。こうした浸出液は速かに無機化され（数分間以内）、自然界の微生物集団には高い分解能力があることを実証している。したがって、微生物群が自然水系中の多様な基質を争奪するので、微生物間に淘汰圧が生じ、速やかに無機化される基質を栄養源にできる日和見微生物種が増殖し、より特殊化した微生物種の増殖は抑えられる。種々の人工物質を分解する能力のある微生物を単離した経験は、こうした微生物はしばしば比較的ゆっくりと増殖し、より増殖の速いバクテリアとの競合のなかで、複雑な炭素源で生存していることを示している。環境中に分解能力を有する微生物が存在している場合、ある人工基質が連続的に放出されて、環境中濃度がその増殖を支えるのに十分になれば、その数は増加するであろう。しかし、水生環境中の有機汚染物質の多くは低い濃度で存在し、二次基質として分解されるだけで、増殖を支えていないであろう。

3.2.3 他方、速やかに無機化される物質が高濃度で存在すれば、共代謝により人工分子の初期変換を促進することがある。そして共代謝された物質は、さらなる分解、および無機化を受けやすくなる。このように他の基質の存在が、物質が分解される可能性を高めることもある。

3.2.4 したがって、自然界の水系中に多様な基質が存在しており、そのなかには速やかに無機化される基質もあることが、一方では微量の汚染物質を分解する能力を有する微生物の増殖を阻害する場合がある、他方、こうした存在は、初期の共代謝によって分解を促進し、次いで、さらに無機化しやすくする場合もある、と結論づけられる。自然条件下でのこれらのプロセスの相対的な重要性は、環境条件および物質の両者によって異なり、一般則はまだ確立できていない。

4. 環境関連要因

4.1 環境についての各要因は、特定の分解プロセスより、むしろ一般的な微生物活動をコントロールしている。しかし、この影響の重要性は、生態系や微生物種の違いによって異なっている（Scow, 1982）。

4.2 酸化還元ポテンシャル

分解性に影響する環境関連要因として最も重要なものの１つは、おそらく酸素の存在であると思われる。酸素含量およびそれに関連して、酸化還元ポテンシャルが、水相中、底質上層部中、および下水処理施設の各部分に存在している好気的生物、および底質中や下水処理施設の各部分に存在している嫌気的生物など、水生環境中における多様な種類の微生物の存在を決定している。ほとんどの水相中では、好気的条件が優先しており、生分解性の予測は好気的試験の結果に基づくべきである。しかし、ある水生環境では、富栄養化および生成した有機物の腐植のために、一年のある期間、酸素含量が極めて低くなることもありうる。こうした期間中は、好気的生物は化学品を分解できないが、もしその化学品が嫌気的条件下で分解されうるならば、嫌気的プロセスがとってかわることもある。

4.3 温度

もう１つ重要なパラメータは温度である。ほとんどの実験試験（標準の好気的易生分解性試験）は 20－25℃で実施されるが、嫌気的試験は、汚泥リアクター内の条件をより適切に模している、35℃で行われることもある。環境中では、微生物活動は 0℃以下から 100℃の温度範囲で見出される。しかし、最適温度は多分 10－30℃の範囲内にあり、おおまかにいえば、この範囲内で温度が 10℃上昇するごとに分解速度は倍増する（de Henau, 1993）。この最適温度範囲の外では、分解者の活動は急激に低下するが、ある特殊化した微生物種（好熱細菌および好冷細菌）は繁殖する。実験条件から外挿する場合、年間のほとんどの期間、氷で覆われており、冬期には分解がほとんど、またはまったく期待されないような水生環境もあることを考慮すべきである。

4.4　　　*pH*

Active micro-organisms are found in the entire pH range found in the environment. However, for bacteria as a group, slightly alkaline conditions favour the activity and the optimum pH range is 6-8. At a pH lower than 5, the metabolic activity in bacteria is significantly decreased. For fungi as a group, slightly acidic conditions favour the activity with an optimum pH range of 5-6 (Scow, 1982). Thus, an optimum for the degrading activity of micro-organisms will probably be within the pH range of 5-8, which is the range most often prevailing in the aquatic environment.

4.5　　　*Presence of nutrients*

The presence of inorganic nutrients (nitrogen and phosphorus) is often required for microbial growth. However, these are only seldom the activity limiting factors in the aquatic environment where growth of micro-organisms is often substrate limited. However, the presence of nutrient influences the growth of primary producers and then again, the availability of readily mineralized exudates.

4.4 *pH*

環境中で見られる、ほぼすべてのpH域で、活性のある微生物が見いだせる。しかし、細菌集団としては、弱アルカリ性の状態が活動に最も適しており、最適pH範囲は6－8である。5より低いpHでは、細菌の代謝活性は著しく低下する。真菌類集団にとっては、弱酸性の状態の方が活動に適切であり、最適pH範囲は5－6である（Scow, 1982）。したがって、微生物の分解活動にとって最適なのは、おそらくpH 5－8の範囲であり、これは水生環境で最も多く見られるpH範囲である。

4.5 *栄養塩の存在*

微生物の増殖には無機栄養塩（窒素およびリン）の存在がしばしば必要になる。しかし、微生物の増殖は基質によって限定されることが多いので、水生環境で無機栄養塩類が活性の限定要因となることは稀である。しかし、栄養塩類の存在は一次生産者の増殖そして、また無機化されやすい浸出物の利用性に影響する。

Annex 9

APPENDIX III

Basic principles of the experimental and estimation methods for determination of BCF and Kow of organic substances

1.   **Bioconcentration factor (BCF)**

1.1   *Definition*

The bioconcentration factor is defined as the ratio between the concentration of the chemical in biota and the concentration in the surrounding medium, here water, at steady state. BCF can be measured experimentally directly under steady-state conditions or calculated by the ratio of the first-order uptake and elimination rate constants, a method that does not require equilibrium conditions.

1.2   *Appropriate methods for experimental determination of BCF*

1.2.1   Different test guidelines for the experimental determination of bioconcentration in fish have been documented and adopted; the most generally applied being the OECD test guideline (OECD 305, 1996) and the ASTM standard guide (ASTM E 1022-94). OECD 305 (1996) was revised and replaced the previous version OECD 305A-E, (1981). Although flow-through test regimes are preferred (OECD 305, 1996), semi-static regimes are allowed (ASTM E 1022-94), provided that the validity criteria on mortality and maintenance of test conditions are fulfilled. For lipophilic substances (log K_{ow} > 3), flow-through methods are preferred.

1.2.2   The principles of the OECD 305 and the ASTM guidelines are similar, but the experimental conditions described are different, especially concerning:

(a)   method of test water supply (static, semi-static or flow through);

(b)   the requirement for carrying out a depuration study;

(c)   the mathematical method for calculating BCF;

(d)   sampling frequency: Number of measurements in water and number of samples of fish;

(e)   requirement for measuring the lipid content of the fish;

(f)   the minimum duration of the uptake phase;

1.2.3   In general, the test consists of two phases: The exposure (uptake) and post-exposure (depuration) phases. During the uptake phase, separate groups of fish of one species are exposed to at least two concentrations of the test substance. A 28-day exposure phase is obligatory unless a steady state has been reached within this period. The time needed for reaching steady-state conditions may be set on the basis of $K_{ow} - k_2$ correlations (e.g. log k_2 = 1.47 – 0.41 log K_{ow} (Spacie and Hamelink, 1982) or log k_2 = 1.69 – 0.53 log K_{ow} (Gobas et al., 1989)). The expected time (d) for e.g. 95 % steady state may thus be calculated by: $-\ln(1-0.95)/k_2$, provided that the bioconcentration follows first order kinetics. During the depuration phase the fish are transferred to a medium free of the test substance. The concentration of the test substance in the fish is followed through both phases of the test. The BCF is expressed as a function of the total wet weight of the fish. As for many organic substances, there is a significant relationship between the potential for bioconcentration and the lipophilicity, and furthermore, there is a corresponding relationship between the lipid content of the test fish and the observed bioconcentration of such substances. Therefore, to reduce this source of variability in the test results for the substances with high lipophilicity, bioconcentration should be expressed in relation to the lipid content in addition to whole body weight (OECD 305 (1996), ECETOC (1995)). The guidelines mentioned are based on the assumption that bioconcentration may be approximated by a first-order process (one-compartment model) and thus that BCF = k_1/k_2 (k_1: first-order uptake rate, k_2: first-order depuration rate, described by a log-linear approximation). If the depuration follows biphasic kinetics, i.e. two distinct depuration rates can be identified, the approximation k_1/k_2 may significantly underestimate BCF. If a second order kinetic has been indicated, BCF may be estimated from the relation: C_{Fish}/C_{Water}, provided that "steady-state" for the fish-water system has been reached.

附属書　9
付録Ⅲ

有機物質の BCF および K_{ow} 測定のための実験法および推定法の基本原理

1. 生物濃縮係数（BCF）

1.1 *定義*

　生物濃縮係数は、定常状態における化学品の生物体内濃度と周囲の媒体、この場合には水中の濃度の比と定義されている。BCF は定常状態において直接、実験的に測定でき、また定常状態である必要なしに、取り込みと排出の一次速度定数の比から計算できる。

1.2 *BCF の実験による測定のための適切な方法*

1.2.1　魚類における生物濃縮の実験的測定のために種々のテストガイドラインが作成され、採用されている。最も一般的に適用されているのは、OECD テストガイドライン（OECD 305, 1996）および ASTM 標準ガイド（ASTM E 1022-94）である。OECD 305 (1996)は改訂され、その前の版である OECD 305A－E (1981)から差し替えられた。流水試験法（OECD 305, 1996）が望ましいが、半止水試験法（ASTM E 1022-94）も認められている。ただし、死亡率および試験条件の維持に関する有意性判定基準が充足されていることを前提とする。親油性物質（$\log K_{ow} > 3$）では、流水法の方が望ましい。

1.2.2　OECD 305 および ASTM ガイドラインの原則は同様ではあるが、記載されている実験条件は、特に以下の点で異なっている：

(a)　試験水の供給方法（止水、半止水、または流水）；

(b)　排泄試験を実施する必要性；

(c)　BCF 算出の数学的方法；

(d)　サンプリング回数：水中濃度の測定回数および魚サンプル数；

(e)　魚の脂質含量測定の必要性；

(f)　取り込み相の最少時間。

1.2.3　一般に、この試験法は 2 つの段階からなっている。すなわち、ばく露（取り込み）段階とばく露後（排泄）段階である。取り込み段階では、1 種類の魚種の各群を、最低 2 種類の濃度の被験物質にばく露する。28 日以内に定常状態に達しなければ、28 日のばく露期間が必要とされている。定常状態に達するのに必要な時間は、$K_{ow} - k_2$ 相関をもとに設定してもよい（例えば $\log k_2 = 1.47 - 0.41 \log K_{ow}$（Spacie と Hamelink, 1982）、または $\log k_2 = 1.69 - 0.53 \log K_{ow}$（Gobas ら, 1989））。したがって、例えば 95% 定常状態に予測される時間（d）は $-\ln(1-0.95)/k_2$ によって計算してもよいが、ただし生物濃縮が一次反応速度論に従うことが前提である。排泄段階では、被験物質を含まない媒体中に魚を移す。試験の両段階を通じて、魚体中の被験物質濃度を追跡する。BCF は魚湿体重の関数として表わされる。有機物質の多くについて、生物濃縮性と親油性の間に有意な関係があり、さらに、試験魚体内の脂質含量とそれらの物質の生物濃縮実測値にも、同様な関係がある。したがって、親油性の高い物質について試験結果を変動させる、このような原因を軽減するために、生物濃縮は、体重の他に脂質含量に関連させて表記すべきである（OECD 305（1996），ECETOC（1995））。ここで示したガイドラインは、生物濃縮は一次反応プロセス（1 コンパートメントモデル）によって、したがって BCF=k_1/k_2（k_1：一次取り込み速度、k_2：一次排泄速度（対数線形近似による））で近似できるという仮定に基づいている。排泄が二段階速度論に従う場合、すなわち 2 つの顕著に異なる排泄速度が見られる場合は、k_1/k_2 という近似は BCF を著しく低く推定することになる。二次反応速度論が示されたら、BCF は C_{Fish}/C_{Water} の関係から推定してよいが、これは魚-水系で「定常状態」に達していることを前提とする。

1.2.4   Together with details of sample preparation and storage, an appropriate analytical method of known accuracy, precision, and sensitivity must be available for the quantification of the substance in the test solution and in the biological material. If these are lacking it is impossible to determine a true BCF. The use of radiolabelled test substance can facilitate the analysis of water and fish samples. However, unless combined with a specific analytical method, the total radioactivity measurements potentially reflect the presence of parent substance, possible metabolite(s), and possible metabolized carbon, which have been incorporated in the fish tissue in organic molecules. For the determination of a true BCF it is essential to clearly discriminate the parent substance from possible metabolites. If radiolabelled materials are used in the test, it is possible to analyse for total radio label (i.e. parent and metabolites) or the samples may be purified so that the parent compound can be analysed separately.

1.2.5   In the log K_{ow} range above 6, the measured BCF data tend to decrease with increasing log K_{ow}. Conceptual explanations of non-linearity mainly refer to either biotransformation, reduced membrane permeation kinetics or reduced biotic lipid solubility for large molecules. Other factors consider experimental artefacts, such as equilibrium not being reached, reduced bioavailability due to sorption to organic matter in the aqueous phase, and analytical errors. Moreover, care should be taken when evaluating experimental data on BCF for substances with log K_{ow} above 6, as these data will have a much higher level of uncertainty than BCF values determined for substances with log K_{ow} below 6.

2.   log K_{ow}

2.1   *Definition and general considerations*

2.1.1   The log *n*-octanol-water partition coefficient (log K_{ow}) is a measure of the lipophilicity of a substance. As such, log K_{ow} is a key parameter in the assessment of environmental fate. Many distribution processes are driven by log K_{ow}, e.g. sorption to soil and sediment and bioconcentration in organisms.

2.1.2   The basis for the relationship between bioconcentration and log K_{ow} is the analogy for the partition process between the lipid phase of fish and water and the partition process between n-octanol and water. The reason for using K_{ow} arises from the ability of octanol to act as a satisfactory surrogate for lipids in fish tissue. Highly significant relationships between log K_{ow} and the solubility of substances in cod liver oil and triolin exist (Niimi, 1991). Triolin is one of the most abundant triacylglycerols found in freshwater fish lipids (Henderson and Tocher, 1987).

2.1.3   The determination of the *n*-octanol-water partition coefficient (K_{ow}) is a requirement of the base data set to be submitted for notified new and priority existing substances within the EU. As the experimental determination of the K_{ow} is not always possible, e.g. for very water-soluble and for very lipophilic substances, a QSAR derived K_{ow} may be used. However, extreme caution should be exercized when using QSARs for substances where the experimental determination is not possible (as for e.g. surfactants).

2.2   *Appropriate methods for experimental determination of K_{ow} values*

2.2.1   For experimental determination of K_{ow} values, two different methods, Shake-flask and HPLC, have been described in standard guidelines e.g. OECD 107 (1995); OECD 117 (1983); EEC A.8. (1992); EPA-OTS (1982); EPA-FIFRA (1982); ASTM (1993). Not only data obtained by the employment of the shake-flask or the HPLC method according to standard guidelines are recommended. For highly lipophilic substances, which are slowly soluble in water, data obtained by employing a slow-stirring method are generally more reliable (De Bruijn *et al.*, 1989; Tolls and Sijm, 1993; OECD Guideline 123).

2.2.2   *Shake-flask method*

The basic principle of the method is to measure the dissolution of the substance in two different phases, water and *n*-octanol. In order to determine the partition coefficient, equilibrium between all interacting components of the system must be achieved after which the concentration of the substances dissolved in the two phases is determined. The shake-flask method is applicable when the log K_{ow} value falls within the range from -2 to 4 (OECD 107, 1995). The shake-flask method applies only to essential pure substances soluble in water and *n*-octanol and should be performed at a constant temperature (±1°C) in the range 20-25 °C.

2.2.3   *HPLC method*

HPLC is performed on analytical columns packed with a commercially available solid phase containing long hydrocarbon chains (e.g. C_8, C_{18}) chemically bound onto silica. Chemicals injected onto such a column move along at different rates because of the different degrees of partitioning between the mobile aqueous phase and the stationary hydrocarbon phase. The HPLC method is not applicable to strong acids and bases, metals complexes, surface-active materials, or substances that react with the eluent. The HPLC method is applicable when the log K_{ow} value falls within

1.2.4 サンプル調製および保存の詳細と共に、試験溶液中および生物試料中の物質を定量するために、精度、正確性および感度がわかっている適切な分析方法が利用できなければならない。これらがなければ、正しい BCF の決定は不可能である。放射性標識された被験物質を用いれば、水および魚サンプルの分析を容易にできる。しかし、特異的な分析法と組み合わせなければ、全放射能の測定は潜在的に、親化合物、可能性のある代謝物、および有機分子として魚組織内に組み入れられた、可能性のある代謝された炭素の存在を反映している。真の BCF を決定するためには、親物質を可能性のある代謝物から明確に区別することが不可欠である。放射性標識された物質を試験に用いるならば、全放射能レベル（すなわち親化合物と代謝物）を分析することも可能であるし、または親化合物を別個に分析できるようサンプルを精製してもよい。

1.2.5 log K_{ow} が 6 より大きい範囲では、log K_{ow} が増加するほど測定された BCF データが小さくなる傾向がある。このような非線形の概念的な説明は、生物変換、膜透過速度の低下、あるいは巨大分子の体内脂質への溶解性の低下等によるものとされている。その他の要因として、平衡に達していなかった、水相中の有機物への吸着による生物学的利用性の低下、ならびに分析誤差などの実験技術上の誤りが考えられる。さらに、log K_{ow} が 6 より大きい物質の BCF についての実験データを評価する際は、log K_{ow} が 6 より低い物質について決定された BCF 値より、不確実性レベルがはるかに高くなるので、注意をはらわなければならない。

2. log K_{ow}

2.1 定義と一般的考察

2.1.1 n-オクタノール/水分配係数の対数値（log K_{ow}）は、物質の親油性の指標である。このことから、log K_{ow} は環境中運命の評価において重要なパラメータである。例えば土壌、底質への吸着および生物体内への生物蓄積等、多くの分配プロセスが log K_{ow} により影響される。

2.1.2 生物濃縮と log K_{ow} の関係の根拠は、魚体内の脂質相と水の間の分配プロセスと、n-オクタノールと水間の分配プロセスとの類似である。K_{ow} を用いる理由は、魚組織中にある脂質の満足できる代用物となるオクタノールの能力から生じている。log K_{ow} と、タラ肝油およびトリオレインへの物質の溶解性との間には、高度に有意な関係が存在している（Niimi, 1991）。トリオレインは、淡水魚の脂質に見出される、最も存在量の多いトリアシルグリセロールの 1 つである（Henderson と Tocher, 1987）。

2.1.3 n-オクタノール/水分配係数（K_{ow}）の測定は、EU 圏内で新規物質または優先既存物質の届け出のために提出しなければならない基本データセットの必要条件である。例えば極めて水溶性の高い物質や極めて親油性の高い物質などの場合、実験による K_{ow} の測定は必ずしも可能ではないので、QSAR から求めた K_{ow} を採用してもよい。しかし、実験による測定が可能ではない物質（例えば界面活性剤等）に QSAR を用いるには細心の注意をはらうべきである。

2.2 実験による K_{ow} 値決定のための適切な方法

2.2.1 K_{ow} の実験による測定には、フラスコ振騰法および HPLC 法という 2 つの異なった方法が、例えば OECD 107 (1995)、OECD 117 (1983)、EEC A.8. (1992)、EPA-OTS (1982)、EPA-FIFRA (1982)、ASTM (1993) 等の標準ガイドラインに記載されている。標準ガイドラインにしたがって、フラスコ振騰法または HPLC 法を採用して得られたデータだけが推奨されるわけではない。極めて親油性が高い物質は、水に対する溶解が遅いので、低速攪拌法を用いて得られたデータの方が一般に信頼性が高い（De Bruijn ら, 1989; Tolls と Sijm, 1993; OECD ガイドライン 123）。

2.2.2 フラスコ振騰法

この方法の基本原理は、2 つの異なった相、すなわち水および n-オクタノールの中での物質の溶解性を測定することである。分配係数を測定するためには、この系すべての相互作用している成分間の平衡が達成されて、その後でこれら二相中に溶解している物質の濃度を測定しなければならない。フラスコ振騰法は、log K_{ow} 値が－2 から 4 の範囲にある場合に適用できる（OECD 107, 1995）。フラスコ振騰法は、水および n-オクタノール中に溶解する基本的に純粋な物質にのみ適用され、20－25℃の範囲内の一定温度(±1℃）において実施しなければならない。

2.2.3 HPLC 法

HPLC 法は、長鎖の炭化水素（例：C_8、C_{18}）がシリカに化学的に結合している、市販されている固定相を充填した分析カラムでおこなわれる。こうしたカラムに注入された物質は、液体の移動相と炭化水素固定相間の分配度の違いによって、異なった速度でカラム上を移動する。HPLC 法は、強酸および強塩基、

the range 0 to 6 (OECD 117, 1989). The HPLC method is less sensitive to the presence of impurities in the test compound compared to the shake-flask method.

2.2.4   *Slow stirring method*

With the slow-stirring method a precise and accurate determination of K_{ow} of compounds with log K_{ow} up till 8.2 is allowed (De Bruijn *et al.*, 1989). For highly lipophilic compounds the shake-flask method is prone to produce artefacts (formation of microdroplets), and with the HPLC method K_{ow} needs to be extrapolated beyond the calibration range to obtain estimates of K_{ow}.

In order to determine a partition coefficient, water, n-octanol, and test compound are equilibrated with each other after which the concentration of the test compound in the two phases is determined. The experimental difficulties associated with the formation of microdroplets during the shake-flask experiment can to some degree be overcome in the slow-stirring experiment as water, octanol, and the test compound are equilibrated in a gently stirred reactor. The stirring creates a more or less laminar flow between the octanol and the water, and exchange between the phases is enhanced without microdroplets being formed.

2.2.5   *Generator column method*

Another very versatile method for measuring log K_{ow} is the generator column method. In this method, a generator column method is used to partition the test substance between the octanol and water phases. The column is packed with a solid support and is saturated with a fixed concentration of the test substance in *n*-octanol. The test substance is eluted from the octanol-saturated generator column with water. The aqueous solution exiting the column represents the equilibrium concentration of the test substance that has partitioned from the octanol phase into the water phase. The primary advantage of the generator column method over the shake flask method is that the former completely avoids the formation of micro-emulsions. Therefore, this method is particularly useful for measuring K_{ow} for substances values over 4.5 (Doucette and Andren, 1987 and 1988; Shiu et al., 1988) as well as for substances having log K_{ow} values less than 4.5. A disadvantage of the generator column method is that it requires sophisticated equipment. A detailed description of the generator column method is presented in the "Toxic Substances Control Act Test Guidelines" (USEPA 1985).

2.3   *Use of QSARs for determination of log K_{ow} (see also in A9.6, « Use of QSARs »)*

2.3.1   Numerous QSARs have been and continue to be developed for the estimation of K_{ow}. Commonly used methods are based on fragment constants. The fragmental approaches are based on a simple addition of the lipophilicity of the individual molecular fragments of a given molecule. Three commercially available PC programs are recommended in the European Commission's Technical Guidance Document (European Commission, 1996) for risk assessment, part III, if no experimentally derived data are available.

2.3.2   CLOGP (Daylight Chemical Information Systems, 1995) was initially developed for use in drug design. The model is based on the Hansch and Leo calculation procedure (Hansch and Leo, 1979). The program calculates log K_{ow} for organic compounds containing C, H, N, O, Hal, P, and/or S. Log K_{ow} for salts and for compounds with formal charges cannot be calculated (except for nitro compounds and nitrogen oxides). The calculation results of log K_{ow} for ionizable substances, like phenols, amines, and carboxylic acids, represent the neutral or unionized form and will be pH dependent. In general, the program results in clear estimates in the range of log Kow between 0 and 5 (European Commission, 1996, part III). However a validation study performed by Niemelä (1993), who compared experimental determined log K_{ow} values with estimated values, showed that the program precisely predicts the log K_{ow} for a great number of organic chemicals in the log K_{ow} range from below 0 to above 9 (n = 501, r^2 = 0.967). In a similar validation study on more than 7000 substances the results with the CLOGP-program (PC version 3.32, EPA version 1.2) were r^2 = 0.89, s.d.= 0.58, n = 7221. These validations show that the CLOGP-program may be used for estimating reliable log K_{ow} values when no experimental data are available. For chelating compounds and surfactants, the CLOGP program is stated to be of limited reliability (OECD, 1993). However, as regards anionic surfactants (LAS) a correction method for estimating adjusted CLOGP values has been proposed (Roberts, 1989).

2.3.3   LOGKOW or KOWWIN (Syracuse Research Corporation) uses structural fragments and correction factors. The program calculates log K_{ow} for organic compounds containing the following atoms: C, H, N, O, Hal, Si, P, Se, Li, Na, K, and/or Hg. Log K_{ow} for compounds with formal charges (like nitrogenoxides and nitro compounds) can also be calculated. The calculation of log K_{ow} for ionizable substances, like phenols, amines and carboxylic acids, represent the neutral or unionized form, and the values will thus be pH dependent. Some surfactants (e.g. alcohol ethoxylates (Tolls, 1998), dyestuffs, and dissociated substances may be predicted by the LOGKOW program (Pedersen *et al*, 1995). In general, the program gives clear estimates in the range of log K_{ow} between 0 and 9 (TemaNord 1995:581). Like the CLOGP-program, LOGKOW has been validated (table 2) and is recommended for classification purposes because of its reliability, commercial availability, and convenience of use.

金属錯体、界面活性物質、あるいは溶出液と反応する物質には適用できない。HPLC法はlog K_{ow}値が0－6の範囲にある場合に適用できる（OECD 117, 1989）。HPLC法の方がフラスコ振騰法よりも被験物質中の不純物の存在に対する影響が少ない。

2.2.4 *低速撹拌法*

低速撹拌法では、log K_{ow}が8.2までの化合物のK_{ow}を正確かつ精密に測定できる（De Bruijnら、1989）。親油性の高い化合物では、フラスコ振騰法では実験上の誤差（微小滴の形成）を生じる傾向があり、またHPLC法ではK_{ow}値の推定値を得るために検量範囲を超えてK_{ow}を外挿する必要がある。

分配係数を決定するためには、水、n-オクタノールおよび被験物質が相互に平衡に達して、その後この二相中の被験物質濃度が測定される。フラスコ振騰法での微小滴形成による実験上の困難さは、低速撹拌法では水、オクタノールおよび被験物質がゆるやかに撹拌されるリアクター内で平衡に達するので、ある程度克服される。撹拌によりオクタノールおよび水の間に多少の層流が生じ、微小滴が形成されることなく、二相間の物質の交換が行われる。

2.2.5 *ジェネレータ・カラム法*

log K_{ow}測定のためのもう1つの、非常に汎用性の高い方法は、ジェネレータ・カラム法である。この方法では、オクタノール相と水相間で物質を分配させるのにジェネレータ・カラムが用いられる。カラムには固体担体が充填され、一定濃度の被験物質を加えたn-オクタノールで飽和されている。被験物質は、オクタノール飽和されたジェネレータ・カラムから、水を用いて溶出される。カラム内にある水溶液は、オクタノール相から水相に分配された被験物質の平衡濃度を表している。ジェネレータ・カラム法がフラスコ振騰法より基本的に優れている点は、前者はミクロエマルジョンの生成が完全に防止されていることである。したがって、この方法はlog K_{ow}が4.5より低い物質と同様に、log K_{ow}が4.5を超える物質でのK_{ow}の測定に特に有用である（DoucetteとAndren, 1987および1988, Shiuら, 1988）。ジェネレータ・カラム法の欠点は、精巧な装置が必要なことである。ジェネレータ・カラム法の詳しい説明は"Toxic Substances Control Act Test Guidelines"（USEPA 1985）に示されている。

2.3 *log K_{ow} 決定のためのQSARの使用*（A9.6「QSARの使用」も参照のこと）

2.3.1 K_{ow}を推定するために、多数のQSARが開発され、また現在も開発され続けている。一般的に用いられている方法は、フラグメント定数に基づいている。このフラグメントによるアプローチは、与えられた分子について、個々の分子フラグメントの親油性を単純に加算することに基づいている。リスクアセスメントに関する欧州委員会の技術指針（European Commission, 1996）パートⅢでは、実験的に求められたデータがない場合に、3種類の市販されているPCプログラムが推奨されている。

2.3.2 CLOGP（Daylight Chemical Information Systems, 1995）は最初、ドラッグデザインのために開発された。このモデルはHanschとLeoの計算法（HanschとLeo, 1979）をもとにしている。このプログラムは、C、H、N、O、ハロゲン、PまたはSを含む有機化合物のlog K_{ow}を計算する。塩類および形式電荷のある化合物のlog K_{ow}は計算できない（ただしニトロ化合物および窒素酸化物を除く）。フェノール、アミン、あるいはカルボン酸などのイオン化する物質のlog K_{ow}計算結果は、中性またはイオン化していない形態を表しており、pH依存的である。一般的にこのプログラムでは、log K_{ow}が0－5の範囲で、明確な推定値が得られる（European Commission, 1996, part III）。しかし、Niemelä（1993）が実験的に測定したlog K_{ow}値を、推定値と比較しておこなった有意性評価研究では、このプログラムは多数の有機化合物のlog K_{ow}を、0以下から9以上というlog K_{ow}範囲で正確に推定することが示された（n=501, r^2=0.967）。7000種以上の物質の同様な有意性評価研究では、CLOGPプログラム（PC version 3.32, EPA version 1.2）の結果はr^2 = 0.89, s.d. = 0.58, およびn = 7221であった。これらの有意性は、CLOGPプログラムが、実験データが利用できない場合に、信頼できるlog K_{ow}の推定に用いられることを示している。キレート化合物および界面活性剤では、CLOGPプログラムの信頼性には限界があるとされている（OECD, 1993）。しかし、非イオン系界面活性物質（LAS）については、調整されたCLOGP値を推定するための補正方法が提案されている（Roberts, 1989）。

2.3.3 LOGKOWまたはKOWWIN（Syracuse Research Corporation）は、構造フラグメントと補正係数を採用している。このプログラムは、C、H、N、O、ハロゲン、Si、P、Se、Li、Na、KまたはHgを含む、有機化合物のlog K_{ow}を計算する。（窒素酸化物やニトロ化合物のような）形式電荷を有する化合物のlog K_{ow}も計算できる。フェノール、アミン、カルボン酸などのイオン化する物質のlog K_{ow}計算結果は、中性または非イオン化された形態を表しており、したがってpH依存的である。ある種の界面活性物質（例えばアルコールエトキシレート（Tolls, 1998））、染料および解離物質は、LOGKOWプログラムで予測されよう（Pedersenら, 1995）。一般に、このプログラムは0－9のlog K_{ow}域で明確な予測値を与える（TemaNord 1995: 581）。CLOGPプログラムと同様に、LOGKOWプログラムも有意性が確認されており（表2）、信頼性、市販で入手できること、および使用の簡便さの理由で、分類目的に推奨されている。

2.3.4   AUTOLOGP (Devillers *et al.*, 1995) has been derived from a heterogeneous data set, comprising 800 organic chemicals collected from literature. The program calculates log K_{ow} values for organic chemicals containing C, H, N, O, Hal, P, and S. The log K_{ow} values of salts cannot be calculated. Also, the log K_{ow} of some compounds with formal charges cannot be calculated, with the exception of nitro compounds. The log K_{ow} values of ionizable chemicals like phenols, amines, and corboxylic acids can be calculated although pH-dependencies should be noted. Improvements are in progress in order to extend the applicability of AUTOLOGP. According to the presently available information, AUTOLOGP gives accurate values especially for highly lipophilic substances (log $K_{ow} > 5$) (European Commission, 1996).

2.3.5   SPARC. The SPARC model is still under development by EPA's Environmental Research Laboratory in Athens, Georgia, and is not yet public available. SPARC is a mechanistic model based on chemical thermodynamic principles rather than a deterministic model rooted in knowledge obtained from observational data. Therefore, SPARC differs from models that use QSARs (i.e. KOWWIN, LOGP) in that no measured log K_{ow} data are needed for a training set of chemicals. EPA does occasionally run the model for a list of CAS numbers, if requested. SPARC provides improved results over KOWWIN and CLOGP only for compounds with log K_{ow} values greater than 5. Only SPARC can be employed in a general way for inorganic or organometallic compounds.

In table 1, this appendix, an overview of log K_{ow} estimation methods based on fragmentation methodologies is presented. Also other methods for the estimation of log K_{ow} values exist, but they should only be used on a case-by-case basis and only with appropriate scientific justification.

Table 1: Overview of QSAR methods for estimation of log K_{ow} based on fragmentation methodologies (Howard and Meylan (1997))

Method	Methodology	Statistics
CLOGP Hansch and Leo (1979), CLOGP Daylight (1995)	Fragments + correction factors	Total n = 8942, r^2 = 0,917, sd = 0,482 Validation: n = 501, r^2 = 0,967 Validation: n = 7221, r^2 = 0,89, sd = 0,58
LOGKOW (KOWWIN) Meylan and Howard (1995), SRC	140 fragments 260 correction factors	Calibration: n = 2430, r^2 = 0,981, sd = 0,219, me = 0,161 Validation: n = 8855, r^2 = 0,95, sd = 0,427, me = 0,327
AUTOLOGP Devillers *et al.* (1995)	66 atomic and group contributions from Rekker and Manhold (1992)	Calibration: n = 800, r^2 = 0,96, sd = 0,387
SPARC Under development by EPA, Athens, Georgia.	Based upon fundamental chemical structure algorithm.	No measured log Kow data are needed for a training set of chemicals.
Rekker and De Kort (1979)	Fragments + correction factors	Calibration: n = 1054, r^2 = 0,99 Validation: n = 20, r^2 = 0,917, sd = 0,53, me = 0,40
Niemi et al. (1992)	MCI	Calibration: n = 2039, r^2 = 0,77 Validation: n = 2039, r^2 = 0,49
Klopman et al. (1994)	98 fragments + correction factors	Calibration: n = 1663, r^2 = 0,928, sd = 0,3817
Suzuki and Kudo (1990)	424 fragments	Total: n= 1686, me = 0,35 Validation: n = 221, me = 0,49
Ghose et al. (1988) ATOMLOGP	110 fragments	Calibration: n = 830, r^2 = 0,93, sd = 0,47 Validation: n = 125, r^2 = 0,87, sd = 0,52
Bodor and Huang (1992)	Molecule orbital	Calibration: n = 302, r^2 = 0,96, sd = 0,31, me = 0,24 Validation: n = 128, sd = 0,38
Broto et al. (1984) ProLogP	110 fragments	Calibration: n = 1868, me= ca. 0,4

2.3.4 AUTOLOGP (Devillers ら, 1995) は、文献から収集された 800 種類の有機化学品をまとめた、不均一なデータセットから導かれた。このプログラムは、C、H、N、O、ハロゲン、P および S を含む有機化学品の log K_{ow} 値を計算する。塩の log K_{ow} 値は計算できない。また、ニトロ化合物を除いて、形式電荷を有する化合物には log K_{ow} が計算できないものもある。フェノール、アミン、カルボン酸などのイオン化する物質の log K_{ow} 値を計算できるが、pH への依存性については注意が必要である。AUTOLOGP の適用性を拡大するための改良作業が進行中である。現在利用可能な情報によれば、AUTOLOGP は特に非常に親油性の高い物質（log K_{ow}>5）で、正確な数値が得られる（European Commission, 1996）。

2.3.5 SPARC. SPARC モデルは、まだジョージア州アセンズにある EPA の環境研究所で開発中であり、公に利用できるようになっていない。SPARC は観察データから得られた知識に基づいた決定論的モデルというより、むしろ化学熱力学の原理に基づいたメカニズムモデルである。したがって、SPARC は、QSAR を用いるモデル（すなわち、KOWWIN、LOGP）と違って、化学品の訓練用セットには log K_{ow} の測定値を必要としない。EPA は要請があれば、このモデルを CAS 番号のリストで稼動させることもある。SPARC は log K_{ow} 値が 5 より大きい化合物では、KOWWIN および CLOGP より優れた結果を与える。無機化合物または有機金属化合物に一般的なやり方で採用できるのは SPARC だけである。

　本付録の表 1 に、フラグメント化の方法論に基づいた log K_{ow} 推定方法の概要を示した。log K_{ow} の推定法は他にもあるが、これらはケースバイケースのみで使用されるべきで、また適切な科学的根拠をつけてのみ使用されるべきである。

表 1：フラグメント化の方法論に基づいた log K_{ow} 推定方法の概要（Howard と Meylan, 1997）

方法	方法論	統計
CLOGP Hansch & Leo (1979), CLOGP Daylight (1995)	フラグメント＋補正係数	合計 n=8942, r²=0.917, sd=0.482 バリデーション：n=501, r²=0.967 バリデーション：n=7221, r²=0.89 sd=0.58
LOGKOW (KOWWIN) Meylan & Howard (1995), SRC	140 フラグメント 260 補正係数	キャリブレーション：n=2430, r²=0.981, sd=0.219, me=0.161 バリデーション：n=8855, r²=0.95, sd=0.427, me=0.327
AUTOLOGP Devillers ら(1995)	Rekker & Manhold (1992)による 66 の原子および置換基の関与	キャリブレーション：n=800, r²=0.96, sd =0.387
SPARC EPA（ジョージア州アセンズ）により開発中	基本的な化学品構造アルゴリズムに基づいている	訓練用セットの化学品に log K_{ow} 測定値は必要ない
Rekker & De Kort (1979)	フラグメント＋補正係数	キャリブレーション：n=1054, r²=0.99 バリデーション：n=20, r²=0.917, sd=0.53, me=0.40
Niemi ら(1992)	MCI	キャリブレーション：n=2039, r²=0.77 バリデーション：n=2039, r²=0.49
Klopman ら(1994)	98 フラグメント＋補正係数	キャリブレーション：n=1663, r²=0.928, sd=0.3817
Suzuki & Kudo (1990)	424 フラグメント	合計：n=1686, me=0.35, バリデーション：n=221, me=0.49
Ghose ら(1988) ATOMLOGP	110 フラグメント	キャリブレーション：n=830, r²=0.93, sd=0.47 バリデーション：n=125, r²=0.87, sd=0.52
Bodor & Huang (1992)	分子軌道法	キャリブレーション：n=302, r²=0.96, sd=0.31, me=0.24 バリデーション：n=128, sd=0.38
Broto ら(1984) ProLogP	110 フラグメント	キャリブレーション：n=1868, me=計算値 0.4

Annex 9

APPENDIX IV

Influence of external and internal factors on the bioconcentration potential of organic substances

1. Factors influencing the uptake

The uptake rate for lipophilic compounds is mainly a function of the size of the organism (Sijm and Linde, 1995). External factors such as the molecular size, factors influencing the bioavailability, and different environmental factors are of great importance to the uptake rate as well.

1.1 *Size of organism*

Since larger fish have a relatively lower gill surface to weight ratio, a lower uptake rate constant (k_1) is to be expected for large fish compared to small fish (Sijm and Linde, 1995; Opperhuizen and Sijm, 1990). The uptake of substances in fish is further controlled by the water flow through the gills; the diffusion through aqueous diffusion layers at the gill epithelium; the permeation through the gill epithelium; the rate of blood flow through the gills, and the binding capacity of blood constituents (ECETOC, 1995).

1.2 *Molecular size*

Ionized substances do not readily penetrate membranes; as aqueous pH can influence the substance uptake. Loss of membrane permeability is expected for substances with a considerable cross-sectional area (Opperhuizen *et al.*, 1985; Anliker *et al.*, 1988) or long chain length (> 4.3 nm) (Opperhuizen, 1986). Loss of membrane permeability due to the size of the molecules will thus result in total loss of uptake. The effect of molecular weight on bioconcentration is due to an influence on the diffusion coefficient of the substance, which reduces the uptake rate constants (Gobas *et al.*, 1986).

1.3 *Availability*

Before a substance is able to bioconcentrate in an organism it needs to be present in water and available for transfer across fish gills. Factors, which affect this availability under both natural and test conditions, will alter the actual bioconcentration in comparison to the estimated value for BCF. As fish are fed during bioconcentration studies, relatively high concentrations of dissolved and particulate organic matter may be expected, thus reducing the fraction of chemical that is actually available for direct uptake via the gills. McCarthy and Jimenez (1985) have shown that adsorption of lipophilic substances to dissolved humic materials reduces the availability of the substance, the more lipophilic the substance the larger reduction in availability (Schrap and Opperhuizen, 1990). Furthermore, adsorption to dissolved or particulate organic matter or surfaces in general may interfere during the measurement of BCF (and other physical-chemical properties) and thus make the determination of BCF or appropriate descriptors difficult. As bioconcentration in fish is directly correlated with the available fraction of the chemical in water, it is necessary for highly lipophilic substances to keep the available concentration of the test chemical within relatively narrow limits during the uptake period.

Substances, which are readily biodegradable, may only be present in the test water for a short period, and bioconcentration of these substances may thus be insignificant. Similarly, volatility and hydrolysis will reduce the concentration and time in which the substance is available for bioconcentration.

1.4 *Environmental factors*

Environmental parameters influencing the physiology of the organism may also affect the uptake of substances. For instance, when the oxygen content of the water is lowered, fish have to pass more water over their gills in order to meet respiratory demands (McKim and Goeden, 1982). However, there may be species dependency as indicated by Opperhuizen and Schrap (1987). It has, furthermore, been shown that the temperature may have an influence on the uptake rate constant for lipophilic substances (Sijm *et al.* 1993), whereas other authors have not found any consistent effect of temperature changes (Black *et al.* 1991).

附属書　9

付録IV

有機物質の生物濃縮性に対する体外および体内要因の影響

1. 取り込みに影響する要因

親油性化合物の取り込み速度は、主に生物体の大きさの関数である（SijmとLinde, 1995）。分子サイズ等の外部要因、生物学的利用性に影響する要因、および各種の環境要因も、取り込み速度に非常に重要である。

1.1 *生物体の大きさ*

大きい魚体の方が体重に対する鰓表面積の比が相対的に小さいので、小型魚に比べて大型魚の方がより低い取り込み速度定数（k_1）が予測される（SijmとLinde, 1995; OpperhuizenとSijm, 1990）。魚による物質の取り込みは、鰓を通過する水流、鰓表皮における水性拡散層を通しての拡散、鰓表皮を通る浸透、鰓の血流量、および血液成分の結合力によっても支配される（ECETOC, 1995）。

1.2 *分子サイズ*

イオン化された物質は、水相中のpHが物質取り込みに影響するので、膜を容易に透過することはない。かなりの断面積を持つ物質（Opperhuizenら, 1985; Anlikerら, 1988）、または鎖長の長い（>4.3nm）化合物（Opperhuizen, 1986）では、膜透過性が失われると予測されている。分子のサイズによって膜透過性が失われると、取り込みは完全に失われる結果になる。生物濃縮に対する分子量の影響は、その物質の拡散係数への影響で、取り込み速度定数が減少することによる（Gobasら, 1986）。

1.3 *利用性*

物質が生物体内に生物蓄積できるためには、それが水中に存在して、魚の鰓を通した移動のための利用性をもつことが必要である。自然界および実験の両方の条件下で、この利用性に影響する要因は、BCFの予測値と比べて、実際の生物濃縮を変化させる。生物濃縮試験中には魚に給餌するので、かなり高濃度の溶解している、および粒子状の有機物が予測され、これが、実際に鰓を通して直接取り込まれる化学品のフラクションを減少させる。McCarthyとJimenez（1985）は、溶解している腐植物質に親油物質が吸着することが、物質の利用性を低下させ、物質の親油性の大きいほど利用性が低下することを示した（ShrapとOpperhuizen, 1990）。さらに、溶解している、または粒子状の有機物質または表面に対する吸着は、一般にBCF（およびその他の物理化学的性質）の測定を妨害する場合があり、このためBCFの決定および適切な解釈を困難にする。魚における生物濃縮は、水中にある化学品の利用できるフラクションに直接関係しているので、親油性の高い物質の場合には、被験物質が利用性をもつ濃度を、取り込み期間中、比較的狭い範囲内に維持することが必要である。

易生分解性の物質は試験水中に短期間しか存在せず、したがって、このような物質の生物濃縮性は有意でないかもしれない。同様に、揮発および加水分解は、物質濃度を低下させ、物質が生物濃縮のための利用性をもつ時間を短縮させるであろう。

1.4 *環境因子*

生物の生理学的特性に影響する環境パラメータもまた、物質の取り込みを左右する。例えば、水中の酸素含量が低下すると、魚は呼吸需要を満たせるよう、より多量の水を鰓から通過させなければならない（McKimとGoeden, 1982）。ただし、OpperhuizenとSchrap（1987）が指摘したように、種依存性がある。さらに、親油性物質の取り込み速度定数に、温度が影響することも示されている（Sijmら, 1993）が、温度変化について一貫性のある影響を見出さなかった研究者もある（Blackら, 1991）。

2. **Factors influencing the elimination rate**

The elimination rate is mainly a function of the size of the organism, the lipid content, the biotransformation process of the organism, and the lipophilicity of the test compound.

2.1 *Size of organism*

As for the uptake rate the elimination rate is dependent on the size of the organism. Due to the higher gill surface to weight ratio for small organisms (e.g. fish larvae) than that of large organisms, steady-state and thus "toxic dose equilibrium" has shown to be reached sooner in early life stages than in juvenile/adult stages of fish (Petersen and Kristensen, 1998). As the time needed to reach steady-state conditions is dependent on k_2, the size of fish used in bioconcentration studies has thus an important bearing on the time required for obtaining steady-state conditions.

2.2 *Lipid content*

Due to partitioning relationships, organisms with a high fat content tend to accumulate higher concentrations of lipophilic substances than lean organisms under steady-state conditions. Body burdens are therefore often higher for "fatty" fish such as eel, compared to "lean" fish such as cod. In addition, lipid "pools" may act as storage of highly lipophilic substances. Starvation or other physiological changes may change the lipid balance and release such substances and result in delayed impacts.

2.3 *Metabolism*

2.3.1 In general, metabolism or biotransformation leads to the conversion of the parent compound into more water-soluble metabolites. As a result, the more hydrophilic metabolites may be more easily excreted from the body than the parent compound. When the chemical structure of a compound is altered, many properties of the compound are altered as well. Consequently, the metabolites will behave differently within the organism with respect to tissue distribution, bioaccumulation, persistence, and route and rate of excretion. Biotransformation may also alter the toxicity of a compound. This change in toxicity may either be beneficial or harmful to the organism. Biotransformation may prevent the concentration in the organism from becoming so high that a toxic response is expressed (detoxification). However, a metabolite may be formed which is more toxic than the parent compound (bioactivation) as known for e.g. benzo(a)pyrene.

2.3.2 Terrestrial organisms have a developed biotransformation system, which is generally better than that of organisms living in the aquatic environment. The reason for this difference may be the fact that biotransformation of xenobiotics may be of minor importance in gill breathing organisms as they can relatively easily excrete the compound into the water (Van Den Berg *et al.* 1995). Concerning the biotransformation capacity in aquatic organisms the capacity for biotransformation of xenobiotics increases in general as follows: Molluscs < crustaceans < fish (Wofford *et al.*, 1981).

3. **Lipophilicity of substance**

A negative linear correlation between k_2 (depuration constant) and log K_{ow} (or BCF) has been shown in fish by several authors (e.g. Spacie and Hamelink, 1982; Gobas *et al.*, 1989; Petersen and Kristensen, 1998), whereas k_1 (uptake rate constant) is more or less independent of the lipophilicity of the substance (Connell, 1990). The resultant BCF will thus generally increase with increasing lipophilicity of the substances, i.e. log BCF and log K_{ow} correlate for substances which do not undergo extensive metabolism.

2. 排泄速度に影響する要因

排泄速度は、主に生物体の大きさ、脂質含量、その生物体の生物変換プロセスおよび被験物質の親油性の関数である。

2.1 *生物体の大きさ*

取り込み速度と同様に、排泄速度も生物体の大きさに依存する。小型生物（例えば幼魚）の方が大型生物より鰓表面積の体重に対する比が大きいので、未成熟/成熟段階の魚よりも、幼生段階の魚の方が、定常状態に、ひいては「毒性用量平衡」に早く到達することが示された（PetersonとKristensen, 1998）。定常状態に達するのに必要な時間は k_2 に依存するので、生物濃縮試験に用いる魚のサイズは、定常状態を達成するのに必要な時間に重要な関係を持っている。

2.2 *脂質含量*

分配の関係から定常状態において、脂質含量の多い生物は、脂質の少ない生物よりも高濃度の親油性物質を蓄積する傾向がある。したがって、身体に対する負荷はしばしば、ウナギのような「脂肪の多い」魚の方が、タラのような「脂肪の少ない」魚より大きい。さらに、脂質「プール」が親油性の高い物質の貯蔵場所として作用することもある。絶食またはその他の生理学的変化が脂質バランスを変化させ、こうした物質を放出させて、遅発的な影響が出る結果になることもある。

2.3 *代謝*

2.3.1 一般に、代謝または生物体内変換は、親化合物をより水溶性の高い代謝物への変換に導く。その結果、より親水性の高い代謝物が親化合物よりも容易に身体から排泄されるであろう。化合物の化学構造が変化した場合、その化合物の多くの性質も変化する。結果的に代謝物は生物体内で、組織内分布、生物蓄積性、難分解性、および排泄経路と排泄速度の点で違った挙動をすることになる。生物変換はまた、化合物の毒性を変化させることもある。こうした毒性の変化は、生物体にとって有益であることもあれば有害となることもある。生物変換は、生物体内の濃度が、毒性反応が現れる程度に高くなるのを防止することもある（解毒）。ただし、例えばベンゾ（a）ピレンで知られているように、親化合物よりも毒性の高い代謝物が形成されることもある（生体内活性化）。

2.3.2 陸生生物は進化した生物変換システムを備えており、このシステムは一般に水生環境に棲息する生物よりも優れている。この違いの理由は、鰓呼吸生物では化合物を比較的容易に、水中に排泄できるので、異物質の生物変換がそれほど重要ではないという事実によるものと思われる（Van Dem Bergら, 1995）。水生生物における生物変換の能力に関しては、異物の生物変換能力は一般に、軟体動物＜甲殻類＜魚類の順で増加する（Woffordら, 1981）。

3. 物質の親油性

魚では、k_2（排泄速度定数）と $\log K_{ow}$（またはBCF）の間に負の線形関係が何人かの研究者により示されている（例えばSpacieとHamelink, 1982; Gobasら, 1989; PetersenとKristensen, 1998）が、k_1（取り込み速度定数）の方が多少、物質の親油性との関連性が低い（Connel, 1990）。したがって、結果として得られるBCFは一般的に物質の親油性が高いほど大きくなる、すなわち強い代謝を受けない物質では、$\log$ BCF と $\log K_{ow}$ に相関性がある。

Annex 9

APPENDIX V

Test guidelines

1. Most of the guidelines mentioned are found in compilations from the organisation issuing them. The main references to these are:

    (a) EC guidelines: Commission Regulation (EC) No 440/2008 of 30 May 2008 laying down test methods pursuant to Regulation (EC) No 1907/2006 of the European Parliament and of the Council on the Registration, Evaluation, Authorisation and Restriction of Chemicals (REACH);

    (b) ISO guidelines: Available from the national standardisation organisations or ISO (https://www.iso.org/home.html);

    (c) OECD guidelines for the testing of chemicals. OECD, Paris, 1993 with regular updates (https://www.oecd.org/env/ehs/testing/oecdguidelinesforthetestingofchemicals.htm);

    (d) OPPTS guidelines: US-EPA homepage (https://www.epa.gov/test-guidelines-pesticides-and-toxic-substances);

    (e) ASTM: (https://www.astm.org/Standard/standards-and-publications.html)

2. **Test guidelines for aquatic toxicity**[1]

OECD Test Guideline 201 (1984) (Updated in 2011) Alga, Growth Inhibition Test
OECD Test Guideline 202 (1984) (Updated in 2004) Daphnia sp. Acute Immobilisation Test and Reproduction Test
OECD Test Guideline 203 (1992) (Updated in 2019) Fish, Acute Toxicity Test
OECD Test Guideline 210 (1992) (Updated in 2013) Fish, Early-Life Stage Toxicity Test
OECD Test Guideline 211 (1998) (Updated in 2012) Daphnia magna Reproduction Test. Additional OECD test guidelines include:
    OECD Test Guideline 219 (2004) Sediment-Water Chironomid Toxicity Using Spiked Water
    OECD Test Guideline 233 (2010) Sediment-Water Chironomid Life-Cycle Toxicity Test Using Spiked Water or Spiked Sediment
    OECD Test Guideline 238 (2014) Sediment-Free Myriophyllum Spicatum Toxicity Test
    OECD Test Guideline 240 (2015), Medaka Extended One-generation Test
    OECD Test Guideline 242 (2016) Potamopyrgus antipodarum Reproduction Test
    OECD Test Guideline 243 (2016) Lymnaea stagnalis Reproduction Test
OECD Test Guideline 212 (1998) Fish, Short-term Toxicity Test on Embryo and Sac-Fry Stages
OECD Test Guideline 215 (2000) Fish, Juvenile Growth Test
OECD Test Guideline 221 Lemna sp. Growth inhibition test

EC C.1: Acute Toxicity for Fish (1992)
EC C.2: Acute Toxicity for Daphnia (1992)
EC C.3: Algal Inhibition Test (1992)
EC C.14: Fish Juvenile Growth Test (2001)
EC C.15: Fish, Short-term Toxicity Test on Embryo and Sac-Fry Stages (2001)
EC C.20: Daphnia Magna Reproduction Test (2001)

OPPTS Testing Guidelines for Environmental Effects (850 Series Public Drafts):

850.1000 Special consideration for conducting aquatic laboratory studies
850.1000 Special consideration for conducting aquatic laboratory studies
850.1010 Aquatic invertebrate acute toxicity, test, freshwater daphnids

---

[1] *The list below will need to be regularly updated as new guidelines are adopted or draft guidelines are elaborated.*

附属書 9

付録Ⅴ

テストガイドライン

1. 言及されているガイドラインのほとんどは、これらを発行している団体の編纂文書に示されている。主な参照文書は以下の通り：

(a) EC guidelines: Commission Regulation (EC) No 440/2008 of 30 May 2008 laying down test methods pursuant to Regulation (EC) No 1907/2006 of the European Parliament and of the Council on the Registration, Evaluation, Authorisation and Restriction of Chemicals (REACH);

(b) ISO guidelines: Available from the national standardisation organisations or ISO (http://www.iso.org/home.html);

(c) OECD guidelines for the testing of chemicals. OECD, Paris, 1993 with regular updates (http://www.oecd.org/env/ehs/testing/oecdguidelinesforthetestingofchemicals.htm);

(d) OPPTS guidelines: US-EPA homepage (https://www.epa.gov/test-guidelines-pesticides-and-toxic-substances)

(e) ASTM: (https://www.astm.org/Standard/standards-and-publications.html)"

2. 水性毒性に関するテストガイドライン [1]

OECD Test Guideline 201 (1984) (Updated in 2011) Alga, Growth Inhibition Test
OECD Test Guideline 202 (1984) (Updated in 2004) Daphnia sp. Acute Immobilisation Test and Reproduction Test
OECD Test Guideline 203 (1992) (Updated in 2019) Fish, Acute Toxicity Test
OECD Test Guideline 210 (1992) (Updated in 2013) Fish, Early-Life Stage Toxicity Test
OECD Test Guideline 211 (1998) (Updated in 2012) Daphnia magna Reproduction Test
    OECD Test Guideline 219 (2004) Sediment-Water Chironomid Toxicity Using Spiked Water
    OECD Test Guideline 233 (2010) Sediment-Water Chironomid Life-Cycle Toxicity Test Using Spiked Water or Spiked Sediment
    OECD Test Guideline 238 (2014) Sediment-Free Myriophyllum Spicatum Toxicity Test
    OECD Test Guideline 240 (2015), Medaka Extended One-generation Test
    OECD Test Guideline 242 (2016) Potamopyrgus antipodarum Reproduction Test
    OECD Test Guideline 243 (2016) Lymnaea stagnalis Reproduction Test
OECD Test Guideline 212 (1998) Fish, Short-term Toxicity Test on Embryo and Sac-Fry Stages
OECD Test Guideline 215 (2000) Fish, Juvenile Growth Test
OECD Test Guideline 221 Lemna sp. Growth inhibition test

EC C.1: Acute Toxicity for Fish (1992)
EC C.2: Acute Toxicity for Daphnia (1992)
EC C.3: Algal Inhibition Test (1992)
EC C.14: Fish Juvenile Growth Test (2001)
EC C.15: Fish, Short-term Toxicity Test on Embryo and Sac-Fry Stages (2001)
EC C.20: Daphnia Magna Reproduction Test (2001)

OPPTS Testing Guidelines for Environmental Effects (850 Series Public Drafts):

850.1000 Special consideration for conducting aquatic laboratory studies
850.1000 Special consideration for conducting aquatic laboratory studies
850.1010 Aquatic invertebrate acute toxicity, test, freshwater daphnids

---

[1] 以下のリストは、新しいガイドラインの採用、あるいはその原案の推敲に応じて、定期的に更新する必要があろう。

850.1010 Aquatic invertebrate acute toxicity, test, freshwater daphnids
850.1020 Gammarid acute toxicity test
850.1020 Gammarid acute toxicity test
850.1035 Mysid acute toxicity test
850.1035 Mysid acute toxicity test
850.1045 Penaeid acute toxicity test
850.1045 Penaeid acute toxicity test
850.1075 Fish acute toxicity test, freshwater and marine
850.1075 Fish acute toxicity test, freshwater and marine
850.1300 Daphnid chronic toxicity test
850.1300 Daphnid chronic toxicity test
850.1350 Mysid chronic toxicity test
850.1350 Mysid chronic toxicity test
850.1400 Fish early-life stage toxicity test
850.1400 Fish early-life stage toxicity test
850.1500 Fish life cycle toxicity
850.1500 Fish life cycle toxicity
850.1730 Fish BCF
850.1730 Fish BCF
850.4400 Aquatic plant toxicity test using Lemna spp. Tiers I and II
850.4400 Aquatic plant toxicity test using Lemna spp. Tiers I and II
850.4450 Aquatic plants field study, Tier III
850.4450 Aquatic plants field study, Tier III
850.5400 Algal toxicity, Tiers I and II
850.5400 Algal toxicity, Tiers I and II

3. Test guidelines for biotic and abiotic degradation [2]

ASTM E 1196-92
ASTM E 1279-89(95) Standard test method for biodegradation by a shake-flask die-away method
ASTM E 1625-94 Standard test method for determining biodegradability of organic chemicals in semi-continuous activated sludge (SCAS)

EC C.4. A to F: Determination of ready biodegradability. Directive 67/548/EEC, Annex V. (1992)
EC C.5. Degradation: biochemical oxygen demand. Directive 67/548/EEC, Annex V. (1992)
EC C.7. Degradation: abiotic degradation: hydrolysis as a function of pH. Directive 67/548/EEC, Annex V. (1992)
EC C.9. Biodegradation: Zahn-Wellens test. Directive 67/548/EEC, Annex V. (1988)
EC C.10. Biodegradation: Activated sludge simulation tests. Directive 67/548/EEC, Annex V. (1998)
EC C.11. Biodegradation: Activated sludge respiration inhibition test. Directive 67/548/EEC, Annex V. (1988)
EC C.12. Biodegradation: Modified SCAS test. Directive 67/548/EEC, Annex V. (1998)

ISO 9408 (1991). Water quality - Evaluation in an aqueous medium of the "ultimate" biodegradability of organic compounds - Method by determining the oxygen demand in a closed respirometer
ISO 9439 (1990). Water quality - Evaluation in an aqueous medium of the "ultimate" biodegradability of organic compounds - Method by analysis of released carbon dioxide
ISO 9509 (1996). Water quality - Method for assessing the inhibition of nitrification of activated sludge micro-organisms by chemicals and wastewaters
ISO 9887 (1992). Water quality - Evaluation of the aerobic biodegradability of organic compounds in an aqueous medium - Semicontinuous activated sludge method (SCAS)
ISO 9888 (1991). Water quality - Evaluation of the aerobic biodegradability of organic compounds in an aqueous medium - Static test (Zahn-Wellens method)
ISO 10707 (1994). Water quality - Evaluation in an aqueous medium of the "ultimate" biodegradability of organic compounds - Method by analysis of biochemical oxygen demand (closed bottle test)

---

[2] *The list below will need to be regularly updated as new guidelines are adopted or draft guidelines are elaborated.*

850.1010 Aquatic invertebrate acute toxicity, test, freshwater daphnids
850.1020 Gammarid acute toxicity test
850.1020 Gammarid acute toxicity test
850.1035 Mysid acute toxicity test
850.1035 Mysid acute toxicity test
850.1045 Penaeid acute toxicity test
850.1045 Penaeid acute toxicity test
850.1075 Fish acute toxicity test, freshwater and marine
850.1075 Fish acute toxicity test, freshwater and marine
850.1300 Daphnid chronic toxicity test
850.1300 Daphnid chronic toxicity test
850.1350 Mysid chronic toxicity test
850.1350 Mysid chronic toxicity test
850.1400 Fish early-life stage toxicity test
850.1400 Fish early-life stage toxicity test
850.1500 Fish life cycle toxicity
850.1500 Fish life cycle toxicity
850.1730 Fish BCF
850.1730 Fish BCF
850.4400 Aquatic plant toxicity test using Lemna spp. Tiers I and II
850.4400 Aquatic plant toxicity test using Lemna spp. Tiers I and II
850.4450 Aquatic plants field study, Tier III
850.4450 Aquatic plants field study, Tier III
850.5400 Algal toxicity, Tiers I and II
850.5400 Algal toxicity, Tiers I and II

3. 生物的および非生物的分解に関するテストガイドライン[2]

ASTM E 1196-92
ASTM E 1279-89 (95) Standard test method for biodegradation by a shake-flask die-away method
ASTM E 1625-94 Standard test method for determining biodegradability of organic chemicals in semi-continuous activated sludge (SCAS)

EC C.4. A to F: Determination of ready degradability. Directive 67/548/EEC, Annex V. (1992)
EC C.5. Degradation: biochemical oxygen demand. Directive 67/548/EEC, Annex V. (1992)
EC C.7. Degradation: abiotic degradation: hydrolysis as a function of pH. Directive 67/548/EEC, Annex V. (1992)
EC C.9. Biodegradation: Zahn-Wellens test. Directive 67/548/EEC, Annex V. (1988)
EC C.10. Biodegradation: Activated sludge simulation tests. Directive 67/548/EEC, Annex V. (1998)
EC C.11 Biodegradation: Activated sludge respiration inhibition test. Directive 67/548/EEC, Annex V. (1988)
EC C.12. Biodegradation: Modified SCAS test. Directive 67/548/EEC, Annex V.(1998)

ISO 9408 (1991). Water quality – Evaluation in an aqueous medium of the "ultimate" biodegradability of organic compounds – Method by determining the oxygen demand in a closed respirometer
ISO 9439 (1990). Water quality – Evaluation in an aqueous medium of the "ultimate" biodegradability of organic compounds – Method by analysis of released carbon dioxide
ISO 9509 (1996). Water quality – Method for assessing the inhibition of nitrification of activated sludge micro-organisms by chemicals and wastewaters.
ISO 9887 (1992). Water quality – Evaluation of the aerobic biodegradability of organic compounds in an aqueous medium – Semicontinuous activated sludge method (SCAS)
ISO 9888 (1991). Water quality – Evaluation of the aerobic biodegradability of organic compounds in an aqueous medium – Static test (Zahn-Wellens method)
ISO 10707 (1994). Water quality – Evaluation in an aqueous medium of the "ultimate" biodegradability of organic compounds – Method by analysis of biochemical oxygen demand (closed bottle test)

---

[2] 以下のリストは、新しいガイドラインの採用、あるいはその原案の推敲に応じて、定期的に更新する必要があろう。

ISO 11348 (1997). Water quality - Determination of the inhibitory effect of water samples on the light emission of Vibrio fischeri (Luminescent bacteria test)

ISO 11733 (1994). Water quality - Evaluation of the elimination and biodegradability of organic compounds in an aqueous medium - Activated sludge simulation test

ISO 11734 (1995). Water quality - Evaluation of the "ultimate" anaerobic biodegradability of organic compounds in digested sludge - Method by measurement of the biogas production

ISO/DIS 14592. (1999) Water quality - Evaluation of the aerobic biodegradability of organic compounds at low concentrations in water. Part 1: Shake flask batch test with surface water or surface water/sediment suspensions (22.11.1999)

OECD Test Guideline 111 (1981). Hydrolysis as a function of pH. OECD guidelines for testing of chemicals

OECD Test Guideline 209 (1984) (Updated in 2010). Activated sludge, respiration inhibition test. OECD guidelines for testing of chemicals

OECD Test Guideline 301 (1992). Ready biodegradability. OECD guidelines for testing of chemicals

OECD Test Guideline 302A (1981). Inherent biodegradability: Modified SCAS test. OECD guidelines for testing of chemicals

OECD Test Guideline 302B (1992). Zahn-Wellens/EMPA test. OECD guidelines for testing of chemicals

OECD Test Guideline 302C (1981). Inherent biodegradability: Modified MITI test (II). OECD guidelines for testing of chemicals

OECD Test Guideline 303A (1981). Simulation test - aerobic sewage treatment: Coupled units test. OECD guidelines for testing of chemicals. Additional test guidelines include:

> OECD Test Guideline 311 (2006), Anaerobic Biodegradability of Organic Compounds in Digested Sludge: by Measurement of Gas Production
> OECD Test Guideline 314 (2008) Simulation Tests to Assess the Biodegradability of Chemicals Discharged in Wastewater.

OECD Test Guideline 304A (1981). Inherent biodegradability in soil. OECD guidelines for testing of chemicals

OECD Test Guideline 306 (1992). Biodegradability in seawater. OECD guidelines for testing of chemicals

OECD Test Guideline 307 (2002). Aerobic and anaerobic transformation in soil. OECD guidelines for testing of chemicals

OECD Test Guideline 308 (2002). Aerobic and anaerobic transformation in aquatic sediment systems. OECD guidelines for testing of chemicals

OECD Test Guideline 309 (2004). Aerobic mineralisation in surface water – Simulation biodegradation test. OECD guidelines for testing of chemicals. Additional test guidelines include:

> OECD Test Guideline 310 (2014) Ready Biodegradability - CO_2 in sealed vessels (Headspace Test)
> OECD Test Guideline 311 (2006) Anaerobic Biodegradability of Organic Compounds in Digested Sludge: by Measurement of Gas Production
> OECD Test Guideline 316 (2008) Phototransformation of Chemicals in Water – Direct Photolysis

OPPTS 835.2110 Hydrolysis as a function of pH

OPPTS 835.2130 Hydrolysis as a function of pH and temperature

OPPTS 835.2210 Direct photolysis rate in water by sunlight

OPPTS 835.3110 Ready biodegradability

OPPTS 835.3170 Shake flask die-away test

OPPTS 835.3180 Sediment/water microcosm biodegradability test

OPPTS 835.3200 Zahn-Wellens/EMPA test

OPPTS 835.3210 Modified SCAS test

OPPTS 835.3300 Soil biodegradation

OPPTS 835.3400 Anaerobic biodegradability of organic chemicals

OPPTS 835.5270 Indirect photolysis screening test: Sunlight photolysis in waters containing dissolved humic substances

4. Test guidelines for bioaccumulation [3]

ASTM, 1993. ASTM Standards on Aquatic Toxicology and Hazard Evaluation. Sponsored by ASTM Committee E-47 on Biological Effects and Environmental Fate. American Society for Testing and Materials. 1916 Race Street, Philadelphia, PA 19103. ASTM PCN: 03-547093-16., ISBN 0-8032-1778-7

---

[3] *The list below will need to be regularly updated as new guidelines are adopted or draft guidelines are elaborated.*

ISO 11348 (1997). Water quality – Determination of the inhibitory effect of water samples on the light emission of Vibrio fischeri (Luminescent bacteria test)
ISO 11733 (1994). Water quality – Evaluation of the elimination and biodegradability of organic compounds in an aqueous medium – Activated sludge simulation test
ISO 11734 (1995). Water quality – Evaluation of the "ultimate" anaerobic biodegradability of organic compounds in digested sludge – Method by measurement of the biogas production
ISO/DIS 14592 (1999). Water quality – Evaluation of the aerobic biodegradability of organic compounds at low concentrations in water. Part 1: Shake flask batch test with surface water or surface water/sediment suspensions (22.11.1999)

OECD Test Guideline 111 (1981). Hydrolysis as a function of pH. OECD guidelines for testing of chemicals
OECD Test Guideline 209 (1984) (Updated in 2010). Activated sludge, respiration inhibition test. OECD guidelines for testing of chemicals
OECD Test Guideline 301 (1992). Ready biodegradability. OECD guidelines for testing of chemicals
OECD Test Guideline 302A (1981). Inherent biodegradability: Modified SCAS. OECD guidelines for testing of chemicals
OECD Test Guideline 302B (1992). Zahn-Wellens/EMPA test. OECD guidelines for testing of chemicals
OECD Test Guideline 302C (1981). Inherent Biodegradability. Modified MITI test (II). OECD guidelines for testing of chemicals
OECD Test Guideline 303A (1981). Simulation test –aerobic sewage treatment: Coupled units test. OECD guidelines for testing of chemicals.
  OECD Test Guideline 311 (2006), Anaerobic Biodegradability of Organic Compounds in Digested Sludge: by Measurement of Gas Production
  OECD Test Guideline 314 (2008) Simulation Tests to Assess the Biodegradability of Chemicals Discharged in Wastewater.
OECD Test Guideline 304A (1981). Inherent biodegradability in soil. OECD guidelines for testing of chemicals
OECD Test Guideline 306 (1992). Biodegradability in seawater. OECD guidelines for testing of chemicals
OECD Test Guideline 307(2002): Aerobic and Anaerobic Transformation in Soil. OECD guidelines for testing of chemicals.
OECD Test Guideline 308(2002). Aerobic and anaerobic transformation in aquatic sediment Systems. OECD guidelines for testing of chemicals.
OECD Test Guideline 309 (2004). Aerobic mineralisation in surface water – Simulation biodegradation test. OECD guidelines for testing of chemicals.
  OECD Test Guideline 310 (2014) Ready Biodegradability - CO_2 in sealed vessels (Headspace Test)
  OECD Test Guideline 311 (2006) Anaerobic Biodegradability of Organic Compounds in Digested Sludge: by Measurement of Gas Production
  OECD Test Guideline 316 (2008) Phototransformation of Chemicals in Water – Direct Photolysis

OPPTS 835.2110 Hydrolysis as a function of pH
OPPTS 835.2130 Hydrolysis as a function of pH and temperature
OPPTS 835.2210 Direct photolysis rate in water by sunlight
OPPTS 835.3110 Ready biodegradability
OPPTS 835.3170 Shake flask die-away test
OPPTS 835.3180 Sediment/water microcosm biodegradability test
OPPTS 835.3200 Zahn-Wellens/EMPA test
OPPTS 835.3210 Modified SCAS test
OPPTS 835.3300 Soil biodegradation
OPPTS 835.3400 Anaerobic biodegradability of organic chemicals
OPPTS 835.5270 Indirect photolysis screening test: Sunlight photolysis in waters containing dissolved humic substances

4. 生物蓄積に関するテストガイドライン[3]

ASTM, 1993. ASTM Standards on Aquatic Toxicology and Hazard Evaluation. Sponsored by ASTM Committee E-47 on Biological Effects and Environmental Fate. American Society for Testing and Materials. 1916 Race Street, Philadelphia, PA 19103. ASTM PCN: 03-547093-16., ISBN 0-8032-1778-7

---

[3] 以下のリストは、新しいガイドラインの採用、あるいはその原案の推敲に応じて、定期的に更新する必要があろう。

ASTM E 1022-94. 1997. Standard Guide for Conducting Bioconcentration Tests with Fishes and Saltwater Bivalve Molluscs. American Society for Testing and Materials

EC, 1992. EC A.8. Partition coefficient. Annex V (Directive 67/548/EEC). Methods for determination of physico-chemical properties, toxicity and ecotoxicity

EC, 1998. EC.C.13 Bioconcentration: Flow-through Fish Test

EPA-OTS, 1982. Guidelines and support documents for environmental effects testing. Chemical fate test guidelines and support documents. United States Environmental Protection Agency. Office of Pesticides and Toxic Substances, Washington, D.C. 20960. EPA 560/6-82-002. (August 1982 and updates), cf. also Code of Federal Regulations. Protection of the Environment Part 790 to End. Revised as of July 1, 1993. ONLINE information regarding the latest updates of these test guidelines: US National Technical Information System

EPA-FIFRA, 1982. The Federal Insecticide, Fungicide and Rodenticide Act. Pesticide Assessment Guidelines, subdivision N: chemistry: Environmental fate, and subdivision E, J & L: Hazard Evaluation. Office of Pesticide Programs. US Environmental Protection Agency, Washington D.C. (1982 and updates). ONLINE information regarding the latest updates of these test guidelines: US National Technical Information System

OECD Test Guideline 107, 1995. OECD Guidelines for testing of chemicals. Partition Coefficient (n-octanol/water): Shake Flask Method

OECD Test Guideline 117, 1989 (Updated in 2004). OECD Guideline for testing of chemicals. Partition Coefficient (n-octanol/water), High Performance Liquid Chromatography (HPLC) Method

OECD Test Guideline 305, 1996 (Updated in 2012). Bioconcentration: Flow-through Fish Test. OECD Guidelines for testing of Chemicals

OECD Test Guidelines 305 A-E, 1981. Bioaccumulation. OECD Guidelines for testing of chemicals

OECD Test Guideline 123. Partition Coefficient (1-Octanol/Water). Slow-stirring method. OECD Guidelines for testing of chemicals. Additional test guidelines include OECD Test Guideline 315 (2008) Bioaccumulation in Sediment-dwelling Benthic Oligochaetes.

ASTM E 1022-94. 1997. Standard Guide for Conducting Bioconcentration Tests with Fishes and Saltwater Bivalve Molluscs. American Society for Testing and Materials

EC, 1992. EC A.8. Partition coefficient. Annex V (Directive 67/548/EEC). Methods for determination of physico-chemical properties, toxicity and ecotoxicity
EC, 1998. EC.C.13 Bioconcentration: Flow-through Fish Test

EPA-OTS, 1982. Guidelines and support documents for environmental effects testing. Chemical fate test guidelines and support documents. United States Environmental Protection Agency. Office of Pesticides and Toxic Substances, Washington, D.C. 20960. EPA 560/6-82-002. (1982年8月および更新版)。および Code of Federal Regulations も参照。Protection of the Environment Part 790 から最後まで。1993 年 7 月 1 日改訂。これらのテストガイドラインの最新改訂版についての ONLINE 情報：US National Technical Information System

EPA-FIFRA, 1982. The Federal Insecticide, Fungicide and Rodenticide Act. Pesticide Assessment Guidelines, subdivision N: chemistry: Environmental fate, and subdivision E, J & L: Hazard Evaluation. Office of Pesticide Programs. US Environmental Protection Agency, Washington D.C. (1982 および更新版)。これらのテストガイドラインの最新改訂版についての ONLINE 情報: US National Technical Information System

OECD Test Guideline 107, 1995. OECD Guidelines for Testing of Chemicals. Partition Coefficient (n-octanol/water): Shake Flask Method
OECD Test Guideline 117, 1989 (Updated in 2004). OECD Guidelines for Testing of Chemicals. Partition Coefficient (n-octanol/water), High Performance Liquid Chromatography (HPLC) Method
OECD Test Guideline 305, 1996 (Updated in 2012). Bioconcentration: Flow-through Fish Test. OECD Guidelines for Testing of Chemicals
OECD Test Guideline 305 A-E, 1981. Bioaccumulation: OECD Guidelines for Testing of Chemicals
OECD Test Guideline 123: Partition Coefficient (1-Octanol/Water): Slow-Stirring Method. OECD guidelines for testing of chemicals. Additional test guidelines include OECD Test Guideline 315 (2008) Bioaccumulation in Sediment-dwelling Benthic Oligochaetes.

Annex 9

APPENDIX VI

References

1. **Aquatic toxicity**

APHA 1992. Standard Methods for the Examination of Water and Wastewater, 18th edition. American Public Health Association, Washington, DC

ASTM 1999. Annual Book of ASTM standards, Vol. 11.04. American Society for Testing and Materials, Philadelphia, PA

DoE 1996. Guidance on the Aquatic Toxicity Testing of Difficult Substances. United Kingdom Department of the Environment, London

ECETOC 1996. Aquatic Toxicity Testing of Sparingly Soluble, Volatile and Unstable Substances. ECETOC Monograph No. 26, ECETOC, Brussels

Lewis, M. A. 1995. Algae and vascular plant tests. In: Rand, G. M. (ed.) 1995. Fundamentals of Aquatic Toxicology, Second Edition. Taylor & Francis, Washington, DC. pp. 135-169

Mensink, B. J. W. G., M. Montforts, L. Wijkhuizen-Maslankiewicz, H. Tibosch, and J.B.H.J. Linders 1995. Manual for Summarising and Evaluating the Environmental Aspects of Pesticides. Report No. 679101022 RIVM, Bilthoven, The Netherlands

OECD 1998. Harmonized Integrated Hazard Classification System for Human Health and Environmental Effects of Chemical Substances. OECD, Paris. (Document ENV/JM/MONO(2001)6) (Updated in 2001) Series on Testing and Assessment No. 33, OECD, Paris."

OECD 1999. Guidelines for Testing of Chemicals. Organisation for Economic Co-operation and Development, Paris

OECD 2000. Guidance Document on Aquatic Toxicity Testing of Difficult Substances and Mixtures, Series on Testing and Assesment No. 23, OECD, Paris. Updated in 2019: OECD 2019. Second edition - Guidance Document on Aqueous-Phase Aquatic Toxicity Testing of Difficult Test Chemicals, Series on Testing and Assessment No. 23 (second edition). OECD, Paris.

OECD 2006. "Current approaches in the statistical analysis of ecotoxicity data: A guidance to application", OECD Environment Health and Safety Publications Series Testing and Assessment N.54

Pedersen, F., H. Tyle, J. R. Niemeldi, B. Guttmann, L. Lander, and A. Wedebrand 1995. Environmental Hazard Classification – data collection and interpretation guide. TemaNord 1995:581

US EPA 1996. Ecological Effects Test Guidelines – OPPTS 850.1000. Special Considerations for Conducting Aquatic Laboratory Studies. Public Draft, EPA 712-C-96-113. United States Environmental Protection Agency. http://www.epa.gov/opptsfrs/home/testmeth.htm

OECD Monograph 11, Detailed Review Paper on Aquatic Toxicity Testing for Industrial Chemicals and Pesticides

Rand, Gary M., Fundamentals of Aquatic toxicology: Effects, Environmental Fate, and Risk Assessment

2. **Biotic and abiotic degradation**

Boesten J.J.T.I. & A.M.A. van der Linden (1991). Modeling the influence of sorption and transformation on pesticide leaching and persistence. *J. Environ. Qual.* 20, 425-435

Boethling R.S., P.H. Howard, J.A. Beauman & M.E. Larosche (1995). Factors for intermedia extrapolation in biodegradability assessment. *Chemosphere* 30(4), 741-752

de Henau H. (1993). Biodegradation. In: P. Calow. Handbook of Ecotoxicology, vol. I. Blackwell Scientific Publications, London. Chapter 18, pp. 355-377

EC (1996). Technical guidance documents in support of the Commission Directive 93/67/EEC on risk assessment for new notified substances and the Commission Regulation (EC) No. 1488/94 on risk assessment for existing substances. European Commission, Ispra

ECETOC (1998): QSARs in the Assessment of the Environmental Fate and Effects of Chemicals, Technical report No. 74. Brussels, June 1998

附属書　9
付録VI
参考文献

1. 水生毒性

APHA 1992. Standard Methods for the Examination of Water and Wastewater, 18th edition. American Public Health Association, Washington, DC

ASTM 1999. Annual Book of ASTM standards, Vol. 11.04. American Society for Testing and Materials, Philadelphia, PA

DoE 1996. Guidance on the Aquatic Toxicity Testing of Difficult Substances. United Kingdom Department of the Environment, London

ECETOC 1996. Aquatic Toxicity Testing of Sparingly Soluble, Volatile and Unstable Substances. ECETOC Monograph No. 26, ECETOC, Brussels

Lewis, M. A. 1995. Algae and vascular plant tests. In: Rand, G. M. (ed.) 1995. Fundamentals of Aquatic Toxicology, Second Edition. Taylor & Francis, Washington, DC. pp. 135-169

Mensink, B. J. W. G., M. Montforts, L. Wijkhuizen-Maslankiewicz, H. Tibosch, and J.B.H.J. Linders 1995. Manual for Summarising and Evaluating the Environmental Aspects of Pesticides. Report No. 679101022 RIVM, Bilthoven, The Netherlands

OECD 1998. Harmonized Integrated Hazard Classification System for Human Health and Environmental Effects of Chemical Substances. OECD, Paris. (Document ENV/JM/MONO(2001)6). (Updated in 2001) Series on Testing and Assessment No. 33, OECD, Paris.

OECD 1999. Guidelines for Testing of Chemicals. Organisation for Economic Co-operation and Development, Paris

OECD 2000. Guidance Document on Aquatic Toxicity Testing of Difficult Substances and Mixtures, Series on Testing and Assessment No.23, OECD, Paris. Updated in 2019: OECD 2019. Second edition - Guidance Document on Aqueous-Phase Aquatic Toxicity Testing of Difficult Test Chemicals, Series on Testing and Assessment No. 23 (second edition). OECD, Paris.

OECD 2006. "Current approaches in the statistical analysis of ecotoxicity data: A guidance to application", OECD Environmental Health and Safety Publications Series Testing and Assessment N.54

Pedersen, F., H. Tyle, J. R. Niemeldi, B. Guttmann, L. Lander, and A. Wedebrand 1995. Environmental Hazard Classification – data collection and interpretation guide. TemaNord 1995:581

US EPA 1996. Ecological Effects Test Guidelines – OPPTS 850.1000. Special Considerations for Conducting Aquatic Laboratory Studies. Public Draft, EPA 712-C-96-113. United States Environmental Protection Agency. http://www.epa.gov/opptsfrs/home/testmeth.htm

OECD Monograph 11, Detailed Review Paper on Aquatic Toxicity Testing for Industrial Chemicals and Pesticides

Rand, Gary M., Fundamentals of Aquatic toxicology: Effects, Environmental Fate, and Risk Assessment

2. 生物的分解および非生物的分解

Boesten J.J.T.I. & A.M.A. van der Linden (1991). Modeling the influence of sorption and transformation on pesticide leaching and persistence. *J. Environ. Qual.* 20, 425-435

Boethling R.S., P.H. Howard, J.A. Beauman & M.E. Larosche (1995). Factors for intermedia extrapolation in biodegradability assessment. *Chemosphere* 30(4), 741-752

de Henau H. (1993). Biodegradation. In: P. Calow. Handbook of Ecotoxicology, vol. I. Blackwell Scientific Publications, London. Chapter 18, pp. 355-377

EC (1996). Technical guidance documents in support of the Commission Directive 93/67/EEC on risk assessment for new notified substances and the Commission Regulation (EC) No. 1488/94 on risk assessment for existing substances. European Commission, Ispra

ECETOC (1998): QSARs in the Assessment of the Environmental Fate and Effects of Chemicals, Technical report No. 74. Brussels, June 1998

Federle T.W., S.D. Gasior & B.A. Nuck (1997). Extrapolating mineralisation rates from the ready CO_2 screening test to activated sludge, river water, and soil. *Environmental Toxicology and Chemistry* 16, 127-134

Langenberg J.H., W.J.G.M. Peijnenburg & E. Rorije (1996). On the usefulness and reliability of existing QSBRs for risk assessment and priority setting. *SAR and QSAR in Environmental Research* 5, 1-16

Loonen H., F. Lindgren, B. Hansen & W. Karcher (1996). Prediction of biodegradability from chemical structure. In: Peijnenburg W.J.G.M. & J. Damborsky (eds.). Biodegradability Prediction. Kluwer Academic Publishers

MITI (1992). Biodegradation and bioaccumulation data on existing data based on the CSCL Japan. Japan chemical industry, Ecology-toxicology & information center. ISBN 4-89074-101-1

Niemelä J (2000). Personal communication to OECD Environment Directorate, 20 March 2000

Nyholm N., U.T. Berg & F. Ingerslev (1996). Activated sludge biodegradability simulation test. Danish EPA, Environmental Report No. 337

Nyholm N. & F. Ingerslev (1997). Kinetic biodegradation tests with low test substance concentrations: Shake flask test with surface water and short-term rate measurement in activated sludge. In: Hales S.G. (ed.). Biodegradation Kinetics: Generation and use of data for regulatory decision making. From the SETAC-Europe Workshop. Port- Sunlight. September 1996. pp. 101-115. SETAC-Europe, Brussels

Nyholm N. & L. Toräng (1999). Report of 1998/1999 Ring-test: Shalke flask batch test with surface water or surface water / sediment suspensions. ISO/CD 14592-1 Water Quality- Evaluation of the aerobic biodegradability of organic compounds at low concentrations, ISO/TC 147/ SC5/WG4 Biodegradability

OECD (1993). Structure-Activity Relationships for Biodegradation. OECD Environment Monographs No. 68. Paris 1993

OECD (1994): "US EPA/EC Joint Project on the Evaluation of (Quantitative) Structure Activity Relationships." OECD Environment Monograph No. 88. Paris

OECD (1995). Detailed Review Paper on Biodegradability Testing. OECD Environmental Monograph No. 98. Paris

OECD (1997). Guidance document on direct phototransformation of chemical in water. OECD/GD(97)21. Paris

OECD (1998). Harmonized integrated hazard classification system for human health and environmental effects of chemical substances. Paris. OECD, Paris. (Document ENV/JM/MONO(2001)6). (Updated in 2001) Series on Testing and Assessment No. 33, OECD, Paris.

Pedersen F., H. Tyle, J. R. Niemelä, B. Guttmann. L. Lander & A. Wedebrand (1995). Environmental Hazard Classification – data collection and interpretation guide for substances to be evaluated for classification as dangerous for the environment. Nordic Council of Ministers. 2nd edition. TemaNord 1995:581, 166 pp

Schwarzenbach R.P., P.M. Gschwend & D.M. Imboden (1993). Environmental organic chemistry 1st ed. John Wiley & Sons, Inc. New York

Scow K.M. (1982). Rate of biodegradation. In: Lyman W.J., W.F. Reehl & D.H. Rosenblatt (1982): Handbook of Chemical Property Estimation Methods Environmental Behaviour of Organic Compounds. American Chemical Society. Washington DC (ISBN 0-8412-1761-0). Chapter 9

Struijs J. & R. van den Berg (1995). Standardized biodegradability tests: Extrapolation to aerobic environments. *Wat. Res.* 29(1), 255-262

Syracuse Research Corporation. Biodegradation Probability Program (BIOWIN). Syracuse. N.Y. http://esc.syrres.com/~esc1/biodeg.htm

Westermann P., B.K. Ahring & R.A. Mah (1989). Temperature compensation in Methanosarcina barkeri by modulation of hydrogen and acetate affinity. *Applied and Environmental Microbiology* 55(5), 1262-1266

3. Bioaccumulation

Anliker, R., Moser, P., Poppinger, D. 1988. Bioaccumulation of dyestuffs and organic pigments in fish. Relationships to hydrophobicity and steric factors. Chem. 17(8):1631-1644

Bintein, S.; Devillers, J. and Karcher, W. 1993. Nonlinear dependence of fish bioconcentration on *n*-octanol/water partition coefficient. SAR and QSAR in Environmental Research. Vol.1.pp.29-39

Black, M.C., Millsap, D.S., McCarthy, J.F. 1991. Effects of acute temperature change on respiration and toxicant uptake by rainbow trout, *Salmo gairdneri* (Richardson). Physiol. Zool. 64:145-168

Bodor, N., Huang, M.J. 1992. J. Pharm. Sci. 81:272-281

Federle T.W., S.D. Gasior & B.A. Nuck (1997). Extrapolating mineralisation rates from the ready CO_2 screening test to activated sludge, river water, and soil. *Environmental Toxicology and Chemistry* 16, 127-134

Langenberg J.H., W.J.G.M. Peijnenburg & E. Rorije (1996). On the usefulness and reliability of existing QSBRs for risk assessment and priority setting. *SAR and QSAR in Environmental Research* 5, 1-16

Loonen H., F. Lindgren, B. Hansen & W. Karcher (1996). Prediction of biodegradability from chemical structure. In: Peijnenburg W.J.G.M. & J. Damborsky (eds.). Biodegradability Prediction. Kluwer Academic Publishers

MITI (1992). Biodegradation and bioaccumulation data on existing data based on the CSCL Japan. Japan chemical industry, Ecology-toxicology & information center. ISBN 4-89074-101-1

Niemelä J (2000). Personal communication to OECD Environment Directorate, 20 March 2000

Nyholm N., U.T. Berg & F. Ingerslev (1996). Activated sludge biodegradability simulation test. Danish EPA, Environmental Report No. 337

Nyholm N. & F. Ingerslev (1997). Kinetic biodegradation tests with low test substance concentrations: Shake flask test with surface water and short term rate measurement in activated sludge. In: Hales S.G. (ed.). Biodegradation Kinetics: Generation and use of data for regulatory decision making. From the SETAC-Europe Workshop. Port-Sunlight. September 1996. pp. 101-115. SETAC-Europe, Brussels

Nyholm N. & L. Toräng (1999). Report of 1998/1999 Ring-test: Shalke flask batch test with surface water or surface water / sediment suspensions. ISO/CD 14592-1 Water Quality- Evaluation of the aerobic biodegradability of organic compounds at low concentrations, ISO/TC 147/ SC5/WG4 Biodegradability

OECD (1993). Structure-Activity Relationships for Biodegradation. OECD Environment Monographs No. 68. Paris 1993

OECD (1994): "US EPA/EC Joint Project on the Evaluation of (Quantitative) Structure Activity Relationships." OECD Environment Monograph No. 88. Paris

OECD (1995). Detailed Review Paper on Biodegradability Testing. OECD Environmental Monograph No. 98. Paris

OECD (1997). Guidance document on direct phototransformation of chemical in water. OECD/GD(97)21. Paris

OECD (1998). Harmonized integrated hazard classification system for human health and environmental effects of chemical substances. Paris. (Document ENV/JM/MONO(2001)6). (Updated in 2001)

Pedersen F., H. Tyle, J. R. Niemelä, B. Guttmann. L. Lander & A. Wedebrand (1995). Environmental Hazard Classification – data collection and interpretation guide for substances to be evaluated for classification as dangerous for the environment. Nordic Council of Ministers. 2nd edition. TemaNord 1995:581, 166 pp

Schwarzenbach R.P., P.M. Gschwend & D.M. Imboden (1993). Environmental organic chemistry 1st ed. John Wiley & Sons, Inc. New York

Scow K.M. (1982). Rate of biodegradation. In: Lyman W.J., W.F. Reehl & D.H. Rosenblatt (1982): Handbook of Chemical Property Estimation Methods Environmental Behaviour of Organic Compounds. American Chemical Society. Washington DC (ISBN 0-8412-1761-0). Chapter 9

Struijs J. & R. van den Berg (1995). Standardized biodegradability tests: Extrapolation to aerobic environments. *Wat. Res.* 29(1), 255-262

Syracuse Research Corporation. Biodegradation Probability Program (BIOWIN). Syracuse. N.Y. http://esc.syrres.com/~esc1/biodeg.htm

Westermann P., B.K. Ahring & R.A. Mah (1989). Temperature compensation in Methanosarcina barkeri by modulation of hydrogen and acetate affinity. *Applied and Environmental Microbiology* 55(5), 1262-1266

3. 生物蓄積性

Anliker, R., Moser, P., Poppinger, D. 1988. Bioaccumulation of dyestuffs and organic pigments in fish. Relationships to hydrophobicity and steric factors. Chem. 17(8):1631-1644

Bintein, S.; Devillers, J. and Karcher, W. 1993. Nonlinear dependence of fish bioconcentration on n-octanol/water partition coefficient. SAR and QSAR in Environmental Research. Vol.1.pp.29-39

Black, M.C., Millsap, D.S., McCarthy, J.F. 1991. Effects of acute temperature change on respiration and toxicant uptake by rainbow trout, *Salmo gairdneri* (Richardson). Physiol. Zool. 64:145-168

Bodor, N., Huang, M.J. 1992. J. Pharm. Sci. 81:272-281

Broto, P., Moreau, G., Vandycke, C. 1984. Eur. J. Med. Chem. 19:71-78

Chiou, T. 1985. Partition coefficients of organic compounds in lipid-water systems and correlations with fish bioconcentration factors. Environ. Sci. Technol 19:57-62

CLOGP. 1995. Daylight Chemical Information Systems, Inf. Sys. Inc. Irvine, Ca

CSTEE (1999): DG XXIV Scientific Committee for Toxicity and Ecotoxicity and the Environment Opinion on revised proposal for a list of Priority substances in the context of the water framework directive (COMMs Procedure) prepared by the Frauenhofer-Institute, Germany,. Final report opinion adopted at the 11th CSTEE plenary meeting on 28th of September 1999

Comotto, R.M., Kimerle, R.A., Swisher, R.D. 1979. Bioconcentration and metabolism of linear alkylbenzenesulfonate by Daphnids and Fathead minnows. L.L.Marking, R.A. Kimerle, Eds., Aquatic Toxicology (ASTM, 1979), vol. ASTM STP 667

Connell, D.W., Hawker, D.W. 1988. Use of polynomial expressions to describe the bioconcentration of hydrophobic chemicals by fish. Ecotoxicol. Environ. Saf. 16:242-257

Connell, D.W. 1990. Bioaccumulation of xenobiotic compounds, Florida: CRC Press, Inc. pp.1-213

De Bruijn, J., Busser, F., Seinen, W. & Hermens, J. 1989. Determination of octanol/water partition coefficients with the "slow stirring" method. Environ. Toxicol. Chem. 8:499-512

Devillers, J., Bintein, S., Domine, D. 1996. Comparison of BCF models based on log P. Chemosphere 33(6):1047-1065

DoE, 1996. Guidance on the aquatic toxicity testing of difficult substance. Unites Kingdom Department of the Environment, London

Doucette, W.J., Andren, A.W. 1987. Correlation of octanol/water partition coefficients and total molecular surface area for highly hydrophobic aromatic compounds. Environ. Sci. Technol., 21, pages 821-824

Doucette, W.J., Andren, A.W. 1988. Estimation of octanol/water partition coefficients: evaluation of six methods for highly hydrophobic aromatic compounds. Chemosphere, 17, pages 345-359

Driscoll, S.K., McElroy, A.E. 1996. Bioaccumulation and metabolism of benzo(a)pyrene in three species of polychaete worms. Environ. Toxicol. Chem. 15(8):1401-1410

ECETOC, 1995. The role of bioaccumulation in environmental risk assessment: The aquatic environment and related food webs, Brussels, Belgium

ECEOOC, 1996. Aquatic toxicity testing of sparingly soluble, volatile and unstable substances. ECETOC Monograph No. 26, ECETOC, Brussels

European Commission, 1996. Technical Guidance Document in support of Commission Directive 93/96/EEC on Risk Assessment for new notified substances and Commission Regulation (EC) No 1488/94 on Risk Assessment for Existing Substances. Brussels

Ghose, A.K., Prottchet, A., Crippen, G.M. 1988. J. Computational Chem. 9:80-90

Gobas, F.A.P.C., Opperhuizen, A., Hutzinger, O. 1986. Bioconcentration of hydrophobic chemicals in fish: Relationship with membrane permeation. Environ. Toxicol. Chem. 5:637-646

Gobas, F.A.P.C., Clark, K.E., Shiu, W.Y., Mackay, D. 1989. Bioconcentration of polybrominated benzenes and biphenyls and related superhydrophobic chemicals in fish: Role of bioavailability and elimination into feces. Environ. Toxicol. Chem. 8:231-245

Goodrich, M.S., Melancon, M.J., Davis, R.A., Lech J.J. 1991. The toxicity, bioaccumulation, metabolism, and elimination of dioctyl sodium sulfosuccinate DSS in rainbow trout (*Oncorhynchus mykiss*) Water Res. 25: 119-124

Hansch, C., Leo, A. 1979. Substituent constants for correlation analysis in chemistry and biology. Wiley, New York, NY, 1979

Henderson, R.J., Tocher, D.R. 1987. The lipid composition and biochemistry of freshwater fish. Prog. Lipid. Res. 26:281-347

Howard, P.H. and Meyland, W.M., 1997. Prediction of physical properties transport and degradation for environmental fate and exposure assessments, QSAR in environmental science VII. Eds. Chen, F. and Schüürmann, G. pp. 185-205

Kimerle, R.A., Swisher, R.D., Schroeder-Comotto, R.M. 1975. Surfactant structure and aquatic toxicity, Symposium on Structure-Activity correlations in Studies on Toxicity and Bioconcentration with Aquatic Organisms, Burlington, Ontario, Canada, pp. 22-35

Broto, P., Moreau, G., Vandycke, C. 1984. Eur. J. Med. Chem. 19:71-78

Chiou, T. 1985. Partition coefficients of organic compounds in lipid-water systems and correlations with fish bioconcentration factors. Environ. Sci. Technol 19:57-62

CLOGP. 1995. Daylight Chemical Information Systems, Inf. Sys. Inc. Irvine, Ca

CSTEE (1999): DG XXIV Scientific Committee for Toxicity and Ecotoxicity and the Environment Opinion on revised proposal for a list of Priority substances in the context of the water framework directive (COMMs Procedure) prepared by the Frauenhofer-Institute, Germany,. Final report opinion adopted at the 11th CSTEE plenary meeting on 28th of September 1999

Comotto, R.M., Kimerle, R.A., Swisher, R.D. 1979. Bioconcentration and metabolism of linear alkylbenzenesulfonate by Daphnids and Fathead minnows. L.L.Marking, R.A. Kimerle, Eds., Aquatic Toxicology (ASTM, 1979), vol. ASTM STP 667

Connell, D.W., Hawker, D.W. 1988. Use of polynomial expressions to describe the bioconcentration of hydrophobic chemicals by fish. Ecotoxicol. Environ. Saf. 16:242-257

Connell, D.W. 1990. Bioaccumulation of xenobiotic compounds, Florida: CRC Press, Inc. pp.1-213

De Bruijn, J., Busser, F., Seinen, W. & Hermens, J. 1989. Determination of octanol/water partition coefficients with the "slow stirring" method. Environ. Toxicol. Chem. 8:499-512

Devillers, J., Bintein, S., Domine, D. 1996. Comparison of BCF models based on log P. Chemosphere 33(6):1047-1065

DoE, 1996. Guidance on the aquatic toxicity testing of difficult substance. Unites Kingdom Department of the Environment, London

Doucette, W.J., Andren, A.W. 1987. Correlation of octanol/water partition coefficients and total molecular surface area for highly hydrophobic aromatic compounds. Environ. Sci. Technol., 21, pages 821-824

Doucette, W.J., Andren, A.W. 1988. Estimation of octanol/water partition coefficients: evaluation of six methods for highly hydrophobic aromatic compounds. Chemosphere, 17, pages 345-359

Driscoll, S.K., McElroy, A.E. 1996. Bioaccumulation and metabolism of benzo(a)pyrene in three species of polychaete worms. Environ. Toxicol. Chem. 15(8):1401-1410

ECETOC, 1995. The role of bioaccumulation in environmental risk assessment: The aquatic environment and related food webs, Brussels, Belgium

ECEOOC, 1996. Aquatic toxicity testing of sparingly soluble, volatile and unstable substances. ECETOC Monograph No. 26, ECETOC, Brussels

European Commission, 1996. Technical Guidance Document in support of Commission Directive 93/96/EEC on Risk Assessment for new notified substances and Commission Regulation (EC) No 1488/94 on Risk Assessment for Existing Substances. Brussels

Ghose, A.K., Prottchet, A., Crippen, G.M. 1988. J. Computational Chem. 9:80-90

Gobas, F.A.P.C., Opperhuizen, A., Hutzinger, O. 1986. Bioconcentration of hydrophobic chemicals in fish: Relationship with membrane permeation. Environ. Toxicol. Chem. 5:637-646

Gobas, F.A.P.C., Clark, K.E., Shiu, W.Y., Mackay, D. 1989. Bioconcentration of polybrominated benzenes and biphenyls and related superhydrophobic chemicals in fish: Role of bioavailability and elimination into feces. Environ. Toxicol. Chem. 8:231-245

Goodrich, M.S., Melancon, M.J., Davis, R.A., Lech J.J. 1991. The toxicity, bioaccumulation, metabolism, and elimination of dioctyl sodium sulfosuccinate DSS in rainbow trout (*Oncorhynchus mykiss*) Water Res. 25: 119-124

Hansch, C., Leo, A. 1979. Substituent constants for correlation analysis in chemistry and biology. Wiley, New York, NY, 1979

Henderson, R.J., Tocher, D.R. 1987. The lipid composition and biochemistry of freshwater fish. Prog. Lipid. Res. 26:281-347

Howard, P.H. and Meyland, W.M., 1997. Prediction of physical properties transport and degradation for environmental fate and exposure assessments, QSAR in environmental science VII. Eds. Chen, F. and Schüürmann, G. pp. 185-205

Kimerle, R.A., Swisher, R.D., Schroeder-Comotto, R.M. 1975. Surfactant structure and aquatic toxicity, Symposium on Structure-Activity correlations in Studies on Toxicity and Bioconcentration with Aquatic Organisms, Burlington, Ontario, Canada, pp. 22-35

Klopman, G., Li, J.Y., Wang, S., Dimayuga, M. 1994. Computer automated log P calculations based on an extended group contribution approach. J. Chem. Inf. Comput. Sci. 34:752-781

Knezovich, J.P., Lawton, M.P., Inoue, L.S. 1989. Bioaccumulation and tissue distribution of a quaternary ammonium surfactant in three aquatic species. Bull. Environ. Contam. Toxicol. 42:87-93

Knezovich, J.P., Inoue, L.S. 1993. The influence of sediment and colloidal material on the bioavailability of a quaternary ammonium surfactant. Ecotoxicol. Environ. Safety. 26:253-264

Kristensen, P. 1991. Bioconcentration in fish: Comparison of BCFs derived from OECD and ASTM testing methods; influence of particulate matter to the bioavailability of chemicals. Danish Water Quality Institute

Mackay, D. 1982. Correlation of bioconcentration factors. Environ. Sci. Technol. 16:274-278

McCarthy, J.F., Jimenez, B.D. 1985. Reduction in bioavailability to bluegills of polycyclic aromatic hydrocarbons bound to dissolved humic material. Environ. Toxicol. Chem. 4:511-521

McKim, J.M., Goeden, H.M. 1982. A direct measure of the uptake efficiency of a xenobiotic chemical across the gill of brook trout (*Salvelinus fontinalis*) under normoxic and hypoxic conditions. Comp. Biochem. Physiol. 72C:65-74

Meylan, W.M. and Howard, P.H., 1995. Atom/Fragment Contribution Methods for Estimating Octanol-Water Partition Coefficients. J.Pharm.Sci. 84, 83

Niemelä, J.R. 1993. QTOXIN-program (ver 2.0). Danish Environmental Protection Agency

Niemi, G.J., Basak, S.C., Veith, G.D., Grunwald, G. Environ. Toxicol. Chem. 11:893-900

Niimi, A.J. 1991. Solubility of organic chemicals in octanol, triolin and cod liver oil and relationships between solubility and partition coefficients. Wat. Res. 25:1515-1521

OECD, 1993. Application of structure activity relationships to the estimation of properties important in exposure assessment. OECD Environment Directorate. Environment Monograph No. 67

OECD, 1998. Harmonized integrated hazard classification system for human health and environmental effects of chemical substances. As endorsed by the 28th joint meeting of the chemicals committee and the working party on chemicals in November 1998

OECD, 2000. Guidance Document on Aquatic Toxicity Testing of Difficult Substances and Mixtures, OECD, Paris. (Updated in 2019): OECD 2019. Second edition - Guidance Document on Aqueous-Phase Aquatic Toxicity Testing of Difficult Test Chemicals, Series on Testing and Assessment No. 23 (second edition). OECD, Paris.

Opperhuizen, A., Van der Velde, E.W., Gobas, F.A.P.C., Liem, A.K.D., Van der Steen, J.M.D., Hutzinger, O. 1985. Relationship between bioconcentration in fish and steric factors of hydrophobic chemicals. Chemosphere 14:1871-1896

Opperhuizen, A. 1986. Bioconcentration of hydrophobic chemicals in fish. In: Poston T.M., Purdy, R. (eds), Aquatic Toxicology and Environmental Fate: Ninth Volume, ASTM STP 921. American Society for Testing and Materials, Philadelphia, PA, 304-315

Opperhuizen, A., Schrap, S.M. 1987. Relationship between aqueous oxygen concentration and uptake and elimination rates during bioconcentration of hydrophobic chemicals in fish. Environ. Toxicol. Chemosphere 6:335-342

Opperhuizen, A., Sijm, D.T.H.M. 1990. Bioaccumulation and biotransformation of polychlorinated dibenzo-p-dioxins and dibenzofurans in fish. Environ. Toxicol. Chem. 9:175-186

Pedersen, F., Tyle, H., Niemelä, J.R., Guttmann, B., Lander, L. and Wedebrand, A., 1995. Environmental Hazard Classification – data collection and interpretation guide (2nd edition). TemaNord 1995:581

Petersen, G.I., Kristensen, P. 1998. Bioaccumulation of lipophilic substances in fish early life stages. Environ. Toxicol. Chem. 17(7):1385-1395

Rekker, R.F., de Kort, H.M. 1979. The hydrophobic fragmental constant: An extension to a 1000 data point set. Eur. J. Med. Chem. – Chim. Ther. 14:479-488

Roberts, D.W. 1989. Aquatic toxicity of linear alkyl benzene sulphonates (LAS) – a QSAR analysis. Communicaciones Presentadas a las Jornadas del Comite Espanol de la Detergencia, 20 (1989) 35-43. Also in J.E. Turner, M.W. England, T.W. Schultz and N.J. Kwaak (eds.) QSAR 88. Proc. Third International Workshop on Qualitative Structure-Activity Relationships in Environmental Toxicology, 22-26 May 1988, Knoxville, Tennessee, pp. 91-98. Available from the National Technical Information Service, US Dept. of Commerce, Springfield, VA

Schrap, S.M., Opperhuizen, A. 1990. Relationship between bioavailability and hydrophobicity: reduction of the uptake of organic chemicals by fish due to the sorption of particles. Environ. Toxicol. Chem. 9:715-724

Shiu, WY, Doucette, W., Gobas, FAPC., Andren, A., Mackay, D. 1988. Physical-chemical properties of chlorinated dibenzo-p-dioxins. Environ. Sci. Technol. 22: pages 651-658

Klopman, G., Li, J.Y., Wang, S., Dimayuga, M. 1994. Computer automated log P calculations based on an extended group contribution approach. J. Chem. Inf. Comput. Sci. 34:752-781

Knezovich, J.P., Lawton, M.P., Inoue, L.S. 1989. Bioaccumulation and tissue distribution of a quaternary ammonium surfactant in three aquatic species. Bull. Environ. Contam. Toxicol. 42:87-93

Knezovich, J.P., Inoue, L.S. 1993. The influence of sediment and colloidal material on the bioavailability of a quaternary ammonium surfactant. Ecotoxicol. Environ. Safety. 26:253-264

Kristensen, P. 1991. Bioconcentration in fish: Comparison of BCFs derived from OECD and ASTM testing methods; influence particulate matter to the bioavailability of chemicals. Danish Water Quality Institute

Mackay, D. 1982. Correlation of bioconcentration factors. Environ. Sci. Technol. 16:274-278

McCarthy, J.F., Jimenez, B.D. 1985. Reduction in bioavailability to bluegills of polycyclic aromatic hydrocarbons bound to dissolved humic material. Environ. Toxicol. Chem. 4:511-521

McKim, J.M., Goeden, H.M. 1982. A direct measure of the uptake efficiency of a xenobiotic chemical across the gill of brook trout (*Salvelinus fontinalis*) under normoxic and hypoxic conditions. Comp. Biochem. Physiol. 72C:65-74

Meylan, W.M. and Howard, P.H., 1995. Atom/Fragment Contribution Methods for Estimating Octanol-Water Partition Coefficients. J.Pharm.Sci. 84, 83

Niemelä, J.R. 1993. QTOXIN-program (ver 2.0). Danish Environmental Protection Agency

Niemi, G.J., Basak, S.C., Veith, G.D., Grunwald, G. Environ. Toxicol. Chem. 11:893-900

Niimi, A.J. 1991. Solubility of organic chemicals in octanol, triolin and cod liver oil and relationships between solubility and partition coefficients. Wat. Res. 25:1515-1521

OECD, 1993. Application of structure activity relationships to the estimation of properties important in exposure assessment. OECD Environment Directorate. Environment Monograph No. 67

OECD, 1998. Harmonized integrated hazard classification system for human health and environmental effects of chemical substances. As endorsed by the 28th joint meeting of the chemicals committee and the working party on chemicals in November 1998

OECD, 2000. Guidance Document on Aquatic Toxicity Testing of Difficult Substances and Mixtures, OECD, Paris (Updated in 2019)

Opperhuizen, A., Van der Velde, E.W., Gobas, F.A.P.C., Liem, A.K.D., Van der Steen, J.M.D., Hutzinger, O. 1985. Relationship between bioconcentration in fish and steric factors of hydrophobic chemicals. Chemosphere 14:1871-1896

Opperhuizen, A. 1986. Bioconcentration of hydrophobic chemicals in fish. In: Poston T.M., Purdy, R. (eds), Aquatic Toxicology and Environmental Fate : Ninth Volume, ASTM STP 921. American Society for Testing and Materials, Philadelphia, PA, 304-315

Opperhuizen, A., Schrap, S.M. 1987. Relationship between aqueous oxygen concentration and uptake and elimination rates during bioconcentration of hydrophobic chemicals in fish. Environ. Toxicol. Chemosphere 6:335-342

Opperhuizen, A., Sijm, D.T.H.M. 1990. Bioaccumulation and biotransformation of polychlorinated dibenzo-p-dioxins and dibenzofurans in fish. Environ. Toxicol. Chem. 9:175-186

Pedersen, F., Tyle, H., Niemelä, J.R., Guttmann, B., Lander,L. and Wedebrand, A., 1995. Environmental Hazard Classification – data collection and interpretation guide (2nd edition). TemaNord 1995:581

Petersen, G.I., Kristensen, P. 1998. Bioaccumulation of lipophilic substances in fish early life stages. Environ. Toxicol. Chem. 17(7):1385-1395

Rekker, R.F., de Kort, H.M. 1979. The hydrophobic fragmental constant: An extension to a 1000 data point set. Eur. J. Med. Chem. – Chim. Ther. 14:479-488

Roberts, D.W. 1989. Aquatic toxicity of linear alkyl benzene sulphonates (LAS) – a QSAR analysis. Communicaciones Presentadas a las Jornadas del Comite Espanol de la Detergencia, 20 (1989) 35-43.

Also in J.E. Turner, M.W. England, T.W. Schultz and N.J. Kwaak (eds.) QSAR 88. Proc. Third International Workshop on Qualitative Structure-Activity Relationships in Environmental Toxicology, 22-26 May 1988, Knoxville, Tennessee, pp. 91-98. Available from the National Technical Information Service, US Dept. of Commerce, Springfield, VA

Schrap, S.M., Opperhuizen, A. 1990. Relationship between bioavailability and hydrophobicity: reduction of the uptake of organic chemicals by fish due to the sorption of particles. Environ. Toxicol. Chem. 9:715-724

Shiu, WY, Doucette, W., Gobas, FAPC., Andren, A., Mackay, D. 1988. Physical-chemical properties of chlorinated dibenzo-p-dioxins. Environ. Sci. Technol. 22: pages 651-658

Sijm, D.T.H.M., van der Linde, A. 1995. Size-dependent bioconcentration kinetics of hydrophobic organic chemicals in fish based on diffusive mass transfer and allometric relationships. Environ. Sci. Technol. 29:2769-2777

Sijm, D.T.H.M., Pärt, P., Opperhuizen, A. 1993. The influence of temperature on the uptake rate constants of hydrophobic compounds determined by the isolated perfused gill of rainbow trout (*Oncorhynchus mykiss*). Aquat. Toxicol. 25:1-14

Spacie, A., Hamelink, J.L. 1982. Alternative models for describing the bioconcentration of organics in fish. Environ. Toxicol. Chem. 1:309-320

Suzuki, T., Kudo, Y.J. 1990. J. Computer-Aided Molecular Design 4:155-198

Syracuse Research Corporation, 1999.

Tas, J.W., Seinen, W., Opperhuizen, A. 1991. Lethal body burden of triphenyltin chloride in fish: Preliminary results. Comp. Biochem. Physiol. 100C(1/2):59-60

Tolls J. & Sijm, D.T.H.M., 1993. Bioconcentration of surfactants, RITOX, the Netherlands (9. Nov. 1993). Procter and Gamble Report (ed.: M.Stalmans)

Tolls, J. 1998. Bioconcentration of surfactants. Ph.D. Thesis. Utrecht University, Utrecht, The Netherlands

Toshima, S., Moriya, T. Yoshimura, K. 1992. Effects of polyoxyethylene (20) sorbitan monooleate on the acute toxicity of linear alkylbenzenesulfonate (C_{12}-LAS) to fish. Ecotoxicol. Environ. Safety 24: 26-36

USEPA 1985. U.S. Environmental Protection Agency. Office of Toxic Substances. Toxic Substances Control Act Test Guidelines. 50 FR 39252

US EPA/EC, 1993. US EPA/EC Joint Project on the Evaluation of (Quantitative) Structure Activity Relationships

US EPA, 1996. Ecological effects test guidelines – OPPTS 850.1000. Special considerations for conducting aquatic laboratory studies. Public Draft, EPA712-C-96-113. United States Environmental Protection Agency. http://www.epa.gov/opptsfrs/home/testmeth.htm

Van Den Berg, M., Van De Meet, D., Peijnenburg, W.J.G.M., Sijm, D.T.H.M., Struijs, J., Tas, J.W. 1995. Transport, accumulation and transformation processes. In: Risk Assessment of Chemicals: An Introduction. van Leeuwen, C.J., Hermens, J.L.M. (eds). Dordrecht, NL. Kluwer Academic Publishers, 37-102

Wakabayashi, M., Kikuchi, M., Sato, A. Yoshida, T. 1987. Bioconcentration of alcohol ethoxylates in carp (*Cyprinus carpio*), Ecotoxicol. Environ. Safety 13, 148-163

Wofford, H.W., C.D. Wilsey, G.S. Neff, C.S. Giam & J.M. Neff (1981): Bioaccumulation and metabolism of phthalate esters by oysters, brown shrimp and sheepshead minnows. Ecotox.Environ.Safety 5:202-210, 1981

4. Reference for QSAR

Boethling, R.S., Howard, P.H., Meylan, W.M. Stiteler, W.M., Beauman, J.A., and Tirado, N. (1994). Group contribution method for predicting probability and rate of aerobic biodegradation. Envir. Sci. Technol., 28, 459-465

De Bruijn, J, Busser, F., Seinen, W., and Hermens, J. (1989), Determination of octanol/water partition coefficients for hydrophobic organic chemicals with the "slow-stirring method," Environ. Toxicol. Chem., 8, 499-512

ECETOC (1998), QSARs in the Assessment of the Environmental Fate and Effects of Chemicals, Technical report No 74

Hansch, C. and A. Leo (1995), *Exploring QSAR*, American Chemical Society

Hilal, S. H., L. A. Carreira and S. W. Karickhoff (1994), *Quantitative Treatments of Solute/solvent Interactions, Theoretical and Computational Chemistry, Vol. 1*, 291-353, Elsevier Science

Howard, P.H., Boethling, R.S, Stiteler, W.M., Meylan, W.M., Hueber, A.E., Beaumen, J.A. and Larosche, M.E. (1992). Predictive model for aerobic biodegradation developed from a file of evaluated biodegradation data. Envir. Toxicol. Chem. 11, 593-603

Howard, P. And Meylan, W.M. (1992). Biodegradation Probability Program, Version 3, Syracuse Research Corp., NY

Langenberg, J.H., Peijnenburg, W.J.G.M. and Rorije, E. (1996). On the usefulness and reliability of existing QSARs for risk assessment and priority setting. SAR QSAR Environ. Res., 5, 1-16

R.L. Lipnick (1986). Charles Ernest Overton: Narcosis studies and a contribution to general pharmacology. *Trends Pharmacol. Sci.*, 7, 161-164

R.L. Lipnick (1989a). Hans Horst Meyer and the lipoid theory of narcosis, *Trends Pharmacol. Sci.*, 10 (7) July, 265-269; Erratum: 11 (1) Jan (1990), p. 44

Sijm, D.T.H.M., van der Linde, A. 1995. Size-dependent bioconcentration kinetics of hydrophobic organic chemicals in fish based on diffusive mass transfer and allometric relationships. Environ. Sci. Technol. 29:2769-2777

Sijm, D.T.H.M., Pärt, P., Opperhuizen, A. 1993. The influence of temperature on the uptake rate constants of hydrophobic compounds determined by the isolated perfused gill of rainbow trout (*Oncorhynchus mykiss*). Aquat. Toxicol. 25:1-14

Spacie, A., Hamelink, J.L. 1982. Alternative models for describing the bioconcentration of organics in fish. Environ. Toxicol. Chem. 1:309-320

Suzuki, T., Kudo, Y.J. 1990. J. Computer-Aided Molecular Design 4:155-198

Syracuse Research Corporation, 1999.

Tas, J.W., Seinen, W., Opperhuizen, A. 1991. Lethal body burden of triphenyltin chloride in fish: Preliminary results. Comp. Biochem. Physiol. 100C(1/2):59-60

Tolls J. & Sijm, D.T.H.M., 1993. Bioconcentration of surfactants, RITOX, the Netherlands (9. Nov. 1993). Procter and Gamble Report (ed.: M.Stalmans)

Tolls, J. 1998. Bioconcentration of surfactants. Ph.D. Thesis. Utrecht University, Utrecht, The Netherlands

Toshima, S., Moriya, T. Yoshimura, K. 1992. Effects of polyoxyethylene (20) sorbitan monooleate on the acute toxicity of linear alkylbenzenesulfonate (C_{12}-LAS) to fish. Ecotoxicol. Environ. Safety 24: 26-36

USEPA 1985. U.S. Environmental Protection Agency. Office of Toxic Substances. Toxic Substances Control Act Test Guidelines. 50 FR 39252

US EPA/EC, 1993. US EPA/EC Joint Project on the Evaluation of (Quantitative) Structure Activity Relationships

US EPA, 1996. Ecological effects test guidelines – OPPTS 850.1000. Special considerations for conducting aquatic laboratory studies. Public Draft, EPA712-C-96-113. United States Environmental Protection Agency. http://www.epa.gov/opptsfrs/home/testmeth.htm

Van Den Berg, M., Van De Meet, D., Peijnenburg, W.J.G.M., Sijm, D.T.H.M., Struijs, J., Tas, J.W. 1995. Transport, accumulation and transformation processes. In: Risk Assessment of Chemicals: An Introduction. van Leeuwen, C.J., Hermens, J.L.M. (eds). Dordrecht, NL. Kluwer Academic Publishers, 37-102

Wakabayashi, M., Kikuchi, M., Sato, A. Yoshida, T. 1987. Bioconcentration of alcohol ethoxylates in carp (*Cyprinus carpio*), Ecotoxicol. Environ. Safety 13, 148-163

Wofford, H.W., C.D. Wilsey, G.S. Neff, C.S. Giam & J.M. Neff (1981): Bioaccumulation and metabolism of phthalate esters by oysters, brown shrimp and sheepshead minnows. Ecotox.Environ.Safety 5:202-210, 1981

4. QSAR に関する参考文献

Boethling, R.S., Howard, P.H., Meylan, W.M. Stiteler, W.M., Beauman, J.A., and Tirado, N. (1994). Group contribution method for predicting probability and rate of aerobic biodegradation. Envir. Sci. Technol., 28, 459-465

De Bruijn, J, Busser, F., Seinen, W., and Hermens, J. (1989), Determination of octanol/water partition coefficients for hydrophobic organic chemicals with the "slow-stirring method," Environ. Toxicol. Chem., 8, 499-512

ECETOC (1998), QSARs in the Assessment of the Environmental Fate and Effects of Chemicals, Technical report No 74

Hansch, C. and A. Leo (1995), *Exploring QSAR*, American Chemical Society

Hilal, S. H., L. A. Carreira and S. W. Karickhoff (1994), *Quantitative Treatments of Solute/solvent Interactions, Theoretical and Computational Chemistry, Vol. 1,* 291-353, Elsevier Science

Howard, P.H., Boethling, R.S, Stiteler, W.M., Meylan, W.M., Hueber, A.E., Beaumen, J.A. and Larosche, M.E.(1992). Predictive model for aerobic biodegradation developed from a file of evaluated biodegradation data. Envir. Toxicol. Chem. 11, 593-603

Howard, P. And Meylan, W.M. (1992). Biodegradation Probability Program, Version 3, Syracuse Research Corp., NY

Langenberg, J.H., Peijnenburg, W.J.G.M. and Rorije, E. (1996). On the usefulness and reliability of existing QSARs for risk assessment and priority setting. SAR QSAR Environ. Res., 5, 1-16

R.L. Lipnick (1986). Charles Ernest Overton: Narcosis studies and a contribution to general pharmacology. *Trends Pharmacol. Sci.*, 7, 161-164

R.L. Lipnick (1989a). Hans Horst Meyer and the lipoid theory of narcosis, *Trends Pharmacol. Sci.*, 10 (7) July, 265-269; Erratum: 11 (1) Jan (1990), p. 44

R.L. Lipnick (1989b). Narcosis, electrophile, and proelectrophile toxicity mechanisms. Application of SAR and QSAR. *Environ. Toxicol. Chem.*, 8, 1-12

R.L. Lipnick (1990). Narcosis: Fundamental and Baseline Toxicity Mechanism for Nonelectrolyte Organic Chemicals. In: W. Karcher and J. Devillers (eds.) *Practical Applications of Quantitative Structure-Activity Relationships (QSAR) in Environmental Chemistry and Toxicology*, Kluwer Academic Publishers, Dordrecht, The Netherlands, pp. 129-144

R.L. Lipnick (ed.) (1991a). *Charles Ernest Overton: Studies of Narcosis and a Contribution to General Pharmacology*, Chapman and Hall, London, and Wood Library-Museum of Anesthesiology

R.L. Lipnick (1991b). Outliers: their origin and use in the classification of molecular mechanisms of toxicity, *Sci. Tot. Environ.*, 109/110 131-153

R.L. Lipnick (1995). Structure-Activity Relationships. In: Fundamentals of Aquatic Toxicology, 2nd edition, (G.R. Rand, ed.), Taylor & Francis, London, 609-655

Loonen, H., Lindgren, F., Hansen, B., Karcher, W., Niemela, J., Hiromatsu, K., Takatsuki, M., Peijnenburg, W., Rorije, E., and Struijs, J. (1999). Prediction of biodegradability from chemical structure: modeling of ready biodegradation test data. Environ. Toxicol. Chem., 18, 1763-1768

Meylan, W. M. and P. H. Howard (1995), *J. Pharm. Sci.*, 84, 83-92

OECD (1993), Structure-Activity Relationships for Biodegradation. OECD Environment Monograph No. 68 OECD, Paris, France

OECD (1995). Environment Monographs No. 92. Guidance Document for Aquatic Effects Assessment. OECD, Paris

F. Pedersen, H. Tyle, J. R. Niemelä, B. Guttmann, L. Lander, and A. Wedebrand (1995), Environmental Hazard Classification: Data Collection and Interpretation Guide for Substances to be Evaluated for Classification as Dangerous for the Environment, 2nd Edition, TemaNord 1995:581, Nordic Council of Ministers, Copenhagen, January

US EPA (1999) Development of Chemical Categories in the HPV Challenge Program, http://www.epa.gov/HPV/pubs/general/categuid.htm

US EPA (2000a), The Use of Structure-Activity Relationships (SAR) in the High Production Volume Chemicals Challenge Program, http://www.epa.gov/hpv/pubs/general/sarfinl1.htm

US EPA (2000b), ECOSAR, http://www.epa.gov/oppt/newchems/tools/21ecosar.htm

US EPA/EC (1993): US EPA Joint Project on the Evaluation of (Quantitative) Structure Activity Relationships, Commission of European Communities, Final Report, July

G.D. Veith, R.L. Lipnick, and C.L. Russom (1989). The toxicity of acetylenic alcohols to the fathead minnow, Pimephales promelas. Narcosis and proelectrophile activation. *Xenobiotica*, 19(5), 555-565

5. **Metals and metal compounds**

Brown, D.S. and Allison, J.D. (1987). MINTEQA1 Equilibrium Metal Speciation Model: A user's manual. Athens, Georgia, USEPA Environmental Research Laboratory, Office of Research and Development

Garman, E.R., Meyer, J.S., Bergeron, C.M., Blewett, T.A., Clements, W.H., Elias, M.C., Farley, K.J., Gissi, F. and Ryan, A.C. (2020), Validation of Bioavailability-Based Toxicity Models for Metals. Environmental Toxicology & Chemistry, 39: 101-117.

OECD (1998). Harmonized Integrated Hazard Classification System for Human Health and Environmental Effects of Chemical Substances (Document ENV/JM/MONO(2001)6)

OECD (2000). Guidance Document on Aquatic Toxicity Testing of Difficult Substances and Mixtures (Updated in 2019): OECD 2019. Second edition - Guidance Document on Aqueous-Phase Aquatic Toxicity Testing of Difficult Test Chemicals, Series on Testing and Assessment No. 23 (second edition). OECD, Paris.

OECD (2001). Guidance Document on Transformation/Dissolution of Metals and Metals Compounds in Aqueous Media

OECD (2015). Guidance on selecting a strategy for assessing the ecological risk of organometallic and organic metal salt substances based on their environmental fate. OECD Series on Testing and Assessment nr. 212. OECD, Paris, France.

Santore, R.C. and Driscoll, C.T. (1995). The CHESS Model for Calculating Chemical Equilibria in Soils and Solutions, Chemical Equilibrium and Reaction Models. The Soil Society of America, American Society of Agronomy

Santore, R.C. and Di Toro, D.M. et al (1999). A biotic ligand model of the acute toxicity of metals. II. Application to fish and daphnia exposure to copper. Environ. Tox. Chem. Submitted

R.L. Lipnick (1989b). Narcosis, electrophile, and proelectrophile toxicity mechanisms. Application of SAR and QSAR. *Environ. Toxicol. Chem.*, 8, 1-12

R.L. Lipnick (1990). Narcosis: Fundamental and Baseline Toxicity Mechanism for Nonelectrolyte Organic Chemicals. In: W. Karcher and J. Devillers (eds.) *Practical Applications of Quantitative Structure-Activity Relationships (QSAR) in Environmental Chemistry and Toxicology*, Kluwer Academic Publishers, Dordrecht, The Netherlands, pp. 129-144

R.L. Lipnick (ed.) (1991a). *Charles Ernest Overton: Studies of Narcosis and a Contribution to General Pharmacology*, Chapman and Hall, London, and Wood Library-Museum of Anesthesiology

R.L. Lipnick (1991b). Outliers: their origin and use in the classification of molecular mechanisms of toxicity, *Sci. Tot. Environ.*, 109/110 131-153

R.L. Lipnick (1995). Structure-Activity Relationships. In: Fundamentals of Aquatic Toxicology, 2nd edition, (G.R. Rand, ed.), Taylor & Francis, London, 609-655

Loonen, H., Lindgren, F., Hansen, B., Karcher, W., Niemela, J., Hiromatsu, K., Takatsuki, M., Peijnenburg, W., Rorije, E., and Struijs, J. (1999). Prediction of biodegradability from chemical structure: modeling of ready biodegradation test data. Environ. Toxicol. Chem., 18, 1763-1768

Meylan, W. M. and P. H. Howard (1995), *J. Pharm. Sci.*, 84, 83-92

OECD (1993), Structure-Activity Relationships for Biodegradation. OECD Environment Monograph No. 68 OECD, Paris, France

OECD (1995). Environment Monographs No. 92. Guidance Document for Aquatic Effects Assessment. OECD, Paris

F. Pedersen, H. Tyle, J. R. Niemelä, B. Guttmann, L. Lander, and A. Wedebrand (1995), Environmental Hazard Classification: Data Collection and Interpretation Guide for Substances to be Evaluated for Classification as Dangerous for the Environment, 2nd Edition, TemaNord 1995:581, Nordic Council of Ministers, Copenhagen, January

US EPA (1999) Development of Chemical Categories in the HPV Challenge Program, http://www.epa.gov/HPV/pubs/general/categuid.htm

US EPA (2000a), The Use of Structure-Activity Relationships (SAR) in the High Production Volume Chemicals Challenge Program, http://www.epa.gov/hpv/pubs/general/sarfinll.htm

US EPA (2000b), ECOSAR, http://www.epa.gov/oppt/newchems/tools/21ecosar.htm

US EPA/EC (1993): US EPA Joint Project on the Evaluation of (Quantitative) Structure Activity Relationships, Commission of European Communities, Final Report, July

G.D. Veith, R.L. Lipnick, and C.L. Russom (1989). The toxicity of acetylenic alcohols to the fathead minnow, Pimephales promelas. Narcosis and proelectrophile activation. *Xenobiotica*, 19(5), 555-565

5. 金属および金属化合物

Brown, D.S. and Allison, J.D. (1987). MINTEQA1 Equilibrium Metal Speciation Model: A user's manual. Athens, Georgia, USEPA Environmental Research Laboratory, Office of Research and Development

Garman, E.R., Meyer, J.S., Bergeron, C.M., Blewett, T.A., Clements, W.H., Elias, M.C., Farley, K.J., Gissi, F. and Ryan, A.C. (2020), Validation of Bioavailability - Based Toxicity Models for Metals. Environmental Toxicology & Chemistry, 39: 101-117.

OECD (1998). Harmonized Integrated Hazard Classification System for Human Health and Environmental Effects of Chemical Substances (Document ENV/JM/MONO(2001)6)

OECD (2000). Guidance Document on Aquatic Toxicity Testing of Difficult Substances and Mixtures (Updated om 2019)

OECD (2001). Guidance Document on Transformation/Dissolution of Metals and Metals Compounds in Aqueous Media

OECD (2015). Guidance on selecting a strategy for assessing the ecological risk of organometallic and organic metal salt substances based on their environmental fate. OECD Series on Testing and Assessment nr. 212. OECD, Paris, France.

Santore, R.C. and Driscoll, C.T. (1995). The CHESS Model for Calculating Chemical Equilibria in Soils and Solutions, Chemical Equilibrium and Reaction Models. The Soil Society of America, American Society of Agronomy

Santore, R.C. and Di Toro, D.M. et al (1999). A biotic ligand model of the acute toxicity of metals. II. Application to fish and daphnia exposure to copper. Environ. Tox. Chem. Submitted

Skeaff, J., Delbeke, K., Van Assche, F. and Conard, B. (2000) A critical surface are concept for acute hazard classification of relatively insoluble metal-containing powders in aquatic environments. Environ. Tox. Chem. 19:1681-1691

Tipping, E. (1994). WHAM – A computer equilibrium model and computer code for waters, sediments, and soils incorporating discrete site/electrostatic model of ion-binding by humic substances. Computers and Geoscience 20 (6): 073-1023

Tipping, E., Lofts, S., and Sonke, J.E. (2011). Humic Ion-Binding Model VII: a revised parameterisation of cation-binding by humic substances. Environmental Chemistry 8 225—235.

Skeaff, J., Delbeke, K., Van Assche, F. and Conard, B. (2000) A critical surface are concept for acute hazard classification of relatively insoluble metal-containing powders in aquatic environments. Environ. Tox. Chem. 19:1681-1691

Tipping, E. (1994). WHAM – A computer equilibrium model and computer code for waters, sediments, and soils incorporating discrete site/electrostatic model of ion-binding by humic substances. Computers and Geoscience 20 (6): 073-1023

Tipping, E., Lofts, S., and Sonke, J.E. (2011). Humic Ion-Binding Model VII: a revised parameterisation of cation-binding by humic substances. Environmental Chemistry 8 225—235.

ANNEX II

GUIDANCE ON TRANSFORM FLOODPLAIN USING DIGITAL TERRAIN COMPUTER PROGRAM MIKE 11

ANNEX 10

GUIDANCE ON TRANSFORMATION/DISSOLUTION OF METALS AND METAL COMPOUNDS IN AQUEOUS MEDIA

附属書 10

水性媒体中の
金属および金属化合物の
変化/溶解に関する手引き

ANNEX 10

GUIDANCE ON TRANSFORMATION/DISSOLUTION OF METALS AND METAL COMPOUNDS IN AQUEOUS MEDIA[1]

A10.1 Introduction

A10.1.1 This test guidance is designed to determine the rate and extent to which metals and sparingly soluble metal compounds can produce soluble available ionic and other metal-bearing species in aqueous media under a set of standard laboratory conditions representative of those generally occurring in the environment. Once determined, this information can be used to evaluate the short term and long-term aquatic toxicity of the metal or sparingly soluble metal compound from which the soluble species came. This test guidance is the outcome of an international effort under the OECD to develop an approach for the toxicity testing and data interpretation of metals and sparingly soluble inorganic metal compounds (reference 1, this annex and section A9.7 of annex 9). The experimental work on several metals and metal compounds upon which this test guidance is based has been conducted and reported (references 5 to 15, this annex). This test guidance has subsequently also been published as a guidance document by OECD (reference 16).

A10.1.2 The evaluation of the short term and long term aquatic toxicity of metals and sparingly soluble metal compounds is to be accomplished by comparison of (a) the concentration of the metal ion in solution, produced during transformation or dissolution in a standard aqueous medium with (b) appropriate standard ecotoxicity data as determined with the soluble metal salt (acute and chronic values). This document gives guidance for performing the transformation/dissolution tests. The strategy to derive an environmental hazard classification using the results of the Transformation/Dissolution Protocol is not within the scope of this guidance document and can be found in annex 9, section A9.7.

A10.1.3 For this test guidance, the transformations of metals and sparingly soluble metal compounds are, within the context of the test, defined and characterized as follows:

    (a)   metals, M^0, in their elemental state are not soluble in water but may transform to yield the available form. This means that a metal in the elemental state may react with the media to form soluble cationic or anionic products, and in the process the metal will oxidize, or transform, from the neutral or zero oxidation state to a higher one;

    (b)   in a simple metal compound, such as an oxide or sulphide, the metal already exists in an oxidized state, so that further metal oxidation is unlikely to occur when the compound is introduced into an aqueous medium. However, while oxidization state may not change, interaction with the media may yield more soluble forms. A sparingly soluble metal compound can be considered as one for which a solubility product can be calculated, and which will yield small amount of the available form by dissolution. However, it should be recognized that the final solution concentration may be influenced by a number of factors, including the solubility product of some metal compounds precipitated during the Transformation/Dissolution test, e.g. aluminium hydroxide.

A10.1.4 This test guidance is not applicable to organometallic compounds.

A10.2 Principles

A10.2.1 This test guidance is intended to be a standard laboratory transformation/ dissolution protocol based on a simple experimental procedure of agitating various quantities of the test substance in a pH buffered aqueous medium, and sampling and analysing the solutions at specific time intervals to determine the concentrations of dissolved metal ions in the water. Two different types of tests are described in the text below:

A10.2.2 *Screening transformation/dissolution test – sparingly soluble metal compounds*

A10.2.2.1 For sparingly soluble metal compounds, the maximum concentration of total dissolved metal can be determined by the solubility limit of the metal compound or from a screening transformation/dissolution test. The intent of the screening test, performed at a single loading, is to identify those compounds which undergo either dissolution or

---

[1] OECD Environment, Health and Safety Publications, Series on Testing and Assessment, No. 29, Environment Directorate, Organisation for Economic Co-operation and Development, April 2001.

附属書 10

水性媒体中の金属および金属化合物の変化/溶解に関する手引き [1]

A10.1 序文

A10.1.1 この試験の手引きは、一般に環境中に生じる条件を再現した一連の標準的な実験的条件の下で、金属および難水溶性の金属化合物が、利用性のある水溶性イオン性やその他の金属を含んだ物質種を生成する速度と程度を決定するために考案されたものである。これらが決定されれば、その情報を、水溶性物質種が由来する金属または難水溶性金属化合物の、短期および長期の水生毒性を評価するのに用いることができる。この手引きは、金属並びに難水溶性の無機金属化合物の毒性試験およびデータ解釈の確立に向けたOECDの下での国際的な努力の成果である(本附属書の参考文献1および附属書9のA9.7章を参照)。本手引きの根拠となるいくつかの金属及び金属化合物に関する実験が実施され、報告されている(本附属書の参考文献5～15)。本手引きは、その後OECDによってガイダンス文書としても発行された(参考文献 16)。

A10.1.2 金属および難溶性金属化合物の短期および長期の水生毒性に関する評価は、(a) 標準的な水性媒体中における変化または溶解によって生成される、溶液中の金属イオン濃度と、(b) 水溶性金属塩に関して測定された、適切な標準的環境毒性データ(急性および慢性毒性値)とを比較することになっている。本文書は、こうした変化/溶解試験を実施するための指針を提供するものである。こうした変化/溶解プロトコールの結果を用いて、環境有害性の分類を行うための戦略については、この手引きの範囲ではなく、附属書9のA9.7を参照されたい。

A10.1.3 本手引きに関しては、金属および難溶性金属化合物の変換が、試験との関係で以下のように定義、特徴付けられている:

(a) 金属 M^0 は、元素状態では水に溶けないが、変換されて利用性のある形態をとることがある。これは、元素状態の金属が媒体と反応して可溶性の陽イオン性または陰イオン性の生成物を形成し、その過程においてこの金属が中性もしくはゼロ酸化状態からより高い酸化数の状態に酸化、もしくは変換することを意味している;

(b) 酸化物や硫化物などの単純な金属化合物においては、金属は既に酸化された状態で存在するので、そうした化合物が水性媒体に導入されても、さらに金属酸化が起こることはありそうもない。しかし、酸化状態に変化はなくても、媒体との相互作用によって、より可溶性の形態を生成することはある。難溶性の金属化合物は、その溶解度積を算定することが可能であり、かつ溶解によって少量の利用性のある形態を生成する化合物と考えることができる。しかし例えば水酸化アルミニウムのように、最終的な溶液濃度は変換/溶解試験の過程で析出する金属化合物の溶解度積などの、数多くの要因によって影響されることは認識すべきである。

A10.1.4 本手引きは有機金属化合物には適用されない。

A10.2 原則

A10.2.1 この試験の手引きは、pHの緩衝能力を持つ水性媒体中で、様々な量の被験物質を攪拌し、一定時間ごとにサンプルをとって水溶液の分析を行い、水中に溶存する金属イオンの濃度を測定するという、単純な実験手続に基づく、標準的な実験室での変化/溶解プロトコールとして意図されたものである。以下のテキストでは、2つの異なったタイプの試験について述べる。

A10.2.2 *変化/溶解のスクリーニング試験:難溶性金属化合物*

A10.2.2.1 難溶性金属化合物に関しては、金属化合物の溶解限度によって、または変化/溶解のスクリーニング試験によって溶存する金属の最高濃度を決定できる。単回添加で行われるこのスクリーニング試験

---

[1] OECD Environment. Health and Safety Publications, Series on Testing and Assessment, No.29, Environment Directorate, Organisation for Economic Co-operation and Development, April 2001.

rapid transformation such that their ecotoxicity potential is indistinguishable from soluble forms and to verify the pH dependency of the dissolution, in preparation of the full transformation/dissolution test (see A9.7.2.3).

A10.2.2.2　　Sparingly soluble metal compounds, having the smallest representative particle size on the market are introduced into the aqueous medium at a single loading of 100 mg/l. Such dissolution as will occur is achieved by agitation during a 24 hours period. After 24 hours agitation, the dissolved metal ion concentration is measured.

A10.2.3　　*Full Transformation/Dissolution test - metals and sparingly soluble metal compounds*

A10.2.3.1　　The full transformation/dissolution test is intended to determine level of the dissolution or transformation of metals and metal compounds after a certain time period at different loadings of the aqueous phase. Normally massive forms and/or powders are introduced into the aqueous medium at three different loadings: 1, 10 and 100 mg/l. A single loading of 100 mg/l may be used if a significant release of dissolved metal species is not anticipated. transformation/dissolution is accomplished by standardized agitation, without causing abrasion of the particles. The short-term transformation/dissolution endpoints are based on the dissolved metal ion concentrations obtained after a 7 days transformation/dissolution period. The long-term transformation/dissolution endpoint is obtained during a 28 days transformation/dissolution test, using a a loading of 1 mg/l, 0.1 mg/l, or 0.01 mg/l depending on the transformation rate.

A10.2.3.2　　As pH has a significant influence on transformation/dissolution both the screening test and the full test should in principle be carried out at a pH that maximizes the concentration of the dissolved metal ions in solution. With reference to the conditions generally found in the environment a pH range of 6 to 8.5 must be used, except for the 28-day full test where the pH range of 5.5 to 8.5 is recommended if technically feasible to take into consideration possible long term effects on acidic lakes.

A10.2.3.3　　As in addition the surface area of the particles in the test sample has an important influence on the rate and extent of transformation/dissolution, powders are tested at the smallest representative particle size as placed on the market, while massive forms are tested at a particle size representative of normal handling and use. A default diameter value of 1 mm should be used in absence of this information. For massive metals, this default may only be exceeded when sufficiently justified. The specific surface area should be determined in order to characterize and compare similar samples. The tested material should also be free from oxidation/corrosion layers due to storage, given the latter may disturb the transformation rate. Appropriate pre-treatment of the samples is recommended.

A10.3　　Applicability of the test

This test applies to all metals and sparingly soluble inorganic metal compounds. Exceptions, such as certain water reactive metals, should be justified.

A10.4　　Information on the test substance

Substances as placed on the market should be used in the transformation/dissolution tests. In order to allow for correct interpretation of the test results, it is important to obtain the following information on the test substance(s):

(a)　　substance name, formula and use on the market;

(b)　　physical-chemical method of preparation;

(c)　　identification of the batch used for testing;

(d)　　chemical characterization: overall purity (%) and specific impurities (% or ppm);

(e)　　density (g/cm^3) or specific gravity;

(f)　　measured specific surface area (m^2/g)- measured by BET N$_2$ adsorption desorption or equivalent technique, and particle size distribution;

(g)　　storage, expiration date;

(h)　　known solubility data and solubility products;

(i)　　hazard identification and safe handling precautions;

(j)　　Safety Data Sheets (SDS) or equivalent.

の意図は、変化/溶解の詳細試験（A9.7.2.3 参照）に備え、生態毒性の可能性が可溶性と区別できないような溶解または急速な変化を起こす化合物を同定し、溶解のpH依存性を確認することである。

A10.2.2.2　市販されている最も小さな代表粒子サイズを持つ難溶性金属化合物が、100mg/lの単回添加で水性媒体に加えられる。これによる溶解は、24時間の攪拌によって達成される。24時間の攪拌の後、溶解した金属イオンの濃度が測定される。

A10.2.3　変化/溶解の詳細試験：金属および難溶性金属化合物

A10.2.3.1　変化/溶解の詳細試験は、様々な用量での添加が行われた水相において、一定の時間の後に金属および金属化合物の溶解または変換のレベルを決定するためのものである。通常は塊状または粉末状の金属が3種類の異なる添加量（1、10、100mg/l）で水性媒体に加えられる。100mg/lの単回添加は、溶解性の金属物質種の著しい放出が予想されない場合に行われる。変化/溶解は、微粒子の磨耗を起こさない、標準化された攪拌作業によって得られるものである。短期の変化/溶解のエンドポイントは、7日間の変化/溶解期間の後に得られる溶存金属イオン濃度に基づくものである。長期の変化/溶解のエンドポイントは、変化速度に応じて 1 mg/l、0.1 mg/l、0.01 mg/l の負荷量を用いて、28日間の変化/溶解試験中に得られる。

A10.2.3.2　pHは変化/溶解に著しく影響するので、スクリーニング試験も詳細試験も原則的には溶液中の溶存金属イオン濃度が最大になるpH値で行うべきである。技術的に可能な場合、酸性湖に対する長期的な影響の可能性を考慮するために 5.5～8.5 のpH値が推奨される28日間の詳細試験を除き、一般の環境に見出される条件に照らせば、6～8.5 の範囲のpH値を用いなければならない。

A10.2.3.3　加えて、試験サンプル粒子の表面積は変化/溶解の速度と程度に大きく影響するので、粉末は市販されている中で最も小さな代表粒子径のものを試験し、塊状の形態のものは通常の取り扱いおよび使用において代表される粒子径で試験される。こうした情報がない場合は、直径1mmのデフォルト値を用いるべきである。塊状の金属の場合は、十分な根拠がある場合を除いて、このデフォルト値を超えるべきではない。似たようなサンプルを比較し、特徴付けるために、比表面積を測定すべきである。保管によって酸化/腐食層が変化速度を阻害する可能性があるため、試験材料には保管による酸化/腐食層がないことも必要である。サンプルの適切な前処理が推奨される。

A10.3　試験の適用範囲

上記の試験はすべての金属および難溶性の無機金属化合物に適用できる。一部の水反応性金属のような例外は根拠を明らかにすべきである。

A10.4　被験物質に関する情報

変化/溶解試験では市販されている状態の物質を用いるべきである。試験結果を正しく解釈できるようにするために、被験物質に関して次に示す情報を得ることが重要である：

(a)　物質の名称、化学式、市販の用途；
(b)　調製のための物理化学的な方法；
(c)　試験に用いる製品バッチの特定；
(d)　化学的特性：全体の純度(%)および特定の不純物（%またはppm）；
(e)　密度（g/cm^3）または比重；
(f)　比表面積の測定値（m^2/g）－BET N$_2$吸脱着法、または同等の技術、および粒度分布；
(g)　貯蔵期間および品質期限；
(h)　既知の溶解度データおよび溶解生成物；
(i)　危険有害性の特定と安全な取り扱いのための注意点；
(j)　安全データシート（SDS）またはこれに相当するもの。

A10.5   **Description of the test method**

A10.5.1   *Apparatus and reagents*

A10.5.1.1   The following apparatus and reagents are necessary for performing tests:

    (a)   pre-cleaned and acid rinsed closed glass sample bottles (A10.5.1.2);

    (b)   transformation /dissolution medium (ISO 6341) (A10.5.1.3);

    (c)   test solution buffering facilities (A10.5.1.4);

    (d)   agitation equipment: orbital shaker, laboratory shaker or equivalent (A10.5.1.5);

    (e)   appropriate filters (e.g. 0.2 µm Acrodisc) or centrifuge for solids-liquid separation (A10.5.1.10) filter should be flushed at least three times with fresh medium to avoid elevetaed trace metals in sample at time 0;

    (f)   means to control the temperature of the reaction vessels to ± 1.5°C in the range 20-23 °C, such as a temperature-controlled cabinet or a water bath;

    (g)   syringes and/or automatic pipettes;

    (h)   pH meter showing acceptable results within + 0.2 pH units;

    (i)   dissolved oxygen meter, with temperature reading capability;

    (j)   thermometer or thermocouple; and

    (k)   analytical equipment for metal analysis (e.g. atomic adsorption spectrometry, inductively coupled plasma mass spectrometry) of acceptable accuracy, preferably with a limit of quantification (LOQ) five times lower than the lowest chronic ecotoxicity reference value or the lowest acute ecotoxicity reference value if only a 7-day test is conducted;

A10.5.1.2   All glass test vessels must be carefully cleaned by standard laboratory practices, acid-cleaned (e.g. HCl or aqua regia) and subsequently rinsed with de-ionized water. Specific attention to the type of glassware is required for metals that can be released from the glass. The test vessel volume and configuration (e.g. one- or two-litre reaction kettles) should be sufficient to hold 1 or 2 l of aqueous medium without overflow during the agitation specified. If air buffering is used (tests carried out at pH 8), it is advised to increase the air buffering capacity of the medium by increasing the headspace/liquid ratio (e.g. 1 l medium in 2.8 l flasks).

A10.5.1.3   A reconstituted standard water based on ISO 6341 should be used[2], as the standard transformation/dissolution medium. The medium should be sterilized by filtration (0.2 µm) before use in the tests. The chemical composition of the standard transformation/dissolution medium (for tests carried out at pH 8) is as follows:

        $NaHCO_3$:   65.7 mg/l
        KCl   :   5.75 mg/l
        $CaCl_2.2H_2O$:   294 mg/l
        $MgSO_4.7H_2O$:   123 mg/l

    For tests carried out at lower or higher pH values, adjusted chemical compositions are given in A10.5.1.7.

---

[2] *For hazard classification purposes the results of the Transformation/Dissolution Protocol are compared with existing ecotoxicity data for metals and metal compounds. However, for purposes such as data validation, there might be cases where it may be appropriate to use the aqueous medium from a completed transformation test directly in an OECD 202 and 203 daphnia and fish ecotoxicity test. If the $CaCl_2.2H_2O$ and $MgSO_4.7H_2O$ concentrations of the transformation medium are reduced to one-fifth of the ISO 6341 medium, the completed transformation medium can also be used (upon the addition of micronutrients) in an OECD 201 algae ecotoxicity test.*

A10.5 試験方法についての解説

A10.5.1 試験器具と試薬

A10.5.1.1 次に示す器具および試薬は、試験の実施に必要なものである：

(a) 予備洗浄および酸洗いされた密閉式ガラスサンプル瓶（A10.5.1.2）；

(b) 変化/溶解媒体（ISO6341）（A10.5.1.3）；

(c) 緩衝能のある試験溶液（A10.5.1.4）；

(d) 攪拌設備：回転式振騰器、実験室用振騰器、または同等の器具（A10.5.1.5）；

(e) 適切なフィルター（例えば0.2μmのアクロディスク）または固液分離用遠心機（A10.5.1.7）フィルターは、時間0においてサンプル内で微量金属の濃度が高まることを避けるために、未使用の媒体で少なくとも3回洗浄すべきである。；

(f) 温度制御キャビネットやウォーターバスなど、反応器の温度を20℃〜23℃の範囲内で±1.5℃まで制御できる装置；

(g) シリンジまたは自動ピペット；

(h) ＋0.2pH単位の範囲内で容認可能な結果が得られるpHメータ；

(i) 温度表示機能付の溶存酸素メータ；

(j) 温度計または熱電対；

(k) 許容精度を有する金属分析用の分析装置（例えば原子吸光分光分析装置、誘導結合プラズマ質量分析装置）望ましくは定量限界（LOQ）が最小の慢性生態毒性基準値または7日間試験のみの場合は最小の急性生態毒性基準値の5分の1より小さいこと。

A10.5.1.2 すべてのガラス製の試験容器は、標準の実験方法によって注意深く洗浄し、酸洗浄（例；塩酸、王水）を行い、次いで脱イオン水ですすがなければならない。ガラスから放出される可能性のある金属については、ガラス器具の種類に特別な注意が必要である。試験容器の大きさと形状（例えば1または2リットル反応器）は、指定された攪拌によって溢れ出すことなく1〜2リットルの水性媒体を入れるのに十分なものであるべきである。空気緩衝を用いる場合（pH8で行われる試験）は、上部空間/液体比率を増大させることによって（例えば1リットルの媒質に対して2.8リットルのフラスコ）、媒体の空気緩衝能力を向上させることが勧められる。

A10.5.1.3 標準の変換/溶解用の媒体としては、ISO6341に基づく標準人工調整水を用いるべきである[2]。この媒体は、試験に使用する前に、ろ過（0.2μm）によって滅菌処理を行うべきである。（pH8で実施される試験用の）標準の変換/溶解媒体の化学組成は次のとおりである：

$NaHCO_3$: 65.7mg/l
KCl : 5.75mg/l
$CaCl_2 \cdot 2H_2O$: 294mg/l
$MgSO_4 \cdot 7H_2O$: 123mg/l

低いpH値または高いpH値で実施される試験のための、調整された化学組成はA10.5.1.7に示す。

---

[2] 危険有害性分類を目的とする場合、変化/溶解プロトコールの結果が金属および金属化合物に関する既存の環境毒性データと比較される。しかし、データの検証などを目的とする場合は、完了した変化試験の水性媒体を直接、OECD202および203のミジンコ及び魚類の環境毒性試験に用いることが適切なことがあるかもしれない。変化媒体における$CaCl_2 \cdot 2H_2O$及び$MgSO_4 \cdot 7H_2O$の濃度をISO6341の媒体の5分の1にまで下げれば、生成された変化媒体はまた（微量栄養素を添加すれば直ちに）OECD201の藻類環境毒性試験において利用できる。

A10.5.1.4 The concentration of total organic carbon in the medium before adding the substance, should not exceed 2.0 mg/l.

A10.5.1.5 In addition to the fresh water medium, the use of a standardized marine test medium may also be considered when the solubility or transformation of the metal or metal compound is expected to be significantly affected by the high chloride content or other unique chemical characteristics of marine waters and when toxicity test data are available on marine species. When marine waters are considered, the chemical composition of the standard marine medium is as follows:

NaF	:	3mg/l
$SrCl_2 \cdot 6H_2O$	:	20 mg/l
H_3BO_3	:	30 mg/l
KBr	:	100 mg/l
KCl	:	700 mg/l
$CaCl_2 \cdot 2H_2O$	:	1.47g/l
Na_2SO_4	:	4.0 g/l
$MgCl_2 \cdot 6H_2O$	:	10.78 g/l
NaCl	:	23.5 g/l
$Na_2SiO_3 \cdot 9H_2O$	:	20 mg/l
$NaHCO_3$	:	200 mg/l

The salinity should be 34 ± 0.5 g/kg and the pH should be 8.0 ± 0.2. The reconstituted salt water should also be stripped of trace metals (from ASTM E 729-96).

A10.5.1.6 The transformation/dissolution tests are to be carried out at a pH that maximizes the concentration of the dissolved metal ions in solution within the prescribed pH range. A pH-range of 6 to 8.5 must be used for the screening test and the 7-day full test, and a range of 5.5 to 8.5 for the 28 day full test (A10.2.3.2).

A10.5.1.7 Buffering at pH 8 may be established by equilibrium with air, in which the concentration of CO_2 provides a natural buffering capacity sufficient to maintain the pH within an average of ± 0.2 pH units over a period of one week (reference 7, annex 10). An increase in the headspace/liquid ratio can be used to improve the air buffering capacity of the medium.

For pH adjustment and buffering down to pH 7 and 6 and up to pH 8 and 8.5, table A10.1 shows the recommended chemical compositions of the media, as well as the CO_2 concentrations in air to be passed through the headspace, and the calculated pH values under these conditions.

Table A10.1: Recommended chemical composition of testing medium

Chemical composition of medium	$NaHCO_3$	6.5 mg/l	12.6 mg/l	64.75 mg/l	194.25 mg/l
	KCl	0.58 mg/l	2.32 mg/l	5.75 mg/l	5.74 mg/l
	$CaCl_2.2H_2O$	29.4 mg/l	117.6 mg/l	294 mg/l	29.4 mg/l
	$MgSO_4.7H_2O$	12.3 mg/l	49.2 mg/l	123.25 mg/l	123.25 mg/l
CO_2 concentration (balance is air) in test vessel		0.50 %	0.10 %	0.038 % (air)	0.038 %(air)
Calculated pH		6.09	7.07	7.98	8.5

NOTE 1: *The pH values were calculated using the FACT (Facility for the Analysis of Chemical Thermodynamics) System (http://www.crct.polymtl.ca/fact/fact.htm).*

NOTE 2: *While the protocol was only validated for the pH range 6.0-8.0, this table does not prevent attaining pH 5.5. Composition for pH 8.5 has not been verified experimentally in presence of metal.*

NOTE 3: *Equilibration via headspace is recommended given CO_2 gas bubbling does not guarantee equal distribution between different test vessels.*

A10.5.1.8 Alternative equivalent buffering methods may be used if the influence of the applied buffer on the chemical speciation and transformation rate of the dissolved metal fraction would be minimal. pH should not be adjusted during the test using an acid or alkali.

A10.5.1.4　物質添加前の媒体中の全有機炭素濃度は、2.0mg/l を超えるべきではない。

A10.5.1.5　金属または金属化合物の溶解度または変化が高濃度の含有塩素もしくはその他の海水に固有の化学的性質によって著しく影響されると予想され、また海生生物種での毒性データが利用可能な場合は、淡水媒体に加え、基準化された海水試験媒体も考慮されてよい。海水を考慮する際、標準の海水媒体の化学組成は次のとおりである：

NaF	: 3mg/l
$SrCl_2・6H_2O$	: 20mg/l
H_3BO_3	: 30mg/l
KBr	: 100mg/l
KCl	: 700mg/l
$CaCl_2・2H_2O$	: 1.47g/l
Na_2SO_4	: 4.0g/l
$MgCl_2・6H_2O$	: 10.78g/l
NaCl	: 23.5g/l
$Na_2SiO_3・9H_2O$	: 20mg/l
$NaHCO_3$	: 200mg/l

塩分濃度は 34±0.5g/kg、また pH は 8.0±0.2 とすべきである。また人工調整塩水からは微量金属を取り除いておくべきである（ASTM E 729-96）。

A10.5.1.6　変化/溶解試験は、定められた pH 域内において、溶液中の溶存金属イオン濃度が最大になる pH 値において実施すべきである。スクリーニング試験と 7 日間の詳細試験では、6～8.5 の範囲内の pH 域を用い、28 日間の詳細試験では 5.5～8.5 までの pH 域を用いるべきである。（A10.2.3.2）

A10.5.1.7　pH8 における緩衝は空気との平衡によって得ることができ、1 週間の試験期間においては、CO_2 が pH を平均で±0.2pH 単位の範囲内に維持するに十分な自然の緩衝機能を果たす（附属書 10 の参考文献 7）。上部空間/液体の比率を引き上げることで媒体に対する空気の緩衝能力を改善できる。

下は pH7 から 6 まで、上は pH8 から 8.5 までの pH 調整および緩衝については、表 A10.1 に推奨される媒質の化学組成と上部空間に通すべき空気中の CO_2 濃度、ならびに pH の計算値を示す。

表 A10.1：試験溶媒の推薦組成

媒体の化学組成	$NaHCO_3$	6.5 mg/l	12.6mg/l	64.75mg/l	194.25mg/l
	KCl	0.58mg/l	2.32mg/l	5.75mg/l	5.74mg/l
	$CaCl_2・2H_2O$	29.4mg/l	117.6mg/l	294mg/l	29.4mg/l
	$MgSO_4・7H_2O$	12.3mg/l	49.2mg/l	123.25mg/l	123.25mg/l
試験容器内の CO_2 濃度（大気緩衝）		0.50%	0.10%	0.038%（大気）	0.038%（大気）
pH の計算値		6.09	7.07	7.98	8.5

注記 1：pH は FACT (Facility for the Analysis of Chemical Thermodynamics) システム（http://www.crct.polymtl.ca/fact/fact.htm）を用いて計算したものである。

注記 2：プロトコールは pH 範囲 6.0-8.0 についてのみ有用性が確認されているが、本表は pH5.5 を得るのを妨げるものではない。pH8.5 の組成は、金属存在下では実験的には確認されていない。

注記 3：CO_2 ガスのバブリングが異なる試験容器間の均等な分配を保証しないことを考慮し、ヘッドスペースによる平衡化が推奨される。

A10.5.1.8　溶存金属成分の化学種分化および変化の速度に与える影響がわずかであれば、同等の代替緩衝法を使用してもよい。酸やアルカリを使用した試験中は pH 値は変更されるべきではない。

A10.5.1.9   During the full transformation/dissolution tests, agitation should be used which is sufficient to maintain the flow of aqueous medium over the test substance while maintaining the integrity of the surface of the test substance and of any solid reaction product coatings formed during the test. For 1 l of aqueous medium, this may be accomplished by the use of a 1.0 to 3.0 l flask capped with a rubber stopper and placed on an orbital or laboratory shaker set at 100 r.p.m. Other methods of gentle agitation may be used provided they meet the criteria of surface integrity and homogeneous solution.

A10.5.1.10   The choice of solids-liquid separation method depends on whether adsorption of soluble metal ions on filters occurs and whether or not a suspension is generated by the agitation prescribed in A10.5.1.9, which will in turn depend on particle size distributions the shape of the particles and particle density. For solids of density greater than approximately 6 g/cm^3 and particle size ranges as low as 50 % < 8 μm, experience has shown that the gentle agitation methods prescribed in A10.5.1.9 are unlikely to result in suspensions. Alternative techniques may be considered in case of finer particles. If there is concern that particles will remain in suspension, then filtration efficiency should be checked prior to any testing. Options that could be considered to increase filtration efficiency include centrifugation followed by filtration, or waiting for about 5 minutes for the suspension to settle prior to taking a solution sample.

A10.5.2   *Prerequisites*

A10.5.2.1   *Analytical method*

A suitable validated analytical method for the total dissolved metal analysis is essential to the study. The analytical detection limit should preferably be 5 times lower than the appropriate chronic ecotoxicity reference value, or the acute ecotoxicity reference value in case a 7-day test is conducted.

The following analytical validation aspects are at a minimum to be reported:

(a)   detection and quantification limit of the analytical method;

(b)   analytical linearity range within the applicable analytical range;

(c)   a blank run consisting of transformation medium (this can be done during the tests);

(d)   matrix effect of the transformation medium on the measurement of the dissolved metal ion;

(e)   mass balance (%) after completion of the transformation test;

(f)   reproducibility of the analysis;

(g)   adsorptive properties of the soluble metal ions on the filters (if filtration is used for the separation of the soluble from the solid metal ion).

A10.5.2.2   *Determination of the appropriate pH of the dissolution medium*

If no relevant literature data exist, a preliminary screening test may need to be carried out in order to ensure that the test is performed at a pH maximizing transformation/dissolution within the pH range described in A10.2.3.2 and A10.5.1.6.

A10.5.2.3   *Reproducibility of transformation data*

A10.5.2.3.1   For a standard set-up of three replicate test vessels and two replicate samples per test vessel at each sampling time, it is reasonable to anticipate that for a constant loading of a substance, tested in a narrow particle and total surface area range, the within-vessel variation in transformation data should be < 10 % and the between-vessel variation should be < 20 % (reference 5, this annex). This variability may be higher at the lower loadings.

A10.5.2.3.2   To estimate the reproducibility of the transformation test, some Guidance is given in the following. The results can be used to eventually improve on reproducibility by adjusting the final test set-up through varying the number of replica test vessels and/or replica samples or further screening of the particles. The preliminary tests also allow for a first evaluation of the transformation rate of the tested substance and can be used to establish the sampling frequency.

A10.5.2.3.3   In preparing the transformation/dissolution medium, the pH of the medium should be adjusted to the desired pH (air buffering or CO$_2$ buffering) by agitation for about half an hour to bring the aqueous medium into

A10.5.1.9　変化/溶解の詳細試験においては、試験物質の表面および試験中に形成された固体反応生成物の被覆の完全性を維持しながら、試験物質上の水性媒体の流れを維持するのに十分な攪拌を行うものとする。1 l の水性媒体の場合、これはゴム栓で蓋をし、100 r.p.m.に設定したオービタル又は実験室用シェーカーに載せた 1.0～3.0 l のフラスコを使用することで達成できる。その他の穏やかな攪拌方法であっても、表面の完全性と溶液の均一性の基準を満たしていれば、使用することができる。

A10.5.1.10　固体・液体分離方法の選択は、溶存金属イオンのフィルターへの吸着が起こるか、およびA10.5.1.9 に述べた攪拌方法によって懸濁が生じるかに依存し、さらには粒度分布、粒子の形状および粒子密度に依存する。密度が約 6g/cm^3 を超え、50%粒子径の範囲が <8μm と小さい固形物については、A10.5.1.9 に述べた低速攪拌方法によって懸濁が生じる可能性は低いことが経験的に示されている。粒子がより微細な場合は、別の方法を考慮してもよい。粒子が懸濁液中に留まる懸念がある場合は、試験の前にろ過効率を確認する必要がある。ろ過効率を上げるために考えられる選択肢には、遠心分離の後にろ過を行うか、溶液サンプルを採取する前に約 5 分間静置して懸濁が沈殿するのを待つ方法などがある。

A10.5.2　*必須条件*

A10.5.2.1　*分析方法*

　全溶存金属分析には、妥当性が確認された適切な分析法が不可欠である。分析上の検出下限は、適切な慢性生態毒性基準値、または 7 日間試験を実施する場合は急性生態毒性基準値の 5 分の 1 より小さいことが望ましい。

　分析の検証に関する次の事項は、報告すべき最低事項である：

(a)　分析方法の検出限界および定量限界；

(b)　適用される分析範囲内における分析上の直線性が保たれる範囲；

(c)　変化媒体からなるブランクラン（試験中に行うことができる）；

(d)　変化媒体が溶存金属イオンの測定に与えるマトリックス効果；

(e)　変化試験終了後のマスバランス（%）；

(f)　分析の再現性；

(g)　溶存金属イオンのフィルターへの吸着性（固体金属と溶出イオンとを分離するためにろ過法が用いられる場合）。

A10.5.2.2　*適切な溶媒 pH 値の決定*

　文献上の関連データが存在しない場合は、A10.2.3.2 および A10.5.1.6 に示した pH 域内で、変化/溶解が最大になる pH 値で試験を行うために、予備スクリーニング試験を実施する必要があろう。

A10.5.2.3　*変化データの再現性*

A10.5.2.3.1　各サンプル採取時に、3 つの試験容器からそれぞれ 2 回の繰り返しサンプリングを行う標準の試験設定において、分布の狭い粒子径および全表面積範囲で試験された物質を定常的に添加する場合には、容器内の変換データの変動は 10%未満、容器間の変動は 20%未満とするべきである（本附属文書、参考文献 5）。この変動は、低負荷ではより大きくなる可能性がある

A10.5.2.3.2　変化試験の再現性を評価するために、以下の手引きが与えられている。試験結果を利用して、繰り返し試験容器または繰り返しサンプルの数を変えたり、微粒子をさらに選別することによって、最終の試験条件を調整して再現性を高めることができる。予備試験もまた被験物質の変化速度の一次評価に利用できるほか、サンプル採取頻度を定めるのにも利用できる。

A10.5.2.3.3　変化/溶解媒体調製の際は、約 30 分間の攪拌をして水性媒体を緩衝雰囲気と平衡にさせることにより、媒体の pH 値を目的の pH 値（空気緩衝または CO$_2$ 緩衝）に調整すべきである。物質を添加す

equilibrium with the buffering atmosphere. At least three samples (e.g. 10 - 15 ml) are drawn from the test medium prior to addition of the substance, and the dissolved metal concentrations are measured as controls and background.

At least five test vessels, containing the metal or metal compound (e.g.100 mg solid/l medium), are agitated as described in A10.5.1.9 at a temperature ± 1.5°C in the range 20 - 23 °C, and triplicate samples are taken by syringe from each test vessel after 24 hours. The solid and solution are separated by membrane filter as described in A10.5.1.10, the solution is acidified with one or two drops of trace metal grade HNO_3 with the target pH 1 and analysed for total dissolved metal concentration.

A10.5.2.3.4 The within-test vessel and between-test vessel means and coefficients of variation of the measured dissolved metal concentrations are calculated.

A10.5.2.3.5 To ensure reproducibility of transformation data, it is recommended that:

(a) new laboratories use a training set;

(b) one metal powder with specified surface conditions be used as standard control; and

(c) one or two laboratories be responsible for reference chemicals.

It is a requirement to check the specific surface area of powder samples.

A10.5.3 *Test performance*

A10.5.3.1 *Screening transformation/dissolution – sparingly soluble metal compounds*

A10.5.3.1.1 After dissolution medium is prepared, add the medium into at least three test vessels (number of test vessels depend on the reproducibility obtained during the preliminary test). After a half-hour of agitation to bring the aqueous medium into equilibrium with the atmosphere or buffering system (paras. A10.5.1.6 to A10.5.1.8), the pH, temperature and dissolved O_2 concentrations of the medium are measured. Then at least two 10 - 15 ml samples are taken from the test medium (prior to addition of the test material) and the dissolved metal concentration measured as controls and background.

A10.5.3.1.2 The metal compound is added to the test vessels at a loading of 100 mg/l and the test vessels are covered and agitated rapidly and vigorously (e.g. on an orbital shaker at 200 rpm, if feasible). After the 24 hours agitation, the pH, temperature and dissolved O_2 concentrations are measured in each test vessel, and two to three solution samples are drawn by syringe from each test vessel and the solution is passed through a membrane filter as described in A10.5.1.10 above, acidified (e.g. 1 % HNO_3) and analysed for total dissolved metal concentration.

A10.5.3.2 *Full transformation/dissolution test - metals and metal compounds*

A10.5.3.2.1 Repeat A10.5.3.1.1.

A10.5.3.2.2 For 7-day test, substance loadings of 1, 10 and 100 mg/l, respectively, are added to the test vessels (number of which depends on the reproducibility as established in subsection A10.5.2.3), containing the aqueous medium. The test vessels are closed (but allowing for equilibration with air if required) and agitated as described in A10.5.1.9. If a 28-day test is to be conducted, , then the loading may be 0.01 mg/l, 0.1 mg/l or 1 mg/l depending on the transformation rate. The test with 1 mg/l loading may be extended to 28 days, provided that the same pH value is to be chosen for both 7 day and 28-day tests. The 7-day tests are only conducted at pH ranges of 6 up to 8.5, while a somewhat broader pH range of 5.5 and 6 to 8.5 is recommended if technically feasible for the 28-day tests.A concurrent control test with no substance loaded (i.e. a blank test solution) is required. At established time intervals (e.g. 2 hours, 6 hours, 1, 4 and 7 days for the short-term test and additionally at e.g. 14, 21 and 28 days for the long-term test), the temperature, pH and dissolved O_2 concentrations are measured in each test vessel, and at least two samples (e.g. 10 - 15 ml) are drawn by syringe from each test vessel. The solid and dissolved fractions are separated as per A10.5.1.10 above. The solutions are acidified (e.g. 1 % HNO_3) and analysed for dissolved metal concentration. After the first 24 hours, the solution volumes should be replenished with a volume of fresh dissolution medium equal to that already drawn. Repeat after subsequent samplings. The maximum total volume taken from the test solutions should not exceed 20 % of the initial test solution volume. The test can be stopped when three subsequent total dissolved metal concentration data points vary no more than 15 %. The maximum duration for the loadings of 10 and 100 mg/l is seven days (the short-term test) and 28 days for the loading of 1 mg/l test medium (the long-term test).

る前に少なくとも 3 サンプル（例えば 10～15ml）を試験媒体から採取し、コントロールおよびバックグラウンドとしての溶存金属濃度を測定する。

金属または金属化合物（例えば媒体 1 リットル中に 100mg の固形物質）を含む、少なくとも 5 個の試験容器を A10.5.1.9 に述べたように 20～23℃の温度域で±1.5℃に管理しながら攪拌し、24 時間後に各試験容器からシリンジで 3 回ずつサンプルを採取する。固形物質と溶液は A10.5.1.10 に述べたようにメンブレンフィルターで分離し、溶液は微量金属グレードの HNO_3 1～2 滴を滴下して目標 pH1 に酸性化された後に全溶存金属濃度が分析される。

A10.5.2.3.4　同一の試験容器内および異なる試験容器間で溶存金属濃度の測定値の平均値および変動係数を計算する。

A10.5.2.3.5　変化データの再現性を確保するため、以下の点が推奨される：

 (a)　新しい実験室ではトレーニングセットを用いること；

 (b)　特定の表面条件を有する 1 つの金属粉を標準管理に使用すること；

 (c)　関連する化学品に対して責任を担う実験室が 1 つか 2 つあること。

粉末サンプルの比表面積を確認することが要求される。

A10.5.3　*試験の実施*

A10.5.3.1　*スクリーニング変化/溶解試験－難溶性金属化合物*

A10.5.3.1.1　溶解媒体を調製し、この媒体を少なくとも 3 個の試験容器に入れる（試験容器の数は予備試験で得られた再現性に依存する）。水性媒体と空気、あるいは緩衝システム（A10.5.1.6～A10.5.1.8 参照）と平衡にさせるため 30 分間の攪拌の後に、この媒体の pH 値、温度、溶存酸素濃度を測定する。次に少なくとも 2 回、10～15ml のサンプルを（試験物質の添加前の）試験媒体から採取し、コントロールおよびバックグラウンドとしての溶存金属濃度を測定する。

A10.5.3.1.2　試験容器に金属化合物を 100mg/l の用量で添加し、試験容器に蓋をして、急速かつ激しく（例えば、可能であれば 200 rpm のオービタルシェーカーで）攪拌する。24 時間の攪拌後、各試験容器において pH 値、温度、溶存酸素濃度を測定し、各試験容器から 2 ないし 3 回の溶液サンプルをシリンジで採取し、上記 A10.5.1.10 に述べたように、この溶液をメンブレンフィルターに通し、酸性に調整し（例えば 1％の HNO_3）、全溶存金属濃度を分析する。

A10.5.3.2　*変化/溶解の詳細試験－金属および金属化合物*

A10.5.3.2.1　A10.5.3.1.1 を反復する。

A10.5.3.2.2　7 日間の試験では、1、10、100mg/l の物質負荷量を、それぞれ、水性媒体の入ったいくつかの試験容器に添加する（試験容器の数は A10.5.2.3 に述べたようにその再現性に依存する）。試験容器は密閉し（ただし必要であれば空気と平衡化できるようにする）蓋をして、A10.5.1.9 に述べたように攪拌する。28 日間の試験を行う場合は、負荷量は変化速度に応じて 0.01 mg/l、0.1 mg/l または 1 mg/l とする。7 日間と 28 日間の両方の試験で同一の pH 値を選択することを条件に、1 mg/l の負荷による試験を 28 日間まで延長することができる。7 日間試験は pH6 から 8.5 までの範囲で実施されるが、28 日間試験では技術的に可能であれば、pH5.5 と 6 から 8.5 のやや広い pH 範囲が推奨される。物質を添加しない対照試験（すなわち、ブランク試験液）が同時に必要である。定められた時間間隔で（短期試験では 2 時間、6 時間、1 日間、4 日間、7 日間、長期試験では 14 日間、21 日間、28 日間など）で、各試験容器の温度、pH 値、溶存酸素濃度を測定し、各試験容器から少なくとも 2 回、溶液サンプル（例えば 10－15ml）をシリンジで採取する。固形物と溶存成分は上記の A10.5.1.10 に述べた方法で分離する。この溶液は酸性に調整し（例えば 1％の HNO_3）、溶存金属濃度を分析する。当初の 24 時間が経過して後、採取した媒体と同量の新たな溶解媒体を溶液に補充すべきである。以降のサンプル採取ではこの操作を繰り返す。試験溶液から採取する最大量は当初の試験溶液量の 20％を超えるべきではない。3 回続けての全溶存金属濃度データポイントが 15％以下しか変化しなかった場合には、試験を中断することができる。10 および 100mg/l の添加における最長試験期間は 7 日間で（短期試験）、1mg/l の添加における最長試験期間は 28 日間である（長期試験）。

A10.5.4   *Test conditions*

A10.5.4.1   The transformation/dissolution tests should be done at a controlled ambient temperature ± 1.5°C in the range 20 – 23 °C.

A10.5.4.2   The transformation/dissolution tests are to be carried out within the pH range described in A10.2.3.2 and A10.5.1.6. The test solution pH should be recorded at each solution sampling interval. The pH can be expected to remain constant (± 0.2 units) during most tests, although some short-term pH variations have been encountered at 100 mg/l loadings of reactive fine powders (reference 7, this annex), due to the inherent properties of the substance in the finely divided state.

A10.5.4.3   Above the aqueous medium, the head space provided by the reaction vessel should be adequate in most instances to maintain the dissolved oxygen concentration above about 6.0 mg/l, which is 70 % of the saturation level of 8.5 mg/l. However, in certain instances, reaction kinetics may be limited not by the availability of molecular oxygen in the head space above the solution but by the transfer of dissolved oxygen to, and removal of reaction product away from, the solid-solution interface. In this case, little can be done, other than await the restoration of equilibrium.

A10.5.4.4   To reduce chemical and biological contamination as well as evaporation, the transformation/dissolution kinetics must be performed in closed vessels and in the dark, whenever possible.

A10.6   **Treatment of the results**

A10.6.1   *Screening test*

The mean dissolved metal concentrations at 24 hours are calculated (with confidence intervals).

A10.6.2   *Full test: Determination of the extent of transformation/dissolution*

A10.6.2.1   *Short-term test*

The dissolved metal concentrations, measured during the different short-term (7 days) tests, are plotted versus time, and the transformation/dissolution kinetics may be determined, if possible. The following kinetic models could be used to describe the transformation/dissolution curves:

(a)   Linear model:

$$C_t = C_0 + kt, \text{ mg/l}$$

where:

C_0 = initial total dissolved metal concentration (mg/l) at time t = 0;
C_t = total dissolved metal concentration (mg/l) at time t;
k = linear rate constant, mg/l-days.

(b)   First order model:

$$C_t = A(1-e^{-kt}), \text{ mg/l}$$

where:

A = limiting dissolved metal concentration (mg/l) at apparent equilibrium = constant;
Ct = total dissolved metal concentration (mg/l) at time t;
k = first order rate constant, 1/days.

(c)   Second order model:

$$C_t = A(1-e^{-at}) + B(1-e^{-bt}), \text{ mg/l}$$

where:

Ct = total dissolved metal concentration (mg/l), at time t;
a = first order rate constant, 1/days;
b = second order rate constant, 1/days;
C = A + B = limiting dissolved metal concentration (mg/l).

A10.5.4　試験条件

A10.5.4.1　変化/溶解試験は、20－23℃の範囲内で、±1.5℃以内に管理された室温下で実施するべきである。

A10.5.4.2　変化/溶解試験は、A10.2.3.2 および A10.5.1.6 に述べた pH 域内で実施される。試験溶液の pH 値は、各溶液のサンプル採取間隔ごとに記録すべきである。pH 値は、大半の試験では一定（±0.2 単位）に保たれると予想されるが、100mg/l の添加量で行う反応性微粉末の試験では、細かく分散した状態での、物質固有の性質によって、いくらかの短期的な pH 値の変動が見られた（本附属書　参考文献 7）。

A10.5.4.3　反応容器内における水性媒体の上部空間は、大半の事例において溶存酸素濃度を大気飽和状態（8.5mg/l）の 70％である約 6.0mg/l 以上に維持するのに適切な大きさであるべきである。しかし、一部の事例においては、水溶液上部の空間における酸素分子の利用性によってではなく、固体と溶液との界面への溶存酸素の移動、および同界面からの反応生成物の除去によって、反応速度が律速になることもある。この場合、平衡状態の回復を待つ以外できることはほとんどない。

A10.5.4.4　化学的および生物学的な汚染、ならびに蒸発を抑えるために、変化/溶解反応はできる限り、密閉された容器で、かつ暗所で実施しなければならない。

A10.6　試験結果の取り扱い

A10.6.1　スクリーニング試験

24 時間の溶存金属平均濃度を計算する（信頼区間を含む）。

A10.6.2　詳細試験：変化/溶解の程度を測定する

A10.6.2.1　短期試験

様々な短期テスト（7 日間）において測定される溶存金属濃度は、時間に対してプロットされ、可能であれば変化/溶解速度が決定される。次に示す速度論モデルは変化/溶解曲線を解釈する際に使用できる：

(a)　直線モデル：

$$C_t = C_0 + kt、mg/l$$

ここで、
C_0 ＝ 時間 t＝0 における全溶存金属濃度初期値（mg/l）；
C_t ＝ 時間 t における全溶存金属濃度（mg/l）；
k ＝ 一次速度定数、mg/l・日。

(b)　一次モデル：

$$C_t = A(1-e^{(-kt)})、mg/l$$

ここで、
A ＝ 見かけの平衡時における溶存金属濃度限界（mg/l）＝定数；
C_t ＝ 時間 t における全溶存金属濃度（mg/l）；
k ＝ 一次の速度定数、1/日。

(c)　二次モデル：

$$C_t = A(1-e^{(-at)}) + B(1-e^{(-bt)})、mg/l$$

ここで、
C_t ＝ 時間 t における全溶存金属濃度（mg/l）；
a ＝ 一次の速度定数、1/日；
b ＝ 二次の速度定数、1/日；
C ＝ A＋B ＝ 溶存金属濃度限界（mg/l）。

(d) Reaction kinetic equation:

$$C_t = a[1 - e^{-bt} - (c/n)\{1 + (b\,e^{-nt} - n\,e^{-bt})/(n-b)\}], \text{mg/l}$$

where:

C_t = total dissolved metal concentration (mg/l) at time t;
a = regression coefficient (mg/l);
b,c,d = regression coefficients (1/days);
n = c+d.

Other reaction kinetic equations may also apply (reference 7 and 8, this annex).

For each replicate vessel in the transformation test, these model parameters are to be estimated by regression analyses. The approach avoids possible problems of correlation between successive measurements of the same replicate. The mean values of the coefficients can be compared using standard analysis of variance if at least three replicate test vessel were used. The coefficient of determination, r^2, is estimated as a measure of the "goodness of fit" of the model.

The release rate may also be expressed relative to the surface area of the test substance (e.g. $\mu g/mm^2$) to allow for a comparison of the release rates between different surface loadings or particle sizes.

A10.6.2.2 *Long-term test*

The dissolved metal concentrations, measured from the 1 mg/l loading during the 28-day test, are plotted versus time and the transformation/dissolution kinetics determined, if possible, as described in A10.6.2.1.

A10.7 **Test report**

The test report should include (but is not limited to) the following information (see also A10.4 and A10.5.2.1):

(a) Identification of the sponsor and testing facility;

(b) Description of the tested substance;

(c) Description of the reconstituted test medium and metal loadings;

(d) Test medium buffering system used and validation of the pH used (as per paras. A10.2.3.2 and A10.5.1.6 to A10.5.1.8) description of the analytical method;

(e) Detailed descriptions of the test apparatus and procedure;

(f) Preparation of the standard metal solution;

(g) Results of the method validation;

(h) Results from the analyses of metal concentrations, pH, temperature, oxygen;

(i) Dates of tests and analyses at the various time intervals;

(j) Mean dissolved metal concentration at different time intervals (with confidence intervals);

(k) Transformation curves (total dissolved metal as a function of time);

(l) Results from transformation/dissolution kinetics, if determined;

(m) Estimated reaction kinetic equation, if determined;

(n) Deviations from the study plan if any and reasons;

(o) Any circumstances that may have affected the results; and

(p) Reference to the records and raw data.

(d) 反応速度式：

$$C_t = a[1-e^{-bt}-(c/n)\{1+(be^{-nt}-ne^{-bt})/(n-b)\}], \text{ mg/l}$$

ここで、

C_t ＝ 時間 t における全溶存金属濃度（mg/l）；
a ＝ 回帰係数（mg/l）；
b、c、d ＝ 回帰係数（1/日）；
n ＝ c+d 。

この他の反応速度式もまた適用できる（本附属書 参考文献 7、8）。

変化試験における各繰り返しサンプル容器について、これらのモデルパラメータを回帰分析によって推計することができる。この手法は、同一繰り返しの連続測定間で自己相関による問題が起こるのを回避するためのものである。これらの係数の平均値は、少なくとも 3 つの繰り返し試験容器を用いている場合は、標準偏差の分析を用いて比較できる。決定係数、r^2 は、モデルの「適合度」の尺度として評価される。

放出速度は、異なる表面負荷量または粒子径間での放出速度の比較を可能にするために、試験物質の表面積（例：$\mu g/mm^2$）と比較して表すこともできる。

A10.6.2.2 *長期試験*

28 日間の試験中に 1mg/l の添加量で測定された溶存金属濃度は時間に対してプロットされ、可能であれば A10.6.2.1 に述べたように、変化/溶解速度が決定される。

A10.7 試験報告

試験報告には、以下の情報を含むべきである（ただしこれらに限定されるわけではない）。（A10.4 および A10.5.2.1 を参照）：

(a) スポンサーおよび試験機関の明示；

(b) 被験物質の説明；

(c) 試験媒体の組成と金属添加量の説明；

(d) 用いた試験媒体の緩衝方法および pH 値の確認（A10.2.3.2 および A10.5.1.6～A10.5.1.8）、分析方法の説明；

(e) 試験器具および手順に関する詳細な説明；

(f) 標準の金属溶液の調製；

(g) 分析手法の検証結果；

(h) 金属濃度、pH 値、温度、酸素濃度の分析結果；

(i) 様々な時間間隔で行った試験および分析の日時；

(j) さまざまな時間間隔における溶存金属濃度の平均値（信頼区間を含む）；

(k) 変化曲線（時間の関数としての全溶存金属濃度）；

(l) 変化/溶解速度論による結果（解析された場合）；

(m) 推定された反応速度式（解析された場合）；

(n) 実験計画からの逸脱があった場合、その記録、および起こった理由；

(o) 結果に影響を与えた可能性のある状況；および

(p) 記録および生データの参照。

Annex 10

APPENDIX

References

1. "Draft Report of the OECD Workshop on Aquatic Toxicity Testing of Sparingly Soluble Metals, Inorganic Metal Compounds and Minerals", Sept. 5-8, 1995, Ottawa

2. OECD Metals Working Group Meeting, Paris, June 18-19, 1996

3. European Chemicals Bureau. Meeting on Testing Methods for Metals and Metal Compounds, Ispra, February 17-18, 1997

4. OECD Metals Working Group Meeting, Paris, October 14-15, 1997

5. LISEC [1] Staff, "Final report "transformation/dissolution of metals and sparingly soluble metal compounds in aqueous media - zinc", LISEC no. BO-015 (1997)

6. J.M. Skeaff [2] and D. Paktunc, "Development of a Protocol for Measuring the Rate and Extent of Transformations of Metals and Sparingly Soluble Metal Compounds in Aqueous Media. Phase I, Task 1: Study of Agitation Method." Final Report, January 1997. Mining and Mineral Sciences Laboratories Division Report 97-004(CR)/Contract No. 51545

7. Jim Skeaff and Pierrette King, "Development of a Protocol For Measuring the Rate and Extent of Transformations of Metals and Sparingly Soluble Metal Compounds in Aqueous Media. Phase I, Tasks 3 and 4: Study of pH and of Particle Size/Surface Area.", Final Report, December 1997. Mining and Mineral Sciences Laboratories Division Report 97-071(CR)/Contract No. 51590

8. Jim Skeaff and Pierrette King, Development of Data on the Reaction Kinetics of Nickel Metal and Nickel Oxide in Aqueous Media for Hazard Identification, Final Report, January 1998. Mining and Mineral Sciences Laboratories Division Report 97-089(CR)/Contract No. 51605

9. LISEC Staff, "Final report "transformation/dissolution of metals and sparingly soluble metal compounds in aqueous media - zinc oxide", LISEC no. BO-016 (January, 1997)

10. LISEC Staff, "Final report "transformation/dissolution of metals and sparingly soluble metal compounds in aqueous media - cadmium", LISEC no. WE-14-002 (January, 1998)

11. LISEC Staff, "Final report "transformation/dissolution of metals and sparingly soluble metal compounds in aqueous media - cadmium oxide", LISEC no. WE-14-002 (January, 1998)

12. Skeaff, J.M., Hardy, D.J. and King, P. (2008), A new approach to the hazard classification of alloys based on transformation/dissolution. Integr Environ Assess Manag, 4: 75-93. https://doi.org/10.1897/IEAM_2007-050.1

13. Skeaff, J., Adams, W.J., Rodriguez, P., Brouwers, T. and Waeterschoot, H. (2011), Advances in metals classification under the United Nations globally harmonized system of classification and labeling. Integr Environ Assess Manag, 7: 559-576. https://doi.org/10.1002/ieam.194

14. Skeaff, J.M. and Beaudoin, R. (2015), Transformation/dissolution characteristics of a nickel matte and nickel concentrates for acute and chronic hazard classification. Integr Environ Assess Manag, 11: 130-142. https://doi.org/10.1002/ieam.1573

15. Huntsman-Mapila, P., Skeaff, J.M., Pawlak, M. and Beaudoin, R. (2016), Addressing aquatic hazard classification for metals, metal compounds and alloys in marine systems, Marine Pollution Bulletin 109:550-557. https://doi.org/10.1016/j.marpolbul.2016.03.055

16. OECD Environment Health and Safety Publications; Series on Testing and Assessment n° 29. Guidance document on Transformation Dissolution of Metals and Metal Compounds in Aqueous media, July 2001.

---

[1] *LISEC, Craenevenne 140, 3600 Genk, Belgium.*

[2] *CANMET, Natural Resources Canada, 555 Booth St., Ottawa, Canada K1A 0G1.*

附属書 10
付録

参考文献

1. "Draft Report of the OECD Workshop on Aquatic Toxicity Testing of Sparingly Soluble Metals, Inorganic Metal Compounds and Minerals", Sept. 5-8, 1995, Ottawa
2. OECD Metals Working Group Meeting, Paris, June 18-19, 1996
3. European Chemicals Bureau. Meeting on Testing Methods for Metals and Metal Compounds, Ispra, February 17-18, 1997
4. OECD Metals Working Group Meeting, Paris, October 14-15, 1997
5. LISEC[1] Staff, "Final report "transformation/dissolution of metals and sparingly soluble metal compounds in aqueous media - zinc", LISEC no. BO-015 (1997)
6. J.M. Skeaff[2] and D. Paktunc, "Development of a Protocol for Measuring the Rate and Extent of Transformations of Metals and Sparingly Soluble Metal Compounds in Aqueous Media. Phase I, Task 1: Study of Agitation Method." Final Report, January 1997. Mining and Mineral Sciences Laboratories Division Report 97-004(CR)/Contract No. 51545
7. Jim Skeaff and Pierrette King, "Development of a Protocol For Measuring the Rate and Extent of Transformations of Metals and Sparingly Soluble Metal Compounds in Aqueous Media. Phase I, Tasks 3 and 4: Study of pH and of Particle Size/Surface Area.", Final Report, December 1997. Mining and Mineral Sciences Laboratories Division Report 97-071(CR)/Contract No. 51590
8. Jim Skeaff and Pierrette King, Development of Data on the Reaction Kinetics of Nickel Metal and Nickel Oxide in Aqueous Media for Hazard Identification, Final Report, January 1998. Mining and Mineral Sciences Laboratories Division Report 97-089(CR)/Contract No. 51605
9. LISEC Staff, "Final report "transformation/dissolution of metals and sparingly soluble metal compounds in aqueous media - zinc oxide", LISEC no. BO-016 (January, 1997)
10. LISEC Staff, "Final report "transformation/dissolution of metals and sparingly soluble metal compounds in aqueous media - cadmium", LISEC no. WE-14-002 (January, 1998)
11. LISEC Staff, "Final report "transformation/dissolution of metals and sparingly soluble metal compounds in aqueous media - cadmium oxide", LISEC no. WE-14-002 (January, 1998)
12. Skeaff, J.M., Hardy, D.J. and King, P. (2008), A new approach to the hazard classification of alloys based on transformation/dissolution. Integr Environ Assess Manag, 4: 75-93. https://doi.org/10.1897/IEAM_2007-050.1
13. Skeaff, J., Adams, W.J., Rodriguez, P., Brouwers, T. and Waeterschoot, H. (2011), Advances in metals classification under the United Nations globally harmonized system of classification and labeling. Integr Environ Assess Manag, 7: 559-576. https://doi.org/10.1002/ieam.194
14. Skeaff, J.M. and Beaudoin, R. (2015), Transformation/dissolution characteristics of a nickel matte and nickel concentrates for acute and chronic hazard classification. Integr Environ Assess Manag, 11: 130-142. https://doi.org/10.1002/ieam.1573
15. Huntsman-Mapila, P., Skeaff, J.M., Pawlak, M. and Beaudoin, R. (2016), Addressing aquatic hazard classification for metals, metal compounds and alloys in marine systems, Marine Pollution Bulletin 109:550-557. https://doi.org/10.1016/j.marpolbul.2016.03.055
16. OECD Environment Health and Safety Publications; Series on Testing and Assessment n° 29. Guidance document on Transformation Dissolution of Metals and Metal Compounds in Aqueous media, July 2001.

---

[1] *LISEC, Craenevenne 140, 3600 Genk, Belgium.*

[2] *CANMET, Natural Resources Canada, 555 Booth St., Ottawa, Canada K1A 0G1*

Bibliography

1. OECD Guideline for testing of chemicals, Paris (1984). Guideline 201 (Updated in 2011) Alga, Growth Inhibition test
2. OECD Guideline for testing of chemicals, Paris (1984). Guideline 202 (Updated in 2004): Daphnia sp. Acute immobilisation test and Reproduction Test
3. OECD Guideline for testing of chemicals, Paris (1992). Guideline 203 (Updated in 2019): Fish, Acute Toxicity Test
4. OECD Guideline for testing of chemicals, Paris (1992). Guideline 204: Fish, Prolonged Toxicity Test: 14- Day study[3]
5. OECD Guideline for testing of chemicals, Paris (1992). Guideline 210 (Updated in 2013): Fish, Early-Life Stage Toxicity Test
6. International standard ISO 6341 (1989 (E)). Determination of the inhibition of the mobility of Daphnia magna Straus (Cladocera, Crustacea)

---

[3] *This test guideline has been cancelled.*

関連文献

1. OECD Guideline for testing of chemicals, Paris (1984). Guideline 201 (Updated in 2011) Alga, Growth Inhibition test

2. OECD Guideline for testing of chemicals, Paris (1984). Guideline 202 (Updated in 2004) Daphnia sp. Acute immobilisation test and Reproduction Test

3. OECD Guideline for testing of chemicals, Paris (1992). Guideline 203 (Updated in 2019) Fish, Acute Toxicity Test

4. OECD Guideline for testing of chemicals, Paris (1992). Guideline 204 : Fish, Prolonged Toxicity Test : 14- Day study[3]

5. OECD Guideline for testing of chemicals, Paris (1992). Guideline 210 (Updated in 2013) Fish, Early-Life Stage Toxicity Test

6. International standard ISO 6341 (1989 (E)). Determination of the inhibition of the mobility of Daphnia magna Straus (Cladocera, Crustacea)

---

[3] 本テストガイドラインは無効とされた。

ANNEX 11

GUIDANCE ON OTHER HAZARDS NOT RESULTING IN CLASSIFICATION

附属書 11

分類に結び付かない他の危険有害性に関する手引き

ANNEX 11

GUIDANCE ON OTHER HAZARDS NOT RESULTING IN CLASSIFICATION

A11.1 Introduction

This guidance aims to provide information that facilitates the identification of hazards which do not result in classification, but which may need to be assessed and communicated.

A11.2 Dust explosions

This section provides guidance on the factors that contribute to a dust explosion hazard and on hazard identification and the need for risk assessment, prevention, mitigation, and communication.

A11.2.1 *Scope and applicability*

A11.2.1.1   Any solid substance or mixture, which is combustible, may pose a dust explosion risk when in the form of fine particles in an oxidizing atmosphere such as air. A risk assessment may be needed for many substances, mixtures, or solid materials, not just those classified as flammable solids in accordance with chapter 2.7. In addition, dusts may be formed (intentionally or unintentionally) during transfer or movement, or in a facility during handling or mechanical processing (e.g. milling, grinding) of substances/mixtures/solid materials (e.g. agricultural commodities, wood products, pharmaceuticals, dyes, coal, metals, plastics). Thus, the possibility of the formation of small particles and their potential accumulation should also be assessed. Where a dust explosion risk is identified, effective preventive and protective measures should be implemented as required by national legislation, regulations, or standards.

A11.2.1.2   This guidance identifies when combustible dusts may be present and thus, when the risk of a dust explosion should be considered. The guidance:

    (a)   Gives a flow chart specifying the key steps to identify a possible combustible dust;

    (b)   Identifies the factors contributing to a dust explosion;

    (c)   Sets out principles of hazard and risk management; and

    (d)   Indicates where expert knowledge is required.

A11.2.2 *Definitions*

In this annex, the following terms, specific to dust explosion hazards and risks, are used:

Combustible dust: Finely divided solid particles of a substance or mixture that are liable to catch fire or explode on ignition when dispersed in air or other oxidizing media;

Combustion: Energy releasing (exothermic) oxidation reaction of (or with) combustible substances/mixtures/solid materials;

Dispersion: Distribution of fine dust particles in the form of a cloud;

Dust deflagration index (K_{st}): A safety characteristic related to the severity of a dust explosion. The larger the value for K_{st}, the more severe the explosion. K_{st} is dust specific and volume independent, and is calculated using the cubic law equation:

$$\left(d_p/d_t\right)_{max} \cdot V^{1/3} = const. = K_{st}$$

    where:
    $(d_p/d_t)_{max}$ = maximum rate of pressure rise
    V = volume of testing chamber

附属書11

分類に結び付かない他の危険有害性に関する手引き

A11.1 序

この手引きは、分類には結び付かないが評価され情報提供される必要がありうる、危険有害性の特定を支援するための情報提供を目的としている。

A11.2 粉じん爆発

この節では、粉じん爆発に寄与する要因に関して、さらに危険性の確認およびリスクアセスメント、防止、軽減および情報伝達に関する手引きを提供している。

A11.2.1 範囲及び適用

A11.2.1.1 可燃性であるいかなる固体物質または混合物も、空気のような酸化性雰囲気において微小粒子になった場合には、粉じん爆発のリスクをもたらす。第2.7章にしたがって可燃性固体として分類されたものだけではなく、多くの物質、混合物または固体材料に関してはリスクアセスメントが必要であろう。さらに粉じんは、物質/混合物/固体材料（例えば農業製品、木材製品、薬品、染料、石炭、金属、プラスチック）の輸送または移動、または設備内での取り扱いまたは機械的な処理（例えば製粉、粉砕）で（意図されてまたは意図されずに）形成されるであろう。それゆえ小さな粒子の形成に関する可能性およびそれらの潜在的な蓄積も評価されるべきである。粉じん爆発のリスクが明らかにされた場合には、効果的な予防的および防護的対策が、国の法律、規則または基準として施行されなければならない。

A11.2.1.2 この手引きは、いつ可燃性の粉じんが存在し、そしていつ粉じん爆発のリスクが検討されなければならないかを明らかにする。本手引きでは：

  (a) 可能性のある可燃性粉じんを確認するためにカギとなるステップを明示するフローチャートを示している；

  (b) 粉じん爆発に寄与する要因を明らかにしている；

  (c) 危険性およびリスクのマネジメントの原則を提示している；さらに

  (d) 専門家の知識が必要とされるところを示している。

A11.2.2 定義

本附属書では、粉じん爆発の危険性およびリスクに特有な、以下の用語が使用されている：

可燃性粉じん：空気中または他の酸化性媒体において散乱した場合に、発火または着火により爆発しやすくなる物質または混合物の細かい固体粒子；

燃焼：可燃性物質/混合物/固体材料の（または、を伴った）エネルギー放出（発熱）酸化反応；

分散：雲状の細かい粉じん粒子の分布；

粉じん爆燃指数（Kst）：粉じん爆発の重大性に関連した安全特性。K_{st}の値が大きくなるほど、爆発は激しくなる。K_{st}は粉じんに特有で容量には無関係であり、立法式を用いて計算される：

$$(d_p/d_t)_{max} \cdot V^{1/3} = const. = K_{st}$$

ここで：
$(d_p/d_t)_{max}$ = maximum rate of pressure rise
V = volume of testing chamber

Dusts are classified into dust explosion classes *in accordance with their K_{st} value:*

St 1: $0 < K_{st} \leq 200$ bar m s^{-1}
St 2: $200 < K_{st} \leq 300$ bar m s^{-1}
St 3: $K_{st} > 300$ bar m s^{-1}

The K_{st} value and the maximum explosion pressure are used to design appropriate safety measures (e.g. pressure relief venting).

Explosible dust atmosphere: A dispersion of a combustible dust in air which after ignition results in a self-sustaining flame propagation;

Explosion: Abrupt oxidation or decomposition reaction producing an increase in temperature, pressure, or both simultaneously;[1]

Limiting oxygen concentration (LOC): maximum oxygen concentration in a mixture of a combustible dust and air and an inert gas, in which an explosion will not occur, determined under specific test conditions;

Maximum explosion pressure: Highest pressure registered in a closed vessel for a dust explosion at optimum concentration;

Minimum Explosible Concentration (MEC)/Lower Explosible Limit (LEL): The minimum concentration of a combustible dust dispersed in air measured in mass unit per volume that will support an explosion;

Minimum ignition energy (MIE): Lowest electrical energy stored in a capacitor, which upon discharge is sufficient to ignite the most sensitive dust/air mixture under specific test conditions;

Minimum ignition temperature (MIT) of a dust cloud: Lowest temperature of a hot surface on which the most ignitable mixture of a dust with air is ignited under specified test conditions;

Particle size: Smallest sieve aperture through which a particle will pass if presented in the most favourable orientation;[2]

A11.2.3 Identification of combustible dust

A11.2.3.1   The purpose of this section is to identify whether a combustible dust is present. If there is applicable data from a recognized and validated test method that supports a conclusion that the substance or mixture is or is not a combustible dust (see considerations in A11.2.3.2.10) then a decision can be made without the application of figure A11.2.1. Otherwise, figure A11.2.1 presents a flow chart that helps to identify whether a substance or mixture is a combustible dust and hence whether the risk of a dust explosion has to be assessed. section A11.2.3.2 contains detailed explanations and guidance on the interpretation of each box used in the flow chart.

---

[1]   *Explosions are divided into deflagration and detonation depending on whether they propagate with subsonic velocity (deflagration) or supersonic velocity (detonation). The reaction of a combustible dust which is dispersed in air and ignited normally propagates with subsonic speed, i.e. as a deflagration. Whereas explosive substances ("Explosives"; see chapter 2.1) have the intrinsic potential for highly energetic decomposition and react in the condensed phase, combustible dusts need to be dispersed in the presence of an oxidizing atmosphere (generally oxygen) to create an explosible dust atmosphere.*

[2]   *For further information on particle size see A11.2.4.1.*

粉じんは、K_{st}の値にしたがって粉じん爆発のクラスに分類される：

St 1: $0 < K_{st} \leq 200\ bar\ m\ s^{-1}$
St 2: $200 < K_{st} \leq 300\ bar\ m\ s^{-1}$
St 3: $K_{st} > 300\ bar\ m\ s^{-1}$

K_{st}値および最大爆発圧力は適当な安全対策（例えば圧抜きベント）を設計するために使用される。

爆発性粉じん雰囲気：着火後、自己持続的火炎伝搬が起こる可燃性粉じんの空気中への分散；

爆発：温度、圧力、または両方同時の上昇を生じる急激な酸化または分解反応；[1]

限界酸素濃度（LOC）：可燃性粉じんおよび空気および不活性ガスの混合物における、爆発が起こらない、最大酸素濃度、特定の試験条件下で決定される；

最大爆発圧力：最適濃度における、密閉容器での粉じん爆発に関して記録される最大圧力；

最小爆発濃度（MEC）／爆発下限（LEL）：爆発を起こすであろう単位容積当たりの質量で測定された空気中に分散している可燃性粉じんの最小濃度；

最小着火エネルギー（MIE）：特定の試験条件下で最も感度が高い粉じん／空気の混合物に着火するのに十分な最小の電気的エネルギー、電気は蓄電池に蓄えられこれが放電される；

粉じん雲の最低着火温度（MIT）：特定の試験条件下で、空気と粉じんの最も着火しやすい混合物が着火する熱表面の最低温度；

粒子サイズ：　最適の方向性がある場合、粒子が通る最も小さなふるいの開口部；[2]

A11.2.3　可燃性粉じんの同定

A11.2.3.1　この節の目的は、可燃性粉じんが存在するかどうかを同定することである。もし物質または混合物が可燃性粉じんであるか否かの（A11.2.3.2.10の検討を参照）結論を裏付ける、認められ検証されている試験方法からのデータがある場合には、図A11.2.1の適用なしに決定を下すことができる。他の場合には、図A11.2.1に物質または混合物が可燃性粉じんかどうか、そして粉じん爆発のリスクが評価されるべきかどうかの確認を支援するフローチャートが示されている。A11.2.3.2には、フローチャートで使用されているそれぞれのボックスに関する詳細な説明と手引きが含まれる。

---

[1]　爆発は、それらの伝搬が音速以下（爆燃）かまたは超音速（爆轟）かによって、爆燃と爆轟に分けられる。空気中に分散し着火した可燃性粉じんの反応は通常音速以下で、すなわち爆燃として、伝搬する。爆発性物質（「爆発物」、第2章参照）が高エネルギー分解の潜在的な能力を有して濃縮された状態で反応するのに対して、可燃性粉じんは、爆発性粉じんの雰囲気を作り出す酸化性雰囲気（一般には酸素）の中で分散している必要がある。

[2]　粒子サイズに関するさらなる情報はA11.2.4.1参照。

Figure A11.2.1: Flow chart for decision on combustible dusts

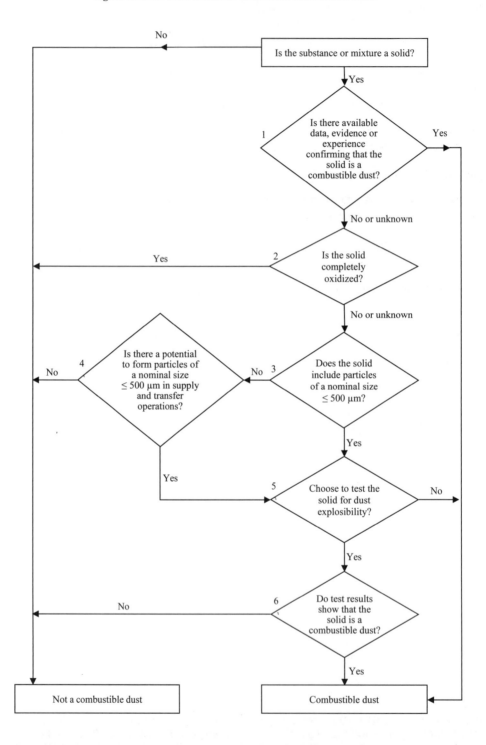

図 A11.2.1：可燃性粉じんの決定に関するフローチャート

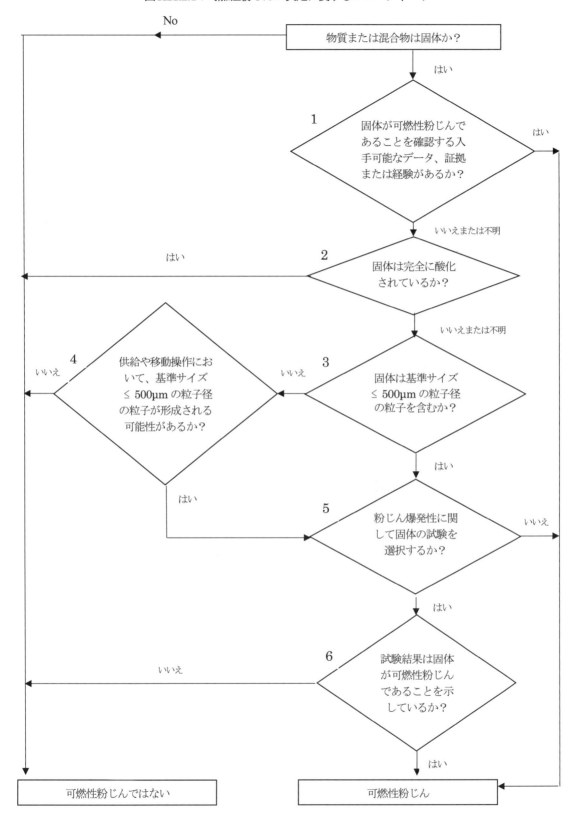

A11.2.3.2   *Explanations to figure A11.2.1*

A11.2.3.2.1   Care has to be taken when using available data, because the behaviour of combustible dusts is very sensitive to conditions such as particle size, moisture content etc. If the conditions under which available data were generated are not known, or are not applicable to the substance, mixture, or solid material under investigation, the data might not be relevant and a conservative approach is recommended when going through the flow chart.

Box 1:   Is there available data, evidence or experience confirming that the solid is a combustible dust?

A11.2.3.2.2   Clear evidence for a combustible dust may be obtained from publicly available incident reports relevant to the substance, mixture, or solid material in question. Similarly, if experience has shown that the substance, mixture, or solid material is combustible in powder form, a dust explosion risk can be assumed. If a substance, mixture, or solid material is not classified as flammable, it may still have the potential to form an explosible dust-air mixture. Specifically, any organic or metallic material handled in powder form or from which a powder may be formed in processing, should be assumed to be a combustible dust unless explicit evidence to the contrary is available.

A11.2.3.2.3   The following are examples of available data indicating a combustible dust:

(a)   Classification of the substance or one of the components of the mixture as pyrophoric or flammable solid.

(b)   Availability of relevant information such as MIE, Kst values, flammability limits, ignition temperatures.

(c)   Results from screening tests (such as Burning index in accordance with VDI 2263, Hartmann tube in accordance with ISO IEC 80079-20-2).

A11.2.3.2.4   In the absence of data, it is common practice to assume the presence of a combustible dust and to apply appropriate risk management measures (see A11.2.6).

Box 2:   Is the solid completely oxidized?

A11.2.3.2.5   When a solid substance or mixture is completely oxidized, e.g. silicon dioxide, further combustion will not occur. Consequently, the solid substance or mixture will not ignite, even if it is exposed to a source of ignition. However, if a solid substance or mixture is not entirely oxidized, combustion of the solid substance or mixture is possible if it is exposed to a source of ignition.

Box 3:   Does the solid include particles of a nominal size ≤ 500 μm?

A11.2.3.2.6   When evaluating materials in relation to box 3, users should consider whether the material includes fine particles which might be released during normal or foreseeable conditions of use.

A11.2.3.2.7   When assessing the particle size with regard to the risk of dust explosions, only the fine particles with a size ≤ 500 μm are relevant[3], even if the median particle size of the whole sample is larger than 500 μm. Hence, only the dust fraction itself, and not the mixture of coarse and fine particles, has to be considered to evaluate the risk of forming explosible dust atmospheres. However, a lower concentration limit for the dust particles in a solid (e.g. by weight percent) that will not lead to such a risk cannot be defined and therefore also small fractions of fine particles are relevant. For further explanation see A11.2.4.1.

Box 4   Is there a potential to form particles of a nominal size ≤ 500 μm in supply and transfer operations?

A11.2.3.2.8   At this stage in the flow chart the solid, as presented, does not include particles smaller than 500 μm. In that form it is not a combustible dust. However, it is not completely oxidized and fine particles could form during supply and transfer operations. Therefore, such conditions should be critically reviewed in detail, especially with respect to foreseeable effects which may lead to the formation of fine particles, e.g. mechanical stress such as abrasion during transport or transfer operations, or desiccation of moisturized material. If such effects cannot be excluded, expert opinion should be sought. See section A11.2.6.2.1 for considerations related to the generation of fine particles during operations and processing.

A11.2.3.2　図A11.2.1の説明

A11.2.3.2.1　可燃性粉じんの挙動は、粒子サイズ、湿度などのような条件に非常に敏感なので、入手可能なデータを使用する際には注意をしなければならない。入手可能なデータがとられた条件が、不明または検討している物質、混合物または固体材料には適用できない場合には、そのデータは関連性がないかもしれない、またフローチャートを進む場合には慎重なアプローチを勧める。

ボックス1：　固体が可燃性粉じんであることを確認する入手可能なデータ、証拠または経験があるか？

A11.2.3.2.2　可燃性粉じんに関する明らかな証拠が、問題となっている物質、混合物または固体材料に関連する公表され入手可能な事故報告から得られるかもしれない。同様に、物質、混合物または固体材料が粉体で可燃性であることが経験からわかっていれば、粉じん爆発のリスクが考えられる。物質、混合物または固体材料が可燃性と分類されていない場合でも、爆発性粉じん－空気混合物を形成する可能性はある。特に、粉体で扱われるまたは工程中に粉体が形成されるいかなる有機物または金属材料も、明確な反対の証拠が得られない限りは、可燃性粉じんとされるべきである。

A11.2.3.2.3　以下は、可燃性粉じんを示す入手可能なデータの例である；

(a) 物質または混合物の成分の1つが自然発火性または可燃性固体として分類されている。

(b) MIE、K_{st}値、可燃限界、着火温度のような関連する情報が入手可能である。

(c) スクリーニング試験（VDI 2263にしたがったBurning index、ISO IEC 80079-20-2にしたがったHartmann tube）の結果がある。

A11.2.3.2.4　データがない場合、可燃性粉じんデータの存在を仮定して、適当なリスクマネジメント対策を適用するのが一般的な方法である（A11.2.6参照）。

ボックス2：　固体は完全に酸化されているか？

A11.2.3.2.5　固体の物質または混合物が完全に酸化されている場合、例えば二酸化ケイ素など、それ以上の燃焼は起きないであろう。結果として、着火源にさらされたとしても、固体の物質または混合物は着火しないであろう。しかし、固体の物質または混合物が完全に酸化されていない場合には、固体の物質または混合物は、着火源にさらされた場合、燃焼の可能性がある。

ボックス3：　固体は基準サイズ≤500μmの粒子を含むか？

A11.2.3.2.6　ボックス3に関係する固体を評価する際、利用者は固体が通常のまたは予期しうる使用状態において放出されるかもしれない細粒子を含むかどうか検討するべきである。

A11.2.3.2.7　粉じん爆発のリスクに関して粒子サイズを評価する場合、全サンプルの粒子サイズ中央値が500μm超であったとしても、サイズ≤500μmの細粒だけが関連する[3]。それゆえ、粗いおよび細かい粒子の混合物ではなく、粉じんの画分だけが、爆発性粉じんの雰囲気を形成するリスクを評価するために検討されなければならない。とはいえ、そのようなリスクに結びつかない固体中の粉じん粒子に関する濃度下限（例えば重量%）は定義することができないし、細粒子の小さい画分は同様に直接的に関連している。さらなる説明はA11.2.4.1を参照すること。

ボックス4：　供給や移動操作において、基準サイズ≤500μmの粒子が形成される可能性があるか？

A11.2.3.2.8　フローチャートにおけるこの段階で、記述されているように、固体は500μm未満の粒子は含んでいない。この形状は可燃性粉じんではない。とはいえ完全に酸化されておらず、供給や移動操作の間に細粒が形成されることがある。したがって、特に細粒の形成に結び付く予測可能な影響に関して、例えば移動または移動操作中の摩耗のような機械的な力または湿った材料の乾燥など、そのような条件は詳細に注意深く検討されなければならない。そのような影響が排除できない場合には、専門家の意見を求めるべきである。操作および工程中の細粒の産生に関する検討はA11.2.6.2.1を参照する。

Box 5   Choose to test the solid for dust explosibility?

A11.2.3.2.9   If testing for dust explosibility is carried out, it should be done in accordance with recognized and validated testing standards, such as those listed in A11.2.8.1. Where a solid is tested, and the solid as presented does not consist of particles ≤ 500 μm, it has to be ground for the purposes of testing for dust explosibility.

Box 6   Do test results show that the solid is a combustible dust?

A11.2.3.2.10   Properties such as particle size, chemistry, moisture content, shape, and surface modification (e.g. oxidation, coating, activation, passivation) can influence the explosion behaviour. Standard tests determine whether a dust is actually able to form explosible mixtures with air.

A11.2.4   *Factors contributing to a dust explosion*

A dust explosion may occur when there is a combustible dust, air or another oxidizing atmosphere, an ignition source, and the concentration of the combustible dust dispersed in air or another oxidizing atmosphere is above the minimum explosible concentration. The relationship between these factors is complex. The following sections give further information on the specific factors that contribute to a dust explosion hazard. In some cases, expert advice may be needed.

A11.2.4.1   *Particle characteristics (size and shape)*

A11.2.4.1.1   The 500 μm size criterion is based on the fact that particles of greater size generally have a surface-to-volume ratio that is too small to pose a deflagration hazard. However, this criterion should be used with care. Flat platelet-shaped particles, flakes, or fibres with lengths that are large compared to their diameter usually do not pass through a 500 μm sieve, yet could still pose a deflagration hazard. In addition, many particles accumulate electrostatic charge in handling, causing them to attract each other, forming agglomerates. Often agglomerates behave as if they were larger particles, yet when they are dispersed they can present a significant hazard. In such cases, a conservative approach is recommended and the material should be treated as a combustible dust.

A11.2.4.1.2   The particle size influences the explosion severity as well as the ignition sensitivity. A decrease in particle size tends to lower the MIE and the MIT of a dust cloud while the maximum explosion pressure and K_{St} value will rise.

A11.2.4.1.3   A concentration limit for the fraction of small dust particles in a combustible solid substance or mixture (e.g. by weight percent) that will not lead to a combustible dust hazard cannot be defined because:

   (a)   Small amounts of dust are sufficient to form an explosible dust-air mixture. Assuming the lower explosion limit of a combustible dust is 30 g/m³, an amount of 0.3 g dispersed in 10 *l* of air would be sufficient to form a hazardous explosible dust atmosphere. Therefore, a (combustible) dust cloud with a volume of 10 *l* has to be considered as hazardous even when unconfined.

   (b)   Dust may not be equally distributed in a substance or mixture and may accumulate and/or separate.

A11.2.4.2   *Concentration of combustible dust*

A11.2.4.2.1   A dust explosion may occur if the concentration of combustible dust dispersed in air reaches a minimum value, the MEC/LEL[4]. This value is specific for each dust.

A11.2.4.2.2   The MEC/LEL of many materials have been measured, varying from 10 to about 500 g/m³. For most combustible dusts it may be assumed that 30 g/m³ is the MEC/LEL (it has to be taken into account that 30 g dispersed in 1 m³ of air, resembles a very dense fog).

---

[3]   *Use of ≤ aligns with NFPA 652, Standard on the Fundamentals of Combustible Dust. However, this notation implies a precision which this parameter does not have in practice.*
[4]   *Although there is an Upper Explosive Limit (UEL) for dusts in air, it is difficult to measure and imprecise. Furthermore, in practice it is not generally possible to consistently maintain a dust-in-air concentration above the UEL; tests in a blender showed dust explosibility even when 75% filled. In consequence, and in contrast to gases and vapours, seeking to maintain safety by operating with dust concentrations above the UEL is not generally a viable approach.*

ボックス5　粉じんの爆発性に関して固体の試験を選択するか？

A11.2.3.2.9　粉じんの爆発性に関する試験を実行する場合、試験はA11.2.8.1に記載されているような、認知され検証されている試験基準にしたがって行われるべきである。固体が試験される場合、現状の固体が500 μm以下の粒子を含まない場合、粉じんの爆発性に関する試験を行うためには、固体はすりつぶされなければならない。

ボックス6　試験結果は固体が可燃性粉じんであることを示しているか？

A11.2.3.2.10　粒子サイズ、化学的性質、水分量、形状および表面の改質（例えば、酸化、コーティング、活性化、非活性化）のような性質が爆発挙動に影響する。標準的な試験が、粉じんが実際に空気と爆発性混合物を形成することができるかどうかを決定する。

A11.2.4　粉じん爆発に寄与する因子

粉じん爆発は、可燃性粉じん、空気、または他の酸化性雰囲気、着火源があり、さらに空気や酸化性雰囲気中に散乱している可燃性粉じんの濃度が最小爆発濃度を超えると起こるであろう。これらの因子の関係は複雑である。以下の節では、粉じん爆発危険性に寄与する特有の因子に関するさらなる情報を伝える。ケースによっては、専門家のアドバイスが必要であろう。

A11.2.4.1　粒子の特性（サイズおよび形状）

A11.2.4.1.1　500 μmサイズの基準は、一般により大きなサイズの粒子は、爆燃危険性をもたらすには小さすぎる表面/体積比であるいう事実に基づいている。しかしこの基準は注意して使用されるべきである。径に比べて大きな長さを持つ、平板状粒子、薄片、繊維は通常500 μmのふるいは通らないが、爆燃危険性の原因となりうる。さらに多くの粒子は取扱い中に静電気を蓄え、互いに引きつけあい塊を形成する。しばしば塊は大きな粒子のようにふるまい、さらに散乱した際には重大な危険性を示す。そのようなケースでは、慎重なアプローチが推薦され、材料は可燃性粉じんとして扱われるべきである。

A11.2.4.1.2　粒子サイズは、着火の感度と同様に爆発の重大性にも影響する。粒子サイズが減少すると紛じん雲のMIEおよびMITは低くなる傾向があり、一方、最大爆発圧力およびKst値は上昇する。

A11.2.4.1.3　粉じん爆発に結び付かない可燃性固体物質または混合物における小さい粉じん粒子の分画に関する濃度限界（例えば重量%）を定義することはできない、なぜなら：

(a) 少しの粉じん量で爆発性粉じん－空気混合物を形成するのに十分である。可燃性粉じんの爆発下限を30 g/m³と仮定すると、10リットルの空気中に0.3gの量の散乱で危険な爆発粉じん雰囲気を形成するのに十分である。したがって10リットル容積の（可燃性）粉じん雲は、閉じ込められていない状態であっても、危険であると考えなければならない。

(b) 粉じんは物質または混合物中に均等に分布してはおらず、溜まりおよび/または分散しているであろう。

A11.2.4.2　可燃性粉じんの濃度

A11.2.4.2.1　空気中に分散している可燃性粉じんの濃度が最小値MEC/LEV[4]に達した場合に、粉じん爆発が起こるかもしれない。この値はそれぞれの粉じんに特有である。

A11.2.4.2.2　多くの材料のMEC/LEVが測定されており、10から約500 g/m³まで変化する。ほとんどの可燃性粉じんについて、30 g/m³がMEC/LEVであると考えられる（空気1m³に30gの分散は非常に濃い霧のようであると考えてよい）。

---

[3] ≤ の使用はNEPA, Standard on the Fundamentals of Combustible Dustにしたがっている。しかしこの表記は、実際には当該パラメータが持たない精度も包含している。

[4] 空気中の粉じんに対して爆発上限（UEL）があるものの、測定するのは困難でしかも不正確である。さらに実際問題として、空気中の粉じん濃度をUELより高く維持するのは一般に可能ではない；配合機内の試験では75%でも粉じん爆発性を示した。つまりガスおよび蒸気と異なり、UELを超えた粉じん濃度において操作によって安全を維持しようとするのは一般に実行可能なことではない。

A11.2.4.3   *Air or other oxidizing atmospheres*

Generally, air is the oxidizing agent in dust explosions, however, if combustible dusts are handled in other oxidizing gases or gas mixtures dust explosions may also occur.

A11.2.4.4   *Ignition sources*

A11.2.4.4.1   Dust explosions will occur when an effective ignition source is present in an explosible dust-air mixture (explosible atmosphere). The effectiveness of a potential source of ignition reflects the ability to ignite an explosible atmosphere. It depends not only on the energy of the ignition source, but also on its interaction with the explosible atmosphere.

A11.2.4.4.2   The assessment of ignition sources is a two-step procedure: First, possible ignition sources are identified. In the second step, each possible ignition source is assessed with respect to its ability to ignite the explosive atmosphere. The ignition sources identified as effective in this procedure then require appropriate preventive measures within the explosion protection concept (see A11.2.6.1).

A11.2.4.4.3   Potential sources of ignition include:

(a)   Hot surfaces;

(b)   Flames and hot gases;

(c)   Mechanically generated sparks;

(d)   Electric apparatus;

(e)   Stray electric currents and cathodic corrosion protection;

(f)   Lightning;

(g)   Static electricity;

(h)   Radio frequency electromagnetic waves (10^4 Hz - 3×10^{12} Hz);

(i)   Electromagnetic waves (3×10^{11} Hz - 3×10^{15} Hz);

(j)   Ionizing radiation;

(k)   Ultrasonics;

(l)   Adiabatic compression and shock waves;

(m)   Exothermic reactions, including self-ignition of dusts, smouldering/glowing particles or dusts, and thermite reactions (e.g. between aluminium and rusty steel).

A11.2.5   *Other factors impacting the severity of a dust explosion*

In addition to the factors explained in A11.2.4, other conditions also influence how severe a dust explosion can be. The more significant of these are environmental factors and confinement, which are explained below. Since the list of factors presented in this section is not complete, expert advice should be sought as appropriate when assessing the risks in a given situation.

A11.2.5.1   *Influence of temperature, pressure, oxygen availability, and humidity*

A11.2.5.1.1   Safety relevant data are frequently given under the tacit assumption of atmospheric conditions and are usually valid in the following range ("standard atmospheric conditions"):

(a)   Temperature –20 °C to +60 °C;

(b)   Pressure 80 kPa (0.8 bar) to 110 kPa (1.1 bar);

(c)   Air with standard oxygen content (21 % v/v).

A11.2.5.1.2   An increase in temperature may have multiple effects such as a decrease in MEC and MIE, thus increasing the likelihood of a dust explosion.

A11.2.4.3　空気または他の酸化性雰囲気

　一般に粉じん爆発においては空気が酸化剤であるが、可燃性粉じんが他の酸化性ガスまたはガス混合物の中で扱われた場合にも粉じん爆発は起こるかもしれない。

A11.2.4.4　着火源

A11.2.4.4.1　効果的な着火源が爆発性粉じん－空気混合物（爆発性雰囲気）の中にあれば、粉じん爆発は起こるであろう。潜在的着火源の効力は爆発性雰囲気を着火する能力である。これは着火源のエネルギーのみならず爆発可能雰囲気との相互作用にもよる。

A11.2.4.4.2　着火源の評価は2段階で行う：最初に可能性のある着火源を特定する。第2段階でそれぞれの可能性のある着火源を爆発性雰囲気に着火する能力に関して評価する。この過程で効力があると特定された着火源には、爆発防護策（A11.2.6.1参照）の中の適当な防止対策が必要となる。

A11.2.4.4.3　潜在的な着火源を以下に示す：

(a) 高温表面；
(b) 炎および高温ガス；
(c) 機械的に発生したスパーク；
(d) 電気機器；
(e) 迷走電流および陰極腐食防護；
(f) 雷；
(g) 静電気；
(h) ラジオ波帯域の電磁波（$10^4\,\mathrm{Hz}-3\times 10^{12}\,\mathrm{Hz}$）；
(i) 電磁波（$3\times 10^{11}\,\mathrm{Hz}-3\times 10^{15}\,\mathrm{Hz}$）；
(j) 電離放射線；
(k) 超音波；
(l) 断熱圧縮および衝撃波；
(m) 発熱反応、以下を含む、粉じんの自己発火、くん焼き/白熱した粒子または粉じん、およびテルミット反応（例えばアルミニウムとさびた鋼）。

A11.2.5　粉じん爆発の重大性に影響を与える他の要因

　A11.2.4で説明されている因子に加えて、他の条件もまた粉じん爆発がどのように激烈になるかに影響を与える。より重要なものは環境因子および閉じ込めであり、これらは以下で説明されている。この節で示されている因子のリストは完全ではないので、ある状況におけるリスクの評価においては、適切に専門家のアドバイスが求められるべきである。

A11.2.5.1　*温度、圧力、酸素の有無および湿度の影響*

A11.2.5.1.1　安全に関連するデータはしばしば大気の状態が暗黙の了解になっており、それらは通常下記の範囲（「標準大気状態」）があてはまる：

(a) 温度−20℃から+60℃；
(b) 圧力80kPa (0.8bar) から110kPa (1.1bar)；
(c) 標準酸素濃度（21%v/v）の空気。

A11.2.5.1.2　温度の上昇は、MECおよびMIEの減少、すなわち粉じん爆発の可能性の増加、のような複合的な効果があるかもしれない。

A11.2.5.1.3　　An increase in pressure tends to lower the MIE and the MIT of a dust cloud while the maximum explosion pressure will rise. The effect is increased sensitivity, thus increasing the likelihood and severity of a dust explosion.

A11.2.5.1.4　　Higher oxygen content can significantly increase the sensitivity of an explosible atmosphere and the severity of an explosion due to higher explosion pressures. Equally lower oxygen concentration can reduce the risk of an explosion. The LEL may also rise. Such a situation can occur when a process is undertaken under an inert atmosphere.

A11.2.5.1.5　　Low or high humidity (of air, gas phase) may influence the occurrence of electrostatic discharges.

A11.2.5.1.6　　Therefore, the risk and severity of dust explosions under non-standard atmospheric conditions should be evaluated by expert consideration of the actual process conditions.

A11.2.5.2　　*Confinement*

　　　　Confinement means the dust is in an enclosed or limited space. A combustible dust (as defined above) can react without confinement or when confined. When confined, the explosion pressure is likely to be higher than when unconfined, as confinement allows pressure to build up, increasing the severity of an explosion. Using suitably sized and located explosion relief allows the burning dust cloud and hot products of a dust explosion to vent to safe places outside the confined area, reducing the potential for the pressure to increase, and so limiting the potential explosion severity. Expert advice may be needed on the possible application and design of explosion relief venting based upon the physical and chemical properties and potential health/physical hazards of the substance, mixture, or solid material.

A11.2.6　　*Hazard prevention, risk assessment and mitigation*

A11.2.6.1　　*General explosion protection concept for dusts*

A11.2.6.1.1　　Table A11.2.1 shows the principles of explosion protection. The table presents both preventive and mitigative measures and identifies which safety characteristics are most relevant to the measures proposed. For guidance on safety characteristics, refer to annex 4, table A4.3.9.3.

A11.2.6.1.2　　The first priority should involve preventive measures such as substitution and application of dust-free processes to avoid where possible the presence of combustible dusts, as shown in the column "Avoidance of combustible dusts".

A11.2.6.1.3　　Where the presence of combustible dusts cannot be avoided, measures such as exhaust ventilation should be taken to prevent the concentration of combustible dusts reaching the explosible range; see the column "Avoidance of reaching the explosion range". Good housekeeping practices are important to prevent the formation of dust clouds or – if that is not achieved - the propagation of pressure waves and fireballs from an initial explosion, e.g. inside equipment or enclosures, dispersing and igniting dust accumulations into a work area. Such secondary explosions can often be more destructive than the primary explosion. A written housekeeping plan with regular inspection for excessive dust levels, including emphasis on priority areas, is strongly recommended. Housekeeping should be conducted concurrently with operations.

A11.2.6.1.4　　Where measures cannot be taken to avoid or reduce explosible dust atmospheres, then, ignition sources should be assessed and avoided where possible (see A11.2.4.4 and table A11.2.2). Ignition sources can include fires and heat caused by the frictional energy of mechanical equipment. Heat or arcing caused by the failure of or the use of improper electrical equipment, such as lighting, motors, and wiring, have also been identified as ignition sources. Improper use of welding and cutting equipment can be a factor. Periodic inspections, lubrication, and adjustment of equipment can be a major tool to prevent ignitions which can lead to explosions. Additional examples of what to consider when evaluating ignition sources are in the column "Avoidance of ignition sources".

A11.2.6.1.5　　Where ignition of an explosible dust atmosphere cannot be excluded, the effects should be mitigated by protective measures. When containment is used as a mechanism to reduce the risk or when the dust is confined, then explosion-proof design or relief venting should be considered. Equipment and buildings with known combustible dusts should be equipped with devices or systems designed to prevent an explosion, minimize its propagation, or limit the damage it causes. Explosion relief venting is one of the most common approaches taken to reduce the explosion pressure. Examples of other mitigating measures are shown in the column "Minimizing effects of a dust explosion".

A11.2.6.1.6　　Section A11.2.8.2 contains a list of regulations and guidance documents on prevention and mitigation of dust explosions, including those discussing explosion prevention systems and the use of deflagration venting.

A11.2.5.1.3　圧力の増加は、最大爆発圧力を上げる一方、粉じん雲のMIEおよびMITを下げる傾向がある。この効果は感度を上げ、粉じん爆発の可能性や重大性を増加させる。

A11.2.5.1.4　より高い酸素含有は、爆発性雰囲気の感度およびより高い爆発圧力により爆発の重大性を増大させることができる。同時により低い酸素濃度は爆発のリスクを減少させることができる。LELもまた上昇するであろう。そのような状況は、非活性雰囲気の下で行われたプロセスでも起こりうる。

A11.2.5.1.5　低いまたは高い湿度（空気、ガス状態の）は静電気放電の発生に影響するかもしれない。

A11.2.5.1.6　それゆえ、非標準状態での粉じん爆発のリスクおよび重大性は、実際のプロセスでの条件を専門家が考慮して評価されるべきである。

A11.2.5.2　*閉じ込め*

　閉じ込めとは、粉じんが封入されているまたは限られた空間にあることを意味する。可燃性粉じん（上記の定義による）は閉じ込められなくてもまたは閉じ込められても反応する。閉じ込められた場合、爆発圧力は、閉じ込めは圧力の上昇を可能にするため、閉じ込められていない場合に比べてより高くなり、爆発の重大性を増大させる。適当なサイズで配置された爆発開放器の使用は、粉じん爆発の燃焼粉じん雲および高温生成物を閉じ込め空間から外部の安全な場所に排出し、圧力の上昇を低下させ、爆発の重大可能性を制限する。物質、混合物または固体材料の物理的および化学的性質さらに潜在的な健康有害性/物理的危険有害性に基づいた、爆発開放用ベントの適用やデザインに関して、専門家のアドバイスが必要かもしれない。

A11.2.6　*危険防止、リスクアセスメントおよび軽減*

A11.2.6.1　*粉じんに関する一般的な爆発防護の考え方*

A11.2.6.1.1　表A11.2.1に爆発防護の原則を示した。表では防止および軽減対策を提示し、提案された対策に対してどの安全特性が最も適しているかを明らかにしている。安全特性に関する手引きは、附属書4、表A4.3.9.3を参照すること。

A11.2.6.1.2　最優先として、「可燃性粉じんの回避」欄に示されているように、可能であれは可燃性粉じんの存在を回避するため、代替および無粉じん工程の適用のような予防的対策を含むべきである。

A11.2.6.1.3　可燃性粉じんの存在が避けられない場合には、可燃性粉じんの濃度が爆発可能範囲に達するのを防ぐために、排気装置のような対策が取られるべきである；「爆発可能範囲への到達回避」欄を参照する。良い清掃の実践は、粉じん雲の形成の防止または－これが達成されない場合には－装置または封入容器の内部での最初の爆発からの圧力および火球の伝搬、作業場への散乱および粉じん堆積物への着火を防止するために重要である。そのような二次爆発はしばしば一次爆発よりも破壊的である。優先順位の高い場所に対する強調も含み、過剰な粉じんレベルに関する定期的な検査を伴った、記載された清掃計画は強く推薦される。清掃は操作とともに並行して実施されるべきである。

A11.2.6.1.4　爆発性粉じん雰囲気を回避するまたは削減するための対策をとることができない場合、可能であれば着火源が評価され、避けられるべきである（A11.2.4.4および表A11.2.2参照）。着火源には、機械装置の摩擦エネルギーによる火および熱も含まれる。照明、モーターおよび配線のような不適切な電気機器の故障または使用によって生じる熱やアークもまた着火源として同定されている。溶接や切断装置の不適切な使用も因子になりうる。定期的な検査、潤滑および装置の調整は爆発につながる着火を防ぐ主な手段となりうる。着火源を評価する際考慮すべき事項の追加例は、「着火源の回避」欄にある。

A11.2.6.1.5　爆発性粉じん雰囲気の着火が排除できない場合、防護対策により影響は軽減されるべきである。閉じ込めがリスクを削減する方法として使用される、すなわち粉じんが封じ込められた場合、防爆仕様または開放ベントが検討されるべきである。既知の可燃性粉じんがある装置および建物は、爆発を防止し、伝搬を最小にしまたはそれによるダメージを制限するために設計された機器またはシステムを備えているべきである。爆発開放ベントは、爆発圧力を低減するためにとられる最も知られた方法の1つである。他の軽減対策の例は「粉じん爆発の影響の最小化」欄に示されている。

A11.2.6.1.6　A11.2.8.2は、爆発防止システムおよび爆燃ベントの使用に関する検討を含む、粉じん爆発の防止および軽減に関する規則および手引きのリストを含む。

- 581 -

A11.2.6.1.7 Every facility where there is a potential for dust explosions should have a safety program and an established emergency action plan. A communication system is needed to notify everyone at the plant when there is an emergency and they might be at risk. A central alarm system, page system or horn can be used to signal the need for evacuation. All workers should be trained in the hazards of combustible dust, the risk of explosions, and proper preventive measures.

Table A11.2.1: General concept to prevent and mitigate dust explosions

Prevention		Mitigation
Avoidance or reduction of explosible dust atmospheres	**Avoidance of ignition sources**	**Minimizing effects of a dust explosion**
Relevant safety characteristics • Dust explosibility **Avoidance of combustible dusts by [examples below]** • Substitution • Passivation • Application of dust-free processes • …	**Identification of relevant ignition sources** • Identification of relevant areas and activities (zoning) • Identification of potential ignition sources • Determination of relevant safety characteristics (see below)	*Relevant safety characteristics* • Maximum explosion pressure • Deflagration index (K_{st}) **Explosion pressure proof design by [examples below]** • Venting (reduction of explosion pressure) • Explosion resistance • …
Relevant safety characteristics • Lower explosible limit (LEL)/ Minimum Explosible Concentration (MEC) **Avoidance of reaching the explosible range by [examples below]** • Good house keeping • Exhaust ventilation • Dust reduced procedures • …	*Relevant safety characteristics* • Minimum ignition energy • Minimum ignition temperatures (dust clouds and dust layers) • Self-ignition behaviour **Prevention of effective ignition sources by [examples below]** • Avoidance of open fire or flames • No smoking • Limitation of surface temperatures • Use of approved electrical and mechanical equipment (in accordance with respective zone) • Prevention of electrostatic discharges (e.g. grounding, dissipative materials) • Prevention of mechanical heating or sparks (e.g. temperature monitoring, misalignment monitoring of moving parts, …) • Spark detection and extinguishing • …	**Explosion suppression by [examples below]** • Explosion detection and dispersion of extinguishing media (powder, water, …) • …
Relevant safety characteristics • Limiting oxygen concentration (LOC) **Oxygen reduction by [examples below]** • Inerting (N_2, CO_2, argon, flue gas, water vapour, …) •		**Explosion isolation by [examples below]** • Ignition and flame resistant components (rotary valves, double acting valves, quick acting gate valves, …) • Extinguishing barriers • …

A11.2.6.1.7 粉じん爆発の可能性のあるすべての施設は安全プログラムおよび確立された緊急行動計画を持つべきである。緊急事態が発生し、リスクにさらされた時に、プラントにいる全員に知らせるための情報伝達システムが必要とされる。中央警告システム、呼び出しシステムおよび警笛は非難の必要性を知らせるために使用することができる。すべての作業者は粉じん爆発、爆発のリスク、および適切な防止対策について訓練されるべきである。

表A11.2.1：粉じん爆発を防止し軽減するための一般的な考えかた

防止		軽減
爆発性粉じん雰囲気の回避または削減	着火源の回避	粉じん爆発の影響の最小化
関連する安全特性 ・粉じんの爆発可能性 **爆発性粉じんの回避** [以下例] ・代替 ・不活性化 ・無粉じん工程の導入 ・…	**関連した着火源の特定** ・関連場所および作業の特定（ゾーニング） ・潜在的着火源の特定 ・関連する安全特性の決定（以下参照）	*関連する安全特性* ・最大爆発圧力 ・爆燃指標 *(Kst)* **耐爆発圧力デザイン** [以下例] ・ベント（爆発圧力の縮小） ・爆発抵抗 ・…
関連する安全特性 ・爆発下限 *(LEL)* /最小爆発濃度 *(MEC)* **爆発可能範囲への到達回避** [以下例] ・清掃 ・排気装置 ・粉じん削減手順 ・…	*関連する安全特性* ・最小着火エネルギー ・最低着火温度 *(粉じん雲および粉じん層)* ・自己発火特性 **効力のある着火源の防止** [以下例] ・裸火および炎の回避 ・禁煙 ・表面温度の制限 ・認可された電気および機器装置の使用（それぞれの場所による） ・静電気放電の防止（例えばアース、散逸性材料） ・機械的過熱または火花の防止（例えば温度監視、可動部分調整不良監視、…） ・火花検知および消火 ・…	**爆発の抑制** [以下例] ・爆発検知および消火剤（粉末、水、…）の分散 ・…
関連する安全特性 ・酸素濃度制限 *(LOC)* **…による酸素削減** [以下例] ・不活性化（N_2、CO_2、アルゴン、排ガス、水蒸気、…）		**爆発の隔離** [以下例] ・着火および炎抵抗性部品（回転バルブ、複動式バルブ、急動扉バルブ、…） ・消火扉 ・…

A11.2.6.2 *Considerations for dust explosion protection during operations and processing*

A11.2.6.2.1   Processing operations may change the physical form of substances, mixtures, and solid materials such that smaller particles are formed (e.g. sieving, milling, grinding). When substances, mixtures, and solid materials that are not completely oxidized are subjected to such operations, this may result in the formation of combustible dusts. In such cases, the principles of this guidance apply equivalently, and the measures for hazard prevention, risk assessment and mitigation described in A11.2.6.1 should be considered. The responsible party (e.g. manufacturer, employer) at a facility performing processing operations has the best knowledge about the operation that is necessary to conduct an appropriate dust explosion risk assessment and determine the proper measures for hazard prevention and risk mitigation.

A11.2.6.2.2   Table A11.2.2 presents potential ignition sources that may be present during operations and that should be considered. The table uses ignition sources as an example when evaluating potential dust explosion protection measures during operations. Expert advice may be needed to develop and apply appropriate preventive and mitigative measures.

A11.2.6.2　*操作および処理中の粉じん爆発防護に関する検討*

A11.2.6.2.1　小さな粒子が形成されるような（例えば、ふるいにかける、粉砕する、すりつぶす）加工処理が物質、混合物および固体材料の物理的な形状を変化させるかもしれない。完全には酸化されていない物質、混合物および固体材料がそのような操作を受けると、これは可燃性粉じんの形成につながるであろう。そのような場合、この手引きの原則が等しく適用され、A11.2.6 に記載されている危険防止、リスクアセスメントおよび軽減対策が検討されるべきである。加工処理操作をしている施設において責任のある関係者（例えば製造者、雇用者）は、適切な粉じん爆発のリスクアセスメントを実行するために必要な操作について熟知し、危険防止およびリスク軽減のための適当な対策を決定する。

A11.2.6.2.2　表 A11.2.2 には、操作中に存在するかもしれない、そして検討されなければならない潜在的着火源を示した。この表は、操作中の潜在的粉じん爆発防護対策を評価する際の例として、着火源を利用している。適切な予防および軽減対策を開発し適用するためには専門家のアドバイスが必要とされるであろう。

Table A11.2.2: Potential ignition sources during operations

Type of ignition source [see A11.2.4.4.3]	Facility management — Construction work, repair, maintenance	Storage	Transfer-operations — Conveying (solids)	Transfer-operations — Pumping (liquids)	Transfer-operations — Other transfer operations	Formulation and packaging — Mixing (no reaction)	Formulation and packaging — Sieving/milling/grinding	Formulation and packaging — Formulation operations	Formulation and packaging — Packaging	Reaction and downstream processing — Reaction	Reaction and downstream processing — Off-gas handling / scrubbing	Reaction and downstream processing — Work-up (phase separation; crystallization; filtration, isolation)	Reaction and downstream processing — Distillation	Reaction and downstream processing — Drying
Hot surfaces	Caused by friction of moving parts at bearings, shaft seals, etc.									Heated equipment, pipes, heat exchangers				
Flames and hot gases	Hot work: welding, cutting, etc.	Generally not relevant								Possible formation of hot gases	Generally not relevant			
Mechanically generated sparks	Sparks generated by use of tools (e.g. hammering, drilling, grinding)	Sparks generated due to grinding, friction or impact (frequently caused by mechanical failures or entrainment of foreign parts into moving equipment or machinery)								Generally not relevant				Sparks generated due to grinding, friction or impact
Electric apparatus	Stray currents, e.g. from welding or faulty equipment	Machines, process control technology installations, motors, switches, cables, lighting												
Stray electric currents and cathodic corrosion protection		Relevant in some cases, e. g.: backflow to electricity generation plants, train tracks, vicinity of electric system with high current												
Lightning	Relevant in some cases, e. g.: thunderstorm even with invisible lightning bolts, activities near lightning protection systems													
Static electricity	Relevant in some cases	Frequently generated by flow or separation processes												
Radio frequency electromagnetic waves	Relevant in some cases, e. g.: radio transmitting station, high frequency generators for heating, curing, welding, cutting													
Electromagnetic waves	Relevant in some cases, e. g.: insolation, powerful light source, laser radiation													
Ionizing radiation	Relevant in some cases, e. g.: X-ray machine, radioactive materials													
Ultrasonics	Relevant in some cases, e. g.: ultrasound scanner, ultrasonic testing, sonic driller													
Adiabatic compression and shock waves	Generally not relevant	Compression of gases, rapidly shutting valves when conveying / pumping material			Generally not relevant					Relevant in some cases, e. g.: relaxation of high-pressure gases in pipelines, hammer blow				
Exothermic reactions	Generally not relevant	Pyrophoric and self-heating substances	Transfer of smouldering nests into other areas			Pyrophoric and self-heating substances		Generally not relevant		Strongly exothermic reaction	Self-heating and ignition of charcoal absorbers	Activated catalysts or residues	Possible decomposition of residue	Self-ignition of dust layers (esp. spray drying)

表 A11.2.2：操作中の潜在的着火源

着火源のタイプ [A11.2.4.3 参照]	装置管理 建設作業、修理、維持管理	保管	移動操作 運搬作業（固体）	移動操作 ポンプ作業（液体）	移動操作 他の移動作業	製剤および包装作業 混合(反応なし)作業	製剤および包装作業 ふるい/粉砕/ろっぱし作業	製剤および包装作業 製剤操作	製剤および包装作業 包装作業	反応	反応および川下工程 気体廃棄物処理/ガス洗浄	反応および川下工程 操作(層分離；晶析；ろ過；分離)	反応および川下工程 蒸留	反応および川下工程 乾燥
高温表面	高温作業：溶接、切断等		ベアリング、軸シール等における摩擦による							高温装置、パイプ、熱交換器	高温ガス生成の可能性	一般には該当しない		
炎および高温ガス						一般には該当しない					一般には該当しない			
機械的火花	道具の使用による火花（例えばスパナー、ドリル、研磨）			研磨、摩擦または衝撃による生じる火花（しばしば機械的故障または装置内の異物の噛み）										研磨、摩擦または衝撃によるよる火花
電気機器					機械、工程管理技術設備、モーター、スイッチ、ケーブル、照明									
迷走電流および陰極腐食防止	迷走電流、例えば、溶接または大陥装置				該当する場合あり、例えば：発電プラントの逆流、鉄道線路、大電流の電気システム近くの活動									
雷	一般には該当しない	該当する場合あり				. 流動または分離工程でよく発生				該当する場合あり、例えば：見えない稲妻を伴った雷雨、カミナリ防護システム近くでの活動				
静電気														
ラジオ波帯電磁波						該当する場合あり、例えば：ラジオ送信所、加熱・治療・溶接・切断用高周波発生装置								
電磁波						該当する場合あり、例えば：日射、強力な光源、レーザー照射								
電離放射線						該当する場合あり、例えば：X-線装置、放射性材料								
超音波									該当する場合あり、例えば：超音波スキャナー、超音波検査、超音波ドリル					
断熱圧縮および衝撃波	一般には該当しない		ガスの加圧、材料の運搬（汲み上げ時の速動閉鎖バルブ）			一般には該当しない				該当する場合あり、例えば：パイプライン中の高圧ガスのリラクゼーション、槌打ち				
発熱反応	一般には該当しない	自然発火性および自己発熱性物質	くん焼巣の他の場所への移動			自然発火性および自己発熱性物質				強い発熱反応	木炭吸着剤の自己発熱および着火	活性化触媒または残渣	残渣の自己分解可能性	粉じん層の自己着火（特にスプレー乾燥）

- 584 -

A11.2.7 *Supplemental information for hazard and risk communication*

A11.2.7.1 As explained in 1.4.6.3, there are many communication elements which have not been standardized in the harmonized system. Some of these clearly need to be communicated to the downstream user. Competent authorities may require additional information, or suppliers may choose to add supplementary information on their own initiative. Each party producing or distributing a product that is determined to be hazardous, including if it becomes hazardous during downstream processing, should create and provide their downstream user with appropriate information, in the form of a Safety Data Sheet (SDS) or another format as appropriate, to alert the user to the hazards and risks.

A11.2.7.2 For substances, mixtures, or solid materials, sections 2, 5, 7, and 9 of the SDS, at a minimum, should provide information on combustible dusts. Annex 4 provides further guidance on each section of the SDS. For example, section 2 (A4.3.2) addresses hazards that do not result in classification; section 5 (A4.3.5) covers requirements for fighting a fire; section 7 (A4.3.7) provides guidance on safe handling practices and section 9 (A4.3.9) describes the physical and chemical properties of a substance, mixture, or solid material.

A11.2.7.3 To communicate combustible dust hazards, and thus a potential risk of dust explosions under the approach described in this annex in a standardized manner, competent authorities may require the use of the following phrases on labels, SDSs and/or in operating instructions or may leave the choice to the manufacturer or supplier:

(a) In the case where a substance or mixture is identified as a combustible dust in accordance with figure A11.2.1: "May form explosible dust-air mixture if dispersed"; or

(b) In the case where a substance, mixture, or solid material is to be further processed in such a manner that the processing creates a combustible dust in accordance with A11.2.6.2.1, in combination with figure A11.2.1: "May form explosible dust-air mixture if small particles are generated during further processing, handling, or by other means.";

(c) In addition, the phrase "Warning" may be used in conjunction with items (a) or (b).

A11.2.8 *References*

A11.2.8.1 *Test methods*

Recognized and scientifically validated testing methods and standards, such as those listed below, should be used when evaluating dust explosibility.

International standards

ISO/IEC 80079-20-2, "Explosive atmospheres - Part 20-2: Material characteristics – Combustible dusts test methods"

National standards

ASTM E1226, "Standard Test Method for Explosibility of Dust Clouds"

VDI 2263-1, "Dust Fires and Dust Explosions; Hazards – Assessment – Protective Measures; Test Methods for the Determination of the Safety Characteristics of Dusts"

A11.2.8.2 *Regulations and guidance on prevention and mitigation*

There are a number of documents available providing guidance on preventive and mitigation measures to minimize or eliminate dust explosions. A partial list is provided below. The use of country-specific documents, including those addressing specific hazards and risks associated with materials such as wood, coal, sulfur, combustible metals, and agricultural and food, is encouraged where available.

(a) Directive 1999/92/EC of the European Parliament and of the Council (ATEX), Annex 1

(b) U.S. OSHA's Combustible Dust Directive (Combustible Dust National Emphasis Program)

(c) Health and Safety Executive, UK, HSG 103, Safe Handling of Combustible Dusts: Precautions Against Explosions

A11.2.7 危険性情報伝達およびリスクコミュニケーションに関する補足情報

A11.2.7.1　1.4.6.3 で説明したように、調和システムにおいては標準化されていない多くの情報伝達要素がある。これらのいくつかは明らかに川下使用者に伝えられる必要がある。所管官庁は追加情報を要求することができるし、また供給者は補足情報を彼ら自身の主導で追加することが選択できる。川下の工程で危険になることも含めて、危険であると決定された製品を生産しまたは供給する者は、危険性およびリスクを警告するために安全データシート（SDS）または他の形式でしかるべく適切な情報を作成し、川下の使用者に対して提供すべきである。

A11.2.7.2　物質、混合物または固体材料に関して、最低限、SDS の 2、5、7 および 9 節では可燃性粉じんに関する情報を提供するべきである。附属書 4 では SDS のそれぞれの節でさらなる手引きを提供している。例えば、2 節（A4.3.2）では分類に結び付かない危険有害性についての取組み；5 節（A4.3.5）では消火に関する必要事項を扱い；7 節（A4.3.7）では安全な取り扱いの実践に関する手引きを提供し、9 節（A4.3.9）では物質、混合物または固体材料の物理的および化学的性質について記載している。

A11.2.7.3　この附属書に標準的な様式で記載されている方法により、可燃性粉じんの危険性、そして粉じん爆発の潜在的なリスクを情報伝達するために、所管官庁はラベル、SDS および/または操作説明書に以下の警句の使用を要求してもよいし、または製造者または供給者に選択を委ねてもよい：

(a) 物質または混合物が、A11.2.1 にしたがって可燃性粉じんとして同定された場合：「分散した場合には、爆発性粉じん－空気混合物を形成する恐れ」；または

(b) 物質、混合物または固体材料が、図 A11.2.1 と共に A11.2.6.2.1 にしたがった可燃性粉じんを作り出す工程などのように、さらに処理された場合；「さらなる処理、取扱いまたは他の方法によって細粒子が発生した場合、爆発性粉じん－空気混合物を形成する恐れ」；

(c) 加えて、注意喚起語「警告」が (a) または (b) とともに使用されてもよい。

A11.2.8　参考文献

A11.2.8.1　試験方法
粉じんの爆発可能性を評価する際には、以下に示されたような認められ科学的に検証された試験方法および規格が使用されるべきである。

国際規格

ISO/IEC 80079-20-2、「爆発性雰囲気－20-2 部：材料の特性－可燃性粉じん試験方法」

各国標準

ASTM E 1226、「粉じん雲の爆発可能性に関する標準試験方法」

VDI2263-1、「粉じん火災および粉じん爆発；危険性－評価－防護対策；粉じんの安全特性の決定に関する試験方法」

A11.2.8.2　予防及び軽減に関する規則および手引き

粉じん爆発を最小化および排除するための防止および軽減対策に関する手引きを提供している、入手可能な文書は多くある。一部のリストを以下に示した。入手可能であれば、木材、石炭、硫黄、可燃性金属および農業や食料に関連した特有の危険性やリスクへの対応も含んだ、国に固有の文書の使用も薦める。

(a) 欧州議会および理事会（ATEX）指令 1999/92/EC、附属書 1

(b) US OSHA 可燃性粉じん指令（可燃性粉じんに関する国家重点プログラム）

(c) 英国 HSE（Health and Safety Executive）、HSG 103、可燃性粉じんの安全な取り扱い：爆発防止措置

(d) U.S. National Fire Protection Association (NFPA)

NFPA 652: Standard on the Fundamentals of Combustible Dust

NFPA 654: Standard for the Prevention of Fire and Dust Explosions from the Manufacturing, Processing, and Handling of Combustible Particulate Solids

NFPA 68: Standard on Explosion Protection by Deflagration Venting

NFPA 69: Standard on Explosion Prevention Systems

(d) U.S. 全米防火協会（NFPA）

NFPA 652: Standard on the Fundamentals of Combustible Dust

NFPA 654: Standard for the Prevention of Fire and Dust Explosions from Manufacturing, Processing, and Handling of Combustible Particulate Solids

NFPA 68: Standard on Explosion Protection by Deflagration Venting

NFPA 69: Standard on Explosion Prevention Systems

英和対訳
化学品の分類および表示に関する
世界調和システム（GHS）
改訂 10 版
GHS 関係省庁連絡会議　仮訳

2025 年 3 月 14 日　第 1 版第 1 刷発行

発行者　　　朝日　　弘
発行所　　　一般財団法人　日本規格協会
　　　　　　〒108-0073　東京都港区三田 3 丁目 11-28　三田 Avanti
　　　　　　　　　　　https://www.jsa.or.jp/
　　　　　　　　　　　振替　00160-2-195146
製作　　　　日本規格協会ソリューションズ株式会社
印刷・製本　株式会社 平文社

Japanese Standards Association, 2025　　　　　　　　　　　　　　Printed in Japan
ISBN 978-4-542-40419-9

●お求めは、下記をご利用ください。
　JSA Webdesk（オンライン注文）：https://webdesk.jsa.or.jp/
　電話：050-1742-6256　E-mail：csd@jsa.or.jp